Vector Quantization

Edited by
Hüseyin Abut
Electrical and Computer Engineering Department
San Diego State University

A volume in the IEEE PRESS Selected Reprint Series, prepared under the sponsorship of the IEEE Acoustics, Speech, and Signal Processing Society.

IEEE PRESS

The Institute of Electrical and Electronics Engineers, Inc., New York

PRINTED IN THE UNITED STATES OF AMERICA

IEEE Order Number: PCO255-0

Library of Congress Cataloging-in-Publication Data

Vector quantization/edited by Hüseyin Abut.
p. cm.—(IEEE Press selected reprint series)
Reprints of works published 1981–1988.
Includes bibliographical references.
ISBN 0-87942-265-3
1. Image processing—Digital techniques. 2. Signal processing—Digital techniques. 3. Data compression (Telecommunication)
I. Abut, Hüseyin, 1945-
TA1632.V43 1990
621.36′7—dc20 90-32686
CIP

Contents

Introduction vi

Part 1: Tutorials on Vector Quantization 1

Vector Quantization: A Pattern-Matching Technique for Speech Coding, *A. Gersho and V. Cuperman* (*IEEE Communications Magazine,* December 1983) 2

Vector Quantization, *R. M. Gray* (*IEEE Acoustics, Speech, and Signal Processing Magazine,* April 1984) 9

Vector Quantization in Speech Coding, *J. Makhoul, S. Roucos, and H. Gish* (*Proceedings of the IEEE,* November 1985) 34

Image Coding Using Vector Quantization: A Review, *N. M. Nasrabadi and R. A. King,* (*IEEE Transactions on Communications,* August 1988) 72

Part 2: Theoretical Studies and Algorithms 87

An Algorithm for Vector Quantizer Design, *Y. Linde, A. Buzo, and R. M. Gray* (*IEEE Transactions on Communications,* January 1980) 88

Asymptotically Optimal Block Quantization, *A. Gersho* (*IEEE Transactions on Information Theory,* July 1979) 100

On the Structure of Vector Quantizers, *A. Gersho* (*IEEE Transactions on Information Theory,* March 1982)... 108

Voronoi Regions of Lattices, Second Moments of Polytopes, and Quantization, *J. H. Conway and N. J. A. Sloane* (*IEEE Transactions on Information Theory,* March 1982) 118

Fast Quantizing and Decoding Algorithms for Lattice Quantizers and Codes, *J. H. Conway and N. J. A. Sloane* (*IEEE Transactions on Information Theory,* March 1982) 134

Finite-State Vector Quantization for Waveform Coding, *J. Foster, R. M. Gray, and M. O. Dunham* (*IEEE Transactions on Information Theory,* March 1985) 140

Using Simulated Annealing to Design Good Codes, *A. A. El Gamal, L. A. Hemachandra, I. Shperling, and V. K. Wei* (*IEEE Transactions on Information Theory,* January 1987) 152

An Algorithm for Uniform Vector Quantizer Design, *K. Sayood, J. D. Gibson, and M. C. Rost* (*IEEE Transactions on Information Theory,* November 1984) 160

Vector Quantizer Design for Memoryless Gaussian, Gamma, and Laplacian Sources, *T. R. Fischer and R. M. Dicharry* (*IEEE Transactions on Communications,* September 1984) 170

A Pyramid Vector Quantizer, *T. R. Fischer* (*IEEE Transactions on Information Theory,* July 1986) 175

Asymptotic Performance of Unrestricted Polar Quantizers, *P. F. Swaszek and T. Ku* (*IEEE Transactions on Information Theory,* March 1986) 191

Entropy-Constrained Vector Quantization, *P. A. Chou, T. Lookabaugh, and R. M. Gray* (*IEEE Transactions on Acoustics, Speech, and Signal Processing,* January 1989) 195

Part 3: Speech Coding 207

Rate Distortion Speech Coding with a Minimum Discrimination Information Distortion Measure, *R. M. Gray, A. H. Gray, Jr., G. Rebolledo, and J. E. Shore* (*IEEE Transactions on Information Theory,* November 1981) 208

An 800 bit/s Vector Quantization LPC Vocoder, *D. Y. Wong, B.-H. Juang, and A. H. Gray, Jr.* (*IEEE Transactions on Acoustics, Speech, and Signal Processing,* October 1982) 222

Very Low Data Rate Speech Compression with LPC Vector and Matrix Quantization, *D. Y. Wong, B.-H. Juang, and D. Y. Cheng* (*Proceedings of the IEEE International Acoustics, Speech, and Signal Processing Conference,* April 1983) 233

Matrix Quantizer Design for LPC Speech Using the Generalized Lloyd Algorithm, *C. Tsao and R. M. Gray* (*IEEE Transactions on Acoustics, Speech, and Signal Processing,* June 1985) 237

A Segment Vocoder at 150 B/S, *S. Roucos, R. M. Schwartz, and J. Makhoul* (*Proceedings of the IEEE International Acoustics, Speech, and Signal Processing Conference,* April 1983) 246

LPC Speech Coding Based on Variable-Length Segment Quantization, *Y. Shiraki and M. Honda* (*IEEE Transactions on Acoustics, Speech, and Signal Processing,* September 1988) 250

Vector Quantization of Speech and Speech-Like Waveforms, *H. Abut, R. M. Gray, and G. Rebolledo* (*IEEE Transactions on Acoustics, Speech, and Signal Processing,* June 1982) 258

Speech and Speaker Independent Codebook Design in VQ Coding Schemes, *H. Reininger and D. Wolf* (*Proceedings of the IEEE International Acoustics, Speech, and Signal Processing Conference,* March 1985) 271

Product Code Vector Quantizers for Waveform and Voice Coding, *M. J. Sabin and R. M. Gray* (*IEEE Transactions on Acoustics, Speech, and Signal Processing,* June 1984) 274

Efficient Bit Allocation for an Arbitrary Set of Quantizers, *Y. Shoham and A. Gersho* (*IEEE Transactions on Acoustics, Speech, and Signal Processing,* September 1988) 289
Embedded Coding of Speech: A Vector Quantization Approach, *A. Haoui and D. G. Messerschmitt* (*Proceedings of the IEEE International Acoustics, Speech, and Signal Processing Conference,* March 1985) 297
Vector Predictive Coding of Speech at 16 kbits/s, *V. Cuperman and A. Gersho* (*IEEE Transactions on Communications,* July 1985) 300
Low-Rate Speech Encoding Using Vector Quantization and Subband Coding, *H. Abut and S. Ergezinger* (*Proceedings of the IEEE International Acoustics, Speech, and Signal Processing Conference,* April 1986) 312
A High Quality Subband Speech Coder with Backward Adaptive Predictor and Optimal Time-Frequency Bit Assignment, *F. K. Soong, R. V. Cox, and N. S. Jayant* (*Proceedings of the IEEE International Acoustics, Speech, and Signal Processing Conference,* April 1986) 316
Quantization Procedures for the Excitation in CELP Coders, *P. Kroon and B. S. Atal* (*Proceedings of IEEE International Acoustics, Speech, and Signal Processing Conference,* April 1987) 320
CELP Coding for High Quality Speech at 8 kbits/s, *M. Copperi and D. Sereno* (*Proceedings of IEEE International Acoustics, Speech, and Signal Processing Conference,* April 1986) 324
An Expandable Error-Protected 4800 BPS CELP Coder (U.S. Federal Standard 4800 BPS Voice Coder), *J. P. Campbell, Jr., V. C. Welch, and T. E. Tremain* (*Proceedings of IEEE International Acoustics, Speech, and Signal Processing Conference,* May 1983) 328

Part 4: Image Coding 332
Image Compression Using Non-Adaptive Spatial Vector Quantization, *R. L. Baker and R. M. Gray* (*Record of the Sixteenth Asilomar Conference on Circuits, Systems, and Computers,* October 1982) 333
Image Coding Using Vector Quantization, *A. Gersho and B. Ramamurthi* (*Proceedings of the IEEE International Acoustics, Speech, and Signal Processing Conference,* May 1982) 338
Vector Quantiser of Video Signals, *T. Murakami, K. Asai, and E. Yamazaki* (*Electronics Letters, vol. 7,* November 1982) 342
Image Vector Quantization with a Perceptually-Based Cell Classifier, *B. Ramamurthi and A. Gersho* (*Proceedings of the IEEE International Acoustics, Speech, and Signal Processing Conference,* March 1984) 344
A Matrix Quantizer Incorporating the Human Visual Model, *K. S. Thyagarajan, S. Parthasarathy, and H. Abut* (*Proceedings of the IEEE International Acoustics, Speech, and Signal Processing Conference,* March 1985) 348
High Quality Image Coding with a Model-Testing Vector Quantizer and a Human Visual System Model, *V. Ramamoorthy and N. S. Jayant* (*Proceedings of the IEEE International Acoustics, Speech, and Signal Processing Conference,* April 1988) 352
A New Vector Quantization Clustering Algorithm, *W. H. Equitz* (*IEEE Transactions on Acoustics, Speech, and Signal Processing,* October 1989) 356
Vector Quantization of Color Images, *T. Murakami, K. Asai, and A. Itoh* (*Proceedings of the IEEE International Acoustics, Speech, and Signal Processing Conference,* April 1986) 364
Image Compression Using Adaptive Vector Quantization, *M. Goldberg, P. R. Boucher, and S. Shlien* (*IEEE Transactions on Communications,* February 1986) 368
Variable Block Vector DPCM: Vector Predictive Coding of Color Images, *C. W. Rutledge* (*Proceedings of the IEEE International Communications Conference,* June 1987) 376
Linear Block Prediction with Source Classification for Image Encoding Applications, *K. S. Thyagarajan and S. Bhatt* (*Proceedings of the IEEE International Acoustics, Speech, and Signal Processing Conference,* April 1988) 382
Adaptive Discrete Cosine Transform Coding with Vector Quantization for Color Images, *K. Aizawa, H. Harashima, and H. Miyakawa* (*Proceedings of the IEEE International Acoustics, Speech, and Signal Processing Conference,* April 1986) 386
Image Sequence Coding Using Vector Quantization, *M. Goldberg and H. Sun* (*IEEE Transactions on Communications,* July 1986) 390
A Finite State/Frame Difference Interpolative Vector Quantizer for Low Rate Image Sequence Coding, *H.-H. Shen and R. L. Baker* (*Proceedings of the IEEE International Acoustics, Speech, and Signal Processing Conference,* April 1988) 398
Image Coding Based on Segmentation Using Region Growing, *K. S. Thyagarajan, H. Bohlmann, and H. Abut* (*Proceedings of the IEEE International Acoustics, Speech, and Signal Processing Conference,* April 1987) 402

Part 5: Segmentation, Classification, and Recognition 406
Vector Quantization and Markov Source Models Applied to Speech Recognition, *R. Billi* (*Proceedings of the IEEE Acoustics, Speech, and Signal Processing Conference,* May 1982) 407

Discrete Utterance Recognition Based upon Source Coding Techniques, *A. Buzo, H. Martinez, and C. Rivera* (*Proceedings of the IEEE International Acoustics, Speech, and Signal Processing Conference,* May 1982) 411
Discrete Utterance Speech Recognition without Time Alignment, *J. E. Shore and D. K. Burton* (*IEEE Transactions on Information Theory,* July 1983) . . . 415
Network-Based Connected Digit Recognition Using Vector Quantization, *M. A. Bush and G. E. Kopec* (*Proceedings of the IEEE International Acoustics, Speech, and Signal Processing Conference,* March 1985) . . . 434
Recent Developments in the Application of Hidden Markov Models to Speaker-Independent Isolated Word Recognition, *B.-H. Juang, L. R. Rabiner, S. E. Levinson, and M. M. Sondhi* (*Proceedings of the IEEE International Acoustics, Speech, and Signal Processing Conference,* March 1985) . . . 438
A Vector-Quantization-Based Preprocessor for Speaker-Independent Isolated Word Recognition, *K.-C. Pan, F. K. Soong, and L. R. Rabiner* (*IEEE Transactions on Acoustics, Speech, and Signal Processing,* June 1985) 442
Connected Digit Recognition Using Vector Quantization, *H. Bourlard, C. J. Wellekens, and H. Ney* (*Proceedings of the IEEE International Acoustics, Speech, and Signal Processing Conference,* March 1984) . . . 457
BYBLOS: The BBN Continuous Speech Recognition System, *Y.-L. Chow, M. O. Dunham, O. A. Kimball, M. A. Krasner, G. F. Kubula, J. Makhoul, P. J. Price, S. Roucos, and R. M. Schwartz* (*Proceedings of the IEEE International Acoustics, Speech, and Signal Processing Conference,* April 1987) . . . 461
A VQ-Based Preprocessor Using Cepstral Dynamic Features for Large Vocabulary Word Recognition, *S. Furui* (*Proceedings of the IEEE International Acoustics, Speech, and Signal Processing Conference,* April 1987) 465
A Vector Quantization Approach to Speaker Recognition, *F. K. Soong, A. E. Rosenberg, L. R. Rabiner, and B.-H. Juang* (*Proceedings of the IEEE International Acoustics, Speech, and Signal Processing Conference,* March 1985) . . . 469
Text-Dependent Speaker Verification Using Vector Quantization Source Coding, *D. K. Burton* (*IEEE Transactions on Acoustics, Speech, and Signal Processing,* February 1987) . . . 473

Part 6: Real-Time Implementation . . . 484
Hardware Realization of Waveform Vector Quantizers, *B. P. M. Tao, H. Abut, and R. M. Gray* (*IEEE Journal of Selected Areas in Communications,* March 1984) . . . 485
Real-Time Speech Compression with a VLSI Vector Quantization Processor, *G. Davidson, T. Stanhope, R. Aravind, and A. Gersho* (*Proceedings of the IEEE International Acoustics, Speech, and Signal Processing Conference,* March 1985) . . . 495
Systolic Architectures for Vector Quantization, *G. A. Davidson, P. R. Cappello, and A. Gersho* (*IEEE Transactions on Acoustics, Speech, and Signal Processing,* October 1988) . . . 499
Vector Quantizer Architectures for Speech and Image Encoding, *H. Abut, B. P. M. Tao, and J. L. Smith* (*Proceedings of the IEEE International Acoustics, Speech, and Signal Processing Conference,* April 1987) 513
Real-Time Vector APC Speech Coding at 4800 bps with Adaptive Postfiltering, *J. H. Chen and A. Gersho* (*Proceedings of the IEEE International Acoustics, Speech, and Signal Processing Conference,* April 1987) 517
A Multiple Rate Low Rate Voice Codec, *J. Rothweiler and J. Carmody* (*Proceedings of the IEEE International Acoustics, Speech, and Signal Processing Conference,* March 1985) . . . 521
A VLSI Chip Set for Real-Time Vector Quantization of Image Sequences, *R. Dianysian and R. L. Baker* (*Proceedings of the International Symposium on Circuits and Systems,* May 1987) . . . 525

Bibliography . . . 529

Author's Index to Bibliography . . . 553

Author Index . . . 559

Subject Index . . . 561

Editor's Biography . . . 566

Introduction

THIS collection of reprints on *vector quantization* (VQ) presents, under one cover, many aspects of this recently emerging branch of data compression and signal processing. Our goal is to provide a balanced exposure to the work of a large number of investigators. The topics include: (1) the development and theory of vector quantization; (2) applications to medium-band through narrow-band speech coding; (3) applications to speech/speaker recognition systems; (4) monochromatic/color still/moving frame image coding at rates under one bit per pixel; and finally, (5) state-of-the-art implementations of a number of systems.

A common objective of all data compression techniques is to reduce the bit rate of digitized signals, such as speech or image sample values, for communication over a digital channel or for storage in a digital memory. Numerous coding techniques have been developed during the last thirty years, many of which are widely used for encoding arbitrary waveforms while others are application-specific and are usually employed in conjunction with a transform operation.

Pulse-code modulation (PCM), differential pulse-code modulation (DPCM), delta modulation (DM), adaptive-predictive coding (APC), and various hybrid, embedded, and adaptive versions of these techniques are widely used for speech and image waveform encoding problems. If system requirements can tolerate medium-to-high data rates (one bit or more per sample) waveform coding techniques are preferred because of their simplicity and good-to-transparent quality. These coders attempt to produce a replica of the original waveform and have proven to be cost effective.

Another important class of coders uses parametric coding techniques, in which the original signal is subjected to a coordinate transformation that concentrates the signal energy along a few dimensions. Efficient coding schemes exist to exploit the lower redundancy transformed signal. Hadamard (HT), Fourier (FFT), discrete-cosine (DCT), and Karhunen–Loeve (KLT) transforms are examples of useful coordinate transforms. As in the case of waveform samples, the transformed coefficients are quantized by means of a standard coding technique. Image coding at one bit per pixel per color on each of sixty four coefficients obtained from an 8×8 DCT is typical of this class of encoders. Another group of encoding and recognition applications obtains a set of model parameters from the analysis of meaningful and manageable segments of the signal. These parameters are then encoded for transmission or storage using appropriate coding techniques. For example, a speech recognizer might use linear prediction (LPC) parameters or LPC derivatives as templates. In this application an LPC analysis of order 10–16 would be performed on short segments of speech and the LPC coefficients (or reflection coefficients, or the log area ratios, or line spectral pairs) are quantized to achieve the desired data compression ratio.

Often individual real-valued data points (or transformed points) are quantized point-at-a-time even though the data are available in block form. Possible correlations among the block's elements are not exploited. A fundamental result of Shannon's rate-distortion theory states that better performance can always be achieved by coding vectors instead of scalars, even if the data source consists of a sequence of independent random variables [237, 332]. Vector quantization was introduced in the late 1970s as a scheme for effectively mapping a sequence of vectors into a digital sequence of numbers. The technique employed uses a multidimensional extension of a simple algorithm by Lloyd [333] due to Linde, Buzo, and Gray [325].

A vector quantizer (block quantizer, K-dimensional quantizer, block source code, or matrix quantizer) consists of a reproduction alphabet or codebook $\boldsymbol{c} = \{y_i;\ i = 1, 2, \cdots, N\}$ of N codewords together with a mapping $q(\cdot)$ that assigns to each K-dimensional real-valued input vector $\boldsymbol{x}$ a codeword $y_i \in \boldsymbol{c}$. For waveform coding, the codewords $\boldsymbol{y}$ should be similar to the source vectors $\boldsymbol{x}$—that is, the codewords are also K-dimensional real-valued vectors. In the pattern recognition literature, the codewords are called the reference patterns or templates. The size N of the codebook is also called the number of levels and it determines the encoding rate.

A vector quantizer can also be seen as a combination of two functions: an *encoder*, which observes a particular input vector $\boldsymbol{x}$ and generates the address i of the codeword y_i specified by $q(\boldsymbol{x})$, and a matching *decoder*, which uses this address to generate the reproduction vector y_i. When $\boldsymbol{x}$ is quantized as y_i, a quantization error results with respect to a distortion measure $d(\boldsymbol{x}, y_i)$. The distortion measure represents the penalty or the cost associated with reproducing vector $\boldsymbol{x}$ by y_i. Then the best mapping $q(\cdot)$ is the one that minimizes $d(\boldsymbol{x}, y_i)$.

Given a distortion measure, the performance of a vector quantizer is measured by the average distortion per sample. To design a codebook or to populate the templates, the K-dimensional space of the vector $\boldsymbol{x}$ is partitioned into N disjoint regions or cells. Associated with each cell s_i, the quantizer assigns a vector y_i as the codeword if $\boldsymbol{x}$ belongs to s_i.

Lloyd proposed an iterative nonvariational technique known as Lloyd's Method I for designing scalar quantizers based on a long training sequence of actual data [331]. Recently, Linde, Buzo, and Gray [325] extended Lloyd's basic approach to the general case of vector quantizers by incorporating the concepts of the k-means clustering technique of MacQueen [337] and the generalized centroid. This algorithm and its variations are presented later in this book and hence they will not be repeated here.

Choice of Papers

The papers reprinted in this book have been selected in an attempt to provide an equitable exposure to all aspects of vector quantization and to key works by leading researchers in the field. In addition, we have expanded the list of papers that we have found to be useful in our own work into a rather extensive bibliography, which appears at the end of the book. Before proceeding with the outline of the book a few words about the bibliography are in order.

It should be noted that the list is by no means complete. Since the field is still quite active it is not possible to be exhaustive. It is certain that several papers published in languages other than English have been overlooked. It was quite beyond our reach to search through the vast Soviet and other European literature. We have included a number of papers which appeared in German, French, and Japanese journals and proceedings. One major class of works that has been omitted intentionally due to the difficulties involved in performing a comprehensive search are the numerous MS and Ph.D. thesis reports prepared in universities throughout the world. In order to be consistent, all dissertation and thesis reports have been excluded. Papers that appeared in abstract form only are also omitted from the bibliography. They include the works presented at the International Conference on Circuits and Systems and Information Theory and Acoustical Society of America meetings.

Due to space limitations we were faced with many difficult choices in making our selections. We assume the responsibility for the choices and hope the authors whose papers were omitted will understand that we intended to produce the most useful and comprehensive collection of papers, consistent with economic limitations.

Section Specific Notes

The four tutorial papers in Part 1 present the theory, algorithms, and applications of vector quantization to speech and image encoding more eloquently than can be made by any editorial comments. Nonetheless, readers may find some of the following remarks useful.

In the first paper, Gersho and Cuperman introduce the vector quantization (VQ) as a pattern-matching technique for speech-coding problems. Readers unfamiliar with VQ will find the paper easily accessible. The second review paper, by Gray, formally introduces and studies the subject of vector quantization and its numerous extensions, from its inception through 1984. In the next paper, Makhoul, Roucos, and Gish present an extensive review on applications of VQ to speech coding. Here, in addition to offering a unique model for VQ, the authors attempt to answer the following three questions on the subject: (1) how it works, (2) why it works, and (3) how it compares with scalar quantizers. Finally, the Nasrabadi and King paper gives recent review of VQ applications to image coding.

Part 2 is devoted to analytical and algorithmic papers. The Linde, Buzo, and Gray paper is not only the traditional reference paper, it is the fundamental guide for newcomers. In this paper, the VQ design algorithm, frequently referred to as the LBG or generalized Lloyd algorithm, is presented. The authors walk the reader through the VQ design algorithm with a numerical example. We would like to point out that there were two earlier works presented in 1978 at two different conferences [223, 11]. Although Gray *et al.* explicitly studied the extensions of Lloyd's Method I to voice coding (vocoding) in the first paper, Adoul, Collin, and Dalle's approach to block coding for speech waveforms has close resemblance. The next two papers by Gersho attempt to establish the analytical framework for this field while in its infancy. Conway and Sloane's papers present the mathematical background to multidimensional quantizers. In the next paper, Foster, Gray, and Dunham combine the concepts of finite-state machines and VQ for waveform coding applications. El Gamel *et al.* discuss the applications of simulated annealing to the design of "good" codes. In the Sayood *et al.*, Fischer *et al.*, and Fischer papers, lattice and pyramid vector quantizers are designed beginning with stochastic processes rather than resorting to the iterative techniques of the LBG algorithm. The final two papers in this section are Swaszek *et al.* and Chou *et al.* works on constrained vector quantizers.

In Part 3, seventeen papers are included covering all aspects of narrow- and medium-band speech coding. In the first paper the authors present, in a very precise fashion, the concepts of source modeling for linear predictive vocoders, Itakura–Saito-type distortion measures for nearest-neighbor distance computations, voicing and pitch modeling, and analysis–synthesis filter codebooks from an information-theoretic viewpoint. The first of two papers by Wong, Juang *et al.* presents an 800-bits-per-second vocoder design having separate codebooks for gain and filter parameters. They have presented their synthesized speech to a number of scientists working in the field in an informal manner, including the editor, and the quality of the reconstructed speech was roughly comparable to the U.S. federal standard LPC-10 vocoder operating at 2400 bits per second. In the next paper the same authors introduce the concept of matrix quantization to design segment vocoders. Tsao and Gray extend the VQ design algorithm to the design of matrix quantizers. Roucos *et al.* and Shiraki and Honda couple VQ with speech segmentation methods to design systems operating under 300 bits per second. In this last group of papers the objective has been to obtain intelligible speech.

In the Abut, Gray, and Rebolledo study, the authors design vector quantizers for speech waveforms and speech-like signals with encoding rates 1.0 and 2.0 bits per sample. According to similar informal listening tests, the synthesized speech had acceptable quality—much better than vocoders but significantly lower than broadcast quality or transparent speech. Speech and speaker-independent codebook design procedures are discussed in the Reininger and Wolf study. In the next paper, Sabin and Gray attempt to overcome the problems of exponential growth in codebook size and the search time by designing product codebooks in which every codevector has a separate gain component and a shape component. The bit allocation for ordered codewords is investigated in the Shoham and Gersho study.

Remaining papers in this group can be classified as embedded vector quantizers—a term first used by Haoui and Messerchmitt. As the name implies, vector quantization is not the underlying process, but rather is implicitly or explicitly embedded in speech-specific encoding techniques. In Cuperman and Gersho, it is coupled with the notion of vector prediction, similar to that of vector DPCM. Chang and Gray have developed gradient algorithms for designing predictive vector quantizers along the lines of Cuperman and Gersho's work [103]. Subband coding, source classification, and LPC ideas are used together with VQ to bring the encoding rate under 4800 bits per second with acceptable quality in the Abut and Ergezinger paper. VQ is used in optimal-bit assignment for subband coders by Soong, Cox, and Jayant. In the Atal and Kroon and Copperi and Sereno studies, we find vector quantizers embedded in code-excited linear prediction (CELP) vocoders. An adaptive vector quantizer designed to perform in the transform domain is presented in the final study of this section.

Fifteen works on applications of vector quantization to image coding, with rates less than one bit per picture element (pixel), are presented in Part 4. The first group to apply VQ to image coding was Yamada, Tazaki, and Fujita of Japan. Their works were written in Japanese and appeared in Japanese journals [598, 599, 600, 601]. The first set of papers chosen in this section are three similar papers from three different groups, which appeared in different journals or proceedings in 1982, and are included for the most part for historical reasons. The codebooks of these quantizers are designed from monochromatic still images and the block sizes are normally 4×4 or smaller. Spatial intensity values from k rows and m columns are re-indexed to vectors of size $K = km \times 1$. The vectors formed in this manner are then presented to the vector quantizer design algorithm just as in the speech case.

Separate codebooks for shade and edge segments of monochromatic images are designed in the Ramamurthi and Gersho paper. The Thyagarajan *et al.* and the Ramamoorthy and Jayant studies incorporate the human visual perception model in the distortion measure to design improved codebooks. In the majority of these works, the quality of reconstructed images has been informally classified as very good. In the next paper, Equitz introduces the concept of K-D quad tree search for savings in computational complexity.

Murakami *et al.* investigate quantization of color imagery in the Y-I-Q domain to better exploit the redundancy in image components. Adaptive VQ is used to improve the quality and/or reduce the bit rate in the next paper by Goldberg, Boucher, and Shlien. Ruthledge and Thyagarajan and Bhatt study the vector DPCM systems in the subsequent two papers. Goldberg and Sun's study extends these concepts to image-sequence coding. Shen and Baker also extend the concept of finite-state VQ to low-rate image-sequence coding. In the final paper of this section, Thyagarajan, Abut, and Bohlmann incorporate a classifier based on region growing and merging to design improved codebooks for still images.

Part 5 is devoted to works on applications of VQ to speech and speaker recognition, speech segmentation, and classification problems. In this application area, the vector quantizer is implicitly used as a clustering technique to generate "good and reliable" reference templates. In the first paper Billi introduces VQ and hidden Markov modeling into the techniques of isolated utterance recognition. The Buzo *et al.* and Shore and Burton studies are devoted to isolated word recognition using VQ codebooks. In the Shore and Burton approach there is no need for time alignment between the reference templates and test vectors, which eliminates the costly problem of dynamic time warping (DTW).

Vector quantizers are used in speaker-independent discrete-utterance recognition tasks in the Bush and Kopec and the subsequent two papers. Although vector quantizers are employed in a connected digit recognition task in the Bourlard *et al.* study, the next two papers implicitly use them in large-vocabulary speech recognition environments. The last two works by Soong *et al.* and Burton investigate the speaker recognition problem using VQ techniques.

Finally, Part 6 is devoted to real-time implementation of VQ-based systems. The Tao, Abut, and Gray paper is the first published work attempting to demonstrate the real-time implementation feasibility of a vector quantizer for speech waveforms. Prior to this publication, the critics of VQ claimed real-time implementation would present a severe roadblock. After this initial LSI and MSI broad-level realization, a number of VLSI chip and board-level VQ systems were built by Gersho and his colleagues, Baker and his students, Fischer and his group, and Rothweiler, in addition to our efforts at San Diego State University. At the time of this writing, there are several chip sets developed at a number of university research laboratories for vector distortion computations. Several systolic, pyramid, hyperplane, and other parallel-processing architectures for incorporating vector quantization in speech, image, and image-sequence-coding applications are currently in place.

Acknowledgments

We hope readers of this book will find the paper selections to be comprehensive, objective, and relevant. It is a great privilege to extend our appreciations to Professors R. M. Gray, M. Ostendorf, D. Wolf, A. Gersho, K. S. Thyagarajan, F. J. Harris, R. L. Baker, J. Adoul, J. D. Gibson, T. R. Fischer, K. Sayood and M. Goldberg, Drs. F. K. Soong, J. Makhoul, L. R. Rabiner, B. S. Atal, G. Davidson and N. J. A. Sloane, and Mr. B. P. M. Tao for their comments and for supplying us with extensive reference lists. The editor is grateful to his family for their help in assembling the extensive bibliography and the author's index to the bibliography, which are presented in the Appendices A and B, respectively. We would also like to thank Professor M. Kaveh, the IEEE Acoustics, Speech, and Signal Processing Society, and the IEEE PRESS for sponsoring this work.

Part 1
Tutorials on Vector Quantization

Vector Quantization: A Pattern-Matching Technique for Speech Coding

ALLEN GERSHO AND VLADIMIR CUPERMAN

VECTOR QUANTIZATION (VQ), a new direction in source coding, has recently emerged as a powerful and widely applicable coding technique. It was first applied to analysis/synthesis of speech, and has allowed Linear Predictive Coding (LPC) rates to be dramatically reduced to 800 b/s with very slight reduction in quality, and further compressed to rates as low as 150 b/s while retaining intelligibility [1,2]. More recently, the technique has found its way to waveform coding [3–5], where its applicability and effectiveness is less obvious and not widely known.

There is currently a great need for a low-complexity speech coder at the rate of 16 kb/s which attains essentially "toll" quality, roughly equivalent to that of standard 64-kb/s log PCM codecs. Adaptive DPCM schemes can attain this quality with low complexity for the proposed 32 kb/s CCITT standard, but at 16 kb/s the quality of ADPCM or adaptive delta modulation schemes is inadequate. More powerful methods, such as subband coding or transform coding, are capable of producing acceptable speech quality at 16kb/s but have a much higher implementation complexity.

The difficulty is further compounded by the need for a scheme that can handle both speech and voiceband data at the 16 kb/s rate. These two types of waveforms occupy the same bandwidth in the subscriber loop part of the telephone network, yet they have a widely different statistical character. Effective speech coding at this rate must be geared to the specific character of speech and must exploit our knowledge of human hearing. On the other hand, a waveform that carries data must be coded and later reconstructed so that a modem can still extract the data with an acceptably low error rate. This is purely a signal processing operation not involving human perception. Vector quantization appears to be a suitable coding technique which caters to this dual requirement. VQ may become the key to 16 kb/s coding; it may also lead to improved quality waveform coding at 8 or 9.6 kb/s.

In this paper, we review recent results obtained in waveform coding of speech with vector quantization and describe some new developments not previously published. Although VQ has not yet become competitive, extrapolation of some of the results reviewed here suggests that it will indeed become a technique of major importance in waveform coding of speech. For a broader and more comprehensive review of VQ, see Gray [19].

This work was supported in part by the Air Force Office of Scientific Research under grant AFOSR 82-008.

This is an expanded version of a paper presented by the authors at GLOBECOM '83.

Vector PCM and Pattern Matching

The starting point and benchmark for all speech coding schemes is conventional PCM, the direct digitization of the speech waveform, sample by sample. PCM is based on the separate digital coding of each sample with linear, logarithm, or perhaps some optimized spacing of quantization levels. Vector quantization, when applied to a block of samples of a waveform, offers a direct generalization of PCM to what might be called "Vector PCM" or VPCM. In VPCM, a block of consecutive samples forms a vector that is treated as one entity. The vector is encoded with a binary word, and an approximation to the original vector is generated using only this binary word. The essential operation here is the quantization of a random vector. In other applications of VQ, the components of the vector may be parameters extracted from a signal rather than the signal samples themselves. In VPCM, the vector specifically represents a block of consecutive samples.

Regardless of the nature of the vector, the VQ coding process is a pattern matching technique. Each vector is encoded by comparison with a set of stored reference vectors, known as codevectors or *patterns*. Each pattern will be used to represent input vectors that are somehow identified as "similar" to this pattern. The best matching pattern in the *codebook*, the set of stored reference patterns, is selected by the encoding process according to a suitable fidelity measure, and a binary word is used to identify this pattern in the "codebook" of patterns. An example of a VPCM codebook for speech waveform coding is illustrated in Fig. 1, where each of 64 patterns consists of six samples joined by straight line segments and the various patterns are superimposed. Note that the patterns describe a variety of possible shapes for a waveform segment of this length. The patterns were optimized to allow the lowest mean-square error matching to

Reprinted from *IEEE Commun. Mag.*, vol. 21, pp. 15–21, Dec. 1983.

speech waveforms subject to the limitation that only 64 patterns could be included in the codebook. The codebook size (number of patterns) is a critical parameter which determines the encoding complexity needed for searching through the codebook, the memory required to store the codebook in both transmitter and receiver, and the bit rate for digital transmission.

Let N be the number of patterns in the codebook, k the dimension of the vector, and r the (coded) transmission rate in bits/sample. A sampling rate of 8 kHz is assumed throughout this article. These three parameters are related by:

$$r = \frac{\log_2 N}{k}$$

Thus, for the codebook of Fig. 1, $k=6$, $N=64$, and $r=1$ bit/sample. The signal-to-quantization noise ratio (SNR) for VPCM in the high rate region is given by the relation:

$$\mathrm{SNR} = 6\,\frac{\log_2 N}{k} + C_k \qquad (1)$$

where C_k is a constant depending on the dimension k, and SNR and C_k are expressed in dB units. This relation is based on the asymptotic theory when the number of patterns, N, is sufficiently large [6].

According to (1), the SNR for VPCM increases approximately at the rate of $6/k$ dB for each doubling of the codebook size (for each additional bit used to code the entire vector), where k is the dimension of the vector. Note that (1) includes as a special case the well-known formula for scalar (one-dimensional) quantization ($k=1$) [7]. By considering SNR as a function of bit-rate, we see that the same trade-off occurs as in scalar PCM, with a 6 dB increase in SNR per unit increase in rate (bits/sample). However, the advantage of VPCM is that its constant term, C_k, is higher since the *correlation* between components of the vector is now exploited. This, in fact, is the chief virtue of VQ: it exploits, in an optimal manner, correlations within a vector of data values. *The constant term C_k increases with the dimension*, approaching the ultimate rate-distortion limit for a given source. The fact that VQ can, in principle, approach the ultimate performance limit is a consequence of Shannon's coding theorem, which is based on the concept of a block source code. Of more practical interest is the fact that a significant gain over scalar PCM is obtained in low dimensions. For example, from experiments with speech signals, we have found that C_2 is larger than C_1 by more than 3 dB and C_8 is larger than C_1 by more than 7 dB. These observations agree with earlier results reported in [8]. VPCM is clearly an attractive improvement over standard PCM. It appears to offer the potential for robustness to changes in signal statistics including power level, but it needs a more complex implementation than scalar PCM.

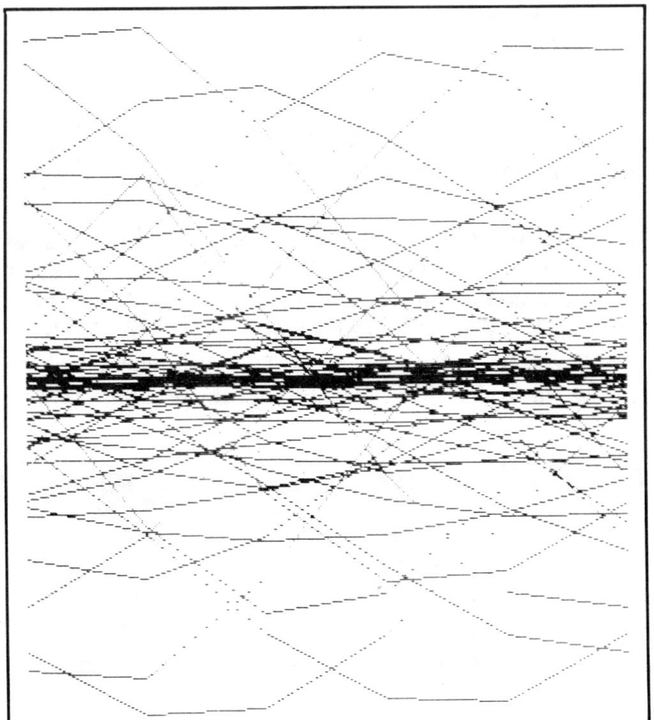

Fig. 1. Superimposed patterns for six-dimensional vectors at rate 1 bit/sample.

The Complexity Issue in Pattern Matching

Implementation complexity in signal processing hardware is typically estimated by the amount of computation or "number crunching" needed per unit of time. Computational complexity is highly processor dependent and is difficult to estimate without consideration of particular architectures. Nevertheless, we can make some preliminary assessments by assuming that multiplications are the most demanding basic operations needed. Although additions can often require equal execution times, the count of multiplications still give a useful order of magnitude estimate of complexity.

Another aspect of implementation complexity is memory. Although the capacity of read-only memories (ROM's) is increasing and their cost decreasing, the codebook storage requirements in VQ can still be very demanding. Thus, in addition to computational complexity, an important consideration is the memory size needed for a particular implementation.

The main obstacle to using VPCM for high dimensional vectors is the exponential growth of encoding complexity with the dimension for a given rate. A vector quantizer of dimension k for a rate of r bits/sample requires a number of computations per sample of the order 2^{kr} and a memory of $k\,2^{kr}$ words if a full search through the codebook is performed for each input vector. (Compare Figs. 2 and 3.)

For these reasons, the direct use of VQ of speech waveforms with optimal codebooks has been limited to dimension-rate products (kr) of 8 or less. For example, at 2 bits/sample and dimension 4, Abut et al. [1] obtained an SNR in the range of 12–14 dB, and at 1 bit/sample and dimension 8, SNR values in the range of 8–10 dB have been obtained. This performance is certainly better than what can be offered by scalar PCM at the same rate. Nevertheless, VPCM is not competitive with the performance achieved by

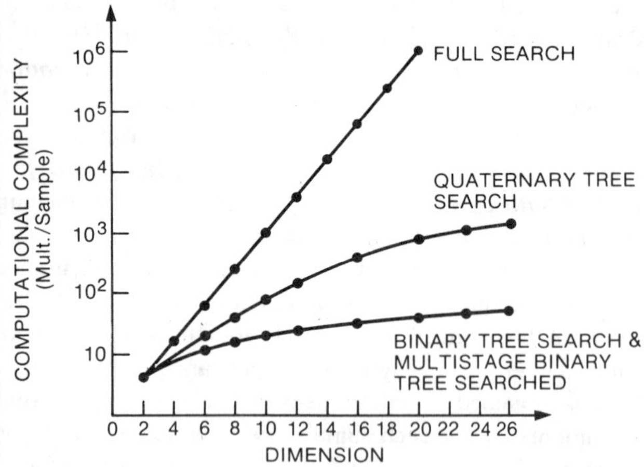

Fig. 2. Computational complexity at rate 1 bit/sample.

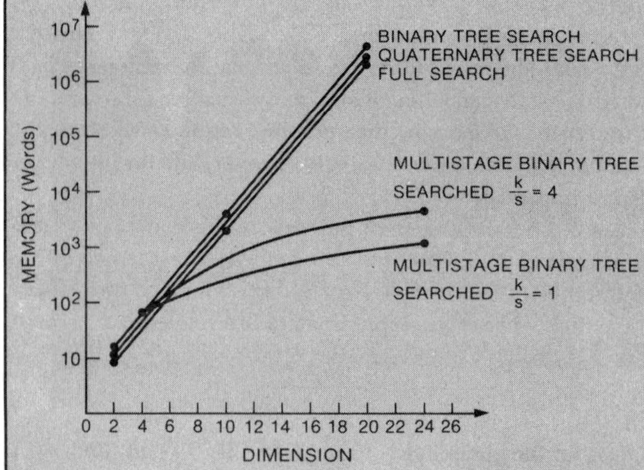

Fig. 3. Memory requirement at rate 1 bit/sample.

more sophisticated scalar speech compression methods such as subband coding, transform coding, or even ADPCM.

Several attempts have been made to circumvent the complexity problem. All are very recent developments and, based on the increasing activity in this area, we may expect to see important advances in the next few years. The approaches being taken fall into two general categories: The first is to use specially structured codebooks which reduce the search and/or memory complexity to make higher dimensional coding feasible. In the second approach, VQ is imbedded as a building block in some coding system which uses other redundancy reducing techniques that have proven effective prior to the advent of vector coding.

Reduced Complexity Vector Quantization Techniques

Several techniques have been proposed to circumvent the complexity problem yet retain the basic VPCM feature that each input vector is separately encoded, independently of prior or subsequent vectors. They generally allow higher dimensionality vector coding for the same complexity, but usually yield a performance inferior to what would be achieved with a direct optimal codebook design for the given dimension and bit rate.

Tree-Structured Codebooks

Tree-structured vector quantizers, first used for LPC [1], greatly reduce the encoding complexity at the expense of a need for more memory and a lower SNR performance than obtained with an optimal "full-search" codebook. The encoding process in a binary tree search is similar to the "successive approximation" method of analog-to-digital conversion which is routinely used in PCM encoders. Each intermediate decision supplies information which determines what test is performed at the next step.

In tree-structured VQ, a large part of the set of stored patterns in the codebook are used only to guide the search. For example, in a binary tree codebook of size 2^{kr}, the search consists of kr binary decisions; only at the last decision stage is the input vector compared with two patterns which are actual candidate codevectors to represent the input vector. The computational complexity for tree-structured codebooks increases only linearly with the dimension. On the other hand, the memory required may be up to twice as much as for standard full-search VQ.

The tree structure is not optimal because of the structural constraints imposed on the location of the pattern vectors in the final stage of the tree. Gray and Abut [8] compared the performance of standard and tree-structured VQ for speech waveforms and found a degradation of about 1 dB in SNR as a result of the tree structure. Increased memory is a major limitation. For example, in dimension 6 at 2 bits/sample, the memory is 48 000 words; in dimension 7 at the same rate, the memory increases to 224 000 words (assuming a binary tree in both cases).

Multistage Vector Quantizers

Multistage vector quantization, first proposed for quantization of LPC coefficients by Juang [9], is a technique that reduces memory as well as encoding complexity. Here, the encoding error of a vector quantizer is formed by taking the difference between the original and quantized vectors and then feeding it into a second-stage vector quantizer. This is illustrated in Fig. 4. Note that the actual binary word that must be transmitted consists of the concatenation of the binary words needed to identify the correct pattern in each

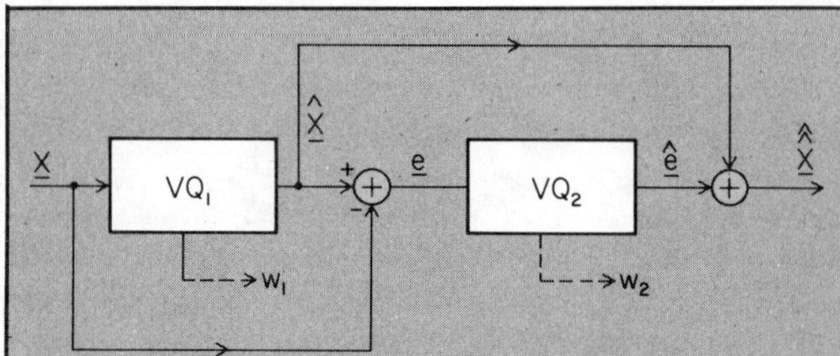

Fig. 4. Two-stage vector quantizer.

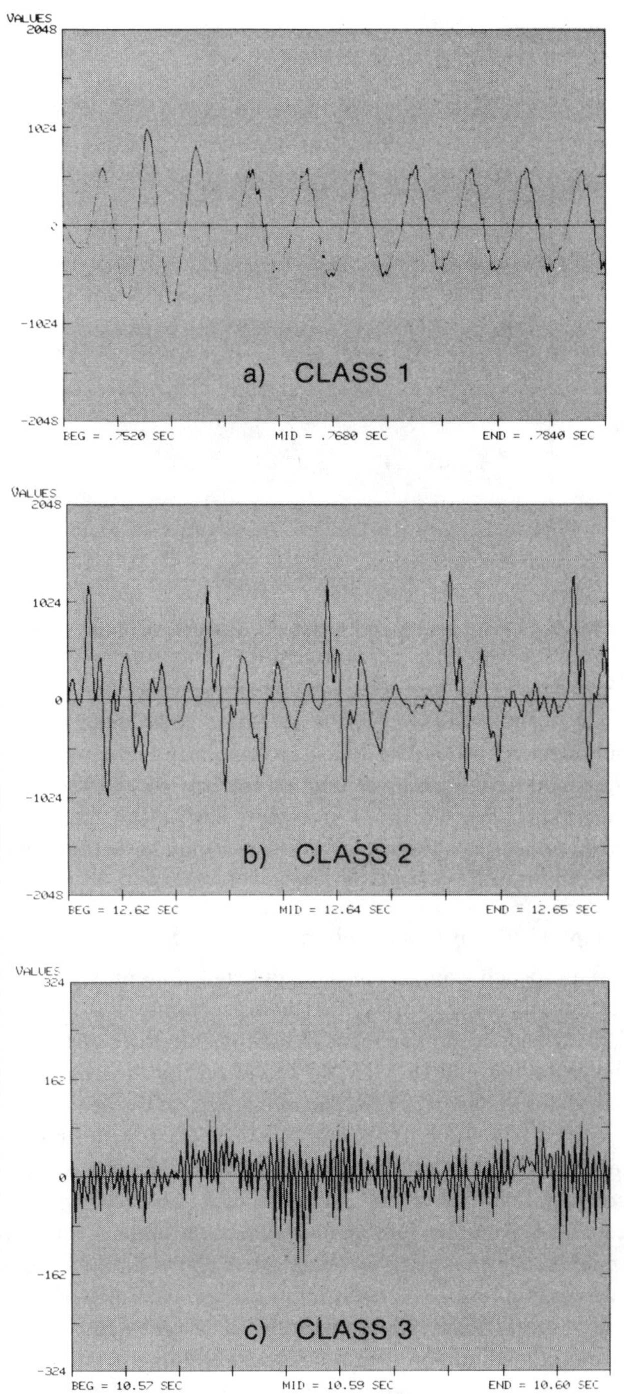

Fig. 5. Typical examples of speech waveforms for the three classes.

codebook. The process can be repeated by feeding the second stage error into a third stage vector quantizer, and so on. Note that this technique is equivalent to a single stage VQ whose codevectors are specially structured. For example, a two-stage VQ with N_1 and N_2 patterns in the first and second stages, respectively, is equivalent to a particular single stage VQ with a codebook of N_1N_2 patterns.

With this approach, both the complexity and memory can be greatly reduced in comparison with single stage vector quantization at the same rate. If s stages are used and the vector dimension is increased in proportion to the number of stages, thus keeping k/s constant, the complexity only increases quadratically with the dimension (that is, increasing as k^2) rather than exponentially as in one-stage VQ. The encoding complexity can be further reduced by using tree-structured codebooks for each stage.

Figures 2 and 3 present, respectively, the computational complexity and required memory versus dimension for different versions of suboptimal vector quantization and for one-stage optimal vector quantization, at a rate of 1 bit/sample. Note that the combination of multistage and binary tree searched coding requires the same encoding complexity but much less memory than the binary tree searched coder without use of multiple stages. One can see from these figures that binary-structured multistage VQ is the only procedure which achieves affordable complexity at higher dimensions.

We have compared the use of standard full-search vector quantizers with multistage vector quantizers for coding speech waveforms. The results show that performance may be improved by increasing the dimension for a given rate while holding the ratio k/s constant and hence keeping the complexity manageable. For example, at a rate of 1 bit/sample the SNR increases from 9.6 dB for standard full-search VQ with dimension 4 to 10.7 dB for multistage VQ with dimension 12.

The performance of a full-search vector quantizer can be improved via the multistage technique at the expense of only a moderate increase in complexity. However, the attainable improvement is quite limited since the increase in SNR with dimension saturates very quickly. This effect is due to a combination of two opposing forces. On the one hand, the performance of multistage VQ tends to decrease as more stages are used, due to greater structural constraints on the equivalent single-stage codebook. On the other hand, increasing the dimensionality tends to improve performance since more correlation in the waveform is being exploited to reduce redundancy.

In summary, tree-structured and multistage codebooks make it possible to implement higher dimensional vector quantizers at relatively low complexity. However, because of the constraints imposed on the codebook structures, they achieve only very limited improvements in performance for a given level of complexity.

Gain-Shape Vector Quantizers

Another way to reduce both memory and encoding complexity in waveform coding is to normalize the energy level ("gain") of a vector prior to encoding and to separately encode the gain with a scalar quantizer. The energy-normalized vector is called the "shape." This technique was studied by Sabin and Gray [5], who found an optimal way to jointly design the gain and shape quantizers. The effectiveness of gain-shape VQ depends on the degree to which the gain of a randomly selected speech vector is statistically independent of its shape. With this technique, higher dimensional vectors can be used with tolerable complexity but the optimality is compromised. For example, at a rate of 1 bit/sample (8 kb/s) the highest "affordable" dimension for standard VQ is $k = 8$ and the SNR obtained is 9.7 dB. By using gain-shape VQ with dimension 12, an improvement of 0.7 dB was obtained at the same level of complexity. Again, only a very limited improvement is achieved. The complexity problem is a tough nut to crack!

Fast Nearest-Neighbor Search Algorithms

In standard full-search VQ, the optimal encoding procedure for a given codebook requires a search process through the codebook to determine the nearest codevector to the given input vector. Since the codebook is designed by a clustering algorithm such as the LBG algorithm [10], the codevectors are at irregularly located points in k-dimensional Euclidean space. It has generally been assumed that an *exhaustive* search through such an unstructured codebook is necessary to find the best matching pattern or so-called "nearest neighbor" to a given input vector. The resulting exponential growth in computational complexity with the dimension for a fixed rate is the main obstacle to approaching the Shannon limits in source coding. For a fixed rate, this implies that computation grows exponentially with k, thus preventing a close approach to Shannon's source coding bound.

In recent work, we have found a way to greatly reduce encoding complexity without degrading performance by generating a tree-structured search algorithm specifically tailored for a given optimal codebook. This gives a successive approximation method that will always lead to the best matching pattern, the so-called "nearest neighbor" to the input vector [11]. The approach, first introduced in [12], is based on an intensive one-time precomputation which results in an efficient nonuniform tree structure. This allows simple binary tree searching that eliminates large groups of codewords in each step and rapidly finds the nearest-neighbor codeword. Preliminary computational results confirm the efficiency of this approach. It is important to note that this approach does not compromise the optimality of the original codebook. An optimal "full-search" codebook is first designed and then the precomputation is used to find an efficient way to speed up the search process in performing the encoding operation. In contrast, the tree-structured codebook designs previously studied do, in fact, lead to suboptimal codebooks [1].

This work does not imply that we can now achieve Shannon limits in practice. A major remaining obstacle is the exponential growth in memory (with dimensionality) required for codebook storage. Some additional memory is also needed to store the binary decision tree for the encoding algorithm.

Imbedded Vector Quantization Methods

Just as other waveform coding systems such as DPCM use one dimensional quantizers as building blocks, so also can we use vector quantization in a coding system that exploits other methods for removing redundancy. The idea of using a vector quantizer as a building block in waveform coding systems may lead to new and powerful waveform coding devices for rates such as 16 kb/s and 9.6 kb/s with important applications to digital voice systems including secure voice and voice-mail.

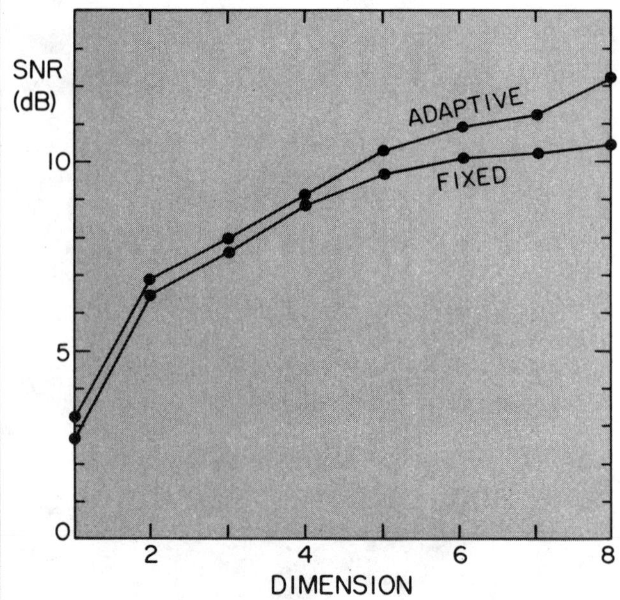

Fig. 6. Performance of fixed vs. adaptive vector quantization.

Adaptive Vector Quantization

The speech waveform is generally not stationary but local stationarity may be assumed within "frames" of duration 10–20 ms. Ideally, one would like to adapt the codebook to the changing statistics of the waveform by "learning" the particular character of the incoming speech waveform, and updating the patterns in the codebook to suit the current talker. Recently, one study was reported using this extremely complex approach for very-low-rate analysis/synthesis coding [13]. The problems related to use of adaptive codebooks for waveform coding are the large amount of side information needed to transmit the new codebook vectors to the receiver, and the complexity and processing delays related to estimation of the new codevectors.

We have experimented with a simpler adaptive scheme using a speech classifier which computes some simple statistics for each frame of speech and selects one of K classes to characterize the frame. The vectors that make up the frame are then coded with one of K codebooks corresponding to the selected classification. The class identification is transmitted to the receiver as side information once for each frame of speech. Hence, for reasonable frame lengths and a moderate number of classes, the increase in bit rate due to the side information is negligible.

We used a specific three-way classifier, described in [4], which computes the energy and the unit-lag autocorrelation and selects the appropriate class based on these two values. In the first class, the waveform is highly correlated, has high

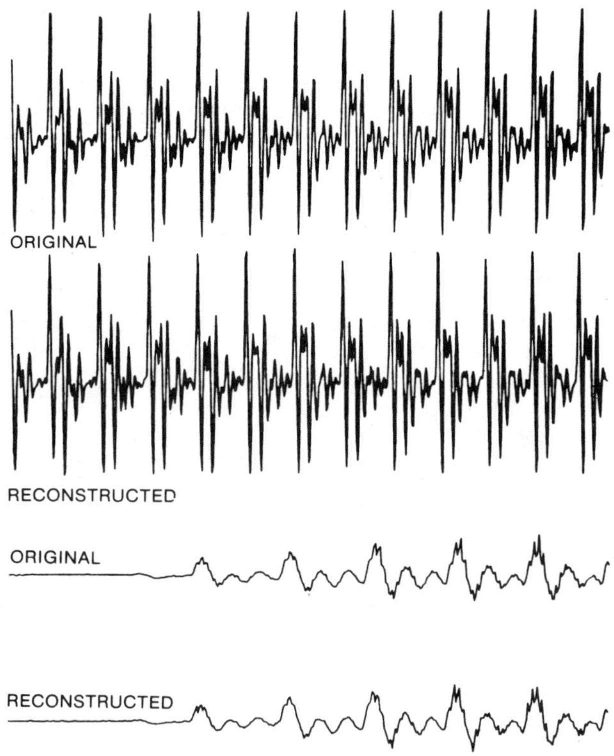

Fig. 7. Examples of original and reconstructed waveforms for adaptive vector predictive coding at 2 bits/sample.

energy, and generally contains voiced speech; it may also contain certain types of unvoiced sounds such as plosives. The second class consists of moderately correlated and high-energy segments. Voiced speech with a strong first formant is a good example of this class. Finally, the third class is made up of low-correlation or low-energy frames and generally consists of unvoiced sounds. Examples of waveforms for the three classes are shown in Fig. 5.

Some useful improvements in SNR (in one case as high as 2.5 dB) were obtained with a rate increase of only 0.03 bit/sample by using this form of adaptive VQ rather than standard VQ with the same dimension and rate. Figure 6 presents the performance of this simple adaptive VQ versus dimension at a rate of 1 bit/sample compared with standard VQ where a fixed codebook is used for every vector.

Adaptive Vector Predictive Coding

Another imbedded approach, AVPC, begins with a generalized DPCM coder by using vector prediction to remove the redundant information from each new input vector, and VQ to encode the unpredictable residual vector [4,14]. To improve performance, the system is made adaptive by classifying input frames of speech into categories and using an appropriate predictor and quantizer for each class. Examples of original speech segments and the reconstructed waveforms after AVPC coding at 2 kb/s are shown in Fig. 7. SNR performance in the range of 18–20 dB has been obtained with this technique for dimension 5 and rate 2 bits/sample (16 kb/s). These SNR values are substantially higher than those achieved by other VQ approaches and are competitive with the best known results for waveform speech coding at a similar level of complexity.

Subband Coding Using Vector Quantization

Subband coding has proven a powerful technique for medium rate speech coding. Its effectiveness is based on a strategy of allocating bits to the different frequency bands according to perceptual criteria. Performance can be further improved by adaptive bit allocation based on the instantaneous power in the subbands [15]. In one recent study, VQ was used to code the LPC spectral information allowing an efficient description of the side information for adaptive bit allocation [1]. Another recent development is a fully vector-quantized subband coder where the side information is generated by vector coding the short-term subband power levels, and VQ is used on each subband with codebooks of appropriate size allocated by the side information [17]. Vector quantization was found to cut in half the bit rate needed for side information, and to distinctly improve perceptual quality by vector coding the subband signals without increasing the bit rate of the main information. With the addition of Time Domain Harmonic Scaling (TDHS), a technique used effectively in reducing the rate of scalar quantized subband coding [18], this approach may lead to improved coders at 9.6 kb/s.

Conclusion

Vector quantization is an exciting new development that is still in its infancy as a coding technique for speech waveforms. We expect to see an increased research effort in the next few years in speech waveform coding based on VQ. Hardware development of competitive speech codecs at 16 kb/s employing this fundamental and versatile technique should not be very far away.

References

[1] A. Buzo, A. H. Gray, Jr., R. M. Gray, and J. D. Markel, "Speech coding based upon vector quantization," *IEEE Trans. Acoust., Speech, and Sig. Process.*, ASSP-28, no. 5, pp. 562–574, Oct. 1980.
[2] S. Roucos, R. Schwartz, and J. Makhoul, "Vector quantization for very-low-rate coding of speech," *Conf. Rec.*, 1982 IEEE Global Communications Conf., pp. E6.2.1-E6.2.5, Miami, FL, Nov. 29-Dec. 2, 1982.
[3] H. Abut, R. M. Gray, and G. Rebolledo, "Vector quantization of speech and speech-like waveforms," *IEEE Trans. Acoust., Speech, and Sig. Process.*, ASSP-30, pp. 423-436, June 1982.
[4] V. Cuperman and A. Gersho, "Adaptive differential vector coding of speech," *Conf. Rec.*, 1982 IEEE Global Communications Conf., pp. E6.6.1-E6.6.5, Miami, FL, Nov. 29-Dec. 2, 1982.
[5] M. J. Sabin and R. M. Gray, "Product code vector quantizers for speech waveform coding," *Conf. Rec.*, 1982 IEEE Global Communications Conf., pp. E6.5.1-E6.5.5, Miami, FL, Nov. 29-Dec. 2, 1982.
[6] A. Gersho, "Asymptotically optimal block quantization," *IEEE Trans. Inform. Theory*, IT-25, no. 4, pp. 373–380, July 1979.
[7] A. Gersho, "Quantization," *IEEE Communications Magazine*, vol. 15, pp. 16–29, Sept. 1977. (Reprinted as "Principles of quantization," *IEEE Trans. Circuits Syst.*, CAS-25, no. 7, July 1978.)
[8] R. M. Gray and H. Abut, "Full-search and tree-searched vector quant. of speech waveforms," *Proc. Intl. Conf. on Acoust., Speech, and Sig. Process.*, pp. 593-596, Paris, 1982.

[9] B. H. Juang and A. H. Gray, Jr., "Multiple stage vector quantization for speech coding," *Proc. Intl. Conf. on Acoust., Speech, and Sig. Process.*, pp. 597–600, Paris, 1982.

[10] Y. Linde, A. Buzo, and R. M. Gray, "An algorithm for vector quantizer design," *IEEE Trans. Commun.*, COM-28, pp. 84–95, Jan. 1980.

[11] A. Gersho and D.-Y. Cheng, "Fast nearest-neighbor search for Euclidean codes," Abstract presented at IEEE Intl. Symp. on Inform. Theory, Ste. Jovite, Canada, Sept. 1983.

[12] A Gersho, "On the structure of vector quantizers," *IEEE Trans. Inform. Theory*, IT-28, no. 2, pp. 157–166, March 1982.

[13] D. Paul, "A 500–800 b/s adaptive vector quantization vocoder using a perceptually motivated distance measure," *Conf. Rec.*, 1982 IEEE Global Communications Conf., pp. E6.3.1–E6.3.5, Miami, FL, Nov. 29–Dec. 2, 1982.

[14] V. Cuperman and A. Gersho, "Vector predictive coding of speech at 16 kb/s," to be published.

[15] T. Ramstad, "Subband coder with a simple adaptive bit-allocation algorithm: a possible candidate for digital mobile telephony?," *Proc. Int. Conf. on Acoust., Speech, and Sig. Process.*, pp. 203–207, Paris, May 1982.

[16] C. D. Heron, R. E. Crochiere, and R. V. Cox, "A 32-band subband/transform coder incorporating vector quantization for dynamic bit allocataion," *Proc. Intl. Conf. on Acoust., Speech, and Sig. Process.*, pp. 1276–1279, Boston, MA, April 1983.

[17] A. Gersho, T. Ramstad, and I. Versvik, "Fully vector-quantized subband coding with adaptive codebook allocation," to be published.

[18] R. E. Crochiere, R. V. Cox, J. D. Johnston, and L. A. Seltzer, "A 9.6-kb/s DSP speech coder," *Bell Syst. Tech. J.*, vol. 61, no. 9, November 1982.

[19] R. M. Gray, "Vector quantization," to be published in *IEEE ASSP Mag.*, 1984.

Vector Quantization

Robert M. Gray

A vector quantizer is a system for mapping a sequence of continuous or discrete vectors into a digital sequence suitable for communication over or storage in a digital channel. The goal of such a system is data compression: to reduce the bit rate so as to minimize communication channel capacity or digital storage memory requirements while maintaining the necessary fidelity of the data. The mapping for each vector may or may not have memory in the sense of depending on past actions of the coder, just as in well established scalar techniques such as PCM, which has no memory, and predictive quantization, which does. Even though information theory implies that one can always obtain better performance by coding vectors instead of scalars, scalar quantizers have remained by far the most common data compression system because of their simplicity and good performance when the communication rate is sufficiently large. In addition, relatively few design techniques have existed for vector quantizers.

During the past few years several design algorithms have been developed for a variety of vector quantizers and the performance of these codes has been studied for speech waveforms, speech linear predictive parameter vectors, images, and several simulated random processes. It is the purpose of this article to survey some of these design techniques and their applications.

DATA compression is the conversion of a stream of analog or very high rate discrete data into a stream of relatively low rate data for communication over a digital communication link or storage in a digital memory. As digital communication and secure communication have become increasingly important, the theory and practice of data compression have received increased attention. While it is true that in many systems bandwidth is relatively inexpensive, e.g., fiber optic and cable TV links, in most systems the growing amount of information that users wish to communicate or store necessitates some form of compression for efficient, secure, and reliable use of the communication or storage medium.

A prime example arises with image data, where simple schemes require bit rates too large for many communication links or storage devices. Another example where compression is required results from the fact that if speech is digitized using a simple PCM system consisting of a sampler followed by scalar quantization, the resulting signal will no longer have a small enough bandwidth to fit on ordinary telephone channels. That is, digitization (which may be desirable for security or reliability) causes bandwidth expansion. Hence data compression will be required if the original communication channel is to be used.

The author is with the Information Systems Laboratory, Department of Electrical Engineering, Stanford University, Stanford, CA 94305.

The two examples of image compression and speech compression or, as they are often called, image coding and speech coding, are probably the currently most important applications of data compression. They are also among the most interesting for study because experience has shown that both types of data exhibit sufficient structure to permit considerable compression with sufficiently sophisticated codes.

Such conversion of relatively high rate data to lower rate data virtually always entails a loss of fidelity or an increase in distortion. Hence a fundamental goal of data compression is to obtain the best possible fidelity for the given rate or, equivalently, to minimize the rate required for a given fidelity. If a system has a sufficiently high rate constraint, then good fidelity is relatively easy to achieve and techniques such as PCM, transform coding, predictive coding, and adaptive versions of these techniques have become quite popular because of their simplicity and good performance [1, 2, 3]. All of these techniques share a fundamental property: The actual quantization or coding or conversion of continuous quantities into discrete quantities is done on scalars, e.g., on individual real-valued samples of waveforms or pixels of images. PCM does this in a memoryless fashion; that is, each successive input is encoded using a rule that does not depend on any past inputs or outputs of the encoder. Transform coding does it by first taking block transforms of a vector and then scalar coding the coordinates of the transformed vector. Predictive coding does it by quantizing an error term formed as the difference between the new sample and a prediction of the new sample based on past coded outputs.

A fundamental result of Shannon's rate-distortion theory, the branch of information theory devoted to data compression, is that better performance can always be achieved by coding vectors instead of scalars, even if the data source is memoryless, e.g., consists of a sequence of independent random variables, or if the data compression system can have memory, i.e., the action of an encoder at each time is permitted to depend on past encoder inputs or outputs [4, 5, 6, 7, 8]. While some traditional compression schemes such as transform coding operate on vectors and achieve significant improvement over PCM, the quantization is still accomplished on scalars and hence these systems are, in a Shannon sense, inherently suboptimal: better performance is always achievable *in theory* by coding vectors instead of scalars, even if the scalars have been produced by preprocessing the original input data so as to make them uncorrelated or independent!

Reprinted from *IEEE ASSP Mag.*, vol. 1, pp. 4–29, April 1984.

This theory had a limited impact on actual system design because 1) the Shannon theory does not provide constructive design techniques for vector coders, and 2) traditional scalar coders often yield satisfactory performance with enough adaptation and fine tuning. As a result, few design techniques for vector quantizers were considered in the literature prior to the late 1970's when it was found that a simple algorithm of Lloyd [9] for the iterative design of scalar quantization or PCM systems extended in a straightforward way to the design of memoryless vector quantizers, that is, of vector quantizers which encode successive input vectors in a manner not depending on previous encoder input vectors or their coded outputs. Variations of the basic algorithm have since proved useful for the design of vector quantizers with and without memory for a variety of data sources including speech waveforms, speech parameter vectors, images, and several random process models, the latter being useful for gauging the performance of the resulting codes with the optimal performance bounds of information theory.

This paper is intended as a survey of the basic design algorithm and many of its variations and applications. We begin with the simplest example of a memoryless vector quantizer, a vector generalization of PCM. For convenience we use the shorthand VQ for both vector quantization and vector quantizer. Necessary properties of optimal quantizers are described and an algorithm given which uses these properties to iteratively improve a code. For concreteness, we focus on two examples of distortion measures: the ubiquitous mean-squared error and the Itakura-Saito distortion. The first example, which is popular in waveform coding applications, provides a geometric flavor to the development; the second example, which is useful in voice coding applications, helps to demonstrate the generality and power of the technique.

Next, various techniques are described for designing the initial codes required by the algorithm. These techniques also indicate some useful structure that can be imposed on vector quantizers to make them more implementable. Several variations of the basic VQ are described which permit reduced complexity or memory or both at the expense of a hopefully tolerable loss of performance. These include tree-searched codes, product codes, and multistep codes.

We then turn from memoryless vector quantizers to those with memory: feedback vector quantizers such as vector predictive quantizers and finite-state vector quantizers. These codes are not yet well understood, but they possess a structure highly suited to VLSI implementation and initial studies suggest that they offer significant performance gains.

For comparison, we also briefly describe trellis encoding systems or "lookahead" or "delayed decision" or "multipath search" codes which use the same decoder as a feedback vector quantizer but which permit the encoder to base its decision on a longer input data sequence.

A final general code structure is described which uses vector quantization to adapt a waveform coder, which may be another VQ.

We next present a variety of simulation results describing the performance of various VQ systems on various data sources. Examples of all of the above VQ varieties are tested for waveform coding applications on two common data sources: a Gauss Markov source and real sampled speech. One bit per sample coders for these sources are compared on the basis of performance, memory requirements, and computational complexity. Both memoryless and simple feedback vector quantizers are studied for voice coding applications at a rate of 0.062 bits/sample and less and for image coding at a rate of 0.5 bit per sample. One example is given of a simple adaptive predictive vector quantizer for speech waveform coding.

By studying a variety of coding systems on common data sources, the results yield some general comparisons and trends among the various vector quantization techniques. The reader should, however, keep two caveats in mind when interpreting such quantitative results: First, the emphasis here is on low bit rate systems, e.g., speech coders using 1 bit per sample or less and image coders ½ bit per pixel. Comparisons favoring certain systems at such low rates may not be valid for the same systems at higher rates. Second, the numbers reported here are intended to provide comparisons for different systems used on common data sources; they can be compared with other numbers reported in the literature only with great care: the input data and the system design parameters such as sampling rate and pre- or post-filtering may be quite different.

Applications of vector quantization to real data sources such as sampled speech waveforms and images are still young and the algorithms do not yet incorporate the sophisticated "bells and whistles" of many well-established scalar quantization schemes. The preliminary experiments described here, using fairly simple vector quantizers with and without memory, demonstrate that the general approach holds considerable promise for some applications. For example, good quality vocoding systems using VQ and the Itakura-Saito distortion have been developed at 800 bits per second, a significant reduction in the bit rate previously required for comparable quality [10]. While the compression achieved so far in waveform coding and image coding applications using the squared-error distortion has not yet been as significant, we believe that it has yielded comparable or better performance at low rates than traditional scalar schemes of greater complexity. The quality of the ½ bit per pixel images shown here is promising given the simplicity of the coding scheme used.

We attempt to use the minimum of mathematics and a maximum of English in the presentation so as to focus on the intuitive ideas underlying the design and operation of vector quantizers. The detailed descriptions of the various algorithms can be found in the cited references. The reader is also referred to a recent tutorial by Gersho and Cuperman [11] which presents a brief overview of VQ applied to speech waveform coding.

MEMORYLESS VECTOR QUANTIZERS

In this section we introduce the basic definition of memoryless vector quantizers, their properties, and an algorithm for their design.

Quantization

Mathematically, a k-dimensional memoryless vector quantizer or, simply, a VQ (without modifying adjectives) consists of two mappings: an encoder γ which assigns to each input vector $\mathbf{x} = (x_0, x_1, \cdots, x_{k-1})$ a channel symbol $\gamma(\mathbf{x})$ in some channel symbol set $\mathbf{M}$, and a decoder β assigning to each channel symbol v in $\mathbf{M}$ a value in a reproduction alphabet $\hat{\mathbf{A}}$. The channel symbol set is often assumed to be a space of binary vectors for convenience, e.g., $\mathbf{M}$ may be the set of all 2^R binary R-dimensional vectors. The reproduction alphabet may or may not be the same as the input vector space; in particular, it may consist of real vectors of a different dimension.

If $\mathbf{M}$ has M elements, then the quantity $R = \log_2 M$ is called the *rate* of the quantizer in bits per vector and $r = R/k$ is the rate in bits per symbol or, when the input is a sampled waveform, bits per sample.

The application of a quantizer to data compression is depicted in the standard Fig. 1. The input data vectors might be consecutive samples of a waveform, consecutive parameter vectors in a voice coding system, or consecutive rasters or subrasters in an image coding system. For integer values of R it is useful to think of the channel symbols, the encoded input vectors, as binary R-dimensional vectors. As is commonly done in information and communication theory, we assume that the channel is noiseless, that is, that $U_n = \hat{U}_n$. While real channels are rarely noiseless, the joint source and channel coding theorem of information theory implies that a good data compression system designed for a noiseless channel can be combined with a good error correction coding system for a noisy channel in order to produce a complete system. In other words, the assumption of a noiseless channel is made simply to focus on the problem of data compression system design and not to reflect any practical model.

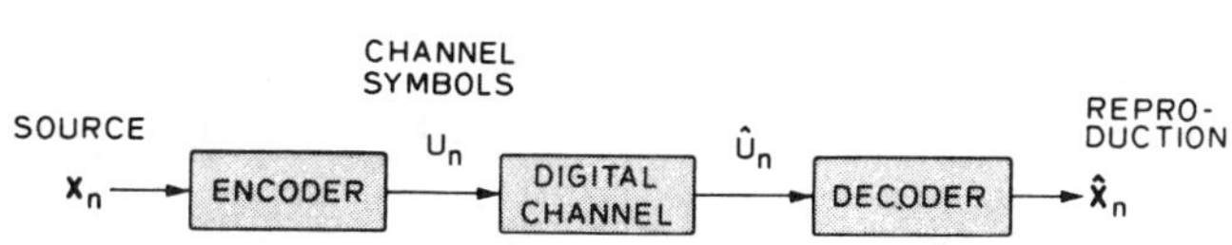

Figure 1. Data Compression System. The data or information source $\{X_n; n = 0, 1, \ldots\}$ is a sequence of random vectors. The encoder produces a sequence of channel symbols $\{U_n; n = 0, 1, 2, \ldots\}$. The sequence $\{\hat{U}_n; n = 0, 1, 2, \ldots\}$ is delivered to the receiver by the digital channel. The decoder then maps this sequence into the final reproduction sequence of vectors $\{\hat{X}_n; n = 0, 1, 2, \ldots\}$.

Observe that unlike scalar quantization, general VQ permits fractional rates in bits per sample. For example, scalar PCM must have a bit rate of at least 1 bit per sample while a k dimensional VQ can have a bit rate of only $1/k$ bits per sample by having only a single binary channel symbol for k-dimensional input vectors.

The goal of such a quantization system is to produce the "best" possible reproduction sequence for a given rate R. To quantify this idea, to define the performance of a quantizer, and to complete the definition of a quantizer, we require the idea of a distortion measure.

Distortion

A distortion measure d is an assignment of a cost $d(\mathbf{x}, \hat{\mathbf{x}})$ of reproducing any input vector $\mathbf{x}$ as a reproduction vector $\hat{\mathbf{x}}$. Given such a distortion measure, we can quantify the performance of a system by an average distortion $Ed(\mathbf{X}, \hat{\mathbf{X}})$ between the input and the final reproduction: A system will be good if it yields a small average distortion. In practice, the important average is the long term sample average or time average

$$\lim_{n\to\infty} \frac{1}{n} \sum_{i=0}^{n-1} d(\mathbf{X}_i, \hat{\mathbf{X}}_i) \tag{1}$$

provided, of course, that the limit makes sense. If the vector process is stationary and ergodic, then, with probability one, the limit exists and equals an expectation $E(d(\mathbf{X}, \hat{\mathbf{X}}))$. For the moment we will assume that such conditions are met and that such long term sample averages are given by expectations. Later remarks will focus on the general assumptions required and their implications for practice.

Ideally a distortion measure should be tractable to permit analysis, computable so that it can be evaluated in real time and used in minimum distortion systems, and subjectively meaningful so that large or small quantitative distortion measures correlate with bad and good subjective quality. Here we do not consider the difficult and controversial issues of selecting a distortion measure; we assume that one has been selected and consider means of designing systems which yield small average distortion. For simplicity and to ease exposition, we focus on two important specific examples:

(1) *The squared error distortion measure:* Here the input and reproduction spaces are k-dimensional Euclidean space

$$d(\mathbf{x}, \hat{\mathbf{x}}) = \|\mathbf{x} - \hat{\mathbf{x}}\|^2 = \sum_{i=0}^{k-1} (x_i - \hat{x}_i)^2,$$

the square of the Euclidean distance between the vectors. This is the simplest distortion measure and the most common for waveform coding. While not subjectively meaningful in many cases, generalizations permitting input-dependent weightings have proved useful and only slightly more complicated. For the squared-error distortion it is common practice to measure the performance of a system by the signal-to-noise ratio (or signal-to-quantization-noise ratio)

$$SNR = 10 \log_{10} \frac{E(\|\mathbf{X}\|^2)}{E[d(\mathbf{X},\hat{\mathbf{X}})]}.$$

This corresponds to normalizing the average distortion by the average energy and plotting it on a logarithmic scale: Large (small) SNR corresponds to small (large) average distortion.

(2) *The (modified) Itakura-Saito distortion:* This distortion measure is useful in voice coding applications where the receiver is sent a linear model of the underlying voice production process. The distortion measure is based on the "error matching measure" developed in the pioneering work of Itakura and Saito on the PARCOR or LPC approach to voice coding [12]. More generally, this distortion measure is a special case of a minimum relative entropy or discrimination measure; VQ using such distortion measures can be viewed as an application of the minimum relative entropy pattern classification technique introduced by Kullback [13] as an application of information theory to statistical pattern classification. (See also [14,15].)

We here introduce a minimum of notation to present a definition of the Itakura-Saito distortion measure. Details and generalizations may be found in [16,17,14,15]. Here the input vector can again be considered as a collection of consecutive waveform samples. Now, however, the output vectors have the form $\hat{\mathbf{x}} = (\alpha, a_1, a_2, \cdots, a_p)$, where α is a positive gain or residual energy term and where the a_i with $a_0 = 1$ are inverse filter coefficients in the sense that if

$$A(z) = \sum_{i=0}^{p} a_i z^{-i}$$

then the all-pole filter with z-transform $1/A(z)$ is a stable filter. Here the reproduction vectors may be thought of as all-pole models for synthesizing the reproduction at the receiver using a locally generated noise or periodic source, in other words, as the filter portion of a linear predictive coding (LPC) model in a vocoding (voice coding) system. The Itakura-Saito distortion between the input vector and the model can be defined in the time domain as

$$d(\mathbf{x}, \hat{\mathbf{x}}) = \frac{\mathbf{a}^t \mathbf{R}(\mathbf{x}) \mathbf{a}}{\alpha} - \ln \frac{\alpha_p(\mathbf{x})}{\alpha} - 1,$$

where $\mathbf{a}^t = (1, a_1, \cdots, a_p)$, $\mathbf{R}(\mathbf{x})$ is the $(p+1) \times (p+1)$ sample autocorrelation matrix of the input vector $\mathbf{x}$, and where $\alpha_p(\mathbf{x})$ is an input gain (residual energy) term defined as the minimum value of $\mathbf{b}^t\mathbf{R}(\mathbf{x})\mathbf{b}$, where the minimum is taken over all vectors $\mathbf{b}$ with first component equal to 1. There are many equivalent forms of the distortion measure, some useful for theory and some for computation. Frequency domain forms show that minimizing the above distortion can be interpreted as trying to match the sample spectrum of the input vector to the power spectral density of the linear all-pole model formed by driving the filter with z-transform $1/A(z)$ by white noise with constant power spectral density $\sqrt{\alpha}$.

The above formula for the distortion is one of the simplest, yet it demonstrates that the distortion measure is indeed complicated—it is not a simple function of an error vector, it is not symmetric in its input and output arguments, and it is not a metric or distance. Because of the intimate connection of this distortion measure with LPC vocoding techniques, we will refer to VQ's designed using this distortion measure as LPC VQ's.

Average distortion

As the average distortion quantifies the performance of a system and since we will be trying to minimize this quantity using good codes, we pause to consider what the average means in theory and in practice.

As previously noted, in practice it is the long term sample average of (1) that we actually measure and which we would like to be small. If the process is stationary and ergodic, then this limiting time average is the same as the mathematical expectation. The mathematical expectation is useful for developing information theoretic performance bounds, but it is often impossible to calculate in practice because the required probability distributions are not known, e.g., there are no noncontroversial generally accepted accurate probability distributions for real speech and image data. Hence a pragmatic approach to system design is to take long sequences of training data, estimate the "true" but unknown expected distortion by the sample average, and attempt to design a code that minimizes the sample average distortion for the training sequence. If the input source is indeed stationary and ergodic, the resulting sample average should be nearly the expected value and the same code used on future data should yield approximately the same averages [18].

The above motivates a training sequence based design for stationary and ergodic data sources. In fact, even if the "true" probability distributions are known as in the case of a Gauss Markov source, the training sequence approach reduces to a standard Monte Carlo approach.

An immediate objection to the above approach, however, is whether or not it makes sense for real sources which may be neither stationary nor ergodic. The answer is an emphatic "yes" in the following sense: The desired property is that if we design a code based on a *sufficiently long* training sequence and then use the code on future data produced by the same source, then the performance of the code on the new data should be roughly that achieved on the training data. The theoretical issue is to provide conditions under which this statement can be made rigorous. For reasonable distortion measures, a sufficient condition for this to be true for memoryless VQ design is that the source be asymptotically mean stationary, it need not be either stationary nor ergodic [19,20,21,22,23]. Asymptotically mean stationary sources include all stationary sources, block (or cyclo) stationary sources, and asymptotically stationary sources. Processes such as speech which exhibit distinct short term and long term stationarity properties are well modeled by asymptotically mean stationary sources [21].

The key point here is that the general design approach using long training sequences does not require either ergodicity nor stationarity to have a solid mathematical foundation. In fact, the mathematics suggest the following pragmatic approach: Try to design a code which minimizes the sample average distortion for a very long training sequence. Then use the code on test sequences produced by the same source, but not in the training sequence. If the performance is reasonably close to the design values, then one can have a certain amount of confidence that the code will continue to yield roughly the same performance in the future. If the training and test performance are significantly different, then probably the training sequence is not sufficiently long. In other words, do not try to prove mathematically that a source is asymptotically mean stationary, instead try to design codes for it and then see if they work on new data.

Henceforth for brevity we will write expectations with the assumption that they are to be interpreted as shorthand for long term sample averages. (A sample average $L^{-1}\sum_{i=0}^{L-1} d(\mathbf{X}_i, \hat{\mathbf{X}}_i)$ is, in fact, an expectation with respect to the sample distribution which assigns a probability of $1/L$ to each vector in the training sequence.)

Properties of optimal quantizers

A VQ is optimal if it minimizes an average distortion $Ed\{\mathbf{X}, \beta[\gamma(\mathbf{X})]\}$. Two necessary conditions for a VQ to be optimal follow easily using the same logic as in Lloyd's [9] classical development for optimal PCM with a mean-squared error distortion measure. The following definition is useful for stating these properties: The collection of possible reproduction vectors $\mathbf{C} = \{\text{all } \mathbf{y} : \mathbf{y} = \beta(v), \text{ some } v \text{ in } \mathbf{M}\}$ is called the *reproduction codebook* or, simply, *codebook* of the quantizer and its members called *codewords* (or templates). The encoder knows the structure of the decoder and hence all of the possible final output codewords.

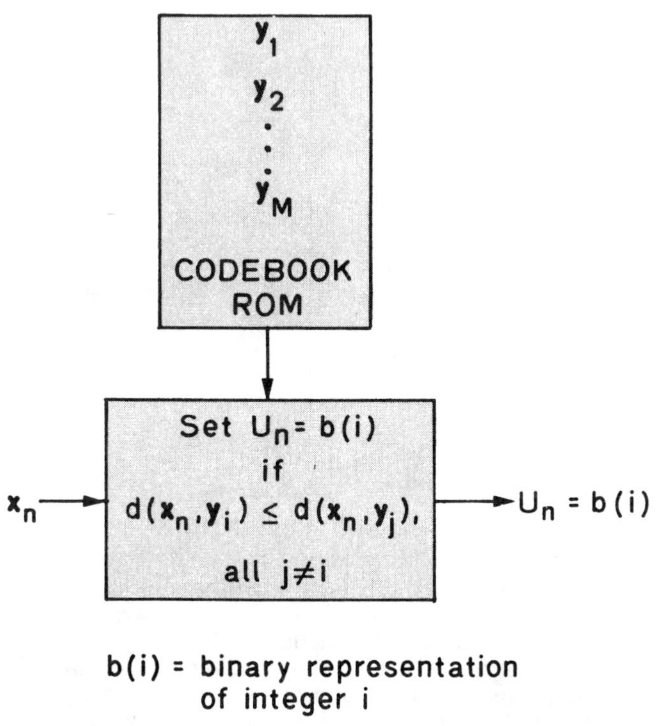

Figure 2. VQ Encoder. The distortion between the input vector and each stored codeword is computed. The encoded output is then the binary representation of the index of the minimum distortion codeword.

Property 1: Given the goal of minimizing the average distortion and given a specific decoder β, no memoryless quantizer encoder can do better than select the codeword v in $\mathbf{M}$ that will yield the minimum possible distortion at the output, that is, to select the channel symbol v yielding the minimum

$$d\{\mathbf{x}, \beta[\gamma(\mathbf{x})]\} = \min_{v \in \mathbf{M}} d[\mathbf{x}, \beta(v)] = \min_{y \in C} d(\mathbf{x}, y). \quad (2)$$

That is, for a given decoder in a memoryless vector quantizer the best encoder is a minimum distortion or nearest neighbor mapping

$$\gamma(\mathbf{x}) = \min_{v \in \mathbf{M}}^{-1} d[\mathbf{x}, \beta(v)], \quad (3)$$

where the inverse minimum notation means that we select the v giving the minimum of (2).

Gersho [24] calls a quantizer with a minimum distortion encoder a Voronoi quantizer since the Voronoi regions about a set of points in a space correspond to a partition of that space according to the nearest-neighbor rule. The word quantizer, however, is practically always associated with such a minimum distortion mapping. We observe that such a vector quantizer with such a minimum distortion encoder is exactly the Shannon model for a block source code subject to a fidelity criterion which is used in information theory to develop optimal performance bounds for data compression systems.

An encoder γ can be thought of as a partition of the input space into cells where all input vectors yielding a common reproduction are grouped together. Such a partition according to a minimum distortion rule is called a Voronoi or Dirichlet partition. A general minimum distance VQ encoder is depicted In Fig. 2.

A simple example of such a partition and hence of an encoder is depicted in Fig. 3 (a more interesting example follows shortly). Observe that this vector quantizer is just two uses of a scalar quantizer in disguise.

As the minimum distortion rule optimizes the encoder of a memoryless VQ for a decoder, we can also optimize the decoder for a given encoder.

Property 2: Given an encoder γ, then no decoder can do better than that which assigns to each channel symbol v the generalized centroid (or center of gravity or barycenter) of all source vectors encoded into v, that is,

$$\beta(v) = cent(v) = \min_{\hat{\mathbf{x}} \in \hat{A}}^{-1} E(d(\mathbf{X}, \hat{\mathbf{x}}) \mid \gamma(\mathbf{X}) = v), \quad (4)$$

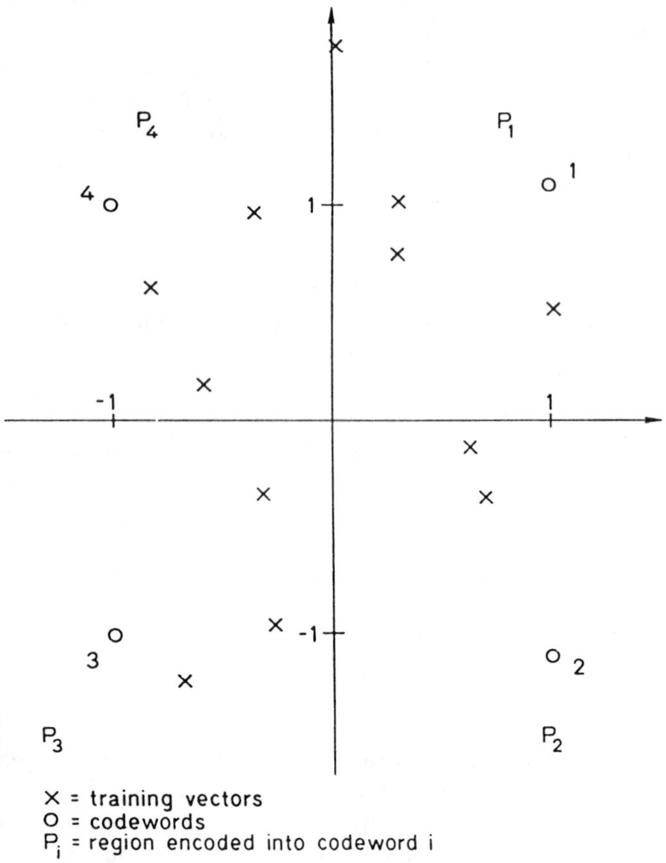

Figure 3. Two-Dimensional Minimum Distortion Partition. The four circles are the codewords of a two-dimensional codebook. The Voronoi regions are the quadrants containing the circles. The x's were produced by a training sequence of twelve two-dimensional Gaussian vectors. Each input vector is mapped into the nearest-neighbor codeword, that is, the circle in the same quadrant.

that is, $\beta(v)$ is the vector yielding the minimum conditional average distortion given that the input vector was mapped into v.

While minimizing such a conditional average may be quite difficult for an arbitrary random process and distortion measure, it is often easy to find for a sample distribution and a nice distortion measure. For example, the centroid in the case of a sample distribution and a squared-error distortion measure is simply the ordinary Euclidean centroid or the vector sum of all input vectors encoded into the given channel symbol, that is, given the sample distribution defined by a training sequence $\{\mathbf{x}_i;\ i = 0, 1, \ldots, L-1\}$, then

$$cent(v) = \frac{1}{i(v)} \sum_{\mathbf{x}_i:\,\gamma(\mathbf{x}_i)=v} x_i\,,$$

where $i(v)$ is the number of indices i for which $\gamma(\mathbf{x}_i) = v$.

The Euclidean centroids of the example of Fig. 3 are depicted in Fig. 4. (The numerical values may be found in [25].) The new codewords better represent the training vectors mapping into the old codewords, but they yield a different minimum distortion partition of the input alphabet, as indicated by the broken line in Fig. 3. This is the key of the algorithm: iteratively optimize the codebook for the old encoder and then use a minimum distortion encoder for the new codebook.

The Itakura-Saito distortion example is somewhat more complicated, but still easily computable. As with the squared error distortion, one groups all input vectors yielding a common channel symbol. Instead of averaging the vectors, however, the sample autocorrelation matrices for all of the vectors are averaged. The centroid is then given by the standard LPC all-pole model for this average autocorrelation, that is, the centroid is found by a standard Levinson's recursion run on the average autocorrelation.

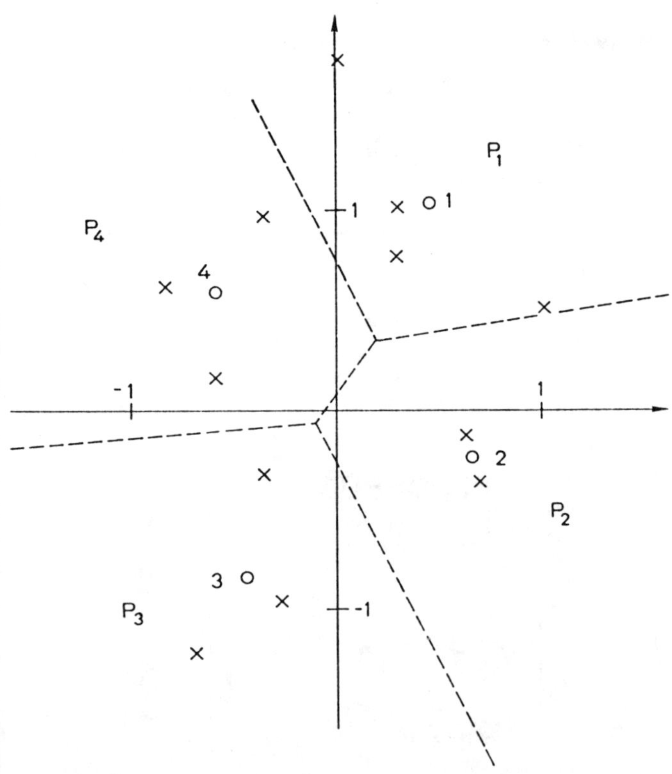

Figure 4. Centroids of Figure 3. The new centroids of the old Voronoi regions of Fig. 3 are drawn as circles. Note that the centroid computation has moved the codewords to better represent the input vectors which yielded those codewords, that is, if one used the same encoder (as in Fig. 3), but replaced the reproduction codewords produced at the decoder by these new centroids, the average distortion would decrease. The broken line delineates the new Voronoi regions for these codewords.

The generalized Lloyd algorithm

The fact that the encoder can be optimized for the decoder and vice versa formed the basis of Lloyd's original optimal PCM design algorithm for a scalar random variable with a known probability density function and a squared error distortion. The general VQ design algorithms considered here are based on the simple observation that Lloyd's basic development is valid for vectors, for sample distributions, and for a variety of distortion measures. The only requirement on the distortion measure is that one can compute the centroids. The basic algorithm is the following:

Step 0. Given: A training sequence and an initial decoder.

Step 1. Encode the training sequence into a sequence of channel symbols using the given decoder minimum distortion rule. If the average distortion is small enough, quit.

Step 2. Replace the old reproduction codeword of the decoder for each channel symbol v by the centroid of all training vectors which mapped into v in Step 1. Go to Step 1.

Means of generating initial decoders will be considered in the next section. Each step of the algorithm must either reduce average distortion or leave it unchanged. The algorithm is usually stopped when the relative distortion decrease falls below some small threshold. The algorithm was developed for vector quantizers, training sequences, and general distortion measures by Linde, Buzo, and Gray [25] and it is sometimes referred to as the LBG algorithm. Previously Lloyd's algorithm had been considered for vectors and difference distortion measures in cluster analysis and pattern recognition problems (e.g., MacQueen [26] and Diday and Simon [27]) and in two-dimensional quantization (e.g., Chen [28] and Adoul *et al.* [29]). Only recently, however, has it been extensively studied for vector quantization applications using several different distortion measures.

Before continuing, it should be emphasized that such iterative improvement algorithms need not in general yield truly optimum codes. It is known that subject to some mathematical conditions the algorithm will yield locally optimum quantizers, but in general there may be numerous such codes and many may yield poor performance. (See, e.g., [30].) It is often useful, therefore, to enhance the algorithm's potential by providing it with good initial codebooks and perhaps by trying it on several different initial codebooks.

INITIAL CODEBOOKS

The basic design algorithm of the previous section is an iterative improvement algorithm and requires an initial code to improve. Two basic approaches have been developed: One can start with some simple codebook of the correct size or one can start with a simple small codebook and recursively construct larger ones.

"Random" codes

Perhaps the simplest example of the first technique is that used in the k-means variation of the algorithm [26]: Use the first 2^R vectors in the training sequence as the initial codebook. An obvious modification more natural for highly correlated data is to select several widely spaced words from the training sequence. This approach is sometimes called random code generation, but we avoid this nomenclature because of its confusion with the random code techniques of information theory which are used to prove the performance bounds.

Product codes

Another example of the first approach is to use a scalar code such as a uniform quantizer k times in succession and then prune the resulting vector codebook down to the correct size. The mathematical model for such a code is a product code, which we pause to define for current and later use: Say we have a collection of codebooks C_i, $i = 0, 1, \ldots, m-1$, each consisting of M_i vectors of dimension k_i and having rate $R_i = \log_2 M_i$ bits per vector. Then the *product codebook* C is defined as the collection of all $M = \Pi_i M_i$ possible concatenations of m words drawn successively from the m codebooks C_i. The dimension of the product codebook is $k = \sum_{i=0}^{m-1} k_i$, the sum of the dimensions of the component codebooks. The product code is denoted mathematically as a Cartesian product:

$$C = \mathop{\times}_{i=0}^{m-1} C_i = \{\textit{all vectors of the form } (\hat{\mathbf{x}}_0, \hat{\mathbf{x}}_1, \cdots, \hat{\mathbf{x}}_{m-1});\ \hat{\mathbf{x}}_i \textit{ in } C_i;\ i = 0, 1, \ldots, m-1\}$$

Thus, for example, using a scalar quantizer with rate R/k k times in succession yields a product k-dimensional vector quantizer of rate R bits per vector. This product code can be used as an initial code for the design algorithm. The scalar quantizers may be identical uniform quantizers with a range selected to match the source, or they may be different, e.g., a positive codebook for a gain and uniform quantizers for $[-1, 1]$ for reflection coefficients in an LPC VQ system.

In waveform coding applications where the reproduction and input alphabets are the same—k-dimensional Euclidean space—an alternative product code provides a means of growing better initial guesses from smaller dimensional codes [31]. Begin with a scalar quantizer C_0 and use a two-dimensional product code $C_0 \times C_0$ as an initial guess for designing a two-dimensional VQ. On completion of the design we have a two-dimensional code, say C^2. Form an initial guess for a three dimensional code as all possible pairs from C^2 and scalars from C_0, that is, use the product code $C^2 \times C_0$ as an initial guess. Continuing in this way, given a good $k-1$ dimensional VQ described by a codebook C^{k-1}, an initial guess for a k-dimensional code design is the product code $C^{k-1} \times C_0$. One can also use such product code constructions with a different initial scalar code C_0, such as those produced by the scalar version of the next algorithm.

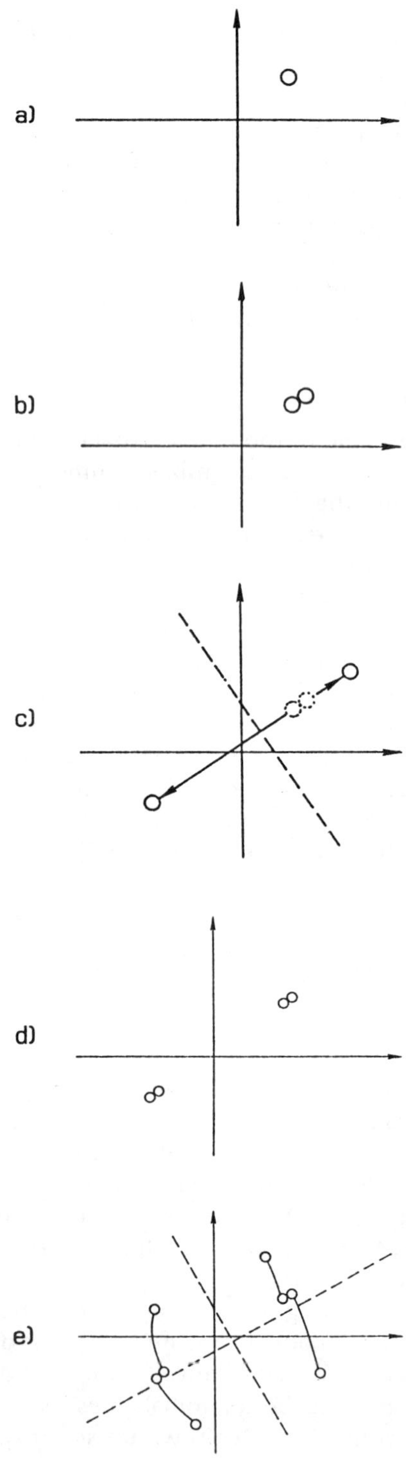

Figure 5. Splitting. A large code is defined in stages: at each stage each codeword of a small code is split into two new codewords, giving an initial codebook of twice the size. The algorithm is run to get a new better codebook. (a) Rate 0: The centroid of the entire training sequence. (b) Initial Rate 1: The one codeword is split to form an initial guess for a two word code. (c) Final Rate 1: The algorithm produces a good code with two words. The dotted line indicates the Voronoi regions. (d) Initial Rate 2: The two words are split to form an initial guess for a four word code. (e) Final Rate 2: The algorithm is run to produce a final four word code.

Splitting

Instead of constructing long codes from smaller dimensional codes, we can construct a sequence of bigger codes having a fixed dimension using a "splitting" technique [25, 16]. This method can be used for any fixed dimension, including scalar codes. Here one first finds the optimum 0 rate code—the centroid of the entire training sequence, as depicted in Fig. 5a for a two-dimensional input alphabet. This single codeword is then split to form two codewords (Fig. 5b). For example, the energy can be perturbed slightly to form a second distinct word or one might purposefully find a word distant from the first. It is convenient to have the original codeword a member of the new pair to ensure that the distortion will not increase. The algorithm is then run to get a good rate 1 bit per vector code as indicated in Fig. 5c. The design continues in this way in stages as shown: the final code of one stage is split to form an initial code for the next.

VARIATIONS OF MEMORYLESS VECTOR QUANTIZERS

In this section we consider some of the variations of memoryless vector quantization aimed at reducing the computation or memory requirements of a full search memoryless VQ.

Tree-searched VQ

Tree-searched vector quantizers were first proposed by Buzo *et al.* [16] and are a natural byproduct of the splitting algorithm for generating initial code guesses. We focus on the case of a binary tree for simplicity, but more general trees will provide better performance while retaining a significant reduction in complexity.

Say that we have a good rate 1 code as in Fig. 5c and we form a new rate two code by splitting the two codewords as in Fig. 5d. Instead of running a full search VQ design on the resulting 4-word codebook, however, we divide the training sequence into two pieces, collecting together all those vectors encoded into a common word in the 1 bit codebook, that is, all of the training sequence vectors in a common cell of the Voronoi partition. For each of these subsequences of training vectors, we then find a good 1-bit code using the algorithm. The final codebook (so far) consists of the four codewords in the two 1-bit codebooks designed for the two subsequences. A tree-searched encoder selects one of the words not by an ordinary full search of this codebook, but instead it uses the first one bit codebook designed on the whole sequence to select a second code and it then picks the best word in the second code. This encoder can then be used to further subdivide the training sequence and construct even better codebooks for the subsequences. The encoder operation can be depicted as a tree in Fig. 6.

The tree is designed one layer at a time; each new layer being designed so that the new codebook available from each node is good for the vectors encoded into the node. Observe that there are 2^R possible reproduction vectors as in the full search VQ, but now R binary searches are made instead of a single 2^R-ary search. In addition, the encoder

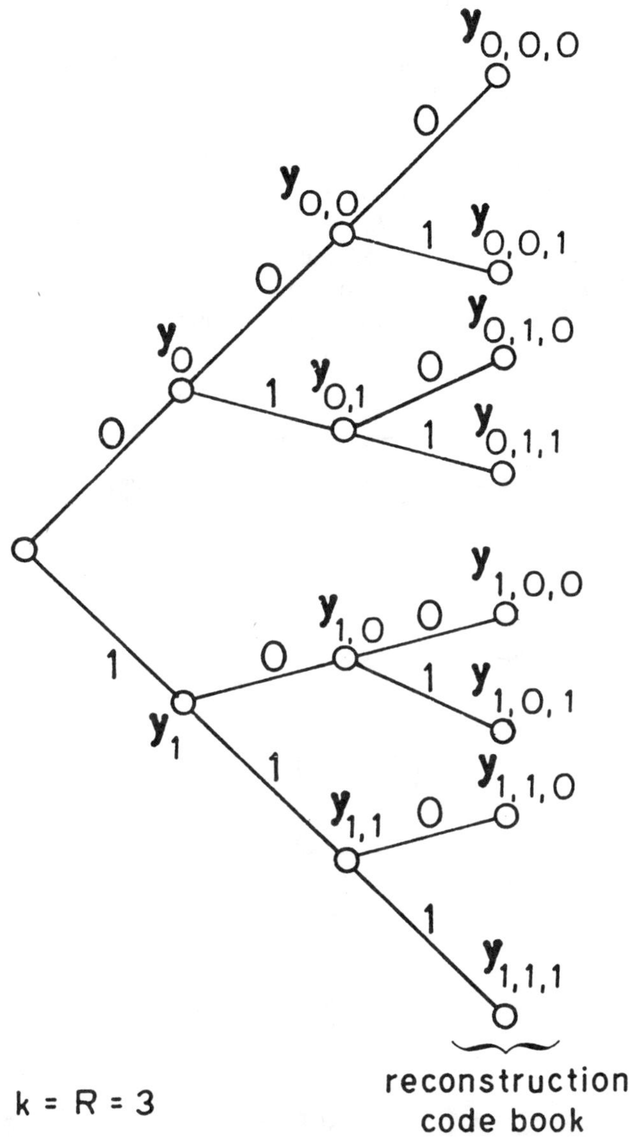

Figure 6. Tree-Searched VQ. A binary encoder tree is shown for a three-dimensional one bit per sample VQ. The encoder makes a succession of R minimum distortion choices from binary codebooks, where the available codebook at each level consists of labels of the nodes in the next level. The labels of the nodes of the final layer are the actual reproduction codewords. At each node the encoder chooses the minimum distortion available label and, if the new index is a 0 (1), sends a channel symbol of 0 (1) and advances up (down) to the next node. After R binary selections the complete channel codeword has been sent and the reproduction codeword specified to the decoder.

storage requirements have doubled. The encoder is no longer optimal for the decoder in the sense of Property 1 since it no longer can perform an exhaustive search of the codebook. The search, however, is much more efficient if done sequentially than is a full search. Thus one may trade performance for efficiency of implementation.

Nonbinary trees can also be used where at the ith layer codebooks of rate R_i are used and the overall rate is then $\Sigma_i R_i$. For example, a depth three tree for VQ of LPC parameter vectors using successive rates of 4, 4, and 2 bits per vector yields performance nearly as good as a full search VQ of the same total rate of 10 bits per vector, yet for the tree search one need only compute $2^4 + 2^4 + 2^2 = 36$ distortions instead of $2^{10} = 1028$ distortions [10].

Other techniques can be used to design tree-searched codes. For example, Adoul *et al.* [32] use a separating hyperplane approach. Another approach is to begin with a full search codebook and to design a tree-search into the codebook. One technique for accomplishing this is to first group the codewords into close disjoint pairs and then form the centroids of the pairs as the node label of the immediate ancestor of the pair. One then works backwards through the tree, always grouping close pairs. Ideally, one would like a general design technique for obtaining a tree search into an arbitrary VQ codebook with only a small loss of average distortion. Gersho and Cheng [33] have reported preliminary results for designing a variable-length tree search for an arbitrary codebook and have demonstrated its implementability for several small dimensional examples.

Multistep VQ

A multistep VQ is a tree-searched VQ where only a single small codebook is stored for each layer of the tree instead of a different codebook for each node of each layer. Such codes provide the computation reduction of tree-searched codes while reducing the storage requirements below that of even ordinary VQ's. The first example of such a code was the multistage codebook [34]. For simplicity we again confine interest to codes which make a sequence of binary decisions. The first layer binary code is designed as in the tree-searched case. This codebook is used to encode the training sequence and then a training sequence of error or residual vectors is formed. For waveform coding applications the error vectors are simply the difference of the input vectors and their codewords. For vocoding applications, the error vectors are residuals formed by passing the input waveform through the inverse filter $A(z)/\alpha$. The algorithm is then run to design a binary VQ for this vector training sequence of coding errors. The reconstruction for these two bits is then formed by combining the two codewords: For waveform coding this is accomplished by adding the first codeword to the error codeword. For voice coding this is accomplished by using the cascade of two all-pole filters for synthesis. This reproduction can then be used to form a "finer" error vector and a code designed for it. Thus an input vector is encoded in stages as with the tree-searched code, but now only R binary codebooks and hence $2R$ total codewords need to be stored. Observe that there are still 2^R possible final codewords, but we have not needed this much storage because the code can be constructed by adding different combinations of a smaller set of words. A multistage VQ is depicted in Fig. 7.

Product codes

Another useful structure for a memoryless VQ is a prod-

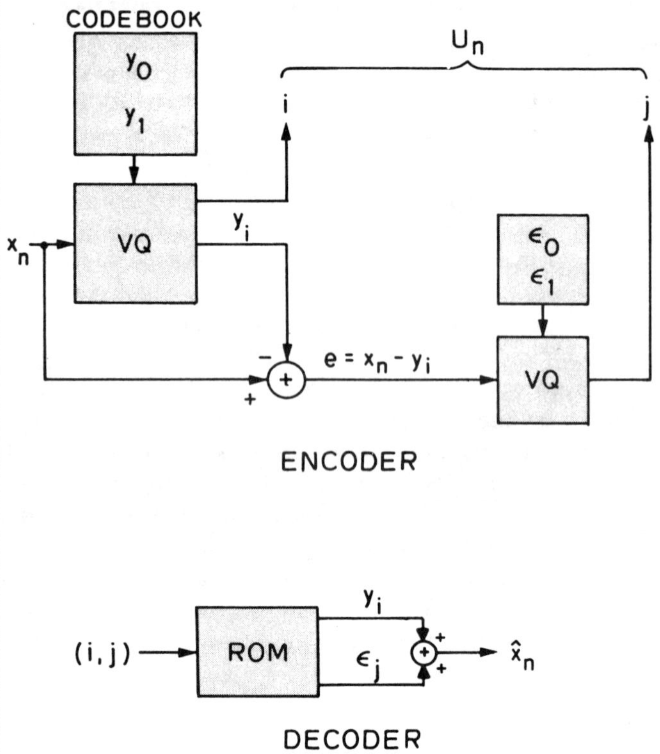

Figure 7. Multistage VQ with 2 Stages. The input vector is first encoded by one VQ and an error vector is formed. The second VQ then encodes the error vector. The two channel symbols from the two VQ's together form the complete channel symbol for the entire encoder. The decoder adds together the corresponding reproduction vectors.

uct code. In one extreme, multiple use of scalar quantizers is equivalent to product VQ's and are obviously simple to implement. More general product VQ's, however, may permit one to take advantage of the performance achievable by VQ's while still being able to achieve the higher rates required for good fidelity. In addition, such codes may yield a smaller computational complexity than an ordinary VQ of the same rate and performance (but different dimension). The basic technique is useful when there are differing aspects of the input vector that one might wish to code separately because of different effects, e.g., on dynamic range or finite word length implementation.

Gain/shape VQ

One example of a product code is a gain/shape VQ where separate, but interdependent, codes are used to code the "shape" and "gain" of the waveform, where the "shape" is defined as the original input vector normalized by removal of a "gain" term such as energy in a waveform coder or LPC residual energy in a vocoder. Gain/shape encoders were introduced by Buzo *et al.* [16] and were subsequently extended and optimized by Sabin and Gray [35, 36]. A gain/shape VQ for waveform coding with a squared-error distortion is illustrated in Fig. 8.

Figure 8 sketches the surprising fact that for the squared error case considered, the two-step selection of the product codeword is an optimal encoding for the given product codebook. We emphasize that here the encoder is optimal for the given product codebook or decoder, but the codebook itself is in general suboptimal because of the constrained product form. A similar property holds for the Itakura-Saito distortion gain/shape VQ. Thus in this case if one devotes R_s bits to the shape and R_g bits to the gain, where $R_s + R_g = R$, then one need only compute 2^{R_s} vector distortions and an easy scalar quantization. The full search encoder would require 2^R vector distortions, yet both encoders yield the same minimum distortion codeword!

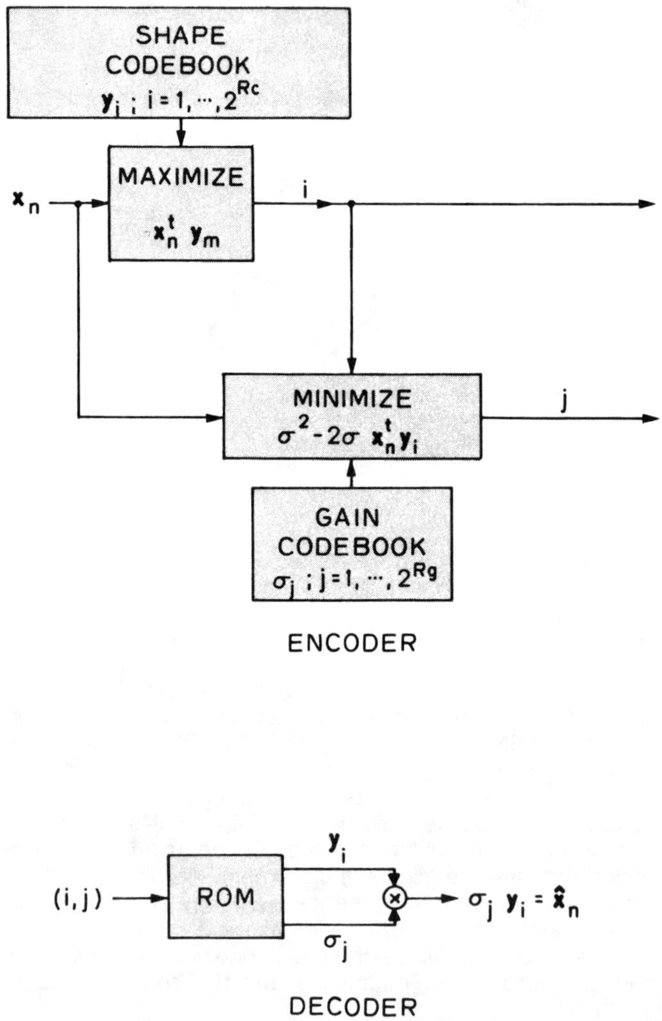

Figure 8. Gain/Shape VQ. First a unit energy shape vector is chosen to match the input vector by maximizing the inner product over the codewords. Given the resulting shape vector, a scalar gain codeword is selected so as to minimize the indicated quantity. The encoder yields the product codeword $\sigma_j y_i$ with the minimum possible squared error distortion from the input vector. Thus this multistep encoder is optimum for the product codebook.

Variations of the basic VQ algorithm can be used to iteratively improve a gain shape code by alternately optimizing the shape for the gain and vice versa. The resulting conditional centroids are easy to compute. The centroid updates can be made either simultaneously or alternately after each iteration [36].

One can experimentally determine the optimal bit allocation between the gain and the shape codebooks.

Separating mean VQ

Another example of a multistep product code is the separating mean VQ where a sample mean instead of an energy term is removed [37]. Define the sample mean $\langle \mathbf{x} \rangle$ of a k-dimensional vector by $k^{-1} \sum_{i=0}^{k-1} x_i$. In a separated mean VQ one first uses a scalar quantizer to code the sample mean of a vector, then the coded sample mean is subtracted from all of the components of the input vector to form a new vector with approximately zero sample mean. This new vector is then vector quantized. Such a system is depicted in Fig. 9. The basic motivation here is that in image coding the sample mean of pixel intensities in a small rectangular block represents a relatively slowly varying average background value of pixel intensity around which there are variations.

To design such a VQ, first use the algorithm to design a scalar quantizer for the sample mean sequence $\langle \mathbf{x}_j \rangle$, $j = 0, 1, \ldots, L-1$. Let $q(\langle \mathbf{x} \rangle)$ denote the reproduction for $\langle \mathbf{x} \rangle$ using the quantizer. Then use the vector training sequence $\mathbf{x}_j - q(\langle \mathbf{x}_j \rangle)\mathbf{1}$, where $\mathbf{1} = (1, 1, \ldots, 1)$, to design a VQ for the difference. Like the gain/shape VQ, a product codebook and a multistep encoder are used, but unlike the gain/shape VQ it can be shown that the multistep encoder here does not select the best possible mean, shape pair, that is, the multistep encoder is not equivalent to a full search encoder.

Lattice VQ

A final VQ structure capable of efficient searches and memory usage is the lattice quantizer, a k-dimensional generalization of the scalar uniform quantizer. A lattice in k-dimensional space is a collection of all vectors of the form $\mathbf{y} = \sum_{i=0}^{n-1} a_i \mathbf{e}_i$, where $n \leq k$, where $\mathbf{e}_0, \ldots, \mathbf{e}_{n-1}$ are a set of linearly independent vectors in $\mathbf{R}^k$, and where the a_i are arbitrary integers. A lattice quantizer is a quantizer whose codewords form a subset of a lattice. Lattice quantizers were introduced by Gersho [38] and the performance and efficient coding algorithms were developed for many particular lattices by Conway and Sloane [39, 40, 41] and Barnes and Sloane [42]. The disadvantage of lattice quantizers is that they cannot be improved by a variation of the Lloyd algorithm without losing their structure and good quantizers produced by the Lloyd algorithm cannot generally be well approximated by lattices. Lattice codes can work well on source distributions that are approximately uniform over a bounded region of space. In fact, lattices that are asymptotically optimal in the limit of large rate are known for this case in two and three dimensions and good lattices are known for dimensions up to 16.

Ideally, one would like to take a full search, unconstrained VQ and find some fast means of encoding having complexity more like the above techniques than that of the full search. For example, some form of multidimensional companding followed by a lattice quantizer as suggested by Gersho [24] would provide both good performance and efficient implementation. Unfortunately, however, no design methods accomplishing this goal have yet been found.

FEEDBACK VECTOR QUANTIZERS

Memory can be incorporated into a vector quantizer in a simple manner by using different codebooks for each input vector, where the codebooks are chosen based on past input vectors. The decoder must know which codebook is being used by the encoder in order to decode the channel symbols. This can be accomplished in two ways: 1) The encoder can use a codebook selection procedure that depends only on past encoder outputs and hence the codebook sequence can be tracked by the decoder. 2) The decoder is informed of the selected codebook via a special low-rate side channel. The first approach is called feedback vector quantization and is the

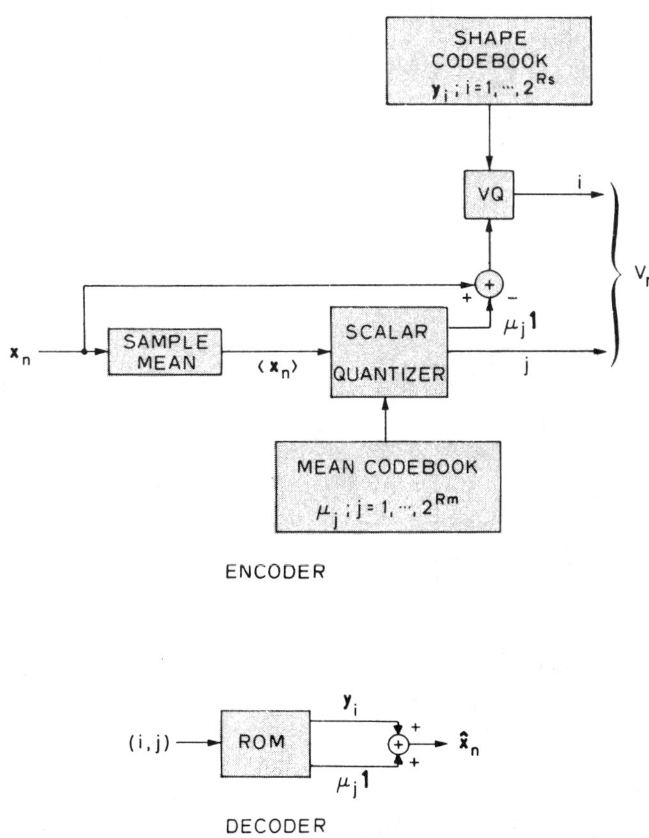

Figure 9. Separating Mean VQ. The sample mean of the input vector is computed, scalar quantized, and then subtracted from each component of the input vector. The resulting vector with approximately zero sample mean is then vector quantized. The decoder adds the coded sample mean to all components of the coded shape vector.

topic of this section. The name follows because the encoder output is "fed back" for use in selecting the new codebook. A feedback vector quantizer can be viewed as the vector extension of a scalar adaptive quantizer with backward estimation (AQB) [3]. The second approach is the vector extension of a scalar adaptive quantizer with forward estimation (AQF) and is called simply adaptive vector quantization. Adaptive VQ will be considered in a later section. Observe that systems can combine the two techniques and use both feedback and side information. We also point out that unlike most scalar AQB and AQF systems, the vector analogs considered here involve no explicit estimation of the underlying densities.

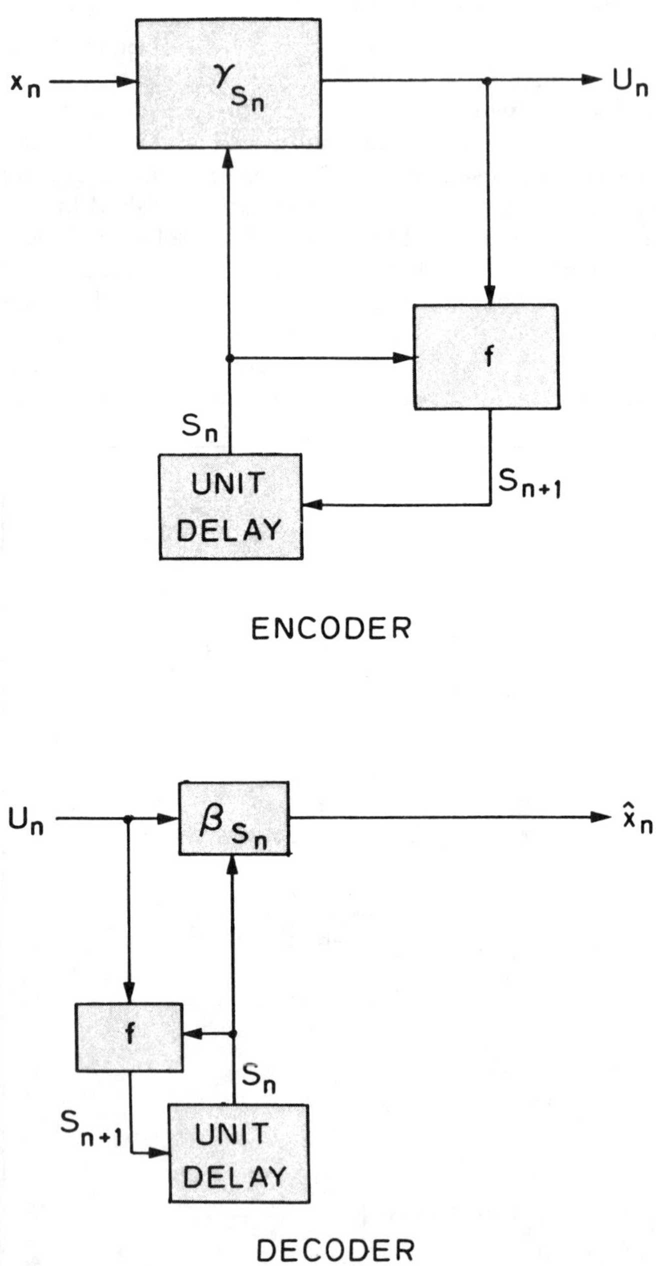

Figure 10. Feedback VQ. At time n both encoder and decoder are in a common state S_n. The encoder uses a state VQ γ_{S_n} to encode the input vector and then selects a new state for the next input vector. Knowing the VQ used and the resulting channel symbol, the decoder can produce the correct reproduction. Note that the state VQ's may be computed at each time from some rule or, if they are small in number, simply stored separately.

It should be emphasized that the results of information theory imply that VQ's with memory can do no better than memoryless VQ's in the sense of minimizing average distortion for a given rate constraint. In fact, the basic mathematical model for a data compression system in information theory is exactly a memoryless VQ and such codes can perform arbitrarily close to the optimal performance achievable using any data compression system. The exponential growth of computation and memory with rate, however, may result in nonimplementable VQ's. A VQ with memory may yield the desired distortion with practicable complexity.

A general feedback VQ can be described as follows [22]: Suppose now that we have a space **S** whose members we shall call states and that for each state s in **S** we have a separate quantizer: an encoder γ_s, decoder β_s, and codebook C_s. The channel codeword space **M** is assumed to be the same for all of the VQ's. Consider a data compression system consisting of a sequential machine such that if the machine is in state s, then it uses the quantizer with encoder γ_s and decoder β_s. It then selects its next state by a mapping called a next-state function or state-transition function f such that given a state s and a channel symbol v, then $f(v,s)$ is the new state of the machine. More precisely, given a sequence of input vectors $\{\mathbf{x}_n;\ n = 0,1,2,\ldots\}$ and an initial state s_0, then the subsequent state sequence s_n, channel symbol sequence v_n, and reproduction sequence $\hat{\mathbf{x}}_n$ are defined recursively for $n = 0,1,2,\ldots$ as

$$v_n = \gamma_{s_n}(\mathbf{x}_n), \qquad \hat{\mathbf{x}}_n = \beta_{s_n}(v_n), \qquad s_{n+1} = f(v_n, s_n). \quad (5)$$

Since the next state depends only on the current state and the channel codeword, the decoder can track the state if it knows the initial state and the channel sequence. A general feedback vector quantizer is depicted in Fig. 10. The freedom to use different quantizers based on the past without increasing the rate should permit the code to perform better than a memoryless quantizer of the same dimension and rate.

An important drawback of all feedback quantizers is that channel errors can accumulate and cause disastrous reconstruction errors. As with scalar feedback quantizer systems, this must be handled by periodic resetting or by error control or by a combination of the two.

If the state space is finite, then we shall call the resulting system a finite-state vector quantizer or FSVQ. For an FSVQ, all of the codebooks and the next-state transition table can all be stored in ROM, making the general FSVQ structure amenable to LSI or VLSI implementation [43].

Observe that a memoryless vector quantizer is simply a feedback vector quantizer or finite-state vector quantizer with only a single state. The general FSVQ is a special case of a tracking finite state source coding system [44] where the encoder is a minimum distortion mapping.

Three design algorithms for feedback vector quantizers using variations on the generalized Lloyd algorithm have been recently developed. The remainder of this section is devoted to brief descriptions of these techniques.

Vector predictive quantization

Cuperman and Gersho [45, 46] proposed a vector predictive coder or vector predictive quantizer (VPQ) which is a vector generalization of DPCM or predictive quantization. A VPQ is sketched in Fig. 11. For a fixed predictor, the VQ design algorithm is used to design a VQ for the prediction error sequence. Cuperman and Gersho considered several variations on the basic algorithm, some of which will be later mentioned.

Chang [47] developed an extension to Cuperman and Gersho's algorithm which begins with their system and then uses a stochastic gradient algorithm to iteratively improve the vector linear predictor coefficients, that is, to better match the predictor to the quantizer. The stochastic gradient algorithm is used only in the design of the system, not as an on line adaptation mechanism as in the adaptive gradient algorithms of, e.g., Gibson *et al.* [48] and Dunn [49]. A scalar version of this algorithm for improving the predictor for the quantizer was developed in unpublished work of Y. Linde.

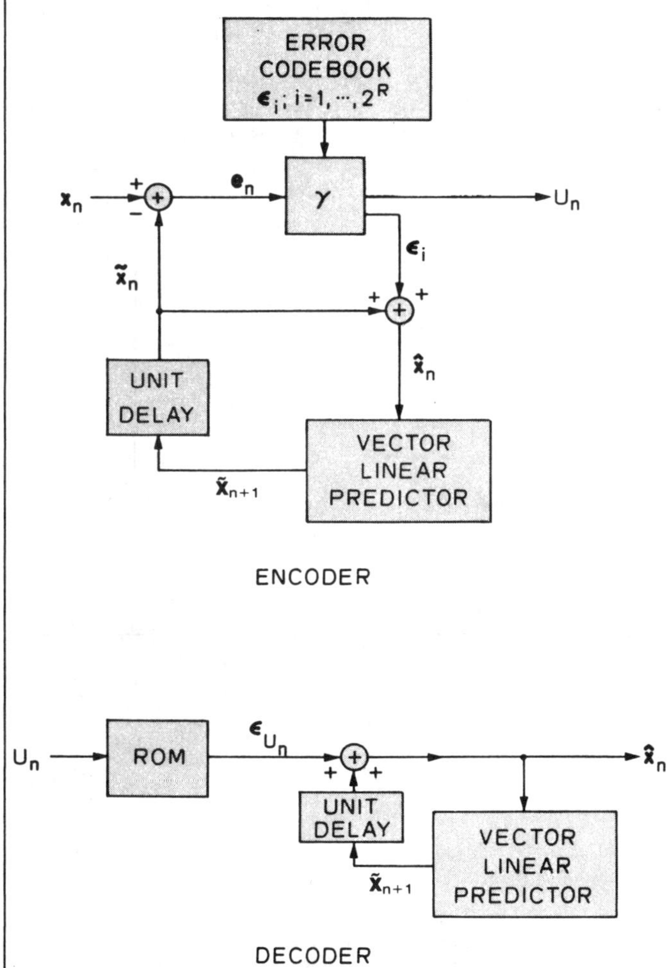

Figure 11. Vector Predictive Quantization. A linear vector predictor for the next input vector of a process given the previous input vector is applied to the previous reproduction of the input vector. The resulting prediction is subtracted from the current input vector to form an error vector which is vector quantized. The decoder uses a copy of the encoder and the received encoded error vectors to construct the reproduction.

Product/multistep FVQ

A second basic approach for designing feedback vector quantizers which is quite simple and works quite well is to use a product multistep VQ such as the gain/shape VQ or the separating mean VQ and use a simple feedback quantizer on the scalar portion and an ordinary memoryless VQ on the remaining vector. This approach was developed in [10] for gain/shape VQ of LPC parameters and in [37] for separating mean VQ of images. Both efforts used simple scalar predictive quantization for the feedback quantization of the scalar terms.

FSVQ

The first general design technique for finite-state vector quantizers was reported by Foster and Gray [50, 51]. There are two principal design components: 1. Design an initial set of state codebooks and a next-state function using an *ad hoc* algorithm. 2. Given the next-state function, use a variation of the basic algorithm to attempt to improve the state codebooks. The second component is accomplished by a slight extension of the basic algorithm that is similar to the extension of [52] for the design of trellis encoders: Encode the data using the FSVQ and then replace all of the reproduction vectors by the centroids of the training vectors which map into those vectors; now, however, the centroids are conditioned on both the channel symbol and the state. While such conditional averages are likely impossible to compute analytically, they are easily computed for a training sequence. For example, in the case of a squared error distance one simply forms the Euclidean centroid of all input vectors which correspond to the state s and channel symbol v in an encoding of the training sequence.

As with ordinary VQ, replacing the old decoder or codebook by centroids cannot yield a code with larger distortion. Unlike memoryless VQ, however, replacing the old encoder by a minimum distortion rule for the new decoder can in principal cause an increase in distortion and hence now the iteration is somewhat different: Replace the old encoder (which is a minimum distortion rule for the old decoder) by a minimum distortion rule for the new decoder. If the distortion goes down, then continue the iteration and find the new centroids. If the distortion goes up, then quit with the encoder being a quantizer for the previous codebook and the decoder being the centroids for the encoder. By construction this algorithm can only improve performance. It turns out, however, that in practice it is a good idea to not stop the algorithm if the distortion increases slightly, but to let it continue: it will almost always eventually drop back down in distortion and converge to something better.

The first design component is more complicated. We here describe one of the more promising approaches of [51] called the omniscient design approach. Say that we wish to design an FSVQ with K states and rate R bits per vector. For simplicity we label the states as 0 through K-1. First use the training sequence to design a memoryless VQ with K codewords, one for each state. We shall call these codewords state labels and this VQ the state quantizer. We call the output of the state VQ the "ideal next state" instead of a channel symbol. Next break up the training sequence into subsequences as follows: Encode the training sequence using the state VQ and for each state s collect all of training vectors which *follow* the occurrence of this state label. Thus for s the corresponding training subsequence consists of all input vectors that occur when the *current* ideal state is s. Use the basic algorithm to design a rate R codebook C_s for the corresponding training sequence for each s.

The resulting state VQ and the collection of codebooks for each state have been designed to yield good performance in the following communication system: The encoder is in an ideal state s chosen by using the state VQ on the last input vector. The encoder uses the corresponding VQ encoder γ_s described by the codebook C_s. The output of γ_s is the channel symbol. In order to decode the channel symbol, the decoder must also know the ideal state. Unfortunately, however, this ideal state cannot be determined from knowledge of the initial state and all of the received channel symbols. Thus the decoder must be omniscient in the sense of knowing this additional side information in order to be able to decode. In particular, this system is not an FSVQ by our definition. We can use the state quantizer and the various codebooks, however, to construct an FSVQ by approximating the omniscient system: Instead of forming the ideal next state by using the state VQ on the actual input vector (as we did in the design procedure), use the state VQ on the current reproduction vector in order to choose the next state. This will yield a state sequence depending only on encoder outputs and the original state and hence will be trackable by the decoder. This is analogous to the scalar practice of building a predictive coder and choosing the predictor as if it knew the past inputs, but in fact applying it to past reproductions.

Combining the previously described steps of (1) initial (state label) codebook design, (2) state codebooks and next-state function design, and (3) iterative improvement of code for given next-state function, provides a complete design algorithm.

In addition to the above design approach, techniques have been developed for iterating on (2) and (3) above in the sense of optimizing the next-state function for a given collection of codebooks. These algorithms, however, are more complicated and require ideas from the theory of adaptive stochastic automata. The reader is referred to [53] for a discussion of these improvement algorithms.

VECTOR TREE AND TRELLIS ENCODERS

As with scalar feedback quantizers, the actions of the decoder of a feedback VQ can be depicted as a directed graph or tree. A simple example is depicted in Fig. 12, where a merged tree or trellis can be drawn since the feedback VQ has only a finite number of states.

Instead of using the ordinary VQ encoder which is only permitted to look at the current input vector in order to decide on a channel symbol, one could use algorithms such as the Viterbi algorithm, M-algorithm or M,L-algorithm, Fano algorithm, or stack algorithm for a minimum cost search through a directed graph and search several levels ahead into the tree or trellis before choosing a channel symbol. This introduces an additional delay into the encoding of several vectors, but it ensures better long run average distortion behavior. This technique is called tree or trellis encoding and is also referred to as look-ahead coding, delayed decision coding, and multipath search coding. (See, e.g., [54, 52] for surveys.) We point out that a tree encoding system uses a tree to denote the

s	u	β_s (u)
0	0	(1,1)
0	1	(1,0)
1	0	(0,-1)
1	1	(-1,-1)

(a) DECODER

s	u	f (u,s)
0	0	0
0	1	1
1	0	0
1	1	1

(b) NEXT-STATE FUNCTION

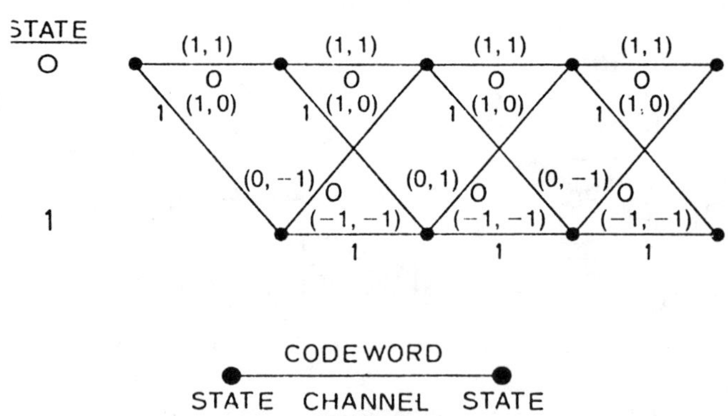

(c) TRELLIS

Figure 12. Decoder trellis for a two state 1 bit per vector two dimensional waveform coder. The trellis depicts the possible state transitions for the given next-state function. The transitions are labeled by the corresponding decoder output (in parentheses) and channel symbol produced by the encoder.

operation on successive vectors by the decoder at successive times while a tree-searched VQ uses a tree to construct a fast search for a single vector at a single time.

A natural variation of the basic algorithm for designing FSVQ's can be used to design trellis encoding systems: Simply replace the FSVQ encoder which finds the minimum distortion reproduction for a single input vector by a Viterbi or other search algorithm which searches the decoder trellis to some fixed depth to find a good long term minimum distortion path. The centroid computation is accomplished exactly as with an FSVQ: each branch or transition label is replaced by the centroid of all training vectors causing that transition, that is, the centroid conditioned on the decoder state and channel symbol. Scalar and simple two dimensional vector trellis encoding systems were designed in [52] using this approach.

Trellis encoding systems are not really vector quantization systems as we have defined them since the encoder is permitted to search ahead to determine the effect on the decoder output of several input vectors while a vector quantizer is restricted to search only a single vector ahead. The two systems are intimately related, however, and a trellis encoder can always be used to improve the performance of a feedback vector quantizer. Very little work has yet been done on vector trellis encoding systems.

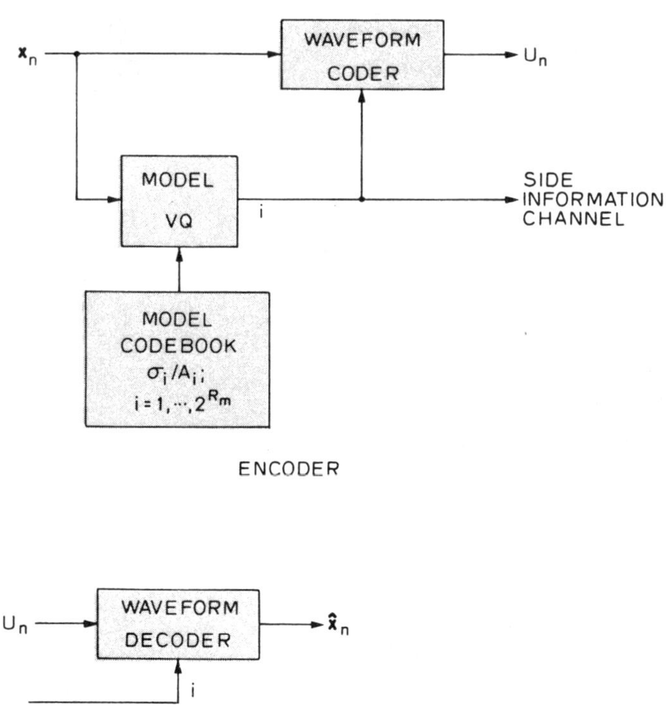

Figure 13. Adaptive VQ. The model VQ uses the Itakura-Saito distortion to select an LPC model to fit the input frame of many sample vectors. This selection in turn determines the waveform coder used to digitize the sample vectors. A side channel then informs the receiver which decoder to use on the channel symbols produced by the waveform coder.

ADAPTIVE VQ

As a final class of VQ we consider systems that use one VQ to adapt a waveform coder, which might be another VQ. The adaptation information is communicated to the receiver via a low rate side information channel.

The various forms of vector quantization using the Itakura-Saito family of distortion measures can be considered as model classifiers, that is, they fit an all-pole model to an observed sequence of sampled speech. When used alone in an LPC VQ system, the model is used to synthesize the speech at the receiver. Alternatively, one could use the model selected to choose a waveform coder designed to be good for sampled waveforms that produce that model. For example, analogous to the omniscient design of FSVQ one could design separate VQ's for the subsequences of the training sequence encoding into common models. Both the model index and the waveform coding index are then sent to the receiver. Thus LPC VQ can be used to adapt a waveform coder, possibly also a VQ or related system. This will yield a system typically of much higher rate, but potentially of much better quality since the codebooks can be matched to local behavior of the data. The general structure is shown in Fig. 13. The model VQ typically operates on a much larger vector of samples and at a much lower rate in bits per sample than does the waveform coder and hence the bits spent on specifying the model through the side channel are typically much fewer than those devoted to the waveform coder.

There are a variety of such possible systems since both the model quantizer and the waveform quantizer can take on many of the structures so far considered. In addition, as in speech recognition applications [55] the gain-independent variations of the Itakura-Saito distortion measure which either normalize or optimize gain may be better suited for the model quantization than the usual form. Few such systems have yet been studied in detail. We here briefly describe some systems of this type that have appeared in the literature to exemplify some typical combinations. All of them use some form of memoryless VQ for the model quantization, but a variety of waveform coders are used.

The first application of VQ to adaptive coding was by Adoul, Debray, and Dalle [32] who used an LPC VQ to choose a predictor for use in a scalar predictive waveform coder. Vector quantization was used only for the adaptation and not for the waveform coding. An adaptive VQ generalization of this system was later developed by Cuperman and Gersho [45,46] who used an alternative classification technique to pick one of three vector predictors and then used those predictors in a predictive vector quantizer. The predictive vector quantizer design algorithm previously described was used, except now the training sequence was broken up into subsequences corresponding to the selected predictor and a quantizer was designed for each resulting error sequence. Chang [47] used a similar scheme with an ordinary LPC VQ as the classifier and with a stochastic gradient algorithm run on each of the vector predictive quantizers in order to im-

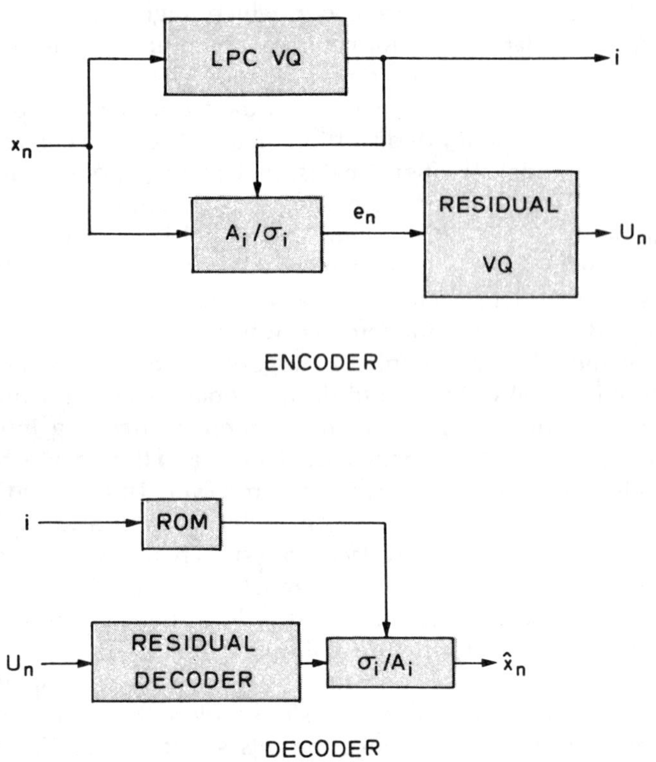

Figure 14. RELP VQ. An LPC VQ is used for model selection and a single VQ to waveform encode the residuals formed by passing the original waveform through the inverse filter $A/\sqrt{\alpha}$. The side information specifies to the decoder which of the model filters $\sqrt{\alpha}/A$ should be used for synthesis.

prove the prediction coefficients for the corresponding codebooks.

Rebolledo *et al.* [56] and Adoul and Mabilleau [57] developed vector residual excited linear predictive (RELP) systems. (See Fig. 14.) A similar system employing either a scalar or a simple vector trellis encoder for the waveform coder was developed by Stewart *et al.* [52]. Both of these systems used the basic algorithm to design both the model VQ and the waveform coders.

The RELP VQ systems yielded disappointingly poor performance at low bit rates. Significantly better performance was achieved by using the residual codebooks produced in the RELP design to construct codebooks for the original waveform, that is, instead of coding the model and the residual, code the model and use the selected model to construct a waveform coder for the original waveform as depicted in Fig. 15 [52]. For lack of a better name, this system might be called an inverted RELP because it uses residual codebooks to drive an inverse model filter in order to get a codebook for the original waveform.

Yet another use of LPC VQ to adapt a waveform coder was reported by Heron, Crochiere, and Cox [58] who used a subband/transform coder for the waveform coding and used the side information to adapt the bit allocation for the scalar parameter quantizers.

Many other variations on the general theme are possible and the structure is a promising one for processes such as speech that exhibit local stationarity, that is, slowly varying short term statistical behavior. The use of one VQ to partition a training sequence in order to design good codes for the resulting distinct subsequences is an intuitive approach to the computer-aided design of adaptive data compression systems.

EXAMPLES

We next consider the performance of various forms of vector quantizers on three popular guinea pigs: Gauss Markov sources, speech waveforms, and images. For the speech coding example we consider both waveform coders using the squared error distortion measure and vocoders using the Itakura-Saito distortion. The caveats of the introduction should be kept in mind when interpreting the results.

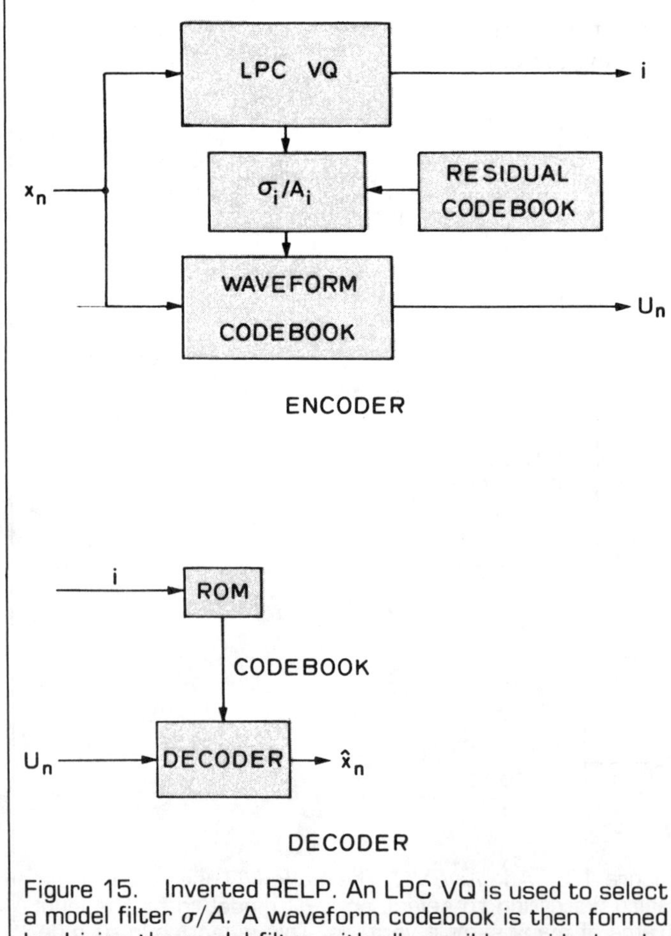

Figure 15. Inverted RELP. An LPC VQ is used to select a model filter σ/A. A waveform codebook is then formed by driving the model filter with all possible residual codewords from a RELP VQ design. Thus, unlike a RELP system, the original waveform (and not a residual) is matched by possible reproduction codewords.

The performance of the systems are given by SNR's for squared error and by an analogous quantity for the Itakura-Saito distortion: In both cases we measure normalized average distortion on a logarithmic scale, where the normalization is by the average distortion of the optimum zero rate code—the average distortion between the input sequence and the centroid of the entire input sequence. This quantity reduces to an SNR in the squared error case and provides a useful dimensionless normalized average distortion in general. We call this quantity the SNR in both cases. The SNR is given in tables instead of graphs in order to facilitate quantitative comparisons among the coding schemes.

Gauss Markov sources

We first consider the popular guinea pig of a Gauss Markov source. This source is useful as a mathematical model for some real data sources and its information theoretic optimal performance bounds as described by the distortion-rate function are known. For this example we consider only the squared error distortion. A Gauss Markov source or a first order Gauss autoregressive source $\{X_n\}$ is defined by the difference equation $X_{n+1} = aX_n + W_n$, where $\{W_n\}$ is a zero mean, unit variance, independent and identically distributed Gaussian source. We here consider the highly correlated case of $a = 0.9$ and vector quantizers of 1 bit/sample. The maximum achievable SNR as given by Shannon's distortion-rate function for this source and rate is 13.2 dB [7].

Various design algorithms were used to design vector quantizers for several dimensions for this source. Table I describes the results of designing several memoryless vector quantizers for a training sequence of 60,000 samples. Given are the design SNR (code performance on the training sequence), the number of multiplications per sample required by the encoder, and the number of real scalars that must be stored for the encoder codebook. The number of multiplications is used as a measure of encoder complexity because it is usually the dominant computation and because the number of additions required is usually comparable. It is given by n = (the number of codewords searched) × (dimension)/(dimension) = the number of codewords searched. The actual storage required depends on the number of bytes used to store each floating point number. Many (but not all) of the final codes were subsequently tested on different test sequences of 60,000 samples. In all cases the open test SNR's were within .25 dB of the design distortion. The systems considered are full search VQ's [25], binary tree-searched VQ's [59], binary multistage VQ's [47], and gain/shape VQ's [36]. The gain and codebook sizes for the gain/shape codes were experimentally optimized.

TABLE I

MEMORYLESS VQ FOR A GAUSS MARKOV SOURCE.

	VQ			TSVQ			MVQ			G/SVQ		
k	SNR	n	M	SNR	n	M	SNR	n	M	SNR	n	M
1	4.4	2	2	4.4	2	2	4.4	2	2			
2	7.9	4	8	7.9	4	12	7.6	4	8	7.9	1	3
3	9.2	8	24	9.2	6	42	8.6	6	18	9.3	1	5
4	10.2	16	64	10.2	8	120	8.4	8	32	9.4	2	10
5	10.6	32	160	10.4	10	310	9.3	10	50	9.8	3	17
6	10.9	64	384	10.7	12	756	9.1	12	72	9.9	4	26
7	11.2	128	896	11.0	14	1778	9.4	14	98	10.2	4	31
8							9.9	16	128	10.6	5	43
9										10.9	6	57

Signal to Noise Ratios (SNR), number of multiplications per sample (n), and storage requirements of memoryless vector quantizers: full search memoryless VQ (VQ), binary tree-searched (TSVQ), binary multistage VQ (MVQ), and gain/shape VQ (G/SVQ). Rate = 1 bit/sample. k = vector dimension. Training Sequence = 60000 samples from a Gauss Markov Source with correlation coefficient 0.9.

TABLE II

FEEDBACK VQ OF A GAUSS MARKOV SOURCE.

	FSVQ1				FSVQ2				VPQ		
k	SNR	K	n	M	SNR	K	n	M	SNR	n	M
1	10.0	64	2	64	9.5	16	2	16	10.0	2	2
2	10.8	256	4	512	10.8	32	4	64	11.2	4	8
3	11.4	512	8	1536	11.1	64	8	192	11.6	8	24
4	12.1	512	16	2048	11.3	128	16	512	11.6	16	64

Signal to Noise Ratios (SNR), number of states (K), number of multiplications per sample (n), and storage (M) for feedback quantizers: FSVQ with number of states increased until negligible change (FSVQ1), FSVQ with fewer states (FSVQ2), VPQ. Rate = 1 bit/sample. k = vector dimension. Training Sequence = 60000 samples from a Gauss Markov Source with correlation coefficient 0.9.

As expected, the full search VQ yields the best performance for each dimension, but the tree-searched VQ is not much worse and has a much lower complexity. The multistage VQ is noticeably inferior, losing more than 1 dB at the higher dimensions, but its memory requirements are small. The gain/shape VQ compares poorly on the basis of performance vs. rate for a fixed dimension, but it is the best code in the sense of providing the minimum distortion for a fixed complexity and rate.

For larger rates and lower distortion the relative merits may be quite different. For example, the multistage VQ is then capable of better performance relative to the ordinary VQ since the quantization errors in the various stages do not accumulate so rapidly. (See, e.g., [34].) Thus in this

TABLE III
MEMORYLESS VQ OF SAMPLED SPEECH.

	VQ				TSVQ			
k	SNRin	SNRout	n	M	SNRin	SNRout	n	M
1	2.0	2.1	2	2	2.0	2.1	2	2
2	5.2	5.3	4	8	5.1	5.1	4	12
3	6.1	6.0	8	24	5.5	5.5	6	42
4	7.1	7.0	16	64	6.4	6.4	8	120
5	7.9	7.6	32	160	7.1	6.9	10	310
6	8.5	8.1	64	384	7.9	7.5	12	756
7	9.1	8.4	128	896	8.3	7.8	14	1778
8	9.7	8.8	256	2048	8.9	8.0	16	4080

	MVQ				G/SVQ			
k	SNRin	SNRout	n	M	SNRin	SNRout	n	M
1	2.0	2.1	2	2				
2	4.3	4.4	4	8				
3	4.3	4.4	6	18	4.5	4.6	4	14
4	4.4	4.5	8	32	6.0	6.1	4	20
5	5.0	5.0	10	50	7.2	6.9	8	44
6	5.0	4.9	12	72	7.7	7.4	16	100
7	5.3	5.1	14	98	8.2	7.7	16	120
8	5.6	5.5	16	128	8.8	8.1	32	264
9					9.3	8.5	64	584
10					9.8	8.9	128	1288
11					10.4	9.3	256	2824

Signal to Noise Ratios inside training sequence (SNRin) of 640000 speech samples, Signal to Noise Ratios outside training sequence (SNRout) of 76800 speech samples, number of multiplications per sample (n), and storage requirements of memoryless vector quantizers: full search memoryless VQ (VQ), binary tree-searched (TSVQ), binary multistage VQ (MVQ), and gain/shape VQ (G/SVQ). Rate = 1 bit/sample. k = vector dimension.

case multistage VQ may be far better because if its much smaller computational requirements.

Table II presents results for three feedback VQ's for the same source. In addition to the parameters of Table I, the number of states for the FSVQ's are given. The first FSVQ and the VPQ were designed for the same training sequence of 60,000 samples. Because of the extensive computation required and the shortness of the training sequence for a feedback quantizer, only dimensions 1 through 4 were considered. The first FSVQ was designed using the omniscient design approach for 1 bit per sample, dimensions 1 through 4, and a variety of numbers of states. For the first example, the number of states was chosen by designing FSVQ's for more and more states until firther increases yielded negligible improvements [51]. It was found, however, that the performance outside of the training sequence for these codes was significantly inferior, by 1 to 2 dB for the larger dimensions. From the discussion of average distortion, this suggests that the training sequence was too short. Hence the second FSVQ design (FSVQ2) was run with a larger training sequence of 128,000 samples and fewer states. The test sequence for these codes always yielded performance within .3 dB of the design value. The VPQ test performance with within .1 dB of the design performance. The scalar predictive quantizer performance and the codebook for the prediction error quantizer are the same as the analytically optimized predictive quantization system of Arnstein [60] run on the same data.

Observe that the scalar FSVQ in the first experiment with 64 states yielded performance quite close to that of the scalar VPQ, which does not have a finite number of states. Intuitively the FSVQ is trying to approximate the infinite state machine by using a large number of states. The VPQ, however, is less complex and requires less memory and hence for this application is superior.

For comparison, the best 1 bit/sample scalar trellis encoding system for this source yields 11.25 dB for this source [52]. The trellis encoding system uses a block Viterbi algorithm with a search depth of 1000 samples for the encoder. It is perhaps surprising that in this example the VPQ and the FSVQ with the short delay of only 4 samples can outperform a Viterbi algorithm with a delay of 1000 samples. It points out, however, two advantages of feedback VQ over scalar trellis encoding systems: 1. The decoder is permitted to be a more general form of finite-state machine than the shift-register based nonlinear filter usually used in trellis encoding systems; and 2. the encoder performs a single full search of a small vector codebook instead of a Viterbi algorithm consisting of a tree search of a sequence of scalar codebooks. In other words, single short vector searches may yield better performance than a "look ahead" sequence of searches of scalar codebooks.

Speech waveform coding

The second set of results considers a training sequence of 640,000 samples of ordinary speech from four different male speakers sampled at 6.5 kHz. The reader is reminded that squared error is not generally a subjectively good distortion measure for speech. Better subjective quality may be obtained by using more complicated distortion measures such as the general quadratic distortion measures with input dependent weightings such as the arithmetic segmented distortions. The VQ design techniques extend to such distortion measures, but the centroid computations are more complicated. (See [30] for the theory and [45, 46] for the application of input-weighted quadratic distortion measures.)

Tables III and IV are the counterparts of Tables I and II for this source. Now, however, the SNR's of the codes on test sequences of samples outside of the training sequence (and by a different speaker) are presented for comparison. In addition, some larger dimensions are con-

TABLE IV
FEEDBACK VQ OF SAMPLED SPEECH.

	FSVQ					VPQ			
k	SNRin	SNRout	K	n	M	SNRin	SNRout	n	M
1	2.0	2.0	2	2	2	2.1	2.6	2	2
2	7.8	7.5	32	4	64	6.4	6.2	4	8
3	9.0	8.3	64	10	192	7.3	6.8	8	24
4	10.9	9.4	512	16	2048	8.0	7.6	16	64
5	12.2	10.8	512	32	2560				

Signal to Noise Ratios inside training sequence (SNRin) of 640000 speech samples, Signal to Noise Ratios outside training sequence (SNRout) of 76800 speech samples, number of states (K), number of multiplications per sample (n), and storage (M) for feedback quantizers: Rate = 1 bit/sample. k = vector dimension.

sidered because the longer training sequence made them more trustworthy. Again for comparison, the best known (nonadaptive) scalar trellis encoding system for this source yields a performance of 9 dB [52]. Here the trellis encoder uses the M-algorithm with a search depth of 31 samples. The general comparisons are similar to those of the previous source, but there are several differences. The tree-searched VQ is now more degraded in comparison to the full search VQ and the multistage VQ is even worse, about 3 dB below the full search at the largest dimension in comparison to about 1 dB for the Gauss Markov case. The complexity and storage requirements are the same except for the shape/gain VQ where different optimum selections of gain and shape codebook size yield different complexity and storage requirements. The VPQ of dimension 4 is inferior to the trellis encoder and the FSVQ of the same dimension. The four dimensional FSVQ, however, still outperforms the scalar trellis encoder.

Observe that an FSVQ of dimension 4 provides better performance inside and outside the training sequence than does a full search memoryless vector quantizer of dimension 8, achieving better performance with 16 4-dimensional distortion evaluations than with 512 8-dimensional distortion computations. The cost, of course, is a large increase in memory. This, however, is a basic point of FSVQ design—to use more memory but less computation.

LPC VQ (vocoding)

Table V presents a comparison of VQ and FSVQ for vector quantization of speech using the Itakura-Saito distortion measure or, equivalently, vector quantization of LPC speech models [16, 14, 53]. The training sequence and test sequence are as above, but now the input dimension is 128 samples and the output vectors are 10th order all-pole models. The training sequence is now effectively shorter since it contains only 5000 input vectors of this dimension. As a result the test results are noticeably different than the design results. Because of the shortness of the training sequence, only FSVQ's of small dimension and few states were considered.

The table summarizes memoryless VQ and two FSVQ designs: the first FSVQ design used was a straightforward application of the design technique outlined previously and the second used the stochastic iteration next-state improvement algorithm of [53]. Observe that the next-state function improvement yields codes that perform better outside of the training sequence then do the ordinary FSVQ codes.

TABLE V
LPC VQ AND FSVQ WITH AND WITHOUT NEXT STATE FUNCTION IMPROVEMENT.

		VQ		FSVQ1		FSVQ2		
R	r	SNRin	SNRout	SNRin	SNRout	SNRin	SNRout	K
1	.008	3.7	2.9					
2	.016	6.1	5.2	7.2	4.3	7.5	6.1	16
3	.023	7.3	6.2	8.4	5.9	9.0	7.5	16
4	.031	8.8	7.9	9.5	7.8	9.6	8.7	4
5	.039	9.7	8.8	10.6	8.9	10.7	9.3	4
6	.047	10.5	9.5					
7	.055	11.6	10.1					
8	.062	12.6	10.7					

Signal to Noise Ratios inside training sequence (SNRin) of 5000 vectors of 128 samples each, Signal to Noise Ratios outside training sequence (SNRout) of 600 vectors of 128 samples each: memoryless VQ, omniscient FSVQ design (FSVQ1), and for omnisicient FSVQ design with next-state function improvement (FSVQ2). K = number of states in FSVQ, R = rate in bits/vector, r = rate in bits/sample. Itakura-Saito distortion measure.

TABLE VI
ADAPTIVE VPQ.

	VPQ	
k	SNRin	SNRout
1	4.12	4.34
2	7.47	7.17
3	8.10	7.67
4	8.87	8.30

Signal to Noise Ratios inside training sequence (SNRin) of 5000 vectors, and Signal to Noise Ratios in test sequence (SNRout) of 600 vectors, rate = 1.023 bits/sample.

a)
b)
c)
d)
e)

Figure 16. Image Training Sequence. The training sequence consisted of the sequence of 3 × 4 subblocks of the five 256 × 256 images shown.

Gain/shape VQ's for this application are developed in [16] and [36]. Tree-searched LPC VQ is considered for binary and nonbinary trees in combination with gain/shape codes in [16] and [10].

Adaptive coding

Table VI presents the results of a simple example of an adaptive VQ, here consisting of an LPC VQ with 8 codewords every 128 samples combined with VPQs of dimensions 1–4. Each of the 8 VPQs is designed for the subsequence of training vectors mapping into the corresponding LPC VQ model [47]. The rate of this system is $1 + 3/128 = 1.023$ bits/sample. The performance is significantly worse than the 10 dB achieved by a hybrid scalar trellis encoder of the same rate [52], but it improves on the nonadaptive VPQ by about ¾ dB. Adaptive vector quantizers are still quite new, however, and relatively little work on the wide variety of possible systems has yet been done.

Image coding

In 1980–1982 four separate groups developed successful applications of VQ techniques to image coding [61, 62, 63, 64, 65, 66, 67, 37]. The only real difference from waveform coding is that now the VQ operates on small rectangular blocks of from 9 to 16 pixels, that is, the vectors are really 2-dimensional subblocks of images, typically squares with 3 or 4 pixels on a side or 3 by 4 rectangles. We here consider both the basic technique and one variation. We consider only small codebooks of 6 bits per 4 × 3 block of 12 pixels for purposes of demonstration. Better quality pictures could be obtained at the same rate of ½ bit per pixel by using larger block sizes and hence larger rates of, say, 8 to 10 bits per block. Better quality could also likely be achieved with more complicated distortion measures than the simple squared error used.

Fig. 16 gives the training sequence of five images. Fig. 17a shows a small portion of the fifth image, an eye, magnified. Fig. 17b is a picture of the $2^6 = 64$ codewords. Fig. 17c shows the decoded eye. Fig. 18 shows the original, decoded image, and error image for the complete picture. The error image is useful for highlighting the problems encountered with the ordinary memoryless VQ. In particular, edges are poorly reproduced and the codeword edges make the picture appear "blocky." This problem was attacked by Ramamurthi and Gersho [62, 67] by constructing segmented (or union or composite) codes—separate codebooks for the edge information and the texture information where a simple classifier was used to distinguish the two in design. In [37] a feedback vector quantizer was developed by using a separating mean VQ with a predictive scalar quantizer to track the mean. Fig. 19 shows the original eye, ordinary VQ, and the feedback VQ. The improved ability to track edges is clearly discernible. Fig. 20 shows the full decoded image for feedback VQ together with the error pattern.

Although image coding using VQ is still in its infancy,

a)

b)

c)

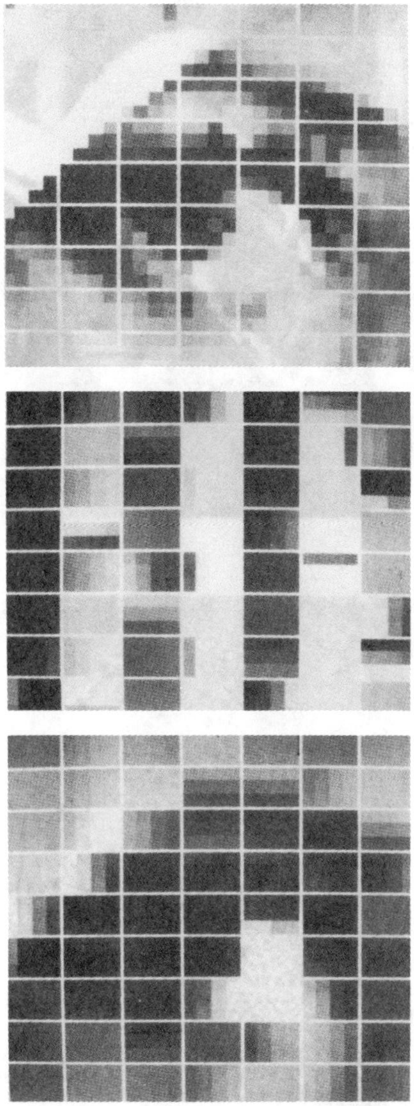

Figure 17. Basic Image VQ Example at ½ bit per pixel. (a) Original Eye Magnified (b) 6 bit codebook VQ codebook for 4 × 3 blocks (c) Decoded Image.

these preliminary experiments using only fairly simple memoryless and feedback VQ techniques with small codebooks demonstrate that the general approach holds considerable promise for such applications.

COMMENTS

We have described Lloyd's basic iterative algorithm and how it can be used to improve the performance of a variety of vector quantization systems, ranging from the fundamental memoryless full search VQ that serves as the basic model for data compression in information theory to a variety of feedback and adaptive systems that can be viewed as vector extensions of popular scalar compression systems. By a variety of examples of systems and code design simulations we have tried to illustrate some of the tradeoffs among performance, rate, complexity, and storage for these codes.

The basic structure of all of the VQ systems is well suited to VLSI implementation: a minimum distortion search algorithm on a chip communicating with off-board storage for codebooks and next-state transition functions. As new and better design algorithms are developed, the chips can be updated by simply reburning the codebook and transition ROM's.

The basic approach can also be incorporated into the design of some traditional scalar data compression schemes, an approach which Gersho calls "imbedded

a)

b)

c)

Figure 18. Full Image for Basic Example (a) Original (b) Decoded Image (c) Error Image.

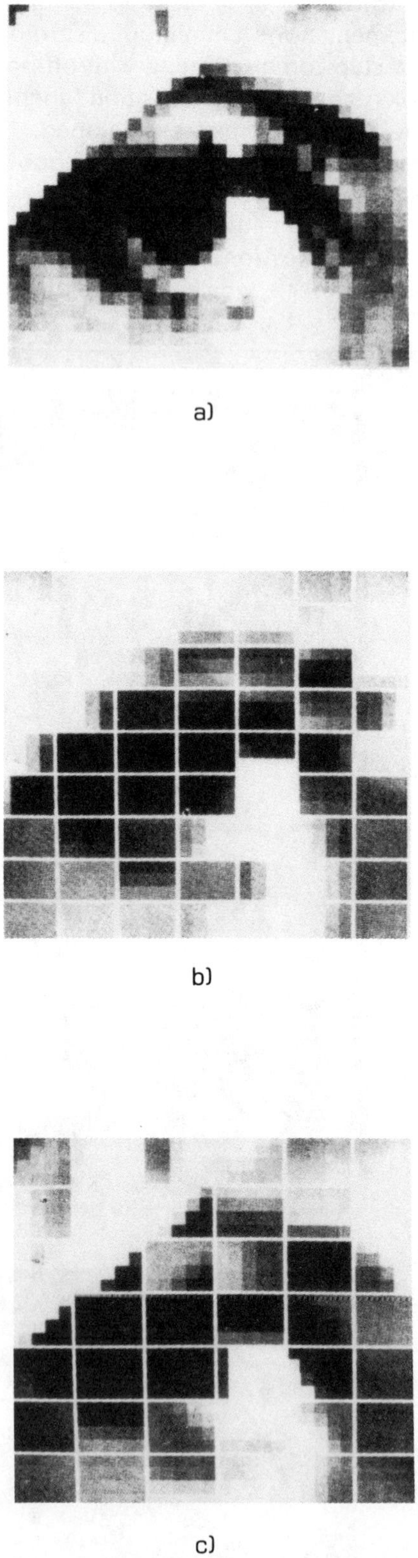

Figure 19. VQ vs. Separating Mean VQ at Rate ½ bit per pixel (a) Original Eye Magnified (b) VQ Decoded Image (c) Separating Mean VQ with DPCM Mean Coding Decoded Image.

Figure 20. Full Image for Separating Mean Example (a) Decoded Image using Separating Mean VQ with DPCM Mean Coding (b) Error Image.

VQ" [11]. Such schemes typically enforce additional structure on the code such as preprocessing, transforming, splitting into subbands, and scalar quantization, however, and hence the algorithms may not have the freedom to do as well as the more unconstrained structures considered here. Even if the traditional schemes prove more useful because of existing DSP chips or intuitive variations well matched to particular data sources, the vector quantization systems can prove a useful benchmark for comparison.

Recently VQ has also been successfully used in isolated word recognition systems without dynamic time warping by using either separate codebooks for each utterance or by mapping trajectories through one or more codebooks [68, 69, 70, 71, 55, 72]. Vector quantization has also been used as a front end acoustic processor to isolated utterance and continuous speech recognition systems which then do approximately maximum likelihood linguistic decoding based on probabilities estimated using "hidden Markov" models for the VQ output data. [73, 74, 75].

Variations of the basic VQ design algorithm have been tried for several distortion measures, including the squared error, weighted squared error, the Itakura-Saito distortion, and an (arithmetic) segmented signal to noise ratio. (See, e.g., [30, 45, 46]). Other distortion measures are currently under study.

The algorithm has not yet been extended to some of the more complicated distortion measures implicit in noise masking techniques for enhancing the subjective performance of scalar quantization speech coding systems. Whether scalar systems designed by sophisticated techniques matched to subjective distortion measures will sound or look better than vector systems designed for mathematically tractable distortion measures remains to be seen. Whenever the subjective distortion measures can be quantified and a means found to compute centroids, however, the vector systems will yield better quantitative performance. Since the centroid computation is only done in design and not in implementation, it can be quite complicated and still yield useful results.

The generalized Lloyd algorithm is essentially a clustering algorithm and we have attempted to demonstrate its applicability to the design of a variety of data compression systems. Other clustering algorithms may yield better codes in some applications. For example, Freeman [76] proposed a design algorithm for scalar trellis encoding systems using the squared error distortion measure which replaced the Lloyd procedure by a conjugate gradient procedure for minimizing the average distortion for a long training sequence. He found that for a memoryless Gaussian source the resulting codes were superior to those obtained by the Lloyd procedure. It would be interesting to characterize the reasons for this superiority, e.g., the procedure may find a better local minimum or it may simply be numerically better suited for finding a continuous local minimum on a digital computer. It would also be interesting to consider variations of this approach for the design of some of the other systems considered here.

A survey article with many topics cannot provide complete descriptions or exhaustive studies of any of systems sketched. It is hoped, however, that these examples impart the flavor of vector quantizer design algorithms and that they may interest some readers to further delve into the recent and current work in the area.

ACKNOWLEDGMENT

The author gratefully acknowledges the many helpful comments from students and colleagues that aided the preparation of this paper.

Portions of the research described here were supported by the Army Research Office, the Air Force Office of Scientific Research, the National Science Foundation, the John Simon Guggenheim Memorial Foundation, and the Joint Services Electronics Program at Stanford University.

REFERENCES

[1] Davisson, L. D. and Gray, R. M., *Data Compression*, Dowden, Hutchinson, & Ross, Inc., Stroudsbug, PA (1976). Benchmark Papers in Electrical Engineering and Computer Science, Volume 14.

[2] Jayant, N. S., Editor, *Waveform coding quantization and Coding*, IEEE Press, NY (1976).

[3] Jayant, N. S. and Noll, P., *Digital Coding of Waveforms*, Prentice-Hall, Englewood Cliffs, NJ (1984).

[4] Shannon, C. E., "A mathematical theory of communication," *Bell Systems Technical Journal* 27 pp. 379–423, 623–656 (1948).

[5] Shannon, C. E., "Coding theorems for a discrete source with a fidelity criterion," *IRE National Convention Record, Part 4*, pp. 142–163 (1959).

[6] Gallager, R. G., *Information theory and reliable communication*, John Wiley & Sons, NY (1968).

[7] Berger, T., *Rate Distortion Theory*, Prentice-Hall Inc., Englewood Cliffs, NJ (1971).

[8] Viterbi, A. J. and Omura, J. K., *Principles of Digital Communication and Coding*, McGraw-Hill Book Company, New York (1979).

[9] Lloyd, S. P., *Least squares quantization in PCM*, Bell Laboratories Technical Note (1957). (Published in the March 1982 special issue on quantization).

[10] Wong, D., Juang, B.-H., and Gray, A. H., Jr., "An 800 bit/s vector quantization LPC vocoder," *IEEE Transactions on Acoustics Speech and Signal Processing* ASSP-30 pp. 770–779 (October 1982).

[11] Gersho, A. and Cuperman, V., "Vector Quantization: A pattern-matching technique for speech coding," *IEEE Communications Magazine*, (December 1983).

[12] Itakura, F. and Saito, S., "Analysis synthesis telephony based on the maximum liklihood method," *Proceedings of the 6th International Congress of Acoustics*, pp. C-17–C-20 (August 1968).

[13] Kullback, S., *Information Theory and Statistics*, Dover, New York (1969).

[14] Gray, R. M., Gray, A. H., Jr., Rebolledo, G., and Shore, J. E., "Rate distortion speech coding with a minimum discrimination information distortion measure," *IEEE Transactions on Information Theory* IT-27 (6) pp. 708–721 (Nov. 1981).

[15] Shore, J. E. and Gray, R. M., "Minimum-cross-entropy pattern classification and cluster analysis," *IEEE Transactions on Pattern Analysis and Machine Intelligence* PAMI-4 pp. 11–17 (Jan. 1982).

[16] Buzo, A., Gray, A. H., Jr., and Gray, R. M., and Markel, J. D., "Speech coding based upon vector quantization," *IEEE Transactions on Acoustics Speech and Signal Processing* ASSP-28 pp. 562–574. (October 1980).

[17] Gray, R. M., Buzo, A., Gray, A. H., Jr., and Matsuyama, Y., "Distortion measures for speech processing," *IEEE Transactions on Acoustics, Speech, and Signal Processing*, ASSP-28 pp. 367–376 (August 1980).

[18] Gray, R. M., Kieffer, J. C., and Linde, Y., "Locally opti-

mal block quantizer design," *Information and Control* 45 pp. 178–198 (May 1980).

[19] Gray, R. M. and Kieffer, J. C., "Asymptotically mean stationary measures," *Annals of Probability* 8 pp. 962–973 (Oct. 1980).

[20] Kieffer, J. C. and Rahe, M., "Markov channels are asymptotically mean stationary," *Siam Journal of Mathematical Analysis* 12 pp. 293–305 (1980).

[21] Fontana, R. J., Gray, R. M., and Kieffer, J. C., "Asymptotically mean stationary channels," *IEEE Transactions on Information Theory* IT-27 pp. 308–316 (May 1981).

[22] Kieffer, J. C., "Stochastic stability for feedback quantization schemes," *IEEE Transactions on Information Theory* IT-28 pp. 248–254 (March 1982).

[23] Sabin, M. J. and Gray, R. M., *Asymptotic properties of the generalized Lloyd algorithm,* Submitted for publication 1983.

[24] Gersho, A., "On the structure of vector quantizers," *IEEE Transactions on Information Theory* IT-28 pp. 157–166 (March 1982).

[25] Linde, Y., Buzo, A., and Gray, R. M., "An algorithm for vector quantizer design," *IEEE Transactions on Communications* COM-28 pp. 84–95 (January 1980).

[26] MacQueen, J., "Some methods for classification and analysis of multivariate observations," *Proc. of the Fifth Berkeley Symposium on Math. Stat. and Prob.* 1 pp. 281–296 (1967).

[27] Diday, E. and Simon, J. C., "Clustering analysis," in *Digital Pattern Recognition,* ed. K. S. Fu, Springer-Verlag, NY (1976).

[28] Chen, D. T. S., "On two or more dimensional optimum quantizers," *Proceedings, 1977 International Conference on Acoustics, Speech, and Signal Processing,* pp. 640–643 (1977).

[29] Adoul, J.-P., Morissette, S., and Rudko, M., "Bit-rate-halving algorithm for PCM-encoded speech using a new bidimensional data compression scheme," *Record of the 1979 IEEE International Conference on Acoustics Speech and Signal Processing,* pp. 432–435 (April 1979).

[30] Gray, R. M. and Karnin, E., "Multiple local optima in vector quantizers," *IEEE Transactions on Information Theory* IT-28 pp. 256–261 (March 1982).

[31] Abut, H., Gray, R. M., and Rebolledo, G., "Vector quantization of speech and speech-like waveforms," *IEEE Transactions on Acoustics Speech and Signal Processing* ASSP-30 pp. 423–435 (June 1982).

[32] Adoul, J.-P., Debray, J.-L., and Dalle, D., "Spectral distance measure applied to the optimum design of DPCM coders with L predictors," *Proceedings of the 1980 IEEE International Conference on Acoustics Speech and Signal Processing,* pp. 512–515 (April 1980).

[33] Gersho, A. and Cheng, D., "Fast nearest neighbor search for nonstructured Euclidean codes," *Abstracts of the 1983 IEEE International Symposium on Information Theory,* p. 88 (September 1983).

[34] Juang, B.-H. and Gray, A. H., Jr., "Multiple stage vector quantization for speech coding," *Proceedings of the IEEE International Conference on Acoustics Speech and Signal Processing* 1 pp. 597–600 (April 1982).

[35] Sabin, M. J. and Gray, R. M., "Product code vector quantizers for speech waveform coding," *Conference Record Globecom '82,* pp. 1087–1091 (December 1982).

[36] Sabin, M. J. and Gray, R. M., *Product code vector quantizers for waveform and voice coding, IEEE Trans. ASSP,* to appear, (April 1984).

[37] Baker, R. L. and Gray, R. M., "Differential vector quantization of achromatic imagery," *Proceedings of the International Picture Coding Symposium,* (March 1983).

[38] Gersho, A., "Asymptotically optimal block quantization," *IEEE Transactions on Information Theory* IT-25 pp. 373–380 (July 1979).

[39] Conway, J. H. and Sloane, N. J. A., "Voronoi regions of lattices, second moments of polytopes, and quantization," *IEEE Transactions on Information Theory* IT-28 pp. 211–226 (March 1982).

[40] Conway, J. H. and Sloane, N. J. A., "Fast quantizing and decoding algorithms for lattice quantizers and codes," *IEEE Transactions on Information Theory* IT-28 pp. 227–232 (March 1982).

[41] Conway, J. H. and Sloane, N. J. A., "On the Voronoi regions of certain lattices," *SIAM Journal of Alg. Disc. Math.,* (1983). in press

[42] Barnes, E. S. and Sloane, N. J. A., "The optimal lattice quantizer in three dimensions," *SIAM Journal of Alg. Disc. Math.* 4 pp. 30–41 (1983).

[43] Foster, J., Newkirk, J., and Gray, R. M., *VLSI implementation of a finite-state vector quantization waveform encoder,* submitted for publication 1983.

[44] Gaarder, N. T. and Slepian, D., "On optimal finite-state digital transmission systems," *IEEE Transactions on Information Theory* IT-28 pp. 167–186 (March 1982).

[45] Cuperman, V. and Gersho, A., "Adaptive differential vector coding of speech," *Conference Record, GlobeCom 82,* pp. 1092–1096 (December 1982).

[46] Cuperman, V. and Gersho, A., *Vector predictive coding of speech at 16 Kb/s,* Submitted for possible publication 1983.

[47] Chang, P. C., Ph.D. Research, Stanford University 1983.

[48] Gibson, J. D., Jones, S. K., and Melsa, J. L., "Sequentially adaptive prediction and coding of speech signals," *IEEE Transactions on Communications* COM-22 pp. 1789–1797 (November 1974).

[49] Dunn, J. G., "An experimental 9600-bit/s voice digitizer employing adaptive prediction," *IEEE Transactions on Communication Technology* COM-19 pp. 1021–1032 (December 1971).

[50] Foster, J. and Gray, R. M., "Finite-state vector quantization," *Abstracts of the 1982 International Sym-*

posium on Information Theory, (June 1982).
[51] Foster, J., Gray, R. M., and Ostendouf, M., *Finite-state vector quantization for waveform coding,* IEEE Trans. Info. Theory, to appear.
[52] Stewart, L. C., Gray, R. M., and Linde, Y., "The design of trellis waveform coders," *IEEE Transactions on Communications* COM-30 pp. 702–710 (April 1982).
[53] Ostendorf, M. and Gray, R. M., *An algorithm for the design of labeled-transition finite-state vector quantizers,* submitted for publication 1983.
[54] Fehn, H. G. and Noll, P., "Multipath search coding of stationary signals with applications to speech," *IEEE Transactions on Communications* COM-30 pp. 687–701 (April 1982).
[55] Shore, J. E. and Burton, D. K., "Discrete utterance speech recognition without time alignment," *IEEE Transactions on Information Theory* IT-29 pp. 473–491 (July 1983).
[56] Rebolledo, G., Gray, R. M., and Burg, J. P., "A multirate voice digitizer based upon vector quantization," *IEEE Transactions on Communications* COM-30 pp. 721–727 (April 1982).
[57] Adoul, J.-P. and Mabilleau, P., "4800 bps RELP vocoder using vector quantization for both filter and residual representation," *Proceedings of the IEEE International Conference on Acoustics Speech and Signal Processing* 1 p. 601 (April 1982).
[58] Heron, C. D., Crochiere, R. E., and Cox, R. V., "A 32-band subband/transform coder incorporating vector quantization for dynamic bit allocation," *Proceedings ICASSP,* pp. 1276–1279 (April 1983).
[59] Gray, R. M. and Linde, Y., "Vector quantizers and predictive quantizers for Gauss-Markov sources," *IEEE Transactions on Communications* COM-30 pp. 381–389 (Feb. 1982).
[60] Arnstein, D. S., "Quantization error in predictive coders," *IEEE Transactions on Communications* COM-23 pp. 423–429 (April 1975).
[61] Yamada, Y., Fujita, K., and Tazaki, S., "Vector quantization of video signals," *Proceedings of Annual Conference of IECE,* p. 1031 (1980).
[62] Gersho, A. and Ramamurthi, B., "Image coding using vector quantization," *Proceedings of the IEEE International Conference on Acoustics Speech and Signal Processing* 1 pp. 428–431 (April 1982).
[63] Baker, R. L. and Gray, R. M., "Image compression using non-adaptive spatial vector quantization," *Conference Record of the Sixteenth Asilomar Conference on Circuits Systems and Computers,* (October 1982).
[64] Murakami, T., Asai, K., and Yamazaki, E., "Vector quantizer of video signals," *Electronic Letters* 7 pp. 1005–1006 (Nov. 1982).
[65] Yamada, Y. and Tazaki, S., "Vector quantizer design for video signals," *IECE Transactions* J66-B pp. 965–972 (1983). (in Japanese)
[66] Yamada, Y. and Tazaki, S., "A method for constructing successive approximation vector quantizers for video signals," *Proceedings of the Annual Conference of the Institute of Television Engineers of Japan,* pp. 6–2 (1983).
[67] Ramamurthi, B. and Gersho, A., "Image coding using segmented codebooks," *Proceedings International Picture Coding Symposium,* (Mar. 1983).
[68] Hamabe, R., Yamada, Y., Murata, M., and Namekawa, T., "A speech recognition system using inverse filter matching technique," *Proceedings of the Ann. Conf. Inst. of Television Engineers,* (June 1981). (in Japanese)
[69] Shore, J. E. and Burton, D. K., "Discrete utterance speech recognition without time alignment," *Proceedings 1982 IEEE International Conference on Acoustics Speech and Signal Processing,* p. 907 (May 1982).
[70] Martinez, H. G., Riviera, C., and Buzo, A., "Discrete utterance recognition based upon source coding techniques," *Proceedings IEEE International Conference on Acoustics Speech and Signal Processing,* pp. 539–542 (May 1982).
[71] Rabiner, L., Levinson, S. E., and Sondhi, M. M., "On the application of vector quantization and hidden Markov models to speaker-independent isolated word recognition," *Bell System Technical Journal* 62 pp. 1075–1106 (April 1983).
[72] Shore, J. E., Burton, D., and Buck, J., "A generalization of isolated word recognition using vector quantization," *Proceedings 1983 International Conference on Acoustics Speech and Signal Processing,* pp. 1021–1024 (April 1983).
[73] Jelinek, F., Mercer, R. L., and Bahl, L. R., "Continuous speech recognition: statistical methods," in *Handbook of Statistics,* Vol. 2, P. R. Khrishaieh and L. N. Kanal, eds., North-Holland, pp. 549–573. (1982).
[74] Billi, R., "Vector quantization and Markov models applied to speech recognition," *Proc. ICASSP 82,* pp. 574–577, Paris (May 1982).
[75] Rabiner, L. R., Levinson, S. E., and Sondhi, M. M., "On the application of vector quantization and hidden Markov models to speaker-independent isolated word recognition," *BSTJ,* Vol. 62, pp. 1075–1105, (April 1983).
[76] Freeman, G. H., "The design of time-invariant trellis source codes," *Abstracts of the 1983 IEEE International Symposium on Information Theory,* pp. 42–43 (September 1983).
[77] Haoui, A. and Messerschmidt, D., "Predictive Vector Quantization," Proceedings of the IEEE International Conference on Acoustics, Speech, and Signal Processing (1984).

Notes added in proof: A similar design technique for FSVQ was independently developed by Haoui and Messerschmidt [77]. It should be pointed out that the FSVQ design algorithm described here is incomplete in that it does not describe the methods used to avoid non-communicating collection of states and wasted states. These issues are discussed in [51].

Vector Quantization in Speech Coding

JOHN MAKHOUL, FELLOW, IEEE, SALIM ROUCOS, MEMBER, IEEE, AND HERBERT GISH, MEMBER, IEEE

Invited Paper

***Quantization**, the process of approximating continuous-amplitude signals by digital (discrete-amplitude) signals, is an important aspect of **data compression** or **coding**, the field concerned with the reduction of the number of bits necessary to transmit or store analog data, subject to a distortion or fidelity criterion. The independent quantization of each signal value or parameter is termed **scalar** quantization, while the joint quantization of a block of parameters is termed block or **vector** quantization. This tutorial review presents the basic concepts employed in vector quantization and gives a realistic assessment of its benefits and costs when compared to scalar quantization. Vector quantization is presented as a process of redundancy removal that makes effective use of four interrelated properties of vector parameters: linear dependency (correlation), nonlinear dependency, shape of the probability density function (pdf), and vector dimensionality itself. In contrast, scalar quantization can utilize effectively only linear dependency and pdf shape. The basic concepts are illustrated by means of simple examples and the theoretical limits of vector quantizer performance are reviewed, based on results from rate-distortion theory. Practical issues relating to quantizer design, implementation, and performance in actual applications are explored. While many of the methods presented are quite general and can be used for the coding of arbitrary signals, this paper focuses primarily on the coding of speech signals and parameters.*

I. Introduction

Current projections for world-wide communications in the 1990s and beyond, point to a proliferation of digital transmission as a dominant means of communication for voice and data. Digital transmission is expected to provide flexibility, reliability, and cost effectiveness, with the added potential for communication privacy and security through encryption. The costs of digital storage and transmission media are generally proportional to the amount of digital data that can be stored or transmitted. While the cost of such media decreases every year, the demand for their use increases at an even higher rate. Therefore, there is a continuing need to minimize the number of bits necessary to transmit signals while maintaining acceptable signal fidelity or quality. The branch of electrical engineering that deals with the latter problem is termed *data compression* or *coding*. When applied to speech, it is known as *speech compression* or *speech coding*.

The conversion of an analog (continuous-time, continuous-amplitude) source into a digital (discrete-time, discrete-amplitude) source, consists of two parts: *sampling* and *quantization*. Sampling converts a continuous-time signal into a discrete-time signal by measuring the signal value at regular intervals of time. Quantization converts a continuous-amplitude signal into one of a set of discrete amplitudes, thus resulting in a discrete-amplitude signal that is different from the continuous-amplitude signal by the quantization error or noise. In this paper, we shall assume that our signals are adequately sampled (see, for example, [107]) so that the only loss in fidelity is attributable to quantization.

When each of a set of parameters (or a sequence of signal values) is quantized separately, the process is known as *scalar quantization*. When the set of parameters is quantized jointly as a single vector, the process is known as *vector quantization* (also known as block quantization or pattern-matching quantization). We shall often abbreviate vector quantization in this paper as VQ.

A. Purpose and Scope

The main purpose of this paper is to present the reader with information that can be used in making a realistic assessment of the benefits and costs of vector quantization relative to scalar quantization, especially in speech coding applications. The emphasis is on the exposition of basic principles rather than the elaboration of various techniques and their variations for which references to the literature are provided. Vector quantization is presented as a process of redundancy removal that makes effective use of four interrelated properties of vector parameters: linear dependency (correlation), nonlinear dependency, shape of the probability density function (pdf), and vector dimensionality itself. We shall see that linear dependency and pdf shape can be employed quite effectively with scalar quantization while the other two properties cannot. Nonlinear dependency plays a significant role in the quantization of speech spectral parameters, while dimensionality is important for waveform quantization. Because of the relatively large cost of vector quantization (generally exponential in the number of dimensions and the number of bits per dimension), given today's computation and storage technologies, the major benefits of vector quantization are

Manuscript received January 15, 1985; revised July 2, 1985.
The authors are with BBN Laboratories Inc., Cambridge, MA 02238, USA.

Reprinted from *Proc. IEEE,* vol. 73, pp. 1551–1558, Nov. 1985.

realized largely at transmission rates of about 1 bit per parameter or less, which is exactly the range where the performance of scalar quantizers degrades sharply. While the concepts presented are quite general, we shall focus in this paper on the low-rate coding of speech (below 8 kbits/s) as an application.

Vector quantization for the purpose of speech coding was used by Dudley [31] in the 1950s and Smith [127] in the 1960s. However, it was not until the introduction of linear predictive coding (LPC) [8], [71], [86], [90] to speech coding that VQ has had significant activity, starting with the work of Kang and Coulter [77], but spurred on mainly by the work of Buzo *et al.* [18], [81]. Until recently, the main purpose for the use of VQ in speech coding has been to reduce the transmission rate of 2400-bit/s vocoders (voice coders) to operate at much lower rates while maintaining acceptable speech intelligibility and quality. Speech coding at very low rates, in the range of 200–800 bits/s, has attracted substantial interest [18], [32], [53], [56], [76], [77], [98], [100], [111], [119], [137], [138] for use in both government and commercial applications. At such low rates, it is important to maximize the cost effectiveness of every bit that is transmitted. Vector quantization has been instrumental in retaining sufficient speech intelligibility to make such systems of actual utility. Today, very-low-rate coding of speech remains one of the major successful applications of VQ. More recently, a mushrooming research activity in the application of VQ to speech waveform coding at somewhat higher data rates has been taking place. While much of the activity has focused on the 8–16-kbit/s range [1], [24], [25], [34], [37], [42], [49], [50], [56], [63], [83], [109], [124], some work has started at data rates below 8 kbits/s [7], [117], which points to exciting possibilities for high-quality speech coding at low rates.

We should point out that, in a sense, VQ has been used regularly and effectively in pattern-recognition type of speech applications, such as in speech and speaker recognition (see, for example, [27], [80], [95], [104], [106], [118], [126]). After all, the VQ problem is part of the general pattern-recognition problem of the classification of data into a discrete number of categories that optimize some fidelity criterion. Indeed, in the design of vector quantizers, one often employs well-known techniques from pattern recognition. However, the basic theory underlying VQ stems from information theory and has wider implications for the transmission of information.

The theoretical foundations of data compression and vector quantization lie in a branch of information theory known as rate-distortion theory, originally set forth by Shannon [122]. Also, most of the theoretical developments since Shannon have taken place as part of the information theory discipline. Of particular relevance is the book by Berger [14] on rate-distortion theory, as well as other information theory texts, such as [46], [92]. Because a full development of vector quantization theory would be highly mathematical, we have chosen in this paper to concentrate on presenting the basic notions and the major results with just enough mathematics that would allow us to be complete without being obscure, we hope.

For further reading, we list a few key references which also contain other references to the literature. The books of collected papers edited by Jayant [73] and Davisson and Gray [26] are devoted to data compression and cover aspects of speech compression. Two relatively recent special issues, the IEEE TRANSACTIONS ON INFORMATION THEORY of March 1982 and the IEEE TRANSACTIONS ON COMMUNICATIONS of April 1982, are devoted to quantization and to speech coding, respectively. The most comprehensive treatment of waveform coding, with applications to speech and video, is the recent book by Jayant and Noll [74]. The review articles by Gold [52], Flanagan *et al.* [39], and Makhoul [87] cover various aspects of speech coding, while the review articles by Gersho and Cuperman [49] and Gray [56] describe some recent work in vector quantization.

B. Paper Outline

In Section II we present the basic VQ problem and several distortion measures that are utilized, along with the basic system design and associated computational and storage costs. The section ends with a VQ model that is introduced, with examples, to aid the reader in visualizing the different processes at work in VQ and in assessing the relative merits of vector and scalar quantization. The remainder of the paper can be viewed as an elaboration of the basic model, supported by theoretical and practical results. Section III contains some of the major theoretical results known about VQ performance from rate-distortion theory. Section IV is devoted to the design of scalar quantizers for vector sources; it includes a comparison between scalar and vector quantization for low-rate speech coding. In Section V we present important practical considerations for vector quantizer design, including methods for reducing computational and storage costs at some loss in performance, and issues of robustness in terms of expected differences between design performance and operational performance. One can take advantage of long-term time-related signal dependencies to reduce the bit rate further without sacrificing signal fidelity; several time-dependent VQ methods are summarized in Section VI. The main paper ends in Section VII with a brief discussion of VQ in speech waveform coding and an outlook to the future.

C. Speech Coding

Before we start the main presentation, we describe the main components of a general speech coding system and two paradigms that we shall use in this paper as a basis for our examples from speech coding.

Fig. 1 shows the basic components of a data compression system appropriate for speech coding. The first component analyzes the discrete-time signal $s(n)$[1] and extracts a vector of unquantized parameters $\boldsymbol{x}(n)$. The set of parameters $\boldsymbol{x}(n)$ is quantized into the vector $\boldsymbol{y}(n)$, which is then encoded into a sequence of bits $\boldsymbol{c}(n)$ and transmitted through the transmission channel or stored in some storage medium. (The quantizer includes any prediction and feedback loops that are an integral part of the quantization process.) In general, the output of the channel $\boldsymbol{c}'(n)$ will be different from $\boldsymbol{c}(n)$ if there are channel errors. At the receiver, the decoder converts the sequence of bits $\boldsymbol{c}'(n)$ into parameter values $\boldsymbol{y}'(n)$, which are then used as input

[1]The dependence of the discrete-time signals on the sampling period will be assumed but not shown explicitly.

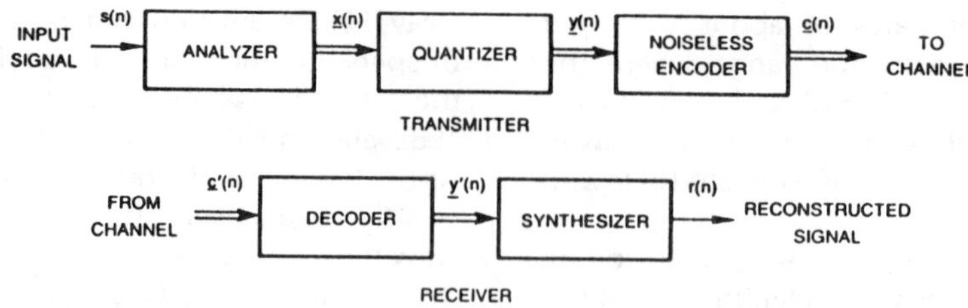

Fig. 1. Basic components of a data compression system for speech coding.

to the synthesizer. The output $r(n)$ is the reconstructed signal which will be an approximation to the input signal $s(n)$.

If there are no channel errors, then $c'(n) = c(n)$ and $y'(n) = y(n)$. The subject of channel errors is important, but will be treated only briefly in this paper. Unless otherwise noted, we shall assume no channel errors.

The nature of the synthesizer in Fig. 1 determines the type of voice coder and dictates the type of analysis to be performed. Fig. 2 shows an example of a synthesis model that is in general use. The model has two major components: an excitation (or source) and a spectral shaping filter.

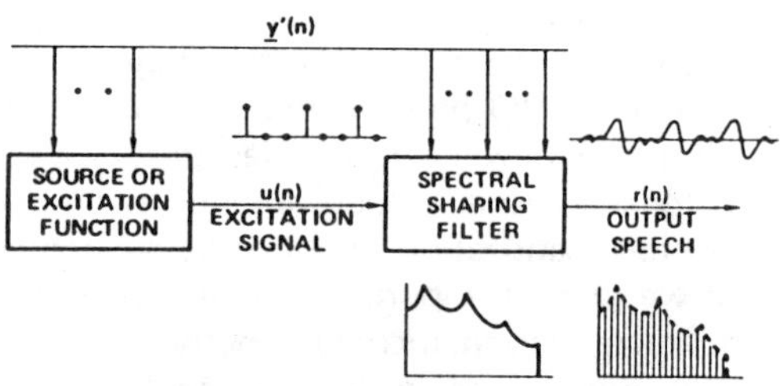

Fig. 2. Major components of the synthesizer in many speech coding systems.

Having chosen a particular synthesis model, any reduction in transmission rate is accomplished by the quantizer and the encoder in Fig. 1. The encoder[2] is assumed to be noiseless, i.e., it does not introduce any additional noise or loss in fidelity. It assigns bits to $y(n)$ in such a way as to minimize the transmission rate, without any loss in fidelity, and may include additional bits to protect the transmitted bit stream against channel errors.

The distortion in the output $r(n)$ relative to the input $s(n)$ may be the result of two processes: modeling and quantization. The modeling effected by the analysis/synthesis system in Fig. 1 may introduce a certain amount of distortion, even in the absence of any quantization (see, for example, the pitch-excited model described below). In this paper, we shall focus only on the distortion caused by the quantization process.

The speech coding task then is to design a system that minimizes the transmission rate while maintaining a certain speech quality, or conversely to maximize speech quality (minimize distortion) for a given transmission rate, subject to certain system cost constraints. For a given choice of analysis/synthesis system, the distortion and transmission rates are determined by the quantizer and the encoder.

[2]The terms "encoder" and "coding" are often used to refer to the whole compression process, as in speech coding. The noiseless encoder in Fig. 1 refers to a very specific part of the compression process. It is hoped that it will be clear from the context which of the two usages is meant.

Speech coding paradigms: We shall employ two basic paradigms as representative of the applications in speech coding, a low-rate coding paradigm and a medium-rate paradigm. The terms low-rate (or narrow-band) and medium-rate (or medium-band) have been used in the literature to denote a wide variety of systems. We shall use the term *medium-rate* for systems operating in the range 8–16 kbits/s, and *low-rate* for systems operating below that range, typically at or below 2400 bits/s. The term *very-low-rate* is often used to denote low-rate systems operating below about 1000 bits/s. Below, we give the basic paradigms that are used in this paper as examples of current systems that operate in the two main ranges.

In our paradigms, the synthesis model shown in Fig. 2 is based on a short-term spectral analysis of speech, where the speech signal is modeled as the output of an all-pole spectral shaping filter

$$H(z) = \frac{G}{A(z)} = \frac{G}{1 + \sum_{k=1}^{N} a(k) z^{-k}} \tag{1}$$

which is excited by a source with a flat spectral envelope. This is the well-known linear-predictive coding (LPC) model for speech. The gain G and the predictor coefficients $\{a(k), 1 \leqslant k \leqslant N\}$ are computed on a short-term basis over a *frame* of about 20–30 ms in which the speech signal can be considered to be approximately stationary. The coefficients are obtained as a result of minimizing the energy of the *prediction residual* obtained by filtering the input signal $s(n)$ through the all-zero filter $A(z)$. Ideally, the *excitation* signal $u(n)$ in Fig. 2 should match the residual signal so that the reconstructed signal $r(n)$ will match the input $s(n)$. In many medium-rate systems, the excitation used is exactly a quantized version of the prediction residual. This is true of many predictive waveform coding systems such as adaptive predictive coding (APC) [10], [89] and certain implementations of adaptive transform coding (ATC) [16]. We shall refer to such systems as *residual-excited* systems. (In Section VII, we shall include another filter that models the periodicity in the speech signal.)

At low rates, without using VQ it becomes necessary to have a simple model of the residual to maintain adequate speech intelligibility and quality. The most popular model is the *pitch-excited* model, where the speech in each frame is declared as either *voiced* or *unvoiced*. (Vowels and nasals are examples of voiced sounds, while consonants such as p, t, k, f, s, are unvoiced.) If a voiced determination is made, i.e., the sound is quasi-periodic at that point, the *pitch*, or fundamental frequency is measured and transmitted as well.

At the receiver, voiced sounds are synthesized by exciting the spectral shaping filter by a sequence of pulses separated by the period (Fig. 2 depicts that situation). For unvoiced sounds, a white random noise source is used as excitation. In either case, the gain G of the filter $H(z)$ is set such that the short-term energy in the output is equal to that of the input speech. Note that the pitch-excited model causes a certain amount of modeling distortion in the output, which can be heard even with no quantization of the model parameters.

II. Vector Quantization

This section begins with a formulation of the vector quantization problem, followed by a discussion of the more common distortion measures that are employed. Next we present the basic VQ system design and its associated computational and storage costs. The section ends with the introduction of a VQ model that is aimed at giving the reader a view of the various processes at work in vector quantization.

A. *Problem Formulation*

We assume that $\boldsymbol{x} = [x_1\ x_2\ \cdots\ x_N]^T$ is an N-dimensional vector whose components $\{x_k, 1 \leqslant k \leqslant N\}$ are real-valued, continuous-amplitude random variables. (The superscript T denotes transpose.) In vector quantization, the vector $\boldsymbol{x}$ is mapped onto another real-valued, discrete-amplitude, N-dimensional vector $\boldsymbol{y}$. We say that $\boldsymbol{x}$ is quantized as $\boldsymbol{y}$, and $\boldsymbol{y}$ is the quantized value of $\boldsymbol{x}$. We write

$$\boldsymbol{y} = q(\boldsymbol{x})$$

where $q(\cdot)$ is the quantization operator. $\boldsymbol{y}$ is also called the *reconstruction vector* or the *output vector* corresponding to $\boldsymbol{x}$. Typically, $\boldsymbol{y}$ takes on one of a finite set of values $Y = \{\boldsymbol{y}_i, 1 \leqslant i \leqslant L\}$, where $\boldsymbol{y}_i = [y_{i1}\ y_{i2}\ \cdots\ y_{iN}]^T$. The set Y is referred to as the reconstruction *codebook*, or simply the codebook, L is the *size* of the codebook, and $\{\boldsymbol{y}_i\}$ are the set of *code vectors*. The vectors $\boldsymbol{y}_i$ are also known in the pattern-recognition literature as the reference patterns or *templates*. The size L of the codebook is also called the *number of levels*, a term borrowed from scalar quantization terminology. Thus one talks about an L-level codebook or L-level quantizer. To design such a codebook, we partition the N-dimensional space of the random vector $\boldsymbol{x}$ into L regions or *cells* $\{C_i, 1 \leqslant i \leqslant L\}$ and associate with each cell C_i a vector $\boldsymbol{y}_i$. The quantizer then assigns the code vector $\boldsymbol{y}_i$ if $\boldsymbol{x}$ is in C_i

$$q(\boldsymbol{x}) = \boldsymbol{y}_i, \qquad \text{if } \boldsymbol{x} \in C_i. \tag{2}$$

The codebook design process is also known as *training* or *populating the codebook*. A method for designing the codebook will be presented in Section II-C.

Fig. 3 shows an example of a partitioning of two-dimensional space ($N = 2$) for the purpose of vector quantization. The region enclosed by the bold lines is the cell C_i. Any input vector $\boldsymbol{x}$ that lies in the cell C_i is quantized as $\boldsymbol{y}_i$. The positions of the code vectors corresponding to the other cells are shown by dots. The total number of code vectors in the example of Fig. 3 is $L = 18$.

For $N = 1$, vector quantization reduces to scalar quantization. Fig. 4 shows an example of a partitioning of the real line for scalar quantization. The code values (output or reconstruction levels) are shown by dots. Here, also, any input value x that lies in the interval C_i is quantized as y_i. The number of levels in Fig. 4 is $L = 10$. Scalar quantization has the very special property that while cells may have different sizes, they all have the *same shape*, namely, they are all intervals on the real line. In comparison, note how in Fig. 3 the cells in two dimensions actually have different shapes. This freedom of having various cell shapes in multidimensional space gives vector quantization an advantage over scalar quantization, as we shall see below.

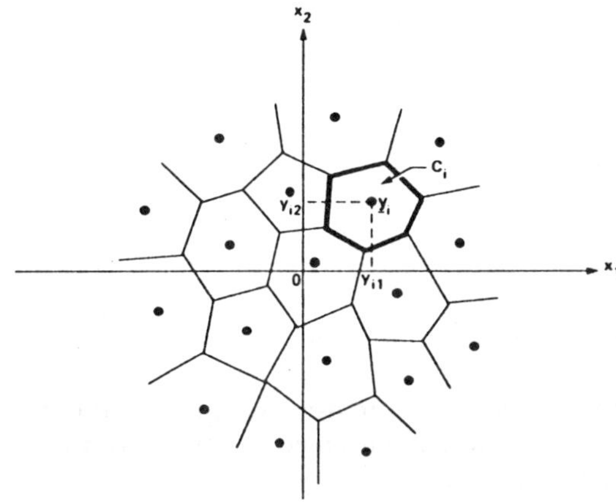

Fig. 3. Partitioning of two-dimensional space ($N = 2$) into $L = 18$ cells. All input vectors in cell C_i will be quantized as the code vector $\boldsymbol{y}_i$. The shapes of the various cells can be very different.

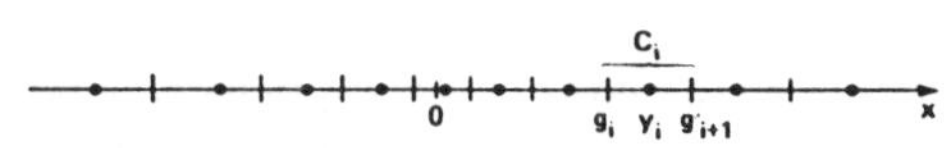

Fig. 4. Partitioning of the real line into $L = 10$ cells or intervals for scalar quantization ($N = 1$).

When $\boldsymbol{x}$ is quantized as $\boldsymbol{y}$, a quantization error results and a *distortion measure* $d(\boldsymbol{x}, \boldsymbol{y})$ can be defined between $\boldsymbol{x}$ and $\boldsymbol{y}$. $d(\boldsymbol{x}, \boldsymbol{y})$ is also known as a dissimilarity measure or distance measure. As the vectors $\boldsymbol{y}(n)$ at different times n are transmitted, one can define an overall average distortion

$$D = \lim_{M \to \infty} \frac{1}{M} \sum_{n=1}^{M} d[\boldsymbol{x}(n), \boldsymbol{y}(n)]. \tag{3}$$

If the vector process $\boldsymbol{x}(n)$ is stationary and ergodic,[3] the sample average in (3) tends in the limit to the expectation

$$\begin{aligned} D &= \mathscr{E}[d(\boldsymbol{x}, \boldsymbol{y})] \\ &= \sum_{i=1}^{L} P(\boldsymbol{x} \in C_i) \mathscr{E}[d(\boldsymbol{x}, \boldsymbol{y}_i) | \boldsymbol{x} \in C_i] \\ &= \sum_{i=1}^{L} P(\boldsymbol{x} \in C_i) \int_{\boldsymbol{x} \in C_i} d(\boldsymbol{x}, \boldsymbol{y}_i) p(\boldsymbol{x})\, d\boldsymbol{x} \end{aligned} \tag{4}$$

where $P(\boldsymbol{x} \in C_i)$ is the discrete probability that $\boldsymbol{x}$ is in C_i, $p(\boldsymbol{x})$ is the multidimensional probability density function (pdf) of $\boldsymbol{x}$, and the integral is taken over all components of the vector $\boldsymbol{x}$.

For purposes of transmission, each vector $\boldsymbol{y}_i$ is encoded

[3] Ergodicity allows us to substitute sample (or time) averages for ensemble averages [28].

into a codeword of binary digits (bits) c_i of length B_i bits. In general, the different codewords will have different lengths. The transmission rate T is then given by

$$T = BF_c \text{ bits/s} \tag{5}$$

where

$$B = \lim_{M \to \infty} \frac{1}{M} \sum_{n=1}^{M} B(n) \quad \text{bits/vector} \tag{6}$$

is the average codeword length, $B(n)$ is the number of bits used to code the vector $x(n)$ at time n, and F_c is the number of codewords transmitted per second. It will also be useful to define the average number of bits per parameter or per dimension[4]

$$R = \frac{B}{N} \text{ bits/dimension.} \tag{7}$$

For a codebook of size L, the maximum number of bits needed to code each vector is

$$B_{\max} = \log_2 L. \tag{8}$$

In designing a data compression system, one attempts to design the quantizer such that the distortion in the output is minimized for a given transmission rate. One major decision in designing a quantizer is what distortion measure to use. This is discussed next.

B. Distortion Measures

To be useful, a distortion measure must be tractable, so that one can analyze it and compute it, and be subjectively relevant, so that differences in distortion values can be used as indicating similar differences in speech quality. Most distortion measures in use today are certainly tractable and, to some extent, subjectively relevant. However, many researchers have experienced the frustration which accompanies their discovery that a few decibels of decrease in the distortion is quite perceivable by the ear in one situation but not in another. The careful researcher has learned that, while objective distortion measures are necessary and useful tools in the design of speech coding systems, periodic subjective quality testing is indispensable to making an informed decision on directions for improving system performance. Below we list some of the major distortion measures in use today.

1) Mean-Square Error: By far the most common distortion measure is the mean-square error (mse)

$$d_2(x,y) = \frac{1}{N}(x-y)^T(x-y) = \frac{1}{N}\sum_{k=1}^{N}(x_k - y_k)^2 \tag{9}$$

where the distortion is defined per dimension. The popularity of the mse lies mainly in its simplicity and mathematical tractability. A more general distortion measure based on the L_r norm is defined by

$$d_r(x,y) = \frac{1}{N}\sum_{k=1}^{N}|x_k - y_k|^r. \tag{10}$$

[4]Note that R is the bit rate per dimension, B is the bit rate per vector, and T is the bit rate per second. We shall often use the generic term *bit rate* to refer to any or all of the three meanings. We trust the intended meaning will be clear from the context.

Note that (10) is equal to (9) for $r = 2$. The two other most popular values of r are $r = 1$ and $r = \infty$. d_1 represents the average absolute error and d_∞ tends towards the maximum error. In fact, one can show that

$$\lim_{r \to \infty} [d_r(x,y)]^{1/r} = \max\{|x_k - y_k|, 1 \leq k \leq N\}. \tag{11}$$

Minimizing D for $r = \infty$ would be equivalent to minimizing the maximum quantization error.

For speech coding, d_2 has been the most popular distortion measure, with d_1 and d_∞ being used occasionally.

2) Weighted Mean-Square Error: In the mse d_2 we assumed that the distortions contributed by quantizing the different parameters $\{x_k\}$ were weighted equally. In general, unequal weights can be introduced to render certain contributions to the distortion more important than others. A general weighted mse is then defined by

$$d_W(x,y) = (x-y)^T W (x-y) \tag{12}$$

where W is a positive-definite weighting matrix. $W = N^{-1}I$, where I is the identity matrix, results in $d_W = d_2$.

One choice for W that is popular in many pattern classification applications is $W = \Gamma^{-1}$, where Γ is the covariance matrix of the random vector x

$$\Gamma = E\left[(x-\bar{x})(x-\bar{x})^T\right], \qquad \bar{x} = \mathcal{E}[x]. \tag{13}$$

In this case, d_W reduces to

$$d_W(x,y) = (x-y)^T \Gamma^{-1} (x-y) \tag{14}$$

and is known as the Mahalanobis distance [85].

If, in addition to being positive definite, the weighting matrix W is symmetric (as in the Mahalanobis distance), one can factor W as follows:

$$W = P^T P. \tag{15}$$

The vectors x and y can be transformed into a new set of vectors $\tilde{x}$ and $\tilde{y}$

$$\tilde{x} = Px \qquad \tilde{y} = Py \tag{16}$$

and

$$\begin{aligned} d_W(x,y) &= (Px - Py)^T(Px - Py) \\ &= (\tilde{x} - \tilde{y})^T(\tilde{x} - \tilde{y}) \\ &= d_2(\tilde{x}, \tilde{y}). \end{aligned} \tag{17}$$

Thus the weighted mse between the original vectors is equal to the mse between the transformed vectors. Therefore, for computational purposes, it may be advantageous to perform the transformation in (16) on all the data before vector quantization is performed.

3) Linear Prediction Distortion Measures: In LPC analysis, the predictor coefficients $\{a(k)\}$ are obtained as a result of minimizing the energy of the prediction residual. One can show [86] that the solution for the optimum $A(z)$ in (1) is unique and is computed from the set of simultaneous linear equations

$$\sum_{k=1}^{N} a(k)\phi(i-k) = -\phi(i), \qquad 1 \leq i \leq N \tag{18}$$

where $\{\phi(i), 0 \leq i \leq N\}$ are the short-term autocorrelation coefficients of the speech signal over a single frame. (For a speech signal band-limited to 5 kHz, for example, the number of coefficients N is typically set to 12–14.) The gain G of the filter $H(z)$ is set so that when excited by a unit-variance source the output energy will be equal to $\phi(0)$. That can be accomplished by setting

$$G^2 = \phi(0) + \sum_{k=1}^{N} a(k)\phi(k) \tag{19}$$

which is equal to the minimum residual energy. The parameters of the filter $H(z)$ are computed every frame, quantized, and transmitted.

The gain G is usually quantized on a logarithmic scale and transmitted separately. (Some work has been done in joint quantization of the gain and the LPC parameters [108].) Because the quantization of predictor coefficients can lead to instability of the resulting all-pole filter, they are usually transformed to another set of parameters known as the reflection coefficients $\{K_k, 1 \leqslant k \leqslant N\}$ or partial correlation (PARCOR) coefficients [69]. Reflection coefficients result as a byproduct of solving (18) or can be computed recursively from the predictor coefficients (see, for example [86], [90]). For a stable $H(z)$, the reflection coefficients have the property

$$|K_k| < 1, \qquad 1 \leqslant k \leqslant N. \tag{20}$$

Because for values of $|K_k|$ approaching 1 the poles approach the unit circle, small changes in K_k can result in large changes in the spectrum. Therefore, for quantization purposes, the reflection coefficients are usually transformed to another set of coefficients that exhibit lower spectral sensitivity as K approaches 1. Two popular transformations are [78]

$$S_k = \frac{2}{\pi}\sin^{-1} K_k, \qquad 1 \leqslant k \leqslant N \tag{21}$$

$$G_k = \frac{1}{2}\log\frac{1+K_k}{1-K_k} = \tanh^{-1} K_k, \qquad 1 \leqslant k \leqslant N. \tag{22}$$

The parameters G_k are known as log-area-ratios (LARs) from the acoustic tube analogy of the vocal tract [8], [90] and possess the property that small changes in G_k are approximately proportional to corresponding changes in the log spectrum of $H(z)$ [131]. The mse d_2 as well as the minimax error d_∞ have been used to quantize S_k and G_k. The quantization properties of K_k, S_k, and G_k have been studied by several researchers [61], [72], [131],

An alternative distortion measure used in quantizing predictor coefficients was proposed by Itakura and Saito [68], [70]; it derives from maximum-likelihood principles. A modified form of the Itakura–Saito distortion between one vector of predictor coefficients $\boldsymbol{x} = [a(1)\ a(2) \cdots a(N)]^T$ and another vector of predictor coefficients $\boldsymbol{y}$ is given by (see [57] for variations on the Itakura–Saito distortion)

$$d_I(\boldsymbol{x},\boldsymbol{y}) = (\boldsymbol{x}-\boldsymbol{y})^T \boldsymbol{\Phi}_x (\boldsymbol{x}-\boldsymbol{y}) \tag{23}$$

where

$$\boldsymbol{\Phi}_x = \{\phi(i-k)/\phi(0), 0 \leqslant i,k \leqslant N-1\} \tag{24}$$

is the normalized autocorrelation matrix whose coefficients $\phi(i-k)$ were used in computing the vector of predictor coefficients $\boldsymbol{x}$ in (18). Since the autocorrelation coefficients in (24) are normalized by $\phi(0)$, one can show that the matrix $\boldsymbol{\Phi}_x$ and the vector $\boldsymbol{x}$ uniquely determine each other. It is important to note that $\boldsymbol{\Phi}_x$ in (23) is effectively a weighting matrix but, unlike in (12) where $\boldsymbol{W}$ is fixed, here $\boldsymbol{\Phi}_x$ changes value as $\boldsymbol{x}$ changes. Since $\boldsymbol{\Phi}_x \neq \boldsymbol{\Phi}_y$ for $\boldsymbol{x} \neq \boldsymbol{y}$, the Itakura–Saito distortion is not symmetric with respect to its arguments, i.e., $d_I(\boldsymbol{x},\boldsymbol{y}) \neq d_I(\boldsymbol{y},\boldsymbol{x})$. The distortion measure is not a distance or a metric. By contrast, the weighted mse distortion is a symmetric distance and metric.

Even though the computation of d_I in (23) implies a matrix multiply, the computation can be simplified considerably and reduced to a scalar (dot) product [68].

4) Perceptually Motivated Distortion Measures: For very small distortions, and therefore high bit rates, most reasonable distortion measures, including those mentioned above, all exhibit similar behavior, by linearity arguments. Furthermore, they would all be expected to correlate well with subjective judgements of speech quality. However, as bit rate decreases and distortion increases, simple distortion measures have not always correlated well with perceptual judgements. Since VQ is expected to be especially useful at low bit rates, it becomes more important to develop and use distortion measures that are correlated better with human auditory behavior. A number of perceptually based distortion measures, and others that correlate well with subjective judgements, have been used in speech coding (see, for example, [74, Appendix E], [11], [12], [94], [100], [101], [134]). If high speech quality at a given bit rate is the most important consideration in a coder design, then one would do well to consider the use of a distortion measure that correlates well with human perception.

C. Codebook Design

As mentioned above, to design our L-level codebook, we partition N-dimensional space into L cells $\{C_i, 1 \leqslant i \leqslant L\}$ and associate with each cell C_i a vector $\boldsymbol{y}_i$. The quantizer then assigns the code vector $\boldsymbol{y}_i$ if $\boldsymbol{x}$ is in C_i. A quantizer is said to be an optimal (minimum-distortion) quantizer if the distortion in (4) is minimized over all L-level quantizers. There are two necessary conditions for optimality. The first condition is that the optimal quantizer is realized by using a minimum-distortion or nearest neighbor selection rule

$$q(\boldsymbol{x}) = \boldsymbol{y}_i, \qquad \text{iff } d(\boldsymbol{x},\boldsymbol{y}_i) \leqslant d(\boldsymbol{x},\boldsymbol{y}_j), \ \ j \neq i, \ \ 1 \leqslant j \leqslant L. \tag{25}$$

That is, the quantizer chooses the code vector that results in the minimum distortion with respect to $\boldsymbol{x}$. (Ties are decided by some rule.) The second necessary condition for optimality is that each code vector $\boldsymbol{y}_i$ is chosen to minimize the average distortion in cell C_i. That is, $\boldsymbol{y}_i$ is that vector $\boldsymbol{y}$ which minimizes

$$D_i = \mathscr{E}[d(\boldsymbol{x},\boldsymbol{y}) | \boldsymbol{x} \in C_i] = \int_{\boldsymbol{x} \in C_i} d(\boldsymbol{x},\boldsymbol{y}) p(\boldsymbol{x})\, d\boldsymbol{x}. \tag{26}$$

We call such a vector the *centroid* of the cell C_i, and we write

$$\boldsymbol{y}_i = \text{cent}(C_i). \tag{27}$$

Computing the centroid for a particular region will depend on the definition of the distortion measure. (The cells thus defined are known as nearest neighbor cells, Voronoi cells, or Dirichlet regions [48].)

In practice, we are given a set of training vectors $\{x(n), 1 \leq n \leq M\}$. A subset M_i of those vectors will be in cell C_i. The average distortion D_i is then given by

$$D_i = \frac{1}{M_i} \sum_{x \in C_i} d(x, y_i). \tag{28}$$

For either the mse or the weighted mse criterion, one can show that D_i is minimized by

$$y_i = \frac{1}{M_i} \sum_{x \in C_i} x(n) \tag{29}$$

or y_i is simply the sample mean of all the training vectors contained in C_i. For the Itakura–Saito distortion d_I, one can show that y_i is computed by first averaging the normalized autocorrelations corresponding to the sample vectors

$$\phi_{y_i}(k) = \frac{1}{M_i} \sum_{x \in C_i} \phi_x(k), \quad 0 \leq k \leq N \tag{30}$$

where $\phi_x(k)$ are normalized such that $\phi_x(0) = 1$. The vector y_i is then obtained as the solution to (18) with $\phi_{y_i}(k)$ as the autocorrelation coefficients.

One method for codebook design is an iterative clustering algorithm known in the pattern-recognition literature as the *K-means algorithm.*[5] In our problem here, $K = L$. The algorithm divides the set of training vectors $\{x(n)\}$ into L clusters[6] C_i in such a way that the two necessary conditions for optimality are satisfied. Below, m is the iteration index and $C_i(m)$ is the ith cluster at iteration m, with $y_i(m)$ its centroid. The algorithm is as follows:

K-Means Algorithm

Step 1: *Initialization:* Set $m = 0$. Choose by an adequate method a set of initial code vectors $y_i(0)$, $1 \leq i \leq L$. (See Section V-E.)

Step 2: *Classification:* Classify the set of training vectors $\{x(n), 1 \leq n \leq M\}$ into the clusters C_i by the nearest neighbor rule

$$x \in C_i(m), \quad \text{iff } d[x, y_i(m)] \leq d[x, y_j(m)], \quad \text{all } j \neq i.$$

Step 3: *Code Vector Updating:* $m \leftarrow m + 1$. Update the code vector of every cluster by computing the centroid of the training vectors in each cluster

$$y_i(m) = \text{cent}(C_i(m)), \quad 1 \leq i \leq L.$$

Step 4: *Termination Test:* If the decrease in the overall distortion $D(m)$ at iteration m relative to $D(m-1)$ is below a certain threshold, stop; otherwise go to Step 2.

Any other reasonable termination test may be substituted for Step 4 above.

The above algorithm can be shown to converge to a local optimum (see, for example, [5], [81]). Furthermore, any such solution is, in general, not unique [33], [58]. Global optimality may be achieved approximately by initializing the code vectors to different values and repeating the above algorithm for several sets of initializations and then choosing the codebook that results in the minimum overall distortion. In Section V-E we shall discuss methods for performing initialization.

D. Computational and Storage Costs

Having designed a codebook as described above, one can then use it to quantize each input vector $x(n)$. The quantization is performed as shown in (25) by computing the distortion between $x(n)$ and each of the code vectors, then choosing the code vector with the minimum distortion as the quantized value of $x(n)$. This type of quantization is known as a *full search* since all code vectors are tested for quantizing each input vector. For an L-level quantizer, the number of distortion computations needed to quantize a single input vector is L. While a distortion computation can be arbitrarily complex, we shall assume here that each distortion computation requires a total of N multiply-adds (this is true for the mse and one version of the Itakura–Saito distortion). Therefore, the *computational cost* for quantizing each input vector is

$$\mathscr{C} = NL. \tag{31}$$

If we encode each code vector into $B = RN = \log_2 L$ bits for transmission, then

$$\mathscr{C} = N\,2^{RN}. \tag{32}$$

Thus computation cost grows exponentially with the number of dimensions and the number of bits per dimension.

Another important part of the quantization cost is *memory* or *storage cost*, i.e., how much storage is needed to store the code vectors. We shall measure storage assuming one storage location per vector parameter. It is clear that storage cost is then given by

$$\mathscr{M} = NL = N\,2^{RN}. \tag{33}$$

Like computational cost, storage cost is exponential in the number of dimensions and the number of bits per dimension.

The costs given above are for the full search algorithm. In Section V we shall present methods that reduce computational costs substantially at the cost of relatively small loss in performance and/or increase in storage.

While we place heavy emphasis on the costs associated with the quantization process itself, it is also important to

[5] The algorithm presented here was described by Forgy in 1965 [41] and is the clustering algorithm described most in the pattern-recognition literature [5], [30], [64], [99], [129]. The name *K-means* comes from MacQueen [84], who actually describes a different algorithm. In an unpublished paper in 1957, Lloyd had independently developed the same algorithm as Forgy's but for the scalar quantization problem and a known distribution (Lloyd's paper has recently been published [82].) The application of this algorithm to a training sequence and the VQ case has been termed in some of the information theory literature as the generalized Lloyd algorithm [56]. Linde, Buzo, and Gray [81] have shown that the algorithm works with a large class of distortion measures, including measures that are not metric, and so the algorithm has also been called the LBG algorithm.

[6] We use the same symbol C_i to represent both the cluster and the cell corresponding to code vector y_i. Cell C_i is that region of N-dimensional space that is closest to y_i based on the nearest neighbor rule, while cluster C_i is that subset of training vectors which are closest to y_i based on the same rule, i.e., cluster C_i is the set of training vectors contained in cell C_i.

keep in mind the costs associated with design of the codebook in the first place. In the K-means algorithm described above, most of the computations result from the classification step; code vector updating presents a negligible amount of computation by comparison. For an L-level quantizer, M training vectors, and I iterations, the computational cost for training is

$$\mathscr{C}_T = NLMI = N2^{NR}MI. \tag{34}$$

The storage cost, including the storage needed to store all the training vectors, is

$$\mathscr{M}_T = N(L + M). \tag{35}$$

For reliable design of the codebook, it has been our experience that one needs at least 10 and preferably about 50 training vectors per code vector, so that M is on the order of $10L$ or more (see Section V-E). So, storage cost for training is largely dominated by the amount of training vectors needed.

E. *Vector Quantization Model*

Before we delve into more mathematics and examine the detailed workings of vector quantization, we shall present a simple model of vector quantization, which we hope will give the reader a basic understanding of the various processes at work. The concepts presented will be illustrated by simplified examples.

1) Basic Model: Our VQ model identifies four properties of vector components which, when utilized appropriately in codebook design, result in optimal performance. The four properties of vector components are: *linear dependency*, *nonlinear dependency*, *pdf shape*, and *dimensionality*. These four properties are interrelated to a certain extent. For example, even though the multidimensional pdf shape completely specifies any dependencies among vector components, we shall see that pdf shape still plays an important role in determining optimal performance, even when all dependencies among components are removed. For a given codebook size, VQ takes advantage of the four properties above by proper *placement* of the code vectors in N-dimensional space. By code vector placement we mean having the freedom to place code vectors where they are needed most so as to minimize the given distortion measure. For example, one would not place code vectors in regions of zero probability. There are two aspects of code vector placement that are of interest: code vector *spacing* or density and *cell shape*. Code vector spacing refers to how close the code vectors are to each other. In general, one would expect closer spacing (higher density) in high pdf regions (i.e., where $p(\mathbf{x})$ is large) and wider spacing of code vectors in regions of low pdf. Once the code vectors are specified, then the cell shapes are automatically determined by the distortion measure. Conversely, once all the cell shapes and positions are specified, then the code vectors are automatically determined as the cell centroids. We shall see below that it will be beneficial to think in terms of cell shapes to gain a better understanding of the workings of VQ, for it is the freedom we have to choose different cell shapes in higher dimensions ($N > 1$) that allows us to exploit dimensionality to minimize distortion in a way that is not possible with scalar quantization. Below, we demonstrate how a vector quantizer chooses its code vector placements and cell shapes, taking advantage of the four vector properties to optimize performance. The examples are designed to expose the processes at work in VQ in a simple way and to show how they may differ from scalar quantization.

2) Dependency: Data compression is largely a process of redundancy removal; it is not necessary to waste bits in transmitting redundant information. Redundancy usually implies some sort of dependency among transmission parameters. We shall classify statistical dependency into two types: *linear dependency* and *nonlinear dependency*. These terms are explained below.

Linear dependency is what we normally think of as *correlation*. Two random variables that are correlated are linearly dependent. If the variables are uncorrelated, they are no longer linearly dependent, but they may still be statistically dependent. The latter "residual" dependency we call nonlinear dependency; it is whatever dependency remains after the linear dependency is removed. Two (zero-mean) variables x_1 and x_2 are uncorrelated if the expected value of their product is zero

$$\mathscr{E}[x_1 x_2] = 0 \quad \text{(uncorrelated)}. \tag{36}$$

But x_1 and x_2 are independent if and only if their joint pdf is equal to the product of the individual (marginal) densities of x_1 and x_2

$$p(x_1, x_2) = p(x_1)p(x_2), \qquad \text{for all } x_1, x_2 \quad \text{(independent)}. \tag{37}$$

If x_1 and x_2 are uncorrelated but dependent (i.e., (36) applies but not (37)), then that dependency we call nonlinear. We shall now demonstrate how one can take advantage of both types of dependency in reducing the necessary bit rate for transmission. In the examples below we shall use the mse as our distortion measure.

EXAMPLE 1

x_1 and x_2 are two random variables whose joint pdf $p(x_1, x_2)$ is shown in Fig. 5. The density is uniform and nonzero inside the rectangle and zero outside the rectan-

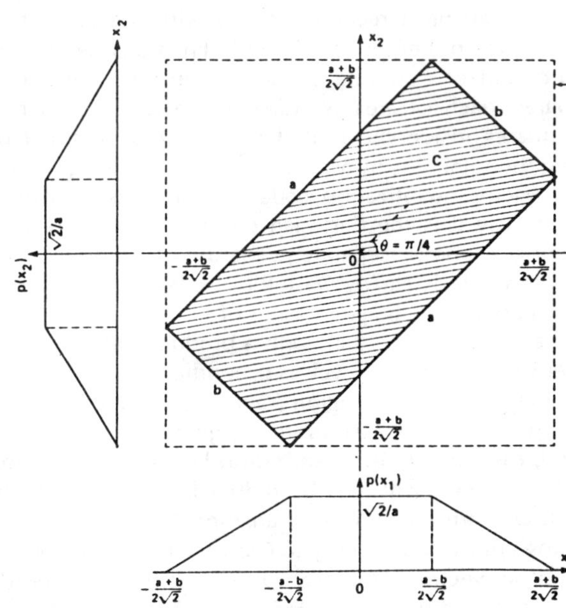

Fig. 5. An example of a uniform pdf $p(x_1, x_2)$ in two dimensions; the density is zero outside the region C. Shown also are the marginal densities $p(x_1)$ and $p(x_2)$. x_1 and x_2 in this example are correlated.

gle. Let the region inside the rectangle be denoted by C, then

$$p(x_1, x_2) = p(\mathbf{x}) = \begin{cases} \dfrac{1}{ab}, & \mathbf{x} \in C \\ 0, & \text{otherwise.} \end{cases} \tag{38}$$

The rectangle has sides a and b, and is oriented at an angle $\theta = \pi/4$. Shown also in Fig. 5 are the marginal densities $p(x_1)$ and $p(x_2)$, which in this example happen to be equal. It is clear from (38) and the marginal densities in Fig. 5 that (37) does not apply, and so x_1 and x_2 are dependent. One can also show that x_1 and x_2 are correlated, so that (36) is not true either.

Now, let us use a scalar *uniform quantizer* to quantize x_1 and x_2 independently. A uniform quantizer is one whose quantizing intervals C_i are of equal length. (Such a quantizer minimizes the mse only for a uniform density and, therefore, would not be optimal for quantizing x_1 and x_2 in this example.) We shall quantize x_1 and x_2 using quantizing intervals equal to Δ. Since x_1 and x_2 have values that range between $-(a+b)/2\sqrt{2}$ and $(a+b)/2\sqrt{2}$, the number of levels needed to quantize each of x_1 and x_2 is equal to

$$L_1 = L_2 = \frac{a+b}{\sqrt{2}\,\Delta}. \tag{39}$$

x_1 and x_2 can be coded using $R_1 = \log_2 L_1$ bits and $R_2 = \log_2 L_2$ bits, respectively. The vector $\mathbf{x}$ can be coded, then, using

$$B_x = R_1 + R_2 = \log_2 L_1 L_2 = \log_2 \frac{(a+b)^2}{2\Delta^2} \text{ bits.} \tag{40}$$

The two scalar quantizers correspond to using a vector quantizer with a total number of levels

$$L_x = L_1 L_2 = \frac{(a+b)^2}{2\Delta^2}. \tag{41}$$

Indeed, such a quantizer can be obtained by first demarking the extreme values of x_1 and x_2 by the dashed rectangle (it is a square in this example) enclosing the region Q in Fig. 5, and then drawing a rectangular grid with the separation between grid lines equal to Δ in both directions (see Fig. 8). Such a quantizer would have quantization cells in the form of squares, each of area Δ^2. Clearly, the total number of such squares inside the dotted region Q is obtained by dividing the total area by Δ^2. The result is equal to L_x in (41). Such a vector quantization code is known as a *product code* because it is the Cartesian product of the codes for quantizing x_1 and x_2 separately. The use of a product code for this example is clearly wasteful of bits since regions of zero probability are assigned some of the bits.

We have seen above that, with a product code, the total number of quantization levels is proportional to the area of the whole quantization region (region Q in Fig. 5). Therefore, if the area of the quantization region can somehow be reduced, the number of quantization levels and the corresponding bit rate will also be reduced. For our example, one can perform a coordinate transformation via a rotation that transforms Fig. 5 into Fig. 6. The vector $\mathbf{x}$ is transformed into another vector $\mathbf{u}$. One can show that the new coordinates u_1 and u_2 are, in fact, uncorrelated. With the proper rotation, any set of random variables can be rendered uncorrelated, as we shall see in Section IV. In the special case of the example in Fig. 6, we see from the marginal

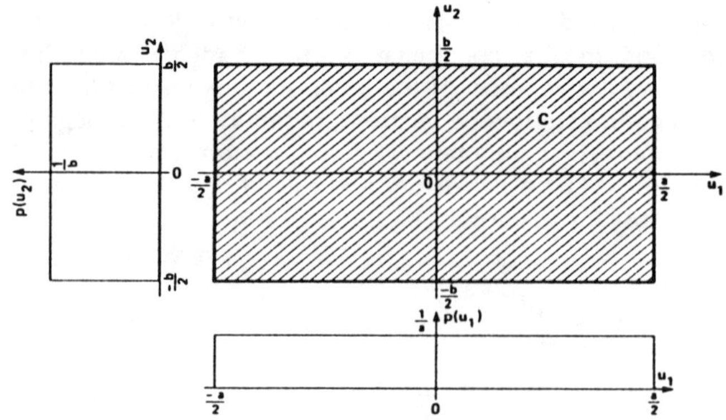

Fig. 6. The pdf in Fig. 5 after a rotation of coordinates. u_1 and u_2 are now uncorrelated and independent.

densities shown in the figure that

$$p(u_1, u_2) = p(u_1)p(u_2), \qquad \text{for all } u_1, u_2$$

and, therefore, u_1 and u_2 are also statistically independent. Performing uniform scalar quantization in this case with a quantizing interval of length Δ yields a number of levels equal to

$$L_1 = \frac{a}{\Delta} \qquad L_2 = \frac{b}{\Delta}$$

$$L_u = L_1 L_2 = \frac{ab}{\Delta^2}. \tag{42}$$

The corresponding number of bits is

$$B_u = \log_2 \frac{ab}{\Delta^2}. \tag{43}$$

The difference in the number of bits needed to code $\mathbf{x}$ and $\mathbf{u}$ can be seen from (40) and (43) to be

$$B_x - B_u = \log_2 \frac{(a+b)^2}{2ab}. \tag{44}$$

For example, for $a = 2b$

$$B_x - B_u = 1.17 \text{ bits} \qquad (a = 2b). \tag{45}$$

Therefore, the rotation saves us in excess of 1 bit per transmitted vector. Such a difference can become significant at low data rates.

In comparing bit rates in (44) we made an implicit assumption that the total quantization distortion is the same for both examples of Figs. 5 and 6. Otherwise, comparing bit rates as such is not meaningful. In fact, for small Δ and, hence, large B_x and large B_u, one can show from (4) that the total distortion in both cases is about the same. Differences in distortion arise as Δ increases and boundary effects near the edges of the rectangle become significant. Under such circumstances one must be careful to equalize the distortions before comparing bit rates. At such low bit rates, (45) would not be expected to hold; it would most likely decrease.

EXAMPLE 2

Example 1 demonstrated how one can take advantage of decorrelation through rotation to reduce the bit rate in scalar quantization of a vector. In this example, we shall demonstrate how, through vector quantization, one can take advantage of nonlinear dependencies to reduce the bit rate.

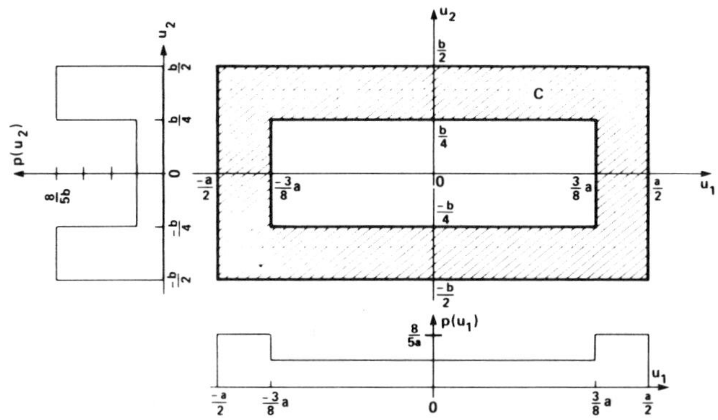

Fig. 7. An example where u_1 and u_2 are uncorrelated but dependent; this type of dependency is termed *nonlinear.*

Fig. 7 shows a different example of a pdf that is nonzero and uniform inside the hatched region C and zero otherwise. Since the area of the hatched region is $5ab/8$, we have

$$p(\mathbf{u}) = \begin{cases} \dfrac{8}{5ab}, & \mathbf{u} \in C \\ 0, & \text{otherwise.} \end{cases} \tag{46}$$

One can show that, like the example in Fig. 6, u_1 and u_2 are uncorrelated but, unlike in Fig. 6, u_1 and u_2 here are not independent, as is clear from (46) and the marginal densities in Fig. 7. Such statistical dependency is termed nonlinear; it cannot be removed by a process of decorrelation. A scalar quantizer designed for u_1 and u_2 in Fig. 7 will yield the same bit rate B_u in (43). To take advantage of the nonlinear dependency we must use a different vector quantizer that partitions only the hatched area in Fig. 7 and does not waste bits inside the small rectangle. One such quantizer would divide only the hatched area into squares of equal area Δ^2. (Such a quantizer would yield the same distortion as the scalar quantizer for small Δ.) The number of levels and bits for this vector quantizer would be

$$L'_u = \frac{5}{8}\frac{ab}{\Delta^2} \qquad B'_u = \log_2 \frac{5ab}{8\Delta^2}. \tag{47}$$

The reduction in bit rate between a scalar quantizer and a vector quantizer in this case is the difference between (43) and (47)

$$B_u - B'_u = \log_2 \frac{8}{5} = 0.68 \text{ bits}. \tag{48}$$

Therefore, taking advantage of nonlinear dependency in this case saved us 0.68 bits/vector.

We have seen how, with proper cell or code vector placement, a vector quantizer was able to take advantage of nonlinear dependencies to reduce the bit rate. It should be clear that any vector rotation will not affect the vector quantizer's ability to locate its cells properly. Therefore, VQ can make effective use of all dependencies (linear or nonlinear) simply by proper code vector placement.

3) Dimensionality:

EXAMPLE 3

The vector quantizer in Example 2 employed the square as the shape of all its cells. The square shape was inspired from the scalar quantizers used earlier. But one property of vector quantizers in higher dimensions is that one has the freedom to choose other cell shapes and not be restricted to the N-dimensional cubes suggested by scalar quantization. Let us examine the effect of using a different shape, such as the *hexagon*, to partition our two-dimensional space. Fig. 8 shows a covering of space by squares and Fig. 9 shows a space covering by regular hexagons. Let each

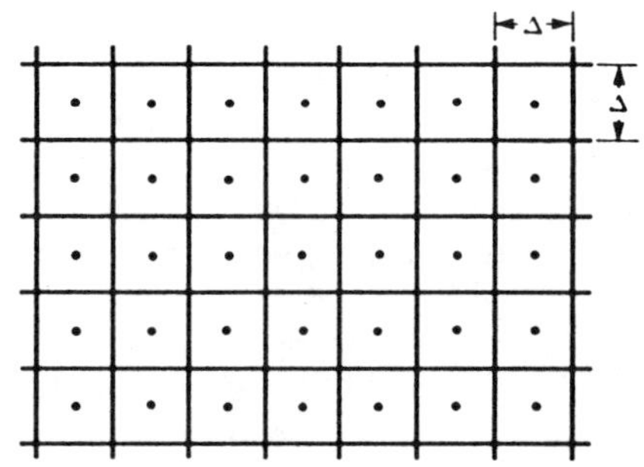

Fig. 8. Packing of two-dimensional space with squares.

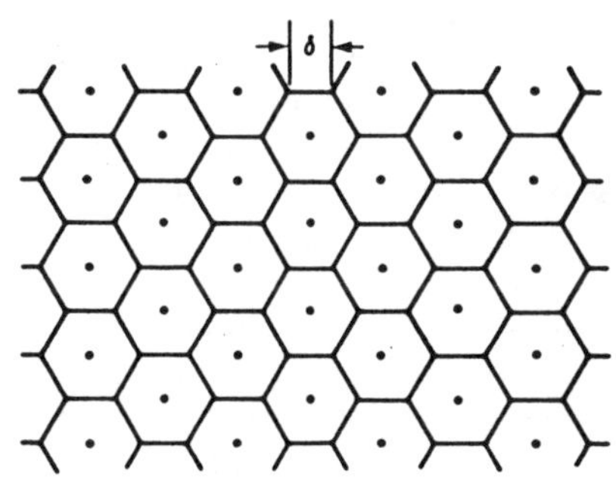

Fig. 9. Packing of two-dimensional space with regular hexagons. For the same number of cells in a given area with a uniform pdf, hexagons have a lower mse than squares.

hexagon have sides of size δ. The area of the hexagon is then

$$A_H = \frac{3\sqrt{3}}{2}\delta^2. \tag{49}$$

To compare the performance of the square quantizer to the hexagon quantizer, we must compare the quantization mse for both quantizers. If we assume that the code vectors are located in the center of the square and the hexagon in each case (as shown by the dots in Figs. 8 and 9), one can show that the mse for each cell is given by

$$E_S = \frac{\Delta^4}{6} \qquad \text{(square)} \tag{50}$$

$$E_H = \frac{5\sqrt{3}}{8}\delta^4 \quad \text{(hexagon)} \tag{51}$$

where the subscripts denote the respective quantizers. The total mse is then obtained by multiplying (50) and (51) by the number of cells (levels). If we make the area of the hexagon equal to the area of the square, i.e., set $A_S = \Delta^2 = A_H$ in (49), then, neglecting edge effects, both quantizers will have the same number of cells covering any given area and, therefore, they will have the same bit rate. However, the ratio of the distortions can be shown to be

$$\frac{E_H}{E_S} = \frac{5\sqrt{3}}{9} = 0.962 \qquad (A_S = A_H) \tag{52}$$

which is equivalent to -0.167 dB. That is, the hexagon quantizer will yield a lower mse than the square quantizer by a fraction of a decibel. Conversely, if we equalize the distortions, i.e., set $E_S = E_H$, then the ratio of the areas will be

$$\frac{A_H}{A_S} = \frac{3}{\sqrt[4]{3}\sqrt{5}} = 1.0194 \qquad (E_S = E_H). \tag{53}$$

This means that, for the same distortion, the hexagon has a slightly larger area than the square and, therefore, will require proportionately fewer cells to cover the same space. Therefore, the bit rate for the hexagon will be lower by an amount equal to $\log_2 A_H/A_S$, which from (53) is equal to 0.028 bits.

The gain of 0.028 bits resulting from the use of a hexagon quantizer can be obtained in the examples of Figs. 6 and 7 in addition to any other gains obtained by taking advantage of linear and nonlinear dependencies. This means that, even in the case when two random variables are independent, as in Fig. 6, vector quantization can squeeze out an additional small fraction of a bit by taking advantage of the higher dimensionality using an appropriate cell shape.

4) PDF Shape: In the examples above, the cells always had not only the same shape throughout the quantization region but also the same size. Effectively, the cell *spacing* was uniform. Uniform cell spacing is certainly a reasonable choice with a uniform pdf. However, for nonuniform pdf shapes, one would expect that some form of nonuniform cell spacing would be needed for optimal performance. We shall see in Section III how scalar and vector quantizers make effective use of pdf shape to reduce the bit rate.

In this section, we presented a model for vector quantization and gave a few examples to illustrate some of the basic principles. In the following sections, we shall review some of the theoretical foundations of vector quantization and give practical examples from speech coding. However, the basic concepts presented in the vector quantization model above and in the simplified examples will still hold when we deal with real-world data.

III. Theoretical Vector Quantizer Performance

In this section we review some of the important *theoretical* tools that are used to help estimate the performance of vector quantizers, along with some of the known results. The presentation is under three headings: rate-distortion theory, scalar quantization, and asymptotic vector quantization. Rate-distortion theory and scalar quantization will give us lower and upper bounds, respectively, on the minimum bit rate achievable by vector quantizers for any given distortion.

This section is theoretical in nature and is rather long. At first reading, we recommend that the reader take only a cursory look at the contents of this section to gain familiarity with the basic definitions and concepts presented, and move quickly to Section IV and the remainder of the paper.

A. Rate-Distortion Theory

Since a major purpose in performing data compression is to minimize the bit rate for a desired level of distortion, it is important in a particular situation to know the theoretical lower bound on the bit rate for any quantizer. By knowing such a bound, one can compare the performance of different quantizers to that bound and decide whether to search for other possibly more complex quantizers that might approach that bound more closely. Rate-distortion theory, a branch of information theory, deals with obtaining such lower bounds without requiring the design of actual quantizers. For a given distortion D, one can compute either $R(D)$, the *rate-distortion function*, defined as the minimum achievable rate (per dimension) for a given distortion D, or its inverse $D(R)$, the *distortion-rate function*, defined as the minimum achievable distortion for a given rate R. The performance limit provided by $D(R)$ or $R(D)$ applies to all methods of source coding, not just vector quantization. It specifically allows for coders that incorporate arbitrarily long delays. In light of the complexity of encoding permitted by this theory, not being able to come very close to the optimal performance when investigating practical vector quantizers may not be indicative of a meaningful disparity. It is when the performance of practical coders is relatively close to the rate-distortion limits that the theory is most useful. Below, we present a summary of the relevant results from rate-distortion theory, preceded by an introduction to the concept of *entropy* from information theory.

1) Coding of Quantizer Output: We mentioned in Section II that one can code the quantizer output vectors $\{y_i, 1 \leq i \leq L\}$ with $\log_2 L$ bits each. If L is a power of 2, the coding is very simple, but if L is not a power of 2, then one can group several vectors together and then perform the coding. For example, if $L = 5$, then to code a single vector would require 3 bits instead of the desired $\log_2 5 = 2.32$ bits, since one cannot transmit a fractional number of bits as such. However, if we group three vectors together, the total number of levels is equal to $L^3 = 125 \cong 2^7$, then each triplet can be coded using 7 bits, for an average of 2.33 bits which is close to the desired average. Henceforth, we shall disregard the problem of fractional bits and simply assume that values can be grouped together, if desired, to achieve the needed rate.

In general, coding L vectors with $\log_2 L$ bits each is actually the maximum rate needed for coding. The minimum average achievable rate to code the vectors $\{y_i\}$ is given by the *entropy* of $\{y_i\}$ [46], defined as

$$H(y) = -\sum_{i=1}^{L} P(y_i)\log_2 P(y_i) \tag{54}$$

where $H(y)$ is the entropy of the discrete-amplitude variable y and $P(y_i)$ is the discrete probability of y_i. $H(y)$ is also called *self-information*, and it is a measure in bits of the information in $\{y_i\}$. Since all discrete probabilities must obey

$$0 \leq P(y_i) \leq 1 \qquad \sum_{i=1}^{L} P(y_i) = 1 \tag{55}$$

one can show [46] that entropy is bounded by

$$0 \leq H(y) \leq \log_2 L. \tag{56}$$

Each vector y_i is coded using

$$B_i = -\log_2 P(y_i) \quad \text{bits} \tag{57}$$

so that vectors with different probabilities will have different wordlengths. The resulting code will be a *variable-length* code, with an average rate equal to the entropy $H(y)$ in (54). This type of coding is known as *entropy coding*. The Huffman code [66] is a well-known, straightforward method for performing entropy coding. With variable-length coding, appropriate buffering schemes would need to be implemented if transmission is over a fixed-rate channel. Any such buffering would, of course, introduce a delay in the system. Also, variable-length coding is particularly sensitive to channel errors. *Permutation codes* [15] offer an alternative for entropy coding using fixed-length codes; however, the codewords can be very long, which also results in delays. In practice, variable-length codes with buffering are used.

One can show that maximum entropy is achieved when all vectors y_i have equal probability [46], i.e., $P(y_i) = 1/L$, in which case

$$H_{max}(y) = B_{max} = \log_2 L \quad \text{bits}. \tag{58}$$

For this special case, a fixed-length code is, in fact, optimal.

Entropy coding is one form of *noiseless coding* in that the coding does not introduce any additional noise or distortion beyond that introduced by the quantization process; it merely takes advantage of the probability distribution to minimize the bit rate. (The purpose of the noiseless encoding box in Fig. 1 is to minimize the bit rate without introducing extra distortion.)

One important and useful property of entropy coding is that, even if the number of levels L is infinite, the entropy can still be finite. (Values with very small probability contribute very little to the entropy since the limit as $P \rightarrow 0$ of $-P\log_2 P$ is zero.) Therefore, one can partition N-dimensional space into a countably infinite number of finite (nonzero) cells, and the entropy of the resultant codebook will be finite. In practice, L must be finite, but it could be made large without substantially increasing the bit rate, provided entropy coding is used.

As the cells in the quantization process are made smaller so that their size goes to zero, the quantizer discrete outputs $\{y_i\}$ tend to the original continuous-amplitude source x; the quantization distortion goes to zero; and the entropy becomes infinite. In other words, it takes infinitely many bits to transmit a continuous-amplitude source. For such a source, it will prove useful to define its *differential entropy* [14]

$$h(x) = -\int_x p(x) \log_2 p(x).dx \tag{59}$$

where $h(x)$ is the differential entropy of the vector x and the integral is over all the components of x. Unlike *absolute* entropy, which was defined in (54), differential entropy may be positive or negative. As such, differential entropy values are not meaningful in and of themselves; they are meaningful only relative to other differential entropies. Thus the difference between two differential entropies is a meaningful measure of the difference in information (in bits) between the corresponding sources.

For a random variable x with a given variance σ^2, one can show that $h(x)$ is bounded above by [122]

$$h(x) \leqslant h_G(x) = \tfrac{1}{2} \log_2 (2\pi e \sigma^2) \tag{60}$$

where $h_G(x)$ is the differential entropy of a Gaussian pdf with variance σ^2. We shall see that Gaussian sources play a special role in bounding the performance of coding systems.

2) Rate-Distortion Theory Results: Most of the results of rate-distortion theory have been obtained for a scalar source that is defined as a function of a discrete variable (such as sampled time). Let $x(n)$ be a discrete-time, continuous-amplitude random source, i.e., a stochastic sequence. If all samples of $x(n)$ are independent and identically distributed, we say that the source is *memoryless* and is completely specified by a single pdf, $p(x)$. It would appear that in quantizing a memoryless source, vector quantization should not have any advantages over scalar quantization. However, we have already seen in Example 3 that VQ indeed can achieve better performance even in the case of a memoryless source. Therefore, in investigating rate-distortion limits we group samples of the source in blocks or vectors and determine the conditions that achieve those limits.

We shall group N consecutive values of $x(n)$ into a single vector x, so that x is a random vector. The vector x is then quantized into a vector $y = q(x)$, where y is one of the vectors in the set $\{y_i, 1 \leqslant i \leqslant L\}$, with L possibly infinite. The average distortion D in representing x as y is given by $\mathscr{E}[d(x,y)]$, where $d(x,y)$ is the distortion per dimension (i.e., per sample of $x(n)$). The vectors y can be transmitted at an average bit rate of $R = H(y)/N$ bits per sample, where $H(y)$ is the entropy of y as defined in (54). The minimum achievable distortion $D_N(R)$ for a given rate R is given by

$$D_N(R) = \min_{q(x)} \mathscr{E}[d(x,y)], \qquad \text{with } \frac{1}{N} H(y) \leqslant R \tag{61}$$

where the minimum is taken over all possible mappings $q(x)$. The distortion-rate function $D(R)$ is obtained in the limit as $N \rightarrow \infty$

$$D(R) = \lim_{N \rightarrow \infty} D_N(R). \tag{62}$$

$D(R)$ is the minimum attainable distortion in coding the source $x(n)$ at a rate R; it is a lower bound on the performance of any quantization scheme. The rate-distortion function $R(D)$ is the inverse of $D(R)$ and is defined similarly.[7] We shall have occasion to use both functions in this paper, although $D(R)$ is used more in practice since we are typically given the rate R and we design our quantizer to minimize the distortion.

The main result of rate-distortion theory that relates to VQ is that, *by using a vector quantizer, one can in principle approach the distortion-rate function arbitrarily closely by increasing the vector size N.* This is essentially implied in the definition of $D(R)$ in (62). Therefore, $D(R)$ is not merely a lower bound for any quantizer, it is actually achievable, in theory, by a vector quantizer of high dimension.

The distortion-rate function $D(R)$ has two important properties: it is *monotone decreasing* with R and it is *convex*. Furthermore, for the mse distortion, $D(R)$ decreases at the rate of about 6 dB/bit for large R. We shall exhibit these properties in specific examples below.

While defining $R(D)$ or $D(R)$ is straightforward, neither is simple to evaluate analytically, except for a few special cases. For a memoryless (zero-mean) Gaussian source with variance σ^2 and a mse distortion, $R(D)$ and $D(R)$ are

[7] The general definition of $R(D)$ in rate-distortion theory is given in terms of the concept of *mutual information* [14]. We have chosen here a somewhat narrower definition that is sufficient for our purposes.

given by [14]

$$R_G(D) = \max\left\{0, \tfrac{1}{2}\log_2 \frac{\sigma^2}{D}\right\}$$

$$= \begin{cases} \tfrac{1}{2}\log_2 \sigma^2/D, & 0 \leq D \leq \sigma^2 \\ 0, & D > \sigma^2 \end{cases} \quad (63)$$

$$D_G(R) = 2^{-2R}\sigma^2. \quad (64)$$

If $D > \sigma^2$ is given, then we need not transmit any information ($R = 0$) because we can always obtain $D = \sigma^2$ by using zeros for the quantized values of $x(n)$, since the quantization error is then equal to $x(n)$. By dividing (64) by σ^2 we obtain the *normalized distortion* which can be measured in decibels

$$\mathscr{D}_G(R) = 10\log_{10}\frac{D_G(R)}{\sigma^2} = -6.02R \text{ dB} \quad (65)$$

where *script* $\mathscr{D}$ is the normalized distortion in decibels. The negative of $\mathscr{D}$ is just the signal-to-noise ratio (SNR) in decibels of the quantizer

$$\text{SNR} = -\mathscr{D} = 10\log_{10}\frac{\sigma^2}{D} \text{ dB}. \quad (66)$$

For much of this paper we shall use $\mathscr{D}$ instead of SNR to measure quantizer mse distortion.

There are scant explicit $D(R)$ results for memoryless non-Gaussian sources (see the result for Laplacian densities by Berger [14]). However, lower and upper bounds exist

$$D^*(R) \leq D(R) \leq D_G(R). \quad (67)$$

Thus $D(R)$ is upper bounded by the distortion-rate function for a Gaussian source. The lower bound $D^*(R)$, known as the *Shannon lower bound*, for a mse distortion is given by

$$D^*(R) = \frac{1}{2\pi e} 2^{2h(x)} 2^{-2R} \quad (68)$$

$$R^*(D) = h(x) - \tfrac{1}{2}\log_2 2\pi e D \quad (69)$$

where $h(x)$ is the differential entropy of the memoryless source. The Shannon lower bound is achievable for many sources only as $R \to \infty$. From (60) and (68), we can write $\mathscr{D}^*(R)$ as a normalized distortion in decibels

$$\mathscr{D}^*(R) = -6.02R - 6.02[h_G(x) - h(x)] \text{ dB} \quad (70)$$

or

$$\mathscr{D}_G(R) - \mathscr{D}^*(R) = 6.02[h_G(x) - h(x)] \text{ dB}$$
$$= 6.02[R_G(D) - R^*(D)] \text{ dB}. \quad (71)$$

Since $h_G(x) > h(x)$, $\mathscr{D}^*(R)$ is less than the Gaussian distortion-rate function by an amount equal to the difference between the source and Gaussian differential entropies (in bits) multiplied by 6.02 dB/bit. Equation (70) makes it very clear that the asymptotic behavior of the distortion-rate function for many sources is expected to decrease at a rate of -6.02 dB/bit as $R \to \infty$.

Table 1 shows four pdfs that are common models used for certain signal distributions. Shown are the differential entropies, the difference $R_G(D) - R^*(D)$, and the corresponding difference in distortion. The Gamma pdf shows the greatest deviation from the Gaussian; it is by far the sharpest or most peaked of the four pdfs, and it becomes unbounded at $x = 0$. The Gamma density is often mentioned as a good model for the first-order pdf of speech [74, p. 32]. However, this model is only good for long-term statistics. Medium-term statistics (on the order of 100 ms), where the speech amplitudes are normalized with respect to the medium-term energy, show that the Laplacian pdf becomes a better model of speech [96]. Short-term statistics (on the order of 20 ms) show the Gaussian pdf to be a good first-order model [96]. The Gaussian pdf also appears to be a good short-term first-order model for the prediction residual in adaptive predictive coding [7], [74]. Since any speech coding system operating at 2 bits/sample or less would need to be normalized with respect to short-term energy to minimize distortion, we shall assume in this paper that the first-order pdf for speech and the prediction residual is essentially Gaussian.

Table 1 Four Common PDFs and their Differential Entropies $h(x)$. Column 4 gives the difference in bits between the Gaussian rate-distortion function $R_G(D)$ and the Shannon lower bound $R^*(D)$ for each pdf. Column 5 shows the corresponding difference in distortion in decibels; it is obtained by multiplying column 4 by 6.02. Column 6 is the asymptotic (high bit rate) difference in bits between the rate of a Lloyd–Max quantizer R_{LM} and the Shannon lower bound, and column 7 is the corresponding difference for the distortion. (From Jayant and Noll [74].)

pdf	$p(x)$	$h(x)$	$R_G(D) - R^*(D) = h_G(x) - h(x)$ (bits)	$\mathscr{D}_G(R) - \mathscr{D}^*(R)$ (dB)	$R_{LM} - R^*$ (bits)	$\mathscr{D}_{LM} - \mathscr{D}^*$ (dB)
Gaussian	$\frac{1}{\sqrt{2\pi}\,\sigma}\exp[-x^2/2\sigma^2]$	$\frac{1}{2}\log_2(2\pi e\sigma^2)$	0	0	0.722	4.35
Uniform	$\frac{1}{2\sqrt{3}\,\sigma}$, $\lvert x\rvert \leq \sqrt{3}\,\sigma$; 0, otherwise	$\frac{1}{2}\log_2(12\sigma^2)$	0.255	1.53	0.255	1.53
Laplacian	$\frac{1}{\sqrt{2}\,\sigma}\exp[-\sqrt{2}\lvert x\rvert/\sigma]$	$\frac{1}{2}\log_2(2e^2\sigma^2)$	0.104	0.63	1.190	7.17
Gamma	$\frac{\sqrt[4]{3}}{\sqrt{8\pi\sigma\lvert x\rvert}}\exp[-\sqrt{3}\lvert x\rvert/2\sigma]$	$\frac{1}{2}\log_2(4\pi e^{1-C}\sigma^2/3)$; C = Euler's constant = 0.5772	0.709	4.27	1.963	11.82

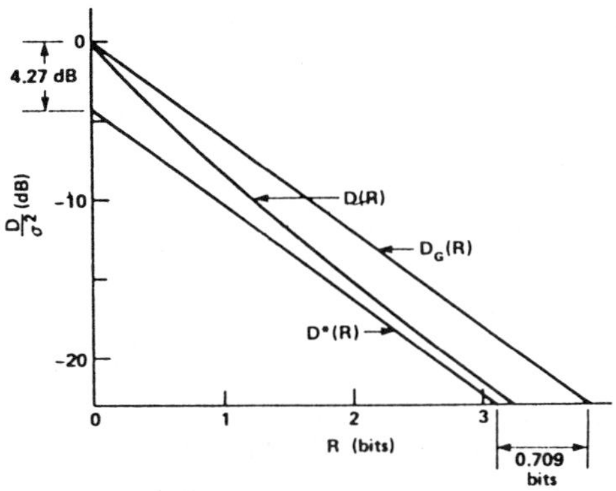

Fig. 10. Distortion-rate function $D(R)$ for a memoryless Gamma source and a mse distortion measure, bounded above by the Gaussian distortion-rate function $D_G(R)$ and below by the Shannon lower bound $D^*(R)$, from Noll and Zelinski [97]. $D_G(R)$ and $D^*(R)$ are always parallel straight lines in the decibel scale with a slope of -6.02 dB/bit. $D(R)$ tends toward a slope of -6.02 dB/bit at high bit rates (usually $R > 3$ bits).

Fig. 10 shows a plot of $D(R)$ for the Gamma pdf along with the Gaussian upper bound $D_G(R)$ and the Shannon lower bound $D^*(R)$. (The reason for choosing the Gamma pdf is because it shows very clearly the departure from the Gaussian.) The $D(R)$ plot was obtained using Blahut's algorithm [17]. Note that the $D(R)$ curve is monotone decreasing and convex. Also, as R increases beyond a few bits, $D(R)$ decreases at the rate of about 6.02 dB/bit which is the slope of the upper and lower bounds. The slope of -6.02 dB/bit will recur over and over again for many types of quantizers as R increases.

Thus far we have considered only memoryless sources. For sources with memory, where the samples are dependent, explicit $D(R)$ results are even more meager. What little is known is for the Gaussian case where the dependence is linear and can be specified completely by the spectral density $\Phi(\omega)$ of the stochastic sequence $x(n)$. For this special case, $D(R)$ is given parametrically by [14]

$$D_G(\theta) = \frac{1}{2\pi}\int_{-\pi}^{\pi} \min\{\theta, \Phi(\omega)\}\, d\omega \tag{72}$$

$$R_G(\theta) = \frac{1}{2\pi}\int_{-\pi}^{\pi} \max\left\{0, \frac{1}{2}\log_2 \frac{\Phi(\omega)}{\theta}\right\} d\omega. \tag{73}$$

For the case of small distortions defined by

$$\theta \leqslant \min_{\omega}\{\Phi(\omega)\} \tag{74}$$

$D_G(R)$ is given by

$$D_G(R) = \gamma 2^{-2R}\sigma^2 \tag{75}$$

where

$$\gamma = \frac{1}{\sigma^2}\exp\left[\frac{1}{2\pi}\int_{-\pi}^{\pi}\log\Phi(\omega)\, d\omega\right]$$
$$= \frac{\mathrm{GM}\{\Phi(\omega)\}}{\mathrm{AM}\{\Phi(\omega)\}} \leqslant 1. \tag{76}$$

GM and AM are the geometric mean and arithmetic mean, respectively. (σ^2 = arithmetic mean of the spectrum.) γ is a *spectral flatness measure* which is a nonnegative quantity that is equal to one if and only if the spectrum is flat [86], [90]. For a flat spectrum, i.e., a white source, the Gaussian signal values are independent and $D_G(R)$ in (75) becomes equal to (64). Therefore,

$$D_G(R)\big|_{\text{correlated source}} = \gamma D_G(R)\big|_{\text{white source}} \tag{77}$$

for distortions obeying (74). Note that for this case of small distortions, the distortion in (72) is equal to θ for all frequencies, which means that the reconstruction error will have a flat spectral density: a result that we shall meet again below.

Equation (77) states that $D(R)$ for a correlated Gaussian source is less than that for the corresponding white (memoryless) source. The difference for small distortions (or high rates) is equal to $-10\log_{10}\gamma$ decibels. Equation (77) quantifies the intuitive notion that one can transmit dependent sources at a lower distortion than independent sources. This general notion applies to non-Gaussian sources as well. The $D(R)$ function for such sources is still bounded above and below as in (67), where the Shannon lower bound is still defined by (68). However, $h(x)$ is now the differential entropy for the dependent source, defined as

$$h(x) = \lim_{N\to\infty} \frac{1}{N} h(\boldsymbol{x})$$

which will be smaller than the differential entropy for the corresponding memoryless source.

B. Scalar Quantization

We showed above how the distortion-rate function $D(R)$ provides a *lower bound* on the minimum distortion achievable by a vector quantizer at a given bit rate. Scalar quantizers provide us with an *upper bound* on the minimum achievable distortion. The difference between the two bounds gives the reduction in distortion that is potentially attainable by vector quantization. In this section we shall consider the scalar quantization of memoryless sources. The scalar quantization of correlated sources is discussed in Section IV.

1) Lloyd–Max Quantization: A scalar quantizer may be designed using the K-means algorithm described in Section II-C. However, because in one dimension the cells are restricted to be adjacent line segments, as shown in Fig. 4, one can instead use the well-known *Lloyd–Max quantizer*[8] [82], [91]. Given a pdf $p(x)$ and a number of levels L, this quantizer determines the intervals C_i and the reconstruction values y_i that minimize the average mse. The necessary conditions for the minimum are obtained by straightforward differentiation with respect to y_i and the interval boundary values g_i. The necessary conditions for optimality can be shown to be

$$g_i = \tfrac{1}{2}(y_i + y_{i-1}), \qquad 2 \leqslant i \leqslant L \tag{78}$$
$$g_1 = -\infty, \qquad g_{L+1} = \infty$$
$$y_i = \mathrm{cent}(C_i), \qquad 1 \leqslant i \leqslant L \tag{79}$$

where the centroid of C_i is simply the mean value of x in that interval. Equation (78) states that the interval boundaries must lie halfway between the reconstruction values. Note that (78) corresponds to (25) in the general case and (79) is the same as (27). For $L > 3$, (78) and (79) are solved itera-

[8] Also known as the Lloyd quantizer or the Max quantizer.

tively to obtain a set of optimal values. Those values may indicate only a local optimum. For the case when $p(x)$ is log-concave, the above necessary conditions are also sufficient for optimality [40] and the optimum will be global. In general, the optimum quantizer will not be uniform, i.e., the interval lengths will not be the same.[9] The interval spacing tends to be smaller where the pdf is larger. Performance, typically, must be evaluated by numerical methods, except when the number of quantization levels is large there is the asymptotic formula given by Algazi [4] for the rth-power distortion

$$D_r = \frac{2^{-r}}{r+1} L^{-r}\left[\int_{-\infty}^{\infty} [p(x)]^{1/(r+1)}\, dx\right]^{r+1} \quad \text{(Lloyd–Max)}. \tag{80}$$

For $R = \log_2 L$ and $r = 2$, (80) can be written in decibels as

$$\mathscr{D}_2 = 10\log_{10} D_2 = -6.02R + F_{LM}(p) \quad \text{dB} \tag{81}$$

where script $\mathscr{D}_2$ is the distortion in decibels and $F_{LM}(p)$ is a constant that depends on the pdf shape. Note again the -6.02-dB/bit behavior for large R.

Fig. 11 shows the normalized mse obtained by the Lloyd–Max quantizer as a function of the rate $R = \log_2 L$ for four pdfs. While the Gamma pdf has the lowest $D(R)$ of

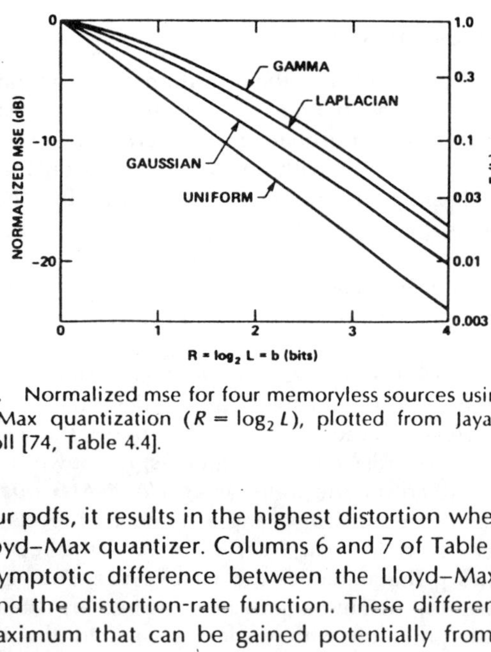

Fig. 11. Normalized mse for four memoryless sources using Lloyd–Max quantization ($R = \log_2 L$), plotted from Jayant and Noll [74, Table 4.4].

the four pdfs, it results in the highest distortion when using the Lloyd–Max quantizer. Columns 6 and 7 of Table 1 show the asymptotic difference between the Lloyd–Max quantizer and the distortion-rate function. These differences are the maximum that can be gained potentially from vector quantization.

2) Constrained-Entropy Quantization: In Lloyd–Max quantization, the number of levels L is fixed and the bit rate is defined simply as $\log_2 L$. However, if the reconstruction values y_i are not equally probable, we can use entropy coding to reduce the rate below $\log_2 L$ to $H(y)$. One can go even further and restate the optimization problem to minimize the distortion *subject to a given entropy* $H(y) = R$. We shall call the resulting quantizer a *constrained-entropy quantizer.* The number of levels L can now be any desired value including infinity. Gish and Pierce [51] have shown that, *at high bit rates, the uniform quantizer is the optimum constrained-entropy quantizer.* With very mild restrictions, this result is true for all pdfs and distortion measures. More recently, Farvardin and Modestino [33] showed experimentally that, with mse distortion, the uniform quantizer is very close to optimal even at very low rates. The obvious conclusion is that, if one is willing to perform variable-rate entropy coding, then the uniform quantizer is always the scalar quantizer of choice.

The asymptotic (high bit rate) performance for the constrained-entropy quantizer is given in [51], and can be written for an rth-power distortion measure as

$$D_r = \frac{2^{-r}}{r+1} 2^{rh(x)} 2^{-rH_{\min}} \quad \text{(constrained entropy)} \tag{82}$$

where $H_{\min}$ is the entropy of the constrained-entropy quantizer outputs and $h(x)$ is the differential entropy of the source. Equation (82) can be written in decibels for $r = 2$

$$\mathscr{D}_2 = -6.02H_{\min} + F_{CE}(p) \quad \text{dB} \tag{83}$$

where F_{CE} is another constant that depends on the pdf shape.

It is interesting that, for the mse distortion, there is a simple relationship between $\log_2 L$ of the Lloyd–Max quantizer and $H_{\min}$. In particular, we have from [51] that, for large L

$$H_{LM} - H_{\min} = \tfrac{1}{3}(\log_2 L - H_{\min}) \tag{84}$$

where H_{LM} is the entropy of the Lloyd–Max quantizer outputs. This relation shows the additional benefit of constrained-entropy quantization over entropy coding the outputs of the Lloyd–Max quantizer.

Another significant result obtained by Gish and Pierce [51] is that the constrained-entropy quantizer approaches the rate-distortion bound within a fixed constant that is dependent only on the distortion measure and not on the pdf. This constant, derived for an rth-power distortion measure, is given by

$$H_{\min} - R(D_r) = \frac{1}{r}\log_2 \frac{re}{1+r} + \log_2 \Gamma\left(1 + \frac{1}{r}\right) \tag{85}$$

where Γ is the Gamma (factorial) function. For the mse distortion, (85) reduces to

$$H_{\min} - R(D_2) = \frac{1}{2}\log_2 \frac{\pi e}{6} = 0.255 \text{ bits}. \tag{86}$$

Fig. 12 shows a plot of (85) as a function of r. The horizontal scale has been chosen to illustrate the important result that for $r = \infty$, which corresponds to the minimax criterion, the curve goes to zero, and $H_{\min} = R(D_\infty)$. Therefore, *a uniform scalar quantizer with entropy coding can achieve the rate-distortion function for the minimax criterion.* For $r = 1$, the constrained-entropy quantizer can approach $R(D)$ within 0.443 bits, and within 0.255 bits for $r = 2$.

The low-rate region ($R < 3$ bits) has been explored by a number of researchers (see [74]), with the results of Farvardin and Modestino [33] being the most useful for our purposes. For the mse distortion, the constrained-entropy quantizer at low rates is even closer to $D(R)$ than 0.255 bits, especially for the more peaked pdfs such as the Laplacian and the Gamma densities. In fact, of all the nonuniform pdfs tested, the Gaussian pdf registers the least gain and

[9]Note that a uniform quantizer is one whose interval lengths are equal. However, depending on the pdf shape and the distortion measure, the output levels of a minimum-distortion uniform quantizer will not be equally spaced, in general.

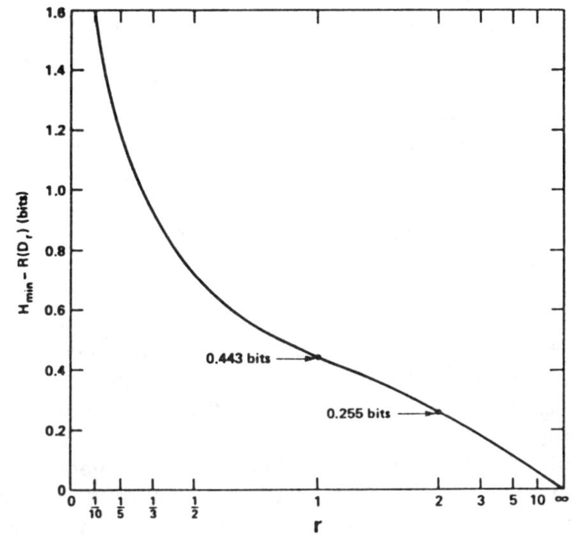

Fig. 12. The difference between the entropy of a constrained-entropy (uniform) quantizer and the rate-distortion function at high bit rates for an rth-power distortion function. This difference is independent of the pdf of the memoryless source; it depends only on r. For $r = \infty$, the minimax distortion, that difference is zero. The data for the plot were taken from Gish and Pierce [51].

lies the furthest away from its $D(R)$ function at low rates. Figs. 13 and 14 show plots of the mse distortion with the constrained-entropy quantizer (solid curves) compared to Lloyd–Max quantization and the distortion-rate functions for the Gaussian (Fig. 13) and the Laplacian (Fig. 14) memoryless sources. The constrained-entropy curve for the

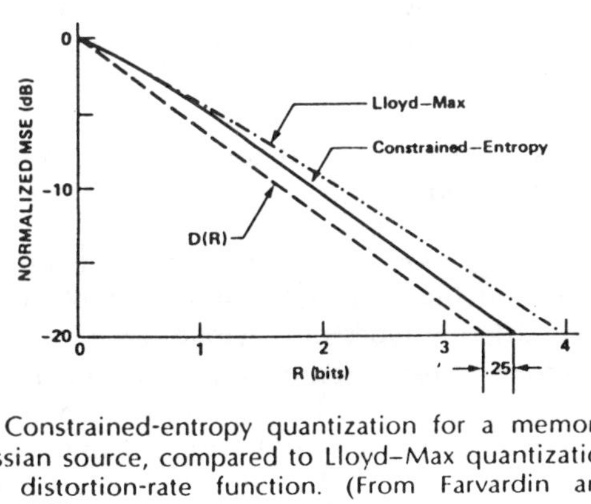

Fig. 13. Constrained-entropy quantization for a memoryless Gaussian source, compared to Lloyd–Max quantization and the distortion-rate function. (From Farvardin and Modestino [33].)

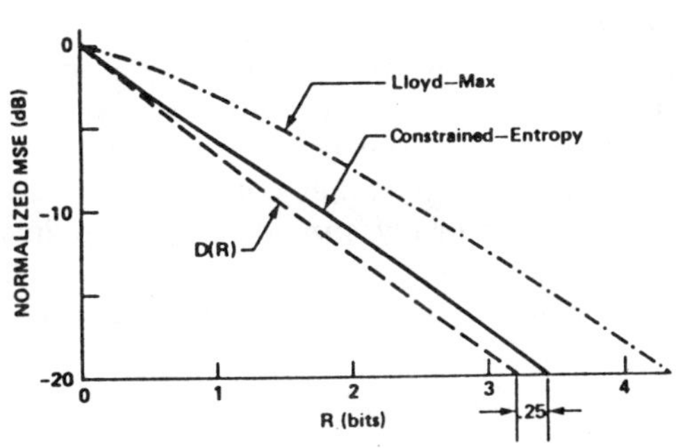

Fig. 14. Same as Fig. 13 for a memoryless Laplacian source. (From Farvardin and Modestino [33].)

Laplacian pdf is much closer to its distortion-rate function than the Gaussian curve is to its $D(R)$, while the opposite is true for the Lloyd–Max curves.

We saw above how constrained-entropy quantization (with entropy coding) takes advantage of the pdf shape to minimize the bit rate at a given distortion. The difference between the solid curves and the $D(R)$ curves in Figs. 13 and 14 is what is not achievable by scalar quantization. Vector quantization with very large N is capable of closing that gap without the need for entropy coding. At the low data rate of $R = 1$ bit in Fig. 13, being able to reach the distortion-rate bound for a Gaussian source would mean a reduction of about 25 percent in bit rate over the scalar quantizers. This result will be important for the coding of the speech prediction residual since it is modeled well as a memoryless Gaussian source.

C. Asymptotic Vector Quantization

We now present some of the known theoretical results for vector quantization. Most of the results below were derived under asymptotic conditions: either L is large (i.e., small distortion) and the vector size N is arbitrary, or N is large and L is arbitrary. We begin by discussing the optimum L-level quantizer in N dimensions, which is analogous to Lloyd–Max quantization in the scalar case, followed by constrained-entropy quantization in the vector case. It is important to notice that, unlike the results of the previous section which applied only to memoryless sources, *the VQ results below apply to sources with arbitrary pdfs, including linear and nonlinear dependencies*, unless noted otherwise.

The asymptotic formulas for scalar quantizer performance mentioned in Section III-B have analogues for vector quantizers. Zador [141] showed that, for large L, the optimum (minimum rth-power distortion) L-level quantizer in N dimensions has a distortion

$$D_r = A(r, N)\, L^{-r/N} \left[\int_{-\infty}^{\infty} [p(\mathbf{x})]^{N/(N+r)}\, d\mathbf{x} \right]^{(N+r)/N} \tag{87}$$

where $p(\mathbf{x})$ is the N-dimensional pdf of the vector process $\mathbf{x}$, and $A(r, N)$ is a term that is independent of the pdf; it represents how well cells can be packed in N-dimensional space for the rth-power distortion measure. Equation (87) is analogous to (80) in the scalar case, but it is important to note that (80) was derived for a memoryless source while (87) is true for an arbitrary source. For the mse distortion, (87) may be written in decibels as

$$\begin{aligned} \mathscr{D}_2 &= -\frac{6.02}{N} B + F_{VQ}(p, N) \quad \text{dB} \\ &= -6.02R + F_{VQ}(p, N) \quad \text{dB} \end{aligned} \tag{88}$$

where $B = \log_2 L$ is the number of bits per vector, R is the number of bits per dimension, and $F_{VQ}(p, N)$ is a term that depends on the pdf $p(\mathbf{x})$ and the number of dimensions.

The difficulty in utilizing the above expressions is that $A(r, N)$ is known exactly for very few cases. For other than $N = 1$, only $A(2, 2)$ is known exactly. For two dimensions and the mse criterion [35], [47], optimal vector quantizer performance (without entropy coding) is obtained with hexagonal regions, for large L. Note that, except for a uniform pdf, these hexagons will not be regular and will not have the same size in general. For higher dimensions, there are a number of upper and lower bounds on $A(r, N)$, but such bounds, though theoretically interesting, appear to be of little practical value. Recently, Conway and Sloane [23] conjectured a lower bound on $A(2, N)$ that seems interesting and useful (see Fig. 15). Other useful information on performance bounds is contained in [47], [139].

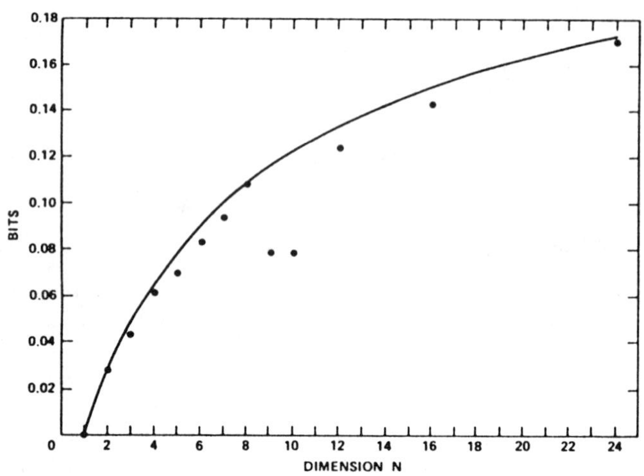

Fig. 15. The savings in bits when using certain N-dimensional lattice quantizers compared to quantization with N-dimensional cubes for a uniform pdf and high bit rate. The data were computed from Conway and Sloane [23, Table I] as follows:

$$\text{BITS} = [10 \log_{10} 0.08333/G_n]/6.02$$

where G_n is as defined in [23]. The solid curve is a conjectured bound given by Conway and Sloane [23] and the dots are the savings for the best known lattice quantizer for each N.

Although the optimal N-dimensional quantizer and its performance are not known in most cases, some understanding has been obtained about its structure. Zador [141] has shown that the optimum density of output values to use for random quantizer selection is proportional to $p^{N/(N+r)}$, where p is the pdf of the vector $\boldsymbol{x}$. (A related result is also obtained by Gersho [47].) For very large N ($N \gg r$), therefore, the output values should have a density similar to $p(\boldsymbol{x})$ for optimum performance.

In a manner similar to scalar quantizer design, one can constrain the entropy rather than the number of levels for the vector quantizer; we call the resulting quantizer the *constrained-entropy vector quantizer.* For such a quantizer, Zador [141] gives the asymptotic rth-power distortion as

$$D_r = B(r, N) 2^{rh(\boldsymbol{x})/N} 2^{-rH_{\min}/N} \tag{89}$$

where $B(r, N)$ is a term similar to $A(r, N)$ in (87). Equation (89) is analogous to (82) in the scalar case. One could write (89) for $r = 2$ to show the -6.02-dB/bit behavior.

Again, values of $B(r, N)$ are known for only a few cases. However, if entropy coding is to be used, there are suboptimal solutions for which the performance is known for large L. One solution, considered by Gish and Pierce [51], is analogous to the solution in the scalar case; namely, apply a uniform scalar quantizer to each of the vector dimensions and perform *joint entropy coding* on the vector outputs. The quantization interval should be the same in all dimensions, which is equivalent to specifying an N-dimensional cube for the cells of the vector quantizer. The result for this coding procedure is that the bit rate will differ from the rate-distortion function by an amount equal to that shown in Fig. 12 for an rth-power distortion measure, independent of the shape of the multidimensional pdf. For the mse, this difference is again 0.255 bits per dimension. This result is true at high bit rates; experiments similar to those performed by Farvardin and Modestino [33] for the scalar case will need to be done for the vector case to test the performance at low bit rates. The 0.255-bit differential that exists for the mse criterion when quantizing with cubes can be reduced by using different cell shapes for different dimensions, as we shall see below in the discussion under space packing.

While the VQ method outlined above is very simple indeed and requires a minimum amount of design and computations, it requires a very large amount of storage. Note that the entropy coding needs to be performed in N-dimensional space. Now, when we use entropy coding we typically use a large number of levels per dimension; therefore, the total number of levels in N-dimensional space will be very large, and the entropy coding will require a storage location for each of the codewords. In essence, we are substituting storage for computation and the amount of storage needed may be excessive for many applications.

Another important result from rate-distortion theory is that, for fixed L and for very large N, the entropy of the vector quantizer approaches $\log_2 L$ [123]. This implies that the output values $\boldsymbol{y}_i$ must have equal probability. (Note that the output values may have a *density* proportional to $p(\boldsymbol{x})$ but still have *discrete* probabilities that are equal.) This result is satisfying in that it confirms the fact that a vector quantizer can achieve its optimal performance for high dimensions without the need for entropy coding. In other words, the fixed-L and constrained-entropy quantizers achieve the same performance for high dimensions. This statement is in sharp contrast to the scalar case ($N = 1$) where the two types of quantizers give different performance.

Space packing: An interesting set of results have been obtained in studying the benefits of packing N-dimensional space with different types of polytopes (volumes bounded by planes) for the mse criterion. In fact, in a manner similar to the hexagon result mentioned above for $N = 2$, it has been conjectured by Gersho [47] that, asymptotically for large L and for *any pdf,* the optimum fixed-L vector quantizer partitions the space into cells whose shapes are derived from a specific polytope for each N.

A special set of polytopes, known as parallelohedra, have been studied by Gersho [47] and by Conway and Sloane [21] for the purpose of quantizing a uniform pdf in N dimensions. The centers of these parallelohedra lie on a *lattice* in N-dimensional space.[10] The lattice points are the output values of the quantizer (i.e., they constitute the codebook), and they form a regularly spaced array of points in N-dimensional space. Such a quantizer is termed a *lattice quantizer.* One very attractive feature of lattice quantizers is

[10] The concept of a lattice is of importance in a number of diverse areas, including the geometry of numbers [19], solid-state physics [143], and image processing [29].

that, given an input vector $\boldsymbol{x}$, it is relatively straightforward to compute the nearest lattice output value [22], without having to compare $\boldsymbol{x}$ against all codebook values and without having to store all codebook values. Conway and Sloane [23] have investigated the use of a number of lattices in the vector quantization of a uniform pdf for large L and using the mse criterion. Fig. 15 shows the saving in bits when those lattices are used relative to using cubes as quantization regions. For each dimension, the lattice used is different. For $N = 1$ we have, of course, the uniform cube quantizer. For $N = 2$, it is the hexagonal lattice, and for $N = 3$, the polyhedron is known as the truncated octahedron. Note that the saving in bit rate increases with N for the different lattices except for $N = 9$ and $N = 10$ for which no lattices have been found that perform better than $N = 8$. The value of 0.028 bits for $N = 2$ is the same as that obtained in Example 3 earlier.

Although the results in Fig. 15 were obtained for a uniform pdf, they have wider applicability. One can show that, if entropy coding is used, then for *any pdf*, the lattice quantizers with large L achieve an entropy less than the cube quantizer by an amount equal to the rates shown in Fig. 15. Since from the results of Gish and Pierce the bit rate of the cube quantizer is 0.255 bits above the rate-distortion function, the bit rate for the lattice quantizers is even closer to $R(D)$ by an amount equal to that shown in Fig. 15 for different dimensions. For example, for $N = 8$, the lattice quantizer with entropy coding will have a bit rate that is only $0.255 - 0.109 = 0.146$ bits greater than $R(D)$ for large L. However, because of the potentially excessive amount of storage needed for the codewords, this method may not be practical. In the remainder of the paper we shall discuss methods that do not require entropy coding.

Experimental results: Much of this section has been devoted to theoretical asymptotic vector quantization results. Few attempts have been made to see how these results may apply in practice and for relatively small values of L and N. For a memoryless Gaussian source, the best VQ results with $R = 1$ bit/sample show practically no improvement over scalar quantization for $N = 2$, a reduction in mse of 0.09 dB (about 0.015 bits) for $N = 3$, and a reduction of about 0.52 dB (about 0.087 bits) for $N = 6$ [36], [58], [81], which is a good improvement for $R = 1$ bit. These results were obtained with no entropy coding.

Much greater gains are achievable for certain non-Gaussian sources. For example, for a memoryless Gamma source and $R = 1$ bit/sample, Fischer and Dicharry [36] obtained a substantial reduction in mse for $N = 6$. However, the resulting mse was still approximately 2 dB higher than $D(R)$ and well above the distortion of the constrained-entropy uniform quantizer, which was found by Granzow and Noll [55] to be only 0.7 dB above $D(R)$. However, this result serves to illustrate that, for cases where Lloyd–Max quantization performs poorly relative to $D(R)$, and entropy coding is not desired or cannot be used, one can make good use of vector quantization to improve performance substantially.

Results with a single-pole Gaussian source have also been obtained at $R = 1$ bit/sample, which show that VQ could outperform differential PCM (DPCM) coding with vector lengths $N > 3$ [60]. But, for $N = 7$, the VQ mse was still 1.6 dB higher than $D(R)$.

IV. Scalar versus Vector Quantization of Vector Sources

We indicated in Section II that the redundancy owing to linear dependency (correlation) can be utilized effectively by vector as well as scalar quantization, provided the latter is performed after the appropriate coordinate transformation. In this section we show how best to use scalar quantizers in the quantization of vector components. The purpose for this presentation is twofold. We wish to give the reader tools to help assess how much of the vector quantization performance in a particular application takes advantage of linear dependencies, and to help evaluate whether a scalar or a vector quantizer is more cost effective in that application, given other real-world constraints.

Throughout this section we use the mse as the distortion measure. The first two subsections are devoted to the scalar quantization of vector sources and the last subsection compares scalar to vector quantization. First, we show that, given a total number of bits to allocate to scalar coding of the various vector components, it is best (minimum mse) to allocate different number of bits to components with different variances. A bit-allocation procedure is described which yields the minimum mse. We then show that, if the vector components are correlated, one can reduce the mse further by first performing a decorrelating transformation or rotation on the vector, followed by independent (scalar) quantization of the components of the rotated vector, utilizing the bit-allocation procedure. Finally, we compare the performance of scalar and vector quantizers in a particular application.

The subject matter of this section is very applicable to the coding of linear prediction parameters (such as log-area ratios), which are generally correlated and have different probability distributions and different variances (see Section IV-C). It is also applicable to the transform coding of speech and other waveforms.

A. *Bit Allocation*

We are given a random vector $\boldsymbol{x}$ whose components x_k have identical pdf shapes but generally unequal variances σ_k^2. (The case of different pdf shapes will be treated below.) We wish to allocate a given number of bits B among the components and use scalar quantizers to quantize each component x_k independently, in such a way as to minimize the average vector distortion. Let R_k be the number of bits allocated to component x_k and let the distortion in quantizing that component be

$$D_k = \mathscr{E}(x_k - y_k)^2 \tag{90}$$

where y_k is the quantized value of x_k. The problem then is to determine the set $\{R_k\}$ so as to minimize the average distortion

$$D = \frac{1}{N}\sum_{k=1}^{N} D_k \tag{91}$$

subject to

$$\frac{1}{N}\sum_{k=1}^{N} R_k = \frac{B}{N} = R \tag{92}$$

where R is the average bit rate per component.

Let us assume for the present that we are operating in the high bit-rate region of the quantization distortion curve for each of the components (usually $R > 3$). Then we can write from the results of Section III-B

$$R_k = \rho + \frac{1}{2}\log_2 \frac{\sigma_k^2}{D_k} \tag{93}$$

where ρ is a constant value that depends on the pdf shape and on the particular scalar quantization method used. (For example, for a Lloyd–Max quantizer, we see from Fig. 11 that $\rho = 0$ for a uniform pdf and $\rho = 0.722$ for a Gaussian pdf.) From (91) and (93) we have

$$D = \frac{1}{N}\sum_{k=1}^{N} \sigma_k^2 2^{-2(R_k-\rho)}. \tag{94}$$

(Note that D in (94) has a slope of $-6.02/N$ decibels/bit, as expected for high bit rates.) The problem now is to find the R_k values that minimize D in (94) subject to (92). Using Lagrange multipliers, one can show that the distortions D_k must be equal

$$D_{\min} = D_k = 2^{-2(R-\rho)}\left[\prod_{k=1}^{N}\sigma_k^2\right]^{1/N}, \quad \text{for all } k \tag{95}$$

and the bit allocation is given by

$$R_k = R + \frac{1}{2}\log_2 \frac{\sigma_k^2}{\left[\prod_{k=1}^{N}\sigma_k^2\right]^{1/N}}. \tag{96}$$

If the components have different variances, the result implies that the number of bits used to quantize each component will be different.

By allocating a different number of bits to different components we realize a lower distortion than we would, had we used an equal number of bits R to quantize each component. In the latter case the total distortion would be

$$D_0 = 2^{-2(R-\rho)}\frac{1}{N}\sum_{k=1}^{N}\sigma_k^2. \tag{97}$$

Therefore, the ratio of the minimum distortion $D_{\min}$ to the equal allocation distortion D_0 is

$$\frac{D_{\min}}{D_0} = \frac{\left[\prod_{k=1}^{N}\sigma_k^2\right]^{1/N}}{\frac{1}{N}\sum_{k=1}^{N}\sigma_k^2} = \frac{\mathrm{GM}\{\sigma_k^2\}}{\mathrm{AM}\{\sigma_k^2\}} = \gamma \leq 1 \tag{98}$$

which is the ratio of the geometric mean to the arithmetic mean of the component variances, with equality iff all variances are equal. Thus with bit allocation, one would have an increase in SNR of $-10\log_{10}\gamma$ decibels. (Note that (98) was derived assuming the same pdf and high bit rates for all components.)

If $\gamma \ll 1$, implying that the component variances have a large dynamic range, then the bit-allocation scheme in (96) would result in a wide range of bit rates R_k. Some of these may be very low (even though the average may be high) and some may even be negative. Low rates would violate our assumption of high rates that allowed us to write (93), and negative rates are, of course, unallowable. The latter condition results for components whose variances are less than $D_{\min}$ in (95). For such cases, one would not transmit anything because by simply using the mean value at the receiver one would have at worst a distortion equal to the signal variance. If such negative bit rates are obtained from the bit-allocation scheme, the corresponding components are allocated zero bits. When negative values of R_k are thus increased to zero, the rates of other components will have to be modified to maintain the same average rate R. Several procedures have been developed which reoptimize the allocation. Segall [120] has solved the above optimal *continuous* bit allocation problem, assuming independent Gaussian components. Fox [43] had earlier presented an algorithm for optimal *integer* bit allocation in the general case where the pdfs of the components may be different. Below we give a brief description of the algorithm and the conditions for its applicability.

Optimum Integer Bit Allocation: Let us assume that we wish to have a bit allocation where all rates R_k are integers. We associate with each component x_k a normalized (unit-variance) quantization distortion function $E_k(b)$, which gives the distortion as a function of the number of bits b, assuming x_k has unit variance. (We assume that each x_k may have a different pdf.) Clearly then, the distortion of x_k is given by

$$D_k(b) = \sigma_k^2 E_k(b). \tag{99}$$

$E_k(b)$ is an actual curve obtained by applying a specific quantizer to the random variable x_k with its specific pdf. The quantizer can be a Lloyd–Max quantizer or a constrained-entrópy quantizer or any other quantizer of interest. The main thing is to generate the curve $E_k(b)$ for each component. Of course, if $E_k(b)$ is not optimized in any way to the component x_k, then the results may be suboptimal. (Fig. 11 shows $E(b)$ for a Lloyd–Max quantizer for four different pdfs.)

The bit-allocation procedure simply assigns the next bit to the component that causes the maximum incremental decrease in the overall distortion. The procedure is as follows:

Step 1: For each bit rate b, calculate the incremental decrease in the distortion when a bit is added to each component:

$$\Delta_k(b) = \sigma_k^2[E_k(b-1) - E_k(b)], \quad 1 \leq k \leq N, \quad b = 1, 2, \cdots. \tag{100}$$

Step 2: Sort the values $\Delta_k(b)$ in decreasing order.

Step 3: Assign the given bits one by one according to the resulting order.

(In actual implementation, the values $\Delta_k(b)$ are computed sequentially as they are needed in allocating the bits.) This algorithm can also be applied to assigning an integral number of *levels* to each component. The variable b would then refer to the number of levels.

The above algorithm gives the optimal solution when all quantization distortion functions $E_k(b)$ are monotone decreasing and convex, i.e., successive bits decrease the quantization error by an amount smaller than earlier bits allocated to that component. For the general case, where the quantization distortion functions are nonconvex or even nondecreasing, Shoham and Gersho [125] present an algorithm that finds the optimum bit allocation.

Note that in the case where the distortions do not obey (94), the minimum distortions for the different components after bit allocation will generally not be equal.

B. Vector Rotation for Correlated Sources

We saw above how one can take advantage of differences in component variances to decrease the distortion, for a given rate, over that of equal bit allocation. The development did not make any assumptions about possible dependence among components. We present below a procedure that takes advantage of any correlation that may exist among the vector components.

Fig. 16 shows a block diagram for the scalar quantization of a vector source, which includes a vector rotation $\boldsymbol{A}$ before quantization, followed by an inverse rotation $\boldsymbol{B}$ at the receiver. $\{q_k,\ 1 \leq k \leq N\}$ are N independent scalar quantizers that quantize the components of the rotated vector $\boldsymbol{u}$. In Section IV-A we considered the case of $\boldsymbol{A} = \boldsymbol{B} = \boldsymbol{I}$. Here we ask the question: What are the rotations $\boldsymbol{A}$ and $\boldsymbol{B}$ and the quantizers $\{q_k\}$ that minimize the distortion at the output, subject to a given bit rate? The answer to this question is not known for general distributions. However, for a source $\boldsymbol{x}$ whose components are jointly Gaussian, Huang and Schultheiss [65] and then Segall [120] showed that the optimal rotation $\boldsymbol{A}$ is given by a matrix $\boldsymbol{S}$ whose rows are the normalized eigenvectors of $\boldsymbol{\Gamma}_x$, the covariance matrix of the vector $\boldsymbol{x}$ (see (13)), and $\boldsymbol{B}$ is the inverse of $\boldsymbol{A}$

$$\boldsymbol{A} = \boldsymbol{S} \qquad \boldsymbol{B} = \boldsymbol{S}^{-1} \tag{101}$$

with

$$\boldsymbol{S}^{-1} = \boldsymbol{S}^T \tag{102}$$

and

$$\boldsymbol{S}\boldsymbol{\Gamma}_x\boldsymbol{S}^T = \boldsymbol{\lambda} = \mathrm{diag}[\lambda_1\lambda_2 \cdots \lambda_N] \tag{103}$$

where $\boldsymbol{\lambda}$ is a diagonal matrix whose elements are the eigenvalues of the covariance matrix $\boldsymbol{\Gamma}_x$. Equation (103) is a well-known property of a symmetric matrix [13]. The eigenmatrix $\boldsymbol{S}$ is known to be an orthogonal matrix. The rotated vector $\boldsymbol{u}$

$$\boldsymbol{u} = \boldsymbol{A}\boldsymbol{x} = \boldsymbol{S}\boldsymbol{x} \tag{104}$$

will then have a covariance matrix

$$\begin{aligned}\boldsymbol{\Gamma}_u &= \mathscr{E}\left[(\boldsymbol{u} - \bar{\boldsymbol{u}})(\boldsymbol{u} - \bar{\boldsymbol{u}})^T\right]\\ &= \boldsymbol{S}\boldsymbol{\Gamma}_x\boldsymbol{S}^T = \boldsymbol{\lambda}.\end{aligned} \tag{105}$$

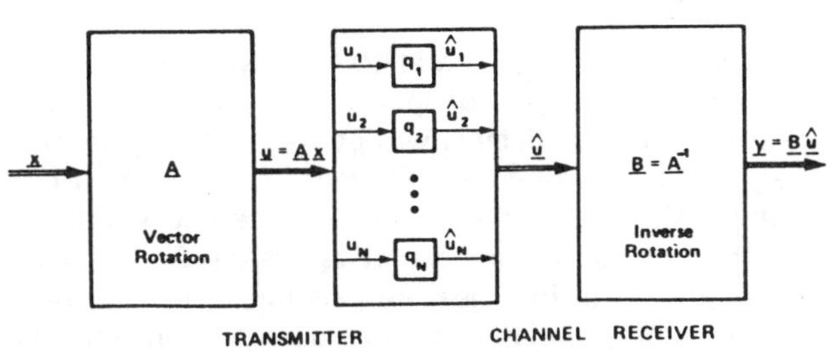

Fig. 16. Scalar quantization of a vector source. The eigenvector rotation $\boldsymbol{A}$ converts the input $\boldsymbol{x}$ into a vector $\boldsymbol{u}$ with uncorrelated components. An inverse rotation after the scalar quantizers gives $\boldsymbol{y}$, the quantized value of $\boldsymbol{x}$.

Therefore, the covariance of $\boldsymbol{u}$ is a diagonal matrix whose elements are the eigenvalues of the covariance $\boldsymbol{\Gamma}_x$. Since $\boldsymbol{\Gamma}_u$ is diagonal, the vector $\boldsymbol{u}$ has uncorrelated components, and since $\boldsymbol{x}$ is Gaussian, $\boldsymbol{u}$ will be Gaussian with independent components. The variance of each component u_k is then equal to the corresponding eigenvalue λ_k.

Using property (102) of an orthogonal matrix, one can show that the distortion at the receiver is equal to the distortion in quantizing the vector $\boldsymbol{u}$. Since

$$\boldsymbol{y} = \boldsymbol{S}^{-1}\hat{\boldsymbol{u}} \tag{106}$$

we have from (104), (106), and (102)

$$D = \frac{1}{N}\mathscr{E}\left[(\boldsymbol{x} - \boldsymbol{y})^T(\boldsymbol{x} - \boldsymbol{y})\right] = \frac{1}{N}\mathscr{E}\left[(\boldsymbol{u} - \hat{\boldsymbol{u}})^T(\boldsymbol{u} - \hat{\boldsymbol{u}})\right]. \tag{107}$$

Therefore, minimum distortion is obtained by designing quantizers that will minimize the average distortion in quantizing $\boldsymbol{u}$. The optimal scalar quantization scheme is then to quantize the components u_k using the bit-allocation scheme described in Section IV-A.

For the special case of high bit rates, the minimum distortion and the bit allocation are given by (95) and (96), respectively, with λ_k substituted for σ_k^2. The ratio of the minimum distortion after rotation $D'_{\min}$ to the equal bit allocation distortion D_0 is then given by

$$\frac{D'_{\min}}{D_0} = \frac{\left[\prod_{k=1}^{N}\lambda_k\right]^{1/N}}{\frac{1}{N}\sum_{k=1}^{N}\sigma_k^2}. \tag{108}$$

Note that, since [74]

$$\mathrm{GM}\{\lambda_k\} \leq \mathrm{GM}\{\sigma_k^2\} \tag{109}$$

we have from (98) and (108)

$$D'_{\min} \leq D_{\min} \tag{110}$$

with equality iff the components of the Gaussian vector $\boldsymbol{x}$ are uncorrelated and, hence, the variances are equal to the eigenvalues. In fact, the eigenvector rotation is the transform that minimizes the geometric mean of the resulting component variances. The same transformation, therefore, minimizes the quantization distortion. From (104) and (102), we can also write

$$\begin{aligned}\frac{1}{N}\sum_{k=1}^{N}\sigma_k^2 &= \frac{1}{N}\mathscr{E}\left[(\boldsymbol{x} - \bar{\boldsymbol{x}})^T(\boldsymbol{x} - \bar{\boldsymbol{x}})\right]\\ &= \frac{1}{N}\mathscr{E}\left[(\boldsymbol{u} - \bar{\boldsymbol{u}})^T(\boldsymbol{u} - \bar{\boldsymbol{u}})\right]\\ &= \frac{1}{N}\sum_{k=1}^{N}\lambda_k.\end{aligned} \tag{111}$$

Substituting in (108), we have

$$\frac{D'_{\min}}{D_0} = \frac{\mathrm{GM}\{\lambda_k\}}{\mathrm{AM}\{\lambda_k\}} \leq 1. \tag{112}$$

We note that there are many possible decorrelating rotations that can be applied to $\boldsymbol{x}$ to obtain a vector $\boldsymbol{u}$ whose components are uncorrelated. The eigenmatrix $\boldsymbol{S}$ is the one decorrelating matrix that results in the minimum distortion.

Optimal Scalar Quantization: Therefore, for a Gaussian vector process, the scalar quantization procedure that minimizes the mse for a given bit rate is

Step 1: Using the eigenmatrix $\boldsymbol{S}$ derived from the covariance of the vector process, transform (rotate) the input vector $\boldsymbol{x}$ to form another vector $\boldsymbol{u}$.

Step 2: Quantize the components of $\boldsymbol{u}$ using some optimal scalar quantizer and the bit-allocation procedure in Section IV-A. The result is the quantized vector $\hat{\boldsymbol{u}}$.

Step 3: Perform an inverse transform (rotation) on $\hat{\boldsymbol{u}}$ with the matrix $\boldsymbol{S}^T$, resulting in the output vector $\boldsymbol{y}$.

The above procedure is known to be optimal for Gaussian sources, but experience shows that the procedure works well for other distributions. In Example 1, we transformed the vector $\boldsymbol{x}$ in Fig. 5 into the vector $\boldsymbol{u}$ in Fig. 6. One can show that the rotation used is in fact an eigenvector rotation. Note how the marginal densities for the vector components change shape as we go from $\boldsymbol{x}$ to $\boldsymbol{u}$, and the components u_1 and u_2 in this case become independent. With some computation, one can show the interesting result in this example that the reduction in distortion achieved by the eigenvector rotation and the bit allocation is far greater than that predicted by (112). The reason, we believe, is the change in the shape of the marginal densities effected by the rotation.

Transform Coding: The transformation introduced by the eigenvector rotation is sometimes referred to as the *Karhunen–Loéve Transform* (KLT) (see [74]), or simply the *eigenvector transform*. The transform involves the computation of the eigenvectors and eigenvalues of the covariance matrix. For a relatively small number of dimensions N, the computation of the eigenvectors can be accomplished via any of a number of standard routines [67]. The eigenvector transform has been used, for example, in quantizing the outputs of a filter bank in a speech vocoder [79] and in quantizing LPC log-area-ratio parameters [45], [112], [116] (see Section IV-C). However, as N increases to values in the hundreds, the amount of computation may make computing the eigenvectors prohibitively expensive. In such cases, one often resorts to simpler orthogonal transforms, such as the discrete cosine transform [3], which, even if not optimal, has been shown to produce very good results for speech and other data [74]. Adaptive transform coding (ATC) [16], [130], [142] is one of the well-known methods for the coding of speech waveforms in the range 8–16 kbits/s.

Distortion Weighting: In Section II-B we mentioned the possibility of using a weighting matrix $\boldsymbol{W}$ to weight the errors in specific components differently. The reason for utilizing such weighting is often guided by perceptual considerations. Minimizing the weighted mse in quantizing $\boldsymbol{x}$ is equivalent to minimizing the mse in quantizing the transformed vector $\tilde{\boldsymbol{x}}$ given in (16). The quantization procedure then follows Fig. 16 with $\tilde{\boldsymbol{x}}$ instead of $\boldsymbol{x}$, and $\boldsymbol{A}$ is the eigenmatrix corresponding to the covariance of $\tilde{\boldsymbol{x}}$, which can be shown to be

$$\boldsymbol{\Gamma}_{\tilde{x}} = \boldsymbol{P}\,\boldsymbol{\Gamma}_x\,\boldsymbol{P}^T \tag{113}$$

where $\boldsymbol{P}$ is given by (15).

C. *Comparisons with Vector Quantization*

The performance of VQ in any specific application can be compared to scalar quantization to help assess the benefits accrued by VQ. Such comparisons are particularly important because of the substantial storage and computational costs typically associated with VQ. In this section we present an example to help illustrate the concepts presented and to see the specific benefits in an important application in speech coding.

EXAMPLE 4

In very-low-rate speech coding employing LPC analysis, one generally transmits for each frame a set of LPC coefficients, a gain term, a voiced/unvoiced decision, and a pitch value if the sound is voiced. At an analysis rate of 40 frames/s (i.e., every 25 ms) and 20 bits/frame, for example, the total bit rate will be 800 bits/s. In this paper we shall focus only on the quantization of spectral (LPC) parameters, since that is where vector quantization is mostly used. Typically, about 10–13 bits/frame are used to code the spectrum. Coincidentally, computational and storage costs tend to become unacceptable for applications requiring significantly more bits per frame to code the spectrum.

Fig. 17 shows the relative performance of scalar and vector quantizers for LPC parameters. Speech, low-pass filtered at 5 kHz and sampled at 10 kHz, was recorded from ten male speakers, 1 min of speech from each speaker. LPC analysis with 14 coefficients was performed at a rate of 100 frames/s. The resulting 60 000 frames of data were used to train the scalar and vector quantizers. The LPC coefficients were represented by log-area-ratios (LARs) G_k defined in (22) and the distortion measure was the mse. The mse in Fig. 17 is plotted relative to the distortion at zero bits, which is equal to the average squared value of the LARs for all the data. Fig. 17 shows three plots of mse as a function of bit rate. Plot (a) was obtained using a separate Lloyd–Max

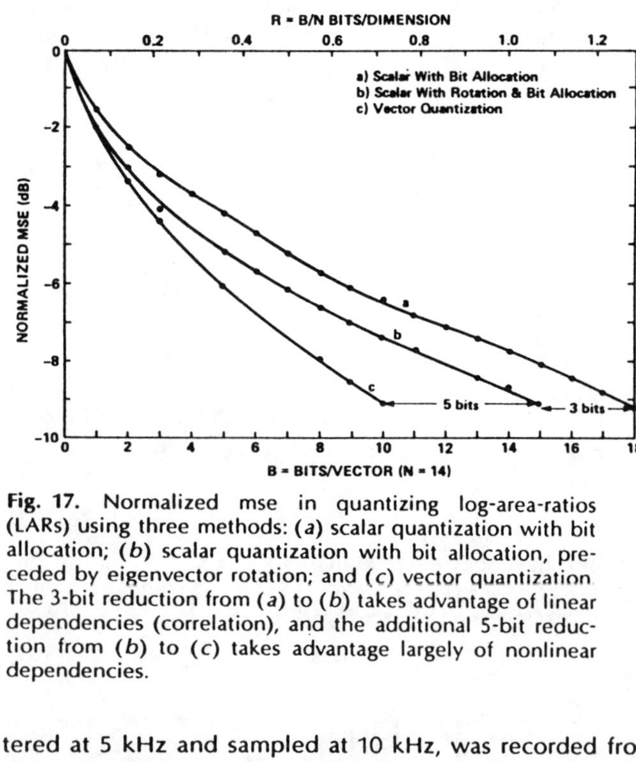

Fig. 17. Normalized mse in quantizing log-area-ratios (LARs) using three methods: (a) scalar quantization with bit allocation; (b) scalar quantization with bit allocation, preceded by eigenvector rotation; and (c) vector quantization. The 3-bit reduction from (a) to (b) takes advantage of linear dependencies (correlation), and the additional 5-bit reduction from (b) to (c) takes advantage largely of nonlinear dependencies.

scalar quantizer for each LAR with the integer bit-allocation scheme described in Section IV-A. Plot (*b*) was obtained by first performing a fixed eigenvector rotation, as described in Section IV-B, followed by bit allocation and Lloyd–Max scalar quantization. Plot (*c*) shows the performance of a vector quantizer whose codebook was designed using the *K*-means algorithm described in Section II-C. For the same mse obtained for the vector quantizer with 10 bits, the scalar quantizer with rotation requires 15 bits, while the scalar quantizer without rotation requires 18 bits. The benefit of 5–8 bits with vector quantization at $B = 10$ bits would be expected to increase as the number of bits is increased, until the slope of the vector quantization curve reaches -0.43 dB/bit ($6.02/14 = 0.43$). However, in terms of percentage of total bit rate, the relative benefit of vector over scalar quantization decreases as the bit rate increases.

The 3-bit reduction in bit rate with eigenvector rotation shows the benefit of taking advantage of linear dependency. Much of the additional 5-bit reduction effected by vector quantization takes advantage of nonlinear dependencies among LARs. The existence of such significant nonlinear dependencies in the space of LARs (or other spectral parameters) for speech is what makes vector quantization truly attractive in this application.

The 10-bit vector quantization system and the 15-bit scalar quantization system with rotation were also compared using subjective listening tests. The analysis/synthesis employed variable-frame-rate transmission (see Section VI-A) at an average of 30 frames/s. Pitch, gain, and duration were unquantized. Upon informal listening, no clear difference could be detected between the two systems.

The bit allocation for the scalar quantizers utilized the pdfs of the different components. Fig. 18 shows the estimated (unnormalized) pdfs for G_1, G_2, and G_{14}. The pdfs for G_1 and G_2 are quite asymmetric and their variances are significantly larger than the other LARs. All pdfs for G_4 to G_{14} tend to be Gaussian in shape, as in the pdf of G_{14} shown in the figure. Table 2 shows the bit allocation before and after eigenvector rotation for the first 15 bits. The table shows the parameter index to which each additional bit is allocated. The parameters are the LARs for the case with no rotation and the transformed parameters after rotation (the index in the latter case is that of the corresponding eigenvector with the eigenvalues ordered in decreasing value). For example, the seventh bit was allocated to the sixth LAR in one case and to the parameter corresponding to the fourth eigenvector in the other. Table 3 shows the total number of bits allocated to each parameter for the 15-bit case. Generally, the parameters with larger variance are allocated more bits. Note in both cases how not all parame-

Table 2 Bit Allocation Before and After the Eigenvector Rotation of LARs. The table shows the parameter index to which each bit is allocated.

Bit Number	Parameter Index (k) No Rotation	Parameter Index (k) With Rotation
1	1	1
2	2	2
3	1	1
4	2	2
5	4	3
6	3	1
7	6	4
8	1	5
9	5	6
10	3	3
11	8	7
12	4	8
13	2	4
14	10	2
15	6	9

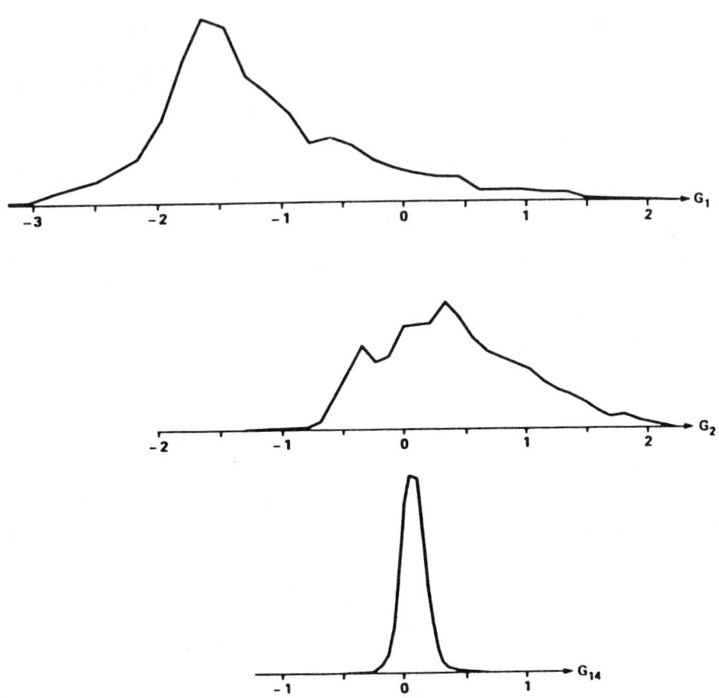

Fig. 18. Histograms of LARs G_1, G_2, and G_{14}. Histograms for G_4 through G_{14} can be modeled well by Gaussian pdfs.

Table 3 Total Number of Bits Allocated to Each Parameter for the 15-Bit Case

Parameter Index (k)	Number of Bits	
	No Rotation	With Rotation
1	3	3
2	3	3
3	2	2
4	2	2
5	1	1
6	2	1
7	0	1
8	1	1
9	0	1
10	1	0
11–14	0	0
	15 bits	15 bits

ters are allocated bits. Parameters whose variances are smaller than the expected distortion are allocated zero bits.

Distortion Measure: In the example above, the mse was used in measuring the distortion of quantizing LARs. The same distortion measure was used for the vector and the scalar quantizers. Using the *same* distortion measure to assess the relative performance of quantizers is very important; otherwise, one can obtain misleading results. However, one freedom we have in vector quantization is to choose distortion measures that are not usually defined for the scalar case, such as the Itakura–Saito distortion in (23). It would be unfair, for example, to use the Itakura–Saito distortion to compare the performance of vector and scalar quantizers unless they were both designed using that distortion measure. Unfortunately, it is not simple to design a scalar quantizer using the Itakura–Saito distortion.[11] One, of course, could always compare the performance of any two quantizers fairly by subjective listening tests. We have designed LPC vector quantizers using the Itakura–Saito distortion as well as the mse on LARs and found no clear difference in subjective speech quality between the two methods.

Dimensionality and Bit Allocation: It was clear in the example above that if sufficient bits are not available, then certain vector components may not be allocated any bits with scalar quantization. For cases where vector components are uncorrelated and of equal variance, and $R < 1$ bit, the bit-allocation procedure will have to assign 1 bit to each of several components in some arbitrary fashion and assign zero bits to the other components, such that the average bit rate is R. The major limitation we are witnessing here is that a scalar quantizer cannot assign a fraction of a bit to a component; it must assign either 1 bit or none corresponding to either a two-level or a one-level quantizer (a fractional number of levels is not possible). (The only exception is if entropy coding is used, in which case fractional bits are possible.) A major advantage of the dimensionality of vectors is that fractional bits come in a very natural way, and all vector components are allocated bits in an equitable fashion. It is this aspect of dimensionality, in addition to the ability to specify arbitrary cell shapes, that makes VQ especially important at low rates. These properties of dimensionality are what makes it possible to code the prediction residual at $R < 1$ bit/sample and maintain good speech quality (see Section VII-B).

Vector Sources: The sequence of LPC-parameter vectors in Example 4 constitutes what one might consider as a "naturally occurring" vector source. Each vector is not merely a collection of scalars from a scalar source, but somehow represents a single entity, namely, the short-term spectrum. The VQ of this vector source as described above takes advantage of the dependencies among vector parameters *in parameter space*. VQ theory allows us, however, to form longer vectors from sequences of LPC vectors and perform VQ on the longer vectors, thereby taking advantage of dependencies *in time* between adjacent LPC vectors, and benefiting from the higher dimensionality. (This is precisely what is done in segment quantization, as described in Section VI.) However, the basic VQ process and its properties remain the same whether it is performed on an inherently scalar or vector source.

[11]Strictly speaking, one cannot design an optimal scalar quantizer using the Itakura–Saito distortion, because scalar quantization implies the independent quantization of vector parameters and the Itakura–Saito distortion is not separable into distortions on the individual parameters. However, one could design a product code instead (see Section V-C).

V. Codebook Design

It should be clear now that VQ can offer substantial performance advantages over scalar quantization at very low rates, especially for sources that exhibit significant nonlinear dependencies. Unfortunately, these advantages are obtained at considerable computational and storage costs, as we saw in Section II-D. For full-search coding, the costs are exponential in the number of bits per vector. Computational and storage costs double for each increase of 1 bit in the rate. In Example 4, to perform 10-bit quantization of vectors of 14 LARs, it takes $14 \times 2^{10} = 14\,336$ multiply-adds to quantize *each* input vector. The codebook requires an equal number of storage locations. This number doubles if 11-bit quantization is desired. Very quickly, computational and storage costs become prohibitively expensive as the number of bits/vector increases.

A number of fast-search algorithms have been proposed in the pattern-recognition literature [44], [121], [140] and more recently in VQ [20], [48], which are designed to reduce the computations in a full search. Most of the algorithms are based on geometrical notions in Euclidean spaces, they require preprocessing of the codebook, and tend to trade off multiplications with comparisons and with increased storage requirements. The number of multiplications can be reduced by as much as an order of magnitude.

In this section we present variations on the basic VQ scheme which are intended to reduce computational costs in a very significant manner (linear rather than exponential growth with the number of bits per vector), but at some reduction in performance. Other methods that are designed to reduce storage costs result in relatively greater reduction in performance. The section ends with a discussion of training and testing issues and codebook robustness.

A. Binary Search

With the K-means algorithm (with $K = L$ levels), a *full search* of the L code vectors is required to quantize each input vector. *Binary search* [18], [112], known in the pattern-recognition literature as hierarchical clustering [5], [64], is a method for partitioning space in such a way that the search for the minimum-distortion code vector is proportional to $\log_2 L$ rather than L. Specifically, N-dimensional space is first divided into two regions (using the K-means algorithm with $K = 2$), then each of the two regions is divided further into two subregions, and so on, until the space is divided into L regions or cells. Here, L is restricted to be a power of 2, $L = 2^B$, where B is an integral number of bits. (A relaxation of this condition will be discussed below.) Associated with each region at each binary division is its centroid. Fig. 19 is a schematic of binary division of space into $L = 8$ cells. At the first binary

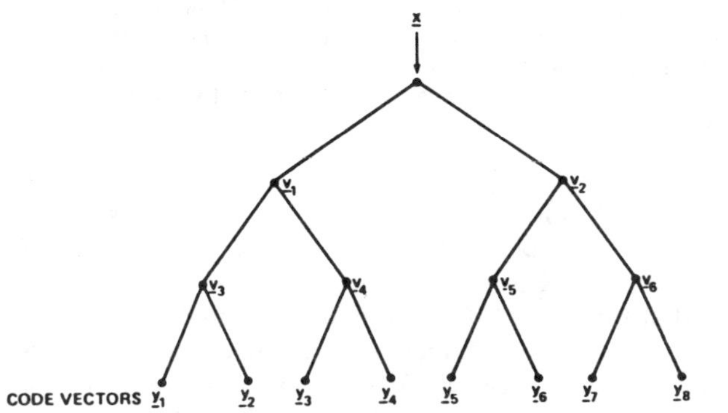

Fig. 19. Uniform tree for a binary-search vector quantizer. The vectors $\boldsymbol{v}_i$ are intermediate code vectors that are compared with the input vector $\boldsymbol{x}$. The code vectors are the vectors $\boldsymbol{y}_i$. The codebook size L is restricted to be a power of 2 in a uniform binary search.

division, $\boldsymbol{v}_1$ and $\boldsymbol{v}_2$ are the region centroids. At the second binary division, there are four regions with centroids $\boldsymbol{v}_3$ through $\boldsymbol{v}_6$. The centroids of the regions after the third binary division are the code vectors $\boldsymbol{y}_i$. An input vector $\boldsymbol{x}$ is quantized by traversing (searching) the tree in Fig. 19 along a path that gives the minimum distortion at each node in the path. Thus $\boldsymbol{x}$ is compared to $\boldsymbol{v}_1$ and $\boldsymbol{v}_2$. If $d(\boldsymbol{x}, \boldsymbol{v}_2) < d(\boldsymbol{x}, \boldsymbol{v}_1)$, for example, then the path leading to $\boldsymbol{v}_2$ is taken. Next $\boldsymbol{x}$ is compared to $\boldsymbol{v}_5$ and $\boldsymbol{v}_6$. If, for example, $d(\boldsymbol{x}, \boldsymbol{v}_5) < d(\boldsymbol{x}, \boldsymbol{v}_6)$, then the path to $\boldsymbol{v}_5$ is taken and $\boldsymbol{x}$ is finally compared to $\boldsymbol{y}_5$ and $\boldsymbol{y}_6$. If $d(\boldsymbol{x}, \boldsymbol{y}_6) < d(\boldsymbol{x}, \boldsymbol{y}_5)$, then $\boldsymbol{y}_6$ is chosen as the quantized value of $\boldsymbol{x}$. Clearly, the total number of distortion computations is, in general, equal to $2 \log_2 L$. Again assuming N multiply-adds for each distortion computation, we have a total computational cost of

$$\mathscr{C} = 2N\log_2 L = 2NB \qquad \text{(binary search)} \qquad (114)$$

which is only *linear* with the number of bits. In the example of 10-bit quantization of 14 LARs, the number of computations is now only 280, as compared to 14336 with full search. (The 280 figure is close to the cost of scalar quantization when rotation is used.) Thus a vast reduction in computation has been effected. The storage cost, however, has increased. Note in Fig. 19 that, in addition to storing the code vectors $\boldsymbol{y}_i$, one must also store all the intermediate vectors $\boldsymbol{v}$. The total storage cost can be shown to have approximately doubled:

$$\mathscr{M} = 2N(L - 2) \qquad \text{(binary search)}. \qquad (115)$$

Also, we shall see below that an additional price is that a certain loss in performance takes place.

The cost requirements in (114) and (115) can be cut in half if the mse is used as the distortion measure. In that case, instead of comparing $\boldsymbol{x}$ to two vectors, one can test where $\boldsymbol{x}$ lies relative to the hyperplane that separates the two regions. Such a test involves the computation of a single scalar dot product of two vectors.

We call the quantization tree depicted in Fig. 19 *uniform*, in that *all* regions at any stage are each divided into two subregions. When performing the actual training to obtain the quantization tree, there may result one or more clusters at certain points in the subdivision process which contain very few training samples, perhaps even one sample. Such clusters would be essentially wasting bits because any further subdivision of those clusters would not reduce the distortion. To ensure a lower average distortion and maximize the utilization of bits, we do not branch the tree uniformly. Specifically, in the training phase, at each point in the subdivision process, the total distortion contributed by each cluster is examined. The cluster that contributes the largest amount of distortion is subdivided next, and the process is repeated. The result is typically a *nonuniform* tree, as depicted in Fig. 20. Note that, in this case, the

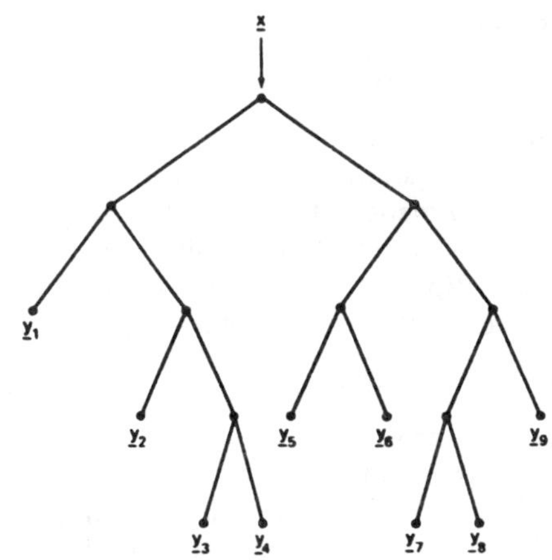

Fig. 20. Nonuniform tree for a binary-search vector quantizer. The codebook size in this case can be any integer.

number of levels can be any desired integer; it is not restricted to be an integral power of 2. A nonuniform tree with $L = 9$ is shown in Fig. 20.

To evaluate the relative performance of binary search (uniform and nonuniform) compared to the full search we give a specific example from very-low-rate speech coding. Fig. 21 shows the average distortion as a function of the number of bits per vector for the quantization of LPC frames of 14 LARs each. As expected, for a given distortion, full search has the lowest rate, followed by nonuniform binary, followed by uniform binary. For the highest rates shown in Fig. 21, the difference between full search and nonuniform binary is approximately 0.5 bits only. For many applications, this small difference in performance is well worth the savings in computation effected by binary search. The plots in Fig. 21 were obtained from the speech of 15 male speakers. For a single speaker, the difference between nonuniform binary and full search is typically larger (about 0.7 bits at $B = 8$ bits).

Binary search is a special case of a class of VQ methods known as *tree-searched* VQ, with the binary search being the simplest. In general, one could divide the space at each node in the tree into more than two subregions (using

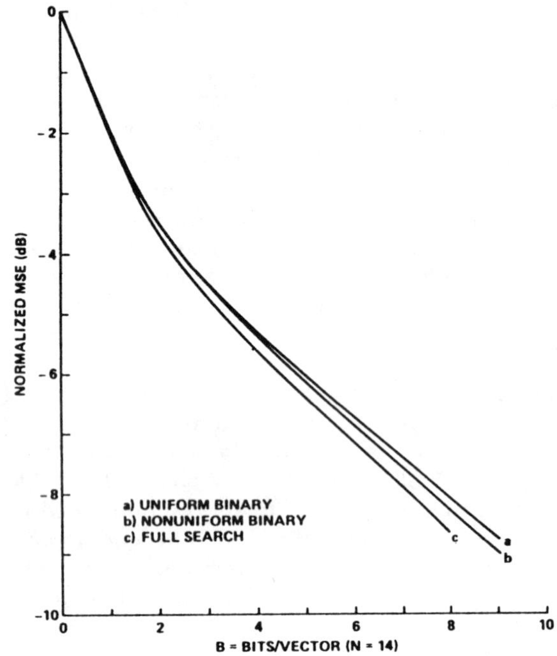

Fig. 21. Comparison of mse when quantizing LARs with three types of vector quantization: (*a*) uniform binary search; (*b*) nonuniform binary search; and (*c*) full search.

K-means with $K > 2$). Such a method would result in an increase in computation over binary search, with some increase in performance. However, because the performance of binary search is usually quite close to full search for many applications, it is typically most cost effective to use binary search with a nonuniform tree.

B. Cascaded Quantization

The major advantage of binary search was the substantial decrease in computational cost relative to full search, accompanied by a relatively small decrease in performance. However, the storage cost was not reduced; in fact, for non-mse distortions the memory cost doubled relative to full search. *Cascaded VQ* is a method intended to reduce storage as well as computational cost [76], [112]. As the name implies, cascaded (also known as multistage) VQ consists of a sequence of VQ stages, each operating on the "residual" of the previous stage. A two-stage cascade is depicted in Fig. 22. The input vector $\boldsymbol{x}$ is first quantized using a B_1-bit (L_1-level) vector quantizer (the quantizer can use full search or binary search, as desired). The residual or "error" $\boldsymbol{e}$ between $\boldsymbol{x}$ and its quantized value $\boldsymbol{z}_i$ is then used as the input to a B_2-bit (L_2-level) second VQ stage with output $\boldsymbol{w}_j$. (The matrix $\boldsymbol{A}_i$ plays a useful role that is explained below; here, we assume that $\boldsymbol{A}_i = \boldsymbol{I}$ and $\boldsymbol{u} = \boldsymbol{e}$.) The final quantized value of $\boldsymbol{x}$ is then simply the sum of the two vectors $\boldsymbol{z}_i$ and $\boldsymbol{w}_j$

$$q(\boldsymbol{x}) = \boldsymbol{y} = \boldsymbol{z}_i + \boldsymbol{w}_j \qquad (\boldsymbol{A}_i = \boldsymbol{I}). \tag{116}$$

During the training phase, the residuals $\boldsymbol{e}$ from the first stage are pooled together and treated as a new random vector that is to be quantized. The process can be repeated for as many stages as desired.

There can be confusion between tree-searched VQ and cascaded VQ. While both processes take place in stages, what one does at each stage is different in the two methods. In tree-searched VQ, one is trying to find the desired

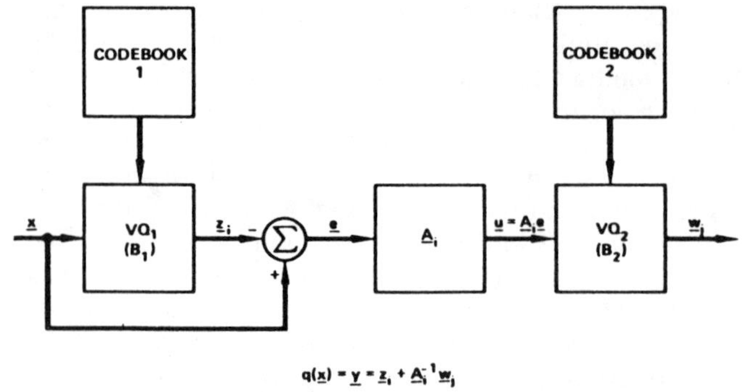

Fig. 22. A two-stage cascaded vector quantizer. In the first stage, a B_1-bit vector quantizer quantizes $\boldsymbol{x}$ into a particular vector $\boldsymbol{z}_i$. The "residual" vector $\boldsymbol{e} = \boldsymbol{x} - \boldsymbol{z}_i$ is then quantized using a B_2-bit quantizer. (The rotation $\boldsymbol{A}_i$ is used to "realign" all the residual vectors to reduce the distortion.) The quantized value of $\boldsymbol{x}$ is then given as the sum of the two code vectors, as shown in the figure.

code vector by searching the space in a systematic manner; at each step, one gets closer and closer to the desired code vector. The role played by the intermediate vectors $\boldsymbol{v}$ in Fig. 19 is simply to guide the search; the desired code vector is found at the end of the search. In cascaded VQ, each stage is a complete VQ search of the space which yields a separate code vector; the final quantized value of the input $\boldsymbol{x}$ is then the *sum* of the code vectors found at each of the stages. The input to each stage is the difference (residual) between the chosen code vector and the input for the previous stage.

The computational and storage costs for a two-stage cascaded vector quantizer (assuming K-means for each stage) are simply

$$\mathscr{C} = N(L_1 + L_2) \qquad \text{(cascade)} \tag{117}$$

$$\mathscr{M} = N(L_1 + L_2) \tag{118}$$

instead of NL_1L_2 for a single-stage full-search quantization. For the example of 10-bit quantization of 14 LARs, the cost of two-stage cascaded VQ with $B_1 = B_2 = 5$ bits is $14(2^5 + 2^5) = 896$ multiply-adds as compared to 14 336 multiply-adds for single-stage VQ. Also, storage cost is reduced in the same proportion. What we are sacrificing for this significant reduction in cost is a loss in performance, as we shall see below.

We use as our example 14-LAR quantization in speech coding. Fig. 23 shows the mse distortion as a function of bit rate for several VQ schemes. Plot (*d*) shows the performance of a nonuniform binary quantizer. Plot (*a*) shows the performance of a cascaded vector quantizer with each stage quantizing to only 1 bit. Clearly, this constitutes the most extreme form of cascaded VQ, with a maximum reduction in cost. As can be seen from plots (*a*) and (*d*), the performance of this cascaded quantizer is significantly worse than the single-stage binary vector quantizer for each bit rate.

What causes this vast reduction in performance for the cascaded quantizer? Recall that during the training phase, after each stage, the residuals are pooled together to form the input to the next stage. If all the clusters in the first stage possess the same relative pdf structure within the cluster, the sets of residuals from the different clusters will have the same pdfs and pooling them together should not reduce performance. However, in general, the residual pdfs from the different clusters will be different and pooling them together will result in a single pdf that will lose many

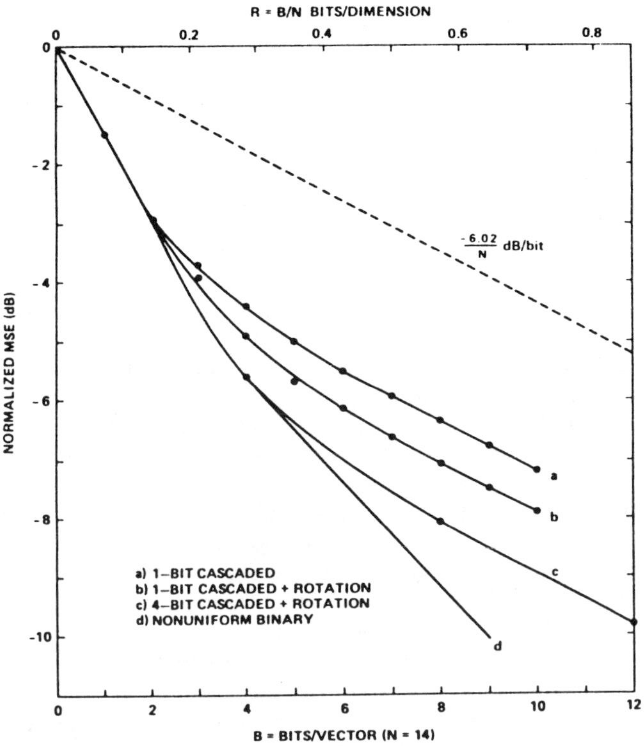

Fig. 23. A comparison of the mse in quantizing LARs using four vector quantizers: (a) multistage (10 stages shown) 1-bit cascaded vector quantization with no rotation between stages; (b) same as (a) but with rotation between stages; (c) 3-stage, 4-bit cascaded vector quantization with rotation (each stage used a nonuniform binary search); and (d) one-stage vector quantization with a nonuniform binary search.

of the dependencies that existed in the initial clusters. As a result, the distortion-rate curve will reach its asymptotic behavior of $-6.02/N$ decibels per bit at a higher distortion level. We see from plot (a) that it reaches its asymptotic behavior at 7 bits and a distortion of -6 dB.

In an effort to forestall the loss of dependence among residual components, we have inserted a vector rotation $\boldsymbol{A}_i$ after each stage [112], as shown in Fig. 22. The idea is to attempt to at least preserve some of the linear dependencies that exist among residual components before they are pooled together. The matrix $\boldsymbol{A}_i$ is set equal to the eigenmatrix corresponding to the covariance matrix of the residuals from the ith cluster. In this manner, the residuals from the different clusters are rotated so that their covariance matrices will have the same structure. Therefore, the covariance of $\boldsymbol{u} = \boldsymbol{A}_i \boldsymbol{e}$ resulting from each cluster will have the same structure. So, pooling the vectors $\boldsymbol{u}$ from the different clusters together will keep the same covariance structure. Fig. 23, plot (b), shows the performance of a 1-bit cascaded quantizer with vector rotation. Note how the asymptotic behavior begins at a lower distortion level. Plot (b) shows about a 1.8-bit advantage over plot (a) for the same distortion at the higher rates. This advantage with using rotation would be expected to be smaller if each stage in the cascade employed a larger number of bits.

Plot (c) in Fig. 23 shows the performance of a 4-bit, 3-stage, cascaded vector quantizer with eigenvector rotation after each stage. The 4-bit VQ in each stage utilized a binary tree search. (That is why plots (c) and (d) coincide up till $B = 4$ bits.) The asymptotic behavior is reached after the second stage at a distortion level of about -8 dB. The sharp departure of plot (c) from plot (d) at 4 bits is a clear indicator of the devastating effect that pooling residuals has, even if eigenvector rotation is utilized. The nonlinear dependencies among LARs are so important and they are virtually destroyed by the pooling of residuals. In fact, the performance for $B > 4$ bits in plot (c) is comparable to the performance of a scalar quantizer at that point. So, there may not be any need to go through the expense of additional vector quantizers after the first stage; a scalar quantizer (with rotation) could effectively replace all stages beyond the first.

The improvement in performance afforded by the vector rotation between the two stages of the cascade is obtained at a cost of additional computation and storage. The additional cost can be seen to be

$$\mathscr{C} = 2N^2 \quad \text{(rotation)} \tag{119}$$

$$\mathscr{M} = N^2 L_1. \tag{120}$$

The storage cost is especially large relative to that for the cascade in (118). Because of these additional costs, the inclusion of the rotation becomes less attractive as a method for improving performance.

It appears, therefore, that the maximum that one can effectively utilize in practice is about two stages of cascaded quantization. The first stage would employ VQ and should be allocated the maximum number of bits that one can afford in terms of computations and storage. The second stage would then perform scalar quantization on the residuals from the first stage. However, if the bit rate for the second stage is less than 1 bit/parameter ($R < 1$), then VQ would be advisable for the second stage as well. Also, if the distortion measure is not conveniently defined in the scalar case (such as the Itakura–Saito distortion), then VQ would have to be used for all stages.

Storage Reduction: Cascaded quantization was designed chiefly for the purpose of reducing storage (though computations are reduced as well), for in tree-searched VQ we already have a very effective method for reducing computations with relatively little loss in performance and a modest increase in storage. In fact, with binary search, for example, one could claim with some justification that, for many applications, computational cost is no longer a major limitation in VQ. If storage cannot be reduced without a concomitant major reduction in performance, then one would have to conclude that *storage cost is ultimately the major limitation in VQ*, for it would prevent us from taking advantage of the power of very large codebooks. With binary search, many would not hesitate to use a 30-bit codebook, but the required storage would tax the addressing capabilities of most of today's computers. Furthermore, the amount of training data needed to design such a codebook can be a serious limitation as well. While much research has been devoted to reducing computational costs, relatively little effort has been expended specifically in reducing storage costs. And little or no work appears to have been done on reducing storage at the expense of increased computation instead of decreased performance. Below, we present another approach to reducing storage costs, namely, product codes. This approach also reduces computations and results in reduced performance, but it could serve as a more attractive solution for many applications.

C. Product Codes

Assume that we structure our parameter space such that two or more component vectors, each with its own codebook, jointly represent all points in the space. The Cartesian product of the component codebooks is known as a *product code*. If, for example, we structure parameter space in terms of two component vectors of dimensions N_1 and N_2, and let the corresponding codebooks be of sizes L_1 and L_2, respectively, then the product code of dimension N and size L, where

$$N = N_1 + N_2 \qquad L = L_1 L_2 \tag{121}$$

will need memory storage equal to the sum of the storage requirements of the two component codebooks

$$\mathcal{M} = N_1 L_1 + N_2 L_2 \qquad \text{(product code).} \tag{122}$$

Compared to a storage cost of NL for an N-dimensional codebook of size L, we see that the storage costs for a product code can be substantially lower.[12]

While storage costs are reduced through the use of a product code, computational costs are reduced only by using certain types of distortion measures or by making simplifying assumptions and approximations. To illustrate the problem, assume we want to use a product code for LARs, i.e., have a separate codebook for each LAR (see Example 4), but instead of the mse use the Itakura–Saito distortion[13] to find the nearest code vector. Then the only way to quantize an input LAR vector would be to perform a *full search* of the product code for the nearest code vector, because unlike the mse, the Itakura–Saito distortion is not separable into distortions on the individual LARs. The result is that there is no reduction in computation relative to the full-search codebook designed by K-means.

There are at least two conditions under which computational costs for quantization with a product code can be reduced in a comparable manner to the storage cost reduction in (122): *independent* quantization and *sequential* quantization. Under independent quantization, the component vectors are quantized independently using their respective codebooks. If the distortion measure is *separable* into independent distortions on the individual vectors, then independent quantization will, in fact, give the nearest product code vector. The mse is one example of a distortion measure that is separable in this fashion.

Under sequential quantization, one of the component vectors is quantized first, then the quantized value of that vector is used in the quantization of a second component vector, and so on. (Clearly, sequential quantization includes independent quantization as a special case.) The LPC "gain–shape" product code [18], [114], using a different definition of the Itakura–Saito distortion, is one example where the nearest product code vector can be obtained by quantizing the LPC parameters (shape) first, then using the quantized LPC parameters to quantize the gain such that the overall distortion is minimized.

In practice, sequential quantization is often used as an approximate suboptimal solution so as to minimize computations. One example is in a residual-excited speech coder, where the product code is the product of the spectrum codebook and the residual codebook. In one of the known implementations, the spectral envelope is quantized first, then the residual is computed using the values of the quantized spectrum, and finally the residual is quantized. This sequential quantization procedure does not minimize any known overall distortion on the reconstructed speech waveform. (See Section VII-B for a different example of a residual-based product code.)

Thus far we have considered only the quantization process in using a product code. Another important consideration is the design of the component codebooks in the first place. If an optimal (minimum-distortion) product code is desired, then, in general, the component codebooks cannot be designed independently unless the component vectors are statistically independent *and* the distortion measure is separable. In practice, independent or sequential quantization is employed in the design process to reduce computations even if the final product code is suboptimal.

In structuring our parameter space into two or more component vectors, a question arises as how best to define the component vectors and allocate the bits among the respective codebooks. Ideally, one would want to attempt to *structure parameter space such that the component vectors are statistically independent*, as much as possible. For if the component vectors were independent, then the only reduction in performance would be owing to the reduced dimensionality of the vectors. Otherwise, we would expect an additional reduction in performance in proportion to the dependence between the component vectors. While it is possible to remove linear dependencies by proper vector rotations, as we saw in Section IV, removing nonlinear dependencies is a difficult task, in general.

To maximize performance, it should be clear that, as a general design criterion, the number of component codebooks should be as low as possible and the size of each codebook as large as can be afforded in storage. The problem of how to choose the sizes of the component codebooks is a bit-allocation problem similar to the one discussed in Section IV-A.

In comparing a two-vector product code and two-stage cascaded quantization, we note that if L_1 and L_2 in (122) are equal to L_1 and L_2 in (118), then the cascade storage would be greater than the product code storage. But a fair comparison between the performance of the two methods would be to design the component codebooks in each case such that the memory costs are equal, then compare the resulting distortion. From (118) and (122), that would mean that the sizes of the component codebooks in each case would have to be different. Which of the two methods would be expected to perform better in practice depends on the particular problem and the structure of the statistical dependencies.

Below we mention briefly two specific methods of using product codes for coding LPC spectra, which can be used as alternatives to cascaded quantization. In Section VII-B we present an effective method for utilizing product codes in speech waveform coding.

[12] It is possible to structure parameter space such that $N_1 + N_2$ is larger than the dimension of the original space. (One example is structuring a speech waveform into two vectors: one representing the spectral envelope and another representing the residual.) In that case one has to compare (122) with $N'L$, where N' is the dimension of the original space.

[13] Computing the Itakura–Saito distortion between two LAR vectors would necessitate converting the LAR vectors to predictor coefficients first and then using (23).

Split-Vector Codes: In this method, a vector (such as a LAR vector) is split into two or more subvectors and each subvector is coded independently using a codebook designed for that part of the vector. Consider again the coding of LAR vectors in Example 4 above using a total of 15 bits/vector for 14 LARs, and assume we are interested in comparing a product code with a two-stage cascade for a given storage cost. In particular, assume that the storage cost is the minimum possible for the cascade case, which, from (118) and (121), can be seen to occur for $L_1 = L_2 = 2^{7.5}$, i.e., 7.5 bits for each of the two codebooks, so that $\mathcal{M} = 5068$ for a 14-dimensional vector. A product code can be designed using the bit-allocation table in Table 3 so that the storage is similar. A good solution would be to split the 14-LAR vector into two subvectors and use two codebooks: a 4-dimensional, 10-bit codebook corresponding to parameters 1–4, and a 10-dimensional, 5-bit codebook corresponding to parameters 5–14. The resulting storage cost from (122) is 4416. One would then compare the distortions for the two separate cases. (We note from Table 3 that a similar product code could be designed for the rotated parameters, which may perform better than a product code on the LARs.) To our knowledge, the suggestion to use split-vector codes for LPC parameters is novel. As yet no comparisons have been made with cascaded quantization.

Split-Band Codes: In this method, the spectral envelope is split into two or more bands and each band is vector quantized separately. Copperi and Sereno [24] designed a residual-excited coder in which the signal was split into two equal frequency bands, and separate LPC model parameters for each band were vector quantized separately. Using the split-band approach allows the designer to choose lower order spectral models and to make use of the fact that human perception is generally less sensitive to high-frequency distortions than to low-frequency distortions. This method is reminiscent of sub-band coding [74] except that here we attempt to keep the number of bands to a minimum (preferably not more than two) and the spectrum in each band is modeled explicitly.

D. *Random Codebooks*

While it is important to reduce the costs associated with the vector *quantization* process, there are times when reducing the costs in the *training* process is of interest. One simple method to design a codebook with essentially no computational training cost is to choose the code vectors at random from a given set of training data. We call the resulting codebook a *random codebook*.

At first glance it would appear that a random codebook would not perform well for quantization purposes. However, results on the asymptotic (large L and N) optimality of the expected performance of random quantizers [115] and the use of random codebooks in proving the basic rate-distortion theorems is a strong motivation for the use of such codebooks. While the use of a random codebook is a reasonable choice for large L and N, it is nevertheless used in practice when these conditions are not satisfied. Random codebook selection and performance under these non-asymptotic conditions is a research area of current interest.

In Fig. 24 we show the relative performance of random and nonuniform binary codebooks for the quantization of 14-LAR vectors using the mse distortion measure. The training data consisted of 15 min of speech (1 min from each of 15 speakers), which comprised approximately 90 000 training vectors The random codebook was obtained by selecting a set of vectors from the training data at equal intervals. Thus if a 100-level random codebook was desired, we chose every 900th vector from the training data. The test data consisted of sentences from five male speakers not used in the training. Note how at low bit rates the random codebook performs significantly worse than the binary codebook. However, the two curves begin to converge as the bit rate increases. We would expect the curves to get even closer at higher rates, but asymptotically they would become parallel to each other with some separation in bits. The difference in performance of about 1.3 bits at 10-bit quantization is smaller than one might expect.

We must bear in mind that, although a random codebook is simple to design, it is a full-search codebook and so

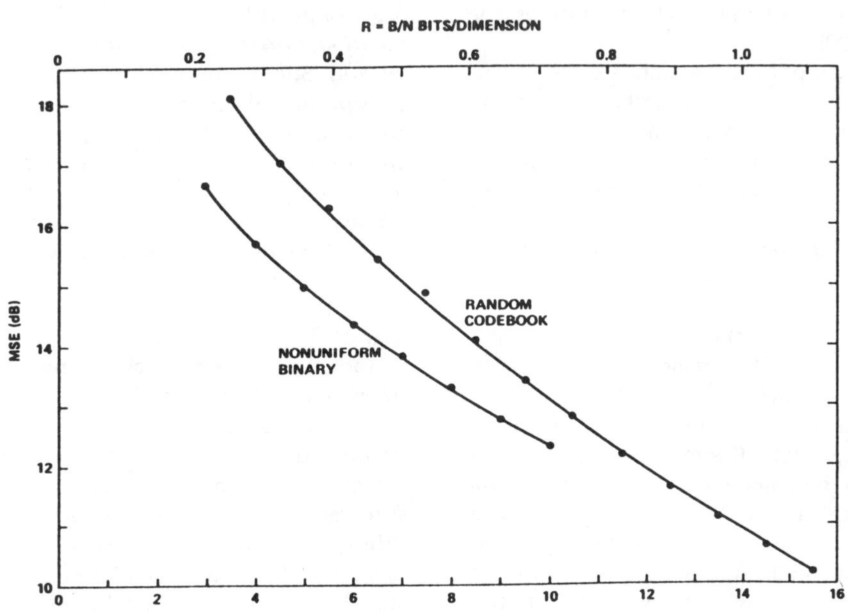

Fig. 24. The mse in quantizing LARs using a random codebook, compared with a nonuniform binary-structured codebook.

requires the same computation and storage. We find random codebooks to be useful in exploring how the VQ distortion decreases as a function of bit rate for higher data rates. In Section VI we mention another useful application of random codebooks.

E. Training and Testing

An important aspect of the design of any codebook is the training procedure used to populate the codebook, i.e., to fill its entries. In the *K*-means algorithm, the training procedure was given in Section II-C, except for a crucial initialization step where one chooses an initial set of code vectors. Below we describe initialization procedures that we have found useful for the VQ of LPC parameters in speech coding. The testing of the resulting codebook and its robustness are discussed next.

1) Training:

Binary clustering: In binary search we need at each stage a method to initialize the partitioning of space into two regions. Ideally, what we would like to do is to draw a hyperplane through the mean of the training data and orthogonal to the largest eigenvector (we assume a mse distortion). Such a hyperplane would divide space into two initial regions from which to start the iterative training procedure. For data where one dimension has a larger variance than the other dimensions, we approximate the eigenvector direction by the dimension with the largest variance. The centroids of the two clusters separated by the orthogonal hyperplane are used as the initial two code vectors. The *K*-means algorithm is then continued with $K = 2$ to determine the final code vectors. (We have found 5 to 10 iterations sufficient to obtain convergence.) After the first partition is completed, each of the subregions is divided into two using the same procedure, etc.

K-means initialization: Because the *K*-means is not guaranteed to result in a codebook that is globally optimum, it is often suggested that one repeat the algorithm with a number of different initial sets of code vectors. The codebook that results in the minimum distortion is then chosen for actual use. The pattern-recognition and VQ literatures contain a number of initialization methods for *K*-means [30], [56], [64], [99].

Because the performance of binary search is close to the optimal *K*-means (see Fig. 21), we have found the codebook generated by nonuniform binary search as a good initial codebook to start the *K*-means iteration. This initialization method has produced better results than using a random initial codebook.

2) Testing: After a codebook is designed with a given set of training data, it is important to test the performance of the codebook on independent data that were not used in the training. Testing only on the training data always presents an overly optimistic view of how the codebook will perform on operational data.

Fig. 25 shows the mse distortion for a 6-bit codebook ($L = 64$ levels) of 14-LAR vectors. The mse is plotted against the number of vectors in the training data. The training data were taken from the speech of 15 male speakers. The resulting codebook for each set of training data was tested on the same training data and on an independent set of test data taken from 5 male speakers who were not used in the training. With increased training, the convergence of the two plots increases. However, the convergence is very slow beyond 20 training vectors per codebook level (i.e., per code vector). As a rule of thumb, using 50 training data per code vector is considered sufficient for most applications. If sufficient data are not available or if the computation or storage becomes excessive, then using as few as 10 training data per code vector may be adequate. The remaining difference between the two plots in Fig. 25 is due to the fact that the test speakers were different from the training speakers. This issue relates to codebook *robustness*, which is discussed below.

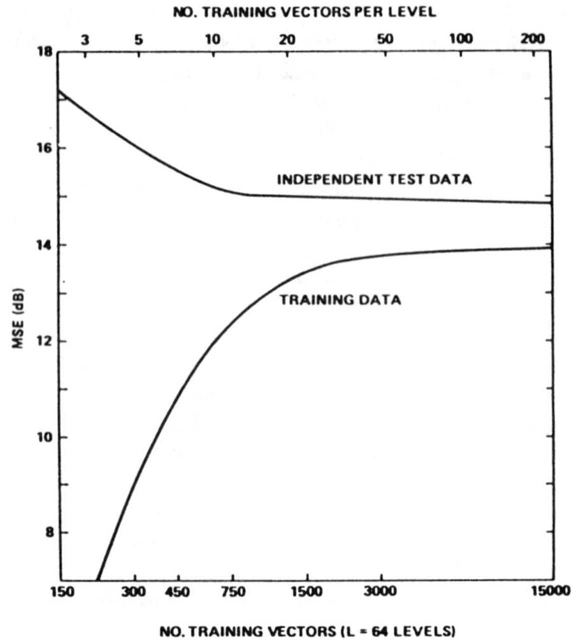

Fig. 25. The mse in quantizing LARs for a 6-bit codebook ($L = 64$ levels) as a function of the number of training data vectors, when tested on the training data and on an independent data set. The smaller the gap between the two curves, the more robust is the codebook.

In general, convergence of training and test performance curves with increased training set size is not guaranteed for arbitrary sources. One theoretical result of interest here is that, for reasonable distortion measures, such convergence is guaranteed for a class of sources known as *asymptotically mean stationary* [59], in which the stationarity condition is relaxed. Speech from a single speaker under a fixed set of environmental conditions, for example, exhibits short-term and long-term stationary properties and is well modeled as an asymptotically mean stationary source. However, the model may not apply as well under more variable, nonstationary conditions.

3) Codebook Robustness: Codebook robustness refers to the resistance of a codebook to degraded performance when tested on data whose distribution is different from that of the training data. Under operational conditions, one cannot usually predict all of the situations under which a quantizer will be used, and so the distribution of the operational data will in general be different from that of the training data. (Fischer and Dicharry [36] have studied the effects of designing codebooks for particular memoryless sources and testing on others with different pdfs.) We differentiate two major types of variabilities that impact the design and operational performance of a codebook: input signal variability and digital transmission channel errors. We discuss each of these below and compare VQ to scalar quantization.

For speech, signal variability can be classified further into speaker variability and environmental variability. Intraspeaker variability is that due to changes in each speaker's voice: normal everyday changes, changes due to different health conditions, and changes brought about by various emotional states. Interspeaker variability refers to voice differences across speakers. Environmental variability refers to the level and type of background noise that surrounds the speaker (e.g., office environment, outdoors, helicopter, etc.) and signal pickup characteristics, including the types of microphones and transmission facilities (e.g., telephone, radio). The performance of any codebook would be expected to deteriorate if used with types of signals for which it had not been designed. Thus if a codebook is designed (trained) for the voice of one speaker, it would not be expected to perform as well for other speakers. Likewise, a codebook trained in an office environment would not be expected to perform as well in a helicopter environment, for example. It is clear that, for optimal performance, one should train the codebook using data that are representative of what the VQ system will meet in actual operation.

The two plots of Fig. 25 would have converged even more had the test data been taken from the same speakers used in the training. The remaining difference of about 1 dB between training and test data in Fig. 25 is significant. Increasing the amount of training data from the same 15 male speakers will not reduce the difference appreciably, it will simply render the codebook better at quantizing the data from those same 15 speakers. If what is desired is the best possible speaker-independent performance, then the gap between training and test shown in Fig. 25 can be bridged further only by increasing the number of speakers in the training rather than by increasing the amount of speech from each speaker. If the final system is to be used with female speech as well, then including female speakers in the training becomes important.

For a given bit rate, a speaker-independent codebook cannot possibly perform as well as a speaker-dependent codebook. (We have seen differences in performance between 1 to 2 bits per vector for LAR quantization around 10 bits.) One interesting possibility for maximizing performance of a VQ system is to design a speaker-independent codebook initially. Then, as the system is used, it adapts to the speech of the new speaker. Such a system would have the extra advantage of also automatically adapting to the acoustic environment of the speaker. Codebook adaptation means that the code vectors will change in time, which necessitates transmitting the new code vectors to the receiver as well. Two such systems were designed by Paul [100] for real-time speech coding at 800 bits/s. In one system, a full-search codebook minimizes the maximum distortion between the input vectors and the code vectors. When the quantization distortion exceeds a certain threshold, the input vector is taken as a new code vector and the least used code vector up to that point is discarded. The new code vector is then transmitted to the receiver.

Scalar quantizers are generally more robust to signal variability than vector quantizers. For the same number of bits per vector, a scalar quantizer usually has fewer output levels than the vector quantizer. (For example, a 10-bit vector quantizer has 1024 levels, while 10 1-bit scalar quantizers have a total of only 20 output levels.) Therefore, given the same amount of training data, the scalar quantizer will have more training data per level and, hence, the resulting quantizer will be more robust. Another more important reason for the greater robustness of the scalar quantizer is exactly the same reason that renders its performance lower than the vector quantizer. The scalar quantizer, by "averaging" or "projecting" the data on one dimension at a time, is less able to model the detailed multidimensional dependencies in the data and, therefore, will be less susceptible to changes in those detailed dependencies. In other words, the one-dimensional pdfs are expected to change less between training and test than the corresponding multidimensional pdfs. Because vector quantizers are not in general as robust as scalar quantizers, it is very important to test vector quantizers extensively on independent data. Also, whenever changes in the algorithm are made, a new set of test data should be used. Otherwise, there might be a tendency to unconsciously change the algorithm to improve performance on the same test set, effectively rendering the test set part of the training data.

Transmission channel errors present a different type of problem to system robustness. Channel errors translate directly into distortions in the output; the higher the error rate, the greater the distortion. Again, in general, VQ systems tend to be less robust to random channel errors than scalar quantizers. (Burst errors may affect both types of quantizers equally.) A simple example will illustrate the difference. Let us take the example of the 10-bit vector quantizer versus the ten 1-bit scalar quantizers, and assume a channel error rate of 1 percent. Then, in 100 bits there will be 1 bit that is in error, on the average. In the scalar quantizer, that 1-bit error causes one value in one dimension to be wrong, while in the vector quantizer, the same 1-bit error causes a whole 10-bit vector to be wrong, which would result in larger distortions on the average.

By using extra transmission bits, standard error-correcting techniques can be employed to help correct for channel errors [103]. The use of such techniques is strongly recommended for high error rates (> 0.1-percent errors). In an unstructured codebook, all bits in the code vector would have to be protected in some fashion, while in structured codebooks (binary-structured or scalar), one can choose to protect only those bits that are most important to the fidelity of the output.

Channel errors can play havoc with any system that employs variable-length coding (such as entropy coding). A single bit error affects not only the one value in which it occurs but can also affect many succeeding values. The error causes the receiver to lose synchronization with the input and succeeding bits may be erroneously decoded. The use of "self-synchronizing codes" [105] reduces the magnitude of the problem significantly. However, independent synchronization schemes, involving the transmission of additional synchronization bits, would still be needed to avoid loss of synchronization.

It should be clear by now that simply optimizing system performance over a set of training data is only one aspect of choosing the best system for some application. Overall robustness of the system to the operational environment must be included in system performance assessment.

VI. Time-Dependent Vector Quantization

Thus far, we have assumed that each input vector is quantized independently of any other input vector. While we have considered dependencies among vector components (in short-term spectral space, for example), we have not yet taken advantage of longer term vector dependencies in time. Certainly, any further dependencies we can uncover should allow us to transmit the data at even lower rates for a given average distortion. Speech, of course, is a time-varying process that is rich in time-related dependencies. The statistical structures of the spectral sequences for each sound and of the sequences of sounds in each utterance are fertile sources of dependency. In this section, we survey briefly several methods in which time dependencies (beyond a single frame) are exploited to reduce the bit rate. We begin first with frame transmission schemes that select the frames to be transmitted judiciously, followed by time-dependent VQ methods. Most of the methods described below incorporate additional delays.

A. *Selective Frame Transmission*

Perhaps the simplest, and yet effective, method for selective frame transmission is known as *frame repeat* or *frame fill* [93], [101], [102]. In this method, only every other frame is transmitted; for the intervening frames, a 1-bit code is transmitted which specifies whether the missing frame should be replaced by the previous frame or the following frame. Which frame to repeat is determined by a nearest neighbor rule. A variation on this method is to send 1 or 2 bits which specify values for the missing frame which are a linear interpolation of the transmitted frames [119].

Another simple and very effective method for selective frame transmission is *variable-frame-rate* transmission [100], [119], [132]. In this method, information is transmitted only when the properties of the speech signal have changed sufficiently (in some predetermined sense) since the preceding transmission. The parameters of the untransmitted frames are regenerated at the receiver through linear interpolation between the parameters of the two adjacent transmitted frames. Thus parameter transmissions occur more frequently when speech characteristics are changing rapidly, as in transitions between sounds, and less frequently when speech characteristics are relatively constant, as in steady-state sounds.

B. *Segment Quantization*

Variable-frame-rate VQ of spectral parameters is a useful procedure for coding at rates as low as 400 bits/s. For data rates of 300 bits/s or lower, the linear interpolation assumption breaks down and the method is no longer adequate in maintaining reasonable intelligibility. At these lower rates, the method of choice is to perform VQ on a sequence of frames comprising a larger segment of speech. The method is known as *segment quantization* [111], [137], in contrast with frame quantization where each frame is quantized separately. VQ is now employed over a larger vector $\boldsymbol{X} = [\boldsymbol{x}(1)^T \boldsymbol{x}(2)^T \cdots \boldsymbol{x}(J)^T]^T$ comprising the elements of J consecutive frame vectors. In this manner, the dependence between consecutive frames is included implicitly in the larger vectors.[14] The duration J of the segment can be either fixed or variable. Clearly, a fixed J results in a simpler quantization problem. However, utilizing variable-duration segments has led to better speech quality.

Segment quantization is a good example where a random codebook has been especially useful. Because of the relatively high-dimensional space ($10 \times 14 = 140$ if $J = 10$), one can argue that a random codebook should give good results. The overriding reason for using a random codebook, however, comes from perceptual considerations. We have found that the clustering and centroid computations in the K-means algorithm produce code vectors (segments) that sound muffled upon listening. So, even if a random codebook produces a larger mse distortion than the corresponding K-means codebook, the synthesized speech from the random codebook yields higher quality and intelligibility. This is yet another instance where an objective measure of distortion is not a good measure of perceptual distance. In the experiments performed thus far, a 13-bit codebook is used ($L = 8192$ segments). Because a random codebook is not structured, quantization of the speech input requires a substantial amount of computation.

Segment quantization is truly in the spirit of VQ theory. For, in principle, one should be able to approach the performance of the optimal coding system by simply using longer blocks for quantization. Unfortunately, if signal fidelity is to be maintained, exponentially larger codebooks will need to be used, which very quickly becomes impractical. An alternative solution is to include time dependence explicitly in the modeling (as opposed to implicitly, as in segment quantization). In this manner, the complexity of the vector quantizer can remain manageable. We have already described two such methods above, frame repeat and variable-frame-rate transmission. Below, we describe other models that adapt to the changing properties of the signal.

C. *Adaptive VQ*

Adaptive VQ implies a change in the codebook as a function of time. There are two types of adaptive vector quantizers: forward-adaptive and backward-adaptive. In forward-adaptive schemes, the system examines *future* data and decides whether a change in the codebook is necessary or not, and transmits to the receiver the desired changes using additional bits. Typically, such schemes incorporate some delay to allow for the inspection of upcoming data. Backward-adaptive schemes examine only *past* transmitted data and change the codebook accordingly. Since past information is available at both the transmitter and the receiver, the necessary codebook updating computation can be performed at both ends simultaneously, thus obviating the need for transmitting additional bits. Backward-adaptive systems have the added advantage of having no extra delay, but are quite susceptible to degraded performance under channel errors. If in forward-adaptive systems the additional bits are not counted, then such systems will result in lower quantization distortion than backward-adaptive systems. However, for an overall fixed data rate, which of the two types of systems performs better depends on the relative number of additional bits to the overall data rate. The principles of forward and backward adaptation are similar to those used in scalar quantization [74].

We should mention at this point that VQ, by its very nature, implies some delay to allow for the input vectors to be specified. Even a backward-adaptive vector quantizer

[14] Segment quantization has also been termed *matrix quantization* [137]. We prefer to use the former term because matrix quantization could be taken to imply another method beyond scalar and vector quantization, while in reality it is simply VQ applied over a longer vector that comprises a larger segment of speech.

will have this inherent delay. A forward-adaptive vector quantizer may have additional delays above and beyond the delays inherent in specifying the vectors in the first place.

An example of a forward-adaptive system is Paul's adaptive VQ system [100] described above. Most adaptive VQ schemes to date, however, are backward-adaptive and most such schemes are *feedback* VQ systems [56]. Two major classes of feedback VQ systems are *vector predictive* systems and *finite-state VQ* systems. Vector predictive systems [25] are a generalization of scalar predictive systems to the vector case and, thus far, they have been used for medium-rate coding of speech waveforms.

In finite-state VQ, we have a finite-state network with transitions among states and we associate with each state a codebook. If one is in a particular state, then the codebook associated with that state is used to quantize the input vector at that time. Depending on which code vector was used, a transition is made to another state (which could also be the same state). The codebook at that state is now used to quantize the next input vector, and so on. The union of all codebooks from all states is usually very large. By subsetting the larger codebook into smaller codebooks at each state, a smaller number of bits is used to quantize each input vector, so that the average bit rate is reduced from the case where the large codebook is used for every input vector. If the time-dependent structure of the data is extensive, substantial reduction in bit rate without loss in fidelity can be accomplished. Finite-state VQ schemes have been employed with segment quantization [113], frame quantization [32], [110], and waveform coding [42], [62].

VII. Speech Waveform Coding

Our understanding of the vector quantization of speech *spectral parameters* has advanced and its usefulness and practicability have been demonstrated to the point where real-time hardware implementation of very-low-rate speech coders is becoming a reality. The application of VQ to speech *waveform* coding is a more recent but intense research activity, and lends itself well to the development of a large number of techniques and variations. We believe it will be some time before the relative merits of the various techniques are worked out and fully appreciated. While some waveform VQ techniques are becoming practical now, especially for coding at about 16 kbits/s, other techniques for low-rate coding require significant computational resources that are not widely available. With additional research effort, the role of waveform VQ in speech coding will become clearer and the possibilities of achieving toll quality at low data rates (below 8 kbits/s) may become a reality.

Below, we give the reader a view of the difficulties inherent in waveform VQ and present a framework that we hope will help sharpen our understanding of the waveform VQ process. It is instructive first to take a brief look at the state of the art in the scalar quantization of speech waveforms, especially those methods that will have a bearing on our VQ discussion below.

A. Scalar Waveform Quantization

It is generally accepted that at 16 kbits/s, adaptive predictive coding (APC) and adaptive transform coding (ATC) result in speech that has approximately toll quality [74] (i.e., quality equal to a high-grade telephone line). In a particular configuration for both methods, one computes and transmits: 1) a representation of the short-term spectrum (typically in the form of LPC coefficients); 2) the short-term gain; 3) the pitch and parameters of a pitch filter; and 4) the prediction residual waveform.[15] The pitch filter is usually an all-pole filter with a maximum of three parameters; it is of the form $1/C(z)$, where

$$C(z) = 1 + c(\tau - 1)z^{-(\tau-1)} + c(\tau)z^{-\tau} + c(\tau + 1)z^{-(\tau+1)} \tag{123}$$

and τ is the estimated pitch period in samples. Pitch periods for adults vary from less than 3 ms for high-pitched females to over 12 ms for low-pitched males. (In certain system implementations, the inclusion of a pitch filter is found to be unnecessary or undesirable; see, for example [89], [133].) Parameter sets 1–3 above are computed and transmitted every frame as *side information* with the residual. At the receiver, the residual excites a filter of the form

$$H(z) = \frac{G}{A(z)\,C(z)} \tag{124}$$

to result in the output speech. The speech signal is usually sampled at either 6.67 or 8 kHz. At 6.67-kHz sampling, the residual is coded with 2 bits/sample, which leaves about 2.7 kbits/s to transmit the side information. At 8-kHz sampling, a 3-level quantizer is used to code the residual (or about 1.6 bits/sample), which leaves over 3 kbits/s to code the side information.

At 16 kbits/s, the short-term SNR achievable by APC and ATC is approximately equal to the gains predicted by rate-distortion theory. The SNR gain over simple quantization of the speech waveform itself is approximately equal to $-10 \log_{10} \gamma$, where γ is the ratio of the geometric mean to the arithmetic mean of the short-term speech spectrum as represented by $H(z)$ (see (76)). The short-term SNR, therefore, changes as the spectrum changes and, in speech, a typical range is 5–20 dB, with higher SNR values associated with vowel sounds and lower values with unvoiced sounds.

As the number of bits used to quantize the residual is reduced to about 1 bit/sample or less, speech quality deteriorates rapidly. A number of methods have been employed to improve the quality at transmission rates in the range 8–9.6 kbits/s. One technique known as *multipulse coding* [9] models the residual as a sequence of a relatively small number of pulses (smaller than the number of residual samples) whose amplitudes and time locations are optimized and transmitted each frame. Another class of coders, with a long history in speech coding, is known as *baseband-excited coders* [89] or *voice-excited coders* [38]. This is a frequency-selective method where only a low-pass band (known as the baseband) of the residual, for example, is transmitted. At the receiver, the high frequencies are obtained from the low frequencies by nonlinear high-frequency regeneration techniques. Though quite effective, these methods also break down as the bit rate goes much

[15] Most ATC systems operate directly on the speech waveform and not on the prediction residual [74]. However, if done appropriately, ATC on the prediction residual can produce similar results [16].

below 1 bit/sample. Vector quantization promises to carry speech coding with high quality into these lower rates.

B. Vector Waveform Quantization

Direct VQ of the speech waveform can be accomplished by simply dividing the waveform into consecutive blocks of N samples each and designing a codebook whose vectors are obtained by one of the clustering methods described earlier. At high data rates, such a method would give good results at relatively small values of N. However, as the bit rate is decreased, it becomes more and more important to increase the block length to minimize the distortion. As N increases, two problems arise. The first problem is the one we already know, and that is the exponential increase of needed resources. This problem can be dealt with partly by proper structuring of the codebook, as we shall see below. The second problem is less obvious; it concerns the concatenation of waveforms for synthesis. As we quantize each block independently to its closest code vector, we minimize the distortion over the whole block so that, for a long block, there is no guarantee that the end of one code vector will match the beginning of the next code vector in time. The result will be a discontinuity in the speech waveform, which may be heard as roughness in the speech. The problem is aggravated at low data rates since the choices for appropriate code vectors are relatively few. One solution to the concatenation problem is to quantize blocks that overlap by a small amount. Then, the quantized code vectors are overlapped and added with a decreasing weighting during the overlap region which goes to zero in both directions, so that the transition from one quantized block to the next is made smoother. This method is used effectively in ATC of the speech waveform to prevent waveform discontinuities from one block to the next [130]. Although the weighted-overlap method just mentioned would be expected to work well, it has the drawback that the overlap actually wastes bits since the bits for the overlapped regions are in effect transmitted twice. Another solution to the waveform discontinuity problem is to quantize the residual instead of the speech waveform itself, as will be described below. ATC of the residual has been used effectively in eliminating the need for overlap [16].

Much of the recent activity in applying VQ to the coding of speech waveforms has concentrated on the medium-rate range of 8–16 kbits/s (see, for example, [1], [24], [25], [34], [37], [42], [49], [50], [56], [63], [83], [109], [124]). In that range, scalar quantization methods already produce approximately toll quality speech at 16 kbits/s and communications quality [39] at 8–9.6 kbits/s. Of course, with the use of VQ, we have the potential of moving the toll quality boundary further down. An exposition of the various waveform VQ techniques under development is beyond the intended scope of this paper. Instead, we have chosen in this brief section to discuss a particular approach to speech coding at low rates (below 8 kbits/s), for it is at those rates that scalar methods tend to break down and VQ offers the potential of achieving toll quality. This approach was chosen because it forms a natural extension to existing well understood scalar speech coding techniques, and appears to have a potential for high-quality low-rate coding of speech.

The main problem in speech waveform coding is how to structure and design our codebook such that signal fidelity is maintained but computational and storage costs are reduced to manageable proportions. One relatively natural manner to structure the codebook is to factor out the short-term spectral and pitch dependencies and code them separately, in effect use a product code. By removing the short-term linear dependencies, one is left with a residual that is relatively white. Fig. 26 shows three waveforms: plot (a) is the speech waveform; plot (b) is the prediction residual after filtering the speech waveform with the spectral inverse filter $A(z)$; and plot (c) is the residual after filtering the speech waveform with the combined spectral and pitch inverse filter $A(z)C(z)$. (Note that in Fig. 26, waveform (b)

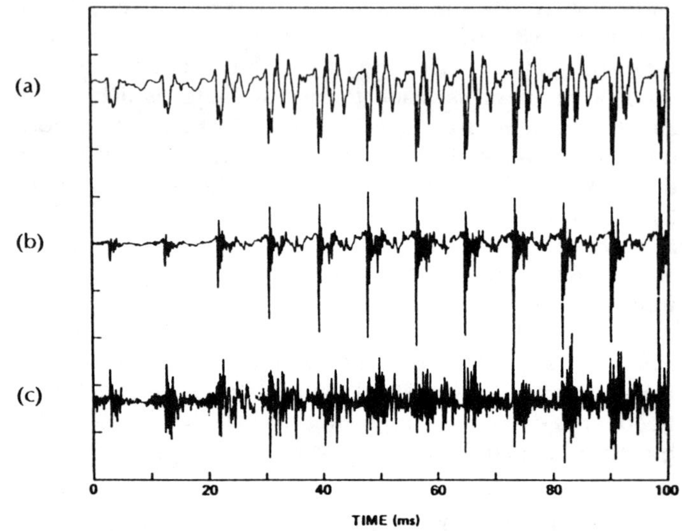

Fig. 26. (a) A speech waveform. (b) The prediction residual after filtering the speech with the all-zero filter $A(z)$. (The plot is amplified 10 dB relative to (a).) (c) The residual after filtering the speech with the combined filter $A(z)C(z)$, thus removing much of the pitch-related structure. (The plot is amplified 20 dB relative to (a).) (From Atal [7].)

was amplified 10 dB and waveform (c) 20 dB relative to the speech waveform.) The residual waveform (c) has most of its short-term linear dependencies removed; it certainly looks much more random than (a) or (b). The statistics of waveform (c) tend to be Gaussian [7]. Our codebook can now be structured in four parts corresponding to the spectral filter $A(z)$, the pitch filter $C(z)$, the gain G, and the residual (shown as waveform (c) in Fig. 26). The first three sets of parameters can be coded and transmitted as side information, exactly as in APC, except that here we can perform VQ on each set of parameters [2], [24] to reduce the bit rate. Also as in APC, the residual is *not quantized separately*, but rather as part of a synthesis procedure. If performed separately, time-domain quantization of the residual will result in no SNR gain over PCM coding of the waveform.

There is a compelling reason why the product code suggested above would be expected to perform well. We mentioned in Section V-C that a product code would be expected to yield good results if the component vectors are chosen to be independent. One could argue with some justification that the short-term spectral envelope, the pitch, and the whitened residual are to some extent independent,

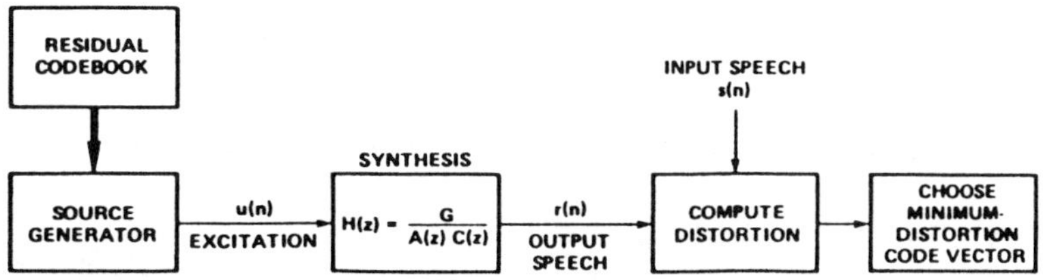

Fig. 27. A block diagram of a system for low-rate speech waveform coding which employs vector quantization. The spectral and pitch filter parameters are quantized and transmitted separately. The codebook shown is used to find the excitation that results in the minimum distortion. As shown in the figure, the excitation is white. If desired, $C(z)$ can be eliminated, in which case the excitation should have some pitch structure, as in the residual waveform (b) in Fig. 26.

so that the use of a product code in this case should reduce computation and storage at only a relatively small reduction in performance over a single large codebook obtained by K-means.

Fig. 27 shows a basic block diagram of the coding system. The first step is to compute and quantize the parameters of $A(z)$, $C(z)$, and G, then substitute the *quantized* values in the synthesis filter $H(z)$ in Fig. 27. Then, the source generator takes one code vector at a time from the residual codebook and sends the samples from the code vector sequentially in time as excitation to the filter $H(z)$. The distortion between the output speech and the input speech in the block to be coded is computed. The process is repeated for all code vectors in the residual codebook, each time comparing the output to the same block of input speech. The code vector that results in the minimum distortion is chosen for transmission. The whole process is then repeated for the next input block. Note that nowhere is there an explicit quantization of the actual prediction residual; the quantization takes place implicitly by searching all vectors in the residual codebook for the vector that minimizes the distortion in the output speech. (The distortion computation may benefit from perceptually based operations such as the use of spectral noise shaping [11], [88].) In product code terminology, the above is a sequential quantization procedure, whereby the side information is quantized first, then the quantized parameter values are used in a full-search of the residual codebook such that an overall distortion criterion is minimized.

The only remaining issues are to choose the block length N and the bit rate R, and to decide how to populate the residual codebook for a given N and R. There is evidence from experiments by Atal [7] that block lengths of about at least 4 ms are needed for good results. This is equivalent to $N = 32$ samples at 8-kHz sampling. At $R = 1$ bit/sample, the codebook contains 32×2^{32} residual samples or about one-half year of speech! At $R = 0.5$ bits/sample, the equivalent amount of speech is just over 4 min. However, the size of the codebook is still $L = 2^{16} = 65\,536$ vectors, which means that the process in Fig. 27 will need to be repeated that many times for each block of 32 samples, if a full search is desired. These illustrative numbers place in perspective the magnitude of the computational problem in VQ. To render the computations more manageable, one will need to either go to lower rates, use shorter blocks, or structure the residual codebook in some fashion.

Because the number of dimensions N here is relatively large, a random codebook would be expected to do well. This is especially so because the residual is relatively white. Therefore, one could populate the codebook from a random sample of residual vectors. Since the residual is relatively Gaussian and white, another possibility would be to simply use a random Gaussian number generator to populate the codebook. Both methods have been used by Atal and Schroeder with good results. In recent simulations on a Cray-1 computer, with $N = 40$ and $R = 0.25$ bits/sample for the residual (excitation), good speech quality was reported with no quantization of the side information [117]. It is important to note that even if the excitation consists of random numbers, each code vector will have a different detailed spectrum and time-domain structure. The quantization process chooses the particular code vector that minimizes the distortion between the original and reconstructed signals.

The use of a pitch filter $1/C(z)$ may not always be desirable. If a pitch filter is not used, then the residual codebook will need to contain wave shapes similar to waveform (b) in Fig. 26. Because of the need for the "pitch pulses," one can no longer use a random Gaussian source to populate the residual codebook. A codebook will have to be designed from a set of residual training data. One could, of course, use any of the other methods we mentioned in Section V. No one, to our knowledge, has attempted to use a more optimized residual codebook for large block lengths.

Tree and trellis coding: Another approach to structuring the codebook to reduce computations and storage is to have code vectors share some of their elements, i.e., some of the elements will have the same values. The two most well-known methods in this class are *tree coding* and *trellis coding* [74], [75], [135]. In tree coding, the vector elements are on a tree, while in trellis coding, the elements are on a trellis. Both types of coding permit the use of longer blocks because of the reduced computational requirements. However, the computational and memory savings are obtained at the cost of reduction in performance, with trellis coding achieving the greatest computational savings but at reduced performance relative to tree coding and to VQ. A number of studies have utilized tree and trellis coding of speech waveforms [6], [7], [34], [49], [54], [56], [74], [128], [136], but there has not been a systematic study comparing these two methods against VQ at lower data rates.

VIII. Conclusions

In the coding of signals at a given rate, it is always possible to reduce the distortion further by using vector quantization instead of scalar quantization. However, because of the substantial computational and memory costs associated with vector quantization, it is most advantageous to use it at low data rates of less than 1 bit/parameter or signal value to be transmitted. In addition, vector quantizers, on the whole, may not be as robust as scalar quantizers when used in an operational environment, especially under conditions of channel errors. However, at low data rates, vector quantization may still be the method of choice when all factors are considered.

Speech presents a fertile signal for the application of vector quantization. Speech parameters are rich with linear and nonlinear dependencies that are utilized effectively by vector quantizers to optimize performance. Much of the success in applying vector quantization to speech coding has been in the coding of spectral parameters for low rate coding of intelligible speech below 800 bits/s. Recent results in the vector quantization of speech waveforms point to exciting possibilities for high-quality speech coding at 8 kbits/s and less.

Acknowledgment

The authors wish to thank the following friends and colleagues who have read and commented on an earlier version of this paper: B. Atal, M. Ostendorf Dunham, A. Gersho, N. Jayant, B-H. Juang, R. Mercer, P. Noll, D. Paul, and Y. Shoham. Their comments have been very valuable in enhancing the accuracy of the presentation. The authors also thank L. Willson for her assistance in the preparation of the manuscript.

References

[1] H. Abut and S. A. Luse, "Vector quantizers for subband coded waveforms," in *Proc. IEEE Int. Conf. Acoust., Speech, Signal Processing* (San Diego, CA, Mar. 1984), paper 10.6.

[2] J-P. Adoul, J-L. Debray, and D. Dalle, "Spectral distance measure applied to the optimum design of DPCM coders with L predictors," in *Proc. IEEE Int. Conf. Acoust., Speech, Signal Processing* (Denver, CO, Apr. 1980), pp. 512–515.

[3] N. Ahmed, T. Natarajan, and K. R. Rao, "Discrete cosine transform," *IEEE Trans. Comput.*, vol. C-23, no. 1, pp. 90–93, Jan. 1974.

[4] V. R. Algazi, "Useful approximations to optimum quantization," *IEEE Trans. Commun. Technol.*, vol. COM-14, pp. 297–301, 1966.

[5] M. R. Anderberg, *Cluster Analysis for Applications.* New York, NY: Academic Press, 1973.

[6] J. B. Anderson and J. B. Bodie, "Tree encoding of speech," *IEEE Trans. Inform. Theory*, vol. IT-21, pp. 379–381, July 1975.

[7] B. S. Atal, "Predictive coding of speech at low bit rates," *IEEE Trans. Commun.*, vol. COM-30, no. 4, pp. 600–614, Apr. 1982.

[8] B. S. Atal and S. L. Hanauer, "Speech analysis and synthesis by linear prediction of the speech wave," *J. Acoust. Soc. Amer.*, vol. 50, no. 2, pp. 637–655, Aug. 1971.

[9] B. S. Atal and J. R. Remde, "A new model of LPC excitation for producing natural-sounding speech at low bit rates," in *Proc. IEEE Int. Conf. Acoust., Speech, Signal Processing* (Paris, France, May 1982), pp. 614–617.

[10] B. S. Atal and M. R. Schroeder, "Adaptive predictive coding of speech signals," *Bell Syst. Tech. J.*, vol. 49, pp. 1973–1986, Oct. 1970.

[11] ——, "Predictive coding a speech signals and subjective error criteria," *IEEE Trans. Acoust., Speech, Signal Processing*, vol. ASSP-27, no. 3, pp. 247–254, June 1979.

[12] T. P. Barnwell, III, "Correlation analysis of subjective and objective measures for speech quality," in *Proc. IEEE Int. Conf. Acoust., Speech, Signal Processing* (Denver, CO, Apr. 1980), pp. 706–709.

[13] R. E. Bellman, *Introduction to Matrix Analysis.* New York, NY: McGraw-Hill, 1960.

[14] T. Berger, *Rate-Distortion Theory.* Englewood Cliffs, NJ: Prentice-Hall, 1971.

[15] T. Berger, "Minimum entropy quantizers and permutation codes," *IEEE Trans. Inform. Theory*, vol. IT-28, no. 2, pp. 149–157, Mar. 1982.

[16] M. Berouti and J. Makhoul, "An adaptive-transform baseband coder," in *Speech Communication Papers: 97th Meeting of the Acoustical Society of America*, J. J. Wolf and D. H. Klatt, Eds. (Cambridge, MA, June 1979), pp. 377–380.

[17] R. E. Blahut, "Computation of channel capacity and rate-distortion functions," *IEEE Trans. Inform. Theory*, vol. IT-18, pp. 460–473, July 1972.

[18] A. Buzo, A. H. Gray Jr., R. M. Gray, and J. D. Markel, "Speech coding based upon vector quantization," *IEEE Trans. Acoust., Speech, Signal Processing*, vol. ASSP-28, no. 5, pp. 562–574, Oct. 1980.

[19] J. W. S. Cassels, *An Introduction to the Geometry of Numbers.* Berlin, Germany: Springer-Verlag, 1959.

[20] D. Y. Cheng, A. Gersho, B. Ramamurthi, and Y. Shoham, "Fast search algorithms for vector quantization and pattern matching," in *Proc. IEEE Int. Conf. Acoust., Speech, Signal Processing* (San Diego, CA, Mar. 1984), paper 9.11.

[21] J. H. Conway and N. J. A. Sloane, "Voronoi regions of lattices, second moments of polytopes, and quantization," *IEEE Trans. Inform. Theory*, vol. IT-28, no. 2, pp. 211–226, Mar. 1982.

[22] ——, "Fast quantizing and decoding algorithms for lattice quantizers and codes," *IEEE Trans. Inform. Theory*, vol. IT-28, no. 2, pp. 227–232, Mar. 1982.

[23] ——, "A lower bound on the average error of vector quantizers," *IEEE Trans. Inform. Theory*, vol. IT-31, no. 1, pp. 106–109, Jan. 1985.

[24] M. Copperi and D. Sereno, "9.6 kbit/s piecewise LPC residual excited coder using multiple-stage vector quantization," in *Proc. IEEE Int. Conf. Acoust., Speech, Signal Processing* (San Diego, CA, Mar. 1984), paper 10.5.

[25] V. Cuperman and A. Gersho, "Vector predictive coding of speech at 16 kbits/s," *IEEE Trans. Commun.*, vol. COM-33, pp. 685–696, July 1985.

[26] L. D. Davisson and R. M. Gray, Eds., *Data Compression.* Stroudsburg, PA: Dowden Hutchinson & Ross, 1976.

[27] N. R. Dixon and T. B. Martin, Eds., *Automatic Speech and Speaker Recognition.* New York, NY: IEEE PRESS, 1979.

[28] J. L. Doob, *Stochastic Processes.* New York, NY: Wiley, 1953.

[29] E. Dubois, "The sampling and reconstruction of time-varying imagery with application in video systems," *Proc. IEEE*, vol. 73, no. 4, pp. 502–522, Apr. 1985.

[30] R. O. Duda and R. E. Hart, *Pattern Classification and Scene Analysis.* New York, NY: Wiley, 1973.

[31] H. Dudley, "Phonetic pattern recognition vocoder for narrow-band speech transmission," *J. Acoust. Soc. Amer.*, vol. 30, no. 8, pp. 733–739, Aug. 1958.

[32] M. O. Dunham and R. M. Gray, "An algorithm for design of labeled-transition finite-state vector quantizers," *IEEE Trans. Commun.*, vol. COM-33, no. 1, pp. 83–89, Jan. 1985.

[33] N. Farvardin and J. W. Modestino, "Optimum quantizer performance for a class of non-Gaussian memoryless sources," *IEEE Trans. Inform. Theory*, vol. IT-30, no. 3, pp. 485–497, May 1984.

[34] H. G. Fehn and P. Noll, "Multipath search coding of stationary signals with applications to speech," *IEEE Trans. Commun.*, vol. COM-30, no. 4, pp. 687–701, Apr. 1982.

[35] L. Fejes Toth, "Sur la representation d'une population infinie par un nombre fini d'elements," *Acta Math. Acad. Scient. Hung.*, vol. 10, pp. 299–304, 1959.

[36] T. R. Fischer and R. M. Dicharry, "Vector quantizer design for memoryless Gaussian, Gamma, and Laplacian sources," *IEEE Trans. Commun.*, vol. COM-32, no. 9, pp. 1065–1069, Sept. 1984.

[37] T. R. Fischer and K. T. Malone, "Contour vector quantization and waveform coding," in *Proc. IEEE Int. Conf. Acoust., Speech, Signal Processing* (Tampa, FL, Mar. 1985), pp. 1707–1710.

[38] J. L. Flanagan, *Speech Analysis Synthesis and Perception*, 2nd ed. New York: Academic Press, 1972.

[39] J. L. Flanagan, M. R. Schroeder, B. S. Atal, R. E. Crochiere, N. S. Jayant, and J. M. Tribolet, "Speech coding," *IEEE Trans. Commun.*, vol. COM-27, no. 4, pp. 710–737, Apr. 1979.

[40] P. E. Fleischer, "Sufficient conditions for achieving minimum distortion in a quantizer," in *1964 IEEE Int. Conv. Rec.*, pt. 1, pp. 104–111, 1964.

[41] E. W. Forgy, "Cluster analysis of multivariate data: Efficiency vs. interpretability of classifications," *Biometrics*, vol. 21, p. 768, abstract, 1965.

[42] J. Foster, R. M. Gray, and M. O. Dunham, "Finite-state vector quantization for waveform coding," *IEEE Trans. Inform. Theory*, vol. IT-31, pp. 348–359, May 1985.

[43] B. Fox, "Discrete optimization via marginal analysis," *Manag. Sci.*, vol. 13, no. 3, pp. 210–216, Nov. 1966.

[44] J. H. Friedman, F. Bashett, and L. J. Shustek, "An algorithm for finding nearest neighbors," *IEEE Trans. Comput.*, vol. C-24, pp. 1000–1006, Oct. 1975.

[45] J. W. Fussell, "The Karhunen-Loéve transform applied to the log area ratios of a linear predictive speech coder," in *Proc. IEEE Int. Conf. Acoust., Speech, Signal Processing* (Denver, CO, Apr. 1980), pp. 36–39.

[46] R. G. Gallager, *Information Theory and Reliable Communication*. New York, NY: Wiley, 1968.

[47] A. Gersho, "Asymptotically optimal block quantization," *IEEE Trans. Inform. Theory*, vol. IT-25, no. 4, pp. 373–380, July 1979.

[48] ——, "On the structure of vector quantizers," *IEEE Trans. Inform. Theory*, vol. IT-28, pp. 157–166, Mar. 1982.

[49] A. Gersho and V. Cuperman, "Vector quantization: A pattern-matching technique for speech coding," *IEEE Commun. Mag.*, vol. 21, pp. 15–21, Dec. 1983.

[50] A. Gersho, T. Ramstad, and I. Versvik, "Fully vector-quantized subband coding with adaptive codebook allocation," in *Proc. IEEE Int. Conf. Acoust., Speech, Signal Processing* (San Diego, CA, Mar. 1984), paper 10.7.

[51] H. Gish and J. N. Pierce, "Asymptotically efficient quantizing," *IEEE Trans. Inform. Theory*, vol. IT-14, no. 5, pp. 676–683, Sept. 1968.

[52] B. Gold, "Digital speech networks," *Proc. IEEE*, vol. 65, no. 12, pp. 1636–1658, Dec. 1977.

[53] ——, "Experiments with a pattern-matching channel vocoder," in *Proc. IEEE Int. Conf. Acoust., Speech, Signal Processing* (Atlanta, GA, Apr. 1981), pp. 32–35.

[54] A. J. Goldberg, "Predictive coding with delayed decision," in *Proc. IEEE Int. Conf. Acoust., Speech, Signal Processing* (Hartford, CT, May 1977), pp. 405–408.

[55] W. Granzow and P. Noll, "Quantization of a memoryless gamma source," submitted to *IEEE Trans. Commun.*

[56] R. M. Gray, "Vector quantization," *IEEE ASSP Mag.*, vol. 1, pp. 4–29, Apr. 1984.

[57] R. M. Gray, A. Buzo, A. H. Gray, Jr., and Y. Matsuyama, "Distortion measures for speech processing," *IEEE Trans. Acoust., Speech, Signal Processing*, vol. ASSP-28, no. 4, pp. 367–376, Aug. 1980.

[58] R. M. Gray and E. D. Karnin, "Multiple local optima in vector quantizers," *IEEE Trans. Inform. Theory*, vol. IT-28, no. 2, pp. 256–261, Mar. 1982.

[59] R. M. Gray and J. C. Kieffer, "Asymptotically mean stationary measures," *Ann. Probability*, vol. 8, pp. 962–973, Oct. 1980.

[60] R. M. Gray and Y. Linde, "Vector quantizers and predictive quantizers for Gauss-Markov sources," *IEEE Trans. Commun.*, vol. COM-30, no. 2, pp. 381–389, Feb. 1982.

[61] A. H. Gray, Jr. and J. D. Markel, "Distance measures for speech processing," *IEEE Trans. Acoust., Speech, Signal Processing*, vol. ASSP-24, no. 5, pp. 380–391, Oct. 1976.

[62] A. Haoui and D. G. Messerschmitt, "Predictive vector quantization," in *Proc. IEEE Int. Conf. Acoust., Speech, Signal Processing* (San Diego, CA, Mar. 1984), paper 10.10.

[63] ——, "Embedded coding of speech: A vector quantization approach," in *Proc. IEEE Int. Conf. Acoust., Speech, Signal Processing* (Tampa, FL, Mar. 1985), pp. 1703–1706.

[64] J. A. Hartigan, *Clustering Algorithms*. New York: Wiley, 1975.

[65] J. Huang and P. Schultheiss, "Block quantization of correlated Gaussian random variables," *IEEE Trans. Commun. Syst.*, vol. CS-11, pp. 289–296, Sept. 1963.

[66] D. A. Huffman, "A method for the construction of minimum-redundancy codes," *Proc. IRE*, vol. 40, no. 9, pp. 1098–1101, Sept. 1952.

[67] IMSL Library, International Mathematical and Statistical Libraries, Inc., Houston, TX.

[68] F. Itakura, "Minimum prediction residual principle applied to speech recognition," *IEEE Trans. Acoust., Speech, Signal Processing*, vol. ASSP-23, no. 1, pp. 67–72, Feb. 1975.

[69] F. Itakura, S. Saito, T. Koike, H. Sawabe, and M. Nishikawa, "An audio response unit based on partial autocorrelation," *IEEE Trans. Commun.*, vol. COM-20, pp. 792–797, Aug. 1972.

[70] F. Itakura and S. Saito, "Analysis synthesis telephony based on the maximum likelihood method," in *Proc. 6th Int. Congr. Acoust.* (Tokyo, Japan, 1968), pp. C17–C20.

[71] ——, "A statistical method for estimation of speech spectral density and formant frequencies," *Electron. Commun. Japan*, vol. 53-A, pp. 36–43, 1970.

[72] ——, "On the optimum quantization of feature parameters in the PARCOR speech synthesizer," in *Conf. Rec., IEEE Conf. Speech Communication and Processing* (Newton, MA, Apr. 1972), pp. 434–437.

[73] N. S. Jayant, Ed., *Waveform Quantization and Coding*. New York, NY: IEEE PRESS, 1976.

[74] N. S. Jayant and P. Noll, *Digital Coding of Waveforms: Principles and Applications to Speech and Video*. Englewood Cliffs, NJ: Prentice-Hall, 1984.

[75] F. Jelinek, "Tree encoding of memoryless time-discrete sources with a fidelity criterion," *IEEE Trans. Inform. Theory*, vol. IT-15, pp. 584–590, Sept. 1969.

[76] B-H. Juang and A. H. Gray, Jr., "Multiple stage vector quantization for speech coding," in *Proc. IEEE Int. Conf. Acoust., Speech, Signal Processing* (Paris, France, May 1982), pp. 597–600.

[77] G. S. Kang and D. C. Coulter, "600 bps voice digitizer," in *Proc. IEEE Int. Conf. Acoust., Speech, Signal Processing* (Philadelphia, PA, Apr. 1976), pp. 91–94.

[78] N. Kitawaki and F. Itakura, "Nonlinear coding of PARCOR coefficients," in *Meeting of the Acoustical Society of Japan* (in Japanese), pp. 449–450, Oct. 1973.

[79] H. P. Kramer and M. V. Mathews, "A linear encoding for transmitting a set of correlated signals," *IRE Trans. Inform. Theory*, vol. IT-2, pp. 41–46, Sept. 1956.

[80] S. E. Levinson, "Structural methods in automatic speech recognition," this issue, pp. 1625–1650.

[81] Y. Linde, A. Buzo, and R. M. Gray, "An algorithm for vector quantizer design," *IEEE Trans. Commun.*, vol. COM-28, no. 1, pp. 84–95, Jan. 1980.

[82] S. P. Lloyd, "Least squares quantization in PCM," *IEEE Trans. Inform. Theory*, vol. IT-28, no. 2, pp. 129–137, Mar. 1982.

[83] Ph. Mabilleau and J-P. Adoul, "Medium band speech coding using a dictionary of waveforms," in *Proc. IEEE Int. Conf. Acoust., Speech, Signal Processing* (Atlanta, GA, Mar. 1981), pp. 804–807.

[84] J. MacQueen, "Some methods for classification and analysis of multivariate observations," in *Proc. 5th Berkeley Symp. on Math., Statist., and Prob.* Berkeley, CA: Univ. of California Press, 1967, pp. 281–297.

[85] P. C. Mahalanobis, "On the generalized distance in statistics," *Proc. Indian Nat. Inst. Sci.* (Calcutta), vol. 2, pp. 49–55, 1936.

[86] J. Makhoul, "Linear prediction: A tutorial review," *Proc. IEEE*, vol. 63, no. 4, pp. 561–580, Apr. 1975.

[87] ——, "Speech coding and processing," in *Modern Signal Processing*, T. Kailath, Ed. New York, NY: Hemisphere/Springer-Verlag, 1985, pp. 211–248.

[88] J. Makhoul and M. Berouti, "Adaptive noise spectral shaping and entropy coding in predictive coding of speech," *IEEE Trans. Acoust., Speech, Signal Processing*, vol. ASSP-27, no. 1, pp. 63–73, Feb. 1979.

[89] ——, "Predictive and residual encoding of speech," *J. Acoust. Soc. Amer.*, vol. 66, no. 6, pp. 1633–1641, Dec. 1979.

[90] J. D. Markel and A. H. Gray, Jr., *Linear Prediction of Speech*. New York, NY: Springer-Verlag, 1976.

[91] J. Max, "Quantizing for minimum distortion," *IRE Trans. Inform. Theory*, vol. IT-6, no. 2, pp. 7–12, Mar. 1960.

[92] R. J. McEliece, *The Theory of Information and Coding: A Mathematical Framework for Communication*. Reading, MA: Addison-Wesley, 1977.

[93] E. McLarnon, "A method for reducing the transmission rate of a channel vocoder by using frame interpolation," in *Proc.*

IEEE Int. Conf. Acoust., Speech, Signal Processing (Tulsa, OK, Apr. 1978), pp. 458–461.

[94] P. Mermelstein, "Evaluation of a segmental SNR measure as an indicator of the quality of ADPCM coded speech," *J. Acoust. Soc. Amer.*, vol. 66, no. 6, pp. 1664–1667, Dec. 1979.

[95] A. Nadas, R. L. Mercer, L. R. Bahl, R. Bakis, P. S. Cohen, A. G. Cole, F. Jelinek, and B. L. Lewis, "Continuous speech recognition with automatically selected acoustic prototypes obtained by either bootstrapping or clustering," in *Proc. IEEE Int. Conf. Acoust., Speech, Signal Processing* (Atlanta, GA, Apr. 1981), pp. 1153–1155.

[96] P. Noll, "Adaptive quantizing in speech coding systems," in *Proc. Int. Zurich Seminar on Digital Commun.* (Zurich, Switzerland, Mar. 1974), paper B3.

[97] P. Noll and R. Zelinski, "Bounds on quantizer performance in the low bit-rate region," *IEEE Trans. Commun.*, vol. COM-26, no. 2, pp. 300–304, Feb. 1978.

[98] J. J. O'Donnell, "A system for very low data rate speech communication," in *Proc. IEEE Int. Conf. Acoust., Speech, Signal Processing* (Atlanta, GA, Apr. 1981), pp. 8–11.

[99] E. A. Patrick, *Fundamentals of Pattern Recognition.* Englewood Cliffs, NJ: Prentice-Hall, 1972.

[100] D. B. Paul, "An 800 bps adaptive vector quantization vocoder using a perceptual distance measure," in *Proc. IEEE Int. Conf. Acoust., Speech, Signal Processing* (Boston, MA, Apr. 1983), pp. 73–76.

[101] ——, "The Lincoln low-rate vocoder: A 1200/2400 bps LPC-10 voice terminal," Tech. Rep. 676, Lincoln Lab., Lexington, MA, Mar. 1984, DCC AD-A141291/5.

[102] D. B. Paul and P. E. Blankenship, "Two distance measure-based vocoder quantization algorithms for very-low-data rate applications: Frame-fill and spectral vector quantization," in *Proc. IEEE Int. Conf. Commun.* (June 1982), pp. 1–6.

[103] W. W. Peterson and E. J. Weldon, Jr., *Error-Correcting Codes*, 2nd ed. Cambridge, MA: M.I.T. Press, 1972.

[104] R. Pieraccini and R. Billi, "Experimental comparison among data compression techniques in isolated word recognition," in *Proc. IEEE Int. Conf. Acoust., Speech, Signal Processing* (Boston, MA, Apr. 1983), pp. 1025–1028.

[105] S. U. H. Qureshi and G. D. Forney, Jr., "Adaptive residual coder—An experimental 9.6/16 kb/s speech digitizer," in *EASCON Rec.*, pp. 29A–29E, 1975.

[106] L. Rabiner, S. E. Levinson, and M. M. Sondhi, "On the application of vector quantization and hidden Markov models to speaker-independent isolated word recognition," *Bell Syst. Tech. J.*, vol. 62, pp. 1075–1105, Apr. 1983.

[107] L. R. Rabiner and R. W. Schafer, *Digital Processing of Speech Signals.* Englewood Cliffs, NJ: Prentice-Hall, 1978.

[108] L. R. Rabiner, M. M. Sondhi, and S. E. Levinson, "A vector quantizer incorporating both LPC shape and energy," in *Proc. IEEE Int. Conf. Acoust., Speech, Signal Processing* (San Diego, CA, Mar. 1984), paper 17.1.

[109] H. Reininger and D. Wolf, "Speech and speaker independent codebook design in VQ coding schemes," in *Proc. IEEE Int. Conf. Acoust., Speech, Signal Processing* (Tampa, FL, Mar. 1985), pp. 1700–1702.

[110] S. Roucos, J. Makhoul, and R. Schwartz, "A variable-order Markov chain for coding of speech spectra," in *Proc. IEEE Int. Conf. Acoust., Speech, Signal Processing* (Paris, France, May 1982), pp. 582–585.

[111] S. Roucos, R. Schwartz, and J. Makhoul, "Segment quantization for very-low-rate speech coding," in *Proc. IEEE Int. Conf. Acoust., Speech, Signal Processing* (Paris, France, May 1982), pp. 1565–1569.

[112] ——, "Vector quantization for very-low-rate coding of speech," in *Proc. IEEE Global Telecommunications Conf.* (Miami, FL, Nov. 29–Dec. 2, 1982), pp. 1074–1078.

[113] ——, "A segment vocoder at 150 B/S," in *Proc. IEEE Int. Conf. Acoust., Speech, Signal Processing* (Boston, MA, Apr. 1983), pp. 61–64.

[114] M. J. Sabin and R. M. Gray, "Product code vector quantizers for waveform and voice coding," *IEEE Trans. Acoust., Speech, Signal Processing*, vol. ASSP-32, no. 3, pp. 474–488, June 1984.

[115] D. J. Sakrison, "A geometric treatment of the source encoding of a Gaussian random variable," *IEEE Trans. Inform. Theory*, vol. IT-14, no. 3, pp. 96–101, May 1968.

[116] M. R. Sambur, "An efficient linear-prediction vocoder," *Bell Syst. Tech. J.*, vol. 54, pp. 1693–1723, Dec. 1975.

[117] M. R. Schroeder and B. S. Atal, "Code-excited linear prediction (CELP): High-quality speech at very low bit rates," in *Proc. IEEE Int. Conf. Acoust., Speech, Signal Processing* (Tampa, FL, Mar. 1985), pp. 937–940.

[118] R. Schwartz, Y. Chow, S. Roucos, M. Krasner, and J. Makhoul, "Improved hidden Markov modeling of phonemes for continuous speech recognition," in *Proc. IEEE Int. Conf. Acoust., Speech, Signal Processing* (San Diego, CA, Mar. 1984), paper 35.6.

[119] R. Schwartz and S. Roucos, "A comparison of methods for 300–400 b/s vocoders," in *Proc. IEEE Int. Conf. Acoust., Speech, Signal Processing* (Boston, MA, Apr. 1983), pp. 69–72.

[120] A. Segall, "Bit allocation and encoding for vector sources," *IEEE Trans. Inform. Theory*, vol. IT-22, no. 2, pp. 162–169, Mar. 1976.

[121] I. K. Sethi, "A fast algorithm for recognizing nearest neighbors," *IEEE Trans. Syst., Man, Cybern.*, vol. SMC-11, no. 3, pp. 245–248, Mar. 1981.

[122] C. E. Shannon, "A mathematical theory of communication," *Bell Syst. Tech. J.*, vol. 27, pp. 379–423, 623–656, 1948.

[123] ——, "Coding theorems for a discrete source with a fidelity criterion," in *IRE Nat. Conv. Rec.* (pt. 4), pp. 142–163, 1959. (Also in *Information and Decision Processes*, R. E. Machol, Ed. New York, NY: McGraw-Hill, 1960, pp. 93–126.)

[124] Y. Shoham and A. Gersho, "Pitch synchronous transform coding of speech at 9.6 kb/s based on vector quantization," in *Proc. IEEE Int. Conf. Commun.* (Amsterdam, The Netherlands, May 1984), pp. 1179–1182.

[125] ——, "Efficient codebook for an arbitrary set of vector quantizers," in *Proc. IEEE Int. Conf. Acoust., Speech, Signal Processing* (Tampa, FL, Mar. 1985), pp. 1696–1699.

[126] J. E. Shore, D. Burton, and J. Buck, "A generalization of isolated word recognition using vector quantization," in *Proc. IEEE Int. Conf. Acoust., Speech, Signal Processing* (Boston, MA, Apr. 1983), pp. 1021–1024.

[127] C. P. Smith, "Perception of vocoder speech processed by pattern matching," *J. Acoust. Soc. Amer.*, vol. 46, no. 6 (pt. 2), pp. 1562–1571, June 1969.

[128] L. C. Stewart, R. M. Gray, and Y. Linde, "The design of trellis waveform coders," *IEEE Trans. Commun.*, vol. COM-30, no. 4, pp. 702–710, Apr. 1982.

[129] J. T. Tou and R. C. Gonzales, *Pattern Recognition Principles.* Reading, MA: Addison-Wesley, 1974.

[130] J. M. Tribolet and R. E. Crochiere, "Frequency domain coding of speech," *IEEE Trans. Acoust., Speech, Signal Processing*, vol. ASSP-27, no. 5, pp. 512–530, Oct. 1979.

[131] R. Viswanathan and J. Makhoul, "Quantization properties of transmission parameters in linear predictive systems," *IEEE Trans. Acoust., Speech, Signal Processing*, vol. ASSP-23, no. 3, pp. 309–321, June 1975.

[132] V. Viswanathan, J. Makhoul, R. Schwartz, and A. W. F. Huggins, "Variable-frame-rate transmission: A review of methodology and application to narrow-band LPC speech coding," *IEEE Trans. Commun.*, vol. COM-30, no. 4, pp. 674–686, Apr. 1982.

[133] V. Viswanathan, W. Russell, A. Higgins, M. Berouti, and J. Makhoul, "Speech-quality optimization of 16 kb/s adaptive predictive coders," in *Proc. IEEE Int. Conf. Acoust., Speech, Signal Processing* (Denver, CO, Apr. 1980), pp. 520–525.

[134] V. Viswanathan, W. Russell, and A. W. F. Huggins, "Objective speech quality evaluation of mediumband and narrowband real-time speech coders," in *Proc. IEEE Int. Conf. Acoust., Speech, Signal Processing* (Boston, MA, Apr. 1983), pp. 543–546.

[135] A. J. Viterbi and J. K. Omura, "Trellis encoding of memoryless discrete-time sources with a fidelity criterion," *IEEE Trans. Inform. Theory*, vol. IT-20, pp. 325–332, May 1974.

[136] S. G. Wilson and S. Hussain, "Adaptive tree encoding of rion," *IEEE Trans. Commun.*, vol. COM-27, pp. 165–170, Jan. 1979.

[137] D. Y. Wong, B. H. Juang, and D. Y. Cheng, "Very low data rate speech compression with LPC vector and matrix quanti-

zation," in *Proc. IEEE Int. Conf. Acoust., Speech, Signal Processing* (Boston, MA, Apr. 1983), pp. 65–68.

[138] D. Y. Wong, B. H. Juang, and A. H. Gray, Jr., "An 800 bit/s vector quantization LPC vocoder," *IEEE Trans. Acoust., Speech, Signal Processing*, vol. ASSP-30, no. 5, pp. 770–780, Oct. 1982.

[139] Y. Yamada, S. Tazaki, and R. M. Gray, "Asymptotic performance of block quantizers with difference distortion measures," *IEEE Trans. Inform. Theory*, vol. IT-26, no. 1, pp. 6–14, Jan. 1980.

[140] T. P. Yunck, "A technique to identify nearest neighbors," *IEEE Trans. Syst., Man, Cybern.*, vol. SMC-6, pp. 678–683, Oct. 1976.

[141] P. L. Zador, "Asymptotic quantization error of continuous signals and the quantization dimension," *IEEE Trans. Inform. Theory*, vol. IT-28, no. 2, pp. 139–149, Mar. 1982.

[142] R. Zelinski and P. Noll, "Adaptive transform coding of speech signals," *IEEE Trans. Acoust., Speech, Signal Processing*, vol. ASSP-25, no. 4, pp. 299–309, Aug. 1977.

[143] J. M. Ziman, *Principles of the Theory of Solids.* Cambridge, UK: Cambridge Univ. Press, 1972.

Image Coding Using Vector Quantization: A Review

NASSER M. NASRABADI, MEMBER, IEEE, AND ROBERT A. KING

Abstract—**This paper presents a review of vector quantization techniques used for encoding digital images. First the concept of vector quantization is introduced, then its application to digital images is explained. Spatial, predictive, transform, hybrid, binary, and subband vector quantizers are reviewed. The emphasis here is on the usefulness of the vector quantization when it is combined with conventional image coding techniques or when it is used in different domains.**

I. Introduction

IMAGE compression is essential for applications such as TV transmission, video conferencing, facsimile transmission of printed material, graphics images, or transmission of remote sensing images obtained from satellites and reconnaissance aircraft. Another area for the application of efficient coding is where pictures are stored in a database, such as archiving medical images, multispectral images, finger prints, and drawings [1] and [2].

The vector quantization algorithms for reducing the transmission bit rate or the storage have recently been extensively investigated for speech and image signals [8] and [11]. In this paper, only the applications of vector quantization to image coding are discussed; the contents are thus restricted to coding techniques that employ vector quantizers (VQ's). Recently, there have been several review articles on general technique of image compression [3]–[6]. In particular, vector quantization has been applied to speech coding for a number of years [7]–[10], and Gray has recently published an excellent review paper on this subject [11].

A fundamental goal of data compression is to reduce the bit rate for transmission or data storage while maintaining an acceptable fidelity or image quality. Numerous bandwidth compression techniques have been developed, such as differential pulse code modulation, transform coding, hybrid coding, and adaptive versions of these techniques in response to the growth of image-processing methods. These techniques usually exploit the psychovisual as well as statistical redundancies in the image data to reduce the bit rate [12]. One deficiency with all of the conventional coding techniques is that quantization is performed on individual real-valued samples of waveforms or pixels of images. For example, transform coding does this by first taking block transforms of a vector and then scalar quantizing the transformed samples within the vector. Predictive coding does it by coding an error term formed as the difference between the current sample and its prediction. These techniques are not optimal since the processed samples are still somehow correlated or dependent. According to Shannon's rate-distortion theory, a better performance is always achievable in theory by coding vectors instead of scalars, even though the data source is memoryless.

This paper is not intended as a survey of the basic vector quantization design algorithms or the searching techniques, but rather as a demonstration of the application of vector quantizers to image coding. A definition of the vector quantizer is given followed by its applications to images in the spatial domain, the predictive domain, the transform domain, and combinations of these known as hybrid domains.

II. Definition of Vector Quantization

Extensive studies of vector quantizers or multidimensional quantizers has recently been performed by many researchers [13]–[17]. The design of optimal vector quantizers from empirical data were proposed and extensively studied by Linde, Buzo, and Gray [9] using a clustering approach which is discussed in Section II-A. This algorithm is now commonly referred to as the LBG algorithm. A vector quantizer can be defined as a mapping Q of K-dimensional Euclidean space R^k into a finite subset Y of R^K. Thus,

$$Q : R^K \rightarrow Y \tag{1}$$

where $Y = (\hat{x}_i;\ i = 1, 2, \cdots, N)$ is the set of reproduction vectors and N the number of vectors in Y. It can also be seen as a combination of two functions: an encoder, which views the input vector x and generates the address of the reproduction vector specified by $Q(x)$, and a decoder, which uses this address to generate the reproduction vector $\hat{x}$. If a distortion measure $d(x, \hat{x})$ which represents the penalty or cost associated with reproducing vectors x by $\hat{x}$ is defined, then the best mapping Q is the one which minimizes $d(x, \hat{x})$ [13]. The LBG algorithm and other variations of this algorithm are based upon this minimization, using a training set as the signal.

One simple distortion measure for waveform coding is the square error distortion given by

$$d(x, \hat{x}) = \|x - \hat{x}\|^2 = \sum_{j=0}^{K-1} (x_j - \hat{x}_j^2). \tag{2}$$

A weighted mean square error (WMSE) distortion can also be used [37]. Other error distortion measures have also been suggested, but they are too expensive computationally [18] for practical implementation. One problem with vector quantization is that a large effort is required by the encoder to search the whole codebook in order to identify the nearest matching vector template to an input vector. Recently, Buzo *et al.* [7] have proposed a tree searched (TSVQ) encoder in order to reduce the search effort of a vector quantizer. A TSVQ encoder can be explained as an encoder that searches a sequence of small codebooks instead of one large codebook. The encoder structure can be depicted as a tree and each search and decision corresponds to advancing one level or stage in the tree, starting from the root of the tree. A detailed description of TSVQ's and TSVQ design algorithms may be found in [19] and [20]. The disadvantages of TSVQ encoders are that their

Paper approved by the Editor for Signal Processing and Communication Electronics of the IEEE Communications Society. Manuscript received October 12, 1984; revised June 16, 1987. This paper was presented at the IEEE Acoustics, Speech, and Signal Processing Conference 1985, Tampa, FL, March 26–29, 1985.

N. M. Nasrabadi is with the Department of Electrical Engineering, Atwater Kent Laboratories, Worcester Polytechnic Institute, MA 01609.

R. A. King is with the Department of Electrical Engineering, Imperial College of Science and Technology, London, SW7, England.

IEEE Log Number 8822470.

Reprinted from *IEEE Trans. Commun.*, vol. COM-36, pp. 957–971, Aug. 1988.

codebook storage requirement is greater than full-search VQ's, and the codewords they select are not, in general, optimum in the sense of minimizing $d(x, \hat{x})$. Other vector quantization techniques that reduce memory as well as encoding complexity can be found in [21]–[22] and [109]–[110] where Fischer designed a quantizer (Pyramid Vector Quantizer) based on the implicit geometry of an independent and identically distributed Laplacian source.

A. Vector Quantizer Design

The goal in designing an optimal vector quantizer is to obtain a quantizer consisting of N reproduction vectors, such that it minimizes the expected distortion. Optimality is said to be achieved if there is no other quantizer that can achieve the minimum expected distortion. Lloyd [17] proposed an iterative nonvariational technique known as his "Method I" for the design of scalar quantizers. Recently, Linde *et al.* [9] extended Lloyds' basic approach to the general case of vector quantizers.

Let the expected distortion be approximated by the time-averaged square error distortion given by the expression

$$D(x, q(x)) = \frac{1}{N} \sum_{i=0}^{N-1} d(x_i, \hat{x}_i). \tag{3}$$

The algorithm for an unknown distribution training sequence is given in [9].

1) Let N = number of levels; distortion threshold $\epsilon \geq 0$. Assume an initial N level reproduction alphabet $\hat{A}_0$, and a training sequence $(x_j;\ j = 0, 1, \cdots, n-1)$, and m = number of iterations, set to zero.

2) Given $\hat{A}_m = (y_i;\ i = 1, \cdots, N)$, find the minimum distortion partition $P(\hat{A}_m) = (S_i;\ i = 1, \cdots, N)$ of the training sequence: $x_j \in S_i$ if $d(x_j, y_i) \leq d(x_j, y_l)$, for all l. Compute the average distortion

$$D_m = D[(\hat{A}_m, P(\hat{A}_m))] = n - 1 \sum_{j=0}^{n-1} \min_{y \in A_m} d(x_j, y). \tag{4}$$

3) If $(D_{m-1} - D_m)/D_m \leq \epsilon$, stop the iteration with $\hat{A}_m$ as the final reproduction alphabet; otherwise continue.

4) Find the optimal reproduction alphabet $\hat{x}(P(\hat{A}_m)) = (\hat{x}(S_i);\ i = 1, \cdots, N)$ for $P(\hat{A}_m)$ where

$$\hat{x}(S_i) = \frac{1}{\|S_i\|} \sum_{j: x_j \in S_i}^{m} x_j. \tag{5}$$

5) Set $\hat{A}_{m+1} = \hat{x}(S_i)$, increment m to $m + 1$, and go to (2).

In the above iterative algorithm an initial reproduction alphabet $\hat{A}_0$ was assumed in order to start the algorithm. There are a number of techniques to obtain the initial codebook. The simplest technique is to use the first widely spaced words from the training sequence. Linde *et al.* [9] used a splitting technique where the centroid for the training sequence was calculated and split into two close vectors. The centroids or the reproduction vectors for the two partitions were then calculated. Each resulting vector was then split into two vectors and the above procedure was repeated until an N level initial reproduction vector was created. Splitting was performed by adding a fixed perturbation vector ϵ to each vector y_i producing two vectors $y_i + \epsilon$, $y_i - \epsilon$. Another approach for designing initial codebooks is to employ product code techniques [22]. In this approach, a scalar quantizer is used K times successively to yield a K-dimensional vector quantizer.

There are a number of other techniques for designing codebooks, such as Kohonen self-organization feature maps [111]–[112] and simulated annealing (Boltzmann Machine) [113]–[114]. Kohonen introduced a neural network clustering technique where weights (synaptic strength) between neurons are adaptively sensitized to the prevailing input patterns. The asymptotic values of the weights define the codebook entries. Simulated annealing is an stochastic optimization technique that can be used to find the optimal codebook patterns with respect to the training sequence. It will avoid local minimums by introducing noise in to the system.

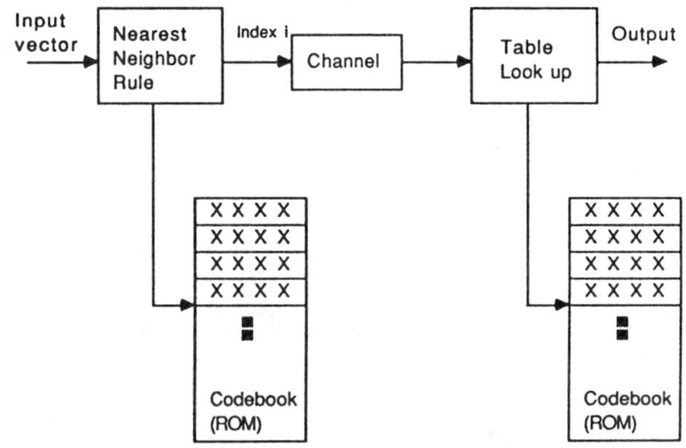

Fig. 1. Block diagram of a simple vector quantizer [37].

III. Spatial Vector Quantizers (SVQ)

Recently, the performance of vector quantizers on digital images has been investigated in the spatial domain [23]–[45]. Adaptive and nonadaptive vector quantizers have been designed to code single frame images. The basic concept involves partitioning an image into two-dimensional vectors. Each vector is then compared to a set of standard vector templates stored in a ROM, and a codeword identifying the best match is then transmitted. The receiver reconstructs the image using the corresponding templates in place of the original vectors. Fig. 1 shows the block diagram of a simple vector quantization coding system. Murakami *et al.* [30] designed a vector quantizer system by converting each vector (block) of the video signal into a vector with zero mean and unit standard deviation as shown below for a vector of input video samples $S = [s_1, s_2, \cdots, s_k]_l$ of dimension K,

$$m_l = \frac{1}{K} \sum_{i=1}^{K} s_i \tag{6}$$

$$\sigma_l = \left[\frac{1}{K} \sum_{i=1}^{K} (s_i - m_l)^2 \right]^{1/2} \tag{7}$$

$$x_i = \frac{(s_i - m_l)}{\sigma_l} \tag{8}$$

$$X_l = [x_1, x_2, \cdots, x_k]_l \tag{9}$$

where m_l, and σ_l are mean and standard deviations of block l. By the above normalization the probability function $P(X)$ of input vectors X is approximately similar for video signals from different scenes. Each vector X_l is then coded by a codebook designed by the LBG algorithms using a mini-max distortion criterion. The mean and the variance are quantized by a scalar quantizer and then transmitted. A two-dimensional quantizer can also be employed to code these statistical measures.

A. Mean/Shape VQ

Baker and Gray [23]–[26] also designed a similar technique called mean/shape vector quantizer (M/SVQ) this is shown in Fig. 2. The sample mean was scalar quantized and the resulting error vector, obtained by subtracting the sample

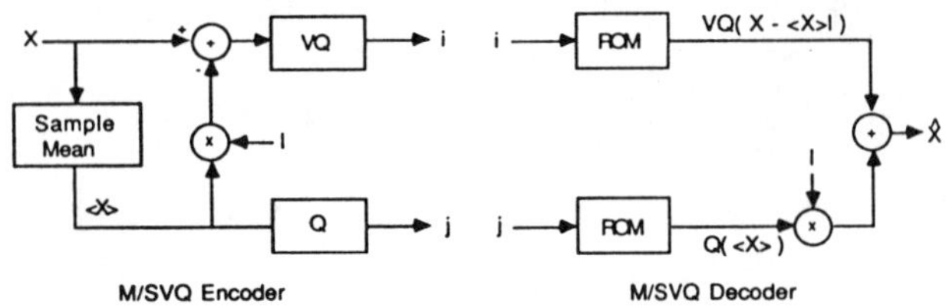

Fig. 2. Mean/shape vector quantizer system [26].

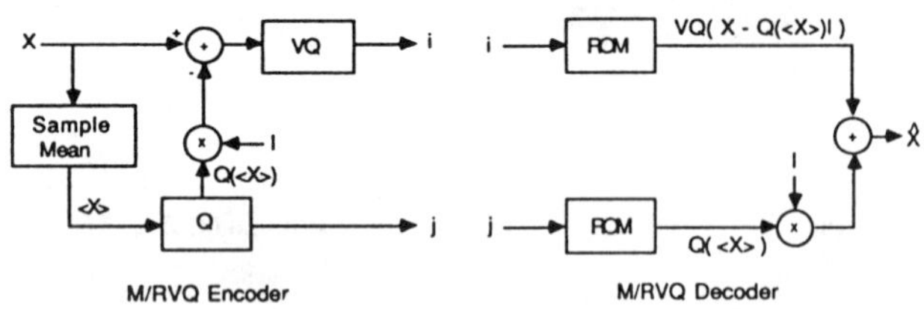

Fig. 3. Mean/residual vector quantizer system [26].

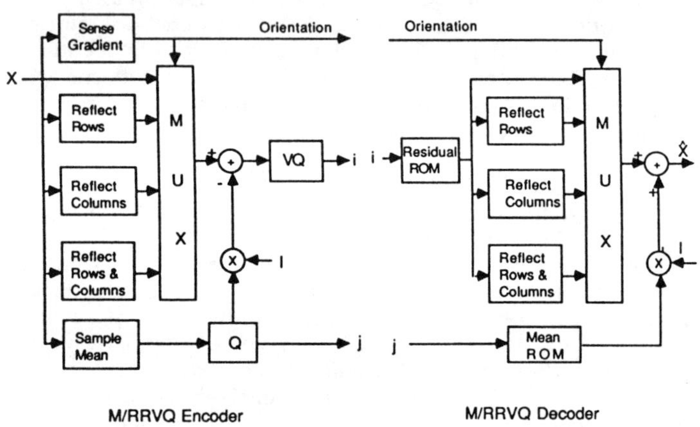

Fig. 4. Mean/reflected residual vector quantizer system [26].

mean from the input vector, is vector quantized. A variation of M/SVQ was also introduced, and is called mean/residual vector quantizer (M/RVQ) as shown in Fig. 3. In this coding system, the sample mean is first scalar quantized and then subtracted from the vector so that any error in quantization of the mean is incorporated into the error vector. The M/RVQ system reduces the blocking distortion that is caused by coarse quantization of the sample mean in the M/SVQ system [25]. To exploit the interblock correlation, Baker suggested the use a scalar quantizer with memory. A differential pulse code modulation (DPCM) was used in his system.

Baker also extended his coding system M/RVQ to include the symmetrical properties of the square vectors [26]. Since some square vectors differ only be reflections about their horizontal or vertical axes, it is possible to transmit this orientation as well as the corresponding codebook vector as shown in Fig. 4 this is called mean/reflected residual VQ. The advantage of this approach is that by sending the corresponding reflection information, we do not have to store all the possible code-vectors. An increase in bit rate is expected because information about the symmetry of the vector block has to be sent. If the block size is small 3 × 3, this increase in side information is significant.

B. Classified VQ

The vector coded images at very low bit rates usually suffer from edge distortion since edges cannot be reproduced perfectly by a small sized codebook. Gersho and Ramamurthi tackled this problem by classifying each vector of the image into several categories [33] and [36]. A simple classification of the training vectors into edge and shade types, followed by a

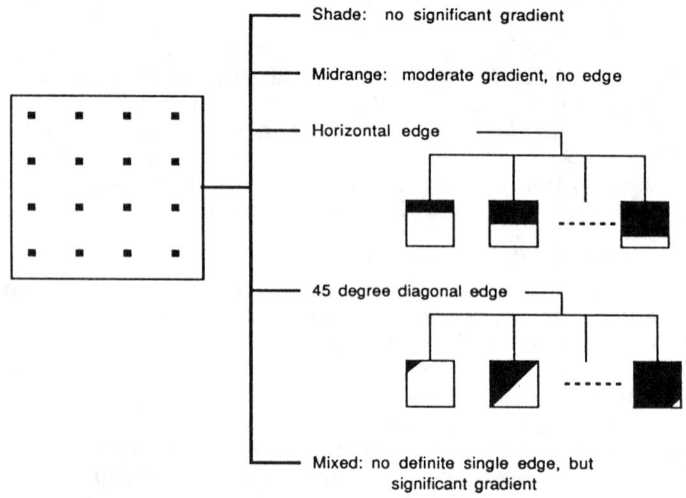

Fig. 5. Each vector is categorized into several classes such as shade and edge blocks [37].

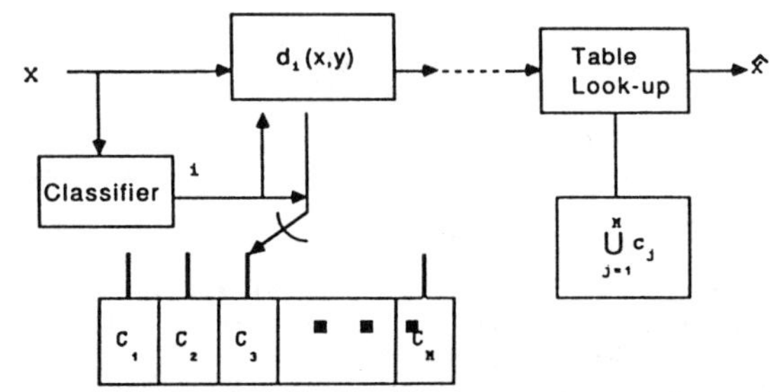

Fig. 6. Classified vector quantizer coding system [37].

separate codebook design for each class was proposed. They called this approach a classified vector quantizer (CVQ). Each vector was then coded by the appropriate vector quantizer. The computational complexity was also reduced since only one appropriate subclass of the codebook is checked for each input vector. However, since it is of great perceptual importance to preserve the fidelity of edges in location and angular orientation, the authors extended their earlier classification idea to include a larger number of classes [37]. Each edge block was then subdivided into more classes depending upon the orientation and location of the edge. Fig. 5 shows the classification that they used. Subcodebooks were designed separately for each class by using the LBG algorithm. For example, let C represent the overall codebook and C_i represent the codebook for each class i so $C = \cup_{i=1}^{M} C_i$ assuming there are M classes. Codebooks were designed by using the LBG algorithm for each class, allowing the training sequence $[x_j; j = 1, \cdots, n]$ to be classified into the appropriate class, $[x_j^i; j = 1, \cdots, d, i = 1, \cdots, M]$; each codebook C_i was designed by using the training sequence that belonged to that class x_j^i. The LBG algorithm was used to minimize the class distortion $d_i(X, Y)$ where x and y were restricted to the corresponding training sequence x^i and the codebook C_i. Fig. 6 shows the coding procedure. Gradient edge operators were employed on the edge enhanced images for classification of each vector image. By increasing the size of the edge codebook, they noticed the quality of the images was seen to be greatly improved.

Finally, Ramamurthi and Gersho extended the above CVQ technique to include different codebooks and appropriate coding processes for each class. For example, the blocks that fall within the midrange class, which represents moderate intensity variation but excluding any definite edges, are two-dimensional transformed by using a discrete cosine transform

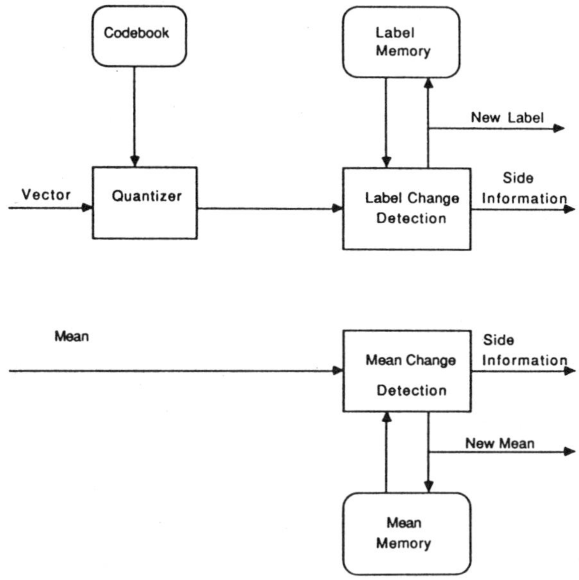

Fig. 7. The block diagram of a codebook replenishment system [40].

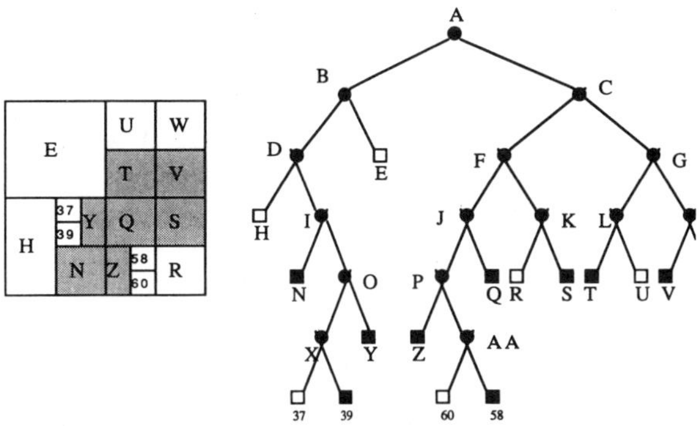

Fig. 8. A typical quad-tree representation of an image [43].

(DCT). A number of high-frequency coefficients are discarded (zonal sampling) [38] thereby reducing the VQ computational complexity in exchange for a small increase in distortion. The distortion is negligible because most of the energy is carried by the low frequency coefficients, especially for the blocks that belong to the midrange class. For the blocks belonging to the edge class, the separate-mean VQ coder of Baker and Gray [23] is employed because the edge blocks are similar in edge orientation and location and differ only in the average intensity level; these can be represented by codebook vectors that are normalized by the mean. These vector quantization techniques can be extended to multidimensional vector quantizers for coding moving images, but the storage for the codebooks as well as the computational complexity will be exponentially increased.

C. Codebook Replenishment VQ

Goldberg *et al.* [39]–[40] introduced an adaptive vector quantizer for the coding of monochromatic and color pictures. In their system, adaptivity was based upon computing and transmitting a small codebook that matched the local statistics of the image to be coded. Fig. 7 shows the block diagram of the proposed adaptive vector quantization system. The image is subdivided into nonoverlapping subimages, and for each subimage, a separate codebook consisting of about 16–64 representative vectors is formed. For each subimage, the locally generated representative vectors of the codebook are transmitted, followed by the codewords or labels for the vectors of the image. Images coded by this technique produce visually acceptable pictures falling between 1–1.5 bit-pixel. The block size of 2×2 was employed in their algorithm. One problem with this method is that the redundancy between different codebooks used in the different subimages is not removed as indicated by the authors [40]. The extension of this technique to color images is discussed in Section VIII. The major problem with this technique is that the new codebooks must be sent to the receiver involving the transformation of a large amount of side information and also needing a large computational effort for designing the new codebooks. Gersho and Yano [41] developed a similar algorithm in which adaptation was applied to change only a small part of the codebook rather than the entire codebook. The algorithm monitors the distortion for each input vector and if it is larger than a predetermined threshold, that input vector is added to the codebook as a new vector.

D. Hierarchical VQ

An important point that has been ignored in adaptive SVQ is that the block size is constant throughout the encoding process. A better technique which could lower the bit rate significantly is to use variable block length. One such technique was introduced by Nasrabadi [42] called adaptive hierarchical VC (AHVQ). In this coding technique, a quad-tree algorithm [43] is first used to partition the image into blocks of size 2×2, 4×4, 8×8, and 16×16. This information about partitioning the image is represented by a quad-tree and is counted as the side information to be transmitted. Fig. 8 shows a typical quad-tree representation. In the coding process, the small blocks, 2×2 and 4×4, are coded by using a typical SVQ. The larger blocks representing constant patterns are encoded by applying VQ in the transform domain. Many of the high-frequency coefficients can be discarded and thus the effective dimension of the blocks is reduced for computational purposes. A similar but one-dimensional algorithm was also proposed for speech [44].

Recently, Vaisey and Gersho [45] implemented such a variable block-size vector quantizer by using a standard split-merging segmentation technique to partition the image into subblocks. Two approaches were studied for vector quantization of images. In the first approach, vector quantizer was used as well as a transform VQ, and in the second approach an adaptive vector quantizer was used where the adaptivity was based upon updating or replenishing the codebook by vector patterns that represent the statistics of the current image more efficiently.

E. Interframe VQ

In an image sequence, successive frames are usually very highly correlated. A simple interframe coding system that only transmits the intensity of the pixels that change between successive frames is known as a frame replenishment coding system. Recently, Goldberg and Sun [39]–[40] introduced several interframe coding systems where the ideas of label replenishment and codebook replenishment were incorporated in their coding system. Fig. 9 shows the schematic diagram for a label replenishment VQ system. In this system, each frame is first divided into two-dimensional blocks, and vectors are then extracted from the resulting blocks either directly in the spatial domain or indirectly through the transform domain. The vectors corresponding to the first frame are taken as the training set for generating the first codebook. The first frame is quantized, and the labels are stored in the label memory and transmitted as indicated by dashed line 2. Each subsequent frame is subdivided into two-dimensional blocks and each vector is quantized. The label is then compared to the one in the label memory, and if it differs, the memory is replenished.

In an image sequence, usually some areas within the image sequence get occluded or are uncovered by the movement of

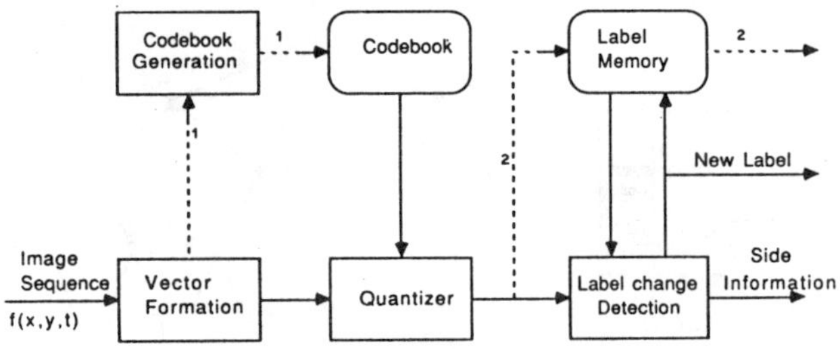

Fig. 9. A schematic diagram for a label replenishment vector quantization system [40].

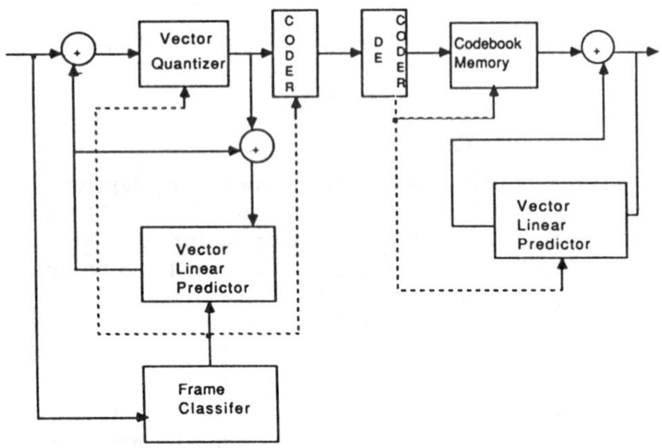

Fig. 10. Adaptive differential vector quantizer [51].

objects. In order for the codebook to follow the statistics of the image, a codebook replenishment VQ system was suggested in [40]. For each vector from the jth frame, the best representative vector pattern is found from the codebook that was generated using the $j - 1$th frame. If the distance between these vectors is above a threshold, that vector is added to the codebook àt the transmitter and the receiver, thereby replenishing the codebooks. If the scene changes drastically, then a large number of new vector patterns has to be transmitted. A better approach would be to use the new frame and form a totally a new codebook as suggested in [40].

IV. Predictive Vector Quantizer (PVQ)

In the previous section, we considered zero memory vector quantization of scalar random variables. However, since the consecutive vectors are statistically dependent, better performance can be achieved if intervector correlation is incorporated in the encoder. A memory vector quantizer can be represented by a finite state machine [46]–[48] where each state s represents a separate quantizer with its own codebook C_s. At each state s_i, the encoder uses the codebook C_{s_i} to produce the channel symbols. It selects its next state by a mapping called the state transition function F such that, given a state s_i and a channel symbol v_i then $F(v_i, s_i)$ is the new state of the machine. Feedback vector quantizers are known as finite state vector quantizers (FSVQ) [47]–[50] or predictive vector quantizers [46]. The application of FSVQ to speech and image coding is still not fully developed. A design procedure was developed by Foster *et al.* [48]. In FSVQ, the codebooks as well as the state transition table are stored in ROM. Recently, Aravind and Gersho [49] and [108] applied a FSVQ coding system to encode digital images. A state transition function F was developed based upon a classifier where intensity and geometric correlations between neighboring blocks were used. The application of FSVQ to image sequences was reported in [50].

A coding system has been introduced by Cuperman and Gersho [51] involving a vector generalization of DPCM and is shown in Fig. 10. The coder consists of a vector predictor and

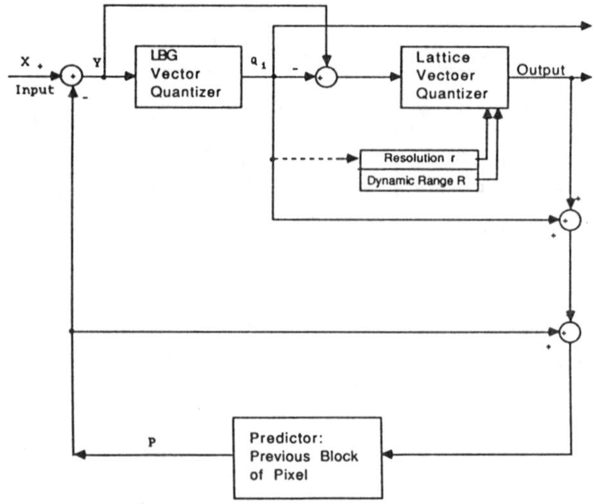

Fig. 11. A two stage hybrid vector quantizer [65].

an error vector quantizer. An error vector signal e is formed by subtracting the predicted vector $\hat{x}$ estimated from previous vectors from the actual present sample vector x. This error vector is then coded by using an error codebook. To improve performance, the system is made adaptive by classifying input frames of speech into categories, and using an appropriate predictor and quantizer for each class.

Hang and Woods developed a predictive vector quantizer in [52]–[53]. In their coding system, the input consists of a vector having the current pixel as the first element and the previous inputs as the remaining elements in the vector. The system is like a mapping or a recursive filter predicting the next sample. The mapping is implemented by a VQ lookup table which produces the error signal directly.

The idea of predictive vector quantization can easily be extended to interframe DPCM coders. In conventional interframe coders, usually two frames of an image sequence are stored, and an interframe difference signal is obtained where the moving areas are detected and their resulting difference signal is scalar quantized [54]–[57]. Murakami *et al.* [58] recently introduced a coding system where vector quantization was used to code the interframe difference signal obtained from the moving areas.

Adaptive predictors, such as motion compensated predictors, have recently been studied extensively [59]–[63]. These adaptive predictors can be incorporated into the interframe predictive vector quantizer. Finally, it should be noted that vector quantization schemes can be employed to code the error signal obtained from conventional hybrid coders [64]; we call these coders hybrid vector quantizers (HVQ).

An interframe predictive vector quantizer was designed by Bage [65]. In his first coding system, the prediction vector P consists of the same spatial block of pixels in the previous frame. An error vector Y is formed by subtracting from an input vector X a prediction vector P. This error vector is then vector quantized by the symbols Q in the codebook. A two-stage hybrid VQ was also proposed as shown in Fig. 11. In this coding system, the first stage is an LBG VQ, in which the difference between the input vector Y and the ith output vector Q of the first stage is Lattice-quantized [65]. This hybrid vector quantization system successively approximates the input vectors by the error vectors in each stage. It uses initially the LBG VQ for coarse quantization in the region where the training set respectively is highest. Then it follows with a Lattice VQ in the region where LBG VQ is less representative. At the decoder, each output vector is formed by component wise addition of the representative vector Q at the output of the first stage and the vector L at the output of the second stage.

V. Transform Vector Quantizer (TVQ)

The purpose of transform coding is to convert statistically dependent or correlated picture elements into independent or uncorrelated coefficients [67]–[73]. Because of the computational complexity an image is usually divided into subimages of reasonable size and then a one- or two-dimensional unitary transform is performed on each subimage. The transformed coefficients are then nonuniformly quantized by a scalar quantizer. The quantization levels are given by a bit assignment matrix in which some of the high-frequency coefficients are discarded. The same bit assignment matrix is stored at the receiver or is transmitted for each image [70]. In adaptive transform coders, several bit assignment matrices are employed where each image block is classified into one of several categories and coded by the corresponding bit assignment matrix [70]. In the following subsections, we introduce the application of the vector quantizer instead of the scalar quantizer to code the transformed coefficients. Transformation has the effect of compressing most of the energy within the vector (block) into some low-frequency coefficients. Thus, if a vector quantizer for an image is designed by the LBG algorithm, using the low-frequency coefficients as the training set, a very low bit rate can be achieved. The codebook designed in the transform domain is believed to be more nearly optimal than the one designed in the space domain because the transformed coefficients have better defined distributions than the image pixels. Also some of the high-frequency coefficients are discarded, and the computational cost is therefore reduced. The correlation between the transformed samples is exploited by the vector quantizer which makes the technique more nearly optimal compared to scalar quantizers.

A. Adaptive Transform Coding Using Vector Quantization

In adaptive transform coders several bit assignment matrices are designed. Each subblock of the image is then classified into one of several classes according to the activity content of the block and coded by the appropriate bit assignment matrix. Typical bit assignment matrices for an adaptive coder with four different classes are shown in Fig. 12. A large number of the high-frequency coefficients are usually discarded. However, these high-frequency coefficients can be transmitted by a vector quantization scheme with a small increase in the bit rate [82]. In this coding system, the bit assignment matrix, as shown in Fig. 13 is now in two parts where the low-frequency coefficients are coded by a scalar quantizer, and the high-frequency coefficients by a vector quantizer. It is shown in [74] that with a small increase in bit rate the high-frequency coefficients are transmitted, and the reconstructed image has a better quality around the edges especially at very low bit rate.

Saito *et al.* [75] designed a TVQ where images were divided into blocks of 16 × 16 pixels. Each block was transformed by a two-dimensional discrete cosine transform (DCT). The dc coefficient was scalar quantized by an uniform scalar quantizer of 256 levels. The ac coefficients were grouped into several different zones by grouping the coefficients in a zig-zig manner. Each zone was then vector quantized by the appropriate codebook. Vector quantization of transform coefficients has recently received some attention by a number of researchers [76]–[80]. In most of these techniques, a conventional adaptive transform coder is combined with vector quantization. An image is usually divided into blocks of 4 × 4 and then categorized into several classes; each class is then vector quantized by the appropriate codebook. There are several problems with this approach: small blocks (4 × 4) are only feasible with the current computational power so in turn the high-frequency coefficients cannot be discarded. Performing VQ in the transform domain in the above manner where each sub-block is considered as a vector has only one advantage, that is, that ac coefficients have a well-behaved Laplacian distribution which could be exploited in the designed of the codebook. However, in the following subsections, several TVQ techniques are reviewed. These techniques use VQ to exploit the intra- and interframe correlation between the blocks.

class 1

```
8 5 4 3 3 2 2 2 1 1 1 1 0 0 0 0
5 4 4 3 3 2 2 1 1 1 0 0 0 0 0 0
4 4 3 3 3 2 2 1 1 0 0 0 0 0 0 0
3 3 3 3 3 2 2 1 1 0 0 0 0 0 0 0
3 3 3 3 2 2 2 1 1 0 0 0 0 0 0 0
3 3 3 2 2 2 1 1 1 0 0 0 0 0 0 0
2 2 2 2 2 1 1 1 0 0 0 0 0 0 0 0
2 2 2 1 1 1 1 1 0 0 0 0 0 0 0 0
2 1 1 1 1 1 1 0 0 0 0 0 0 0 0 0
1 1 1 1 1 1 0 0 0 0 0 0 0 0 0 0
1 1 1 1 1 1 0 0 0 0 0 0 0 0 0 0
1 0 0 0 0 0 0 0 0 0 0 0 0 0 0 0
0 0 0 0 0 0 0 0 0 0 0 0 0 0 0 0
0 0 0 0 0 0 0 0 0 0 0 0 0 0 0 0
0 0 0 0 0 0 0 0 0 0 0 0 0 0 0 0
0 0 0 0 0 0 0 0 0 0 0 0 0 0 0 0
```

class 3

```
8 3 2 2 1 1 1 1 0 0 0 0 0 0 0 0
3 3 2 2 1 1 1 0 0 0 0 0 0 0 0 0
3 2 2 1 1 1 1 0 0 0 0 0 0 0 0 0
2 2 1 1 1 1 1 0 0 0 0 0 0 0 0 0
2 1 1 1 1 1 0 0 0 0 0 0 0 0 0 0
2 1 1 1 1 0 0 0 0 0 0 0 0 0 0 0
2 1 1 1 0 0 0 0 0 0 0 0 0 0 0 0
1 1 1 1 0 0 0 0 0 0 0 0 0 0 0 0
1 0 0 0 0 0 0 0 0 0 0 0 0 0 0 0
1 0 0 0 0 0 0 0 0 0 0 0 0 0 0 0
1 1 0 0 0 0 0 0 0 0 0 0 0 0 0 0
0 0 0 0 0 0 0 0 0 0 0 0 0 0 0 0
0 0 0 0 0 0 0 0 0 0 0 0 0 0 0 0
0 0 0 0 0 0 0 0 0 0 0 0 0 0 0 0
0 0 0 0 0 0 0 0 0 0 0 0 0 0 0 0
0 0 0 0 0 0 0 0 0 0 0 0 0 0 0 0
```

class 2

```
8 4 3 3 2 2 2 2 1 1 1 0 0 0 0 0
4 4 3 3 2 2 2 1 1 1 0 0 0 0 0 0
4 3 3 3 2 2 2 1 1 1 0 0 0 0 0 0
3 3 3 2 2 2 2 1 1 1 0 0 0 0 0 0
3 3 3 2 2 2 2 1 1 1 0 0 0 0 0 0
3 2 2 2 2 2 1 1 1 0 0 0 0 0 0 0
2 2 2 2 2 1 1 1 0 0 0 0 0 0 0 0
2 2 2 1 1 1 1 1 0 0 0 0 0 0 0 0
1 1 1 1 1 1 1 0 0 0 0 0 0 0 0 0
1 1 1 1 1 1 1 1 0 0 0 0 0 0 0 0
1 1 1 1 1 1 0 0 0 0 0 0 0 0 0 0
0 0 0 0 0 0 0 0 0 0 0 0 0 0 0 0
0 0 0 0 0 0 0 0 0 0 0 0 0 0 0 0
0 0 0 0 0 0 0 0 0 0 0 0 0 0 0 0
0 0 0 0 0 0 0 0 0 0 0 0 0 0 0 0
0 0 0 0 0 0 0 0 0 0 0 0 0 0 0 0
```

class 4

```
8 2 1 0 0 0 0 0 0 0 0 0 0 0 0 0
2 1 0 0 0 0 0 0 0 0 0 0 0 0 0 0
1 0 0 0 0 0 0 0 0 0 0 0 0 0 0 0
0 0 0 0 0 0 0 0 0 0 0 0 0 0 0 0
0 0 0 0 0 0 0 0 0 0 0 0 0 0 0 0
0 0 0 0 0 0 0 0 0 0 0 0 0 0 0 0
0 0 0 0 0 0 0 0 0 0 0 0 0 0 0 0
0 0 0 0 0 0 0 0 0 0 0 0 0 0 0 0
0 0 0 0 0 0 0 0 0 0 0 0 0 0 0 0
0 0 0 0 0 0 0 0 0 0 0 0 0 0 0 0
0 0 0 0 0 0 0 0 0 0 0 0 0 0 0 0
0 0 0 0 0 0 0 0 0 0 0 0 0 0 0 0
0 0 0 0 0 0 0 0 0 0 0 0 0 0 0 0
0 0 0 0 0 0 0 0 0 0 0 0 0 0 0 0
0 0 0 0 0 0 0 0 0 0 0 0 0 0 0 0
0 0 0 0 0 0 0 0 0 0 0 0 0 0 0 0
```

Fig. 12. Typical bit assignment matrices for several different classes.

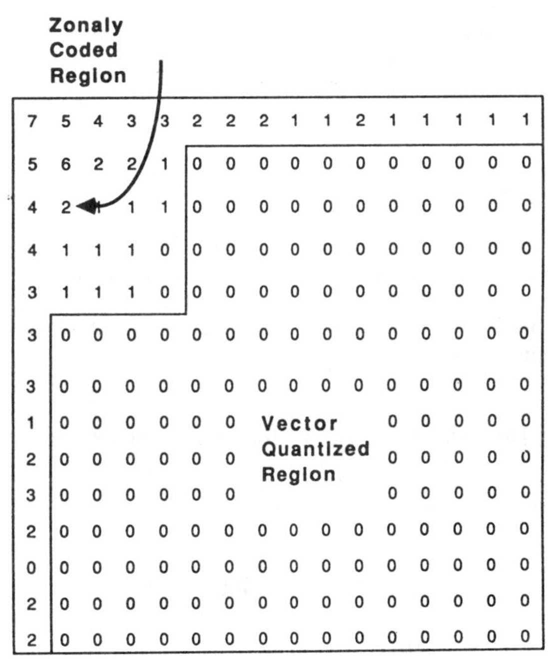

Fig. 13. An adaptive bit assignment matrix.

B. One-Dimensional Transform Vector Quantizer

In one-dimensional transform coders, the image lines are transformed by a one-dimensional unitary transform and a bit

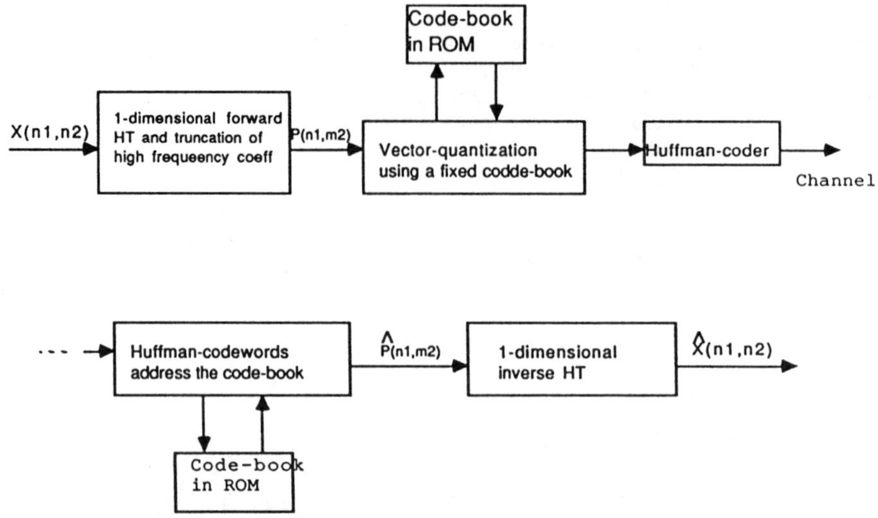

Fig. 14. One-dimensional transform VQ system [81].

assignment matrix is then used to code the normalized coefficients in each transformed line. The quantizer is a scalar where the correlation between the coefficients in each line or previous lines are not incorporated into the quantizer. The correlation within the transformed coefficients in each image line is not very significant since the linear transform is believed to decorrelate the samples within each line. However, the interline correlation is very significant and this can be exploited by a vector quantization scheme. One such technique was proposed by King and Nasrabadi [81], as shown in Fig. 14. A one-dimensional unitary transform is taken along each image line of $N_1 \times N_2$ image data $x(n_1, n_2)$ yielding a sequence of transform coefficients.

$$P(n_1, m_2) = \sum_{n_2=0}^{N_2-1} x(n_1, n_2) A_F(n_2, m_2) \tag{10}$$

for $m_2 = 0, 1, \cdots, N_2 - 1$ where $A_F(n_2, m_2)$ is the one-dimensional forward transform kernel and m_2 is the transform-domain row coordinate. Each transformed coefficient $P(n_1, m_2)$ is then normalized by its respective standard deviation $\sigma(m_2)$ given by

$$[\sigma(m_2)]^2 = \frac{1}{N_1} \sum_{n_1=0}^{N_1-1} [P(n_1, m_2)]^2 - [\mu(m_2)]^2 \tag{11}$$

where the mean $\mu(m_2)$ is given by

$$\mu(m_2) = \frac{1}{N_1} \sum_{n_1=0}^{N_1-1} P(n_1, m_2) \tag{12}$$

for $m_2 = 0, 1, \cdots, N_2 - 1$.

Since most of the energy is contained in the low-frequency coefficients a number of high-frequency coefficients are discarded by a zonal sampling selection process [71] and [72]. The neighboring rows of the transformed image are highly correlated, and this interrow correlation can be exploited by constructed a set of K-dimensional vectors where each vector of the K vector is subsequently compared to an optimized codebook of standard vector templates and is represented by its nearest matching vector template, with the mean square error (MSE) used as a matching criterion. A binary codeword (Huffman codeword) is assigned to each permissible vector template which is then transmitted. At the receiver, these codewords will address a similar codebook to that of the transmitter to reconstruct the normalized transformed samples. In practice, this reconstruction can be done very rapidly by using the codewords to address a read only memory (ROM) in which the standard vector templates are stored as shown in Fig. 15. The reconstructed samples are then multiplied by the corresponding variances which are either transmitted or estimated at the receiver [73]. A number of the high-frequency coefficients are discarded at the transmitter and each line is padded with zeros at the receiver. Then an inverse unitary transform is performed on each line given by

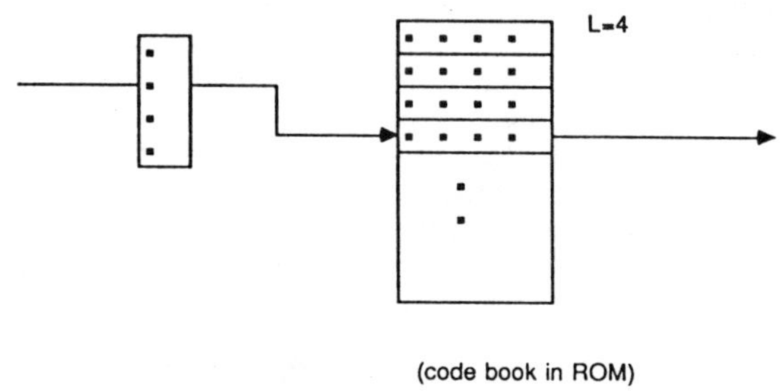

Fig. 15. A codeword addressing a read only ROM.

$$x(n_1, n_2) = \sum_{m_2=0}^{N_2-1} P(n_1, m_2) B_I(n_2, m_2) \tag{13}$$

where $B_I(n_2, m_2)$ is the one-dimensional inverse transform kernal. The bit rate R achieved is given by

$$R = C \frac{\log_2 M}{K} \tag{14}$$

where M is the size of the codebook of vectors of dimension, K and C are the ratio of the number of samples retained over the total number of samples in each line (block).

In the above system, some of the high-frequency coefficients are discarded. However, it is possible to design two codebooks, one to code the low-frequency coefficients and one for the high-frequency coefficients. In this system, the low-frequency coefficients with variances above a given threshold are coded by the first codebook and the high-frequency coefficients by the second codebook. The overhead information representing the codebook is also transmitted.

The codebooks are designed by using the LBG algorithm in

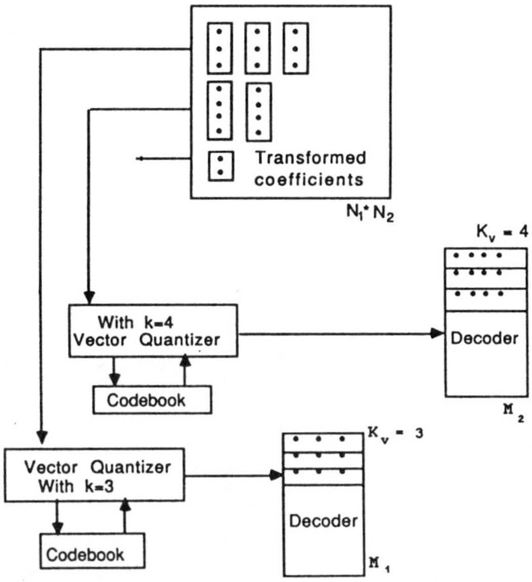

Fig. 16. Adaptive transform vector quantizer where M represents the number of vector patterns in the codebook and K the vector dimension.

the transform domain for vectors of dimension K. A decision on the value K is highly dependent on the spatial resolution and the correlation factor of the images to be coded. One technique would be to find a value of K that gives the distortion error between the original image and the coded image by a fixed optimized codebook of dimension K. However, since the activities within an image could be significantly different from one another, K cannot remain fixed.

An adaptive system was proposed by Nasrabadi *et al.* [83] where several codebooks of different vector lengths K were employed in the vector quantization scheme as shown in Fig. 16. The adaptivity was accomplished by varying the vector mapping window ($K_v \times 1$) in accordance with the statistical variation in the vertical direction of the row-transformed picture. The statistical variation is described by the Euclidean distance between transformed rows. Thus, the adaptive mapping is such that if there is sufficient intercorrelation between $K_v = b$ neighboring rows, then the rows are encoded with a codebook of vector lengths $K_v = b$. Previously developed codebooks for different template lengths are switched to and addressed in accordance with precomputed thresholds of row interseparation. The system is now a variable bit rate coder. The average bit rate is given by

$$R = A_1 C_1 \frac{\log_2 M}{K_1} + A_2 C_2 \frac{\log_2 M_2}{K_2} + \cdots + A_i C_i \frac{\log_2 M_i}{K_i} \tag{15}$$

where A_i is the number of times the codebook i has been used, M_i is the size of the codebook, and C_i is the percentage of the retained coefficients. Some extra bits are also required to identify each codebook as well as the used adaptivities, such as the zonal discarding of the high-frequency coefficients.

C. Two-Dimensional Transform Vector Quantization

As a consequence of the computational complexity involved in two-dimensional transform coders, an image array $x(n_1, n_2)$ is usually divided into small blocks of $f_i(n_1, n_2)$ for $i = 1, 2, \cdots, Q$ where Q is the number of blocks. Each block is then coded as a unit independent of all other blocks; unfortunately, this reduces the efficiency of the coder since the elements of the various blocks remain correlated in the transform domain. In order to exploit this interblock correlation, Habibi [64] introduced a two-dimensional hybrid coder where the interblock correlation is used by employing a DPCM coder on the corresponding coefficients of the transformed blocks.

A new two-dimensional hybrid coder was introduced by Nasrabadi and King [81] where a vector quantization scheme was used instead of DPCM. In this coding system, a two-dimensional transform is performed on each subblock of the image as given by

$$F_i(m_1, m_2) = \sum_{n_1=0}^{N_1-1} \sum_{n_2=0}^{N_2-1} f_i(n_1, n_2) A_F(n_1, n_2) \tag{16}$$

for $m_1 = 0, 1, \cdots, N_1 - 1$, $m_2 = 0, 1, \cdots, N_2 - 1$ and $i = 1, 2, \cdots, Q$ where $A_F(n_1, n_2)$ is the forward transform kernel. Since most of the energy is compacted within the low-frequency coefficients, a large number of the high-frequency coefficients are discarded as in the one-dimensional transform coders. However, since the elements of various neighboring blocks are still correlated, this interblock correlation can be exploited by clustering the corresponding retained and normalized transformed samples at m_1 and m_2 of each K neighboring blocks into vectors of dimension K. The vector quantization scheme illustrated in the Section (V-B) is then applied on the vectors, and the corresponding codewords transmitted. At the receiver, these codewords will address the codebook stored in the ROM to reconstruct the transformed blocks by substituting the corresponding vector templates in place of the original vectors. A two-dimensional inverse transform is then applied on each block as given by

$$f_i(n_1, n_2) = \sum_{m_1=0}^{N_1-1} \sum_{m_2=0}^{N_2-1} F_i(m_1, m_2) B_I(m_1, m_2) \tag{17}$$

for $n_1 = 0, 1, \cdots, N_1 - 1$, $n_2 = 0, 1, \cdots, N_2 - 1$ where $B_I(m_1, m_2)$ is the inverse transform kernel. The vector quantizer employed in this system can be made adaptive as was illustrated for the one-dimensional transform vector quantizer in Section (V-B).

D. Interframe Transform Vector Quantization (ITVQ)

In transform coders, as discussed in Section V, a transformation is performed on a subblock of an image and then coded using a bit assignment matrix. In three-dimensional coders a transformation is applied in each spatial direction as well as in the temporal direction. The resulting coefficients are usually normalized and coded by a three-dimensional bit assignment matrix. However, it is possible to vector quantize these transform coefficients to reduce the bit rate even further. The computational complexity of such a three-dimensional transform coder is very large. Nasrabadi and King [84] introduced a coding system in which a transformation was performed in each spatial direction and a vector quantizer was used along the temporal direction. In the coding system shown in Fig. 17, a two-dimensional unitary transform is performed on each $j \times k$ pixel subblock of L frames. The two-dimensional transform is given by

$$F_i(u, v, l) = \sum_{j=0}^{J-1} \sum_{k=0}^{K-1} f_i(j, k, l) A_F(j, k; u, v) \tag{18}$$

for $u = 0, 1, \cdots, J - 1$, $v = 0, 1, \cdots, K - 1$, $l = 0, 1, \cdots, L$, $i = 1, 2, \cdots, Q^2$ where $f_i(j, k, l)$ denotes the i three-dimensional block amplitude values for a digital image sequence of L frames; $A_F(j, k; u, v)$ represents the forward two-dimensional transform kernel and Q the number of subblocks in each direction. The transformed coefficients are

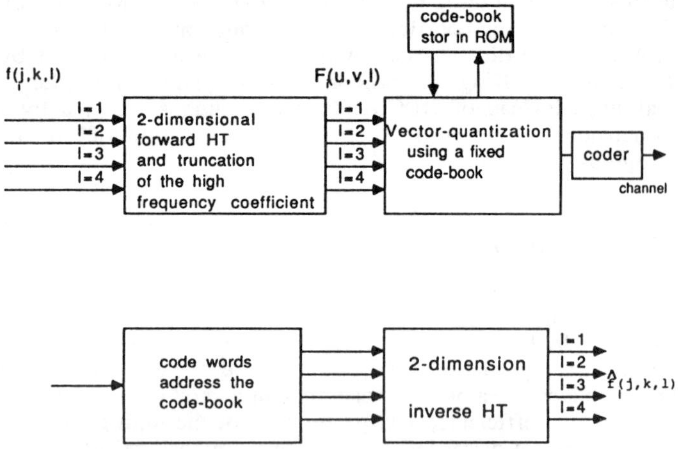

Fig. 17. A block diagram of an interframe transform vector quantizer [81].

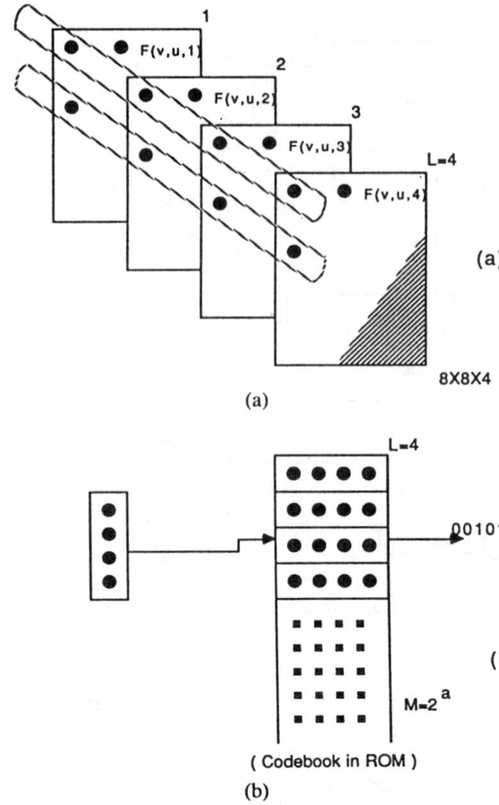

Fig. 18. (a) Shows the clustering of four frames where shaded area represents the high-frequency coefficients that are discarded. (b) Lookup table at the receiver.

then normalized by their corresponding mean and standard deviations given by the following expressions, respectively:

$$\mu(u, v)=\frac{1}{Q^2}\sum_{m=0}^{Q-1}\sum_{n=0}^{Q-1} F_{m,n}(u, v) \tag{19}$$

and

$$\sigma^2(u, v)=\frac{1}{Q^2}\sum_{m=0}^{Q-1}\sum_{n=0}^{Q-1} \{[F_{m,n}(u, v)]^2-[\mu(u, v)]^2\} \tag{20}$$

for $u = 0, 1, \cdots, J - 1$, $v = 0, 1, \cdots, K - 1$.

A number of the high-frequency coefficients are zonally discarded. Each spatial frequency (u, v) of L temporally adjacent blocks are clustered into vectors of dimension L. Each vector is then compared to a codebook of standard vector templates and is represented by its nearest (in a mean square error sense) matching vector template as shown in Fig. 18. A binary codeword is then assigned to each permissible vector template and transmitted. At the receiver the transmitted codewords will address the same lookup table codebook where L temporarily adjacent, transformed blocks are reconstructed. The coefficients are rescaled by their corresponding standard deviation and a two-dimensional inverse transform is applied on each block of the frames, as given by expression

$$f_i(j, k, l)=\sum_{u=0}^{J-1}\sum_{v=0}^{K-1} F_i(u, v, l)B_l(u, v; j, k) \tag{21}$$

for $j = 0, 1, \cdots, J - 1$, $k = 0, 1, \cdots, K - 1$, $l = 0, 1, \cdots, L$, $i=1, 2, \cdots, Q^2$ where $B_l(u, v, j, k)$ is the two-dimensional inverse transform kernel. The bit rate R is obtained from the relation

$$R=C\frac{\log_2 M}{L} \tag{22}$$

where M is the number of patterns (vector templates) in the codebook, L the dimension of the vector (or the number of frames clustered), and C is the number of the retained coefficients divided by the total number of samples in the block.

So far the dimension of the vector templates in the codebook has been kept fixed. However, the number of frames that can be clustered depends on the temporal variation within the image sequence. Thus, L should be made variable, changing to exploit the temporal correlation between the frames. Thus, the vector quantizer has to be made adaptive with respect to the movement of the objects within the image sequence, such that in a sequence of frames with little movement a large number of frames are clustered; this is a consequence of the high temporal correlation of frames. However, with a high density of moving objects within the image sequence, a smaller number of frames have to be clustered since the scene changes considerably from one frame to another. Thus, to adapt to the motion of the moving objects within the image sequence several codebooks of different dimensions are employed. The image sequence is then classified, and coded by switching to the appropriate codebook as shown in Fig. 19.

Fig. 20 shows a coding system with only two codebooks of dimension L and $L/2$ which has been simulated on the computer. After performing the two-dimensional transformation, the adaptive quantization scheme first classifies each three-dimensional block as belonging to a stationary area or an area of rapid motion. The classification is based upon a temporal activity index which is a measure given by the difference between the sum of the square of the ac coefficients of each subblock sequence. The three-dimensional block, after being classified, is coded by the appropriate vector quantizer by specifying a suitable threshold indexed for the two classes. The overhead information (identifying the codebook) is minimized by coding each block as a packet where each packet will have a header which specifies the vector quantizer that has been used, any adaptivities, and including any error correcting redundancies.

The above algorithm has been performed on a typical image sequence of frame size (136, 200) quantized to 8 bits/pixel and

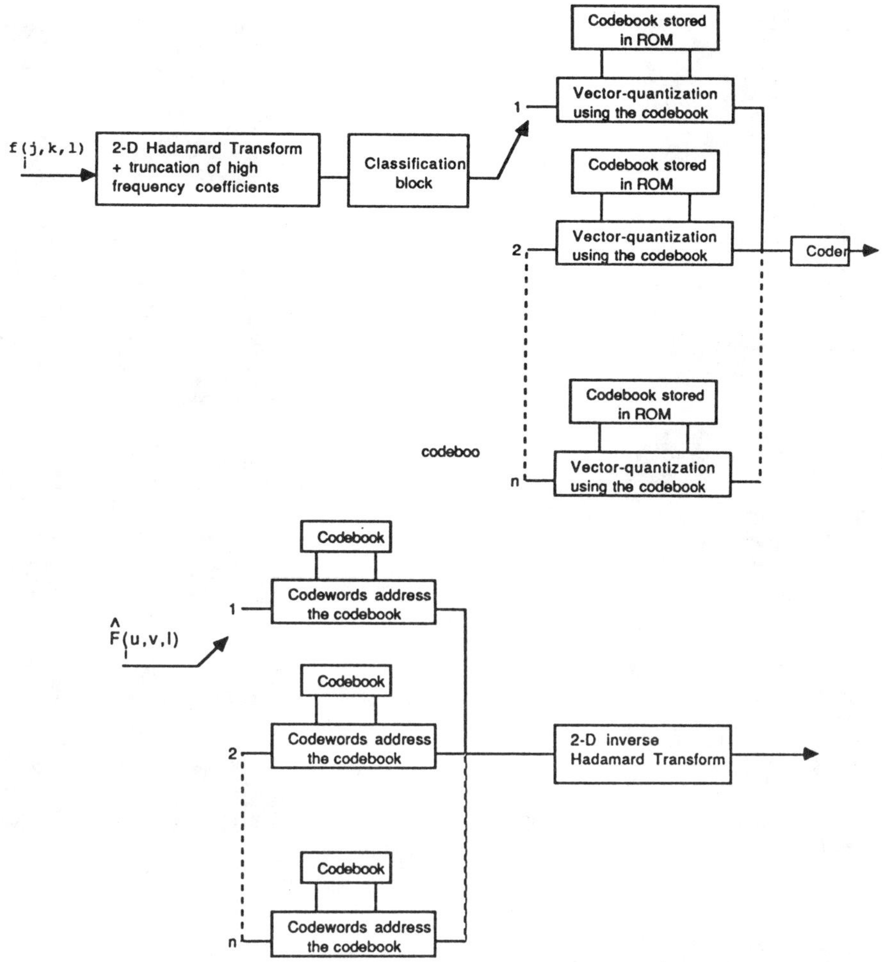

Fig. 19. Adaptive interframe transform vector quantizer system [84].

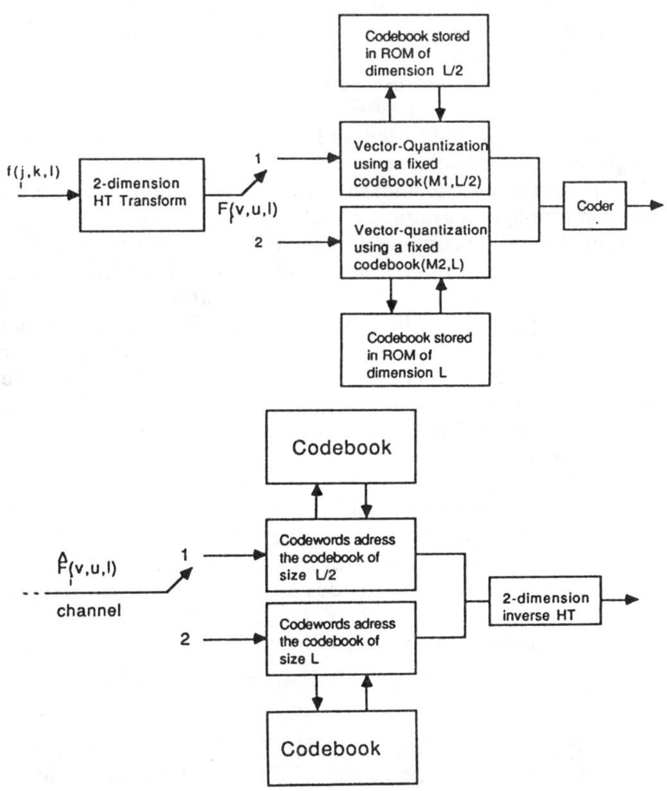

Fig. 20. The simulated adaptive interframe transform VQ coder with only two codebooks [84].

scaled to 5 bits for display as shown in Fig. 21. The codebook was designed using the LBG algorithm. The initial codebook was taken from the training sequence by retaining the most probable patterns with a given distance between them. Processes images are shown in Figs. 22 and 23.

The system shown in Fig. 20 is only adaptive with respect to variation of correlation in the temporal direction. However, other adaptivities such as the use of adaptive zonal discarding of the high-frequency coefficients as well as using specially optimized codebooks to code the high-frequency coefficients are still to be incorporated to provide a fully adaptive system. The bit rates achieved for the processing were obtained by using only two codebooks. A significantly greater reduction is possible if several codebooks or multidimensional vectors are employed. Using a number of codebooks will result in optimized vector quantizers for each particular class as well as the reduction of complexity of searching procedure [11].

VI. Binary Vector Quantizer (BVQ)

A. Facsimile Coding Using BVQ

Digital images such as business letters and documents, weather maps, engineering drawings, geographical maps, and newspaper pages are nominally two-tone (black and white). Efficient coding for the digital transmission or storage of two-tone images has been accomplished by a number of techniques such as run length coding techniques and entropy coding of the original data [1], [2].

One such technique employs entropy coding of vectors obtained by dividing the original image into blocks of N pixels, transmitting short codewords for more frequent vectors and long ones for less frequent vectors. An efficient way to

Fig. 21. Original input image sequence of resolution (136, 200) with amplitude resolution of 8 bits [82].

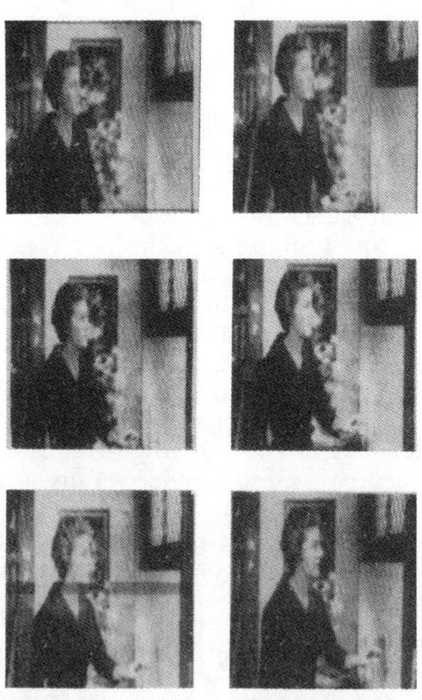

Fig. 22. Processed image sequence at bit rate of $R = 1.6$ bits/pixel/frame with codebooks of size $CB1 = (285, 3)$ and $CB2 = (111, 6)$ with NMSE $= 9.2 - 9.5 \times 10^{-3}$ [82].

construct such codes is to use Huffman codes which are based upon the joint probability of the pel patterns of the vectors. The bit rate per vector R will satisfy the inequality

$$H_N \leq R \leq H_N + 1 \tag{23}$$

and the bit rate per pixel r is given by

$$\frac{H_N}{N} \leq r \leq \frac{H_N}{N} + \frac{1}{N} \tag{24}$$

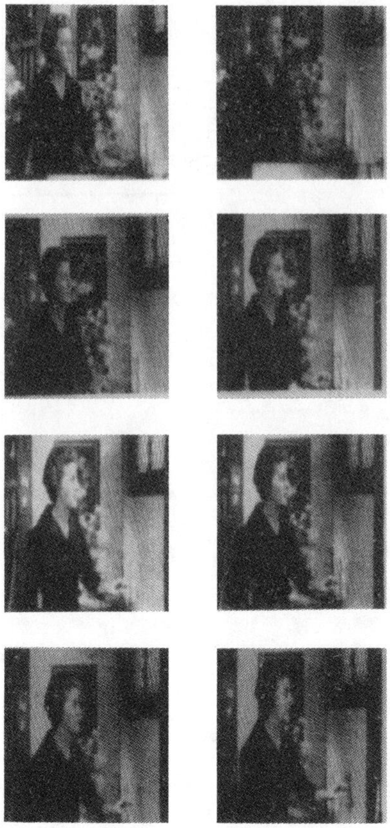

Fig. 23. Processed image sequence at bit rate of $R = 0.6$ bits/pixel/frame with codebooks of size $CB1 = (313, 4)$ and $CB2 = (203, 8)$ with NMSE $= 1.8 - 2.0 \times 10^{-3}$ [82].

where H_N is the entropy of the vectors of length N pixels. It is thus obvious that we can reduce r by making N larger. However, for large N the implementation of the encoder and decoder becomes complicated. A lower bit rate can obviously be achieved by using the conditional entropy [1], but the design and implementation of the Huffman codes will now be very complicated since they require the evaluation of the joint and conditional probabilities and a large lookup table for the storage of the codewords.

One possible way to increase the compression factor is to allow some image degradation in the coding process. One such approximate coding scheme was studied by Knudson [85]. For a given vector size, only a small subset of all possible pixel patterns of the block is allowed. Any pattern which is not allowed is replaced by the pattern in the allowed subset that matches it most closely. This is known as a binary vector quantizer and the codebook could be designed by using the LBG algorithm on a set of training sequences, but there is a high possibility of getting trapped into a local minima. Knudson carried out some experiments using newspaper texts and graphics using blocks of 8×8. It was found that a subset of 62 patterns (out of a possible 2^{64}) was sufficient to obtain satisfactory image quality. It is also possible to use several binary vector quantizers to adapt to the information variations within each block. Huffman codes generated from the statistical distribution measures can be used to represent the permissible block patterns in each codebook.

B. Block Truncation Coding Using Binary Vector Quantization

Recently, a new technique called block truncation coding (BTC) has been developed [86], [87]. This technique uses a one-bit nonparametric quantizer, adaptive over local regions of the image. In general, the picture is divided into $N \times N$

blocks which are coded individually, each into a two-level signal. The levels, a and b, for each block are chosen such that the first two sample moments are preserved. These are given by

$$a = \bar{X} - \bar{\sigma}\sqrt{\left[\frac{q}{m-q}\right]} \quad (25)$$

$$b = \bar{X} - \bar{\sigma}\sqrt{\left[\frac{m-q}{q}\right]} \quad (26)$$

where $\bar{X}$ and $\bar{\sigma}$ are mean and standard deviations of the block, respectively, q is the number of samples greater than $\bar{X}$, and $m = N^2$ is the number of samples in the block. Each block is then described by the values of $\bar{X}$, $\bar{\sigma}$, and an $M \times N$ bit plane consisting of 1's or 0's indicating whether the pixels have values above or below $\bar{X}$. Assuming $\bar{X}$ and $\bar{\sigma}$ are quantized to 8 bits, a data rate of 2 bits/pixel is achieved for blocks of 4×4 pixels. In order to decrease this bit rate, an efficient way to code the bit planes must be devised. One approach is to use an entropy coder on the patterns discussed in Section VI-A) above where several small binary codebooks are employed to code each classified block. These suggestions were recently reported in [88].

VII. Subband Vector Quantizers (Subband VQ)

One feature of the human visual system is that the early stages of visual information processing carried out by the brain consist of analyzing the retinal information over multiple, independent channels each of which responds selectively to a narrow-band of spatial frequencies [89].

Subband coding of speech signals was first proposed by Crochiere *et al.* [90] for medium-to-low rate speech coding systems. In conventional subband coding systems a bank of quadrature mirror filters [91] is employed to produce several replicas of the original signal at different frequency bands; a varying bit assignment strategy for each of these bands is used. Traditionally, fewer numbers of bits are assigned to high-frequency bands than are assigned to low-frequency bands. Recently, vector quantization has been applied to subband speech signals by constructing vectors of dimension K where one sample from successive K subbands was taken [92]–[93]. Recently, Woods [106] designed a subband image coding system. In his system, each band was DPCM coded. In [107], vector quantization was applied to Woods' subband coder in order to exploit the correlation or the redundancies between the spectral bands.

The above techniques can obviously be extended to natural multiband images such as multispectral images [94]–[97] or components of color TV images [98]–[105]. Boucher and Goldberg have recently studied the application of vector quantization to color TV images [98]. They extended their codebook replenishment technique discussed in Section III to color images. The three color signals R, G, and B are first transformed to the *NTSC YIQ* coordinate system by using the standard transformation. The reason for this mapping is that the transformation tends to decorrelate the information and to compact most of the energy in the luminance component Y. Both chrominance components I and Q are severely band limited and there is little correlation between the *NTSC* components Y, I, and Q. In their color coding system, each spectral band, the luminance component Y, and the chrominance components I and Q which were subsampled by a factor of 2 were vector quantized by the codebook replenishment technique.

Yamaguchi [101] designed a coding system where highly correlated R, G, and B, components of color images were vector quantized with bit rates of about 2–3 bits/pixel for each color component. The R, G, and B predictor errors denoted by e_R, e_G, e_B, were then calculated. For the G component signals (intracomponent prediction) the prediction error is $e_G = G - \hat{G}$. For the R and B components the predictor is an intercomponent prediction, for example, $e_R = R - \hat{R}(G')$ where the prediction signal is based on the R component and the decoded G' component. The prediction errors e_R, e_G, e_B are then vector quantized because there exist some correlation between the error components.

Murakami *et al.* extended his algorithm [103] to code color images. In their technique, a multistage VQ was applied to Y, I, Q components of the color images. A combined component VQ was also designed where a multidimensional vector was used. Budge and Baker used a mean/residual vector quantizer to encode each color component [102] individually where the I and Q components were subsampled by a factor of 4.

VIII. Conclusion

We have presented a survey of vector quantization techniques for image coding. The basic concept of vector quantization was given in Section II. Spatial vector quantizers were discussed in Section III. As digital images are made of nonstationary signals, a variety of SVQ's have been developed such as subtracting the mean and the variance of each vector before using the VQ, or using several codebooks to match the statistics of the image blocks. One major problem with VQ's is that they do not reconstruct the edge vectors efficiently, and this is because the codebook cannot reproduce all possible patterns. A simple way to solve this is to use a codebook replenishment VQ (dynamic VQ). The effect that channel noise has on a spatial VQ coding system has not yet been investigated. Obviously, in a simply VQ system if an error occurs during transmission of a codeword, the vector represented by that codeword is corrupted, therefore, the whole block of 4×4 is not correctly reconstructed. A comparison of spatial VQ systems is a difficult task because computational complexity, codebook storage, and overhead information must be considered for each system, as well as the fact that more computational power and large memory storage are becoming more feasible. Since each spatial VQ has its own merit perhaps a combination of them would be the best idea. One major advantage of spatial VQ systems over, for example, transform VQ systems is that the decoder is very simple as it is just a lookup table.

The combination of VQ with conventional coders will result in a more efficient coder than when a conventional coder is used with a scalar quantizer. For example, in an interframe VQ the error signal obtained from two consecutive frmaes are correlated and dependent, but the VQ will exploit this redundancy. Predictive VQ's continue to be developed because designing the codebook and the effect that the channel noise will have on the feedback VQ systems are still major problems.

Several transform VQ systems were also discussed in this paper. The TVQ is computationally complex because of the one- or two-dimensional transformation that has to be performed at the encoder and decoder. TVQ's have several advantages over SVQ's and PVQ's. For example, some of the high-frequency coefficients can be discarded which will reduce the vector size or the bit rate depending on the TVQ system reported in Section V. It can also exploit the intra- and interframe correlation between the vectors to reduce the bit rate. A TVQ will perform better than the conventional hybrid transform/DPCM coders because DPCM coders perform significantly less well in the presence of noise, and the correlation and the dependency between error signal is not fully exploited by scalar quantization. The effect of the channel noise on TVQ will not be so severe because the error is distributed over the whole transform region (block size).

The TVQ should always perform better than conventional transform coders because scalar quantizers are forced to allocate integral numbers of bits to each transform coefficient. However, VQ permits a more distributed allocation.

ACKNOWLEDGMENT

The authors would like to thank Mr. Y. Feng for drawing the block diagrams.

REFERENCES

[1] T. S. Huang, "Coding of two-tone images," *IEEE Trans. Commun.*, vol. COM-25, pp. 1406–1428, Nov. 1977.
[2] A. N. Netravali and F. W. Mounts, "Ordering techniques for facsimile coding: A review," *Proc. IEEE*, vol. 68, pp. 796–807, July 1980.
[3] A. N. Netravali and J. O. Limb, "Picture Coding: A review," *Proc. IEEE*, vol. 68, pp. 366–406, Mar. 1980.
[4] A. K. Jain, "Image data compression: A review," *Proc. IEEE*, vol. 69, pp. 349–389, Mar. 1981.
[5] A. Habibi, "Survey of adaptive image coding techniques," *IEEE Trans. Commun.*, vol. COM-25, pp. 1275–1284, Nov. 1977.
[6] M. Kunt, A. Ikonomopoulos, and M. Kocher, "Second-generation image-coding techniques," *Proc. IEEE*, vol. 73, pp. 549–574, Apr. 1985.
[7] A. Buzo, A. H. Gray, R. M. Gray, and J. D. Markel, "Speech coding based upon vector quantization," *IEEE Trans. Acoust. Speech, Signal Processing*, vol. ASSP-28, pp. 562–574, Oct. 1980.
[8] A. Gersho and V. Cuperman, "Vector quantization: A pattern-matching technique for speech coding," *IEEE Commun. Mag.*, pp. 15–21, Dec. 1983.
[9] Y. Linde, A. Buzo, and R. M. Gray, "An algorithm for vector quantizer design," *IEEE Trans. Commun.*, vol. COM-28, pp. 84–95, Jan. 1980.
[10] H. Abut, R. M. Gray, and G. Rebolledo, "Vector quantization of speech and speech-like waveforms," *IEEE Trans. Acoust., Speech, Signal Processing.*, vol. ASSP-30, pp. 423–435, June 1982.
[11] R. M. Gray, "Vector quantization," *IEEE ASSP Mag.*, pp. 4–29, Apr. 1984.
[12] W. K. Pratt, *Image Transmission Techniques*. New York: Academic, 1979.
[13] A. Gersho, "On the structure of vector quantizers," *IEEE Trans. Inform. Theory*, vol. IT-28, pp. 157–166, Mar. 1982.
[14] J. P. Adoul, C. Collin, and D. Dalle, "Block encoding and its application to data compression of PCM speech," in *Proc. Canadian Commun. EHV Conf.*, Montreal, P.Q., Canada, 1978, pp. 145–148.
[15] D. T. S. Chen, "On two- or more dimensional optimum quantizers," in *Proc. IEEE Int. Conf. Acoust., Speech, Signal Processing*, 1977, pp. 640–643.
[16] A. Gersho, "Asymptotically optimal block quantization," *IEEE Trans. Inform. Theory*, vol. IT-25, pp. 373–380, July 1979.
[17] S. P. Lloyd, "Least-squares quantization in PCM," *IEEE Trans. Inform. Theory*, vol. IT-28, pp. 129–137, Mar. 1982.
[18] R. M. Gray, A. Buzo, A. H. Gray, and Y. Matsuyama, "Distortion measures for speech processing," *IEEE Trans. Acoust., Speech, Signal Processing*, vol. ASSP-28, pp. 367–376, Aug. 1980.
[19] R. M. Gray and H. Abut, "Full search and tree searched vector quantization of speech waveforms," in *Proc. IEEE Int. Conf. Acoust., Speech, Signal Processing*, May 1982, pp. 593–596.
[20] R. M. Gray and Y. Linde, "Vector quantization and predictive quantizers for Gauss Markov source," *IEEE Trans. Commun.*, vol. COM-30, pp. 381–389, Feb. 1982.
[21] B. H. Juang, Jr., and A. H. Gray, "Multiple stage vector quantization for speech coding," in *Proc. IEEE Int. Conf. Acoust., Speech, Signal Processing*, Apr. 1982, pp. 597–600.
[22] M. J. Sabin and R. M. Gray, "Product code vector quantizers for waveform and voice coding," *IEEE Trans. Acoust., Speech, Signal Processing*, vol. ASSP-32, pp. 474–488, June 1984.
[23] R. L. Baker and R. M. Gray, "Differential vector quantization of achromatic imagery," presented at Proc. Int. Picture Coding Symp., Mar. 1983.
[24] S. E. Budge and R. L. Baker, "Compression of color digital images using vector quantization in product codes," in *Proc. IEEE Int. Conf. Acoust., Speech, Signal Procesing*, Apr. 1985, pp. 129–132.
[25] R. L. Baker and R. M. Gray, "Image compression using non-adaptive spatial vector quantization," in *Proc. Conf. Rec. Sixteenth Asilomar Conf. Circuits, Syst., Comput.*, Oct. 1982, pp. 55–61.
[26] R. L. Baker, "Vector quantization of digital images," Ph.D. dissertation, Stanford Univ., Stanford, CA, June 1984.
[27] Y. Yamada, K. Fujita, and S. Tazaki, "Vector quantization of video signals," in *Proc. Ann. Conf. IECE*, 1980, p. 1031.
[28] Y. Yamada and S. Tazaki, "Vector quantizer design for video signals," *IECE Trans. J66-B*, pp. 965–972, 1983 (in Japanese).
[29] ——, "A method for constructing successive approximation vector quantizers for video signals," in *Proc. Ann. Conf. Inst. Television Eng. Japan*, 1983, pp. 6-2.
[30] T. Murakami, K. Asai, and E. Yamazaki, "Vector quantizer of video signals," *Electron. Lett.*, vol. 7, pp. 1005–1006, Nov. 1982.
[31] T. Saito, H. Takeo, H. Harashima, and H. Miyakawa, "A scheme of diminishing performance deterioration due to mismatch in vector quantization of picture signals," presented at Proc. Int. Picture Coding Symp., July 1984.
[32] H. Van Helden, F. Booman, and J. Biemond, "A generalized tree structured algorithm for vector quantization of images," presented at Proc. Int. Picture Coding Symp., July 1984.
[33] A. Gersho and B. Ramamurthi, "Image coding using vector quantization," in *Proc. IEEE Int. Conf. Acoust., Speech, Signal Processing*, Apr. 1982, pp. 428–431.
[34] B. Ramamurthi and A. Gersho, "Image coding using segmented codebooks," presented at Proc. Int. Picture Coding Symp., Mar. 1983.
[35] B. Ramamurthi, A. Gersho, and A. Sekey, "Low-rate image coding using vector quantization," in *Proc. IEEE Global Commum. Conf. Rec.*, Nov. 1983, pp. 184–187.
[36] B. Ramamurthi and A. Gersho, "Image vector quantization with a perceptually-based cell classifier," presented at IEEE Proc. Int. Conf. Acoust., Speech, Signal Processing, Mar. 1984.
[37] ——, "Classified vector quantization of images," *IEEE Trans. Commun.*, vol. COM-34, pp. 1105–1115, Nov. 1986.
[38] B. Ramamurthi, "Vector quantizer of images based on a composite source model," Ph.D. dissertation, Univ. California, Santa Barbara, CA, Apr. 1985.
[39] H. F. Sun and M. Goldberg, "Adaptive vector quantization for image sequence coding," in *Proc. IEEE Int. Conf. Acoust., Speech, Signal Processing*, Mar. 1985, pp. 339–342.
[40] M. Goldberg and H. F. Sun, "Image sequence coding using vector quantization," *IEEE Trans. Commun.*, vol. COM-34, pp. 703–710, July 1986.
[41] A. Gersho and M. Yano, "Adaptive vector quantization by progressive code-vector replacement," in *Proc. IEEE Int. Conf. Acoust., Speech, Signal Processing*, 1985, pp. 133–136.
[42] N. M. Nasrabadi, "Use of vector quantizers in image coding," in *Proc. IEEE Int. Conf. Acoust., Speech, Signal Processing*, Mar. 1985, pp. 125–128.
[43] H. Samet, "The quad-tree and related hierarchical data structures," *ACM Comput. Surveys*, vol. 16, no. 2, pp. 188–260, June 1984.
[44] A. Gersho and Y. Shoham, "Hierarchical vector quantization of speech with dynamic codebook allocation," in *Proc. IEEE Int. Conf. Acoust., Speech, Signal Processing*, 1984, pp. 10.9.1–10.9.4.
[45] D. J. Vaisey and A. Gersho, "Variable block-size image coding," in *Proc. IEEE Int. Conf. Acoust., Speech, Signal Processing*, Apr. 1987, pp. 1051–1054.
[46] A. Haoui and D. Messerschmitt, "Predictive vector quantization," in *Proc. IEEE Int. Conf. Acoust., Speech, Signal Processing*, Mar. 1984, pp. 10.10.1–10.10.4.
[47] M. Dunham and R. Gray, "An algorithm for the design of labelled-transition finite-state vector quantizers," *IEEE Trans. Commun.*, vol. COM-33, pp. 83–89, Jan. 1985.
[48] J. Foster, R. M. Gray, and M. Dunham, "Finite-state vector quantization of waveform coding," *IEEE Trans. Inform. Theory*, vol. IT-31, pp. 348–355, May 1985.
[49] A. Aravind and A. Gersho, "Low-rate image coding with finite-state vector quantization," in *Proc. IEEE Int. Conf. Acoust., Speech, Signal Processing*, Mar. 1986, pp. 137–140.
[50] R. L. Baker and H. H. Shen, "A finite-state vector quantizer for low-rate image sequence coding," in *Proc. IEEE Int. Conf. Acoust., Speech, Signal Processing*, Apr. 1987, pp. 760–763.
[51] V. Cuperman and A. Gersho, "Adaptive differential vector coding of speech," in *Proc. Conf. Rec., GLOBECOM '82*, Dec. 1982, pp. 1092–1096.

[52] H. M. Hang, "Predictive coding of images," Ph.D. dissertation, Elect., Comput., Syst. Dep., Rensselaer Poly. Inst., Troy, NY, Aug. 1984.

[53] H. M. Hang and J. W. Woods, "Predictive vector quantization of images," *IEEE Trans. Commun.*, vol. COM-33, pp. 1208–1219, Nov. 1985.

[54] J. O. Limb and R. F. W. Pease, "A simple interframe coder for video telephony," *Bell Syst. Tech. J.*, vol. 50, pp. 1877–1888, Aug. 1971.

[55] B. G. Haskell *et al.*, "Interframe coding of video-telephone pictures," *Proc. IEEE*, vol. 60, pp. 792–800, July 1972.

[56] B. G. Haskell *et al.*, "Interframe coding of 525-line, monochrome television at 1.5 M bit/s," *IEEE Trans. Commun.*, vol. COM-25, pp. 1339–1348, Nov. 1977.

[57] H. Yasuda *et al.*, "Transmitting 4-MHz TV signals by combinational difference coding," *IEEE Trans. Commun.*, vol. COM-25, pp. 508–516, May 1977.

[58] T. Murakami, K. Asai, A. Itoh, and E. Yamazaki, "Interframe vector coding of color video signals," presented at Proc. Int. Picture Coding Symp., July 1984, also 1983.

[59] C. Cafforio and F. Rocca, "Methods for measuring small displacements of television images," *IEEE Trans. Inform. Theory*, vol. IT-22, pp. 573–579, Sept. 1976.

[60] A. N. Netravali and J. A. Stuller, "Motion compensation transform coding," *Bell Syst. Tech. J.*, pp. 1703–1718, Sept. 1974.

[61] J. R. Jain and A. K. Jain, "Displacement measurement and its application in interframe image coding," *IEEE Trans. Commun.*, vol. COM-29, pp. 1799–1808, Dec. 1981.

[62] E. Dubois, B. Prasada, and S. Sabri, "Image sequence coding," in *Image Sequence Analysis*, T. S. Huang, Ed. New York: Springer-Verlag, 1981, ch. 3.

[63] A. N. Netravali and J. D. Robbins, "Motion-compensated television coding: Part I," *Bell Syst. Tech. J.*, vol. 58, pp. 631–670, Mar. 1979.

[64] A. Habibi, "Hybrid coding of pictorial data," *IEEE Trans. Commun.*, vol. COM-22, pp. 614–624, May 1974.

[65] M. J. Bage, "Interframe predictive coding of images using hybrid vector quantization," *IEEE Trans. Commun.*, vol. COM-34, pp. 411–415, Apr. 1986.

[66] J. A. Roese, W. K. Pratt, and G. S. Robinson, "Interframe cosine transform image coding," *IEEE Trans. Commun.*, vol. COM-25, pp. 1329–1338, Nov. 1977.

[67] H. C. Andrews, J. Kane, and W. K. Pratt, "Hadamard transform image coding," *Proc. IEEE*, vol. 57, pp. 58–68, Jan. 1969.

[68] P. A. Wintz, "Transform picture coding," *Proc. IEEE*, vol. 60, pp. 809–823, July 1972.

[69] W. K. Pratt, W. H. Chen, and L. R. Welch, "Slant transform image coding," *IEEE Trans. Commun.*, vol. COM-22, pp. 1075–1093, Aug. 1974.

[70] W. H. Chen and C. H. Smith, "Adaptive coding of monochrome and color images," *IEEE Trans. Commun.*, vol. COM-25, pp. 1285–1292, Nov. 1977.

[71] G. B. Anderson and T. S. Huang, "Piecewise Fourier transformation for picture bandwidth compression," *IEEE Trans. Commun.*, vol. COM-20, pp. 388–491, June 1972.

[72] M. Tasto and P. A. Wintz, "Image coding by adaptive block quantization," *IEEE Trans. Commun. Technol.*, vol. COM-19, pp. 956–972, Dec. 1971.

[73] K. N. Nagan, "Adaptive transform coding of video signals," *Proc. IEE*, vol. 129, Pt.F, Feb. 1982.

[74] N. M. Nasrabadi and R. A. King, "Computationally efficient adaptive block-transform coding," presented at Proc. EUSIPCO-83, 2nd European Conf. Sig., Sept. 1983.

[75] T. Saito, H. Takeo, K. Aizawa, H. Harashima, and H. Miyakawa, "Adaptive discrete cosine transform image coding using gain/shape vector quantization," in *Proc. IEEE Int. Conf. Acoust., Speech, Signal Processing*, Apr. 1986, pp. 129–132.

[76] J. P. Mareseq and C. Labit, "Vector quantization in transformed image coding," in *Proc. IEEE Int. Conf. Acoust., Speech, Signal Processing*, Apr. 1986, pp. 145–147.

[77] A. A. Abdelwahab and S. C. Kwatra, "Image data compression with vector quantization in the transform domain," in *Proc. IEEE Int. Conf. Commun.*, 1986, pp. 1285–1288.

[78] B. Hammer and M. Schielein, "Image coding by vector quantization of *M*-Hadamard transform coefficients," *SPIE*, vol. 594, Image Coding, pp. 72–78, 1985.

[79] M. Haqhiri and C. Renius, "An adaptive block-quantizer approach to transform coding of pictures," *SPIE*, vol. 594, Image Coding, pp. 66–71, 1985.

[80] K. S. Thyagarajan, H. Abut, and H. Bheda, "A matrix quantizer for image processing," in *Proc. IEEE Int. Conf. Acoust., Speech, Signal Processing*, Mar. 1984, pp. 48.2.1–48.2.3.

[81] N. M. Nasrabadi and R. A. King, "Image coding using vector quantization in the transform domain," *Pat. Recog. Lett.*, pp. 323–329, 1983.

[82] N. M. Nasrabadi, "Orthogonal transforms and their applications to image coding," Ph.D. dissertation, Imperial College, Univ. London, England, 1984.

[83] N. M. Nasrabadi, J. O. Ibikunle, and R. A. King, "Design of a new codec system for video conferencing," presented at Int. Conf. Digital Signal Proc., Florence, Italy, Sept. 1984.

[84] N. M. Nasrabadi and R. A. King, "A new image coding technique using transform vector quantization," presented at 1984 IEEE Int. Conf. Acoust., Speech, Signal Processing, Mar. 1984.

[85] D. R. Knudson, "Digital encoding of newspaper graphics," Electron. Syst. Lab., Mass. Inst. Technol., Rep. ESL-616, Aug. 1975.

[86] E. J. Delp and D. R. Mitchell, "Image compression using block truncation coding," *IEEE Trans. Commun.*, vol. COM-27, Sept. 1979.

[87] D. J. Healy and D. R. Mitchell, "Digital video bandwidth compression using block truncation coding," *IEEE Trans. Commun.*, vol. COM-29, pp. 1809–1817, Dec. 1981.

[88] V. R. Udpikar and J. P. Raina, "BTC image coding using vector quantization," *IEEE Trans. Commun.*, vol. COM-35, pp. 352–356, Mar. 1987.

[89] G. Westheimer and F. W. Campbell, "Light distribution in the image formed by the living human eye," *J. Opt. Soc. Amer.*, vol. 67, no. 2, pp. 207–212, Feb. 1977.

[90] R. E. Crochiere, S. A. Webber, and J. L. Flanagan, "Digital coding of speech in sub-bands," *Bell Syst. Tech. J.*, vol. 55, Oct. 1986.

[91] J. D. Johnston, "A filter family designed for use in quadrature mirror filter banks," presented at Proc. IEEE Int. Conf. Acoust., Speech, Signal Processing, 1980.

[92] H. Abut and S. A. Luse, "Vector quantization for sub-band coded waveforms," presented at Proc. IEEE Int. Conf. Acoust., Speech, Signal Processing, 1984.

[93] A. Gersho, T. Ramstad, and I. Versvik, "Fully vector-quantized sub-band coding with adaptive codebook allocation," presented at Proc. IEEE Int. Conf. Acoust., Speech, Signal Processing, 1980.

[94] E. E. Hilbert, "Cluster compression algorithm: A joint clustering/data compression concept," *Jet Prop. Lab.*, Pub. 77-43, Dec. 1977.

[95] A. Habibi and A. S. Samulon, "Bandwidth compression of multispectral data," *SPIE*, vol. 66, pp. 23–35, 1975.

[96] A. Habibi, "A review of image bandwidth compressed techniques," *SPIE*, vol. 528, Digital Image Processing, pp. 138–144, 1985.

[97] G. E. Lowitz, "Image data reduction," in *Proc. Int. Conf. Spacecraft On-Board Data Manage.*, Nice 24-27, Oct. 1978, pp. 291–294.

[98] P. Boucher, "Adaptive vector quantization of pictorial data," Masters thesis, Univ. Ottawa, Appl. Sci. Elec. Eng., 1984.

[99] P. Boucher and M. Goldberg, "Color image compression by adaptive vector quantization," presented at Proc. IEEE Int. Conf. Acoust., Speech, Signal Processing, 1984.

[100] M. Goldberg, P. R. Boucher, and S. Shlien, "Image compression using adaptive vector quantization," *IEEE Trans. Commun.*, vol. COM-34, Feb. 1986.

[101] H. Yamaguchi, "Efficient encoding of colored pictures in *R*, *G*, *B* components," *IEEE Trans. Commun.*, vol. COM-32, pp. 1201–1209, Nov. 1984.

[102] S. E. Budge and R. L. Baker, "Compression of color digital image using vector quantization in product codes," in *Proc. IEEE Int. Conf. Acoust., Speech, Signal Processing*, Mar. 1985, pp. 129–132.

[103] T. Murakami, K. Asai, and A. Itoh, "Vector quantization of color images," in *Proc. IEEE Int. Conf. Acoust., Speech, Signal Processing*, 1986, pp. 133–135.

[104] J. Barrilleaux, R. Hinkle, and S. Wells, "Efficient vector quantization for color image coding," in *Proc. IEEE Int. Conf. Acoust., Speech, Signal Processing*, 1987, pp. 740–743.

[105] C. L. Yeh, "Color image-sequence compression using adaptive binary-tree vector quantization with codebook replenishment," in *Proc. IEEE Int. Conf. Acoust., Speech, Signal Processing*, Apr. 1987, pp. 1059–1062.

[106] J. W. Woods and S. D. O'Neil, "Sub-band coding of images," *IEEE Trans. Acoust., Speech, Signal Procesing*, vol. ASSP-34, pp. 1278–1289, Oct. 1986.

[107] P. H. Westerink, J. Biemond and D. E. Boekee, "Sub-band coding of images using predictive vector quantizer," in *Proc. IEEE Int. Conf. Acoust., Speech, Signal Processing*, Apr. 1987, pp. 1378–1381.

[108] A. Aravind and A. Gersho, "Image compression based on vector quantization with finite memory," *Opt. Eng.*, vol. 26, no. 7, pp. 570–580, July 1987.

[109] T. R. Fischer, "A pyramid vector quantizer," *IEEE Trans. Inform. Theory*, vol. IT-32, pp. 568–583, July 1986.

[110] H. Tseng and T. R. Fischer, "Transform and hybrid transform/DPCM coding of images using pyramid vector quantization," *IEEE Trans. Commun.*, vol. COM-35, pp. 79–86, Jan. 1987.

[111] T. Kohonen, *Self-Organization and Associative Memory.* New York: Springer-Verlag, 1984.

[112] ——, "The neural phonetic typewriter," *IEEE Comput.*, vol. 21, pp. 11–21, Mar. 1988.

[113] N. Metropolis, A. Rosenbluth, M. Rosenbluth, A. Teller, and E. Teller, "Equation of state calculations by fast computing machines," *J. Chem. Phys.*, vol. 21, no. 6, pp. 1087–1092, June 1953.

[114] S. Kirkpartrick, C. D. Gelatt, Jr., and M. P. Vecchi, "Optimization by simulated annealing," *Science*, vol. 220, no. 4598, pp. 671–680, May 1983.

Part 2
Theoretical Studies and Algorithms

An Algorithm for Vector Quantizer Design

YOSEPH LINDE, MEMBER, IEEE, ANDRÉS BUZO, MEMBER, IEEE, AND ROBERT M. GRAY, SENIOR MEMBER, IEEE

***Abstract*—An efficient and intuitive algorithm is presented for the design of vector quantizers based either on a known probabilistic model or on a long training sequence of data. The basic properties of the algorithm are discussed and demonstrated by examples. Quite general distortion measures and long blocklengths are allowed, as exemplified by the design of parameter vector quantizers of ten-dimensional vectors arising in Linear Predictive Coded (LPC) speech compression with a complicated distortion measure arising in LPC analysis that does not depend only on the error vector.**

INTRODUCTION

An efficient and intuitive algorithm for the design of good block or vector quantizers with quite general distortion measures is developed for use on either known probabilistic source descriptions or on a long training sequence of data. The algorithm is based on an approach of Lloyd [1], is not a variational technique, and involves no differentiation; hence it works well even when the distribution has discrete components, as is the case when a sample distribution obtained from a training sequence is used. As with the common variational techniques, the algorithm produces a quantizer meeting necessary but not sufficient conditions for optimality. Usually, however, at least local optimality is assured in both approaches.

We here motivate and describe the algorithm and relate it to a number of similar algorithms for special cases that have appeared in both the quantization and cluster analysis literature. The basic operation of the algorithm is simple and intuitive in the general case considered here and it is clear that variational techniques are not required to develop nor to apply the algorithm.

Several of the algorithm's basic properties are developed using heuristic arguments and demonstrated by example. In a companion theoretical paper [2], these properties are given precise mathematical statements and are proved using arguments from optimization theory and ergodic theory. Those results will occasionally be quoted here to characterize the generality of certain properties.

Paper approved by the Editor for Data Communication Systems of the IEEE Communications Society for publication after presentation in part at the 1979 International Telemetering Conference, Los Angeles, CA, November 1978 and the 1979 International Symposium on Information Theory, Gringano, Italy, June 1979. Manuscript received May 22, 1978; revised August 21, 1979. This work was supported by Air Force Contract F44620-73-0065, F49620-78-C-0087, and F49620-79-C-0058 and by the Joint Services Electronics Program at Stanford University, Stanford, CA.

Y. Linde is with the Codex Corporation, Mansfield, MA.

A. Buzo was with Stanford University, Stanford, CA and Signal Technology Inc., Santa Barbara, CA. He is now with the Instituto de Ingenieria, National University of Mexico, Mexico City, Mexico.

R. M. Gray is with the Information Systems Laboratory, Stanford University, Stanford, CA 94305.

In particular, the algorithm's convergence properties are demonstrated herein by several examples. We consider the usual test case for such algorithms—quantizer design for memoryless Gaussian sources with a mean-squared error distortion measure, but we design and evaluate block quantizers with a rate of one bit per symbol and with blocklengths of 1 through 6. Comparison with recently developed lower bounds to the optimal distortion of such block quantizers (which provide strict improvement over the traditional bounds of rate-distortion theory) indicate that the resulting quantizers are indeed nearly optimal and not simply locally optimal. We also consider a scalar case where local optima arise and show how a variation of the algorithm yields a global optimum.

The algorithm is also used to design a quantizer for 10-dimensional vectors arising in speech compression systems. A complicated distortion measure is used that does not simply depend on the error vector. No probabilistic model is assumed, and hence the quantizer must be designed based on a training sequence of real speech. Here the convergence properties for both length of the training sequence and the number of iterations of the algorithm are demonstrated experimentally. No theoretical optimum is known for this case, but our system was used to compress the output of a traditional 6000 bit/s Linear Predictive Coded (LPC) speech system down to a rate of 1400 bits/s with only a slight loss in quality as judged by untrained listeners in informal subjective tests. To the authors' knowledge, direct application of variational techniques have not succeeded in designing block quantizers for such large block lengths and such complicated distortion measures.

BLOCK QUANTIZERS

An N-level k-dimensional quantizer is a mapping, q, that assigns to each input vector, $x = (x_0, \cdots, x_{k-1})$, a reproduction vector, $\hat{x} = q(x)$, drawn from a finite reproduction alphabet, $\hat{A} = \{y_i; i = 1, \cdots, N\}$. The quantizer q is completely described by the reproduction alphabet (or codebook) $\hat{A}$ together with the partition, $S = \{S_i; i = 1, \cdots, N\}$, of the input vector space into the sets $S_i = \{x: q(x) = y_i\}$ of input vectors mapping into the ith reproduction vector (or codeword). Such quantizers are also called block quantizers, vector quantizers, and block source codes.

DISTORTION MEASURES

We assume the distortion caused by reproducing an input vector x by a reproduction vector $\hat{x}$ is given by a nonnegative distortion measure $d(x, \hat{x})$. Many such distortion measures have been proposed in the literature. The most common for

Reprinted from *IEEE Trans. Commun.*, vol. COM-28, pp. 84–95, Jan. 1980.

reasons of mathematical convenience is the squared-error distortion.

$$d(x,\hat{x}) = \sum_{i=0}^{k-1} |x_i - \hat{x}_i|^2 \tag{1}$$

Other common distortion measures are the l_ν, or Holder norm,

$$d(x,\hat{x}) = \left\{ \sum_{i=0}^{k-1} |x_i - \hat{x}_i|^\nu \right\}^{1/\nu} \triangleq \|x - \hat{x}\|_\nu, \tag{2}$$

and its νth power, the νth-law distortion:

$$d(x,\hat{x}) = \sum_{i=0}^{k-1} |x_i - \hat{x}_i|^\nu = \|x - \hat{x}\|_\nu^\nu. \tag{3}$$

While both distortion measures (2) and (3) depend on the νth power of the errors in the separate coordinates, the measure of (2) is often more useful since it is a distance or metric and hence satisfies the triangle inequality, $d(x, \hat{x}) \leq d(x, y) + d(y, \hat{x})$, for all y. The triangle inequality allows one to bound the overall distortion easily in a multi-step system by the sum of the individual distortions incurred in each step. The usual ν^{th}-law distortion of (3) does not have this property. Other distortion measures are the l_∞, or Minkowski norm,

$$d(x,\hat{x}) = \max_{0 \leq i \leq k-1} |x_i - \hat{x}_i|, \tag{4}$$

the weighted-squares distortion,

$$d(x,\hat{x}) = \sum_{i=0}^{k-1} w_i |x_i - \hat{x}_i|^2, \tag{5}$$

where $w_i \geq 0$, $i = 0, \cdots, k-1$, and the more general quadratic distortion

$$\begin{aligned} d(x,\hat{x}) &= (x - \hat{x})B(x - \hat{x})^t \\ &= \sum_{i=0}^{k-1} \sum_{j=0}^{k-1} B_{i,j}(x_i - \hat{x}_i)(x_j - \hat{x}_j), \end{aligned} \tag{6}$$

where $B = \{B_{i,j}\}$ is a $k \times k$ positive definite symmetric matrix.

All of the previously described distortion measures have the property that they depend on the vectors x and $\hat{x}$ only through the error vector $x - \hat{x}$. Such distortion measures having the form $d(x, \hat{x}) = L(x - \hat{x})$ are called difference distortion measures. Distortion measures not having this form but depending on x and $\hat{x}$ in a more complicated fashion have also been proposed for data compression systems. Of interest here is a distortion measure of Itakura and Saito [3, 4] and Chaffee [5, 32] which arises in speech compression systems and has the form

$$d(x,\hat{x}) = (x - \hat{x})R(x)(x - \hat{x})^t, \tag{7}$$

where for each x, $R(x)$ is a positive definite $k \times k$ symmetric matrix. This distortion resembles the quadratic distortion of (6), but here the weighting matrix depends on the input vector x.

We are here concerned with the particular form and application of this distortion measure rather than its origins, which are treated in depth in [3-9] and in a paper in preparation. For motivation, however, we briefly describe the context in which this distortion measure is used in speech systems. In the LPC approach to speech compression [10], each frame of sampled speech is modeled as the output of a finite-order all-pole filter driven by either white noise (unvoiced sounds) or a periodic pulse train (voiced sounds). LPC analysis has, as input, a frame of speech and produces parameters describing the model. These parameters are then quantized and transmitted. One collection of such parameters consists of a voiced/unvoiced decision together with a pitch estimate for voiced sounds, a gain term σ (related to volume), and the sample response of the normalized inverse filter $(1, a_1, a_2, \cdots, a_K)$, that is, the normalized all-pole model has transfer function or z-transform $\{\Sigma_{k=0}^{K} a_k z^{-k}\}^{-1}$. Other parameter descriptions such as the reflection coefficients are also possible [10].

In traditional LPC systems, the various parameters are quantized separately, but such systems have effectively reached their theoretical performance limits [11]. Hence it is natural to consider block quantization of these parameters and compare the performance with the traditional scalar quantization techniques. Here we consider the case where the pitch and gain are (as usual) quantized separately, but the parameters describing the normalized model are to be quantized together as a vector. Since the lead term is 1, we wish to quantize a vector $(a_1, a_2, \cdots, a_K) \triangleq x = (x_0, \cdots, x_{K-1})$. A distortion measure, $d(x, \hat{x})$, between x and a reproduction x, can then be viewed as a distortion measure between two normalized (unit gain) inverse filters or models. A distortion measure for such a case has been proposed by Itakura and Saito [3, 4] and by Chaffee [5, 32] and it has the form of (7) with $R(x)$ the autocorrelation matrix $\{r_x(k-j); k = 0, 1, \cdots, K-1; j = 0, 1, \cdots, K-1\}$ defined by

$$r_x(k) = \int_{-\pi}^{\pi} \left| \sum_{m=0}^{K} a_i e^{-im\theta} \right|^{-2} e^{ik\theta} d\theta/2\pi, \tag{8}$$

described by x when the input has a flat unit amplitude spectrum.

Many properties and alternative forms for this particular distortion measure are developed in [3-9], where it is also shown that standard LPC systems implicitly minimize this distortion, which suggests that it is also an appropriate distortion measure for subsequent quantization. Here, however, the important fact is that it is not a difference distortion measure; it is one for which the dependence on x and $\hat{x}$ is quite complicated.

We also observe that various functions of the previously defined distortion measures have been proposed in the literature, for example, distortion measures of the forms $\|x - \hat{x}\|^r$ and $\rho(\|x - \hat{x}\|)$, where ρ is a convex function and the norm is any of the previously defined norms. The techniques to be developed here are applicable to all of these distortion measures.

PERFORMANCE

Let $X = (X_0, \cdots, X_{k-1})$ be a real random vector described by a cumulative distribution function $F(x) = \Pr\{X_i \leqslant x_i; i = 0, 1, \cdots, k-1\}$. A measure of the performance of a quantizer q applied to the random vector X is given by the expected distortion

$$D(q) = Ed(X, q(X)), \tag{9}$$

where E denotes the expectation with respect to the underlying distribution F. This performance measure is physically meaningful if the quantizer q is to be used to quantize a sequence of vectors $X_n = (X_{nK}, \cdots, X_{nK+K-1})$ that are stationary and ergodic, since then the time-averaged distortion,

$$n^{-1} \sum_{i=0}^{n-1} d(X_i, q(X_i)),$$

converges with probability one to $D(q)$ as $n \to \infty$ (from the ergodic theorem), that is, $D(q)$ describes the long-run time-averaged distortion.

An alternative performance measure is the maximum of $d(x, q(x))$ over all x in A, but we use only the expected distortion (9) since, in most problems of interest (to us), it is the average distortion and not the peak distortion that determines subjective quality. In addition, the expected distortion is more easily dealt with mathematically.

OPTIMAL QUANTIZATION

An N-level quantizer will be said to be optimal (or globally optimal) if it minimizes the expected distortion, that is, q^* is optimal if for all other quantizers q having N reproduction vectors $D(q^*) \leqslant D(q)$. A quantizer is said to be locally optimum if $D(q)$ is only a local minimum, that is, slight changes in q cause an increase in distortion. The goal of block quantizer design is to obtain an optimal quantizer if possible and, if not, to obtain a locally optimal and hopefully "good" quantizer. Several such algorithms have been proposed in the literature for the computer-aided design of locally optimal quantizers. In a few special cases, it has been possible to demonstrate global optimality either analytically or by exhausting all local optima. In 1957, in a classic but unfortunately unpublished Bell Laboratories' paper, S. Lloyd [1] proposed two methods for quantizer design for the scalar case ($k = 1$) with a squared-error distortion criterion. His "Method II" was a straightforward variational approach wherein he took derivatives with respect to the reproduction symbols, y_i, and with respect to the boundary points defining the S_i and set these derivatives to zero. This in general yields only a "stationary-point" quantizer (a multidimensional zero derivative) that satisfies necessary but not sufficient conditions for optimality. By second derivative arguments, however, it is easy to establish that such stationary-point quantizers are at least locally optimum for ν^{th}-power law distortion measures. In addition, Lloyd also demonstrated global optimality for certain distributions by a technique of exhaustively searching all local optima. Essentially the same technique was also proposed and used in the parallel problem of cluster analysis by Dalenius [12] in 1950, Fisher [13] in 1953, and Cox [14] in 1957. The technique was also independently developed by Max [15] in 1960 and the resulting quantizer is commonly known as the Lloyd-Max quantizer. This approach has proved quite useful for designing scalar quantizers, with power-law distortion criteria and with known distributions that were sufficiently well behaved to ensure the existence of the derivatives in question. In addition, for this case, Fleischer [16] was able to demonstrate analytically that the resulting quantizers were globally optimum for several interesting probability densities.

In some situations, however, the direct variational approach has not proved successful. First, if k is not equal to 1 or 2, the computational requirements become too complex. Simple combinations of one-dimensional differentiation will not work because of the possibly complicated surface shapes of the boundaries of the cells of the partition. In fact, the only successful applications of a direct variational approach to multidimensional quantization are for quantizers where the partition cells are required to have a particular simple form such as multidimensional "cubes" or, in two dimensions, "pie slices," each described only by a radius and two angles. These shapes are amenable to differentiation techniques, but only yield a local optimum within the constrained class. Secondly, if, in addition, more complex distortion measures such as those of (4)-(7) are desired, the required computation associated with the variational equations can become exorbitant. Thirdly, if the underlying probability distribution has discrete components, then the required derivatives may not exist, causing further computational problems. Lastly, if one lacks a precise probabilistic description of the random vector X and must base the design instead on an observed long training sequence of data, then there is no obvious way to apply the variational approach. If the underlying unknown process is stationary and ergodic, then hopefully a system designed by using a sufficiently long training sequence should also work well on future data. To directly apply the variational technique in this case, one would first have to estimate the underlying continuous distribution based on the observations and then take the appropriate derivatives. Unfortunately, however, most statistical techniques for density estimation require an underlying assumption on the class of allowed densities, e.g., exponential families. Thus these techniques are inappropriate when no such knowledge is available. Furthermore, a good fit of a continuous model to a finite-sample histogram may have ill-behaved differential behavior and hence may not produce a good quantizer. To our knowledge, no one has successfully used such an approach nor has anyone demonstrated that this approach will yield the correct quantizer in the limit of a long training sequence.

Lloyd [1] also proposed an alternative nonvariational approach as his "Method I." Not surprisingly, both approaches yield the same quantizer for the special cases he considered, but we shall argue that a natural and intuitive extension of his Method I provides an efficient algorithm for the design of good vector quantizers that overcomes the problems of the variational approach. In fact, variations of Lloyd's Method I have been "discovered" several times in the literature for squared-error and magnitude-error distortion criteria for both scalar and multidimensional cases (e.g., [22], [23], [24], [31]). Lloyd's basic development, however, remains the simplest, yet it extends easily to the general case considered here.

To describe Lloyd's Method I in the general case, we first assume that the distribution is known, but we allow it to be either continuous or discrete and make no assumptions requiring the existence of derivatives. Given a quantizer q described by a reproduction alphabet $\hat{A} = \{y_i;\, i = 1, \cdots, N\}$ and partition $S = \{S_i;\, i = 1, \cdots, N\}$, then the expected distortion $D(\{\hat{A}, S\}) \triangleq D(q)$ can be written as

$$D(\{\hat{A}, S\}) = Ed(X, q(X))$$

$$= \sum_{i=1}^{N} E(d(X, y_i) \mid X \in S_i) \Pr(X \in S_i), \tag{10}$$

where $E(d(X, y_i) \mid X \in S_i)$ is the conditional expected distortion, given $X \in S_i$, or, equivalently, given $q(X) = y_i$.

Suppose that we are given a particular reproduction alphabet $\hat{A}$, but a partition is not specified. A partition that is optimum for $\hat{A}$ is easily constructed by mapping each x into the $y_i \in \hat{A}$ minimizing the distortion $d(x, y_i)$, that is, by choosing the minimum distortion or nearest-neighbor codeword for each input. A tie-breaking rule such as choosing the reproduction with the lowest index is required if more than one codeword minimizes the distortion. The partition, say $P(\hat{A}) = \{P_i;\, i = 1, \cdots, N\}$ constructed in this manner is such that $x \in P_i$ (or $q(x) = y_i$) only if $d(x, y_i) \leq d(x, y_j)$, all j, and hence

$$D(\{\hat{A}, P(\hat{A})\}) = E\,(\min_{y \in \hat{A}} d(X, y)) \tag{11}$$

which, in turn, implies for any partition S that

$$D(\{\hat{A}, S\}) \geq D(\{\hat{A}, P(\hat{A})\}). \tag{12}$$

Thus for a fixed reproduction alphabet $\hat{A}$, the best possible partition is $P(\hat{A})$.

Conversely, assume we are given a partition $S = \{S_i;\, i = 1, \cdots, N\}$ describing a quantizer. For the moment, assume also that the distortion measure and distribution are such that, for each set S with nonzero probability in k-dimensional Euclidean space, there exists a minimum distortion vector $\hat{x}(S)$ for which

$$E(d(X, \hat{x}(S)) \mid X \in S) = \min_{u} E(d(X, u) \mid X \in S). \tag{13}$$

Analogous to the case of a squared-error distortion measure and a uniform probability distribution, we call the vector $\hat{x}(S)$ the centroid or center of gravity of the set S. If such points exist, then clearly for a fixed partition $S = \{S_i;\, i = 1, \cdots, N\}$, no reproduction alphabet $\hat{A} = \{y_i; i = 1, \cdots, N\}$ can yield a smaller average distortion then the reproduction alphabet $\hat{x}(S) \triangleq \{\hat{x}(S_i); i = 1, \cdots, N\}$ containing the centroids of the sets in S since

$$D(\{\hat{A}, S\}) = \sum_{i=1}^{N} E(d(X, y_i) \mid X \in S_i) \Pr(X \in S_i)$$

$$\geq \sum_{i=1}^{N} \min_{u} E(d(X, u) \mid X \in S_i) \Pr(X \in S_i)$$

$$= D(\{\hat{x}(S), S\}). \tag{14}$$

It is shown in [2] that the centroids of (13) exist for all sets S with nonzero probability for quite general distortion measures including all of those considered here. In particular, if $d(x, y)$ is convex in y, then centroids can be computed using standard convex programming techniques as described, e.g., in Luenberger [17, 18] or Rockafellar [19]. In certain cases, they can be found easily using variational techniques. If the probability of a set S is zero, then the centroid can be defined in an arbitrary manner since then the conditional expectation given that S in (13) has no unique definition.

Equations (12) and (14) suggest a natural algorithm for designing a good quantizer by taking any given quantizer and iteratively improving it:

Algorithm (Known Distribution)

(0) Initialization: Given N = number of levels, a distortion threshold $\epsilon \geq 0$, and an initial N-level reproduction alphabet $\hat{A}_0$ and a distribution F. Set $m = 0$ and $D_{-1} = \infty$.

(1) Given $\hat{A}_m = \{y_i;\, i = 1, \cdots, N\}$, find its minimum distortion partition $P(\hat{A}_m) = \{S_i;\, i = 1, \cdots, N\}$: $x \in S_i$ if $d(x, y_i) \leq d(x, y_j)$ for all j. Compute the resulting average distortion, $D_m = D(\{\hat{A}_m, P(\hat{A}_m)\}) = E \min_{y \in \hat{A}_m} d(X, y)$.

(2) If $(D_{m-1} - D_m)/D_m \leq \epsilon$, halt with $\hat{A}_m$ and $P(\hat{A}_m)$ describing final quantizer. Otherwise continue.

(3) Find the optimal reproduction alphabet $\hat{x}(P(\hat{A}_m)) = \{\hat{x}(S_i);\, i = 1, \cdots, N\}$ for $P(\hat{A}_m)$. Set $\hat{A}_{m+1} \triangleq \hat{x}(P(\hat{A}_m))$. Replace m by $m + 1$ and go to (1).

If, at some point, the optimal partition $P(\hat{A}_m)$ has a cell S_i such that $\Pr(X \in S_i) = 0$, then the algorithm, as stated, assigns an arbitrary vector as centroid and continues. Clearly, alternative rules are possible and may perform better in practice. For example, one can simply remove the cell S_i and the corresponding reproduction symbol from the quantizer without affecting performance, and then continue with an $(N - 1)$ level quantizer. Alternatively, one could assign to S_i its Euclidean center of gravity or the ith centroid from the previous iteration. One could also simply reassign the reproduction vector corresponding to S_i to another cell S_j and continue the algorithm. The stated technique is given simply for convenience, since zero probability cells were not a problem in the examples considered here. They can, however, occur and in such situations alternative techniques such as those described may well work better. In practice, a simple alternative is that, if the final quantizer produced by the algorithm has a zero probability (hence useless) cell, simply rerun the algorithm with a different initial guess.

From (12) and (14), $D_m \leqslant D_{m-1}$ and hence each iteration of the algorithm must either reduce the distortion or leave it unchanged. We shall later mention some minor additional details of the algorithm and discuss techniques for choosing an initial guess, but the previously given description contains the essential ideas.

Since D_m is nonincreasing and nonnegative, it must have a limit, say D_∞, as $m \to \infty$. It is shown in [2] that if a limiting quantizer $\hat{A}_\infty$ exists in the sense $\hat{A}_m \to \hat{A}_\infty$ as $m \to \infty$ in the usual Euclidean sense, then $D(\{\hat{A}_\infty, P(\hat{A}_\infty)\}) = D_\infty$ and $\hat{A}_\infty$ has the property that $\hat{A}_\infty = \hat{x}(P(\hat{A}_\infty))$, that is, $\hat{A}_\infty$ is exactly the centroid of its own optimal partition. In the language of optimization theory, $\{\hat{A}_\infty, P(\hat{A}_\infty)\}$ is a *fixed point* under further iterations of the algorithm [17, 18]. Hence the limit quantizer (if it exists) is called a fixed-point quantizer (in contrast to a stationary-point quantizer obtained by a variational approach). In this light, the algorithm is simply a standard technique for finding a fixed point via the method of successive approximation (see, e.g., Luenberger [17, p. 272]). If $\epsilon = 0$ and the algorithm halts for finite m, then such a fixed point has been attained [2].

It is shown in [2] that a necessary condition for a quantizer to be optimal is that it be a fixed-point quantizer. It is also shown in [2] that, as in Lloyd's case, if a fixed-point quantizer is such that there is no probability on the boundary of the partition cells, that is if $\Pr(d(X, y_i) = d(X, y_j)$, for some $i \neq j) = 0$, then the quantizer is locally optimum. This is always the case with continuous distributions, but can in principle be violated for discrete distributions. It was never found to occur in our experiments, however. As Lloyd suggests, the algorithm can easily be modified to test a fixed point for this condition and if there is nonzero probability of a vector on a boundary, the strategy would be to reassign the vector to another cell of the partition and continue the iteration.

For the $N = 1$ case with a squared-error distortion criterion, the algorithm is simply Lloyd's Method I, and his arguments apply immediately in the more general case considered herein. A similar technique was earlier proposed in 1953 by Fisher [13] in a cluster analysis problem using Bayes decisions with a squared-error cost. For larger dimensions and distortion measures of the form $d(x, \hat{x}) = \|x - \hat{x}\|_2^r, r \geqslant 1$, the relations (12) and (14) were observed by Zador [20] and Gersho [21] in their work on the asymptotic performance of optimal quantizers, and hence the algorithm is certainly implicit in their work. They did not, however, actually propose or apply the technique to design a quantizer for fixed N. In 1965, Forgy [31] proposed the algorithm for cluster analysis for the multidimensional squared-error distortion case and a sample distribution (see the discussion in MacQueen [25]). In 1977, Chen [22] proposed essentially the same algorithm for the multidimensional case with the squared-error distortion measure and used it to design two-dimensional quantizers for vectors uniformly distributed in a circle.

Since the algorithm has no differentiability requirements, it is valid for purely discrete distributions. This has an important application to the case where one does not possess *a priori* a probabilistic description of the source to be compressed, and hence must base his design on an observed long training sequence of the data to be compressed. One approach would be to use standard density estimation techniques of statistics to obtain a "smooth" distribution of F and to then apply variational techniques. As previously discussed, we do not adopt this approach as it requires additional assumptions on the allowed densities. Instead we consider the following approach: Use the training sequence, say $\{x_k; k = 0, \cdots, n-1\}$ to form the time-average distortion

$$\frac{1}{n} \sum_{i=0}^{n-1} d(x_i, q(x_i))$$

and observe that this is exactly the expected distortion $E_{G_n} d(X, q(X))$ with respect to the sample distribution G_n determined by the training sequence, i.e., the distribution that assigns probability m/n to a vector x that occurs in the training sequence m times. Thus we can design a quantizer that minimizes the time-average distortion for the training sequence by running the algorithm on the sample distribution G_n (1,2). This yields the following variation of the algorithm:

Algorithm (Unknown Distribution)

(0) Initialization: Given N = number of levels, distortion threshold $\epsilon \geqslant 0$, an initial N-level reproduction alphabet $\hat{A}_0$, and a training sequence $\{x_j; j = 0, \cdots, n-1\}$. Set $m = 0$ and $D_{-1} = \infty$.

(1) Given $\hat{A}_m = \{y_i; i = 1, \cdots, N\}$, find the minimum distortion partition $P(\hat{A}_m) = \{S_i;\ i = 1, \cdots, N\}$ of the training sequence: $x_j \in S_i$ if $d(x_j, y_i) \leqslant d(x_j, y_l)$, for all l. Compute the average distortion

$$D_m = D(\{\hat{A}_m, P(\hat{A}_m)\}) = n^{-1} \sum_{j=0}^{n-1} \min_{y \in \hat{A}_m} d(x_j, y).$$

(2) If $(D_{m-1} - D_m)/D_m \leqslant \epsilon$, halt with $\hat{A}_m$ final reproduction alphabet. Otherwise continue.

(3) Find the optimal reproduction alphabet $\hat{x}(P(\hat{A}_m)) = \{\hat{x}(S_i);\ i = 1, \cdots, N\}$ for $P(\hat{A}_m)$. Set $\hat{A}_{m+1} \triangleq \hat{x}(P(\hat{A}_m))$. Replace m by $m + 1$ and go to (1).

Observe above that while designing the quantizer, only partitions of the training sequence (the input alphabet) are considered. Once the final codebook $\hat{A}_m$ is obtained, however, it is used on new data outside the training sequence with the optimum nearest-neighbor rule, that is, an optimum partition of k-dimensional Euclidean space.

(1) It was observed by a reviewer that application of the algorithm to the sample distribution provides a "Monte Carlo" design of the quantizer for a vector with a known distribution, that is, the design is based on samples of the random vectors rather than on an explicit distribution.

(2) During the period this paper was being reviewed for publication, two similar techniques were reported for special cases. In 1978, Capria, Westin, and Esposito [23] presented a similar technique for the scalar case using dynamic programming arguments. Their approach was for average squared-error distortion and for maximum distortion over the training sequence. In 1979, Menez, Boeri, and Esteban [24] proposed a similar technique for scalar quantization using squared-error and magnitude-error distortion measures.

If the sequence of random vectors is stationary and ergodic, then it follows from the ergodic theorem that, with probability one, G_n goes to the true underlying distribution F as $n \to \infty$. Thus if the training sequence is sufficiently long, hopefully a good quantizer for the sample distribution G_n should also be good for the true distribution F, and hence should yield good performance for future data produced by the source. All of these ideas are made precise in [2] where it is shown that, subject to suitable mathematical assumptions, the quantizer produced by applying the algorithm to G_n converges, as $n \to \infty$, to the quantizer produced by applying the algorithm to the true underlying distribution F. We observe that no analogous results are known to the authors for the density estimation/variational approach, and that independence of successive blocks is not required for these results, only block stationarity and ergodicity.

We also point out that, for finite alphabet distributions such as sample distributions, the algorithm always converges to a fixed-point quantizer in a finite number of steps [2].

A similar technique used in cluster analysis with squared-error cost functions was developed by MacQueen in 1967 [25] and has been called the k-means approach. A more involved technique using the k-means approach is the "ISODATA" approach of Ball and Hall [26]. The basic idea of finding minimum distortion partitions and centroids is the same, but the training sequence data is used in a different manner and the resulting quantizers will, in general, be different. Their sequential technique incorporates the training vectors one at a time and ends when the last vector is incorporated. This is in contrast to the previous algorithm which considers all of the training vectors at each iteration. The k-means method can be described as follows: The goal is to produce a partition $S_0 = \{S_0, \cdots, S_{N-1}\}$ of the training alphabet, $A = \{x_i; i = 0, \cdots, n-1\}$ consisting of all vectors in the training sequence. The corresponding reproduction alphabet $\hat{A}$ will then be the collection of the Euclidean centroids of the sets S_i, that is, the final reproduction alphabet will be optimal for the final partition (but the final partition may not be optimal for the final reproduction alphabet, except as $n \to \infty$). To obtain S, we first think of the each S_i as a bin in which to place training sequence vectors until all are placed. Initially, we start by placing the first N vectors in separate bins, i.e., $x_i \in S_i$, $i = 0, \cdots, N-1$. We then proceed as follows: at each iteration, a new training vector x_m is observed. We find the set S_i for which the distortion between x_m and the centroid $\hat{x}(S_i)$ is minimized and then add x_m to this bin. Thus, at each iteration, the new vector is added to the bin with the closest centroid, and hence the next time, this bin will have a new centroid. This operation is continued until all sample vectors are incorporated.

Although similar in philosophy, the k-means algorithm has some crucial differences. In particular, it is suited for the case where *only* the training sequence is to be classified, that is, where a long sequence of vectors is to be grouped in a low distortion manner. The sequential procedure is computationally efficient for grouping, but a "quantizer" is not produced until the procedure is stopped. In other words, in cluster analysis, one wishes to group things and the groups can change with time, but in quantization, one wishes to fix the groups (to get a time-invariant quantizer), and then use these groups (or the quantizer) on future data outside of the training sequence.

An additional problem is that the only theorems which guarantee convergence, in the limit of a long training sequence, require the assumption that successive vectors be independent [25], unlike the more general case for the proposed algorithm [2].

Recently Levenson *et al.* used a variation of the k-means and ISODATA algorithms with a distortion measure proposed by Itakura [4] to determine reference templates for speaker-independent word recognition [27]. They used, as a distortion measure, the logarithm of the distortion of (7) (which is a gain-optimized Itakura-Saito distortion [7]—our use of the distortion measure with unit-gain-normalized models results in no such logarithmic function). In their technique, however, a minimax rule was used to select the reproduction vectors (or cluster points) rather than finding the "optimum" centroid vector. If instead, the distortion measure of (7) is used, then the centroids are easily found, as will be seen.

CHOICE OF $\hat{A}_0$

There are several ways to choose the initial reproduction alphabet $\hat{A}_0$ required by the algorithm. One method for use on sample distributions is that of the k-means method, namely choosing the first N vectors in the training sequence. We did not try this approach as, intuitively, one would like these vectors to be well-separated, and N consecutive samples may not be. Two other methods were found to be useful in our examples. The first is to use a uniform quantizer over all or most of the source alphabet (if it is bounded). For example, if used on a sample distribution, one uses a k-dimensional uniform quantizer on a k-dimensional Euclidean cube including all or most of the points in the training sequence. This technique was used in the Gaussian examples described later.

The second technique is useful when one wishes to design quantizers of successively higher rates until achieving an acceptable level of distortion. Here we consider M-level quantizers with $M = 2^R$, $R = 0, 1, \cdots$, and continue until we achieve an initial guess for an N-level quantizer as follows:

INITIAL GUESS BY "SPLITTING"

(0) Initialization: Set $M = 1$ and define $\hat{A}_0(1) = \hat{x}(A)$, the centroid of the entire alphabet (the centroid of the training sequence, if a sample distribution is used).

(1) Given the reproduction alphabet $\hat{A}_0(M)$ containing M vectors $\{y_i; i = 1, \cdots, M\}$, "split" each vector y_i into two close vectors $y_i + \epsilon$ and $y_i - \epsilon$, where ϵ is a fixed perturbation vector. The collection $\tilde{A}$ of $\{y_i + \epsilon, y_i - \epsilon, i = 1, \cdots, M\}$ has $2M$ vectors. Replace M by $2M$.

(2) Is $M = N$? If so, set $\hat{A}_0 = \tilde{A}(M)$ and halt. $\hat{A}_0$ is then the initial reproduction alphabet for the N-level quantization algorithm. If not, run the algorithm for an M-level quantizer on $\tilde{A}(M)$ to produce a good reproduction alphabet $\hat{A}_0(M)$, and then return to step (1).

Using the splitting algorithm on a training sequence, one starts with a one-level quantizer consisting of the centroid of the training sequence. This vector is then split into two vectors and the two-level quantizer algorithm is run on this pair to obtain a good (fixed-point) two-level quantizer. Each of these two vectors is then split and the algorithm is run to produce a good four-level quantizer. At the conclusion, one has fixed-point quantizers for $1, 2, 4, 8, \cdots, N$ levels.

EXAMPLES

Gaussian Sources

The algorithm was used initially to design quantizers for the classical example of memoryless Gaussian random variables with the squared-error distortion criterion of (1), based on a training sequence of data. The training and sample data were produced by a zero-mean, unit-variance memoryless sequence of Gaussian random variables. The initial guess was a unit quantizer on the k-dimensional cube, $\{x\colon |x_i| \leqslant 4; i = 0, \cdots, k-1\}$. A distortion threshold of 0.1% was used. The overall algorithm can be described as follows:

(0) Initialization: Fix N = number of levels, k = block length, n = length of training sequence, $\epsilon = .001$. Given a training sequence $\{x_j; j = 0, \cdots, n-1\}$. Let $\hat{A}_0$ be an N-level uniform quantizer reproduction alphabet for the k-dimensional cube, $\{u\colon |u_i| \leqslant 4, i = 0, \cdots, k-1\}$. Set $m = 0$ and $D_{-1} = \infty$.

(1) Given $\hat{A}_m = \{y_i; i = 1, \cdots, N\}$, find the minimum-distortion partition $P(\hat{A}_m) = \{S_i; i = 1, \cdots, N\}$. For example, for each $j = 0, \cdots, n-1$, compute $d(x_j, y_i)$ for $i = 1, \cdots, N$. If $d(x_j, y_i) \leqslant d(x_j, y_l)$ for all l, then $x_j \in S_i$. Compute:

$$D_m = D(\{\hat{A}_m, P(\hat{A}_m)\}) = n^{-1} \sum_{j=0}^{n-1} \min_{y \in \hat{A}_m} d(x_j, y).$$

(2) If $(D_{m-1} - D_m)/D_m \leqslant \epsilon = .001$, halt with final quantizer described by $\hat{A}_m$. Otherwise continue.

(3) Find the optimal reproduction alphabet $\hat{x}(P(\hat{A}_m)) = \{\hat{x}(S_i); i = 1, \cdots, N\}$ for $P(\hat{A}_m)$. For the squared-error criterion, $\hat{x}(S_i)$ is the Euclidean center of gravity or centroid given by

$$x(S_i) = \frac{1}{\| S_i \|} \sum_{j: x_j \in S_i} x_j,$$

where $\|S_i\|$ denotes the number of training vectors in the cell S_i. If $\|S_i\| = 0$, set $\hat{x}(S_i) = y_i$, the old codeword. Define $\hat{A}_{m+1} = \hat{x}(P(A_m))$, replace m by $m+1$, and go to (1).

Table 1 presents a simple but nontrivial example intended to demonstrate the basic operation of the algorithm. A two-dimensional quantizer with four levels is designed, based on a short training sequence of twelve training vectors. Because of the short training sequence in this case, the final distortion is lower than one would expect and the final quantizer may not work well on new data outside of the training sequence. The tradeoffs between the length of the training sequence and the performance inside and outside the training sequence are developed more carefully in the speech example.

Observe that, in the example of Table 1, the algorithm could actually halt in step (1) of the $m = 1$ iteration since, if $P(\hat{A}_m) = P(\hat{A}_{m-1})$, it follows that $\hat{A}_{m+1} = \hat{x}(P(\hat{A}_m)) = \hat{x}(P(\hat{A}_{m-1})) = \hat{A}_m$, and hence $\hat{A}_m$ is the desired fixed point.

TABLE 1
A SIMPLE EXAMPLE

(0) <u>Initialization</u>: $N = 4$, $k = 2$, $\epsilon = .001$, $n = 12$.

Training Sequence:

$\underset{\sim}{x}_1$ = (-.37449, .98719)	$\underset{\sim}{x}_7$ = (-.59161, .17968)
$\underset{\sim}{x}_2$ = (.63919,-.11875)	$\underset{\sim}{x}_8$ = (.14093,1.76413)
$\underset{\sim}{x}_3$ = (-.83293, .60645)	$\underset{\sim}{x}_9$ = (.70898,-.35017)
$\underset{\sim}{x}_4$ = (-.70534,-1.21856)	$\underset{\sim}{x}_{10}$ = (.30038, .79836)
$\underset{\sim}{x}_5$ = (-.28952,-.94821)	$\underset{\sim}{x}_{11}$ = (.30165,1.06552)
$\underset{\sim}{x}_6$ = (1.09924, .516)	$\underset{\sim}{x}_{12}$ = (.37801,-.32708)

$\hat{A}_0 = \{(2,2), (2,-2), (-2,2), (-2,-2)\}$
$= \{\underset{\sim}{y}_1, \underset{\sim}{y}_2, \underset{\sim}{y}_3, \underset{\sim}{y}_4\}$

D_{-1} = 9.99E + 62 (∞ on a microcomputer)

Set $m = 0$.

<u>m=0</u> (1) Find $P(\hat{A}_0) = \{S_1, S_2, S_3, S_4\}$:

$\underset{\sim}{x}_j \in S_i$ if $d(\underset{\sim}{x}_j, \underset{\sim}{y}_i) \le d(\underset{\sim}{x}_j, \underset{\sim}{y}_m)$, all m.

$S_1 = \{\underset{\sim}{x}_6, \underset{\sim}{x}_8, \underset{\sim}{x}_{10}, \underset{\sim}{x}_{11}\}$

$S_2 = \{\underset{\sim}{x}_2, \underset{\sim}{x}_9\}$

$S_3 = \{\underset{\sim}{x}_1, \underset{\sim}{x}_3, \underset{\sim}{x}_7\}$

$S_4 = \{\underset{\sim}{x}_4, \underset{\sim}{x}_5, \underset{\sim}{x}_{12}\}$

Compute D_0:

$$D_0 = \frac{1}{12} \sum_{j=1}^{12} \min_{\underset{\sim}{y} \in \hat{A}_0} d(\underset{\sim}{x}_j, \underset{\sim}{y}) = 2.0172 .$$

(2) $(D_{-1} - D_0)/D_0 > .001$, continue.

(3) Find the optimal reproduction alphabet $\hat{A}_1 \triangleq \hat{\underset{\sim}{x}}(P(\hat{A}_0)) = \{\hat{\underset{\sim}{x}}(S_i), i = 1, \ldots, 4\}$:

$\hat{\underset{\sim}{x}}(S_1) = (\underset{\sim}{x}_6 + \underset{\sim}{x}_8 + \underset{\sim}{x}_{10} + \underset{\sim}{x}_{11})/4 = (.46055, 1.036)$

$\hat{\underset{\sim}{x}}(S_2) = (\underset{\sim}{x}_2 + \underset{\sim}{x}_9)/2 = (.674085, -.23446)$

$\hat{\underset{\sim}{x}}(S_3) = (\underset{\sim}{x}_1 + \underset{\sim}{x}_3 + \underset{\sim}{x}_7)/3 = (-.599676, .591106)$

$\hat{\underset{\sim}{x}}(S_4) = (\underset{\sim}{x}_4 + \underset{\sim}{x}_5 + \underset{\sim}{x}_{12})/3 = (-.457623, -.831283)$

Set $m = 1$. Go to (1).

<u>m=1</u> (1) Find $P(\hat{A}_1)$:

Evaluating distortions shows $P(\hat{A}_1) = P(\hat{A}_0)$ (no change in partition)

Compute D_1:

$$D_1 = \frac{1}{12} \sum_{j=1}^{12} \min_{\underset{\sim}{y} \in \hat{A}_1} d(\underset{\sim}{x}_j, \underset{\sim}{y}) = .0997308 .$$

(2) $(D_0 - D_1)/D_1 \cong 19 > .001$

(3) $\hat{A}_2 \triangleq \hat{\underset{\sim}{x}}(P(\hat{A}_1)) = \hat{A}_1$, since $P(\hat{A}_1) = P(\hat{A}_0)$ and hence $\hat{\underset{\sim}{x}}(P(\hat{A}_1)) = \hat{\underset{\sim}{x}}(P(\hat{\underset{\sim}{A}}_0)) = \hat{A}_1$. Thus $\hat{A}_1$ is a fixed point. Set m=2. Go to (1).

<u>m=2</u> (1) $P(\hat{A}_1) = P(\hat{A}_0)$ and hence $D_2 = D_1$ and hence $(D_1 - D_2)/D_2 = 0 < .001$.

Halt with final quantizer described by $\{\hat{A}_1, P(\hat{A}_1)\}$.

Note: Characters with tildes underneath appear boldface in text.

In other words, if the quantizer stays the same for two iterations, then the two distortions are equal and an "$\epsilon = 0$" threshold is satisfied.

As a more realistic example, the algorithm was run for the scalar ($k = 1$) case with $N = 2, 3, 4, 6$ and 8, using a training sequence of 10,000 samples per quantizer output from a zero-mean, unit-variance memoryless Gaussian source. The resulting quantizer outputs and distortion were within 1% of the optimal values reported by Max [15]. No more than 20

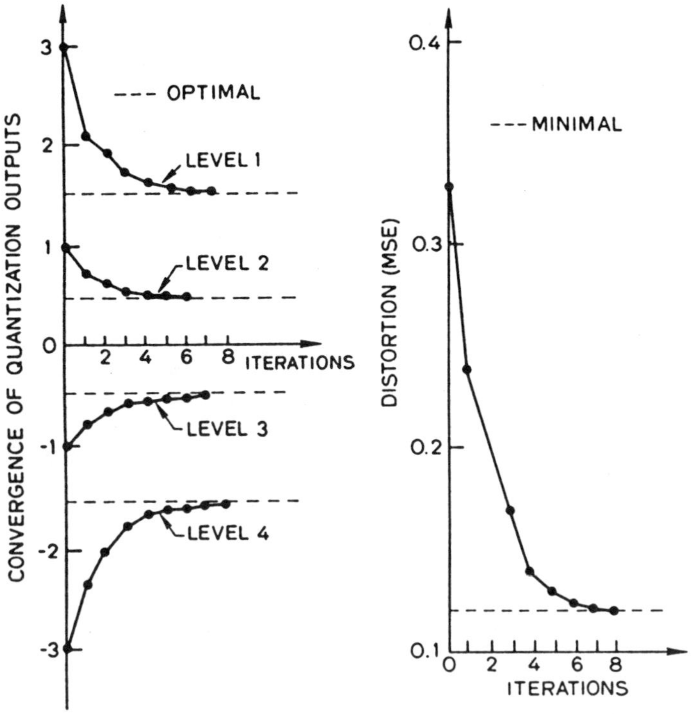

Fig. 1. The Basic Algorithm: Gaussian Source $N = 4$.

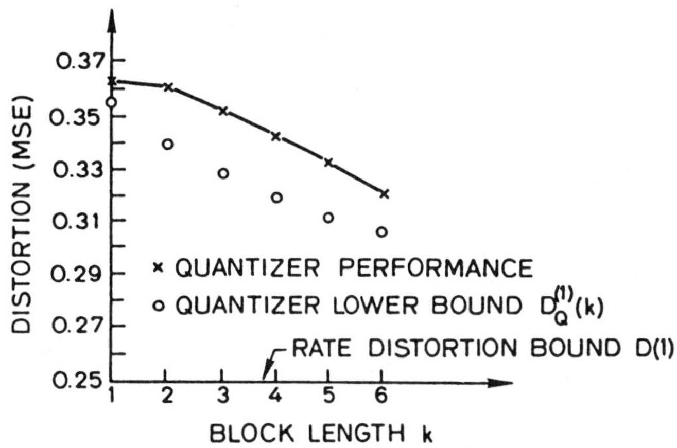

Fig. 2. Block Quantization Rate 1 bit/symbol Gaussian Source.

iterations were required for $N = 8$ and, for smaller N, the number of iterations was considerably smaller. Figure 1 describes the convergence rate of one of the tests for the case $N = 4$.

The algorithm was then tried for block quantizers for memoryless Gaussian variables with block lengths k equal to 1, 2, 3, 4, 5 and 6 and a rate of one bit per sample, so that $N = 2^k$. The distortion criterion was again the squared-error distortion measure of (1). The algorithm used a training sequence of 100,000 samples. In each case, the algorithm converged in fewer than 50 iterations and the resulting distortion is plotted in Fig. 2, together with the one bit-per-symbol scalar case as a function of block length. For comparison, the rate-distortion bound [28, p. 99] $D(R) = 2^{-2R}$ for $R = 1$ bit-per-symbol is also plotted. As expected and as shown in Fig. 2, the block quantizers outperform the scalar quantizer, but for these block lengths, the performance is still far from the rate distortion bound (which is achievable, in principle, only in the limit as $k \to \infty$). A more favorable comparison is obtained using a recent result of Yamada, Tazaki, and Gray [29] which provides a lower bound to the performance of an optimal N-level k-dimensional quantizer with a difference distortion measure when N is large. This bound provides strict improvement over the rate-distortion bound for fixed k and tends to the rate-distortion bound as $k \to \infty$. In the current case, the bound has the form

$$D_Q{}^{(k)}(R) = D(R) \cdot \left\{ \left(\frac{e}{1 + k/2} \right) \Gamma(1 + k/2)^{2/k} \right\},$$

where Γ is the gamma function. This bound is theoretically inappropriate for small k, yet it is surprisingly close for the $k = 1$ result, which is known to be almost optimal. For $k = 6$, $N = 2^6 = 64$ is moderately large and the closeness of the actual performance to the lower bound, compared to optimal performance provided by $D_Q{}^{(k)}(1)$, suggests that the algorithm is indeed providing a quantizer with block length six and rate one bit-per-symbol that is nearly optimal (within 6% of the optimal).

LLOYD'S EXAMPLE

Lloyd [1] provides an example where both variational and fixed-point approaches can yield locally optimal quantizers instead of a globally optimum quantizer. We next propose a slight modification of the fixed-point algorithm that indeed finds a globally optimum quantizer in Lloyd's example. We conjecture that this technique will work more generally, but we have been unable to prove this theoretically. A similar technique can be used with the stationary-point algorithm.

Instead of using samples from the source that we wish to quantize, we use samples corrupted by additive independent noise, where the marginal distribution of the noise is such that only one locally optimum quantizer exists for it. As an example, for scalar quantization with the squared-error distortion measure, we use Gaussian noise. In this case, any locally optimal quantizer is also globally optimum. Other distributions, such as the uniform or a discrete amplitude noise with an alphabet size equal to the number of quantizer output levels, can also be used.

When the noise power is much greater than the source power, the distribution of their sum is essentially the distribution of the noise. We assume that, initially, the noise power is so large that only one locally optimum quantizer exists for the sum; hence, regardless of the initial guess, the algorithm will converge to this optimum. On the next step, the noise power is reduced slightly and the quantizer resulting from the previous run is used as the initial guess. Intuitively, since the noise has been reduced by a small amount, the global optimum for the new sum should be close to that of the previous sum (we use the same source and noise samples with reduced noise power). Thus we expect that the algorithm will converge to the global optimum even though new local optimum points might have been introduced. We continue in the same manner reducing the noise gradually to zero.

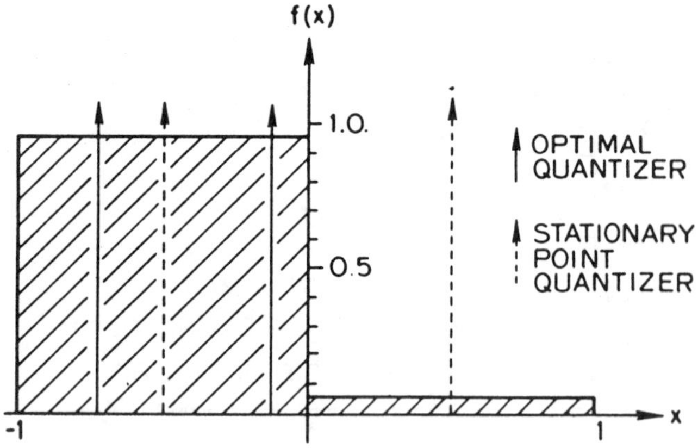

Fig. 3. The Probability Density Function.

To illustrate how the algorithm works, we use a source with a probability density function as shown in Fig. 3. In this case, there are two locally optimum two-level quantizers. One has the output levels +0.5 and −0.5 and yields a mean-squared error of 0.083, the second (which is the global optimum) has output levels −0.71 and 0.13 and yields a mean-squared error 0.048. (This example is essentially the same as one of Lloyd's [1].)

The modified algorithm was tested on a sequence of 2,000 samples chosen according to the probability density shown in Fig. 3. Gaussian noise was added starting at unity variance and reducing the variance by approximately 50% on each successive run. The initial guess was + 0.5 and − 0.5 which is the non-global optimum. Each run was stopped when the distortion was changed by less than 0.1% from its previous value.

The results are given in Fig. 4 and it is seen that, in spite of the bad initial guess, the modified algorithm converges to the globally optimum quantizer.

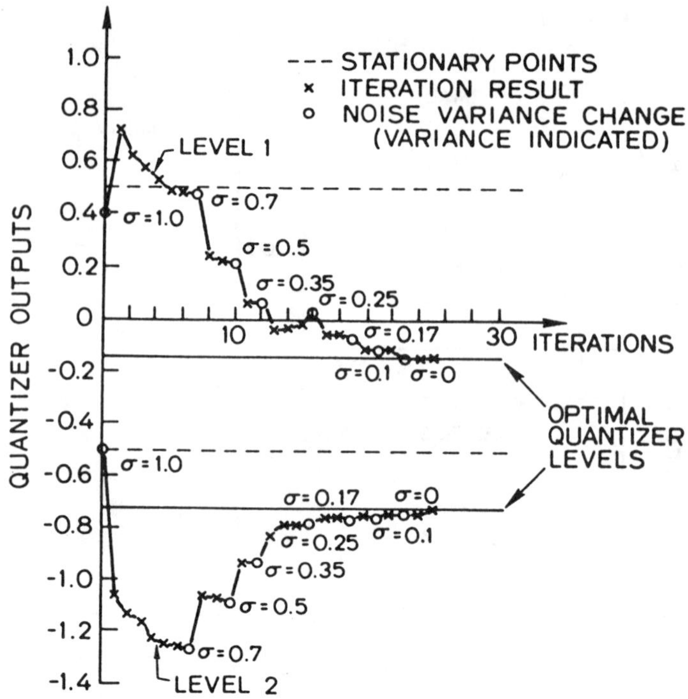

Fig. 4. The Modified Algorithm.

SPEECH EXAMPLE

In the next example, we consider the case of a speech compression system consisting of an LPC analysis of 20 ms-long speech frames producing a voiced/unvoiced decision and a pitch, a gain, and a normalized inverse filter as previously described, followed by quantization where the pitch and gain are separately quantized as usual, but the normalized filter coefficients $(a_1, \cdots, a_K) = (x_0, x_1, \cdots, x_{K-1})$ are quantized as a vector with $K = 10$, using the distortion measure of (7)-(8). The training sequence consisted of a sequence of normalized inverse filter parameter vectors(3). The LPC analysis was digital, and hence the training sequence used was already "finely quantized" to 10 bits per sample or 100 bits for each vector. The original gain required 12 bits per speech frame and the pitch used 8 bits per speech frame. The total rate of the LPC output (which we wish to further compress by block quantization) is 6000 bits/s. No further compression of gain or pitch was attempted in these experiments as our goal was to study only potential improvement when block quantizing the normalized filter parameters. The more complete problem including gain and pitch involves many other issues and is the subject of a paper still in preparation.

(3) The training sequence and additional test data of LPC reflection coefficients were provided by Signal Technology Inc. of Santa Barbara and were produced using standard LPC techniques on a single male speaker.

For the distortion measure of (7)-(8), the centroid of a subset S of a training sequence $\{x_j, j = 0, 1, \cdots, n-1\}$ is the vector u minimizing

$$\sum_{j:x_j \in S} (x_j - u)R(x_j)(x_j - u)^t.$$

We observe that the autocorrelation matrix $R(x)$ is a natural byproduct of the LPC analysis and need not be recomputed. This minimization, however, is a minimum-energy-residual minimization problem in LPC analysis and it can be solved by standard LPC algorithms such as Levinson's algorithm [10]. Alternatively, it is a much studied minimization problem in Toeplitz matrix theory [30] and the centroid can be shown via variational techniques to be

$$\hat{x}(S) = \left\{ \sum_{j:x_j \in S} R(x_j) \right\}^{-1} \sum_{j:x_j \in S} R(x_j) x_j^t. \tag{15}$$

The splitting technique for the initial guess and a distortion threshold of 0.5% were used. The complete algorithm for this example can thus be described as follows:

(0) Initialization: Fix $N = 2^R$, R an integer, where N is the largest number of levels desired. Fix $K = 10$, n = length of training sequence, $\epsilon = .005$. Set $M = 1$.

Given a training sequence $\{x_j; j = 0, \cdots, n-1\}$, set $A = \{x_j; j = 0, \cdots, n-1\}$, the training sequence alphabet. Define $\hat{A}(1) = \hat{x}(A)$, the centroid of the entire training sequence using (15) or Levinson's algorithm.

(1) (Splitting): Given $\hat{A}(M) = \{y_i, i = 1, \cdots, M\}$, split each reproduction vector y_i into $y_i + \epsilon$ and $y_i - \epsilon$, where ϵ is a fixed perturbation vector. Set $\hat{A}_0(2M) = \{y_i + \epsilon, y_i - \epsilon, i = 1, \cdots, M\}$ and then replace M by $2M$.

(2) Set $m = 0$ and $D_{-1} = \infty$.

(3) Given $\hat{A}_m(M) = \{y_1, \cdots, y_M\}$, find its optimum partition $P(\hat{A}_m(M)) = \{S_i; i = 1, \cdots, M\}$, that is, $x_j \in S_i$ if $d(x_j, y_i) \leqslant d(x_j, y_l)$, all l. Compute the resulting distortion

$$D_m = D(\{\hat{A}_m(M), P(\hat{A}_m(M))\})$$

$$= n^{-1} \sum_{j=0}^{n-1} \min_{y \in \hat{A}_m} d(x_j, y).$$

(4) If $(D_{m-1} - D_m)/D_m \leqslant \epsilon = .005$, then go to step (6). Otherwise continue.

(5) Find the optimal reproduction alphabet $\hat{A}_{m+1}(M) = \hat{x}(P(\hat{A}_m(M)) = \{\hat{x}(S_i); i = 1, \cdots, 1\}$ for $P(\hat{A}_m(M))$. Replace m by $m + 1$ and go to (3).

(6) Set $\hat{A}(M) = \hat{A}_m(M)$. The final M-level quantizer is described by $\hat{A}(M)$. If $M = N$, halt with final quantizer described by $\hat{A}(N)$. Otherwise go to step (1).

Table 2 describes the results of the algorithm for $N = 64$, and hence for one- to eight-bit quantizers trained on $n = 19{,}000$ frames of LPC speech produced by a single speaker. The distortion at the end of each iteration is given and, in all cases, the algorithm converged in fewer than 14 iterations. When the resulting quantizers were applied to data from the same speaker outside of the training sequence, the resulting distortion was within 1% of that within the training sequence. A total of three and one-half hours of computer time on a PDP 11/35 was required to obtain all of these codebooks.

Figure 5 depicts the rate of convergence of the algorithm with a training sequence length for a 16-level quantizer. Note the marked difference between the distortion for 2400 frames inside the training sequence and outside the training sequence for short training sequences. For a long training sequence of over 12,000 frames, however, the distortion is nearly the same.

Tapes of the synthesized speech at 8 bits per frame for the normalized model sounded similar to those of the original LPC speech with 100 bits per frame for the normalized model (the gain and the pitch were both left at the original LPC rate of 12 and 8 bits per frame, respectively). While extensive subjective tests were not attempted, all informal listening tests judged the synthesized speech perfectly intelligible (when heard *before* the original LPC!) and the quality only slightly inferior when the two were compared. The overall compression was from 6000 bits/s to 1400 bits/s. This is not startling as existing scalar quantizers that optimally allocate bits among the parameters and optimally quantize each parameter using a spectral deviation distortion measures [11] also perform well in this range. It is, however, promising as these were preliminary results with no attempt to further compress pitch and gain (which, taken together in our system, had more than twice the bit rate of the normalized model vector quantizer). Further results on applications of the algorithm to the overall speech compression system will be the subject of a forthcoming paper [33].

TABLE 2
ITAKURA-SAITO DISTORTION VS. NUMBER OF ITERATIONS. TRAINING SEQUENCE LENGTH = 19,000 FRAMES.

NUMBER OF LEVELS	DISTORTION	ITERATION NUMBER
2	10.33476	1
	1.98925	2
	1.78301	3
	1.67244	4
	1.55983	5
	1.49814	6
	1.48493	7
	1.48249	8
4	1.38765	1
	1.07906	2
	1.04223	3
	1.03252	4
	1.02709	5
8	0.96210	1
	0.85183	2
	0.81353	3
	0.79191	4
	0.77472	5
	0.76188	6
	0.75130	7
	0.74383	8
	0.73341	9
	0.71999	10
	0.71346	11
	0.70908	12
	0.70578	13
	0.70347	14
16	0.64653	1
	0.55665	2
	0.51810	3
	0.50146	4
	0.49235	5
	0.48761	6
	0.48507	7
32	0.44277	1
	0.40452	2
	0.39388	3
	0.38667	4
	0.38128	5
	0.37778	6
	0.37574	7
	0.37448	8
64	0.34579	1
	0.31766	2
	0.30850	3
	0.30366	4
	0.30086	5
	0.29891	6
	0.29746	7
128	0.27587	1
	0.25628	2
	0.24928	3
	0.24550	4
	0.24309	5
	0.24142	6
	0.24021	7
	0.23933	8
256	0.22458	1
	0.20830	2
	0.20228	3
	0.19849	4
	0.19623	5
	0.19479	6
	0.19386	7
	0.19319	8

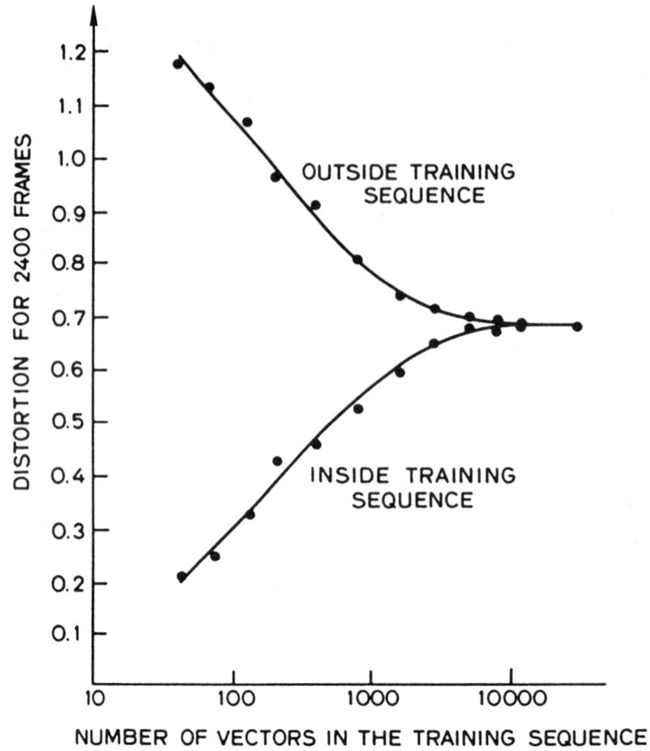

Fig. 5. Convergence with Training Sequence.

EPILOGUE

The Gaussian example of Figure 2, Lloyd's example, and the speech example were run on a PDP 11/34 minicomputer at the Stanford University Information Systems Laboratory. The simple example of Table 1 was run in BASIC on a Cromemco System 3 microcomputer. As a check, the microcomputer program was also used to design quantizers for the Gaussian case of Figure 2 using the splitting method, $k = 1, 2$, and 3, and a training sequence of 10,000 vectors. The results agreed with the PDP 11/34 run to within one percent.

ACKNOWLEDGMENT

The authors would like to acknowledge the help of J. Markel of Signal Technology, Inc. of Santa Barbara and A. H. Gray, Jr., of the University of California at Santa Barbara in both the analysis and synthesis of the speech example.

REFERENCES

[1] Lloyd, S. P., "Least Squares Quantization in PCM's," Bell Telephone Laboratories Paper, Murray Hill, NJ, 1957.
[2] Gray, R. M., J. C. Kieffer and Y. Linde, "Locally Optimal Block Quantization for Sources without a Statistical Model," Stanford University Information Systems Lab Technical Report No. L-904-1, Stanford, CA, May 1979 (submitted for publication).
[3] Itakura, F. and S. Saito, "Analysis Synthesis Telephony Based Upon Maximum Likelihood Method," *Repts. of the 6th Internat'l. Cong. Acoust.*, Y. Kohasi, ed., Tokyo, C-5-5, C17-20, 1968.
[4] Itakura, F., "Maximum Prediction Residual Principle Applied to Speech Recognition," *IEEE Trans. ASSP*, 23, pp. 67-72, Feb. 1975.
[5] Chaffee, D. L., "Applications of Rate Distortion Theory to the Bandwidth Compression of Speech Signals," Ph.D. Dissertation, Univ. of Calif. at Los Angeles, 1975.
[6] Gray, R. M., A. Buzo, A. H. Gray, Jr., and J. D. Markel, "Source Coding and Speech Compression," *Proc. of the 1978 Internat'l. Telemetering Conf.*, pp. 371-878, 1978.
[7] Matsuyama, Y., A. Buzo and R. M. Gray, "Spectral Distortion Measures for Speech Compression," Stanford Univ. Inform. Systems Lab. Tech. Rept. 6504-3, Stanford, CA, April, 1978.
[8] Buzo, A., "Optimal Vector Quantization for Linear Predicted Coded Speech," Stanford Univ., Ph.D. Dissertation, Dept. of Elec. Engrg., August, 1978.
[9] Matsuyama, Y. A., "Process Distortion Measures and Signal Processing," Ph.D. Dissertation, Dept. of Elec. Engrg., Stanford Univ., 1978.
[10] Markel, J. D. and A. H. Gray, Jr., *Linear Prediction of Speech*, Springer-Verlag, NY 1976.
[11] Gray, A. H., Jr., R. M. Gray and J. D. Markel, "Comparison of Optimal Quantizations of Speech Reflection Coefficients," *IEEE Trans. ASSP*, Vol. 24, pp. 4-23, Feb. 1977.
[12] Dalenius, T., "The Problem of Optimum Stratification," *Skandinavisk Aktuuvielidskrift*, Vol. 33, pp. 203-213, 1950.
[13] Fisher, W. D., "On a Pooling Problem from the Statistical Decision Viewpoint," *Econometrica*, Vol. 21, pp. 567-585, 1953.
[14] Cox, D. R., "Note on Grouping," *J. of the Amer. Statis. Assoc.*, Vol. 52, pp. 543-547, 1957.
[15] Max, J., "Quantizing for Minimum Distortion," *IRE Trans. on Inform. Theory*, IT-6, pp. 7-12, March 1960.
[16] Fleischer, P., "Sufficient Conditions for Achieving Minimum Distortion in a Quantizer," *IEEE Int. Conv. Rec.*, pp. 104-111, 1964.
[17] Luenberger, D. G., *Optimization by Vector Space Methods*, John Wiley & Sons, NY, 1969.
[18] Luenberger, D. G., *Introduction to Linear and Nonlinear Programming*, Addison-Wesley, Reading, MA, 1973.
[19] Rockafellar, R. T., *Convex Analysis*, Princeton Univ. Press, Princeton, NJ, 1970.
[20] Zador, P., "Topics in the Asymptotic Quantization of Continuous Random Variables," Bell Telephone Laboratories Technical Memorandum, Feb. 1966.
[21] Gersho, A., "Asymptotically Optimal Block Quantization," *IEEE Trans. on Inform. Theory*, Vol. IT-25, pp. 373-380, 1979.
[22] Chen, D. T. S., "On Two or More Dimensional Optimum Quantizers," *Proc. 1977 IEEE Internat'l. Conf. on Acoustics, Speech, & Signal Processing*, pp. 640-643, 1977.
[23] Caprio, J. R., N. Westin and J. Esposito, "Optimum Quantization for Minimum Distortion," *Proc. of the Internat'l. Telemetering Conf.*, pp. 315-323, Nov. 1978.

[24] Menez, J., F. Boeri, and D. J. Esteban, "Optimum Quantizer Algorithm for Real-Time Block Quantizing," *Proc. of the 1979 IEEE Internat'l. Conf. on Acoustics, Speech, & Signal Processing*, pp. 980-984, 1979.

[25] MacQueen, J., "Some Methods for Classification and Analysis of Multivariate Observations," *Proc. of the Fifth Berkeley Symposium on Math., Stat. and Prob.*, Vol. 1, pp. 281-296, 1967.

[26] Ball, G. H. and D. J. Hall, "Isodata—An Iterative Method of Multivariate Analysis and Pattern Classification," in *Proc. IFIPS Congr.*, 1965.

[27] Levinson, S. E., L. R. Rabiner, A. E. Rosenberg and J. G. Wilson, "Interactive Clustering Techniques for Selecting Speaker-Independent Techniques for Selecting Speaker-Independent Reference Templates for Isolated Word Recognition," *IEEE Trans. ASSP*, Vol. 27, pp. 134-141, 1979.

[28] Berger, T., *Rate Distortion Theory*, Prentice-Hall, Englewood Cliffs, NJ, 1971.

[29] Yamada, Y., S. Tazaki and R. M. Gray, "Asymptotic Performance of Block Quantizers with Difference Distortion Measures," to appear, *IEEE Trans. on Inform. Theory*.

[30] Grenander, U. and G. Szego, *Toeplitz Forms and Their Applications*, Univ. of Calif. Press, Berkeley, 1958.

[31] Forgy, E., "Cluster Analysis of Multivarate Data: Efficiency vs. Interpretability of Classifications," Abstract, *Biometrics*, Vol. 21, p. 768, 1965.

[32] Chaffee, D. L. and J. K. Omura, "A Very Low Rate Voice Compression System," Abstract in *Abstracts of Papers, 1974, IEEE Intern. Symp. on Inform. Theory*, Notre Dame, Oct. 28-31, IEEE, 1974.

[33] Buzo, A., A. H. Gray, Jr., R. M. Gray, and J. D. Markel, "Speech Coding Based on Vector Quantization," submitted for publication.

Asymptotically Optimal Block Quantization

ALLEN GERSHO, SENIOR MEMBER, IEEE

***Abstract*—In 1948 W. R. Bennett used a companding model for nonuniform quantization and proposed the formula**

$$D = \frac{1}{12N^2} \int p(x)[E'(x)]^{-2}\, dx$$

for the mean-square quantizing error where N is the number of levels, $p(x)$ is the probability density of the input, and $E'(x)$ is the slope of the compressor curve. The formula, an approximation based on the assumption that the number of levels is large and overload distortion is negligible, is a useful tool for analytical studies of quantization. This paper gives a heuristic argument generalizing Bennett's formula to block quantization where a vector of random variables is quantized. The approach is again based on the asymptotic situation where N, the number of quantized output vectors, is very large. Using the resulting heuristic formula, an optimization is performed leading to an expression for the minimum quantizing noise attainable for any block quantizer of a given block size k. The results are consistent with Zador's results and specialize to known results for the one- and two-dimensional cases and for the case of infinite block length ($k \to \infty$). The same heuristic approach also gives an alternate derivation of a bound of Elias for multidimensional quantization. Our approach leads to a rigorous method for obtaining upper bounds on the minimum distortion for block quantizers. In particular, for $k=3$ we give a tight upper bound that may in fact be exact. The idea of representing a block quantizer by a block "compressor" mapping followed with an optimal quantizer for uniformly distributed random vectors is also explored. It is not always possible to represent an optimal quantizer with this block companding model.

I. Introduction

DIGITAL CODING of analog sources is today a subject of considerable importance, yet very little is understood about optimal block quantization. On the one hand, extensive results are available for the one-dimensional (or zero-memory) quantizer. On the other hand, useful bounds are available in the limiting case where the block length approaches infinity. What is needed is a theory of quantization for finite block lengths of arbitrary size. In this note an attempt is made to apply some of the appealing features of the one-dimensional theory to the study of block quantization. A heuristically derived formula is found for the asymptotic case of high-quality quantization. This formula specializes to known results for the one- and two-dimensional cases and for the limiting case of infinite block length.

Manuscript received March 3, 1978; revised January 9, 1979. This work was supported in part by the Electronics Program of the Office of Naval Research while the author was visiting the Department of System Science at the University of California, Los Angeles. This paper was presented at the Information Theory International Symposium, Cornell University, Ithaca, NY, October 10–14, 1977.

The author is with Bell Laboratories, Murray Hill, NJ 07974.

II. Formulation

Let X be a k dimensional random vector with joint density $p(x) = p(x_1, x_2, \cdots, x_k)$. An N point "block" quantizer is a function $Q(x)$ which maps x in R_k into one of N output vectors or "output points" $y_1, y_2, \cdots, y_N$ each in R_k. The quantizer is specified by the values of the output points and by a partition of the space R_k into N disjoint and exhaustive regions $S_1, S_2, \cdots, S_N$, where $S_i = Q^{-1}(y_i) \subset R_k$. Then

$$Q(x) = y_i, \qquad \text{if } x \text{ is in } S_i$$

for $i = 1, 2, \cdots, N$. The term "block" quantizer is used to indicate that the quantizer operates on a "block" of k random variables, i.e., a k-dimensional random vector.

The performance of such a quantizer can be measured by the distortion:

$$D = \frac{1}{k} E\|X - Q(X)\|^r$$

where $\|\cdot\|$ denotes the usual l_2 norm. We assume that $E\|X\|^r$ is finite. Note that for $r=2$, D is the familiar mean-square "per-letter" distortion measure and for $k=1$ it is the usual rth absolute moment of the quantizing error.

We wish to determine a) the minimum distortion $D_1(N)$ attainable over the set of all N point quantizers and b) the minimum distortion $D_2(H_Q)$ attainable over the set of all quantizers having a fixed output entropy H_Q where

$$H_Q = -\sum_{1}^{N} p_i \log p_i \qquad p_i = P\{X \in S_i\}.$$

We consider only the asymptotic case of high quality quantization where N is very large in problem a) or H_Q is very large in problem b).

III. Previous Work

In 1948 W. R. Bennett [1] modeled the one-dimensional quantizer using a memoryless monotonically increasing nonlinearity $E(x)$ (called the compressor) followed by a uniform N point quantizer. This model is completely general since any finite partition of the real line into intervals can be obtained in this way using a suitable continuous compressor curve. He showed that the distortion could be approximated by the integral

$$D \cong \frac{1}{12N^2} \int_{L_1}^{L_2} \frac{p(x)}{[\lambda(x)]^2}\, dx \tag{1}$$

where $\lambda(x) = E'(x)/(L_2 - L_1)$. The result assumes that the $N-2$ finite regions S cover the interval (L_1, L_2) and that L_1 and L_2 are appropriately chosen so that the contribu-

Reprinted from *IEEE Trans. Inform. Theory,* vol. IT-25, pp. 373–380, July 1979.

tion to the distortion due to the tail or "overload" regions can be neglected. The integral is based on some implicit regularity conditions on the density $p(x)$ and on the assumption that N is very large. Bennett's formula is a convenient analytical tool for optimization studies of one dimension quantization.

Several authors have pursued the problem of minimizing the distortion in one-dimensional quantization. Panter and Dite [2] found an expression for the minimum mean-square distortion ($r=2$) for large N. Lloyd [3] found optimality conditions and an algorithmic approach for finding the optimum quantizer valid for each N. Smith [4] was the first to use Bennett's formula to find the best compressor curve and the minimum distortion for optimum quantization for large N. Algazi [5] used the rth power distortion measure and showed that for large N the minimum distortion is

$$D_1(N)=\frac{1}{r+1}(2N)^{-r}\|p_1(x_1)\|_{1/(1+r)} \tag{2}$$

where $p_1(x_1)$ is the (one-dimensional) density of the random variable X_1 and

$$\|p(x)\|_\alpha=\left[\int[p(x)]^\alpha\,dx\right]^{1/\alpha}. \tag{3}$$

For the fixed entropy constraint, Gish and Pierce [6] solved the one-dimensional problem. Their result can be expressed as

$$D_2(H_Q)=\frac{1}{r+1}2^{-r}e^{-r[H_Q-H(p_1)]} \tag{4}$$

where $H(p_1)$ is the differential entropy of X_1, $H(p_1)=-\int p_1(x_1)\log p_1(x_1)\,dx_1$. Both (2) and (4) are valid only as asymptotic results for high quality quantization.

Extensions to block quantization have been studied by Zador [7], Schutzenberger [8], and Elias [9]. Schutzenberger derived an inequality bounding the distortion D for given H_Q, namely,

$$D\geqslant Ke^{-H_Q r/k}$$

where K is an unspecified constant depending on k, r, and $p(x)$. Elias defined the quantizer distortion measure $D^*=\sum_i^N P(S_i)[V(S_i)]^{r/k}$ where $V(S_i)$ is the k-dimensional volume of the region S_i. He showed that

$$D^*\geqslant N^{-r/k}\|p(x)\|_{k/k+r} \tag{5}$$

where $\|p(x)\|_\alpha$, the L_α norm of $p(x)$, is defined in (3) except that the integration is now k-dimensional. He also showed that for N sufficiently large there exists a quantizer with D^* arbitrarily close to this bound. Elias assumed the input vector $\boldsymbol{x}$ to be bounded so that each region S_i has finite volume. Zador, in a lengthy unpublished manuscript, found for the asymptotic case of high quality quantization that

$$D_1(N)=A(k,r)N^{-r/k}\|p(x)\|_{k/(k+r)} \tag{6}$$

and that

$$D_2(H_Q)=B(k,r)e^{-r/k[H_Q-H(p)]} \tag{7}$$

where $H(p)$ is the differential entropy of the random vector $\boldsymbol{X}$. Zador did not obtain $A(k,r)$ or $B(k,r)$ explicitly, but he showed that

$$\frac{1}{1+\beta}V_k^{-\beta}\leqslant kB(k,r)\leqslant kA(k,r)\leqslant\Gamma(1+\beta)V_k^{-\beta} \tag{8}$$

where $\beta=r/k$, $\Gamma(x)$ is the gamma function, and V_k is the volume of a unit sphere in k dimensions. A key feature of Zador's result is that $A(k,r)$ and $B(k,r)$ are independent of the density $p(\boldsymbol{x})$. Hence these functions can be studied in the simpler context of the uniform density on a unit cube in k dimensions. For $k=2$ and $r=2$, Fejes Toth [10] found the (asymptotically) minimum distortion having the form (6) with the explicit value $A(2,2)=5\sqrt{3}/108$.

Recently Gray and Gray [11] gave a simple derivation of the one-dimensional results (2) and (4) using Bennett's integral. Here we explore a similar approach but generalized to the k-dimensional case.

IV. Admissable and Optimal Polytopes

In the one-dimensional case, Bennett's integral is derived by separating the description of a quantizer into two aspects: a) a uniform quantizer that is optimal for the uniform density function and b) the compressor slope function which determines how the output points of the uniform quantizer must be redistributed to take into account the probability density function of the random variable to be quantized. Zador's expression for the minimal distortion also has the striking feature that the factor $A(k,r)$ is independent of the probability density of the random variable. Since $\|p(\boldsymbol{x})\|_\alpha=1$ if $p(\boldsymbol{x})$ is unity in a bounded region of unit volume and zero elsewhere, it follows that $A(k,r)$ is determined by the optimal quantizer for a *uniformly* distributed random variable. These observations suggest that Bennett's integral can be generalized by first considering the optimum quantizer for a uniformly distributed k-dimensional random variable and then considering the effect of a nonuniform distribution of output points on the distortion of a quantizer. We begin by exploring some relevant geometrical features of partitions in R_k.

For every finite (or countably infinite) set of points $\boldsymbol{y}_1,\boldsymbol{y}_2,\cdots,\boldsymbol{y}_N$ in R_k a Dirichlet partition is defined with each point in S_i closer to $\boldsymbol{y}_i$ than to any other $\boldsymbol{y}_j$ for $j\neq i$. That is,

$$S_i=\{\boldsymbol{x}:\ \|\boldsymbol{x}-\boldsymbol{y}_i\|\leqslant\|\boldsymbol{x}-\boldsymbol{y}_j\|\text{ for each }j\neq i\}.$$

An optimal N-point quantizer that minimizes distortion will clearly have a Dirichlet partition. An example of a Dirichlet partition in the plane is shown in Fig. 1. In general, each bounded Dirichlet region is a polytope (bounded by segments of $k-1$ dimensional hyperplanes) and is convex. For a quantizer an effective partition would have the property that the unbounded regions or "overload" regions would make a sufficiently small contribution to the distortion. This is always possible when $E\|X\|^r<\infty$.

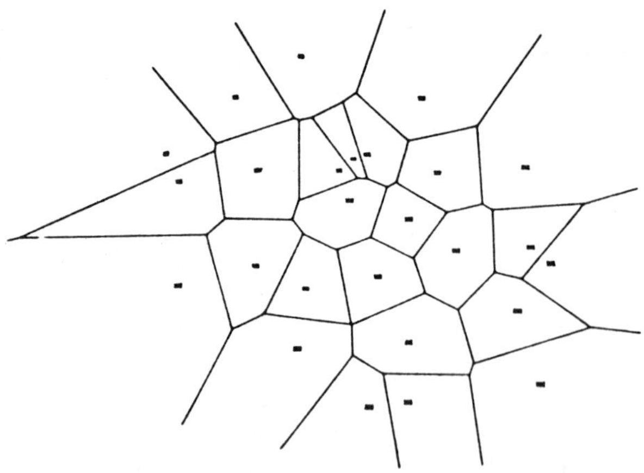

Fig. 1. Direchlet partition for Cambridge, Massachusetts, schools (from H. L. Loeb [17]).

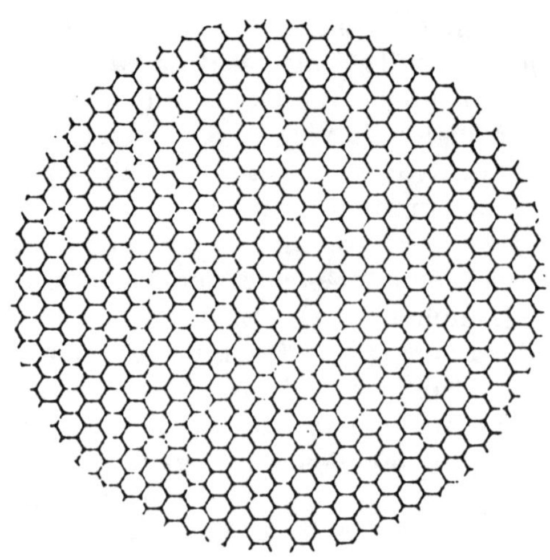

Fig. 2. Tesselation of regular hexagons.

The *centroid* $\hat{y}$ of a convex polytope H in R_k is the value of y that minimizes $\int_H \|x-y\|^r\, dx$. For $r=2$, $\hat{y}$ coincides with the usual definition for the centroid of a body with uniform mass distribution. It should be noted (see Fig. 1) that in general the points generating a Dirichlet partition are not necessarily the centroids of their respective regions. For a uniformly distributed random vector x, a quantizer will have a Dirichlet partition defined on the bounded set in R_k where $p(x)$ is positive. For the quantizer that minimizes distortion it is clearly necessary that each output point will be the centroid of the region in which it lies. The two necessary conditions for optimality, i.e., that the partition be a Dirichlet partition and that the output points be centroids, were noted for $k=1$ by Lloyd [3].

A convex polytope H is said to generate a tessellation if there exists a partition of R_k whose regions are all congruent to H. For example in the plane all triangles, quadrilaterals, and hexagons generate tessellations.

We now make the basic conjecture that for N sufficiently large the optimal (distortion-minimizing) quantizer for a random vector uniformly distributed on some convex set S will have a partition whose regions are all congruent to some polytope H, with the possible exception of regions touching the boundary of S. In other words, the optimal partition is essentially a tessellation of S. This conjecture plays a key role in the heuristic approach which follows.

We define H_k, the class of admissible polytopes in R_k as follows. A convex polytope H in R_k is in H_k if a)H generates a tessellation that is a Dirichlet partition with respect to the centroids of each region in the partition. For example, the equilateral triangle, the rectangle, and the regular hexagon are the admissible polygons in H_2. Fig. 2 illustrates a tessellation of the regular hexagon. Now we define the *normalized inertia* $I(H)$ of a polytope H as

$$I(H)=\int_H \|x-\hat{x}\|^r\, dx/[V(H)]^{1+r/k} \qquad (9)$$

where $\hat{x}$ is the centroid of H and $V(H)$ is the k-dimensional volume of H. The normalization has the property that $I(\alpha H)=I(H)$ for $\alpha>0$ where the polytope $\alpha H=\{\alpha x: x\in H\}$. In other words, when the size of H is scaled, its normalized inertia remains unchanged.

We define the *coefficient of quantization*

$$C(k,r) \triangleq \frac{1}{k} \inf_{H\in H_k} I(H).$$

An *optimal polytope* H^* is an admissible polytope which attains the minimum inertia of all admissable polytopes with the same volume. Hence

$$I(H^*)=kC(k,r).$$

By calculating the normalized inertia of each admissible polygon, it can be shown that for $k=2$ and $r=2$, the optimal polytope is the regular hexagon. We conjecture that an optimal polytope exists for each k.

It is a classic isoperimetric result that every convex polytope has a greater moment of inertia with respect to its centroid than a k-dimensional sphere with the same volume. For the unit radius sphere B centered at the origin it is known that

$$\int_B \|x\|^r\, dx=\frac{k}{k+r}V_k$$

where V_k is the volume of B. Hence we have

$$I(B)=\frac{k}{k+r}V_k^{-r/k}$$

so that we have the lower bound

$$C(k,r) \geqslant \frac{1}{k+r}V_k^{-r/k}. \qquad (10)$$

An upper bound on $kC(k,r)$ can be found by calculating the normalized inertia for any admissible polytope in H_k. One such choice is the k-dimensional cube (centered at the origin), which is clearly admissible. The cube has normalized inertia $k/[(r+1)2^r]$ so that

$$C(k,r) \leqslant \frac{1}{1+r}2^{-r}. \qquad (11)$$

Note that this bound is independent of the dimension k.

V. Heuristic Derivation of the Distortion Integral

Generalizing the concept of "asymptotic fractional density of quanta" introduced by Lloyd in a classic paper [3] on one-dimensional quantization, define the output point density function of a k-dimensional quantizer as

$$g_N(x)=\frac{1}{NV(S_1)}, \qquad \text{if } x\in S_i, \text{ for } i=1,2,\cdots,N.$$

where $V(S_1)$ denotes the volume of S_1. Note that $g_N(x)=0$ if x is in a region of the partition having infinite volume. In the asymptotic situation where N is very large, $g_N(x)$ can be expected to approximate closely a continuous density function $\lambda(x)$ having unit volume. Then $\lambda(x)\Delta V(x)$ may be taken as the fraction of output points located in an incremental volume element $\Delta V(x)$ containing x. Thus the volume of the quantizing region S_i associated with the output point y_i is given approximately by

$$V(S_i)\approx\frac{1}{N\lambda(y_i)} \tag{12}$$

for every bounded region S_i. Note that $N\lambda(y_i)$ is the number of points per unit volume in the neighborhood of y_i so that its reciprocal (12) is the volume per output point.

The distortion (1) can be expressed as

$$D=\frac{1}{k}\sum_{i=1}^{N}\int_{S_i}\|x-y_i\|^r p(x)\,dx. \tag{13}$$

For N large it is reasonable to assume that most of the regions S_i will be bounded sets, and the "overload" regions S_i will correspond to the tail region of the density $p(x)$. Assume the partition has been suitably chosen so that the overload distortion is negligible, treat N as the number of bounded regions, and for N large make the approximation

$$p(x)\backsim p(y_i), \qquad \text{for } x\in S_i.$$

Then we obtain

$$D=\frac{1}{k}\sum_{i=1}^{N}p(y_i)\int_{S_i}\|x-y_i\|^r\,dx. \tag{14}$$

As N becomes large the partition for any bounded region should look more and more like the partition for a uniform density, assuming $\lambda(x)$ is smoothly varying. Thus we approximate S_i by a suitably rotated, translated, and scaled optimal polytope H^*. Then

$$\int_{S_i}\|x-y_i\|^r\,dx=I(H^*)[V(S_i)]^{1+r/k} \tag{15}$$

using (9). We then have

$$D=\frac{1}{k}\sum_{r=1}^{N}p(y_i)I(H^*)[V(S_i)]^{1+r/k}, \tag{16}$$

and from (12) we obtain

$$D=N^{-\beta}C(k,r)\sum_{i=1}^{N}p(y_i)[\lambda(y_i)]^{-\beta}V(S_i). \tag{17}$$

The summation can be approximated by an integral yielding

$$D=N^{-\beta}C(k,r)\int\frac{p(y)}{[\lambda(y)]^{\beta}}\,dy. \tag{18}$$

The region of integration is actually the union of all bounded regions of the partition but may be taken to be the entire k-dimensional space since the contribution to the distortion of the overload regions will be negligible for any reasonable quantizer with sufficiently large N.

Equation (18) may be recognized as the k-dimensional version of Bennett's formula (1) for one-dimensional quantization with mean-square distortion.

VI. Minimization of the Distortion Integral

The distortion integral (18) allows the minimization of the distortion by optimizing the choice of $\lambda(x)$, the asymptotic output point density function. No reference is needed to the explicit quantizer characteristics (the output points and partition regions).

For problem a), D is to be minimized over all quantizers with N fixed. Hölder's inequality gives

$$\int p\lambda^{-\beta}\,dy\{\int\lambda\,dy\}^{\beta} \geqslant \{\int(p\lambda^{-\beta})1/(1+\beta)\lambda\beta/(1+\beta)\,dy\}^{1+\beta}.$$

Noting that $\int\lambda\,dy=1$, we obtain the result

$$\int p\lambda^{-\beta}dy\geqslant\|p\|_{1/(1+\beta)}$$

with equality attained only when λ is proportional to $p^{1/(1+\beta)}$. Hence the minimum value of D, referring to (18), is

$$D_1(N)=C(k,r)N^{-\beta}\|p(x)\|_{k/(k+r)}. \tag{19}$$

This is the desired result. Note that (19) coincides with Zador's result (6) when we take $A(k,r)=C(k,r)$. Furthermore using (10) we obtain a lower bound for $D_1(N)$ that coincides with Zador's lower bound.

A significant property of the optimum quantizer can now be demonstrated. Since the optimum point density λ is proportional to $p^{1/(1+\beta)}$, we observe that each term in the sum (16) reduces to a constant independent of the index i. *Therefore each region S_i of the partition makes an equal contribution to the distortion for an optimal quantizer.* This property was observed by Panter and Dite [2] for $k=1$ and by Fejes Toth [10] for $k=2$.

In problem b), D is to be minimized subject to a constraint on the quantizer output entropy H_Q. Since $p_i\backsim p(y_i)V(S_i)$ for each bounded set S_i and for large N,

$$H_Q=-\Sigma p(y_i)\frac{1}{N\lambda(y_i)}\log[p(y_i)/N\lambda(y_i)]$$

$$=-\Sigma p(y_i)\log p(y_i)\Delta V(y_i)-\Sigma p(y)\log\frac{1}{N\lambda(y_i)}\Delta V(y_i)$$

where $\Delta V(y_i)=1/N\lambda(y_i)$. As in the derivation of (18), the sums can be approximated by integrals, yielding

$$H_Q = H(p) - \int p(y) \log \frac{1}{N\lambda(y)} \, dy \tag{20}$$

where $H(p)$ is the differential entropy of the random vector $\boldsymbol{X}$. Equation (20) reduces for $k=1$ to the corresponding one-dimensional result given by Gish and Pierce [6]. From (18) we have

$$D = C(k,r) \int e^{-\beta \log [N\lambda(y)]} p(y) \, dy. \tag{21}$$

Now applying Jensen's inequality we get

$$D \geqslant C(k,r) e^{-\beta \int p(y) \log [N\lambda(y)] \, dy}. \tag{22}$$

Applying (20) we see that

$$D \geqslant C(k,r) e^{-\beta [H_Q - H(p)]}. \tag{23}$$

The application of Jensen's inequality yields an equality when $\lambda(y)$ is a constant corresponding to a *uniform* distribution of output points. Hence the solution to problem b) is

$$D_2(H_Q) = C(k,r) e^{-\beta [H_Q - H(p)]}. \tag{24}$$

Note that (24) coincides with Zador's result (7) when we take $B(k,r) = C(k,r)$. Furthermore applying the lower bound (10) to (24) gives a bound for $D_2(H_Q)$ which coincides with Zador's lower bound (8). *It is significant to observe that for large N the optimal quantizer for a constrained entropy is very nearly, the uniform quantizer.* For $k=1$ this was noted by Gish and Pierce [6].

As an additional illustration of the use of the function $\lambda(\boldsymbol{x})$ we give a heuristic derivation of Elias' result [9]. Since

$$P(S_i) \approx p(y_i) V(S_i),$$

$$D^* \approx \sum_{i=1}^{N} p(y_i) V(S_i)^{r/k} V(S_i).$$

Using (11) gives

$$D^* \approx \sum_{i=1}^{N} p(y_i) \left[\frac{1}{N\lambda(y_i)} \right]^{r/k} \Delta V(y_i),$$

and approximating the sum for N large by an integral yields

$$D^* = N^{-\beta} \int \frac{p(y)}{[\lambda(y)]^{\beta}} \, dy. \tag{25}$$

The minimization of this integral as shown above leads to the result that

$$D^* = N^{-\beta} \| p(\boldsymbol{x}) \|_{k/k+r} \tag{26}$$

which is Elias' lower bound.

Finally, it should be noted that the formulas (2), (4), (6), (7), (18), (19), (24), and (26), which have been written as equalities, should more correctly be taken as lower bounds on attainable distortion for any finite N. Since the minimum distortion attainable is nonincreasing as N (or H_Q) increases, the actual distortion can only be greater than these asymptotic values for any quantizer with a finite number of quantizing regions N.

VII. Special Cases and Bounds

For $k=1$, a finite interval on the real line is the only admissible polytope. The interval is therefore the optimal polytope for $k=1$. Calculating its normalized inertia gives

$$C(1,r) = \frac{1}{r+1} 2^{-r}.$$

For $r=2$, we have $C(1,2) = 1/12$ and hence our generalized Bennett integral (18) reduces to the original Bennett integral (1). For $k=1$, the minimum distortion formula (19) coincides with the known result (2) as given by Algazi [5]. Also for $k=1$, our constrained-entropy minimum distortion formula (24) reduces to the known result (4) due to Gish and Pierce [6].

For $k=2$ we have already noted that the regular hexagon is the optimal polytope. This yields the coefficient of quantization

$$C(2,2) = \frac{5}{36\sqrt{3}}.$$

A theorem by Fejes Toth [12] shows in effect that for a uniformly distributed random variable the minimum distortion for each r is obtained by a tessellation of regular hexagons. Newman [13] independently found a proof of this result for $r=2$. Their results imply that Zador's coefficient $A(2,2)$ has the value $5/(36\sqrt{3})$. Hence the complete solution for nonuniform densities $p(x_1,x_2)$ is in fact given by

$$D \sim \frac{5}{36\sqrt{3}} N^{-1} \left[\int \int \sqrt{p(x_1,x_2)} \, dx_1 \, dx_2 \right]^2$$

asymptotically as $N \to \infty$, when $k=2$ and $r=2$. Unknown to Newman and Zador, Fejes Toth [10] had given a complete proof of this result. Hence our minimum distortion formula reduces to this known result for $k=2$ and $r=2$.

Fejes Toth [10] noted that the optimal partition for a given probability density $p(x_1,x_2)$ in the plane consists of "approximately" regular hexagons with the centroids distributed with a nonuniform density over the plane. An example of a hexagonal partition whose centroids are distributed nonuniformly in the plane is shown in Fig. 3. These results for $k=2$ help to clarify the role of the output point density function $\lambda(\boldsymbol{x})$ in characterizing a quantizer as used in this paper.

For $k \geqslant 3$, the minimum distortion attainable for a quantizer is not known. However we can obtain upper bounds on the quantization coefficient $C(k,r)$ as noted in Section IV by calculating the normalized inertia for any admissible polytope. Any admissible polytope generates a tessellation that can be used for the quantization of a random vector that is uniformly distributed on a unit volume region. Hence neglecting the boundary regions when N is large, the normalized inertia I' of that polytope gives the attainable distortion

$$D = \left(\frac{1}{k} I' \right) N^{-r/k}.$$

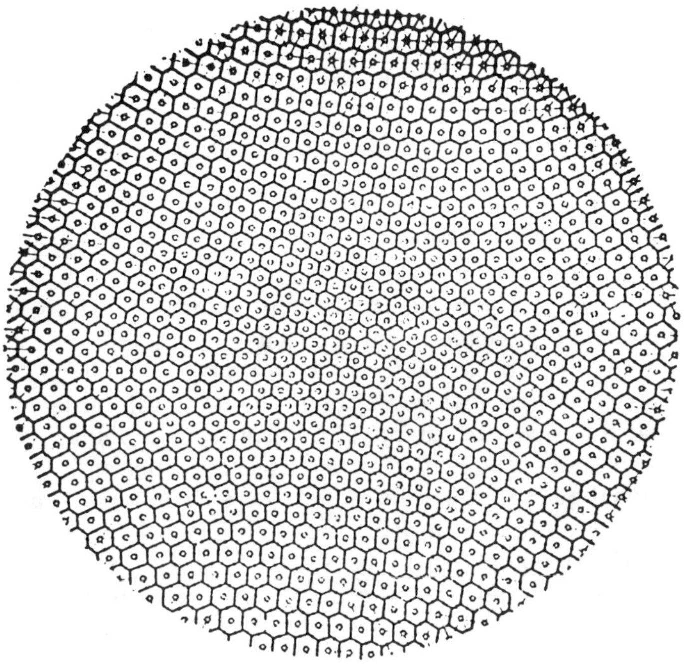

Fig. 3. Hexagonal partition for nonuniform density of points (from Fejes Toth [10]).

Hence

$$A(k,r) \leq \frac{1}{k} I'.$$

Therefore any upper bounds we obtain for $C(k,r)$ are in fact upper bounds for Zador's $A(k,r)$. Even though our derivation of (24) is not rigorous, these upper bounds are rigorously valid.

For $k=3$ the admissible polyhedra include the five principal parallelohedra: cube, hexagonal prism, rhombic dodecahedron, elongated dodecahedron, and the truncated octahedron. Of these five, the truncated octahedron shown in Fig. 4 and specified by the set

$$\{(x_1,x_2,x_3): |x_1|+|x_2|+|x_3|<1.5, |x_1|<1, |x_2|<1, |x_3|<1\}$$

has by direct calculation the smallest normalized inertia with

$$I = \frac{19}{64\sqrt[3]{3}} = 0.23563\cdots$$

which gives the upper bound

$$C(3,2) \leq 0.0785445\cdots$$

which is surprisingly close to the sphere lower bound (0.07697) discussed in Section IV. The truncated octahedron is clearly the best parallelohedron and is very likely the optimal polyhedron.

For $k=4$, the analog of the truncated octahedron is the admissible polytope

$$\{x: |x_1|+|x_2|+|x_3|+|x_4| \leq 2; |x_i| \leq 1, i=1,2,3,4\}$$

which by a crude Monte Carlo integration gives the bound

$$C(4,2) \leq 0.0766\cdots.$$

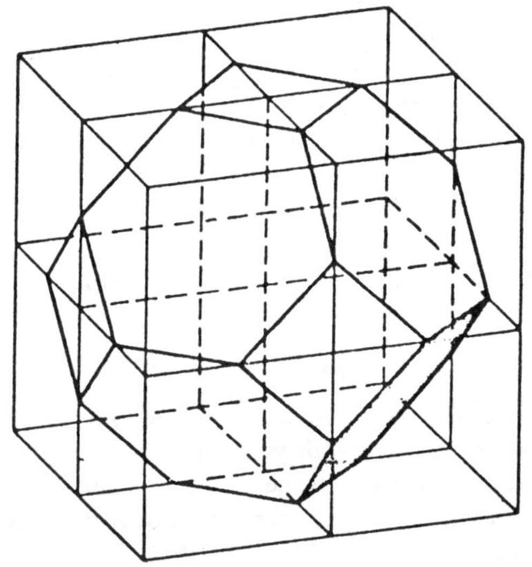

Fig. 4. Truncated octahedron imbedded in cube of side length 2, corresponding to analytical description given in text.

TABLE I
VALUES AND BOUNDS FOR QUANTIZATION COEFFICIENT $C(k,2)$

K	SPHERE LOWER BOUND	ACTUAL VALUE	POLYTOPE UPPER BOUND	CUBE UPPER BOUND	ZADOR UPPER BOUND
1	.08333	.08333(a)		.08333	.5
2	.07958	.08019(b)		.08333	.159
3	.07697		.07854(c)	.08333	.1157
4	.07503		.0766(d)	.08333	.09974
5	.07352		.0779(e)	.08333	.09133
100	.0574		.0766(f)	.08333	.05789

(a) interval
(b) regular hexagon
(c) truncated octahedron
(d) four dimensional analog of (c)
(e) cross product of truncated octahedron with an interval
(f) cross product of 25 type (d) polytopes

One technique for obtaining upper bounds is to select admissible polytopes in a higher dimensional space by forming "cross-products" or prisms using lower dimensional polytopes. For example, the cube in k-space is simply the kth cross-product of the interval, the hexagonal prism in $k=3$ is the cross product of the regular hexagon ($k=2$) with the interval ($k=1$), and the cross product of the regular hexagon with the truncated octahedron gives an admissible polytope for $k=5$. For such cross-product polytopes the normalized moment of inertia when $r=2$ is trivially obtained by summing the normalized moments of inertia of each lower dimensional polytope.

In Table I values of $C(k,2)$ are given when known together with the available lower and upper bounds.

VIII. INFINITE BLOCK LENGTH

All of the preceding results have been based on a fixed (but arbitrary) block length k with $1 \leq k < \infty$. The results can be compared with the known performance bounds of rate distortion theory by examining the limiting case for mean-square distortion ($r=2$) letting $k\to\infty$. Since Zador's upper and lower bounds coincide asymptotically as $k\to\infty$, we use the lower bound to study the limiting behavior of $D_1(N)$. As noted by Zador, Stirling's formula gives $V_k^{-\beta}\sim (2\pi e)^{-1}k$ as $k\to\infty$. Zador also conjectured that in general

$$\|p_k(x)\|_{k/k+2}\to e^{2\bar{H}} \qquad \text{as } k\to\infty \tag{27}$$

where $\bar{H}$ is the differential entropy rate of the source

$$\bar{H}=-\lim_{n\to\infty}\frac{1}{n}\int p_n(x)\log p_n(x)\,dx$$

and a subscript has been added to the joint density $p(x)$ to identify its dimensionality. In the particular case of a stationary Gaussian process, Zador showed that (27) holds. The result (27) appears to be of some fundamental theoretical interest. A proof that it holds for any stationary ergodic process is given in the Appendix. Without assuming ergodicity, the weaker result

$$\lim_{k\to\infty}\|p_k\|_{k/k+2}\geq e^{2\bar{H}} \tag{28}$$

will now be shown

$$\begin{aligned}\log\|p_k\|_{k/k+2}&=\frac{k+2}{k}\log\int p_k(x)e^{-[2/(k+2)]\log p_k(x)}\,dx\\ &\geq\frac{k+2}{k}\int p_k(x)\log e^{-[2/(k+2)]\log p_k(x)}\,dx\\ &=-\frac{2}{k}\int p_k(x)\log p_k(x)\,dx,\end{aligned}$$

using Jensen's inequality. Thus

$$\|p_k\|_{k/k+r}\geq e^{2H(p_k)/k} \tag{29}$$

from which (28) follows.

Now using (19) and the lower bound (10) on $C(k,r)$ gives the result for very large block length k that

$$D\geq\frac{1}{2\pi e}e^{-2(R-\bar{H})} \tag{30}$$

where $R=(\log N)/k$ is the *rate* or average number of nats per component of X needed to identify (or transmit) the quantizer output approximation $Q(x)$. Inverting (30) gives a more familiar result

$$R\geq\bar{H}-\frac{1}{2}\log(2\pi eD), \tag{31}$$

which is the generalized Shannon lower bound on the rate-distortion function (see Berger [14]). Note that (31) as derived here is only valid for the asymptotic case of small distortion (corresponding to large N).

IX. Companding Revisited

For $k=1$, Bennett introduced the "companding" model of a quantizer as a monotonically increasing nonlinear mapping $E(x)$ the compressor, followed by a uniform quantizer, and by the inverse mapping, E^{-1}. Lloyd's point density function $\lambda(x)$ is then the slope of the mapping $E(x)$. For block quantization we have introduced the concept of a point density function and derived a generalization of Bennett's integral without any reference to a mapping. To complete the connection with the one-dimensional case we can define a continuously differentiable and invertible mapping f which maps a point x in R_k into another point $f(x)$ in R_k. Then a family of k dimensional quantizers can be modeled as shown in Fig. 5 as the cascade of such a mapping with an optimal N-point uniform quantizer, let us say on the unit cube in R_k followed by the inverse mapping f^{-1}. The overall quantizer is described by

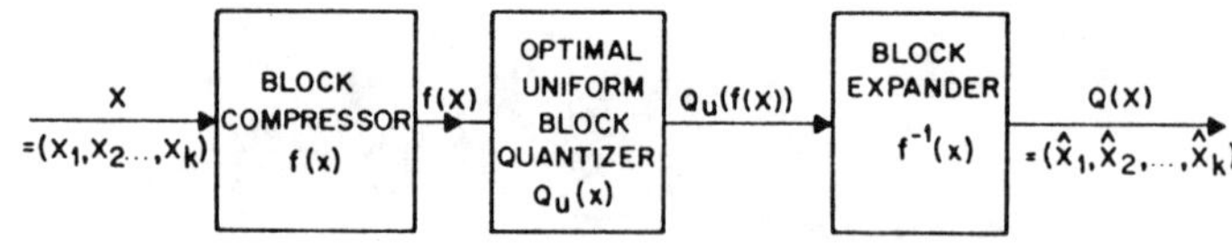

Fig. 5. Companding model of block quantization.

$$Q(x)=f^{-1}\circ Q_u\circ f(x)$$

where $Q_u(x)$ denotes the uniform quantizer. The point density function for the overall quantizer is then given by the Jacobian determinant of f. The Dirichlet partition for the uniform quantizer with regions S_i, will induce a new partition for the overall quantizer with regions $S_i^*=f^{-1}(S_i)$. In general the new partition will not be a Dirichlet partition.

The question then arises, for a given probability density $p(x)$, does there exist a mapping $f(x)$ which makes $Q(x)$ the optimum quantizer? In order to preserve the Dirichlet property it is necessary that the mapping be conformal since the line joining the centroids of two adjacent regions in a Dirichlet partition must always be perpendicular to the hyperplane separating the two regions. For $k=2$, Heppes and Szusz [15] have noted, in effect, that a necessary and sufficient condition for the existence of $f(x)$ is that the logarithm of the point density function be a harmonic function, i.e., that $\log\lambda(x)$ satisfy Laplace's equation. Since for the optimum quantizer we have shown that λ must be proportional to a power of $p(x)$, it follows that the condition is equivalent to having $\log p(x)$ satisfy Laplace's equation. This condition eliminates the joint normal density as well as any other density whose curves of constant density close. Hence there appears to be a fundamental limitation to the possibility of generalizing Bennett's companding approach to the multidimensional case.

X. Concluding Remarks

In this paper we have shown that the point density function of a quantizer, first conceptualized by Lloyd, can be generalized to the multidimensional case to provide a fruitful and intuitively satisfying way to develop the theory of optimal quantization for random vectors. With it we have heuristically generalized the classic integral of Bennett for the distortion of a quantizer which we hope will be useful for future studies of block quantization. In deriving the results of Zador, albeit heuristically, we have gone further toward a constructive theory of optimal quantization by introducing some of the salient geometrical features of the partition of space defined by a quantizer. Finally, we have pointed out the possibilities and limitations of the companding approach to modeling a quantizer for random vectors, an approach that has been of great practical importance in the one-dimensional case.

Acknowledgment

I would like to thank Thomas Liggett of the University of California, Los Angeles, who provided the proof given in the Appendix, Hans Witsenhausen, who introduced me to the work of Fejes Toth, and Neil Sloane, who introduced me to the polyhedra associated with sphere packings.

Appendix[1]

Theorem 1:

$$\lim_{k\to\infty} \|p_k(x)\|_{k/k+r} = e^{r\bar{H}}, \qquad 0<r<r_0$$

if $p_k(x)$ is the joint density of $X^k=(X_1,X_2,\cdots,X_k)$ where $\{X_k\}$ is a stationary ergodic process and $\|p_1(x_1)\|_{1/(1+r_0)}<\infty$.

Proof: Define the random variables

$$Z_k = -\frac{1}{k}\log p_k(x^k).$$

Then

$$\log \|p_k(x)\|_{k/k+r} = \frac{k+r}{k}\log Ee^{[kr/(k+r)]Z_k}. \tag{A.1}$$

From the Shannon-McMillan theorem for abstract alphabets [16], Z_k converges in probability to $\bar{H}$. Hence $e^{a_kZ_k}$ converges in distribution to $e^{r\bar{H}}$ where $a_k=rk/(k+r)$. If also

$$Ee^{dZ_k}<C<\infty \tag{A.2}$$

where C is independent of k, and $r<d$, then convergence in the mean would follow, namely,

$$Ee^{a_kZ_k}\to e^{r\bar{H}} \qquad \text{as } k\to\infty$$

which would complete the proof. To prove (A.2), define

$$\rho_k(X^k) \triangleq \frac{k+r}{r}\log Ee^{a_kZ_k}.$$

Then (A.2) will follow from the fact that $\rho_k(X^k)$ is subadditive, that is,

$$\rho_{k+l}(X^{k+l}) \leqslant \rho_k(X^k)+\rho_l(X^l), \tag{A.3}$$

which is analogous to the subadditivity of the entropy. Denote $s=k+l+r$, $u=(x_1,x_2,\cdots,x_k)$, $v=(x_{k+1},x_{k+2},\cdots,x_{k+l})$, and let $f=p_{k+l}(u,v)$, $p_k=p_k(u)$, $p_l=p_l(v)$. Then

$$\iint f^{(k+l)/s}\,du\,dv = \iint \left[fp_k^{-r/(k+r)}\right]^{k/s}\left[fp_l^{-r/(l+r)}\right]^{l/s} \cdot \left[p_k^{k/(k+r)}p_l^{l/(l+r)}\right]^{r/s} du\,dv$$

$$\leqslant \left[\iint fp_k^{-r/(k+r)}\,du\,dv\right]^{k/s}\left[\iint fp_l^{-r/(l+r)}\,du\,dv\right]^{l/s} \cdot \left[\iint p_k^{k/(k+r)}p_l^{l/(l+r)}\,du\,dv\right]^{r/s}$$

$$= \left[\int p_k^{k/(k+r)}\,du\right]^{(k+r)/s}\left[\int p_l^{l/(l+r)}\,dv\right]^{(l+r)/s}.$$

using the triple Hölder inequality. Hence $k^{-1}\rho_k(X^k)$ is bounded by $\rho_1(X_1)$ so that

$$\frac{k+r}{k}\log\{Ee^{a_kZ_k}\} \leqslant K\log\|p_1(x_1)\|_{1/(1+r_0)}<\infty \tag{A.4}$$

which holds, in particular, for $r=r_0$. Given d, with $0<d<r_0$, for sufficiently large k,

$$d<\frac{kr_0}{k+r_0}.$$

Hence

$$\log Ee^{dZ_k}<\log Ee^{[kr_0/(k+r_0)]Z_k},$$

so that (A.4) implies (A.2), completing the proof of the theorem.

[1]This proof is due to Thomas Liggett.

References

[1] W. R. Bennett, "Spectra of quantized signals," *Bell Syst. Tech. J.*, vol. 27, pp. 446–472, July, 1948.

[2] P. F. Panter and W. Dite, "Quantization in pulse-count modulation with nonuniform spacing of levels," *Proc. IRE*, vol. 39, pp. 44–48, 1951.

[3] S. P. Lloyd, "Least squares quantization in PCM," unpublished memorandum, Bell Laboratories, 1957.

[4] B. Smith, "Instantaneous companding of quantized signals," *Bell Syst. Tech. J.*, vol. 52, pp. 1037–1076, Sept. 1973.

[5] V. R. Algazi, "Useful approximations to optimum quantization," *IEEE Trans. Commun. Tech.*, vol. COM-14, pp. 297–301, 1966.

[6] H. Gish and J. N. Pierce, "Asymptotically efficient quantizing," *IEEE Trans. Inform. Theory*, vol. IT-14, pp. 676–683, Sept. 1968.

[7] P. Zador, "Asymptotic quantization of continuous random variables," unpublished memorandum, Bell Laboratories, 1966.

[8] M. P. Schutzenberger, "On the quantization of finite dimensional messages," *Inform. Contr.*, vol. 1, pp. 153–158, 1958.

[9] P. Elias, "Bounds and asymptotes for the performance of multivariate quantizers," *Ann. Math. Stat.*, vol. 41, pp. 1249–1259, 1970.

[10] L. Fejes Toth, "Sur la representation d'une population infinie par un nombre fini d'éléments," *Acta Math. Acad. Scient. Hung.*, vol. 10, pp. 299–304, 1959.

[11] R. M. Gray and A. H. Gray, Jr., "Asymptotically optimal quantizers," *IEEE Trans. Inform. Theory*, vol. IT-23, pp. 143–144, Jan. 1977.

[12] L. Fejes Toth, *Lagerungen in der Ebene auf der Kugel und im Raum. Springer-Verlag* p. 81, 1953.

[13] D. J. Newman, "The hexagon theorem," unpublished memorandum, Bell Laboratories, 1964.

[14] T. Berger, *Rate-Distortion Theory.* Englewood Cliffs, N.J.: Prentice-Hall, 1971.

[15] A. Heppes and P. Szüsz, "Bemerkung zu einer Arbeit von L. Fejes Toth, " *El. Math.*, vol. 15, pp. 134–136, 1960.

[16] A. Perez, "On the theory of information in the case of an abstract alphabet," in *Trans. First Prague Conf. on Inform. Theory, Statistical Decisions Functions, and Random Processes,* (held at Prague, Nov. 28–30, 1956). Prague: Publishing House of the Czechoslovak Academy of Sciences, 1957, pp. 209–243.

[17] H. L. Loeb, *Space Structures*, Addison-Wesley, 1975.

On the Structure of Vector Quantizers

ALLEN GERSHO, FELLOW, IEEE

Abstract—Vector quantization is intrinsically superior to predictive coding, transform coding, and other suboptimal and *ad hoc* procedures since it achieves optimal rate distortion performance subject only to a constraint on memory or block length of the observable signal segment being encoded. The key limitation of existing techniques is the very large randomly generated code books which must be stored, and the computational complexity of the associated encoding procedures. The quantization operation is decomposed into its rudimentary structural components. This leads to a simple and elegant approach to derive analytical properties of optimal quantizers. Some useful properties of quantizers and algorithmic approaches are given, which are relevant to the complexity of both storage and processing in the encoding operation. Highly disordered quantizers, which have been designed using a clustering algorithm, are considered. Finally, lattice quantizers are examined which circumvent the need for a code book by using a highly structured code based on lattices. The code vectors are algorithmically generated in a simple manner rather than stored in a code book, and fast algorithms perform the encoding algorithm with negligible complexity.

I. Introduction

ORDINARY one-dimensional scalar quantization is an operation on a single sample value of an analog signal. The quantizer replaces the sample by one of a finite set of representative values. Thereby a digital codeword can be used to identify which representative value best approximates the original analog value.

In vector quantization (also known as block or multidimensional quantization), an ordered set of k samples (i.e., a k-dimensional vector in Euclidean space) is mapped into one of a finite set of representative or "output" vectors. A binary word can then be used to identify which of the standard set of output vectors represents the original input vector. After digital storage or transmission of this codeword, a table lookup or "code book" can be used to reproduce the corresponding output vector, which approximately reconstructs the original input vector.

Manuscript received October 27, 1981; revised November 3, 1981. This work was supported in part by the Air Force Office of Scientific Research under Grant AFOSR-82-0008.

The author is with the Department of Electrical and Computer Engineering, University of California, Santa Barbara, CA 93106.

A quantizer may be viewed as the cascade of a coder and a decoder. The coder identifies in which region of the space the input vector lies and assigns a corresponding codeword. The decoder, which can be implemented as a code book, takes the codeword and generates the corresponding output vector. Note that the coder is essentially defined by a partition of the k-dimensional space of vectors or "points" into a finite number of regions. The geometry of this partition is of fundamental interest in the theory of optimal quantization.

A. Summary of Relevant Prior Work

The basic theory of one-dimensional quantization is reviewed by Gersho [1]. A survey with an extensive bibliography is given by Morris [2]. Although vector quantization is not a new subject, only a few scattered papers appeared on this topic before 1978. Noteworthy are the following papers: Steinhaus [3] first formulated the problem of optimal vector quantization; Schutzenberger [4] found the asymptotic behavior of the mean-square error as a function of the number of representative points; Fejes Toth [5] found the exact asymptotical expression for the two-dimensional case; Newman [6] showed the optimal property of the regular hexagonal array; Dunn [7] found that permutation codes can be used to obtain good quantizers for spherically symmetrical Gaussian vectors; Zador [8] studied the asymptotic properties of multidimensional quantizers in depth and obtained several important results; Elias [9] treated the theory of vector quantization using a performance measure that depended only on the partition and not on the representative points; Chen [10] computed two-dimensional quantizers for the Gaussian density function.

The study of vector quantization has accelerated in the past three years. Gersho [11] introduced the companding

Reprinted from *IEEE Trans. Inform. Theory,* vol. IT-28, pp. 157–166, March 1982.

approach to vector quantization and developed a theoretical approach for asymptotically optimal quantization consistent with Zador's results and based on an extension of Lloyd's point density function. The companding approach has been pursued by Bucklew [12], and the analytical approach was extended to more general performance measures by Yamada *et al.* [13]. Bucklew and Wise [14] recently developed new proofs of the asymptotic theory of vector quantization.

The design of optimal vector quantizers from empirical data was proposed and extensively studied by Linde, Buzo, and Gray [15] using a clustering approach that is closely related to an algorithm used by Lloyd [16] for one-dimensional quantization with statistically specified random variables. The algorithm they have developed has turned out to be of major importance in vector quantization and is referred to here as the LBG algorithm. Adoul *et al.* [17] also proposed the use of a clustering algorithm for two-dimensional quantizer design. A clustering algorithm was also used for one-dimensional quantizer design by Menez *et al.* [18].

Quantizers, whose points lie on a lattice, were defined for two, three, and four dimensions by Gersho [11] and for higher dimensions in unpublished work at Bell Laboratories [31]. Some properties of lattices relevant to quantization are treated by Conway and Sloane in this issue [32]. The basic theory of lattices, a branch of geometric number theory, is presented by Cassels [25]. A comprehensive review of the literature on lattices, sphere packings, and the use of binary codes to generate lattices is given by Sloane [26]. Sloane [27] also gives a useful set of tables of several important lattices.

A major contribution to speech coding based on vector quantization is given by Buzo *et al.* [28]. Subsequent advances in speech coding based on vector quantization are given by Abut *et al.* [29] and Juang *et al.* [30].

B. Plan of this Paper

We introduce a systematic formulation of the structural components of a quantizer. Specifically, the basic structure of a quantizer is described in terms of the partition and output set. Using a convenient set of binary-valued mappings, an analytically useful model of quantization is found and applied to very simply obtain some optimality properties. A quantizer whose partition has the optimal nearest neighbor character is then further decomposed into the fundamental building blocks common to all such quantizers. The specific characteristics of random quantizers, which have been designed using a clustering algorithm, is considered next. We provide some basic tools for determining the faces bounding each region of the partition of a random quantizer. Finally, the special case of lattice quantizers is examined and contrasted with random quantizers whose output points are disordered. The intent of this paper is to develop basic tools for subsequent studies of encoding complexity and fast encoding algorithm design.

II. Basic Structure of Quantization

A quantizer is a mapping of k-dimensional Euclidean space R^k into a finite subset Y of R^k. Thus,

$$Q: R^k \to Y, \tag{1}$$

where $Y = \{y_1, y_2, \cdots, y_N\}$ and y_i, called an output point, is in R^k for each i. A quantizer is refered to as a vector quantizer to distinguish from the special case of a one-dimensional quantizer where $k = 1$.

Associated with every N point quantizer in R^k is a *partition*

$$R_1, R_2, \cdots, R_n,$$

where

$$R_i = Q^{-1}(y_i) = \{x \in R^k: Q(x) = y_i\}. \tag{2}$$

With this definition, it follows that

$$\bigcup_{i=1}^{N} R_i = R^k \quad \text{and} \quad R_i \cap R_j = 0, \qquad \text{for } i \neq j. \tag{3}$$

The quantizer Q is thus uniquely defined by jointly specifying the output set Y and the corresponding partition $\{R_i\}$.

In practice, a quantizer does not exist as a single entity. It is actually the composite of two separate functions, the *coder* and the *decoder*. A coder C is defined as a mapping of R^k into the index set J, and the decoder D is defined as the mapping of J into the output set Y where

$$J = \{1, 2, 3, \cdots, N\}. \tag{4}$$

Thus,

$$C: R^k \to J \quad \text{and} \quad D: J \to R^k. \tag{5}$$

Hence, a quantizer Q is defined by the composition of C and D,

$$Q = D \cdot C. \tag{6}$$

An effective structural characterization of quantizers should provide a compact and elegant description of the fundamental theory of optimal quantization. Also, in order to study the complexity associated with the implementation of a quantizer and the complexity of the encoding operation, it is necessary to decompose the quantizer operation into fundamental and canonical building blocks. We begin by defining the *cell assignment function*, S_i, as the binary valued function

$$S_i: R^k \to \{1, 0\}$$

according to

$$S_i(x) = \begin{cases} 1, & \text{if } x \in R_i, \\ 0, & \text{otherwise.} \end{cases} \tag{7}$$

Thus S_i is the indicator function for the set R_i.

Let x denote an input vector, the input to the coder. Then the binary valued variable a_i is defined by

$$a_i = S_i(x) \tag{8}$$

and the binary-valued vector $\boldsymbol{a}$ is defined by

$$\boldsymbol{a} = \{a_1, a_2, \cdots, a_N\}. \tag{9}$$

Note that one and only one component of the vector $\boldsymbol{a}$ is nonzero. Of course the vector $\boldsymbol{a}$ has a very restricted set of possible values. Nevertheless, it turns out to be worthwhile introducing the binary variables a_i for later analytical convenience as well as for its conceptual value in describing the structure of a quantizer. With these definitions we have a very convenient and compact way to characterize the quantizer function. Thus, an N-point quantizer can be expressed in the form

$$Q(\boldsymbol{x}) = \sum_{i=1}^{N} y_i S_i(\boldsymbol{x}). \tag{10}$$

For the specific vector $\boldsymbol{x}$ as input to the quantizer, the output can be described in the form

$$Q(\boldsymbol{x}) = \sum_{i=1}^{N} y_i a_i. \tag{11}$$

This expression is the first stage of a general decomposition of the quantizer function into structural building blocks.

In order to separately characterize the structure of the coder and decoder, it is convenient to define an *index generator* F as the mapping

$$F\colon \{1,0\}^N \to J,$$

that is, a mapping from the set of binary N-vectors to the index set J of integers from 1 to N. Specifically,

$$F(\boldsymbol{a}) = j \qquad \text{if } j \text{ is the largest index } i \text{ with } a_i = 1. \tag{12}$$

This definition is meaningful for an arbitrary binary vector. An index generator can readily be implemented with combinational logic to produce a binary word that specifies the index. If N is a power of two, the index generator would then map an N bit word into a $\log_2 N$ bit word. Finally, we define an *address generator* as the mapping

$$G\colon J \to \{1,0\}^N,$$

which is defined by

$$G(j) = (\delta_{1j}, \delta_{2j}, \cdots, \delta_{Nj}), \tag{13}$$

where

$$\delta_{ij} = \begin{cases} 1, & \text{if } i = j, \\ 0, & \text{if } i \neq j. \end{cases} \tag{14}$$

The address generator corresponds to what is known as an address decoder used in a computer to enable a particular memory chip. With these definitions, the encoder C can be represented as

$$C = F \cdot S \tag{15}$$

so that

$$C(\boldsymbol{x}) = F(\boldsymbol{a}) = F(S_1(\boldsymbol{x}), S_2(\boldsymbol{x}), \cdots, S_N(\boldsymbol{x})), \tag{16}$$

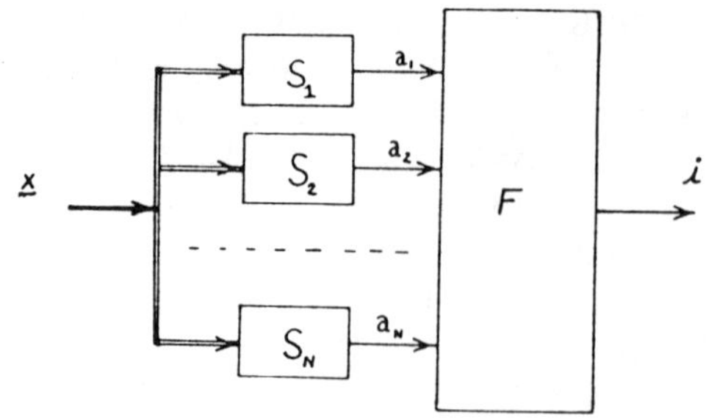

Fig. 1. Coder.

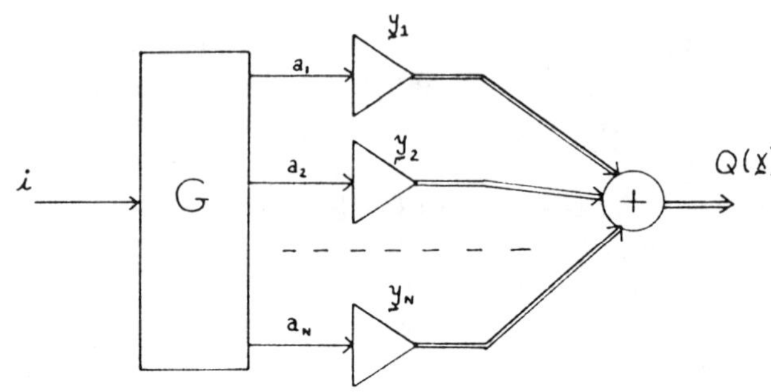

Fig. 2. Decoder.

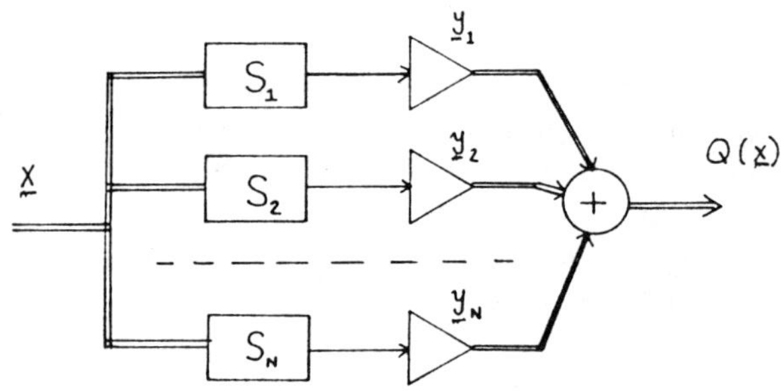

Fig. 3. Quantizer.

and the decoder D is represented as

$$D = \sum_{i=1}^{N} y_i G_i, \tag{17}$$

and so

$$D(j) = y_j. \tag{18}$$

The basic structure of the coder, decoder, and the quantizer are illustrated in Figs. 1, 2, and 3, respectively. Note that any quantizer as depicted in Fig. 3, can be viewed as the composite of a nonlinear and a linear operation. Thus the mapping from $\boldsymbol{x}$ to the binary vector $\boldsymbol{a}$ is a nonlinear operation, while the mapping from $\boldsymbol{a}$ to the output $Q(\boldsymbol{x})$ is a linear operation. This observation is helpful to clearly understand the basic results on optimal quantization and to appreciate the basic mechanism of the algorithmic approach to vector quantizer design.

Without adding further restrictions to the class of quantizers of interest, no further decomposition of the structure of a quantizer is possible.

III. Optimal Quantization

Let X denote a continuously distributed random vector in R^k with a specified joint probability density function, $f(x)$. We focus on the usual Euclidean distance measure and define and N-point quantizer as *optimal* with respect to the random variable X if it minimizes the mean-square Euclidean distance e between X and $Q(X)$. Thus, Q^* is optimal if

$$e = E\|X - Q^*(X)\|^2 \le E\|X - Q(X)\|^2,$$

for any N-point quantizer Q.

We first consider the linear portion of a quantizer as depicted in Fig. 3. From (11), it is clear that $Q(X)$ is a linear weighted sum of the random variables A_i, where

$$A_i = S_i(X)$$

Specifically, we have the representation of the output of the quantizer in the form:

$$Q(X) = \sum_{i=1}^{N} y_i A_i. \tag{19}$$

Thus, for a given partition the random variables A_i can be regarded as the observable data that is correlated with the input vector. The quantizer output is intended to estimate X as accurately as possible by forming a weighted sum of these observables with vector-valued weights. This is a standard problem in linear estimation. From the orthogonality principle of linear estimation, it follows that a necessary condition for optimality of the linear weights y_i is that

$$E(X - Q(X))A_i = \mathbf{0}, \qquad \text{for } i = 1,2,3,\cdots,N, \tag{20}$$

where $\mathbf{0}$ denotes the zero vector in R^k. Thus, we have

$$E\{XA_i\} = E\{Q(X)A_i\}. \tag{21}$$

But A_i is zero or one so that (21) simplifies to

$$E(X \mid A_i = 1)p_i = y_i p_i, \tag{22}$$

where

$$p_i = P(A_i = 1) = P(X \in R_i). \tag{23}$$

We thus have the following result.

Condition A: For an optimal quantizer, the output points must be given by

$$y_i = E(X \mid X \in R_i). \tag{24}$$

In words, the output point y_i must be the conditional mean (or the centroid) of X given that X is in the region R_i.

A simple but important and well-known theorem in quantization theory (see, for example, [35]) applies whenever Condition A holds and can be easily proved at this point.

Theorem: Given an N point quantizer Q with any suboptimal partition and optimal output points, the mean-square quantization error e, is given by

$$e = E\|X\|^2 - E\|Q(X)\|^2. \tag{25}$$

Thus, the quantizer output $Q(X)$ always has lower energy than the input X to the quantizer.

Note that this optimality condition holds regardless of the choice of partition $\{R_i\}$ for the quantizer. In the past, this property has often been assumed to hold for an optimal quantizer (one whose partition is also optimal).

Proof: From the orthogonality condition (20), any quantizer with optimal output points must satisfy

$$e = E(X - Q(X))^{tr}X, \tag{26}$$

since $Q(X)$ is a linear combination of the A_i. Thus,

$$e = E\|X\|^2 - EX^{tr}Q(X). \tag{27}$$

But

$$E\|Q(X)\|^2 = \sum_i y_i^{tr} E\{A_i Q(X)\}$$

and using (21)

$$\sum_i y_i^{tr} E\{A_i Q(X)\} = \sum_i y_i^{tr} E\{XA_i\} = EX^{tr}Q(X).$$

Hence,

$$EX^{tr}Q(X) = E\|Q(X)\|^2, \tag{28}$$

and we obtain (25) and the theorem is proved.

A second necessary condition for an optimal quantizer is obtained by taking the linear portion (the output points) as given and finding the best partition to minimize mean-square distance. The result is well-known and does not require any particular use of the structural decomposition of the quantizer function. For completeness, the result is included here. Assume again that $p_i > 0$ for each i. This means that each output point y_i occurs with positive probability and that each region R_i has positive volume in k-dimensional space. Let x be an internal point in some region R_i and suppose that

$$\|x - y_i\| > \|x - y_j\|, \qquad \text{for some } j \neq i. \tag{29}$$

Then there exists a sphere W contained in R_i with center x such that for each point x in W,

$$\|x - y_i\| > \|x - y_j\|.$$

Then it may readily be seen that the mean-square distance is not increased if the partition is altered by removing the set W from R_i and including it in R_j. Hence, we have the necessary condition for an optimal partition.

Condition B: For any optimal quantizer, the partition must satisfy

$$R_i \supset \left\{x \in R^k\colon \|x - y_i\| < \|x - y_j\|, \text{ for all } j \neq i\right\}.$$

This determines the partition except for the boundaries of dimensionality $k-1$ or less. For x continuously distributed, the boundary regions have zero probability and can be arbitrarily assigned. Thus, for any boundary point x that is equidistant from the output points $\{y_i\}$ for i in some set I, assign x to R_m where m is the smallest index in I. Thus, we have the so-called "nearest neighbor" partition, also known as the Voronoi partition. The regions or cells R_i are known as Dirichlet or Voronoi cells.

IV. Secondary Structural Decomposition

To proceed further we restrict our focus to an important special class of quantizers. A *Voronoi quantizer* is a quantizer whose partition regions, R_i, have the property of being "nearest neighbor" regions. Note that this is the standard Shannon model for a block source code. See, for example, Berger [36]. The closure, R_i^c of each region, R_i, is said to be a *Voronoi* cell if it is equal to the region, V_i, determined from the output point y_i according to

$$V_i = \{x: \|x - y_i\| \le \|x - y_j\|, \text{ for all } j \neq i\}. \quad (30)$$

We shall say that a quantizer with a given output set Y has Voronoi cells $R_i^c = V_i$ where V_i is defined with respect to $y_i \in Y$. With this construction a quantizer is entirely defined by specifying the set Y of output points and adopting a consistent convention for the unambiguous assignment of the boundaries of the Voronoi cells to particular regions of the partition. Thus, if a point x is equidistant from two or more output points, simply assign it to the region R_i of lowest index i.

Let H_{ij} denote the *half-space* defined by the relation

$$H_{ij} = \{x: \|x - y_i\| \le \|x - y_j\|\}, \quad (31)$$

and let L_{ij} denote the hyperplane

$$L_{ij} = \{x: \|x - y_i\| = \|x - y_j\|\}, \quad (32)$$

for $j = 1, 2, 3, \cdots, N\ j \neq i$. Then the Voronoi cell V_i can be described as the intersection of half-spaces

$$V_i = \bigcap_{j \neq i} H_{ij}, \quad (33)$$

and each face of the polytope V_i must lie in L_{ij} for some j.

Note also that the half-space H_{ij} can also be expressed in the form

$$H_{ij} = \{x: y_{ij} \cdot x + b_{ij} \ge 0\}, \quad (34)$$

where

$$y_{ij} = y_i - y_j \quad (35)$$

and

$$b_{ij} = \frac{1}{2}(\|y_j\|^2 - \|y_i\|^2). \quad (36)$$

Let

$$T_{ij}(x) = \begin{cases} 1, & \text{if } x \in H_{ij}, \\ 0, & \text{otherwise.} \end{cases} \quad (37)$$

We may now decompose the structure of the nonlinear operation S_i into more elementary building blocks.

Let $u(v)$ denote the unit step function defined by

$$u(v) = \begin{cases} 1, & \text{if } v \ge 0, \\ 0, & \text{if } v < 0. \end{cases} \quad (38)$$

Then,

$$T_{ij}(x) = u(y_{ij} \cdot x + b_{ij}) \quad (39)$$

so that

$$S_i(x) = \prod_{j \neq i} T_{ij}(x). \quad (40)$$

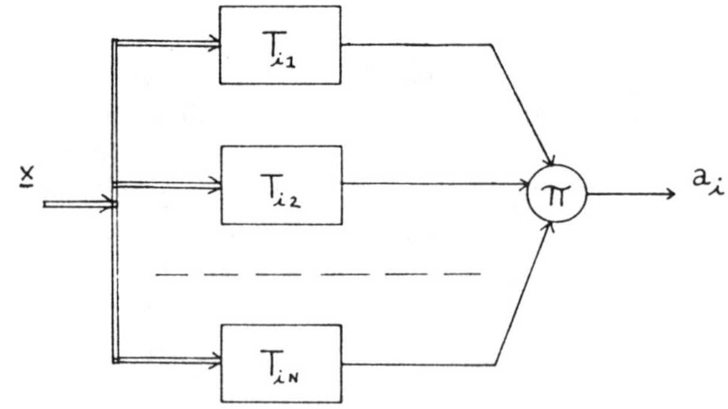

Fig. 4. Cell assignment function.

We therefore have a decomposition of the cell assignment function S_i into a product of $N - 1$ more elementary operations T_{ij}. Each function T_{ij} identifies whether or not the input vector is contained in the particular half-space H_{ij} defined by the pair of output points (y_i, y_j). In the case of a Euclidean distance measure, the boundary of this half-space is a hyperplane and the function is evaluated by a linear inequality involving the input vector. Note that the function T_{ij} reduces to the familiar binary comparator when $k = 1$, which is the standard building block of any analog to digital conversion system.

The encoding operation is in essence equivalent to the evaluation of T_{ij} for each $i, j \in J$ which amounts to $N(N - 1)/2$ evaluations since, for $T_{ij}(x) \neq 0$,

$$T_{ij}(x) = 1 - T_{ji}(x). \quad (41)$$

Fig. 4 illustrates the secondary structure of a quantizer by showing the decomposition of the cell assignment function.

Typically, in applications of vector quantization, the *rate*, defined by

$$\beta = \frac{\log_2 N}{k}, \quad (42)$$

is constrained. The dimension, k, is chosen sufficiently large to obtain adequate performance. Consequently, N increases exponentially with k and usually $N \gg k$. As a result, the quantizer will consist of a very large number of Voronoi cells each of which has a number of $(k - 1)$-dimensional faces that is much less than N. This means that many of the half-spaces H_{ij} used in (33) to define the cell V_i are redundant. Only a subset of the functions T_{ij}, for $j \in J$ and $j \neq i$, need be evaluated to precisely determine if any given input point x is contained in the cell V_i. For example, in the one-dimensional case the Voronoi cells are intervals and have only two "faces" so that only two comparisons, rather than N, are needed to determine if a given input point lies in a particular interval.

In general, a half-space H_{ij} is redundant with respect to the cell V_i if L_{ij}, the boundary of H_{ij} does not contain a face of V_i. Specifically, this means that $L_{ij} \cap V_i$ has dimension less than $k - 1$.

To characterize the structure of a quantizer fully, it is necessary to identify the faces of each cell. For this purpose it is convenient to define the adjacent points of a cell as follows: call y_j an *adjacent point* of the Voronoi cell V_i if

y_j is in Y and L_{ij} contains a face of V_i. Let $k(i)$ denote the *adjacency number*, the number of faces of cell V_i, and define the set of indices

$$J_i = \{j_1^i, j_2^i, \cdots, j_{k(i)}^i\}. \tag{43}$$

We then have

$$V_i = \bigcap_{j \in J_i} H_{ij} \tag{44}$$

and

$$S_i(x) = \prod_{j \in J_i} T_{ij}(x). \tag{45}$$

As a further aid to characterizing the nonredundant half-spaces of the cell V_i, define, for each i, the binary vector b^i according to

$$b_j^i = \begin{cases} 1, & \text{if } j \in J_i, \\ 0, & \text{otherwise.} \end{cases} \tag{46}$$

Thus, $b_j^i = 1$ implies that y_j is an adjacent point of V_i.

An important task in reducing encoding complexity using precomputation is to find for each cell V_i the binary vector b^i that determines the set of adjacent points of the cell. This task itself can be of a very substantial computational complexity. In the next section, we describe a probabilistic algorithm that can be appended to the quantizer design algorithm and leads to at least a nearly optimal solution. A probabilistic algorithm is one that has a positive (but controllably small) probability of failing. In our case, a failure can be limited to two modes. The first is when a redundant half-space is not recognized as redundant and the second occurs when a nonredundant half-space is incorrectly said to be redundant. The first case may increase the encoding computational cost of quantization but will not alter the performance of the quantizer. The second case will degrade the performance of the quantizer since there will be a positive probability that an input point will be incorrectly encoded.

V. Random Quantizers

The only effective method for the design of multidimensional quantizers is the use of a clustering algorithm. This approach was thoroughly developed by Linde, Buzo, and Gray [15]. Their design algorithm, which we refer to as the LBG algorithm, is based on the use of a training set of random vectors generated from a source for which the quantizer is to be optimized. The LBG algorithm in rudimentary form has been used by Chen [10], Menez [18], and Adoul [17] for one- and two-dimensional quantizers and is similar to some standard statistical clustering algorithms (see for example, Anderberg [19] or Hartigan [20]). Although the details of the LBG algorithm are not relevant here, the resulting character of the quantizer, i.e., the code book, obtained in this way is important for its implications on the structural properties and implementation complexity of the resulting quantizer.

A key issue in code book design is the complexity of the coder implementation resulting from a cluster design. Since the output vectors in the code book have no natural structure, an exhaustive search is needed to locate the nearest output vector for a given input vector. The costly processing time resulting from such a procedure is greatly reduced by a binary tree-structured code book at the price of suboptimality and increased code book storage requirements [28]. A more general M-ary tree-structured code book design can offer a better trade-off between performance and processing time, but the resulting quantizer is always inferior to the "full-search" code book [30].

It is not necessary to compromise the optimality of the full-search code book in order to reduce the encoding complexity. Suitable preprocessing of the optimal code book can result in novel data structures which offer some of the processing speed possessed by fast sorting algorithms. It is for this reason that we focus attention on the characterization and structure of random quantizers with optimal code books obtained from clustering algorithms.

Let

$$\text{TS} = \{x_1, x_2, \cdots, x_M\}$$

denote M independent observations of the continuously distributed random variable X with a specified joint density function. We call TS the *training set*, and assume that $M \gg N$, where, as before, N is the number of points of the quantizer of interest. A quantizer design algorithm is a procedure which determines a specific N-point quantizer from the given TS.

Suppose Q is an N-point quantizer with an output set Y and a partition $\{R_j\}$. The LBJ algorithm and other variations of this algorithm produce a quantizer design, which is (at least locally) optimal in the following sense. We shall say that Q is *optimal with respect to* TS, if the following two conditions are satisfied.

Condition I: For each output point y_j of Q,

$$y_j = \frac{\sum_{i=1}^{M} x_i S_j(x_i)}{\sum_{i=1}^{M} S_j(x_i)}. \tag{47}$$

Condition II: For all i and j, with $i = 1, 2, 3, \cdots, M$, and $j = 1, 2, 3, \cdots, N$,

$$\|Q(x_i) - x_i\| \le \|y_j - x_i\|. \tag{48}$$

Condition I means that y_j is the centroid of all training vectors in cell R_j. Condition II means that every point in TS that is contained in R_i is also contained in the Voronoi cell V_i defined with respect to the output set Y.

Note that Condition II is weaker than the requirement that the partition be Voronoi. Thus, an output set Y together with a partition $\{R_i\}$ may not define an optimal quantizer; however, it can be optimal with respect to TS.

We define a quantizer design algorithm as *admissible* if it operates on a given TS and produces an N-point quantizer Q, which is optimal with respect to TS. In particular, the LBG algorithm is admissable. Since the training set is random, the quantizer Q obtained by an admissible algorithm is also random. We make the reasonable assumption that the (random) code book Y obtained from such an algorithm satisfies the following condition.

Weak Condition of Disorder: Every subset of n output points from Y with $n \leq k$, forms an $n \times k$ matrix which is of rank n with probability one.

In other words, the output points of the random quantizer are linearly independent. In particular, we will have the property that no two distinct hyperplanes L_{ij} and $L_{i'j'}$ of dimensions $k-1$ will be parallel. It is also convenient to define the stronger condition.

Strong Condition of Disorder: Each subset of $n \leq k$ output points of Y is continuously distributed in nk dimensions with probability one.

An immediate consequence of the strong condition is that no two output points can lie on a given surface of dimension $k-1$. The natural disorder in the location of output points of a random quantizer places a lower bound on the computational complexity of the encoding operation. This character is in marked contrast with the highly ordered lattice quantizers to be examined in Section VII.

VI. Code Book Analysis

Given a code book Y obtained by an admissible design algorithm, an extensive degree of precomputation on this code book may be justified if it can lead to a reduction in the processing complexity needed to perform the encoding operation. It should be apparent that a fast encoding algorithm based on a random quantizer can make use of the fact that each Voronoi cell determined by the code book will usually have an adjacency number, ($k(i)$, for cell i) that is much less than N. To uniquely determine that a given vector $\boldsymbol{x}$ is contained in cell V_i the encoder must in one way or another evaluate the nonlinear mapping S_i for the argument $\boldsymbol{x}$. Thus, the first step of any precomputation is to determine the subset J_i of indices needed to specify the half-spaces H_{ij} which determine the Voronoi cell V_i. The computation of J_i, or equivalently of the binary vectors $\boldsymbol{b}^i$, is an essential step of code book analysis.

A. Eliminating Remote Hyperplanes

We first note a simple theorem which can greatly reduce the number of half-spaces that need be considered in searching for a minimal specification of a Voronoi cell. For convenience of notation, consider the cell V_1 and suppose the output points $y_2, y_3, \cdots, y_N$ are so numbered that they have monotonically nondecreasing distance from y_1. That is, we have

$$\| y_i - y_1 \| \leq \| y_{i+1} - y_1 \|. \tag{49}$$

For any random quantizer satisfying the strong condition of disorder, the output point y_3 will have zero probability of lying on the surface of a sphere with center y_1 and radius equal to the distance between y_1 and y_2. Hence, the ordered output points will than have strictly monotonically increasing distance from y_1.

For $1 < \tau < N$, define the region R_1^τ:

$$R_1^\tau = \bigcap_{j=1}^{\tau} H_{1j}. \tag{50}$$

This set necessarily contains the Voronoi cell V_1. We wish to establish whether or not it contains any points that are not in $\cap_{\tau+1}^{N} H_{ij}$. In this way we can determine if the Voronoi cell has any faces contained in the hyperplanes L_{1j} for $j \geq \tau + 1$. In other words, we can determine if J_1 contains any index j with $j > \tau$. Let

$$r_j = \| y_j - y_1 \|. \tag{51}$$

We then have the basic result:

Theorem: If

$$\| \boldsymbol{x} - y_1 \| \leq \frac{1}{2} r_{\tau+1}, \tag{52}$$

for all $\boldsymbol{x} \in R_1^\tau$, then

$$R_1^\tau = V_1. \tag{53}$$

Proof: For any $\boldsymbol{x} \in R_1^\tau$, (52) holds. Hence,

$$(\boldsymbol{x} - y_1) \cdot (y_{\tau+1} - y_1) \leq \| \boldsymbol{x} - y_1 \| \| y_{\tau+1} - y_1 \| \leq \frac{1}{2} r_{\tau+1} r_{\tau+1} \tag{54}$$

so that

$$(\boldsymbol{x} - y_1) \cdot (y_{\tau+1} - y_1) \leq \frac{1}{2} r_i^2, \tag{55}$$

for $\tau + 1 \leq i \leq N$, which is equivalent to the condition that

$$\| \boldsymbol{x} - y_1 \|^2 \leq \| \boldsymbol{x} - y_i \|^2, \qquad \text{for } \tau + 1 \leq i \leq N, \tag{56}$$

which proves the theorem.

Although this is a very simple result, it provides a useful way to reduce the computational task of identifying the bounded faces of the Voronoi cells. We shall see that it is also useful for characterizing the Voronoi cells of a lattice quantizer.

B. An Algorithm for Finding Adjacent Points

We next describe a general approach for determining systematically all of the faces of each cell for a partition that is optimal with respect to a given TS. An admissable design algorithm operating on a *TS* produces a specific output set Y. Given the code book Y it may be computationally infeasible to determine all the faces of each Voronoi cell of Y. With the following procedure, however, we obtain all the faces of a partition which together with Y defines a quantizer that is optimal with respect to TS. The procedure is adapted from an approach of Boneh [33] for eliminating redundant constraints in nonlinear programming problems. The method generalizes easily to more general distortion measures than the mean-square error.

For each point $\boldsymbol{x}_\sigma$ in TS, compute

$$w_{\sigma j}^i = 1 - T_{ij}(\boldsymbol{x}_\sigma). \tag{57}$$

Thus, $w_{\sigma j}^i = 1$ if $\boldsymbol{x}_\sigma$ is not contained in the half-space H_{ij}. Note that this computation is performed anyhow in the last iteration of the LBG algorithm in order to perform the final partition of the training vectors [15].

If $\boldsymbol{x}_\sigma$ is not in V_i, then it is not in

$$\bigcap_{j \in J_i} H_{ij}$$

so that for at least one j in J_i, we will have

$$w_{\sigma j}^i = 1 \quad \text{and} \quad b_j^i = 1. \tag{58}$$

Hence,

$$\sum_{\substack{j=1 \\ j \neq i}}^{N} w_{\sigma j}^i b_j^i \geq 1, \qquad \text{for all } \sigma \text{ with } S_i(\boldsymbol{x}_\sigma) = 0, \tag{59}$$

which holds for each Voronoi cell V_i.

Equation (59) gives a nontrivial necessary condition on $\boldsymbol{b}^i$ for each σ with x_σ not in V_i. For each i, (59) defines a set of M inequality constraints that must be satisfied by the binary vector $\boldsymbol{b}^i$. It is possible, however, to find an extraneous solution $\boldsymbol{c}^i$ to (59) that is not a correct specification of the Voronoi cell V_i. This can happen in two ways.

One possibility is that a solution may retain some redundant hyperplanes as in the trivial solution $c_j^i = 1$, for all j. Clearly, any vector obtained from $\boldsymbol{b}^i$ by replacing some of the zero components by unity will be a solution of (59). On the other hand, it is also possible to have a solution with $c_j^i = 0$ when in fact $b_j^i = 1$. This situation can occur in the following way. Suppose that y_2 is an adjacent point of V_1, and let

$$Z = \bigcap_{\substack{j \in J_1 \\ j \neq 2}} H_{1j} - V_1,$$

where we have used relative complement notation. Then the set Z has positive k-dimensional volume. Suppose further that there is no point in TS that lies in Z. In this case, (59) will be satisfied by the binary vector

$$c_j^1 = \begin{cases} 0, & \text{if } j = 2, \\ b_j^1, & \text{if } j \neq 2. \end{cases} \tag{60}$$

Thus, the hyperplane H_{12} would be incorrectly identified as redundant with respect to V_1. But since Z contains no training vectors, the partition boundaries obtained by the solution $\boldsymbol{c}^i$ will still be optimal with respect to the TS. Hence, the solution of (59) will still correctly characterize a quantizer that is optimal with respect to the training set.

Since the most useful solution vector is the one with the most zeros, the best solution is that which minimizes

$$\sum_{j=1}^{N} c_j^i \tag{61}$$

subject to the inequality constraints (59). This is a standard zero-one integer programming problem for which many algorithms are available. The computation is not trivial due to the large dimensionality of the $M \times N$ matrix characterizing (59) and an efficient algorithm should be used. Note that the dimensionality of the problem may be reduced by prior elimination of remote redundant hyperplanes, as discussed in Section VI-B.

It should also be noted that a solution to (59) is needed for each of the N cells. This allows a reduction in the computation of $\boldsymbol{b}^i$ for each successive cell i after the first by making use of the fact that each nonredundant hyperplane L_{ij} of cell i also determines a face of cell j. Hence, if

$$b_j^i = 1,$$

then

$$b_i^j = 1. \tag{62}$$

These techniques provide the basis for combining a quantizer design algorithm, such as the LBG algorithm, with a face-determining algorithm. There remains the task of effectively using the knowledge of the faces of each cell to develop a fast encoding algorithm for a given code book.

VII. Lattice Quantizers

We have seen that a random quantizer requires a computationally significant computational complexity for encoding as well as a substantial memory for code book storage. This complexity in processing and storage is inevitable because of the condition of disorder. The lack of structure in the location of the output points prohibits a truly fast and simple encoder.

A natural and elegant way to achieve a structured code book is to use a set of output points which lie in a bounded region of a lattice. A *lattice* in k dimensions is defined by any nonsingular $k \times k$ matrix U so that if $\boldsymbol{m}$ is any k-dimensional vector (column matrix) of integers, the lattice Λ is the set of all vectors of the form $U\boldsymbol{m}$. Note that the columns of U are points of the lattice and all other lattice points are formed by taking a linear combination of these basis vectors with integer-valued coefficients.

The lattice points form a regularly spaced array of points in k-dimensional space. The origin is always a lattice point. Any translation of the lattice which shifts the point at the origin to another lattice point position produces the identical lattice. In other words, if you sit on one lattice point and view the surrounding set of lattice points, you will see the identical environment regardless of which point you are sitting on. The Voronoi cell containing a given lattice point $\boldsymbol{x}$ is the set of all points closer to $\boldsymbol{x}$ than to any other point of the lattice. Since each lattice point has an identical environment, the Voronoi cells for all lattice points are congruent. These cells, called parallelohedra, are translations of one another, and collectively they fill the space without overlapping. (The boundaries are not consequential and may be arbitrarily defined.) Properties of lattices were thoroughly studied in a classical treatise by Voronoi [34].

A *lattice quantizer* may be defined as a quantizer whose output set Y is a subset of a lattice Λ. In one-dimensional quantization, the only lattice quantizer is the uniform quantizer and the Voronoi cells are intervals in R^1. The companding approach to one-dimensional quantization shows that any nonuniform quantizer may be modeled using memoryless nonlinear functions and a uniform (lattice) quantizer [1]. More recently, the companding approach was extended to higher dimensions and a similar representation of a nonuniform quantizer was proposed using memoryless nonlinear mappings on R^k and a uniform k-dimensional quantizer [11]. The notion of a uniform quantizer in k dimensions was defined as one whose cells (apart from boundary or overload cells) are congruent translates of one another. In other words, a lattice quantizer is a uniform quantizer.

For the characterization of lattice quantizers, it is necessary to understand some basic features of lattices. Three useful properties of a lattice are the density, the kissing number, and the normalized moment of inertia. The density of the lattice is the largest fraction of space that can be filled with nonoverlapping spheres centered about the lattice points. The kissing number is the number of spheres that touch a given sphere, where the spheres are centered about the lattice points and have the maximum diameter without overlapping. The normalized moment of inertia is the moment of inertia of the parallelohedron (Voronoi cell) taken about the lattice point contained in the cell when the lattice coordinates have been scaled so that the cell has unit volume. The first two properties give some indication of the quality of a particular lattice for quantization. The third property directly determines the performance of a lattice quantizer when the mean-square error distortion measure is used. In [11] it was found that the densest lattice in three dimensions is suboptimal for quantization.

The beauty of lattices is the ease with which an arbitrary encoding can be performed. In other words, given an arbitrary point in k-space, we wish to identify to which lattice point it is nearest. For a random array of points, the encoding complexity grows with the number of points in the array. With a lattice, however, the encoding operation becomes trivially simple.

For example, consider the two-dimensional lattice generated by the matrix

$$U = \begin{bmatrix} \sqrt{3} & 0 \\ 1 & 2 \end{bmatrix}.$$

It may easily be seen that this generates a lattice whose Voronoi cells are regular hexagons. The optimal properties of this lattice for quantization are discussed in Gersho [11]. This lattice is equivalent to the A_2 lattice in Sloane's tabulation of common lattices [27]. Since any hexagonal lattice is the union of two rectangular lattices, decoding is simply achieved by finding the nearest point in each rectangular sublattice and selecting the nearer of these two points.

In four dimensions, the candidate cell proposed in [11] is generated by the matrix

$$U = \begin{bmatrix} 2 & 0 & 0 & 1 \\ 0 & 2 & 0 & 1 \\ 0 & 0 & 2 & 1 \\ 0 & 0 & 0 & 1 \end{bmatrix}.$$

This lattice is denoted as D_4 in Sloane's tabulation [27]. It is interesting to note that this lattice can be defined as the set of points whose coordinates are all even integers or all odd integers. From this observation a simple decoding method can be deduced.

Specifically, to encode an arbitrary vector x, first find the nearest point to x in the sublattice whose coordinates are all even. This can be done by finding the nearest point to $x/2$ in the sublattice whose coordinates are arbitrary integers and then doubling the values of each coordinate. Next, find the nearest point to x in the sublattice whose coordinates are all odd. This can be done by finding the nearest point with integer coordinates to $(x - 1)/2$ and then doubling each coordinate value and adding one to the coordinates. Finally, select the closer of the two candidate points.

With this limited space it is not possible to develop the fascinating theory of lattices in any depth. However, it is apparent that there is a natural and elegant structure to the partition induced by a quantizer whose points lie on a lattice. For example, in the four-dimensional lattice described above, the Voronoi cell is bounded by 12 parallel pairs of hyperplanes. This can be shown using the theorem of Section VI-A, where R_0^{24} is the cell formed by the 24 output points nearest to the origin. The theorem shows that in this case the adjacency number is equal to the kissing number and all the adjacent points to the origin lie on the surface of a sphere of radius two. Consequently, to encode a given point in four-space only 12 scalar products are needed. Hence, 48 multiplications is an upper bound on the encoding complexity (if we neglect additions). The encoding is really easier if we recognize that the 12 hyperplanes are defined by 12 vectors with *integer* coordinates. Hence, only very simple arithmetic operations are needed.

Of course, this use of lattices as discussed here is only directly useful for quantizing uniformly distributed random vectors and for the mean-square error distortion measure. However, the lattice structure serves as a natural starting point for design of more general quantizers. The elegant structural properties still remain if we consider non-Euclidean spaces. One approach for extending the results for more practical applications is the multidimensional generalization of companding introduced in [11].

VIII. Concluding Remarks

The preceding presentation has offered a foundation for a structural characterization of vector quantizers. In order to provide such a foundation it is necessary to devote considerable space to definitions. Such a presentation is clearly intended as a foundation for future studies rather than as a self-contained theory. Future work will treat the computational complexity of encoding in greater depth. Novel fast encoding algorithms for random quantizers are also being explored. The subject of vector quantization is becoming increasingly important and a deeper understanding of the structural properties should play an important role in future studies of complexity, algorithm design, and performance capabilities.

References

[1] A. Gersho, "Quantization," *IEEE Commun. Soc. Magazine*, vol. 15, pp. 16–29, Sept. 1977. reprinted with minor corrections as: "Principles of quantization, *IEEE Trans. Circuits Syst.*, vol. CAS-25, July 1978.

[2] J. M. Morris, "Quantization and source encoding with a fidelity criterion: A survey," Naval Res. Lab. Rep. 8082, AD A038650, Mar. 25, 1977.

[3] H. Steinhaus, "Sur la division des corps materiels en parties," *Bull. Acad. Polon. Sci.*, Cl. III, vol. IV, no. 12, pp. 801–804, 1956.

[4] M. P. Schutzenberger, "On the quantization of finite dimensional messages," *Inform. Contr.*, vol. 1, pp. 153–158, 1958.

[5] J. Fejes Toth, "Sur la representation d'une population infinie par un nombre fini d'elements," *Acta Math. Acad. Sci. Hung.*, vol. 10, pp. 299–304, 1959.

[6] D. J. Newman, "The hexagon theorem," this issue, pp. 137–139.

[7] J. G. Dunn, "The performance of a class of *n*-dimensional quantizers for a Gaussian source," in *Proc. Columbia Symp. Signal Transmission Processing*, Columbia Univ., New York, NY, 1965, pp. 76–81.

[8] P. Zador, "Asymptotic quantization of continuous random variables," this issue, pp. 139–149.

[9] P. Elias, "Bounds and asymptotes for the performance of mutivariate quantizers," *Ann. Math. Statist.*, vol. 41, pp. 1249–1259, 1970.

[10] D. T. S. Chen, "On two or more dimensional optimum quantizers," in *Proc. 1977 IEEE Int. Conf. Acoust., Speech Signal Processing*, 1977, pp. 640–643.

[11] A. Gersho, "Asymptotically optimal block quantization," *IEEE Trans. Inform. Theory*, vol. IT-25, pp. 373–380, July 1979.

[12] J. A. Bucklew, "Companding and random quantization in several dimensions," to be published.

[13] Y. Yamada, S. Tazaki, and R. M. Gray, "Asymptotic performance of a block quantizer with difference distortion measure," *IEEE Trans. Inform. Theory*, vol. IT-26, pp. 6–14, Jan. 1980.

[14] J. A. Bucklew and G. L. Wise, "Multidimensional asymptotic quantization theory with *r*th power distortion measures," this issue, pp. 239–247.

[15] Y. Linde, A. Buzo, and R. M. Gray, "An algorithm for vector quantizer design," *IEEE Trans. Commun. Technol.*, vol. COM-28, pp. 84–95, Jan. 1980.

[16] S. P. Lloyd, "Least-squares quantization in PCM," this issue, pp. 129–137.

[17] J. P. Adoul, C. Collin, and D. Dalle, "Block encoding and its application to data compression of PCM speech," in *Proc. Canadian Commun. and EHV Conf.*, Montreal, 1978, pp. 145–148.

[18] J. Menez, F. Boeri, and D. J. Esteban, "Optimum quantizer algorithm for real-time block quantizing," in *Proc. Int. Conf. Acoust., Speech, Signal Processing*, pp. 980–984, 1979.

[19] M. Anderberg, *Cluster Analysis for Applications*. New York: Academic, 1973.

[20] J. A. Hartigan, *Clustering Algorithms*. New York: Wiley, 1975.

[21] J. B. MacQueen, "Some methods for classification and analysis of multivariate observations," in *Proc. 5th Berkeley Symp. Math. Statist. Prob.*, 1967, pp. 281–197.

[22] G. H. Ball and D. J. Hall, "A clustering technique for summarizing mulativariate data," *Behavioral Sci.*, vol. 12, pp. 153–155, Mar. 1967.

[23] R. C. Jancey, "Multidimensional group analysis," *Aust. J. Bot.*, vol. 14, no. 1, pp. 127–130, 1966.

[24] E. Forgey, "Cluster analysis of multivariate data: Efficiency vs. interpretability of classifications" (Abstract), *Biometrics*, vol. 21, p. 768, 1965.

[25] J. W. S. Cassels, "An introduction to the geometry of numbers," in *Grundleheren. Math. Wissenschaften*, vol. 99. New York: Springer-Verlag, 1959.

[26] N. J. A. Sloane, "Binary codes, lattices, and sphere packings," in *Combinatorial Surveys (Proc. 6th British Combinatorial Conf.)*, P. J. Cameron, Ed. New York: Academic, 1977.

[27] N. J. A. Sloane, "Tables of sphere packings and spherical codes," *IEEE Trans. Inform. Theory*, vol. IT-27, pp. 327–338, May 1981.

[28] A. Buzo, A. H. Gray, Jr., R. M. Gray, and J. D. Markel, "Speech coding based upon vector quantization," *IEEE Trans. Acoust., Speech Signal Processing*, vol. ASSP-28, pp. 562–574, Oct. 1980.

[29] H. Abut, R. M. Gray, and G. Rebolledo, "Vector quantization of speech and speech-like waveforms," submitted for publication.

[30] B. Juang, D. Y. Wong, and A. H. Gray, Jr., "Recent developments in vector quantization for speech processing," in *Proc. Int. Conf. Acoust., Speech, Signal Processing*, Apr. 1981.

[31] A. Gersho, Bell Labs., unpublished work, 1980.

[32] J. H. Conway and N. J. A. Sloane, "Voronoi regions of lattices, 2nd moments of polytopes, and quantization," this issue, pp. 211–226.

[33] A. Boneh, "Minimal representation of nonlinear inequalities by a probabilistic set covering (SC) equivalence," to be published.

[34] G. Voronoi, "Recherches sur les paralleloedres primitives," *J. Reine Angew. Math.*, vol. 134, pp. 198–287, 1908.

[35] N. C. Gallagher and J. A. Bucklew, "Properties of minimum mean-square error block quantizers," *IEEE Trans. Inform. Theory*, vol. IT-28, no. 1, pp. 105–107, 1982.

[36] T. Berger, *Rate Distortion Theory*. Englewood Cliffs, NJ: Prentice-Hall, 1971.

Voronoi Regions of Lattices, Second Moments of Polytopes, and Quantization

J. H. CONWAY AND N. J. A. SLOANE, FELLOW, IEEE

***Abstract*—If a point is picked at random inside a regular simplex, octahedron, 600-cell, or other polytope, what is its average squared distance from the centroid? In n-dimensional space, what is the average squared distance of a random point from the closest point of the lattice A_n (or D_n, E_n, A_n^* or D_n^*)? The answers are given here, together with a description of the Voronoi (or nearest neighbor) regions of these lattices. The results have applications to quantization and to the design of signals for the Gaussian channel. For example, a quantizer based on the eight-dimensional lattice E_8 has a mean-squared error per symbol of $0.0717\cdots$ when applied to uniformly distributed data, compared with $0.08333\cdots$ for the best one-dimensional quantizer.**

I. Quantization; Codes for Gaussian Channel

A. Introduction

THE MOTIVATION for this work comes from block quantization and from the design of signals for the Gaussian channel. Let us call a finite set of points $y_1,\cdots,y_M$ in n-dimensional Euclidean space $\mathbb{R}^n$ a *Euclidean code*. An n-dimensional *quantizer* with outputs $y_1,\cdots,y_M$ is the function Q: $\mathbb{R}^n \to \mathbb{R}^n$ which sends each point $x \in \mathbb{R}^n$ into $Q(x) =$ closest codepoint y_i (in case of a tie, pick that y_i with the smallest subscript). If x has probability density function $p(x)$, the mean-squared error per symbol of this quantizer is

$$E(n, M, p, \{y_i\}) = \frac{1}{n}\int_{\mathbb{R}^n} \|x - Q(x)\|^2 p(x)\,dx,$$

where $\|x\| = (x \cdot x)^{1/2}$. Around each codepoint y_i is its *Voronoi region* $V(y_i)$ (see [49]), consisting of all points of the underlying space which are closer to that codepoint than to any other. More precisely, we define $V(y_i)$ to be the closed set

$$V(y_i) = \{x \in \mathbb{R}^n: \|x - y_i\| \le \|x - y_j\| \text{ for all } j \ne i\}.$$

(Voronoi regions are also called Dirichlet regions, Brillouin zones, Wigner–Seitz cells, or nearest neighbor regions.) If x is an interior point of $V(y_i)$, the quantizer replaces x by $Q(x) = y_i$. Then we may write

$$E(n, M, p, \{y_i\}) = \frac{1}{n}\sum_{i=1}^{M}\int_{V(y_i)} \|x - y_i\|^2 p(x)\,dx. \quad (1)$$

Given n, M, and $p(x)$ one wishes to find the infimum

$$E(n, M, p) = \inf_{\{y_i\}} E(n, M, p, \{y_i\})$$

over all choices of $y_1,\cdots,y_M$. Zador ([53]; see also [6], [7], [24], [52]) showed under quite general assumptions about $p(x)$ that

$$\lim_{M\to\infty} M^{2/n}E(n, M, p) = G_n\left(\int_{\mathbb{R}^n} p(x)^{n/(n+2)}\,dx\right)^{(n+2)/n}, \quad (2)$$

where G_n depends only on n. Zador also showed that

$$\frac{1}{(n+2)\pi}\Gamma\left(\frac{n}{2}+1\right)^{2/n} \le G_n \le \frac{1}{n\pi}\Gamma\left(\frac{n}{2}+1\right)^{2/n}\Gamma\left(1+\frac{2}{n}\right). \quad (3)$$

Asymptotically the upper and lower bounds in (3) agree, giving

$$G_n \to \frac{1}{2\pi e} = 0.0585498\cdots \quad \text{as } n \to \infty. \quad (4)$$

Since the probability density function $p(x)$ only appears in the last term of (2), we may choose any convenient $p(x)$ when attempting to find G_n. From now on we assume that the input x is uniformly distributed over a large region in n-dimensional space, and we can usually avoid edge effects by passing to a limiting situation with infinitely many y_i. With this assumption the mean-squared error is minimized if each codepoint y_i lies at the centroid of the corresponding Voronoi region $V(y_i)$ (see [24]). It is known that, for an optimal one-dimensional quantizer with a uniform input distribution, the points y_i should be uniformly spaced along the real line; correspondingly

$$G_1 = \frac{1}{12} = 0.08333\cdots. \quad (5)$$

Similarly for the optimal two-dimensional quantizer it is known that the points y_i should form the hexagonal lattice

Manuscript received December 7, 1980; revised November 4, 1981.
J. H. Conway is with the Department of Pure Mathematics and Mathematical Statistics, University of Cambridge, 16 Mill Lane, Cambridge CB2 1SB, England.
N. J. A. Sloane is with the Mathematics and Statistics Research Center, Bell Laboratories, Murray Hill, NJ 07974.

Reprinted from *IEEE Trans. Inform. Theory,* vol. IT-28, pp. 211–226, March 1982.

A_2 (described in Section III-A); correspondingly

$$G_2 = \frac{5}{36\sqrt{3}} = 0.0801875 \cdots, \tag{6}$$

(see [21], [23], [24], [37]).

In three dimensions Gersho [24] conjectures that the optimal quantizer is based on the body-centered cubic lattice A_3^*, and that

$$G_3 = \frac{19}{192\sqrt[3]{2}} = 0.0785433 \cdots. \tag{7}$$

Similarly in four dimensions he conjectures that the optimal quantizer is based on the lattice D_4, and that

$$G_4 = 0.076602 \tag{8}$$

(obtained by Monte Carlo integration). Furthermore, he conjectures that, in all dimensions, any optimal quantizer is such that for large M the Voronoi regions $V(y_i)$ are all congruent, to some polytope P say. For such a quantizer we obtain, from (1) and (2),

$$G_n = \frac{1}{n} \frac{\int_P \|x - \hat{x}\|^2 \, dx}{\left(\int_P dx\right)^{(n+2)/n}}, \tag{9}$$

where $\hat{x}$ is the centroid of P. The expression on the right makes sense for any polytope and will be denoted by $G(P)$: we refer to it as the *dimensionless second moment of* P. It is also convenient to have symbols for the *volume*, *unnormalized second moment*, and *normalized second moment* of P: these are

$$\mathrm{vol}(P) = \int_P dx,$$

$$U(P) = \int_P \|x - \hat{x}\|^2 \, dx,$$

and

$$I(P) = \frac{U(P)}{\mathrm{vol}(P)},$$

respectively. Then

$$G(P) = \frac{1}{n} \frac{U(P)}{\mathrm{vol}(P)^{1+2/n}} = \frac{1}{n} \frac{I(P)}{\mathrm{vol}(P)^{2/n}}.$$

If Gersho's conjecture is correct then G_n may be determined from

$$G_n = \min_P G(P), \tag{10}$$

taken over all n-dimensional space-filling polytopes. Whether or not the conjecture is true, any value of $G(P)$ for a space-filling polytope is an upper bound to G_n. Furthermore (1), (2) and (9) allow us to interpret G_n and $G(P)$ as *mean-squared quantization errors per symbol*, assuming a uniform input distribution to the quantizer.

In the second application, the same Euclidean code $y_1, \cdots, y_M$ is used as a code for the Gaussian channel. Now the Voronoi regions are the decoding regions: all points x in the interior of $V(y_i)$ are decoded as y_i. If the codewords are equally likely and all the Voronoi regions $V(y_i)$ are congruent to a polytope P, the probability of correct decoding is proportional to

$$\int_P e^{-x \cdot x} \, dx.$$

The description of the Voronoi regions given in Section III thus makes it possible to calculate this probability exactly for many Euclidean codes. These results will be described elsewhere.

B. Summary of Results

In Sections II and III we compute $G(P)$ for a number of important polytopes (not just space-filling ones), including all regular polytopes (see Theorem 4). The three- and four-dimensional polytopes are compared in Tables I and II. The chief tools are Dirichlet's integral (Theorem 1), an explicit formula for the second moment of an n-simplex (Theorem 2), and a recursion formula giving the second moment of a polytope in terms of its cells (Theorem 3).

In Section III we study lattices, in particular the root lattices A_n, D_n, E_6, E_7, E_8, and their duals (defined in Section III-A). We determine the Voronoi regions for these lattices, and their second moments. The second moment gives the average squared distance of a point from the lattice. The *maximum* distance of any point of the underlying space from the lattice is its *covering radius*. The covering radii of these lattices were mostly already known (see for example [2], [4], [31]), but for completeness we rederive them. The final section (Section IV) compares the quantization errors of the different lattices—see Table V and Fig. 20. E_8 is the clear winner.

It is worth mentioning that for most of these lattices there are very fast algorithms for finding the closest lattice point to an arbitrary point; these are described in a companion paper [12]. The sizes of the spherical codes obtained from these lattices have been tabulated in [47].

Although we have tried to keep this paper as self-contained as possible, some familiarity with Coxeter's book [16] will be helpful to the reader. $\mathbb{Z}$ denotes the integers, $\mathbb{Q}$ the rationals and $\mathbb{R}$ the reals.

II. Second Moments of Polytopes

In this section we compute the second moments of a number of fairly simple polytopes; many others will be analyzed in Section III. The methods used are described in Theorems 1–3. A *polytope* in this paper means a convex region of $\mathbb{R}^n$ enclosed by a finite number of hyperplanes (cf. [16, p. 126]). The part of the polytope that lies in one of hyperplanes is called a *cell*. We usually denote the edge-length of our polytopes by $2l$. The main source for information about polytopes is Coxeter [16], but there is an extensive literature, particularly for low-dimensional figures (see for example [1], [13]–[20], [22], [23], [26], [28], [29], [33], [34], [36], [40], [48], and [50]). Although second moments *about an axis* are tabulated for many simple polyhedra in standard engineering handbooks (see also [43]), the results given here appear to be new.

A. Dirichlet's Integral

A few special figures can be handled using Dirichlet's integral.

Theorem 1 ([51, §12.5]): Let f be continuous and $\alpha_1,\cdots,\alpha_n > 0$. Then

$$\int f(x_1+\cdots+x_n)x_1^{\alpha_1}\cdots x_n^{\alpha_n}\frac{dx_1}{x_1}\cdots\frac{dx_n}{x_n} = \frac{\Gamma(\alpha_1)\cdots\Gamma(\alpha_n)}{\Gamma(\alpha_1+\cdots+\alpha_n)}\int_0^1 f(\tau)\tau^{\Sigma\alpha_i-1}\,d\tau,$$

where the integral on the left is taken over the region bounded by $x_1 \ge 0,\cdots,x_n \ge 0$ and $x_1+\cdots x_n \le 1$.

B. Generalized Octahedron or Crosspolytope

Consider for example the n-dimensional generalized octahedron or crosspolytope β_n [16, p. 121] of edge-length $2l$. Taking $f = 1$, $\alpha_1 = 1$ (to get the volume) or $\alpha_1 = 3$ (to get the second moment) and $\alpha_i = 1$ for $i \ge 2$ in Theorem 1 we find

$$\frac{\operatorname{vol}(\beta_n)}{(2l)^n} = \frac{2^{n/2}}{n!}, \quad \frac{I(\beta_n)}{(2l)^2} = \frac{n}{(n+1)(n+2)},$$
$$G(\beta_n) = \frac{(n!)^{2/n}}{2(n+1)(n+2)} \to \frac{1}{2e^2} = 0.0676676\cdots \quad \text{as } n\to\infty. \tag{11}$$

C. The n-Sphere

As a second application, for the n-dimensional (solid) sphere S_n of radius ρ we find

$$\frac{\operatorname{vol}(S_n)}{\rho^n} = \frac{\pi^{n/2}}{\Gamma(\frac{1}{2}n+1)}, \quad \frac{I(S_n)}{\rho^2} = \frac{n}{n+2},$$
$$G(S_n) = \frac{\Gamma(\frac{1}{2}n+1)^{2/n}}{(n+2)\pi} \to \frac{1}{2\pi e} = 0.0585498\cdots \quad \text{as } n\to\infty. \tag{12}$$

It is clear from the definition that the sphere has the smallest value of $G(P)$ of any figure [53]. Thus $G_n \ge G(S_n)$ for all n, which is the sphere lower bound of (3) (see Fig. 20). Unfortunately quantizers cannot be built with either generalized octahedra (for $n \ge 3$) or spheres (for $n \ge 2$) as Voronoi regions, since these objects do not fill space.

D. n-Dimensional Simplexes

The next result makes it possible to find the second moment of any figure provided it can be decomposed into simplexes.

Theorem 2: Let P be an arbitrary simplex in $\mathbb{R}^n$ with vertices $v_i = (v_{i1},\cdots,v_{in})$ for $0 \le i \le n$. Then

a) the centroid of P is at the barycenter

$$\hat{v} = \frac{1}{n+1}(v_0+\cdots+v_n) \tag{13}$$

of the vertices,

b)

$$\operatorname{vol}(P) = \frac{1}{n!}\det\begin{vmatrix} 1 & v_{01} & \cdots & v_{0n} \\ 1 & v_{11} & \cdots & v_{1n} \\ & \cdots & \cdots & \\ 1 & v_{n1} & \cdots & v_{nn}\end{vmatrix}, \tag{14}$$

and

c) the normalized second moment about the origin 0 is

$$I_0 = \frac{n+1}{n+2}\|\hat{v}\|^2 + \frac{1}{(n+1)(n+2)}\sum_{i=0}^{n}\|v_i\|^2. \tag{15}$$

In other words I_0 is equal to the second moment of a system of $n+1$ particles each of mass $1/(n+1)(n+2)$ placed at the vertices and one particle of mass $(n+1)/(n+2)$ placed at the barycenter.

Proof: a) is elementary, b) is well known (c.f. [25, p. 349]), and c) follows from [25, eq. (24), the case $n = 2$].

E. Regular Simplex

For example if P is a regular n-simplex of edge-length $2l$ then

$$\frac{\operatorname{vol}(P)}{(\sqrt{2}\,l)^n} = \frac{\sqrt{n+1}}{n!}, \quad \frac{I(P)}{(\sqrt{2}\,l)^2} = \frac{n}{(n+1)(n+2)},$$
$$G(P) = \frac{(n!)^{2/n}}{(n+1)^{1+(1/n)}(n+2)} \to e^{-2} = 0.135335\cdots \quad \text{as } n\to\infty. \tag{16}$$

For $n = 1$, 2, and 3 the values of $G(P)$ are $1/12$, $1/6\sqrt{3} = 0.0962250\cdots$, and $3^{2/3}/20 = 0.104004\cdots$, and $G(P)$ *increases* monotonically with n.

F. Volume and Second Moment of a Polytope in Terms of its Cells

Instead of decomposing a figure into simplexes one may proceed by induction, expressing the volume and second moment of a polytope in terms of the volume and second moment of its cells, then in terms of its $(n-2)$-dimensional faces, and so on. Theorem 3 is the basis for this procedure.

Suppose P is an n-dimensional polytope with N_1 congruent cells $F_1, F_1', F_1'',\cdots, N_2$ congruent cells $F_2, F_2', F_2'',\cdots$, and so on. Suppose also that P contains a point 0 such that all of the generalized pyramids $0F_1, 0F_1', \cdots$ are congruent, all of $0F_2, 0F_2', \cdots$ are congruent, $\cdots$. Let $\boldsymbol{a}_i \in F_i$ be the foot of the perpendicular from 0 to F_i, let $h_i = \|0\boldsymbol{a}_i\|$, and let $V_{n-1}(i)$ be of volume of F_i and $U_{n-1}(i)$ the unnormalized second moment of F_i about $\boldsymbol{a}_i$.

Theorem 3: The volume and unnormalized second moment about 0 of P are given by

$$\operatorname{vol}(P) = \sum_i \frac{N_i h_i}{n} V_{n-1}(i),$$

$$U(P) = \sum_i \frac{N_i h_i}{n+2}\left[h_i^2 V_{n-1}(i) + U_{n-1}(i)\right].$$

Proof: Follows from elementary calculus by dividing each generalized pyramid $0F_i$ into slabs parallel to the cell F_i.

G. Truncated Octahedron

For example let P be the truncated octahedron with vertices consisting of all permutations of $\sqrt{2}\,l(0, \pm 1, \pm 2)$. P has $N_1 = 6$ square cells and $N_2 = 8$ cells which are regular hexagons, all with edge-length $2l$. The second moments of these cells can be calculated directly, or else found in Section II-I below. Then from the theorem we find that

$$\operatorname{vol}(P) = \frac{6 \cdot l\sqrt{8}}{3} 4l^2 + \frac{8 \cdot l\sqrt{6}}{3} 6\sqrt{3}\,l^2 = 64\sqrt{2}\,l^3,$$

$$U(P) = \frac{6 \cdot l\sqrt{8}}{5}\left[8l^2 \cdot 4l^2 + \frac{8l^4}{3}\right] + \frac{8 \cdot l\sqrt{6}}{5}\left[6l^2 \cdot 6\sqrt{3}\,l^2 + 10\sqrt{3}\,l^4\right] = 304\sqrt{2}\,l^5,$$

hence

$$I(P) = \frac{U(P)}{\operatorname{vol}(P)} = \frac{19l^4}{4},$$

$$G(P) = \frac{1}{3}\frac{I(P)}{\operatorname{vol}(P)^{2/3}} = \frac{19}{192^3\sqrt{2}} = 0.0785433 \cdots. \tag{17}$$

H. Second Moment of Regular Polytopes

The next theorem gives an explicit formula for the second moment of any regular polytope. Suppose P is an n-dimensional regular polytope [16]. For $0 \le j \le n$ choose a j-dimensional face F_j of P so that $F_0 \subseteq F_1 \cdots \subseteq F_n = P$, and let 0_j be the center of F_j, $R_j = \|0_n 0_j\|$, and for $j \ge 1$ let $r_j = \|0_{j-1}0_j\|$. Thus r_j is the inradius of F_j measured from 0_j, and $r_j^2 = R_{j-1}^2 - R_j^2$. Let $N_{j,j-1}$ be the number of $((j-1)$-dimensional) cells of F_j. Then it is known that the symmetry group of P has order

$$g = N_{n,n-1}N_{n-1,n-2}\cdots N_{2,1}N_{1,0}$$

(see [16, p. 130]), and that the volume of P is

$$\operatorname{vol}(P) = N_{n,n-1}\cdots N_{2,1}N_{1,0}\cdot\frac{r_1 r_2 \cdots r_n}{n!} \tag{18}$$

(see [16, p. 137]).

Theorem 4: The second moment of any n-dimensional regular polytope P about its center 0_n is given by

$$I(P) = \frac{2}{(n+1)(n+2)} \cdot \left(R_0^2 + 2R_1^2 + 3R_2^2 + \cdots + nR_{n-1}^2\right) \tag{19}$$

or equivalently

$$I(P) = \frac{2}{(n+1)(n+2)} \cdot \left(r_1^2 + 3r_2^2 + 6r_3^2 + \cdots + \frac{n(n+1)}{2}r_n^2\right). \tag{20}$$

Proof: The proof is by induction, the one-dimensional case being immediate. From Theorem 3 we have

$$U(P) = \frac{N_{n,n-1}r_n}{n+2}\left[r_n^2 V_{n-1}(P) + U_{n-1}(P)\right],$$

where (from (18) and the induction hypothesis)

$$V_{n-1}(P) = N_{n-1,n-2}\cdots N_{2,1}N_{1,0}\cdot\frac{r_1 r_2\cdots r_{n-1}}{(n-1)!},$$

$$U_{n-1}(P) = V_{n-1}(P) \cdot \frac{2}{n(n+1)}\left[r_1^2 + \cdots + \frac{n(n-1)}{2}r_{n-1}^2\right].$$

Then

$$I(P) = \frac{U(P)}{\operatorname{vol}(P)} = \frac{n}{n+2}\left[r_n^2 + \frac{2}{n(n+1)} \cdot \left(r_1^2 + \cdots + \frac{n(n-1)}{2}r_{n-1}^2\right)\right]$$

which simplifies to (20).

The values of g, $\operatorname{vol}(P)$ and R_j are tabulated for all regular polytopes in [16, table I, pp. 292–295]. We have already dealt with the simplex and generalized octahedron. For an n-dimensional cube $G(P) = 1/12$ for all n (since the cube is a direct product of line segments). We now treat the remaining regular polytopes.

I. Regular Polygons

If P is a regular p-gon of edge-length $2l$ then from Theorem 4 we find

$$\operatorname{vol}(P) = pl^2\cot\frac{\pi}{p},$$

$$I(P) = \frac{l^2}{6}\left(1 + 3\cot^2\frac{\pi}{p}\right),$$

$$G(P) = \frac{1}{6p}\left(\operatorname{cosec}\frac{2\pi}{p} + \cot\frac{\pi}{p}\right). \tag{21}$$

For $p = 3, 4$, and 6, $G(P) = 1/6\sqrt{3}$, $1/12$, and $5/36\sqrt{3}$.

J. Icosahedron and Dodecahedron

For the icosahedron

$$\mathrm{vol} = \frac{20 l^3 \tau^2}{3}, \quad I = \frac{3 l^2 \tau^2}{5},$$

$$G = \frac{1}{20}\left(\frac{6\tau}{5}\right)^{2/3} = 0.0778185 \cdots, \tag{22}$$

where $\tau = (\sqrt{5}+1)/2$, and for the dodecahedron

$$\mathrm{vol} = 4\sqrt{5}\, l^3 \tau^4, \quad I = l^2 \cdot \frac{39\tau + 28}{25},$$

$$G = \frac{11\tau + 17}{300}\left(\frac{2}{\tau\sqrt{5}}\right)^{2/3} = 0.0781285 \cdots. \tag{23}$$

K. The Exceptional Four-Dimensional Polytopes

There are three "exceptional" regular polytopes in four dimensions, the 24-cell, the 120-cell, and the 600-cell (the prefix giving the number of cells—see [16]). For the 24-cell

$$\mathrm{vol} = 32 l^4, \quad I = 26 l^2/15,$$

$$G = \frac{13}{120\sqrt{2}} = 0.0766032 \cdots; \tag{24}$$

for the 120-cell

$$\mathrm{vol} = 120\sqrt{5}\, l^4 \tau^8, \quad I = \frac{2 l^2}{15\sqrt{5}}\left(282 + 127\sqrt{5}\right),$$

$$G = \frac{43\tau + 13}{300\sqrt{6}\, 5^{1/4}} = 0.0751470 \cdots; \tag{25}$$

and for the 600-cell

$$\mathrm{vol} = 100 l^4 \tau^3, \quad I = \frac{4 l^2}{15}\left(12 + 5\sqrt{5}\right),$$

$$G = \frac{(3\tau + 4)\tau^{1/2}}{150} = 0.0750839 \cdots. \tag{26}$$

L. Comparisons

The three- and four-dimensional polytopes that have been considered are compared in Tables I and II.

TABLE I
COMPARISON OF DIMENSIONLESS SECOND MOMENT $G(P)$ FOR VARIOUS THREE-DIMENSIONAL POLYHEDRA P

P	G(P)
tetrahedron	.1040042 ...
cube*	.0833333 ...
octahedron	.0825482 ...
hexagonal prism*	.0812227 ...
rhombic dodecahedron*	.0787451 ...
truncated octahedron*	.0785433 ...
dodecahedron	.0781285 ...
icosahedron	.0778185 ...
sphere	.0769670 ...

*A space-filling polyhedron.

TABLE II
COMPARISON OF DIMENSIONLESS SECOND MOMENT $G(P)$ FOR VARIOUS FOUR-DIMENSIONAL POLYTOPES P

P	G(P)
simplex	.1092048 ...
cube*	.0833333 ...
generalized octahedron	.0816497 ...
24-cell*	.0766032 ...
120-cell	.0751470 ...
600-cell	.0750839 ...
sphere	.0750264 ...

*A space-filling polytope.

III. VORONOI REGIONS OF LATTICES AND THE MEAN-SQUARE ERROR OF LATTICE QUANTIZERS

In this section we determine the Voronoi regions of various n-dimensional lattices, and their volumes and second moments. Since these lattices can be used to construct n-dimensional quantizers for uniformly distributed inputs, the dimensionless second moments give upper bounds to G_n (see (10) and Section IV). The lattices considered are the *root lattices*[1] and their duals, namely A_n, A_n^*, D_n, D_n^*, E_6, E_7, and E_8 (we shall not consider E_6^* or E_7^* here, while $E_8^* = E_8$).

[1]A root lattice is a lattice spanned by the root system of a Lie algebra—see [5], [30].

For general information about lattices see for example [4], [5], [8], [11], [14]–[17], [28], [32], [38], [39], and [45]–[47]. In particular if Λ is a lattice in $\mathbb{R}^n$ the *dual* (or *reciprocal* or *polar*) lattice Λ^* consists of all points $\boldsymbol{x}$ in the span $\mathbb{Q}\Lambda$ such that $\boldsymbol{x} \cdot \boldsymbol{y} \in \mathbb{Z}$ for all $\boldsymbol{y} \in \Lambda$. Since all the points in a lattice are equivalent, it is enough to find the Voronoi region around the origin, i.e., the closed set

$$V(0) = \{\boldsymbol{x} \in \mathbb{R}^n \mid \|\boldsymbol{x}\| \le \|\boldsymbol{x} - \boldsymbol{u}\| \text{ for all nonzero } \boldsymbol{u} \in \Lambda\}.$$

The volume of the Voronoi region can be written down immediately from the other standard parameters of the lattice:

$$\mathrm{vol}\, V(0) = \frac{V_n \rho^n}{\Delta} = \frac{\rho^n}{\delta} = \sqrt{d}, \tag{27}$$

where V_n is the volume of a unit sphere in $\mathbb{R}^n$, 2ρ is the minimum distance between the points of Λ, and Δ, δ and d are respectively the density, center density, and determinant of Λ. The covering radius will be denoted by R_c.

A. Definition of the Root Lattices

For $n \geq 1$, A_n is the n-dimensional[2] lattice consisting of the points $(x_0, x_1, \cdots, x_n)$ in $\mathbb{Z}^{n+1}$ with $\Sigma x_i = 0$. The dual A_n^* consists of the union of $n+1$ cosets of A_n:

$$A_n^* = \bigcup_{i=0}^{n} ([i] + A_n),$$

where

$$[i] = \left(\frac{-j}{n+1}, \frac{-j}{n+1}, \cdots, \frac{-j}{n+1}, \frac{i}{n+1}, \cdots, \frac{i}{n+1}\right)$$
$$= \left(\left(\frac{-j}{n+1}\right)^i, \left(\frac{i}{n+1}\right)^j\right) \tag{28}$$

and $i + j = n + 1$. For $n = 1$ and 2, $A_n^* \cong A_n$ (i.e., they differ only by a rotation and change of scale).

For $n \geq 2$, D_n consists of the points $(x_1, x_2, \cdots, x_n)$ in $\mathbb{Z}^n$ with Σx_i even. In other words, if we color the integer lattice points alternately red and blue in a checkerboard coloring, D_n consists of the red points. The dual D_n^* is the union of four cosets of D_n:

$$D_n^* = \bigcup_{j=0}^{3} ([i] + D_n),$$

where

$$[0] = (0^n), \quad [1] = (\tfrac{1}{2}^n),$$
$$[2] = (0^{n-1}, 1), \quad [3] = (\tfrac{1}{2}^{n-1}, -\tfrac{1}{2}).$$

Also $D_2 \cong A_1 \oplus A_1$, $D_3 \cong A_3$, and $D_4^* \cong D_4$. Equivalently, D_n may be obtained by applying Construction A of [32] or [45] to the even weight code of length n. Similarly D_n^* is obtained by applying Construction A to the dual code $\{0^n, 1^n\}$.

There are many possible definitions of the lattices E_6, E_7, and E_8 (see the references given at the beginning of this section). We shall use the following: E_8 is the union of D_8 and the coset

$$(\tfrac{1}{2}, \tfrac{1}{2}, \tfrac{1}{2}, \tfrac{1}{2}, \tfrac{1}{2}, \tfrac{1}{2}, \tfrac{1}{2}, \tfrac{1}{2}) + D_8.$$

In other words E_8 consists of the points $(x_1, \cdots, x_8)$ with $x_i \in \mathbb{Z}$ and Σx_i even, together with the points $(y_1, \cdots, y_8)$ with $y_i \in \mathbb{Z} + \frac{1}{2}$ and Σy_i even. E_7 is a subspace of dimension 7 in E_8, consisting of the points $(u_1, \cdots, u_8) \in E_8$ with $u_7 = -u_8$. E_6 is a subspace of dimension 6 in E_8, consisting of the points $(u_1, \cdots, u_8) \in E_8$ with $u_6 = u_7 = -u_8$.

B. Voronoi Region of a Root Lattice

In this section we give a uniform method for finding the Voronoi region of any root lattice Λ (the dual lattices must be handled differently). The method is based on finding a fundamental simplex for the affine Weyl group of the lattice (cf. [3], [5], [16]).

[2] The subscript gives the dimension of the lattice.

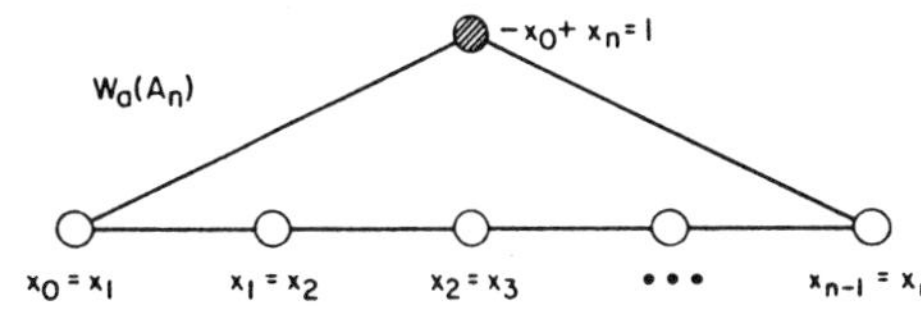

Fig. 1. Extended Coxeter–Dynkin diagram for $W_a(A_n)$. The extending node is indicated by a solid circle. The $n+1$ nodes are labeled with the equations to the hyperplanes which are the walls of the fundamental simplex. The labelings in Fig. 1-3 are based on [5, pp. 250–2100] and [10].

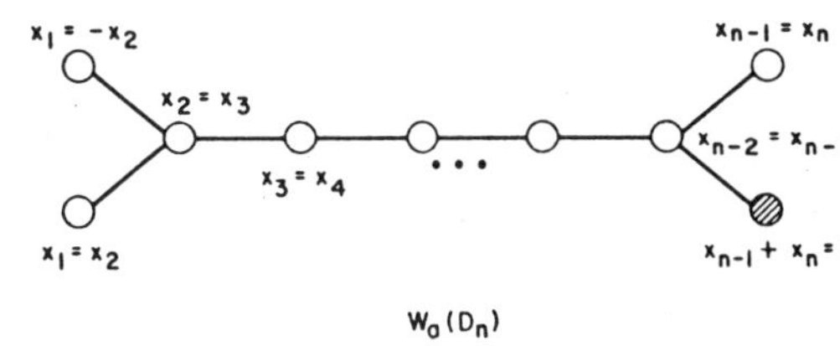

Fig. 2. Extended Coxeter–Dynkin diagram for $W_a(D_n)$. There are $n+1$ nodes.

The (ordinary) *Weyl group* $W(\Lambda)$ of an n-dimensional root lattice Λ is a certain finite group of orthogonal transformations of $\mathbb{R}^n$ which sends Λ to itself (for the precise definition see [5, p. 143] or [30, p. 43]).[3] Similarly the *affine Weyl group* $W_a(\Lambda)$ is a certain infinite group of isometries of $\mathbb{R}^n$ which sends Λ to itself (see [5, p. 173]); and $W(\Lambda)$ is the subgroup of $W_a(\Lambda)$ fixing the origin.[4] The affine Weyl group is described by the extended Coxeter–Dynkin diagram shown in Figs. 1–3.

This diagram can be read in at least three different ways (see [3], [5], [16], [27]). First, it provides a presentation for $W_a(\Lambda)$, defining the group in terms of generators and relations; however we shall not make use of this interpretation here. Second, it can be used to specify a *fundamental simplex* S for $W_a(\Lambda)$. This is an n-dimensional closed simplex whose images under the action of $W_a(\Lambda)$ are distinct and tile $\mathbb{R}^n$. In other words we can write

$$\mathbb{R}^n = \bigcup_{g \in W_a(\Lambda)} g(S), \tag{29}$$

where (except for the boundaries of $g(S)$, a set of measure zero) each point $x \in \mathbb{R}^n$ belongs to a unique $g(S)$. In this interpretation the nodes of the diagram represent the hyperplanes which are the walls of the fundamental simplex [16, p. 191]. The angle between two walls or hyperplanes is indicated by the branch of the diagram joining the corresponding nodes. If the hyperplanes are at an angle of $\pi/3$ the nodes are joined by a single branch, if the angle is $\pi/4$ they are joined by a double branch (see Fig. 4), if the angle is π/p with $p > 4$ they are joined by a branch labeled p, and finally if the hyperplanes are perpendicular the nodes

[3] $W(\Lambda)$ is generated by the reflections in the hyperplanes through the origin perpendicular to the minimal vectors of the lattice. Alternatively, in the terminology of [11], $W(\Lambda)$ is the group $G_0(\Lambda)$, while the full group of orthogonal transformations sending Λ to itself is a split extension of $G_0(\Lambda)$ by $G_1(\Lambda)$.

[4] We may think of Λ itself as being an Abelian group of translations of $\mathbb{R}^n$, which sends Λ to Λ. Then $W_a(\Lambda)$ is a split extension of Λ by $W(\Lambda)$.

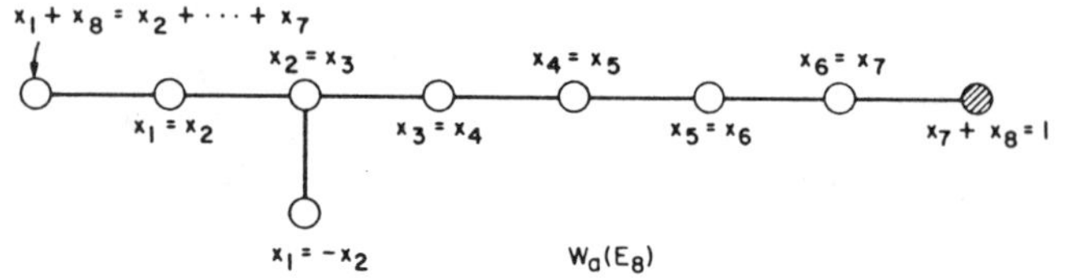

(a)

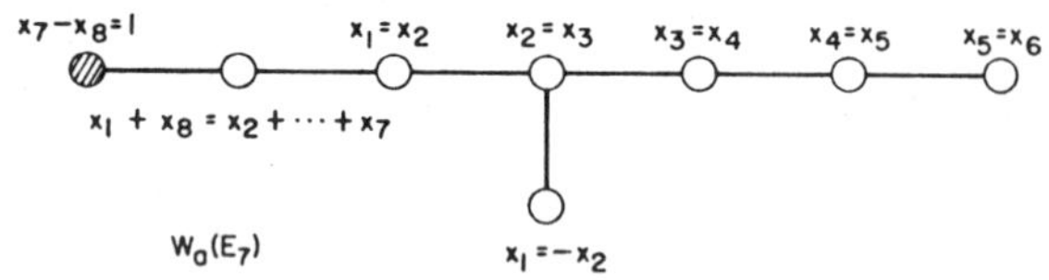

(b)

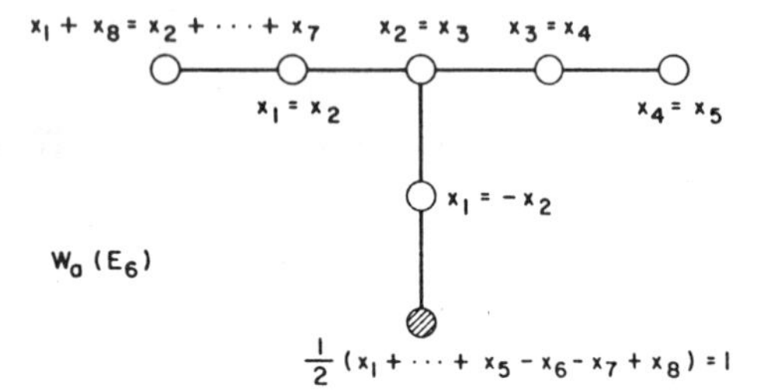

(c)

Fig. 3. Extended Coxeter–Dynkin diagrams for (a) $W_a(E_8)$, (b) $W_a(E_7)$, and (c) $W_a(E_6)$.

are not joined by a branch. The nodes in Figs. 1–3 have been labeled with the equations to the corresponding hyperplanes.

In the third interpretation the nodes in the extended Coxeter–Dynkin diagram are taken to represent the *vertices* of a fundamental simplex, rather than the bounding hyperplanes. Each node represents the vertex opposite to the corresponding hyperplane (some examples are shown in Figs. 6, 8, and 9)—see [16, p. 196].

One of the nodes in the diagram is indicated by a solid circle. This is the *extending node*; removing it leaves a Coxeter–Dynkin diagram for the Weyl group $W(\Lambda)$. Of the $n+1$ hyperplanes represented by the extended diagram, all except that corresponding to the extending node pass through the origin. It is helpful to think of the latter hyperplane as forming the *roof* of the fundamental simplex. The vertex of the fundamental simplex opposite the roof is the origin.

For later use we remark that the finite Weyl group $W(\Lambda)$ also has a fundamental domain, consisting of an infinite cone centered at the origin. A fundamental simplex for $W_a(\Lambda)$ is obtained by taking the finite part of the cone beneath the roof. The intersection of this cone with the roof, or more precisely with a unit sphere centered at the origin, is a *spherical simplex*. The Coxeter–Dynkin diagram for $W(\Lambda)$—i.e., with the extending node deleted—describes this spherical simplex in the same way as the extended diagram describes the fundamental simplex for $W_a(\Lambda)$. These spherical simplexes and (unextended) Coxeter–Dynkin diagrams can be used to define the Weyl

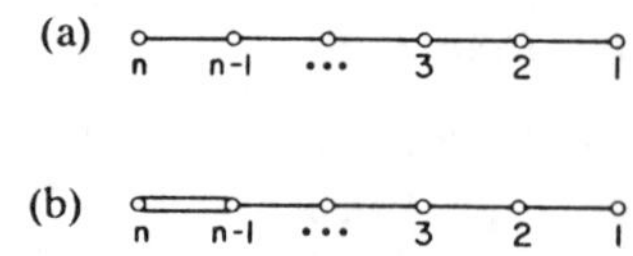

Fig. 4. Coxeter–Dynkin diagrams for the spherical simplexes of (a) $W(A_n)$ and (b) $W(C_n)$. (The labeling of the nodes is for convenience only and has no geometrical significance.)

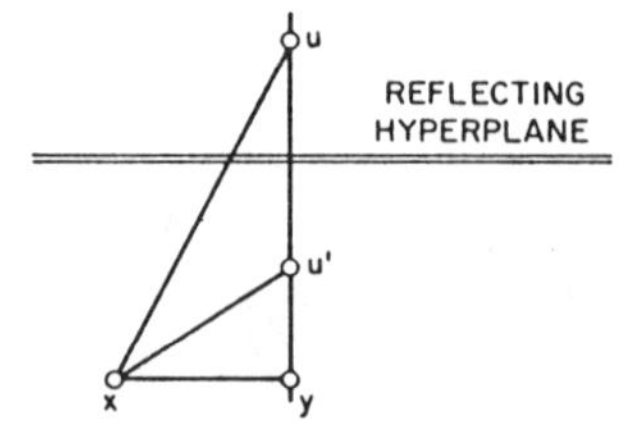

Fig. 5.

groups of all the root *systems* (and not just the root *lattices* A_n, D_n, and E_n). In Sections F and G we shall require the spherical simplexes corresponding to $W(A_n)$ and $W(C_n)$, shown in Fig. 4. The Weyl group $W(A_n)$ is usually written as $[3^{n-1}]$ and is isomorphic to the symmetric group on $n+1$ letters. $W(C_n)$ is written $[3^{n-2}, 4]$ and has order $2^n n!$. (See [3], [5], [16], [18], and [30].)

Lemma: The origin is the closest lattice point to any interior point of the fundamental simplex.

Proof: Let $\boldsymbol{u}$ be the closest lattice point to $\boldsymbol{x} \in S$. Suppose $\boldsymbol{u} \neq 0$. Then $\boldsymbol{u} \notin S$, and $\boldsymbol{u}$ and $\boldsymbol{x}$ are on opposite sides of a reflecting hyperplane of $W_a(\Lambda)$. Let $\boldsymbol{u}' \in \Lambda$ be the image of $\boldsymbol{u}$ in this hyperplane, and let $\boldsymbol{y}$ be the foot of the perpendicular from $\boldsymbol{x}$ to the line $\boldsymbol{uu}'$ (see Fig. 5). Then $\|\boldsymbol{xu}'\|^2 = \|\boldsymbol{xy}\|^2 + \|\boldsymbol{yu}'\|^2 < \|\boldsymbol{xy}\|^2 + \|\boldsymbol{yu}\|^2 = \|\boldsymbol{xu}\|^2$, and $\boldsymbol{x}$ is closer to $\boldsymbol{u}'$ than to $\boldsymbol{u}$, a contradiction. Therefore $\boldsymbol{u} = 0$.

The connection between the fundamental simplex and the Voronoi region is given by the following basic theorem.

Theorem 5: For any root lattice Λ, the Voronoi region around the origin is the union of the images of the fundamental simplex under the Weyl group $W(\Lambda)$.

Proof: Let $\boldsymbol{x}$ be any point of the Voronoi region around the origin. From (29), $\boldsymbol{x} \in g(S)$ for some $g \in W_a(\Lambda)$. Suppose $\boldsymbol{x}$ is an interior point of $g(S)$. By the lemma, the closest lattice point to $\boldsymbol{x}$ is $g(0)$. Therefore $g(0) = 0$, $g \in W(\Lambda)$, and

$$\boldsymbol{x} \in \bigcup_{g \in W(\Lambda)} g(S).$$

We omit the discussion of the case when $\boldsymbol{x}$ is a boundary point of $g(S)$. The converse statement, that $\boldsymbol{x} \in \cup g(S)$ implies $\boldsymbol{x}$ is in the Voronoi region, follows by reversing the steps.

It follows from Theorem 5 that the Voronoi region is the union of $|W(\Lambda)|$ copies of the fundamental simplex S. Furthermore the cells of the Voronoi region are the images of the roof of the fundamental simplex under $W(\Lambda)$. Thus the Voronoi region is bounded by hyperplanes which are the perpendicular bisectors of the lines joining 0 to its nearest neighbors in the lattice.

Corollary: The number of $(n-1)$-dimensional cells of the Voronoi region of a root lattice is equal to the contact number of the lattice (the number of nearest neighbors of any lattice point).

This is not true for all lattices, as we shall see in Section III-H.

The second moment of the Voronoi region can now be obtained from that of the fundamental simplex. The results are given in the following sections.

C. Voronoi Region for A_n

We first find the vertices $v_0, v_1, \cdots, v_n$ of the fundamental simplex S. These are found by omitting each of the hyperplanes of Fig. 1 in turn and calculating the point of intersection of the remaining n hyperplanes. The results are shown in Fig. 6, where each node is labeled with the coordinates of the vertex opposite the corresponding hyperplane. The ith vertex is

$$v_i = \left(\left(-\frac{j}{n+1} \right)^i \left(\frac{i}{n+1} \right)^j \right)$$

where $i + j = n + 1$, for $0 \le i \le n$, and is the same as the coset representative $[i]$ for A_n in A_n^* (see (28)). Also

$$\|v_i\|^2 = \frac{ij}{n+1}. \tag{30}$$

The barycenter of S (13) is

$$\begin{aligned} \hat{v} &= \frac{1}{n+1} \sum_{i=0}^{n} v_i \\ &= \left(\frac{-n}{2n+2}, \frac{-n+2}{2n+2}, \cdots, \frac{n-2}{2n+2}, \frac{n}{2n+2} \right) \end{aligned}$$

and from (15) the normalized second moment about the origin is

$$\begin{aligned} I(S) &= \frac{n+1}{n+2} \|\hat{v}\|^2 + \frac{1}{(n+1)(n+2)} \sum_{i=0}^{n} \|v_i\|^2 \\ &= \frac{n+1}{n+2} \cdot \frac{n(n+2)}{12(n+1)} + \frac{1}{(n+1)(n+2)} \cdot \frac{n(n+2)}{6} \\ &= \frac{n}{12} + \frac{n}{6(n+1)}. \end{aligned}$$

Now $|W(A_n)| = (n+1)!$ and the determinant d is $n+1$ [5, pp. 250–251]. By Theorem 5 the Voronoi region around the origin, $V(0)$, is the union of $|W(A_n)|$ copies of S, so

$$\begin{aligned} I(V(0)) &= \frac{U(V(0))}{\text{vol}(V(0))} \\ &= \frac{|W(A_n)| \cdot U(S)}{|W(A_n)| \cdot \text{vol}(S)} \\ &= \frac{U(S)}{\text{vol}(S)} = I(S). \end{aligned} \tag{31}$$

Also $\text{vol}(V(0)) = \sqrt{n+1}$ from (27). Therefore the dimensionless second moment of the Voronoi region of A_n is

$$\begin{aligned} G(A_n) &= \frac{1}{n} \frac{I(V(0))}{\text{vol}(V(0))^{2/n}} \\ &= \frac{1}{(n+1)^{1/n}} \left(\frac{1}{12} + \frac{1}{6(n+1)} \right) \\ &\to \frac{1}{12} \qquad \text{as } n \to \infty. \end{aligned} \tag{32}$$

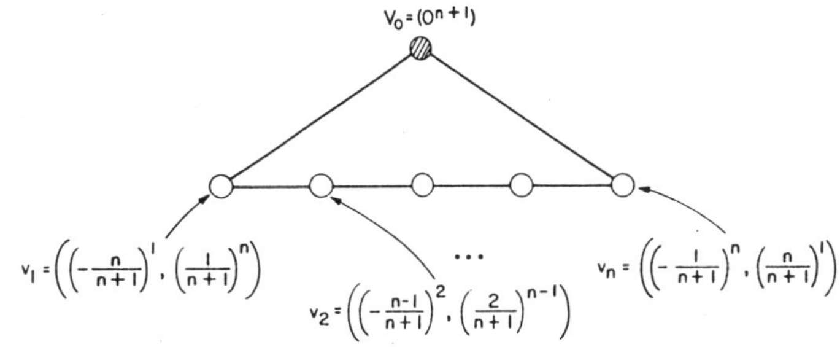

Fig. 6. Vertices of fundamental simplex for A_n.

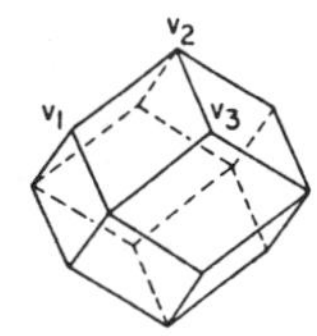

Fig. 7. A rhombic dodecahedron, the Voronoi region $V(0)$ for the face-centered cubic lattice A_3. The points v_0 (not shown, but at the center of the figure), v_1, v_2, and v_3 form a fundamental simplex S, and the rhombic dodecahedron is the union of 24 copies of S.

Once the Voronoi region has been found we can also determine the points in $\mathbb{R}^n$ at maximum distance from the lattice, since these are necessarily vertices of the Voronoi regions. From (30) it follows that the covering radius of A_n (the maximum distance of any point in $\mathbb{R}^n$ from A_n) is

$$R_c = \sqrt{\frac{ab}{n+1}} = \rho \sqrt{\frac{2ab}{n+1}},$$

where ρ is the packing radius (see (27)) and $a = [(n+1)/2]$, $b = n + 1 - a$. Typical points at this distance from A_n are the vertex v_a of $V(0)$ and its images under $W(A_n)$.

The lattice A_1 consists of equally spaced points on the real line, and $G(A_1) = 1/12$. The lattice A_2 is the hexagonal lattice, the fundamental simplex is an equilateral triangle, the Voronoi region is a hexagon, and $G(A_2) = 5/36\sqrt{3}$ (compare Section II-I). The lattice A_3 is the face-centered cubic lattice, the densest known sphere packing in $\mathbb{R}^3$, the Voronoi region is a rhombic dodecahedron (Fig. 7; see also [33, p. 130] and [22, p. 294 and anaglyph XI]), and $G(A_3) = 2^{-11/3} = 0.0787451 \cdots$.

For $n = 1$ and 2 it is known that A_n is the optimal quantizer (see (5) and (6)), but for $n = 3$ the dual lattice A_3^* is better. The values of $G(A_n)$ for $n \le 9$ are plotted in Fig. 20. $G(A_n)$ decreases to its minimum value of $0.0773907 \cdots$ at $n = 8$ and then slowly increases to $1/12$ as $n \to \infty$.

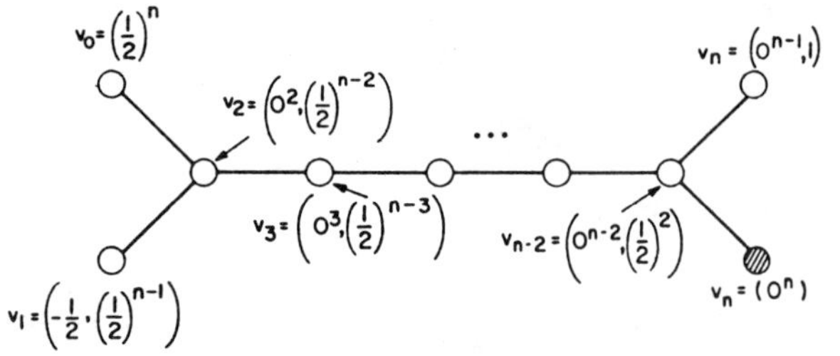

Fig. 8. Vertices of fundamental simplex for D_n.

D. Voronoi Region for D_n ($n \ge 4$)

D_n (for $n \ge 4$), E_6, E_7, and E_8 are handled in the same way as A_n and our treatment will be brief. The vertices $v_0, \cdots, v_n$ of a fundamental simplex for D_n, $n \ge 4$, are shown in Fig. 8. Their barycenter is

$$\hat{v} = \frac{1}{2(n+1)}(0, 2, 3, \cdots, n-2, n-1, n+1),$$

$|W(D_n)| = 2^{n-1} \cdot n!$ and the determinant $d = 4$. For the Voronoi region $V(0)$ we find

$$I(V(0)) = \frac{n}{12} + \frac{1}{2(n+1)},$$

$$\mathrm{vol}(V(0)) = 2,$$

and

$$G(D_n) = \frac{1}{2^{2/n}}\left(\frac{1}{12} + \frac{1}{2n(n+1)}\right)$$

$$\to \frac{1}{12} \quad \text{as } n \to \infty. \tag{33}$$

The covering radius of D_n (for $n \ge 4$) is

$$R_c = \sqrt{\frac{n}{4}} = \rho\sqrt{\frac{n}{2}},$$

as illustrated by the vertex $v_0 = (\frac{1}{2}^n)$. For $n = 4$ and 5, D_n is the densest known sphere packing in $\mathbb{R}^n$ (cf. [32]). For $n = 4$ the Voronoi region is a 24-cell (see for example [16, p. 156]), and $G(D_4) = 13/120\sqrt{2} = 0.0766032 \cdots$, in agreement with (24). This number also agrees closely with the value (8) that Gersho obtained for this region by Monte Carlo integration. The values of $G(D_n)$ for $n \le 9$ are plotted in Fig. 20. $G(D_n)$ takes its minimum value of $0.0755905 \cdots$ at $n = 6$ and then slowly increases to $1/12$ as $n \to \infty$.

E. Voronoi Regions for E_6, E_7, E_8

The vertices of fundamental simplexes for E_6, E_7, and E_8 are shown in Fig. 9. For E_8

$$|W(E_8)| = 2^{14} \cdot 3^5 \cdot 5^2 \cdot 7 = 696729600, \qquad d = 1,$$

$$\hat{v} = \frac{1}{1080}(5, 35, 55, 79, 109, 149, 209, 751),$$

$$I(V(0)) = \frac{929}{1620},$$

$$\mathrm{vol}(V(0)) = 1,$$

$$G(E_8) = \frac{929}{12960} = 0.0716821 \cdots. \tag{34}$$

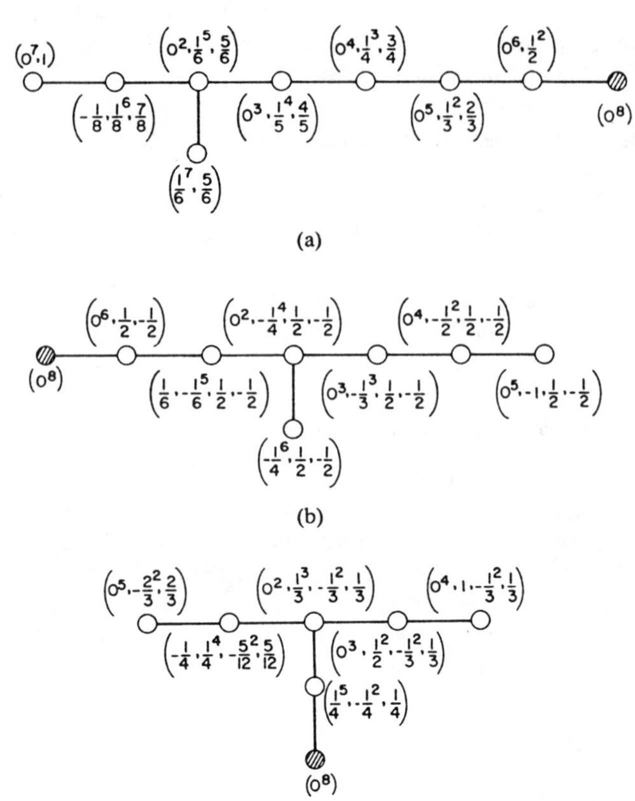

Fig. 9. Vertices of fundamental simplexes for (a) E_8, (b) E_7, and (c) E_6.

The Voronoi region $V(0)$ is an eight-dimensional polytope which is the reciprocal[5] to the Gosset polytope 4_{21} [16, p. 204].

Let N_i denote the number of i-dimensional faces of $V(0)$. Then from [16, p. 204] we have $N_0 = 19440$, $N_1 = 207360$, $N_2 = 483840$, $N_3 = 483840$, $N_4 = 241920$, $N_5 = 60480$, $N_6 = 6720$, and $N_7 = 240$. The 19440 vertices consist of 2160 at distance one from $\mathbf{0}$ and 17280 at distance $2\sqrt{2}/3$. The former are the images of the vertex $(0^7, 1)$ of the fundamental simplex under $W(E_8)$, and are at the maximum possible distance from E_8, while the latter are the images of the vertex $((1/6)^7, 5/6)$ under $W(E_8)$. Thus $R_c = 1 = \rho\sqrt{2}$. The other seven vertices of the fundamental simplex are not vertices of the Voronoi region.

For E_7, $|W(E_7)| = 2^{10} \cdot 3^4 \cdot 5 \cdot 7 = 2903040$, $d = 2$,

$$\hat{v} = -\frac{1}{96}(1, 5, 8, 12, 18, 30, -42, 42),$$

$$I(V(0)) = \frac{163}{288},$$

$$\mathrm{vol}(V(0)) = \sqrt{2},$$

$$G(E_7) = \frac{163}{2016} \cdot 2^{-1/7} = 0.0732306 \cdots. \tag{35}$$

The covering radius of E_7 is $R_c = \sqrt{3/2} = \rho\sqrt{3}$, as illustrated by the vertex $(0^5, -1, \frac{1}{2}, -\frac{1}{2})$.

[5]It is not difficult to give a direct proof of this statement; it also follows from Theorem 8 below.

For E_6, $|W(E_6)| = 2^7 \cdot 3^4 \cdot 5 = 51840$, $d = 3$,

$$\hat{v} = \frac{1}{42}(0,3,5,8,14,-14,-14,14),$$

$$I(V(0)) = \frac{15}{28},$$

$$\mathrm{vol}(V(0)) = \sqrt{3},$$

$$G(E_6) = \frac{5}{56 \cdot 3^{1/6}} = 0.0743467 \cdots. \quad (36)$$

The Voronoi regions for E_7 and E_6 are the reciprocals of the Gosset polytopes 2_{31} and 1_{22} described in [16, pp. 202–203]. The covering radius of E_6 is $R_c = 2/\sqrt{3} = \rho\sqrt{8/3}$, as illustrated by the vertices $(0^5, -2/3, -2/3, 2/3)$ and $(0^4, 1, -1/3, -1/3, 1/3)$.

F. Voronoi Region for D_n^*

In order to determine the Voronoi regions for the dual lattices A_n^* and D_n^* we shall use *Wythoff's construction*, as described in [13] and [16, § 11.6]. The idea is to construct new polytopes out of the spherical simplexes described in Section B, the vertices of the new polytope being indicated by drawing rings around certain nodes in the Coxeter–Dynkin diagram. More precisely, let $v_1, \cdots, v_n$ be the vertices of a spherical simplex for a Weyl group $W(\Lambda)$. If a single node of the diagram is ringed, say that corresponding to v_i, the vertices of the new polytope are the images of v_i under the Weyl group. If two or more nodes are ringed, say those corresponding to $v_i, v_j, \cdots$, the symbol represents a polytope whose vertices are the images under $W(\Lambda)$ of some interior point of the spherical subsimplex with vertices $v_i, v_j, \cdots$. We can adjust the metrical properties of the polytope (for example, equalize its edge lengths) by choosing this interior point suitably. Some one-, two-, and three-dimensional examples are shown in Fig. 10; others may be found in [13] and [16].

We now use this construction to find the Voronoi region for D_n^*, $n \geq 3$. In Section III-A we saw (using the second definition) that D_n^* is the union of the sets $(2\mathbb{Z})^n$ and $(1^n) + (2\mathbb{Z})^n$. The closest points to the origin from the first set consist of $2n$ points of the form $(\pm 2, 0^{n-1})$, and the closest from the second set consist of 2^n points of the form $(\pm 1^n)$. The Voronoi region $V(0)$ is the intersection of the Voronoi regions determined by these two sets. The first of these, P say, is a cube centered at zero with vertices $(\pm 1^n)$. The second, Q say, is a generalized octahedron with vertices $(\pm n/2, 0^{n-1})$. Furthermore Q can be obtained by reciprocating P in a sphere of radius $\rho = \sqrt{n/2}$ centered at the origin.

Thus the Voronoi region $V(0)$ is the intersection of P and a reciprocal polytope $Q = P^*$. In other words $V(0)$ is obtained by *truncating* P in the manner described in [16, p. 147], and is therefore specified by ringing one or two nodes of the Coxeter—Dynkin diagram (Fig. 4(b)) for the spherical simplex of P ([13], [16, § 8.1 and § 11.7]).

The radii R_j (defined in Section II-H) for the cube P are given by $R_j = \sqrt{n-j}$ ([16, p. 295]). If n is even the radius ρ of the sphere of reciprocation is equal to $R_{n/2}$, and we must ring the node labeled $n/2$ in Fig. 4(b). If n is odd ρ lies between $R_{(n-1)/2}$ and $R_{(n+1)/2}$ and both nodes $(n-1)/2$ and $(n+1)/2$ must be ringed. We have therefore established the following theorem.

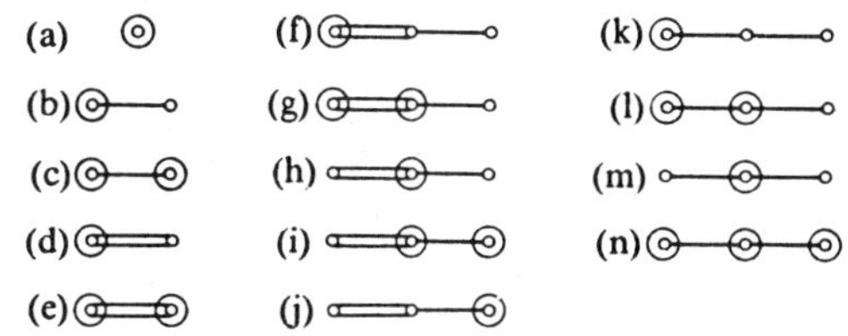

Fig. 10. Examples of polytopes obtained by Wythoff's construction: (a) edge, (b) triangle, (c) hexagon, (d) square, (e) octagon, (f) cube, (g) truncated cube, (h) cuboctahedron, (i) truncated octahedron, (j) octahedron, (k) tetrahedron, (l) truncated tetrahedron, (m) octahedron (again), and (n) truncated octahedron (again).

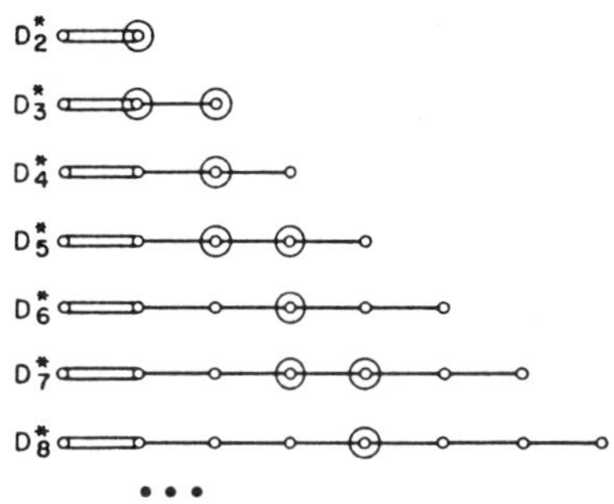

Fig. 11. Voronoi regions for the lattices D_n^*.

Theorem 6: The Voronoi region around the origin of the lattice D_n^* is the polytope defined by the diagrams in Fig. 11.

The coordinates for $\beta(n,k)$ and $\delta(n,k)$ given below show that the edge-lengths of the Voronoi regions are all equal. In Coxeter's notation [16, p. 146] the Voronoi region for D_{2t}^* is

$$\begin{Bmatrix} 3 & 3 & \cdots & 3 & \\ 3 & 3 & \cdots & 3 & 4 \end{Bmatrix}$$

with $t-1$ threes in each row, and for D_{2t+1}^* it is

$$\left\{ 3 \begin{matrix} 3 & 3 & \cdots & 3 & \\ 3 & 3 & \cdots & 3 & 4 \end{matrix} \right\}$$

with $t-1$ threes in the top and bottom rows.

We shall determine the second moments of these Voronoi regions recursively, using Theorem 3. In order to do this it will be necessary to find the second moments of all the polytopes $\alpha(n,k)$, $\beta(n,k)$, $\gamma(n,k)$, and $\delta(n,k)$ defined in Fig. 12. In this notation the Voronoi region of D_n^* is (up to a scale factor) equal to $\beta(n, n/2)$ if n is even and to $\delta(n,(n-1)/2)$ if n is odd. Let $R_\alpha(n,k)$, $V_\alpha(n,k)$, and $U_\alpha(n,k)$ denote respectively the circumradius, volume, and unnormalized second moment about the center of $\alpha(n,k)$, with a similar notation for $\beta(n,k)$, $\gamma(n,k)$, and $\delta(n,k)$.

For the vertices of the polytope $\alpha(n,k)$ it is convenient to take the points in $\mathbb{R}^{n+1}$ whose coordinates are all

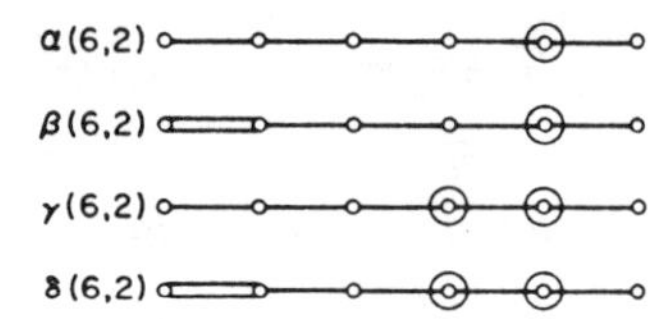

Fig. 12. Polytopes $\alpha(n,k)$, $\beta(n,k)$, $\gamma(n,k)$, and $\delta(n,k)$. In general $\alpha(n,k)$ has n nodes with the kth node from the right ringed, and $\gamma(n,k)$ has n nodes with the kth and $(k+1)$st ringed (except that $\gamma(n,0)=\alpha(n,1)$ and $\gamma(n,n)=\alpha(n,n)$). $\beta(n,k)$ and $\delta(n,k)$ are the same as $\alpha(n,k)$ and $\gamma(n,k)$, respectively, except that the left branch is a double bond. By convention $\alpha(0,0)$ and $\gamma(0,0)$ represent a point.

permutations of $(0^{n-k+1}, 1^k)$ [16, pp. 157–158]. The centroid of $\alpha(n,k)$ is the point

$$\frac{k}{n+1}(1^n),$$

and so the circumradius is

$$R_\alpha(n,k)=\sqrt{\frac{k(n-k+1)}{n+1}}.$$

Similarly

$\beta(n,k)$ has vertices $(0^{n-k+1}, \pm 1^k)$, $R_\beta(n,k)=\sqrt{k}$,

$\gamma(n,k)$ has vertices $(0^{n-k}, 1, 2^k)$,

$$R_\gamma(n,k)=\sqrt{\frac{4k(n-k)+n}{n+1}},$$

$\delta(n,k)$ has vertices $(0^{n-k}, \pm 1, \pm 2^k)$,

$$R_\delta(n,k)=\sqrt{4k+1}.$$

(These polytopes appear with different names in [17].) Each of these polytopes has two kinds of cells obtained by deleting either the left or the right node of its diagram [16, § § 7.6, 11.6, 11.7]. For example deleting the left node of the $\alpha(n,k)$ diagram produces an $\alpha(n-1,k-1)$, while deleting the right node produces an $\alpha(n-1,k)$. Thus in general $\alpha(n,k)$ has cells of type $\alpha(n-1,k-1)$ and $\alpha(n-1,k)$. (If $k=0$ the first type is absent, while if $k=n$ the second type is absent.) The number of cells of each type is given by the ratio of the orders of the underlying Weyl groups (obtained by ignoring the rings on the diagram). Thus the number of $\alpha(n-1,k-1)$-type cells of an $\alpha(n,k)$ is

$$\frac{|[3^{n-1}]|}{|[3^{n-2}]|}=\frac{(n+1)!}{n!}=n+1.$$

This is also the number of $\alpha(n-1,k)$-type cells. We represent this process of finding the cells by the graph shown in Fig. 13.

We can now apply Theorem 3 to $\alpha(n,k)$, obtaining

$$V_\alpha(n,k)=\frac{(n+1)h_L}{n}V_\alpha(n-1,k-1)+\frac{(n+1)h_R}{n}V_\alpha(n-1,k),$$

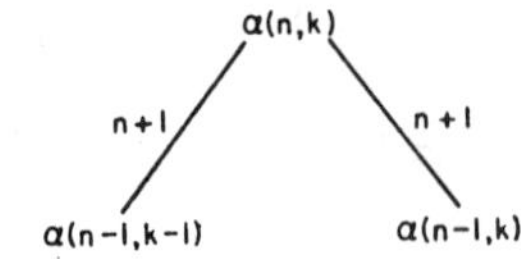

Fig. 13. The polytope $\alpha(n,k)$ has $n+1$ cells of type $\alpha(n-1,k-1)$ and $n+1$ of type $\alpha(n-1,k)$.

$$U_\alpha(n,k)=\frac{(n+1)h_L}{n+2}\left(h_L^2V_\alpha(n-1,k-1)+U_\alpha(n-1,k-1)\right)+\frac{(n+1)h_R}{n+2}\left(h_R^2V_\alpha(n-1,k)+U_\alpha(n-1,k)\right),$$

where

$$h_L^2=R_\alpha(n,k)^2-R_\alpha(n-1,k-1)^2=\frac{(n-k+1)^2}{n(n+1)},$$

$$h_R^2=R_\alpha(n,k)^2-R_\alpha(n-1,k)^2=\frac{k^2}{n(n+1)},$$

the subscripts on h standing for left and right. If we write

$$V_\alpha(n,k)=v_\alpha(n,k)\frac{\sqrt{n+1}}{n!}, \tag{37}$$

$$U_\alpha(n,k)=u_\alpha(n,k)\frac{\sqrt{n+1}}{(n+2)!}, \tag{38}$$

then v_α and u_α are integers satisfying the recurrences

$$v_\alpha(n,k)=(n-k+1)v_\alpha(n-1,k-1)+kv_\alpha(n-1,k), \tag{39}$$

for $n\ge 2$ and $1\le k\le n$, with $v_\alpha(n,0)=v_\alpha(n,n+1)=0$ for $n\ge 1$, and $v_\alpha(1,1)=1$, and

$$u_\alpha(n,k)=(n-k+1)^3v_\alpha(n-1,k-1)+k^3v_\alpha(n-1,k)+(n-k+1)u_\alpha(n-1,k-1)+ku_\alpha(n-1,k), \tag{40}$$

for $n\ge 2$ and $1\le k\le n$, with $u_\alpha(n,0)=u_\alpha(n,n+1)=0$ for $n\ge 1$, and $u_\alpha(1,1)=1$. The first few values of v_α and u_α are shown in Table III. With the help of [44] the v_α may be identified as the Eulerian numbers [41, p. 215], and are given by

$$v_\alpha(n,k)=\sum_{j=0}^{k}(-1)^j\binom{n+1}{j}(k-j)^n. \tag{41}$$

There is a more complicated formula for $u_\alpha(n,k)$ which we omit.

Similarly for the polytope $\beta(n,k)$ we have the graph shown in Fig. 14, and writing

$$V_\beta(n,k)=v_\beta(n,k)\frac{2^n}{n!}, \tag{42}$$

$$U_\beta(n,k)=u_\beta(n,k)\frac{2^n}{(n+2)!}, \tag{43}$$

TABLE III
THE FIRST FEW VALUES OF $v_\alpha(n,k)$, $u_\alpha(n,k)$, $v_\beta(n,k)$, AND $u_\beta(n,k)$. THE DIAGONALS CORRESPOND TO $k = 1, 2, \cdots$.

$v_\alpha(n,k)$

n						
6	1	57	302	302	57	1
5	1	26	66	26	1	
4	1	11	11	1		
3	1	4	1			
2	1	1				
1	1					

$u_\alpha(n,k)$

n						
6	6	1158	8916	8916	1158	6
5	5	400	1290	400	5	
4	4	116	116	4		
3	3	24	3			
2	2	2				
1	1					

$v_\beta(n,k)$

n						
6	1	58	360	662	719	720
5	1	27	93	119	120	
4	1	12	23	24		
3	1	5	6			
2	1	2				
1	1					

$u_\beta(n,k)$

n						
6	12	2568	28848	69624	80388	80640
5	10	950	5490	8250	8400	
4	8	312	880	960		
3	6	84	120			
2	4	16				
1	2					

we obtain the recurrences

$$v_\beta(n,k) = n v_\beta(n-1,k-1) + k v_\alpha(n-1,k), \tag{44}$$

for $n \ge 2$ and $1 \le k \le n$, with $v_\beta(n,0) = 0$ for $n \ge 1$, and $v_\beta(1,1) = 1$, and

$$u_\beta(n,k) = k^3(n+1)v_\alpha(n-1,k) + k u_\alpha(n-1,k) + n^2(n+1)v_\beta(n-1,k-1) + n u_\beta(n-1,k-1), \tag{45}$$

for $n \ge 2$ and $1 \le k \le n$, with $u_\beta(n,0) = 0$ for $n \ge 1$, $u_\beta(1,1) = 2$ (see Table III). Furthermore one can show by induction that

$$v_\beta(n,k) = \sum_{i=1}^{k} v_\alpha(n,i), \tag{46}$$

which implies $v_\beta(n,n) = n!$. Since the $v_\alpha(n,k)$ satisfy $v_\alpha(n,k) = v_\alpha(n, n-k+1)$ it follows that

$$v_\beta(2t,t) = \frac{1}{2}(2t)! \tag{47}$$

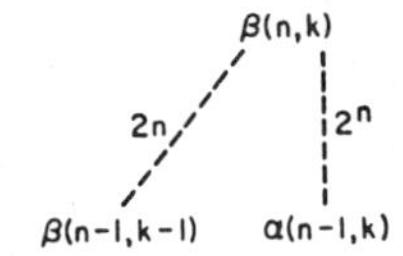

Fig. 14. $\beta(n,k)$ has $2n$ cells of type $\beta(n-1,k-1)$ and 2^n of type $\alpha(n-1,k)$. (We use broken lines here to make the structure of Figs. 15 and 16 more visible.)

For $\gamma(n,k)$ and $\delta(n,k)$ we have a pair of graphs similar to Figs. 13 and 14 (simply replace α by γ and β by δ in Figs. 13 and 14). As before we set

$$V_\gamma(n,k) = v_\gamma(n,k)\frac{\sqrt{n+1}}{n!},$$

$$U_\gamma(n,k) = u_\gamma(n,k)\frac{\sqrt{n+1}}{(n+2)!},$$

$$V_\delta(n,k) = v_\delta(n,k)\frac{2^n}{n!},$$

$$U_\delta(n,k) = u_\delta(n,k)\frac{2^n}{(n+2)!},$$

and obtain the recurrences

$$v_\gamma(n,k) = (2n-2k+1)v_\gamma(n-1,k-1) + (2k+1)v_\gamma(n-1,k), \tag{48}$$

for $n \ge 1$ and $0 \le k \le n$, with $v_\gamma(n,-1) = v_\gamma(n,n+1) = 0$ for $n \ge 0$, and $v_\gamma(0,0) = 1$;

$$u_\gamma(n,k) = (2n-2k+1)^3 v_\gamma(n-1,k-1) + (2k+1)^3 v_\gamma(n-1,k) + (2n-2k+1)u_\gamma(n-1,k-1) + (2k+1)u_\gamma(n-1,k), \tag{49}$$

for $n \ge 1$ and $0 \le k \le n$, with $u_\gamma(n,-1) = u_\gamma(n,n+1) = 0$ for $n \ge 0$, and $u_\gamma(0,0) = 0$;

$$v_\delta(n,k) = 2n v_\delta(n-1,k-1) + (2k+1)v_\gamma(n-1,k), \tag{50}$$

for $n \ge 1$ and $0 \le k \le n$, with $v_\delta(n,-1) = 0$ for $n \ge 0$, and $v_\delta(0,0) = 1$;

$$u_\delta(n,k) = (2k+1)^3(n+1)v_\gamma(n-1,k) + (2k+1)u_\gamma(n-1,k) + 8n^2(n+1)v_\delta(n-1,k-1) + 2n u_\delta(n-1,k-1), \tag{51}$$

for $n \ge 1$ and $0 \le k \le n$, with $u_\delta(n,-1) = 0$ for $n \ge 0$, and $u_\delta(0,0) = 0$ (see Table IV). Also

$$v_\gamma(n,k) = \sum_{j=0}^{k} (-1)^j \binom{n+1}{j} (2k+1-2j)^n,$$

$$v_\delta(n,k) = \sum_{i=0}^{k} v_\gamma(n,i),$$

$$v_\delta(n,n) = 2^n n!, \quad v_\delta(2t+1,t) = 2^{2t}(2t+1)!. \tag{52}$$

TABLE IV
$v_\gamma(n, k)$, $u_\gamma(n, k)$, $v_\delta(n, k)$, AND $u_\delta(n, k)$. THE DIAGONALS CORRESPOND TO $k = 0, 1, \cdots$.

$v_\gamma(n,k)$

n						
5	1	237	1682	1682	237	1
4	1	76	230	76	1	
3	1	23	23	1		
2	1	6	1			
1	1	1				
0	1					

$u_\gamma(n,k)$

n						
5	5	10065	124330	124330	10065	5
4	4	2416	10520	2416	4	
3	3	477	477	3		
2	2	60	2			
1	1	1				
0	0					

$v_\delta(n,k)$

n						
5	1	238	1920	3602	3839	3840
4	1	77	307	383	384	
3	1	24	47	48		
2	1	7	8			
1	1	2				
0	1					

$u_\delta(n,k)$

n						
5	10	20840	369740	954120	1074490	1075200
4	8	5224	41240	61048	61440	
3	6	1140	3654	3840		
2	4	188	256			
1	2	16				
0	0					

The j-dimensional faces of $\alpha(n, k), \cdots, \delta(n, k)$ for any j can be found from Figs. 15 and 16.

The special cases we are most interested in are $\beta(2t, t)$ and $\delta(2t+1, t)$, the Voronoi regions for D_{2t}^* and D_{2t+1}^* respectively. For D_{2t}^* we have

$$\operatorname{vol}(V(0)) = v_\beta(2t, t)\frac{2^{2t}}{(2t)!} = 2^{2t-1}, \quad \text{from (47)},$$

$$U(V(0)) = u_\beta(2t, t)\frac{2^{2t}}{(2t+2)!},$$

$$G(D_{2t}^*) = \frac{u_\beta(2t, t)}{2^{2-1/t}t(2t+2)!},$$

$$\rho = 1, \tag{53}$$

and covering radius

$$R_c = R_\beta(2t, t) = \sqrt{t} = \rho\sqrt{t}.$$

For this version of D_{2t+1}^* (which differs by a scale factor

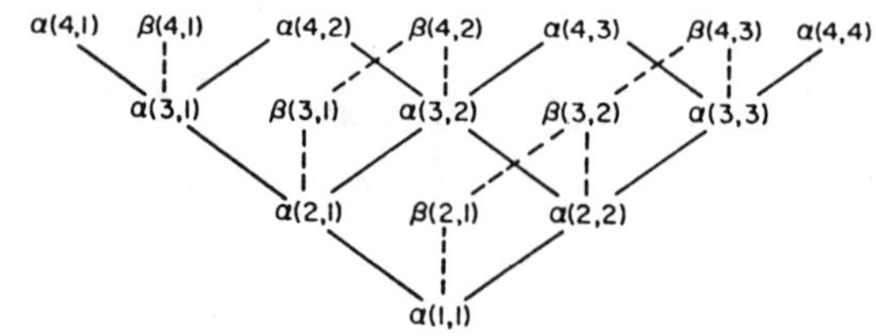

Fig. 15. Interconnections between the $\alpha(n, k)$ and $\beta(n, k)$.

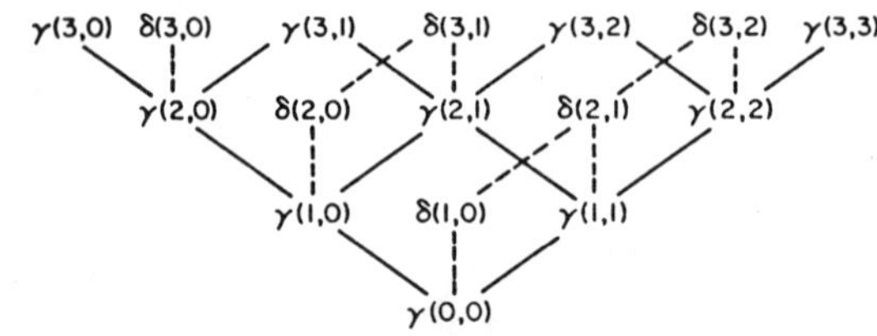

Fig. 16. Interconnections between the $\gamma(n, k)$ and $\delta(n, k)$.

from the definitions given in Section III-A) we have

$$\operatorname{vol}(V(0)) = v_\delta(2t+1, t)\frac{2^{2t+1}}{(2t+1)!} = 2^{4t+1}, \quad \text{from (52)},$$

$$U(V(0)) = u_\delta(2t+1, t)\frac{2^{2t+1}}{(2t+3)!},$$

$$G(D_{2t+1}^*) = \frac{u_\delta(2t+1, t)}{(2t+1)(2t+3)!2^{f(t)}}, \tag{54}$$

where

$$f(t) = \frac{2(2t^2 + 5t + 1)}{2t+1},$$

$$\rho = \sqrt{3} \quad (\text{if } t = 1), \qquad \rho = 2 \quad (\text{if } t > 1),$$

$$R_c = R_\delta(2t+1, t) = \sqrt{4t+1}$$

$$= \rho\sqrt{\frac{5}{3}} \quad (\text{if } t = 1), \qquad \rho\sqrt{t + \frac{1}{4}} \quad (\text{if } t > 1).$$

For example $D_3^* \cong A_3^*$ is the body-centered cubic lattice, the Voronoi region is a truncated octahedron (see [22, page 294 and anaglyph XI], [24, fig. 4], and [33, p. 129]), and $G(D_3^*) = 19/192\sqrt[3]{2}$ (see (7) and Section II-G). Also

$$G(D_4^*) = 13/120\sqrt{2} = G(D_4),$$

$$G(D_5^*) = \frac{2641}{23040 \cdot 2^{3/5}} = 0.0756254 \cdots, \tag{55}$$

$$G(D_6^*) = \frac{601 \cdot 2^{1/3}}{10080} = 0.0751203 \cdots. \tag{56}$$

The values of $G(D_n^*)$ are plotted in Fig. 20. The minimum value is $0.0746931 \cdots$ at $n = 9$.

G. Voronoi Region for A_n^*

Theorem 7: The Voronoi region for the lattice A_n^* is the polytope P_n defined in Fig. 17. If we rescale A_n^* by multiplying it by $n + 1$, the vertices of the Voronoi region may be taken to be the images of the point

$$\sigma = \left(-\frac{n}{2}, -\frac{n-2}{2}, -\frac{n-4}{2}, \cdots, \frac{n-2}{2}, \frac{n}{2}\right)$$

under the action of the Weyl group $W(A_n)$.

Since σ is sometimes referred to as the Weyl vector for A_n (see [10]), it is appropriate to call P_n the *Weyl polytope* of A_n. The case $n = 4$ of this theorem may be found in [15, pp. 72–73].

Sketch of Proof: It is easy to check that σ is equidistant from the walls of the fundamental simplex S for A_n; i.e. that σ is the incenter of S. Let P be the convex hull of the images of σ under $W(A_n)$. Since the walls of S are reflecting hyperplanes for $W_a(A_n)$, P and its images under $W_n(A_n)$ tile $\mathbb{R}^n$. Thus P is the Voronoi region for some lattice $\Lambda \subseteq A_n^*$. But Λ must contain all the points (28), since these are the images of $\mathbf{0}$ in the walls of P. Since these points span A_n^*, $\Lambda = A_n^*$.

The second moment of P_n may be found as follows. First, the covering radius of A_n^*, $R_c(n)$ say, is the circumradius of P_n, which is

$$R_c(n) = \sqrt{\sigma \cdot \sigma} = \left\{\frac{1}{2}\binom{n+2}{3}\right\}^{1/2}$$

$$= \rho\sqrt{\frac{n+2}{3}}, \tag{57}$$

since now $\rho = \sqrt{n(n+1)/4}$, and the volume of P_n is

$$V_n = (n+1)^{n-1/2} \text{ from (27)}. \tag{58}$$

Let $I_n = I(P_n)$ be the normalized second moment. A typical cell of P_n is obtained by deleting say the rth node from the left in Fig. 17, and is a prism $P_r \times P_s$ with $r + s = n - 1$ (see Fig. 18). The number of such faces is

$$\frac{|W(A_n)|}{|W(A_r)||W(A_S)|} = \binom{n+1}{r+1}. \tag{59}$$

Furthermore $I(P_r \times P_s) = I_r + I_s$.

Let h_{rs} be the height of the perpendicular from the center of P_n to a typical face $P_r \times P_s$. Then (see Fig. 19)

$$h_{rs}^2 = R_c(r+s+1)^2 - R_c(r)^2 - R_c(s)^2$$

$$= \frac{(r+1)(s+1)(n+1)}{4}, \quad \text{using (57)}.$$

We may now apply Theorem 3, to obtain

$$I_nV_n = \frac{1}{n+2}\sum_{r=0}^{n-1}\binom{n+1}{r+1}h_{rs}V_rV_s\left(h_{rs}^2 + I_r + I_s\right)$$

(with $r + s = n - 1$), which, if we write $J_n = I_{n-1}/n$, becomes

$$J_n = \frac{1}{2(n+1)n^{n-1}}\sum_{r=1}^{n-1}\binom{n}{r}r^r s^s\left(\frac{n}{4} + \frac{J_r}{s} + \frac{J_s}{r}\right),$$

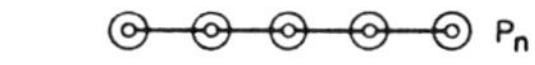

Fig. 17. Voronoi region $P_n = V(0)$ for A_n^*. There are n nodes, all ringed.

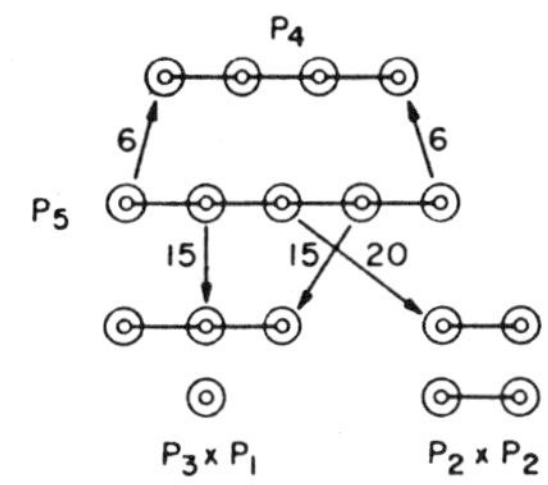

Fig. 18. The three types of cells of P_5.

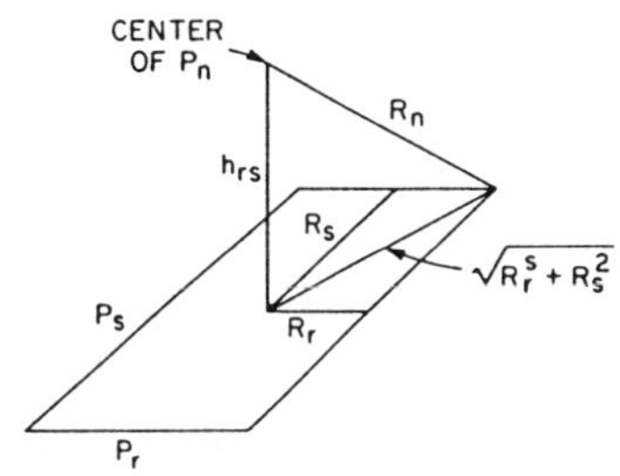

Fig. 19. Calculation of height h_{rs} of perpendicular from center of P_n to cell $P_r \times P_s$.

where $r + s = n - 2$. Using Abel's identity [42, section 1.5] to simplify the first term, this becomes

$$J_n = \frac{n!}{8(n+1)n^{n-2}}\sum_{k=0}^{n}\frac{n^k}{k!} - \frac{n^2}{4(n+1)}$$

$$+ \frac{1}{n+1}\sum_{r=1}^{n-1}\binom{n}{r}\left(\frac{r}{n}\right)^r\left(\frac{n-r}{n}\right)^{n-r-1}J_r \tag{60}$$

for $n \geq 2$, with $J_1 = 0$. The first few values are as follows:

n:	1	2	3	4	5	6	7
J_n:	0	$\frac{1}{12}$	$\frac{5}{18}$	$\frac{19}{32}$	$\frac{389}{375}$	$\frac{1045}{648}$	$\frac{78077}{33614}$.

Finally, the dimensionless second moment of the Voronoi region of A_n^* is

$$G(A_n^*) = \frac{J_{n+1}}{n(n+1)^{1-(1/n)}}. \tag{61}$$

The values for $n \leq 9$ are plotted in Fig. 20. The curve is extremely flat, the minimum value of $0.0754913 \cdots$ occurring at $n = 16$.

H. When is the Voronoi Region Determined by the First Layer of the Lattice?

We have seen in the Corollary to Theorem 5 that for a root lattice the walls of the Voronoi region are determined solely by the minimum vectors of the lattice. To give a precise statement of this property for an arbitrary lattice Λ, let us write $\Lambda = \Lambda_0 \cup \Lambda_1 \cup \Lambda_2 \cup \cdots$, where Λ_i, the ith *layer* from the origin, is chosen so that $\boldsymbol{u} \cdot \boldsymbol{u}$ is a constant λ_i (say) for all $\boldsymbol{u} \in \Lambda_i$, and $0 = \lambda_0 < \lambda_1 < \lambda_2 < \cdots$. We say that the Voronoi region is determined by the first layer of the lattice if the walls of the Voronoi region around the origin, $V(0)$, are bounded by the hyperplanes

$$x \cdot u = \tfrac{1}{2}\lambda_1, \quad \text{for } u \in \Lambda_1. \tag{62}$$

If this property holds then there is a simple description of the Voronoi region.

Theorem 8: If the Voronoi region $V(0)$ is determined by the first layer of the lattice, then $V(0)$ is the reciprocal of the vertex figure of Λ at the origin. Equivalently, $V(0)$ is (on a suitable scale) the reciprocal of the polytope with vertices Λ_1.

Proof: This follows immediately from the definitions of vertex figure and reciprocal polytope—see [16].

The final two theorems give sufficient conditions for this property to hold.

Theorem 9: Suppose that (i) $\Lambda_r \subseteq \Lambda_1 + \Lambda_1 + \cdots + \Lambda_1$ (r times) and (ii) $r\lambda_1 \le \lambda_r$, for all $r = 1, 2, \cdots$. Then the Voronoi region is determined by the first layer.

Condition (i) states that Λ_1 spans Λ, and moreover does it economically in the sense that any vector in Λ_r is the sum of not more than r vectors of Λ_1. In practice this condition is very easily checked by induction. An important class of lattices satisfying (i) are those obtained by applying Construction A of [32] or [45] to a linear binary code with minimum distance ≤ 4, which is spanned by the codewords of minimum weight, and which if $d < 4$ has the additional property that no coordinate of the code is always **0**.

Proof of Theorem 9: Suppose the contrary, so that there is a point $u \in \Lambda_r$ with $r > 1$ and a point $x \in \mathbb{R}^n$ such that

$$x \cdot u > \tfrac{1}{2}\lambda_r$$

but

$$x \cdot v \le \tfrac{1}{2}\lambda_1, \quad \text{for all } v \in \Lambda_1.$$

From (i), $u = \Sigma n_i v_i$ with $v_i \in \Lambda_1$, $n_i > 0$ and $\Sigma n_i \le r$. Then

$$x \cdot u = \Sigma n_i (x \cdot v_i) \le \tfrac{1}{2}\lambda_1 \Sigma n_i \le \tfrac{1}{2} r \lambda_1 \le \tfrac{1}{2}\lambda_r,$$

a contradiction.

Theorem 9 can be used to give an alternative proof that any root lattice has the property. On the other hand the dual lattices A_n^* for $n \ge 3$ and D_n^* for $n \ge 5$ do not, as the previous section demonstrated, nor does the Leech lattice in 24 dimensions [9], [32]. In fact one can show that the Voronoi region of the Leech lattice is determined just by the first two layers: in other words the Voronoi region has 196560 + 16773120 = 16969680 cells.

A second test is the following.

Theorem 10 (A. Gersho, private communication): Let V_1 be the intersection of the half-planes defined by (62). If $x \cdot x \le \frac{1}{4}\lambda_2$ holds for all $x \in V_1$, then V_1 is the Voronoi region.

TABLE V
DIMENSIONLESS SECOND MOMENT $G(\Lambda)$.

n	sphere bound	best lattice known Λ	$G(\Lambda)$	Zador bound
1	.0833	A_1	.0833	.500
2	.0796	A_2	.0802	.159
3	.0770	A_3^*	.0785	.116
4	.0750	D_4	.0766	.100
5	.0735	D_5^*	.0756	.091
6	.0723	E_6	.0743	.086
7	.0713	E_7	.0732	.082
8	.0704	E_8	.0717	.080
9	.0697	D_9^*	.0747	.078
10	.0691	D_{10}^*	.0747	.076

Proof: Take any $x \in V_1$, $v \in \Lambda_2 \cup \Lambda_3 \cup \cdots$. Then $x \cdot v \le \|x\| \, \|v\| \le \frac{1}{2}\sqrt{\lambda_2}\sqrt{\lambda_2} = \frac{1}{2}\lambda_2$, so $x \in V(0)$.

IV. COMPARISON OF QUANTIZERS

Comparing the different lattices analyzed in this section we see that the best quantizers found so far in dimensions 1–10 are the following:

dimension	lattice	dimension	lattice
1	$A_1(\cong A_1^*)$	6	E_6
2	$A_2(\cong A_2^*)$	7	E_7
3	$A_3^*(\cong D_3^*)$	8	$E_8(= E_8^*)$
4	$D_4(\cong D_4^*)$	9	D_9^*
5	D_5^*	10	D_{10}^* .

The values of the dimensionless second moment $G(\Lambda)$, which is our measure of the mean-squared quantization error per symbol, are shown in Table V and Fig. 20, together with Zador's bounds (3). It is known that A_1 and A_2 are optimal, and it is tempting to make the following conjecture.

Conjecture

The best *lattice* quantizer in $\mathbb{R}^n$—that with the lowest $G(\Lambda)$—is the dual of the densest lattice packing.

Certainly E_6^*, E_7^*, and the Leech lattice should be investigated.

It is worth drawing attention to the remarkably low value of the mean-squared error for E_8 (see Fig. 20). Furthermore there is a fast algorithm [12] available for performing the quantization with this lattice (and in fact for any of the lattices described here).

Note added in proof: It has recently been shown that the body-centered cubic lattice A_3^* is the optimal three-dimensional lattice quantizer for uniformly distributed data: see E. S. Barnes and N. J. A. Sloane, "The optimal lattice

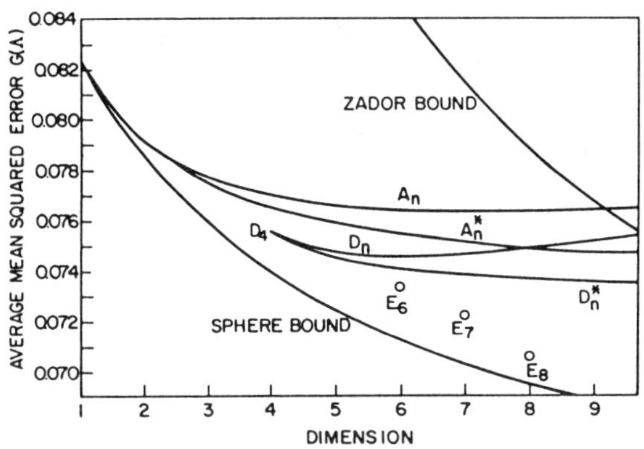

Fig. 20. Comparison of mean-squared quantization error per symbol, $G(\Lambda)$, for different lattices Λ in dimensions 1–9.

quantizer in three dimensions," *SIAM J. Discrete and Algebraic Methods*, to appear.

Acknowledgment

During the early stages of this work we were greatly helped by several discussions with Allen Gersho. Some of the calculations were performed on the MACSYMA system [35]. We should also like to thank E. S. Barnes and H. S. M. Coxeter for their comments on the manuscript.

References

[1] A. D. Alexandrow, *Konvexe Polyeder*. Berlin: Akademie-Verlag, 1958.

[2] E. S. Barnes, "Construction of perfect and extreme forms II," *Acta Arithmetica*, vol. 5, pp. 205–222, 1959.

[3] C. T. Benson and L. C. Grove, *Finite Reflection Groups*. Tarrytown-on-Hudson, NY: Bogden and Quigley, 1971.

[4] M. N. Bleicher, "Lattice coverings of *n*-space by spheres," *Can. J. Math.*, vol. 14, pp. 632–650, 1962.

[5] N. Bourbaki, *Groupes et Algèbres de Lie, Chapitres 4, 5 et 6*. Paris: Hermann, 1968.

[6] J. A. Bucklew, "Upper bounds to the asymptotic performance of block quantizers," *IEEE Trans. Inform. Theory*, vol. IT-27, no. 5, pp. 577–581, Sept. 1981.

[7] J. A. Bucklew and G. L. Wise, "A note on multidimensional asymptotic quantization theory," *Proc. 18th Allerton Conf. on Commun., Control, and Computing*, 1980, pp. 454–463.

[8] J. W. S. Cassels, *An Introduction to the Geometry of Numbers*. New York: Springer-Verlag, 1971.

[9] J. H. Conway, R. A. Parker, and N. J. A. Sloane, "The covering radius of the Leech lattice," *Proc. Royal Soc.*, 1982, to be published.

[10] J. H. Conway and L. Queen, "On computing the characters of Lie groups," in preparation.

[11] J. H. Conway and N. J. A. Sloane, "On the enumeration of lattices of determinant one," *J. Number Theory*, to be published.

[12] ——, "Fast quantizing and decoding algorithms for lattice quantizers and codes," *IEEE Trans. Inform. Theory*, this issue, pp. 227–232.

[13] H. S. M. Coxeter, "Wythoff's construction for uniform polytopes," *Proc. London Math. Soc.*, Ser. 2, vol. 38, pp. 327–339, 1935; reprinted in [15].

[14] ——, "Extreme forms," *Can. J. Math.*, vol. 3, pp. 391–441, 1951.

[15] ——, *Twelve Geometric Essays*. Carbondale: Southern Illinois Univ., 1968.

[16] ——, *Regular Polytopes*. New York: Dover, 3rd ed., 1973.

[17] ——, "Polytopes in the Netherlands," *Nieuw Archief voor Wiskunde (3)*, vol. 26, pp. 116–141, 1978.

[18] H. S. M. Coxeter and W. O. J. Moser, *Generators and Relations for Discrete Groups*. New York: Springer-Verlag, 4th ed., 1980.

[19] H. M. Cundy and A. P. Rollett, *Mathematical Models*. Oxford: The Univ., 1961.

[20] E. L. Elte, "The semiregular polytopes of the hyperspaces," Doctor of Math. and Science Thesis, State-Univ. of Gronigen, The Netherlands, 13 May 1912.

[21] L. Fejes Tóth, "Sur la représentation d'une population infinie par un nombre fini d'éléments," *Acta Math. Acad. Sci. Hungar.*, vol. 10, pp. 299–304, 1959.

[22] ——, *Regular Figures*. Oxford: Pergamon, 1964.

[23] ——, *Lagerungen in der Ebene, auf der Kugel und im Raum*. New York: Springer-Verlag, 2nd ed., 1972.

[24] A. Gersho, "Asymptotically optimal block quantization," *IEEE Trans. Inform. Theory*, vol. IT-25, no. 4, pp. 373–380, July 1979.

[25] I. J. Good and R. A. Gaskins, "The centroid method of numerical integration," *Numer. Math.*, vol. 16, pp. 343–359, 1971.

[26] B. Grünbaum, *Convex Polytopes*. New York: Wiley, 1967.

[27] M. Hazewinkel *et al.*, "The ubiquity of Coxeter–Dynkin diagrams (an introduction to the A-D-E problem)," *Nieuw Archief voor Wiskunde* (3), vol. 25, pp. 257–307, 1977.

[28] D. Hilbert and S. Cohn-Vossen, *Geometry and The Imagination*. New York: Chelsea, 1952.

[29] A. Holden, *Shapes, Space, and Symmetry*. New York: Columbia Univ., 1971.

[30] J. E. Humphreys, *Introduction to Lie Algebras and Representation Theory*. New York: Springer-Verlag, 1972.

[31] G. Kaur, "Extreme quadratic forms for coverings in four variables," *Proc. Nat. Inst. Sci. India*, vol. 32A, pp. 414–417, 1966.

[32] J. Leech and N. J. A Sloane, "Sphere packing and error-correcting codes," *Can. J. Math.* vol. 23, pp. 718–745, 1971.

[33] A. L. Loeb, *Space Structures: Their Harmony and Counterpoint*. Reading, MA: Addison-Wesley, 1976.

[34] L. A. Lyusternik, *Convex Figures and Polyhedra*. New York: Dover, 1963.

[35] Mathlab Group, *MACSYMA Reference Manual*. Cambridge, MA: Lab. Comp. Sci., M.I.T., version 9, 1977.

[36] P. McMullen and G. C. Shephard, *Convex Polytopes and the Upper Bound Conjecture*, London Math. Soc. Lecture Note Series 3. Cambridge: The Univ., 1971.

[37] D. J. Newman, "The hexagon theorem," *IEEE Trans. Inform. Theory*, this issue, pp. 137–139.

[38] H. V. Niemeier, "Definite quadratische Formen der Dimension 24 und Diskriminante 1," *J. Number Theory*, vol. 5, pp. 142–178, 1973.

[39] O. T. O'Meara, *Introduction to Quadratic Forms*. New York: Springer-Verlag, 1971.

[40] A. Pugh, *Polyhedra: A Visual Approach*. Berkeley: Univ. of Calif., 1976.

[41] J. Riordan, *An Introduction to Combinatorial Analysis*. New York: Wiley, 1958.

[42] J. Riordan, *Combinatorial Identities*. New York: Wiley, 1968.

[43] J. Satterly, "The moments of inertia of some polyhedra," *Math. Gazette*, vol. 42, pp. 11–13, 1958.

[44] N. J. A. Sloane, *A Handbook of Integer Sequences*. New York: Academic, 1973.

[45] N. J. A. Sloane, "Binary codes, lattices and sphere packings," in *Combinatorial Surveys* (Proc. 6th British Combinatorial Conf.), P. J. Cameron, Ed. London and New York: Academic, 1977, pp. 117–164.

[46] N. J. A. Sloane, "Self-dual codes and lattices," in *Relations Between Combinatorics and Other Parts of Mathematics*, Proc. Symp. Pure Math., vol. 34, Providence RI: Amer. Math. Soc., 1979, pp. 273–308.

[47] N. J. A. Sloane, "Tables of sphere packings and spherical codes," *IEEE Trans. Inform. Theory*, vol. IT-27, no. 3, pp. 327–338, May 1981.

[48] B. M. Stewart, *Adventures among the Toroids*. Published by the author, 4494 Wausau Rd., Okemos, MI, 1970.

[49] G. Voronoi, "Récherches sur les parallèloèdres primitifs I, II," *J. reine angew. Math.*, vol. 134, pp. 198–287, 1908, and vol. 136, pp. 67–181, 1909.

[50] M. J. Wenninger, *Polyhedron Models*. Cambridge: The Univ., 1971.

[51] E. T. Whittaker and G. N. Watson, *A Course of Modern Analysis*. Cambridge: The Univ., 4th ed., 1963.

[52] Y. Yamada, S. Tazaki, and R. M. Gray, "Asymptotic performance of block quantizers with difference distortion measures," *IEEE Trans. Inform. Theory*, vol. IT-26, no. 1, pp. 6–14, Jan. 1980.

[53] P. Zador, "Topics in the asymptotic quantization of continuous random variables," *IEEE Trans. Inform. Theory*, this issue, pp. 139–149.

Fast Quantizing and Decoding Algorithms for Lattice Quantizers and Codes

J. H. CONWAY AND N. J. A. SLOANE, FELLOW, IEEE

Abstract—**For each of the lattices $A_n (n \geq 1)$, $D_n (n \geq 2)$, E_6, E_7, E_8, and their duals a very fast algorithm is given for finding the closest lattice point to an arbitrary point. If these lattices are used for vector quantizing of uniformly distributed data, the algorithm finds the minimum distortion lattice point. If the lattices are used as codes for a Gaussian channel, the algorithm performs maximum likelihood decoding.**

I. Introduction

THE SO-CALLED *root lattices* are the n-dimensional lattices A_n $(n \geq 1)$, $D_n (n \geq 2)$, and $E_n (n = 6, 7, 8)$ defined in Section II. These lattices and their duals give rise to the densest known sphere packing and coverings in dimensions $n \leq 8$, and they can be used as the basis for efficient block quantizers for uniformly distributed inputs and to construct codes for a band-limited channel with Gaussian noise (see [6], [9], [11], [16]). Around each lattice point is its *Voronoi region*, consisting of all points of the underlying space which are closer to that lattice point than to any other. (Voronoi regions are also called Dirichlet regions, Brillouin zones, Wigner–Seitz cells, or nearest neighbor regions.) If the lattice is used as a quantizer, all the points in the Voronoi region around the lattice point x are represented by x; while if the lattice is used as a code for a Gaussian channel, all the points in the Voronoi region around x are decoded as x. In the preceding paper [6] we found the Voronoi regions for most of the root lattices and their duals, as well as the mean-squared quantization error when these lattices are used to quantize uniformly distributed data.

In the present paper we give very fast and simple algorithms which, for any of the lattices A_n, D_n, E_n, and their duals (as well as many other lattices), find the closest lattice point to an arbitrary point of the space. In other words the algorithms find which Voronoi region the given point belongs to. The algorithms can therefore be used either for vector quantizing or for channel decoding. The running time of the algorithms for D_n, D_n^*, E_n, and E_n^* is proportional to n, while for A_n and A_n^* it is proportional to $n \log n$ and $n^2 \log n$, respectively.

Although considerable work has been done in the past on n-dimensional quantizers and codes (see [1], [3], [8], [9], [14], [16]–[18], and the references given there), these algorithms appear to be new.

Manuscript received May 7, 1981; revised November 3, 1981.

J. H. Conway is with the Department of Pure Mathematics and Mathematical Statistics, University of Cambridge, 16 Mill Lane, Cambridge CB2 1SB, England.

N. J. A. Sloane is with the Mathematics and Statistics Research Center, Bell Laboratories, Murray Hill, NJ 07974.

II. Definition of Root Lattices and their Duals

If $a_1, \cdots, a_n$ are linearly independent vectors in m-dimensional real Euclidean space $\mathbb{R}^m$ with $m \geq n$, the set of all vectors

$$x = u_1 a_1 + \cdots + u_n a_n,$$

where $u_1, \cdots, u_n$ are arbitrary integers, is called an n-dimensional *lattice* Λ. The *dual* (or *reciprocal*) lattice Λ^* consists of all points y in the subspace of $\mathbb{R}^m$ spanned by $a_1, \cdots, a_n$ such that the inner product $x \cdot y = x_1 y_1 + \cdots + x_m y_m$ is an integer for all $x \in \Lambda$. Most of the lattices considered here are contained in their duals, so that there are coset representatives $r_0, \cdots, r_{d-1}$ such that

$$\Lambda^* = \bigcup_{i=0}^{d-1} (r_i + \Lambda). \tag{1}$$

The number d is called the *determinant* of Λ. The *norm* (or energy) of a vector $x = (x_1, \cdots, x_m) \in \mathbb{R}^m$ is

$$\|x\| = \sqrt{x \cdot x} = \left(\sum x_i^2\right)^{1/2}.$$

For further information about lattices see for example [2], [4]–[7], [11], [12], [15], [16].

For $n \geq 1$, A_n is the n-dimensional[1] lattice consisting of the points $(x_0, x_1, \cdots, x_n)$ having integer coordinates that sum to zero. (Thus A_n is an n-dimensional lattice described by $(n + 1)$-dimensional coordinates.) If two lattices A and B differ only by a rotation and change of scale, we say they are equivalent and write $A \cong B$. Then A_1 is equivalent to the one-dimensional lattice of integer points, $\mathbb{Z}$, and A_2 is equivalent to the familiar two-dimensional hexagonal lattice (see Fig. 1).

The dual lattice A_n^* consists of the union of $n + 1$ cosets of A_n:

$$A_n^* = \bigcup_{i=0}^{n} (r_i + A_n), \tag{2}$$

[1] The subscript gives the dimension of the lattice.

Reprinted from *IEEE Trans. Inform. Theory,* vol. IT-28, pp. 227–232, March 1982.

where, for $i = 0, 1, \cdots, n$ and $j = n + 1 - i$,

$$r_i = \left(\frac{-j}{n+1}, \frac{-j}{n+1}, \cdots, \frac{-j}{n+1}, \frac{i}{n+1}, \cdots, \frac{i}{n+1}\right) = \left(\left(\frac{-j}{n+1}\right)^i, \left(\frac{i}{n+1}\right)^j\right). \quad (3)$$

Then $A_1^* \cong A_1 \cong \mathbb{Z}$ and $A_2^* \cong A_2$. Also A_3 ($\cong D_3$) is the face-centered cubic lattice and A_3^* ($\cong D_3^*$) is the body-centered cubic lattice. It is known that A_2 is the optimal two-dimensional quantizer for a uniformly distributed input, and it is conjectured that A_3^* is optimal in three dimensions.

For $N \geq 2$, D_n consists of the points $(x_1, x_2, \cdots, x_n)$ having integer coordinates with an even sum. The dual D_n^* is the union of four cosets of D_n:

$$D_n^* = \bigcup_{i=0}^{3} (r_i + D_n), \quad (4)$$

where

$$r_0 = (0^n), \quad r_1 = \left(\tfrac{1}{2}^n\right),$$
$$r_2 = (0^{n-1}, 1), \quad r_3 = \left(\tfrac{1}{2}^{n-1}, -\tfrac{1}{2}\right).$$

Also $D_2 \cong A_1 \oplus A_1 \cong D_2^*$, $D_3 \cong A_3$, and $D_4^* \cong D_4$.

For example the vectors of small norm in D_4 consist of all permutations and sign changes of the following:

type	number	norm^2
(0, 0, 0, 0)	1	0
(±1, ±1, 0, 0)	24	2
(±2, 0, 0, 0)	8	4
(±1, ±1, ±1, ±1)	16	4
(±2, ±1, ±1, 0)	96	6
(±2, ±2, 0, 0)	24	8
...	...	...

(5)

(See also [16, Table V].)

There is a second, equivalent definition of D_n^* which is sometimes easier to use: D_n^* consists of the points of the n-dimensional integer lattice $\mathbb{Z}^n$, together with the translate of $\mathbb{Z}^n$ by the vector $(\frac{1}{2}, \frac{1}{2}, \cdots, \frac{1}{2})$, i.e.,

$$D_n^* = \mathbb{Z}^n \cup \left(\left(\tfrac{1}{2}^n\right) + \mathbb{Z}^n\right). \quad (6)$$

The most convenient definitions of E_6, E_7, and E_8 are the following.[2] E_8 is the union of D_8 and the coset

$$\left(\tfrac{1}{2}, \tfrac{1}{2}, \tfrac{1}{2}, \tfrac{1}{2}, \tfrac{1}{2}, \tfrac{1}{2}, \tfrac{1}{2}, \tfrac{1}{2}\right) + D_8.$$

In other words E_8 consists of the points $(x_1, \cdots, x_8)$ with $x_i \in \mathbb{Z}$ and Σx_i even, together with the points $(y_1, \cdots, y_8)$ with $y_i \in \mathbb{Z} + \frac{1}{2}$ and Σy_8 even. Also $E_8^* = E_8$.

[2]A slightly different (although equivalent) definition of E_7 was used in [6]. The present definition leads to a simpler algorithm in Section VII.

E_7 is a subspace of dimension 7 in E_8, consisting of the points $(u_1, \cdots, u_8)$ in E_8 with $\Sigma u_i = 0$. Equivalently

$$E_7 = A_7 \cup \left(\left(-\tfrac{1}{2}^4, \tfrac{1}{2}^4\right) + A_7\right). \quad (7)$$

The dual E_7^* is given by

$$E_7^* = E_7 \cup \left(\left(-\tfrac{3}{4}^2, \tfrac{1}{4}^6\right) + E_7\right) = \bigcup_{i=0}^{3} (s_i + A_7), \quad (8)$$

where

$$s_i = \left(\left(\frac{-j}{4}\right)^{2i}, \left(\frac{i}{4}\right)^{2j}\right), \quad i + j = 4. \quad (9)$$

Finally E_6 is a subspace of dimension 6 in E_8, which we may take for example to consist of the points $(u_1, \cdots, u_8)$ in E_8 with $\Sigma u_i = 0$ and $u_1 + u_8 = 0$. However this lattice and its dual do not appear to be as important as the others for the applications considered here, and we shall not discuss them in as much detail.

III. Finding the Closest Point of the n-Dimensional Integer Lattice $\mathbb{Z}^n$

The algorithm for finding the closest point of the integer lattice $\mathbb{Z}^n$ to an arbitrary point $x \in \mathbb{R}^n$ is particularly simple, and serves to introduce the notation. For a real number x, let

$$f(x) = \text{closest integer to } x.$$

In case of a tie, choose the integer with the smallest absolute value. For $\boldsymbol{x} = (x_1, \cdots, x_n) \in \mathbb{R}^n$, let

$$f(\boldsymbol{x}) = (f(x_1), \cdots, f(x_n)).$$

For later use we also define $g(\boldsymbol{x})$, which is the same as $f(\boldsymbol{x})$ except that the *worst* component of $\boldsymbol{x}$—that furthest from an integer—is rounded the *wrong way*. In case of a tie, the component with the lowest subscript is rounded the wrong way.

More formally, for $x \in \mathbb{R}$ we define $f(x)$ and the function $w(x)$ which rounds the wrong way as follows. (Here m is an integer.)

If $x = 0$,	then $f(x) = 0$, $w(x) = 1$.
If $0 < m \leq x \leq m + \frac{1}{2}$,	then $f(x) = m$, $w(x) = m + 1$.
If $0 < m + \frac{1}{2} < x < m + 1$,	then $f(x) = m + 1$, $w(x) = m$.
If $-m - \frac{1}{2} \leq x \leq -m < 0$,	then $f(x) = -m$, $w(x) = -m - 1$.
If $-m - 1 < x < -m - \frac{1}{2}$,	then $f(x) = -m - 1$, $w(x) = -m$.

(10)

(Ties are handled so as to give preference to points of

smaller norm.) We also write

$$x = f(x) + \delta(x),$$

so that $|\delta(x)| \leq 1/2$ is the distance from x to the nearest integer.

Given $x = (x_1, \cdots, x_n) \in \mathbb{R}^n$, let k $(1 \leq k \leq n)$ be such that

$$|\delta(x_k)| \leq |\delta(x_i)| \quad \text{for all } 1 \leq i \leq n$$

and

$$|\delta(x_k)| = |\delta(x_i)| \Rightarrow k \leq i.$$

Then $g(x)$ is defined by

$$g(x) = (f(x_1), f(x_2), \cdots, w(x_k), \cdots, f(x_n)).$$

Algorithm 1—To Find the Closest Point of $\mathbb{Z}^n$ to x: Given $x \in \mathbb{R}^n$, the closest point of $\mathbb{Z}^n$ is $f(x)$. (If x is equidistant from two or more points of $\mathbb{Z}^n$, this procedure finds the one with the smallest norm.)

To see that the procedure works, let $u = (u_1, \cdots, u_n)$ be any point of $\mathbb{Z}_n$. Then

$$\|u - x\|^2 = \sum_{i=1}^{n} (u_i - x_i)^2,$$

which is minimized by choosing $u_i = f(x_i)$ for $i = 1, \cdots, n$. Because of (10) ties are broken correctly, favoring the point with the smallest norm.

IV. Finding the Closest Point of D_n

Algorithm 2—To Find the Closest Point of D_n to x: Given $x \in \mathbb{R}^n$, the closest point of D_n is whichever of $f(x)$ and $g(x)$ has an even sum of components (one will have an even sum, the other an odd sum). If x is equidistant from two or more points of D_n this procedure produces a nearest point having the smallest norm.

This procedure works because $f(x)$ is the closest point of $\mathbb{Z}^n$ to x and $g(x)$ is the next closest. $f(x)$ and $g(x)$ differ by one in exactly one coordinate, and so precisely one of $\Sigma f(x_i)$ and $\Sigma g(x_i)$ is even and the other is odd. Again (10) implies that ties are broken correctly.

Example: Find the closest point of D_4 to $x = (0.6, -1.1, 1.7, 0.1)$. We compute

$$f(x) = (1, -1, 2, 0)$$

and

$$g(x) = (0, -1, 2, 0),$$

since the first component of x is the furthest from an integer. The sum of the components of $f(x)$ is $1 - 1 + 2 + 0 = 2$, which is even, while that of $g(x)$ is $0 - 1 + 2 + 0 = 1$, which is odd. Therefore $f(x)$ is the point of D_4 closest to x.

To illustrate how ties are handled, suppose

$$x = (\tfrac{1}{2}, \tfrac{1}{2}, \tfrac{1}{2}, \tfrac{1}{2}).$$

In fact (see (5)) x is now equidistant from eight points of D_4, namely $(0,0,0,0)$, any permutation of $(1,1,0,0)$, and $(1,1,1,1)$. The algorithm computes

$$f(x) = (0,0,0,0), \quad \text{sum} = 0, \quad \text{even},$$
$$g(x) = (1,0,0,0), \quad \text{sum} = 1, \quad \text{odd},$$

and selects $f(x)$. Indeed $f(x)$ does have the smallest norm of the eight neighboring points.

V. Finding the Closest Point of a Coset or of a Dual Lattice

A procedure Φ for finding the closest point of a lattice Λ to a given point x can be easily converted to a procedure for finding the closest point of a coset $r + \Lambda$ to x. For if $\Phi(x)$ is the closest point of Λ to x,

$$\Phi(x - r) + r$$

is the closest point of $r + \Lambda$ to x.

Suppose further that L is a lattice (or in fact any set of points) which is a union of cosets of Λ:

$$L = \bigcup_{i=0}^{d-1} (r_i + \Lambda).$$

Then Φ can be used as the basis for the following procedure for finding the closest point of L.

Algorithm 3—To Find the Closest Point of L (A Union of Cosets of a Lattice) to a Given Point x: Given x, compute

$$y_i = \Phi(x - r_i) + r_i$$

for $i = 0, 1, \cdots, d-1$. Compare each of $y_0, \cdots, y_{d-1}$ with x and choose the closest.

In view of (1), (2), (4), (8) this algorithm reduces the problem of finding the closest point of the dual lattices A_n^*, D_n^* and E_6^*, E_7^*, E_8^* to that of finding the closest point of the original lattices. Alternatively, a faster algorithm for D_n^* is obtained from (6) and Algorithm 1. We illustrate Algorithm 3 by applying it first to D_5^* using definition (4), and then to D_3^* using definition (6).

For the first example we observe from (4) that D_5^* is the union of four cosets of D_5 with coset representatives

$$r_0 = (0, 0, 0, 0, \ 0),$$
$$r_1 = (\tfrac{1}{2}, \tfrac{1}{2}, \tfrac{1}{2}, \tfrac{1}{2}, \ \tfrac{1}{2}),$$
$$r_2 = (0, 0, 0, 0, \ 1),$$
$$r_3 = (\tfrac{1}{2}, \tfrac{1}{2}, \tfrac{1}{2}, \tfrac{1}{2}, -\tfrac{1}{2}).$$

To find the closest point of D_5^* to

$$x = (0.1, 0, -0.3, 0.4, 0.8)$$

we proceed as follows:

$$f(x) = (0,0,0,0,1), \quad \text{sum} = 1, \quad \text{odd},$$
$$g(x) = (0,0,0,1,1), \quad \text{sum} = 2, \quad \text{even},$$

therefore

$$y_0 = (0,0,0,1,1).$$

Also

$$x - r_1 = (-0.4, -0.5, -0.8, -0.1, 0.3),$$
$$f(x - r_1) = (0,0,-1,0,0), \quad \text{sum} = -1, \quad \text{odd},$$
$$g(x - r_1) = (0,-1,-1,0,0), \quad \text{sum} = -2, \quad \text{even},$$

therefore

$$y_1 = g(x - r_1) + r_1 = (0.5, -0.5, -0.5, 0.5, 0.5);$$
$$x - r_2 = (0.1, 0, -0.3, 0.4, -0.2),$$
$$f(x - r_2) = (0,0,0,0,0), \quad \text{sum} = 0, \quad \text{even},$$
$$g(x - r_2) = (0,0,0,1,0), \quad \text{sum} = 1, \quad \text{odd},$$

therefore

$$y_2 = f(x - r_2) + r_2 = (0,0,0,0,1);$$

and

$$x - r_3 = (-0.4, -0.5, -0.8, -0.1, 1.3),$$
$$f(x - r_3) = (0,0,-1,0,1), \quad \text{sum} = 0, \quad \text{even},$$
$$g(x - r_3) = (0,-1,-1,0,1), \quad \text{sum} = -1, \quad \text{odd},$$

therefore

$$y_3 = f(x - r_3) + r_3 = (0.5, 0.5, -0.5, 0.5, 0.5).$$

The final step is to see which of $y_0, \cdots, y_3$ is closest to x. We compute

$$\|x - y_0\|^2 = 0.5,$$
$$\|x - y_1\|^2 = 0.55,$$
$$\|x - y_2\|^2 = 0.3,$$
$$\|x - y_3\|^2 = 0.55.$$

Thus

$$y_2 = (0,0,0,0,1)$$

is the point of D_5^* closest to x.

For the second example we use (6) to define the body-centered cubic lattice D_3^* as the union of two cosets of $\mathbb{Z}^3$ with coset representatives

$$r_0 = (0,0,0),$$
$$r_1 = (\tfrac{1}{2}, \tfrac{1}{2}, \tfrac{1}{2}).$$

To find the closest point of D_3^* to

$$x = (0.2, 0.5, 0.8),$$

we compute

$$y_0 = f(x) = (0,0,1),$$
$$x - r_1 = (-0.3, 0, 0.3),$$
$$f(x - r_1) = (0,0,0),$$
$$y_1 = r_1 + f(x - r_1) = (0.5, 0.5, 0.5).$$

The final step is to find which of y_0 and y_1 is closer to x:

$$\|x - y_0\|^2 = 0.33, \quad \|x - y_1\|^2 = 0.18.$$

Thus $y_1 = (0.5, 0.5, 0.5)$ is the closest point of D_3^* to x.

Obviously the second definition of D_n^* leads to the faster algorithm. For any n it is only necessary to compute $f(x)$, $f(x - r_1)$, and to calculate the squared norms $\|x - y_0\|^2$ and $\|x - y_1\|^2$.

VI. Finding the Closest Point of E_8

Since E_8 is the union of two cosets of D_8 the discussion in the previous section leads to the following procedure.

Algorithm 4—To Find the Closest Point of E_8 to x: Given $x = (x_1, \cdots, x_8) \in \mathbb{R}^8$.

Compute $f(x)$ and $g(x)$, and select whichever has an even sum of components; call it y_0.

Compute $f(x - \tfrac{1}{2})$ and $g(x - \tfrac{1}{2})$, where

$$\tfrac{1}{2} = (\tfrac{1}{2}, \tfrac{1}{2}, \tfrac{1}{2}, \tfrac{1}{2}, \tfrac{1}{2}, \tfrac{1}{2}, \tfrac{1}{2}, \tfrac{1}{2}),$$

and select whichever has an even sum of components; add $\tfrac{1}{2}$ and call the result y_1.

Compare y_0 and y_1 with x and choose the closest.

For example, to find the closest point of E_8 to

$$x = (0.1, 0.1, 0.8, 1.3, 2.2, -0.6, -0.7, 0.9),$$

we compute

$$f(x) = (0,0,1,1,2,-1,-1,1), \quad \text{sum} = 3, \quad \text{odd},$$
$$g(x) = (0,0,1,1,2,0,-1,1), \quad \text{sum} = 4, \quad \text{even},$$

and take $y_0 = g(x)$. Also

$$x - \tfrac{1}{2} = (-0.4, -0.4, 0.3, 0.8, 1.7, -1.1, -1.2, 0.4),$$
$$f(x - \tfrac{1}{2}) = (0,0,0,1,2,-1,-1,0), \quad \text{sum} = 1, \quad \text{odd},$$
$$g(x - \tfrac{1}{2}) = (-1,0,0,1,2,-1,-1,0), \quad \text{sum} = 0, \quad \text{even},$$

and so

$$y_1 = g(x - \tfrac{1}{2}) + \tfrac{1}{2} = (-0.5, 0.5, 0.5, 1.5, 2.5, -0.5, -0.5, 0.5).$$

Finally,

$$\|x - y_0\|^2 = 0.65, \quad \|x - y_1\|^2 = 0.95$$

and we conclude that $y_0 = g(x)$ is the closest point to x.

VII. Finding the Closest Point of A_n, A_n^*, E_7, and E_7^*

Algorithm 5—To Find the Closest Point of A_n to x:

Step 1: Given $x \in \mathbb{R}^{n+1}$, compute $s = \Sigma x_i$ and replace x by

$$x' = x - \frac{s}{n+1}(1, 1, \cdots, 1).$$

Step 2: Calculate $f(x') = (f(x'_0), \cdots, f(x'_n))$ and the *deficiency* $\Delta = \Sigma f(x'_i)$.

Step 3: Sort the x'_i in order of increasing value of $\delta(x'_i)$ (defined in Section III). We obtain a rearrangement of the numbers $0, 1, \cdots, n$, say $i_0, i_1, \cdots, i_n$, such that

$$-\tfrac{1}{2} \le \delta(x'_{i_0}) \le \delta(x'_{i_1}) \cdots \le \delta(x'_{i_n}) \le \tfrac{1}{2}.$$

Step 4: If $\Delta = 0$, $f(x')$ is the closest point of A_n to x.
If $\Delta > 0$, the closest point is obtained by subtracting 1 from the components $f(x'_{i_0}), \cdots, f(x'_{i_{\Delta-1}})$.
If $\Delta < 0$, the closest point is obtained by adding 1 to the components $f(x'_{i_n}), f(x'_{i_{n-1}}), \cdots, f(x'_{i_{n-\Delta+1}})$.

Remarks: We know from Section III that $f(x)$ is the closest point of $\mathbb{Z}^{n+1}$ to x. The procedure described here finds the closest point of A_n because it makes the smallest changes to the norm of $f(x')$ needed to make $\Sigma f(x'_i)$ vanish.

Step 1 projects x onto x', the closest point of the hyperplane $\Sigma x_i = 0$. Since A_n is by definition contained in this hyperplane it may be possible to assume that x already lies there, in which case Step 1 can be omitted.

The only substantial amount of computation needed is for the sort in Step 3, which takes $O(n \log n)$ steps [10], [13]. However Step 3 can be omitted if x is expected to be close to A_n. In this case Δ will be small, and Steps 3 and 4 can be replaced by the following:

Step 3': If $\Delta = 0$, $f(x')$ is the closest point of A_n to x.

If $\Delta > 0$, find the Δ components of x', say $x'_{i_0}, \cdots, x'_{i_{\Delta-1}}$, for which $\delta(x'_i)$ is as small (i.e., as close to $-\frac{1}{2}$) as possible. The closest point of A_n is obtained by subtracting one from the components $f(x'_{i_0}), \cdots, f(x'_{i_{\Delta-1}})$ of $f(x)$.

If $\Delta < 0$, find the $|\Delta|$ components of x', say $x'_{i_n}, x'_{i_{n-1}}, \cdots, x'_{i_{n-\Delta+1}}$, for which $\delta(x'_i)$ is as large (i.e., as close to $\frac{1}{2}$) as possible. The closest point of A_n is obtained by adding 1 to the components of $f(x'_{i_n}), \cdots, f(x'_{i_{n-\Delta+1}})$ of $f(x)$.

In any case $|\Delta|$ cannot exceed $n/2$. However if Δ is expected to be large the first version of the algorithm is preferable.

The closest point of A_n^* can be found by Algorithm 3, using the fact that A_n^* is the union of $n+1$ cosets of A_n (see (2)).

For example the hexagonal lattice A_2 is shown in Fig. 1, together with ordinary two-dimensional coordinates (u_1, u_2) for the points. The three-dimensional coordinates (x_0, x_1, x_2) with $x_0 + x_1 + x_2 = 0$ that we have used are obtained by multiplying (u_1, u_2) on the right by the matrix

$$M = \begin{pmatrix} 1 & 0 & -1 \\ \frac{1}{\sqrt{3}} & \frac{-2}{\sqrt{3}} & \frac{1}{\sqrt{3}} \end{pmatrix}.$$

Conversely the u-coordinates may be obtained from the x-coordinates by

$$(u_1, u_2) = (x_0, x_1, x_2) \cdot \tfrac{1}{2} M^{\text{tr}}.$$

For example the points $(0,0)$, $(1,0)$, $(1/2, \sqrt{3}/2)$, $(-1/2, \sqrt{3}\,2)$ have x-coordinates $(0,0,0)$, $(1,0,-1)$, $(1,-1,0)$, $(0,-1,1)$, respectively. To find the closest point of A_n to the point P with coordinates

$$(u_1, u_2) = (0.4, -0.4)$$

we first find the x-coordinates of P, which are

$$x = (x_0, x_1, x_2) = (0.169, 0.462, -0.631).$$

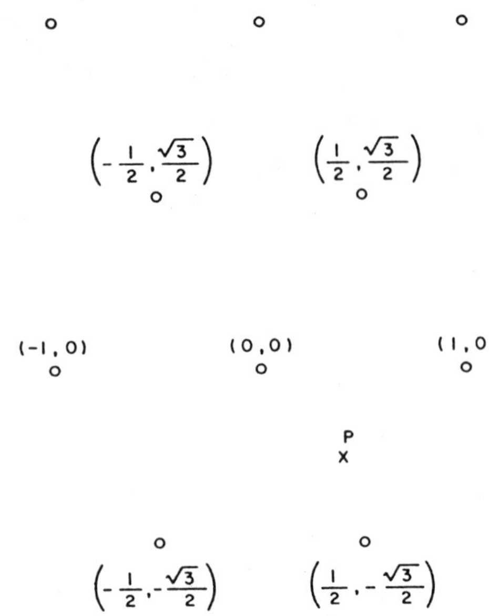

Fig. 1. The hexagonal lattice A_2.

Step 1 of the algorithm can be omitted, since $x_0 + x_1 + x_2 = 0$ holds automatically. Step 2 produces

$$f(x) = (0, 0, -1),$$

with deficiency $\Delta = -1$. At Step 3 we find

$$\delta(x_0) = 0.169 < \delta(x_2) = 0.369 < \delta(x_1) = 0.462.$$

At Step 4 we add 1 to $f(x_1)$, obtaining

$$(0, 1, -1)$$

which is the closest point of A_2. The u-coordinates for this point are

$$(0, 1, -1) \cdot \frac{1}{2} M^{\text{tr}} = \left(\frac{1}{2}, -\frac{\sqrt{3}}{2} \right)$$

(see Fig. 1).

Since $A_3 \cong D_3$, Algorithm 2 is preferable to Algorithm 5 for finding the closest point of the face-centered cubic lattice. Finally E_7 and E_7^* can be handled via the algorithm for A_7, using (7), (8), and Algorithm 3.

Acknowledgment

During the early stages of this work we had some helpful discussions with Allen Gersho.

References

[1] T. Berger, "Optimum quantizers and permutation codes," *IEEE Trans. Inform. Theory*, vol. IT-18, no. 6, 759–765, Nov. 1972.
[2] N. Bourbaki, *Groupes et Algèbres de Lie, Chapitres 4, 5 et 6*. Paris: Hermann, 1968.
[3] J. A. Bucklew, "Companding and random quantization in several dimensions," *IEEE Trans. Inform. Theory*, vol. IT-27, no. 2, 207–211, March 1981.
[4] J. W. S. Cassels, *An Introduction to the Geometry of Numbers*. New York: Springer-Verlag, 1971.
[5] J. H. Conway and N. J. A. Sloane, "On the enumeration of lattices of determinant one," *J. Number Theory*, to be published.
[6] J. H. Conway and N. J. A. Sloane, "Voronoi regions of lattices, second moments of polytopes, and quantization," *IEEE Trans. Inform. Theory*, this issue on pp. 211–226.

[7] H. S. M. Coxeter, *Regular Polytopes*, 3rd ed. New York: Dover, 1973.

[8] N. C. Gallagher, Jr., and J. A. Bucklew, "Some recent developments in quantization theory," in *Proc. 12th Annu. Sympos. System Theory*, Virginia Beach, VA, May 19–20, 1980.

[9] A. Gersho, "Asymptotically optimal block quantization," *IEEE Trans. Inform. Theory*, vol. IT-25, no. 4, 373–380, July 1979.

[10] D. E. Knuth, *The Art of Computer Programming*, vol. 3, *Sorting and Searching*. Reading, MA: Addison-Wesley, 1973.

[11] J. Leech and N. J. A. Sloane, "Sphere packing and error-correcting codes," *Canad. J. Math.*, 23, 718–745, 1971.

[12] H. V. Niemeier, "Definite quadratische Formen der Dimension 24 und Diskriminante 1," *J. Number Theory*, 5, 142–178, 1973.

[13] E. M. Reingold, J. Nievergelt, and N. Deo, *Combinatorial Algorithms: Theory and Practice*. Englewood Cliffs, NJ: Prentice-Hall, 1977.

[14] K. D. Rines and N. C. Gallagher, Jr., "The design of multidimensional quantizers using prequantization," in *Proc. 18th Allerton Conf. Commun., Control, and Computing*, Univ. of Illinois, Monticello, IL, Oct. 8–10, 1980, pp. 446–453.

[15] N. J. A. Sloane, "Binary codes, lattices and sphere packings," in *Combinatorial Surveys* (Proc. 6th British Combinatorial Conference), P. J. Cameron, Ed. London and New York: Academic, 1977, pp. 117–164.

[16] ——, "Tables of sphere packings and spherical codes," *IEEE Trans. Inform. Theory*, vol. IT-27, no. 3, 327–338, May 1981.

[17] Y. Yamada, S. Tazaki, and R. M. Gray, "Asymptotic performance of block quantizers with difference distortion measures," *IEEE Trans. Inform. Theory*, vol. IT-26, no. 1, 5–14, Jan. 1980.

[18] P. Zador, "Topics in the asymptotic quantization of continuous random variables," *IEEE Trans. Inform. Theory*, this issue on pp. 139–149.

Finite-State Vector Quantization for Waveform Coding

JOHN FOSTER, MEMBER, IEEE, ROBERT M. GRAY, FELLOW, IEEE, AND MARI OSTENDORF DUNHAM

Abstract—**A finite-state vector quantizer is a finite-state machine used for data compression: Each successive source vector is encoded into a codeword using a minimum distortion rule, and into a code book, depending on the encoder state. The current state and the selected codeword then determine the next encoder state. A finite-state vector quantizer is capable of making better use of the memory in a source than is an ordinary memoryless vector quantizer of the same dimension or blocklength. Design techniques are introduced for finite-state vector quantizers that combine ad hoc algorithms with an algorithm for the design of memoryless vector quantizers. Finite-state vector quantizers are designed and simulated for Gauss–Markov sources and sampled speech data, and the resulting performance and storage requirements are compared with ordinary memoryless vector quantization.**

I. Introduction

A FINITE-STATE vector quantizer (FSVQ) is an example of a finite-state machine for data compression. It can be viewed as a finite collection of ordinary vector quantizers (VQ's) or block source codes, where each successive source vector is encoded using a VQ determined by the current encoder state. The current state and the codeword selected then determine the next encoder state. Unlike an ordinary memoryless VQ, an FSVQ can take advantage of the memory or correlation between successive source vectors by choosing the appropriate code or VQ, given the past behavior.

Finite-state machines for vector quantization are not new: ordinary VQ's and sliding-block or sliding-window codes are special cases of finite-state machines. Trellis encoders are also examples of finite-state machines where the encoder is permitted to look ahead of the current source vector. More general finite-state machine models were considered by Ziv [1] and Gaardner and Slepian [2]. Ziv focused on rate-distortion theory for such codes, and Gaarder and Slepian developed several optimality properties and some coding theorems for such codes. Gaarder and Slepian also considered several properties of the example of a sliding-block or sliding-window code for a Gauss–Markov source. Unlike such general finite-state machines, finite-state vector quantizers assume an encoder that uses the minimum distortion rule to select a codeword from a state code book, that is, it operates as a quantizer, but at each time the quantizer may be different. Thus, finite-state vector quantizers are a special case of feedback quantizers, as introduced by Kieffer [3], as a model for predictive quantization schemes.

We introduce a class of finite-state machines for data compression called finite-state vector quantizers or FSVQs. These codes are not as general as the basic models of [1] and [2], but they have an intuitive structure and their similarity to ordinary VQ's motivates a design algorithm that is the principle contribution of this work. In the language of [2], our finite-state machine codes are of the "tracking finite-state system" variety, this is, the new encoder state is a function of the previous encoder state and the selected channel codeword rather than a more general function of the previous state and the current input vector. This permits the decoder to track the encoder state if it knows the initial condition (and the channel is noiseless).

General FSVQ's have two principal differences from ordinary memoryless VQ's. First, the code book may vary with time as a function of the encoder state. This resembles a universal source code or a finite-choice adaptive code, but here no side information is transmitted to specify the code book selected; such information must be inferred from the received sequence of channel symbols and an initial state. Second, the minimum distortion selection rule is not necessarily optimum for a given decoder since a low distortion codeword may lead to a bad state and hence to poor long-term behavior. Thus the encoder mapping is not necessarily optimal for the decoder in any long-run sense. It is, however, a reasonable heuristic. In addition, the problem of selecting a bad state by this encoder can be minimized by good design.

No new Shannon theory is required by FSVQ's. Since FSVQ's contain ordinary block source codes (memoryless VQ's) as a special case, the usual positive source coding theorem relative to a distortion measure applies. Since they are special cases of finite-state machines, the negative coding theorems of [1] and [2] apply. In addition, since asymptotically mean stationary (ams) sources driving finite-state machines yield ams input/output processes

Manuscript received June 27, 1983; revised June 18, 1984. This work was supported in part by a Bell Laboratories Cooperative Research Fellowship, the National Science Foundation, the U.S. Army Research Office, and the Joint Services Electronics Program at Stanford University. The material in this paper was presented in part at the 1982 IEEE International Symposium on Information Theory in Les Arcs, France, June 1982.

J. Foster was with Bell Laboratories, Holmdel, NJ, and the Information Systems Laboratory, Stanford University, Stanford, CA. He is now with the Tuskegee Institute, Tuskegee, AL 36088, USA.

R. M. Gray is with the Information Systems Laboratory, Stanford University, Stanford, CA 94305, USA.

M.O. Dunham is with BBN Laboratories, Inc., Cambridge, MA, 02238, USA.

Reprinted from *IEEE Trans. Inform. Theory*, vol. IT-31, pp. 348–359, May 1985.

[4]–[7], the converse for ams sources [7] applies. Regardless, the optimal performance that is theoretically achievable is given by the Shannon rate-distortion function of the source and distortion measure, given an ams ergodic source and a sufficiently well-behaved distortion measure. In contrast to [1] and [2], the goal of this work is to present the development and simulation of design algorithms for the class of codes considered.

A finite-state vector quantizer is not the same as the finite-state Markov models now popular for use in isolated utterance and continuous speech recognition systems. (See, e.g., [8]–[11]). We pause to discuss these systems since we are aware of some confusion concerning their similarities and differences. In these recognitions systems, a memoryless VQ is first used as a front end (or acoustic processor) to encode speech parameters, e.g., selected spectral coefficients or LPC parameter vectors. An algorithm such as the forward-backward algorithm is then used to fit a "hidden Markov model" of a given structure to the observed VQ outputs, that is, to construct a Markov model for the observed compressed data [8]. The estimation algorithm used to find the parameters of the Markov source that best matches the observed training data is approximately a maximum likelihood algorithm if one assumes the given model. In particular, the estimation algorithm tries to match the observed relative frequencies to those that would be produced by the Markov model. Once the model is found (in the design or training stage), it is then used in a linguistic decoder to perform approximate maximum likelihood recognition on the speech data. In the recognition problem the quantization is in the memoryless VQ; the modeling is inherently a continuous estimation problem in which one tries to find transition probabilities that best explain the VQ output observations. No compression is involved in the hidden Markov modeling; the compression is performed prior to the modeling in the front end.

On the other hand, an FSVQ is a compression device and hence a possible alternative to the VQ as a front end of such a recognition system. No Markovian assumptions are made, nor are transition probabilities estimated; the machine is designed to minimize average distortion between its input and output. Waveforms are matched in the sense of minimizing distortion, not in the sense of matching relative frequencies to assumed model distributions. In short, FSVQ codewords are designed given a finite-state quantizer structure, whereas the hidden Markov model structure is designed given a sequence of VQ codewords. It should also be observed that the output of the finite-state machine is not a Markov source in general (unless the process being compressed is memoryless).

There is a tempting parallel between modeling a source and building a good FSVQ for the source in the sense that both the model outputs and the possible FSVQ outputs should "look like" the typical source outputs. However, this view has not yet yielded any design methods for FSVQ, probably because the notions of "look like" and the fundamental goals are quite different. In addition, the hidden Markov techniques have so far concentrated on coding LPC or spectral parameter vectors; our goal is to code waveforms. (FSVQ for LPC parameter vectors is developed in [12].)

In the next section we define and discuss tracking finite-state source codes and FSVQ's. Then two different structures are considered for FSVQ, the difference being in how the reproduction codewords relate to the states of the decoder. The decoder finite-state machine can either be a Moore machine [13] with its outputs, the reproduction codewords, associated with the states (or nodes), or it can be a Mealy machine [13] with its outputs associated with the transitions (or branches). We shall refer to these two structures as labeled-state and labeled-transition FSVQ's, respectively. The two structures are equivalent in the sense that Mealy and Moore finite-state machines are equivalent [13]. That is, given an FSVQ of one form, one can find an FSVQ of the other form that is equivalent in the sense that a given initial state and input sequence will yield the same output sequences, except possibly for the initial state output. Thus the two forms of FSVQ are capable of the same rate-distortion performance. The complexity and storage requirements of the two forms may be quite different in different applications, however, and one form or the other may be preferable in certain cases. In addition, iterative design algorithms such as those developed here may produce quite different final codes even if run on equivalent initial codes.

In the subsequent section we introduce several design algorithms for FSVQ's. These algorithms combine a basic VQ design algorithm (see, e.g., [14]) with various *ad hoc* algorithms for selecting the next-state functions or state transition functions. The basic approach was motivated by the preliminary tests of Rebolledo [15], who observed that in voice coding (LPC) VQ systems, each codeword was virtually always followed by one of a small subset of codewords, that is, all codewords were almost certainly followed by a small subcode. Thus, a lower rate should suffice for the same performance if successors to a codeword are restricted to lie in a small subcode. The problem, of course, is that in the rare instances where the source changes in an unusual way, the input vector may not be well reproduced by the available subcode. In this case there may be no transition to an appropriate state and the coder can become derailed. A useful design technique must produce codes that can find their way back to good reproductions if so derailed.

The next section is devoted to simulation studies for two different sources: a Gauss–Markov source and a sampled speech source. The first is useful because it is a popular test source for data compression systems and since its rate-distortion performance bounds are known. The second is of more practical interest because of the increasing importance of digital voice systems. The FSVQ's are compared to ordinary VQ's on the basis of performance as measured by both signal-to-(quantization)-noise ratio (SNR) and complexity. Further comparisons with other feedback quantizers, such as Cuperman and Gersho's vector predictive quantizer [16], [17], may be found in the general survey paper [18]. The closing section provides some conclusions and some suggestions for future research.

II. Tracking Finite-State Source Codes

Define a *state space* $\boldsymbol{S}$ as a finite collection of symbols $\boldsymbol{S} = \{\sigma_i;\ i = 0, 1, 2, \cdots, K-1\}$ called states. Fix a dimension k, called the code block length, and let $\boldsymbol{R}^k$ denote k-dimensional Euclidean space. Each input source vector $\boldsymbol{x} \in \boldsymbol{R}^k$ is to be encoded into a channel symbol for communication or storage. For convenience we denote the alphabet of channel symbols as $\boldsymbol{N} = \{0, 1, 2, \cdots, N-1\}$. The rate of the coding is $\log N$ bits per input vector or $k^{-1}\log N$ bits per source symbol (or sample), where all logarithms are to base 2. Often it is convenient to consider only integer rates in bits per vector and hence to consider $\boldsymbol{N} = \{0, 1\}^R$, where R is the rate in bits per vector. In this case the channel symbols are simply binary vectors.

A *finite-state encoder* is a mapping α: $\boldsymbol{R}^k \times \boldsymbol{S} \to \boldsymbol{N}$, that is, given an input source vector $\boldsymbol{x} \in \boldsymbol{R}^k$ and a state $s \in \boldsymbol{S}$, then the vector $\boldsymbol{x}$ is encoded into the channel symbol or channel codeword $\alpha(\boldsymbol{x}, s) \in \boldsymbol{N}$. The encoder works in conjunction with a *next-state function*, a mapping f: $\boldsymbol{N} \times \boldsymbol{S} \to \boldsymbol{S}$. If the encoder has a current state $s \in \boldsymbol{S}$, views an input vector $\boldsymbol{x} \in \boldsymbol{R}^k$, and produces a codeword $u = \alpha(\boldsymbol{x}, s)$, then the next encoder state is $f(u, s)$. The encoder and next-state function together describe a finite-state machine, which we will call the encoding machine.

A *finite-state decoder* is a mapping β: $\boldsymbol{N} \times \boldsymbol{S} \to \boldsymbol{R}^k$, that is, given a channel symbol $u \in \boldsymbol{N}$ and a state s, the channel symbol u is decoded into a reproduction vector

$$\hat{\boldsymbol{x}} = \beta(u, s) = \beta(\alpha(\boldsymbol{x}, s), s).$$

The decoder and next-state function together form a finite-state machine that we will refer to as the decoding machine.

One constraint on the finite-state machines considered here is already implicit: we assume that the encoder and decoder share the same state space. By restricting the next-state mapping to depend only on the current state and the encoder output rather than on the current state and the source vector, we are assured that the decoder will be able to track the state sequence given a common initial state, and hence we have a tracking finite-state system in the terminology of Gaarder and Slepian [2]. We shall refer to such codes as tracking finite-state source codes or, simply, tracking finite-state codes.

Associated with each state $s \in \boldsymbol{S}$ is a *state code book* $C_s = \{\beta(u, s),\ u \in \boldsymbol{N}\}$ of possible reproduction vectors obtainable by an encoder and decoder in that state. The collection of state code books is simply the decoder outputs indexed by the pairs of states and channel symbols. We define the *super code book* $\boldsymbol{C} = \cup_{s \in \boldsymbol{S}} C_s$ as the collection of all of the reproduction codewords in all of the state code books. The maximum size of the super code book is $K2^R$. It need not actually be this large, however, as the state code books may overlap, that is, contain common reproduction words.

A finite-state code is used to encode a sequence of vectors or a random process $\{\boldsymbol{X}_n;\ n = 0, 1, 2, \cdots\}$ as follows. Given an initial state $S_0 = \sigma_0 \in \boldsymbol{S}$, the channel symbol sequence $\{U_n;\ n = 0, 1, 2, \cdots\}$, the state sequence $\{S_n;\ n = 0, 1, 2, \cdots\}$, and the reproduction sequence $\{\hat{\boldsymbol{X}}_n;\ n = 0, 1, 2, \cdots\}$ are defined recursively for $n = 0, 1, \cdots$ as

$$U_n = \alpha(\boldsymbol{X}_n, S_n) \qquad S_{n+1} = f(U_n, S_n) \qquad \hat{\boldsymbol{X}}_n = \beta(U_n, S_n).$$

III. Finite-State Vector Quantizers

What makes a quantizer a special case of a finite-state machine is the use of a minimum distortion encoding rule. Hence we next introduce a distortion measure. Given a source vector $\boldsymbol{x} = (x_1, x_2, \cdots, x_k)$ and a reproduction vector $\hat{\boldsymbol{x}} = (\hat{x}_1, \hat{x}_2, \cdots, \hat{x}_k)$, the distortion $d(\boldsymbol{x}, \hat{\boldsymbol{x}})$ between $\boldsymbol{x}$ and $\hat{\boldsymbol{x}}$ is defined as the mean-squared error

$$d(\boldsymbol{x}, \hat{\boldsymbol{x}}) = \|\boldsymbol{x} - \hat{\boldsymbol{x}}\|^2 = \sum_{i=1}^{k} (x_i - \hat{x}_i)^2,$$

where $\|\cdot\|$ denotes the Euclidean norm. As with ordinary VQ's, other distortion measures can be considered, but we shall see that we do need more than the principal requirement of ordinary VQ, the existence of a generalized centroid with respect to the distortion measure.

Given the above terminology, a *finite-state vector quantizer* (FSVQ) is a tracking finite-state code described as above by an initial state, say $\sigma_0 \in \boldsymbol{S}$; a decoder mapping β; a next state mapping f; and a minimum distortion encoder mapping defined in terms of β by

$$\alpha(\boldsymbol{x}, s) = \min_u{}^{-1}\, d(\boldsymbol{x}, \beta(u, s)), \tag{3.1}$$

where the inverse minimum notation means that $\alpha(\boldsymbol{x}, s)$ is the index u for which the reproduction codeword $\beta(u, s)$ yields the minimum possible distortion over all possible reproduction codewords in the state code book C_s. Here and throughout we assume an arbitrary tie-breaking rule. Thus

$$d(\boldsymbol{x}, \beta(\alpha(\boldsymbol{x}, s), s)) = \min_{\hat{\boldsymbol{x}} \in C_s} d(\boldsymbol{x}, \hat{\boldsymbol{x}}).$$

As do Gaarder and Slepian [2], we shall measure the performance of an FSVQ on a given source by the limit

$$\Delta = \lim_{n \to \infty} n^{-1} \sum_{i=0}^{n-1} Ed(\boldsymbol{X}_i, \hat{\boldsymbol{X}}_i)$$

if the limit exists, where E denotes expectation. This limit will exist, for example, if the joint source-reproduction process is ams [4] and the sequence $\{d(\boldsymbol{X}_i, \hat{\boldsymbol{X}}_i);\ i = 0, 1, 2, \cdots\}$ is uniformly integrable. If the input process is ams, then the joint input/output process using a finite-state machine code is also ams [5], [6]. If the sample average energy of the source also converges, say to $\overline{E}$, then the SNR is defined as $\overline{E}/\Delta$. The ergodic theorem for ams sources [4] implies that the sample average distortion and energy will converge with probability one. If the input/output pair process of vectors is also ergodic, it will converge to the above asymptotic expectations, that is,

$$\lim_{n \to \infty} n^{-1} \sum_{i=0}^{n-1} d(X_i, \hat{X}_i) = \Delta$$

$$\lim_{n \to \infty} n^{-1} \sum_{i=0}^{n-1} \|X_i\|^2 = \bar{E}; \qquad \text{almost everywhere.}$$

Even if the process is not ergodic, however, the ergodic decomposition theorem for ams sources [7] implies that the sample averages will converge with probability one to their expectation with respect to an ergodic component of the stationary mean of the ams process. The point is simply that if the source is ams, then the joint input/output process of an FSVQ driven by the source will have convergent sample average distortion and energy, and hence the expectations under the true (but possibly unknown) ergodic component can be estimated by long term sample averages.

An FSVQ is optimal if Δ is minimized (or the SNR is maximized) over all decoder mappings and next-state mappings. As in the case of ordinary VQ design, when the underlying source probability is unknown or when it is known but the expectations are computed by Monte Carlo integration, we will attempt to design codes that minimize the long-term sample average distortion for a long training sequence of data. Also as in ordinary VQ design, if the appropriate stationarity assumptions are satisfied, then this long-term sample average should well approximate the expectation, and hence similar performance should be achieved on future data from the same source. As with ordinary VQ design (and as observed by Gaarder and Slepian [2]), there seems little hope of designing globally optimum codes. The best one can hope for, given current techniques, is the design of reasonably good codes, e.g., locally optimum codes. Although we will consider design algorithms similar to those for ordinary VQ design, unlike the VQ case the difficulty of the theory of such finite-state machines has thus far precluded any demonstration that the resulting codes are locally optimal.

As a final observation, if we were to permit the encoder to search several nodes into the future and find the path minimizing the sum of the distortions encountered, then the system would become an ordinary trellis encoding system with a vector alphabet. In other words, an FSVQ can be considered to a be a trellis source encoding system with a vector alphabet, a search depth of one, and a decoder that is a finite-state machine, but not necessarily a sliding-block code structure, as usually assumed. If the encoder is allowed to look ahead and use a longer search depth or delay, then such an encoder would be nearly optimal for the decoder. In an FSVQ, however, only a search of one node ahead is permitted.

IV. Labeled-State and Labeled-Transition FSVQ's

As stated in the introduction, the outputs of the encoding and decoding machines can either be associated with the transitions between states (a Mealy finite-state machine) or with the states themselves (a Moore finite-state machine). While these two structures are mathematically equivalent in the sense that one can always find a machine of one kind with essentially the same output behavior as a machine of the other kind, it is of interest to consider both types of machines since the resulting complexity may differ for a particular application. In addition, running iterative improvement algorithms on equivalent codes need not produce equivalent codes. There are also distinct properties associated with the two structures that may make a difference in some applications. For example, given a labeled-transition FSVQ, one can construct an equivalent labeled-state FSVQ by simply defining a new state space consisting of old state and channel symbol pairs. The new FSVQ possesses many more states and hence has greater transition storage requirements, but it also possesses a certain additional freedom for the designer that may more than compensate for this apparent disadvantage: by changing the next-state function, one can cause a given reproduction codeword to have as successors any N of the super codewords. On the other hand, changing the next-state function of a labeled-transition FSVQ only permits one to choose which state code book, that is, which of a fixed collection of codewords, can succeed the given codeword. To be more precise, in a labeled-transition system there are K^{NK} ways to choose a next-state function since each of the N transitions from a given state may connect to any of the K states. Note that duplicate transitions are permitted and may be useful since they can bear different labels. The corresponding labeled-state system has KN states, but now we do not permit duplicate transitions, as they are a waste. Thus there are $\binom{KN}{N}^{KN}$ possible choices of the next-state function in the equivalent labeled-state code. This freedom can lead to quite different code books when the initial code is altered using an iterative improvement technique.

Intuitively, in both systems the state itself can be considered to correspond to a coarse approximation of the last source vector and hence a form of prediction of the next source vector to be encoded. In the labeled-state case the approximation is simply the last reproduction codeword chosen, the label of the current state or node. In the labeled-transition case, the approximation is given not by a specific codeword but by a specific code book, the state code book consisting of the labels of the branches leaving the node. Thus, the labeled-state code can use the extra states to perhaps produce a better coarse approximation of the source through the state sequence than can a labeled-transition code.

These observations suggest that for a fixed rate, the labeled-state FSVQ may be capable of using its additional storage to provide better performance than can an equivalent labeled-transition FSVQ. In addition, several of the design techniques for next-state functions to be considered are most natural for the labeled-state case. Neither structure is, however, *a priori* superior.

To describe the design algorithms in a common fashion for both kinds of machines, we introduce a common notation. A labeled-state FSVQ has one output label for each state in $\boldsymbol{S}$. A labeled-transition FSVQ has one output label for each transition, and we can consider the transitions to be the members of the Cartesian product space $N \times S$ consisting of channel symbol/next-state pairs. Define the *label index set* $\boldsymbol{I}$ to be the space $\boldsymbol{S}$ for a labeled-state FSVQ and $\boldsymbol{N} \times \boldsymbol{S}$ for a labeled transition FSVQ. Define the *label code book* as the super code book indexed by the label index set, that is $\boldsymbol{C} = \{\boldsymbol{c}(i);\ i \in \boldsymbol{I}\}$.

In both the labeled-state and labeled-transition case, each transition or channel symbol/next-state pair is associated with a reproduction vector in the label code book; define the label index function ϕ: $\boldsymbol{N} \times \boldsymbol{S} \to \boldsymbol{I}$, which gives the index of the label of the super codeword implied by the transition. For a labeled-transition, FSVQ $\phi(u,s) = (u,s)$, that is, the transition index and label index are identical. For a labeled-state FSVQ, $\phi(u,s) = f(u,s)$, and the label index function gives the next state. With this terminology an FSVQ is completely specified by an initial state σ_0, a label code book $\boldsymbol{C} = \{\boldsymbol{c}(i);\ i \in \boldsymbol{I}\}$, a next-state function f, and a label index function ϕ: $\boldsymbol{N} \times \boldsymbol{S} \to \boldsymbol{I}$. The operation of either a labeled-state or labeled-transition FSVQ can then be described as

$$\begin{aligned} S_0 &= \sigma_0, \\ U_n &= \min_{u \in N}{}^{-1}\, d(\boldsymbol{X}_n, \boldsymbol{c}(\phi(u, S_n))) \\ \hat{\boldsymbol{X}}_n &= \boldsymbol{c}(\phi(U_n, S_n)) \\ S_{n+1} &= f(U_n, S_n). \end{aligned} \tag{4.1}$$

V. FSVQ Design Algorithms

In this section we present design algorithms for FSVQ's. The design is divided into three steps and specific algorithms for each step are presented. The basic steps in the design of a dimension k rate R bits per vector (or $r = R/k$ bits per symbol) FSVQ having K states and $N = 2^R$ transitions from each state and hence N reproduction words in each state code book are the following:

1) initial code book design
2) state code books and next-state function design
3) iterative improvement of the label code book.

The first step is to design an ordinary VQ having K codewords. This code book can be viewed as a collection of tentative state labels even if the FSVQ is to have labeled transitions. The code book is simply an initial super code book if the FSVQ has labeled states. If the FSVQ is to have labeled transitions, then this initial code book can be viewed as a set of state labels from which the state code books—the transition labels—will be developed. In both cases we can say that the state represents a coarse approximation to the previous input vector, and hence a form of prediction of the current vector. Next a family of state code books together with a next-state function is designed using the initial code book. Lastly, for the fixed next-state function, an iterative improvement algorithm is run on the label code book to better match it to the connection rule.

This design approach is *ad hoc* and no results are obtained demonstrating even local optimality. It has, however, several useful features. It is simple and reasonably intuitive; it provides the first (to our knowledge) design methodology for general finite-state vector quantizers; and the separation into these steps permits almost direct application of the ordinary VQ design algorithm of [14] to the first and third steps, and a variation of the algorithm together with several heuristics to the second step.

The VQ design algorithm of [14] and principles of optimization theory suggest that better performance could be obtained from a more involved algorithm that followed Step 3 by redoing the next-state function to better match the improved decoder. One could then iterate, optimizing the next-state function for the label code book and vice versa until convergence. Algorithms to accomplish this are reported in [12], but the simpler single pass design algorithm above and its performance are the focus of this work. We now develop algorithms for each design step in detail.

A. Initial Code Book Design

The initial super code book design is accomplished using an ordinary VQ design algorithm. A brief summary of this procedure is presented in the Appendix for reference. Details may be found in [14], [19]–[21].

Having completed the above design, one has a complete code book of reproduction values that can be considered as initial state labels. For later use we denote this final code book, which is the initial code book of the next step as $\boldsymbol{S} = \{\boldsymbol{c}(s);\ s \in \boldsymbol{S}\}$, where we now use the state names as indices to emphasize the fact that these reproduction codewords are associated with the states whether the FSVQ has labeled states or labeled transitions.

B. State Code Books and Next-State Function Design

Three basic techniques are introduced to find state code books and next-state functions for a given initial code book. The first two are appropriate for the design of labeled-state FSVQ's and have the advantages of being intuitive and simple. They focus on choosing a reasonable next-state function for the given initial code and attempt no further reproduction codeword design at this stage. The third technique uses a variation of the VQ design algorithm to design a collection of state code books by conditioning on the minimum distortion state label in the code book of the previous step. Thus, candidate state code books are obtained based on the assumption that the encoder knows the best word in the initial code book. A next-state function is then formed by trying to approximate this omniscient encoder as well as possible by an FSVQ. This is accomplished in different but similar ways for the cases of labeled states and labeled transitions.

The Conditional Histogram Technique: A simple and, historically, the first technique for finding a next-state function for a labeled-state FSVQ is to keep the initial code book and to form a conditional codeword histogram using this super code book on the training sequence; that is, to determine for each state the relative frequencies of all of the successor states. Then for each of the N states σ having the largest relative frequencies of following state s, set $f(u, s) = \sigma$. In other words, constrain the labeled-state FSVQ to have as next states the N most likely next states when the same super code book is used as an ordinary VQ. This approach was motivated by the results of Rebolledo [15], who found that in voice coding VQ systems codewords in the initial code book were almost always followed by one of a very small subset of codewords.

Nearest Neighbor Design: A second and even simpler approach is to assume that the input vectors are highly correlated and hence the available state codewords should well approximate the current reproduction code word. As in the conditional histogram design, the initial code book is kept and we simply determine a next-state function for that code book. In this case, for each state s find the N nearest neighbor state labels in the entire super code book and then assign these values to the next-state function. This method depends only on the distortion between the codewords, while the previous method depends only on the relative frequency of successor codewords. The resulting code can be viewed as a form of multidimensional delta modulation since the allowed next reproductions are always close to the current reproduction. This method is clearly inappropriate if the number of codewords K is much larger than the number of possible transitions N, since it will likely produce codes that are unable to track rapid variations, a condition analogous to slope overload in delta modulators. The method is considered, however, because it is simple and useful for comparison.

Omniscient Design: The third technique is the most complicated one considered but has the potential to make better use of the memory of the resulting FSVQ. The name chosen reflects the approach of initially designing the state code books under the assumption that the encoder and decoder are omniscient in the sense of knowing a sequence of idealized states. The omniscient encoder is then approximated by a more constrained FSVQ that can be tracked by a nonomniscient decoder. If the approximation is reasonably good, then the code books designed assuming the extra knowledge should also be fairly good. Labeled-state and labeled-transition designs are identical except for the final step, but in both cases this step is different in a crucial way that yields different interpretations of the operation of the two codes.

Consider the state s to be associated with the state label $\boldsymbol{c}(s) \in \boldsymbol{S}$. We wish to design, for each such state, a state code book and a next-state selection rule. If the states were not constrained to be selected according to a specific next-state function, but instead could be chosen in a memoryless fashion, then a reasonable heuristic would be to select the state using a minimum distortion rule with the state label VQ $\boldsymbol{S}$. This selection would then partition the training sequence according to the K different possible state labels, thus producing K separate training subsequences with the property that the vectors in a given subsequence all followed vectors in the original sequence that were clustered about the initial reproduction words. These new training sequences could then be used to design K good memoryless VQ's of N words each using the ordinary VQ design technique; this would produce, for each s, a state code book $\boldsymbol{C}_s$ that is good on the average for all of the input vectors in the training sequence that follow vectors that are closer to $\boldsymbol{c}(s)$ than to any other initial state label. This would be a good code if the encoder and decoder were omniscient in the sense of knowing the state label having minimum distortion to the previous input vector. While an encoder could easily find this, the decoder cannot know it without increasing the communication rate. Hence this does not immediately provive the desired FSVQ. To approximate the omniscient code by an FSVQ, two approaches seem reasonable. If the FSVQ has labeled states, then the new super code book is too large, having KN words instead of the allowed K; in addition, we already have a candidate super code book, the initial code book $\boldsymbol{S}$. If instead of sending the best codeword in the omniscient state code book for the given state, the encoder sends the best possible reproduction codeword in the initial code book, that is, the codeword in the initial code book closest to the ideal codeword in the omniscient state code book, then the decoder can track the encoder and should produce an output reproduction close to that of the omniscient decoder. This can be accomplished as follows. For each state s and for each reproduction codeword $\hat{\boldsymbol{c}}(u, s)$, $u \in \boldsymbol{N}$, in the omniscient state code book $\boldsymbol{C}_s$, find the state σ for which $d(\hat{\boldsymbol{c}}(u, s), \boldsymbol{c}(\sigma))$ is minimized. Set $f(u, s) = \sigma$. Doing this for each of the N values of u will provide the required next-state function. Since a possible transition would be wasted if fewer than N connections were made, if the minimum distortion selection yields a redundant connection (one already selected), then the connection yielding the next smallest distortion is selected instead (and so on if necessary). In this way N distinct connections are made. Doing this for all s completes the design of an initial labeled-state FSVQ. Note that the state code books for the omniscient code are discarded; they are simply used to obtain a next-state function for the original code book.

In the case of a labeled-transition FSVQ, the omniscient state code books provide a set of transition labels, but now we need a next-state rule. The transition label just selected by an encoder provides an estimate of the state label of the next state to be entered, but there may be no state with exactly that label. Hence, analogous to the above case, we do the best we can: pick as the next state the one with the label closest to the reproduction label just chosen from the state code book. This implies a next-stage function as follows. For each state s and each $u \in \boldsymbol{N}$, find the state σ for which $d(\hat{\boldsymbol{c}}(u, s), \boldsymbol{c}(\sigma))$ is minimized, then set $f(u, s) = \sigma$. Doing this for each u and for each state s will yield the required next-state function and hence, in combination with the state code books, the design of the labeled-transition FSVQ. Observe that in contrast to the labeled-state design, here the initial code book is discarded and the state code books are retained for the final code.

Both the omniscient techniques assume that the distortion measure is such that one can measure the distortion between reproduction codewords as well as between an input vector and a reproduction codeword. This is clearly the case for symmetric distortion measures such as the squared error distortion considered here. The technique must be modified, however, for nonsymmetric distortion measures such as the Itakura–Saito distortion. (See [12] for details.)

The omniscient design algorithm is summarized in the Appendix.

The final labeled-state FSVQ is described by the super code book or label code book $\boldsymbol{S}$ and the next-state function f. The final labeled-transition FSVQ is described by the state code books $\{C_s;\ s \in \boldsymbol{S}\}$ and the next-state function f. Alternatively, the FSVQ of either kind is specified as in (4.1) by label code book $\boldsymbol{C} = \{c(i);\ i \in \boldsymbol{I}\}$, where $\boldsymbol{I}$ is the label index set, a label index function ϕ: $\boldsymbol{N} \times \boldsymbol{S} \to \boldsymbol{I}$ giving the index of the label produced by each transition or channel symbol-state pair, and the next-state function f: $\boldsymbol{N} \times \boldsymbol{S} \to \boldsymbol{S}$. This completes the design of the initial FSVQ.

C. Iterative Improvement of State Code Books

In the final stage of FSVQ design, the initial FSVQ produced by the previous step is "fine tuned" to the next-state function by replacing state or transition labels by centroids. The algorithm is a variation on the ordinary VQ algorithm with the only difference being that the centroids are conditioned on states in the labeled-state case or channel symbol-state pairs (or transitions) in the labeled-transition case rather than on channel symbols as in the usual case. These two cases are considered together using the label code book and label index function as in (4.1). The algorithm is summarized in the Appendix.

Unlike ordinary VQ and trellis encoder cases, it is possible for a code book to be made worse by the centroid replacement. To guarantee a final code book at least as good as the initial code book, we can halt the iterative process if an increase in distortion occurs. Alternatively, we can allow increases in distortion and continue the process in hopes of improvement in a later iteration. Experimentally, we find that allowing increases in distortion almost always results in improved code books. Therefore, we halt the improvement algorithm only when the relative decrease in distortion is less than 0.005.

We conclude this section with a note on the choice of the initial state. In most of the experiments presented, the initial state is determined by a full search encoding. This technique adds (negligbly) to the rate but is used to avoid initial derailing effects in the test sequence encoding. It is likely that the appropriate fixed state, such as a silence frame, will provide equivalent performance to full search encoding in some applications. However, this may not be true in general and we leave the issue to specific applications.

VI. Simulations

Codes were designed and tested for two sources, a Gauss–Markov source and a sampled speech sequence. The first experiments focused on a comparison of the next-state function design techniques and the labeled-state versus labeled-transition structures. We also compared the performance of these codes with the rate-distortion function. The remaining experiments focused on comparison of FSVQ to VQ for various vector lengths.

A. Gauss–Markov Sources

A Gauss–Markov source or first order Gauss autoregressive source $\{X_n\}$ is defined by the difference equation

$$X_{n+1} = aX_n + W_n,$$

where $\{W_n\}$ is a zero mean unit variance independent, identically distributed Gaussian time series. This source is of interest since it is a common model of real data, since it is a common guinea pig for comparing data compression systems, and since its optimal performance—the rate-distortion function—can be expressed parametrically and either evaluated, approximated, or bounded. (See, e.g., [22]–[24] and [19]). We here consider only the case $a = 0.9$ of a highly correlated source, the correlation coefficient being typical of speech and scanned image data.

To begin with, we present experimental results for the Gauss–Markov source based on a training sequence of 128,000 samples and a separate test sequence of the same length. The first experiment considered the various next-state function design methods. The conditional histogram, nearest-neighbor, and omniscient designs for labeled-state FSVQ's were all made for a dimension $k = 4$ with $K = 128$ states. FSVQ's with rates of $R = 0, 1, \cdots, 4$ bits per vector or $R/k = 0, 1/4, \cdots, 1$ bits per sample (bps) were designed using each method. Labeled-transition FSVQ's of the same rates, dimension, and super code-book size were designed using the omniscient next-state function design algorithm. Thus the labeled-state FSVQ in each case had the same number of states as would the labeled-state FSVQ equivalent to the labeled-transition FSVQ. The resulting average distortions for encoding the design and test sequences are depicted in Table I, together with the Shannon lower bound to the rate-distortion function (which actually yields the rate-distortion function for the $R = 1$ bit per vector case).

TABLE I
Comparison of Transition Picking Techniques for Gauss–Markov Source, Vector Length 4 and 128 Codewords[a]

N	Distortion (dB) in Train					
	VQ	NN	CH	OLS	OLT	$D(r)$
2	-3.59	-3.59	-2.73	-4.30	-5.06	8.7
4	-6.51	-6.60	-6.61	-6.81	-7.89	-10.2
8	-8.24	-8.32	-9.09	-8.67	-10.11	11.8
16	-10.15	-9.88	-10.90	-10.24	-11.34	-13.2

N	Distortion (dB) in Test				
	VQ	NN	CH	OLS	OLT
2	-3.55	-3.55	-2.68	-4.25	4.99
4	-6.46	-6.34	-6.47	-6.72	-7.76
8	-8.15	-8.18	-8.91	-8.61	-9.91
16	-10.07	-9.74	-10.78	-10.14	-11.12

[a] $N = 2^R$ = number of transitions. NN: nearest-neighbor. CH: conditional histogram. OLS: omniscient labeled-state. OLT: omniscient labeled transition. VQ: full search VQ with N codewords. $D(r)$: rate-distortion function ($r = R/k = \log_2 N/4$).

In these experiments, the labeled-transition FSVQ performs consistently better than any of the labeled-state FSVQ's. This is probably because the labeled-transition FSVQ's had fewer empty states than the labeled-state FSVQ's. For the labeled-state case alone, no one algorithm was consistently superior to the others. In other experiments [25] where an attempt was made to relabel empty states, the omniscient labeled-state FSVQ proved to be superior. Although these results are quite optimistic because the vector lengths and code book sizes were large relative to the training sequence length, they demonstrate the potential for better performance than is here reported. For comparison, Arnstein's optimal one-bps predictive quantizer [22] and an ordinary VQ of dimension 4 and 1 bps achieve an SNR of about 10 dB for this source. The dimension-4 labeled-transition FSVQ achieves this at 3/4 bps, and with *ad hoc* empty state relabeling algorithms such as the techniques described in [25], we expect that an FSVQ would achieve this performance at lower rates.

The next experiment compared the SNR obtainable using labeled-transition FSVQ's and ordinary VQ's for a fixed rate and varying blocklength of from one to five samples. The number of states in each case was fixed at eight so as not to exhaust the training sequence. Greater improvement in the FSVQ results could certainly be obtained by increasing the number of states. The results are depicted in Table II.

TABLE II
DISTORTION OF VQ AND 8 STATE LABELED-TRANSITION FSVQ FOR GAUSS–MARKOV SOURCE[a]

k	Distortion (dB)			
	Training Sequence		Test Sequence	
	VQ	FSVQ	VQ	FSVQ
1	-4.42	-9.21	-4.40	-9.14
2	-7.90	-11.04	-7.86	-10.90
3	-9.24	-11.22	-9.17	-11.08
4	-10.15	-11.34	-10.07	-11.12
5	-10.56	-11.61	-10.47	-11.26

[a]Vector length = k, rate = 1 bps. $D(r) = 13.2$.

At dimension 1, a rate one-bps FSVQ outperforms a scalar quantizer by almost 5 dB and comes close to the performance of Arnstein's predictive quantizer [22]. This is interesting since, like an FSVQ, a predictive quantizer is a sequential machine and uses a minimum distortion rule; but unlike an FSVQ it is not a finite-state machine. In fact, a finite-state vector quantizer can be viewed as a form of a predictive vector quantizer allowed to have only a finite number of states. Here we expect that 16 states would suffice to match the performance of the infinite-state machine.

The gap between FSVQ and VQ rapidly closes to about 1 dB and then remains at that level. It is interesting, however, that the memoryless VQ requires a dimension of 3 before it can match a dimension-1 FSVQ of the same rate. At the highest dimension of 5, the eight-state FSVQ yielded performance approximately 2 dB below that of the rate-distortion bound, 11.3 dB in comparison with 13.2 dB. For reference, the best rate one trellis encoder for this source with constraint length 5 and 32 codewords yields a performance of 11.43 dB [24, 26].

Memory requirements of the VQ, labeled-transition FSVQ and labeled-state FSVQ are determined by the vector dimension k, the number of bytes needed to represent an element of the vector b, the number of transitions N ($N = 2^k$ for a rate of 1 bps), and the number of states K for the labeled-transition FSVQ or KN for the equivalent labeled-state FSVQ. Assuming that b_t bytes are used to represent a transition, the memory requirements are respectively kNb, $kbKN + KNb_t$, and $kbKN + KN^2b_t$. Memory requirements for VQ, eight-state labeled-transition FSVQ and the equivalent labeled-state FSVQ are depicted in Table III. In this table, we use 1 byte for representing a transition when $K < 2^8$ and 2 bytes otherwise. We also use 32 bits to represent an element of a vector, noting that it is possible to represent the elements more efficiently. Although the labeled-state structure requires significantly more memory than the labeled-transition structure, especially for large numbers of states, both structures have memory requirements well within current hardware technology.

TABLE III
STORAGE REQUIREMENTS FOR VQ (kNb), LT-FSVQ ($kKNb + KN$), LS-FSVQ ($kKNb + KN^2$)[a]

k	Storage (bytes)		
	VQ	LT-FSVQ	LS-FSVQ
1	8	80	96
2	32	288	384
3	96	832	1280
4	256	2176	4096
5	640	5376	13312

[a]$N = 2^k$ (rate 1 bps), $K = 8$, $b = 4$, and 1 byte is used to represent each transition.

Next, we briefly mention the results of earlier FSVQ experiments [25], using Gauss–Markov training sequence of length 60,000 samples. In these experiments, the state size is increased until further improvement is negligible. In addition, an *ad hoc* algorithm for relabeling empty states was used to improve quantizer performance. This involved relabeling any empty states with the labels of popular states and repeating the transition design procedure. Table IV shows the results of a comparison between VQ and

TABLE IV
DISTORTION OF VQ AND LABELED-STATE FSVQ FOR GAUSS–MARKOV SOURCE[a]

k	Distortion (dB)		K	Memory (bytes)	
	VQ	FSVQ		VQ	FSVQ
1	-4.3	-10.0	64	8	384
2	-7.9	-10.8	256	32	3k
3	-9.2	-11.4	512	96	14k
4	-10.1	-12.1	512	256	24k
5	-10.5	-12.8	512	640	42k
6	-11.3	-13.1	1024	1536	152k
7	-11.6	-13.6	1024	3584	284k

[a]Vector length = k, rate = 1 bps. Corresponding storage requirements given K states, 4 bytes/scalar, 1 byte for transition storage when $K \leq 256$ and otherwise 2 bytes to represent each transition. $D(r) = 13.2$. 1 kbyte-1024 bytes.

labeled-state FSVQ at a rate of 1 bps with the corresponding storage requirements. Here the performance of FSVQ is as much as 5.7 dB better than VQ for dimension 1 and levels off to 2 dB at higher dimensions. The results are rather optimistic due to the shortness of the training sequence and the large number of states used, as evidenced by the fact that they exceed the Shannon bound to the rate-distortion function.

Table V depicts more realistic labeled-state FSVQ results, using a training sequence of 120,000 samples and a separate test sequence of 60,000 samples. For dimension 6, 64-state FSVQ's were designed for rates of 1–6 bits per block. An arbitrary initial state was fixed and hence the design algorithm was not permitted an initial full search. In addition, the algorithm for replacing isolated states was not used. These results seem to indicate that full search, initial state determination may not be necessary for a FSVQ for a Gauss–Markov source, as the test results are quite close to the training sequence results.

TABLE V
DISTORTION OF LABELED-STATE FSVQ FOR GAUSS–MARKOV SOURCE[a]

R	r	$D(r)$	Distortion (dB) Training Sequence	Distortion (dB) Test Sequence
1	1/6	-8.2	-4.3	-4.2
2	1/3	-9.2	-6.6	-6.5
3	1/2	-10.2	-8.0	-7.9
4	2/3	-11.2	-9.2	-9.1
5	5/6	-12.2	-10.4	-10.3
6	1	-13.2	-11.0	-10.8

[a]Vector length = 6,64 states, $r = R/k$.

B. Sampled Speech

The experiments were repeated for a training sequence of sampled speech waveforms composed of 640,000 samples from five male speakers sampled at 6.5 kHz (about 20 seconds of speech for each speaker). The resulting codes were then tested on a test sequence of 76,800 samples from a male speaker not in the training sequence. The test and training sequences were the same as those used in [21] and [24].

The comparison of next-state function designs assumed, as before, a dimension of $k = 4$ and a label bode book size of 128 with rates of $R = 1$–4 bits per vector. The results for the training sequence are shown in Table VI. As with the Gaussian source, the labeled-transition FSVQ consistently yields better codes, and no one of the three labeled-state design algorithm is clearly superior.

Next labeled-transition FSVQ's were designed for rate 1 bps and dimensions 1–5. The design technique and storage requirements were as in the Gaussian case. The average distortion is depicted in Table VII for the resulting codes used both inside and out of the training sequence. The gain over ordinary VQ is more marked in this case than in the Gauss–Markov results, approximately 1.5 dB. As in the Gauss–Markov case, more positive results can be achieved by increasing the number of states further and relabeling empty states. Here the gains were as much as 6 dB. The results and storage requirements for these labeled-state FSVQs are given in Table VIII.

TABLE VI
COMPARISON OF TRANSITION PICKING TECHNIQUES FOR SPEECH, VECTOR LENGTH 4 AND 128 CODEWORDS[a]

N	Distortion (dB) in Train VQ	NN	CH	OLS	OLT
2	-1.31	0.00	-1.17	-1.06	-2.48
4	-3.33	-3.44	-3.19	-3.75	-4.02
8	-5.10	-4.88	-5.62	-5.57	-6.81
16	-7.08	-6.65	-8.16	-7.58	-8.84

N	Distortion (dB) in Test VQ	NN	CH	OLS	OLT
2	-1.41	0.00	-1.09	-0.92	-2.50
4	-3.57	-3.58	-2.44	-3.84	-3.96
8	-5.11	-4.83	-5.47	-5.52	-6.33
16	-6.94	-6.55	-7.69	-7.37	-8.17

[a]$N = 2^R$ = number of transitions. NN: nearest-neighbor. CH: conditional histogram. OLS: omniscient labeled-state. OLT: omniscient labeled transition. VQ: full search VQ with N codewords.

TABLE VII
DISTORTION OF VQ AND 8 STATE LABELED-TRANSITION FSVQ FOR SPEECH[a]

k	Distortion (dB) Training Sequence		Test Sequence	
	VQ	FSVQ	VQ	FSVQ
1	-2.01	-2.29	-2.05	-2.47
2	-5.25	-6.74	-5.27	-6.68
3	-5.76	-7.56	-5.69	-7.01
4	-7.08	-8.84	-6.94	-8.17
5	-7.86	-9.60	-7.49	-8.54

[a]Vector length = k, rate = 1 bps.

TABLE VIII
DISTORTION OF VQ AND LABELED-STATE FSVQ FOR SPEECH[a]

k	Distortion (dB)		K	Memory (bytes)	
	VQ	FSVQ		VQ	FSVQ
1	-2.0	-2.0	2	8	12
2	-5.9	-7.8	32	32	384
3	-6.4	-9.0	64	96	1280
4	-7.3	-10.9	512	256	25k
5	-8.1	-12.2	512	640	43k
6	-8.6	-13.2	1024	1536	156k
7	-9.3	-14.0	2048	3584	582k
8	-9.5	-15.3	2048	8192	1114k

[a]Vector length = k, rate = 1 bps. Corresponding storage requirements given K states, 4 bytes/scalar, 1 byte for transition storage when $K \leq 256$ and 2 bytes otherwise.

Outside the training sequence the eight-state, dimension-5, labeled-transition FSVQ compares favorably (8.54 dB) with the constraint length 5 trellis encoding system of [26] (7.35 dB) at a rate of 1 bps. In addition its performance is close to that of the constraint length 8 trellis encoding system (8.67 dB).

Both the speech and the Gaussian experiments indicate that vector trellis encoders with a short search (one vector) can outperform scalar trellis encoders with long searches. The former have the advantage of having a more general finite-state machine as a decoder rather than the sliding-block code structure or nonlinear filter usually used by a scalar trellis encoder. This provides evidence in confirmation of Stewart's [26] conjecture that more general finite-state machine decoders (or more freedom in next-state function design) should improve trellis encoding performance and that vector alphabets should also yield noticeable improvements.

The coded speech was synthesized, informal listening tests were consistent with the measured distortion, and the FSVQ codes were substantially better than the ordinary VQ codes of the same dimension and rate. The storage requirements were, however, significantly greater.

The differences between the design distortion and the test sequence distortion were reasonably typical of speech waveform coding designs based on the VQ design algorithm, about 1 to 3 dB.

VII. Comments

The design techniques presented have been shown to yield FSVQ's that provide good performance on both Gauss–Markov and sampled speech sequences. In addition, the use of *ad hoc* algorithms to deal with the problem of empty states should result in even better performance than reported here. The good performance of FSVQ's indicates that a finite-state machine operating on vectors is able to take advantage of the memory of a process, even with the relatively small dimensions considered. Clearly, FSVQ's should outperform ordinary VQ's of the same dimension, but it is perhaps surprising that they also outperform scalar trellis encoding systems, since trellis encoders use a finite-state machine decoder and are permitted to search into the future before deciding. We conjecture that the improvement is due to the fact that FSVQ's operate on vectors and not scalars, that more general decoders are permitted, and that the design algorithms produce good codes within the allowed class.

FSVQ's offer an alternative to trellis encoding for achieving performance near the rate-distortion bound, and they do not require the delays of trellis encoders, since they have a search depth of only a single vector. The same FSVQ design techniques could, however, be used to design vector alphabet trellis encoders with even better performance by simply allowing the encoder in the FSVQ to look ahead before making its decision.

Unlike ordinary VQ's, no optimality or convergence properties have yet been developed for the design algorithms developed here. A theory of design of similar codes for Markov sources has been proposed by Dunham [27], and there appear to be similarities with the approach taken here. His approach assumes knowledge of various conditional distributions, and here such distributions are estimated using the training sequence. Unlike Dunham's approach, our approach leads to specific codes for specific sources. Perhaps a combination of the two will help fill the theoretical gaps in this work.

The FSVQ's designed here are amenable to implementation on a single very large scale integration (VLSI) chip. The development of such a chip, which permits user supplied code books and next-state functions, is described in [25].

A major problem with FSVQs is that of many source codes with feedback—Channel errors can have disastrous effects. The degree of the effect and means to combat it remain a subject for future research. Several options seem possible, however. The code can be periodically updated by a full search VQ, perhaps simultaneously with the transmission synchronization information. The distortion measure could be modified to be the conditional expected distortion over a noisy channel instead of the original distortion. This results in a joint source and channel code [28]. The design techniques would be essentially unchanged save for the computation of the distortion and the centroids. For simple channel models such as a binary symmetric channel, these computations should be reasonable.

Although labeled-state FSVQ's yield slightly inferior results to labeled-transition FSVQs in the examples considered here, it seems likely that each will have its uses.

As a final note, during the revision of this paper, an FSVQ design technique quite similar to the omniscient design approach described here for labeled-transition FSVQ's was independently developed by Haoui and Messerschmidt [29].

Acknowledgment

The authors gratefully acknowledge the help of Mr. Tom Flynn of the Information Systems Laboratory, Stanford University, Stanford, CA, for numerous valuable comments and for his help with the simulations.

Appendix
Design Algorithms

Initial Code Book Design Algorithm

Step 1) Initialization:

Given

training sequence $\{X_i = x_i;\ i = 0, 1, 2, \cdots, L-1\}$,
dimension (blocksize) $= k$,
rate $= R$ bits per vector $= R/k$ bits per sample,
convergence threshold $\epsilon > 0$,
an initial super code book $C_0 = \{c(i);\ i = 0, 1, \cdots, K-1\}$; the initial code may be obtained by the VQ design algorithm and the splitting techniques of [14].

Set $m = 0$ and $d_{-1} = \infty$.

Step 2) Minimum Distortion Encoding:

Set $m = m + 1$.
Encode training sequence using minimum distortion encoder for given code book and save running count centroid for each code-book index: Set $n(u) = 0$; $u = 0, 1, \cdots, K-1$, $\text{cen}(u) = 0$; $u = 0, 1, \cdots, K-1$, $\Delta = 0$.

For $l = 0, 1, \cdots, L-1$

$$U_l = \min_u{}^{-1}\, d(X_l, c(u))$$

$$\Delta = \Delta + d(X_l, c(U_l)) = \Delta + \min_{c \in C} d(X_l, c)$$

$$n(U_l) = n(U_l) + 1$$

$$\text{cen}(U_l) = \text{cen}(U_l) + X_l.$$

Set

$$d_m = \frac{\Delta}{L}.$$

For $u = 0, 1, \cdots, K-1$ such that $n(u) \neq 0$ set

$$\text{cen}(u) = \frac{\text{cen}(u)}{n(u)},$$

the Euclidean centroid of all source vectors mapping into the codeword with index u. If $n(u) = 0$, the centroid can be defined in an arbitrary manner (e.g., split the most popular centroid).

Step 3) Reproduction Codeword Update:

Compute the distortion increment

$$\delta = \frac{d_{m-1} - d_m}{d_m}.$$

If $\delta \leq \epsilon$, then quite with final code book C_m. Otherwise form a new code book $C_{m+1} = \{\text{cen}(u);\ u = 0, 1, \cdots, K-1\}$ and go to Step 2.

Omniscient State Code Book / Next-State Function Design Algorithm

Step 1) Initialization:

Given

an initial state label code book $S = \{c(s);\ s \in \boldsymbol{S}\}$, $N = 2^R$,
a threshold $\epsilon \geq 0$,
a training sequence $\boldsymbol{L} = \{x_i;\ i = 0, 1, \cdots, L-1\}$.

Step 2) State Code-Book Design:

For each $s \in \boldsymbol{S}$ use the ordinary VQ algorithm to design a code book $C_s = \{C(u, s);\ u \in \boldsymbol{N}\}$ for the training sequence

$$\boldsymbol{L}_s = \left\{x_i \in \boldsymbol{L}:\ d(x_{i-1}, c(s)) = \min_{\sigma \in \boldsymbol{S}} d(x_{i-1}, c(\sigma))\right\}.$$

This can be accomplished either by partitioning the training sequence using the minimum distortion rule and separately running K VQ designs or by doing the designs in parallel; that is, each iteration first classifies an input vector to find the minimum distortion state and then does the VQ algorithm update step for the next input vector and the appropriate state code book. The threshold ϵ is either used on the separate designs or on the overall parallel design and total distortion.

Step 3) Next-State Function Design:

For each $s \in \boldsymbol{S}$ and each $u \in \boldsymbol{N}$, set

$$f(u, s) = \min_{\sigma \in \boldsymbol{S}}{}^{-1}\, d(\hat{c}(u, s), c(\sigma)).$$

Iterative Label Improvement Algorithm

Step 1) Initialization:

Given

training sequence $\{\boldsymbol{X}_i;\ i = 0, 1, 2, \cdots, L-1\}$,
dimension (blockwise) $= k$,
rate $= R$ bits per vector $= R/k$ bits per sample,
a label code book $\boldsymbol{C} = \{c(i);\ i \in \boldsymbol{I}\}$, where $\boldsymbol{I}$ is the label index set $\boldsymbol{I} = \boldsymbol{S}$ for a labeled-state FSVQ and $\boldsymbol{I} = \boldsymbol{N} \times \boldsymbol{S}$ for a labeled-transition FSVQ,
an initial state $\sigma_0 \in \boldsymbol{S}$,
a label index function ϕ: $\boldsymbol{N} \times \boldsymbol{S} \to \boldsymbol{I}$,
a next-state function f: $\boldsymbol{N} \times \boldsymbol{S} \to \boldsymbol{I}$,
a convergence threshold $\epsilon \geq 0$.

Set $m = 0$, $d_{-1} = -\infty$.

Step 2) FSVQ Encoding:

Set $m = m + 1$.
Encode training sequence using FSVQ encoder for given code book and save running count centroid for each codebook index as follows.

Set $n(i) = 0$; $\text{cen}(i) = 0$; $i \in \boldsymbol{I}$, $\Delta = 0$, $S_0 = \sigma_0$.
For $l = 0, 1, \cdots, L-1$

$$U_l = \min_{u \in \boldsymbol{N}}{}^{-1}\, d(x_l, c(\phi(u, S_l)))$$

$$\Delta = \Delta + d(x_l, c(\phi(U_l, S_l)))$$

$$= \Delta + \min_{c \in C_{S_l}} d(x_l, c)$$

$$n(\phi(U_l, S_l)) = n(\phi(U_l, S_l)) + 1$$

$$\text{cen}(\phi(U_l, S_l)) = \text{cen}(\phi(U_l, S_l)) + X_l$$

$$S_{l+1} = f(U_l, S_l).$$

Set

$$d_m = \frac{\Delta}{L}.$$

For i such that $n(i) \neq 0$ set

$$\text{cen}(i) = \frac{\text{cen}(i)}{n(i)},$$

the Euclidean centroid of all source vectors mapping into the codeword with label index i. If $n(i) = 0$, the centroid can be defined in an arbitrary manner (e.g., split the most popular centroid).

Step 3) Reproduction Codeword Update:

Compute the distortion increment

$$\delta = \frac{d_{m-1} - d_m}{d_m}.$$

If $0 \leq \delta \leq \epsilon$, then quit with final label code book C_m. Otherwise form a new code book

$$C_{m+1} = \{\text{cen}(i);\ i \in \boldsymbol{I}\}$$

and go to Step 2.

References

[1] J. Ziv, "Distortion-rate theory for individual sequences," *IEEE Trans. Inform. Theory*, vol. IT-26, pp. 137–143, Mar. 1980.

[2] N. T. Gaarder and D. Slepian, "On optimal finite-state digital transmission systems," *IEEE Trans. Inform. Theory*, vol. IT-28, pp. 167–186, Mar. 1982.

[3] J. C. Kieffer, "Stochastic stability for feedback quantization schemes," *IEEE Trans. Inform. Theory*, vol. IT-28, pp. 248–254, Mar. 1982.

[4] R. M. Gray and J. C. Kieffer, "Asymptotically mean stationary measures," *Ann. Prob.*, vol. 8, pp. 962–973, Oct. 1980.

[5] J. C. Kieffer and M. Rahe, "Markov channels are asymptotically mean stationary," *SIAM J. Math. Anal.*, vol. 12, pp. 293–305, 1980.

[6] J. C. Kieffer and J. G. Dunham, "On a type of stochastic stability for a class of encoding schemes," *IEEE Trans. Inform. Theory*, vol. IT-29, pp. 793–797, Nov. 1983.

[7] R. M. Gray and F. Saadat, "Block source coding theory for asymptotically mean stationary sources," *IEEE Trans. Inform. Theory*, vol. IT-30, pp. 54–68, Jan. 1984.

[8] F. Jelinek, R. L. Mercer, and L. R. Bahl, "Continuous speech recognition: statistical methods," in *Handbook of Statistics*, vol. 2, P. R. Krishnaiah and L. N. Kanal, Eds. New York: North Holland, 1982, pp. 549–573.

[9] R. Billi, "Vector quantization and Markov source models applied to speech recognition," in *Proc. IEEE Int. Conf. Acoustics, Speech, Signal Processing*, Paris, May 3–5, 1982.

[10] L. Rabiner, S. E. Levinson, and M. M. Sondhi, "On the application of vector quantization and hidden Markov models to speaker-independent isolated word recognition," *Bell Syst. Tech. J.*, vol. 62, pp. 1075–1106, Apr. 1983.

[11] J. E. Shore, D. Burton, and J. Buck, "A generalization of isolated word recognition using vector quantization," in *Proc. IEEE Int. Conf. Acoust., Speech, Signal Proc.*, Boston, MA, Apr. 14–16, 1983.

[12] M. O. Dunham and R. M. Gray, "An algorithm for the design of labeled-transition finite-state vector quantizers," *IEEE Trans. Commun.*, vol. COM-33, Jan. 1985.

[13] J. E. Hopcraft and J. D. Ullman, *Introduction to Automata Theory, Languages, and Computation*. Reading, MA: Addison-Wesley, 1979.

[14] Y. Linde, A. Buzo, and R. M. Gray, "An algorithm for vector quantizer design," *IEEE Trans. Commun.*, vol. COM-28, pp. 84–95, Jan. 1980.

[15] G. Rebolledo, "Speech and waveform coding based on vector quantizations," Ph.D. dissertation, Inform. Syst. Lab., Stanford Univ., Stanford, CA, Aug. 1982.

[16] V. Cuperman and A. Gersho, "Adaptive differential vector coding of speech," in *Proc. IEEE Conf. Global Communications*, Miami, FL, Dec. 1982.

[17] ——, "Vector predictive coding of speech at 16 kb/s," *IEEE Trans. Commun.*, to appear.

[18] R. M. Gray, "Vector quantization," *IEEE ASSP Mag.*, pp. 4–29, Apr. 1984.

[19] R. M. Gray and Y. Linde, "Vector quantizers and predictive quantizers for Gauss–Markov sources," *IEEE Trans. Commun.*, vol. COM-30, pp. 381–389, Feb. 1982.

[20] R. M. Gray and E. Karnin, "Multiple local optima in vector quantizers," *IEEE Trans. Inform. Theory*, vol. IT-28, pp. 256–261, Mar. 1982.

[21] H. Abut, R. M. Gray, and G. Rebolledo, "Vector quantization of speech and speech-like waveforms," *IEEE Trans. Acoust., Speech, Signal Processing*, vol. ASSP-30, pp. 423–435, June 1982.

[22] D. S. Arnstein, "Quantization error in predictive coders," *IEEE Trans. Commun.*, vol. COM-23, pp. 423–429, Apr. 1975.

[23] J. Y. Huang and P. M. Schultheiss, "Block quantization of correlated Gaussian random variables," *IEEE Trans. Commun. Syst.*, vol. CS-11, pp. 289–296, Sept. 1963.

[24] L. C. Stewart, R. M. Gray, and Y. Linde, "The design of trellis waverform coders," *IEEE Trans. Commun.*, vol. COM-30, pp. 702–710, Apr. 1982.

[25] J. Foster, "Finite-state vector quantization for waveform coding," Ph.D. dissertation, Inform. Syst. Lab., Stanford Univ., Stanford, CA, Nov. 1982.

[26] L. C. Stewart, "Trellis data compression," Inform. Syst. Lab, Stanford Univ., Stanford, CA, Tech. rep. L905-1, July 1981.

[27] J. G. Dunham, "An iterative theory of code design," *IEEE Trans. Inform. Theory*, submitted for publication.

[28] J. G. Dunham and R. M. Gray, "Joint source and noisy channel trellis encoding," *IEEE Trans. Inform. Theory*, vol. IT-27, pp. 516–519, July 1981.

[29] A. Haoui and D. G. Messerschmitt, "Predictive vector quantization," in *Proc. IEEE Int. Conf. Acoustics, Speech, Signal Processing*, San Diego, CA, Mar. 19–21, 1984.

Using Simulated Annealing to Design Good Codes

ABBAS A. EL GAMAL, SENIOR MEMBER, IEEE, LANE A. HEMACHANDRA, ITZHAK SHPERLING, AND VICTOR K. WEI, MEMBER, IEEE

Abstract—**Simulated annealing is a computational heuristic for obtaining approximate solutions to combinatorial optimization problems. It is used to construct good source codes, error-correcting codes, and spherical codes. For certain sets of parameters codes that are better than any other known in the literature are found.**

I. Introduction

ANNEAL, as defined in *Webster's New Collegiate Dictionary*, means: 1) to heat (as glass) in order to fix laid-on colors, and 2) to heat and then cool (as steel or glass) usually for softening and making less brittle.

The cooling stage of annealing is usually done slowly to reduce the number of defects in the crystal structure (as steel) and to minimize the potential energy stored in the molecular configuration (as glass). A fundamental question in statistical mechanics concerns whether the system solidifies in the limit of low temperature, and if it does, whether it forms a crystalline solid or a glass. Motivated by a subtlety in numerical integration, the Metropolis algorithm [1] has often been used to simulate numerically the annealing process to gain an understanding of the ground-state configuration.

A typical combinatorial optimization problem seeks the minimum of a given objective (or cost) function of many variables. The variables are subject to intertwining constraints, and they interact with each other in complicated ways not unlike the molecules in a physical system. By appropriately defining an effective temperature for the multivariable system and imitating the physical annealing process, researchers have sought to solve a diverse collection of problems, with varying degrees of success.

Kirkpatrick [2] first investigated the use of simulated annealing in connection with the physical design of computers. He found the technique useful for solving VLSI layout and partitioning problems. More recently, researchers have compared simulated annealing with other heuristic techniques for *NP*-complete problems [5]. At present, simulated annealing provides the best heuristic algorithm for the graph partitioning problem, but extensive fine tuning has failed to make it the method of choice for graph coloring or the traveling salesman problem [3]. Johnson *et al.* [4] have investigated the interplay among local search, neighborhood structure, and combinatorial optimization problems.

The successes of simulated annealing have resulted in a surge of interest in the method. Many evaluations of the technique and its application to diverse areas are in progress [6]–[8]. In this paper we use simulated annealing to construct good source codes, error-correcting codes, and spherical codes. We have rediscovered many optimal or near-optimal source codes. For several parameter sets we have found constant-weight error-correcting codes that are better than any previous such code. We have also obtained spherical codes better than those constructed by an *ad hoc* method analogous to apple peeling. If we conceptualize codewords as repelling molecules, then the analogy between a code with good distance properties and a molecule structure with low potential energy is apparent. This may account for our success here. A more sophisticated attempt to explain the success of simulated annealing in a diverse variety of problems can be found in [2].

The rest of the paper is organized into four sections. In Section II we describe the coding problems we investigate. In Section III we present our simulated annealing algorithm. In Section IV we summarize the results obtained from computer experiments with the algorithm, including a few record-breaking codes. In Section V we conclude the paper by remarking on the advantages and disadvantages of the algorithm.

A detailed explanation of the intuition behind the simulated annealing method can be found in [2] or in many recent papers. We shall not include one here.

II. The Design of Good Codes

In a communication system (Fig. 1) the design of good codes is of fundamental importance. Source information, such as English text and audio and video data, is represented efficiently via a *source code* [10]. American Na-

Manuscript received October 9, 1985; revised January 1, 1986. This work was supported in part by the National Science Foundation under NSF Grant ECS-83-00988. This paper was presented in part at the Simulated Annealing Conference, Yorktown Heights, NY, 1984; at the 22nd Annual Allerton Conference on Communication, Control, and Computing, Monticello, IL, October 3–5, 1984; and at the International Symposium on Information Theory, Brighton, England, June 1985.

A. A. El Gamal and I. Shperling are with the Information Systems Laboratory, Department of Electrical Engineering, Stanford University, Stanford, CA 94305, USA.

L. A. Hemachandra was with Bell Communications Research, Morristown, NJ. He is now with the Department of Computer Science, Cornell University, Ithaca, NY 14853, USA.

V. K. Wei is with Bell Communications Research, 435 South Street, Morristown, NJ 07960, USA.

IEEE Log Number 8610106.

Reprinted from *IEEE Trans. Inform. Theory,* vol. IT-33, pp. 116–123, Jan. 1987.

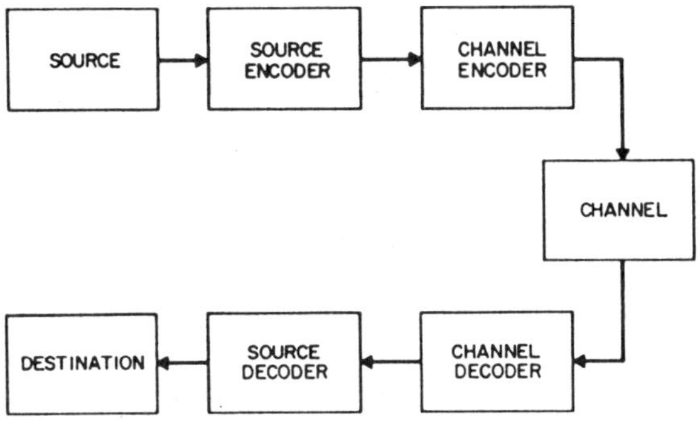

Fig. 1. Communication system.

tional Standard Code for Information Interchange (ASCII) codes, the Huffman code, Morse code, and various voice and picture encoding schemes fall into this category. The source-encoded information is then sent over a channel (transmission media), such as a telephone line, microwave link, or optical fiber, where the signals are often corrupted by noise. To ensure reliable transmission, the data are further encoded via a *channel code* (*error-correcting code*). This enables the detection of errors and their eradication. Prominent channel codes include Hamming codes, Reed–Solomon codes, convolutional codes, and trellis codes [11], [12], [26].

To be precise, the source code we study here is a set of M binary n-tuples

$$S = \{x_1, x_2, \cdots, x_M\} \subset \{0,1\}^n.$$

The rate of the code S is $R = (1/n)\log M$. (All logarithms in this paper are base two.) Let $B^n = \{0,1\}^n$ denote the set of all n-tuples. The *distortion* of a source code is the following quantity:

$$\delta(S) = \frac{1}{2^n} \sum_{y \in B^n} \min_{x \in S} d_H(x, y)$$

where the Hamming distance $d_H(x, y)$ between two binary n-tuples is the number of bit positions where they differ. A binary n-tuple is encoded to its nearest neighbor in S. The distortion is the average number of erroneous bits in encoding a random n-tuple. When S is used in a communication system, it can encode a random binary source with rate R and average distortion δ. A desirable property of a source code is a small distortion.

Given the length n and the size M, the construction of a source code with minimum distortion is a very difficult problem. The Lloyd–Max algorithm [25] can be used, but it often gets stuck in a local optimum and fails to find a minimum-distortion code. An exhaustive search algorithm would have to check all $\binom{2^n}{M}$ possible codes, a practical impossibility even for small values of n and M. For example, if $n = 7$ and $M = 16$, then about $\binom{128}{16} \approx 10^{20}$ codes would need to be checked. However, we shall see in the next section how simulated annealing can be used to find an optimum or near-optimum source code without doing an exhaustive search.

The channel code (error-correcting code) we study here is also a set of binary n-tuples

$$C = \{x_1, x_2, \cdots, x_M\}$$

with rate $R = (1/n)\log M$. Each member $x \in C$ is called a codeword. The minimum distance of C is

$$d_{\min} = \min_{\substack{x, y \in C \\ x \neq y}} d_H(x, y).$$

When the channel code is used in a communication system, up to $d_{\min}/2$ transmission errors can be eradicated. A desirable property of a channel code is a large minimum distance $d_{\min}$. The Hamming weight of a binary n-tuple x is the number of ones in x. In this paper we restrict ourselves to *constant-weight codes* where all codewords have the same Hamming weight. An important number in the theory of error-correcting codes is $A(n, d, w)$—the maximum size of a constant-weight code of length n, minimum distance d or more, and constant codeword weight w.

The constants $A(n, d, w)$ are extremely difficult to determine. Sophisticated mathematical theories, such as the Leech lattice [12], have been used to calculate these constants. Only a handful of values have been exactly determined [12], [13], notably $A(24, 8, 8) = 759$ and $A(24, 8, 12) = 2576$. For other values the gaps between upper and lower bounds are often sizable [12], [13]. One of the smallest undetermined values is $17 \le A(23, 10, 7) \le 23$. In the next section we shall see how simulated annealing is used to raise this lower bound to 18 and to help narrow other gaps.

Another channel code we study is the spherical code. A *spherical code* is a set of real vectors $\Phi = \{x_1, x_2, \cdots, x_m\}$ on the surface of a unit sphere in n-dimensional Euclidean space R^n. The minimum separating angle of a spherical code is

$$\theta_{\min} = \min_{\substack{x, y \in \Phi \\ x \neq y}} \cos^{-1}(x \cdot y)$$

where $x \cdot y$ is the inner (dot) product of x and y. Spherical codes have important applications to transmission over the Gaussian channel [15] and to many other areas [16], [17]. A desirable property of a spherical code is to have a large minimum separating angle. An important number is $M(n, \theta)$—the maximum number of vectors of a spherical code in n dimensions having a minimum separating angle greater than or equal to θ.

The case $\theta = \pi/3$ is particularly interesting. The number $M(n, \pi/3)$, called the *kissing number*, is the maximum number of nonoverlapping unit balls in n-dimensional space that can touch a given unit ball simultaneously. The determination of the kissing numbers has a long and fascinating history [18]–[20], culminating in the resolution of $M(8, \pi/3) = 240$ and $M(24, \pi/3) = 196\,560$. Other kissing numbers beyond three dimensions remain undecided.

Generally, $M(n, \theta)$ is very difficult to determine, save for a few trivial cases. In the next section we shall see how simulated annealing can be used to produce interesting lower bounds for $M(n, \theta)$. These results may not improve upon existing lower bounds for particular values of n and θ. For example, simulated annealing has not yielded any improvement on existing lower bounds on the kissing numbers. However, the technique can be used to generate reasonable lower bounds for general values of n and θ rather quickly.

III. The Simulated Annealing Algorithm

We use the simulated annealing algorithm outlined in Fig. 2 to find good source codes, constant-weight codes, and spherical codes in our experiments. Our algorithm is similar to those simulated annealing algorithms used in other applications, such as those in [2].

```
CHOOSE CODE C, TEMPERATURE T
DO
  {
   DO
     {
     CHOOSE C', A PERTURBATION OF C
     LET ΔE=ENERGY(C')-ENERGY(C)
     IF ΔE<0 THEN C←C'
     ELSE WITH PROBABILITY EXP(-ΔE/T) C←C'
     }
   UNTIL (SEVERAL ENERGY DROPS OR TOO MANY ITERATIONS)
   LOWER TEMPERATURE
  }
UNTIL(STABLE CODE CONFIGURATION)
```

Fig. 2. Simulated annealing algorithm.

Initially, we choose a random code and a sufficiently high temperature. The outer loop decreases the temperatures until a series of stable code configurations has been seen; the inner loop perturbs the code until a prescribed number of energy drops occurs (not necessarily in a row), or too many iterations. After each perturbation, if the energy is decreased, we adopt the new code; if the energy is increased, we adopt the new code with probability $\exp(-\Delta E/T)$.

The temperature T is a control parameter of the algorithm. At high temperatures, $\exp(-\Delta E/T)$ is invariably close to one for positive ΔE. Therefore, we almost always adopt the new code. At low temperatures $\exp(-\Delta E/T)$ is close to zero for large ΔE, so we are not likely to adopt a new code which greatly increases the energy function. The initial temperature is set sufficiently high. For source codes we used $T_{\text{initial}} = 10 \cdot \text{length}$ in our experiments. For constant-weight codes and spherical codes we use $T_{\text{initial}} = 1000$. The energy function E is chosen to steer the code transfiguration in the general direction of reduced objective function. For source codes the energy function E is set to be the distortion $\delta(S)$, the objective function. For constant-weight codes we choose

$$E = \sum_{\substack{x,\, y \in C \\ x \neq y}} [d_H(x, y)]^{-k}$$

where k is a constant. As the simulated annealing process proceeds, the codewords have a tendency to repel each other, thus increasing the minimum distance. We do not choose the minimum distance as the energy function because it depends on only a pair of codewords; changes not involving any minimum-distance codeword pair would not be reflected in the energy function. The choice of an energy function different from the objective function is not an uncommon practice in simulated annealing [2]. We will comment on this later. We often use $k = 2$ in the computation. Other values of k and even other energy functions have been examined experimentally without noticeable improvement.

To perturb a code, we randomly select a codeword and randomly "jiggle" it, i.e., change one or two bits. The initial code has M codewords, each of which is an all-zero n-tuple (a very bad code indeed). We have tried other random initial codes with similar results because the all-zero initial code gets randomized very quickly at high temperatures.

The temperature is lowered in geometric series, i.e., $T \leftarrow \alpha T$. Typically, we choose the constant $\alpha > 0.9$. For source codes the temperature is lowered when the number of acceptances or rejections exceeds code size $\cdot$ (1 + 0.1 $\cdot$ code size/T). The initial temperature is set at around 10 $\cdot$ length. The algorithm terminates when the temperature drops below 0.01 $\cdot$ length, or when five consecutive temperature stages produce no change in the code. For constant-weight codes and spherical codes we decrease the temperature when several (three to five) energy drops have occurred, or several hundred iterations. The initial temperature is chosen to be 1000, and we terminate the algorithm manually when we see a stable code configuration.

If we had chosen the minimum distance as the energy function, perturbations not involving any minimum-distance codeword pair would have resulted in no change in the energy function and hence would be adopted with probability one. This would have resembled a random walk. We feel that distance variations between every codeword pair should be reflected in the energy function. Experimental success seems to vindicate our decision. We have also tried the energy function $E = \sum f[d_H(x, y)]$, where $f(\cdot)$ is a concave function with a single minimum at the desired minimum distance. We attempted this energy function because in the cases where $A(n, d, w)$ is determined, the distance profile of the optimum codes tends to be a convex function with a single maximum at the desired distance. However, this energy function converged more slowly than our original choice in the experiments we have performed.

For the spherical code the simulation algorithm is essentially identical. The energy function is chosen to be

$$E = \sum_{x \neq y \in C} [\cos^{-1}(x \cdot y)]^{-k}.$$

Note that $\cos^{-1}(x \cdot y)$ is the separating angle between the two vectors. For each perturbation we jiggle a random codeword on the spherical surface.

TABLE I
SOURCE CODES VIA SIMULATED ANNEALING

Length n	Rate R	Distortion by Simulated Annealing	Hamming Lower Bound on Distortion	Number of Codes Checked
6	1/2	1/6	1/6	300
7	4/7	1/8	1/8	1782
10	0.3322	0.2371	0.2359	1674
10	0.5	0.16396	0.16250	3000

IV. RESULTS

Preliminary simulations have been performed on source codes, constant-weight codes, and spherical codes. The results are summarized in the following. Due to the nature of the simulated annealing algorithm, the codes in each parameter set are not exhaustively searched. Only a small portion of all possible codes are examined. We have also limited the CPU time spent in a simulation run to a few hours. The full power of the algorithm is not explored. The results obtained here demonstrate the potential power, not the ultimate limit, of the simulated annealing algorithm.

Source Codes

Simulations were performed on source codes with lengths ranging from six to ten and sizes varying from eight to 32. The resulting distortion is equal to or near the Hamming lower bound (Table I). The total number of iterations run ranges from 300 to 3000 for each set of parameters. Thus simulated annealing proves useful in finding good source codes, at least for the ranges of parameters tested.

For critical comparison we ran a greedy version of the algorithm in which we start with a low initial temperature. This modification has the effect of taking only the code transfigurations which reduce distortion. For $n = 7$, $m =$ 16, and about 200 iterations we got stuck with a code having a distortion 10–20 percent above minimum in eight out of ten runs.

Fig. 3 illustrates the history of a single run of our algorithm which successfully finds a source code of length seven, rate 4/7, and distortion 1/8 which meets the Hamming lower bound. (This gives the result indicated in the second row of Table I.) The horizontal axis corresponds to the temperature, and the vertical axis corresponds to the distortion per bit, i.e., $\delta(C)/\text{length}$. The figure should be read from right to left because we proceed from high temperatures to low temperatures. Although we started out with the code consisting of 16 all-zero codewords at temperature $10 \cdot \text{length} = 70$, only the part of the history corresponding to temperatures below ten is shown. We stay at each temperature stage for a total of 10*(1 + 2*code size/T) iterations. The temperature is lowered in a geometric series by a factor of 0.9 at each reduction. The total number of iterations in this run is 3206. The per-bit distortions of the source codes found by the algorithm at each temperature stage is shown in circles. More codes may be found than different values of distortion because different codes may have identical distortion. The complete figure shows that the distortion has a decreasing tendency, resulting in the minimum at low-enough temperatures. Notice that, occasionally, a code with larger distortion than any code found at the previous (higher) temperature may occur. This is the particular characteristic of the simulated annealing algorithm that enables it to avoid the trap of a local minimum in its search for the global minimum.

The histories of other runs of our algorithm which have generated the remaining codes listed in Table I are not included here. The figures for all histories exhibit a decreasing tendency for the distortion as the temperature drops, resulting in minimum or near-minimum distortions.

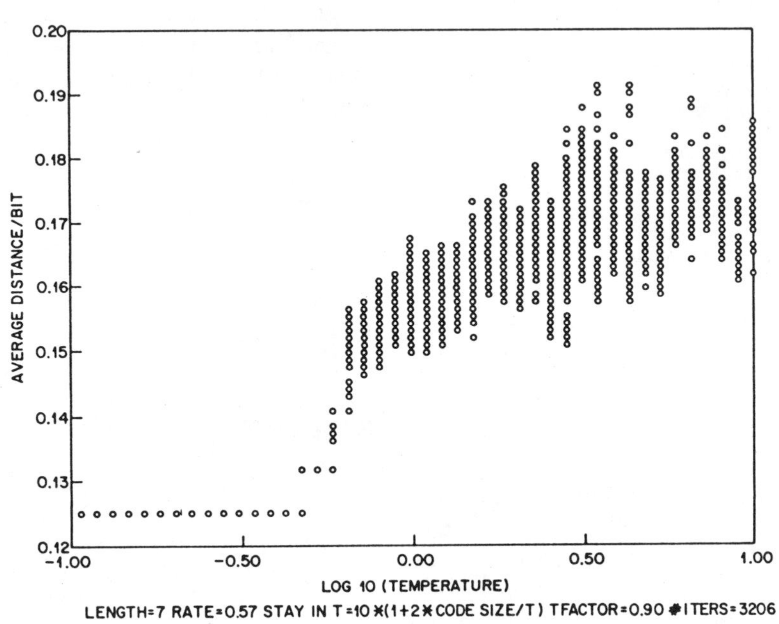

Fig. 3. History of simulated annealing run.

TABLE II
CONSTANT-WEIGHT CODES VIA SIMULATED ANNEALING

Length	Weight	Distance	Code Size	Upper Bound
21	9	10	22	41
22	9	10	23	57
23	7	10	18[a]	23
23	8	10	28[a]	50
23	9	10	24	87
23	10	10	39	117
23	11	10	39	135
24	8	10	33[a]	68
24	9	10	24	119
24	11	10	57	223
24	12	10	60	247

[a] Best-known result in the literature.

TABLE III
SIZE 18 CONSTANT-WEIGHT CODE FOR $A(23, 10, 7)$

1.	00000010010000001111 0001
2.	00000101011010010001000
3.	00001101000000100010101
4.	00010000101000111000100
5.	00010001000101010000011
6.	00011000000011100101000
7.	00101010101001000000001
8.	00110010000010000010110
9.	01000000110000000001111
10.	01001000001110001000010
11.	01010110010100100000000
12.	01100001000100000111000
13.	10000000110111000010000
14.	10000011100000100100010
15.	10001010000100010001100
16.	10111001010000001000000
17.	11000100000001001100100
18.	11100000000010110000001

TABLE IV
SIZE 28 CONSTANT-WEIGHT CODE FOR $A(23, 10, 8)$

1.	00000101000110100001110
2.	00100101100000010100011
3.	10000100110100010011000
4.	10001110000100001100010
5.	00001001101111010000000
6.	01100000101010100010100
7.	11000011000101100010000
8.	01001101010010000110000
9.	11001100100001000000101
10.	11010010100010000001010
11.	10011000000000110010110
12.	00100100011010001001001
13.	00010001011000010101100
14.	10000000001001000111011
15.	00000011100000001011101
16.	00101011010001000001010
17.	00010100100001101101000
18.	00110110000011010010000
19.	10100010001100010000101
20.	01000110010000111000100
21.	10000001111000101000010
22.	01010100011101000000010
23.	00111000110100001000100
24.	00011111001000100000001
25.	01101000000100110101000
26.	10010000010110100100001
27.	01000000000110011010011
28.	10100001000011001100100

TABLE V
SIZE 33 CONSTANT-WEIGHT CODE FOR $A(24, 10, 8)$

1.	000000000101100011101010
2.	000000100100111001010100
3.	000000110110000100101100
4.	000001000011111000001001
5.	000001010100010110000011
6.	000010001110011010100000
7.	000010101000000001100111
8.	000100001010010000011110
9.	000100110001011000100010
10.	000101010001000001010101
11.	001000000011001110000110
12.	001001101100001000001010
13.	001011000010000011011000
14.	001011011000100100000100
15.	001100101000010111000000
16.	001110000100100000010011
17.	010000000010010101110001
18.	010001110010100000010010
19.	010100001100001100000101
20.	010110010100010001001000
21.	010111001001000010000010
22.	011000001001101000110000
23.	100000101111000000010001
24.	100001100000001110110000
25.	100010001001001101001000
26.	100010010001110010010000
27.	100100100000100010001101
28.	100101000110100101000000
29.	101000010100001001100001
30.	110000011010000011000100
31.	110001000101010000100100
32.	111000000000110100001010
33.	111110100010000000100000

Constant Weight Codes

Using the simulated annealing algorithm, we have discovered many good constant-weight codes whose parameters are listed in Table II. They provide improvements to the previous results in Graham and Sloane [13]. During the course of this research, the authors became aware of the newer results of Conway and Sloane [14] which supersede our findings. After more computational effort we were able to find three codes that are better than those reported in [14]. They are a (23, 10, 7) code with 18 codewords, a (23, 10, 8) code with 28 codewords, and a (24, 10, 8) code with 33 codewords. The numbers in the parentheses are the length, the minimum distance, and the codeword weight, respectively. The codewords of these codes are listed in Tables III–V. The upper bounds in Table II are taken from Graham and Sloane [13].

Spherical Codes

We have also found many good spherical codes. They provide lower bounds on the constants $M(n, \theta)$. Using a method analogous to apple peeling, we have constructed a class of spherical codes. The details of our construction are contained in the Appendix. The results of the apple-peeling construction and simulated annealing in three-dimensional space are compared in Table VI. Simulated annealing found better codes than the apple-peeling construction. One of these codes is listed in Table VII. It is possible that bounds exist rivaling or even superseding the apple-peeling bound in the literature. The comparison made in Table VI is preliminary.

The fourth column in Table VI contains the Wyner lower bound [27], which is similar to the one derived by Shannon [28] but slightly sharper. The bound is

$$M(n, \theta) \geq \frac{n\sqrt{\pi}\,\Gamma\left(\dfrac{n+1}{2}\right)}{(n-1)\Gamma\left(\dfrac{n+2}{2}\right)} \left[\int_0^\theta \sin^{n-2}\phi \, d\phi\right]^{-1}$$

where $\Gamma(\cdot)$ is the Gamma function. In three dimensions it

TABLE VI
SPHERICAL CODES VIA SIMULATED ANNEALING

Dimension	Angle $\cos\theta$	θ	Lower Bound for $M(3,\theta)$ Wyner	Lattice	Apple Peel	Annealing	Upper Bound Rankin
3	~ 2/3	0.841	6		14	18	24
3	13/16	~ 0.62	10	12	30	31	42
3	5/6	~ 0.586	12	24	34	35	48
3		0.524	14		40	45	59
3	~ 9/10	0.45	20		56	59	80
3	13/14	~ 0.38	28	48	82	83	112
3		$\pi/9$	32		98	99	132

TABLE VII
SIZE 35 SPHERICAL CODE FOR $M(3, \text{arccos}\,(5/6))$

0.	−0.183291	0.022164	0.982809
1.	0.100781	−0.959698	−0.262341
2.	0.790523	−0.462605	0.401335
3.	−0.810027	−0.387179	−0.440396
4.	0.765679	0.579538	0.279056
5.	−0.269579	0.954104	0.130435
6.	0.302644	−0.846476	0.438047
7.	0.623330	−0.776523	−0.092038
8.	−0.066041	0.897189	−0.436681
9.	−0.208099	0.471576	−0.856919
10.	−0.129810	−0.588486	0.798018
11.	−0.255299	0.666100	0.700809
12.	0.761180	0.103834	0.640175
13.	−0.973947	−0.001453	0.226773
14.	0.874298	−0.322242	−0.362993
15.	0.258973	0.858889	0.441863
16.	−0.615145	0.670210	−0.415228
17.	−0.803612	−0.577832	0.142539
18.	−0.233786	−0.571646	−0.786489
19.	−0.652256	−0.345887	0.674481
20.	0.840418	0.417846	−0.345113
21.	0.291185	0.399403	0.869304
22.	0.620219	0.023453	−0.784078
23.	0.397097	−0.587405	−0.705173
24.	0.995169	0.034284	0.092000
25.	−0.678789	0.259884	0.686809
26.	−0.924050	0.182779	−0.335743
27.	−0.775339	0.603816	0.185083
28.	−0.294027	−0.908184	0.297910
29.	0.378628	0.576793	−0.723845
30.	0.447649	0.881887	−0.147941
31.	0.384192	−0.297955	0.873853
32.	0.069441	−0.068884	−0.995205
33.	−0.555841	−0.007377	−0.831256
34.	−0.466423	−0.836125	−0.288695

reduces to

$$M(3,\theta) \geq \frac{2}{1-\cos\theta}.$$

The second lower bound (fifth column of Table VI) is obtained from well-known lattice structures as described in Sloane [19]. The upper bound (column eight of Table VI) is due to Rankin [29] as follows:

$$M(n,\theta) \leq \frac{\sqrt{\pi}\,\Gamma\left(\dfrac{n-1}{2}\right)\sin\psi\tan\psi}{2\Gamma\left(\dfrac{n}{2}\right)\displaystyle\int_0^{\psi}(\sin\phi)^{n-2}(\cos\phi-\cos\psi)\,d\phi}$$

where $\psi = \sin^{-1}[\sqrt{2}\sin(\theta/2)]$. This bound is tighter than the standard volume bound. In three dimensions it reduces to

$$M(3,\theta) \leq \frac{\sin\psi\tan\psi}{(1-\cos 2\psi)/4-\cos\psi(1-\cos\psi)}.$$

Among the lower bounds included in Table VI, simulated annealing provided the best results.[1] The preliminary investigations in four dimensions also look promising. Simulated annealing shows that $M(4, \cos^{-1}(5/6)) \geq 97$, which betters the lattice bound of 96, and $M(4, \cos^{-1}(9/10)) \geq 145$, which betters the lattice bound of 144.

Implementation Issues

The simulation programs for constant-weight codes and spherical codes are implemented on a VAX 780 with a floating-point accelerator. In each application the *C* program consists of less than a thousand statements. Most of the statements are for bookkeeping, debugging, and input/output. The core of the program, which executes the annealing process, is less than two hundred statements long.

Each of the three constant-weight codes listed in Tables III–V are found after a few runs of the algorithm. The CPU time spent on a typical run ranges from 15 min for the code in Table III to a maximum of about 2 h for the codes in Tables IV and V. Unsuccessful runs are terminated manually after about 1–2 h. Once a code has been found, we have been able to reproduce the success in almost every repeat run using the same cooling rate, although the code found each time is different.

The simulation runs for three-dimensional spherical codes typically take less than 30 min. The test runs for higher dimensions are allowed to consume up to several hours of CPU time.

We have used the basic algorithm as shown in Fig. 2 throughout our experiments. No major speed-up modification has been attempted. If large-scale experiments are to be embarked upon, methods such as described in [21] can be incorporated to reduce the running time. Jiggling several codewords simultaneously in each iteration of the algorithm in Fig. 2 may also help.

V. CONCLUDING REMARK

Simulated annealing promises to be a useful tool in designing good codes. We have found new codes better than any code previously known in the literature for many

[1] During the writing of this paper, the excellent constructions of three-dimensional spherical codes for several specific minimum separating angles [30] have been brought to the authors' attention. It appears unlikely that the simulated annealing algorithm can better these codes. However, the algorithm remains useful in the automatic generation of codes with general parameters.

sets of parameters. The method of simulated annealing has already proven useful in many combinatorial optimization problems. Our experiment here indicates that simulated annealing is also useful in providing lower bounds to combinatorial constants, such as the coding theoretical constants $A(n, d, w)$ and $M(n, \theta)$. Improvement to other constants, such as the Ramsey numbers, may be obtainable. Several related source-coding problems, such as vector quantization [32] and discrete Markov source codes [10], are also subject to analysis by simulated annealing. Essentially, the same algorithm can be used with very little modification.

The algorithm itself seems robust with respect to modifications of the choice of energy function, perturbation pattern, and annealing schedule (consisting of the initial temperature, the temperature decrement, and the termination conditions) as long as they stay reasonable. Though some interest exists in the fine tuning of annealing schedules [9], we have adhered to one kind of schedule in most of our experiments. In our experience simulated annealing appears robust for reasonable changes in the schedule. Better results are usually produced by the devotion of more CPU time. An important issue in simulated annealing is the convergence of the algorithm. We have largely ignored this issue in our studies. However, in all our experiments the results are converging. For more details on this issue see [33] and [34]. The method is easily applicable to new areas, and the results may be exciting.

One disadvantage of the simulated annealing algorithm is its liberal consumption of CPU time. In our experiments the required computation time goes up very rapidly with increasing parameter values. The choice of simulated annealing for a given problem must be weighed against other computational methods. For certain parameter values the simulated annealing algorithm finds it hard to contend with tight code constructions which make ingenuous use of advanced mathematical structures such as those contained in [12], [19], and [30]. The usefulness of the algorithm lies in its ability to generate reasonably good results automatically. The algorithm is relatively simple to implement and to adapt to particular problem constraints. In places where computation power is abundant and where specific mathematical constructions are lacking or unsatisfactory, the simulated annealing algorithm can be used effectively to generate reasonably good results mechanically. The codes generated in this paper are all nonsystematic codes, which could limit their applicability in practical situations.

Acknowledgment

Fruitful discussions with N. J. A. Sloane and D. S. Johnson are heartily acknowledged.

Appendix

In this Appendix we prove the apple-peeling lower bound on the numbers $M(n, \theta)$. We use a method of code construction which is similar to apple peeling in three dimensions, thus the name. We also present another construction method based on certain lattices in multiple dimensions.

Theorem (The Apple-Peeling Bound): For $0 < \theta \leq \pi$, $n \geq 3$, we have

$$M(n, \theta) \geq 2 \cdot \sum_{i=0}^{k} M\left(n-1, \arccos^*\left[\frac{\cos\theta - \sin^2((i+1/2)\theta)}{\cos^2((i+1/2)\theta)}\right]\right)$$

where $k = \lceil \pi/2\theta - 1/2 \rceil$ and

$$\arccos^*(x) = \begin{cases} \arccos(x), & -1 \leq x < 1 \\ 2\pi, & x < -1. \end{cases}$$

Remark: In any dimension $n \geq 1$, $M(n, \pi + \epsilon) = 1$ for $0 < \epsilon \leq \pi$. In particular, $M(n, 2\pi) = 1$.

Proof: Let $x = (\sin\alpha, z_1 \cos\alpha, z_2 \cos\alpha, \cdots, z_{n-1}\cos\alpha)$ and $y = (\sin\alpha, w_1\cos\alpha, w_2\cos\alpha, \cdots, w_{n-1}\cos\alpha)$ be two points on the surface of the unit sphere with identical first coordinates. The two points $z = (z_1, z_2, \cdots, z_{n-1})$ and $w = (w_1, w_2, \cdots, w_{n-1})$ are on the surface of the unit sphere in $n-1$ dimensions. The points x and y are separated by an angle $\cos^{-1}(x \cdot y) = \cos^{-1}(\sin^2\alpha + z \cdot w \cos^{22}\alpha)$, while z and w are separated by $\cos^{-1}(z \cdot w)$.

For each i and $\alpha = \pm(i + 1/2)\theta$ we have $M(n - 1, \arccos^*(\cos\theta - \sin^2\alpha)/\cos^2\alpha))$ code points in n dimensions with first coordinates equal to $\cos\alpha$ and a minimum separating angle $\geq \theta$. Note that when $(\cos\theta - \sin^2\alpha)/\cos^2\alpha < -1$, we have precisely one point with a first coordinate equal to $\sin\alpha$ because $M(n-1, 2\pi) = 1$. Two code points x and y with different first coordinates have their inner product $x \cdot y \leq \sin\alpha\sin\beta + \cos\alpha\cos\beta = \cos(\alpha - \beta) \leq \cos\theta$. Hence $\cos^{-1}(x \cdot y) \geq \theta$. By summing over all legal values of i, we obtain the apple-peeling bound.

In two dimensions it is easy to show that $M(2, \theta) = \lceil 2\pi/\theta \rceil$ for $0 < \theta \leq 2\pi$. This, together with the apple-peeling theorem, gives us the sixth column of Table VI. Note that when $\cos((i + 1/2)\theta) = 0$, then $\cos\theta - \sin^2((i + 1/2)\theta) < 0$, and we take $\arccos^*[(\cos\theta - \sin^2((i+1/2)\theta))/\cos^2((i+1/2)\theta)] = 2\pi$.

The apple-peeling construction provides large codes in many dimensions. Although lattices provide the best-known lower bounds for kissing numbers ($\theta = \pi/3$) and excellent lower bounds in eight and 24 dimensions [16], [19], [20], [21], [31], the apple-peeling bound works well in other cases [32]. An additional advantage of the apple-peeling bound is its ease of implementation in any dimension. Equivalent or superior bounds may exist in the literature. The studies made here are preliminary.

We have also investigated another construction method based on certain lattices in n-dimensional space. First, choose k real numbers $x_1, x_2, \cdots, x_k$ and k integers $n_1, n_2, \cdots, n_k$ which add up to n. The collection of points with n_i coordinates being x_i and with the n_i satisfying certain constraints form a spherical code. The minimum separating angle and the size of the code can be calculated for specific choices of the x_i and the constraints on the n_i. For example, the set of n-dimensional vectors with coordinates taken from $\{0, 1/\sqrt{m}, -1/\sqrt{m}\}$ and with exactly m nonzero coordinates forms a spherical code with $2^m\binom{n}{m}$ code points and a minimum separating angle $= \cos^{-1}(1 - (1/m))$. The lattice construction presented earlier can be used in combination with the apple-peeling bound to produce good lower bounds to the size of spherical codes.

REFERENCES

[1] N. Metropolis, A. W. Rosenbluth, M. N. Rosenbluth, A. H. Teller, and E. Teller, "Equation of state calculations by fasting computing machines," *J. Chem. Phys.*, vol. 21, June 1953.

[2] S. Kirkpatrick, C. D. Gelatt, and M. P. Vecchi, "Optimization by simulated annealing," *Science*, vol. 220, pp. 671–680, May 1983.

[3] D. S. Johnson, L. McGeoch, C. Rodriguez, and C. Schevon, "Optimization by simulated annealing: An experimental evaluation," presented at the *Workshop on Statistical Physics in Engineering and Biology*, Apr. 1984.

[4] D. S. Johnson, C. H. Papadimitriou, and M. Yannakakis, "How easy is local search?" in *Proc. 26th Symp. Foundations of Computer Science*, Oct 1985, pp. 39–42.

[5] M. R. Garey and D. S. Johnson, *Computers and Intractability: A Guide to the Theory of NP-Completeness*. San Francisco, CA: Freeman, 1979.

[6] F. Romeo, C. Sechen, and A. Sangiovanni-Vincentelli, "Simulated annealing research at Berkeley," in *Proc. IEEE Int. Conf. Computer Design*, 1984, pp. 652–657.

[7] M. P. Vecchi and S. Kirkpatrick, "Global wiring by simulated annealing," *IEEE Trans. Comput. Aided Des. Integrated Circuits Syst.*, vol. CAD-2, pp. 215–222, Oct. 1983.

[8] S. White, "Concepts of scale in simulated annealing," in *Proc. IEEE Int. Conf. Computer Design*, 1984, pp. 646–651.

[9] M. Lundy "Applications of the annealing algorithm to combinatorial problems in statistics," *Biometrika*, vol. 72, pp. 191–198, 1985.

[10] R. G. Gallager, *Information Theory and Reliable Communication*. New York: Wiley, 1968.

[11] S. Lin and D. J. Costello, Jr., *Error Control Coding: Fundamentals and Applications*. Englewood Cliffs, NJ: Prentice-Hall, 1983.

[12] F. J. MacWilliams and N. J. A. Sloane, *The Theory of Error-Correcting Codes*. New York: North-Holland, 1977.

[13] R. L. Graham and N. J. A. Sloane, "Lower bounds for constant weight codes," *IEEE Trans. Inform. Theory*, vol IT-26, pp. 31–43, Jan. 1980.

[14] J. H. Conway and N. J. A. Sloane, "Lexicographic codes: Error-correcting codes from game theory," to appear.

[15] I. F. Blake, "The Leech lattice as a code for the Gaussian channel," *Inform. Contr.*, vol. 19, pp. 66–74, 1971.

[16] A. Gersho, "Asymptotically optimal block quantization," *IEEE Trans. Inform. Theory*, vol. IT-25, pp. 373–380, 1979.

[17] L. Shepp, "Computerized tomography and nuclear magnetic resonance," *J. Comput. Assisted Tomog.*, vol. 4, pp. 94–107, Feb. 1980.

[18] N. J. A. Sloane, "Tables of sphere packings and spherical codes," *IEEE Trans. Inform. Theory*, vol. IT-27, pp. 327–338, May 1981.

[19] ——, "Recent bounds for codes, sphere packings, and related problems obtained by linear programming and other methods," to appear.

[20] H. S. M. Coxeter, "The problem of packing a number of equal nonoverlapping circles on a sphere," *Trans. N.Y. Acad. Sci.*, series II, vol. 24, pp. 320–331, Jan. 1962.

[21] J. W. Greene and K. J. Supowit, "Simulated annealing without rejected moves," in *Proc. IEEE Int. Conf. Computer Design*, 1984, pp. 658–664.

[22] J. L. Lutton and E. Bonomi, "An efficient non-deterministic heuristic for the minimum weighted perfect Euclidean matching problem: The Metropolis algorithm," preprint.

[23] A. El Gamal and I. Shperling, "Design of good codes via simulated annealing," presented at the Simulated Annealing Conf., Yorktown Heights, NY, 1984.

[24] L. A. Hemachandra and V. K. Wei, "Using simulated annealing to calculate combinatorial constants," in *Proc. 22nd Annual Allerton Conf. Communication, Control, and Computing*, Oct. 3–5, 1984, pp. 545–552.

[25] S. P. Lloyd, "Least squares quantization in PCM," *IEEE Trans. Inform. Theory*, vol. IT-28, pp. 129–137, 1982.

[26] A. R. Calderbank, J. E. Mazo, and V. K. Wei, "Asymptotic upper bounds on the minimum distance of trellis codes," *IEEE Trans. Commun.*, vol. COM-33, pp. 305–309, Apr. 1985.

[27] A. D. Wyner, "Capabilities of bounded discrepancy decoding," *Bell Syst. Tech. J.*, vol. 44, pp. 1061–1122, July/Aug. 1965.

[28] C. E. Shannon, "Probability of error for optimal codes in a Gaussian channel," *Bell Syst. Tech. J.*, vol. 38, pp. 611–656, May 1959.

[29] R. A. Rankin, "The closest packing of spherical caps in *n*-dimensions," *Proc. Glasgow Math. Ass.*, vol. 2, pp. 139–144, 1955.

[30] T. Tarnai and Z. S. Gáspár, "Improved packing of equal circles on a sphere and rigidity of its graph," *Math. Proc. Cambridge Phil. Soc.*, vol. 93, parit 2, pp. 191–218, Mar. 1983.

[31] L. Hemachandra and V. K. Wei, "A note on the size of spherical codes," Bell Commun. Res., Tech. Memo., 1984.

[32] A. Gersho, "Quantization," *IEEE Commun. Soc. Mag.*, vol. COM-15, pp. 16–29, Sept. 1977.

[33] S. Romeo and S. Sangiovanni-Vincentelli, "Probabilistic hill-climbing algorithms: Properties and applications," in *Proc. Chapel Hill Conf. VLSI*, 1985, pp. 393–418.

[34] B. Hajek, private communication.

An Algorithm for Uniform Vector Quantizer Design

KHALID SAYOOD, MEMBER, IEEE, JERRY D. GIBSON, SENIOR MEMBER, IEEE, AND MARTIN C. ROST

Abstract—A vector quantizer maps a k-dimensional vector into one of a finite set of output vectors or "points". Although certain lattices have been shown to have desirable properties for vector quantization applications, there are as yet no algorithms available in the quantization literature for building quantizers based on these lattices. An algorithm for designing vector quantizers based on the root lattices A_n, D_n, and E_n and their duals is presented. Also, a coding scheme that has general applicability to all vector quantizers is presented. A four-dimensional uniform vector quantizer is used to encode Laplacian and gamma-distributed sources at entropy rates of one and two bits / sample and is demonstrated to achieve performance that compares favorably with the rate distortion bound and other scalar and vector quantizers. Finally, an application using uniform four- and eight-dimensional vector quantizers for encoding the discrete cosine transform coefficients of an image at 0.5 bit / pel is presented, which visibly illustrates the performance advantage of vector quantization over scalar quantization.

I. Introduction

THE problem of optimal vector or block quantizer design currently is receiving substantial research interest. For uniform quantization, a quantization region must be able to tile space. In the scalar or one-dimensional case, this is not a problem since the interval is the only kind of region that tiles space. Thus, the only requirement for uniform scalar quantization is to find the size of the interval, given the source probability density function (pdf) and the distortion measure. Unfortunately, for higher dimensions many shapes (polytopes) tile space. Furthermore, the interaction between the source pdf and the distortion measure is so complex that the approach used for scalar quantizer design [1] is generally inapplicable.

Fortunately, Linde, Buzo, and Gray [2] have built upon Lloyd's Method I [1] to develop a vector quantizer design algorithm that uses the training sequences of the random variable to be quantized to generate locally optimal multidimensional quantizer designs. The theoretical properties of this algorithm are presented in [3], the applications of the algorithm to speech coding are reported in [2], [4]–[6]. For the case in which no training sequence is available, Conway and Sloane [7], [8] use the root lattices and their duals as the basis for designing efficient block quantizers for uniformly distributed inputs. By finding the Voronoi regions for these lattices and evaluating the mean squared error (mse) when these lattices are used to quantize uniformly distributed data, they are able to list the lattices that achieve the minimum mse for dimensions one through ten. These results constitute the best known uniform vector quantizers for the given block lengths. Additional important results on vector quantization are available in the very readable paper by Gersho [9] and the special issue of this Transactions edited by Gray and published in March 1982 (Part I). Conway and Sloane [7] present the properties of various lattices and measures of their efficiency when used as quantizers, and they also develop techniques for coding points based upon the special properties of particular lattices [8].

In this paper we present an explicit, step by step algorithm for generating and coding all of the root lattices—along with their duals—mentioned by Conway and Sloane—independent of their special symmetries. As an example, we have used the A_n^* lattice. To determine how these lattices perform as quantizers for a clearly nonasymptotic case and for nonuniformly distributed inputs, a four-dimensional quantizer is used to quantize Laplacian and Gamma distributed random variables at 1 and 2 bits/sample. The results are compared to the rate distortion bound, to scalar quantization, and to vector quantizers designed using the Linde, Buzo, and Gray algorithm. Finally, we present an application of these vector quantizers to image data compression at 0.5 bit/pixel.

II. Notation

Let X be an n-dimensional source vector with joint pdf $p(X) = p(x_1, x_2, \cdots, x_n)$. An n-dimensional vector quantizer is a function $Q(X)$ that maps $X \in \mathscr{R}^n$ into one of N output points with each output point corresponding to an output vector $Y_1, Y_2, \cdots, Y_N$ belonging to $\mathscr{R}^n$. The quantizer is completely specified by listing the N output vectors and their corresponding partitions of the space $\mathscr{R}^n$ into N disjoint and exhaustive regions. Following [2] and [9], we let these regions be denoted by $\mathscr{S}_1, \mathscr{S}_2, \cdots, \mathscr{S}_N$ so that $Q(X) = Y_i$, if $X \in \mathscr{S}_i$, for $i = 1, 2, \cdots, N$. Such a vector

Manuscript received December 6, 1982; revised May 4, 1984. This research was supported in part by the Office of Naval Research under Contract N00014-81-K-0299 and in part by the National Science Foundation under Grant ECS-8020304.

K. Sayood and M. C. Rost are with the Department of Electrical Engineering, University of Nebraska—Lincoln, Lincoln, NE 68588, U.S.A.

J. D. Gibson is with the Department of Electrical Engineering, Texas A & M University, College Station, TX 77843, U.S.A.

Reprinted from *IEEE Trans. Inform. Theory,* vol. IT-30, pp. 805–814, Nov. 1984.

quantizer may also be called a block quantizer with block length n. If $n = 1$, we have the familiar scalar quantization problem.

The distortion measure selected for this work is the MSE per symbol (letter) given by

$$d = \frac{1}{n} E\{d(X - Y)\}$$
$$= \frac{1}{n} E\|X - Q(X)\|^2$$
$$= \frac{1}{n} \int_{\mathscr{R}^n} \|X - Q(X)\|^2 p(X)\, dX, \tag{1}$$

where $\|\cdot\|$ denotes the usual l_2 norm.

III. Lattice Quantizer Design

In this section we develop a method for designing a lattice vector quantizer. A lattice is defined as a set of vectors

$$L = \{y: y = u_1 a_1 + u_2 a_2 + \cdots + u_n a_n\}, \tag{2}$$

where u_i are integers and the a_i, $i = 1, \cdots, n$ are the basis vectors for the lattice. The vector quantizer encodes the source vector X by mapping it into the lattice set

$$Y_i = Q(X), \qquad Y_i \in L, \text{ if } X \in \mathscr{S}_i.$$

The quantization regions $\mathscr{S}_i$ are Voronoi regions with centroid Y_i. The quantizer output can be uniquely represented by the vector

$$U_i = \{u_{i1}, u_{i2}, \cdots, u_{in}\}, \tag{3}$$

where

$$Y_i = \{u_{i1} a_1 + u_{i2} a_2 + \cdots + u_{in} a_n\}. \tag{4}$$

The decoding of U_i is accomplished by projecting its components u_{ij} back into the source space.

The mapping Q is implemented in two steps. First the lattice basis $\{a_i\}$ must be found and imbedded into $\mathscr{R}^n$. This will give explicit meaning to the components of U_i as it spans the lattice set. This imbedding is done by generating each of the basis vector's components from the defining lattice metric [12]–[14]. For the root lattices it is possible to construct a basis from the symmetry of their reflection properties. This approach is taken in [8], [12], and [14]. We start with a metric and require only that it be symmetric and invertible. Then a method for matching the source vector to the output vector that minimizes the distortion measure must be derived. A general method that works for any lattice is presented in Section IV.

In general a lattice can be represented in a quadratic form [11]

$$\sum_{i=1}^{n} \sum_{i=1}^{n} g_{ij} x_i x_j, \tag{5}$$

where

$$g_{ij} = a_i \cdot a_j \tag{6}$$

and the centered dot represents the usual vector inner product. The metric coefficients g_{ij} can be written as a symmetric matrix

$$G = \begin{bmatrix} g_{11} & g_{12} & \cdots & g_{1n} \\ g_{21} & \ddots & & \\ \vdots & & \ddots & \vdots \\ g_{n1} & & \cdots & g_{nn} \end{bmatrix}. \tag{7}$$

As $g_{ij} = g_{ji}$, there are $(n/2)(n+1)$ distinct coefficients in G, each an inner product of two basis vectors. These coefficients will be used to determine each a_i.

Let the components of a_i be $[a_{i1}, a_{i2}, \cdots, a_{in}]$. The n-basis vectors can also be written as a matrix

$$A = \begin{bmatrix} a_1 \\ a_2 \\ \vdots \\ a_n \end{bmatrix} = \begin{bmatrix} a_{11} & a_{12} & \cdots & a_{1n} \\ \vdots & & & \\ a_{n1} & & & a_{nn} \end{bmatrix}.$$

These n^2 unknowns can be reduced to $(n/2)(n+1)$ by requiring a lower triangular form for the matrix of basis vectors

$$A = \begin{bmatrix} a_{11} & & & 0 \\ a_{21} & a_{22} & & \\ \vdots & & \ddots & \\ a_{n1} & \cdots & & a_{nn} \end{bmatrix}. \tag{8}$$

The triangular form forces a specific orientation of the lattice in which a_1 is parallel to e_1, the first basis vector of $\mathscr{R}^n$.

Now that the number of unknowns matches the number of independent entries in G, the basis vectors can be obtained by using the following sequential method.

Notice that

$$a_1 \cdot a_1 = g_{11}$$

but

$$a_1 \cdot a_1 = a_{11}^2.$$

Therefore

$$a_{11} = \sqrt{g_{11}}. \tag{9}$$

Thus a_1 is completely determined. Now for a_2, the two equations

$$a_2 \cdot a_1 = g_{21}$$
$$a_2 \cdot a_2 = g_{22}$$

imply that

$$a_{11} a_{21} = g_{21}$$
$$a_{21}^2 + a_{22}^2 = g_{22}.$$

Using (9) gives

$$a_{21} = g_{21}/\sqrt{g_{11}}, \tag{10}$$

so that

$$a_{22} = \sqrt{g_{22} - g_{12}^2/g_{11}}. \tag{11}$$

The method for finding the components of the basis vectors can be generalized in the following manner.

Step 1: Find the lattice metric $\boldsymbol{G}$.
Step 2: $j = 1$

$$a_{11} = \sqrt{g_{11}}. \tag{12}$$

Step 3: $j = j + 1$

$$a_{ji} = \left(g_{ji} - \sum_{k=1}^{i-1} a_{jk} a_{ik} \right) / a_{ii}; \qquad i = 1, 2, \cdots, j-1 \tag{13}$$

$$a_{jj} = \left(g_{jj} - \sum_{k=1}^{j-1} (a_{jk})^2 \right)^{1/2}. \tag{14}$$

We assume that $\sum_{k=a}^{b} = 0$ when $b < a$.
Step 4: Repeat Step 3 for $j = 2, 3, \cdots, n$.

This algorithm generates the basis vectors of all of the root lattices of Conway and Sloane [7]. The metrics for the various lattices can be obtained from Humphreys [12].

Obtaining the basis vectors for the dual lattice L^* to the lattice L becomes easy once we realize that the basis vectors for the dual lattice are simply the reciprocal basis vectors of the root lattice [11]. Thus the metric matrix of the dual lattice is simply the inverse of the metric matrix of the root lattice. Once the metric matrix is obtained, the same algorithm can be used to generate the basis vectors of the dual lattice.

The A_n^* Lattice

The A_n^* lattices were chosen for the example in this paper. The metric for these lattices can be obtained from its quadratic form, which is given by Gameckii [13] as

$$\sum_{i=1}^{n} n x_i^2 - \sum_{\substack{i,j=1 \\ i \neq j}}^{n} x_i x_j. \tag{15}$$

Therefore for $n = 3$

$$\boldsymbol{G}^* = \begin{bmatrix} 3 & -1 & -1 \\ -1 & 3 & -1 \\ -1 & -1 & 3 \end{bmatrix} = 3 \begin{bmatrix} 1 & -1/3 & -1/3 \\ -1/3 & 1 & -1/3 \\ -1/3 & -1/3 & 1 \end{bmatrix}. \tag{16}$$

Solving for $\boldsymbol{A}$, we get

$$\boldsymbol{A} = \begin{bmatrix} \boldsymbol{a}_1 \\ \boldsymbol{a}_2 \\ \boldsymbol{a}_3 \end{bmatrix} = \begin{bmatrix} \sqrt{3} & 0 & 0 \\ \frac{-\sqrt{3}}{3} & \frac{2\sqrt{6}}{3} & 0 \\ \frac{-\sqrt{3}}{3} & \frac{-\sqrt{6}}{3} & \sqrt{2} \end{bmatrix}. \tag{17}$$

In the Soviet literature the A_n^* lattice is denoted by L_1^n. Also, depending on the form of the quadratic used, the enumeration of the basis vectors may be different. Confusion can be avoided by transforming the matrix of basis vectors into the minimal canonical form that is unique (within a scale factor) for a given lattice [15]. Let a_{ij} be the elements of the matrix of minimal basis vectors where the minimal basis set is a set of basis vectors that uniquely determine the lattice points. Then a minimal canonical basis is defined such that [15]

$$0 \le a_{ji} < a_{ii}, \qquad j > i$$
$$a_{ji} = 0, \qquad j < i.$$

For example, for the A_3 lattice the metric matrix obtained from the metric given by Humphreys [12] is

$$\boldsymbol{G} = \begin{bmatrix} 1 & -1/2 & 0 \\ -1/2 & 1 & -1/2 \\ 0 & -1/2 & 1 \end{bmatrix}. \tag{18}$$

Therefore, the metric matrix of A_3^* is

$$\boldsymbol{G}^* = \boldsymbol{G}^{-1} = \begin{bmatrix} 3/4 & 1/2 & 1/4 \\ 1/2 & 1 & 1/2 \\ 1/4 & 1/2 & 3/4 \end{bmatrix}. \tag{19}$$

By using the algorithm $\boldsymbol{A}$ is then found to be

$$\boldsymbol{A} = \begin{bmatrix} \boldsymbol{a}_1 \\ \boldsymbol{a}_2 \\ \boldsymbol{a}_3 \end{bmatrix} = \frac{1}{2} \begin{bmatrix} \sqrt{3} & 0 & 0 \\ \frac{2}{3}\sqrt{3} & \frac{2\sqrt{6}}{3} & 0 \\ \frac{\sqrt{3}}{3} & \frac{\sqrt{6}}{3} & \sqrt{2} \end{bmatrix}. \tag{20}$$

This is already in minimal canonical form. From (17), which was obtained by using the metric given by Gameckii, replacing $\boldsymbol{a}_2$ by $\boldsymbol{a}_1 + \boldsymbol{a}_2$, and then replacing $\boldsymbol{a}_3$ by $\boldsymbol{a}_2 + \boldsymbol{a}_3$, we obtain

$$\boldsymbol{A} = \begin{bmatrix} \sqrt{3} & 0 & 0 \\ \frac{2\sqrt{3}}{3} & \frac{2\sqrt{6}}{3} & 0 \\ \frac{\sqrt{3}}{3} & \frac{\sqrt{6}}{3} & \sqrt{2} \end{bmatrix}, \tag{21}$$

which is the same as (20) within a scale factor of 1/2.

Once the basis vectors are determined, the remaining lattice points can be obtained by taking integral combinations of these basis vectors. The distortion achieved by a given quantizer is related to the density of its points in space. Scaling the metric or basis vectors will change the density of the resulting lattice proportional to α^n, where α is proportional to the distance change between two points of the lattice, and n is the lattice dimension. By changing the density of the lattice, the distortion of its resulting quantizer can be changed. How this scaling affects other parameters of the lattice can be found in Sloane [14]. For code books of limited size, the source pdf and edge effects of the outermost code points weigh heavily in the determination of the expected distortion. In this case the optimum point density for a given code book must be determined experimentally.

For this paper the code book size was chosen based upon a given data transmission rate. Once the desired rate is chosen—for example, 2 bits/sample—the number of code points is fixed depending upon the dimension of the particular lattice to be used. A 2 bits/sample rate with a four-dimensional quantizer requires a code book size of

$2^4 = 16$ points. The code book consists of the 16 closest points to the origin. The lattice density is then chosen to minimize the distortion by experimentation.

IV. Coding

One of the major problems with general vector quantizers is the size of the code book and hence the number of operations required to find the closest output point to a given input point. The brute force approach of comparing the input point to each output point and thus finding the output point that minimizes the distortion measure may be time-consuming and cause large delays. However, the regular structure of the lattice quantizers affords ways of substantially reducing this delay. Conway and Sloane [8] have developed very efficient methods of coding for a certain class of lattice quantizers. The quantizers we use in this paper (A_n^*) fall into that class. The coding techniques of Conway and Sloane—though highly efficient—are restrictive in their use. In this section we present coding techniques that have a general applicability to all vector quantizers, although they are not as efficient. Two general strategies are presented: one for the case where the codebook is small enough to store, another for the case where the code book is too large to store.

The first strategy involves partitioning the range space of the input random variable into more convenient regions than those partitioned by the quantizer. This idea is similar in some sense to the prequantization for two dimensions presented by Rines and Gallagher [16]. The partitioning is done as follows.

Let r_1 be the length of the shortest vector of the basis set. Let $r_{\max}$ be the maximum distance of any output point from the origin. Pick $\Delta r > 0$ such that $r_{\max} - r_1$ is an integral multiple of Δr; that is, $r_{\max} - r_1 = M\Delta r$, $M \in Z$, where Z is the set of integers. The choice of M depends upon the spacing of the output points, as will be seen later on. $\mathscr{R}^n$ can now be partitioned into shells $\{S_i\}_{i=0}^{M+1}$ for which

$$S_i = \{X \in \mathscr{R}^n: r_1 + (i-1)\Delta r \le \|X\| < r_1 + i\Delta r\} \quad i = 1,\cdots, M \tag{22}$$

$$S_0 = \{X \in \mathscr{R}^n: \|X\| < r_1\} = \{(0,0\cdots,0)\} \tag{23}$$

$$S_{M+1} = \{X \in \mathscr{R}^n: \|X\| \ge r_{\max}\}. \tag{24}$$

There is another more "natural" subdivision of space that can be utilized. This is the partitioning of space by its coordinate axes, the n-dimensional equivalent of quadrants. They are referred to as axis-partitions. Using this subdivision, $\mathscr{R}^n$ is partitioned into regions S_{ij}: $i = 0,\cdots, M+1$, $j = 1,\cdots,2^n$. The axes are included in all partitions that they abut. Then an $M + 1 \times 2^{n+1}$ look-up table is constructed where the rows correspond to shells, and two columns are assigned to each axis-partition. The S_0 shell is not included in the table because it only contains the point at the origin. The output points are ordered so that the first column assigned to each axis-partition contains the location of the first point in that axis-partition and the second column contains the number of output points in the axis-partition of the particular shell. An example of this ordering is given in the Appendix.

TABLE I
Look-Up Table for Axis Partitions

Axis-Partitions	Neighboring Axis-Partitions	
1	2	4
2	1	3
3	2	4
4	1	3

A $2^n \times (n+1)$ look-up table contains an enumeration of the neighboring axis-partitions to each axis-partition. Table I shows how this look-up table would appear in two dimensions. For the sake of brevity, define a function $QL(k)$, where $QL(k)$ is simply the enumeration of the axis-partitions neighboring axis-partition k. In the two-dimensiónal case

$$QL(1) = \{2,4\}$$
$$QL(2) = \{1,3\}. \tag{25}$$

When an input point is obtained, it is classified according to shell and axis partition. (If any of the coordinates is zero, the situation is treated as a separate case and will be discussed later). If the input point falls in the kth shell and mth axis-partition, then it belongs to region $S_{k,m}$. It is then compared to the output points in $S_{k,m}$. The location and number of output points can be found by looking in locations $(k, 2m-1)$ and $(k, 2m)$ of the look-up table. The input point is also compared to the output points in $S_{k,QL(m)}$, $S_{k-1,m}$, $S_{k-1,QL(m)}$, $S_{k+1,m}$, and $S_{k+1,QL(m)}$ to find the output point that minimizes the distortion measure. The subscripts $QL(m)$ denote the neighboring axis-partitions. If any of the coordinates of the input point is zero, then copies of the input point are made depending on the number of zero coordinates. A zero coordinate implies that the point lies on the boundary between two axis-partitions. If there are k zero coordinates, then the point lies on the boundary of 2^k axis-partitions. Therefore, this point should be included in all 2^k axis-partitions. This is accomplished by making 2^k copies of this point and assigning a copy to each axis partition. Each of these copies is treated in the manner described above, and the output that minimizes the distortion measure over all copies is selected.

The preceding strategy presupposes that the code book itself can be stored. This might not always be possible because the size of the code book increases almost exponentially with increasing bit allocation and increasing blocklength. For example, to quantize a sample with eight bits using an eight-dimensional quantizer, about 1.8×10^7 trillion vectors are needed! To avoid this exhaustion of resources, a different coding strategy is in order.

Again the regularity of the lattice can be utilized. As Gersho [9] points out, any translation of the lattice that shifts the point at the origin to any other lattice point position produces the identical lattice. Quoting Gersho, "if you sit on a lattice point and view the surrounding lattice points, you will see the identical environment regardless of the point you are sitting on". This observation, combined with the fact that a relatively small number of points are required to generate a Voronoi region, provides an attractively simple coding strategy. Let X be an input point with magnitude m_x. Let C be a group of points that define a Voronoi region at the origin, and let $S \in C$ be a basis set for the lattice. Let m_1 be the distance of the farthest point in the group C. If m_x is less than or equal to m_1, then the required output point lies in C. If m_x is greater than m_1, then define

$$M = [m_x/m_1], \tag{26}$$

where $[\cdot]$ implies the smallest integer greater than m_x/m_1 and let

$$\overline{X} = \frac{1}{M} X; \tag{27}$$

then $|\overline{X}| \leq m_1$. Find $X_0 \in S$ that minimizes the distortion measure $d(\overline{X}, X_i)$, $X_i \in S$. Let θ be the origin; then

$$\theta' = \theta + MX_0 \tag{28}$$

is also a lattice point. Find

$$d_1 = \|X - \theta'\|. \tag{29}$$

If d_1 is less than or equal to m_1, then translate C by θ' to obtain

$$C' = C + \theta'.$$

Because of the lattice structure, C' is a subset of the lattice, and the point in C' closest to X is a valid output point.

If d_1 is greater than m_1, then define

$$X' = X - \theta'.$$

Continue the process until d_n is less than m_1. Note that each time the translation is preserved in $\theta^{(n)}$.

V. Performance Evaluation

To obtain performance data, we designed uniform four-dimensional vector quantizers based on the A_4^* lattice to be used on Laplacian and gamma-distributed sources at entropy rates of 1 and 2 bits/sample. These quantizers were then used to quantize 50 000 independent and identically distributed variates with the appropriate distribution. The resulting signal-to-noise ratios (defined as $10 \log_{10} [1/\text{mse}]$) are listed in Table II along with the rate distortion bound; the scalar quantization results of Noll and Zelinski [17] and Farvardin and Modestino [18]; and four-dimensional vector quantizers designed by Fischer and Dicharry [19] using the Linde, Buzo, and Gray (LBG) algorithm [2]. In Table II, "Ent" indicates that the outputs are entropy-coded, while "Uncod" means that the rate is $\log_2 l$, where l is the number of output points. For the algorithm developed in this paper, it is necessary to adjust the quantization cell size by simulation. Furthermore, to obtain the entropy-coded results in column (b), a "cut-and-try" approach is used to achieve the specified entropy rates of 1 and 2 bits/sample. Because these searches were not exhaustive, the signal-to-noise ratios in columns (b) and (e) may not be the maximum ratios achievable using the proposed algorithm.

Before proceeding, it must be noted that if entropy coding of the LBG algorithm quantizer were to be performed, a cut-and-try method would also be required to produce a given entropy rate. For scalar quantization at a specified entropy rate, a search must be performed among all quantizers having the desired output entropy to find the one with the minimum distortion (maximum signal-to-noise ratio).

The results in column (b) are particularly satisfying for a Laplacian source at 2 bits/sample and for a gamma-distributed source at 1 bit/sample. For the Laplacian source at 1 bit/sample, the optimum nonuniform scalar quantizer designed by Farvardin and Modestino [18] outperforms our uniform four-dimensional quantizer. It is possible that we have not found the best four-dimensional quantizer that our algorithm can produce for this source and entropy rate.

VI. Image Transform Coding Application

As stated earlier in this paper, the vector quantizers generated by the algorithm presented here were used in a transform domain image compression scheme. A brief introduction to image compression via transform coding is presented in the following. For a more detailed introduction see Hunt [20].

Image compression schemes usually consist of three steps [20]. The first step is to remove as much as possible of the data correlation, which is called redundancy removal. The processed data are then quantized, and the quantizer outputs are coded. There are various methods used for the removal of data correlation; some operate in the spatial domain, and others operate in the transform domain.

The method used in this application was a transform domain method using the discrete cosine transform (DCT). To best understand transform domain image coding, it is instructive to look at the Karhunen–Loève (KL) transform. In this method the image is represented by an $N \times N$ matrix of the samples of the image intensity values. Let $\mathbf{g}$ represent a vector of length N^2 constructed by lexicographically ordering the elements of the $N \times N$ image intensity matrix. Let ψ be the covariance matrix of $\mathbf{g}$; then ψ can be written as

$$\psi = P \Lambda P^T,$$

where P is a matrix of eigenvectors, and Λ is a diagonal matrix of eigenvalues of ψ. The eigenvalues of ψ are arranged in decreasing order, and P is ordered accordingly. Then the transformed vector G obtained by

$$G = P\mathbf{g}$$

has the property that all of its elements are uncorrelated and they are arranged in order of decreasing statistical variance. Now image compression is simply a matter of retaining the first M components ($M < N^2$) of G to form

TABLE II
SIGNAL-TO-NOISE RATIO COMPARISONS

Distribution	Rate bits/ sample	(a) RDF	(b) Uniform, 4-dim. Vector (Ent)	(c) Optimum Scalar (Ent)	(d) Scalar (Ent)	(e) Uniform, 4-dim. Vector (Uncod)	(f) Nonuniform, 4-dim. Vector (Uncod)	(g) Scalar (Uncod)
		Signal-to-Noise Ratios (dB)						
Laplacian	1	6.62	4.92	5.76	3.01	3.97	4.62	3.01
	2	12.66	12.25	11.31	10.85	9.81		7.53
Gamma	1	8.53	8.35		1.77	4.15	5.65	1.77
	2	15.21	13.79		11.60	7.55		6.33

Column Descriptions: (a) Rate distortion bound (Ref. [26]);
(b) Uniform, four-dimensional quantizer with entropy coding (Sec. V);
(c) Optimum, nonuniform scalar quantizer with entropy coding (Ref. [30]);
(d) Nonuniform, scalar quantizer (symmetric, even no. of levels, Ref. [26]);
(e) Uniform, four-dimensional quantizer without entropy coding (Sec. V);
(f) Nonuniform, four-dimensional quantizer without entropy coding (Ref. [31]);
(g) Nonuniform, scalar quantizer without entropy coding (Ref. [32]).

a compressed vector $\hat{G}$. The magnitude of M depends upon the amount of compression desired. As the elements of G are arranged in order of decresing statistical importance, the statistically more important M elements are retained, while the $N^2 - M$ elements discarded have less statistical importance. The vector $\hat{G}$ can now be transmitted, and at the receiver it can be transformed back into the spatial domain by

$$\hat{g} = P^T\hat{G}.$$

This method, known as the Karhunen-Loève transform, has the smallest mean squared error among all transform coding methods.

While the KL transform obtains total decorrelation of the KL components, it requires the computation of the eigenvectors of the covariance matrix of the data. Because of this, it is computationally very cumbersome. The Fourier transform is a good approximation to the KL transform, but it suffers because the discrete Fourier series assumes that the data are periodic. Hence the last sample is assumed to be a neighbor of the first sample. This in turn creates a discontinuity that gives rise to the Gibbs phenomenon. This can be avoided by forcing symmetry onto the data in the following manner. If the image block looks like Fig. 1(a), then the Fourier transform is taken of the larger symmetric block, shown in Fig. 1(b). This larger block can now be taken to be periodic without introducing discontinuities at the boundary. However, because of the symmetry, the discrete Fourier transform of this larger block is precisely the DCT of the original image block. Because of this, the DCT is popular as a transform technique.

Let $f(x, y)$ represent the image intensity value at coordinate (x, y), and let $C(u, v)$ represent the DCT value at coordinate (u, v). Then the DCT pair is given by

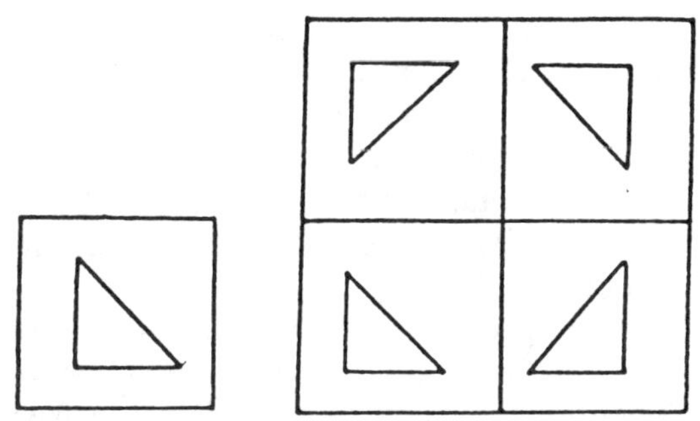

Fig. 1. Symmetrizing an image.

$$C(u,v) = \frac{1}{N^2}\sum_{x=0}^{N-1}\sum_{y=0}^{N-1} f(x,y)\Big[\cos\frac{\pi}{2N}((2x+1)u + (2y+1)v) + \cos\frac{\pi}{2N} \cdot ((2x+1)u - (2y+1)v)\Big] \tag{30}$$

and

$$f(x,y) = \sum_{u=0}^{N-1}\sum_{v=0}^{N-1} C(u,v) \cdot\Big[\cos\frac{\pi}{2N}((2x+1)u + (2y+1)v) + \cos\frac{\pi}{2N}((2x+1)u - (2y+1)v)\Big]. \tag{31}$$

In matrix notation the elements of the transform matrix Υ

are given by

$$\Upsilon(k,m) = \begin{cases} \frac{2}{N} \cos \frac{(2m-1)(k-1)\pi}{2N}, \\ \quad 2 \le k \le N,\ 1 \le m \le N \\ \frac{1}{N}, \quad k = 1,\ 1 \le m \le N, \end{cases} \tag{32}$$

and the transform pair is given by

$$[C] = \Upsilon[f]\Upsilon^T \tag{33}$$

$$[f] = \Upsilon^T[C]\Upsilon. \tag{34}$$

Because manipulating 256 × 256 or 512 × 512 matrices is cumbersome, the image is partitioned into smaller subblocks, usually 16 × 16 or 32 × 32. Another reason often proffered for this partitioning is that pixel correlation will not usually exceed 16 or 32 pixels; hence it is pointless to take transforms of larger blocks. However, as pointed out by Tescher [21], this reasoning is not valid. Even if all the pixels within the subblock are decorrelated, the pixels on the border of the subblock remain correlated with the pixels on the border of the adjacent subblock. Regardless of this, the computational advantages gained from partitioning are sufficient to outweigh this concern.

To summarize, the image is partitioned into subblocks. Each subblock is then transformed using (33). Depending on the number of bits available, some of the lower energy coefficients are discarded. The remaining coefficients are allocated bits according to their variance under the constraint that the total number of bits allocated not exceed the number of bits available. At the receiver, the image is recovered using (34).

This approach was used to code the DCT coefficients of the USC 256 GIRL image, using four-dimensional and eight-dimensional quantizers. The original image was composed of a 256 × 256 matrix of pixels, eight bits being used to represent the gray level of each pixel. The image was partitioned into 16 × 16 blocks, which resulted in 16 rows, or stripes, of 16 blocks each. The DCT was applied to each block separately resulting in 256 coefficients for each block. The Wintz–Kurtenbach algorithm [22] was used to decide which coefficients should be discarded and what the bit allocation for the remaining coefficients should be. Because the coefficients were allocated with a differing number of bits and hence required quantizers with a differing number of levels, the input vectors to the quantizer consisted of the same coefficients from different blocks. Initially, the coefficients were taken from neighboring blocks. Because the image is composed of locally stationary areas along with areas of high fluctuation, some areas were quantized with greater clarity and others had a blurred appearance. To eliminate this difficulty, the following technique was used. For the four-dimensional quantizer, coefficients were selected from every fourth block. Hence the first vector contained the DCT coefficients from the first, fifth, ninth, and thirteenth blocks in the stripe. The second vector contained the DCT coefficients from the second, sixth, tenth, and fourteenth blocks, and so on. This method gave a much more even quality to the picture and thus was much more perceptually pleasing. In a similar manner for the eight dimensional quantizer, a coefficient from every other block was picked, resulting in similar improvements. The quantizer for the DCT coefficient was designed assuming a zero mean unit variance Gaussian distribution. The quantizers for the other coefficients were designed assuming a zero-mean unit variance Laplacian input. The quantizer inputs were normalized by subtracting the mean and dividing by the standard deviation. At the output of the quantizer the normalization procedure was reversed.

Fig. 2. USC girl (original).

Fig. 3. Scalar quantizer.

The results are presented in Figs. 2–6. Fig. 2 shows the original image coded at eight bits per pixel. Fig. 3 is the image coded at a half bit per pixel using a scalar quantizer. The distortion is obvious. Fig. 4 is the image coded at a half bit per pixel using a four-dimensional quantizer. The improvement over Fig. 3 is noticeable. The most striking improvement is apparent in Fig. 5, which was coded at a half bit per pixel with an eight-dimensional quantizer. Notice that this is very close to Fig. 6, which is the ideal quantization case. The ideal quantization case was ob-

Fig. 4. Four-dimensional quantizer.

Fig. 5. Eight-dimensional quantizer.

Fig. 6. Ideal quantizer.

tained in the following manner. The coefficients assigned zero bits by the bit allocation procedure were discarded. The coefficients that had been assigned a nonzero number of bits were left unquantized. This was done to see how much of the perceived distortion was due to the absence of certain coefficient values and how much was due to quantization error. This is the best that can be hoped for from any quantization scheme, without reallocating bits among coefficients.

These pictures show very clearly the advantage of vector quantization in reducing distortion. The price that has been paid is in processing time. While coding with the four-dimensional quantizer took only twice as long as coding with the scalar quantizer, coding with the eight-dimensional quantizer took 30 times as long. This shows that the coding problem for larger blocklengths for general vector quantizers is far from solved, and until a satisfactory resolution of this problem is obtained, the cost in processing time must be weighed against the improvement in the distortion level.

VII. Conclusions

A general design algorithm has been presented that can be used to generate all of the root lattices and their duals presented by Conway and Sloane. As an example, the A_n^* lattice was investigated for encoding Laplacian or gamma-distributed sources at entropy rates of 1 and 2 bits/sample. A four-dimensional vector quantizer designed using our algorithm performs extremely well, achieving a signal-to-noise ratio within 0.18 dB of the rate distortion bound for 1 bit/sample encoding of a gamma-distributed source and within 0.41 dB of the rate distortion bound for 2 bits/sample encoding of a Laplacian source. Additionally, an application using four- and eight-dimensional vector quantizers for encoding the two-dimensional DCT coefficients of an image at 0.5 bit/pel visibly demonstrates the improvement available using vector quantization compared to the usual scalar quantization approach.

Appendix
Example of Small Code Book Design

This is an example of code book design for a short code book. For ease of understanding, a two-dimensional example is presented, and the number of points in the code book is very small. Because of this, the method's saving is not very striking. It would greatly increase if the dimensionality and/or the size of the code book were to increase.

The original set of codewords is presented in Table III. First, each codeword is assigned to a particular shell and axis-partition. (In two dimensions the axis-partitions are simply the quadrants.) For this example the radius of the innermost circle r_1 was taken to be one and the width of the annular regions Δr was taken to be 0.5.

The separation of codewords into shells is easily accomplished by finding the magnitude of each codeword and then assigning it to shell S_n by using the following rule:

$$X_i \in S_n \text{ iff } r_1 + (n-1)\Delta r \leq |X_i| < r_1 + (n\Delta r), \quad n = 2, 3, \cdots$$

Using this rule, the code book shown in Table III can be partitioned into shells in the following manner:

$$S_1 = \{(0.5, 0.866), (1.0, 0.0), (-0.5, 0.866), (-1.0, 0.0), (-0.5, -0.866), (0.5, -0.866)\}$$

$$S_2 = \{(0.0, 1.732), (-1.5, 0.866), (-1.5, -0.866), (0.0, -1.732), (1.5, -0.866), (1.5, 0.866)\}$$

$$S_3 = \{(1.0, 1.732), (-1.0, 1.732), (-1.0, -1.732), (1.0, -1.732), (2.0, 0.0), (-2.0, 0.0)\}.$$

Now each of the sets S_i can be divided into subsets according to the quadrants in which the codewords lie. After numbering the

TABLE III
ORIGINAL CODE BOOK

-.5,	.866	-1.0,	0.0
-.5,	-.866	1.5,	.866
.5,	.866	-1.5,	-.866
1.0,	1.732	1.0,	-1.732
-1.0	-1.732	-1.0,	1.732
.5,	-.866	1.5,	-.866
1.0,	0.0	-1.5,	.866
0.0,	-1.732	2.0,	0.0
0.0,	1.732	-2.0,	0.0

Quadrant 2	Quadrant 1
Quadrant 3	Quadrant 4

Fig. 7. Quadrant numbering.

quadrants as in Fig. 7, each set S_n is divided into four subsets: S_{n1}, S_{n2}, S_{n3}, and S_{n4}. Because the quadrants are assumed to include their axes, any codeword lying on an axis would be included in two subsets. For example a point $(a,0)$, $a>0$, which is in shell S_n would be an element of the set S_{n1} and the set S_{n4}. Using these rules, the subsets would be

$$\begin{aligned}
S_{01} &= S_{02} = S_{03} = S_{04} = (0.0,0.0)\\
S_{11} &= \{(0.5,0.866),(1.0,0.0)\}\\
S_{12} &= \{(-0.5,0.866),(-1,0,0.0)\}\\
S_{13} &= \{(-1.0,0.0),(-0.5,-0.866)\}\\
S_{14} &= \{(1.0,0.0),(0.5,-0.866)\}\\
S_{21} &= \{(1.5,0.866),(0.0,1.732)\}\\
S_{22} &= \{(-1.5,0.866),(0.0,1.732)\}\\
S_{23} &= \{(-1.5,-0.866),(0.0,-1.732)\}\\
S_{24} &= \{(0.0,-1.732),(1.5,-0.866)\}\\
S_{31} &= \{(1.0,1.732),(2.0,0.0)\}\\
S_{32} &= \{(-1.0,1.732),(-2.0,0.0)\}\\
S_{33} &= \{(-1.0,-1.732),(-2.0,0.0)\}\\
S_{34} &= \{(1.0,-1.732),(2.0,0.0)\}
\end{aligned}$$

The rearranged code book can now be tabulated in a two-dimensional array as in Table IV.

What remains is to construct a look-up table, so that after a received input vector is assigned to a particular shell and quadrant, the quantizer knows where to look for the associated codewords. The look-up table is constructed in the form of a matrix $L=[l_{ij}]$. Each row of the matrix is associated with a particular shell, and two columns are assigned to each quadrant. Therefore, to each subset S_{nm} there are two corresponding locations in the matrix l_{n2m-1}, l_{n2m}. In the first location the position of the first codeword in the code book belonging to the set S_{nm} is listed, and in the second location the number of codewords in S_{nm} is listed. For example, if the received vector is in S_{32}, then $l_{33}=19$ and $l_{34}=2$. This way, instead of looking through the entire code book, the quantizing routine can go to location 19 of the code book and compare the received vector with the two codewords starting from codeword number 19. The look-up table for this example is shown in Table V.

TABLE IV
REARRANGED CODE BOOK

1	0.5	0.866
2	1.0	0.0
3	-0.5	0.866
4	-1.0	0.0
5	-1.0	0.0
6	-0.5	-0.866
7	1.0	0.0
8	0.5	-0.866
9	1.5	0.866
10	0.0	1.732
11	-1.5	0.866
12	0.0	1.732
13	-1.5	-0.866
14	0.0	-1.732
15	0.0	-1.732
16	1.5	-0.866
17	2.0	0.0
18	1.0	1.732
19	-2.0	0.0
20	-1.0	1.732
21	-2.0	0.0
22	-1.0	-1.732
23	2.0	0.0
24	1.0	-1.732

To see how this algorithm functions, suppose that an input point falls in the third quadrant of the third shell. The routine looks at the elements (3,5) and (3,6) of the look-up table and finds that there are two points in this region and that these are the twenty-first and twenty-second elements in Table IV. The routine then calculates the distance from these points and stores the value of the shortest distance and the corresponding code point. It then looks at the neighboring quadrants (found by using the quadrant look-up table shown in Table II) and finds two points in each of them, in the nineteenth, twentieth, twenty-third, and twenty-fourth locations in Table IV. It calculates the distance from these points and retains the location of the point having the least distance. The routine then moves to the neighboring shell (shell 2) and repeats the procedure with the second, third, and fourth quadrants in that shell, each time keeping the location of the point with the least distance from the input point. Because there are two points each in these axis-partitions in the second shell, a total of twelve comparisons is made. While the savings in the number of comparisons is somewhat modest in this example, they improve as the dimensionality of the quantizer and the size of the code book increase.

TABLE V
CODE BOOK LOOK-UP TABLE

	Quadrant 1		Quadrant 2		Quadrant 3		Quadrant 4	
	location	# of words	location	# of words	location	# of words	location	# of words
Shell 1	1	2	3	2	5	2	7	2
Shell 2	9	2	11	2	13	2	15	2
Shell 3	17	2	19	2	21	2	23	2

REFERENCES

[1] S. P. Lloyd, "Least squares quantization in PCM," *IEEE Trans. Inform. Theory*, vol. IT-28, pp. 129–137, Mar. 1982.

[2] Y. Linde, A. Buzo, and R. M. Gray, "An algorithm for vector quantizer design," *IEEE Trans. Commun.*, vol. COM-28, pp. 84–95, Jan. 1980.

[3] R. M. Gray, J. C. Kieffer, and Y. Linde, "Locally optimal block quantizer design," *Inform. Contr.*, vol. 45, pp. 178–198, 1980.

[4] A. Buzo, A. H. Gray, Jr., R. M. Gray, and J. D. Markel, "Speech coding based upon vector quantization," *IEEE Trans. Acoust., Speech, Signal Process.*, vol. ASSP-28, pp. 562–574, Oct. 1980.

[5] G. Rebolledo, R. M. Gray, and J. P. Burg, "A multirate voice digitizer based upon vector quantization," *IEEE Trans. Commun.*, vol. COM-30, pp. 721–727, Apr. 1982.

[6] H. Abut, R. M. Gray, and G. Rebolledo, "Vector quantization of speech and speech-like waveforms," *IEEE Trans. Acoust., Speech, and Signal Processing*, vol. ASSP-30, pp. 423–435, Jun. 1982.

[7] J. H. Conway and N. J. A. Sloane, "Voronoi regions of lattices, second moments of polytopes, and quantization," *IEEE Trans. Inform. Theory*, vol. IT-28, pp. 211–226, Mar. 1982.

[8] J. H. Conway and N. J. A. Sloane, "Fast quantizing and decoding algorithms for lattice quantizers and codes," *IEEE Trans. Inform. Theory*, vol. IT-28, pp. 227–232, Mar. 1982.

[9] A. Gersho, "Asymptotically optimal block quantization," *IEEE Trans. Inform. Theory*, vol. IT-25, pp. 373–380, Jul. 1979.

[10] R. M. Gray, ed., "Special Issue on Quantization," *IEEE Trans. Inform. Theory*, Part I, vol. IT-28, March 1982.

[11] H. M. S. Coexter, *Regular Polytopes*, 3rd ed. New York: Dover, 1973.

[12] J. E. Humphreys, *Introduction to Lie Algebras and Representation Theory*. New York: Springer Verlag, 1972.

[13] A. F. Gameckii, "On the theory of covering an n-dimensional Euclidean space with equal spheres," *Soviet Math.*, vol. 3, pp. 1410–1414, 1962.

[14] N. J. A. Sloane, "Tables of sphere packing and spherical codes," *IEEE Trans. Inform. Theory*, vol. IT-27, pp. 327–338, May 1981.

[15] H. Cohn, *Advanced Number Theory*. New York: Dover, 1962.

[16] K. D. Rines and N. C. Gallagher, Jr., "The design of two-dimensional quantizers using prequantization," *IEEE Trans. Inform. Theory*, vol. IT-28, pp. 232–239, Mar. 1982.

[17] P. Noll and R. Zelinski, "Bonds on quantizer performance in the low bit-rate region," *IEEE Trans. Commun.*, vol. COM-26, pp. 300–304, Feb. 1978.

[18] N. Farvardin and J. W. Modestino, "Optimum quantizer performance for a class of non-Gaussian memoryless sources," in *Proc. 20th Ann. Allerton Conf. Commun., Contr., Comput.*, Monticello, IL, 1982.

[19] T. R. Fischer and R. M. Dicharry, "Vector quantizer design for gaussian, gamma, and laplacian sources," Dept. Elec. Eng., Texas A & M Univ., TCSL Res. Rep. 82-03-04, Oct. 7, 1982; also *IEEE Trans. Commun.*, COM-32, 1984.

[20] B. R. Hunt, "Digital image processing," in *Applications of Digital Signal Processing*, A. V. Oppenheim, Ed. Englewood Cliffs, NJ: Prentice-Hall, 1978, pp. 169–237.

[21] A. G. Tescher, "Transform Image Coding," in *Image Transmission Techniques*, W. K. Pratt, Ed. New York: Academic, 1979, pp. 113–155.

[22] A. J. Kurtenbach and P. A. Wintz, "Quantizing for noisy channels," *IEEE Trans. Commun. Technol.*, vol. COM-17, pp. 291–302, Apr. 1969.

Vector Quantizer Design for Memoryless Gaussian, Gamma, and Laplacian Sources

THOMAS R. FISCHER AND ROY M. DICHARRY

Abstract—**Locally optimum vector quantizer (VQ) designs are presented for memoryless Gaussian, gamma, and Laplacian sources. For Gaussian sources, low (2–6) dimensional vector quantization provides relatively little improvement in mean-squared error (MSE) compared to the minimum mean-squared error (MMSE) scalar quantizer. For Laplacian or gamma sources, however, significant improvement in MSE is available with vector quantization. The Laplacian and gamma 6 bit, six-dimensional vector quantizers achieve, respectively, improvements of 2 and 4.5 dB over the corresponding scalar MMSE quantizer distortions.**

I. Introduction

Ever since Shannon [1] first demonstrated the superiority of block source encoding methods, there has been considerable interest in finding good source codes or vector quantizers. Uniform vector quantizers have been examined and several useful asymptotic properties developed (e.g., [2], [3]). The optimum uniform quantizer partitions have been found for two- and three-dimensional quantizers [4], [5], and several classes of rapidly encodable and decodable quantizers developed based on the properties of lattices [6], [7]. A significant advance in the design of nonuniform quantizers was provided by the iterative algorithm of Linde, Buzo, and Gray [8], and denoted here as the LBG algorithm. This approach has been used to calculate low bit-rate quantizers for vector Gaussian sources [8] and for real speech and autoregressive Laplacian sources [9]. This iterative design approach suffers from some drawbacks, most notably the large computational burden of the design algorithm [10] and the large number of locally optimum vector quantizers [11].

For the 1 bit/dimension vector quantization of a memoryless Gaussian source described in [8], relatively small decrease in mean-squared error (MSE) was achieved (for two-to-six-dimensional quantizers), compared to the Lloyd-Max [12], [13] scalar quantization result. This is not especially surprising since the optimum scalar (1 bit) quantizer has a corresponding MSE that is only 50 percent larger than the rate-distortion bound. Other sources, such as Laplacian and gamma, have (1 bit) optimum quantizer MSE's of more than double the corresponding rate-distortion bound [14]. The results presented here for the Laplacian and gamma 1 bit/dimension vector quantizers show significant improvement over the scalar quantizer MSE for relatively low (2–6) dimensions.

The main purpose of this correspondence is to provide selected vector quantizer designs and the corresponding mean-squared errors for vector Gaussian, gamma, and Laplacian sources. No entropy coding of the quantizer outputs is considered; however, entropy considerations have been shown to reduce bit rate requirements in the scalar case [15], and similar improvements may be expected for entropy coding in the vector case.

Paper approved by the Editor for Signal Processing and Communication Electronics of the IEEE Communications Society for publication after presentation in part at the International Conference on Communications, Boston, MA, June 1983. Manuscript received October 29, 1982; revised August 4, 1983. This work was supported in part by the National Science Foundation under Grant ECS 8106699.

T. R. Fischer is with the Department of Electrical Engineering, Texas A & M University, College Station, TX 77843.

R. M. Dicharry is with Teledyne Geotech, Garland, TX 75041.

II. The Quantization Problem

The basic quantization problem considered in the sequel is the following. Let x be an L-dimensional random vector with independent and identically distributed vector components. Let Q denote the quantizer and $y = Q(x)$ the quantizer output with y_i, $i = 1, \cdots, n$, the (n possible) quantizer output vectors. The quantizer Q is to be selected to minimize the mean-squared error per dimension,

$$E\left\{\frac{1}{L}\|x - y\|^2\right\} = E\left\{\frac{1}{L}(x - y)^T(x - y)\right\}. \tag{1}$$

Necessary conditions for an optimum vector quantizer may be easily derived [8]. In summary, associated with each quantizer output y_i is a (nearest neighbor) partition cell q_i, $i = 1, \cdots, n$, satisfying

$$q_i = \{x: \|x - y_i\|^2 < \|x - y_j\|^2, \forall j \neq i\}. \tag{2}$$

For each partition cell q_i, y_i is the conditional mean,

$$y_i = E[x \mid x \in q_i]. \tag{3}$$

Conditions (2) and (3) are also sufficient for optimality in the scalar case, for log-convex source pdf's [16]. General sufficiency conditions for optimality in the vector case are not known, and there may exist many locally optimum quantizers that satisfy (2) and (3).

The necessary conditions (2) and (3) provide the basis of the iterative LBG quantizer design algorithm. The details of the algorithm are presented in [8] and will not be repeated here. The basic idea of the algorithm is quite simple and can be summarized as follows. Since analytical or numerical computation of (3) is not feasible, a random number generator is used to produce a training sequence $\{x_i\}_{i=1}^N$, representing the distribution of random vector x (or, for a physical source of interest such as speech, an appropriate training sequence is selected from available data). Given an initial set of vector outputs $\{y_j\}_{j=1}^n$, (2) can be applied to the training sequence and the various x assigned to cells q_j of a partition. Equation (3) can then be solved for a new set of quantizer output vectors. These output vectors are in general different from the initial values and can be used again in (2) with the training sequence to reassign each x_i to a new partition. Iterating between (2) and (3) provides a nonincreasing distortion, and the algorithm eventually converges to a locally optimum design.

The LBG algorithm has the overwhelming advantage of being the *only* general (locally optimum) vector quantizer design algorithm. There are, however, two difficulties with the approach. First, the resulting design is only locally optimum, and there is no way (other than searching) to determine if better designs exist. Second, a large amount of computer time may be required to obtain a single design. For reasonable representation of the source random vector, a large training sequence is required. Obviously the computational burden increases with vector dimension (for a fixed number of bits per dimension). The 32-output vector, five-dimensional Laplacian quantizer design presented in the next section used a training sequence of length 320 000 vectors and required in excess of

Reprinted from *IEEE Trans. Commun.*, vol. COM-32, pp. 1065–1069, Sept. 1984.

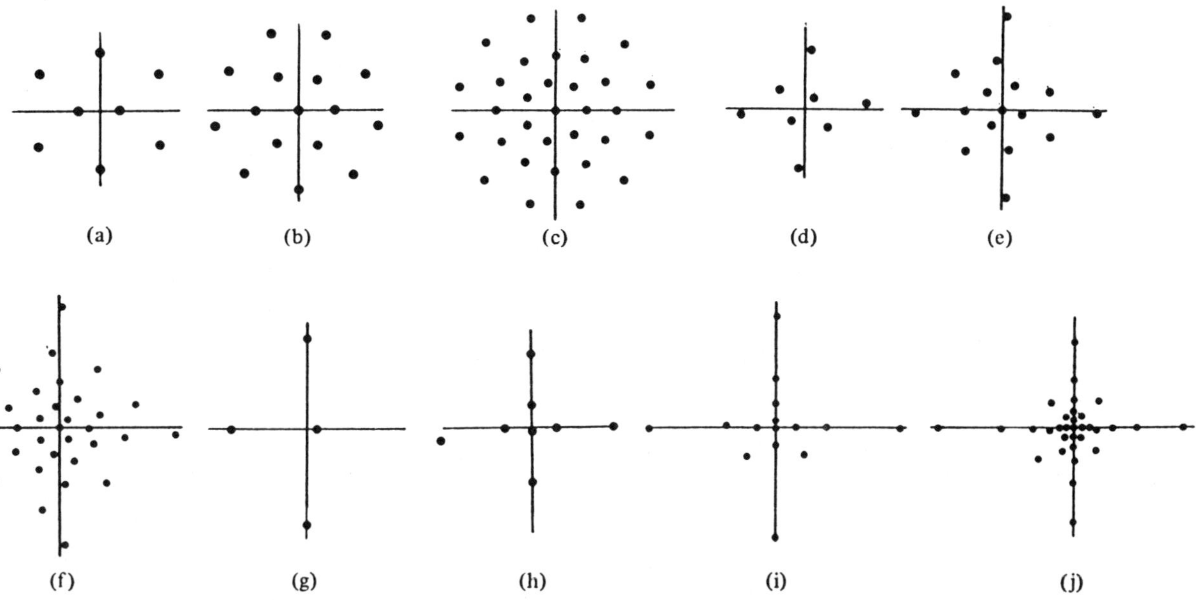

Fig. 1. Vector quantizer designs. (a) (*GS*, 2, 8), MSE = 0.203. (b) (*GS*, 2, 16), MSE = 0.108. (c) (*GS*, 2, 32), MSE = 0.0569. (d) (*L*, 2, 8), MSE = 0.235. (e) (*L*, 2, 16), MSE = 0.132. (f) (*L*, 2, 32), MSE = 0.0717. (g) (*GM*, 2, 4), MSE = 0.474. (h) (*GM*, 2, 8), MSE = 0.236. (i) (*GM*, 2, 16), MSE = 0.123. (j) (*GM*, 2, 32), MSE = 0.0627.

25 CPU hours on a VAX 11/780 computer (without floating point accelerator).

III. Vector Quantizer Designs and Performance

The LBG algorithm was used to generate most of the VQ designs presented. The algorithm in turn depends on an adequate training sequence representation of the source pdf. As suggested in [8], a training sequence of 10 000 vectors per output vector was used for most of the designs and a convergence threshold of $\epsilon = 0.001$ was chosen. All MSE values presented are based on the training sequence used to design the quantizer. Mean-squared error results for data outside the training sequence are typically slightly larger.

In order to simplify the description of the various vector quantizer designs, the following notation is adopted:

(distribution, dimension, number of output vectors).

For example, (*GS*, 3, 16) would correspond to a Gaussian quantizer in three dimensions with 16 output vectors. The other abbreviations used will be *L* for Laplacian and *GM* for gamma. All designs are based on a distribution with zero mean and unit variance in each dimension.

A. Gaussian Source

The LBG algorithm was used to design vector quantizers (VQ) for a unit variance Gaussian source. The performance of 1 bit/dimension VQ's for one to six dimensions was reported in [8]. These designs were repeated and distortions very similar to those in [8] were achieved.

Two-dimensional quantizer designs were also generated for 8, 16, and 32 quantizer output vectors. These designs are shown in Fig. 1. Because of the circular symmetry of the Gaussian pdf, an arbitrary rotation of each of these designs is also locally optimum. The vector quantizer MSE's are summarized in Table I, along with the corresponding values of MSE for the rectangular quantizer and Wilson's polar quantizer [17].

TABLE I
MEAN-SQUARED ERROR FOR LOCALLY OPTIMUM, RECTANGULAR, AND POLAR TWO-DIMENSIONAL GAUSSIAN QUANTIZERS

No. of Output Values	MSE
8 - (GS,2,8)	.2027
8 - Rectangular (1 bit x 2 bit)	.2405
8 - Wilson	.204
16 - (GS,2,16)	.1078
16 - Rectangular (2 bit x 2 bit)	.118
16 - Wilson	.109
32 - (GS,2,32)	.0569
32 - Rectangular (3 bit x 2 bit)	.0760
32 - Wilson	.0584

B. Laplacian Source

A locally optimum four-output vector, two-dimensional Laplacian quantizer can be derived analytically. The Laplacian pdf is given by

$$p(x_1, x_2) = \frac{\lambda^2}{4} e^{-\lambda(|x_1|+|x_2|)} \tag{4}$$

where $\lambda = \sqrt{2}$ for unit variance in each dimension. The four-output vector, two-dimensional rectangular quantizer (1 bit scalar quantization in each dimension) achieves an MSE of 0.50 per dimension, and has output points at $y_1 = (\alpha, \alpha)$, $y_2 = (\alpha, -\alpha)$, $y_3 = (-\alpha, \alpha)$, and $y_4 = (-\alpha, -\alpha)$, where $\alpha = \sqrt{\frac{1}{2}}$. These output points lie on a circle of radius 1 in the $x_1 - x_2$ plane. However, it is clear that these points are at exactly the minimum of $p(x_1, x_2)$ for all points on the unit circle. Hence, a better quantizer is obtained by a 45° rotation of the output points and corresponding partition so that $p(x_1, x_2)$ is maximum, subject to $x_1^2 + x_2^2 = 1$, at each output vector (however, this rotated quantizer is not locally optimum).

This intuitive argument leads to the consideration of the quantizer partition shown in Fig. 2, with parameters m and Δ. For such a quantizer the output points lie at $(\pm y_1^*, 0)$ and

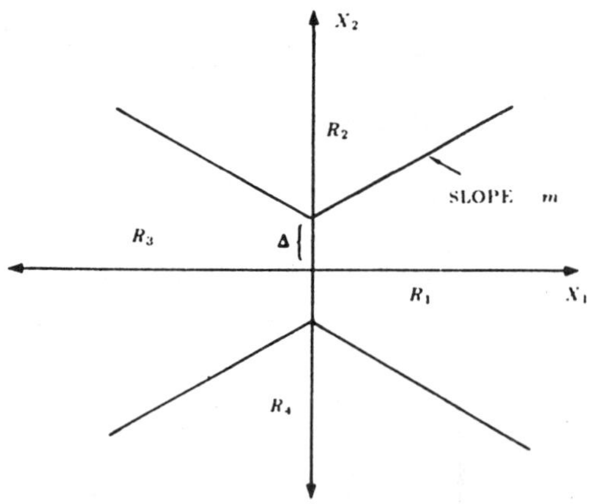

Fig. 2. Parameters m and Δ, and quantization regions for the four-output vector, two-dimensional Laplacian quantizer.

$(0, \pm y_2^*)$, where y_1^* and y_2^* may be calculated to be

$$y_1^* = E[x_1 \mid (x_1, x_2) \in R_1] = \frac{(m+1)^2 - e^{-\lambda\Delta}}{\lambda(m+1)(m+1-e^{-\lambda\Delta})} \tag{5}$$

and

$$y_2^* = E[x_2 \mid (x_1, x_2) \in R_2] = \Delta + \frac{2m+1}{\lambda(m+1)} \tag{6}$$

where R_1 and R_2 are the quantization regions shown in Fig. 2. The probability of (x_1, x_2) falling in either of the four quantization regions (R_1, R_2, R_3, R_4) may also be calculated as

$$P(R_1) = P(R_3) = \frac{1}{2}\left[1 - \frac{e^{-\lambda\Delta}}{m+1}\right] \tag{7}$$

and

$$P(R_2) = P(R_4) = \frac{e^{-\lambda\Delta}}{2(m+1)}. \tag{8}$$

The MSE of any locally optimum vector quantizer is given by

$$\text{MSE} = \text{trace}(\text{cov}[\mathbf{x}] - \text{cov}[\mathbf{y}])$$

and thus, in the present case, the MSE per dimension for the design of (5) and (6) is

$$\text{MSE} = 1 - \frac{1}{2}\left[\frac{(m+1)^2 - e^{-\lambda\Delta}}{(m+1)(m+1-e^{-\lambda\Delta})}\right]^2 \cdot \left[1 - \frac{e^{-\lambda\Delta}}{m+1}\right] + \left[\Delta + \frac{2m+1}{\lambda(m+1)}\right]^2 \frac{e^{-\lambda\Delta}}{2(m+1)}. \tag{9}$$

Equation (9) may be minimized with respect to m and Δ, and the resulting locally optimum vector quantizer parameters are

$$\Delta^* = 0.5667$$
$$m^* = 0.4948$$
$$y_1^* = 0.8258$$
$$y_2^* = 1.4577$$
$$\text{MSE}^* = 0.4303. \tag{10}$$

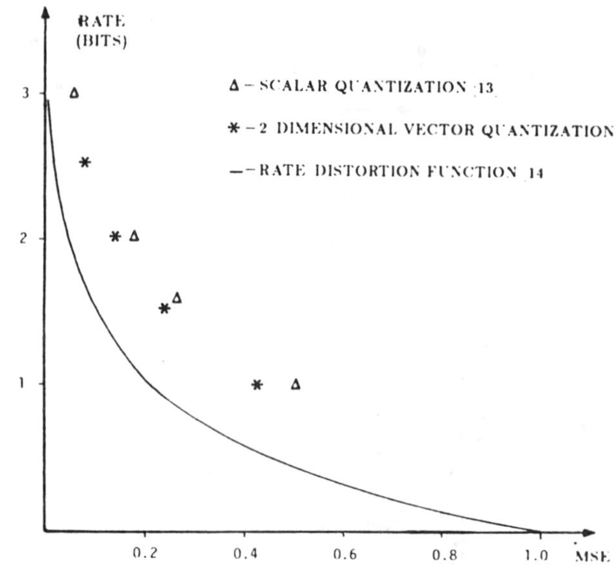

Fig. 3. Rate versus MSE for a Laplacian source.

This four-output vector, two-dimensional Laplacian quantizer provides a substantial improvement over the 1 bit/dimension rectangular (scalar) quantizer. Clearly a $\pm 90°$ rotation of the design in (10) provides an equivalent vector quantizer.

The development of (5)–(8) is easily generalized to the case of an independent, but not identically distributed Laplacian source with pdf

$$p(x_1, x_2) = \frac{\lambda_1\lambda_2}{4} e^{-(\lambda_1|x_1| + \lambda_2|x_2|)}.$$

In this case for Δ, m as in Fig. 2 and $\lambda_1 \leqslant \lambda_2$, (5)–(8) become

$$y_1^* = \frac{(\lambda_1 + m\lambda_2)^2 - \lambda_1^2 e^{-\lambda_2\Delta}}{\lambda_1(\lambda_1 + m\lambda_2)(\lambda_1(1 - e^{-\lambda_2\Delta}) + m\lambda_2)}$$

$$y_2^* = \Delta + \frac{2m\lambda_2 + \lambda_1}{\lambda_2(m\lambda_2 + \lambda_1)}$$

$$P(R_1) = P(R_3) = \frac{\lambda_1(1 - e^{-\lambda_2\Delta}) + m\lambda_2}{2(\lambda_1 + m\lambda_2)}$$

$$P(R_2) = P(R_4) = \frac{\lambda_1 e^{-\lambda_2\Delta}}{2(\lambda_1 + m\lambda_2)}$$

and the m and Δ that minimize the MSE can be solved numerically for specified parameters λ_1 and λ_2.

The LBG algorithm was used to design two-dimensional vector quantizers with 4, 8, 16, and 32 output vectors, and 1 bit/dimension quantizers for two to six dimensions. In all but the six-dimensional, 6 bit quantizer, a training sequence with 10 000 vectors per quantizer output vector was used. In the 6 bit, six-dimensional case, a smaller training sequence of only 1000 vectors per quantizer output was used to reduce the computational burden.

The four-vector, two-dimensional design obtained from the LBG algorithm was very similar to (10). The 8-, 16-, and 32-output vector quantizers are shown in Fig. 1 and the corresponding values of MSE are shown in Fig. 3 along with the

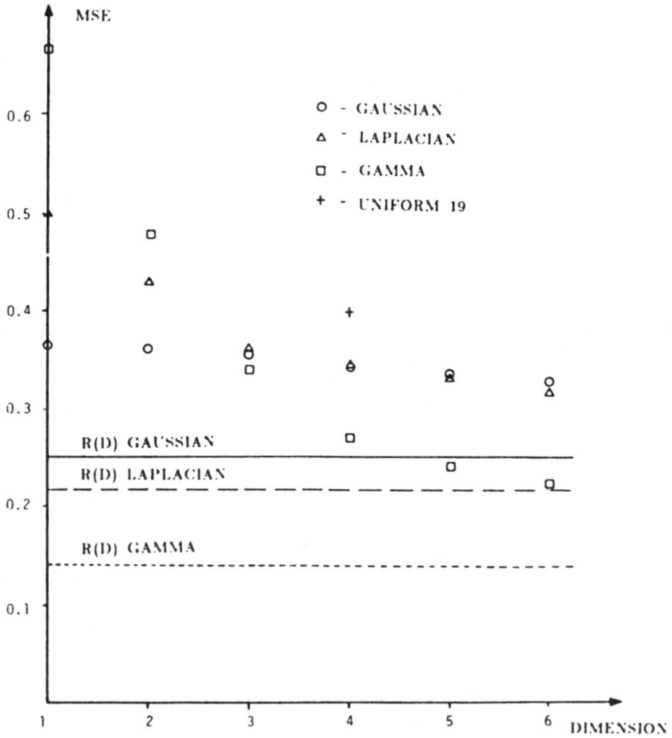

Fig. 4. MSE versus dimension for 1 bit/dimension vector quantizers.

scalar quantizer distortion and the rate-distortion bound. As is easily seen, two-dimensional vector quantization of the Laplacian source offers significant decrease in MSE compared to scalar quantization, even for the four-output vector quantizer. This is in marked contrast with the Gaussian case.

The performance of the 1 bit/dimension vector Laplacian quantizer is shown in Fig. 4. Initially the distortion decreases rapidly as the quantizer dimension is increased. However, for five or six dimensions, far less improvement in MSE is achieved with each increase in dimension. By five dimensions the MSE has dropped roughly two-thirds of the difference between the scalar quantizer distortion and the rate-distortion bound.

Also shown in Fig. 4 is the performance of the uniform four-dimensional quantizer of Sayood *et al.* [18], which performed quite well in an image coding application. This design, although significantly better than scalar quantization, has approximately 18 percent larger MSE than the locally optimum design.

C. Gamma Source

The LBG algorithm was used to design 1 bit/dimension vector gamma quantizers for two to six dimensions and 8-, 16-, and 32-output vector gamma quantizers for two dimensions. The four-output vector, two-dimensional design is shown in Fig. 1, and is similar to a design in [19]. This design is quite asymmetric and is initially rather surprising. However, because the gamma pdf is highly peaked at the origin, a quantizer output point close to the origin is strongly suggested. The remaining two-dimensional quantizers are also shown in Fig. 1.

The gamma VQ distortion for one to six dimensions is shown in Fig. 4. By five dimensions the MSE has been reduced by more than 75 percent of the difference between the scalar quantizer MSE and the rate-distortion bound. For 1 bit/dimension data compression, the gamma rate-distortion bound is less than that of the Laplacian, which in turn is less than the Gaussian case. Conversely, the corresponding scalar quantizer MSE's are exactly the opposite. By four dimensions the VQ distortions have adjusted to the same relative order as the rate-distortion bounds, and this continues for larger dimensions. At six dimensions, the highly peaked nature of the gamma pdf allows considerably less distortion than either of the Gaussian or Laplacian cases.

TABLE II
MEAN-SQUARED ERROR FOR GAUSSIAN, LAPLACIAN, AND GAMMA SOURCES (10 000 VECTORS PER QUANTIZER OUTPUT VECTOR†) AND MISMATCHED QUANTIZER DESIGNS

	MSE (per dimension)		
Design	Gaussian Source	Laplacian Source	Gamma Source
(GS, 2, 4)	0.361*/0.361**	0.431*/0.501**	0.631*/0.776**
(L, 2, 4)	0.374	0.422	0.565
(GM, 2, 4)	0.520	0.486	0.480
(GS, 2, 8)	0.200	0.271	0.416
(L, 2, 8)	0.231	0.235	0.321
(GM, 2, 8)	0.338	0.281	0.235
(GS, 2, 16)	0.107	0.159	0.2883
(L, 2, 16)	0.128	0.132	0.189
(GM, 2, 16)	0.187	0.160	0.127
(GS, 2, 32)	0.0571	0.0960	0.2215
(L, 2, 32)	0.0689	0.0717	0.1150
(GM, 2, 32)	0.0950	0.0843	0.0651
(GS, 3, 8)	0.355	0.434	0.627
(L, 3, 8)	0.386	0.359	0.387
(GM, 3, 8)	0.493	0.404	0.353
(GS, 4, 16)	0.342	0.391	0.470
(L, 4, 16)	0.393	0.345	0.358
(GM, 4, 16)	0.477	0.373	0.289

* - Rectangular quantizer rotated by 45°.

** - Rectangular quantizer.

† - An exact expression has not been obtained for the accuracy of the MSE values; however, numerical experience indicates that the values are accurate only to a few percent. The important feature of the data is the relative MSE for the different quantizer designs, rather than an absolute accuracy.

D. Quantizer–Source Mismatch

In many applications the source distribution may not be well matched to the quantizer design. This may occur because the source pdf is not known, or because the quantizer must be used for a variety of sources (or a nonstationary source). In the scalar case, quantizer–source mismatch was thoroughly examined by Mauersberger [20], with the basic conclusion that the Laplacian scalar quantizer is considerably more robust to pdf changes than is the Gaussian quantizer.

In order to get an indication of the effect of quantizer–source mismatch in the vector case, the MSE distortions were calculated for the 1 bit/dimension (two to four dimensions) and 8-, 16-, and 32-output vector, two-dimensional, Gaussian, Laplacian, and gamma quantizers for each of the Gaussian, Laplacian, and gamma sources. The MSE calculations are based on the respective random number generator, with 10 000 random vectors per quantizer output value. The results are presented in Table II. Certainly the cases considered are far from exhaustive, but do indicate that the Laplacian design appears to be the more robust of the three designs for the Gaussian, Laplacian, and gamma sources. Note that (as expected) the four-level, two-dimensional Gaussian quantizer design produces an MSE distortion for the Gaussian source that is invariant to quantizer rotation. However, from a robustness viewpoint, the design version with quantizer outputs along the coordinate axes is desirable, providing smaller MSE for the Laplacian and gamma sources.

A detailed examination of the robustness features of vector quantizers would be desirable. Note, however, that a locally optimum robust quantizer for the Gaussian, Laplacian and gamma sources (or for any other combination of known sources) may be easily obtained from the LBG algorithm. Using a training sequence with an equal number of vectors from the three sources in the LBG algorithm will generate the desired robust quantizer.

References

[1] C. E. Shannon, "A mathematical theory of communication," *Bell Syst. Tech. J.*, vol. 27, pp. 379–423, 623–656, 1948.

[2] J. A. Bucklew and G. L. Wise, "Multidimensional asymptotic quantization theory with rth power distortion measures," *IEEE Trans. Inform. Theory*, vol. IT-28, pp. 239–247, Mar. 1982.

[3] A. Gersho, "Asymptotically optimal block quantization," *IEEE Trans. Inform. Theory*, vol. IT-25, pp. 373–380, July 1979.

[4] D. J. Newman, "The hexagon theorem," *IEEE Trans. Inform. Theory*, vol. IT-28, pp. 137–139, Mar. 1982.

[5] E. S. Barnes and N. J. A. Sloane, "The optimal lattice quantizer in three dimensions," *SIAM J. Discrete and Algebraic Methods*, to be published.

[6] J. H. Conway and N. J. A. Sloane, "Voronoi regions of lattices, second moments of polytopes, and quantization," *IEEE Trans. Inform. Theory*, vol. IT-28, pp. 211–226, Mar. 1982.

[7] ——, "Fast quantizing and decoding algorithms for lattice quantizers and codes," *IEEE Trans. Inform. Theory*, vol. IT-28, pp. 227–232, Mar. 1982.

[8] Y. L. Linde, A. Buzo, and R. M. Gray, "An algorithm for vector quantizer design," *IEEE Trans. Commun.*, vol. COM-28, pp. 84–95, Jan. 1980.

[9] H. Abut, R. M. Gray, and G. Rebolledo, "Vector quantization of speech and speech-like waveforms," *IEEE Trans. Acoust., Speech, Signal Processing*, vol. ASSP-30, pp. 423–435, June 1982.

[10] M. R. Garey, D. S. Johnson, and H. S. Witsenhausen, "The complexity of the generalized Lloyd–Max problem," *IEEE Trans. Inform. Theory*, vol. IT-28, pp. 255–256, Mar. 1982.

[11] R. M. Gray and E. D. Karnin, "Multiple local optima in vector quantizers," *IEEE Trans. Inform. Theory*, vol. IT-28, pp. 256–261, Mar. 1982.

[12] S. P. Lloyd, "Least squared quantization in PCM," unpublished memo., Bell Lab., 1957; see *IEEE Trans. Inform. Theory*, vol. IT-28, pp. 129–137, 1982.

[13] J. Max, "Quantizing for minimum distortion," *IRE Trans. Inform. Theory*, vol. IT-6, pp. 7–12, Mar. 1960.

[14] P. Noll and R. Zelinski, "Bounds on quantizer performance in the low bit-rate region," *IEEE Trans. Commun.*, vol. COM-26, pp. 300–304, Feb. 1978.

[15] N. Farvardin and J. W. Modestino, "Optimum quantizer performance for a class of non-Gaussian memoryless sources," in *Proc. 20th Annu. Allerton Conf.*, Univ. Illinois, Urbana, 1982.

[16] A. V. Trushkin, "Sufficient conditions for uniqueness of a locally optimal quantizer for a class of convex error weighting functions," *IEEE Trans. Inform. Theory*, vol. IT-28, pp. 187–198, Mar. 1982.

[17] S. G. Wilson, "Magnitude/phase quantization of independent Gaussian variates," *IEEE Trans. Commun.*, vol. COM-28, pp. 1924–1929, Nov. 1980.

[18] H. Sayood and J. D. Gibson, "An algorithm for vector quantizer design," in *Proc. 20th Annu. Allerton Conf.*, Univ. Illinois, Urbana, Oct. 6–8, 1982.

[19] P. Noll, "Source coding with vector quantization schemes" (in German), *Frequenz*, vol. 36, pp. 187–193, 1982.

[20] W. Mauersberger, "Experimental results on the performance of mismatched quantizers," *IEEE Trans. Inform. Theory*, vol. IT-25, pp. 381–386, July 1979.

A Pyramid Vector Quantizer

THOMAS R. FISCHER, MEMBER, IEEE

Abstract—The geometric properties of a memoryless Laplacian source are presented and used to establish a source coding theorem. Motivated by this geometric structure, a pyramid vector quantizer (PVQ) is developed for arbitrary vector dimension. The PVQ is based on the cubic lattice points that lie on the surface of an L-dimensional pyramid and has simple encoding and decoding algorithms. A product code version of the PVQ is developed and generalized to apply to a variety of sources. Analytical expressions are derived for the PVQ mean square error (mse), and simulation results are presented for PVQ encoding of several memoryless sources. For large rate and dimension, PVQ encoding of memoryless Laplacian, gamma, and Gaussian sources provides mse improvements of 5.64, 8.40, and 2.39 dB, respectively, over the corresponding optimum scalar quantizer. Although suboptimum in a rate-distortion sense, because the PVQ can encode large-dimensional vectors, it offers significant reduction in mse distortion compared with the optimum Lloyd–Max scalar quantizer, and provides an attractive alternative to currently available vector quantizers.

I. Introduction

THE DATA COMPRESSION or vector quantization of a continuous-valued source has a rich and varied history, dating to the original work of Shannon [1]. An algorithm for the design of scalar quantizers based on the necessary conditions for optimality is well-known [2], [3], and the resulting Lloyd–Max quantizer is routinely used in data compression applications. Despite the relative simplicity of the optimum scalar quantizer design, the performance fails to achieve a distortion close to the rate-distortion bound, particularly for sources with memory. For certain memoryless sources, however, including entropy coding in the design process has yielded very good performing scalar quantizers [4].

In an attempt to obtain quantization performance closer to that promised by rate-distortion theory, there has recently been considerable interest in vector quantization. In this case the average statistical properties of a block or vector of data values can be used to advantage. A theoretical basis for asymptotically optimum vector quantization has been provided by Zador [5] and Gersho [6], [7], and asymptotic vector quantizer (VQ) performance bounds were developed by Yamada, Tazaki, and Gray [8]. The necessary conditions for an optimum scalar quantizer have been generalized to the vector case, and Linde, Buzo, and Gray (LBG) [9] developed an iterative vector quantizer design algorithm based on a training sequence. This LBG algorithm has been rigorously studied [10] and applied to the vector quantization of a number of sources (see, e.g., [11]–[16]). Although the LBG approach is quite general, it suffers from two basic drawbacks. First, the training sequence approach is essentially a Monte Carlo method and requires considerable computer time for the design of each vector quantizer. Second and more importantly, the design that is produced has no general structure, and the implementation of the VQ design requires considerable computation. In the worst case the implementation requires a distance calculation between the input vector and every output vector, so that the best representation vector can be determined. Such a brute force search for the nearest representation point becomes infeasible for modest rates and dimensions.

The principal alternative to quantizers based on the LBG algorithm is the lattice quantizer [7], [17]–[20]. By using a regular set of points in space, lattice quantizers offer the potential of rapid encoding and decoding algorithms [18], [19]. Sayood, Gibson, and Rost [20] have used a lattice quantizer in transform image coding, but so far lattice quantizers have been primarily oriented toward memoryless uniform sources.

The purpose of the present paper is twofold; first, the geometric properties of a Laplacian source are developed and used to prove a source coding theorem; second, motivated by the geometric approach, an instrumentable vector quantizer is developed for arbitrary vector dimension. The general approach is strongly influenced by Sakrison's [21] clever development of the geometric properties of a Gaussian source. As noted by a perceptive reviewer, the source coding theorem presented in Section III basically replaces the Gaussian source model, mean square error (mse) criterion, and representation sphere of Sakrison's treatment, with a Laplacian source model, mean absolute error (mae) criterion, and representation pyramid. That such an extension of Sakrison's approach is possible, is interesting in its own right, and a general geometric formulation for source coding is presented elsewhere [22]. Unfortunately, as in [21], the proof of the source coding theorem is based on a random coding argument, which is not instrumentable [23]. Unlike Sakrison's treatment of the Gaussian source, however, the geometric approach is then used to develop an implementable vector quantizer, termed a pyramid vector quantizer (PVQ). The PVQ uses codewords corresponding to points in the cubic (i.e., all integer) lattice, that also lie

Manuscript received November 1, 1983; revised December 20, 1984. This work was supported in part by the Air Force Office of Scientific Research, under Grant AFOSR 84-0003. A preliminary version of this paper was presented at the 21st Annual Allerton Conference on Communication, Control, and Computing, October 5–7, 1983.

The author is with the Telecommunications and Control Systems Laboratory, Department of Electrical Engineering, Texas A & M University, College Station, TX 77843, U.S.A.

IEEE Log Number 8608405.

Reprinted from *IEEE Trans. Inform. Theory,* vol. IT-32, pp. 568–583, July 1986.

on a particular pyramid. As such, the PVQ is a type of lattice quantizer, but interestingly, not one based on a uniformly distributed source.

The paper is organized as follows. Sections II and III develop the geometric properties of a memoryless Laplacian source and prove a source coding theorem, verifying the asymptotic optimality (in a rate-distortion sense) of the geometric approach. In Section IV the PVQ is developed and generalized to a product code [24] version appropriate for moderate sizes of vector dimension. Analytical expressions approximating the large-dimensional mse performance of the PVQ and product code PVQ are then developed through use of Shannon's entropy power [1]. For the product code PVQ, the optimum rate allocation is determined for the "gain" and "shape" (see [24]) code books. In Section V the PVQ is generalized to apply to a wide variety of other sources, with memoryless Gaussian and gamma sources treated in detail. Section VI provides the results of Monte Carlo simulations of PVQ and product code PVQ performance for Laplacian, Gaussian, and gamma sources, at a variety of rates and dimensions. These results are compared with the (large-dimensional) performance expressions, the rate-distortion bounds, optimum scalar quantizer performance, the Sayood, Gibson, and Rost VQ, and several LBG algorithm-based results in the literature. It is demonstrated that because the PVQ can take advantage of large block sizes, for memoryless sources the PVQ and product code PVQ offer an attractive alternative to scalar quantizers and LBG algorithm-based vector quantizers.

II. The Geometric Structure of a Laplacian Source

Let X_i be a sequence of independent and identically distributed (i.i.d.) Laplacian random variables with probability density function

$$p_X(x_i) = \frac{\lambda}{2} e^{-\lambda |x_i|}. \tag{1}$$

The sequence X_i is assembled into vectors X of length L, with a resulting density

$$f_X(x) = \prod_{i=1}^{L} p_X(x_i).$$

A contour of constant probability density is specified by the condition

$$r \triangleq \sum_{i=1}^{L} |X_i| = \|X\|_1 = \text{constant}, \tag{2}$$

where $\|X\|_1$ is the familiar l_1 norm. Let $S(L, K)$ be defined as

$$S(L, K) = \left\{ X\colon \sum_{i=1}^{L} |X_i| = K \right\}. \tag{3}$$

Geometrically, $S(L, K)$ is the surface of a hyperpyramid in L-dimensional space. The pyramid $S(L, K)$ encloses an L-dimensional volume $V(L, K)$ and has an L-dimensional surface area ($(L-1)$-dimensional volume), $A(L, K)$. This volume and surface area can be calculated to be [31, p. 620]

$$V(L, K) = \frac{2^L K^L}{\Gamma(L+1)}$$

$$A(L, K) = \frac{2^L \sqrt{L}\, K^{L-1}}{\Gamma(L)} \tag{4}$$

where $\Gamma(L) = (L-1)!$.

The scalar random variable r may be thought of as a radial parameter (in the l_1 sense) that indexes a particular contour of constant density $f_X(x)$ or, equivalently, the pyramid $S(L, r)$. The probability density function for r is easily calculated (using the moment generating function) as

$$p_r(r) = \frac{\lambda^L r^{L-1} e^{-\lambda r}}{\Gamma(L)}.$$

Defining

$$\rho = \frac{r}{L}$$

as the per dimension l_1 norm of X, it is easily verified that

$$E[r] = \frac{L}{\lambda}$$

$$\text{var}[r] = \frac{L}{\lambda^2}$$

$$E[\rho] = \frac{1}{\lambda}$$

$$\text{var}[\rho] = \frac{1}{L\lambda^2}. \tag{5}$$

With respect to the (per dimension) l_1 distance measure, it is clear that for large L, vector X becomes highly localized around the particular contour of constant density indexed by $r = L/\lambda$. That is, the per dimension l_1 distance between X and a point on pyramid $S(L, L/\lambda)$ goes to zero as L becomes arbitrarily large. More generally, for any of the norms

$$\|X\|_\nu = \left(\sum_{i=1}^{L} |X_i|^\nu \right)^{1/\nu}, \qquad \nu \geq 1, \tag{6}$$

the per dimension (norm) distance between X and the closest point on pyramid $S(L, L/\lambda)$ goes to zero as dimension L gets large. This may be argued intuitively based on the differential entropy of the source, as follows.

The vector X has entropy per degree of freedom [1],

$$h = -\frac{1}{L} \int \cdots \int f_X(x) \log f_X(x)\, dx$$

$$= \log\left(\frac{2e}{\lambda}\right).$$

Further, it was observed by Shannon [1] that

$$\frac{1}{L} \log f_X(x) \overset{\text{a.s.}}{\to} -h \text{ as } L \to \infty. \tag{7}$$

Substituting (1) and (2) into (7) yields

$$\frac{\lambda}{L} r \to 1,$$

so that the important pyramid for source coding is (from (3)) $S(L, L/\lambda)$.

For each vector X let $\hat{X}$ be a closest vector on $S(L, L/\lambda)$, in the norm sense of (6). Then, asymptotic in dimension, negligible distortion is introduced by approximating X by $\hat{X}$. This is summarized by the following theorem.

Theorem 1: For each i.i.d. Laplacian vector X, let $\hat{X} \in S(L, L/\lambda)$ be a vector closest to X (in the $\|X - \hat{X}\|_\nu$ sense). Then, for arbitrary $\epsilon > 0$, $\delta > 0$, and any fixed α satisfying $0 < \alpha < \infty$, for large enough dimension L,

$$\frac{1}{L}\|X - \hat{X}\|_\nu^\alpha < \epsilon$$

for all X except on a set of total probability δ. Thus, for each X a corresponding $\hat{X} \in S(L, L/\lambda)$ exists such that

$$\frac{1}{L}\|X - \hat{X}\|_\nu^\alpha \xrightarrow{p} 0.$$

Proof: See Appendix I.

The norm distance described in the theorem can be computed explicitly if $\nu = 1$ (and $\alpha = 1$). In this case the mae distance is computed (with $\hat{X} = LE[|X|]X/\|X\|_1$) as

$$E\left\{\frac{1}{L}\|X - \hat{X}\|_1\right\} = E\{|\rho - E(\rho)|\} = \frac{2L^L e^{-L}}{\lambda\Gamma(L+1)}$$

which, using Stirling's approximation for large L, is

$$E\left\{\frac{1}{L}\|X - \hat{X}\|_1\right\} \simeq \frac{2}{\lambda\sqrt{2\pi L}}.$$

For $\nu = 2$ the distortion measure is mse and the vector $\hat{X}$ can be computed from the well-known projection theorem. Let vector $S(X)$ have components

$$S_i = \begin{cases} 1, & \text{if } X_i > 0 \\ 0, & \text{if } X_i = 0, \\ -1, & \text{if } X_i < 0 \end{cases} \quad \text{if } \|X\|_1 \geq L/\lambda, \tag{8a}$$

or

$$S_i = \begin{cases} 1, & \text{if } X_i \geq 0 \\ -1, & \text{if } X_i < 0 \end{cases}, \quad \text{if } \|X\|_1 < L/\lambda. \tag{8b}$$

The two cases correspond to X being either on or outside (8a) or inside (8b) the pyramid, respectively. Using the projection theorem, the vector on the pyramid $S(L, L/\lambda)$ that is closest to X can be computed as

$$\hat{X} = X - \left[\left(X, \frac{S}{\|S\|_2}\right) - \frac{L\|S\|_2}{\lambda\|S\|_1}\right]\frac{S}{\|S\|_2} \tag{9}$$

where $(\cdot, \cdot)$ is the usual inner product and provided

$$S(X) = S(\hat{X}). \tag{10}$$

If application of (9) yields a vector $\hat{X}$ that does not satisfy (10) (corresponding to X being "above" a corner or edge of the pyramid), then $\hat{X} \notin S(L, L/\lambda)$ and all components of X that fail (10) should be set to zero and (8) and (9) reapplied. This condition (10) can easily be checked using a computer algorithm. Setting the indicated components of X to zero allows (9) to generate the closest point on the pyramid.

For the norm-based difference distortion measures of (6), nothing is lost (for large L) by first representing X by an appropriate point on the pyramid and then quantizing this point with an optimum vector quantizer. Selection of the optimum vector quantizer for points on the pyramid will naturally depend on the particular distortion measure. The essential concept is that only a single contour of constant probability density is important for designing the optimum source code. Further (somewhat obviously), since the probability density is constant along this important geometric surface, quantizer representation regions should tend to be distributed uniformly on the surface of the pyramid. This intuitive observation is formalized in the following section.

III. A Source Coding Theorem

For the norm-based distortion measures of (6), an asymptotically (in dimension) optimum source code or vector quantizer for a memoryless Laplacian source can be designed based solely on the pyramid $S(L, L/\lambda)$. That is, from Theorem 1 it follows that an optimum source code can be decomposed into the two steps of 1) finding a point on $S(L, L/\lambda)$ that is closest to X in the distortion sense, and 2) quantizing this point with an optimum vector quantizer. Actual construction of the vector quantizer depends, however, on the geometric structure of the hyperpyramid.

Based on Sakrison's development of an optimum source code for a memoryless Gaussian source and an mse distortion criterion [21], a source encoding procedure for an i.i.d. Laplacian vector X and the mae distortion measure is as follows.

1) Form

$$X = X \cdot \frac{E[\|X\|_1]}{\|X\|_1}.$$

2) Let a rate $R \geq 0$ be selected and choose 2^{RL} points at random on $S(L, L/\lambda)$ according to a uniform distribution. About each point place a representation region of geometric shape $S^*(L, \epsilon)$ defined by

$$S^*(L, \epsilon) = \{x\colon \|x\|_1 \leq \epsilon\}.$$

Adjust the location of each region (i.e., adjust the centroid) so that the region both contains the original point and has maximum area of intersection with pyramid $S(L, L/\lambda)$. Let the centroid of region i be denoted as y_i.

3) If X falls in the ith region, then X is encoded as y_i.

The performance of the source code generated by the construction of 1)–3) is summarized by the following theorem.

Theorem 2: Let $D(R) = E\{(1/L)\|X - y\|_1\}$ be the mae distortion resulting from the construction of 1)–3). Then, for arbitrary $\delta > 0$ and $R \geq 0$, a positive integer L_δ exists such that for $L > L_\delta$

$$D^*(R) \leq D(R) < (1 + \delta)D^*(R),$$

where $D^*(R) = 2^{-R}/\lambda$ is the distortion-rate function for an i.i.d. Laplacian source.

Proof: See Appendix II.

In the case of the mean-square error performance criterion, a construction analogous to 1)–3) (using (9) for $\hat{X}$ and selecting the representation regions as spheres instead of pyramids) can be shown to provide asymptotically (in L) optimum performance in the small distortion case. For either performance measure, however, the construction is based on a random assignment of points to $S(L, L/\lambda)$ and leads to designs which are extremely difficult to implement. An alternative design approach is to look for points on $S(L, L/\lambda)$ that have useful structural properties. Although such an approach does not necessarily lead to designs that achieve optimum (rate-distortion) performance, the implementation can be quite simple and provide performance significantly better than that of the optimum scalar quantizer. This approach is developed in the following sections.

IV. A Pyramid Vector Quantizer

A PVQ can be constructed based on a subset of the points in the cubic lattice (that is, the set of all vectors with integer components). As such, the PVQ is a type of lattice quantizer but, significantly, a lattice quantizer that is not based on a uniform source pdf.

The construction of the PVQ will proceed in three distinct steps. First, the number of cubic lattice points on the pyramid is evaluated. Second, an encoding/decoding algorithm is developed, and the resulting source code is denoted the pyramid vector quantizer. The PVQ codewords are scaled versions of the lattice points that lie on a specified pyramid. Enumeration encoding and decoding algorithms are provided for representing the pyramid codewords as binary codewords for transmission or storage. Although asymptotic in dimension it is sufficient to base an optimum VQ design on a single $(S(L, L/\lambda))$ pyramid, for moderate sizes of dimension significant distortion may be introduced in approximating X by $\hat{X}$. As a final modification to the PVQ design, concentric pyramids are used to generate a hybrid scalar-vector quantizer in product code [24] form. In this design a fraction of the total bit rate is allocated to the scalar quantization of the radial random variable r with the remaining bit rate allocated to the PVQ for a normalized pyramid. The VQ construction is carried out for the mse distortion measure but may be readily extended to any of the norm-based distortion criteria of (6).

A. Integer Points on the Pyramid

If pyramid $S(L, L/\lambda)$ has $L/\lambda = K$ where K is a positive integer, then it is possible to calculate the number of vectors with integer components that lie on $S(L, K)$. Let $N(L, K)$ be the number of such vectors, defined explicitly as

$$N(L,K) = \left\{ \begin{array}{l} \text{the number of vectors } x \text{ such that} \\ \sum_{i=1}^{L} |x_i| = K, \quad \text{and} \quad x_i \text{ an integer, for } i = 1, \cdots, L \end{array} \right\}. \tag{11}$$

For $K = 1$, only one of the x_i in (11) is nonzero, with value either 1 or -1. Hence $N(L, 1) = 2L$. Conversely, if $L = 1$, then x_i in (11) is either K or $-K$, so that $N(1, K) = 2$. Using a combinatorial argument, $N(L, K)$ is shown in Appendix III to satisfy the recursive formula

$$N(L,K) = N(L-1,K) + N(L-1,K-1) + N(L,K-1),$$

for $L \geq 2$ and $K \geq 2$. Since $N(1, K)$ and $N(L, 1)$ are known, it is simple to compute $N(L, K)$. The integer coordinate points on $S(3, 4)$ are illustrated in Fig. 1.

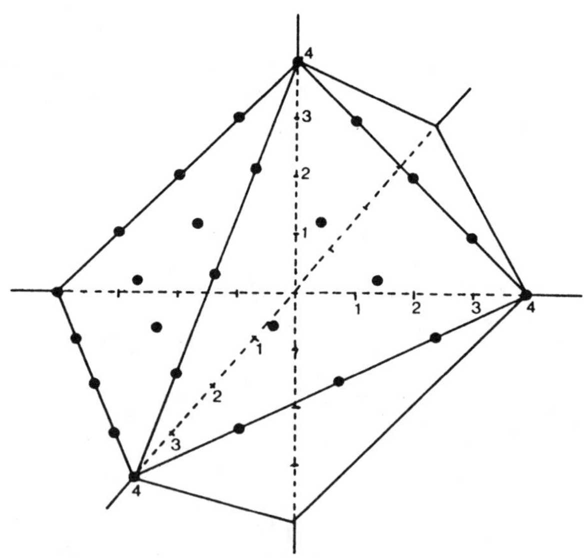

Fig. 1. Integer coordinate points on $S(3, 4)$.

B. An Encoding Algorithm

To encode L-dimensional vectors at a specified rate per dimension R, the largest value of K should be found such that

$$N(L,K) \leq 2^{RL}. \tag{12}$$

The $N(L, K)$ points of $S(L, K)$ tend to be uniformly spread over the pyramid surface and make a natural set of representation points for the source code. The basic encoding algorithm is as follows.

1) Form $\hat{X} \in S(L, L/\lambda)$ from (9).

2) Scale $\hat{X}$ by $K\lambda/L$ to obtain a corresponding point, $\tilde{X}$, on $S(L, K)$.

3) Find the closest of the $N(L, K)$ integer component vectors on $S(L, K)$ to $\tilde{X}$. This may be done componentwise as follows.

a) Round each component of $\tilde{X}$ to the nearest integer; call the resulting vector $\tilde{y}$.

b) Compute $\|\tilde{y}\|_1$. If $\|\tilde{y}\|_1 = K$, then the closest point on $S(L, K)$ has been found. If $\|\hat{y}\|_1 < K$, then increase by one in magnitude the $K - \|\tilde{y}\|_1$ components of $\tilde{y}$ that both contribute the largest distortion and were previously rounded down. If $\|\tilde{y}\|_1 > K$, then decrease by one in magnitude the $\|\tilde{y}\|_1 - K$ nonzero components of $\tilde{y}$ that contribute the most distortion and were previously rounded up. In the case of a tie for which component is to be modified, choose either one arbitrarily.

Let the resulting integer component vector be denoted $\hat{y}$.

4) The vector quantizer representation point is $y = \alpha\hat{y}/K$, where α is a scale parameter selected to minimize the distortion.

Selecting parameter α as

$$\alpha = \|\hat{X}\|_1 = \frac{L}{\lambda} \tag{13}$$

provides quantizer output points on pyramid $S(L, L/\lambda)$. The VQ representation regions should be adjusted, however, to provide the maximum volume of intersection with the pyramid surface. For the mse criterion the appropriate representation region is the sphere, while for the mae criterion the representation region is the pyramid. (The VQ encoding truncates these shapes to account for neighboring output points.) If one of the scaled $N(L, K)$ points of $S(L, K)$ lies in the middle of a face of the pyramid, then this point should be used as the representation sphere centroid. This follows because the pyramid face acts as an $(L-1)$-dimensional hyperplane slicing through the sphere, and the maximum volume of intersection occurs if the hyperplane passes through the sphere centroid. For low rates, however, K may be much less than L so that all of the $N(L, K)$ points lie along edges of the pyramid. The representation sphere centroid should then be located somewhat inside the pyramid. That is, the scaling factor α should be chosen as

$$\alpha = \gamma\frac{L}{\lambda} \tag{14}$$

where $\gamma \leq 1$ is a parameter that can be adjusted experimentally.

The source encoding algorithm is quite easy to implement, since it involves only rounding of the vector components and (if $\|\tilde{y}\|_1 \neq K$) a search over the vector components for the dimensions contributing the largest distortion. Implementation of the algorithm requires addition, multiplication, comparison, and rounding operations. The number of multiplications is a standard measure of implementation complexity, and steps 1)–4) of the algorithm show that no more than $4L$ multiplications are required. In contrast, the LBG algorithm has a design complexity that grows exponentially with the rate-dimension product (for a training sequence length that is proportional to the number of VQ output vectors), and a full-search encoding complexity that also depends exponentially on the rate-dimension product.

The performance of the basic pyramid VQ can be improved in two ways. First, the encoding algorithm of 1)–4) may be improved (for low rates) by inserting a step 1′) prior to 1).

1′) Use a threshold comparison on the components of X so that

$$X_i' = \begin{cases} X_i, & \text{if } |X_i| > \text{threshold} \\ 0, & \text{if } |X_i| \leq \text{threshold} \end{cases}, \tag{15}$$

for $i = 1, \cdots, L$. The modified X' is then used in 1)–4). Modification 1′) is justified by observing that if $|X_i|$ is small, then it will likely be rounded to zero in step 3) (for low rates and hence small K). Setting small X_i to zero in step 1′) reduces the size of the normalizing factor $\|\hat{X}\|_1$ used in step 2). Since K is substantially smaller than L for low rates, many of the components of the possible $\hat{y}$ are zero. Although the algorithm (step 3)) correctly finds the closest integer coordinate point on the pyramid, the VQ output is scaled by α, and the *scaled* point may no longer be closest to the original. The thresholding reduces this effect.

Using step 1′) requires that a new value of α be selected in step 4). For a high rate VQ, parameter α should be selected as

$$\alpha = E\left[\|\hat{X}\|_1\right] = L\left(T + \frac{1}{\lambda}\right)e^{-\lambda T}, \tag{16}$$

where $T =$ threshold used in (15). For low rates, α may be selected as

$$\alpha = \gamma L\left(T + \frac{1}{\lambda}\right)e^{-\lambda T}, \tag{17}$$

again with $\gamma \leq 1$. In practice, the threshold parameter may be adjusted experimentally for a given R and L. Obviously, however, the threshold parameter T should be no larger than the value that satisfies

$$E\left[X^2 \| X| \leq T\right] = D$$

where D is the mse to be obtained by the quantizer.

The second way in which the PVQ performance can be improved is related to the distortion introduced in approximating X by $\hat{X}$. Although for large dimension this distortion (per dimension) is negligible, for moderate sizes of dimension the distortion may be significant. Improved PVQ performance may be obtained by using concentric pyramids as the basis for the VQ output points. This approach is developed in Section IV-E. First, however, we develop a binary codeword representation for the PVQ codewords and an asymptotic approximation for the mse of the PVQ.

C. Binary Codeword Representation

For the PVQ to be useful in a digital communication or storage system, the $N(L, K)$ pyramid codewords corresponding to the lattice points on $S(L, K)$ must be uniquely represented as binary sequences of RL bits. Since the number of pyramid codewords can be counted, there exists an enumeration (encoding) procedure that assigns to each lattice point a unique integer in $\{0,\cdots, N(L, K) - 1\}$, which can be represented in the usual binary format (as a binary codeword) by RL bits. Inverting this procedure (enumeration decoding) assigns to each binary codeword the corresponding pyramid (lattice point) codeword. Obviously, if the rate-dimension product is small, then the enumeration encoding/decoding can be accomplished with a simple table lookup operation. For larger rate-dimension products the memory requirements of a table lookup become prohibitive, and efficient enumeration encoding and decoding algorithms are necessary.

A general enumeration algorithm may be based on partitioning the pyramid lattice points as follows. If $\boldsymbol{x}^T = (x_1,\cdots, x_L)$, then for $x_1 = 0$, $N(L-1, K)$ pyramid codewords are possible, corresponding to all permissible combinations of $(x_2,\cdots, x_L)$. Index these as $\{0,\cdots, N(L-1, K) - 1\}$. If $x_1 = 1$, there are $N(L-1, K-1)$ possible pyramid codewords, and index these as $\{N(L-1, K),\cdots, N(L-1, K) + N(L-1, K-1) - 1\}$. If $x_1 = -1$, there are also $N(L-1, K-1)$ possible combinations of $(x_2,\cdots, x_L)$, and so index these codewords as $\{N(L-1, K) + N(L-1, K-1),\cdots, N(L-1, K) + 2N(L-1, K-1) - 1\}$. If $x_1 = 2$, there are $N(L-1, K-2)$ possible codewords, and these are indexed as $\{N(L-1, K) + 2N(L-1, K-1),\cdots, N(L-1, K) + 2N(L-1, K-1) + N(L-1, K-2) - 1\}$. The partitioning continues until $x_1 = K$, with index $N(L, K) - 2$, and $x_1 = -K$, with the final index $N(L, K) - 1$.

Next, for each value of x_1 the respective range of indices is further partitioned based on x_2. For example, assume that $x_1 = 0$. Then for $x_2 = 0$ there are $N(L-2, K)$ pyramid codewords corresponding to all permissible values of $(x_3,\cdots, x_L)$. These codewords are indexed by $\{0,\cdots, N(L-2, K) - 1\}$. If $x_2 = 1$, then the indices are $\{N(L-2, K),\cdots, N(L-2, K) + N(L-2, K-1) - 1\}$, etc. The partitioning continues through the vector component x_L, at which point there is a one-to-one correspondence between the pyramid codewords and the integers $\{0,\cdots, N(L, K) - 1\}$. For any specified pyramid codeword the corresponding index can be determined, represented as an RL bit binary sequence, and transmitted or stored. To decode, the binary sequence is interpreted as an integer and the pyramid codeword values are extracted sequentially, beginning with x_1. The PVQ enumeration encoding and decoding procedures can be summarized by the following algorithms.

Encoding Algorithm:

0) Set index $b = 0$, $i = 1$, $k = K$, $l = L$. Define $N(l, 0) = 1$ for all $l \geq 0$, $N(0, k) = 0$ for all $k \geq 1$, and

$$\operatorname{sgn}(x) = \begin{cases} 1, & \text{if } x > 0 \\ 0, & \text{if } x = 0. \\ -1, & \text{if } x < 0 \end{cases}$$

1)

if $x_i = 0$, then $b = b + 0$,

if $|x_i| = 1$, then $b = b + N(l-1, k)$

$$+ \left[\frac{1 - \operatorname{sgn}(x_i)}{2}\right] N(l-1, k-1),$$

if $|x_i| > 1$, then $b = b + N(l-1, k)$

$$+ 2 \sum_{j=1}^{|x_i|-1} N(l-1, k-j)$$

$$+ \left[\frac{1 - \operatorname{sgn}(x_i)}{2}\right] N(l-1, k-|x_i|).$$

2) Replace

$$k \leftarrow k - |x_i|$$
$$l \leftarrow l - 1$$
$$i \leftarrow i + 1.$$

If $(k = 0)$, then stop and transmit b; otherwise, return to 1).

Decoding Algorithm:

0) Assume that the binary codeword is interpreted as an integer $b \in \{0,\cdots, N(L, K) - 1\}$ and is to be decoded as $\hat{\boldsymbol{x}}^T = (\hat{x}_1,\cdots, \hat{x}_L)$. Set $\hat{\boldsymbol{x}} = 0$; $i = 1$; $xb = 0$; $k = K$; $l = L$. Define $N(l, 0)$, $N(0, k)$ as in the encoding algorithm.

1) If $(b = xb)$, then $\hat{x}_i = 0$; go to 5).

2) If $(b < xb + N(l-1, k))$, then $\hat{x}_i = 0$; go to 4). Otherwise, $xb = xb + N(l-1, k)$; set $j = 1$.

3) If $(b < xb + 2N(l-1, k-j))$, then

$$\hat{x}_i = \begin{cases} j, & \text{if } xb \leq b < xb + N(l-1, k-j) \\ -j, & \text{if } b \geq xb + N(l-1, k-j) \end{cases}.$$

Otherwise, $xb = xb + 2N(l-1, k-j)$; $j = j + 1$; go to 3).

4) $k = k - |\hat{x}_i|$; $l = l - 1$; $i = i + 1$. If $(k > 0)$, go to 1).

5) If $(k > 0)$, then $\hat{x}_L = k - |\hat{x}_i|$. Vector $\hat{\boldsymbol{x}}$ is the pyramid codeword.

The algorithms are most simply implemented if the values for $N(l, k)$ are stored in memory for $l = 1,\cdots, L$ and $k = 1,\cdots, K$. Since K is roughly proportional to the product RL, the memory requirement is proportional to RL^2. Both the enumeration encoding and decoding algorithms require at most one loop through the algorithm for each PVQ vector component.

Example: Using the pyramid $S(4,2)$, binary encode the PVQ codeword $x = (0,1,0,-1)$. By direct computation $N(4,2) = 32$, $N(3,2) = 18$, $N(2,2) = 8$.

0) Set $b = 0$; $i = 1$; $k = 2$.
1) $x_1 = 0$, so $b = 0$.
2) $i = 2$, $k = 2$.
Repeat.
1) $|x_2| = 1$, so $b = 0 + N(2,2) + [(1 - \text{sgn}(1))/2]N(2,1) = 8$.
2) $i = 3$, $k = 1$.
Repeat.
1) $x_3 = 0$, so $b = b + 0 = 8$.
2) $i = 4$, $k = 1$.
Repeat.
1) $x_4 = -1$, so $b = 8 + N(0,1) + [(1 - \text{sgn}(-1))/2]N(0,0) = 9$.
2) $i = 5$, $k = 0$, so stop with $b = 9$.

D. Asymptotic Performance

Let $D_{\text{PVQ}}(R)$ be the normalized mse for pyramid quantization at rate R. This distortion is certainly lower bounded by the distortion in the Shannon lower bound (SLB) [23] to the rate-distortion function,

$$D_{\text{PVQ}}(R) \geq D_{\text{SLB}}(R) = \frac{2e}{\pi\lambda^2}2^{-2R}. \tag{18}$$

However, a better approximation is possible. Since the PVQ output vectors are a regular subset of points in the cubic lattice, the normalized mse per quantization cell can asymptotically (in dimension) approach a value no smaller than 1/12. (Interestingly, this is also true for several other lattices (such as A_n and D_n) as well [17], and so basing a PVQ on such lattices would provide similar asymptotic performance.) This normalized mse is precisely that obtained from optimum scalar quantization of a uniformly distributed source, or equivalently, from optimum scalar quantization of each component (i.e., rectangular vector quantization) of a vector of i.i.d. uniform random variables. The resulting mse per dimension is then

$$\text{mse}_{\text{uniform}} = \sigma_u^2 2^{-2R}. \tag{19}$$

It remains to relate σ_u^2 to the parameters of the Laplacian source, and this can be done using Shannon's notion of entropy power [1]. All vectors $\hat{X}$ to be PVQ encoded have "sample differential entropy" of exactly $\log 2e/\lambda$. That is,

$$-\frac{1}{L}\log f_X(\hat{X}) = \log\frac{2e}{\lambda} \tag{20}$$

for each $\hat{X}$. For an i.i.d. uniform source, say Y, every realization satisfies

$$-\frac{1}{L}\log f_Y(Y) = \log 2\sqrt{3}\,\sigma_u, \tag{21}$$

since the density, $f_Y(y)$, is constant. Hence equating the entropies of (20) and (21) (equivalently, measuring the volume of the pyramid with a hypercube), the equivalent variance is

$$\sigma_u^2 = \frac{e^2}{3\lambda^2} \tag{22}$$

so that, using (19), the PVQ distortion is approximated as

$$D_{\text{PVQ}}(R) \simeq \frac{e^2}{3\lambda^2}2^{-2R} = \frac{e^2}{3}(E[|X|])^2 2^{-2R} \tag{23}$$

for large enough dimension. Implicit in the development of (23) are the assumptions that the PVQ lattice points are distributed roughly uniformly on the pyramid and that the distribution of $\hat{X}$ is roughly uniform on the pyramid surface.

For large rates the PVQ performance can be compared with the (Shannon lower bound to the) rate-distortion function in another way. To PVQ encode a memoryless source and achieve an mse distortion D requires, from (23), a rate

$$R_{\text{PVQ}}(D) = \frac{1}{2}\log_2\frac{e^2}{3\lambda^2 D}.$$

Comparing this with the rate R_{SLB} required in the Shannon lower bound to the rate-distortion function (18) indicates that the PVQ requires an increased rate of

$$R_{\text{PVQ}} - R_{\text{SLB}} = \frac{1}{2}\log_2\frac{e\pi}{6} \sim 0.255 \quad \text{bits}.$$

This is precisely the rate increase required for optimum scalar quantization with entropy coding [4], [28]. Significantly, however, the PVQ does not require the variable length codewords typically used in entropy coding which can cause synchronization problems with noisy channels. Further, the entropy coding approach can only be improved by using vector, instead of scalar, quantization. The PVQ performance can be improved by using a better lattice. For example, if the PVQ codewords were taken from the Leech lattice, then replacing the cubic lattice value for mse per quantization cell of 1/12 with the known [17] Leech lattice value of $0.06577\cdots$ (i.e., multiply (23) by $12 \times 0.06577\cdots$) yields an expected PVQ distortion of

$$D_{\text{LeechPVQ}} \sim \frac{0.263 \times e^2}{\lambda^2}2^{-2R}.$$

This would reduce the required rate to only 0.084 bits above the rate-distortion bound.

If the thresholding described in the previous section is included in the encoding algorithm, then the pyramid $S(L, L/\lambda)$ is modified to $S(L, L[T + 1/\lambda]e^{-\lambda T})$. This follows since, if X' is the source vector after thresholding, namely,

$$X_i' = \begin{cases} X_i, & \text{if } |X_i| > T \\ 0, & \text{if } |X_i| \leq T, \end{cases}$$

then $E[\|X'\|_1] = L[T + 1/\lambda]e^{-\lambda T}$. Parameter σ_u^2 must then be modified, so that by replacing $1/\lambda$ with $(T +$

$1/\lambda)e^{-\lambda T}$, (22) becomes

$$\sigma_u^2 = \frac{e^2}{3}\left(T + \frac{1}{\lambda}\right)^2 e^{-2T\lambda}. \tag{24}$$

The thresholding operation introduces the additional distortion (per dimension)

$$E\left[X^2 \,|\, |X| \le T\right] \cdot P_r(|X| \le T),$$

so that the overall mse distortion for the PVQ with thresholding is approximated as

$$D_{\text{PVQ}T}(R) \simeq \frac{e^2}{3}\left(T + \frac{1}{\lambda}\right)^2 e^{-2\lambda T} 2^{-2R} + \frac{2}{\lambda^2} - \left(T^2 + \frac{2T}{\lambda} + \frac{2}{\lambda^2}\right)e^{-\lambda T}. \tag{25}$$

The threshold value T can be selected to minimize (25), and this minimizing value is easily computed from the necessary condition

$$\left(1 + \frac{1}{\lambda T}\right)e^{-\lambda T} = \frac{3}{2e^2}2^{2R}. \tag{26}$$

Equations (25) and (26) must be approached with some caution. Since the thresholded vector X' has components that can assume a value $x' = 0$ with positive probability $1 - e^{-\lambda T}$, the entropy power argument breaks down. For the optimizing T in (26), $D_{\text{PVQ}T}(R)$ can actually drop below the Shannon lower bound for very small rates ($R < 1$). For larger rates, however, (25) provides a useful approximation to the reduction in mse that can be achieved through thresholding. The thresholding operation is particularly useful at rate $R = 1$, and this is borne out by the Monte Carlo simulations presented in Section VI.

E. A Product Code PVQ

For moderate sizes of dimension, significant distortion may be introduced in approximating X by $\hat{X} \in S(L, L/\lambda)$. Equivalently, the relative variance of r in (2) may be appreciable. A general VQ design can be based on concentric pyramids, with the number and location of output vectors on each pyramid selected to minimize the average distortion. To simplify design complexity, a product code [24] PVQ is designed with identical relative orientation of output vectors on each pyramid, and the pyramids are indexed by quantized versions of r. That is, a single pyramid VQ is designed for normalized vector $\hat{X}$ (with $2^{R_p L}$ output points), and a scalar quantizer designed for r (with 2^{R_r} output levels). The (PVQ) average rate per dimension R_p and the scalar quantizer rate R_r must be selected to minimize the overall distortion but are constrained to satisfy

$$R_p L + R_r = RL. \tag{27}$$

If $\hat{Y} = \text{PVQ}(\hat{X})$ is the pyramid quantizer output and $\hat{r} = Q(r)$ the scalar quantizer output, then the product code output is $Y = \hat{r}\hat{Y}$.

Since the PVQ was specified in the previous section, it remains to design an optimum quantizer for r. The mse encountered in representing X by $\hat{X} \in S(L, L/\lambda)$ is approximated (using (8) and (9)) as

$$\frac{1}{L}\|X - \hat{X}\|_2^2 \simeq \frac{1}{L}\left\|\left[\left(X, \frac{S}{\|S\|_2}\right) - \frac{L\|S\|_2}{\lambda\|S\|_1}\right]\frac{S}{\|S\|_2}\right\|_2^2. \tag{28}$$

The approximation in (28) is exact if $\hat{X}$ is the orthogonal projection of X onto the appropriate face of the pyramid. Assuming that the set of vectors X for which (28) is not exact contribute negligible additional distortion, then disregarding this unimportant set, and noting that from (8) $\|S\|_1 = L$ and $\|S\|_2 = \sqrt{L}$, the expected value of (28) becomes

$$E\left\{\frac{1}{L}\|X - X\|_2^2\right\} \simeq E\left\{\frac{1}{L}\left|\frac{\sum_{i=1}^{L}|X_i|}{\sqrt{L}} - \frac{\sqrt{L}}{\lambda}\right|^2\right\} = \frac{1}{L}\operatorname{var}[|X_i|]. \tag{29}$$

From the central limit theorem, the random variable inside the absolute value signs in (29) converges in distribution to a Gaussian random variable. Provided the approximation in (29) is valid, it then follows that for large dimension it is appropriate to scalar quantize r with the Max [3] (Gaussian) quantizer. Numerical experience has shown that the Max quantizer is appropriate for small dimension as well.

Rates R_p and R_r can be selected in an optimum way by approximating the PVQ performance by (23) or (25) and by approximating the scalar quantizer distortion by the Huang and Schultheiss [25] model for the Max quantizer. Explicitly, the Max quantizer distortion is modeled as

$$D_{\text{Max}}(R_r) = \sigma^2 K(R_r)2^{-2R_r}, \tag{30}$$

with $K(R_r) \ge 1$ and $K(R_r)$ selected to fit the Max mse data [3].

Using (3), (25), and (30), the overall product code PVQ (PCPVQ) distortion is

$$D_{\text{PCPVQ}}(R) \doteq D_{\text{PVQ}T}(R_p) + \frac{1}{L\lambda^2}K(R_r)2^{-2R_r}, \tag{31}$$

and rates R_p and R_r must be selected to minimize (31) subject to (27). This optimum rate allocation is easily computed as

$$R_p = \frac{L}{L+1}R - \frac{1}{2(L+1)}G(R_p),$$
$$R_r = \frac{L}{L+1}\left(R + \frac{1}{2}G(R_p)\right) \tag{32}$$

where

$$G(R_p) = \log_2\left\{\frac{3e^{2\lambda T}}{e^2(1+T\lambda)^2} \cdot \left[\frac{K'(L(R-R_p))}{2L\ln 2} + K(L(R-R_p))\right]\right\}$$

and $K'(L(R - R_p)) = dK(L(R - R_p))/dR_p$.

V. Other Sources

The basic PVQ approach may be extended to other memoryless sources (or stationary sources with memory, but treated as memoryless) by modifying parameter λ according to

$$\frac{1}{\lambda} = E[|X_i|]. \tag{33}$$

That is, for an arbitrary source (with finite variance) that satisfies the ergodic property

$$\frac{\|X\|_1}{L} \to E[|X_i|],$$

the pyramid $S(L, L/\lambda)$, with $1/\lambda$ as in (33), is appropriate for source coding. Further, provided that the approximation in (28) is valid, (29) demonstrates that asymptotic in dimension there is negligible distortion introduced in approximating $\boldsymbol{X}$ by the vector $\hat{\boldsymbol{X}} \in S(L, L/\lambda)$ that is closest to $\boldsymbol{X}$. Significantly, however, the distribution of $\hat{\boldsymbol{X}}$ on $S(L, L/\lambda)$ is generally nonuniform, so that finding a good codeword assignment for $\hat{\boldsymbol{X}}$ may be extremely difficult. Despite this difficulty, if (29) is valid, then it still follows that for a wide variety of memoryless sources any product code based on the pyramid structure should treat $\|\boldsymbol{X}\|_1/\sqrt{L}$ as a Gaussian random variable (for large L), and consequently, Max quantization is still appropriate for radial parameter $\boldsymbol{r}$.

Regardless of the validity of (28) and (29), for a class of distortion criteria (including the mse distortion measure), simply scaling $\boldsymbol{X}$ to a point on the pyramid (rather than finding the closest point) is a sufficient first step in constructing a PVQ source code. This is established by the following theorem.

Theorem 3: Let X_n be a stationary ergodic source satisfying $E[|X_n|^\nu] < \infty$ for some fixed value of ν, $1 \le \nu < \infty$, and

$$\lim_{L\to\infty} \frac{1}{L}\sum_{i=1}^{L}|X_i|^\beta = E\left[|X_1|^\beta\right]$$

with probability 1 for $\beta = 1, \nu$. For any α satisfying $0 < \alpha \le \nu$, if $\hat{\boldsymbol{X}} = LE[|X|]\boldsymbol{X}/\|\boldsymbol{X}\|_1$, then

$$\frac{1}{L}\|\boldsymbol{X} - \hat{\boldsymbol{X}}\|_\nu^\alpha \to 0 \quad \text{as} \quad L \to \infty.$$

Proof: Let $\hat{\boldsymbol{X}} = \boldsymbol{X}LE[|X|]/\|\boldsymbol{X}\|_1$. Then

$$\begin{aligned}
&\frac{1}{L}\|\boldsymbol{X} - \hat{\boldsymbol{X}}\|_\nu^\alpha \\
&= \frac{1}{L}\left\|\frac{\|\boldsymbol{X}\|_1 \cdot \hat{\boldsymbol{X}}}{LE[|X|]} - \hat{\boldsymbol{X}}\right\|_\nu^\alpha \\
&= \frac{1}{E[|X|]^\alpha}\left|\frac{\|\boldsymbol{X}\|_1}{L} - E[|X|]\right|^\alpha \cdot \frac{1}{L}\left(\sum_{i=1}^{L}|\hat{X}_i|^\nu\right)^{\alpha/\nu} \\
&\le \frac{1}{E[|X|]^\alpha}\left|\frac{\|\boldsymbol{X}\|_1}{L} - E[|X|]\right|^\alpha \left(\frac{1}{L}\sum_{i=1}^{L}|\hat{X}_i|^\nu\right)^{\alpha/\nu}.
\end{aligned}$$

The latter term converges to $E[|X|^\nu]^{\alpha/\nu} < \infty$, the first term is clearly bounded, and by the ergodic assumption the middle term converges to zero as L increases.

The development of Section IV is now extended to several sources of interest. The product code PVQ may then be easily determined by minimizing the equivalent of (31).

A. Gaussian

For the Gaussian source with variance σ^2,

$$\frac{1}{\lambda} = E[|X|] = \sigma\sqrt{\frac{2}{\pi}}$$

or, with thresholding,

$$\begin{aligned}
\frac{1}{\lambda} &= E\left[|X| \mid |X| > T\right] \cdot P_r[|X| > T] \\
&= \sigma\sqrt{\frac{2}{\pi}}\exp\left\{\frac{-T^2}{2\sigma^2}\right\}.
\end{aligned}$$

The PVQ distortion of (22) is then extended to

$$\begin{aligned}
D_{\mathrm{PVQ}T}(R) &\simeq \frac{2e^2\sigma^2}{3\pi}e^{-T^2/\sigma^2}2^{-2R} \\
&\quad + \sigma^2 P_r(|X| \le T) - \sqrt{\frac{2}{\pi}}\,\sigma T e^{-T^2/2\sigma^2},
\end{aligned} \tag{34}$$

and the minimizing threshold value satisfies

$$\sqrt{\frac{2}{\pi}}\,\frac{2e^2}{3}\exp\left\{\frac{-T^2}{2\sigma^2}\right\}2^{-2R} - \frac{T}{\sigma} = 0. \tag{35}$$

As before, if $T > 0$, then (34) is not very accurate for very small rates but is reasonably accurate for $R \ge 1$.

An alternative approach to VQ design, particularly for a Gaussian source, would be to design a spherical VQ (SVQ) with output points corresponding to the lattice points that also lie on a sphere in L-dimensional space. Sloane's sphere packing results [27] would be particularly useful for such a design. If the cubic lattice (or the A_n or D_n lattices) were used for the design, then by replicating the analysis of (18)–(23), for large dimension the expected distortion of the SVQ would be

$$D_{\mathrm{SVQ}}(R) \simeq \frac{2\pi e}{12}\sigma^2 2^{-2R}. \tag{36}$$

Comparing (36) to (34) with $T = 0$, the spherical approach is superior to the pyramid approach by a factor of $\pi^2/4e$ (about 0.42 dB). Note, however, that (36) will remain the same for SVQ encoding of *any* other source (of the same variance). In particular, for the Laplacian source, comparing (36) to (23), the PVQ approach is superior by a factor of e/π (about 0.63 dB).

B. Uniform

Let X be uniformly distributed on $(-\Delta/2, \Delta/2)$, with variance $\sigma^2 = \Delta^2/12$. Then $E[|X|] = \Delta/4$, and (23) becomes

$$D_{PVQ}(R) = \frac{e^2}{4}\sigma^2 2^{-2R}. \tag{37}$$

With thresholding, analogous to (25) the distortion is

$$D_{PVQT}(R) = \frac{e^2}{e}\left[\frac{\sqrt{e}\,\sigma}{2} - \frac{T^2}{2\sqrt{3}\,\sigma}\right]^2 2^{-2R} + \frac{T^3}{3\sqrt{3}\,\sigma} \tag{38}$$

where the minimizing value of T satisfies

$$T = \frac{e^2}{3}\left[\sqrt{3}\,\sigma - \frac{T^2}{\sqrt{3}\,\sigma}\right] 2^{-2R}.$$

From (37) the PVQ obviously provides mse distortion considerably larger than that of the optimum scalar quantizer and so is an inappropriate choice for encoding a uniform source. The difficulty is that the distribution of $\hat{X}$ is decidedly nonuniform on the pyramid. Comparing (37) with (36), the SVQ approach is superior (but also considerably worse than optimum scalar quantization) by a factor of $3e/2\pi$ (about 1.1 dB).

C. Gamma

The memoryless gamma source has density [28]

$$p_X(x) = \frac{3^{\frac{1}{4}}}{\sqrt{8\pi\sigma|x|}} \exp\left[-\frac{\sqrt{3}\,|x|}{2\sigma}\right]$$

and mean absolute value

$$\frac{1}{\lambda} = E[|X|] = \frac{\sigma}{\sqrt{3}}.$$

Hence even without the thresholding operation, (23) becomes

$$D_{PVQ}(R) = \frac{e^2}{9}\sigma^2 2^{-2R}, \tag{39}$$

which is significantly better than the distortion of the optimum scalar quantizer. Analogous to (25) and (26), the optimum threshold parameter can be selected numerically.

As with the Gaussian source, comparing the PVQ performance with that of an SVQ (compare (39) with (36)), the PVQ is superior by a factor of $2e/3\pi$ (about 2.4 dB).

D. Generalized Gaussian

The generalized Gaussian density is [26]

$$p_X(x) = C_1 e^{-C_2|x|^\nu}, \qquad \nu > 0$$

where

$$C_1 = \nu\eta(\sigma,\nu)/2\Gamma(1/\nu)$$

$$C_2 = [\eta(\sigma,\nu)]^\nu$$

$$\eta(\sigma,\nu) = \frac{1}{\sigma}\left[\frac{\Gamma(3/\nu)}{\Gamma(1/\nu)}\right]^{1/2}$$

$$\mathrm{var}(X) = \sigma^2.$$

The entropy is easily computed to be

$$h = C_2 E[|X|^\nu] - \ln C_1,$$

and the mean absolute value is

$$E[|X|] = \frac{2C_1\Gamma(2/\nu)}{\nu C_2^{2/\nu}}.$$

For large rate and dimension the expected mse distortion for PVQ encoding is then

$$D_{PVQ}(R,\nu) = \frac{e^2}{3}\left[\frac{2C_1\Gamma(2/\nu)}{\nu C_2^{2/\nu}}\right]^2 2^{-2R}$$

or, in terms of rate,

$$R_{PVQ}(D,\nu) = \frac{1}{2}\log_2 \frac{e^2}{3D}\left[\frac{2C_1\Gamma(2/\nu)}{\nu C_2^{2/\nu}}\right]^2. \tag{40}$$

The Shannon lower bound for the memoryless generalized Gaussian source is given by

$$R_{SLB}(D,\nu) = h - \tfrac{1}{2}\log 2\pi e D. \tag{41}$$

Comparing R_{SLB} to R_{PVQ} indicates that for a given (small) distortion the PVQ requires a rate $R_{PVQ} = R_{SLB} + \Delta R_{PVQ}$ larger than the minimum rate of the Shannon lower bound. Specifically,

$$\Delta R_{PVQ} = \frac{1}{2}\log_2 \frac{2\pi e^3 c_1^2}{3e^{2C_2\rho}}\left(\frac{2C_1\Gamma(2/\nu)}{\nu C_2^{2/\nu}}\right)^2 \tag{42}$$

where $\rho = E[|X|^\nu]$ and is plotted in Fig. 2 as a function of ν. For $\nu = 1$ the generalized Gaussian density coincides with the Laplacian density and $\Delta R_{PVQ} = 0.255$ bits, as was shown earlier. For $\nu \neq 1$ the PVQ requires (for the same distortion) a slightly larger rate than does scalar quantization with entropy coding but does not require variable length codewords.

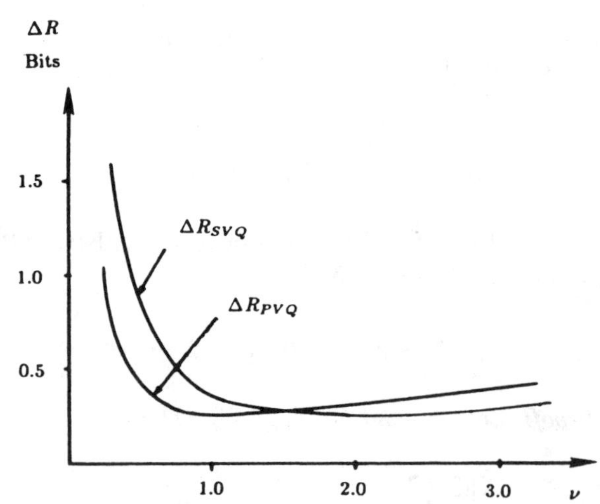

Fig. 2. ΔR_{PVQ}, ΔR_{SVQ} versus ν for generalized Gaussian source.

Equations (40)–(42) can be repeated for the (uniform lattice) SVQ, and the resulting ΔR_{SVQ} is also shown in Fig. 2. For large values of parameter ν the value of ΔR_{SVQ} is smaller than ΔR_{PVQ}, but by no more than 0.188 bits (corresponding to $\nu \to \infty$ and the uniform density). For small ν the PVQ is superior. Interestingly, in transform coding both speech [32] and image [33] sources have transform coefficients that are well-modeled by Laplacian or gamma densities. Fig. 2 then implies that the PVQ is perhaps a better choice than the SVQ for the transform coding of such sources.

VI. Performance

The performance of the PVQ was evaluated by Monte Carlo simulation for memoryless Laplacian, gamma, and Gaussian sources. The average mse was computed for a block of 1000 random (i.i.d., zero-mean, and unit variance component) vectors generated by a random number generator and then averaged over 100 blocks of such vectors. The mse performance is summarized in Tables I–III and Figs. 3 and 4.

The Monte Carlo simulations verified the analytical expressions approximating the PVQ performance that were obtained in Sections IV and V. However, the PVQ encoding of the gamma source was better than anticipated (from (39)) for the low rates simulated. For the gamma source, the nonuniformity of the distribution of $\hat{X}$ on the pyramid surface is apparently well matched to the PVQ encoding procedure. As evidenced by Fig. 4, the PVQ performance is quite close to the rate distortion bound and significantly better than the mse of the optimum scalar quantizer. In the high rate case, comparing (39) to the known optimum scalar quantizer [28] indicates that the PVQ provides an improvement of 8.40 dB.

For the Laplacian source, the Monte Carlo simulations yielded mse values very close to those expected from the asymptotic performance expression (23). At a rate of 1 bit/dimension, the thresholding is very important in reducing the mse. However, for larger rates (3 bits/dimension or above), thresholding offers negligible improvement in performance. For high rates, comparing (23) to the optimum scalar quantizer [28], the PVQ provides a 5.64-dB improvement in mse performance.

TABLE I
MONTE CARLO SIMULATIONS OF THE PYRAMID VECTOR QUANTIZER MSE PERFORMANCE FOR A ZERO-MEAN UNIT-VARIANCE MEMORYLESS LAPLACIAN SOURCE ENCODED AT AN AVERAGE RATE OF R BITS / DIMENSION[a]

	Mean Square Error		
	$R = 1$ ($T = 0.75$, $\gamma = 0.73$)	$R = 2$ ($T = 0.4$, $\gamma = 0.94$)	$R = 3$ ($T = 0.17$, $\gamma = 0.98$)
PVQ ($L = 16$)	0.310 ($K = 4$, $N = 0$)	0.0907 ($K = 10$, $N = 3$)	0.0233 ($K = 22$, $N = 4$)
PVQ ($L = 32$)	0.295 ($K = 7$, $N = 2$)	0.0778 ($K = 21$, $N = 3$)	0.0210 ($K = 45$, $N = 4$)
PVQ ($L = 48$)	0.283 ($K = 11$, $N = 0$)	0.0773 ($K = 31$, $N = 3$)	0.0202 ($K = 68$, $N = 4$)
PVQ ($L = 64$)	0.281 ($K = 14$, $N = 2$)	0.0773 ($K = 41$, $N = 3$)	0.0198 ($K = 91$, $N = 4$)
Distortion-rate bound [30]	0.2178	0.0542	0.0136
Optimum scalar quantizer [29]	0.5	0.1765	0.0514
$D_{PVQ}(R)$	0.3079	0.07697	0.01924
$D_{PVQT}(R)$	0.2464	0.0750	0.01921

[a]For each product code PVQ, the parameters are L, K, N, T, and γ, where L is the dimension, K the pyramid index in (11), N the number of scalar (Max) quantizer bits for radial parameter r, T the threshold in (15), and γ the scale factor in (14).

TABLE II
MONTE CARLO SIMULATIONS OF THE PYRAMID VECTOR QUANTIZER MSE PERFORMANCE FOR A ZERO-MEAN UNIT-VARIANCE MEMORYLESS GAMMA SOURCE ENCODED AT AN AVERAGE RATE OF R BITS / DIMENSION[a]

	Mean Square Error		
	$R = 1$ ($T = 0.45$, $\gamma = 0.82$)	$R = 2$ ($T = 0.2$, $\gamma = 0.92$)	$R = 3$ ($T = 0$, $\gamma = 0.98$)
PVQ ($L = 16$)	0.218 ($K = 3$, $N = 3$)	0.0503 ($K = 10$, $N = 3$)	0.0130 ($K = 22$, $N = 4$)
PVQ ($L = 32$)	0.170 ($K = 7$, $N = 2$)	0.0397 ($K = 21$, $N = 3$)	0.0105 ($K = 45$, $N = 4$)
PVQ ($L = 48$)	0.157 ($K = 10$, $N = 3$)	0.0374 ($K = 31$, $N = 3$)	0.00988 ($K = 68$, $N = 4$)
PVQ ($L = 64$)	0.148 ($K = 14$, $N = 2$)	0.0370 ($K = 41$, $N = 3$)	0.00959 ($K = 91$, $N = 4$)
Distortion-rate bound [30]	0.140	0.030	0.00693
Optimum scalar quantizer [30]	0.665	0.233	0.0713
$D_{PVQ}(R)$	0.205	0.0513	0.0128

[a]The product code parameters L, K, N, T, and γ are as in Table I.

TABLE III
MONTE CARLO SIMULATIONS OF THE PYRAMID VECTOR QUANTIZER MSE PERFORMANCE FOR A ZERO-MEAN UNIT-VARIANCE MEMORYLESS GAUSSIAN SOURCE ENCODED AT AN AVERAGE RATE OF R BITS / DIMENSION[a]

	Mean Square Error		
	$R = 1$ ($T = 0.75$, $\gamma = 0.64$)	$R = 2$ ($T = 0.28$, $\gamma = 0.92$)	$R = 3$ ($T = 0$, $\gamma = 0.98$)
PVQ ($L = 16$)	0.358 ($L = 15$, $K = 4$, $N = 0$)	0.111 ($K = 11$, $N = 1$)	0.0257 ($K = 25$, $N = 2$)
PVQ ($L = 32$)	0.353 ($L = 24$, $K = 6$, $N = 0$)	0.107 ($K = 21$, $N = 3$)	0.0263 ($K = 47$, $N = 2$)
PVQ ($L = 48$)	0.355 ($L = 47$, $K = 11$, $N = 0$)	0.104 ($K = 32$, $N = 1$)	0.0251 ($K = 71$, $N = 2$)
PVQ ($L = 64$)	0.357 ($L = 61$, $K = 14$, $N = 0$)	0.1048 ($K = 43$, $N = 0$)	0.0250 ($K = 94$, $N = 2$)
Distortion-rate bound [1]	0.25	0.0625	0.016525
Optimum scalar quantizer [3]	0.3634	0.1175	0.03454
$D_{PVQ}(R)$	0.392	0.0980	0.0245
$D_{PVQT}(R)$	0.301	0.0960	0.0244

[a] The product code parameters L, K, N, T, and γ are as in Table I.

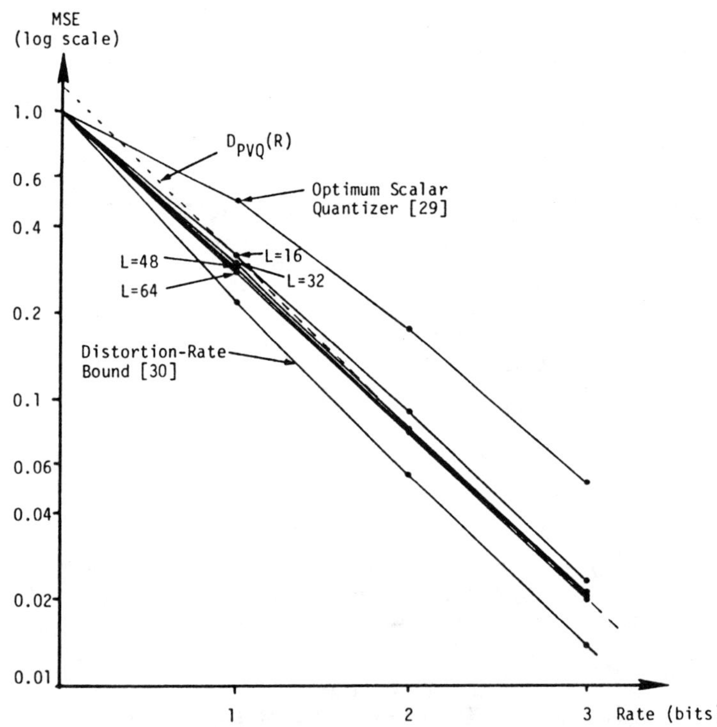

Fig. 3. Normalized mse versus rate for L-dimensional PVQ encoding of memoryless Laplacian source.

For the Gaussian source, the PVQ offers little improvement over the optimum scalar quantizer at a rate of 1 bit/sample, but more significant reduction in mse for larger rates. For large rates, the Max scalar quantizer provides an mse distortion that is a factor of 2.72 larger than the rate distortion function [25], while for large dimension the PVQ provides an mse (from (34) with $T = 0$) of $2e^2/3\pi \simeq 1.568$ times the rate-distortion function. Thus for large rates the PVQ provides an improvement of 2.39 dB over the optimum scalar quantizer.

In Fischer and Dicharry [16], LBG [9] algorithm-based VQ performance is reported for memoryless Gaussian, Laplacian, and gamma sources. The results in [16] are consistent with those of Tables I–III, but the former results were limited (because of the computational burden of the LBG method) to either a rate of 1 bit/dimension and six or fewer dimensions, or two dimensions and 5 bits or less. For Laplacian and gamma sources, the 1-bit/dimension six-dimensional VQ's of [16] provide mse distortions only slightly (less than ten percent) larger than those of the 16-dimensional PVQ. Although the PVQ is based on the simple (and asymptotically suboptimum) cubic lattice, as evidence by Figs. 3 and 4 the advantage of using large dimensions is considerable.

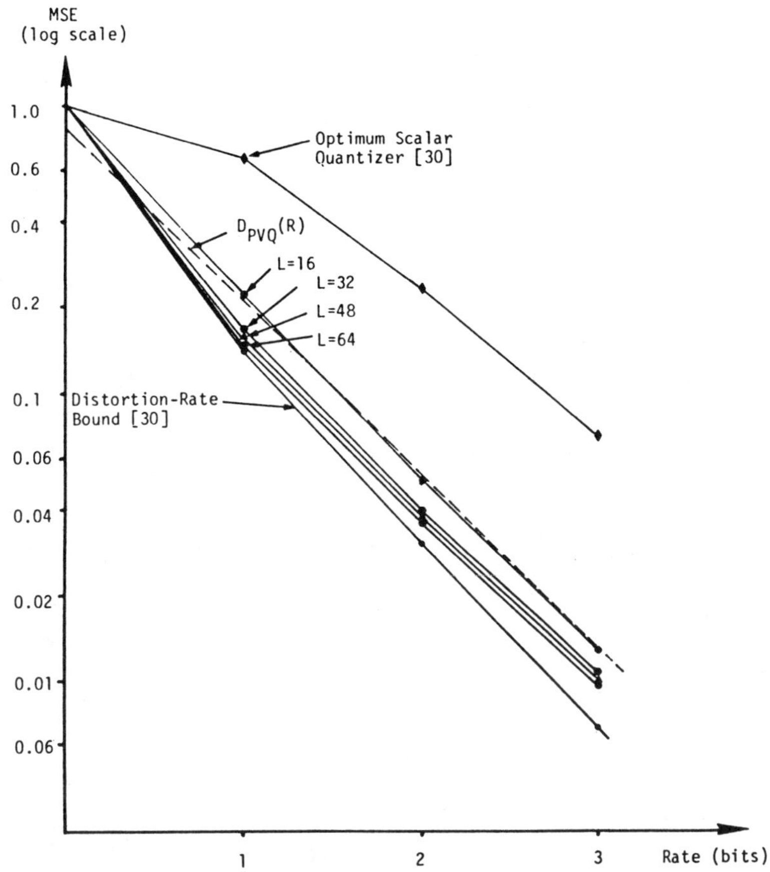

Fig. 4. Normalized mse versus rate for L-dimensional PVQ encoding of memoryless source with gamma pdf.

In Sayood, Gibson, and Rost (SGR) [20], four-dimensional A_n^* lattice quantizers were designed for rate 1 and 2 bits/dimension encoding of memoryless Laplacian and gamma sources. The PVQ mse performance in Tables I and II is uniformly better than the results in [20], due to the larger dimensions considered. For example, in [20] a 1-bit/dimension Laplacian VQ provided an SNR of 3.97 dB, while for the PVQ the SNR is 5.09 dB (for $L = 16$) or 5.51 dB (for $L = 64$). From the results in [17] it is clear that for moderate sizes of dimension (say, $L \sim 16$) the A_n^* lattice is superior to the uniform lattice, so that if the SGR quantizer could be extended to 16 dimensions the performance should be slightly better than that of the PVQ.

The SGR lattice quantizer (and many other lattice quantizers) differs from the PVQ in two important respects. First, the SGR approach uses a lattice centered at the origin and adjusts the lattice density (i.e., radially scales all the lattice points) to "match" the probability density function (pdf) of the source to be encoded. The PVQ is based on Shannon's notion of a region of high probability [1] and the sphere hardening property [34] and assigns the (lattice point) codewords to a specific surface (the pyramid) in L-dimensional space, with no codewords either inside or outside this surface. The product code PVQ uses a radial parameter (the l_1 norm) to radially scale the codewords on a nominal pyramid, and this is different (and simpler) than classifying points in a lattice as lying on concentric (pyramid) shells. Second, the SGR approach provides a general encoding algorithm based on the symmetry of the lattice, but for large code book sizes does not address either how the lattice can be effectively truncated to maintain a fixed coding rate, nor how the lattice codewords can be enumerated to form binary codewords for digital transmission. The PVQ approach solved these problems be finding an explicit formula for computing the number of lattice points on a pyramid and by developing simple enumeration encoding and decoding algorithms for representing each lattice point codeword as a binary codeword.

In Abut *et al.* [12], LBG algorithm-based vector quantization is reported for autoregressive (AR) sources with an i.i.d. Laplacian driving process. These results are not directly comparable with the present PVQ performance because of the source AR structure. For example, at a rate of 1 bit/dimension and scalar quantization, the SNR in [12, Fig. 1] is approximately 3.8 dB. However, the optimum 1-bit scalar quantizer for a memoryless Laplacian source is known [29] to provide an SNR value of only 3.0 dB. Similarly, for two dimensions, [12] has an SNR value of about 4.3 dB, while the best known two-dimensional 2-bit Laplacian VQ [16] provides an SNR of 3.67 dB. Basically, the AR model used in [12] causes the resulting source to be rather unlike the Laplacian driving process (and presumably "more Gaussian"), so that the VQ performance is better than that for the corresponding memoryless Laplacian source. Interestingly, by comparing [12] and [16] for the Laplacian source, and [11] and [9] for the Gaussian source, it appears that as the vector dimension increases, the mse provided by VQ's for sources with memory approaches the rate-distortion bound at a faster rate than does the mse provided by VQ's for the corresponding memoryless source.

Most of the LBG algorithm-based VQ's discussed in the literature [9], [11]–[16] consider a rate-dimension product of less than eight, with product code VQ's [24] extending the rate-dimension product to 12. For large rate-dimension products, the LBG method is computationally prohibitive because of the exponential growth in required VQ design computations (for a training sequence length proportional to the number of VQ output vectors) and the complexity of encoding. The encoding complexity can be reduced significantly by using tree search techniques, but the resulting VQ performance is generally somewhat degraded [15]. Particularly for 2 or 3 bits per dimension encoding (and larger rates), the PVQ offers significant advantages over the LBG algorithm VQ's in simplifying both the VQ design process and the encoding complexity.

VII. Conclusion

By extending the geometric approach of Sakrison [21], the geometric properties of i.i.d. Laplacian source were developed and used to prove a source coding theorem. Analogous to entropy coding of discrete ensembles, the geometric approach attempts to locate VQ output vectors solely in the region of L-dimensional space of high probability. For the memoryless Laplacian source this implies placing the VQ output vectors near the surface of an L-dimensional hyperpyramid. This pyramid is indexed by the differential entropy of the source, or, equivalently, by the mean absolute value of the source.

Motivated by the source coding theorem and geometric structure, a pyramid vector quantizer was developed. The PVQ is based on the cubic lattice points that also lie on the appropriate pyramid. A simple PVQ encoding algorithm was derived that is implementable for arbitrary vector dimension. The PVQ was generalized to a product code version and analytical expressions developed for both the PVQ and product code PVQ to approximate the large-dimensional mse performance. The pyramid vector quantizer's mse performance was evaluated by Monte Carlo simulation and compared with the analytical performance expressions and several VQ results in the literature. Particularly, for rates of 2 bits/dimension or larger, because the PVQ can take advantage of encoding large-dimensional vectors (dimension sizes currently beyond the computational feasibility of the LBG [9] method), the PVQ and product code PVQ offer attractive alternatives to the LBG algorithm-based vector quantizers for the memoryless sources considered.

Acknowledgment

The author would like to thank Prof. J. D. Gibson for many useful discussions, Prof. P. F. Dahm for a critical reading of the manuscript, and the anonymous reviewers for their helpful comments.

Appendix I
Proof of Theorem 1

Let $\epsilon_\alpha \triangleq \epsilon^{1/\alpha}$, so that

$$\frac{1}{L}\|X - \hat{X}\|_\nu < \epsilon_\alpha \Leftrightarrow \frac{1}{L}\|X - \hat{X}\|_\nu^\alpha < \epsilon. \tag{A1}$$

Certainly, (A1) is satisfied if $|X_i - \hat{X}_i| < \epsilon_\alpha$ for each $i = 1,\cdots, L$. For each vector X define set Υ as

$$\Upsilon = \{\hat{x}: |X_i - \hat{x}_i| < \epsilon_\alpha, \quad \text{for } i = 1,\cdots, L\}.$$

Since $p_X(x_i)$ is positive for $-\infty < x_i < \infty$, define α_i^S and α_i^I, $i = 1,\cdots, L$, as

$$\alpha_i^S = \min\left(\sup_{\{\hat{x}_i:\, |X_i - \hat{x}_i| < \epsilon_\alpha\}} \frac{p_X(\hat{x}_i)}{p_X(X_i)}, \beta \right)$$

and

$$\alpha_i^I = \max\left(\inf_{\{\hat{x}_i:\, |X_i - \hat{x}_i| < \epsilon_\alpha\}} \frac{p_X(\hat{x}_i)}{p_X(X_i)}, \frac{1}{\beta} \right)$$

for some fixed real number β satisfying $1 < \beta < \infty$. Clearly,

$$0 < \frac{1}{\beta} \le \alpha_i^I \le 1 \le \alpha_i^S \le \beta < \infty,$$

so that $E[\alpha_i^I]$, $E[\alpha_i^S]$, var $[\alpha_i^I]$, and var $[\alpha_i^S]$ are bounded.

From the continuity of $p_X(\cdot)$, it follows that as $\hat{x}_i$ varies over the set $\{\hat{x}_i: |X_i - \hat{x}_i| < \epsilon_\alpha\}$, variable $\alpha_i = p_X(\hat{x}_i)/p_X(X_i)$ can assume any value in the range (α_i^I, α_i^S). For each $i = 1,\cdots, L$, an acceptable $\hat{x}_i$ exists such that $p(\hat{x}_i)$ can assume any value in the range

$$\alpha_i^I p(X_i) < p(\hat{x}_i) < \alpha_i^S p(X_i).$$

Hence an $\hat{X} \in \Upsilon$ exists such that $(1/L)\log f_X(x)$ can assume any value in the range

$$\frac{\log f_X(X)}{L} - \frac{1}{L}\sum_{i=1}^{L} \log\left(\frac{1}{\alpha_i^I}\right) < \frac{\log f_X(\hat{X})}{L} < \frac{\log f_X(X)}{L} + \frac{1}{L}\sum_{i=1}^{L} \log(\alpha_i^S).$$

Asymptotically, both sums converge as

$$\frac{1}{L}\sum_{i=1}^{L} \log\frac{1}{\alpha_i^I} \xrightarrow{\text{a.s.}} E\left[\log\frac{1}{\alpha_i^I}\right] > 0 \tag{A2}$$

and

$$\frac{1}{L}\sum_{i=1}^{L} \log \alpha_i^S \xrightarrow{\text{a.s.}} E\left[\log \alpha_i^S\right] > 0,$$

so that there exists a positive integer, say L_1, such that if $L > L_1$, then each sum is within ϵ_1 of its limit for all X, except on a set of total probability Δ_1. Further, choose L_1 large enough so that ϵ_2, defined as

$$\epsilon_2 = \min\left(E\left[\log\frac{1}{\alpha_i^I}\right], E\left[\log \alpha_i^S\right] \right) - \epsilon_1$$

is positive. It follows that $\hat{X} \in \Upsilon$ exists such that $(1/L)\log f_X(\hat{X})$ can assume any value in

$$\left(\frac{\log f_X(X)}{L} - \epsilon_2, \frac{\log f_X(X)}{L} + \epsilon_2\right).$$

However, for any $\epsilon_2 > 0$ and $\Delta_2 > 0$, the convergence in (7) implies that a positive integer L_2 exists such that if $L > L_2$, then

$$\frac{\log f_X(X)}{L} - \epsilon_2 < -h(x) < \frac{\log f_X(X)}{L} + \epsilon_2$$

except on a set of total probability Δ_2. Hence (for $L > \max(L_1, L_2)$) for all X (except on a set of probability $\delta \le \Delta_1 + \Delta_2$), an $\hat{X} \in \Upsilon$ (so that $(1/L)\|X - \hat{X}\|_\nu < \epsilon_\alpha$) exists such that $(1/L)\log f_X(\hat{X}) = -h(x)$, implying $\hat{X} \in S(L, L/\lambda)$.

Appendix II
Proof of Theorem II

The proof parallels the argument of Sakrison [21]. Using the source code construction of 1)–3) in Section III, the mae (per dimension) may be upper-bounded as

$$E\left[\frac{1}{L}\|X - y\|_1\right] \le \frac{\epsilon}{L}(1 - P(\epsilon)) + \frac{2}{\lambda}P(\epsilon) + \frac{1}{L}E[\|X - \hat{X}\|_1]$$

where $P(\epsilon)$ is the probability that $\hat{X}$ is not within one of the representation regions. The middle term follows since $\hat{X}$ is certainly within a distance $2L/\lambda$ from some representation point. The last term has already been shown to asymptotically decrease to zero and so is dropped.

Probability $P(\epsilon)$ may be calculated as

$$P(\epsilon) = \prod_{i=1}^{2^{RL}}\left[1 - \frac{\text{area of intersection of region } i \text{ and } S(L, L/\lambda)}{\text{total area of } S(L, L/\lambda)}\right],$$

or, since the points are selected at random in 2)

$$P(\epsilon) = \left[1 - \frac{\text{area of intersection}}{\text{area of } S(L, L/\lambda)}\right]^{2^{RL}}. \tag{A3}$$

Since $S^*(L, \epsilon)$ is the same pyramid shape as $S(L, L/\lambda)$, the smallest area of maximum intersection between $S^*(L, \epsilon)$ and $S(L, L/\lambda)$ occurs if a single face of $S(L, L/\lambda)$ acts as a hyperplane slicing through the centroid of $S^*(L, \epsilon)$. This can be seen by the following argument. The pyramid $S(L, L/\lambda)$ has 2^L faces, $2L$ vertices, and at each vertex 2^{L-1} faces meet. The vertices may be covered by $2L$ representation pyramids, each centered at a point with a single nonzero coordinate of magnitude $L/\lambda - \epsilon$. Any randomly chosen point on the pyramid surface within an l_1 norm distance of 2ϵ from a vertex may then be represented with such a region, and the resulting area of intersection between the representation pyramid and $S(L, L/\lambda)$ is $A(L, \epsilon)/2$ (for $\epsilon < L/\lambda$). Any other point selected at random on $S(L, L/\lambda)$ must be at least an l_1 norm distance of 2ϵ from a vertex, and thus within 2ϵ of at most two faces. Such a point can then be covered by an appropriately placed representation pyramid that intersects with only one face of $S(L, L/\lambda)$. Since this latter area of intersection is smaller than $A(L, \epsilon)/2$, the area of intersection in (A3) is then bounded from below by the $(L-1)$-dimensional volume of the intersection of $S^*(L, \epsilon)$ and an $(L-1)$-dimensional hyperplane parallel to a face of $S^*(L, \epsilon)$ and passing through the centroid. By direct calculation for $2 \le L \le 6$, we determine a bound for this volume as

$$\text{area of intersection} \ge \frac{1}{\sqrt{L}}V(L-1, \epsilon), \tag{A4}$$

where equality occurs if $L = 2$ and the bound becomes looser as L increases. Using (4) and (A4) in (A3) yields the bound for $P(\epsilon)$

$$P(\epsilon) \le \left[1 - \frac{\dfrac{2^{L-1}\epsilon^{L-1}}{\sqrt{L}\,\Gamma(L)}}{\dfrac{2^L\sqrt{L}\left(\dfrac{L}{\lambda}\right)^{L-1}}{\Gamma(L)}}\right]^{2^{RL}}. \tag{A5}$$

The rate-distortion function for an i.i.d. Laplacian source is well-known [23] to be

$$R(D) = \log\frac{1/\lambda}{D},$$

so that

$$D = \frac{1}{\lambda}2^{-R}. \tag{A6}$$

The distortion in L dimensions ϵ may then be expressed as

$$\epsilon = \frac{L(1+\delta)}{\lambda}2^{-R} \tag{A7}$$

for some $\delta > 0$. Substituting (A7) into (A5) yields

$$P(\epsilon) \le \left[1 - \frac{(1+\delta)^{L-1}}{2L}\cdot 2^{-R(L-1)}\right]^{2^{RL}}. \tag{A8}$$

For large L, (A6) becomes approximately

$$P(\epsilon) \simeq \exp\left[-\frac{(1+\delta)^{L-1}2^R}{2L}\right],$$

so that for arbitrary $\delta > 0$ the probability that $\hat{X}$ is not within an mae distance of ϵ from some representation point y_i goes to zero as dimension L increases. Hence the encoding scheme of 1)–3) is asymptotically (in L) optimum and achieves a distortion arbitrarily close to the rate-distortion bound.

Appendix III

The number of integer coordinate vectors on pyramid $S(L, K)$ is denoted $N(L, K)$ and derived as follows. All of the $N(L, K)$ possible vectors on $S(L, K)$ are of the form

$$x = \begin{cases} i \\ x_2 \\ \vdots \\ x_L \end{cases}, \qquad \text{for } i = 0, \pm 1, \cdots, \pm K.$$

The portion of x corresponding to $(x_2, \cdots, x_L)$ has $N(L-1, K - |i|)$ possible permutations. This implies

$$N(L, K) = 2(N(L-1, K-1) + N(L-1, K-2) + \cdots + N(L-1, 1) + 1) + N(L-1, K). \tag{A9}$$

Similarly,

$$N(L, K-1) = 2(N(L-1, K-2) + \cdots + N(L-1, 1) + 1) + N(L-1, K-1). \tag{A10}$$

Combining (A9) and (A10) yields the desired result,

$$N(L, K) = N(L, K-1) + N(L-1, K-1) + N(L-1, K).$$

REFERENCES

[1] C. E. Shannon, "A mathematical theory of communication," *Bell Syst. Tech. J.*, vol. 27, pp. 379–423, 623–656, 1948.

[2] S. P. Lloyd, "Least squares quantization in PCM," unpublished memorandum, Bell Laboratories, 1957; see *IEEE Trans. Inform. Theory*, vol. IT-28, pp. 129–137, 1982.

[3] J. Max, "Quantizing for minimum distortion," *IRE Trans. Inform. Theory*, vol. IT-6, pp. 7–12, Mar. 1960.

[4] N. Farvardin and J. W. Modestino, "Optimum quantizer performance for a class of non-Gaussian memoryless sources," *IEEE Trans. Inform. Theory*, vol. IT-30, pp. 485–497, 1984.

[5] P. L. Zador, "Asymptotic quantization error of continuous signals and quantization dimension," *IEEE Trans. Inform. Theory*, vol. IT-28, pp. 139–149, Mar. 1982. (The paper first appeared as an unpublished Bell Laboratories memorandum in 1966.)

[6] A. Gersho, "Asymptotically optimal block quantization," *IEEE Trans. Inform. Theory*, vol. IT-25, pp. 373–380, July 1979.

[7] ——, "On the structure of vector quantizers," *IEEE Trans. Inform. Theory*, vol. IT-28, pp. 157–166, Mar. 1982.

[8] Y. Yamada, S. Tazaki, and R. M. Gray, "Asymptotic performance of block quantizers with difference distortion measures," *IEEE Trans. Inform. Theory*, vol. IT-26, pp. 6–14, Jan. 1980.

[9] Y. L. Linde, A. Buzo, and R. M. Gray, "An algorithm for vector quantizer design," *IEEE Trans. Commun.*, vol. COM-28, pp. 84–95, Jan. 1980.

[10] R. M. Gray, J. C. Kieffer, and Y. Linde, "Locally optimal block quantizer design," *Inform. Contr.*, vol. 45, pp. 178–198, 1980.

[11] R. M. Gray and Y. Linde, "Vector quantizers and predictive quantizers for Gauss-Markov sources," *IEEE Trans. Commun.*, vol. COM-30, pp. 381–389, Feb. 1982.

[12] H. Abut, R. M. Gray, and G. Rebolledo, "Vector quantization of speech and speech-like waveforms," *IEEE Trans. Acoust., Speech, Signal Processing*, vol. ASSP-30, pp. 423–435, June 1982.

[13] V. Cuperman and A. Gersho, "Adaptive differential vector coding of speech," in *Proc. Globecom '82*, Nov. 1982, pp. 1092–1096.

[14] R. M. Gray, "Vector quantization," *IEEE ASSP Mag.*, vol. 1, pp. 4–29, Apr. 1984.

[15] A. Gersho and V. Cuperman, "Vector quantization: A pattern-matching technique for speech coding," *IEEE Commun. Mag.*, vol. 21, pp. 15–21, Dec. 1983.

[16] T. R. Fischer and R. M. Dicharry, "Vector quantizer design for memoryless Gaussian, gamma, and Laplacian sources," *IEEE Trans. Commun.*, vol. COM-32, pp. 1065–1069, Sep. 1984.

[17] J. H. Conway and N. J. A. Sloane, "Voronoi regions of lattices, second moments of polytopes, and quantization," *IEEE Trans. Inform. Theory*, vol. IT-28, pp. 211–226, Mar. 1982.

[18] ——, "Fast quantizing and decoding algorithms for lattice quantizers and codes," *IEEE Trans. Inform. Theory*, vol. IT-28, pp. 227–232, Mar. 1982.

[19] ——, "A fast encoding method for lattice codes and quantizers," *IEEE Trans. Inform. Theory*, vol. IT-29, pp. 820–824, Nov. 1983.

[20] K. Sayood, J. D. Gibson, and M. C. Rost, "An algorithm for uniform vector quantizer design," *IEEE Trans. Inform. Theory*, vol. IT-30, pp. 805–814, Nov. 1984.

[21] D. J. Sakrison, "A geometric treatment of the source encoding of a Gaussian random variable," *IEEE Trans. Inform. Theory*, vol. IT-14, pp. 481–486, May 1968.

[22] T. R. Fischer, "Geometric source coding and vector quantization," *Abstracts of Papers*, Int. Symp. Inform. Theory, Brighton, England, June 24–28, 1985.

[23] T. Berger, *Rate-Distortion Theory*. Englewood Cliffs, NJ: Prentice-Hall, 1971.

[24] M. J. Sabin and R. M. Gray, "Product code vector quantizers for waveform and voice coding," *IEEE Trans. Acoust., Speech, Signal Processing*, vol. ASSP-32, pp. 474–488, June 1984.

[25] J. J. Y. Huang and P. M. Schultheiss, "Block quantization of correlated Gaussian random variables," *IEEE Trans. Commun. Syst.*, vol. CS-11, pp. 289–296, Sept. 1963.

[26] W. Mauersberger, "Experimental results on the performance of mismatched quantizers," *IEEE Trans. Inform. Theory*, vol. IT-25, July 1979, pp. 381–386.

[27] N. J. A. Sloane, "Tables of sphere packing and spherical codes," *IEEE Trans. Inform. Theory*, vol. IT-27, pp. 327–338, May 1981.

[28] N. S. Jayant and P. Noll, *Digital Coding of Waveforms: Principles and Applications to Speech and Video*. Englewood Cliffs, NJ: Prentice-Hall, 1984.

[29] M. D. Paez and T. H. Glisson, "Minimum mean-squared-error quantization in speech PCM and DPCM systems," *IEEE Trans. Commun.*, vol. COM-20, pp. 225–230, Apr. 1972.

[30] P. Noll and R. Zelinski, "Bounds on quantizer performance in the low bit-rate region," *IEEE Trans. Commun.*, vol. COM-26, Feb. 1978, pp. 300–304.

[31] I. S. Gradshteyn and I. M. Ryzhik, *Table of Integrals, Series, and Products*. New York: Academic, 1980.

[32] R. Zelinski and P. Noll, "Adaptive transform coding of speech signals," *IEEE Trans. Acoust. Speech, Signal Processing*, vol. ASSP-25, pp. 299–309, Aug. 1977.

[33] R. C. Reininger and J. D. Gibson, "Distributions of the two-dimensional DCT coefficients for images," *IEEE Trans. Commun.*, vol. COM-31, pp. 835–839, June 1983.

[34] J. M. Wozencraft and I. M. Jacobs, *Principles of Communication Engineering*. New York: Wiley, 1965.

Asymptotic Performance of Unrestricted Polar Quantizers

PETER F. SWASZEK, MEMBER, IEEE, AND TSU W. KU

Abstract—**Unrestricted polar quantizers (UPQ's) were previously introduced to overcome the weak mean-squared error performance of polar coordinate quantizers for the independent bivariate Gaussian source when the number of levels was small (< 3 bits / dimension). An asymptotic mean-squared error analysis of the UPQ's with a general circularly symmetric source is presented. The performance of the UPQ's is shown to be asymptotically within 0.17 dB of that of the optimum bivariate quantizer. The results are extended to dimensions greater than two. Examples for the independent Gaussian source are included.**

Introduction

Optimum vector quantizers, although known to have performance superior to that of standard scalar quantizers [1], [2], have several drawbacks. For example, the available design algorithms have very slow convergence unless the rate–dimension product is small [3], and implementation of the optimum vector quantizer in a k-dimensional space is a computationally burdensome full-search procedure. Previously, for the two-dimensional case, polar coordinate quantizers were introduced with the goal of improving the performance over that of rectangular coordinate (scalar) quantization without the design and implementation complexity of optimum bivariate quantizers [4]–[6]. Polar quantization to N levels consists of transforming the rectangular coordinate representation of the source (x, y) to the polar coordinates of magnitude r and phase ϕ through

$$r = \left(x^2 + y^2\right)^{1/2} \qquad \phi = \tan^{-1}(y/x),$$

and quantizing r and ϕ separately by scalar quantizers with N_r and N_ϕ levels, respectively ($N_r \times N_\phi = N$). With this scheme the bivariate quantization pattern is a partitioning of the plane into N_r concentric rings about the origin, with each divided angularly into N_ϕ pieces.

The mean-squared error (mse) quantization of the independent bivariate Gaussian source is considered in [4] and [5], minimizing the mse by identifying necessary conditions upon the magnitude and phase quantizers' parameters (breakpoints and output values), and specifying the factorization of N into $N_r \times N_\phi$. In this case the gain in performance over rectangular coordinate quantization was modest with the worst performance occurring for small values of N. One reason for this lack of major improvement in performance is that the polar quantizers are still scalar quantizers, although of transformed coordinates. For more substantial gain, block processing is required.

In [7] Wilson overcame the weak performance of small N polar quantizers for the bivariate Gaussian source by designing unrestricted polar quantizers (UPQ's). The UPQ's are very similar to the original polar quantizers except that the number of regions in the uniform quantizer for the phase angle varies depending upon the result of the magnitude quantization. In other words, each concentric ring in the quantization pattern is allowed to have a different number of phase divisions. Let N_i be the number of divisions in the phase quantizer when r is in the ith magnitude ring. To have a total of N regions for the quantizer the UPQ's must satisfy the constraint $\sum_{i=1}^{N_r} N_i = N$.

Note that these UPQ's are block processors, albeit simple ones, in that the phase quantization depends upon both variables. Their implementation remains simple, requiring only two scalar quantizers (one of them uniform) and a lookup table of the N_i. The bivariate Gaussian UPQ examples in [7] demonstrated vast improvement in performance over the strictly polar quantizers for small $N (N < 36)$. The larger N case was not considered, due to the design method's complexity. In particular, the optimization of the mse performance of all combinations of the N_i which sum to N ($2^N - 1$ such cases) is prohibitive for larger N.

This correspondence presents an asymptotic (large N) companding-type approach to the design and analysis of the UPQ's for a circularly symmetric source. The solution includes both the specification of the magnitude quantizer and the N_i as well as an evaluation of the resulting mse. Next, these results are extended to higher dimensions through a brief discussion of unrestricted spherical quantizers. Finally, finite N examples for the Gaussian source with exact values of their mse's are presented to validate the analytical results.

Derivation

For this analysis assume that the input is from a continuously valued circularly symmetric source with unit variance rectangular coordinate marginals and bivariate density function

$$f(x, y) = p\left(\sqrt{x^2 + y^2}\right).$$

Transforming to polar coordinates, the phase ϕ is uniformly distributed on $[0, 2\pi)$, and the magnitude is distributed on $[0, \infty)$ with density function $f(r) = 2\pi r p(r)$. Note that the magnitude and phase are independent random variables. Although this form is restrictive, the important independent bivariate Gaussian source is included.

By definition the UPQ's employ separate scalar quantizers for r and ϕ—a nonuniform quantizer on $[0, \infty)$ for r and a uniform quantizer on $[0, 2\pi)$ for ϕ. Let r_i and r_{i+1} represent the endpoints and $\hat{r}_i$ the output value of the magnitude quantizer's ith interval, $i = 1, \cdots, N_r$ ($r_{N_r+1} = \infty$). The nonuniform spacing of this magnitude quantizer will be described through a compander implementation with compressor $g(r)$; hence

$$r_i = g^{-1}[(i-1)/N_r] \qquad \hat{r}_i = g^{-1}[(2i-1)/2N_r].$$

Likewise, let $\phi_{j,i}$ and $\phi_{j+1,i}$ be the endpoints, and let $\hat{\phi}_{j,i}$ be the output value of the jth interval in the phase quantizer for the ith magnitude ring, $j = 1, \cdots, N_i$ ($\phi_{N_i+1, i} = 2\pi$). Since the number of levels in the phase quantizer is different for each magnitude ring, define the function $N_\phi(r)$, which when evaluated at the ith magnitude ring's output value represents the number of phase divisions for the magnitude ring, $N_\phi(\hat{r}_i) = N_i$. Using this function, the parameters of the uniform quantizer for the phase are

$$\phi_{j,i} = (j-1)2\pi/N_\phi(\hat{r}_i) \qquad \hat{\phi}_{j,i} = (2j-1)\pi/N_\phi(\hat{r}_i).$$

With this notation the mse/dimension D can be written as

Manuscript received June 5, 1984; revised July 3, 1985. This work was supported in part by the National Science Foundation under NEFRI Grant ECS-8206237, and by a Physio-Control Corporation Career Development Award. This work was presented in part at the IEEE International Symposium on Information Theory, Brighton, England, June 1985.

P. F. Swaszek is with the Department of Electrical Engineering, Kelley Annex, University of Rhode Island, Kingston, RI 02881.

T. W. Ku is with the Department of Electrical Engineering, University of Washington, Seattle, WA 98195.

IEEE Log Number 8406624.

Reprinted from *IEEE Trans. Inform. Theory,* vol. IT-32, pp. 330–333, March 1986.

$$D = \frac{1}{2}\sum_{i=1}^{N_r}\sum_{j=1}^{N_\phi(\hat{r}_i)}\int_{\phi_{j,i}}^{\phi_{j+1,i}}\int_{r_i}^{r_{i+1}}\left[r^2+\hat{r}_i^2-2r\hat{r}_i\cos\left(\phi-\hat{\phi}_{j,i}\right)\right] \cdot \frac{f(r)}{2\pi}\,dr\,d\phi.$$

Assuming that N is large, we simplify D through several fine quantization approximations. First, replace $f(r)$ by $f(\hat{r}_i)$ since it varies little over each interval. Also, ϕ differs from $\hat{\phi}$ only slightly on each region; hence one can replace the cosine by the first two terms of its Taylor expansion about $\hat{\phi}_{j,i}$. Applying these approximations yields

$$D \approx \frac{1}{2}\sum_{i=1}^{N_r} f(\hat{r}_i)\sum_{j=1}^{N_\phi(\hat{r}_i)}\int_{\phi_{j,i}}^{\phi_{j+1,i}}\int_{r_i}^{r_{i+1}}\left[(r-\hat{r}_i)^2+r\hat{r}_i\left(\phi-\hat{\phi}_{j,i}\right)^2\right] \cdot \frac{1}{2\pi}\,dr\,d\phi.$$

Integrating over r and ϕ, using the fact that the phase quantizer is uniform with $N_\phi(\hat{r}_i)$ levels, and simplifying yields

$$D \approx \frac{1}{2}\sum_{i=1}^{N_r} f(\hat{r}_i)\left[\frac{(r_{i+1}-\hat{r}_i)^3-(r_i-\hat{r}_i)^3}{3} + \frac{\pi^2}{3N_\phi^2(\hat{r}_i)}\left(r_{i+1}^2-r_i^2\right)\frac{\hat{r}_i}{2}\right].$$

For large N_r the differences of the magnitude quantizer's parameters can be approximated by derivatives:

$$r_{i+1}-\hat{r}_i \approx \hat{r}_i - r_i \approx \frac{1}{2N_r g'(\hat{r}_i)}.$$

The second term in brackets becomes

$$\begin{aligned}\left(r_{i+1}^2-r_i^2\right) &= (r_{i+1}-r_i)(r_{i+1}+r_i)\\ &= (r_{i+1}-\hat{r}_1-r_i+\hat{r}_i)\\ &\quad\cdot(r_{i+1}-\hat{r}_i+r_i-\hat{r}_i+2\hat{r}_i)\\ &\approx \frac{1}{N_r g'(\hat{r}_i)}2\hat{r}_i\end{aligned}$$

with the result that

$$D \approx \frac{1}{2}\sum_{i=1}^{N_r} f(\hat{r}_i)\left[\frac{1}{12N_r^2[g'(\hat{r}_i)]^2}+\frac{\pi^2\hat{r}_i^2}{3N_\phi^2(\hat{r}_i)}\right]\frac{1}{g'(\hat{r}_i)N_r}.$$

From the definition of the magnitude compressor each quantization region $[r_i, r_{i+1})$ maps through g onto an interval of width $1/N_r$ in the range of the uniform quantizer; thus

$$\begin{aligned}1/N_r &= g(r_{i+1}) - g(r_i)\\ &\approx g'(\hat{r}_i)(r_{i+1}-r_i) = g'(\hat{r}_i)\Delta_i\end{aligned}$$

where Δ_i equals the width of the ith magnitude quantization interval. Using this relationship, we replace $[g'(\hat{r}_i)N_r]^{-1}$ by Δ_i and approximate the sums by integrals ($\Delta_i \approx dr$) to yield

$$D \approx \frac{1}{24N_r^2}\int_0^\infty \frac{f(r)}{[g'(r)]^2}\,dr + \frac{\pi^2}{6}\int_0^\infty \frac{r^2 f(r)}{N_\phi^2(r)}\,dr.$$

For a fixed N quantizer the total number of levels in all of the magnitude rings must sum to N:

$$\sum_{i=1}^{N_r} N_\phi(\hat{r}_i) = N.$$

Dividing by N_r, replacing $1/N_r$ on the left side with $g'(\hat{r}_i)\Delta_i$, and approximating the sum by an integral establishes the constraint

$$\int_0^\infty N_\phi(r)g'(r)\,dr = \frac{N}{N_r}.$$

With this constraint D is optimized over the choice of $N_\phi(r)$ by a calculus of variations argument. The resulting performance is next minimized over the value of N_r by setting the derivative with respect to N_r equal to zero. Finally, the minimum mse/dimension for the UPQ's is obtained through an application of Hölder's inequality

$$D_{\text{UPQ}} \geq \frac{\pi}{6N}\left[\int_0^\infty r^{1/2}f^{1/2}(r)\,dr\right]^2 \tag{1}$$

with equality if and only if the magnitude compressor $g(r)$ is

$$g(r) = \int_0^r s^{-1/4}f^{1/4}(s)\,ds \Big/ \int_0^\infty s^{-1/4}f^{1/4}(s)\,ds. \tag{2}$$

With this optimum compressor the expressions for $N_\phi(r)$ and N_r are

$$N_\phi(r) = \left[\sqrt{2\pi}N^{1/2}r^{3/4}f^{1/4}(r)\right]\Big/\left[\int_0^\infty s^{1/2}f^{1/2}(s)\,ds\right]^{1/2} \tag{3}$$

and

$$N_r = \frac{N^{1/2}}{\sqrt{2\pi}}\left[\int_0^\infty s^{-1/4}f^{1/4}(s)\,ds\right]\Big/\left[\int_0^\infty s^{1/2}f^{1/2}(s)\,ds\right]^{1/2} \tag{4}$$

The design of the UPQ for polar coordinate density $(2\pi)^{-1}f(r)$ is specified by (2)–(4) with the mse/dimension in (1).

The optimum bivariate quantizer [2] has asymptotic performance

$$D_{\text{OPT}} \approx \frac{5}{36\sqrt{3}N}\left[\int_{-\infty}^{\infty}\int_{-\infty}^{\infty} f^{1/2}(x,y)\,dx\,dy\right]^2.$$

Changing variables to polar coordinates and integrating over ϕ yields

$$D_{\text{OPT}} \approx \frac{5\pi}{18\sqrt{3}N}\left[\int_0^\infty r^{1/2}f^{1/2}(r)\,dr\right]^2$$

which is similar to (1). Taking the ratio of the UPQ and optimum performances

$$\frac{D_{\text{UPQ}}}{D_{\text{OPT}}} = 1.039 = 0.166\text{ dB} \tag{5}$$

demonstrates that asymptotically the UPQ's are within 0.166 dB of the performance of the optimum bivariate vector quantizer for all circularly symmetric sources.

Dimension > 2

Referring for details to [8] and [9], a k-dimensional spherically symmetric source ($k > 2$) can be represented by the spherical coordinates of magnitude r, longitude ϕ_1, and $k-2$ latitudes ϕ_j, $j = 2, 3, \cdots, k-1$. The generalization of the UPQ's to higher dimension unrestricted spherical quantizers (USQ's) consists of an N_r-level nonuniform quantizer for the magnitude and $k-1$ uniform quantizers for the angles. The number of levels in the quantizer for the last angle ϕ_{k-1} is controlled by the magnitude quantization through the function $N_{k-1}(\hat{r})$. The number of levels for the ϕ_{k-2} quantizer depends upon the last two quantizations through the function $N_{k-2}(\hat{\phi}_{k-1}, \hat{r})$. Generally, working back toward the longitude ϕ_1, the quantizer for ϕ_j has its number of levels set from the previous quantizations through $N_j(\hat{\phi}_{j+1}, \hat{\phi}_{j+2}, \cdots, \hat{\phi}_{k-1}, \hat{r})$. As was true for the UPQ's these USQ's are block processors of a sequential form and should far outperform scalar quantization techniques. For implementation, k scalar quantizers and a lookup table will be required. An example of a three-dimensional Gaussian USQ appears in Table II and is discussed in the next section.

Consider a unit variance marginal spherically symmetric source with rectangular coordinate density $p(\sqrt{x^T x})$ and magnitude density function

$$f(r) = \frac{2\pi^{k/2} r^{k-1}}{\Gamma(k/2)} p(r).$$

The minimum mse/dimension for the USQ's can be computed in a manner similar to that of the previous section and [8] and [9]. Minimization over the choice of N_r, the level functions N_{k-1} through N_1, and the compressor function for r yield

$$D_{\text{USQ}} \approx \frac{\pi 2^{2/k}}{12\Gamma^{2/k}(k/2)} \cdot \left[\int_0^{\infty} r^{2(k-1)/(k+2)} f(r)^{k/(k+2)}\, dr \right]^{(k+2)/k} N^{-2/k}. \quad (6)$$

Manipulation similar to that before (5) yields that for any spherically symmetric source

$$\frac{D_{\text{USQ}}}{D_{\text{OPT}}} = \frac{1/12}{C(k,2)} \quad (7)$$

where $C(k,2)$ is the coefficient of quantization of dimension k described in [1] and [2]. Since in the limit of large dimension $C(k,2)$ approaches $1/2\pi e$, we see that USQ's are at most 1.533 dB from optimum performance. At lower dimension the difference is even smaller ($k = 2$ yielded 0.166 dB; evaluating (7) with $k = 3$ shows a difference of 0.26 dB).

Examples

To demonstrate the usefulness of the aforementioned results we first consider the asymptotic performance of the USQ's for the independent Gaussian source. In particular, (6) simplifies to

$$D_{\text{USQ,Gauss}} \approx \frac{\pi}{6} \left[\frac{k+2}{k} \right]^{(k+2)/2} N^{-2/k}.$$

Fig. 1 contains a plot of the coefficient of quantization $N^{2/k} \times D$ versus dimension k for this source. For comparison the upper and lower bounds on the performance of the optimum vector quantizer from [1] are included. Note that for the USQ's the mse/dimension decreases for increasing dimension.

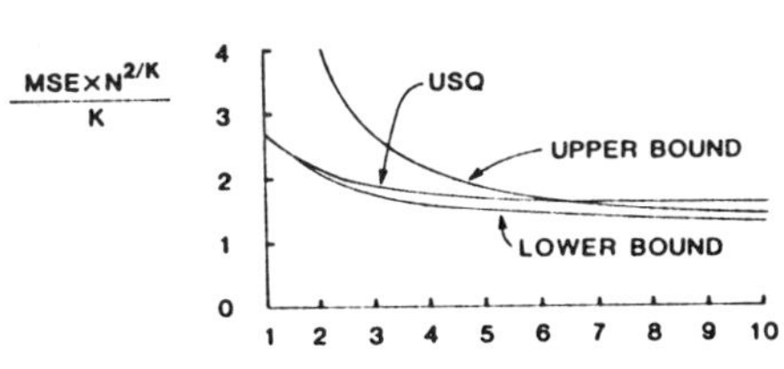

Fig. 1. Coefficient of mse for k-dimensional Gaussian source.

As a more complete example consider the UPQ's for the bivariate Gaussian source with density

$$f(x,y) = \frac{1}{2\pi} e^{-(x^2+y^2)/2}.$$

The magnitude compressor is a function of a Gaussian cumulative distribution function with zero mean and variance four:

$$g(r) = -1 + 2\int_{-\infty}^{r/2} \frac{1}{\sqrt{2\pi}} e^{-x^2/2}\, dx, \qquad r \in [0, \infty).$$

The number of levels in the magnitude quantizer N_r is computed from (4) to be $(N/2)^{1/2}$, and the phase level distribution function $N_\phi(r)$ from (3) is found to be

$$N_\phi(r) = \sqrt{\pi} N^{1/2} r e^{-r^2/8}.$$

Evaluating (1) for this source yields $D \approx 2\pi/3N = 2.09\ N^{-1}$. For comparison, equivalent expressions for the mse/dimension of rectangular, polar, and optimum quantizers are [6]

$$D_{\text{RECT}} \approx 2.72 N^{-1} \qquad D_{\text{POLAR}} \approx 2.48 N^{-1} \qquad D_{\text{OPT}} \approx 2.01 N^{-1}.$$

Note that asymptotically the UPQ's outperform the strictly polar quantizers by quite a large margin, the result being very close to the optimum value.

To validate the assumptions made in the derivation, and hence the aforementioned performance figures, exact calculations for several finite N UPQ's for the bivariate Gaussian source were made. After selecting a value for N, N_r was computed. (Actually, the closest integer value was chosen.) Next, the optimum breakpoint and output values for the N_r-level magnitude quantizer were found by a mapping through g^{-1} of $2N_r$ equally spaced points in $[0,1]$. Finally, the numbers of phase levels in each ring were found by evaluating $N_\phi(r)$ at the N_r magnitude output values, again taking the nearest integer values. Notice that the rounding of the calculated values of N_r and $N_\phi(\hat{r})$ yields a total number of levels slightly different than the original value of N. However, if a specific value of N is desired (as in the forthcoming example), the $N_\phi(r)$ values can be adjusted (rounded up or down) to sum to any N exactly.

Designs of UPQ's were tabulated for several values of N. It was noted that for N larger than 4 bits/dimension, the performance was very close to the predicted value. For smaller N the foregoing design rules lead to clearly suboptimum quantizers. Table I contains data for a UPQ with $N = 256$ (4 bits/dimension). The data tabulated include the $N_r = 11$ output and breakpoint values for the magnitude quantizer along with the number of phase levels in each magnitude ring. Also listed is the exact mse value (mse $= 2 \times D$). For comparison, mse values of strictly polar quantizers [4] and scalar rectangular coordinate quantizers [10] are included.

TABLE I
$N = 256$ Gaussian UPQ Example[a]

Breakpoint	Output	N_i
0.0000	0.1137	4
0.2278	0.3428	10
0.4590	0.5768	16
0.6966	0.8191	22
0.9447	1.0743	27
1.2085	1.3484	31
1.4952	1.6506	34
1.8167	1.9963	35
2.1937	2.4151	34
2.6708	2.9795	28
3.3820	4.0017	15

[a] Exact mse $= 0.016823$, ($D = 2.153\ N^{-1}$), (polar $N = 260$, mse $= 0.017961$), and (scalar $N = 256$, mse $= 0.018994$).

Methods similar to those mentioned above result in the parameters for the USQ's. As an example, Table II presents a 4-bits/di-

TABLE II
DATA FOR THE THREE-DIMENSIONAL $N = 4096$ GAUSSIAN USQ[a]

Shell	$\hat{r}$	$N_\phi(\hat{r})$	$N_\theta(\hat{r}, \hat{\phi})$
1	0.09985	1	2
2	0.3005	5	3, 8, 10, 8, 3
3	0.5038	8	3, 9, 13, 15, 15, 13, 9, 3
4	0.7115	10	3, 9, 14, 18, 20, 20, 18, 14, 9, 3
5	0.9257	13	3, 9, 15, 19, 23, 25, 26, 25, 23, 19, 15, 9, 3
6	1.149	16	3, 9, 15, 20, 25, 28, 30, 32, 32, 30, 28, 25, 20, 15, 9, 3
7	1.384	18	3, 9, 15, 20, 25, 29, 32, 35, 36, 36, 35, 32, 29, 25, 20, 15, 9, 3
8	1.636	19	3, 9, 15, 21, 26, 30, 33, 36, 37, 38, 37, 36, 33, 30, 26, 21, 15, 9, 3
9	1.910	20	3, 9, 15, 21, 26, 30, 34, 37, 39, 40, 40, 39, 37, 34, 30, 26, 21, 15, 9, 3
10	2.217	21	3, 9, 15, 21, 26, 31, 35, 38, 40, 41, 42, 41, 40, 38, 35, 31, 26, 21, 15, 9, 3
11	2.572	21	3, 9, 15, 21, 26, 31, 35, 38, 40, 41, 42, 41, 40, 38, 35, 31, 26, 21, 15, 9, 3
12	3.008	19	3, 9, 15, 21, 26, 30, 33, 36, 37, 38, 37, 36, 33, 30, 26, 21, 15, 9, 3
13	3.603	15	3, 9, 15, 20, 24, 27, 29, 30, 29, 27, 24, 20, 15, 9, 3
14	4.697	8	3, 9, 13, 15, 15, 13, 9, 3

[a] Exact mse = 0.02217; $D = 1.892\ N^{-2/3}$ versus $1.878\ N^{-2/3}$ as predicted.

mension Gaussian USQ with dimension three. In this case $N_r = 14$ nonuniform-thickness spherical shells exist into which the space is partitioned. The table lists their output magnitudes, and nearest-neighbor partitions are assumed for the magnitude breakpoints. Within each shell the latitude is uniformly quantized on $[-\pi/2, \pi/2]$ to $N_\phi(\hat{r})$ levels. Finally, given a shell and latitude partition the longitude is uniformly quantized on $[0, 2\pi)$ to $N_\theta(\hat{r}, \hat{\phi})$ levels. The resulting implementation requires a lookup table of size $\sum_{i=1}^{N_r} N_\phi(\hat{r}_i)$ scalar entries, which is 194 for this example. This is much smaller than the lookup table (containing 4096 three-dimensional entries), which would be required by an optimum trivariate quantizer with full search implementation. Furthermore, quantization of a single length-three input by this USQ requires only three scalar quantizations (each requiring at most $\log_2 42$ scalar comparisons). For comparison the optimum $N = 4096$ trivariate quantizer's implementation would require more than 32 000 scalar arithmetic operations and 4000 scalar comparisons. The USQ's do require that the data be in spherical coordinates, but the complexity of this transformation is low and does not grow with N.

COMMENTS

The results presented, although asymptotic, appear to yield good quantizers for finite N. For small N situations the results can supply a starting point for a local search similar to Wilson's technique in [7] or as starting configurations (to help speed convergence) for the design of optimum quantizers using the iterative Linde, Buzo, and Gray (LBG) algorithm as in [3].

Note that the USQs' mse/dimension for spherically symmetric sources in (6) has the same functional dependence upon the source density as the optimum quantizer. Also, the ratio in (7) can be written as $C(1,2)/C(k,2)$. Together these facts suggest that the USQ's optimally distribute the output points in the input space but do not select the quantization regions optimally; instead they rely on the cross product of scalar interval partitioning. Thus performance could be improved through better uniform quantization of the resulting hypersphere partitions. Such a technique based on the Gosset and Leech lattices in dimensions eight and 24 was recently described [11].

REFERENCES

[1] P. Zador, "Development and evaluation of procedures for quantizing multivariate distributions," Ph.D. dissertation, Dep. Statist., Stanford Univ., Stanford, CA, 1963.
[2] A Gersho, "Asymptotically optimal block quantization," *IEEE Trans. Inform. Theory*, vol. IT-25, pp. 373–380, July 1979.
[3] T. R. Fischer and R. M. Dicharry, "Vector quantizer design for Gaussian, gamma and Laplacian sources," *IEEE Trans. Commun.*, vol. COM-32, pp. 1065–1069, Sept. 1984.
[4] W. A. Pearlman, "Polar quantization of a complex Gaussian random variable," *IEEE Trans. Commun.*, vol. COM-27, pp. 892–899, June 1979.
[5] J. A. Bucklew and N. C. Gallagher, Jr., "Quantization schemes for bivariate Gaussian random variables," *IEEE Trans. Inform. Theory*, vol. IT-25, pp. 537–543, Sept. 1979.
[6] J. A. Bucklew and N. C. Gallagher, Jr., "Two-dimensional quantization of bivariate circularly symmetric densities," *IEEE Trans. Inform. Theory*, vol. IT-25, pp. 667–671, Nov. 1979.
[7] S. G. Wilson, "Magnitude/phase quantization of independent Gaussian variates," *IEEE Trans. Commun.*, vol. COM-28, pp. 1924–1929, Nov. 1980.
[8] P. F. Swaszek and J. B. Thomas, "Multidimensional spherical coordinates quantization," *IEEE Trans. Inform. Theory*, vol. IT-29, pp. 570–576, July 1983.
[9] P. F. Swaszek, "Uniform spherical coordinate quantization of spherically symmetric sources," *IEEE Trans. Commun.*, vol. COM-33, pp. 518–521, June 1985.
[10] J. Max, "Quantizing for minimum distortion," *IEEE Trans. Inform. Theory*, vol. IT-6, pp. 7–12, Mar. 1960.
[11] J. P. Adoul and M. Barth, "Encoding and decoding algorithms for spherical codes and quantizers from the Gosset and the Leech lattices," in *Abs. IEEE Int. Symp. Inform. Theory*, 1985, p. 130.

Entropy-Constrained Vector Quantization

PHILIP A. CHOU, TOM LOOKABAUGH, AND ROBERT M. GRAY, FELLOW, IEEE

Abstract—**An iterative descent algorithm based on a Lagrangian formulation is introduced for designing vector quantizers having minimum distortion subject to an entropy constraint. These entropy-constrained vector quantizers (ECVQ's) can be used in tandem with variable rate noiseless coding systems to provide locally optimal variable rate block source coding with respect to a fidelity criterion. Experiments on sampled speech and on synthetic sources with memory indicate that for waveform coding at low rates (about 1 bit/sample) under the squared error distortion measure, about 1.6 dB improvement in the signal-to-noise ratio can be expected over the best scalar and lattice quantizers when block entropy-coded with blocklength 4. Even greater gains are made over other forms of entropy-coded vector quantizers. For pattern recognition, it is shown that the ECVQ algorithm is a generalization of the *k*-means and related algorithms for estimating cluster means, in that the ECVQ algorithm estimates the prior cluster probabilities as well. Experiments on multivariate Gaussian distributions show that for clustering problems involving classes with widely different priors, the ECVQ outperforms the *k*-means algorithm in both likelihood and probability of error.**

I. Introduction

CONSIDER the quantization or data compression scheme of blocking a discrete-time stationary ergodic source into contiguous vectors of length n, and representing each vector by one of m reproduction vectors. For transmission of the source across a binary channel (or storage in a binary medium), the index of each reproduction vector can be encoded into binary in a straightforward manner requiring $(\log_2 m)/n$ bits per sample. For this encoding scheme, the natural design problem for a fixed block length n is to find the set of m reproduction vectors, known as the reproduction codebook, which minimizes the average distortion between the source and its reproduction, subject to a constraint on the maximum rate of bit transmission. Distortion-rate theory [1]-[3] guarantees that if n is sufficiently large, there exists a finite set of m reproduction vectors with transmission rate $R = (\log_2 m)/n$ bits per sample, such that the average distortion per sample D is arbitrarily close to the minimum distortion $D(R)$, over all possible compression schemes with average transmission rate less than or equal to R bits per sample. This provides a theoretical basis for the use of such vector quantizers; for sufficiently long block lengths, they are theoretically optimal. Unfortunately, practical schemes cannot use arbitrarily long block lengths. For a fixed block length of practical size, a scheme that achieves the same distortion with lower average transmission rate than straightforward binary coding of the indexes is one that "entropy encodes" its index sequence. Entropy encoding reduces the average transmission rate from $(\log_2 m)/n$ bits per sample to the index entropy or, with additional effort, even to the entropy rate of the index sequence. The key word here, however, is "average," since known practical schemes for achieving this lower rate all have a variable rate nature, and hence, an additional amount of complexity. If we are willing to deal with the added complexity, the natural design problem is to find the set of reproduction vectors which minimizes the average distortion between the source and its reproduction, subject to a constraint on the index entropy.

This is a classic problem in the quantization literature, often developing in parallel with the sister problem of quantizer design for minimum distortion with a fixed number of indexes. One broad area of inquiry has been based on calculus approximations that can be made if the probability distribution of the source is smooth and the number of indexes per dimension is large. Investigations in this "asymptotic" or "high resolution" quantization theory began with the scalar case ($n = 1$), in which regular quantizers composed of thresholds and scalar reproductions (the thresholds determining to which reproduction a given source symbol will be mapped) are generally the systems of interest. Gish and Pierce [4] showed, among other things, that the optimal high-resolution entropy-constrained scalar quantizer (ECSQ) i) has uniformly spaced thresholds regardless of the source probability density function and ii) for the Gaussian iid case and squared-error distortion has an index entropy only 0.255 bits/sample greater than the rate-distortion curve. The combination of these results suggests that scalar uniform quantization should be quite effective for this source—a result that had already been noted experimentally by Goblick and Holsinger [5]. Zador [6] developed bounds for the index entropy of a high-resolution entropy-constrained vector quantizer (ECVQ) ($n \geq 1$). Based on the fact that one of these bounds is optimized by a uniform distribution of reproductions, Gersho [7] conjectured the optimal high resolution ECVQ should have the form of a lattice (a lattice in R^n is composed of all integral combinations of a set of linearly independent vectors which span the space). Investigations of the properties of lattices suggest that certain lattices should perform better than others;

Manuscript received September 8, 1987; revised April 8, 1988. This work was supported by the National Science Foundation under NSF Grant IST-8509860; by a Hughes Doctoral Fellowship; by an IBM Doctoral Fellowship; and by ESL, a subsidiary of TRW Inc. This work was performed while all the authors were with the Information Systems Laboratory, Department of Electrical Engineering, Stanford University.

P. A. Chou is with AT&T Bell Laboratories, Murray Hill, NJ 07974.

T. Lookabaugh is with Compression Labs, Inc., San Jose, CA 95134.

R. M. Gray is with the Information Systems Laboratory, Department of Electrical Engineering, Stanford University, Stanford, CA 94305.

IEEE Log Number 8824508.

Reprinted from *IEEE Trans. Acoust., Speech, Signal Processing,* vol. ASSP-37, pp. 31–42, Jan. 1983.

Conway and Sloane [8]–[10] have determined the best known lattices for several dimensions as well as useful encoding and indexing schemes, and Sayood *et al.* have applied lattices to transform image coding [11].

A second avenue of investigation has centered on efforts to determine the optimal ECSQ directly (without high-resolution approximations). Expressions for the entropy and distortion performance of ECSQ typically also include as a parameter the number of reproductions m. Ideally, we would like to find the optimum performance over all m; however, in many analytic approaches and all practical implementations, it is in fact necessary to limit m to some maximum value. Wood [12] provided the first numerical ECSQ design, using a descent algorithm to show that the optimal ECSQ for the Gaussian iid case was only slightly superior to the uniform scalar quantizer even for moderate values of m. Berger [13] described the necessary conditions for an optimal ECSQ under squared-error distortion (and later for the rth power distortion [14]), and first presented the Lagrangian formulation that we will make use of in Section II. Based on this formulation, he developed an iterative algorithm for determining thresholds of the ECSQ. Netravali and Saigal [15] presented another ECSQ design algorithm based on fixed-point considerations. Noll and Zelinski [16] applied Berger's algorithm to a variety of common probability density functions. Farvardin and Modestino [17] extended the necessary conditions for optimality to general distortion measures, and outlined two algorithms for ECSQ design. The first, which they actually used in their experiments, is similar to Berger's algorithm. The second is a fixed-point algorithm which can be considered the scalar version of the ECVQ design algorithm of this paper. Although they reported difficulties with the convergence of the algorithm, we encountered no such difficulties in the present investigation. For the squared-error distortion, Ziv [18] presented an interesting result which shows that a certain dithered uniform scalar quantizer followed by entropy coding of a block of n indexes will have index entropy (conditioned on the dithering variable) within 0.754 bits/sample of the optimal ECVQ of block length n; Gutman [19] extended the result to a variety of other distortions. Unfortunately, this bound is loose at low rates, and tighter bounds from the high-resolution theory are available at high rates.

In this paper we present a descent algorithm for the ECVQ design problem. The algorithm begins with a Lagrangian formulation like Berger's, but in implementation is quite similar to the generalized Lloyd algorithm [20], [21] developed for the constrained number-of-indexes vector quantizer design problem. We conduct a variety of experiments based on the use of a training sequence rather than the integration of probability density functions characteristic of previous investigations. We discover that, generally at the cost of higher complexity, ECVQ outperforms many other entropy-coded quantization schemes including scalar uniform threshold, lattice, constrained number-of-indexes vector quantization, and a recently introduced tree based variable rate vector quantization scheme [22]; performance gains are especially significant for sources with memory such as speech. Finally, we explore applications of the algorithm in clustering and classifier design for pattern recognition.

II. Theoretical Background

Consider the variable rate communications system shown in Fig. 1. In a typical implementation, the encoder blocks the source $\{X_i\}$ into vectors of length n, and maps each vector $X^n \in \mathcal{X}^n$ into a variable-length codeword $c \in \mathcal{C}$. The codeword c is then sent through the channel, which is assumed to be noiseless, and the decoder maps the codeword into a reproduction vector $Y^n \in \mathcal{Y}^n$. The reproduction alphabet $\mathcal{Y}$ is usually, but not necessarily, equal to the source alphabet $\mathcal{X}$, but, in general, the two alphabets are arbitrary and need not be finite or even countable. The channel codeword alphabet Σ is finite, and is usually taken to be binary, $\Sigma = \{0, 1\}$. This shall be assumed throughout the remainder of the paper. The set of channel codewords $\mathcal{C} = \{c_i\}_{i \in \mathcal{I}}$ is known as the channel codebook, and is a subset of Σ^*, the set of all finite length strings from Σ. Thus, $\mathcal{C}$ is finite or countable depending on its index set $\mathcal{I}$, and contains codewords of varying lengths, in general.

Let $\boldsymbol{\alpha}_n: \mathcal{X}^n \to \mathcal{C}$ and $\boldsymbol{\beta}_n: \mathcal{C} \to \mathcal{Y}^n$ denote the encoder and decoder, respectively. It is possible to decompose the encoder into two parts, $\boldsymbol{\alpha}_n = \gamma \circ \alpha$, where $\alpha: \mathcal{X}^n \to \mathcal{I}$ and $\gamma: \mathcal{I} \to \mathcal{C}$, and $\circ$ denotes composition. The first part, α, maps a source vector into an index, and is many-to-one and information lossy, in general. The second part, γ, maps the index into a channel codeword, and is one-to-one and information lossless.

Similarly, the decoder can be decomposed as $\boldsymbol{\beta}_n = \beta \circ \gamma^{-1}$, where $\gamma^{-1}: \mathcal{C} \to \mathcal{I}$ and $\beta: \mathcal{I} \to \mathcal{Y}^n$. The mapping γ^{-1} is the inverse of γ, and maps each channel codeword back to its index. The mapping β outputs a reproduction vector for an index.

The pair $(\boldsymbol{\alpha}_n, \boldsymbol{\beta}_n)$ will be referred to as a variable rate block coder because a block of input is represented by a single codeword in $\mathcal{C}$, and the codewords in $\mathcal{C}$ are generally not all of the same length.

To be information lossless, it is not enough that γ be invertible; it must also be *uniquely decodable*. This means that if c is a string in Σ^*, and c is the concatenation of codewords $\gamma(i_1), \cdots, \gamma(i_m)$ and also the concatenation of codewords $\gamma(i'_1), \cdots, \gamma(i'_{m'})$, then $m = m'$ and $i_1 = i'_1, \cdots, i_m = i'_m$. It can be shown [2, p. 49] that if the mapping $\gamma: \mathcal{I} \to \mathcal{C}$ is uniquely decodable, then there exists a *prefix-free* mapping $\gamma': \mathcal{I} \to \mathcal{C}'$ with the same codeword lengths. That is, there is a codebook $\mathcal{C}' = \{c'_i\}_{i \in \mathcal{I}}$ in which no codeword is the prefix of any other codeword, and in which $|c'_i| = |c_i|$ for all $i \in \mathcal{I}$, where $|c|$ denotes the length of the codeword c. Without loss of generality, then, γ will be assumed to be prefix-free. This places a constraint on the minimum average codeword length, as we shall see.

Assume the source $\{X_i\}_{i=0}^{\infty}$ is a strictly stationary ergodic random process with process distribution P. In what follows, the strict stationarity condition could be replaced

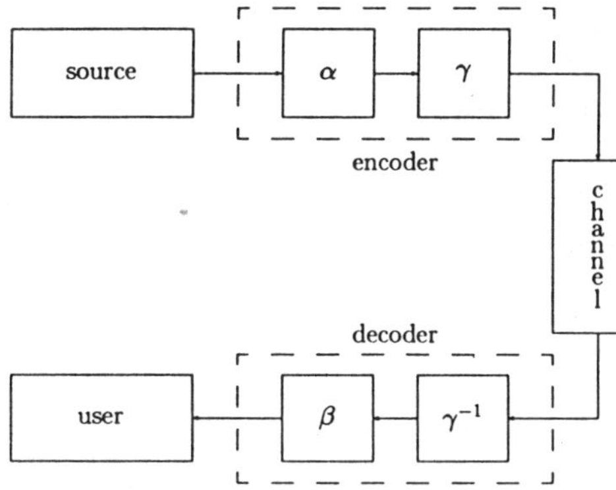

Fig. 1. A variable rate communications system.

by an asymptotic mean stationarity condition, but strict stationarity will be assumed for simplicity.

Let X^n denote the vector $(X_0, \cdots, X_{n-1})$, and let $Y^n = \boldsymbol{\beta}_n(\boldsymbol{\alpha}_n(X^n))$ be its reconstruction. If $l_n(X^n) = |\boldsymbol{\alpha}_n(X^n)|$ is the length of the codeword representing X^n, then

$$R_n = \frac{1}{n} E[l_n(X^n)]$$

is the average *rate*, or average number of channel symbols (here, bits) used to represent each source symbol. If $\rho_n(X^n, Y^n)$ is the distortion between the source vector X^n and its reproduction Y^n, then

$$D_n = \frac{1}{n} E[\rho_n(X^n, Y^n)]$$

is the average distortion between each source letter and its reproduction.

According to distortion-rate theory [2], [3], [23], if the distortion measure $\rho_n(x^n, y^n)$ and the source distribution P_{X^n} are fixed, then the nth-order distortion-rate function

$$D_n(R) = \inf_{P_{Y^n|X^n}} \left\{ \frac{1}{n} E[\rho_n(X^n, Y^n)] \middle| \frac{1}{n} I(X^n; Y^n) \le R \right\}$$

is a lower bound to the nth-order *operational* distortion-rate function

$$\hat{D}_n(R) = \inf_{(\boldsymbol{\alpha}_n, \boldsymbol{\beta}_n)} \left\{ \frac{1}{n} E[\rho_n(X^n, Y^n)] \middle| \frac{1}{n} E[l_n(X^n)] \le R \right\}.$$

If ρ_n is a single-letter fidelity criterion, that is, if $\rho_n(x^n, y^n) = \sum_{i=0}^{n-1} \rho_1(x_i, y_i)$, then as $n \to \infty$, the bound becomes tight, so that the distortion-rate function

$$D(R) = \lim_{n \to \infty} D_n(R)$$

characterizes the region in the distortion-rate plane achievable by some deterministic variable rate block code $(\boldsymbol{\alpha}_n, \boldsymbol{\beta}_n)$.

The nth-order distortion-rate function $D_n(R)$ is nonincreasing and convex, and its computation is a straightforward convex programming problem. In the case where the source and reproduction alphabets are finite, the elegant Blahut algorithm [24] can be used to compute $D_n(R)$.

The Blahut algorithm and more general convex programming algorithms rely on minimizing the Lagrangian

$$J(P_{Y^n|X^n}) = E[\rho_n(X^n, Y^n)] + \lambda I(X^n; Y^n),$$

where the Lagrange multiplier λ has an interpretation as the slope of a line supporting the graph of $D_n(R)$.

The nth-order *operational* distortion-rate function $\hat{D}_n(R)$ is nonincreasing, but it is unfortunately not necessarily convex or even continuous, particularly if P has a discrete component, e.g., if P is a sample distribution from a training sequence. (See Fig. 2.) Hence, Lagrangian methods cannot be used to find $\hat{D}_n(R)$. However, they can be used to find the *convex hull* of $\hat{D}_n(R)$ by minimizing the functional

$$J(\boldsymbol{\alpha}_n, \boldsymbol{\beta}_n) = E[\rho_n(X^n, Y^n)] + \lambda E[l_n(X^n)]. \quad (1)$$

Here, λ has an interpretation as the slope of a line supporting the convex hull.

Fortunately, any point on the convex hull can be achieved by timesharing between two variable rate block coders $(\boldsymbol{\alpha}_n, \boldsymbol{\beta}_n)$ and $(\boldsymbol{\alpha}_n', \boldsymbol{\beta}_n')$, that achieve points on $\hat{D}_n(R)$. In practice, timesharing is usually not necessary because there often exists a single variable rate block coder $(\boldsymbol{\alpha}_n, \boldsymbol{\beta}_n)$ that achieves a point on $\hat{D}_n(R)$ and on its convex hull with average rate (or distortion) sufficiently close to the desired average rate (or distortion).

In any case, our objective is to find the convex hull of $\hat{D}_n(R)$, and to find variable rate block coders $(\boldsymbol{\alpha}_n, \boldsymbol{\beta}_n)$ achieving points on the convex hull. These coders are in some sense optimal, and can be used in practical data compression systems. We solve this problem by explicitly minimizing the Lagrangian functional (1).

III. The ECVQ Algorithm

Recall that $\boldsymbol{\alpha}_n = \gamma \circ \alpha$ and $\boldsymbol{\beta}_n = \beta \circ \gamma^{-1}$, where $\alpha: \mathcal{X}^n \to \mathcal{I}$ quantizes an input vector into an index, $\gamma: \mathcal{I} \to \mathcal{C}$ noiselessly encodes the index into a binary string for transmission across the channel, $\gamma^{-1}: \mathcal{C} \to \mathcal{I}$ decodes the string back into its original index, and $\beta: \mathcal{I} \to \mathcal{Y}^n$ reproduces the index as an output vector which represents the original input vector in a minimum distortion sense.

Hence, we may rewrite the coder $(\boldsymbol{\alpha}_n, \boldsymbol{\beta}_n)$ as (α, γ, β), and reexpress the Lagrangian functional (1) as

$$J_\lambda(\alpha, \gamma, \beta) = E\Big[\rho_n\big(X^n, \beta(\alpha(X^n))\big) + \lambda\big|\gamma(\alpha(X^n))\big|\Big]. \quad (2)$$

Our objective is to find the coder (α, γ, β) which minimizes this functional.

We employ an iterative descent algorithm similar to the generalized Lloyd algorithm. Starting with an arbitrary initial coder $(\alpha^{(0)}, \gamma^{(0)}, \beta^{(0)})$, we repeatedly apply a transformation

$$(\alpha^{(t+1)}, \gamma^{(t+1)}, \beta^{(t+1)}) = T(\alpha^{(t)}, \gamma^{(t)}, \beta^{(t)})$$

such that $J_\lambda(\alpha^{(t)}, \gamma^{(t)}, \beta^{(t)})$ is decreasing in t.

Since $J_\lambda(\alpha, \gamma, \beta)$ is bounded below by zero, the sequence of real numbers $J^{(t)} = J_\lambda(\alpha^{(t)}, \gamma^{(t)}, \beta^{(t)})$ is guar-

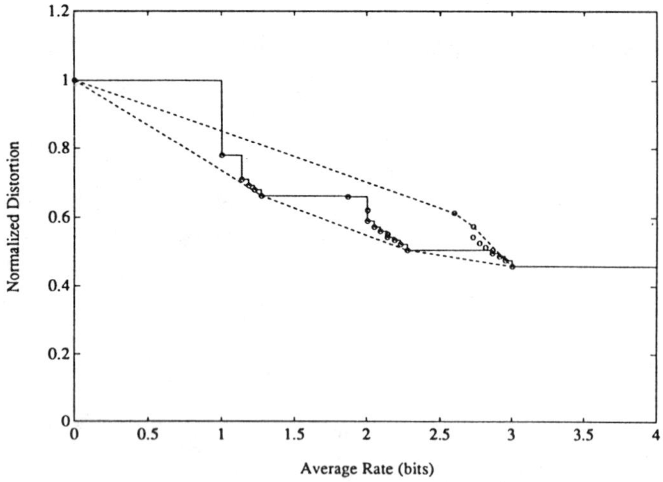

Fig. 2. A typical operational distortion-rate function $\hat{D}(R)$ when the underlying probability measure is discrete. Each circle is a (rate, distortion) pair; the dashed line is their convex hull. The solid staircase line is the operational distortion-rate function.

anteed to converge. As a convergence criterion for the algorithm, we use the simple stopping rule

$$(J^{(t)} - J^{(t+1)})/J^{(t+1)} < \epsilon,$$

where $\epsilon = 0.005$ is typical.

Following the lines of argument of Sabin and Gray [25], it appears that under suitable regularity conditions, the convergence of the coders $(\alpha^{(t)}, \gamma^{(t)}, \beta^{(t)})$ themselves (to a set) is also guaranteed, in a well-defined sense. This aspect, however, will not be investigated in this paper.

The transformation T operates as follows. For fixed $\gamma^{(t)}$ and $\beta^{(t)}$, $\alpha^{(t+1)}$ is chosen to minimize $J_\lambda(\alpha^{(t+1)}, \gamma^{(t)}, \beta^{(t)})$. Then, for fixed $\alpha^{(t+1)}$ and $\beta^{(t)}$, $\gamma^{(t+1)}$ is chosen to minimize $J_\lambda(\alpha^{(t+1)}, \gamma^{(t+1)}, \beta^{(t)})$. Finally, for fixed $\alpha^{(t+1)}$ and $\gamma^{(t+1)}$, $\beta^{(t+1)}$ is chosen to minimize $J_\lambda(\alpha^{(t+1)}, \gamma^{(t+1)}, \beta^{(t+1)})$. This procedure guarantees that $J^{(t)}$ is nonincreasing.

Let us consider the minimizations in more detail.

Fix γ and β, thereby fixing the reproduction codewords and the lengths of their associated channel codewords. A mapping $\alpha: \mathfrak{X}^n \to \mathcal{I}$ that minimizes (2) is one that minimizes the integrand $\rho_n(x^n, \beta(\alpha(x^n))) + \lambda|\gamma(\alpha(x^n))|$ almost everywhere. That is, α maps x^n to the ith cell if the distance from x^n to $\beta(i)$, biased by the length of $\gamma(i)$, is minimized. Thus,

$$\alpha(x^n) = \operatorname*{argmin}_{i \in \mathcal{I}} \left[\rho_n(x^n, \beta(i)) + \lambda\left|\gamma(i)\right|\right]. \quad (3)$$

This α need not be unique since ties may be broken arbitrarily and (3) need hold only for almost all x^n. Equation (3) is analogous to nearest neighbor encoding in standard vector quantization.

Next express (2) as

$$J_\lambda(\alpha, \gamma, \beta) = \sum_{i \in \mathcal{I}} p(i)\, E\big[\rho_n(X^n, \beta(i)) + \lambda\left|\gamma(i)\right|\,\big|\alpha(X^n) = i\big], \quad (4)$$

where $p(i) = P\{\alpha(X^n) = i\}$.

For fixed α and β, and hence for fixed p, a prefix-free code $\gamma: \mathcal{I} \to \mathcal{C}$ that minimizes (4) is one that minimizes the expected codeword length $R = \Sigma_{i \in \mathcal{I}}\, p(i)\,|\gamma(i)|$. A prefix-free code of minimum expected codeword length can be found, in the case of finite $\mathcal{I}$, by the Huffman algorithm [26], for example. The Huffman algorithm could be incorporated into the design algorithm at this stage. However, for simplicity, we allow the fiction that codewords can have noninteger lengths, and assign

$$\left|\gamma(i)\right| = \log_2\left(1/p(i)\right). \quad (5)$$

The "average rate" of the resulting code is exactly equal to its index entropy. Experiments have shown that following this system with a Huffman code produces an overall system nearly identical in performance to a system produced by a procedure that includes the Huffman algorithm within the design loop. Indeed, the average rate of a Huffman code must satisfy bounds such as $H(p) \le R \le H(p) + \max_i p(i) + 0.086$ [27], where $H(p) = \Sigma_{i \in \mathcal{I}}\, p(i) \log_2(1/p(i))$ is the index entropy. On the other hand, experiments have also shown that if the Huffman code is to be followed by a buffering scheme, it is best to incorporate the code and the buffering scheme into the design loop, so that the resulting codebook will be optimized for the buffer. Normally, however, we use (5) so as not to tie our results to a particular entropy code, since there are a number of noiseless codes, e.g., arithmetic codes [28], [29] and Ziv–Lempel codes [30], that also achieve average rates quite close to the codeword entropy.

The last step in the transformation T is to fix α and γ, and hence p. Then the mapping $\beta: \mathcal{I} \to \mathcal{Y}^n$ that minimizes (4) is one that for each $i \in \mathcal{I}$, minimizes $E[\rho_n(X^n, \beta(i)) + \lambda|\gamma(i)|\,|\alpha(X^n) = i]$, i.e.,

$$\beta(i) = \arg\min_{y^n \in \mathcal{Y}^n} E\left[\rho_n(X^n, y^n)\,|\,\alpha(X^n) = i\right].$$

The reproduction codeword $\beta(i)$ is known as the *centroid* of the cell $\{\alpha(X^n) = i\}$, and is the same as the centroid of a cell in standard vector quantization; it need not be unique. Computing the centroid can be performed easily for a number of common distortion measures [21]. For example, if ρ_n is the squared-error distortion $\rho_n(x^n, y^n) = \Sigma_{i=0}^{n-1}(x_i - y_i)^2$, then the centroid is the conditional average $\beta(i) = E[X^n\,|\,\alpha(X^n) = i]$.

A summary of the algorithm is given in Fig. 3.

When the distribution is unknown, it can be estimated from the sample $X_0^n, X_1^n, \cdots, X_{N-1}^n$, where the vector $X_k^n = (X_{nk}, X_{nk+1}, \cdots, X_{n(k+1)-1})$ is the kth block from the original source $\{X_i\}_{i=0}^{\infty}$. The sample distribution

$$P_{X^n}^N(F) = \frac{1}{N}\sum_{k=0}^{N-1} 1_F(X_k^n)$$

converges on events F almost everywhere since P is stationary. Furthermore, if P is n-ergodic, $P_{X^n}^N$ converges to P_{X^n} in distribution. Here, $1_F(x) = 1$ if $x \in F$ and 0 otherwise. The algorithm for this case is identical to the algorithm of Fig. 3, with probabilities and expectations replaced by sample averages in the obvious way. For $\lambda =$

(0) Initialization: Given
a distribution P_{X^n},
a distortion measure ρ_n,
a Lagrange multiplier λ,
a convergence threshold ϵ,
an index set $\mathcal{I}$,
an initial reproduction codebook $\{\beta^{(0)}(i)\}_{i\in\mathcal{I}}$,
and initial codeword lengths $\{\gamma^{(0)}(i)\}_{i\in\mathcal{I}}$,
set $t = 0$ and $J^{(0)} = \infty$.

(1) $\alpha(x^n) = \operatorname{argmin}_{i\in\mathcal{I}} \rho_n(x^n, \beta(i)) + \lambda|\gamma(i)|$.

(2) $|\gamma(i)| = -\log_2 P_{X^n}\{\alpha(X^n) = i\}$.

(3) $\beta(i) = \operatorname{argmin}_{y^n\in\mathcal{Y}^n} E[\rho_n(X^n, y^n)\,|\alpha(X^n) = i]$.

(4) $J^{(t+1)} = E[\rho_n(X^n, \beta^{(t+1)}(\alpha^{(t+1)}(X^n))) + \lambda|\gamma^{(t+1)}(\alpha^{(t+1)}(X^n))|]$.

(5) If $(J^{(t)} - J^{(t+1)})/J^{(t+1)} > \epsilon$,
set $t = t + 1$ and go to (1).
Otherwise, quit.

Fig. 3. Descent algorithm to minimize the Lagrangian functional $J_\lambda(\alpha, \beta, \gamma)$ over all variable rate block coders.

0, the algorithm reduces to the generalized Lloyd algorithm [21].

When the algorithm converges, we obtain the variable rate coder (α, γ, β) that (locally) minimizes the functional

$$\frac{1}{n} J_\lambda(\alpha, \gamma, \beta) = D(\alpha, \gamma, \beta) + \lambda R(\alpha, \gamma, \beta),$$

where $D(\alpha, \gamma, \beta) = (1/n)E[\rho_n(X^n, \beta(\alpha(X^n)))]$ is the average distortion per letter, and $R(\alpha, \gamma, \beta) = (1/n)E[|\gamma(\alpha(X^n))|]$ is the average codeword length per letter. Thus, the line in the distortion-rate plane with slope $-\lambda$ that passes through the point $(R(\alpha, \gamma, \beta), D(\alpha, \gamma, \beta))$ supports the convex hull of the operational distortion-rate function (modulo the assumption that the minimization was global).

Therefore, to find the entire convex hull, it is necessary to repeat the minimization of $J_\lambda(\alpha, \gamma, \beta)$ for various λ's.

Two points on the convex hull are easy to find. The first point (R_∞, D_∞) is obtained by minimizing $J_\lambda(\alpha, \gamma, \beta)$ in the limit of large λ, hence, the notation R_∞ and D_∞. This point corresponds to the rate-zero codebook, in which there is only one channel codeword, the empty string. In this case, $\alpha(X^n) = 0$, $|\gamma(0)| = 0$, and $\beta(0) = \operatorname{argmin}_{y^n\in\mathcal{Y}^n} E[\rho_n(X^n, y^n)]$. Hence, $R_\infty = 0$ and $D_\infty = (1/n)E[\rho_n(X^n, \beta(0))]$, where $\beta(0)$ is the centroid for the entire distribution.

The second point (R_0, D_0), obtained at $\lambda = 0$, corresponds to the full-rate codebook designed by the generalized Lloyd algorithm. For this point to have a finite average rate and positive distortion, we insist that the index set be finite, say $\mathcal{I} = \{0, 1, \cdots, m - 1\}$. This constrains the rate (or resolution) of the code to be at most $1/n \log_2 m$. The average length R_0, or the entropy, of the code will of course be less than this. One could say that our resulting codes are both *rate* and *entropy* constrained. The distortion D_0 for the code is computed by the generalized Lloyd algorithm.

One of many ways to obtain the remainder of the points on the convex hull is to "walk up" the curve starting from the point (R_0, D_0), which corresponds to $\lambda = 0$. The first point visited is the point (R_1, D_1) which minimizes $J_{\lambda_1} = D + \lambda_1 R$, where $-\lambda_1 = \epsilon(D_0 - D_\infty)/(R_0 - R_\infty)$ is some fraction of the slope between (R_∞, D_∞) and (R_0, D_0). The next point visited (R_2, D_2) corresponds to some slope λ_2, etc. Each λ_i in the increasing sequence $\lambda_1, \lambda_2, \cdots$, can be determined *a priori*, e.g., by a geometric sequence, or it can be determined from a prediction of the behavior of the convex hull up to that point. In our experiments, the formula used was

$$\lambda_{n+1} = \frac{\lambda_n - \lambda_{n-1}}{R_{n-1} - R_n} \frac{R_0}{M}, \qquad n \geq 1, \tag{6}$$

with λ_0 and λ_1 initialized as above, and M being the desired number of (equally spaced in R) points on the curve. This procedure generally performed quite well, although in some experiments (on synthetic data such as the Gaussian memoryless source), R was extremely sensitive to λ so that backtracking was necessary. A computational advantage is accrued by using the final codebook for λ_n as the starting codebook for λ_{n+1}: in an experiment, we realized a savings in computation time of a factor of 2.4 over the alternative of using the λ_0 codebook to initialize the algorithm for each λ.

One issue of some importance is the complexity of the algorithm. Each pass through steps 1)–4) of the training sequence based algorithm of Fig. 3 requires roughly $3Nnm$ additions, Nnm multiplications, and m logarithms, with N the training sequence size in vectors, n the vector dimension, and m the codebook size. For M of (6) chosen as 20, we found that in the experiments reported in the next section, convergence for each λ took an average of three passes giving a total effort of about 60 times the "per pass" figures.

IV. Experimental Results

In this section we examine the implementation of the ECVQ design algorithm and compare its performance against several other entropy-coded quantization systems. Implementational issues include the effect of codebook size and the effect of placing specific entropy coding schemes inside and outside the design loop. Experimental comparisons were made against scalar uniform quantizers with entropy coding of single quantizer outputs and blocks of outputs, two- and four-dimensional lattice quantizers, full search, complete tree structured, and entropy-pruned tree structured vector quantizers (all with entropy-coded outputs).

In Fig. 4 we show the signal-to-quantization noise ratio (SQNR) curves for squared-error distortion for an eight-dimensional ECVQ as we varied the codebook size. Not surprisingly, the algorithm performance improves as codebook size increases. The training sequence was 20 000 samples of speech data, and the performance for this example was measured on the training sequence.

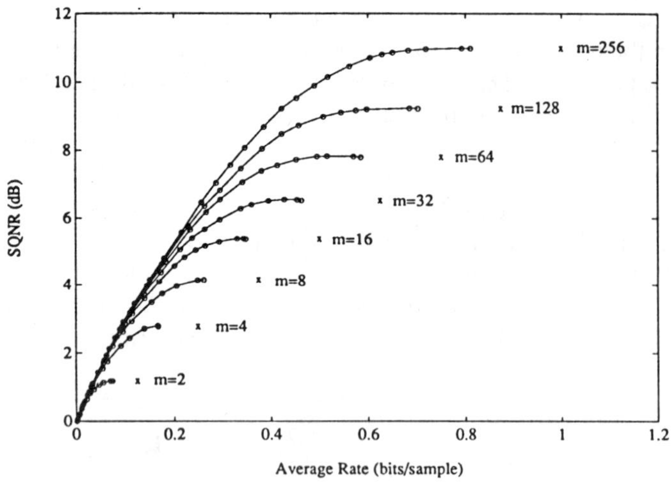

Fig. 4. ECVQ performance versus codebook size. Solid lines and circles are ECVQ performance as λ is varied; the x's are full search VQ performance without entropy coding. Codebook size is given by m.

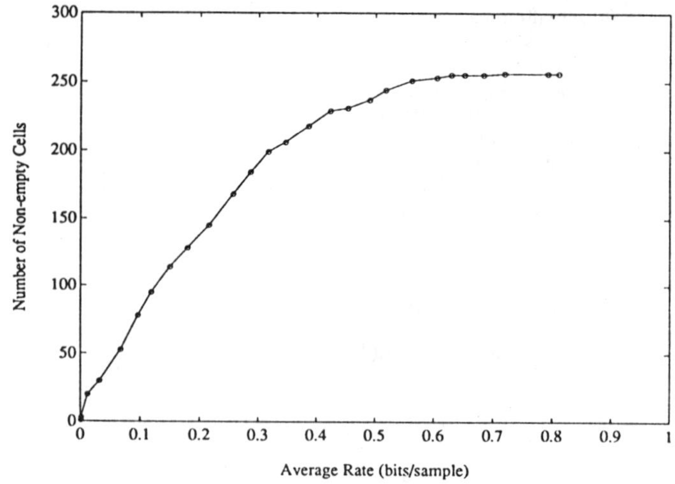

Fig. 5. Effective codebook size for the speech coding experiment of Fig. 4, with initial codebook size $m = 256$.

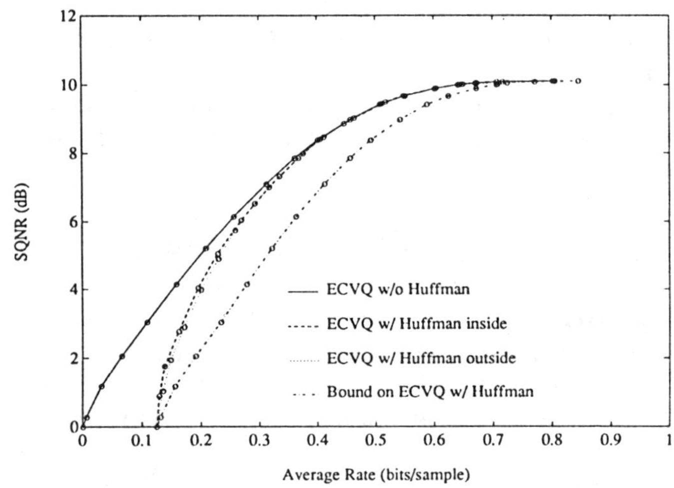

Fig. 6. Effect of Huffman coding inside and outside the ECVQ design algorithm. The solid line is ECVQ performance without Huffman coding (i.e., with noninteger word lengths), the dashed line is performance with Huffman coding inside the design algorithm, the dotted line is performance with a Huffman code applied after the ECVQ algorithm has finished. The dot-dash line shows a bound on Huffman code redundancy due to Gallager [27].

The algorithm itself generally reduces the codebook size as λ is increased, since during the course of the algorithm, a particular cell (say the ith cell) may become unpopulated (i.e., $\alpha(X_k^n)$ never equals i during step 1) of the algorithm). In this case, step 2) will set $|\gamma(i)| = \infty$ so that the cell will never be chosen as the best codeword by an encoder, or equivalently will never become populated as the design algorithm continues. Unlike the generalized Lloyd algorithm, there is no convincing rationale for attempting to repopulate empty cells; splitting a highly populated cell may indeed reduce distortion, but it may also increase the entropy of the quantizer. Hence, we remove empty cells from the codebook and continue the algorithm with the smaller codebook size. Fig. 5 shows the *effective* codebook size versus average rate for the experiment of Fig. 4 with an initial codebook size of 256.

In Fig. 6 we demonstrate the fact (mentioned in Section III) that not integerizing the codeword lengths inside the design loop leads to only a minor penalty in performance compared to the case where we replace step 2) of the design algorithm with an actual Huffman code design to determine the best integral word lengths. Not integerizing the word lengths, which leads to a minimum output entropy for α, is simpler and also advantageous when contemplating other entropy coding algorithms for γ.

On the other hand, including an entropy encoding scheme within the design loop tailors the ECVQ quantizer to that particular scheme. Consider a system in which the variable rate code is used over a fixed rate channel, so that buffering becomes necessary. Although the decoder can track the encoder buffer state if the channel code is prefix-free, buffer overflows and underflows cause a decrease in the effective average bit rate. This causes a decrease in SQNR at the nominal bit rate. By explicitly incorporating the overflow and underflow buffer strategies into the ECVQ design loop, the codewords can be designed to compensate for overflows and underflows. The standard strategy for handling overflows (for a zero mean source) is to output a block of zeros instead of the reproduction codeword whose channel codeword would have overflowed the buffer. A better strategy, due to Berger *et al.* [31] and called buffer adapted Huffman (BAH) coding, is to output the best reproduction codeword possible from among those whose channel codewords fit into the remaining buffer space. The standard strategy for handling underflows is to pad the channel codeword with noninformation carrying bits. A better strategy is to output the best reproduction codeword possible from among those whose channel codewords will underflow the buffer or empty it exactly.

In Fig. 7 we show test sequence SQNR performance versus buffer length on speech data described later in this section, for several 256 codeword, 8-dimensional ECVQ systems with codeword entropies equal to 2.926 bits/vector. All are Huffman coded, buffered, and at-

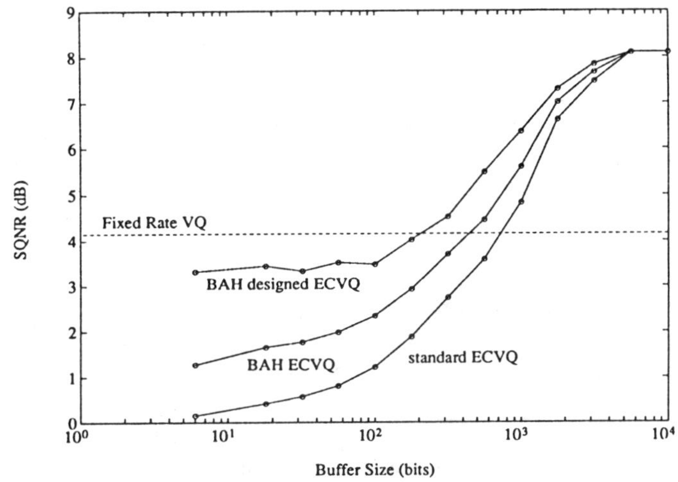

Fig. 7. Effect of buffer length on ECVQ with Huffman coding for a speech source. A fixed rate VQ and an "output zeros during overflow" standard strategy are compared to buffer adapted Huffman coding applied to an independently designed ECVQ (BAH) or included in the ECVQ design loop (BAH designed).

tached to a fixed rate, 3 bits/vector channel. At any buffer size, the ECVQ designed with the BAH overflow strategy inside the design loop outperforms the usual ECVQ system followed by either the standard overflow strategy or the BAH strategy. A fixed rate, 3 bits/vector VQ is shown for comparison; it is outperformed by all the variable rate systems at moderate to high buffer lengths.

We next compare, in a series of experiments, several quantizers in terms of their index entropy versus squared-error distortion. Table I shows several of the systems, giving a qualitative evaluation of their encoding complexity (source-to-index mapping) as well as noting m/n—a measure of the complexity of the required entropy coder for encoding the index. The scalar system considered is a simple uniform threshold quantizer with 17 thresholds and 19 cells, whose centroids serve as the reproduction levels, as in [17, Section VI, B]. The quantizer outputs are individually entropy encoded. The scalar system with block entropy coding is a scalar system with 4 thresholds and 5 cells, whose output indexes have been blocked into four dimensions before entropy encoding the 625 possible index blocks. This system is close in spirit to Ziv's universal quantizer [18], although it is not dithered. The A_2 lattice system employed has as its reproduction levels the innermost 37 points of a two-dimensional hexagonal lattice [32]. For high-resolution entropy-constrained two-dimensional quantization, the hexagonal lattice is optimal [7], [33]. The D_4 lattice system uses the 256 point Voronoi code enumerated by Conway and Sloane [9], [10]. The D_4 lattice, or the set of points in $\boldsymbol{R}^4$ with integral coordinates whose sum is even, is the best-known lattice for high-resolution entropy-constrained four-dimensional quantization [8]. The outputs of both the A_2 and D_4 systems are entropy coded as usual. Performance curves in the distortion-rate plane are traced out by scaling the various lattices. The ECVQ systems of four and eight dimensions were designed using the algorithms of Section III, with an initial codebook size of 256. All of these systems were designed on training sequences rather than on the underlying distributions.

TABLE I
COMPARISON OF ENTROPY-CODED QUANTIZATION SYSTEMS FOR THE EXPERIMENTS OF FIGS. 8, 9, AND 10

System	Block Size (n)	Number of Cells (m)	m/n	Encoding Complexity
Scalar	1	19	19	very simple
Scalar with Block Entropy Coding	4	625	156	simple
A_2 Lattice	2	37	19	moderate
D_4 Lattice	4	256	64	moderate
ECVQ	4	256	64	complex
ECVQ	8	256	32	complex

Results for the Gaussian iid source are shown in Fig. 8. We use training and test sequences of 40 960 samples each. The rate-distortion function for this source is [3, p. 99]

$$R(D) = \frac{1}{2}\log_2\frac{\sigma^2}{D} \qquad \text{bits/sample,}$$

where σ^2 is the source variance and D is the average squared-error distortion. We note that, although ECVQ shows a slight improvement over the other systems, the gain is most likely not worth the additional complexity involved. Indeed, for this source our 8-dimensional ECVQ operating at 0.5 bits/sample with an SQNR of 2.3 dB is outperformed by several easily instrumentable codes of large blocklength such as the Golay code of length 24 evaluated by Adoul and Lamblin [34] as having an SQNR of 2.53 dB, and a length 32 Hadamard scrambling/permutation code due to Schroeder and Sloane [35] with an SQNR of 2.51 dB as measured by Adoul and Lamblin. Fischer's pyramid VQ represents a similar approach to the Laplacian iid source [36]. Observe, however, that in our results the D_4 lattice quantizer is uniformly outperformed by all the other systems. This suggests that we should use high-resolution quantizer conclusions about optimal lattices with care when the number of indexes is small. A similar case, in which an ECSQ outperforms a lattice quantizer based on the A_4^* lattice for the Laplacian memoryless source, was reported in Sayood *et al.* [11]. (The fact that the ECSQ and uniform threshold scalar quantizers have essentially identical performance for this source was demonstrated by Berger [14] and Farvardin and Modestino [17].)

Results for a Gauss–Markov source with correlation coefficient $a = 0.9$ are shown in Fig. 9. Again, the training and test sequences were 40 960 samples each. The rate-distortion function for this source for $R > 0.926$ bits/sample is [3, p. 113]

$$R(D) = \frac{1}{2}\log_2\frac{0.19}{D} \qquad \text{bits/sample.}$$

For $R < 0.926$ bits/sample, this is a lower bound on the rate-distortion curve (and hence the resulting SQNR curve

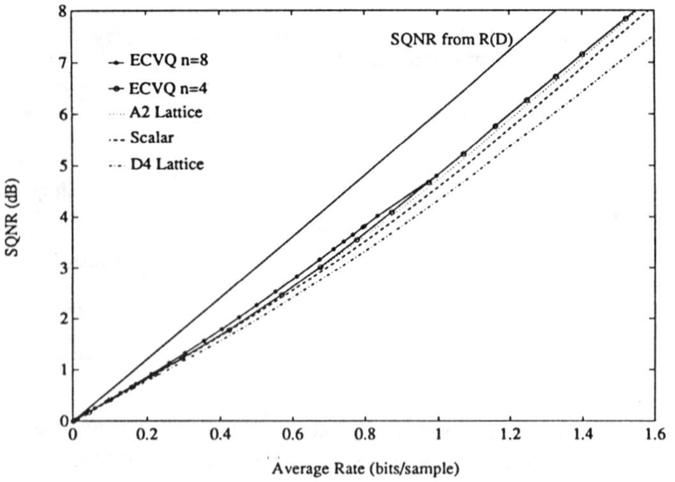

Fig. 8. Performance of some of the systems of Table I on a Gaussian iid source.

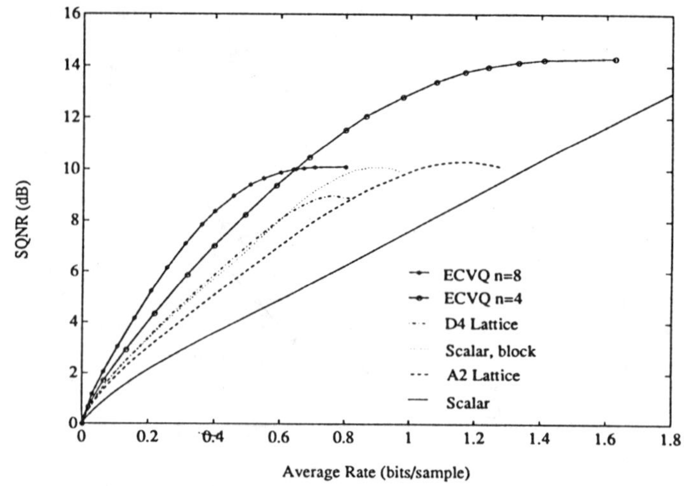

Fig. 10. Performance of the systems of Table I on a speech waveform source.

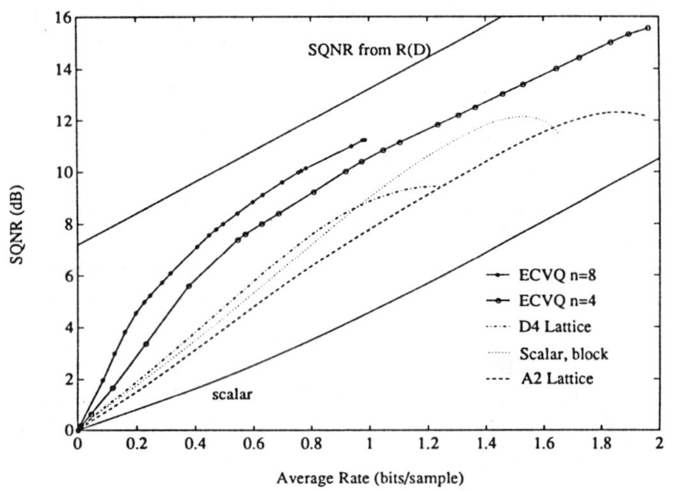

Fig. 9. Performance of the systems of Table I on a Gauss–Markov source with correlation coefficient $a = 0.9$. The $R(D)$ derived SQNR curve is an upper bound for $R < 0.926$ bits/sample.

is an upper bound on the true rate-distortion derived SQNR curve). In this experiment, there is a clear advantage for the ECVQ coders over the other systems; in many cases, rate can be cut in half with no increase in distortion. At 0.75 bits per sample for a blocklength of $n = 4$, ECVQ has a 1.6 dB advantage over the entropy-coded D_4 lattice—its nearest competitor. However, a predictive trellis code with SQNR as high as 12.58 dB at 1 bit/sample has been designed for this source [37]. Note, however, that the memory inherent in both the state and the predictor gives such a code an effective block length much larger than dimension 8.

In Fig. 10 we show results for a speech waveform coding experiment. The training sequence was 20 s of 8 kHz sampled speech (160 000 samples) and the test sequence was 10 s (80 000 samples) from the same (male) speaker. Again, the ECVQ systems have a clear advantage over the other systems. At 0.75 bits per sample for a blocklength of $n = 4$, ECVQ again has a 1.6 dB advantage over its nearest competitor—this time the block entropy-coded scalar system. Admittedly, direct vector quantization of waveform speech has had limited success in producing good subjective quality at such rates. More fruitful applications of ECVQ to real sources will probably come in the form of transform, subband, and image vector quantization. Preliminary results indicate that the technique cannot be used to full advantage on LPC-VQ, however.

Our second set of experiments compares ECVQ to full-search vector quantizers, complete tree-structured vector quantizers, and entropy-pruned tree-structured vector quantization, again for waveform speech coding. The full-search vector quantizers are designed by the generalized Lloyd algorithm [21] to have a minimum distortion subject to a constraint on the number of indexes. However, the indexes are subsequently entropy encoded. Similarly, the binary tree-structured vector quantizers are designed recursively using the generalized Lloyd algorithm, as in [38], but the output indexes are subsequently encoded. The entropy-pruned tree-structured vector quantizers are designed by a recent technique [22] that seeks to prune complete tree-structured quantizers so as to minimize the distortion subject to a constant on the index entropy.

The training sequence for these experiments was locally generated, consisting of 2 min of 8 kHz sampled speech from 3 males and 3 females. The test sequence was 40 s of speech from a male and female not in the training sequence. All vectors had 8 dimensions, so that the training and test sequences consisted of 120 000 and 40 000 vectors, respectively. Generalized Lloyd algorithm and complete tree-structured vector quantizers were designed for a 1.5 bit/sample rate (4096 codewords) and used as initial codebooks ($\lambda = 0$) for the ECVQ and entropy-pruned tree-structured vector quantizers, respectively. Additional generalized Lloyd algorithm and complete tree-structured vector quantizers were designed for lower rates. The results, which again demonstrate a performance advantage for ECVQ, are shown in Fig. 11. At 0.75 bits per sample, ECVQ has a 1.3 dB advantage in SQNR over entropy-pruned tree-structured vector quan-

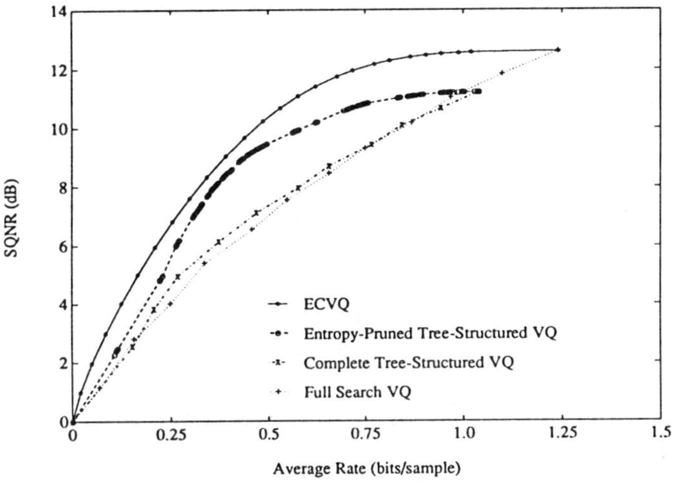

Fig. 11. Performance of entropy-coded vector quantizers in a speech waveform coding experiment.

tization, which in turn has a 1.6 dB advantage over other entropy-coded vector quantization systems. It is important, however, to note the advantage of the pruned tree-structured system in terms of complexity: like ECVQ, it realizes significant gains over the other vector quantizer systems, but it is much easier to search. For instance, the 0.75 bit/sample system requires an average of 6 distortion calculations per input vector compared to 4096 for ECVQ.

V. Applications to Pattern Recognition

The generalized Lloyd algorithm has more than a passing similarity to several clustering algorithms of the statistical pattern recognition community, in particular, an algorithm due to Forgey [39], the k-means algorithm [40], and the ISODATA algorithm [41]. The basic operation of all these algorithms is to assign data to cluster centers on the basis of which center is closest (generally in Euclidean or Mahalanobis distance), next choose new cluster centers by centroiding, and then iterate these two steps until convergence is reached. In the context of clustering, we will refer to these collectively as k-means-type algorithms. In this section, we suggest that the ECVQ algorithm may be useful in applications where it is desired to separate clusters of widely varying population, since it is a k-means-type algorithm that adjusts the distance to a cluster center by a measure of its population.

If the clusters are modeled by a Gaussian mixture, the optimal (minimum risk) classifier is straightforward to design, at least in principle, using Bayes' theorem. Let us assume that n-dimensional observed vectors $x_1, x_2, \cdots$, are drawn independently from a Gaussian mixture:

$$p(x) = \sum_{i=0}^{m-1} p(x \mid \omega_i)\, P(\omega_i),$$

where ω_i, $i = 0, \cdots, m-1$ represents the class, $P(\omega_i)$ is the prior probability of class i, and $p(x \mid \omega_i)$ is Gaussian with mean μ_i and covariance matrix Σ_i. The minimum probability of error rule for determining ω_i from an observation x follows directly from Bayes' theorem [42, ch. 2]:

$$\hat{\imath} = \operatorname*{argmax}_i \left[p(\omega_i | x) \right]$$

$$= \operatorname*{argmax}_i \left[\log p(x | \omega_i) + \log P(\omega_i) \right],$$

which, under the assumption $\Sigma_i = \Sigma\ \forall i$, is equivalent to

$$\hat{\imath} = \operatorname*{argmin}_i \left[\rho_M(x, \mu_i) - \lambda \log P(\omega_i) \right], \quad (7)$$

where $\rho_M(x, \mu_i) = (x - \mu_i)'\Sigma^{-1}(x - \mu_i)$ is the squared Mahalanobis distance, μ' denotes the transpose of μ, and $\lambda = 2$. We recognize (7) as identical to the calculation (3) performed by α, the first stage of the ECVQ encoder of Fig. 1, in mapping a vector to its index, providing we make the identification $\beta(i) = \mu_i$ and $|\gamma(i)| = -\log P(\omega_i)$. (This is precisely the choice of $|\gamma(i)|$ determined by the ECVQ design algorithm in (5) if $P(\omega_i)$ is assumed proportional to the population of cell i.) Hence, the ECVQ encoder first stage corresponds to an optimal classifier for a Gaussian mixture when the covariance matrices are the same for each class. If the prior probabilities are also the same for each class, i.e., the prior probabilities are uniform, then the standard vector quantizer encoder corresponds to the optimal classifier. For $\Sigma = \sigma^2 I$, we can further simplify (7) to write

$$\hat{\imath} = \operatorname*{argmin}_i \left[\rho(x, \mu_i) - \lambda^* \log P(\omega_i) \right], \quad (8)$$

where $\rho(x, \mu_i)$ is squared-error distortion and λ^* is $2\sigma^2$.

The question still remains as to how to determine the parameters of the probability model from a set of training data $x_0, x_1, \cdots, x_{N-1}$. A typical approach is to determine the maximum likelihood estimate of the parameters. Necessary conditions for the estimate $\hat{p} = (\hat{p}_0, \cdots, \hat{p}_{m-1})$ of the prior probabilities and $\hat{\mu} = (\hat{\mu}_0, \cdots, \hat{\mu}_{m-1})$ of the means to be stationary points of the likelihood are [42, p. 193]

$$\hat{p}_i = \frac{1}{N} \sum_{k=0}^{N-1} \hat{p}(\omega_i | x_k, \hat{\mu}) \quad (9)$$

$$\hat{\mu}_i = \frac{\sum_{k=0}^{N-1} \hat{p}(\omega_i | x_k, \hat{\mu})\, x_k}{\sum_{k=0}^{N-1} \hat{p}(\omega_i | x_k, \hat{\mu})} \quad (10)$$

$$\hat{p}(\omega_i | x_k, \hat{\mu}) = \frac{p(x_k | \hat{\mu}_i)\, \hat{p}_i}{\sum_{j=0}^{m-1} p(x_k | \hat{\mu}_j)\, \hat{p}_j}, \quad (11)$$

where $p(x \mid \hat{\mu}_i)$ is the multivariate Gaussian density with mean $\hat{\mu}_i$ and covariance matrix Σ_i.

Equations (9), (10), and (11) form the basis for an iterative algorithm which attempts to find the maximum likelihood estimates of the means and prior probabilities by first fixing $\hat{p}$ and $\hat{\mu}$ and determining the $\hat{p}(\omega_i | x_k, \hat{\mu})$ from (11), then fixing $\hat{p}(\omega_i | x_k, \hat{\mu})$ and determining $\hat{p}$ and

$\hat{\mu}$ from (9) and (10). This algorithm, which we will term the maximum likelihood algorithm, was first proposed by Wolfe [43]. Duda and Hart approximate (11) by

$$\hat{p}(\omega_i | x_k, \hat{\mu}) = \begin{cases} 1 & \text{if } \rho_M(x_k, \hat{\mu}_i) < \rho_M(x_k, \hat{\mu}_j), \quad j \neq i, \\ 0 & \text{otherwise.} \end{cases}$$

This is indeed reasonable if the μ_i are well separated and the $\hat{p}_i$ are all equal. In this case, the resulting iterative algorithm is equivalent to the k-means-type algorithms. However, if we allow the $\hat{p}_i$ to vary, then a more reasonable approximation is

$$\hat{p}(\omega_i | x_k, \hat{\mu}) = \begin{cases} 1 & \text{if } \rho_M(x_k, \hat{\mu}_i) - \lambda \log \hat{p}_i \\ & \quad < \rho_M(x_k, \hat{\mu}_j) - \lambda \log \hat{p}_j, \quad j \neq i, \\ 0 & \text{otherwise.} \end{cases}$$

When applied to (9), (10), and (11), this leads to the ECVQ algorithm. Hence, the ECVQ algorithm is an approximation to the maximum likelihood algorithm for estimating means *and* prior probabilities in the same way that the k-means-type algorithms are an approximation to the maximum likelihood algorithm for estimating means with the classes assumed equiprobable.

To compare the behavior of several of these algorithms, we simulated a 40 000 sample training sequence drawn from a four component bivariate Gaussian mixture with different prior probabilities but all covariance matrices equal to the identity matrix. Several thousand samples from the density are shown in Fig. 12. We used the maximum likelihood algorithm, the generalized Lloyd algorithm, and the ECVQ algorithm (with $\lambda = 2$) to estimate means and prior probabilities. (The ECVQ algorithm and maximum likelihood algorithm used the final estimate of the generalized Lloyd algorithm as the starting point for their iterations.) We then used the Bayes' classifier corresponding to each algorithm to classify data for which we (but not the algorithms) knew the labels. The results of the experiment are shown in Tables II and III. Not surprisingly, the maximum likelihood algorithm provided the best estimates of the prior probabilities and the best classifier performance. It also provided the best estimates of the means (not shown). While not achieving the performance of the maximum likelihood algorithm, the ECVQ algorithm both provided better mean estimates and significantly outperformed the generalized Lloyd algorithm in classification accuracy. Although not shown, we were able to improve the ECVQ algorithm's performance by varying λ from the "optimal" value of 2. In fact, the choice $\lambda = 1$ resulted in a slightly lower misclassification rate than the maximum likelihood algorithm. Coupled with the fact that λ in general depends on knowledge of the cluster variances [see (8)], which may not be readily available, this suggests that a classifier design might consider several different λ's. The curve tracing algorithm of Section III would be useful for selecting such a set. However, dependence of the ECVQ algorithm on the starting codebook, training sequence size, and choice of λ appears to be much more serious in pattern recognition applications than in source coding applications, so that the algorithm should be applied with care, particularly in circumstances where an appeal to maximum likelihood becomes even more difficult.

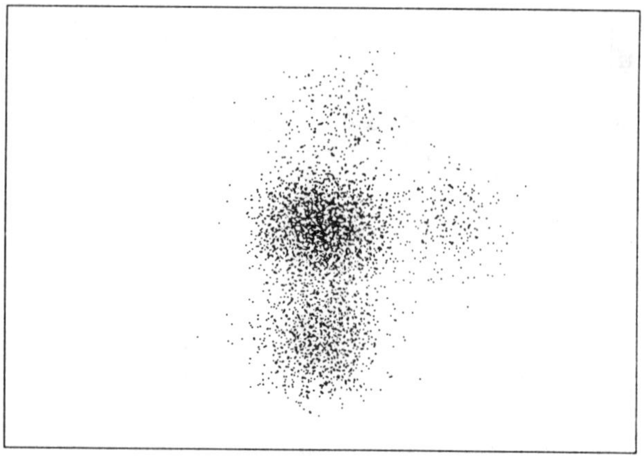

Fig. 12. Four component bivariate Gaussian mixture. The component probabilities are nonuniform.

TABLE II
TRUE AND ESTIMATED CLASS PROBABILITIES FOR A GAUSSIAN MIXTURE

Class	True Probabilities	Maximum Likelihood Estimated Probabilities	ECVQ Estimated Probabilities
1	0.700	0.699	0.725
2	0.050	0.052	0.041
3	0.050	0.050	0.045
4	0.200	0.199	0.189

TABLE III
PERFORMANCE OF VARIOUS ALGORITHMS FOR PARAMETER ESTIMATION AND CLASSIFICATION OF A GAUSSIAN MIXTURE

Algorithm	Maximum Likelihood	Generalized Lloyd Algorithm	ECVQ
Log Likelihood	−72459	−79504	−72633
Misclassification Rate	0.0317	0.1640	0.0356

VI. DISCUSSION AND CONCLUSION

In this paper we have introduced an iterative descent algorithm for designing locally optimal variable rate coders for use in a block quantization or data compression scheme. In essence, the algorithm designs vector quantizers which are optimized to perform well when followed by a variable rate entropy coder, such as a Huffman, arithmetic, or Ziv–Lempel coder. The choice of which entropy coder to use is not made by the algorithm, but is instead left to the implementor. Any of these entropy coders will produce an average rate approximately equal to the codeword entropy. The algorithm simply ensures that the codeword entropy and the distortion are low.

For high rates (high resolution), asymptotic quantization theory predicts that, for the squared error distortion measure and other difference distortion measures expressible as the rth power of a seminorm, the advantage in performance of the optimum ECVQ system over simpler scalar and lattice systems will be small. Our experimental investigations have focused on the low-rate region (on the order of one bit per sample), where it is shown that for memoryless sources, such as the Gaussian iid source, the performance improvement of ECVQ over simple scalar quantization is indeed negligible. However, for sources with memory, such as the first-order Gauss–Markov source with correlation coefficient $a = 0.9$, as well as for real speech, the improvement in mean squared error is shown to be as much as 1.6 dB at 0.75 bits/sample. Gains in the performance of ECVQ over other forms of entropy-coded vector quantization are shown to be even more substantial.

A real-time ECVQ implementation would most likely be no more complex than an entropy-coded VQ implementation. The principal VQ encoder calculation for many interesting distortion measures can be simplified to an inner product. For such distortion measures, the first stage calculation for the ECVQ encoder also has an inner product form (with $-\lambda \log_2 p(i)$ added to the constant); hence, several new inner product based systolic array vector quantizer architectures are also attractive for implementing the first stage of the ECVQ encoder, allowing large codebooks to be searched in real time [44], [45]. The remaining stages of the ECVQ can be implemented by lookup tables. Thus, ECVQ is an attractive option for a number of real-time low rate data compression systems.

The ECVQ algorithm also plays a role in statistical pattern recognition, where it generalizes the "k-means" algorithm and its variants used in clustering, in that the mixture is not assumed to be equiprobable. That is, the ECVQ algorithm estimates both the means *and the prior probabilities* of the unknown mixture. It is shown experimentally that in the case of widely different priors, the ECVQ algorithm outperforms the equivalent k-means-type of algorithm in likelihood as well as in probability of error. In addition, the estimated means and prior probabilities are close to their true values. It is conjectured that the ECVQ algorithm may be useful even in exploratory data analysis if it is suspected that the class probabilities may not be uniform.

References

[1] C. E. Shannon, "Coding theorems for a discrete source with a fidelity criterion," in *IRE Nat. Conv. Rec.*, 1959, pp. 142–163.

[2] R. G. Gallager, *Information Theory and Reliable Communication.* New York: Wiley, 1968.

[3] T. Berger, *Rate Distortion Theory: A Mathematical Basis for Data Compression.* Englewood Cliffs, NJ: Prentice-Hall, 1971.

[4] H. Gish and J. N. Pierce, "Asymptotically efficient quantizing," *IEEE Trans. Inform. Theory*, vol. IT-14, pp. 676–683, Sept. 1968.

[5] T. J. Goblick and J. L. Holsinger, "Analog source digitization: A comparison of theory and practice," *IEEE Trans. Inform. Theory*, vol. IT-13, pp. 323–326, Apr. 1967.

[6] P. Zador, "Asymptotic quantization error of continuous signals and their quantization dimensions," *IEEE Trans. Inform. Theory*, vol. IT-28, pp. 139–149, Mar. 1982. (Previously an unpublished Bell Lab memo, 1966.)

[7] A. Gersho, "Asymptotically optimal block quantization," *IEEE Trans. Inform. Theory*, vol. IT-25, pp. 373–380, July 1979.

[8] J. H. Conway and N. J. A. Sloane, "Voronoi regions of lattices, second moments of polytopes, and quantization," *IEEE Trans. Inform. Theory*, vol. IT-28, pp. 211–226, Mar. 1982.

[9] ——, "Fast quantizing and decoding algorithms for lattice quantizers and codes," *IEEE Trans. Inform. Theory*, vol. IT-28, pp. 227–231, Mar. 1982.

[10] ——, "A fast encoding method for lattice codes and quantizers," *IEEE Trans. Inform. Theory*, vol. IT-29, pp. 820–824, Nov. 1983.

[11] K. Sayood, J. D. Gibson, and M. C. Rost, "An algorithm for uniform vector quantizer design," *IEEE Trans. Inform. Theory*, vol. IT-30, pp. 805–814, Nov. 1984.

[12] R. C. Wood, "On optimum quantization," *IEEE Trans. Inform. Theory*, vol. IT-15, pp. 248–252, Mar. 1969.

[13] T. Berger, "Optimum quantizers and permutation codes," *IEEE Trans. Inform. Theory*, vol. IT-18, pp. 759–765, Nov. 1972.

[14] ——, "Minimum entropy quantizers and permutation codes," *IEEE Trans. Inform. Theory*, vol. IT-28, pp. 149–157, Mar. 1982.

[15] A. N. Netravali and R. Saigal, "Optimum quantizer design using a fixed-point algorithm," *Bell Syst. Tech. J.*, vol. 55, pp. 1423–1435, Nov. 1976.

[16] P. Noll and R. Zelinski, "Bounds on quantizer performance in the low bit-rate region," *IEEE Trans. Commun.*, vol. COM-26, pp. 300–304, Feb. 1978.

[17] N. Farvardin and J. W. Modestino, "Optimum quantizer performance for a class of non-Gaussian memoryless sources," *IEEE Trans. Inform. Theory*, vol. IT-30, pp. 485–497, May 1984.

[18] J. Ziv, "On universal quantization," *IEEE Trans. Inform. Theory*, vol. IT-31, pp. 344–347, May 1985.

[19] M. Gutman, "On uniform quantization with various distortion measures," *IEEE Trans. Inform. Theory*, vol. IT-33, pp. 169–171, Jan. 1987.

[20] S. P. Lloyd, "Least squares quantization in PCM," *IEEE Trans. Inform. Theory*, vol. IT-28, pp. 129–136, Mar. 1982. (Previously an unpublished Bell Lab Tech. Note, 1957.)

[21] Y. Linde, A. Buzo, and R. M. Gray, "An algorithm for vector quantizer design," *IEEE Trans. Commun.*, vol. COM-28, pp. 84–95, Jan. 1980.

[22] P. A. Chou, T. Lookabaugh, and R. M. Gray, "Optimal pruning with applications to tree structured source coding and modeling," *IEEE Trans. Inform. Theory*, to appear.

[23] M. B. Pursley and L. D. Davisson, "Variable rate coding for nonergodic sources and classes of ergodic sources subject to a fidelity constraint," *IEEE Trans. Inform. Theory*, vol. IT-22, pp. 324–337, May 1976.

[24] R. Blahut, "Computation of channel capacity and rate-distortion functions," *IEEE Trans. Inform. Theory*, vol. IT-18, pp. 460–473, July 1972.

[25] M. J. Sabin and R. M. Gray, "Global convergence and empirical consistency of the generalized Lloyd algorithm," *IEEE Trans. Inform. Theory*, vol. IT-32, pp. 148–155, Mar. 1986.

[26] D. A. Huffman, "A method for the construction of minimum-redundancy codes," *Pro. IRE*, vol. 40, pp. 1098–1101, Sept. 1952.

[27] R. G. Gallager, "Variations on a theme by Huffman," *IEEE Trans. Inform. Theory*, vol. IT-24, pp. 668–674, Nov. 1978.

[28] J. Rissanen and G. G. Langdon, Jr., "Arithmetic coding," *IBM J. Res. Develop.*, vol. 23, pp. 149–162, Mar. 1979.

[29] F. Rubin, "Arithmetic stream coding using fixed precision registers," *IEEE Trans. Inform. Theory*, vol. IT-25, pp. 672–675, Nov. 1979.

[30] J. Ziv, "Coding theorems for individual sequences," *IEEE Trans. Inform. Theory*, vol. IT-24, pp. 405–412, July 1978.

[31] T. Berger, M. U. Chang, and S. Y. Tung-Kleinberg, "Quantization-permutation codes and buffer-adapted Huffman codes," in *Proc. 18th Annu. Allerton Conf. Commun., Contr.*, Comput., Monticello, IL, 1980, pp. 443–436.

[32] K. D. Rines and N. C. Gallagher, Jr., "The design of two-dimensional quantizers using pre-quantization," *IEEE Trans. Inform. Theory*, vol. IT-28, pp. 232–238, Mar. 1982.

[33] D. J. Newman, "The hexagon theorem," *IEEE Trans. Inform. The-*

ory, vol. IT-28, pp. 137–139, Mar. 1982. (First appeared as a Bell Lab Tech. Memo, May 1964.)

[34] J. Adoul and C. Lamblin, "A comparison of some algebraic structures for CELP coding of speech," in *Proc. ICAASP*, Dallas, TX, Apr. 1987, pp. 45.8.1–45.8.4.

[35] M. R. Schroeder and N. J. A. Sloane, "New permutation codes using Hadamard unscrambling," *IEEE Trans. Inform. Theory*, vol. IT-33, pp. 144–146, Jan. 1987.

[36] T. R. Fischer, "A pyramid vector quantizer," *IEEE Trans. Inform. Theory*, vol. IT-32, pp. 568–583, July 1986.

[37] E. Ayanoğlu and R. M. Gray, "The design of predictive trellis waveform coders using the generalized Lloyd algorithm," *IEEE Trans. Commun.*, vol. COM-34, pp. 1073–1080, Nov. 1986.

[38] A. Buzo, A. H. Gray, Jr., R. M. Gray, and J. D. Markel, "Speech coding based upon vector quantization," *IEEE Trans. Inform. Theory*, vol. IT-28, pp. 562–574, Oct. 1980.

[39] E. W. Forgey, "Cluster analysis of multivariate data: Efficiency versus interpretability of classifications," *Biometrics*, vol. 21, no. 3, p. 768, 1965.

[40] J. B. MacQueen, "Some methods for classification and analysis of multivariate observations," in *Proc. 5th Berkeley Symp. Math. Stat. Probabil.* Berkeley, CA: University of California Press, 1967, pp. 281–297.

[41] G. H. Ball and D. J. Hall, "ISODATA, A novel method of data analysis and pattern classification," Tech. Rep. AD 699616, Stanford Res. Inst., Menlo Park, CA, 1965.

[42] R. O. Duda and P. E. Hart, *Pattern Classification and Scene Analysis.* New York: Wiley, 1973.

[43] J. H. Wolfe, "Pattern clustering by multivariate mixture analysis," *Multivariate Behavior. Res.*, vol. 5, pp. 329–350, 1970.

[44] P. Cappello, G. Davidson, A. Gersho, C. Koc, and V. Somayazulu, "A systolic vector quantization processor for real-time speech coding," in *Proc. ICASSP*, 1986, pp. 2143–2146.

[45] R. Dianysian and R. L. Baker, "A VLSI chip set for real time vector quantization of image sequences," in *Proc. ISCS*, May 1987.

Part 3
Speech Coding

Rate-Distortion Speech Coding with a Minimum Discrimination Information Distortion Measure

ROBERT M. GRAY, FELLOW, IEEE, AUGUSTINE H. GRAY, JR., SENIOR MEMBER, IEEE, GUILLERMO REBOLLEDO, STUDENT MEMBER, IEEE, AND JOHN E. SHORE, SENIOR MEMBER, IEEE

Abstract—An information theory approach to the theory and practice of linear predictive coded (LPC) speech compression systems is developed. It is shown that a traditional LPC system can be viewed as a minimum distortion or nearest-neighbor system where the distortion measure is a minimum discrimination information between a speech process model and an observed frame of actual speech. This distortion measure is used in an algorithm for computer-aided design of block source codes subject to a fidelity criterion to obtain a 750-bits/s speech compression system that resembles an LPC system but has a much lower rate, a larger memory requirement, and requires no on-line LPC analysis. Quantitative and informal subjective comparisons are made among our system and LPC systems.

I. Introduction

SPEECH COMPRESSION systems can generally be classified as either waveform coders or speech coders. Waveform coders attempt to communicate an approximation to the original sampled speech waveform. Examples are traditional pulse-code modulation (PCM) and differential pulse-code modulation (DPCM) systems, tree coding techniques, and adaptive versions of these systems. Waveform coders typically require from one bit per sample (6500 bits/s for the sampling rate considered here) for adaptive tree coding to 10 bits per sample for PCM.

Speech coders (or voice coders or vocoders) attempt to track the underlying process producing the waveform rather than the waveform itself. For a frame or vector of sampled speech, the encoder first performs a system identification procedure and selects a speech production model from an allowed class for the observed frame. Then a digital approximation to the model is formed and communicated to the receiver, where the reproduction speech is synthesized using the communicated model. At the present time, the best quality low-rate speech coding systems of reasonable complexity are those based on linear predictive coding (LPC) techniques [32,], [36], [22], [23], [2], [31], [39]. These systems typically transmit from 2400 to 6000 bits/s.

Manuscript received March 4, 1980; revised December 12, 1980. This work was supported in part by the Air Force Office of Scientific Research under Contract F 49620-79-0058, in part by the Consejo Nacionale de Ciencia y Technologia de Mexico, and in part by the Naval Research Laboratory under Contract NRL N00039-79-C-0450.

R. M. Gray is with the Information Systems Laboratory, Stanford University, Stanford, CA 94305.

A. H. Gray, Jr., is with Signal Technology, Inc., Santa Barbara, CA.

G. Rebolledo was with the Information Systems Laboratory, Stanford University, Stanford, CA. He is now with Time and Space Processing, Inc., Santa Clara, CA.

J. E. Shore is with the Naval Research Laboratory, Washington, DC 20375.

In an LPC system the speech production or synthesis model is a finite-order autoregressive (all pole) filter driven by either white noise (for unvoiced sounds) or a periodic pulse train (for voiced sounds). Such a model can be described by a random process $\{y_i\}_{i=0}^{\infty}$ defined by the difference equation

$$y_n = -\sum_{j=1}^{M} a_j y_{n-j} + \sigma\left(\beta^{1/2}\epsilon_n + (1-\beta)^{1/2} s_n\right), \qquad n \geq 0$$

$$y_i = u_i, \qquad i = -M,\cdots,-1, \tag{1}$$

where the u_i are the initial conditions for the filter (which are usually determined by the previous frame), σ is the filter grain, and $\beta^{1/2}\epsilon_n + (1-\beta)^{1/2}s_n$ is the driving process, where ϵ_n is a zero-mean unit variance sequence of independent random variables and s_n is a periodic pulse train with period τ, phase φ, and zero time-average mean and unit time-average energy. The parameter $\beta \in [0,1]$ provides a convenient mix of voiced ($\beta = 0$) and unvoiced ($\beta = 1$) sounds [32, p. 247]. We assume for convenience that $\beta > 0$, but it can be very small for voiced sounds and is not known *a priori*.

Intuitively, the process $\{y_i\}$ is produced by driving an autoregressive filter described by the z-transform $\sigma/A(z)$,

$$A(z) = 1 + \sum_{k=1}^{M} a_k z^{-k},$$

by a zero-mean unit power process. We shall refer to $\sigma/A(z)$ or simply σ/A, as the model filter and $1/A(z)$ or $1/A$ as the normalized model filter. We assume that the a_i are chosen so that the filter is stable, and hence $\{y_i\}$ is asymptotically stationary.

The complete model is specified by the parameters $M, \sigma, a_1,\cdots, a_M, \tau, \beta, \varphi$, the initial conditions $\{u_i\}$, and a marginal probability density function for ϵ_n. It has been found experimentally that good quality can be obtained with a variety of densities for ϵ_n, including a simple uniform distribution or a Gaussian distribution. We adopt the

Reprinted from *IEEE Trans. Inform. Theory,* vol. IT-27, pp. 708–721, Nov. 1981.

latter assumption since it is more useful mathematically. The model order M is usually fixed *a priori* as a compromise—a high-order is needed to obtain a good model, but too high an order causes the filter to track spectral properties of the pitch (the driving process component s_n). Furthermore, the typically small amount of speech samples (64–256) used to estimate the parameters make high-order spectral estimates unreliable. Typical orders are 10–20. We here consider M equal to 10.

Strictly speaking, LPC analysis is a method for observing a frame of sampled speech data $x = (x_0, \cdots, x_{n-1})$, where n is the frame length ($n = 128$ here), and producing an estimate of the filter σ/A or some equivalent parameter vector such as $(\sigma, a_1, \cdots, a_M)$. In practice, the speech samples are often prefiltered to enhance desired characteristics before the estimation. In particular, the speech is often bandpass filtered to some desired range (here 200 Hz–3.3 kHz), preemphasized to enhance higher order spectral components (here with a single pole filter with coefficient 0.9), and Hamming windowed. Such prefiltering either has little effect on the speech or is compensated by appropriate postfiltering at the receiver. We do not further consider this aspect of the system; we assume that any such preprocessing has been incorporated into the observed speech frame x and that any necessary post-processing can be added to the synthesizer.

The phase parameter φ is usually ignored and generated locally at the receiver. The voicing parameter β and the pitch period τ can be estimated in a variety of ways that may or may not be related to the model estimation (see, e.g., [32], [36], [39, pt. III]). We focus on the communication of the filter model σ/A and will use straightforward techniques from the literature for estimating and communicating the pitch and voicing parameters.

The procedure described above is an estimation or system identification procedure—genuine data compression has not yet occurred, since LPC analysis produces an analog model σ/A (or a very high-rate digital approximation if the analysis is done digitally). The actual compression occurs in a second digitization or quantization step where σ/A (or some equivalent parameter vector) and (β, τ) are quantized or digitized for communication over a low-rate digital channel. The receiver then uses this digital approximation to synthesize the speech with a locally generated driving function.

The overall communication system for the model filter thus comprises two distinct steps: 1) LPC analysis, wherein an observed frame of speech is mapped into a filter $\sigma/A \in \mathcal{F}_M$, the class of all Mth-order stable autoregressive filters, using standard techniques such as Levinson's algorithm or Burg's algorithm; and 2) quantization, wherein a discrete approximation $\hat{\sigma}/\hat{A}$ to σ/A is selected from a finite code book $\mathcal{C} = \{(\sigma/A)_i;\ i = 1, \cdots, N\}$ of allowed filters in $\mathcal{F}_M$. We shall refer to such two-step speech compression systems as LPC speech compression systems or, simply, LPC systems, since the first step is an LPC analysis and the second step produces LPC models. The pitch and voicing can be handled in a parallel manner.

Virtually all traditional LPC systems perform the digitization step with scalar (single-symbol, independent, memoryless) quantizers applied separately to each parameter in some parameter vector representation of the filter (see, e.g., [47], [16], [33], [39, pt. IV], and the references therein).

Rate distortion theory (see, e.g., [11], [3]) provides alternative mathematical models and possible implementations for speech compression systems with similar synthesis codebooks $\mathcal{C}$ of stable autoregressive filters. One model is an LPC system with an alternative digitization step: if we had a distortion measure $d(\sigma/A, \hat{\sigma}/\hat{A})$ that assigned a distortion or cost to reproducing an LPC model σ/A by an approximation $\hat{\sigma}/\hat{A} \in \mathcal{C}$, then the Shannon rate distortion theory model of a data compression system (for the LPC analysis output) would use a minimum distortion mapping as an encoder. That is, σ/A is mapped into the $\hat{\sigma}/\hat{A} \in \mathcal{C}$, minimizing $d(\sigma/A, \hat{\sigma}/\hat{A})$. As in the similar pattern classification and cluster analysis procedures, we refer to such a mapping as a nearest-neighbor mapping even if the distortion measure is not a metric. (See, e.g., [14], [15].)

A second and more general approach is possible if we have a distortion measure $\rho(x, \sigma/A)$ that measures the distortion between the original frame of sampled speech and any filter $\sigma/A \in \mathcal{F}_M$. In this case we can forego the LPC analysis step and consider a single step system that directly selects a filter $\hat{\sigma}/\hat{A} \in \mathcal{C}$ for an input frame using a nearest-neighbor rule. This system still uses LPC synthesis models, but it is not a standard LPC system since no explicit LPC analysis is performed. Following Chaffee and Omura [7], [8], we refer to such a system as a *rate-distortion speech coder*.

If appropriate distortion measures can be found, these abstract rate distortion theory models could provide a framework for analyzing and comparing LPC systems and could yield a single step implementation of an LPC-like system. This paper provides a preliminary development of these ideas.

Rate-distortion theory yields two immediate observations on LPC systems. First, scalar quantization is, in general, inferior to vector or block quantization, and hence one can always do better by quantizing an entire parameter vector at once than by separately quantizing its components. This is true even if the parameters are probabilistically independent, and it is especially true if the distortion measure couples the parameters. The issue is how much performance improvement is achievable and at what price. Second, traditional LPC speech compression systems are not optimal in a rate-distortion theory sense, since the mapping from the input x or its LPC model σ/A to the codeword $\hat{\sigma}/\hat{A} \in \mathcal{C}$ is not a minimum distortion mapping using any known distortion. The digitization is instead an *ad hoc* two-step mapping consisting of LPC analysis followed by scalar quantizers that are usually designed to separately minimize magnitude error or squared error distortion for each component.

Itakura and Saito [22], [23] showed that the LPC analysis step was mathematically equivalent to a minimum distortion mapping using their so-called "error matching measure" defined by

$$\rho_{IS}(x, \sigma/A) = \int_{-\pi}^{\pi} \frac{d\theta}{2\pi} \{ f_x(\theta) |A(e^{i\theta})|^2/\sigma^2 - \ln (f_x(\theta) |A(e^{i\theta})|^2/\sigma^2) - 1 \} \quad (2)$$

where

$$f_x(\theta) = n^{-1} |X(e^{i\theta})|^2 = n^{-1} \left| \sum_{k=0}^{n-1} x_k z^{-k} \right|^2 \quad (3)$$

is the sample spectrum of the observed speech frame. This same distortion was not, however, minimized in the subsequent digitization step nor in an overall sense.

Observe that we can also define the Itakura–Saito distortion as a distortion measure on spectral densities [18]

$$d_{IS}(f, \hat{f}) = \int_{-\pi}^{\pi} \frac{d\theta}{2\pi} \left\{ \frac{f(\theta)}{\hat{f}(\theta)} - \ln \frac{f(\theta)}{\hat{f}(\theta)} - 1 \right\} \quad (4)$$

in which case $\rho_{IS}(x, \sigma/A) = d_{IS}(f_x, \sigma^2/|A|^2)$, where with a slight abuse of notation we consider $A = A(e^{i\theta})$ as a function of θ. We shall refer to d_{IS} as the spectral Itakura–Saito distortion.

The Itakura–Saito distortion measure has one obvious important attribute: minimizing ρ_{IS} is equivalent to LPC analysis, and high-rate LPC systems sound good, so ρ_{IS} is a logical candidate for a subjectively meaningful distortion measure in speech processing systems. From a rate-distortion theory and intuitive point of view, however, the original development of this distortion has some critical drawbacks. Itakura and Saito developed the LPC analysis equations as an approximate maximum likelihood estimate of the parameters of an Mth order Gaussian autoregressive source ($\{y_n\}$ of (1) with $\beta = 1$) based on an asymptotically long observed frame of samples. Thus the observed speech was assumed to be Gaussian (a controversial assumption), the frame length was assumed to be huge (in practice, it is usually less than 200), and a pure noise driving process with no periodic component was assumed (even when the model might be used to synthesize voiced sounds). Thus any possible effect of a periodic driving component on the filter selection based on the observations was ignored theoretically. Finally, their development was only for continuous LPC analysis (using a variational approach), and hence it is not clear that the same distortion measure is theoretically relevant to digitization.

The first to attempt a speech compression system with digitized LPC models selected by a minimum distortion mapping were Chaffee and Omura [7], [8], Franson [10], and Kang and Coulter [24], [25]. Chaffee and Omura used minimum distortion mappings to find separately the normalized model filter $1/A$ and the gain σ directly from an observed frame $\boldsymbol{x}$. Their distortion measure for the filter selection was the filter distortion measure—the distortion of (2) with $\sigma^2 = \alpha_M = 1$ [18]. Franson used a single-symbol magnitude error distortion on a subset of the PARCOR coefficients and used a combination of scalar and vector quantization. Kang and Coulter used a form of weighted Euclidean distance on formant frequencies to select spectra from a finite set to match the LPC analysis spectrum. All of these efforts used *ad hoc* procedures requiring interactive human selection for choosing the codewords based on a few short standard speech phrases. Hence, even though a minimum distortion rule was used to encode into a given code book, there was no guarantee that the actual code book itself was even locally optimum in the sense of minimizing average distortion. Furthermore, none used the same distortion implicitly used in the LPC analysis step.

Linde *et al.* [29] demonstrated that the filter distortion measure used by Chaffee and Omura was amenable for use in an algorithm for the computer-aided design of locally optimum quantizers and obtained preliminary results for low-rate compression of the normalized model filter alone. This algorithm is a generalization to vectors and general distortion measures of an algorithm of Lloyd [30]. (See [29] for a comparison of this algorithm to related algorithms in quantization and pattern recognition.)

In [18] it was demonstrated that the full (nonnormalized) Itakura–Saito distortion measure is well-suited theoretically and computationally for using the algorithm of [29] to digitize the complete model filter σ/A. It was further shown in [18] that if the Itakura–Saito distortion measure is viewed as a distortion measure between spectral densities as in (3), then if LPC analysis on a speech frame $\boldsymbol{x}$ produces a filter σ_M/A_M and $\hat{\sigma}/\hat{A} \in \mathcal{C} \subset \mathcal{F}_M$,

$$d_{IS}(f_x, \hat{\sigma}^2/|A|^2) = d_{IS}(f_x, \sigma_M^2/|A_M|^2) + d_{IS}(\sigma_M^2/|A_M|^2, \hat{\sigma}^2/|\hat{A}|^2).$$

This "triangle equality" for the Itakura–Saito distortion in a two-step LPC system implies [18], [6] that the algorithm of [29] could be used either to design a two-step LPC system by digitizing the LPC parameters or to design a one-step rate-distortion coder operating directly on the speech frames, and that the two systems would yield equivalent code books and overall distortions.

The first complete speech coding system using the Itakura–Saito distortion and the vector quantization design algorithm on the LPC parameters was developed and simulated by Buzo *et al.* [6]. This system was not, however, a direct and exact application of the design algorithm to the Itakura–Saito distortion: the gain σ and normalized filter $1/A$ were treated separately, forcing the codebook to have a supoptimal product form. The suboptimal separation of gain and normalized filter permitted relatively large rates to be simulated with reasonable memory requirements, but it led to the use of several approximations and bounds in the development, design, and implementation of the system. The approach was successful, however, in that a system based on vector quantization with 10 bits per frame for the filter $1/A$, 5 bits per frame for the gain σ, and 7 bits per frame for the pitch and voicing (at 50 frames per second this yields 1000 bits/s) yielded quantitative and subjective quality close to that of an LPC system with approximately optimal scalar quantization and optimal bit allocation at a rate of 2150 bits/s [6], [33], a significant savings in view of the history of LPC systems.

From an information theory point of view, however, the development of [6] had two fundamental drawbacks. 1) The system was suboptimal, and its design incorporated several approximations. Hence the full performance gains achievable by vector quantization were not obtained. 2) The informal development of [6] shed no light on the theoretical meaning of the Itakura–Saito distortion measure itself or of its appropriateness for vector quantization. The purpose of this paper is to resolve these two problems by providing an information theoretic development of the Itakura–Saito and related distortion measures for estimation and quantization applications and by using such a distortion measure in the design of a rate-distortion speech coder.

The next section presents a brief tutorial description of the basic LPC analysis equations and their various original derivations. We then present our first main result: a development of a means of selecting an LPC model from either a continuous set (LPC analysis) or a discrete set (rate-distortion speech coding) based on a classical technique of Kullback combining information theory and statistics. In particular, we show that the Itakura–Saito style distortion measures are asymptotic minimum discrimination information measures between the model (1) and the sample autocorrelation of the observed speech frame, where the minimum is over all probabilistic descriptions of the input having the prescribed autocorrelation values. Unlike previous stochastic developments, the speech is not assumed to be Gaussian. The Gaussian form of the minimum discrimination information instead arises because of an extremal property of Gaussian processes.

The minimum discrimination information measure was proposed by Kullback [26] as a general measure of resemblance (small values) or dissimilarity (large values) for parameter estimation and pattern classification problems.

The approach in this paper has several theoretical and intuitive advantages absent in previous developments.

1) It provides a single development for the distortion measure in both continuous estimation (e.g., LPC analysis) and quantization or classification applications (e.g., speech coding and classification systems).

2) The original speech is not assumed to be Gaussian. The synthesized speech, however, is assumed to be Gaussian. Thus the distortion measure is a measurement of how well or badly a Gaussian model fits real speech.

3) The number of observed speech samples is not assumed to be asymptotically large. We do, however, view the model as a random process rather than a random vector and hence do take an asymptotic limit.

4) The voicing and pitch parameters β and τ are explicitly included in the model.

As an intermediate result of the development, we obtain a rigorous general statement and solution of the so-called "maximum entropy method" [4], [45], [37], [41], [46] for spectral estimation. In particular, necessary and sufficient conditions are given for the existence of a unique solution to a well-posed maximum differential entropy. To our knowledge, this is the first time that the maximization has been precisely stated without assuming *a priori* the existence of a solution. The maximum entropy approach is also known to formally yield the LPC equations, but it is inferior to the minimum discrimination approach, since differential entropy is mathematically ill-behaved [11, pp. 31–32], since it does not imply a distortion measure, and since the extremum is less generally defined and solutions exist less generally. In addition, minimum discrimination has several more desirable properties for spectral estimation and related problems [41], [43], [44]. The one advantage of considering differential entropy is for the evaluation of the more general minimum information—a role similar to that played by differential entropy in the evaluation of rate-distortion functions [3], [11].

In Sections IV and V our second main result is presented: the development and simulation of a rate-distortion speech coding system based on the above distortion measure. This development has two advantages over that of Buzo *et al.* [6]. First, the gain and normalized model filter are treated as a single vector rather than separately. This obviates the approximations and bounds of [6] in the code book design and yields an exact and more elegant development. Second, the joint quantization of gain and normalized model filter means that the model filter code book is not forced to have the suboptimal product form of [6], and hence lower average distortion for a given rate is theoretically possible. The resulting memory requirements prohibit rates as large as those of [6], but we are able to design a code book with only 9 bits for the model filter and 6 bits for the pitch and voicing (for a total of 750 bits/s) that is only slightly inferior to the 1100-bits/s system of Buzo *et al.* [26], the main difference being a more pronounced "warble" during voiced sounds. Section V contains some concluding remarks.

II. LPC Analysis

There are several derivations of the basic LPC equations [32], [36], [22], [23], [2], [31], [39]. As the name implies, one of these derivations [2], [32] is based on the theory of linear prediction. We collect for reference a few important facts from linear prediction theory and the theory of Toeplitz forms and orthogonal polynomials. Good references are Grenander and Szego [21, ch. 10] and Geronimus [13].

Let $\{x_n\}_{n=-\infty}^{\infty}$ be a stationary zero-mean discrete-time random process with an autocorrelation function $r(k) = Ex_n x_{n-k}$, where E denotes expectation, and power spectral density $f(\theta) = \Sigma_k r(k) e^{-ik\theta}$. We wish to form an Mth-order one-step linear predictor

$$\hat{x}_n = \sum_{k=1}^{M} h_k x_{n-k}$$

of x_n that minimizes the expected squared error $E(e_n^2) = E((x_n - \hat{x}_n)^2)$. If we define the filter with unit sample response a_k, $k = 0, 1, \cdots, M$ by $a_0 = 1$, $a_k = -h_k$, $k =$

$1,\cdots,M$, $a_k = 0$ else, then

$$E(e_n^2) = \sum_{k=0}^{M}\sum_{j=0}^{M} r(k-j)a_k a_j$$
$$= \int_{-\pi}^{\pi}\frac{d\theta}{2\pi} f(\theta)\,|A(e^{i\theta})|^2. \qquad (6)$$

Define the $(M+1)\times(M+1)$ autocorrelation matrix $\boldsymbol{R}^{(M)} = \{r(k-j);\ k=0,\cdots,M;\ j=0,\cdots,M\}$ and the vector $\boldsymbol{a} = (a_0,\cdots,a_M)'$; then $E(e_n^2) = \boldsymbol{a}'\boldsymbol{R}^{(M)}\boldsymbol{a}$. The minimization of the Toeplitz form $\boldsymbol{a}'\boldsymbol{R}^{(M)}\boldsymbol{a}$ over all $\boldsymbol{a}$ with $a_0 = 1$ and a fixed M and f is a much studied problem. The solution can be expressed analytically in terms of orthogonal polynomials and computed numerically via Levinson's and related algorithms (e.g., see [32]). Analytically, the solution can be expressed as the solution $(\sigma_M, \boldsymbol{a})$ the system of equations

$$r(k) = \int_{-\pi}^{\pi}\frac{d\theta}{2\pi} f(\theta)e^{ik\theta}\,d\theta$$
$$= \sigma_M^2\int_{-\pi}^{\pi}\frac{d\theta}{2\pi} e^{ik\theta}\,|A_M(e^{i\theta})|^{-2}, \qquad k=0,\cdots,M. \qquad (7)$$

This may be found, for example, as [13, p. 12, eq. (1.20)]. These equations are called the correlation matching equations in the speech literature [31], [32]. The actual minimum error power is given by (see, e.g., [21, p. 38])

$$\sigma_M^2 = (\det \boldsymbol{R}^{(M)})/(\det \boldsymbol{R}^{(M-1)}). \qquad (8)$$

If the process is such that the one-step prediction error σ_∞^2 defined by

$$\sigma_\infty^2 = \exp\left\{\int_{-\pi}^{\pi}\frac{d\theta}{2\pi}\ln f(\theta)\right\} \qquad (9)$$

is finite, then there exists an A_∞ such that

$$f(\theta) = \sigma_\infty^2/|A_\infty(e^{i\theta})|^2, \qquad (10)$$

and σ_∞/A_∞ is stable.

This provides an exact spectral representation or second-order model of a random process with spectral density $f(\theta)$ as the output of an infinite order stable autoregressive filter σ_∞/A_∞ driven by a zero-mean white (uncorrelated) process. This model is in general only accurate in a second-order moment sense, and the distributions will not be accurate (unless, of course, the original process is Gaussian).

Last, the representation of (10) is a limit of the finite-order linear predictors in the sense that $\sigma_M^2 \downarrow \sigma_\infty^2$ and $A_M \to A_\infty$ as $M\to\infty$ in L_2 [21, sec. 10.8]. Thus the error or residual spectral density $f_e(\theta) = f(\theta)\,|A_M(e^{i\theta})|^2$ becomes increasingly flat, and hence the error process $\{e_n\}$ becomes white in the limit as $M\to\infty$.

The question now arises: what does linear prediction of a known stationary process have to do with LPC analysis —that is, with fitting an autoregressive model to real speech data? Mechanically, in LPC analysis using the so-called autocorrelation method, first the sample autocorrelation

$$r_x(k) = \begin{cases}\sum_{i=0}^{n-|k|-1} x_i x_{i+|k|}, & |k|\le n-1\\ 0, & |k|>n\end{cases} \qquad (11)$$

is found, and then the previously described techniques are used to find the A_M minimizing (6) with $r = n^{-1}r_x$ and resulting minimum squared error σ_M^2. The LPC filter model is then σ_M/A_M. The question remains, however, as to why this is a reasonable procedure. Each of the original developments of the LPC analysis technique provided a different explanation (and we shall provide yet another). The Itakura–Saito development [22] was closest in spirit to the approach proposed here, since it applied a traditional parameter estimation technique to a speech model.

The other two developments [2], [31] adopted the point of view that all-pole models were physiologically good speech models and that a good selection procedure should choose such a model to match well the second-order properties of the observed speech frame. In [31] this was accomplished via choosing σ/A so that when driven by white noise the output autocorrelation matched the observed sample autocorrelation $n^{-1}r_x$, $k=0,\cdots,M$. From (7) this yielded the LPC analysis equations. Observe that this is equivalent to the "matching autoregressive source" used by Anderson and Bodie [1] to design tree encoding systems for speech waveforms. In [2] the model σ_M/A_M was chosen so as to yield a good approximation to the exact input sample spectrum $f_x(\theta) = \sigma_\infty^2/|A_\infty(e^{i\theta})|^2$. A moderate M was selected so that a "smoothed" approximation to f_x would be obtained without trying to model the violent f_x behavior due to unreliable autocorrelation estimates $r_x(k)$ for large k and to effects of the pitch period. Neither of these two justifications are explicit applications of general techniques for system identification. Both are heuristic rather than mathematical developments.

Of the above derivations, only the Itakura–Saito development led to the formulation of LPC analysis as a minimum distortion rule. This formulation is irrelevant to the actual implementation of a traditional LPC system, but it provides the required distortion measure for a rate-distortion coded system.

The Itakura–Saito distortion (2) can be rewritten as [17], [18]

$$\rho_{IS}(x,\sigma/A) = \{\alpha/\sigma^2 + \ln\sigma^2\} - 1 - \ln\sigma_\infty^2, \qquad (12)$$

where σ_∞^2 is given by (9) with $f = f_x$ and where α is the residual energy defined by

$$n\alpha = \boldsymbol{a}'\boldsymbol{R}_x^{(M)}\boldsymbol{a}$$
$$= r_x(0)r_a(0) + 2\sum_{k=1}^{M} r_x(k)r_a(k), \qquad (13)$$

where r_a is the sample autocorrelation of $\boldsymbol{a}$. Observe that given $\boldsymbol{x}$, minimizing $\rho_{IS}(\boldsymbol{x},\sigma/A)$ over σ/A in some class is equivalent to minimizing only the bracketed term in (12). In fact, the computationally difficult term $\ln\sigma_\infty^2$ was incorporated by Itakura and Saito only to ensure that the distortion be nonnegative. This can also be accomplished by subtracting $\ln \sigma_M^2$, and hence the modified Itakura–Saito distortion measure given by

$$\rho(x, \sigma/A) = \{\alpha/\sigma^2 + \ln \sigma^2\} - 1 - \ln \sigma_M^2 \qquad (14)$$

is also a valid distortion measure and is nearest-neighbor equivalent to ρ_{IS} in the sense of [18]; that is, for a fixed x, minimizing $\rho_{IS}(x, \sigma/A)$ and $\rho(x, \sigma/A)$ over any collection of σ/A always yields the same minimizing filter.

III. Minimum Discrimination Information Model Selection

In this section the distortion measures of (12) and (14) are characterized as asymptotic minimum discrimination information measures between the LPC model process of (1) and second-order constraints determined by the speech frame. The proofs of the results are all straightforward applications of results of Csiszár [9] and Kullback [26] and are placed in the Appendix.

Let P and Q be probability distributions (PD's) on k-dimensional Euclidean space $\mathcal{R}^k$. The discrimination information (directed divergence, I-divergence, Kullback–Leibler number, relative entropy, cross entropy) $I(Q\|P)$ is defined by [26], [39], [9] as

$$I(Q\|P) = \begin{cases} \int q_P \ln q_P \, dP, & \text{if } P \ll Q, \\ +\infty, & \text{otherwise,} \end{cases}$$

where $P \ll Q$ means that P is absolutely continuous with respect to Q and where q_P is the Radon–Nikodym derivative of Q with respect to P. If both Q and P are absolutely continuous with respect to Lebesgue measure and hence possess probability density functions (pdf's) q and p, then this assumes the more familiar form $I(Q\|P) = \int dyq(y) \ln (q(y)/p(y))$ where we take $\ln 0 = -\infty$, $\ln (a/0) = +\infty$, $0 \cdot (\pm \log 0) = 0$. The general definition is required to allow possibly singular processes and to ensure that the minimization is well-defined.

Given a measurable vector-valued statistic $T(y)$ and a fixed PD P, define the minimum discrimination information $\vartheta(\theta, P)$ for a vector θ by [26, ch. 3], [9]

$$\vartheta(\theta, P) \triangleq \inf_{Q \in \mathcal{P}(\theta)} I(Q\|P),$$

where $\mathcal{P}(\theta)$ is the set of all PD's that satisfy a vector constraint $\theta = E_Q T = \int T(y)\, dQ(y)$. If the constraint set $\mathcal{P}(\theta)$ is empty, then we define $\vartheta(\theta, P) = +\infty$. If the infimum is actually a minimum, and hence there is a $Q^* \in \mathcal{P}(\theta)$ with $\vartheta(\theta, P) = I(Q^*\|(P))$, then we call Q^* the minimum discrimination PD for θ with respect to P or the I-projection of P onto $\mathcal{P}(\theta)$. Csiszár [9] shows that if such a Q^* exists and $\vartheta(\theta, P) \neq \infty$, then Q^* must be unique.

If one does not actually know the expectation $E_Q T = \theta$ but instead observes a sample data sequence x and forms an estimate $\hat{\theta}(x)$ of θ, then $\vartheta(\hat{\theta}(x), P)$ can be considered as the discrimination information between the observed data x and a PD P. Kullback [26, pp. 37, 82] proposes $\vartheta(\hat{\theta}(x), P)$ as a measure of the "dissimilarity" or "resemblance" of a sample x and a probabilistic model P—the smaller the $\vartheta(\hat{\theta}(x), P)$, the smaller the dissimilarity or the greater the resemblance. Much of the remainder of Kullback's book is then devoted to applications of this measure to statistics in general and classification and hypothesis testing in particular. Of particular interest is the classification problem of Kullback [26] and Kupperman [27] where one has a particular family of models $\mathfrak{M}$, and one classifies a sample x as most resembling a model $P_0 \in \mathfrak{M}$ if P_0 yields the smallest $\vartheta(\hat{\theta}(x), P)$ over all $P \in \mathfrak{M}$.

Observe next that since discrimination is always nonnegative, we can consider $\vartheta(\hat{\theta}(x), P)$ as a distortion measure between the sample x and a model $P \in \mathfrak{M}$ and hence can view Kullback's procedure for minimum discrimination information classification as a minimum distortion or nearest-neighbor rule. If the family $\mathfrak{M}$ is finite, then this is equivalent to the rate distortion theory model with a code book $\mathfrak{M}$. If $\mathfrak{M}$ is a continuum (or approximately so), then this can be viewed as a model estimation or system identification rule as in LPC analysis.

Kupperman [27] and Kullback [26, pp. 83–85] show that if $\mathfrak{M}$ is a class of exponential densities (e.g., Gaussian) and if the unknown source producing the observed sample is itself a member of $\mathfrak{M}$, then the above classification procedure is equivalent to a maximum likelihood rule. In our case this would reduce to the assumption that speech itself is Gaussian and would yield a development similar to that of Itakura and Saito [22]. We emphasize, however, that we do not assume that real speech is actually Gaussian (or anything else) and that we use the more general rule for classifying nearest-neighbor minimum discrimination information.

As a first step, consider a fixed finite dimension k and constraints specified by the general second-order expectations.

$$E_Q y_i = \mu_i, \qquad i = 0, 1, \cdots, k-1$$

$$E_Q\{(y_i - \mu_i)(y_j - \mu_j)\} = R_1(i, j),$$

$$i, j \in \{0, 1, \cdots, k-1\}, \qquad |i-j| \leq J \leq k \qquad (15)$$

for some fixed $J \geq 1$. We confine our attention to a finite set of lags, since the covariance will be estimated from a finite data sample. We assume that R_1 is consistent with some valid covariance matrix; i.e., there exists a symmetric nonnegative definite matrix $\boldsymbol{R} = \{R(i, j)\}$ with $R_1(i, j) = R(i, j)$ for $|i-j| \leq J$. Any $\boldsymbol{R}$ of this form is called an extension of R_1. Only second-order constraints are invoked corresponding to the physical idea that the ear is primarily sensitive to second-order characteristics.

The following theorem and corollary characterize the minimum discrimination information for a general nonsingular Gaussian model p and the given second-order constraints.

Theorem 1: Given a Gaussian pdf p on $\mathcal{R}^k$ with mean $\boldsymbol{m}$ and nonsingular covariance matrix $\boldsymbol{R}_2$ and a constraint set $\mathcal{P}(R_1, \boldsymbol{\mu})$ containing all PD's satisfying, (15) then $\vartheta((R_1, \boldsymbol{\mu}), P)$ is given as follows.

a) If there does not exist any positive definite extension $\boldsymbol{R}$ of R_1, then $I(Q\|P) + \infty$ for all $Q \in \mathcal{P}(R_1, \boldsymbol{\mu})$, and hence $\vartheta((R_1, \boldsymbol{\mu}), P) = \infty$.

b) If there exists any positive definite extension to R_1, then there exists a unique positive definite extension $\boldsymbol{R}^*$ of R_1 called the minimum discrimination information extension of R_1 (with respect to P) such that the pdf

$$q^*(y) = (2\pi)^{-k/2} \det(\boldsymbol{R}^*)^{-1/2} \cdot \exp\{-1/2(y-\boldsymbol{\mu})'\boldsymbol{R}^{*-1}(y-\boldsymbol{\mu})\}$$

yields a PD Q^* with

$$\vartheta((R_1, \mu), P) = I(Q^*\|P) = \frac{1}{2}\ln\{\det \boldsymbol{R}_2/\det \boldsymbol{R}^*\} - \frac{k}{2} + \frac{1}{2}\operatorname{tr} \boldsymbol{R}^*\boldsymbol{R}_2^{-1} + \frac{1}{2}(\boldsymbol{\mu}-\boldsymbol{m})'\boldsymbol{R}_2^{-1}(\boldsymbol{\mu}-\boldsymbol{m}).$$

In particular, the infimum is always a minimum, and the minimum is finite if and only if R_1 has a positive definite extension, in which case Q^* is unique.

Corollary 1: Given P and $\mathcal{P}(R_1, \boldsymbol{\mu})$ as in the theorem, then

$$\vartheta((R_1, \boldsymbol{\mu}), P) = \frac{1}{2}\ln\det \boldsymbol{R}_2 - \frac{k}{2} + \frac{1}{2}(\boldsymbol{\mu}-\boldsymbol{m})'\boldsymbol{R}_2^{-1}(\boldsymbol{\mu}-\boldsymbol{m}) + \min_{\boldsymbol{R}\in\mathcal{Q}_k(R_1)} \frac{1}{2}\{\operatorname{tr} \boldsymbol{R}\boldsymbol{R}_2^{-1} - \ln\det \boldsymbol{R}\},$$

where $\mathcal{Q}_k(R_1)$ is the collection of all covariance matrices that are extensions of R_1. If the minimum of the bracketed term is finite, then the minimizing $\boldsymbol{R}$ is $\boldsymbol{R}^*$ as in the previous theorem.

Since $\mathcal{Q}_k(R_1)$ is a convex set of $k \times k$ matrices, the corollary shows that solving for $\boldsymbol{R}^*$ and hence $\vartheta((R_1, \boldsymbol{\mu}), P)$ is a convex programming problem. We shall see, however that an analytical solution is possible with some additional assumptions.

Next assume that the model mean $\boldsymbol{m}$ and covariance $\boldsymbol{R}_2$ are determined by the LPC model (1). Assume also that the constraint mean $\mu_i = 0$, all i, and that R_1 is Toeplitz with $J \geq M$. These assumptions incorporate into the constraints the assumption that the observed process has zero mean, the fact that the autocorrelation estimate will be Toeplitz since it is based on a sample autocorrelation function, and the requirement that we have at least as many covariance constraints as the order of the autoregressive model.

For this case it follows with some algebra that

$$\boldsymbol{m} = \sigma(1-\beta)^{1/2}\boldsymbol{A}^{-1}\boldsymbol{s} + \boldsymbol{A}^{-1}\boldsymbol{b}$$
$$\boldsymbol{R}_2 = \sigma^2\beta(\boldsymbol{A}'\boldsymbol{A}^{-1})^{-1} \tag{16}$$

where $\boldsymbol{s} = (s_0, \cdots, s_{k-1})$ and

$$b_i = \begin{cases} -\sum_{j=i+1}^{M} a_j u_{i-j}, & 0 \leq i \leq M \\ 0, & \text{else.} \end{cases}$$

Corollary 2: Given the assumptions of Theorem 1, assume also that $\boldsymbol{\mu} = \boldsymbol{0}$, $\boldsymbol{R}_1$ is Toeplitz, $J \geq M$, and that $\boldsymbol{m}$ and $\boldsymbol{R}_2$ are as in (16); then

$$\vartheta((\boldsymbol{R}_1, \boldsymbol{0}), P) = \frac{k}{2}\ln(\sigma^2\beta) - \frac{k}{2} + \frac{1}{2}\|\sigma(1-\beta)^{1/2}\boldsymbol{s} + \boldsymbol{b}\|^2/(\sigma^2\beta) + \frac{1}{2}(\sigma^2\beta)^{-1}\Gamma(R_1, A) - \frac{1}{2}\max_{\boldsymbol{R}\in\mathcal{Q}_k(R_1)} \ln\det \boldsymbol{R},$$

where

$$\Gamma(R_1, A) = \sum_{i=0}^{M-1}\sum_{j=0}^{i}\sum_{m=0}^{i} a_j a_m R_1(j-m) + (k-M)\boldsymbol{a}'\boldsymbol{R}_1^{(M)}\boldsymbol{a},$$

where $\boldsymbol{R}_1^{(M)} = \{R_1(k-j);\ k, j \in \{0, 1, \cdots, M\}\}$, $\|\cdot\|$ denotes Euclidean distance, and the $\boldsymbol{R}$ maximizing $\ln\det \boldsymbol{R}$ over $\mathcal{Q}_k(R_1)$ is $\boldsymbol{R}^*$, the minimum discrimination information extension of R_1 with respect to P. If $\boldsymbol{R}_1$ is positive definite, then $((R_1, \boldsymbol{0}), P)$ is finite and $\boldsymbol{R}^*$ is unique. If $\boldsymbol{R}_1$ is not positive definite, then $\vartheta((\boldsymbol{R}_1, \boldsymbol{0}), P) = \infty$.

Note from the corollary that under the given assumptions $\boldsymbol{R}^*$ does not depend on P and hence is the same for all Gaussian autoregressive P of order M less than J. The following corollary shows that in this case $\boldsymbol{R}^*$ is the so-called "maximum entropy extension" of R_1 [37].

Corollary 3: Let $\mathcal{G}_1$ denote the class of all pdf's q such that the corresponding distribution $\mathcal{Q}$ is in $\mathcal{P}(R_1, \boldsymbol{0})$. Define $\mathcal{G}_2$ as the set of all pdf's q such that the differential entropy $h(q) = -\int dy q(y)\ln q(v)$ exists. (We allow $\pm\infty$, but $q \in \mathcal{G}_2$ only if the integral is well-defined as an appropriate limit.) Define

$$\mathcal{H}(R_1) = \sup_{\mathcal{G}_1\cap\mathcal{G}_2} h(q),$$

where we take $\mathcal{H}(R_1) = -\infty$ if the constraint set is empty.

a) If R_1 is not positive definite, then $\mathcal{H}(R_1) = -\infty$,

b) If R_1 is positive definite, then the supremum is a maximum and $\mathcal{H}(R_1) = h(q^*)$, where q^* is the unique Gaussian density with mean zero and covariance $\boldsymbol{R}^*$, as given in Corollary 2, for any autoregressive Gaussian P or order $M \leq J$. Thus [40, sec. 20],

$$\mathcal{H}(R_1) = \frac{k}{2}\ln(2\pi e) + \frac{1}{2}\max_{\boldsymbol{R}\in\mathcal{Q}_k(R_1)} \ln\det \boldsymbol{R} = \frac{k}{2}\ln(2\pi e) + \frac{1}{2}\ln\det \boldsymbol{R}^*.$$

The connection with the maximum entropy problem yields the following corollary.

Corollary 4: Given the assumptions of Corollary 2 and $k \geq J \geq M$, then

$$\max_{R \in \mathcal{Q}_k(R_1)} \ln \det R = \ln \det R^* = \ln \det R_1^{(J-1)} + (k-J) \ln \left(\det R_1^{(J)} / \det R_1^{(J-1)}\right),$$

and therefore

$$k^{-1}\vartheta((R_1, 0), P) = \frac{1}{2}\Big\{\ln(\sigma^2\beta) - 1 + k^{-1}\|\sigma(1-\beta)^{1/2}s + b\|^2 / (\sigma^2\beta) + (\sigma^2\beta)^{-1}\Gamma(R_1, A) - k^{-1}\ln\det R_1^{(J-1)} - (1 - J/k)\ln\left(\det R_1^{(J)}/\det R_1^{(J-1)}\right)\Big\}.$$

where Γ is given in Corollary 2.

Now let $\bar{P}$ denote the process distribution for the Mth-order autoregressive Gaussian model which implies all kth-order restrictions P^k to $\mathcal{R}^k$. Define the consistent constraint family $\mathcal{P}_k(R_1)$ for all k as the class of all PD's Q on $\mathcal{R}^k$ such that $E_Q y_i = 0$, all i, and $E_Q y_i y_j = R_1(i-j)$, all i, j such that $|i-j| \leq J$. We require that R_1 extend to a valid (nonnegative definite) covariance function. This is the case if and only if the matrix $R_1^{(J)}$ is nonnegative definite [38], [46]. We also again require that $J \geq M$. Define

$$\vartheta_k\left(R_1^{(J)}, P^k\right) = \inf_{Q \in \mathcal{P}_k(R_1)} I(Q \| P^k),$$

and define the asymptotic minimum discrimination information by

$$\vartheta_\infty\left(R_1^{(J)}, \bar{P}\right) = \liminf_{k \to \infty} k^{-1}\vartheta_k\left(R_1^{(J)}, P^k\right).$$

(The limit infimum is used since the limit may not exist.)

Theorem 2: a) If $R_1^{(J)}$ is not positive definite, then $\vartheta_\infty(R_1^{(J)}, \bar{P}) = +\infty$. b) If $R_1^{(J)}$ is positive definite, then the limit infimum defining ϑ_∞ is a limit, and

$$\vartheta_\infty\left(R_1^{(J)}, \bar{P}\right) = \frac{1}{2}\ln(\sigma^2\beta) - \frac{1}{2}\ln\sigma_J^2 - \frac{1}{2} + (2\sigma^2\beta)^{-1}\left\{a' R_1^{(M)} a + \sigma^2(1-\beta)\right\} < \infty,$$

where $\sigma_J^2 = \det R_1^{(J)} / \det R_1^{(J-1)}$.

Next make the dependence of $\bar{P}$ on β explicit as $\bar{P}_\beta$, and recall that $\beta \in (0,1]$. Since β^{-1} is minimized for $\beta \in (0,1]$ as one with $\beta = 1$, and $\beta^{-1} - \ln\beta^{-1}$ is minimized as one with $\beta = 1$,

$$\min_{\beta \in (0,1]} a\vartheta_\infty\left(R_1^{(J)}, \bar{P}_\beta\right) = \frac{1}{2}\left\{\ln\left(\sigma^2/\sigma_J^2\right) + \sigma^{-2}a' R_1^{(M)} a - 1\right\}. \quad (17)$$

Thus, regardless of the input constraints R_1 and the model filter σ/A, the asymptotic minimum discrimination information $\vartheta_\infty(R_1^{(J)}, \bar{P}_\beta)$ is *always* minimized by the choice $\beta = 1$—i.e., by the given filter with an unvoiced driving function.

The last remaining problem is the choice of J. We have seen that $J \geq M$ is mathematically useful, since it yields a simple form for ϑ. Furthermore, it is intuitively reasonable to require knowledge of at least as many second-order constraints as the order of the model. Assume also that the $J+1$ constraints of R_1 are determined from a sample correlation; that is, $R_1(j) = r_x(j)$, $j = 0, \cdots, J$ or $R_1^{(J)} = R_x^{(J)}$.

How large should J be? As we have argued previously, this is simply a manner of convenience, since the same selection rule results regardless of J. Two choices are of main interest. First, since $R_1(j)$ is increasingly unreliable as j gets larger, choose J as small as possible; i.e., set $J = M$. This yields from (13), (14), and (17) that $\vartheta_\infty(R_x^{(M)}, \bar{P}_1) = \frac{1}{2}\rho(x, \sigma/A)$; that is, the modified Itakura–Saito speech distortion measure ρ of (14) is the asymptotic minimum distortion information between the constraints given by $\{r_x(j), j = 0, \cdots, M\}$ and the model filter σ/A. This has the particular advantage of not requiring the computation of a minimum discrimination information extension of the constraints, and it requires the fewest possible constraints. Alternatively, take $J = k-1$ and let $J \to \infty$; then $\sigma_J^2 \to \sigma_\infty^2$, and we have as above that

$$\vartheta_\infty\left(r_x, \bar{P}_1\right) = \lim_{k \to \infty} k^{-1}\vartheta_k\left(R_1^{(k-1)}, P_1^k\right) = \frac{1}{2}\rho_{\mathrm{IS}}(x, \sigma/A);$$

that is, the Itakura–Saito distortion effectively considers $r_x(j)$ as an estimate of the covariance for *all* lags j, including $j \geq n$ where $r_x(j) \triangleq 0$. We believe that it is preferable to place confidence only in the small lag values and, therefore, that $\rho(x, \sigma/A)$ is the better distortion measure. Physically speaking, the values of $\rho_{\mathrm{IS}}(x, \sigma/A)$ will be unrealistically large because it is trying to match a complete autocorrelation function instead of only a few values thereof. Combining the previous observations with the results of the section on linear predictive coding yields the following.

Theorem 3: Let σ/A be in the class of all stable Mth-order stable autoregressive filter models, and let $x = (x_0, x_1, \cdots, x_{n-1})$, $n \geq M$, be an observed frame of speech with sample autocorrelation r_x. Define $R_x^{(M)} = \{r_x(j-m);\ j, m \in \{0, 1, \cdots, M\}\}$ and $\sigma_M^2 = \det R_x^{(M)} / \det R_x^{(M-1)}$. Then

$$\rho(x, \sigma/A) = \ln\left(\sigma^2/\sigma_M^2\right) + \sigma^{-2}a' R_x^{(M)} a - 1 = 2\vartheta_\infty\left(R_x^{(M)}, \bar{P}_1\right) = 2\min_{\beta \in (0,1]} \vartheta_\infty\left(R_x^{(M)}, \bar{P}_\beta\right),$$

Thus the modified Itakura–Saito distortion measure is twice the asymptotic minimum discrimination information between the first M lags of the sample covariance and an Mth-order Gaussian autoregressive model with $\beta = 1$.

Similarly,

$$\rho_{IS}(x, \sigma/A) = \ln(\sigma^2/\sigma_\infty^2) + \sigma^{-2}a'R_x^{(M)}a - 1$$
$$= 2\vartheta_\infty(r_x, \bar{P}_1),$$

where now the constraint set consists of the full sample autocorrelation for all lags (with $R_x(j) = 0$ for $j \geq n$).

As a final observation on these two distortion measures, it follows from (4), (5), (12), (13), and [18, sec. V] that in a two-step LPC system if a frame x is mapped into σ_M/A_M using LPC analysis and $\sigma/A \in \mathcal{C} \subset \mathcal{F}_M$, then

$$\rho(x, \hat{\sigma}/\hat{A}) = d_{IS}(\sigma_M^2/|A_M(e^{i\theta})|^2, \hat{\sigma}^2/|\hat{A}(e^{i\theta})|^2);$$

that is, the modified Itakura–Saito distortion between the speech frame and the filter $\hat{\sigma}/\hat{A}$ is the spectral Itakura–Saito distortion between the LPC model σ_M/A_M and $\hat{\sigma}/\hat{A}$.

IV. Code Book Design

For a fixed rate in bits per frame, we wish to design a code book of model filters $\mathcal{C} = \{\sigma_i/A_i;\ i = 1, 2, \cdots, 2^R\}$ for use in a rate-distortion speech coder using the asymptotic distortion measure of minimum discrimination information $\rho(x, \sigma/A)$ of Theorem 3. The goal is to design a code book minimizing the expected distortion $\rho(\mathcal{C}) = E \min_{\sigma/A \in \mathcal{C}} \rho(x, \sigma/A)$.

Since there is no generally accepted probabilistic model for speech, however, we consider instead a sample distribution and try to minimize the time average distortion for a long training sequence of speech $\{x_i\}_{i=0}^{L-1}$, where $x_i = (x_{in}, \cdots, x_{in+n-1})'$, chosen from a variety of speakers; that is, we attempt to choose $\mathcal{C}$ to minimize

$$\rho(\mathcal{C}) = L^{-1} \sum_{i=0}^{L-1} \min_{\sigma/A \in \mathcal{C}} \rho(x_i, \sigma/A). \tag{18}$$

To describe the algorithm we first require the notion of the centroid of a collection of vectors with respect to ρ. Let $S = \{x_i;\ i = 1, \cdots, \|S\|\}$ denote a collection of $\|S\|$ vectors. The ρ-centroid or center of gravity or minimum distortion point for S is the filter σ/A that minimizes the average distortion

$$\Delta(S, \sigma/A) \triangleq \frac{1}{\|S\|} \sum_{x \in S} \rho(x, \sigma/A).$$

Abbreviate $R_i = R_{x_i}^{(M)}$ and $\sigma_i^2 = \sigma_M^2(i)$, and we have that

$$\Delta(S, \sigma/A) = \sigma^{-2}a'\bar{R}a + \ln \sigma^2 - 1 - \frac{1}{\|S\|} \sum_{i=1}^{\|S\|} \ln \sigma_i^2,$$

where

$$\bar{R} = \frac{1}{\|S\|} \sum_{i=1}^{\|S\|} R_i$$

is the average sample autocorrelation matrix over all $x \in S$. $\Delta(S, \sigma/A)$ is easily minimized using standard LPC analysis: first choose $A = \bar{A}_M$, the filter minimizing $a'\bar{R}a$—that is, the LPC model for the average sample spectrum. If the resulting minimum is $\bar{\sigma}_M^2$, then choose σ^2 to minimize

$$\Delta(S, \sigma/\bar{A}_M) = \frac{\bar{\sigma}_M^2}{\sigma^2} - \frac{1}{\|S\|} \sum_{i=1}^{\|S\|} \ln \sigma_i^2 + \ln \sigma^2 - 1,$$

which is accomplished by $\sigma^2 = \bar{\sigma}_M^2$. Thus the centroid of a set S, which we shall denote by $\text{cen}(S)$, is simply the standard LPC model $\bar{\sigma}_M/\bar{A}_M$ for the average sample autocorrelation over S! If S is empty, the centroid can be defined in an arbitrary fashion. The following algorithm yields a code book giving a local minimum of (17) [29], [20].

Code Book Design Algorithm

Step 0: Initialize—Fix $\epsilon > 0$, R_{max} is the maximum rate in bits per frame. Let $T = \{x_i;\ i = 0, \cdots, L-1\}$ be the training sequence.

Step 1: Set $R = 0$—Find $\text{cen}(T)$. This the optimum zero rate code book $\mathcal{C}_0$.

Step 2: Split—Given a final rate R code book $\mathcal{C}_R = \{\sigma_i/A_i;\ i = 1, \cdots, 2^R\}$, an initial rate $R+1$ code book $\mathcal{C}_{R+1}(0)$ is formed by perturbing each σ_i/A_i by a small amount to obtain two new close codewords. Set $m = 0$, $R = R + 1$, $D_{-1} = \infty$.

Step 3: Encode—Given $\mathcal{C}_R(m) = \{\sigma_i/A_i;\ i = 1, \cdots, 2^R\}$, group all the vectors in the training sequence according to the minimum distortion codeword for each by forming the sets $S_i = \{x\colon \rho(x, \sigma_i/A_i) \leq \rho(x, \sigma_j/A_j), \text{all } j\}$ with some arbitrary tie-breaking rule. Compute the distortion

$$D_m \triangleq \rho(\mathcal{C}_R(m)) = \sum_{i=1}^{L-1} \min_{\sigma/A \in \mathcal{C}_R(m)} \rho(x_i, \sigma/A)$$
$$= \sum_{i=1}^{2^R} \sum_{x \in S_i} \rho(x, \sigma_i/A_i).$$

Step 4: Code Book Update—if $(D_{m-1} - D_m)/D_m > \epsilon$, for a new code book $\mathcal{C}_R(m+1) = \{\text{cen}(S_i);\ i = 1, \cdots, 2^R\}$, set $m = m + 1$ and go to Step 3. Otherwise, go to Step 5.

Step 5: $\mathcal{C}_R = \mathcal{C}_R(m)$ and $D(R) = D_m$. If $R < R_{max}$ then go to Step 2. Otherwise, halt with final code book $\mathcal{C} = \mathcal{C}_R$.

The above description of the algorithm is in the format of [29], but the actual program implementation incorporates several computational shortcuts. For example, in Step 0 the speech frames are converted into vectors of the form $(\ln \sigma_M^2, r_x(0), \cdots, r_x(M))$, and the encoder code book is stored in the form $(\ln \sigma^2, r_a(0)/\sigma^2 \cdots r_x(M)/\sigma^2)$ to avoid repeating the logarithm and division computations. Furthermore, during the encoding step, running averages of the sample autocorrelations mapping into each codeword are kept so that the centroid computation simply requires a Levinson's algorithm on these stored averages.

V. Simulation

Code books were designed and tested for codes with rates of $R = 0$–9 bits per frame for the model filters (0–450 bits/s at 50 frames/s) on a PDP-11/34 at the Stanford University Information Systems Laboratory. The training sequence consisted of 100 s or 5000 frames of preemphasized Hamming windowed bandpass filtered sampled speech from five male speakers (1000 frames each). The speech was normal conversation recorded under good conditions. The design threshold was $\epsilon = 0.01$, and all code books converged in fewer than ten iterations.

The design and quantitative performance of the filter code books is not affected by the pitch/voicing system. In order to facilitate a subjective comparison, however, we adopted a slight variation of the pitch extraction and quantization technique used (but not carefully described) in [6]. The pitch and voicing parameters were extracted and quantized on a DEC VAX at Signal Technology, Inc., Santa Barbara, CA, as follows. The pitch and voicing were obtained using a modified cepstral technique described in [39] and implemented in Version 3 of Signal Technology's Interactive Laboratory System. If the frame was declared to be unvoiced, then the pitch period parameter τ was set to zero. The remaining voiced frames were used as a training sequence for the vector quantizer design algorithm of [29] with a distortion measure $|(\tau - \hat{\tau})/(\tau\hat{\tau})|$ and a codebook of size $2^6 - 1$. This system is a slight modification of that of [6] and [33] and requires six bits per frame, or 300 bits/s, yielding a total bit rate of 750 bits/s.

The code books obtained from the design algorithm were tested both in a "closed" test on the original training sequence and in an "open" test on 600 frames of speech by a male speaker not included in the training sequence. The final design distortion and the distortion resulting from encoding the open test data with the final code books are given in Table I using the modified Itakura–Saito distortion measure [14]. Table II provides a comparison of our final design distortions and those of the suboptimal system of Buzo *et al.* [6], which used five bits per frame for the gain and rates of 0–10 bits for the normalized filter.

TABLE I
Modified Itakura–Saito Distortion for Rate-Distortion Speech Coder

bits/frame	Design Distortion	Open Test
0	4.87	4.41
1	2.10	2.29
2	1.20	1.33
3	.90	1.06
4	.65	.72
5	.52	.58
6	.43	.50
7	.34	.43
8	.27	.38
9	.20	.34

TABLE II
Modified Itakura–Saito Distortion for Combined Filter and Gain Design and Separate Filter and Gain Design.

bits/frame	Design Distortion: Joint Gain & Filter	Design Distortion Separate Gain & Filter
0	4.87	
1	2.10	
2	1.20	
3	.90	
4	.65	
5	.52	
6	.43	.67
7	.34	.56
8	.27	.45
9	.20	.36
10		.30
11		.25
12		.20
13		.15
14		.12
15		.09

Table I shows that there is indeed a quantitative increase in distortion when using a code book on speakers not in the training sequence. In this example the loss is effectively equivalent to 1 or 2 bits in rate. In informal subjective listening tests, the main subjective difference between the two tests was a more pronounced "warble" or yodel-like sound during long voiced sounds in the open test. Both tests produced easily intelligible speech of fair quality at 7–9 bits. The quality was typical of that of higher rate LPC systems, except that the warble was more pronounced.

The code books were also synthesized without using the pitch and voicing data by driving the filter with white noise to produce "whispered" speech. Perhaps surprisingly, many listeners found the whispered speech (at 450 bits/s) to be of better equality than the speech with the voicing and pitch information.

Table II demonstrates that the combined gain and model filter design of the rate-distortion coder indeed provides improved performance over the system of [6]—3 bits of improvement for the highest rate. This is significant as a percentage, but the suboptimal technique requires less memory for a given bit rate (e.g., $2^5 + 2^4$ for nine bits instead of 2^9) and hence is capable of higher rates and better quality given a memory constraint. The rate-distortion coder provides, however, a benchmark for such suboptimal systems.

Our 9 bit per frame (for σ/A) system was quantitatively and subjectively quite close to the 12-bit per frame system of Buzo *et al.* [6] and compared well with the 15-bit per frame system of [6], again except for a more pronounced warble.

The interested reader is referred to [6] for a comparison of the suboptimal vector quantization system of [6] with LPC systems with approximately optimal scalar quantization. We mention, however, that the quantitative distortion achieved by our 9 bit system (750 bit/s) and the Buzo *et al.* 12-bit system (950 bits/s) requires 30 bits each frame for the filter parameter in a scalar system yielding a total rate of 1850 bits/s.

VI. Final Comments

A theoretical development of the Itakura–Saito distortion as an application of classical techniques of information theory and statistics to speech processing has been given and the distortion measure used to design and simulate very low-rate speech coders. Our system inherits several of the difficulties of traditional LPC systems: poor quality on high-pitched female and very low-pitched male voices, possible quality problems owing to interaction between pitch extraction and model filter classification, and inability to reproduce well some sounds such as plosives and "*z*." In addition, the "warble" common to very low-rate speech compression systems is more pronounced in our 750 bits/s than in higher rate LPC systems. On the positive side, our system yields significantly better quality than any LPC system that we are aware of with a comparable rate of 9 bits per frame or 450 bits/s for the model filter.

Two implementations of rate-distortion speech coders seem plausible. The first is a microprocessor based firmwave system that would search the code book sequentially. An obvious problem with such a system is the large memory requirement and the large number of distortion computations required to search the code book—$2^9 = 512$ in our case. The number of distortion computations can be reduced from 2^9 to 18 by using the binary tree searched vector quantizer technique of Buzo *et al.* [6] with approximately a 15-percent increase in distortion for a given rate. Other tree structures provide a range of computation-distortion tradeoffs. Several approaches to trading memory requirements for efficiency have been suggested [6].

An alternative approach is based on the observation that once the code book is designed, the rate-distortion speech encoder consists of the N inverse filters A_i/σ_i in parallel, each followed by the additive fixed bias $\ln \sigma_i$. The observed speech frame is input to each filter, and the minimum distortion codeword σ_i/A_i is specified by the i such that A_i/σ_i has the smallest output energy plus bias. This configuration seems amenable to LSI and VLSI design techniques. The decoder or synthesizer is essentially a standard LPC decoder that first converts the received binary symbols into the appropriate filter and uses the filter to synthesize the speech.

We believe that the 5000-frame training sequence used here is sufficient for preliminary studies and conclusions, but the design of an actual code for implementation would likely require a much larger and richer training sequence.

We believe that some of the techniques developed here may prove useful in other problems combining system identification with classification or quantization. For example, Gersch *et al.* [14], [15] have successfully used a distortion measure equivalent to the Itakura–Saito distortion measure to classify electroencephalograms using the Kullback–Leibler nearest-neighbor rules. They effectively assume that the observed data is Gaussian, as do Itakura and Saito [22], but our work provides more general theoretical support for their technique. It has also been shown recently [42] that many of the useful properties of the Itakura–Saito distortion used here and in [18] and [6] hold more generally for minimum discrimination information measures without the Gaussian model assumption. Hence, some of these techniques may prove useful in other pattern classification and cluster analysis problems.

Acknowledgment

The authors wish to acknowledge helpful and informative discussions with Dr. B. H. Juang of Signal Technology, Inc. on speech processing and with Prof. Bradley Dickinson of Princeton University on positive definite extensions of Toeplitz matrices. We also gratefully acknowledge Dr. Juang's help with the pitch extraction and quantization.

Appendix

In order to evaluate $\vartheta((R_1, \boldsymbol{\mu}), P)$ introduce the $(J+1)k - (J+1)J/2$ multipliers $\psi = \{\psi_{i,j};\ i, j \in \{0, 1, \cdots, k-1\}\ 0 \le i - j \le J\}$ corresponding to the $R_1(i, j)$ and the k multipliers $\gamma = (\gamma_0, \cdots, \gamma_{k-1})$ corresponding to the μ_i. Extend $\psi_{i,j}$ to all values of $i, j \in \{0, 1, \cdots, k-1\}$ by $\psi_{i,j} = 0$ if $|i - j| > J$ and $\psi_{i,j} = \psi_{j,i}$ if $j > i$. Define $\boldsymbol{\Psi} = \{\psi_{i,j};\ i, j \in \{0, 1, \cdots, k-1\}\}$.

Lemma 1: Given $K = (J+1)k - (J+1)J/2 + k$, define the set $\mathcal{G}_P$ of $\mathfrak{R}^K$ by $\mathcal{G}_P = \{\psi, \gamma\colon E_P \exp\{-\frac{1}{2}\boldsymbol{y}'\boldsymbol{\Psi}\boldsymbol{y} - \boldsymbol{\gamma}^T\boldsymbol{y}\} < \infty\}$, where P is a nonsingular k-dimensional Gaussian distribution with mean $\boldsymbol{m}$ and covariance $\boldsymbol{R}_2$. Then $(\psi, \gamma) \in \mathcal{G}_P$ if and only if $\boldsymbol{R}^{-1} + \boldsymbol{\Psi}$ is positive definite. In addition, $\mathcal{G}_P$ is an open set in $\mathfrak{R}^K$.

Proof:

$$E_P \exp\left\{-\frac{1}{2}\boldsymbol{y}'\boldsymbol{\Psi}\boldsymbol{y} - \boldsymbol{\gamma}'\boldsymbol{y}\right\}$$

$$= c_1 \int d\boldsymbol{y} \exp\left\{-\frac{1}{2}\boldsymbol{y}'\{\boldsymbol{R}_2^{-1} + \boldsymbol{\Psi}\}\boldsymbol{y} + \{\boldsymbol{m}'\boldsymbol{R}_2^{-1} - \boldsymbol{\gamma}'\}\boldsymbol{y}\right\}, \tag{A1}$$

where $c_1 < \infty$ depends on $\boldsymbol{m}$ and $\boldsymbol{R}_2$. If $\boldsymbol{R}_2^{-1} + \boldsymbol{\Psi} \triangleq \boldsymbol{\Lambda}$ is positive definite, then it is invertible and we can define $\boldsymbol{\alpha} = \boldsymbol{\Lambda}^{-1}(\boldsymbol{m}^T\boldsymbol{R}_2^{-1} - \boldsymbol{\gamma}')$ and rewrite (A1) as $c_2 \exp\{-\frac{1}{2}(\boldsymbol{y} - \boldsymbol{\alpha})'\boldsymbol{\Lambda}(\boldsymbol{y} - \boldsymbol{\alpha})\}$ where $c_2 < \infty$. This has the form of a Gaussian density with covariance $\boldsymbol{\Lambda}^{-1}$ and integrates to $c_2(2\pi)^{k/2}(\det \boldsymbol{\Lambda}^{-1})^{1/2} < \infty$; hence $(\psi, \gamma) \in \mathcal{G}_P$. Conversely, say that (A1) is finite. By construction, $\boldsymbol{R}_2^{-1} + \boldsymbol{\Psi}$ is symmetric, and hence there exists a unitary matrix $\boldsymbol{u}$ such that $\boldsymbol{R}_2^{-1} + \boldsymbol{\Psi} = \boldsymbol{U}\boldsymbol{D}\boldsymbol{U}'$, where $\boldsymbol{D}$ is a diagonal matrix with diagonal entries $D_{i,i}$, $i = 0, \cdots, k-1$. Set $\boldsymbol{u} = \boldsymbol{U}'\boldsymbol{y}$. Since the Jacobian of $\boldsymbol{U}$ is 1, (A1) can be written

$$c_1 \int d\boldsymbol{u} \exp\left\{-\frac{1}{2}\boldsymbol{u}'\boldsymbol{D}\boldsymbol{u} + \boldsymbol{g}'\boldsymbol{u}\right\}, \tag{A2}$$

where $\boldsymbol{g}' = \{\boldsymbol{m}'\boldsymbol{R}_2^{-1} - \boldsymbol{\gamma}'\}\boldsymbol{U}$, which then becomes

$$c_3 \int d\boldsymbol{u} \exp\left\{-\frac{1}{2}(\boldsymbol{u} - \boldsymbol{g})'\boldsymbol{D}(\boldsymbol{u} - \boldsymbol{g})\right\}$$

$$= 3c_3 \prod_{i=0}^{k-1} \int du_i \exp\left\{-\frac{1}{2}D_{ii}(u_i - g_i)^2\right\}.$$

From Fubini's theorem, the existence and finiteness of the above product integral ensures that for each i, $\int du_i \exp\{-\frac{1}{2}(u_i$

$- g_i)^2 D_{i,i}\} < \infty$, which is true if and only if $D_{i,i} < 0$, all i, and hence if and only if $R_2^{-1} + \Psi$ is positive definite, proving the first statement in the lemma. If the elements of a positive definite matrix are perturbed by a sufficiently small amount, the resulting matrix is also positive definite, since eigenvalues (and hence determinants) are continuous functions of the matrix elements. (See, e.g., Lancaster [28, p. 225].) This implies that all $(\psi, \gamma) \in \mathcal{G}_P$ are interior points and hence that $\mathcal{G}_P$ is open.

Lemma 2: Given k, K, P as in Lemma 1, define $\mathcal{D}_P \subset \mathcal{R}^K$ by $\mathcal{D}_P = \{R_1, \mu$: there exists a PD $Q \in \mathcal{P}(R_1, \mu)$ with $I(Q \| P) < \infty\}$. Then $(R_1, \mu) \in \mathcal{D}_P$ if and only if R_1 has a positive definite extension. In addition, $\mathcal{D}_P$ is an open set in $\mathcal{R}^K$.

Proof: Say that R_1 has a positive definite extension R. Let Q be Gaussian with covariance R and mean μ. From Kullback [26, p. 189],

$$I(Q\|P) = \frac{1}{2}\ln(\det R_2 \det R) - \frac{k}{2} + \frac{1}{2}\operatorname{tr} RR_2^{-1} + \frac{1}{2}\operatorname{tr} R_2^{-1}(\mu - m)(\mu - m)' < \infty,$$

since $\det R > 0$. Thus $(R_1, \mu) \in \mathcal{D}_P$. Conversely, say that no such extension exists, and hence any extension R of R_1 has at least one zero eigenvalue. Since R is symmetric, we can diagonalize it as $R = UDU'$, where U is unitary and D is the diagonal eigenvalue matrix. Let $u = U'y$ so that $E_Q uu' = D$, and hence, for at least one i, $Eu_i^2 = EU_i'y = 0$, where U_i' is the ith row of U'. Define $F = \{y: U_i'y = 0\}$. By construction, $Q(F) = 1$ while $P(F) = 0$, and therefore Q is not absolutely continuous with respect to P. Thus $\mathcal{D}_P$ consists exactly of those (R_1, μ) for which R_1 has a positive definite extension, and hence $\mathcal{D}_P$ is open in $\mathcal{R}^K$ as in the preceding proof.

Remark: Observe that a similar argument shows that if R_1 has no positive definite extension, then Q is not absolutely continuous with respect to Lebesque measure and hence does not possess a pdf.

Proof of Theorem 1: a) Follows from Lemma 2. b) From Lemmas 1 and 2 and Csiszar's Theorem 3.3, $\vartheta((R_1, \mu), P)$ is achieved as a finite minimum by a unique pdf of the form $q^*(y) = cp(y)\exp\{-\frac{1}{2}y'\Psi y - \gamma' y\}$, where $c < \infty$ is a normalization constant where Ψ and γ are as previously described. Since q^* corresponds to a $Q^* \in \mathcal{P}(R_1, \mu)$, some algebra shows that Q^* has covariance

$$R^* = \left(R_2^{-1} + \Psi\right)^{-1}, \tag{A3}$$

where Lemma 1 ensures that the above inverse exists and where R^* is an extension of R_1, and $\gamma' = R^*\mu - m'R_2^{-1}$. In particular, there exists a multiplier matrix Ψ and extension R^* of R_1 satisfying (A3). Thus q^* is Gaussian with covariance R^* and mean μ, and hence b) follows from Kullback [26, p. 189].

Proof of Corollary 1: If R_1 has no positive definite extension, then $\det R = 0$ and $\vartheta((R_1, \mu), P) = +\infty$ as in the theorem. If R_1 has a positive definite extension, then from the theorem $R^* \in \mathcal{Q}_k(R_1)$ and $\vartheta((R_1, \mu), P) = I(Q^* \| P)$, and hence the right-hand side of the equality is a lower bound to $\vartheta((R_1, \mu), P)$. It cannot be less than $I(Q^* \| P)$, however, for this would imply the existence of an $R \in \mathcal{Q}_k(R)$ yielding $I(Q \| P) < \vartheta((R_1, \mu), P)$, a contradiction. Hence the corollary must hold.

Proof of Corollary 2: From the assumptions of Theorem 1 we have

$$\vartheta((R_1, 0), P) = \frac{k}{2}\ln(\sigma^2\beta) - \frac{1}{2}\ln\det R^* - \frac{k}{2} + \frac{1}{2}(\sigma^2\beta)^{-1}\operatorname{tr} R^*A'A + \frac{1}{2}\|\sigma(1-\beta)^{1/2}s + b\|^2/(\sigma^2\beta),$$

where $\|u\| = (u'u)^{1/2}$, and

$$\operatorname{tr} R^*A'A = \operatorname{tr}\left\{E_{Q^*}(yy')A'A\right\} = E_{Q^*}y'A'Ay = E_{Q^*}v'v = \sum_{i=0}^{k-1} E_{Q^*}v_i^2,$$

where $v = Ay$ and hence

$$v_i = \sum_{j=0}^{\min(i,M)} a_j y_{i-j}$$

and

$$E_{Q^*}v_i^2 = \sum_{j=0}^{\min(i,M)} \sum_{m=0}^{\min(i,M)} a_j a_m R^*(i-j, i-m).$$

Since R_1 is Toeplitz and $J \geq M$, $R^*(i-j, i-m) = R_1(j-m)$ for all j and m in the above sum. Observe further that for $i \geq M$ $E_{Q^*}v_i^2 = a'R_1^{(M)}a$; that is, the trace term depends on R^* only through R_1 and hence does not require evaluation of the minimum discrimination information extension of R_1. Thus we have $\operatorname{tr} R^*A'A = \Gamma(R_1, A)$, which with Corollary 1 proves the first part. The remainder of the corollary follows from Theorem 1 and the fact that a Toeplitz matrix $R_1^{(J)}$ has a positive definite extension if and only if it is positive definite [38], [45].

Proof of Corollary 3: a) If R_1 is not positive definite, then it has no positive definite extension, and by the remark following the proof of Lemma 2, if $Q \in \mathcal{P}(R_1, 0)$, then Q does not have a pdf. Thus $\mathcal{G}_1$ is empty, and $\mathcal{H}(R_1) = -\infty$. b) If R_1 is positive definite and therefore has a positive definite extension, then from Theorem 1 and Corollary 2, for any P of the assumed form with $M \leq J$ there exists a unique extension R^* (not depending on P) and a corresponding Gaussian Q^* with

$$I(Q^*\|P) = \inf_{Q \in \mathcal{P}(R_1, 0)} I(Q\|P) = \min_{g \in \mathcal{G}_1} \int dy\, q(y)\ln(q(y)/p(y)). \tag{A4}$$

Since the integral

$$-\int dy\, q(y)\ln p(y) = \frac{k}{2}\ln(2\pi\sigma^2\beta) + \Gamma(R_1, A) + \frac{1}{2}\|\sigma(1-\beta)^{1/2}s + b\|^2/(\sigma^2\beta),$$

where $\Gamma(R_1, A)$ is given in Corollary 2 exists and is finite, then the difference $-h(q)$ between (A4) and the above integral exists, and hence $q \in \mathcal{G}_2$. Thus the minimum in (A4) can be taken over $\mathcal{G}_1 \cap \mathcal{G}_2$, and (A4) can be written as

$$\frac{k}{2}\ln(2\pi\sigma^2\beta) + \Gamma(R_1, A) + \frac{1}{2}\|\sigma(1-\beta)^{1/2}s + b\|^2/(\sigma^2\beta) - \max_{q \in \mathcal{G}_1 \cap \mathcal{G}_2} h(q). \tag{A5}$$

The minimum of (A3) is achieved if and only if the maximum in (A5) is achieved, and the same q yields both. From the theorem the minimizing q^* is unique, hence so is the maximizing q^*. The remainder of the corollary then follows from Corollary 2.

Proof of Corollary 4: Given R^* in Corollaries 2 and 3 and $R^* = \Lambda^{-1}$ in Lemma 1, R^* can be found as in Reid [37] by writing the Λ as a triangular decomposition $\Lambda = L'L$ where L is a lower triangular matrix with J nonzero diagonals and by solving $LR^* = (L')^{-1}$ with $R^* \in Q_k(R_1)$. Let $l_{i,i}$ denote the diagonal elements of L; then

$$\det R^* = \prod_{i=0}^{k-1} (1/l_{i,i})^2. \tag{A6}$$

From [37, eq. (28)], however, $l_{i,i} = l_{J,J}$ for all $i \geq J$. We also have, however, that $R^* = L^{-1}(L')^{-1} = L^{-1}(L^{-1})'$ is a triangular decomposition of R^* and that the ith diagonal of L^{-1} is $1/l_{i,i}$ which by Gaussian elimination must be $\{\det R^{*(i)}/\det R^{*(i-1)}\}^{1/2}$. (See, e.g., Gantmacher [12, ch. 2]). Thus, since R^* is an extension of R_1^J and $R^{*(J)} = R_1^{(J)}$,

$$\det R^* = \prod_{i=1}^{J-1} \left\{\det R_1^{(i)}/\det R_1^{(i-1)}\right\} \cdot \left(\det R_1^{(J)}/\det R_1^{(J-1)}\right)^{k-J}$$

$$= \det R_1^{(J-1)}\left(\det R_1^{(J)}/\det R_1^{(J-1)}\right)^{k-J}.$$

Proof of Theorem 2: a) If $R_1^{(J)}$ is not positive definite and hence has no positive definite extension to a covariance function r, then for some k_0, $R_1^{(J)}$ will have no positive definite extension to a $k \times k$ covariance matrix R for all $k \geq k_0$. The conclusion then follows from Theorem 1. b) By construction, $k^{-1}\|b\| \to 0$, $k^{-1}\|s\| \to 1$, and $k^{-1}b's \to 0$ as $k \to \infty$ and $\Gamma(R_1, A) \to \frac{1}{2}(\sigma^2\beta)^{-1}a'R_1^{(M)}a$ as $k \to \infty$. The remaining term $\to \ln \sigma_J^2$ as $k \to \infty$, completing the proof.

References

[1] J. B. Anderson and J. B. Bodie, "Tree encoding of speech," *IEEE Trans. Inform. Theory*, vol. IT-21, pp. 379–387, 1975.

[2] B. S. Atal and S. L. Hanauer, "Speech Analysis and synthesis by linear prediction of the speech wave," *J. Acoust. Soc. Amer.*, vol. 50, pp. 637–655, 1971.

[3] T. Berger, *Rate Distortion Theory: A Mathematical Basis for Data Compression*. Englewood Cliffs, NJ: Prentice-Hall, 1971.

[4] J. P. Burg, "Maximum entropy spectral analysis," presented at the 37th Annu. Int. Mtg., Soc. of Explor. Geophys., Oklahoma City, OK, 1967.

[5] A. Buzo, A.H. Gray, Jr., R. M. Gray, and J. D. Markel, "A two-step compression system with vector quantizing," in *Proc. 1979 Int. Conf. Acoustics, Speech, and Signal Processing*, pp. 52–55.

[6] ——, "Speech coding based upon vector quantization," *IEEE Trans. Acoust., Sp., Sig. Proc.*, vol. ASSP-28, pp. 562–574, Oct. 1980.

[7] D. L. Chaffee and J. K. Omura, "A very low rate voice compression system," in *Abstracts of Papers 1974 Int. Symp. Inform. Theory* Notre Dame, Oct. 1974.

[8] D. L. Chaffee, "Applications of rate distortion theory to the bandwidth compression of speech," Ph.D. dissertation, University of California, Los Angeles, 1975.

[9] I. Csiszár, "I-divergence geometry of probability distributions and minimization problems," *Ann. Prob.*, Vol. 3, pp. 146–158, 1975.

[10] L. J. Franson, "Applications of pattern matching of linear predictive coding of speech at 1200 bps," NRL Rep. 7931, Oct. 1975.

[11] R. G. Gallager, *Information Theory and Reliable Communication*. New York: Wiley, 1968.

[12] F. R. Gantmacher, *The Theory of Matrices*, vol. 1. New York: Chelsea, 1959.

[13] L. Ya. Geronimus, *Orthogonal Polynomials*. New York: Consultants Bureau, Noble Offset, 1961.

[14] W. Gersh, F. Martinelli, J. Yonemoto, M. D. Low, and J. A. McEwan, "Automatic classification of electroencephalograms: Kullback–Leibler nearest neighbor rules," *Science*, vol. 205, pp. 193–195, 1979.

[15] ——, "Kullback–Leibler nearest neighbor rule classification of EEGS," submitted for publication.

[16] A. H. Gray, Jr., Rm. M. Gray, and J. D. Markel, "Comparison of optimal quantization of speech reflection coefficients," *IEEE Trans.*, vol. ASSP-25, pp. 9–23, 1977.

[17] A. H. Gray, Jr., and J. D. Markel, "Distance measures for speech processing," *IEEE Trans. Acoust., Sp., Sig. Proc.*, Vol. ASSP-24, pp. 380–391, 1976.

[18] R. M. Gray, A. Buzo, A. H. Gray, Jr., and Y. Matsuyama, "Distortion measures for speech processing," *IEEE Trans. Acoust. Sp., Sig. Proc.*, vol. ASSP-28, pp. 367–376, Aug. 1980.

[19] R. M. Gray, A. Buzo, Y. Matsuyama, A. H. Gray, Jr., and J. D. Markel, "Source coding and speech compression," in *Proc. 1978 Int. Telemetering Conf.*, pp. 371–378.

[20] R. M. Gray, J. C. Kieffer, and Y. Linde, "Locally optimal block quantizer design," *Inform. Contr.*, vol. 45, pp. 178–198, May 1981.

[21] U. Grenander and G. Szegö, *Toeplitz Forms and Their Applications*. Berkeley, CA: Univ. of California Press, 1968.

[22] F. Itakura and S. Saito, "Analysis synthesis telephony based on the maximum likelihood method," in *Proc. 6th Int. Congr. Acoust.*, 1968, pp. C-17–C-20.

[23] ——, "A statistical method for estimation of speech spectral density and formant frequencies," *Electron. Commun. Japan*, vol. 53-A, pp. 36–43, 1970.

[24] G. Kang, and D. Coulter, "600 bps voice digitizer," in *Conf. Rec. 1976 IEEE Int. Conf. Acoustics, Speech, and Signal Processing*, pp. 91–94.

[25] G. S. Kang and D. C. Coulter, "600 bit per second voice digitizer," NRL Rep. 8043, Nov. 1976.

[26] S. Kullback, *Information Theory and Statistics*. New York: Dover, 1968.

[27] M. Kupperman, "Probabilities of hypotheses and information-statistics in sampling from exponential class populations," *Ann. Math. Stat.*, vol. 29, pp. 571–574, 1958.

[28] P. Lancaster, *Theory of Matrices*. New York: Academic, 1969.

[29] Y. Linde, A. Buzo, and R. M. Gray, "An algorithm for vector quantizer design," *IEEE Trans. Commun.*, vol. COM-28, pp. 89–95, Jan. 1980.

[30] S. P. Lloyd, "Least squares quantization in PCM," Bell Labs., Murray Hill, NJ, Tech. Paper, 1957.

[31] J. D. Markel and A. H. Gray, Jr., "On autocorrelation with application to speech analysis," *IEEE Trans. Audio Electroacoust.*, vol. AU-21, pp. 69–79, 1973.

[32] ——, *Linear Prediction of Speech*. New York: Springer-Verlag, 1976.

[33] ——, "Implementation of two transformed reflection coefficient scalar quantization methods," *IEEE Trans. Acoust., Sp., Sig. Proc.*, vol. ASSP-28, pp. 575–582, Oct. 1980.

[34] J. D. Markel and F. Juang, "Cepstral based pitch period and voicing estimation with statistical assistance," ARPA NSC Note NO-140, Oct. 1979.

[35] M. Pinsker, *Information and Information Stability of Random Variables, and Processes*. Moscow, USSR: Ize. Akad. Nauk. SSR, 1960 (English Translation—San Francisco, CA: Holden-Day, 1964).

[36] L. R. Rabiner and R. W. Schaefer, *Digital Processing of Speech Signals*. Englewood Cliffs, NJ: Prentice-Hall, 1978.

[37] J. S. Reid, "Confidence limits and maximum entropy spectra," *J. Geophys. Res.*, vol. 84, pp. 5289–5301, 1979.

[38] W. Rudin, "The extension problem for positive definite functions," *Ill. J. Math.*, vol. 7, pp. 532–539, 1963.

[39] R. W. Schafer and J. D. Markel, Eds., *Speech Analysis*. New York: IEEE Press, 1979.

[40] C. E. Shannon, "The mathematical theory of communication," *Bell Syst. Tech. J.*, vol. 27, pp. 379–423, 623–656, 1948.

[41] J. E. Shore, "Minimum cross entropy spectral analysis," *IEEE Trans. Acoust., Sp., Sig. Proc.*, vol. ASSP-29, pp. 230–236, Apr. 1981.

[42] J. E. Shore and R. M. Gray, "Minimum cross entropy pattern classification and cluster analysis," *IEEE Trans. Pattern Anal. Machine Intell.*, to appear.

[43] J. E. Shore and R. W. Johnson, "Properties of cross-entropy mini-

mization," *IEEE Trans. Inform. Theory*, vol. IT-27, no. 4, pp. 472–482, July 1981.

[44] ——, "Axiomatic derivation of the principle of maximum entropy and the principle of minimum cross entropy," *IEEE Trans. Inform. Theory*, vol. IT-26, no. 1, pp. 26–37, Jan. 1980.

[45] T. J. Ulrych and R. W. Clayton, "Time series modelling and maximum entropy," *Phys. Earth Planet. Interiors*, vol. 12, pp. 188–200, 1976.

[46] A. Van den Bos, "Alternative interpretation of maximum entropy spectral analysis," *IEEE Trans. Inform. Theory*, vol. IT-17, pp. 493–494, 1971.

[47] R. Viswanathan and J. Makhoul, "Quantization properties of transmission parameters in linear predictive systems," *IEEE Trans. Acoust., Sp., Sig. Proc.*, vol. ASSP-25, pp. 9–23, 1977.

An 800 bit/s Vector Quantization LPC Vocoder

DAVID Y. WONG, MEMBER, IEEE, BIING-HWANG JUANG, MEMBER, IEEE,
AND AUGUSTINE H. GRAY, JR., FELLOW, IEEE

Abstract—An 800 bit/s vector quantization linear predictive coding (LPC) vocoder has been developed. The recently developed LPC vector quantization theory is applied to reduce the bit rate for LPC coefficients coding by a factor of four. Branch search techniques and separation of voiced and unvoiced codebooks are applied for better algorithm efficiency. Differential coding is applied to reduce the bit rate for the pitch and gain parameters by one third.

Formal subjective evaluation shows that the 800 bit/s vocoder preserves most of the intelligibility of an LPC system. It is also robust under different transmission error and acoustic conditions. Informal listening comparisons show the quality to be acceptable and sometimes very close to 2400 bit/s LPC speech.

The computational cost of the 800 bit/s vocoder is equivalent to or even lower than the 2400 bit/s LPC-10. Compatibility with any LPC-10 vocoder is guaranteed because the 800 bit/s design only differs in the quantization and encoding algorithms. Further bit rate reduction can be achieved by removing frame to frame redundancy in the code.

I. Introduction

THE present lower bound in bit rate for a practical speech coder with acceptable quality is 2400 bits/s. This bit rate is achieved by linear predictive coding (LPC) or channel voice coding. Formant and homomorphic voice coders have also been proposed, but are not fully developed due to implementation, robustness, or quality problems.

Manuscript received September 8, 1981; revised April 14, 1982. This work was supported by the Naval Electronics Systems Command under Contract N00039-79-C-0450.

D. Y. Wong is with Speech Technology Laboratory, Santa Barbara, CA 93105.

B.-H. Juang is with Bell Laboratories, Holmdel, NJ 07733.

A. H. Gray, Jr. is with Signal Technology, Inc., Santa Barbara, CA 93101.

A number of studies have been and are currently being conducted to achieve bit rates lower than 2400 bits/s while minimizing further degradation to the speech quality. This topic of research is one of both fundamental and practical interest.

All of the approaches taken to lower the bit rate of vocoders have exploited one or more of the following three basic ideas.

1) *Time Redundancy Reduction:* Because of the similarity of speech characteristics from frame to frame under steady-state conditions, significant bit rate reduction can be achieved with frame repeat [1], [2] or other differential coding techniques [3], [6].

2) *Parameter Set Reduction:* By adopting a more parsimonious parameter set, the bit rate can be reduced as long as the new parameter set is not any more sensitive to quantization than the original set. A smaller parameter set can be obtained by adopting a better model such as the formant model [4], [5] or by applying some transformation techniques such as Karhunen–Loeve transformation [6].

3) *Improved Quantization:* While speech analysis techniques have been studied extensively for the purpose of low bit rate speech transmission, the quantization problem has received

Reprinted from *IEEE Trans. Acoust., Speech, Signal Processing*, vol. ASSP-30, pp. 770–780, Oct. 1982.

relatively limited attention. Study of the quantization problem has largely been restricted to techniques which quantize each parameter individually. Basic quantization theory [7], [8], a number of studies in pattern matching [4], [5], [9]-[12], [15], and recent studies on LPC vector quantization [13], [14] have pointed to the potentials of dramatic bit rate reduction in quantizing the parameters as a vector.

In this paper the development of an 800 bit/s LPC vocoder by applying vector quantization to the LPC coefficients, with almost no change to the LPC analysis-synthesis algorithms, is discussed. If the first two techniques mentioned above are applied in addition to vector quantization, the bit rate can be reduced even further.

The overall design of the 800 bit/s LPC vocoder is as follows. For a typical 2400 bit/s LPC system, the frame rate is fixed at 44.4 frames/s, and 54 bits are used to encode each set of LPC parameters: 41 bits for filter coefficients, 6 bits for pitch, 1 bit for voicing, 5 bits for gain, and 1 bit for synchronization. The 800 bit/s system encodes three consecutive frames of LPC parameters with 54 bits (i.e., 18 bits/frame): 30 bits (10 bits/frame) for filter coefficients, 12 bits for pitch and voicing, 11 bits for gain, and 1 bit for synchronization. Comparison of the bits allocation for the 2400 bit/s LPC-10 system and the 800 bit/s system on a per frame basis is also summarized in Table I.

An overall functional block diagram of the 800 bit/s vocoder is provided in Fig. 1(a) and (b). The key to attaining the 800 bit/s rate is the reduction from 41 to 10 bits for coding each set of LPC coefficients by vector quantization. The vector quantizer consists of two codebooks, one for voiced speech and one for unvoiced speech (including nonspeech pauses and background noise). Due to the voicing separation, the voicing parameter is also required for quantizing the LPC coefficients. Three frames of LPC coefficients are encoded at a time, resulting in three 10-bit codes. Encoding of pitch and gain is scalar for one of every three frames. For the remaining two frames, the differences are encoded. The voicing parameter is embedded in the pitch code.

Details of the vector quantizer and the differential coding algorithms for pitch, voicing, and gain are discussed in Sections II and III. Formal subjective evaluation results are presented in Section IV.

II. LPC Vector Quantization

A. Theory and Design

The design goal of all speech coders is to minimize some measurement (or perception) of distortion between the original speech signal and that reproduced at the output of the decoder and synthesizer for a specified bit rate. The design features which uniquely define a coder are:

1) the distortion measure, be it quantitative, perceptual, or intuitive,

2) the minimization criterion (such as minimum average or minimax) and codebook generation procedure, and

3) the quantizer configuration or search procedure.

An optimal (in the minimum average distortion sense) LPC vector quantization technique has been developed in [13], [14], and [16]. Its basic concept is schematically depicted in Fig. 2. Its design features are as follows.

1) The distortion measure minimized during quantization is the Itakura-Saito distortion measure, which is also minimized during LPC analysis if the autocorrelation analysis algorithm followed by residual gain matching is performed [20]. Furthermore, the quantization distortion and analysis distortion are additive, so that even if they are separately minimized (during the process of analysis and then quantization), the overall distortion is minimized.

2) The quantization codebook achieves minimum average distortion because it is generated by a non-Euclidean clustering procedure similar to a K-means algorithm [21].

3) The full search vector quantizer is used to match every input parameter vector to the closest codeword in the codebook.

Even though the fully optimal design theoretically achieves locally minimum average distortion, a number of practical problems exist. Due to these practical considerations, an alternate design is proposed. The features of the modified design are discussed below.

1) The Itakura-Saito (IS) distortion measure, which includes both the gain and LPC filter coefficients, is replaced with the likelihood-ratio distortion. The likelihood ratio is a gain normalized spectral distortion measure defined as follows.

Let $X(z)$ be the z transform of a frame of speech and $\sqrt{\alpha_M}/A_M(z)$ be the optimal Mth order LPC model of $X(z)$. The squared gain term α_M is the residual energy resulting from inverse filtering $X(z)$ with $A_M(z)$. Also, let $1/A(z)$ be any Mth order all-pole filter. Inverse filtering $X(z)$ with $A(z)$ will result in a residual energy α, $\alpha \geqslant \alpha_M$ because α is minimized by $A_M(z)$. The residual energy α is given as

$$\alpha = \int_{-\pi}^{\pi} |X(e^{j\theta})|^2 \, |A(e^{j\theta})|^2 \, \frac{d\theta}{2\pi}. \tag{1}$$

Note that $A_M(z)$ and $A(z)$ are both assumed to have all their roots within the unit circle in the z plane.

The likelihood-ratio measure is defined for the two unity gain model spectra as

$$d_{LR}\left(\frac{1}{A_M}, \frac{1}{A}\right) \triangleq \int_{-\pi}^{\pi} |A(e^{j\theta})/A_M(e^{j\theta})|^2 \, \frac{d\theta}{2\pi} - 1$$

$$= \frac{\alpha}{\alpha_M} - 1. \tag{2}$$

TABLE I
Bit Allocation Schemes for 2400 Bit/s and 800 Bit/s LPC Vocoders

2400 BPS LPC-10		
44.4	Frames/Sec	
54	Bits/Frame	
	41 Bits	LPC Coefficients
	13 Bits	Voicing, Pitch, Gain, Sync
800 BPS Vector Quantization LPC		
18	Bits/Frame	
	10 Bits	LPC Coefficients
	8 Bits	Voicing, Pitch, Gain, Sync

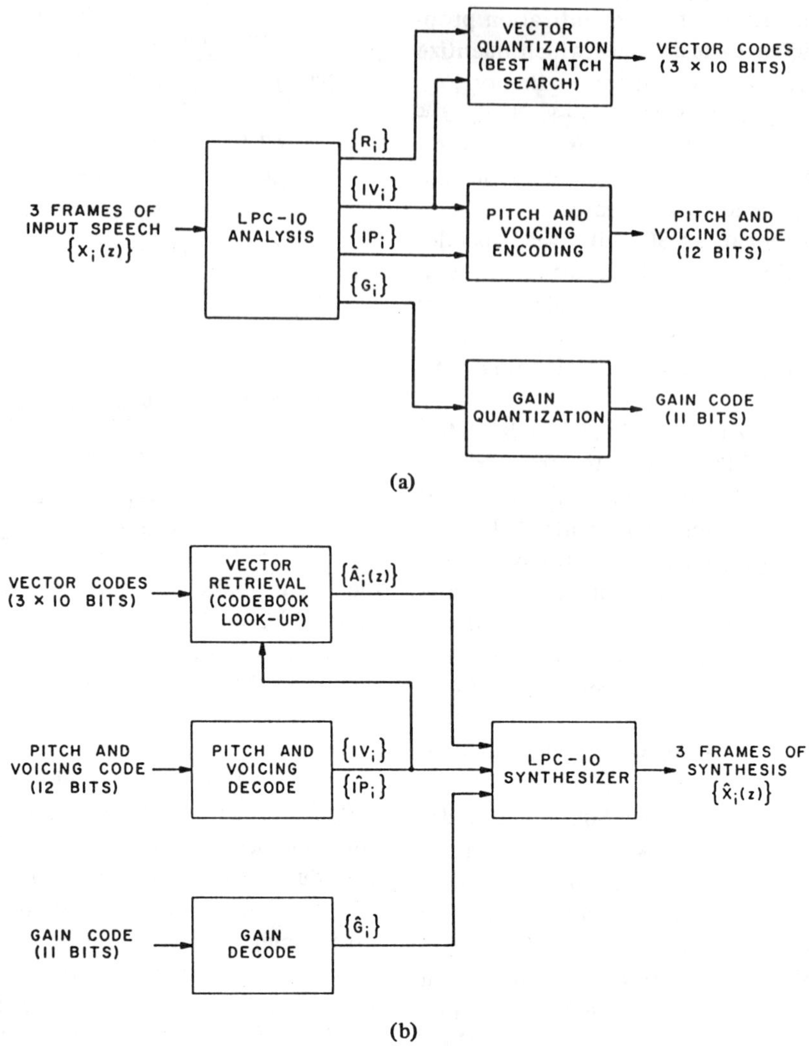

Fig. 1. (a) Overall design of the 800 bit/s LPC analyzer/encoder. (b) Overall design of the 800 bit/s LPC decoder/synthesizer.

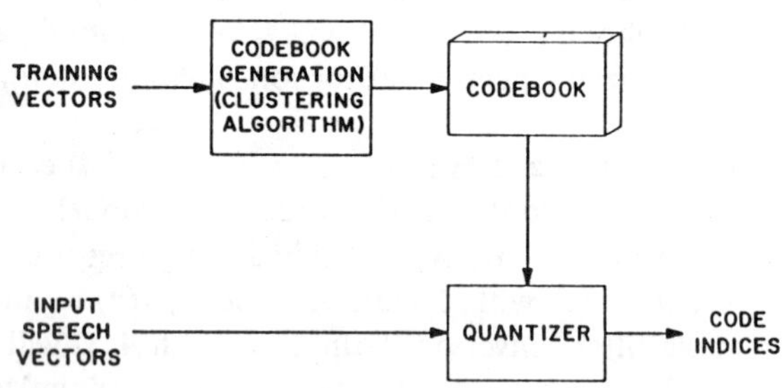

Fig. 2. Basic concept of vector quantization.

Thus, minimizing d_{LR} is equivalent to minimizing the residual energy since the optimal α_M depends on the input only.

The evaluation of d_{LR} can be carried out with a scalar product [14]

$$d_{LR}\left(\frac{1}{A_M}, \frac{1}{A}\right) = \frac{r_x(0)}{\alpha_M} r_a(0) + 2 \sum_{i=1}^{M} \frac{r_x(i)}{\alpha_M} r_a(i) - 1 \qquad (3)$$

where $r_x(i)$ and $r_a(i)$ denote the autocorrelation sequences of the input speech data and the polynomial coefficients of $A(z)$, respectively.

A significant advantage of the likelihood-ratio measure is that the gain term is separated from the LPC filter so it can be quantized separately. This is a very desirable feature because by not having to represent the gain variation of speech in the codebook, the size of the codebook can be kept smaller and more manageable. In practice, leaving the gain term out of the codebook ensures that vocoder performance will not depend on the level of the talker's voice or the gain of the analog input equipment. The separation of the gain (residual energy) from the LPC filter also allows easier tandeming with any 2400 bit/s LPC vocoder.

Although the likelihood ratio lacks the additive property necessary for full optimality in the quantization design as discussed above, consistency between quantization and analysis is maintained because the likelihood ratio is also minimized during LPC analysis.

2) The minimum average distortion criterion is maintained because the K-means clustering process developed for the IS distortion measure is also applicable to the likelihood ratio.

An advantage of a clustering algorithm for training the codebook is that it achieves locally optimal results for a given set of initial codewords. The process is also completely automatic so that very large codebooks can be generated from even larger databases. In some applications, the codebook can be made automatically adaptive to the user's voice for the best performance.

3) Two modifications are made to the search procedure.

First, since voiced and unvoiced LPC vectors are spectrally very dissimilar, and since voicing information is transmitted independent of the LPC vector code, two separate codebooks are searched for the two voicing states. This, of course, requires that two separate codebooks be generated and stored. It also essentially increases the codebook size by one bit without affecting the bit rate of the coder.

Second, in a full search design, the quantizer searches every one of the prestored codewords (vectors) to find the best match. For a 10-bit codebook, this involves computing 1024 likelihood ratio values, which requires 1024 X 11 multiply-adds. If the parameter space is first partitioned into a small number of subspaces, and each subspace into a small number of partitions and so forth until each partition corresponds to just a single codeword, the computation can be reduced using a tree search structure. To illustrate such a tree search, a binary tree for a 3-bit (eight codewords) case is shown in Fig. 3. In the tree illustrated, there is a total of three node levels (including the initial point) and two branches extend from each node. For a 10-bit codebook, a binary tree would contain ten node levels, and the number of searches is reduced from 1024 (for the full search) to only 20. The tradeoff for the tree search is higher average distortion, leading to suboptimal performance. There are also more codewords to be stored. For example, for the binary tree search 1022 additional codewords must be stored in the quantizer for the intermediate nodes.

Numerous other tree structures may be considered. A table of some tree structures and the corresponding number of codebook searches for a 10-bit code is provided in Table II.

For all of the pattern matching vocoders studied previously, only the full search vector quantizer or a hybrid of scalar and vector full search have been considered. As will be shown in Section II-B, the tree search procedure can lead to dramatic computation reduction at a small cost to performance and storage. In addition, a number of interesting interpretations can be drawn from the tree search codebook.

B. *Implementation and Results*

The main effort in the implementation of a 10-bit LPC vector quantization is the generation of an optimal codebook. Without complete *a priori* statistical information on the LPC coefficient vectors for speech, an optimal codebook must be generated from a large collection of coefficient vectors extracted from real speech data. The process is similar to the unsupervised training procedures commonly used in pattern recognition [21]. The training data used in this study are extracted from about 30 min of casual conversational speech collected from ten talkers, seven male, three female. The training data are also separated according to voicing so that two separate codebooks can be generated for voiced and unvoiced speech. Details on the training data and the codebook generation algorithm can be found in [14]. Those results most pertinent to the design of the 800 bit/s vocoder are discussed here.

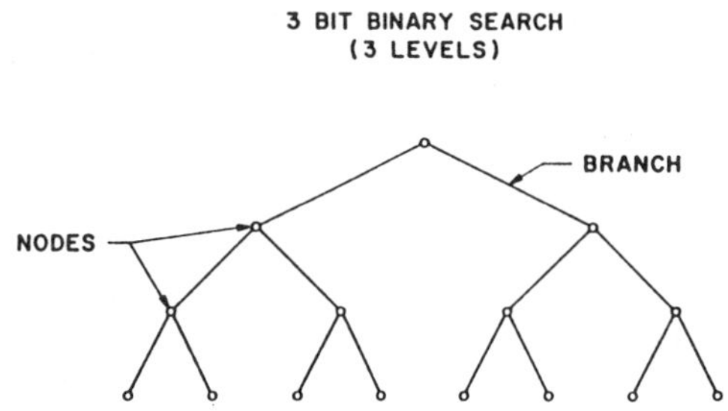

Fig. 3. A 3-bit binary search tree structure.

TABLE II
TREE STRUCTURES FOR A 10-BIT VECTOR CODEBOOK

Number of Branching Levels		Number of Branches at Each Node	Number of Searches
(binary search)	10	all 2's	20
	5	all 4's	20
	4	8,8,8,2	26
	3	16,16,4	36
	2	32,32	64
		64,16	80
		128,8	136
		256,4	260
		512,2	514
(full search)	1	1024	1024

Tree Search: For tree search codebooks, the full search clustering algorithm described in [16] must be modified as follows. Suppose there are K levels in the tree, and at each level there are L_j branches from each node ($j = 1, \cdots, K$).

a) An L_1 size full search codebook is first generated.

b) The training data are partitioned into L_1 subsets, each containing all the training vectors matched to one of the L_1 codewords.

c) For each subset, an L_2 size full search codebook is generated.

d) Steps b) and c) are repeated for L_2 and L_3 and so on, until all of the codewords are generated at the bottom of the tree.

e) The total number of codewords stored for such a codebook is

$$\sum_{i=1}^{K} \prod_{j=1}^{i} L_j. \tag{4}$$

Tree search codebooks for voiced speech from 1-10 bits were generated for all of the tree structures listed in Table II. The average distortion results for the 10-bit codebook are tabulated in Table III. The distortion performance of the binary, 32-32 tree, and the full search codebook from 1-10 bits is also compared in Fig. 4. As expected, the full search procedure consistently outperforms all other tree structures. However, at 10 bits, the 32-32 tree search is better than a 9-bit full search and just slightly inferior to 10-bit full search.

TABLE III
AVERAGE DISTORTION (LIKELIHOOD RATIO) OF 10 DIFFERENT TREE SEARCH 10-BIT CODEBOOKS FOR VOICED SPEECH

Tree Structure (Branches/node)	Average Distortion (Likelihood Ratio)
All 2's (binary)	.234364
All 4's	.214533
8,8,8,2	.211336
16,16,4	.205092
32,32	.189262
64,16	.193070
128,8	.194682
256,4	.191977
512,2	.184769
1024 (full)	.173432

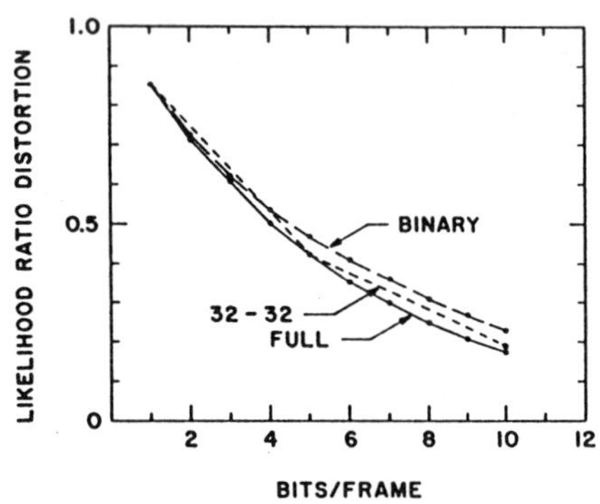

Fig. 4. Performance comparison of full search, 32-32 tree search, and binary tree search vector quantization.

In Table III, it can be seen that the 32-32 tree search performs almost as well as a 512-2 tree. It is also interesting to note that the 32-32 tree search actually outperforms the 64-16, 128-8, and 256-4 tree structures. This result is very significant from an implementation point of view because the computation for a 32-32 codebook search is only 1/16 that of a full (1024) search, while the increase in storage requirements is insignificant (1/32).

Unvoiced Codebook Experiment: It has been shown in [16] that unvoiced codebooks will generally show lower average distortion than voiced codebooks, so that a smaller size codebook for unvoiced speech will yield the same distortion performance. Further experiments were later conducted on the performance of the unvoiced codebook. After obtaining the tree search codebooks for voiced speech, the binary search, 32-32 tree search, and the full search codebooks were also obtained for unvoiced speech. Several test speech samples were then processed by the vector quantizer under three conditions. In the first condition, only the voiced LPC vectors were quantized, using the voiced codebook. In the second condition, both voiced and unvoiced vectors were quantized, using the two different codebooks. In the third condition, all vectors were quantized with the same voiced codebook. The output vectors under all three conditions were then used for LPC synthesis, and informal listening comparison was performed. Perceptually, little difference could be found among all three sets of syntheses, and no preference for any one over the other two could be drawn.

The implication of this short experiment is that a 10-bit unvoiced codebook introduces no perceptible distortion, so it should be reduced to a much smaller size to cut computation and storage cost. Furthermore, it is likely that the unvoiced codebook may be entirely eliminated, using the voiced codebook for all speech. It is also possible that for unvoiced speech the search can be limited to just a subset of the 10-bit voiced codebook to further reduce computation and transmission cost.

Pattern Matching: The input to the quantizer is a set (or vector) of analysis parameters extracted from the speech signal. Regardless of what quantization technique is used, the overall effect is that the input vector of analysis parameters is mapped into a quantized vector. Since each parameter vector represents a spectral pattern, all quantization schemes result in matching the quantizer input spectral pattern to a quantized output pattern. Thus, any quantization can be thought of as a pattern matching process. For LPC vector quantization, each LPC vector corresponds to a spectral pattern (or a vocal tract model).

The pattern matching aspect of vector quantization is particularly well demonstrated by the 32-32 tree search codebook for voiced speech. The 32 nodes (vectors) at the second level of the tree are found to cover the basic spectral patterns for vowels and voiced consonants, whereas the 32 branches of each one of these nodes represent variations of the 32 basic patterns. The first two formants of the top 32 nodes are plotted in Fig. 5 to illustrate this observation. The vowel boundaries proposed by Peterson and Barney (for men, women, and children) [19] are also sketched in the figure as a reference. Note that the formants are centralized because the 32 nodes serve as representative averages for the full 1024 codewords. Note also that children's speech is not represented here, so the high F_1 and F_2 values are not covered. The LPC spectrum for one of the initial 32 nodes is illustrated at the top of Fig. 6. The four spectra below it in the figure are those of selective branches from the node. It can be seen that the locations of the first three formants are quite close for all five spectra. However, considerable variations in the formant amplitudes exist. The codewords, therefore, do not just represent variations in formant values, but also spectral variabilities due to such factors as glottal pulse shape, formant bandwidth and other speaker characteristics.

Algorithm Complexity: It has been shown above that a 32-32 tree search is a good alternative to the full search vector quantizer, leading to considerably reduced computation. Recall from (3) that each comparison between an input LPC vector and codeword requies an evaluation of the likelihood ratio, which involves 11 multiply-add operations. Therefore, a full (1024) search requires about four times the computation needed for loading the covariance matrix [20, p. 222], while a 32-32 tree search requires only a fraction of the computation. Its computation requirement is close to that needed for the Levinson matrix solve algorithm. In fact, with vector quantization, the matrix solve algorithm is no longer needed, so the overall computation at the encoder with vector quantization is equal to LPC analysis with matrix solve. The comparisons are summarized in Table IV.

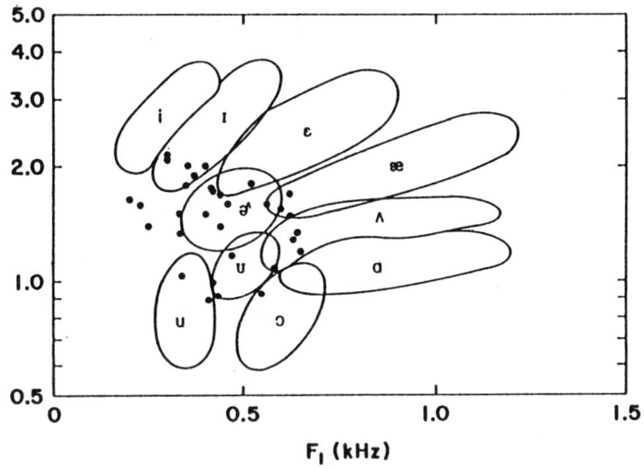

Fig. 5. First and second formant frequencies for the first 32 nodes of a 32-32 tree search codebook.

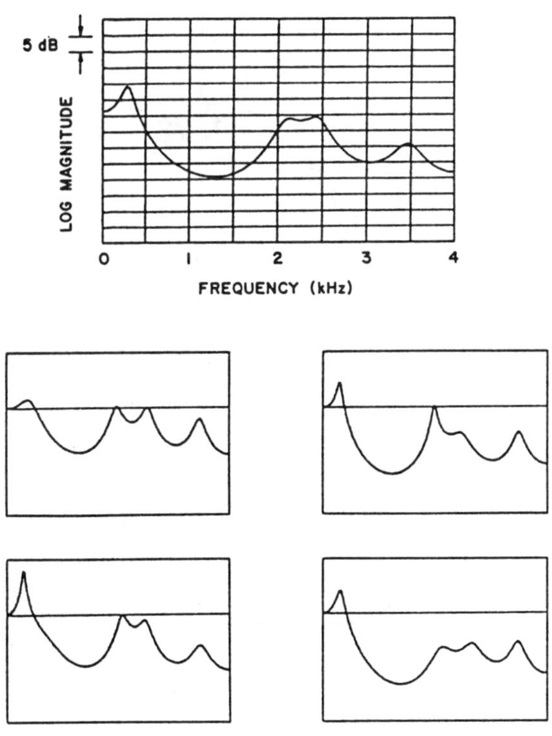

Fig. 6. LPC spectra of a node (Fig. 5) and four of its branches.

TABLE IV
COMPUTATION AND STORAGE REQUIREMENTS FOR VECTOR QUANTIZATION

ALGORITHM COMPLEXITY	MULT/ADD	MAX. MEMORY (16 BITS/COEFF)
FULL SEARCH	1024 × 11 ~ 4 × COVAR MATRIX LOAD	20 K BYTES
32-32 SEARCH	64 × 11 ~ LEVINSON MATRIX SOLVE	20 K + 640 BYTES

Based on an upper bound estimate of using 16 bits to store each coefficient of a codeword vector, the total storage requirement for a 10-bit full search and a 32-32 tree search codebook are also compared in Table IV. At the encoder, storage in the r_a form (autocorrelation of the $A(z)$ coefficients) allows direct computation of the likelihood ratio. At the decoder, the coefficients are stored in whatever form the synthesizer utilizes, typically as reflection coefficients (for lattice form filtering) or polynomial coefficients (for direct form filtering). The upper bound storage requirement is 20 kbytes for each full search codebook and 640 additional bytes for the 32-32 tree search codebook. A more realistic estimate may be about 10 kbytes for the former and 320 additional bytes for the latter. Further reduction in storage is possible if the coefficients are stored in some other equivalent forms such as log area ratios and reflection coefficients at the encoder. However, this would require additional computation to mathematically transform them before the search.

Based on the results discussed above, a 32-32 tree search is adopted for 10-bit LPC vector quantization. Two separate codebooks are generated for voiced and unvoiced speech, using the extensive database collected from ten different talkers.

III. Pitch and Gain Quantization

A. Gain Quantization

The gain term is quantized with a simple differential coding technique. Three frames of gain terms are quantized and encoded into 11 bits. The gain term for the middle frame is quantized to 5 bits using the existing ANDVT LPC-10 quantization table [18]. The difference between frame 1 and 2 and that between frame 3 and 2 are quantized with a 3-bit table. The table is derived from a histogram of gain differences between consecutive frames collected from speech samples of ten different talkers.

B. Pitch and Voicing Coding

For the LPC vocoder used in this study, pitch value quantization is embedded in the pitch detection algorithm, resulting in a 6-bit pitch code for every speech frame. Voicing is updated once per frame. In the 800 bit/s design, three frames of pitch and voicing are encoded to 12 bits total. The voicing decision is updated only once per frame, and it is embedded in the pitch code where possible.

The first bit is assigned to coding the voicing decision for the first frame. Bits 2-7 are assigned to code the pitch value for the first voiced frame, which could be any of the three frames. The functions of bits 8-12 depend on the voicing state of frame 1 as follows.

1) If frame 1 is voiced, bits 8-10 encode the difference in pitch from frame 1-2. A binary code of "111" indicates that frame 2 is unvoiced. Bits 11 and 12 are used to encode the pitch difference from frame 2-3 if frame 2 is voiced; otherwise it encodes the pitch difference from 1-3. A binary code of "11" indicates frame 3 is unvoiced.

2) If frame 1 is unvoiced, then bit 11 (eleven) is used for encoding voicing for frame 2, and bit 12 for frame 3. Bits 8-10 encode the pitch difference from frame 2-3 if both are voiced.

The bit allocation scheme for this 12-bit code is shown in Fig. 7, and a flowchart of the encoder is provided in Fig. 8.

IV. Subjective Evaluation

The 800 bit/s vocoder developed was evaluated by two sets of tests for a number of transmission and acoustic conditions. For intelligibility, the diagnostic rhyme test (DRT) [22] was administered. Four transmission error conditions were tested with the DRT: 0 percent, 1 percent, 2 percent, and 5 percent. To obtain a reference and to separate the effects of vector quantization from pitch and gain coding, two additional vocoder configurations were tested with the DRT: 1) LPC-10 without quantization on any of the vocoder parameters and 2) LPC-10 with vector quantization on the LPC coefficients, but no quantization on pitch and gain. Each DRT involved two male talkers and eight listeners.

For user acceptability, the diagnostic acceptability measure test (DAM) [23] was conducted. The DAM was used to evaluate the vocoder under a quite acoustic environment and four noisy environments: office, carrier, E3A, and helicopter. Two talkers and fourteen listeners were used to test the quiet environment; one talker and fourteen listeners were used to test the noisy environments. The results of both DRT and DAM are discussed below.

In addition to the formal subjective tests, the vocoder has also been tested on a wide variety of talkers (not used in training the quantization codebook), and informal listening found the system to be quite robust for different voices. In some cases, the output quality is quite close to the 2400 bit/s LPC-10. It is important to note that in all of the tests performed, the talkers, vocabulary, and recording conditions used are completely different from those used in collecting the codebook generation data.

A. DRT Results

The DRT scores for the 800 bit/s vocoder (under 0 percent transmission error), those of LPC-10 with no quantization, and LPC-10 with only vector quantization on LPC coefficients are compared in Table V. All six DRT feature scores are tabulated to allow close analysis of the effects of vector quantization on speech intelligibility. The reader is referred to [22] for the definition of the DRT features.

The DRT score of 88.4 for the LPC-10 vocoder is very close to that generally reported for a 2400 bit/s LPC system [18], [24], [27]. The score would be several points higher if the vocoder modifications proposed in [25] were incorporated. The drop of 8.4 DRT points from the unquantized LPC-10 to the 800 bit/s vocoder is primarily due to degradation in the sustention, graveness, and nasality features. LPC vector quantization is responsible for 5.9 DRT points and pitch and gain coding for 2.5 points. The degradation due to LPC vector quantization is primarily in the graveness and sustention features. The two features least well preserved at 800 bits/s in fact are sustention (62) and graveness (57).

The sustention feature at 800 bits/s is almost 40 points biased towards the positive state. [See Fig. 9(c).] This bias is primarily a result of pitch and gain coding, which made most voiced stop consonants (e.g., /b/) sound like continuants

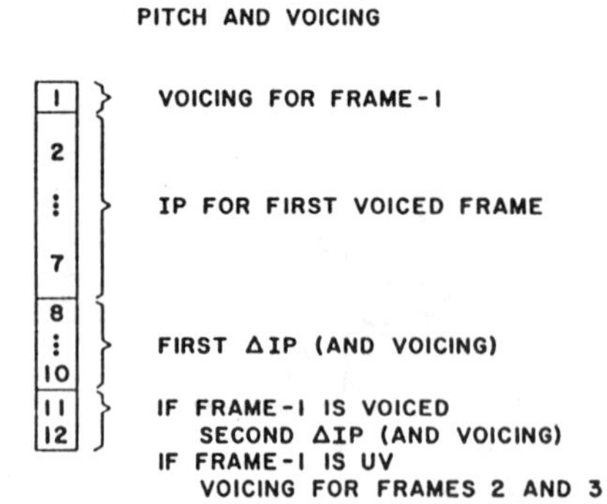

Fig. 7. Bit allocation for the pitch and voicing parameters.

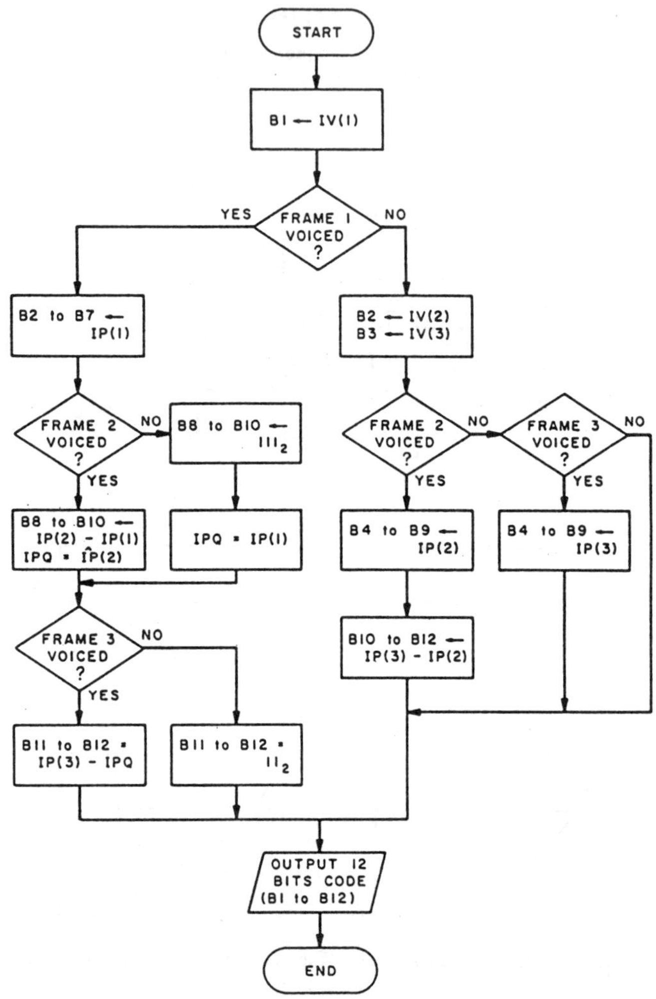

Fig. 8. Flowchart of the pitch and voicing encoder.

(e.g., /v/). This result is expected because the sustention feature is highly dependent on the parameter update rate and the gain parameter as demonstrated in [25]. Significant improvements are possible with faster frame update for unvoiced speech as discussed in [25] and better gain quantization. LPC vector quantization also led to some degradation in the sustention score, causing some continuants (voiced and unvoiced) to be perceived as stops.

The graveness score at 800 bits/s is significantly biased towards the positive state. [See Fig. 9(e).] This means that front position consonants (such as /m/ and /f/) are perceived

TABLE V
DRT SCORES FOR 1) LPC-10 WITHOUT PARAMETER QUANTIZATION, 2) LPC-10 WITH ONLY 10-BIT VECTOR QUANTIZATION ON LPC FILTER COEFFICIENTS, AND 3) 800 BIT/S LPC/VQ VOCODER

DRT Features	LPC-10 No Quan. DRT Scores	LPC-10 VQ Only DRT Scores	800 BPS LP C/VQ DRT Scores
Voicing	90.6	89.8	87.1
Nasality	96.1	92.6	86.7
Sustention	78.9	71.5	62.1
Sibilation	98.0	96.5	94.5
Graveness	73.8	52.7	57.0
Compactness	93.0	91.8	92.6
Overall	88.4	82.5	80.0

as medial positions (/n/ and /θ/). The poor graveness score is almost entirely due to LPC vector quantization as shown in Table V. The bias towards the positive state for the vector quantization LPC may be a result of some degree of centralization of the sounds due to the clustering process as seen in Fig. 5.

The scores for the three features of voicing, compactness, and sibilation at 800 bits/s are very close to the unquantized LPC-10. The compactness score shows no degradation at 800 bits/s, and is consistently one of the highest among all six features. The voicing score is biased towards the negative state, meaning the synthesis sounds devoiced.

The nasality feature is degraded by both LPC vector quantization and pitch and gain quantization. It is biased towards the positive state, meaning the nonnasal consonants (in the word initial position) tend to be perceived as nasal. It should be noted that the consonant pairs (such as /n/-/d/) used for testing the nasality feature also differ in the sustention feature—so the poor nasality score may very well be a side effect of poor sustention discrimination. In fact, the nonnasal consonants used for testing this feature are all stops, and they are perceived as sustained nasals.

Transmission Error Degradation: The overall DRT scores for the 800 bit/s vocoder for the four different transmission errors are compared in Table VI.

The six DRT feature scores for the different transmission error rates are plotted in Figs. 9(a)-(f). All six features, except for sibilation, show consistent degradation with transmission errors. The degradation in the DRT score is approximately linear from 1-5 percent. The drop of 25 points at 5 percent is very close to that for a 2400 bit/s LPC system [24]. The robustness of the vocoder under noisy transmission conditions is remarkable because the 800 bit/s code contains much less redundancy than the 2400 bit/s code.

It is interesting, however, to note that while the sustention score at 0 percent transmission error is much lower for the absent (negative) state, as the transmission error rate increases, the degradation is mostly on the present (positive) state, i.e., smooth transitions begin to sound abrupt. This may be a result of an increasingly choppy quality in the speech as transmission error rate increases.

For the same reason, sustained nasal consonants become more like stop (nonnasal) consonants leading to lower scores for the present state, as shown in Fig. 9(b). The voicing scores for the present state degrade much more with transmission errors, meaning the speech tends to sound more and more devoiced or breathy as transmission errors increase. The other feature scores (except for sibilation) show uniform drop for both the absent and present state.

TABLE VI
DRT SCORES FOR DIFFERENT TRANSMISSION ERROR CONDITIONS

Transmission Error	DRT Score
0%	80.0
1%	75.3
2%	71.7
5%	55.5

Overall, the DRT scores show significant but graceful degradation of the system due to transmission errors. However, it should be pointed out that in the 800 bit/s design, no attempt was made to provide error protection. If the system is to operate under severe error conditions, simple error protection techniques such as that discussed in [26] can be expected to provide significant improvements.

B. DAM Results

The DAM test was used to evaluate the vocoder for a variety of acoustic environments. The results are analyzed for three different acceptability score categories: signal, background, and total effects. The DAM scores for these categories are plotted in Fig. 10(a)-(c). The different noise environments are plotted according to a subjective rank ordering of the severity of the background noise.

For signal quality, six semantic (parametric) scales are used. Of these, the thin (SH), harsh (SD), and nasal (SN) qualities show a consistent trend of degradation with the severity of background noise. The flutter (SF) and muffled (SL) qualities show almost no change throughout. The interrupted (SI) score shows a drop for the carrier and E3A conditions only. From these six scores, one can conclude that acoustic noise contamination causes the speech synthesis to sound thinner (due to high frequency noise) and harsher (due to time discontinuities in the LPC spectral characteristics). The low SI (interrupted) score for the E3A and carrier conditions are perhaps due to bursts of background tones, other talking voices, and engine noise in these environments. The relatively high SI score for the helicopter environment may be due to the constancy in the level and the narrow-band nature of the noise.

The background acceptability scores of Fig. 10(b) show a more consistent trend of degradation with the noise environment. Most of the degradation is heard as buzzing and hissing qualities, due to voicing errors and also discontinuities in the LPC spectral change.

For total effects, four sets of scores are plotted in Fig. 10(c): the metametric pleasantness (TRP) and intelligibility (TRI) scores, the isometric acceptability (CIA) score, and the composite acceptability (CAE) score. The parametric accept-

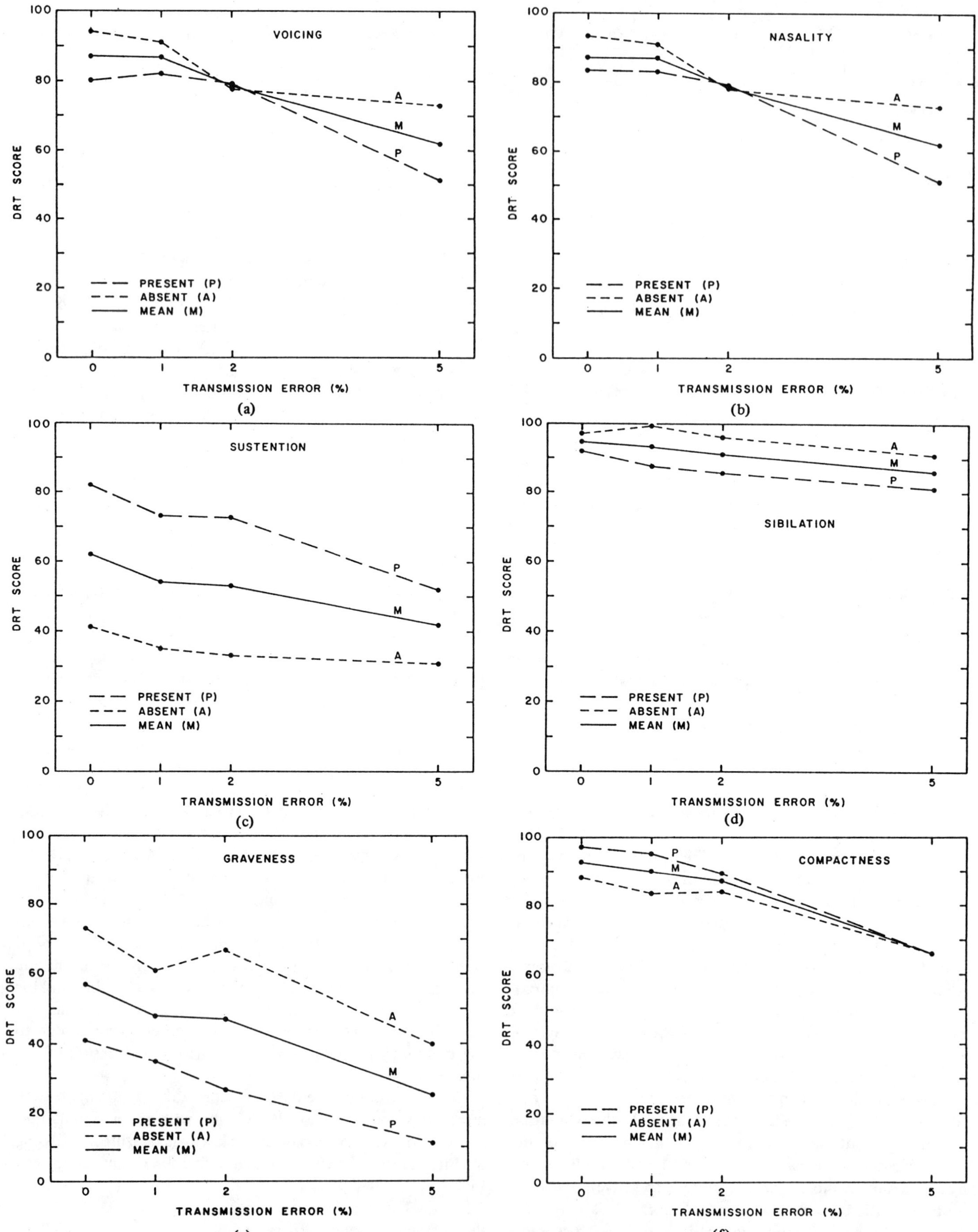

Fig. 9. (a) DRT voicing scores for four different transmission error conditions. (b) DRT nasality scores. (c) DRT sustention scores. (d) DRT sibilation scores. (e) DRT graveness scores. (f) DRT compactness scores.

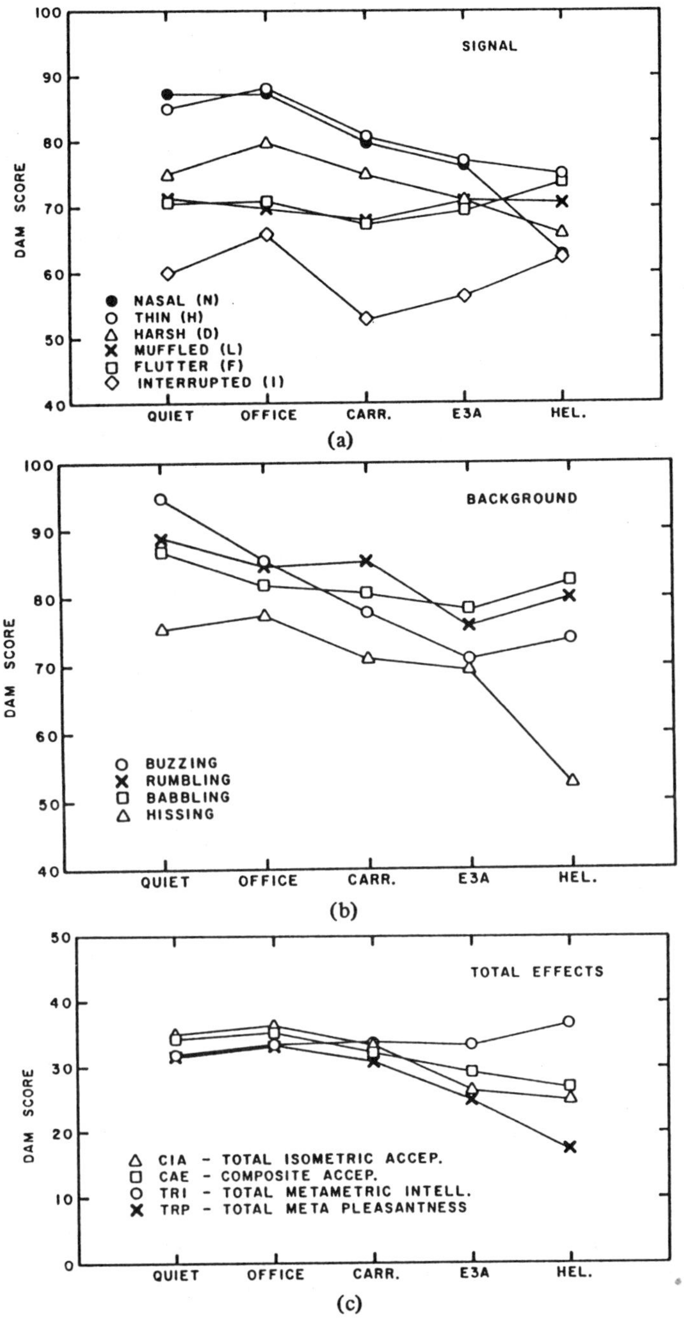

Fig. 10. (a) DAM signal parametric scores for six different acoustic noise environments. (b) DAM background acceptability scores for six different acoustic noise environments. (c) DAM total effects acceptability scores for six different acoustic noise environments.

ability score is omitted because it is a direct derivative of the six parametric scores in Fig. 10(a). The TRI score is completely incompatible with the other three scores. The other three scores, on the other hand, are very consistent with each other. The quiet and office noise environments are practically equal, scoring about 35 points. For comparison, the 2400 bit/s LPC-10 DAM score is 52.6 for the quiet environment. There is a small amount of degradation due to carrier noise, some more for the E3A condition, and severe degradation for the helicopter environment. The helicopter condition, however, can be remedied with noise reduction preprocessing.

Overall, the DAM results indicate that the vector quantization LPC vocoder is relatively insensitive to low levels of noise, and is most sensitive to high-level narrow-band noise typified by the helicopter environment.

V. Discussion

An 800 bit/s vector quantization LPC vocoder has been developed. Subjective evaluation found that overall, the 800 bit/s vocoder compares very well to 2400 bit/s LPC vocoders.

Analysis of the computation and memory requirements needed for the vector quantization LPC showed that with a branch search quantizer, the 800 bit/s vocoder requires minor increase in data memory over a 2400 bit/s LPC-10 vocoder. There is, however, some reduction in computation.

It should also be clear from the design of the 800 bit/s vocoder that it is primarily a better quantization algorithm for LPC-10. Therefore, the design is ideal for tandem operation with any LPC system.

Research is continuing to improve the quality and robustness of the system and to optimize the computation and storage requirements for real time implementation. Research is also underway to further lower the bit rate of the system with frame fill [1], variable frame rate [2], and the new matrix coding [28] techniques.

Acknowledgment

The authors wish to thank C. Leeds for his assistance in implementing the entire software simulation, especially his development of the codebook generation software on the FPS AP-120B array processor. They are also grateful to G. S. Kang of the Naval Research Laboratory for his generous help in coordinating the speech evaluation.

References

[1] P. E. Blankenship and M. L. Malpass, "Frame-fill techniques for reducing vocoder data rates," Lincoln Lab. Tech. Rep. 556, Feb. 1981.

[2] E. McLarnon, "A method for reducing the transmission rate of a channel vocoder using frame interpolation," in *Conf. Rec., 1978 IEEE ICASSP*, 1978, pp. 458–461.

[3] J. W. Fussell, "A differential linear predictive voice coder for 1200 BPS," in *Proc. 1981 IEEE ICASSP*, Mar. 1981, pp. 24–27.

[4] G. S. Kang and D. C. Coulter, "600 BPS voice digitizer (linear predictive formant vocoder)," Naval Res. Lab. Rep. 8043, Nov. 1976.

[5] J. J. O'Donnell, "A system for very low data rate speech communication," in *Proc. 1981 IEEE ICASSP*, Mar. 1981, pp. 8–11.

[6] M. R. Sambur, "An efficient linear prediction vocoder," *Bell. Syst. Tech. J.*, vol. 54, pp. 1693–1723, 1975.

[7] T. Berger, *Rate Distortion Theory*. Englewood Cliffs, NJ: Prentice-Hall, 1971.

[8] L. O. Davisson and R. M. Gray, Ed. *Data Compression*. Stroudsburg, PA: Dowden, Hutchingson and Ross, 1976.

[9] C. P. Smith, "Speech data reduction," AFCRL Rep. TR-57-111 Astia AD117290, Bedford, MA, May 1957.

[10] H. Dudley, "Phonetic pattern recognition vocoder for narrowband speech transmission," *J. Acoust. Soc. Amer.*, no. 30, pp. 733–739, 1958.

[11] D. L. Chaffee, "Applications of rate distortion theory to the bandwidth compression of speech," Ph.D. dissertation, Univ. California, Los Angeles, 1975.

[12] L. J. Fransen, "Application of pattern matching to linear predictive coding of speech at 1200 bps," NRL Rep. 7931, Oct. 1975.

[13] R. M. Gray, A. Buzo, A. H. Gray, Jr., and Y. Matsuyama, "Distortion measures for speech processing," *IEEE Trans. Acoust., Speech, Signal Processing*, vol. ASSP-28, pp. 367–376, Aug. 1980.

[14] A. Buzo, A. H. Gray, Jr., R. M. Gray, and J. D. Markel, "Speech

coding based upon vector quantization," *IEEE Trans. Acoust., Speech, Signal Processing*, vol. ASSP-28, pp. 562–574, Oct. 1980.

[15] B. Gold, "Experiments with a template-matching channel vocoder," in *Proc. 1981 IEEE ICASSP*, Mar. 1981, pp. 32–35.

[16] B. H. Juang, D. Y. Wong, and A. H. Gray, Jr., "Distortion performance of vector quantization for LPC voice coding," *IEEE Trans. Acoust., Speech, Signal Processing*, vol. ASSP-30, pp. 294–304, Apr. 1982.

[17] D. Y. Wong, B. H. Juang, and A. H. Gray, Jr., "An 800 bps LPC vector quantization vocoder," Final Rep., Contr. N00039-79-C-0450, June 1981.

[18] G. S. Kang, "Application of linear prediction encoding to a narrowband voice digitizer," Naval Res. Lab. Rep. 7774, Oct. 1974.

[19] G. E. Peterson and H. L. Barney, "Control methods used in a study of the vowels," *J. Acoust. Soc. Amer.*, vol. 24, pp. 175–184, 1952.

[20] J. D. Markel and A. H. Gray, Jr., *Linear Prediction of Speech.* New York: Springer-Verlag, 1976.

[21] M. R. Anderberg, *Cluster Analysis for Applications.* New York: Academic, 1973.

[22] W. D. Voiers, "Diagnostic evaluation of speech intelligibility," in *Speech Intelligibility and Speaker Recognition*, M. E. Hawley, Ed. Stroudsburg, PA: Dowden, Hutchingson and Ross, 1977.

[23] —, "Diagnostic acceptability measure for speech communication systems," in *Conf. Rec., 1977 IEEE ICASSP*, May 1977, pp. 204–207.

[24] —, "Diagnostic evaluation of speech coding systems," Final Rep., Contr. DCA 100-77-C-0003, Sept. 1978.

[25] D. Y. Wong and J. D. Markel, "An intelligibility evaluation of several linear prediction vocoder modifications," *IEEE Trans. Acoust., Speech, Signal Processing*, vol. ASSP-26, pp. 424–435, Oct. 1978.

[26] J. W. Fussell *et al.*, "Providing channel error protection for a 2400 BPS linear predictive coded voice system," in *Conf. Rec., 1978 IEEE ICASSP*, Apr. 1978, pp. 462–465.

[27] S. Maitra and C. R. Davis, "A speech digitizer at 2400 bits/s," *IEEE Trans. Acoust., Speech, Signal Processing*, vol. ASSP-27, pp. 729–733, Dec. 1979.

[28] D. Y. Wong and B. H. Juang, "Voice coding at 800 bps and lower data rates with LPC vector quantization," in *Conf. Rec., 1982 IEEE ICASSP*, Paris, France, May 1982, pp. 606–609.

VERY LOW DATA RATE SPEECH COMPRESSION WITH LPC VECTOR AND MATRIX QUANTIZATION

D.Y. Wong
Speech Technology Lab.
Santa Barbara, CA 93105

B.H. Juang
Bell Laboratories
Holmdel, NJ 07733

D.Y. Cheng
University of California
Santa Barbara, CA 93106

ABSTRACT

Frame predictive vector quantization is developed to compress the bit rate for coding the LPC filter coefficients to under 250 bits/sec. An innovative LPC compression technique, matrix quantization, is also developed to compress the LPC filter coefficients to a rate under 150 bits/sec. Subjective evaluation with the diagnostic rhyme test (DRT) finds the proposed techniques to be feasible for intelligible speech transmission at bit rates between 400 bits/sec and 200 bits/sec.

I. Introduction

Vector Quantization [1] has been used to reduce the bit rate of an LPC vocoder from 2400 bits/sec to 800 BPS in our earlier study [2]. This study explores the possibility of extending the vector quantization technique for even lower bit rate LPC voice coding. Theoretically, vector quantization achieves optimal distortion performance for encoding LPC coefficients. The vector code is also nearly optimal in its discrete memoryless source code entropy [3]. Higher code efficiency is thus attainable only by exploiting frame to frame redundancy in the LPC coefficient vectors.

To achieve rates below 400 bits/sec, very efficient coding of pitch and gain information is also necessary. That topic, however, will not be addressed in this paper. The reader is referred to [4] for a report on our recent research on the coding of pitch and gain parameters.

II. Frame Repeat LPC Vector Quantization

Design

A vector quantizer maps each input LPC model vector x onto an index j. The index designates the codeword y_j in a codebook $C=\{y_i\}$ which best matches the input vector, i.e., the quantizer output is $\hat{x}=y_j$. Let $x(n)$, $x(n-1)$, $\hat{x}(n)$, and $\hat{x}(n-1)$ denote the input vectors and quantized output vectors for the current and previous frames, respectively. Since the quantizer output vector is always one of the codebook entries, denote $\hat{x}(n-1)=y_j$. In frame repeat coding, the index for frame n-1 is repeated for frame n when at least one of the following conditions is met:

$$d[x(n), \hat{x}(n-1)] < t; \qquad (1.1)$$
$$d[x(n), y_j] = \min_i d[x(n), y_i] \qquad (1.2)$$

where d(.,.) denotes the distortion measure for the vector quantizer, and t is a selected threshold. When frame repetition occurs, no codeword index has to be sent for the new frame. However, a 1-bit/frame repetition flag (repeat/no-repeat) must be transmitted. Due to this overhead bit, for an L-bit codebook, at least 1/(L+1) of the frames must be repeated to achieve any bit rate reduction.

Experimental Results

A test speech sample was processed to study the performance of the frame repeat algorithm. The likelihood ratio measure with full search vector quantization [1,3] was employed. To investigate the tradeoff between frame repeat and codebook size, three sets of codebooks, 6-bit, 8-bit, and 10-bit in size, were tested. Each set consists of two codebooks, one for voiced speech and one for unvoiced speech. The codebooks were obtained using a full scale multi-speaker training speech sequence consisting of over thirty thousand LPC vectors as detailed in [3]. For each codebook size, threshold (t) values ranging from 0.0 to 0.8 in 0.1 increments were tested.

The average distortion for all three codebook sizes is plotted against the average number of bits (including the repetition flag) in Figure 1. As expected, the average distortion is a monotonically decreasing function of the average number of bits per frame for a fixed codebook size, but the incremental

Reprinted from *IEEE Int. Conf. Acoust., Speech, Signal Processing,* vol. 1, pp. 65–68, April 1983.

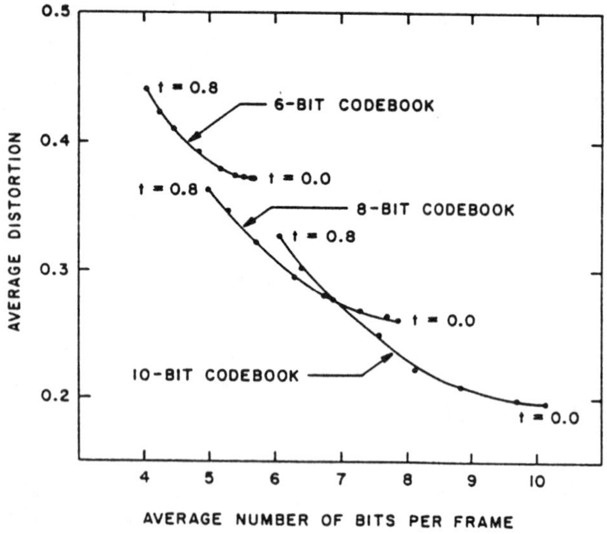

Figure 1: Performance Trade-Off Between Codebook Size and Frame Repeat Threshold

performance gain (i.e. drop in average distortion) decreases with the bit rate. To achieve an average distortion of about 0.3 or less, a 6-bit codebook is simply not adequate; an 8-bit or 10-bit codebook is necessary.

At a frame rate of 44.44 frames/sec, an average rate of about 250 bits/sec (i.e. 5.6 bits/frame) is achieved with a threshold of 0.6 for the 8-bit codebooks. The output speech quality is informally judged to be very close to the 800 bits/sec vector quantized LPC synthesis [2].

III. LPC Matrix Quantization

LPC matrix quantization is based on a speech production model in which only a small set of target phonemes are intended by the speaker, and the articulatory transition from one phoneme to the next always follows the same path. Consequently, a finite number of transition vector sequences are adequate to construct all of the speech sounds produced.

Based on such a model, a natural extension to the vector quantization technique is, therefore, to assemble the LPC vectors into NxM matrices $\underline{X}(n)$, where

$$\underline{X}(n)=(x(n-N+1), x(n-N+2),.., x(n)) \quad (2)$$

$x(i)$=Mx1 vector of LPC coefficients for frame i,
M=order of the LPC filter, and
N=number of vectors in the matrix.

If each frame corresponds to T sec of speech, the duration of a matrix is NT sec. Each matrix will be treated as the basic unit for quantization.

To develop a matrix quantization system, an algorithm for segmenting speech data into LPC matrices must first be developed. A distortion measure between two matrices must then be defined. Then a codebook must be generated. A block diagram of the matrix quantization system is shown in Figure 2.

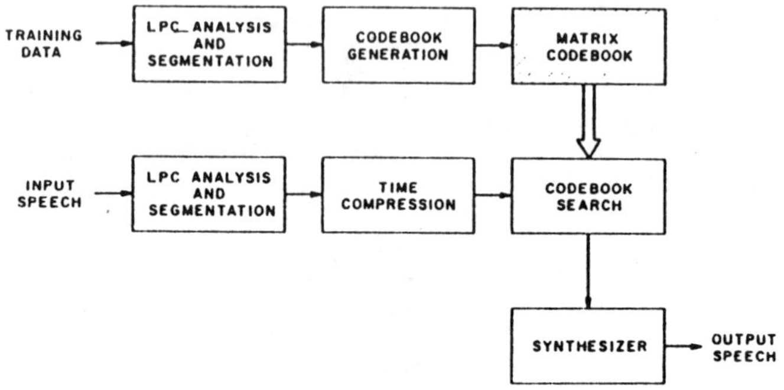

Figure 2: Block Diagram of the LPC Matrix Quantization System

Segmentation

Speech segmentation is defined according to a set of acoustically observable events: speech onset, speech offset, steady states, and transitions. Each segment begins at a speech onset point or steady state center and ends at the next steady state center or speech offset point. The sequence of LPC vectors corresponding to each segment is thus termed a transition matrix. The duration of a transition matrix is found to vary between 50 msec to over 300 msec.

Segmentation as defined above is equivalent to the discrimination between speech and non-speech signals (pauses) and between steady-state and transition speech sounds. The discrimination algorithm developed in this study is based on LPC reflection coefficients, the voicing decision, and the speech energy.

Speech/non-speech discrimination is based on the gain and voicing features. A voiced frame is automatically defined as speech. For unvoiced frames, the s/ns classifier is based on a comparison between the signal energy and an estimate of the background noise. All speech or pause segments shorter than 50 msecs are merged into neighboring segments to eliminate errors due to fluctuations in background noise and speech segments with marginal energy levels.

Steady-state/transition classification is primarily based on a spectral variance measure defined as follows. Denote the LPC vectors extracted from the signal at every T sec by x(n), then the spectral variance σ_d at index n is given by

$$\sigma_d(n) = 1/2L \sum_{j=n-L}^{n+L} d[x(j),x(n)] \quad (3)$$

where L=3 and d[,] is the COSH distortion measure [5].

Heuristics for detecting dips and valleys in the σ_d contour are applied to locate steady state sounds. The center of each steady state segment is defined as a boundary point. Details of the algorithm are discussed in [4].

After the segment boundary points are detected, smoothing algorithms are then applied to alter segments that are too short (<50 msec), too long (>300 msec), or those which contain too many (>3) voicing transitions.

An example of the segmentation results is shown in Figure 3. The onset and offset points are labeled in the plot at the bottom of the figure. The unlabeled vertical bars in the plot mark the location of steady state centers.

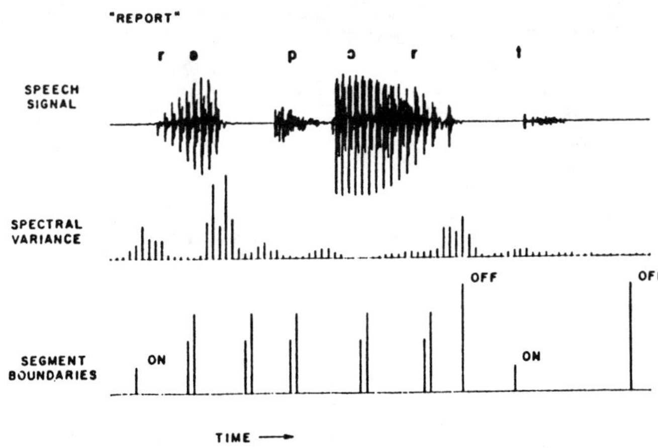

Figure 3: Speech Segmentation Based on Spectral Variance

Distortion Measure

Since an input matrix b(i) and a codeword w(j) may be different in length, the distortion D[w(j), b(i)] must perform time warping between the two matrices. In this study, dynamic time warping is selected so that D[,] is accumulated over the optimal warping path. The continuity condition (slope constraint) selected is the simplified path with slope intensity P=1 as defined in [6]. No global range constraint other than that implied by the continuity and fixed end point conditions are applied. The distance measure between two vectors is the COSH measure. The input matrix is pre-compressed to a minimum length before time warping by eliminating any vector whose COSH distance to the preceding vector is below a specified threshold. With a threshold of 0.4, about 50% of the speech frames are eliminated.

Codebook Generation

A database of sixteen minutes of speech from a single speaker is collected as training data. Standard LPC analysis is performed at a rate of 100 frames/sec. The segmentation and time compression algorithms described above are applied to group the vectors into 3,478 transition matrices. Codebook generation is performed using a minimax criterion [4]. With this approach, the maximum distortion for encoding any matrix in the data base is bounded above by a selected distance t_c. The codebook size thus varies inversely with t_c. Initially, a large value is selected for t_c to prevent memory overflow. Then t_c is reduced until the desired codebook size is obtained. For the given training data of 3,478 matrices, by setting the distance upperbound to t_c = 1.0, 1185 codewords are generated with an average length of 7.46 LPC vectors.

Quantizer Simulation

A matrix quantization system based on the segmentation algorithm and codebook described above has been simulated. Input speech data is transformed into LPC matrices and each matrix is mapped into a codeword index by searching the codebook for the best match according to D[,] defined above. In addition to the matrix code index, timing information must be transmitted so that the codeword can be warped to the right length at the receiver. At a minimum, the duration of the input matrix (50 msec to 300 msec) must be transmitted. This would require 4 bits (in 20 msec steps) or 5 bits (in 10 msec steps) for each matrix. At an average rate of about 8 matrices/sec (assuming no pauses in the speech) the bit rate for duration information is 32 to 40 bits/sec.

At the receiver, the matrix code j is used to retrieve $w(j)$. An output $\hat{B}(i)$ is obtained by time warping $w(j)$ according to the timing information. The quantized LPC matrix $\hat{B}(i)$ is fed to the synthesizer to produce the output speech.

A number of speech samples including words, phrases, and sentences from inside the training data, outside the training data (and vocabulary) by the same speaker and other speakers were used to test the system. The speech is informally judged to be quite intelligible in most cases and the quality is well preserved. Formal test results are presented below.

IV. Subjective Evaluation

A single speaker diagnostic rhyme test (DRT) procedure [7] is used to evaluate the intelligibility of both the frame repeat vector quantization and matrix quantization coding techniques. In these tests, the excitation parameters, pitch, voicing, and gain, are not quantized. Only the LPC filter coefficients are quantized. The DRT results therefore reflect only the effects of quantization on the LPC filters. The results for both the vector and matrix quantization systems are tabulated in Table 1.

	Frame Repeat <u>Vector Quan</u>.	Matrix <u>Quan</u>.
Voicing	93.0	89.1
Nasality	89.1	72.7
Sustention	69.5	53.1
Sibilation	75.8	69.5
Graveness	65.6	44.5
Compactness	80.5	77.3
Total	78.9	67.7

Table 1: DRT scores for the Frame Repeat Vector Quantization and Matrix Quantization Systems

For the frame repeat VQ system, the DRT word list is spoken by a talker outside the training data. Furthermore, almost all of the DRT words do not occur in the training speech data. The test is truly an <u>open test</u>. The total score of 78.9 compares favorably with that of a fixed frame rate (44.4 frames/sec) vector quantization (without pitch and gain coding) which attained a score of 82.5 [2].

The matrix quantization system is tested with the codebook obtained in the experiment discussed in III. Only a few of the DRT words overlap with the training data vocabulary. The speaker for the DRT word list is the same. The overall DRT score obtained is 67.7. Close examination of the individual DRT feature scores indicates that the intelligibility of the matrix quantization system would improve significantly if a larger training data base is used, or if the DRT vocabulary is included in the training data [4].

V. Conclusions

Two different techniques for very low data rate compression of the LPC filter coefficients have been developed and tested. Further research is continuing on improving the matrix quantization technique and on coding the timing, pitch and gain parameters to fully implement speech coders which operate at bit rates between 200 to 400 bits/sec.

This research was supported by the US Air Force under contract F30602-81-C-0054.

<u>References</u>

1. A. Buzo, A.H. Gray Jr., R.M. Gray, and J.D. Markel, "Speech Coding Based on Vector Quantization", <u>IEEE Trans. ASSP</u>, pp. 562-574, October 1980.

2. D.Y. Wong, B.H. Juang, and A.H. Gray Jr., "An 800 bits/sec Vector Quantization LPC Vocoder", <u>IEEE Trans. ASSP</u>., Vol. ASSP-30, pp. 770-780, October 1982.

3. B.H. Juang, D.Y. Wong, and A.H. Gray, "Distortion Performance of Vector Quantization for LPC Voice Coding", <u>IEEE Trans. ASSP</u>, Vol. ASSP-30, pp. 294-304, April 1982.

4. D.Y. Wong, B.H. Juang, and D.Y. Cheng, <u>Vector/Matrix Quantization for Narrow-Band Digital Speech Compression</u>, Signal Technology, Inc., Final Report, Contract F30602-81-C-0054, July 1982.

5. A.H. Gray Jr. and J.D. Markel, "Distance Measures for Speech Processing", <u>IEEE Trans. ASSP</u>. Vol. ASSP-24, pp. 380-391, October 1976.

6. H. Sakoe and S. Chiba, "Dynamic Programming Algorithm Optimization for Spoken Word Recognition", <u>IEEE Trans. ASSP</u>, Vol. ASSP-26, pp. 43-49, February 1978.

7. W.D. Voiers, "Diagnostic Evaluation of Speech Intelligibility", <u>Speech Intelligibility and Speaker Recognition</u>, ed. M.E. Hawley, Dowden, Hutchingson, and Ross, 1977.

Matrix Quantizer Design for LPC Speech Using the Generalized Lloyd Algorithm

CHIEH TSAO, STUDENT MEMBER, IEEE, AND ROBERT M. GRAY, FELLOW, IEEE

***Abstract*—Rate-distortion theory provides the motivation for using data compression techniques on matrices of *N* LPC vectors. This leads to a simple extension of speech coding techniques using vector quantization. The effects of using the generalized Lloyd algorithm on such matrices using a summed Itakura–Saito distortion measure are studied, and an extension of the centroid computation used in vector quantization is presented. The matrix quantizers so obtained offer substantial reductions in bit rates relative to full-search vector quantizers. Bit rates as low as 150 bits/s for the LPC matrix information (inclusive of gain, but without pitch and voicing) have been achieved for a single speaker, having average test sequence and codebook distortions comparable to those in the equivalent full-search vector quantizer operating at 350 bits/s. Preliminary results indicate that higher quality or lower bit rates may be achieved with enough computational resources.**

I. Introduction

VECTOR quantization (VQ) [1], [2] was introduced as a method of compressing vocal tract models obtained by linear predictive coding (LPC). It combines the two steps of linear prediction and scalar quantization into one step by using techniques derived from rate-distortion theory based on the Itakura–Saito distance measure. The method used a generalized form of Lloyd's algorithm to cluster LPC speech frames, resulting in a *codebook* of LPC vectors which is locally optimized for a training sequence of finite length [3]. The LPC frames, however, can themselves be considered as scalars in a large alphabet, and hence rate-distortion theory indicates that better performance for a given rate or lower rate for a given performance can be achieved by encoding groups of LPC frames or matrices of parameters. As with scalar compression schemes, an alternative to memoryless vector quantization is the use of feedback quantizers such as finite-state vector quantizers [4], [5] which also exploit the interframe memory in speech. Applications of finite-state quantizers to LPC vectors are described in [5]. The matrix quantizers considered here compare very favorably in terms of complexity and performance. At 150 bits/s, a three-vector matrix quantizer achieves distortion performance comparable to full-search vector quantizers operating at 350 bits/s, while the finite-state vector quantizer described in [5] at the same rate achieves distortion performance comparable to full-search vector quantizers operating at 250 bits/s. This is further reinforced by subjective testing, where there was overwhelming preference for the sound of the matrix quantizer when compared to the finite-state quantizer at 150 bits/s. Furthermore, matrix quantizers are much simpler and quicker to design, and therefore potentially superior if adaptation is desired.

Current research in very low bit rate transmission of speech lies primarily in using techniques derived from speech recognition. For example, Wong and Juang [6], [7] isolate speech segments between onset/offset and steady-state points to obtain a variable-rate matrix code with very low average rate (the term matrix quantization is due to Wong and Juang in this context) under speaker and vocabulary constraints, while Schwartz and Roucos [8]–[11] isolate speech segments between steady-state points (generally similar to diphones) in their segment quantizer with the same objective. Although the experiments in [10] show that variable-rate matrix codes generally perform better than fixed-rate matrix codes, nevertheless, fixed-rate matrix codes continue to be of interest because they are inherently simpler to design and implement, particularly for small matrices. In this paper, we present the design of locally optimal, fixed-rate codes using the generalized Lloyd algorithm on N-matrices of LPC vectors. The method focuses on small values of N, ($N \leq 4$), uses a simple distortion computation, and offers a continuous tradeoff between codebook size, rate, and quality. We first present a simple extension of LPC speech coding using vector quantization, followed by some experimental results.

II. Matrix Quantization

We define some terms for convenience. The number N is the block length in a matrix of N LPC vectors. The *codebook rate* in bits per frame is given by the logarithm to base 2 of the number of codewords, or reproduction matrices in the codebook, divided by N. We shall call a matrix code using a codebook with 2^M codewords and a block length of N an M-bit N-code, with rate $R = M/N$ bits/frame. Intuitively, increasing N will reduce the rate M/N, but at the expense of higher average codebook distortion. To keep the codebook distortion constant, we also need to increase the number of codewords, which increases M. However, because speech is highly structured, M increases less quickly than N in practice for constant

Manuscript received May 16, 1984; revised September 28, 1984. This work was supported in part by the Singapore Government, the Joint Services Electronics Program at Stanford University, the National Science Foundation, and by the Industrial Affiliated Program of the Stanford University Information Systems Laboratory.

The authors are with the Information Systems Laboratory, Department of Electrical Engineering, Stanford University, Stanford, CA 94305.

Reprinted from *IEEE Trans. Acoust., Speech, Signal Processing,* vol. ASSP-33, pp. 537–545, June 1985.

codebook distortion. The net result is decreasing transmission bit rate with rising N for comparable vocoding quality. Where N is fixed (as in our study here), we also have the advantage of fixed rate encoding.

The heuristic argument above is expressed more precisely in Shannon's source coding theorem [12], [13], applied to the particular case of LPC speech. Looked at another way, a matrix quantizer for LPC speech is analogous to a waveform vector quantizer where the source symbols are LPC vectors instead of individual speech samples.

The technique of speech coding using vector quantization has been dealt with in detail elsewhere [1], [2]. We now present a straightforward extension to the matrix case. Essentially, this consists of taking a training sequence of K speech frames, with K being some (large) integer, and blocking the sequence into LPC N-matrices. The generalized Lloyd algorithm is then applied to this collection of LPC matrices using a summed Itakura–Saito distortion measure, in exactly the manner described in [14]. A codebook of reproduction matrices is then obtained, and this is the matrix quantizer. In the paragraphs below, and in Appendixes I and II, we develop some notation for matrix quantizer design, restate informally the generalized Lloyd algorithm in the matrix notation so developed, and show how centroids using the summed Itakura–Saito distortion measure may be computed.

Let $\boldsymbol{x}$ be a vector of LPC coefficients ($\log \alpha_M$, $r_x(0), \cdots, r_x(M)$), where M is the order of the LPC filter, α_M is the gain of the filter, and $r_x(j)$, $j = 0, \cdots, M$ are the $M + 1$ sample autocorrelations for a speech frame [1]. Then, if N is some integer ≥ 1, define the $(M + 2) \times N$ matrix $\boldsymbol{X}$, where $\boldsymbol{X} = [\boldsymbol{x}_1, \boldsymbol{x}_2, \cdots, \boldsymbol{x}_N]$. Furthermore, define a reproduction matrix $\boldsymbol{Y}$ to be a matrix of LPC vectors $[\boldsymbol{y}_1, \boldsymbol{y}_2, \cdots, \boldsymbol{y}_N]$, where each $\boldsymbol{y} = (\log \sigma^2, r_a(0), \cdots, r_a(M))$, where each $r_a(j)$, $j = 0, \cdots, M$ are the autocorrelations of the inverse filter coefficients and σ is the gain for the LPC model. Consider the distortion measure $D(\boldsymbol{X}, \boldsymbol{Y})$ where

$$D(\boldsymbol{X}, \boldsymbol{Y}) = \frac{1}{N} \sum_{i=1}^{i=N} d_{IS}(\boldsymbol{x}_i, \boldsymbol{y}_i)$$

with

$$d_{IS}(\boldsymbol{x}, \boldsymbol{y}) = \frac{r_a(0) r_x(0) + 2 \sum_{i=1}^{i=M} r_a(i) r_x(i)}{\sigma^2} - \log \alpha_M - 1$$

being the usual modified Itakura–Saito distortion measure [1]. It was shown in [3] that a generalized centroid for $d_{IS}(\cdot, \cdot)$ exists and that the generalized Lloyd algorithm converges to a local minimum for $d_{IS}(\cdot, \cdot)$ in the case of a finite training sequence, since $d_{IS}(\cdot, \cdot)$ is convex and differentiable. It follows that $D(\cdot, \cdot)$, which is a finite sum of $d_{IS}(\cdot, \cdot)$ is also convex, from the properties of convex functions, and differentiable. Hence, by the results in [3], a generalized centroid with respect to $D(\cdot, \cdot)$ exists, and the generalized Lloyd algorithm will converge to a local minimum in the training sequence case. All that needs to be shown now is that the centroid with respect to $D(\cdot, \cdot)$ can be easily computed. This is done in Appendix I. An informal statement of the Lloyd algorithm using the matrix notation above is given in Appendix II.

Some remarks about the use of training sequences with the Lloyd algorithm are appropriate in the matrix case. The training sequence of matrices $\{\boldsymbol{X}\}$ is obtained by sliding the first matrix along the training sequence of speech frames, so that the jth matrix in the sequence is given by $\boldsymbol{X}_j = [\boldsymbol{x}_j, \boldsymbol{x}_{j+1}, \cdots, \boldsymbol{x}_{j+N-1}]$. This sliding block technique ensures that we obtain the greatest number of matrices from a training sequence of any given length, and that all phoneme transitions in speech present in the training sequence are captured.

A test sequence is encoded by blocking the sequence into matrices of N LPC vectors (without sliding) and transmitting the codeword index of the codeword nearest to each block relative to the distortion $D(\cdot, \cdot)$. That is, if the test sequence is given by the sequence of LPC frames $[\boldsymbol{t}_1, \boldsymbol{t}_2, \cdots, \boldsymbol{t}_j \cdots]$, then we block the sequence into matrices $\boldsymbol{T}_j = [\boldsymbol{t}_{N(j-1)+1}, \boldsymbol{t}_{N(j-1)+2}, \cdots, \boldsymbol{t}_{Nj}]$. Then if $\{\boldsymbol{Y}\}$ is the matrix codebook of reproduction vectors obtained above, the transmitted codeword index m for the jth matrix in the test sequence $\boldsymbol{T}_j$ is given by

$$m : D(\boldsymbol{T}_j, \boldsymbol{Y}_m) \leq D(\boldsymbol{T}_j, \boldsymbol{Y}_k) \quad \text{for} \quad \boldsymbol{Y}_k \in \{\boldsymbol{Y}\}$$

where $k = 1, \cdots, B$, B being the number of reproduction matrices in the codebook $\{\boldsymbol{Y}\}$. At the receiver, which will possess a copy of the codebook $\{\boldsymbol{Y}\}$, the matrix $\boldsymbol{Y}_m$ will be synthesized in the normal way. Note that the pitch and voicing parameters are independent of the coding of the LPC information and may be separately transmitted by any previously established method (e.g., using the fake process trellis code as in [6]).

We should add that the whole encode–decode scheme is noted for its simplicity (fixed rate coding, no time-warping required in the distortion computation). Furthermore, no initialization process is required at the start of the transmission, as is required in any comparable finite-state vector quantization scheme.

III. Experimental Results

The experiments were divided into the following two parts.

1) Showing that the generalized Lloyd algorithm converges in practice under the distortion $D(\cdot, \cdot)$. Also, a sequence of codebooks using a relatively short (5000) frame multiple speaker training sequence was generated to show the general trend of bit rate versus distortion obtained via matrix quantization. A short single speaker test sequence (600 frames) was then used on these codebooks to show the effects of a short out-of-train sequence on a codebook generated with a short training sequence. Specific matrix 3-codes and 2-codes were then generated with a longer training sequence (15 000 frames) and these were compared to full-search VQ codebooks of essentially the same average distortion using four test (out-of-train) sequences, each about 2000 frames long. Only a single

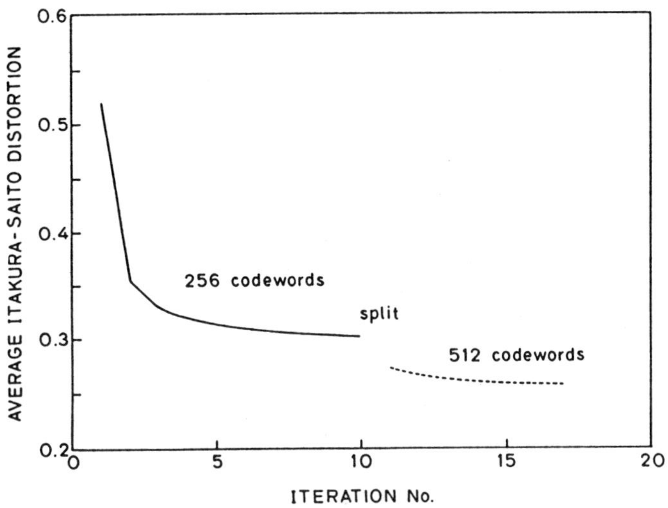

Fig. 1. Convergence of generalized Lloyd algorithm for 9 bit matrix 3-code.

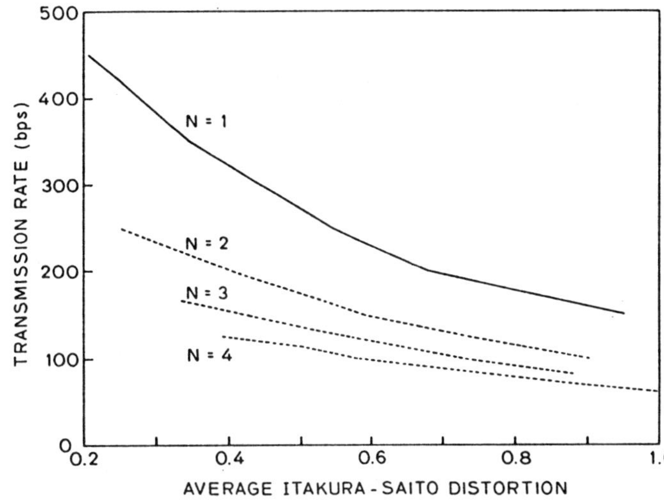

Fig. 2. Transmission rate versus MQ codebook distortion.

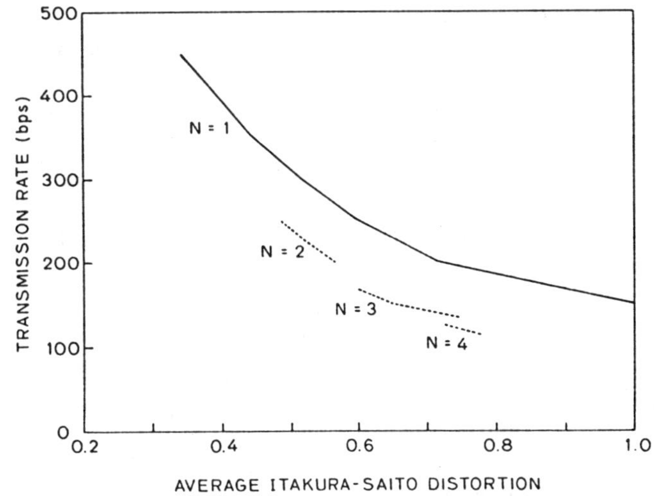

Fig. 3. Transmission rate versus average distortion for test sequence.

speaker was used in both test and training sequences for this latter experiment.

2) Subjective tests comparing the out-of-train single-speaker test sequence vocoded using various matrix and vector quantizers. Intelligibility tests on single-speaker training and test sequences, consisting of reproducing the contents of a 1 min out-of-train test sequence vocoded using a 9-bit 3-code at 150 bits/s for the LPC matrix. These latter tests are not conclusive owing to the fact that the LPC speech used in our experiment was made under difficult conditions, but serve mainly to indicate that matrix quantizers designed under the distortion measure $D(\cdot, \cdot)$ have the potential for producing intelligible speech at lower bit rates.

In all cases, the training sequence had to be kept relatively short (15 000 frames or less) because of the constraints imposed by limited computational resources. This necessarily meant that the codebooks generated would be 'tuned' to the data in the training sequences, making the system speaker and vocabulary dependent. While it is conceivable that true speaker and vocabulary independence may be achieved with a training sequence that is long enough and a codebook that is large enough, such a task is beyond the scope of this study. The details for the test and training sequences are given in Appendix III.

A. Convergence of the Lloyd Algorithm and Distortion Measurements

In Fig. 1, we show the convergence of the generalized Lloyd algorithm for $N = 3$ and a 15 000 frame single speaker training sequence. The initial guess codebook had 256 codewords in it. This converged to a local optimum after 10 iterations (percentage change in distortion ≤ 0.5 percent with the next iteration). The perturbation matrix ϵ was then used to double the number of codewords, and the graph shows convergence to a 512-codeword codebook after the 17th iteration. In Fig. 2, we show plots of transmission bit rate against average codebook distortion, computed as in the previous section, for various block lengths N from 1 (full-search VQ) to 4. We note that for any given value of average distortion, increasing N decreases the bit rate of transmission for the appropriate matrix quantizer. Isolated experiments with longer training sequences indicate that these trends are representative of those obtained with very long training sequences (see the next section).

In Fig. 3 we show plots of transmission bit rate against average distortion obtained with a 600-frame out-of-training test sequence. Because of the shortness of the training and test sequences used, the average distortion with this test sequence is significantly higher than the average codebook distortion. This seems to be particularly true in the matrix case, where the gains obtained with increasing N appear to be substantially less than the codebook distortions would seem to indicate. Nevertheless, we feel that with a substantially longer training sequence, the differences between a test sequence and the training sequence would be considerably smaller. Results from the next section appear to substantiate this claim.

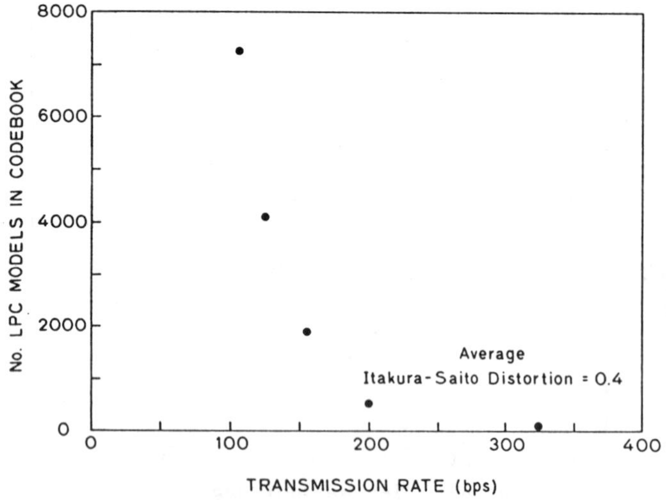

Fig. 4. Codebook size versus transmission rate.

Codebook size, however, also increases very rapidly and at very low bit rates, codebooks with several thousand matrices are required, particularly for $N \geq 4$, if the distortion is to be acceptable. This greater than exponential growth in codebook size is a limitation of the method (vector quantization based methods suffer from exponential codebook growth in general) at present. However, large codebook size seems inherent in current very low-bit rate systems. For example, [6] reports codebook sizes of 3478 matrices comprising matrices with an average size of 13 10-ms frames, giving 45 214 LPC models in the codebook for an overall LPC matrix bit rate of 150 bits/s without gain, pitch, and voicing. Fig. 4 shows the increase in the number of LPC models in the codebook with decreasing transmission bit rate, for a fixed distortion level. Naturally, the number of LPC models for a given rate would increase substantially as the distortion decreases, so the figures implied in the curve cannot be directly compared to those reported in [6]. The trend appears to be approximately proportional to $(1/R)2^{1/R}$ where R is the transmission bit rate. Extensions of shape-gain vector quantizers [1] are possible, however, and the design of shape-gain matrix quantizers is the subject of current research. These could alleviate the problem of very large codebook size considerably.

We now present results obtained from a 15 260-frame single speaker training sequence and four single-speaker test sequences for matrix 3-codes and matrix 2-codes, and compare these against figures obtained for 7 bit and 8 bit VQ codes obtained from the same training and test sequences.

Tables I–IV show the codebook distortions and test sequence distortions obtained for 9 bit and 10 bit matrix 2-codes and 3-codes, respectively. For the matrix code, the average test-sequence distortion is given by

$$\frac{1}{L}\sum_{j=1}^{j=L} \min_{Y \in \{Y\}} D(T_j, Y)$$

where T_j is the jth (nonsliding) block in the test sequence, as defined previously, and L is the number of matrices in the test sequence. The test-sequence distortion computation for the VQ case is exactly the same with $N = 1$. We note that in the case of the 9 bit matrix codes, where we have a compression ratio of about 1:30 (for our purposes, the compression ratio is defined by the number of matrices in the training sequence divided by the number of matrices in the codebook) the increase in distortion for the four test sequences was no worse than 1.14 dB, and was typically significantly less than that (test sequence number 3 seemed to give higher distortion). No part of the

TABLE I
512 Codewords, Matrix 3-Code, Codebook Distortion = 0.257

test seq	distortion	dB increase	no. frames
1	0.287	0.488	1900
2	0.289	0.516	1900
3	0.325	1.02	2420
4	0.271	0.225	2500

mean = 0.293, std dev = 0.020, av dB increase = 0.569

TABLE II
1024 Codewords, Matrix 3-Code, Codebook Distortion = 0.220

test seq	distortion	dB increase	no. frames
1	0.266	0.832	1900
2	0.267	0.844	1900
3	0.296	1.298	2420
4	0.246	0.493	2500

mean = 0.269, std dev = 0.018, av dB increase = 0.873

TABLE III
512 Codewords, Matrix 2-Code, Codebook Distortion = 0.213

test seq	distortion	dB increase	no. frames
1	0.247	0.628	1900
2	0.243	0.573	1900
3	0.277	1.140	2420
4	0.225	0.229	2500

mean = 0.248, std dev = 0.019, av dB increase = 0.661

TABLE IV
1024 Codewords, Matrix 2-Code, Codebook Distortion = 0.179

test seq	distortion	dB increase	no. frames
1	0.221	0.932	1900
2	0.221	0.936	1900
3	0.252	1.407	2420
4	0.206	0.617	2500

mean = 0.225, std dev = 0.017, av dB increase = 0.993

TABLE V
128 CODEWORDS, VQ CODE, CODEBOOK DISTORTION = 0.240

test seq	distortion	dB increase	no. frames
1	0.250	0.176	1900
2	0.244	0.068	1900
3	0.290	0.817	2420
4	0.252	0.216	2500

mean = 0.259, std dev = 0.018, av dB increase = 0.331

TABLE VI
256 CODEWORDS, VQ CODE, CODEBOOK DISTORTION = 0.193

test seq	distortion	dB increase	no. frames
1	0.212	0.420	1900
2	0.205	0.275	1900
3	0.244	1.023	2420
4	0.203	0.237	2500

mean = 0.216, std dev = 0.017, av dB increase = 0.489

mq m - N = m-bit matrix N-code

vq m = m-bit full-search code

TABLE VII

mq9-3 vs vq7			
	average distortion		
	mq9-3	vq7	dB difference
codebook	0.257	0.240	0.294
testseq1	0.287	0.250	0.605
testseq2	0.289	0.244	0.741
testseq3	0.325	0.290	0.496
testseq4	0.271	0.252	0.302

TABLE VIII

mq9-2 vs vq8			
	average distortion		
	mq9-2	vq8	dB difference
codebook	0.213	0.193	0.445
testseq1	0.247	0.212	0.653
testseq2	0.243	0.205	1.074
testseq3	0.277	0.244	0.560
testseq4	0.225	0.203	0.437

test sequences was contained in the training sequence. Furthermore, while some of the vocabulary in the test sequences does inevitably overlap with that in the training sequence, the training sequence was made from random newspaper clippings about medicare and Queen Elizabeth II, while the test sequences were about a day at the zoo and a passage from a fairy tale. In the case of the 10 bit code, where we have a compression ratio of about 1:15, there was a greater increase in the test sequence distortion figures (no worse than 1.41 dB). We should remark that the training and test sequences used are still too short for us to draw precise conclusions about the limiting behavior of matrix quantizers as the training sequence becomes very long, but the distortion figures obtained above do suggest that the matrix codebooks obtained are not very sensitive to out-of-train test sequences which are made under recording conditions identical to the training sequence, for a single speaker.

Tables V and VI show the training and test sequence distortions for 7 bit and 8 bit full-search VQ ($N = 1$, $R = 7$, and $R = 8$). We note that the average codebook distortion is only slightly lower than that for a 9 bit matrix 3-code ($R = 3$). While the difference between the test sequence distortions is a little higher, we attribute this to the higher compression ratios used for the full-search codes (1:119 and 1:60 for 7 bit and 8 bit codes, respectively). Tables VII and VIII compare the distortion figures for matrix and VQ codes. Notice that the differences in codebook and test sequence distortions between the matrix codes and the VQ codes is at most 1 dB, and is typically significantly less. For all practical purposes, this difference is negligible. However, the 9 bit matrix 3-code ($R = 3$) transmits at 150 bits/s, while the 7 bit VQ code ($R = 7$) transmits at 350 bits/s. Similarly, the 9 bit matrix 2-code ($R = 4.5$) transmits at 225 bits/s while the 8 bit VQ code ($R = 8$) transmits at 400 bits/s. This implies that, at least from the point of view of average IS distortion, there is a reduction in bit rate by greater than 50 percent for the 9 bit matrix 3-code, while the 9 bit matrix 2-code gave a 43.8 percent reduction. It is not inconceivable that with 11 or 12 bit 4-codes and 5-codes, we can achieve even lower bit rates (and this seems further strengthened by the trends implicit in Fig. 2). The production of reasonable codes of this size, however, requires training sequences of about 150 000 frames in length, together with very large amounts of CPU time, and we shall leave this exercise to organizations which are better endowed to cope with such burdens. In any case, as the matrices grow larger, variable-rate codes become more efficient, as shown in [10].

B. Subjective Tests

A subjective comparison between vocoded speech using matrix quantizaton and that using full-search VQ and finite-state VQ (designed with stochastic iteration) [5] with varying rates was made with the out-of-train test sequence

TABLE IX

Scores for MQ/VQ/FSVQ Comparison				
Pair		Score		
1.	2.	1 better	2 better	Undecided
mq9-2	vq4	12	0	0
(225 bps)	fsvq3	10	0	2
	vq5	8	0	4
	vq6	7	0	5
	vq7	6	3	3
	vq9	4	0	8
mq9-3	vq4	12	0	0
(150 bps)	fsvq3	12	0	0
	vq5	3	1	8
	vq6	7	1	4
	vq7	5	4	3
	vq9	4	2	6
mq10-4	vq4	4	4	4
(125 bps)	fsvq3	5	4	3
	vq5	0	7	5
	vq6	1	8	3
	vq7	1	10	1
	vq9	0	12	0

Legend:
mqM-N = M-bit matrix codebook with N LPC vectors per matrix
vqM/fsvqM = M-bit full-search/finite state codebook
No. of training matrices used: approx 5000 (sliding blocks on train)

and the multiple-speaker training sequence in the previous part (those used to obtain Figs. 2 and 3). Pitch and voicing information was obtained from the ILS software made by Signal Technology Inc., Santa Barbara, CA. They were not encoded in any way, and were directly used to process the vocoded sequences. Since the issue here is the compression of the LPC information, only this was subjected to compression. Note that pitch and voicing can be compressed using existing methods (e.g., the trellis coder for pitch reported under development in [6] which adds another 25 bits/s to the rate). Results were gathered from 12 listeners, each of which was asked to put down their personal preference for one of a pair of sequences on an *A–B* comparison in terms of intelligibility and general speech quality. The sequences were recorded on cassette tape and played back on loudspeakers from Radio Shack. The results are shown in Table IX. The matrix coded sequences at 225 bits/s clearly out-performed full-search VQ at higher bit rates, and was even judged competitive with full-search VQ at 350 bits/s. At 150 bits/s, the matrix code clearly out-performed both full-search VQ at 200 bits/s and finite state VQ at 150 bits/s. There was some indecision with full-search VQ at 250 bits/s owing to the fact that the order of presentation was reversed for this particular comparison, but those that did decide still preferred the matrix code. The matrix code still seemed competitive at 300 and 350 bits/s. For the matrix code at 125 bits/s, there was overwhelming preference for the VQ codes at the higher bit rates, although it still appeared competitive with finite state at 150 bits/s and full-search at 200 bits/s. The reason for this marked degradation in quality was that this code had 1024 codewords instead of the 512 in the previous matrix codes. Since only 5000 training matrices were used, there was clearly an insufficient number of training matrices to produce a good 10 bit 4-code, and this resulted in a marked degradation in quality. Some remarks should be made regarding the results obtained with the matrix codes at 150 and 225 bits/s and full-search VQ at 450 bits/s. At 450 bits/s, the full-search VQ code began to show more warble than the matrix codes, and this generally pushed some listeners to favor the matrix codes. Listening with headphones, however, indicated that there was degradation in the clarity of the articulation of words, particularly consonants, between 450 and 150 bits/s, but only slight degradation between 450 and 225 bits/s.

The sound of the matrix codes is generally quite different from the kind of speech produced by full-search or finite-state codes. The matrix codes, while retaining intelligibility to a surprising degree at such low bit rates, manifested distortion in the character of the speech reproduced, and this became more marked with lower bit rates. The matrix codes seemed to impart a foreign accent to the speaker (in particular, the "bad" code at 125 bits/s had a very "gargly" quality and gave the speaker a cockney accent). In contrast, the full-search VQ and finite-state VQ codes sounded more muffled at the lower rates, but without much distortion of the character of the speech. It would appear that the matrix codes could somehow approximate the right phonemic units in speech, but not exactly the ones uttered. This seems to be the main effect of the higher average distortions noted in Fig. 3 with the matrix codes.

In order to give some indication of how intelligible speech at 150 bits/s using matrix coding is, we decided to conduct an informal intelligibility test. It should be mentioned at the outset that the test was conducted using LPC speech made under difficult conditions. The training and test sequences were made with an 8-bit D/A converter in a room with air-conditioning noise. To minimize this, the speaker had to move very close to the microphone (about 3 or 4 in), so that there were substantial variations in average gain owing to movements of the speaker's head in some places. Limitations in disk storage and CPU cycles owing to a very heavily used VAX limited the length of the training sequence to about 5 800 20 ms frames, which is just under 2 min of speech. The length of the test sequence was about 1 min and 16 sec. All this resulted in training and test sequences with a low (under 30 dB) signal to noise, and considerable average gain variation in some places, resulting in LPC speech containing some of the problems one might expect in a practical environment. With a 9 bit 3-code (512 codewords, rate 150 bits/s), there were only 11 matrices per codeword in the training sequence. This compression ratio is only just adequate for obtaining a good average to every codeword in the codebook. As usual, the uncoded voicing and pitch parameters from the original test sequence obtained via the ILS package, was used in vocoding the sequence.

The training sequence was made with a single speaker reading a newspaper clipping about Queen Elizabeth II and the salary of the Royal Family. The test sequence was made from a child's essay about a day at the zoo. Care was taken that none of the listeners had significant prior knowledge of the content of the test sequence. The vocoded test sequence was taped and played back to each of the listeners. The listeners were required to write down the contents of the test sequence, dictation-style. They were also allowed to back up to any portion of the sequence they wanted to listen to again as often as they wanted (in practice, the limit seemed to be about four times). Eight listeners correctly identified on average about 81 percent of the 90-word text, with a standard deviation of 3.6 percent.

The ability of the listeners to make sense of the text varied somewhat. The three highest scores were obtained from people who have heard vocoded or synthetic speech before. The lower scores were from those who have not heard such speech. Two of the people approached could not understand the sequence at all. They were discounted from the test. The nature of much of the vocoded speech was akin to heavily accented English, and some people have difficulty with that, especially people whose native language is not English. Most of the errors for all the listeners occurred in two badly garbled places in the text. All seemed to agree that the beginning and ending sentences were clear and easy to comprehend. We attribute most of this to the codebook size, which was dictated by the shortness of the training sequence. There were not enough codewords to adequately capture all transitions present in the test sequence. This is not really surprising, since the quality is comparable to that obtained with 7 bit VQ, which is less than fully intelligible. Some of the problems were also caused by the gain variation mentioned and one or two incorrect voicing decisions.

The point of all this, however, is not to implement a 150 bits/s speech vocoder, but to demonstrate that matrix quantizers designed using the generalized Lloyd algorithm have the potential for synthesizing intelligible speech using fixed-rate codes at very low bit rates.

IV. General Remarks

We should make a few general remarks about the matrix quantizer design technique outlined above. The first thing one observes is that N times as many adds/multiples are required in computing $D(X, Y)$ when compared to standard VQ ($N = 1$). This means (assuming that the total time for all distortion computations is predominant) that a matrix quantizer takes N times as long to train as a standard full-search vector quantizer for the same length of training sequence using the sliding block technique. The other observation is that for a given sequence of speech we want to encode, the number of adds/multiplies required is similar to that for a VQ codebook containing the same number of codewords as the matrix codebook. However, a larger codebook size is typically required to obtain equivalent speech quality for matrix quantization, so this means an increase in the number of adds/multiples in practice. How much this will increase typically depends on N. For a matrix 2-code, only a slight increase is needed (probably at most a factor of 2) while this increases significantly for larger N. From a data-storage viewpoint, a matrix codebook would require N times as much storage as a full-search codebook of the same size. In practice, several times N would be appropriate to obtain equivalent quality.

Up to this point, we have only dealt with the full-search matrix quantizer, in the sense of [2]. Obviously, suboptimal variants are possible, such as tree-search matrix quantizers and shape-gain matrix quantizers [1]. The properties of these variants are subjects for future research. An interesting possibility would be the finite-state matrix quantizer, which might be a good alternative to dealing with huge codebooks with large N.

Within the last one or two years, standard vector quantization techniques have seen growing acceptance as a data reduction method for various speech recognition systems. We listened to individual matrix quantizer codewords with $N = 4$, by breaking down an encoded speech sequence into its constituent matrix codewords, and these codewords sound remarkably similar to groups of phonemes or allophones. Vowel-like sounds seem to be captured especially well by the matrix quantizer, whereas consonants are captured less well. This more general version of the Lloyd vector quantizer for LPC speech might well find some application in the recognition field.

V. Summary

We have presented an extension to the technique of designing vector quantizers for LPC speech motivated by an application of rate-distortion theory. Matrix quantizers designed by the generalized Lloyd algorithm are locally optimal quantizers which generally have substantially lower rates for a given quality of reproduction than standard VQ. They also have the advantage of fixed-rate and use a simple distortion computation. The design technique is a general one which allows a continuous tradeoff between computation time, data storage, and transmission bit-rate for a fixed average distortion by varying the block length. A matrix quantizer operating at 150 bits/s (512 codewords, three-vector matrix) has codebook and test sequence distortions comparable to a full-search vector quantizer operating at 350 bits/s. In contrast, a finite-state vector quantizer at the same rate designed with stochastic iteration has distortion performance comparable to a full-search vector quantizer operating at 250 bits/s. Matrix quantizers are also simpler and quicker to design than finite-state vector quantizers. At 150 bits/s for the LPC matrix on a short training sequence, an average of 81 percent of a 90-word out-of-training text for a single speaker was intelligible based on eight listeners. Subjective MQ/VQ/FSVQ comparison tests conducted among 12 listeners indicated a definite preference for matrix quantizers at comparable rates. Codebook distortion figures tend to confirm this trend.

Appendix I

We show how centroids with respect to the distortion measure $D(\cdot, \cdot)$ may be computed. The centroid is defined as that reproduction matrix Y which minimizes the expected distortion of a cell of matrices X, where X is taken to be random. Since

$$E(D(X, Y)) = \sum_{i=1}^{i=N} E(d_{IS}(x_i, y_i))$$

we may write

$$\min_Y E(D(X, Y)) \leq \sum_{i=1}^{i=N} E(d_{IS}(x_i, y_i))$$

for any given y_i, $i = 1, \cdots, N$. This implies that

$$\min_Y E(D(X, Y)) \leq \sum_{i=1}^{i=N} \min_{y_i} E(d_{IS}(x_i, y_i)).$$

Conversely, we have

$$\sum_{i=1}^{i=N} \min_{y_i} E(d_{IS}(x_i, y_i)) \leq E(D(X, Y))$$

for any given Y, and this implies that

$$\sum_{i=1}^{i=N} \min_{y_i} E(d_{IS}(x_i, y_i)) \leq \min_Y E(D(X, Y)).$$

Hence, we have

$$\min_Y E(D(X, Y)) = \sum_{i=1}^{i=N} \min_{y_i} E(d_{IS}(x_i, y_i)).$$

That is, the centroid of a cell of matrices X is merely the matrix which comprises the centroids of the individual component vectors x_i of X. In practice, we replace the expectation E with the arithmetic average over the cell. From [15] and [1] we know that the centroid of a component vector x_i is merely the arithmetic average of the sample autocorrelations $r_x(j)$, $j = 0, \cdots, M$ for that component.

Appendix II

An informal statement of the generalized Lloyd algorithm using the matrix notation developed in the text is given.

Suppose an initial guess of B reproduction matrices (codewords) $\{Y_j, j = 1, \cdots, B\}$ exists, and we have a training sequence of LPC matrices $\{X_j, j = 1, \cdots, K\}$, where K is some large integer (these will be known as $\{Y\}$ and $\{X\}$ for short). Assign each matrix X in the training sequence to some codeword Y by minimizing $D(X, Y)$ over all the codewords $\{Y\}$, for each X in the training sequence. The training sequence $\{X\}$ is now partitioned into B or fewer cells. Find the centroid $\overline{X}$ of a given cell by averaging each sample autocorrelation coefficient, as follows:

$$\overline{X} = [\overline{x}_1, \cdots, \overline{x}_N]$$

where

$$\overline{x}_j = \frac{1}{L} \sum_{i=1}^{i=L} r_{x_{ji}}(k), \qquad k = 0, \cdots, M \qquad j = 1, \cdots, N$$

with L being the number of matrices in any given cell. The proof that this is the centroid of the matrices for that cell is given in Appendix I. The inverse filter model for each $\overline{x}_j$ is then found by conventional means (e.g., Levinson's algorithm) and the reproduction matrices $\{Y\}$ so obtained from $\{\overline{X}\}$ will constitute the new codebook. This procedure is continued until the percentage change in average distortion given by

$$\frac{1}{K} \sum_{j=1}^{j=K} \min_{Y \in \{Y\}} D(X_j, Y)$$

is smaller than some threshold Δ. At this point, a local minimum is approached. If B is equal to the number of codewords we want, then the algorithm ends. Otherwise, we perturb each reproduction matrix by some perturbation matrix ϵ and repeat the process with an initial codebook of $2B$ reproduction matrices, $\{Y\} \cup \{Y + \epsilon\}$. We perturb the gain term σ^2 by some quantity ϵ for each reproduction vector y in Y in our experiments. Empty cells, that is, codewords around which no matrices cluster, are treated by retaining that codeword for the next iteration. Since the algorithm is only locally optimal, many bad codebooks can exist owing to shallow local optima. However, a good codebook is usually found after one or two different guesses at the initial codebook, or by starting with the centroid of the training sequence and splitting to obtain the desired rate.

Appendix III

The specifications for the training and test sequences used to obtain the experimental results are as follows:

1—Fig. 1: The training sequence was made at Stanford University. It was sampled at 8 kHz, using an 8 bit analog-to-digital converter. Tenth-order LPC analysis using the autocorrelation method was done using a preemphasis of 0.9 and a Hamming window at 160 samples/frame giving standard 20 ms frames. SNR was marginally adequate at about 30 dB. The sequence was 15 200 frames long and contained random extracts from various newspapers read by a single speaker.

2—Fig. 2: The training sequence was obtained from Signal Technology Inc., Santa Barbara, CA. It was sampled at 6.5 kHz, using a 12 bit analog-to-digital converter. Tenth-order LPC analysis using the autocorrelation method was done using a preemphasis of 0.9 and a hamming window at 128 samples/frame giving standard 20 ms frames. SNR was very good at about 50 dB. The sequence was 5000 frames long and contained conversational speech from five speakers.

3—Fig. 3: The test sequence used was also obtained from Signal Technology, Inc., Santa Barbara, CA. Specifications are the same as for (2) except that the sequence was 600 frames long and contained speech from a single speaker. The codebooks were obtained from the training sequence in (2).

4—Tables I–VIII: The four test sequences used have the same specifications as in (1) except that the sequences contained speech based on a child's essay on a day at the zoo and a passage from a fairy tale. The training sequence used for these tests is that in (1).

5—Table IX: The training sequence used was that in (2) and the test sequence that in (3).

REFERENCES

[1] A. Buzo, A. H. Gray, Jr., R. M. Gray, and J. D. Markel, "Speech coding based upon vector quantization," *IEEE Trans. Acoust., Speech, Signal Processing,* vol. ASSP-28, pp. 562–574, Oct. 1980.

[2] R. M. Gray, A. H. Gray, Jr., G. Rebolledo, and J. E. Shore, "Rate distortion speech coding with a minimum discrimination information distortion measure," *IEEE Trans. Inform. Theory,* vol. IT-27, pp. 708–721, Nov. 1981.

[3] R. M. Gray, J. C. Kieffer, and Y. Linde, "Locally optimal block quantizer design," *Inform. Contr.*, vol. 45, pp. 178–198, May 1980.

[4] J. Foster, R. M. Gray, and M. Dunham, "Finite-state vector quantization for waveform coding," *IEEE Trans. Inform. Theory,* to be published.

[5] M. Dunham and R. M. Gray, "An algorithm for the design of labeled-transition finite-state vector quantizers," *IEEE Trans. Commun.*, vol. COM-33, pp. 83–89, Jan. 1985.

[6] D. Y. Wong and B. H. Juang, "Vector/matrix quantization for narrow-bandwidth digital speech compression," Final Rep. Signal Technol., Inc., Contract F30602-81-C-0054, July 1982.

[7] D. Y. Wong, B. H. Juang, and D. Y. Cheng, "Very low data rate speech compression with LPC vector and matrix quantization," in *Proc. ICASSP*, Boston, MA, Apr. 1983, pp. 65–68.

[8] S. Roucos, R. Schwartz, and J. Markhoul, "Vector quantization for very-low-rate coding of speech," *Globecom '82 Conf. Record,* Dec. 1982, pp. 1074–1078.

[9] S. Roucos, J. Makhoul, and R. Schwartz, "A variable-order Markov chain for coding of speech spectra," in *Proc. IEEE Int. Conf. Acoust., Speech, Signal Processing*, Paris, Apr. 1982, vol. 1, pp. 582–585.

[10] S. Roucos, R. Schwartz, and J. Makhoul, "Segment quantization for very-low rate speech coding," in *Proc. IEEE Int. Conf. Acoust. Speech, Signal Processing,* Paris, Apr. 1982, vol. 3, pp. 1565–1568.

[11] ——, "A segment vocoder at 150 B/S," in *Proc. ICASSP,* Boston, MA, Apr. 1983, pp. 61–64.

[12] R. J. McEliece, *The Theory of Information and Coding.* Reading, MA: Addison-Wesley, 1977.

[13] R. G. Gallager, *Information Theory and Reliable Communication.* New York: Wiley, 1968.

[14] Y. Linde, A. Buzo, and R. M. Gray, "An algorithm for vector quantizer design," *IEEE Trans. Commun.*, vol. COM-28, Jan. 1980, pp. 84–95.

[15] R. M. Gray, A. Buzo, A. H. Gray, Jr., and Y. Matsuyama, "Distortion Measures for speech processing," *IEEE Trans. Acoust., Speech,* Signal Processing, vol. ASSP-28, pp. 367–376, Aug. 1980.

A SEGMENT VOCODER AT 150 B/S

S. Roucos, R.M. Schwartz, J. Makhoul

Bolt Beranek and Newman Inc.
Cambridge, MA 02238

ABSTRACT

In this paper we investigate several methods for reducing the bit rate of a segment vocoder [1] by 35% to 150 b/s. In the original vocoder we used a random sample of vectors as a set of templates for vector quantization. We demonstrate in this paper that this random quantizer is near-optimal by comparing it with quantizers that use clustering algorithms for quantizing speech segments. The reduction of the bit rate of the segment vocoder was achieved primarily by using a segment network, i.e., not all segment templates are allowed to follow a given segment template. The spectral continuity of speech is used to determine the subset of templates, that can be used to quantize an input segment. To achieve the low rate of 150 b/s, we also reduced the bit rate for coding pitch, gain, and segment duration.

Finally, we present the bit allocation used for transmitting speech at 150 b/s as a single speaker segment vocoder.

1. INTRODUCTION

In an earlier paper [1], we presented the segment vocoder as a new method for transmitting speech for a single speaker at the low bit rate of 230 b/s. In the segment vocoder, we represent the output of LPC analysis at 100 frames/s as a sequence of segments. Each segment consists of a variable number of consecutive frames and has an average duration of 90 ms. To achieve the low rate of the segment vocoder, vector quantization is used for quantizing all the LPC spectra in a segment as a single unit. The segment templates of the vector quantizer were obtained without using clustering techniques due to the large computational load of segment clustering. Instead, a random set of segments was used as the set of templates. We hypothesized then, that the performance of such a random quantizer is near-optimal. In this paper, we confirm this hypothesis by comparing the above random quantizer to quantizers that use clustering algorithms to determine locally optimal sets of templates.

We also describe in this paper the quantization techniques that were used to reduce the bit rate of the segment vocoder from 230 b/s to 150 b/s, a reduction of 35% in the bit rate. The major decrease in the bit rate was achieved by using a segment network that restricts the number of segment templates that can follow a given template. The segment network is used in the following manner: if the current input segment is quantized to a given template, then only those segment templates that follow this template in the network can be used to quantize the following input segment. We describe below the method used to define the segment network that was used to reduce the segment bit rate from 13 bits to 10 bits (per segment) with minimal increase in the segment quantization error.

Finally, we will describe the bit allocation used for the segment vocoder operating at 150 b/s as a single speaker system. In the following section, a brief description of the original segment vocoder is presented. A more detailed discussion of the design issues of the segment vocoder can be found in [1, 2].

2. DESCRIPTION OF THE SEGMENT VOCODER

The segment vocoder uses a 14th order, unquantized LPC analysis at 100 frames/s to represent the input speech. The LPC input sequence is automatically segmented with an average segmentation rate of 11 segments/s. The segmentation algorithm is a heuristic algorithm that uses a set of thresholds on two spectral derivatives to determine the spectral steady-state regions in the input speech. The segments are defined to begin and end in the middle of consecutive steady-states.

In the original segment vocoder, each segment is quantized as single unit independently from other segments. The spectral trajectory, gain track, pitch track and total duration of a segment are also separately quantized.

The vector quantizer used to quantize the spectral trajectory of a segment uses a distance measure that incorporates the required time alignment of two segments. We describe this distance measure [1] because it will also be used for segment clustering which is described in the following section.

The sequence of LPC spectra in a segment represent a piece-wise linear trajectory in the 14 log-area-ratios (LARs) space. We use LARs to represent an LPC spectrum. The total length (using a Euclidean norm on LARs) of a segment is computed. Then, the spectral trajectory is resampled at M equispaced points in LARs space (usually M=10). The space-sampling representation of a segment consists of M 14-dimensional LARs vector (in our case, this corresponds to a 140 dimensional vector for M=10).

The distance measure used for segment quantization defines the time warping between two segments by the correspondence of the consecutive space-samples of the two segments. The distance measure between two segments X and Y is:

$$d = \sum_{i=1}^{M} w_i \, \| \underline{x}_i - \underline{y}_i \|^2$$

Reprinted from *Proc. IEEE Conf. Acoust., Speech, Signal Processing,* vol. 1, pp. 61–64, April 1983.

where x_i, y_i are 14 LARs vectors that represent the i-th space-samples of the two segments X and Y, || || is the Euclidean norm, and w_i is a weight.

The above distance measure is used to quantize an input segment to the nearest segment template. In the original segment vocoder we use 8000 segment templates which corresponds to 13 bits. In the segment vocoder, voicing is not transmitted. The receiver uses the voicing of the template. Similarly, the gain track of the template is used at the receiver but a level adjustment is transmitted using 2 bits. This adjustment equalizes the average gain (in dB) of the input and template segments. The total duration of a segment is quantized and transmitted with 3 bits. Finally, pitch is modelled as a piece-wise linear trajectory and one pitch value is transmitted per segment using 3 bits. Therefore, we transmit 21 bits/segment which yields an average bit rate of 230 b/s for a segment rate of 11 segments/s. We will describe in Section 5 the bit allocation used for the new segment vocoder operating at 150 b/s.

3. SEGMENT CLUSTERING

The above segment vocoder uses 8000 segment templates (~13 bits) obtained by automatically segmenting 15 minutes of speech at an average segment rate of 11 segments/s. The resulting set of segment templates is considered as a random sample of speech segments. We postulated in [1] that the random quantizer, that uses the above set of segment templates, is near optimal. This was based on the result that for a Gaussian random vector with a large dimensionality and with independent and identically distributed components, a random quantizer is optimal. Since we do not expect segments to be Gaussian, we used clustering techniques to determine if the performance of the random quantizer used in segment quantization is far from optimal.

The database used for the clustering experiments consisted of one hour of speech from a single male speaker reading text. The data was collected in four 15 minute sessions spanning a 6 month interval. The database was automatically segmented with an average segment rate of 11 segments/s. The total number of segments was 34,872 (~15 bits). Clustering was used to determine several codebooks of different sizes which were compared to random quantization. We describe below the clustering algorithm used to determine locally optimal sets of templates.

3.1 Clustering Algorithm

The distance measure used for clustering is the same measure used in segment quantization with unit weights and incorporates the time alignment of two segments with different total durations. Each segment is represented by 10 equally spaced space-samples, where each space-sample is a vector of 14 LARs.

The clustering algorithm is a two-pass procedure: a binary clustering to determine a set of clusters, followed by a template selection process for each cluster. The binary clustering phase uses an 8-th order LAR vector instead of 14 to minimize the storage requirement. Therefore, each segment was an 80-dimensional vector and the Euclidean distance on this vector was used in the binary clustering phase. The template selection phase uses the 14 LARs representation to get a set of segment templates.

The binary clustering algorithm is described in [2, 3]. This algorithm is applied sequentially in the following manner on the training data set of segments. Initially, the binary clustering algorithm divides the training data set into two clusters using the K-means algorithm (with K=2). Each cluster is represented by its mean vector. Then, the quantization error for each cluster is computed (using Euclidean distance on 80 dimensional vector). The cluster that has the largest total quantization error (sum of the quantization errors of all the segments in the cluster) is selected for further subdivision into two clusters. This process is repeated until the desired number of clusters is obtained. The K-means algorithm (with K=2) is always used in dividing a given cluster.

The binary clustering algorithm defines a non-uniform binary tree that can be used to efficiently find the nearest cluster representative to an input vector. The large savings in the computational load of the binary clustering over the K-means algorithm (2 $\log_2 M$ distances instead of M distances for M clusters) is needed for segment clustering. We have also found that the quantization error of binary clustering is not much higher than that of the K-means algorithm both for quantizing 14 dimensional LARs vectors and 140 dimensional segments [2].

Given the M clusters obtained by the non-uniform binary clustering algorithm, the second pass determines a set of segment templates. We have used two methods for selecting these segment templates.

1. i) Mean Template: For each cluster, we recompute the mean segment of all the segments in it using the 14 LARs representation. Since the space-sampling representation specifies a time alignment of the segments, the detailed timing is also averaged to specify the timing of the template. The voicing decision for each space-sample is obtained by a majority vote from the corresponding space-sample of all segments in the cluster.

2. ii) Nearest to the Mean Template: To avoid the smearing due to the averaging process, the template of a cluster was chosen as that segment in the cluster nearest to the mean. This segment was represented using 14 LARs.

Using the training data set of 34,872 segments, we compared two sets of codebooks: random codebooks and clustering based codebooks. The random codebooks were obtained by uniformly sampling the training set, i.e., a 10 bit random codebook is obtained by including the segments at multiples of 32 from the 34,872 (~15 bits) segments.

Figure 1 shows the mean square error in quantizing an independent test set of 6 sentences for both the random quantizer and cluster mean quantizer. The random quantizer requires an additional 2 bits to get the same quantization error as the cluster mean quantizer. This loss in performance is rather small and confirms our earlier assumption. Further, for the same bit rate the synthesized test sentences of the random quantizer are crisper and more intelligible than the cluster-mean quantizer. This effect is probably due to the smearing of the averaging process.

To reduce the smearing of the spectral trajectory of a segment template, we used the nearest to the mean template selection method. The quantization error of this quantizer increases to be almost equal to the quantization error of the random quantizer for the same bit rate. Therefore, due to the small amount of data available for each cluster (for the dimensionality of a segment) the nearest to the mean template selection is equivalent to random template selection, i.e., all segments in a cluster are equally far from the mean of the cluster. Also, in listening tests the quality of the random quantizer and the nearest to the mean quantizer is the same. Therefore, random quantization is an effective method for segment quantization which does not use the computationally expensive clustering algorithm to determine a set of segment templates.

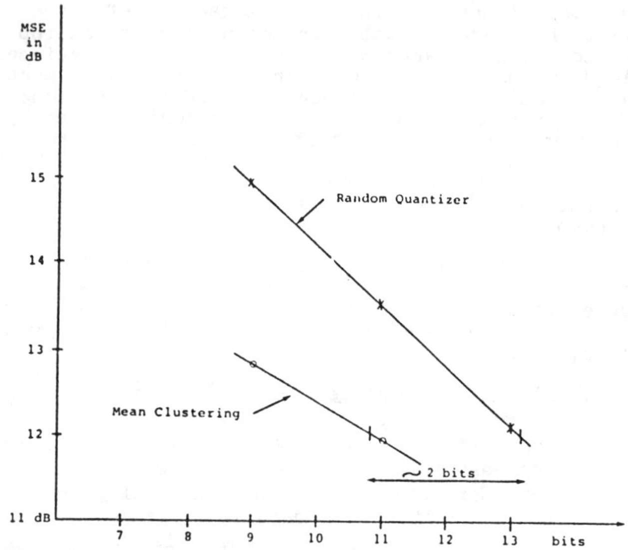

Fig. 1 Comparison of the mean quantization error of a random quantizer with a cluster mean quantizer.

4. SEGMENT NETWORK

In the original segment vocoder the sequence of segment templates that can be used to quantize the sequence of input segments is unconstrained, i.e., any segment template can follow any template. To reduce the bit rate of the segment vocoder we use a segment network to constrain the sequence of segment templates. That is, if the current input segment is quantized to a given segment template, then the following input segment can be quantized only to one of a subset of all available templates. Thus reducing the number of bits used to code the segment template.

A general method for determining the segment network would be to determine statistically which segment templates are most likely to follow a given segment template, i.e., to estimate a Markov chain model for the sequence of speech segment templates. However, this approach would require a prohibitive amount of data to estimate the model. A practical alternative approach is based on an assumption of the spectral continuity of the input speech at the boundary between successive segment templates. If a given segment template matches the current input segment, we assume that the segment template that best matches the following input segment must begin with a spectrum that is close to the last spectrum of the previous segment template. Since the segment templates are generally defined as beginning and ending in relative steady states the assumption stated above is reasonable.

We implemented the above model of the segment network in the following manner. Suppose that the current input segment is quantized to a given template. The last spectrum of this best segment template is used to determine a subset of templates that are allowed in quantizing the following input segment. The templates allowed are those whose first spectrum is nearest to the last spectrum of the template used in quantizing the current input segment. The distance measure used to select the subset of templates is the Euclidean distance using the first 8 LARs. We found that using 8 LARs was slightly better than using all 14, since the high-order LARs vary in time more rapidly than the low-order LARs.

We have tested the segment network idea for a wide range of initial bit rates and bit savings. The effect of using the segment network on the performance of the segment vocoder is shown in Fig. 2. In this figure, we compare the quantization error for the case where all templates were allowed with a segment vocoder that used the segment network. For the unconstrained case, in this figure, we used a total of 1024 segment templates (10 bits). Then for each lower bit rate for the segment network, we restricted the number of templates that could follow a given template to satisfy the bit rate requirement. For example, for the reduced rate of 8 bits/segment, we restricted the choice to the 256 segment templates whose first spectrum was nearest the last spectrum of the previous segment template. The solid line indicates the quantization error for a random quantizer at the same bit rate. When 10 bits of templates are allowed, the segment network allows all templates to follow any template and therefore will have the same quantization error as a 10-bit random quantizer. We can draw two conclusions from the curves in Fig. 2.

1. The segment network achieves a saving of 2.5 to 3 bits over the random quantizer with a negligible increase in quantization error.

2. If the bit rate is reduced by more than 3 bits, the quantization error with the segment network increases much faster, with decreased bit rate, than that of the random quantizer. This indicates that the method used to subset the templates does something consistently worse than a random selection of the templates.

We have found the above two conclusions to hold for several cases. In particular, for the 150 b/s segment vocoder we use a 10-bit network derived from a 13-bit random codebook. During careful listening experiments we were generally not able to distinguish the case of the 10-bit segment network from that of the unconstrained 13-bit segment templates codebook. Careful examination of the templates used in quantization verified that indeed for more than 90% of the input segments, the segment template chosen was the same in both cases.

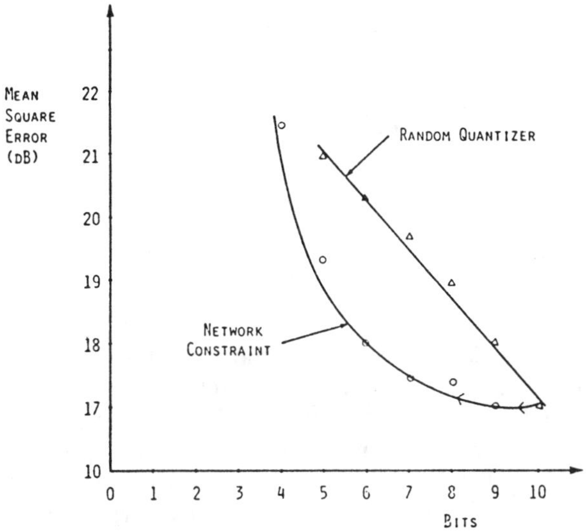

Fig. 2 Mean square quantization error of a random quantizer and a segment network that uses 1024 segment templates.

5. SEGMENT VOCODER AT 150 B/S

We present in this section the quantization methods and bit allocation used for vocoding speech at 150 b/s with the segment vocoder.

The segment templates were obtained by segmenting 15 minutes of speech into 8000 segments (13 bits). A segment network was used to reduce the spectral bit rate to 10 bits/segment as described above.

Pitch was transmitted using 1 bit only for each segment. The input pitch was modelled by a piece-wise linear model (linear over a segment) similarly to the original segment vocoder. The change in pitch from the last transmitted value (quantized value) is quantized using an adaptive 2-level quantizer. The two levels were $\pm 3\sqrt{t}$, where t was the segment duration. The resulting pitch sounded quite similar to the original. However, this coding scheme did occasionally chânge the intended intonation of the input speech.

The gain adjustment and segment duration were quantized using 3 levels each. Therefore we used 14 bits for each segment. At an average segment rate of 11 segments/s, the average bit rate of the segment vocoder is 154 b/s. We used the above segment vocoder to transmit the speech of a single male speaker. In informal listening tests, we have fond that the output speech is quite intelligible (9 words in 10 are intelligible in context).

6. CONCLUSION

We demonstrated in this paper that a random quantizer is effective for segment quantization. The computationally expensive clustering algorithms are not needed to determine a set of segment templates; a random set of segments is equally effective.

We also demonstrated that a segment network based on a spectral continuity model can be used to save 3 bits per segment in quantizing the spectral trajectory of a segment. The segment network was used with quantization techniques for pitch, gain and segment duration to implement a segment vocoder with an average bit rate of 150 b/s. The single speaker vocoder has a good intelligibility to be a useful communication system.

ACKNOWLEDGEMENT

This work was supported by the Advanced Research Projects Agency and monitored by RADC under Contract NO. F19628-80-C-0165.

REFERENCES

1. S. Roucos, R. Schwartz, and J. Makhoul, "Segment Quantization for Very-Low-Rate Speech Coding," ICASSP, Paris, France, May 1982, pp. 1565-1568.

2. S. Roucos, R.M. Schwarts, and J. Makhoul, "Research in Narrowband Communications," Final Report, Contract No. F19628-80-C-0165, Bolt Beranek and Newman Inc., BBN Report No. 5231, November 1982.

3. S. Roucos, R. Schwartz, and J. Makhoul, "Vector Quantization for Very-Low-Rate Coding of Speech," GLOBECOM, Miami, FL, November 1982, pp. 1074-1078.

LPC Speech Coding Based on Variable-Length Segment Quantization

YOSHINAO SHIRAKI, MEMBER, IEEE, AND MASAAKI HONDA, MEMBER, IEEE

Abstract—**This paper presents a low bit rate LPC coder based on variable-length segment quantization. In this segment vocoder, the speech spectral parameter sequence is represented as the concatenation of variable-length spectral segments generated by linearly time-warping fixed-length code segments. Both the sequence of code segments and the segment lengths are efficiently determined using a dynamic programming procedure. This procedure minimizes the spectral distance measure between the original and the coded spectral sequence in a given interval. An iterative algorithm is also developed for designing fixed-length code segments for the training spectral sequence. This iterative algorithm consists of updating segment boundaries of the training spectral sequence using an *a priori* codebook, and updating the codebook using these segment sequences. Convergence of this algorithm is discussed theoretically and experimentally. In experiments, the performance of variable-length segment quantization for voice coding is compared to that of fixed-length segment quantization and vector quantization.**

I. Introduction

VECTOR quantization (VQ) [1]–[3] is well known as a powerful data compression technique in speech processing. Especially, VQ is well fitted to LPC parameter compression, and has been widely used for voice coding in the range of several hundreds bits per second (bits/s). For data rates of about 300 bits/s or lower, however, single frame VQ is no longer sufficient for maintaining fair spectral distortion and speech intelligibility. Several coding methods to make use of multiple frames of spectral parameters, namely, a spectral segment, have been studied to cope with these problems [4]–[8]. These voice coders, referred to as segment vocoders, are classified into two types depending on the segment length, that is, whether the number of frames in the segment is fixed or variable.

Fixed-length segment quantization (SQ), also referred to as matrix quantization (MQ), is a straightforward extension of the ordinary VQ technique [8], [9]. Each segment is successively quantized using a code segment with the same length as that of the input segment so as to minimize a spectral distance measure. Fixed-length SQ is simple, but a large codebook is required to improve SQ performance, because the segment pattern varies significantly due to the segment position of the spectral sequence or the utterance speed.

Manuscript received April 10, 1987; revised August 17, 1987.

The authors are with the Speech Research Group of the Basic Research Laboratories, NTT, 3-9-11, Midori-cho, Musashino-shi, Tokyo, 180 Japan.

IEEE Log Number 8822368.

Unlike fixed-length SQ, variable-length SQ requires additional segmentation of the spectral sequence into variable-length segments and matching between the original and code segments with different segment lengths. Although variable-length SQ is more complicated than fixed-length SQ, it has the potential to reduce codebook size at the expense of the additional information for segment length by designing the code segment to be a more invariant unit related to linguistic patterns like phonemes or demisyllables. Roucos *et al.* have proposed variable-length SQ using segmentation based on the spectral change and segment matching based on variable-to-fixed length conversion [4]. This method offers not only successive segment coding with relatively short coding delay, but also a simple codebook design by extending the ordinal VQ design algorithm. The variable-to-fixed length method warps the input segment to the fixed-length segment by sampling the spectral parameters with equal interval along the parameter trajectory. It then computes the average spectral distance over the fixed number of frames between the input and the code segments. This method is computationally simple, but never minimizes the spectral distortion of the decoded (quantized) segment, which is obtained by inversely warping the fixed-length code segment. In addition, segmentation plays a very important role in designing a better SQ codebook.

In this paper, a method of joint segmentation and variable-length segment quantization is presented. The entire spectral sequence is generated by concatenating the code segments using fixed-to-variable length conversion. Both the sequence of code segments and segment lengths, which are equivalent to segment boundaries, are determined using a dynamic programming procedure. This procedure fully minimizes the spectral distance measure between the original and the coded spectral sequences in a given interval. An iterative algorithm is also presented for designing variable-length SQ based on the joint segmentation and quantization procedure. In the experiments, the distortion performance comparison of variable-length SQ, fixed-length SQ, and ordinary VQ, as well as convergence in the SQ design algorithm and the subjective quality of the coded speech, are discussed.

II. Variable-Length Segment Quantization

SQ in voice coding uses a spectral segment representing multiple frames of spectral parameters as a quantization unit. Denoting the segment by matrix X_j, which has l_j

Reprinted from *IEEE Trans. Acoust., Speech, Signal Processing,* vol. ASSP-36, pp. 1437–1444, Sept. 1988.

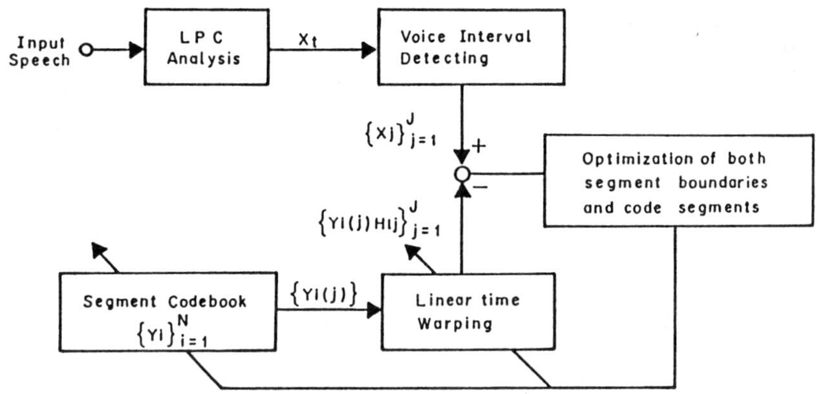

Fig. 1. Block diagram of variable-length segment vocoder based on joint segmentation and quantization.

column vectors consisting of spectral parameters, the original spectral sequence can be represented as the sequence of segments, $X_1, \cdots, X_j, \cdots, X_J$. Here, X_1 is the first segment and X_J is the last segment in a given interval, for example, a voice interval having silence segments at both sides. Each segment length l_j, namely, the column number of the matrix, is variable. On the other hand, the SQ codebook $C = \{Y_i\}_{i=1}^{N}$ has fixed-length segments. If the segment lengths, equivalent to segment boundaries, are obtained using any means like the one based on detecting spectral changes, the variable-length SQ can be extended straightforwardly using ordinary VQ. Thus, the variable-length segment can be successively quantized using a minimum distortion rule as

$$\hat{X}_j = \arg \min_{Y_i \in C} d(X_j, Y_i). \tag{1}$$

Here, $d(X_j, Y_i)$ represents the distance measure between the original and the code segments, and its computation requires a fixed-to-variable segment length conversion unlike the VQ or fixed-length SQ.

However, if the segment lengths are unknown, the variable-length SQ is not an extension of the ordinary VQ. The SQ needs to determine both the sequence of code segments and their length. Such a joint segmentation and quantization procedure is shown in Fig. 1. The entire spectral sequence is generated by concatenating the code segments using fixed-to-variable length conversion and is compared to the original spectral sequence in the voice interval [11]. Thus, a minimum distortion rule can be written as

$$\hat{X}_j \Big|_{j=1}^{J} = \arg \min_{\substack{l_j \\ 1 \leq j \leq J}} \left\{ \min_{Y_{i(j)} \in C} \sum_{j=1}^{J} d(X_j, Y_{i(j)}) \right\}. \tag{2}$$

Here, the sum of the segment lengths l_j is constrained to be equal to the voice interval length. This optimum solution can be obtained using the dynamic programming technique described below in Section II-B.

A. Distortion Measure

Before discussing the joint segmentation and quantization algorithm, it is necessary to describe the distance measure used in the SQ and the codebook design. We use LSP (line spectrum pair) parameters as the spectral parameter [11]. LSP parameters are one of the LPC parameters and offer substantial advantages over other LPC parameters with respect to dynamic behavior and quantizing efficiency. Weighted Euclidean distances of LSP parameters and speech powers were used as the spectral distance measure because of computational simplicity. Denoting the pth-order LSP parameters by $\{\theta_i\}_{i=1}^{p}$ and speech power in decibels by P, the distortion measure between the original and the code spectral parameter vectors is written as

$$d_{\text{frame}}(x, y) = (x - y)' W(x - y), \tag{3}$$

where $x = (\theta_1, \theta_2, \cdots, \theta_p, P)'$, $y = (\hat{\theta}_1, \hat{\theta}_2, \hat{\theta}_p, \hat{P})'$, and W is a diagonal weighting matrix. The diagonal elements $\{w_{ii}\}_{i=1}^{p+1}$, which are weighting factors for the LSP parameters, are determined so that the weighted distance approximates the Euclidean distance between the power-normalized LPC spectra in decibels by the least mean square error criterion. The diagonal element w_{p+1}, which is the weighting factor for speech power, is determined so as to maximize the intelligibility of the coded speech.

Fixed-to-variable length conversion is used to define the distance between two segments with different segment lengths. The L vectors of the code element are linearly interpolated and resampled at equal intervals, so that the number of frames is identical to the original segment length l_t. This procedure is illustrated in Fig. 2. It can be written in matrix form as YH_{lt}, using the L by l_t interpolation or time-warping matrix H_{lt}. The interpolation matrix is specified only by the original segment length l_t for the fixed-length code segment, and each column vector has two nonzero elements

$$\begin{aligned} h_{ij} &= 1 - \alpha \qquad \text{for } i = [\beta] + 1 \qquad j = 1, 2, \cdots, L, \\ &= \alpha \qquad \text{for } i = [\beta] + 2 \end{aligned} \tag{4}$$

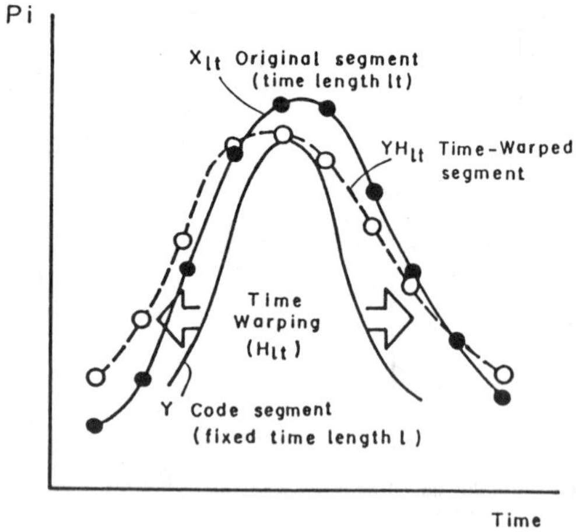

Fig. 2. Illustration of fixed-to-variable segment length conversion.

where $\alpha = \beta - [\beta]$, $\beta = (j - 1)(L - 1)/(l_t - 1)$, and $[\beta]$ denotes the maximum integer not exceeding β. Then, the distance measure between the original and the code segments is defined as the sum of the distances between the original and the interpolated code vectors over the number of frames of the original segment as

$$d(X, Y) = \mathrm{tr}\,(X - YH_{lt})'\, W(X - YH_{lt}), \qquad (5)$$

where tr A denotes the trace of matrix A defined as tr $A = \Sigma\, a_{ii}$.

Note that the distance measure based on variable-to-fixed lengths conversion never ensures minimum distortion for the spectral sequence reconstructed from the code segment using fixed-to-variable length conversion, even if segment quantization is done using the minimum distortion rule. Also, other types of fixed-to-variable length conversion based on nonlinear time warping need more information to transmit the time-warping path, although they may provide better spectral matching between the original and code segments.

B. Joint Segmentation and Quantization

Joint segmentation and quantization determines the sequence of code segments and segment lengths based on the minimum distortion rule given by (2). If the optimum solution needed to satisfy the minimum distortion rule is selected using a full-search algorithm among the spectral sequences generated for all of the combinations of code segments and their lengths, an enormous amount of computation is required. Using a dynamic programming procedure, this computational load, however, can be efficiently reduced [10]. First, we define the following notations:

T	the number of frames contained in the voice interval
J	the number of segments contained in the voice interval
t_j	segment boundary
t_j^0	initial segment boundary
t_j^*	optimum segment boundary
Δ	search range of segment boundary
$X(t_{j-1}, t_j)$	segment having boundary positions t_{j-1} and t_j
$q(X, Y)$	minimum distortion when quantizing X using segment codebook $C = \{Y_i\}_{i=1}^N$.

The accumulated distortion and the optimum boundary position for the end boundary position t_j is obtained using the following forward recursion:

$$\sigma(t_j) = \min_{t_{j-1} \in R_{j-1}} \left\{\sigma(t_{j-1}) + q(X(t_{j-1}, t_j), Y)\right\} \qquad (6)$$

$$s(t_j) = \mathop{\arg\min}_{\substack{t_{j-1} \in R_{j-1} \\ t_j^0 - \Delta \leqq t_j \leqq t_j^0 + \Delta \\ (j = 1, 2, \cdots, J)}} \left\{\sigma(t_{j-1}) + q(X(t_{j-1}, t_j), Y)\right\}, \qquad (7)$$

where

$$\sigma(t_0) = 0, \qquad t_0 = 1, \qquad t_J = T,$$

$$q(X, Y) = \min_{Y_i \in C} d(X, Y_i),$$

$$R_{j-1} = \left\{t \mid t_{j-1}^0 - \Delta \leqq t \leqq \min\,(t_{j-1}^0 + \Delta, t_j)\right\}.$$

After the final accumulated distortion $\sigma(t_J)$ has been computed, the optimum boundary positions are obtained using the following backward recursion:

$$t_{j-1}^* = s(t_j) \qquad j = J, J-1, \cdots, 2. \qquad (8)$$

It is obvious that the performance of the joint segmentation and quantization is at least as good as that of the SQ using the initial segment boundary positions. The search range Δ of the segment boundaries controls the performance and the computational amount of joint segmentation and quantization. As the search range becomes larger, the constraint begins to approach the limit of $t_0 < t_1 < \cdots < t_j \cdots < t_J$. In this case, the proposed algorithm fully satisfies the minimum distortion rule of (2) at the cost of an increased computational amount proportional to Δ^2. A good initial guess of the segment boundaries might be obtained based on detecting spectral changes; this would be useful for reducing the search range with an only slight degradation from the attainable upper bound of the SQ performance. The choice of initial guess of the segment boundaries and the search range will be discussed in the following experiments.

III. Variable-Length SQ Codebook Design

Variable-length SQ can be designed by straightforwardly extending the conventional VQ design algorithm, if the training data are prepared as the variable-length segment data. However, if only the spectral parameter sequence is given and the segment boundaries are unknown, the SQ codebook design is not trivial. Thus, using the training data of the spectral parameter sequence, the variable-length SQ is designed by finding both the segment boundaries and the code segments so as to minimize

$$D = \sum_{j=1}^{J} \min_{Y_i \in C} d\big(X(t_{j-1}, t_j), Y_i\big). \qquad (9)$$

The SQ codebook design algorithm is composed of training segment data updating and SQ codebook updating as shown in Fig. 3 [10].

SQ Codebook Design Algorithm:

Step 0—Initialization: Given vector dimension ($p + 1$), training vector sequence $\{x_t\}_{t=1}^{T}$, total number of training segment data J, segment length l, and thresholds ϵ_1, ϵ_2, choose by an adequate method an initial SQ codebook $C_0 = \{Y_i\}_{i=1}^{Y}$ and initial segment boundaries $\{t_j^0\}_{j=1}^{J}$. Set $k = 0$ and initial distortion $D_0 = \infty$.

Step 1—Training Segment Sequence Updating: Set $k = k + 1$. Update the training segment sequence to $\{X_j(k) = X(t_{j-1}^*, t_j^*\}_{j=1}^{J}$ using the joint segmentation and quantization algorithm with SQ codebook C_{k-1} for the training vector sequence. Update the neutral segment boundaries to $\{t_j^0(k) = t_j^*\}_{j=1}^{J}$. Compute the minimum distortion for the training segment sequence as

$$D_k(0) = \sum_{j=1}^{J} \min_{Y_i \in C_{k-1}} d(X_j, Y_i). \qquad (10)$$

Step 2—Set $m = 0$, $C_k(0) = C_{k-1}$.

Step 3—SQ Codebook Updating: Set $m = m + 1$. Encode the training segment sequence using the minimum distortion rule as

$$S_i = \{ j \mid \hat{X}_j = Y_i \}$$

where

$$\hat{X}_j = \arg \min_{Y_i \in C_k(m-1)} d(X_j, Y_i). \qquad (11)$$

For each cluster, compute the centroid using the training segment data as

$$Y_i = \Big(\sum_{j \in S_i} X_j H_{lj}^t\Big)\Big(\sum_{j \in S_i} H_{lj} H_{lj}^t\Big)^+. \qquad (12)$$

Here A^+ is a generalized inverse matrix of A (see the Appendix). Update the codebook to $C_k(m) = \{Y_i\}_{i=1}^{N}$ and compute the distortion decrease as

$$\delta = \{D_k(m-1) - D_k(m)\}/D_k(m),$$

where

$$D_k(m) = \sum_{Y \in C_k(m)} d(X_j, Y_i). \qquad (13)$$

Step 4—Termination: Go to Step 3 unless $\delta \leqq \epsilon_2$. Set $C_k = C_k(m)$, $D_k = D_k(m)$, and go to Step 1 unless $(D_{k-1} - D_k)/D_k \leqq \epsilon_1$.

This algorithm ensures a nonincreasing sequence of distortions as

$$D_0 \geqq D_1 \geqq \cdots \geqq D_k \geqq D_{k+1} \geqq \cdots. \qquad (14)$$

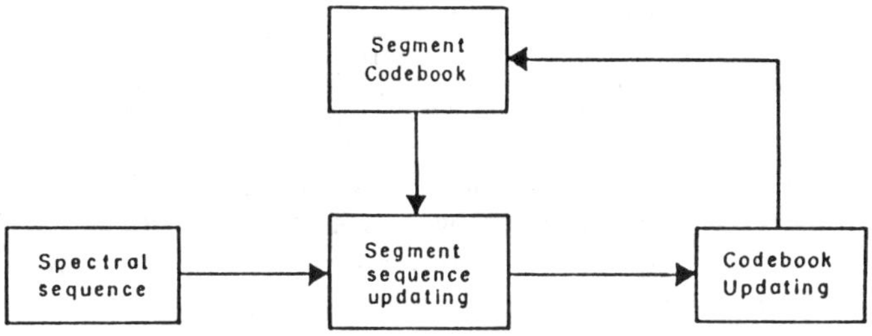

Fig. 3. Iterative algorithm for designing variable-length segment quantizer.

This can be easily proved as follows. First, in the segment updating stage (Step 1), for each voice interval, the distortion for the updated segment boundaries is at least as small as that for their neutral boundaries as described earlier. Thus, summing up the distortions over all of the voice intervals, we obtain $D_{k-1} \geqq D_k(0)$. Next, in the codebook updating stage (Step 3), the centroid calculated by (12) yields the minimum distortion among any code segments for the training segment data in the cluster (see the Appendix). Thus, summing up the distortion over all of the clusters, the inequality $D_k(m-1) \geqq D_k(m)$ holds for every m [12]. Thus, as D_k is the converged value of the sequence of $D_k(m)$ for $m \geqq 1$, we obtain the inequality $D_{k-1} \geqq D_k$ for every k. Since the distortion has a nonnegative value, the convergence of this algorithm is also ensured [13].

IV. Experimental Results

Several experiments were conducted to examine the performance of the proposed SQ method for speech data. Speech data were taken from several sentences uttered at normal speed by a single male speaker. The 12th-order LSP parameters and speech power were calculated using ordinary linear prediction analysis for a frame of 10 ms with a 30 ms-long Hamming window and an 8 kHz sampling rate. The original sequence was divided into a training sequence of 118 500 frames and a test sequence of 14 840 frames, excluding silence frames. Segment boundaries were initially estimated using a procedure based on detecting spectral change. The spectral change was obtained at every frame as the average slope of the regression lines determined by the least mean square error fitting the spectral parameter block. Hence, the segment rate varied according to the spectral change of the input speech. An average rate of 14.3 segments per second was obtained for the training and the test sequences. This indicates that segment rate is approximately equal to phoneme rate. In the following experiments, the segment rate was fixed at this rate. The segment lengths range from 1 to 29 frames (10–290 ms). There were 16 976 total segments for the training sequence, and 2141 for the test sequence. The code segment length was fixed at 10.

A. Convergence of Codebook Design Algorithm

The first experiment concerned convergence of the iterative algorithm of the SQ codebook design. An initial codebook C_0 was obtained using the VQ design algorithm of [13], namely, the splitting method, for the training segment sequence. The 2136 training segments, which are a subset of all the training segments, were used in this experiment. Fig. 4 shows the convergence performance of the iterative algorithm for the 6-bit codebook with a search range of 16 (160 ms). The ordinate represents the relative distortion normalized by the initial distortion, and the abscissa represents the iteration number. The triangular plots in this figure represent the performance achieved at the segment boundary updating stage (Step 1 of the algorithm). It is found that the iterative algorithm converges well after three or four iterations and reduces the distortion to about 80 percent of the initial value after five iterations. Most of the distortion reduction can be achieved by one or two iterations, especially at the first segment boundary updating stage.

Fig. 5 shows the performance comparison of joint SQ between the training and the test sequence for the 6-bit codebook at every iteration. The curve for the training sequence in this figure is the same as that in Fig. 4, except for excluding plots for the segment boundary updating stage. This figure indicates that codebooks designed for the training sequence are also robust for the test sequence and yield nonincreasing distortions for each iteration number, although the performance improvement more rapidly saturates in few iterations compared to that for the training sequence.

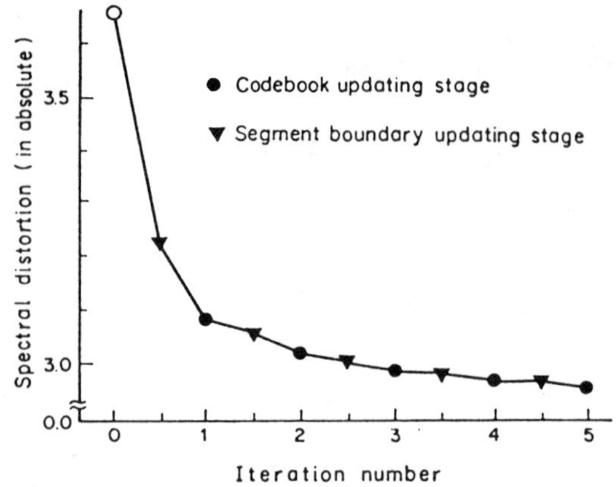

Fig. 4. Distortion convergence of iterative codebook design algorithm.

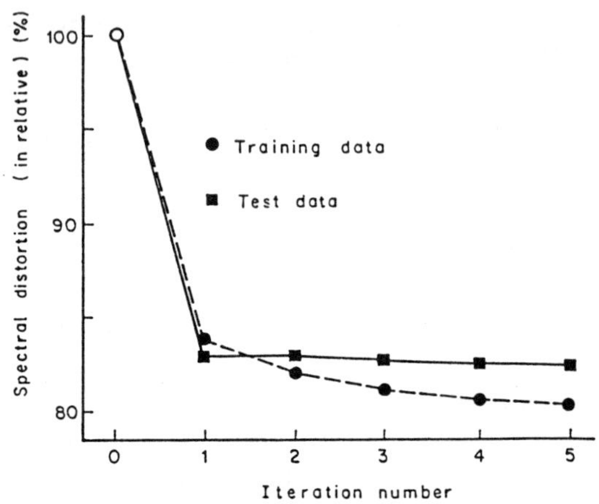

Fig. 5. Distortion of training sequence and test sequence.

B. Performance of Joint Segmentation and Quantization

The performance and computational load of joint segmentation and quantization are dependent on the initial guess of the segment boundaries and the search range as described earlier. Two kinds of initial guesses of segment boundaries were obtained using methods based on detecting the local peak of spectral changes and splitting the voice interval into equal length segments, respectively. The average distortions for the 6-bit codebook designed by the iterative algorithm were computed for several search ranges of segment boundaries. The average distortions normalized by the initial distortion for the latter initial segment boundaries are shown in Fig. 6. The results in Fig. 6 indicate that, as expected, the average distortion converges to a consistent lower bound regardless of the initial segment boundaries as the search range increases, while the initial distortions and the convergence speed is different for both initial guesses. Thus, the choice of the initial guess based on detecting spectral changes and the search range greater than 16 appear to be sufficient to provide SQ performance as good as the attainable bound with fair computational load.

The next experiments compare the performance of the SQ using joint segmentation and quantization, namely, joint SQ, the SQ using separate segmentation, namely, separate SQ, matrix quantization (MQ), or the fixed-length SQ and ordinary VQ. All parameters and the distance measures used in these systems are the same as those described earlier. The codebook size is 6–10 bits for SQ's, 9–13 bits for MQ, and 4–10 bits for VQ. These codebooks were designed using the training sequence of 118 500 frames. Design of the separate SQ corresponds to that of the joint SQ using the initial codebook and the initial segment boundaries. The VQ and MQ were designed using the algorithms of [13] and [8], respectively, except for the distance measure of (3).

Table I shows a comparison of the performance of the joint SQ and the separate SQ. The average distortions were obtained by averaging the weighted Euclidean distance given by (3) over all frames of the sequence. The results in Table I indicate that the joint SQ yields distortions consistently superior to the separate SQ for the training and the test sequence. The average distortion for the 7-bit codebook of the joint SQ is nearly the same as that for the 8-bit and 9-bit codebooks of the separate SQ for the training and the test sequence, respectively.

The MQ was designed for segment lengths of 7 (70 ms) for the same bit rate as the SQ. That is, this MQ has almost the same average segment rate as that of the SQ.

The joint SQ and the MQ are compared in Fig. 7. The two SQ's are indicated by SQ(3) and SQ(4). Their segment length is quantized with 3 bits and 4 bits, respectively. For SQ(3), the segment length was nonuniformly quantized to 2, 4, 5, 6, 7, 9, 11, or 14 considering the

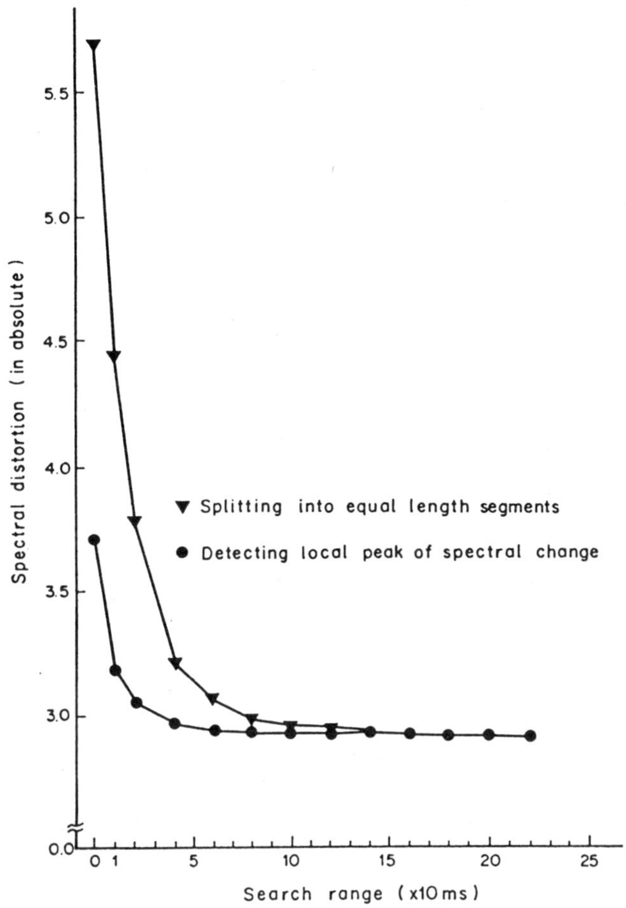

Fig. 6. Distortion for search range of segment boundaries.

TABLE I
DISTORTION OF JOINT SEGMENT QUANTIZATION (SQ) AND SEPARATE SEGMENT QUANTIZATION

	training sequence		test sequence	
codebook size	joint SQ	separate SQ	joint SQ	separate SQ
6	3.44	4.12	3.40	4.19
7	3.00	3.59	3.00	3.67
8	2.59	3.10	2.71	3.26
9	2.26	2.68	2.48	2.97
10	1.96	2.29	2.35	2.77

occurrence frequency of the segment length. For the SQ(4), the segment length was uniformly quantized to 1–16. The bit rates of SQ(3) and SQ(4), including segment length information, are 129–186 bits/s and 143-200 bits/s for a 6- to 10-bit codebook size at 14.3 segments per second. In this figure, the number near each plot indicates codebook size. The results show that the joint SQ performs better than the MQ in the range of 150 bits/s or more. From the standpoints of codebook size and spectral distortion, or speech quality, the variable-length SQ is more efficient than the MQ.

The joint SQ and three VQ's are compared in Fig. 8. The three VQ frame rates are 100 f/s, 50 f/s, and 33.3 f/s, respectively. In the latter two VQ's, spectral distor-

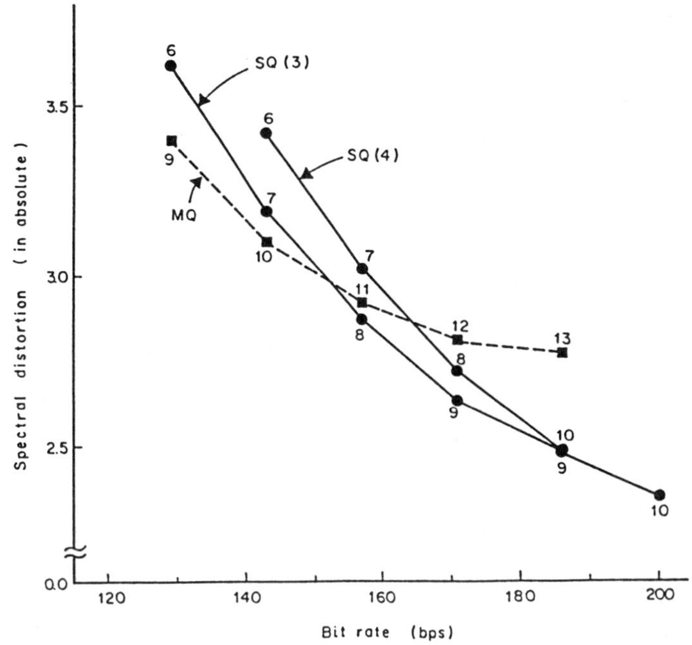

Fig. 7. Comparison of joint SQ's to MQ in bits per second.

tion was computed at every 10-ms frame for the spectral parameters obtained by linearly interpolating adjacent quantized parameters. The result shows that the spectral distortion of the joint SQ at 186 bits/s is compatible with that of the VQ's at 675 bits/s (100 f/s), 350 bits/s (50 f/s), and 290 bits/s (33.3 f/s), respectively. It is found that the joint SQ reduces the bit rate to about 50 percent of that with the 50-f/s VQ.

C. Subjective Listening Test

A sound articulation test was conducted to evaluate the intelligibility of the coded speech using the segment vocoder. Speech samples taken from 100 isolated Japanese CV syllables were used for the test. This intelligibility test was practically open response, since listeners selected every answer independently. A sound articulation score was obtained by averaging the ratio of the number of correctly received phonemes to the total number of phonemes over four listeners and six trials. Speech was encoded using a 10-bit codebook based on data for 20 000 training segments taken from sentence speech in which the test speech data were not included. The pitch period and voice-unvoice decision was assigned once for every segment automatically. The average segment rate for the test speech was 9 per second. Thus, the total bit rate was 150 bits/s ($= (10 + 2.7 + 4) \times 9$) including the information for 2.7 bits/segment for the segment length (using entropy coding) and 4 bits/segment for the pitch period (using differential coding). Articulation scores for the segment vocoder are compared to those for original speech, for the LSP vocoder without parameter quantizer, and for VQ vocoders in Table II. The VQ frame rate is 100 f/s. The total bit rates for the 6- and 7-bit VQ's are 1000 bits/s ($= (6 + 4) \times 100$) bits/s and 1100 bits/s,

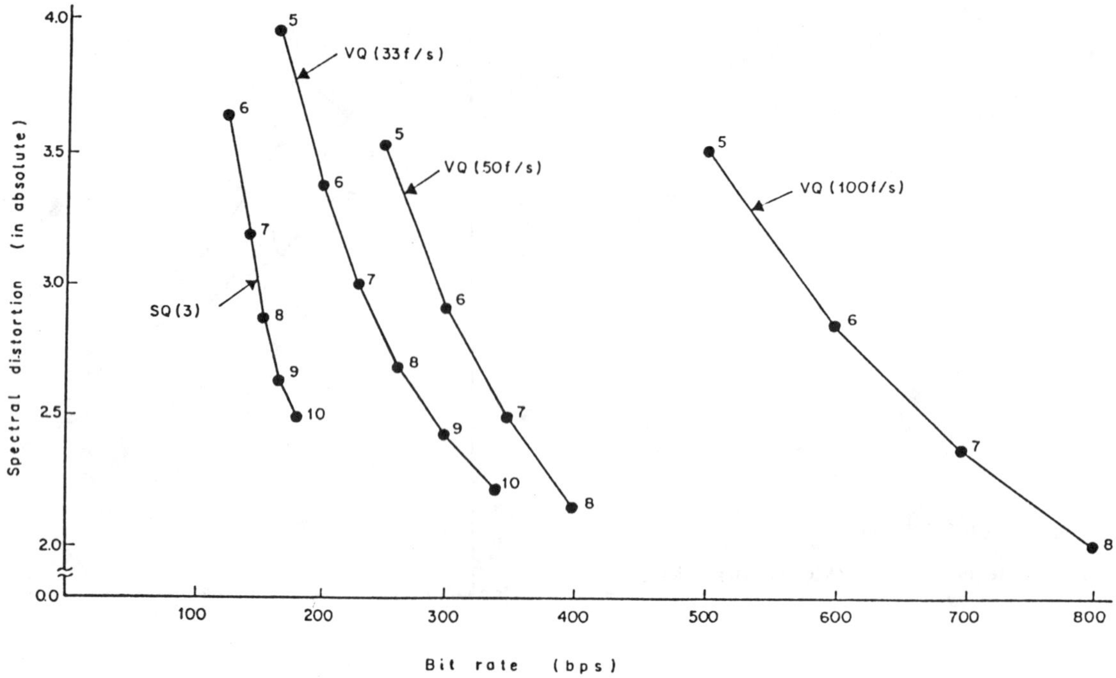

Fig. 8. Comparison of joint SQ(3) to VQ in bits per second.

TABLE II
SOUND ARTICULATION SCORE OF ORIGINAL SPEECH, LSP VOCODER, JOINT SQ VOCODER, AND VQ VOCODERS

	original	LSP	joint-SQ 150bps	6b-VQ 1000bps	7b-VQ 1100bps
vowel	100	100	99	98	99
consonant	88	80	56	56	63
average	94	90	78	77	81

respectively. Here, articulation scores for the LSP vocoder indicate the maximum performance attainable for the SQ and VQ vocoders. It is found that the average articulation score for the SQ vocoder is 78 percent. This is slightly better than that for the VQ vocoder using the 6-bit codebook. Note that articulation scores greater than 75 percent yield intelligibility sufficient to correctly recognize sentence speech.

V. CONCLUSION

A variable-length segment vocoder based on joint segmentation and quantization has been presented for very low bit rate speech coding. The coding problem has been formulated to search for the segment boundaries and the sequence of code segments so as to minimize the spectral distortion measure for a given interval. The problem has been efficiently solved using a dynamic programming procedure. An iterative algorithm for designing the variable-length segment quantizer has been also presented. This iterative algorithm consists of updating the segment boundaries and updating the fixed-length code segments for the training sequence. It has been shown that the iterative algorithm has theoretical convergence and consistently improves the performance of a codebook in several iterations. The performance of the variable-length segment quantization has been investigated experimentally and compared to other voice coding systems. It has been shown that a 150-bits/s segment coder using a 10-bit codebook affords performance comparable to a 1000-bits/s VQ coder in terms of spectral distortion and subjective quality. Thus, the proposed variable segment vocoder offers substantial potential for realizing a very low bit rate voice coding.

APPENDIX
PROOF OF (12) OF SECTION III

For each cluster, the centroid for variable-length segments is obtained so as to minimize

$$D = \sum_{j \in S_i} d(X_j, Y_i)$$
$$= \sum_{j \in S_i} \operatorname{tr}(X_j - Y_i H_{lj})' W (X_j - Y_i H_{lj}). \quad \text{(A1)}$$

By differentiating (A1) with respect to Y_i and setting the matrix equal to zero matrix, we obtain

$$\frac{\partial D}{\partial Y_i} = -2W \sum_{j \in S_i} X_j H_{lj}' + 2WY_i \sum_{j \in S_i} H_{lj} H_{lj}' = 0. \quad \text{(A2)}$$

When $\Sigma\, H_{lj} H_{lj}'$ is nonsingular, we obtain (12) from (A2) immediately. When $\Sigma\, H_{lj} H_{lj}'$ is singular, it is impossible for any Y_i to attain (A2). Letting (A2) be nonzero matrix e_i, we obtain

$$-2 \sum_{j \in S_i} X_j H_{lj}^t + 2Y_i \sum_{j \in S_i} H_{lj} H_{li}^t = W^{-1} e_i. \quad \text{(A3)}$$

In this case, min D is attained at min tr $(W^{-1}e_i)^t W^{-1}e_i$. Thus, if we denote the minimum norm least squares general inverse matrix of A as A^+ [14], then

$$Y_i = \left(\sum_{j \in S_i} X_j H_{lj}^t \right) \left(\sum_{j \in S_i} H_{lj} H_{lj}^t \right)^+. \quad \text{(A4)}$$

Acknowledgment

The authors wish to thank Dr. S. Furui and K. Kakehi for their many stimulating discussions throughout the course of this work; as well as the two anonymous reviewers whose constructive suggestions have improved the quality of this paper.

References

[1] A. Buzo, A. H. Gray, R. M. Gray, and J. D. Markel, "Speech coding based upon vector quantization," *IEEE Trans. Acoust., Speech, Signal Processing*, vol. ASSP-28, pp. 562–574, 1980.

[2] D. Y. Wong, B. H. Juang, and A. H. Gray, "An 800 bit/s vector quantization LPC vocoder," *IEEE Trans. Acoust., Speech, Signal Processing*, vol. ASSP-30, pp. 770–780, 1982.

[3] J. Makhoul, S. Roucos, and H. Gish, "Vector quantization in speech coding," *Proc. IEEE*, vol. 73, pp. 1551–1588, 1985.

[4] S. Roucos, R. Schwarts, and J. Makhoul, "Segment quantization for very low rate speech coding," in *Proc. IEEE Int. Conf. Acoust., Speech, Signal Processing*, 1982, pp. 1565–1568.

[5] ——, "A segment vocoder at 150 bits/s," in *Proc. IEEE Int. Conf. Acoust., Speech, Signal Processing*, 1983, pp. 61–64.

[6] Y. Shiraki and M. Honda, "Very-low-rate speech coding using time space spectrum patterns," *Trans. Tech. Group Speech Acoust. Soc. Japan*, vol. S84-06, Apr. 1984.

[7] D. Y. Wong, B. H. Juang, and D. Y. Cheng, "Very low data rate speech compression with LPC vocoder and matrix quantization," in *Proc. Int. Conf. Acoust., Speech, Signal Processing*, 1983, pp. 65–68.

[8] C. Tsao and R. M. Gray, "Matrix quantizer design for LPC speech using the generalized Lloyd algorithm," *IEEE Trans. Acoust., Speech, Signal Processing*, vol. ASSP-33, pp. 537–545, 1985.

[9] D. K. Burton, "Applying matrix quantization to isolated word recognition," in *Proc. IEEE Int. Conf. Acoust., Speech, Signal Processing*, 1985, pp. 29–32.

[10] Y. Shiraki and M. Honda, "Very low bit rate speech coding based on joint segmentation and variable length segment quantizer," *J. Acoust. Soc. Amer.*, Suppl. 1, vol. 79, 1986.

[11] N. Sugamura and F. Itakura, "Line spectrum representation of linear predictor coefficients of speech signal and its statistical properties," *Trans. Inst. Electron., Commun. Eng. Japan*, vol. J64-A, pp. 323–340, 1981.

[12] S. P. Lloyd, "Least squares quantization in PCM," *IEEE Trans. Inform. Theory*, vol. IT-28, pp. 129–137, 1982.

[13] Y. Linde, A. Buzo, and R. M. Gray, "An algorithm for vector quantizer design," *IEEE Trans. Commun.*, vol. COM-28, pp. 84–95, 1980.

[14] C. R. Rao and S. K. Mitra, *Generalized Inverse of Matrices and Its Applications*. New York: Wiley, 1971.

Vector Quantization of Speech and Speech-Like Waveforms

HÜSEYIN ABUT, SENIOR MEMBER, IEEE, ROBERT M. GRAY, FELLOW, IEEE, AND GUILLERMO REBOLLEDO, MEMBER, IEEE

Abstract—An algorithm for the design of vector quantizers that are locally optimum in the sense of minimizing an average quantitative distortion measure is used to design 1 and 2 bit/sample vector quantizers for both real sampled speech and a simulated speech-like autoregressive random process. Both weighted and unweighted squared-error distortion measures are considered.

Several comparisons are made and discussed based on the average distortions of the vector quantization schemes. The results for the simulated speech are compared to mathematical performance bounds from information theory to provide an indication of how nearly globally optimal vector quantization is for such highly correlated sources. A comparison of the results for the real speech and the simulated speech provides a quantitative measure of the accuracy of such models and, hence, of the applicability of information theory bounds and code designs based on probabilistic models. The signal-to-quantization-noise ratios of vector quantizers designed to minimize squared-error distortion are compared to those of several popular speech waveform coding systems of similar rates.

Manuscript received October 14, 1980; revised August 1, 1981 and November 5, 1981. This work was supported in part by the U.S. Air Force Office of Scientific Research under Contract F49620-79-C-0058 and by the Consejo Nacional de Ciencia y Tecnologia of Mexico.

H. Abut was with the Speech Laboratory, National Semiconductor Corporation, Santa Clara, CA 95053. He is now with the Department of Electrical and Computer Engineering, San Diego State University, San Diego, CA 92182.

R. M. Gray is with the Information Systems Laboratory and the Department of Electrical Engineering, Stanford University, Stanford, CA 94305.

G. Rebolledo was with the Information Systems Laboratory, Stanford University, Stanford, CA 94305, and Time and Space Processing, Inc., Santa Clara, CA 95053. He is now with the Division de Estudios Superiores, Facultad de Ingeniera, University of Mexico, Mexico City, Mexico.

I. Introduction

Waveform coding speech compression systems attempt to digitally communicate a good reproduction of the actual speech waveform. Voice coding systems attempt to estimate and communicate a linear model of the speech production process rather than the specific waveform. Voice coding systems permit communication at lower rates, but waveform coders typically provide better quality and more robustness against speaker variations, multiple speakers, and background noise. It is still important, however, to use the minimum possible bit rate and, hence, the smallest possible bandwidth or storage capacity for a given fidelity.

Waveform coders range from simple fixed high rate scalar quantizers (PCM's) and predictive quantizers (DPCM's), to fairly complex low rate adaptive subband coders, transform coders, DPCM, and tree encoding systems typically operating in the range of 8 kbits/s (kilobits per second)–15 kbits/s. A variety of such systems is discussed, for example, in [1]–[10].

Most people would agree that given a rate constraint, the general goal of such digitization schemes is to minimize the distortion between the original and reconstructed speech, but there is no universally accepted definition of distortion. There are, in fact, two fundamentally different approaches to defining and measuring distortion. The first is a subjective or qualitative notion based on relative distortions perceived by listeners in listening tests. While subjective distortion is clearly of fundamental importance in the evaluation of a voice communication system, it is of little help in the actual design and

Reprinted from *IEEE Trans. Acoust., Speech, Signal Processing,* vol. ASSP-30, pp. 423–435, June 1982.

analysis of a specific system since no analytical techniques exist for the design of a system that is optimized in the sense of minimizing subjective distortion.

The second approach uses a quantitative distortion measure such as the ubiquitous squared-error distortion, weighted versions thereof, or more complicated spectral distortion measures. Such quantitative measures may result from an approximate quantification of a subjective measure or they may be the result of a mathematical development as in the Itakura–Saito distortion [37]. The advantages of a quantitative measure are numerous. 1) It provides a means for the quantitative comparison of different systems. 2) Given a suitably tractable distortion measure, a system can be designed to provide a minimum or, at least, a local minimum of the average or maximum distortion. 3) Using techniques from information theory, one may be able to characterize the optimal performance for a given class of data compression systems on a specified source, where optimal here means yielding the minimum average distortion subject to the system constraints.

In short, a quantitative distortion measure permits the application of tools and techniques of communication and information theory to the design, analysis, and comparison of data compression schemes.

An additional justification for considering a quantitative distortion measure in the design and analysis of data compression systems is that virtually all such systems perform the actual digitization or quantization step using a minimum distortion or nearest neighbor rule to select a digital representation of an analog quantity. This quantization may be applied directly to the waveform, to an error sequence within a feedback loop; to a companded, transformed, or filtered set of samples; or to some combination. The quantization operation, however, invariably involves a minimum distortion selection. Thus, even though a particular class of systems may be selected based on its subjective quality, it is of both theoretical and practical interest to develop the properties of such systems when they are optimized. The resulting performance provides a benchmark for the comparison of similar systems.

When trying to characterize the optimal performance achievable for a given class of data compression systems, it is often of interest to consider the most general possible class. This provides a more fundamental benchmark for comparison with existing systems, and an indication of what additional improvement might be achievable using less structured (and possibly more complex) systems. The original Shannon mathematical model for a data compression system is that of a block source code—a mapping of consecutive nonoverlapping blocks or frames of source data into consecutive nonoverlapping blocks of channel symbols such as binary numbers. Block source codes can be simply viewed as vector or multidimensional quantizers, and include such systems as transform coders, traditional scalar quantization (PCM), permutation codes, and any other schemes that process parsed information. Information theory has also treated sliding-block or sliding-window codes that operate as possibly nonlinear filters (e.g., DPCM) and both classes have been shown to have an optimal achievable performance (in the limit of long block length or constraint length) given by Shannon's distortion rate function (see, e.g., [11], [2]). Here we focus on the design performance of block codes or vector quantizers. The design of trellis waveform coders for speech using an algorithm similar to the vector quantizer design algorithm treated here is considered in [38].

A general goal might be to find (if possible) the optimum vector quantizer of a given rate and block length (dimension) when used on a particular source. Such a result would indicate achievable performance improvements over special cases such as transform coders. Even if one possesses a complete and accurate probabilistic model of the source, however, such theoretical characterizations do not usually exist. Rate distortion theory bounds are achievable only in the limit of long block lengths. Approximations to the optimal performance may be found for a fixed block length using asymptotic quantizer performance theory if the rate is large enough (see, e.g., [12]–[14]). Since theory can characterize only asymptotic optimal performance values (large block length or large rate), the best available means of finding the performance limits of a class of codes of nonasymptotic rate and block length remains the actual design and test of good codes incorporating optimization techniques and a minimum of structural constraints.

An obvious advantage of popular special cases is the fact that their structural constraints usually make simple and efficient implementation possible. For example, most implementations of vector quantizers consist of a simple (often linear) transformation followed by separate scalar quantizers for the transformed samples [15]–[16], [4], [8]. Such codes are suboptimal in a Shannon sense because of the requirement of scalar quantization, but they are reasonable to build, work fairly well, and are field-proven. It remains of interest, however, to determine the potential improvement available by using less structured vector quantizers and whether the improvement merits the possibly increased complexity.

A general vector quantizer consists of a codebook of possible reproduction vectors and a minimum distortion encoding rule. With advances in large-scale integration, cheap memory, and fast search techniques, such a structure will itself become increasingly amenable to real implementation.

In some applications, the potential gains of general vector quantization over simple techniques are known to be small. It is well known, for example, that for large rates, simple PCM performs quite close to the rate distortion bound for Gauss memoryless sources [12], [17]. Based on these rigorous arguments, intuitive developments have been advanced so that DPCM provides nearly optimal performance for large rates and Gauss and Markov sources [18]. These are, however, asymptotic results and should not discourage the investigation of the potential improvements from optimization at more interesting lower (not asymptotically large) rates. We also point out that the Gaussian assumption is a "worst case" assumption in source coding—the Gaussian source is the "least compressible" source, given the source's autocorrelation, in that theoretically more bits are required to achieve a given average distortion [11]. Hence, conclusions based on a Gaussian source may be too conservative for other sources.

An obvious difficulty with the goal of developing general optimal vector quantizers is the fact that there is no general theory for the construction of optimal source codes. For this reason, the standard design technique for vector quantizers comprises two components: 1) choose a code structure that is reasonable to implement and optimize; and 2) develop a probabilistic model of the source for use in the optimization. Almost all traditional data compression schemes use only scalar quantization and the technique of Lloyd [19] (Method II) and Max [20] applied to a model distribution to find the optimal scalar quantizers. Examples are PCM and ADPCM systems [5], [1], [3] and transform coders [4], [8], which are optimized for models and then applied to real speech.

Other model-based techniques have also been incorporated into waveform coder design and analysis. In order to use the transform coder optimization techniques of Kramer and Mathews [15] and Huang and Schultheiss [16] on real speech, the speech is assumed to have a Gaussian distribution [4], [8]. This assumption yields transformed samples that are also Gaussian and provides rate distortion functions in a useful form for determining good bit allocations for the scalar quantizers for the transformed samples.

Thus, in the above examples and in more sophisticated systems based on these examples, the code is optimized for a model of the source and not the source itself. Intuitively, if the model is accurate, the resulting theory and performance should hold approximately for real speech. There is, however, no quantitative measure of the accuracy of a model and the validity of the approximation for real speech.

An alternative to the Lloyd-Max model-based quantizer design technique that has not previously been applied to speech waveform compression is the generalization of Lloyd's Method I [19] to the design of vector quantizers based on a long training sequence of actual data [21]-[24]. Both Lloyd's Method I and Method II, in general, guarantee only locally optimal quantizers in the sense of yielding a local minimum of an average distortion, but Method I has several advantages over the Lloyd (Method II)-Max technique.

First, it requires no explicit modeling of the source. The algorithm is instead driven directly by a long training sequence, and not indirectly through a probabilistic model based on a training sequence. In particular, no assumptions of the source being Gaussian, Laplacian, or autoregressive of a particular order are required. It can be shown mathematically that under certain stationarity assumptions, the design algorithm run on a long training sequence will produce nearly the same code and performance that it would produce if run on the true (but unknown) probabilistic description [22]. No such convergence result is known for the Lloyd-Max algorithm run on a model inferred from training data. Intuitively, this convergence implies that a quantizer designed on a sufficiently rich training sequence will yield similar performance on future data—a property that is experimentally supported for speech waveform coding in a later section.

Second, the technique works for vector quantization as well as scalar quantization to yield a locally optimum quantizer of a given rate and dimension.

Third, quite general distortion measures can be used. The only requirement is the existence of a generalized centroid or center of gravity with respect to an arithmetic average.

Fourth, the design algorithm can be run either on real speech data or as a Monte Carlo design for a speech model. This permits a comparison of the model performance with rate distortion bounds and a new and, we think, informative quantitative comparison between real and simulated speech.

This paper has two distinct but strongly related goals. One is to use the vector quantizer design algorithm of [21] to design, simulate, and evaluate speech waveform coders using vector quantization of 1 and 2 bits/sample and short block lengths of 8 and less. Both a squared error and a weighted squared error are considered. The resulting performance is compared to that of several popular speech waveform coders of comparable rates including optimized PCM and DPCM, transform coding, subband coding, and ADPCM.

The other goal is to apply the same vector quantization algorithm to common probabilistic models of the speech rather than to the speech itself. The models selected are tenth-order autoregressive sources with matching correlations. This is of interest since information-theoretic bounds on optimal performance can be computed for such mathematical models and not for real speech. Thus, the model—which bears at least a resemblance to real speech—provides a means of comparing the performance of the designed vector quantizers with the theoretical optimum, and hence provides an indication of the actual (global) optimality of the algorithm. Such comparisons have not previously been made for waveform coders operating on such highly correlated sources.

While it is generally acknowledged that models have provided a useful tool in the design of speech waveform coders, it is also generally accepted that conclusions regarding the performance using real speech cannot safely be made based only on the models. We plot results of separate experiments on real speech and simulated speech on common graphs to quantify the differences between real speech and mathematical models of speech. The results point out significant differences in both the code structure and the resulting performance.

This paper is concerned with the average quantitative distortion of vector quantizers designed to minimize average distortion. In speech waveform coding systems, however, subjective distortion is also important. For this reason, the digitized speech produced in our experiments was synthesized and listened to by a variety of audiences. We briefly describe these informal subjective tests and their support of the quantitative results.

II. Vector Quantization

Let $X = (X_0, \cdots, X_{k-1})^t$ be a k-dimensional real-valued random vector described by a cumulative distribution function $F(x) = \Pr(X_i \leqslant x_i, i = 0, \cdots, k-1)$. When the dimension k is not clear from context, we will denote X as X^k. We assume for simplicity that X has zero mean, that is, $EX = \mathbf{0}$ where E denotes expectation with respect to F. A vector quantizer (block quantizer, k-dimensional quantizer, block source code) consists of a reproduction alphabet or codebook $\mathcal{C} = \{y_i; i = 1, \cdots, N\}$ of N vectors together with a mapping q that assigns to each x a codeword $y_i \in \mathcal{C}$. For waveform coding, the reproduction vectors or codewords y_i should be similar to the source vectors x, that is, the codewords are also k-dimensional real-valued vectors. It is convenient to define subsets $S_i, i = 1, \cdots, N$, of k-dimensional Euclidean space by $S_i = \{x: q(x) = y_i\}$, that is, S_i, collects together all input vectors mapping into the ith codeword. The vector quantizer is then completely described by the codebook $\mathcal{C}$ and the collection of disjoint sets $\mathcal{S} = \{S_i; i = 1, \cdots, N\}$.

To measure the performance of a vector quantizer, we assume a distortion measure $d(x, \hat{x})$ that assigns a nonnegative distortion value to reproducing x as $\hat{x}$. We here focus on weighted-squared error distrotion measures of the form

$$d(x, \hat{x}) = (x - \hat{x})^t B(x - \hat{x}) = \sum_{i=0}^{k-1} \sum_{j=0}^{k-1} (x_i - \hat{x}_i) B_{i,j} (x_j - \hat{x}_j) \tag{1}$$

where $B = \{B_{i,j}; i = 0, \cdots, k\ 1; j = 0, \cdots, k - 1\}$ is a symmetric, positive definite matrix. In spite of the existence of several more sophisticated and more subjectively meaningful distortion measures for speech coding systems, such as spectral deviation and the Itakura-Saito distortion measure (see, e.g., [23]), various forms of weighted-squared error remain the most popular distortion measure for waveform coding systems because of their simplicity, tractability in analysis and design, and computability in applications.

The mathematical results to be developed are valid for a general weighting matrix B. The actual quantizer design and comparison will, however, focus on two special cases—the traditional squared error with B an identity matrix

$$d_1(x, \hat{x}) = \|x - \hat{x}\|^2 \triangleq (x - \hat{x})^t (x - \hat{x}) \tag{2}$$

and the Mahalanobis distortion measure defined by the weighting $B = R^{-1}$ where $R = E\{(X - EX)(X - EX)^t\}$ is the covariance matrix of the k-dimensional vector X

$$d_2(x, \hat{x}) = (x - \hat{x})^t R^{-1} (x - \hat{x}). \tag{3}$$

We require that R be positive definite so that R^{-1} is well defined. The Mahalanobis distortion has been successfully used in a variety of pattern classification problems (see, e.g., [25]-[29]), including speech classification problems [30], and hence seems a potentially useful distortion measure for vector quantization of speech, which can be viewed as a nearest neighbor pattern classification system [21].

Since R is symmetric and positive definite, so is R^{-1} and we can form a triangular decomposition of the form $R^{-1} = G^t G$ where G is a lower triangular matrix. Thus,

$$d_2(x, \hat{x}) = \|Gx - G\hat{x}\|^2 \tag{4}$$

where Gx is a "whitened" vector since its covariance matrix is the identity matrix. Minimizing the Mahalanobis distortion is equivalent to minimizing the squared-error distortion between the "residual" vectors Gx and $G\hat{x}$ formed by passing x and $\hat{x}$ through a causal time-varying filter described by G.

If the random vector X is quantized to obtain $q(X)$ and an error vector $e = X - q(X)$, then the expected Mahalanobis distortion can be expressed as

$$Ed_2(X, q(X)) = E\{e^t R^{-1} e\} = \mathrm{tr}\,\{R^{-1} R_e\} \tag{5}$$

where "tr" denotes the trace operation and $R_e = E\{(X - q(X))(X - q(X))^t\}$ is the covariance matrix of the error signal. Comparison with

$$Ed_1(X, q(X)) = E(e^t e) = \mathrm{tr}\,\{R_e\} \tag{6}$$

points out that the Mahalanobis distortion couples the components of the error vector, that is, the distortion depends on the cross correlations of the errors as well as on their power.

Given a distortion measure, the performance of a vector quantizer q on a random vector X can be measured by the average or expected distortion per sample

$$\overline{D}(q) = k^{-1} Ed(X, q(X)) = k^{-1} \sum_{i=1}^{N} E\{d(X, \hat{y}_i) | X \in S_i\} \Pr\{X \in S_i\}. \tag{7}$$

The rate of a quantizer is defined by $R = \log_2 N$ bits/block or $r = R/k$ bits/sample.

A vector quantizer is implemented on a discrete-time random process $\{X_i\}_{i=0}^{\infty}$ by successively quantizing nonoverlapping vectors $X_i = (X_{ik}, X_{ik+1}, \cdots, X_{ik+k-1})$ as $q(X_i)$. If the process $\{X_i\}_{i=0}^{\infty}$ is k-stationary and k-ergodic (i.e., the vector process $\{X_i\}_{i=0}^{\infty}$ is stationary and ergodic), then the ergodic theorem implies that

$$\lim_{L \to \infty} L^{-1} \sum_{i=0}^{L-1} d(X_i, q(X_i)) = Ed(X, q(X)) \tag{8}$$

with probability one, and hence the expected distortion yields the long-term time average distortion. Hence, we define a quantizer q to be *optimum* for a given number of levels N and dimension k if it minimizes the expected distortion and, hence also minimizes the long-term sample average distortion over all quantizers of N levels and dimension k.

The goal of vector quantizer design is to produce an optimum vector quantizer. Except for a few special cases, however, the best one can hope for is a locally optimum quantizer, that is, a q such that $\overline{D}(q)$ is a local minimum. In particular, subject to certain technical assumptions, the vector quantizer design algorithm of [21] can be used to design locally optimum quantizers for quite general distributions F and distortion measures d [21]-[24], [37].

In many cases such as speech, however, there is no generally accepted distribution F describing x. If one does possess a training sequence $\{x_i; i = 0, \cdots, L - 1\}$ of data from the source and L is large, then the expectations can be computed based on the sample distribution provided by the data instead of the "true"—but unknown—underlying distribution. If the vector source were stationary and ergodic, the resulting sample averages would be good approximations to the "true" averages. In particular, from (8) one should be able to use the design algorithm to minimize the sample distortion

$$k\overline{D}(q) = L^{-1} \sum_{i=0}^{L-1} d(x_i, q(x_i)) \tag{9}$$

and thereby approximately minimize the true expectation. This approach is made precise in [22] and implies, in particular, that if L is sufficiently large, then a quantizer yielding a given sample distortion on a training sequence will yield nearly the same time average distortion on future data produced by the same source. Furthermore, this fact justifies the use of the algorithm on sample distortions even when F is known, as this is equivalent to a Monte Carlo design for a known source. For this reason, we henceforth focus on sample averages of the form (9).

In order to state the design algorithm, we require the notion of the centroid or center of gravity of a set S of $|S|$ k-dimensional real-valued vectors with respect to the distortion measure d of (1). The centroid of S, cen (S), is defined as the vector $\boldsymbol{u}$ that minimizes the sample average

$$\Delta_S(\boldsymbol{u}) = \sum_{x \in S} d(\boldsymbol{x}, \boldsymbol{u}).$$

It is shown in the Appendix that regardless of the specific positive definite matrix $\boldsymbol{B}$, the centroid is given by

$$\text{cen}\,(S) = |S|^{-1} \sum_{x \in S} \boldsymbol{x}, \tag{10}$$

the usual Euclidean center of gravity of the set S of vectors.

Two versions of the vector quantizer algorithm are considered. The difference is in the manner of generating initial guesses for larger codes from good small codes. Usually, the initial guess for a vector of dimension k is formed as a product code of a dimension $k - 1$ code with a scalar code; thus, higher dimension codes are constructed from good smaller dimension codes. For comparison, some of the codes were designed using the splitting technique of [21] (see [24] for a flowchart), which designs codes for a fixed dimension code by designing codes of rate 0, 1, 2, $\cdots$ bits/block by splitting each word in a rate R code into two close words to form a rate $R + 1$ code. As shall be seen, the two techniques occasionally give distinct results, indicating that genuinely distinct local minima are possible.

The product code form of the design algorithm is next described. It should be pointed out that for the weighted squares distortion being considered, the iteration of Steps 2 and 3 is equivalent to Diday and Simon's Algorithm α for cluster analysis [29].

A. Design Algorithm–(Product Code Initialization)

Step 0–(Initialization): Given are a training sequence of sample data (real numbers) $T = \{x_i; i = 0, \cdots, J - 1\}$, an integer rate r bits/sample, a maximum dimension (vector size) K, a distortion threshold $\epsilon > 0$, and a symmetric positive-definite $K \times K$ weighting matrix $\boldsymbol{B} = \{B(i,j);\ i,j = 0, 1, \cdots, K - 1\}$. Let $\boldsymbol{B}_k$ denote the $k \times k$ submatrices $\{B(i,j);\ i,j = 0, 1, \cdots, k - 1\}$, $k = 1, \cdots, K$ (which are also positive definite), and define for $k = 1, 2, \cdots, K$ the distortion measures $d_k(\boldsymbol{x}, \hat{\boldsymbol{x}}) = (\boldsymbol{x} - \hat{\boldsymbol{x}})^t \boldsymbol{B}_k (\boldsymbol{x} - \hat{\boldsymbol{x}})$ where $\boldsymbol{x} = \boldsymbol{x}^k$. Set $k = 1$.

Step 1–(Initialization for Dimension k): If $k = 1$, let $\mathcal{C}_0(k)$ be an initial scalar code of rate r having 2^r entries. (This can be obtained by the splitting technique of [21] or by an intuitive guess such as a uniform quantizer on some range.) If $k > 1$, then form the product code $\mathcal{C}_0(k) = \mathcal{C}(k - 1) \times \mathcal{C}(1)$ where $\mathcal{C}(k - 1)$ is the final dimension $k - 1$ codebook of Step 3. In other words, each word in $\boldsymbol{y}^k$ in $\mathcal{C}_0(k)$ has the form $(\boldsymbol{u}^{k-1}, z)$ where $\boldsymbol{u}^{k-1} \in \mathcal{C}(k - 1)$ and $z \in \mathcal{C}(1)$. Set $m = 0$, $D_{-1} = \infty$, and $L = J/k$ (which we assume an integer for simplicity). Let T_k denote the k-dimensional vector sequence $\{\boldsymbol{x}_i; i = 1, \cdots, L\}$ where $\boldsymbol{x}_i = \boldsymbol{x}_i^k = (x_{ki}, \cdots, x_{k(i+1)-1})^t$ is formed by blocking the training sequence.

Step 2: Given $\mathcal{C}_m(k) = \{\boldsymbol{y}_i; i = 1, \cdots, 2^{rk}\}$, group all vectors $\boldsymbol{x}_i$ in the training sequence according to the minimum distortion codeword by forming the sets

$$S_i = \{\boldsymbol{x}_j\colon\ d(\boldsymbol{x}_j, \boldsymbol{y}_i) \leqslant d(\boldsymbol{x}_j, \boldsymbol{y}_l), \quad \text{all } l \neq i\}$$

with some arbitrary tie-breaking rule. Compute the distortion resulting when using the quantizer q described by the codebook $\mathcal{C}_m(k)$ and the sets $\mathcal{S} = \{S_i\}$

$$D_m = L^{-1} \sum_{i=0}^{L-1} d(\boldsymbol{x}_i, q(\boldsymbol{x}_i)) = L^{-1} \sum_{i=0}^{L-1} \min_{\boldsymbol{y} \in \mathcal{C}_m(k)} d(\boldsymbol{x}_i, \boldsymbol{y})$$

$$= L^{-1} \sum_{i=1}^{2^{rk}} \sum_{\boldsymbol{x} \in S_i} d(\boldsymbol{x}, \boldsymbol{y}_i).$$

Step 3: If $(D_{m-1} - D_m)/D_m > \epsilon$, then form a new codebook

$$\mathcal{C}_{m+1}(k) = \{\text{cen}\,(S_i); i = 1, \cdots, 2^{rk}\}.$$

(This must reduce distortion or leave it unchanged.) Set $m = m + 1$ and go to Step 2.

If $(D_{m-1} - D_m)/D_m \leqslant \epsilon$, then we are done for dimension k with final codebook $\mathcal{C}(k) = \mathcal{C}_m(k)$.

Step 4: If $k < K$, then set $k = k + 1$ and go to Step 1. If $k = K$, then halt with final codebook $\mathcal{C}(K)$.

III. Simulated Speech

A common probabilistic model for speech that has proved useful in the design of waveform coding systems for real speech is a finite-order autoregressive or all-pole process with a driving process and second-order characteristics estimated from real speech (see, e.g., [4], [6], [8]–[10]). Numerous driving (innovations) densities have been proposed, including Gaussian, Laplacian, gamma, and K_0 (see, e.g., the above references and [31]).

Such autoregressive processes are stationary and ergodic if the autoregressive filter is stable. A standard bromide in the speech literature asserts that speech is only locally or quasi-stationary and "hence" not stationary. Local stationarity and global (usual) stationarity are not, however, incompatible. One can have a model exhibiting both characteristics and have the local stationary properties as the fine structure. (Such models are treated mathematically in Fontana *et al.* [32].) Thus, there is no conflict in considering such autoregressive models as consistent but coarse models of real speech. Such a process models the global behavior of speech, that is, the average over the local properties.

An Mth-order autoregressive or all-pole model for speech is a random process $\{X_n\}_{n=-\infty}^{\infty}$ specified by the difference equation

$$X_n = -\sum_{k=1}^{M} a_k X_{n-k} + \sigma Z_n$$

where $\{Z_n\}_{n=-\infty}^{\infty}$ is an independent, identically distributed sequence of random variables described by a probability density function f_Z with $EZ_n = 0$, $E(Z_n^2) = 1$, and where the a_k are chosen so that the filter with z-transform $1/A(z) = (1 + \sum_{k=1}^{M} a_k z^{-k})^{-1}$ is stable (and hence, $\{X_n\}$ is stationary and ergodic). The parameter σ^2 is the squared gain or one-step prediction error of the process $\{X_n\}$.

If one forms an estimate of the autocorrelation of speech as $r(k)$; $k = 0, 1, \cdots, M$ and chooses σ and the a_k in the model so that $r(k) = EX_nX_{n-k}$, then the resulting process is called the matching autoregressive source (MARS) [6]. If $r(k)$ is given by a sample autocorrelation, then from the correlation matching property of linear prediction (see, e.g., Geronimus [33, p. 12] or Markel and Gray [34]), $\{X_n\}$ is also a linear prediction model for speech. If $r(k)$ is estimated based on a short (~20 ms) speech frame, then the model is exactly the LPC model for that speech frame as used in LPC speech compression systems. We here take $1/A(Z)$ as the zero rate Itakura–Saito centroid of [23], that is, the optimum zero rate LPC model based on a training sequence of 5000 frames of 128 samples each (640 000 samples) normal speech spoken by five male speakers with a sampling rate of 6.5 kHz. This is equivalent to the LPC model for the average of the frame sample autocorrelations. (See [23] for details.) The autocorrelation and inverse filter coefficients are given in Table I. We shall use the same training sequence and portions thereof for direct vector quantizer design.

The driving process pdf $f_Z(z)$ was chosen to approximate the residual marginal pdf produced for the model filter $A(Z)/\sigma$, that is, the real speech is passed through the filter A/σ selected above, and $f_Z(z)$ is chosen to approximate the output. Two pdf's yielding good fits [5], [31], [10] are the unit variance K_0 pdf defined by $f_Z(z) = \pi^{-1}K_0(z)$ where K_0 is a zero-order modified Bessel function, and the unit variance Laplace (or double exponential) density $f_Z(z) = (\sqrt{2})^{-1} \exp(-\sqrt{2}|z|)$. Both pdf's provide good fits to the data and both will be seen to yield similar results. For comparison purposes, we will also evaluate the performance bound for a unit variance Gaussian pdf $f_Z(z) = (2\pi)^{-1/2} \exp(-z^2/2)$ which provides a "worst case" bound for all sources having the same autocorrelation, that is, the largest possible minimum achievable average distortion.

TABLE I
SAMPLE AUTOCORRELATION COEFFICIENTS AND FILTER PARAMETERS

Lag k	r(k)/r(0)	a_k
0	1.0	1.0
1	0.400940	-0.35573
2	-0.047447	0.29671
3	-0.167337	0.09716
4	-0.182098	0.13798
5	-0.198497	0.10661
6	-0.263764	0.18392
7	-0.25939	0.17867
8	-0.129546	0.07244
9	-0.010753	0.11004
10	0.104571	0.03551

$r(0) = 9662.57$

$\sigma^2 = 6561.25$

IV. PERFORMANCE BOUNDS FOR SPEECH-LIKE SOURCES

Given a probabilistic source model, a distortion measure d, a dimension or block length k, and a rate r bits/sample ($R = kr$ bits/block), the optimal achievable performance is defined by

$$\rho_k(r) = \min_q \overline{D}(q),$$

$$\overline{D}(q) = k^{-1}Ed(X, q(X)) \tag{11}$$

where the minimum is taken over all quantizers q with rate r bits/sample and the k^{-1} normalizes the distortion to a per-sample basis. Information-theoretic techniques can be used to obtain lower bounds to $\rho_k(r)$ for the models considered here.

The first bound is given by the generalization of Yamada *et al.* [14] of the Shannon lower bound to the distortion-rate function (see, e.g., [11]). For the distortion measure $(x - \hat{x})^t B(x - \hat{x})$ this bound is given from [14, eq. (52) and (6c)] as

$$\rho_k(r) \geq k^{-1}D_{SLB}^{(k)}(R) \triangleq (\det B)^{1/k}(2e\pi)^{-1}2^{-2(rk-h(X))/k} \tag{12}$$

where $R = kr$ and $h(X)$ is the differential entropy of X, i.e.,

$$h(X) = -\int f_X(x) \log f_X(x)\, dx$$

where $f_X(x)$ is the pdf for X. The differential entropy $h(X)$ of the autoregressive source model is difficult to evaluate exactly for fixed k, but it is shown in the Appendix that

$$h(X) \geq kh(\sigma Z) = k(h(Z) + \log \sigma) \tag{13}$$

where

$$h(Z) = h(Z_0) = -\int f_{Z_0}(z) \log f_{Z_0}(z)\, dz$$

is the differential entropy of the unit variance random variable Z_0. Thus, we have the bound

$$\rho_k(r) \geq D^{(k)}(f) \triangleq \sigma^2 (\det B)^{1/k}(2e\pi)^{-1}2^{-2(r-h(Z))}. \tag{14}$$

We shall refer to this bound simply as the distortion-rate bound.

For the usual squared-error distortion measure, we have that $\det B = 1$, and hence

$$\rho_k(r) \geq D^{(k)}(r) = \sigma^2(2e\pi)^{-1}2^{-2(r-h(Z))} \tag{15}$$

(as in Berger [11]), and for the Mahalanobis distortion measure, we have that $\det B = \det R^{-1}$, and hence

$$\rho_k(r) \geq D^{(k)}(r) = \sigma^2(\det R)^{-1/k}(2e\pi)^{-1}2^{-2(r-h(Z))} \tag{16}$$

which is simply the squared-error bound scaled by $(\det \boldsymbol{R})^{-1/k}$. These bounds are readily evaluated for the sources considered since, for a unit variance Laplace pdf f_Z, we have by direct evaluation that

$$h(Z) = \log \sqrt{2}e \cong 1.347 \text{ nats} = 1.942 \text{ bits} \tag{17}$$

and for the unit variance K_0 pdf f_Z, we have by computer evaluation that

$$h(Z) = \log \pi - (2/\pi) \int_0^\infty K_0(x) \log K_0(x)\, dx$$

$$\cong \log \pi + 0.200 \cong 1.345 \text{ nats} = 1.940 \text{ bits}. \tag{18}$$

To the accuracy of the figures, $h(Z)$ and, hence the distortion-rate and asymptotic quantizer bounds, are the same for both Laplace and K_0 pdf's.

For the Gaussian pdf

$$h(Z) = \tfrac{1}{2} \log 2\pi e \cong 1.419 \text{ nats} \cong 2.047 \text{ bits}. \tag{19}$$

Unfortunately, bounds from distortion-rate theory can be quite poor when the dimension k is not large. If we assume that the codebook size $N = 2^{kr}$ is large, however, then the generalization of Yamada *et al.* [14] of the asymptotic quantizer performance bounds of Gish and Pierce [12] and Gersho [13] can be used to improve the distortion-rate bound. From the above and [14, eq. (55)], if 2^{kr} is large, then

$$\rho_k(r) \geqslant \Delta^{(k)}(r) \triangleq \{e\Gamma(1 + k/2)^{2/k}/(1 + k/2)\}\, D^{(k)}(r) \tag{20}$$

where Γ is the gamma function and the bracketed term is bound from below by 1. We shall refer to this bound as the asymptotic quantizer bound.

It is shown in the Appendix that

$$\lim_{k \to \infty} e\Gamma(1 + k/2)^{2/k}/(1 + k/2) = 1, \tag{21}$$

and hence the asymptotic quantizer bound approaches the distortion-rate bound in the limit of large dimension k. For finite k, however, the asymptotic quantizer always provides a strictly better bound [14].

It is well known from rate-distortion theory that for a given autocorrelation, the distortion rate function is most conservative for the "worst case" of a Gaussian source [11], that is, for a fixed rate, the optimal achievable distortion is larger for the Gaussian than for any other density. From (14), (19), and (20), it is clear that the Gaussian density plays a similar "worst case" role in the bounds developed here. We henceforth focus on the more optimistic bounds for the simulated speech cases of K_0 and Laplacian densities, as even these bounds will be seen to be overly conservative for real speech.

V. Quantizer Performance Bounds

A problem with using the average distortion $\bar{D}(q)$ of (11) as a measure of quantizer performance is its dependence on the input signal amplitude. A common and useful normalization in the case of squared-error distortion is to divide by the input signal variance (or energy, if the signal has zero mean) and use either $\bar{D}(q)/R(0)$, $R(0) = E[(X - EX)^2]$ as a measure of distortion or $R(0)/\bar{D}(q)$—the signal to (quantization) noise ratio (SNR)—as a measure of quality.

As a natural generalization of this normalization, we define

$$\gamma = \min_u k^{-1} E(d(\boldsymbol{X}, \boldsymbol{u})), \tag{22}$$

the smallest average distortion achievable when no information is sent, that is, the average distortion using the optimum zero rate code. We will focus on the normalized average distortion defined by

$$\bar{d}(q) = \bar{D}(q)/\gamma \tag{23}$$

as the measure of performance of a quantizer. The expectations of (11) and (22) will be probabilistic averages if a distribution is assumed and will be estimated by a time average if a training sequence is used.

Analogous to (10), the minimizing $\boldsymbol{u}$ in (22) is simply $E\boldsymbol{X}$, and hence

$$\gamma = k^{-1} E((\boldsymbol{X} - E\boldsymbol{X})^t \boldsymbol{B} (\boldsymbol{X} - E\boldsymbol{X})). \tag{24}$$

If $\boldsymbol{B}$ is the identity matrix, then $\gamma = R(0)$, and hence $1/\bar{d}(q)$ is the usual SNR. In the general case, $1/\bar{d}(q)$ plays a role analogous to that of an SNR.

For the Mahalanobis distortion, $\boldsymbol{B} = \boldsymbol{R}^{-1}$, and hence $\gamma = 1$.

VI. Experimental Results

Quantizers were designed and tested for a variety of rates, dimensions, and data sets. In all of the experiments except for the K_0 simulated speech, duplicate runs were made on separate computers using different programs as a check on the results.

The first series of experiments consisted of the design of vector quantizers of 1 bit/sample and dimensions 1-8 and 2 bits/sample and dimensions 1-4 for both real speech and simulated speech. The training sequence for real speech consisted of the first 128 000 samples or approximately 20 s of the training sequence of [23] described in Section III, and hence contained a single male speaker. The simulated speech training sequences consisted of 60 000 samples produced by a random number generator. The final average normalized distortion is plotted in decibels for the entire training sequence together with the asymptotic quantizer bound and distortion rate bounds in Figs. 1-4—Figs. 1 and 3 are for squared-error distortion and 1 bit/sample and 2 bits/sample, respectively. Figs. 2 and 4 are similar plots for the Mahalanobis distortion. The real speech tests were run on a PDP 11/34 at Stanford University and a PDP 11/70 at Time and Space Processing, Inc. The results for simulated speech were conducted on a Cyber computer at California State University at Long Beach and on a Cromemco System 3 at Stanford.

In this test series, the training sequence was longer for the real speech simply because of greater available computer time. Longer training sequences are always better, but the availability of the Cyber and the speed of the Cromemco necessitated shorter sequences.

The digitized speech was synthesized for all of these experiments. The subjective quality of the squared-error distortion results was almost identical to the corresponding Mahalanobis

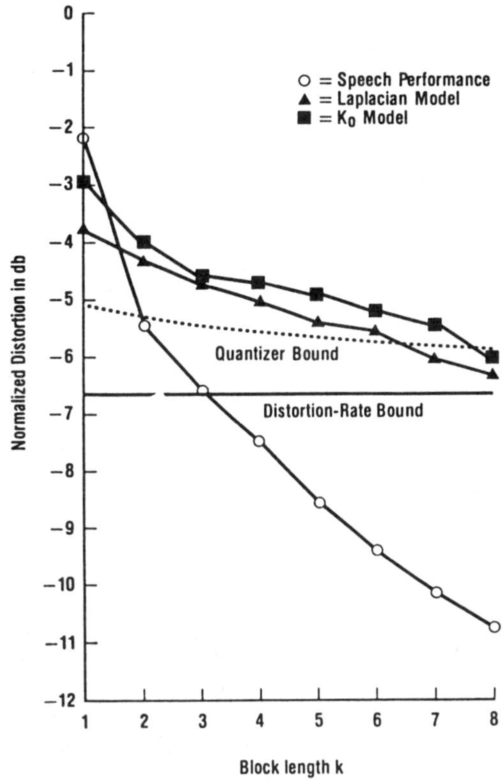

Fig. 1. Design performance of vector quantizers under MSE distortion measure at a rate of 1 bit/sample.

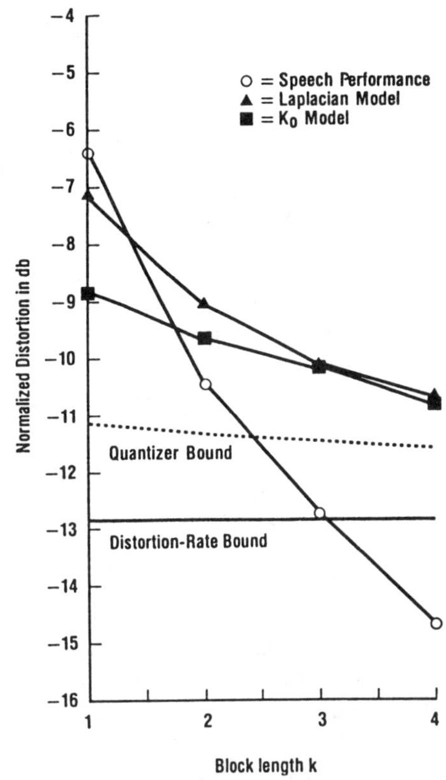

Fig. 3. Design performance of vector quantizers under MSE distortion measure at a rate of 2 bits/sample.

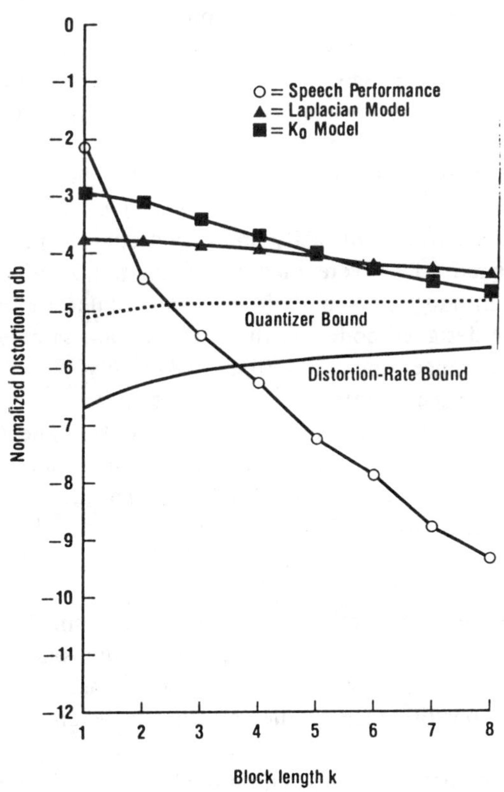

Fig. 2. Design performance of vector quantizers under Mahalanobis distortion measure at a rate of 1 bit/sample.

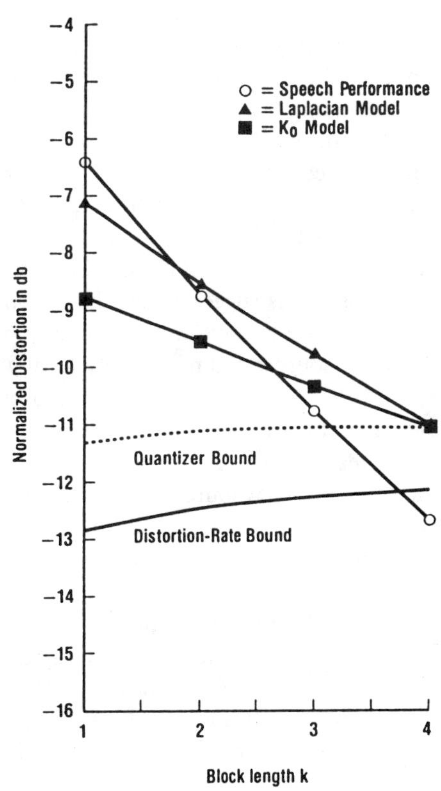

Fig. 4. Design performance of vector quantizers under Mahalanobis distortion measure of a rate of 2 bits/sample.

TABLE II
SNR INSIDE AND OUTSIDE 128K TRAINING SEQUENCE

SNR(dB) \ k	1	2	3	4	5	6	7	8
Inside	2.16	5.44	6.56	7.48	8.55	9.40	10.15	10.75
Outside	2.04	5.26	6.00	6.94	7.37	7.82	8.05	8.16

(a) 1 bit per sample

SNR(dB) \ k	1	2	3	4
Inside	6.41	10.44	12.72	14.68
Outside	6.37	9.71	11.70	12.38

(b) 2 bits per sample

distortion results. As the Mahalanobis distortion is more complicated to implement, further tests were not conducted for it.

Three properties common to all of the figures are apparent. First, the relative performance improvements achievable for the real speech are much larger than those achieved by the model. For example, an improvement of roughly 8 dB is achieved between optimum PCM ($k = 1$) and a rate 8 bits/block ($k = 8$ for 1 bit/sample, $k = 4$ for 2 bits/sample) using MSE vector quantization. The corresponding improvement is only about 3-4 dB on the simulated speech. Second, the performance of the vector quantizers on real speech exceeds that of information theoretic bounds based on the model by several decibels, indicating that model-based theoretical bounds are not appropriate for predicting performance on real speech. Third, the performance of the quantizers designed for the models are quite close to the theoretical quantizer optimal performance bound for the larger k. As the number of levels in the largest rate codes is $2^8 = 256$, the bounds should be good approximations. This suggests that the design algorithm is yielding codes that are approximately globally optimal for the simulated speech sources and not merely locally optimum. Not surprisingly, the differences between the design performance and the theoretical bounds are much smaller for these highly correlated results than in the case of Gauss-Markov sources [24] and memoryless sources [21]—the more memory a source has, the better it can be compressed.

The relative subjective quality of the speech clearly increased with dimension. In the 1 bit/sample case, it went from very noisy, partially, or barely intelligible quality at dimension 1 to an easily intelligible, but somewhat noisy quality at dimension 8. The 2 bit/sample case yielded significantly better quality at all dimensions. The dimension 4, 2 bit/sample example exhibited little audible quantization noise on the recording equipment used.

The next experiments comprised testing the codebooks designed above on the 78 600 sample open test sequence of [23] containing a speaker not in the training sequence. Subjectively there was a clear increase in quantization noise and buzziness of the speech, especially at the smaller dimensions and at 1 bit/sample. The quantitative difference between the design SNR and the test sequence SNR are summarized in Table II for the squared-error distortion (the design values are for the source experiments depicted in Figs. 1 and 3). The quantitative results reinforce the quality loss of the subjective tests. Note that over 2 dB is lost at the larger dimensions by using a quantizer designed for one speaker on another.

The theoretical results of [22], the experimental results of [21], and common sense all suggest that a longer training sequence with more speakers should yield codes with better performance (on the average) outside of the training sequence. Hence the design was repeated on the full 640 000 sample, five speaker training sequence (about 100 s) of [23] on DEC VAX's at Stanford University and at National Semiconductor. The designed codes were then tested on the 78 600 test sequence. In addition, the National Semiconductor experiment tested the designed codes on the first 128 000 samples of the training sequence, that is, the first speaker only.

The two experiments differed in the means of generating initial guesses and in convergence thresholds. The National Semiconductor experiments used the product code initialization introduced here. The Stanford experiments used the splitting technique of [21], [24]. The results of the National Semiconductor tests and the Stanford University tests are given in Tables III and IV, respectively.

Observe that the design SNR is uniformly smaller for the larger training sequence—reflecting the fact that one codebook must now work for five speakers instead of one. On the other hand, the performance of the resulting code on the test sequence is everywhere better—both quantitatively and subjectively—with the codes based on the larger training sequence. Furthermore, the differences between the design and test values are smaller. Somewhat surprisingly, the code based on

TABLE III
SNR INSIDE AND OUTSIDE 640K TRAINING SEQUENCE: PRODUCT CODE INITIALIZATION

SNR(dB) \ k	1	2	3	4	5	6	7	8
Inside (full)	2.01	5.25	6.04	7.11	7.82	8.51	9.13	9.70
Inside (1st 128K samples)	2.16	5.42	6.33	7.48	8.28	8.95	9.54	10.04
Outside	2.05	5.28	6.03	6.99	7.47	8.07	8.47	8.76

(a) 1 bit per sample

SNR(dB) \ k	1	2	3	4
Inside (full)	6.42	9.93	12.18	12.37
Inside (1st 128K samples)	6.33	16.31	14.31	14.57
Outside	6.45	9.68	11.83	11.58

(b) 2 bits per sample

the full sequence is nearly optimal for the first speaker alone (compare the "Inside" results of Table II with the "Inside" (1st 128K samples) of Table III).

Observe also the variation among the SNR's of one code on different speakers. This simply reflects the well-known fact that short-term SNR's can vary widely for a given coding scheme.

In the original experiments, a convergence threshold of $\epsilon = 0.005$ was used for both designs. The 1 bit/sample results were quite close, as indicated in Tables III(a) and IV(a). For 2 bits/sample, however, the product code initial guess resulted in noticeably inferior códes, and hence the threshold for the National Semiconductor 2 bit/sample experiment was lowered to $\epsilon = 0.0025$ to force further iterations. The resulting performance is given in Table III(b) and compares well with Table IV(b) except for the $k = 4$ case. Attempts to improve the $k = 4$ performance of Table III(b) by further iterations yielded no reduction in performance, indicating that the product code had led to a local minimum distinct from and inferior (by over 1 dB) to that of the splitting technique. Note further that the $k = 4$ code of Table III performed poorly on the test sequence, worse than that of the Table III $k = 3$ code! This emphasizes the fact that the choice of initial codes for the algorithm can be crucial to the resulting performance. Observe, however, that the results of Table IV are not all better than those of Table III.

VII. COMPARISONS WITH OTHER SYSTEMS

The vector quantization systems developed in the previous sections were designed to minimize the long run mean squared-error, and hence maximize the long run SNR. As discussed in the Introduction, we therefore use the long run SNR for quantitative comparisons with other waveform coding systems of comparable rates. For these comparisons, we use the results of the 640 000 sample training sequence and the 76 800 test sequence. All of the comparisons are for the 2 bit/sample case as that was the minimum rate considered in the references.

Paez and Glisson [5] developed PCM systems for 2 bits/sample and more using the Lloyd-Max scalar quantizer design procedure on several model densities. Their best system resulted from a gamma model of the marginal density, and yielded an SNR of 6.04 dB when applied to real speech. Lloyd's Method I on the original speech yielded a slightly better 6.4–6.5 dB in both design and open tests. Thus, the model-based design does fairly well for $k = 1$ (as is also suggested by Fig. 3). It is worth observing, however, that the actual codes can be quite different. The code based on the real speech is highly asymmetric, with a value near zero occurring over 80 percent of the time, while the model-based codes have levels symmetric about the origin.

Paez and Glisson also use the Lloyd-Max technique to design quantizers inside the feedback loop of a DPCM system using a model for the error sequence. Their best performance is obtained using a Laplace model which yields an SNR of 11.67 dB at 2 bits/sample. A dimension three 2 bit/sample vector quantizer yields a slightly better performance of 11.85 dB outside the training sequence and a design distortion of 12.13 dB. A dimension 4 code improves these values to 12.70 and 13.48, respectively.

The best low rate transform coder of Campanella and Robinson [4] (using a Karhunen-Loeve expansion as in [15], [16]) yields an SNR of 12.6 dB at 14 kbits/s. Their Fourier transform coder is inferior by about 4 dB. A vector

TABLE IV
SNR INSIDE AND OUTSIDE 640K TRAINING SEQUENCE: SPLITTING INITIALIZATION

SNR(dB) \ k	1	2	3	4	5	6	7	8
Inside	2.01	5.24	6.09	7.09	7.88	8.53	9.08	9.74
Outside	2.05	5.25	6.01	6.99	7.57	8.05	8.41	8.82

(a) 1 bit per sample

SNR(dB) \ k	1	2	3	4
Inside	6.50	9.98	12.13	13.48
Outside	6.41	9.72	11.85	12.70

(b) 2 bits per sample

dimension of $k = 16$ was used. This performance is close to that of our vector quantizer of dimension $k = 4$ and 2 bits/sample (Table IV). This shows that for speech, higher dimension transform coders with an optimum transform perform quite close to optimized vector quantizers of smaller dimensions. Similar comparisons were found for Gauss-Markov sources [24].

Crochiere *et al.* [7] reported performance at 2 bits/sample of 10.9 dB for an adaptive predictive PCM (ADPCM) system and 11.1 dB for their subband coder of the same rate. Our vector quantizers provide at $k = 3$ about 0.7 dB improvement in open tests and about 1 dB in design performance. At $k = 4$, there is roughly 1.5 dB improvement in the open test, and about 2.4 dB in the design performance over the reported subband coder performance.

Of course, the above comparisons must be taken with appropriate caution, especially when the SNR is computed and compared for short segments. In all cases considered, however, a block length of 3 or 4 sufficed for a 2 bit/sample vector quantizer to provide slightly better long term MSE performance than that of the other systems considered.

VIII. COMMENTS

Figs. 1-4 demonstrate the relative performance of vector quantizers designed for real speech and simulated speech. Tables II-IV provide similar results for real speech for longer training sequences and indicate the performance of using the codes on open tests, that is, on speakers not in the training sequence. Comparison of these results with each other and theoretical bounds based on information theory yield several conclusions. Roughly 8 dB in performance improvement is achieved in design performance and 6 dB in open test performance by the 8 bits/block codes over the optimum PCM. This improvement is greater by at least 4 dB than that predicted by design experiments on the simulated speech. In addition, the performance achieved was several decibels better than that predicted by model based information theory bounds. This suggests that except for PCM, theoretical performance bounds based on the model are of little use in predicting performance for real speech—even though the model is of order 10 and the largest block length is 8.

It was also found that longer training sequences with more speakers yield lower design performance, but higher open test performance for the cases considered. Furthermore, the code obtained for the larger sequence gave performance for the first speaker close to that of the code designed only for the first speaker. These results reinforce the theoretical convergence results of [23].

The performance on the simulated speech was close to that of approximate bounds to the optimal achievable distortion at a fixed dimension for the larger dimensions, suggesting that on these sources the vector quantizers were nearly optimal. In the 1 bit/sample MSE case, the results were within 1 dB of the distortion rate bound, the optimal achievable performance for any dimension.

The performances of the vector quantizers were compared with several popular waveform coding systems of 2 bits/sample. The higher rate codes were shown to provide slight improvements in long term SNR over the best of the popular systems. This shows that existing systems are performing fairly near the optimal performance achievable using short vector quantizers. It also shows that short vector quantizers in principle provide an alternative technique for achieving this performance.

The vector quantizers considered here have one clear disadvantage with respect to the other systems in that they have not been implemented in real-time hardware. The quantitative comparisons of this work are based on average distortion and not implementation complexity. The full search vector quantizers are not well suited to software or firmware use because of the large searches required. For example, 1 bit/sample full search quantizer requires the computation of 256 energies of eight-dimensional error vectors, and the selection of a codeword with a minimum energy for each vector consisting of eight consecutive samples. Fast search algorithms using tree-searched vector quantizers have been shown to provide much reduced search effort (16 energies in the above code example) with only small losses of performance for voice coding [23] and waveform coding of Gauss-Markov sources [24]. Other vector quantizer encoder structures are currently being studied that provide other complexity/performance tradeoffs. The full search vector quantizers remain of interest, however, as a benchmark for comparison for other schemes of comparable rates and dimensions. In addition, the general vector quantizer structure has several possible advantages over the popular schemes.

1) Given a code structure, the code is optimized by a straightforward iterative computer procedure without the need for human injected "bells and whistles" in the design.

2) Unlike DPCM and similar systems, there is no feedback loop and no predictor is required. Hence, channel errors cannot propagate past the block length (here of eight or fewer samples).

3) There are no explicit models required for the optimization procedure.

4) The algorithm of [21] can be run for quite general distortion measures. For example, minimizing a distortion measure of the form $d(\boldsymbol{x}, \hat{\boldsymbol{x}}) = \|\boldsymbol{x} - \hat{\boldsymbol{x}}\|^2 / \|\boldsymbol{x}\|^2$ would maximize the average short term or segmented SNR, a currently fashionable fidelity criterion.

5) Shorter block lengths are required for a given performance than for other vector quantizer schemes such as transform coders.

Appendix

Centroids

To find the vector $\boldsymbol{u}$ that minimizes

$$\Delta(\boldsymbol{u}) = \sum_{i=1}^{K} d(\boldsymbol{x}_i, \boldsymbol{u}) = \sum_{i=1}^{K} (\boldsymbol{x}_i - \boldsymbol{u})^t \boldsymbol{B} (\boldsymbol{x}_i - \boldsymbol{u})$$

for a symmetric positive definite $\boldsymbol{B}$, first form the triangular decomposition $\boldsymbol{B} = \boldsymbol{G}^t \boldsymbol{G}$ where $\boldsymbol{G}$ is lower triangular and positive definite. We then have that

$$\Delta(\boldsymbol{u}) = \sum_{i=1}^{K} \|\boldsymbol{G}\boldsymbol{x}_i - \boldsymbol{G}\boldsymbol{u}\|^2.$$

From Linde *et al.* [21], this is minimized if

$$\boldsymbol{G}\boldsymbol{u} = K^{-1} \sum_{i=1}^{K} \boldsymbol{G}\boldsymbol{x}_i,$$

and hence since G is invertible, $\Delta(\boldsymbol{u})$ is minimized by

$$\boldsymbol{u} = K^{-1} \sum_{i=1}^{K} \boldsymbol{x}_i,$$

the Euclidean center of gravity of the $\{\boldsymbol{x}_i\}$.

Differential Entropy of Autoregressive Processes

Given an autoregressive source $\{X_n\}$ as in Section III, we wish to show that for any k and vector $\boldsymbol{X} = (X_0, \cdots, X_{k-1})$, $h(\boldsymbol{X}) \geqslant k(h(Z) + \log \sigma^2)$. First fix initial conditions $X_i = x_i$, $i = -M, \cdots, -1$, and observe that then $\boldsymbol{X} = \sigma A \boldsymbol{Z} + \boldsymbol{b}$ where A is a lower triangular matrix with elements $(A)_{i,j} = a_{i-j}$ where $a_{i-j} = 0$ for $i < j$ or $|i - j| > M$ and $\boldsymbol{b}$ depends on the initial conditions and the a_i. From Problem 1.33 of McEliece [35], the conditional differential entropy

$$h(\boldsymbol{X} | X_{-m} = x_{-m}, \cdots, X_{-1} = x_{-1}) = h(\boldsymbol{Z}) + \log |J|$$

where J is the Jacobian of the transformation $f(\boldsymbol{Z}) = \sigma A \boldsymbol{Z} + \boldsymbol{b}$. The Jacobian is given by $\det(\sigma A) = \sigma^k$. Since the Z_i are independent, $h(\boldsymbol{Z}) = k h(\boldsymbol{Z}_0)$, and hence the conditional differential entropy above is given by $k\{h(Z) + \log \sigma\}$ regardless of the initial conditions, and hence

$$\begin{aligned} h(\boldsymbol{X} | X_{-m}, \cdots, X_{-1}) = - \int & f_{X_{-m}, \cdots, X_{-1}}(x_{-m}, \cdots, x_{-1}) \\ & h(\boldsymbol{X} | X_{-m} = x_{-m}, \cdots, X_{-1} = x_{-1})\, dx_{-m} \cdots dx_{-1} \\ & = k\{h(Z) + \log \sigma\}. \end{aligned} \quad \text{(A1)}$$

From McEliece [35, Sect. 1.3], $h(\boldsymbol{X}) \geqslant h(\boldsymbol{X} | X_{-m}, \cdots, X_{-1})$ and the desired result follows.

Proof of (21)

From Gradshteyn and Ryzhik [36, p. 937, eq. 8.327], as $k \to \infty$,

$$\Gamma(1 + k/2) \cong (1 + k/2)^{(1+k)/2} e^{-(1+k/2)} \sqrt{2\pi}$$

whence

$$e\Gamma(1 + k/2)^{2/k} / (1 + k/2) \cong (1 + k/2)^{1/k} e^{-2/k} (2\pi)^{1/k}$$

$$e\Gamma(1 + k/2)^{2/k} / (1 + k/2) \cong (1 + k/2)^{1/k} \cdot e^{-2/k} (2\pi)^{1/k} \xrightarrow[k \to \infty]{} 1.$$

References

[1] N. S. Jayant, "Digital coding of speech waveforms: PCM, DPCM, and DM quantizers," *Proc. IEEE*, vol. 62, no. 5, 1974.

[2] L. D. Davisson and R. M. Gray, Eds., *Benchmark Papers in Data Compression*. Stroudsburg, PA: Hutchinson and Ross, 1976.

[3] J. L. Flanagan, M. R. Schoeder, B. S. Atal, R. E. Crochiere, N. S. Jayant, and J. M. Tribolet, "Speech coding," *IEEE Trans. Commun.*, vol. COM-27, pp. 710–736, Apr. 1979.

[4] S. J. Campanella and G. S. Robinson, "A comparison of orthogonal transformations for digital speech processing," *IEEE Trans. Commun. Technol.*, vol. COM-19, pp. 1045–1049, 1971.

[5] M. D. Paez and T. H. Glisson, "Minimum mean squared error quantization in speech PCM and DPCM systems," *IEEE Trans. Commun. Technol.*, vol. COM-20, pp. 225–230, Apr. 1972.

[6] J. B. A. Anderson and J. B. Bodie, "Tree encoding of speech," *IEEE Trans. Inform. Theory*, vol. IT-20, pp. 379–387, 1975.

[7] R. E. Crochiere, S. M. Webber, and J. K. L. Flanagan, "Digital coding of speech in sub-bands," *Bell Syst. Tech. J.*, vol. 55, pp. 1069–1086, Oct. 1976.

[8] R. Zelinski and P. Noll, "Adaptive transform coding of speech signals," *IEEE Trans. Acoust., Speech, Signal Processing*, vol. ASSP-25, pp. 299–309, Aug. 1977.

[9] S. G. Wilson and S. Husain, "Adaptive tree encoding of speech at 8000-bits/s with a frequency weighted error criterion," *IEEE Trans. Commun.*, vol. COM-27, pp. 165–170, Jan. 1979.

[10] Y. Matsuyama and R. M. Gray, "Universal tree encoding of speech waveforms," *IEEE Trans. Inform. Theory*, vol. IT-27, pp. 31–40, Jan. 1981.
[11] T. Berger, *Rate Distortion Theory*. Englewood Cliffs, NJ: Prentice-Hall, 1971.
[12] H. Gish and J. N. Pierce, "Asymptotically efficient quantizing," *IEEE Trans. Inform. Theory*, vol. IT-14, pp. 676–683, 1968.
[13] A. Gersho, "Asymptotically optimum block quantization," *IEEE Trans. Inform. Theory*, vol. IT-25, pp. 373–380, July 1979.
[14] Y. Yamada, S. Tazaki, and R. M. Gray, "Asymptotic performance of block quantizers with difference distortion measures," *IEEE Trans. Inform. Theory*, vol. IT-26, pp. 6–14, Jan. 1980.
[15] H. P. Kramer and M. V. Mathews, "A linear coding for transmitting a set of correlated signals," *IRE Trans. Inform. Theory*, vol. IT-2, pp. 41–46, Sept. 1956.
[16] J. Y. Huang and P. M. Schultheiss, "Block quantization of correlated Gaussian random variables," *IEEE Trans. Commun. Syst.*, vol. CS-11, pp. 289–296, Sept. 1963.
[17] T. J. Goblick, Jr. and J. L. Halsinger, "Analog source digitization: A comparison of theory and practice," *IEEE Trans. Inform. Theory*, vol. IT-13, pp. 323–326, 1967.
[18] J. B. O'Neal, Jr., "Differential pulse-code modulation (PCM) with entropy coding," *IEEE Trans. Inform. Theory*, vol. IT-22, pp. 169–174, Mar. 1970.
[19] S. P. Lloyd, "Least squares quantization in PCM," Bell Lab. Tech. Paper, Murray Hill, NJ, 1957; also, *IEEE Trans. Inform. Theory*, Mar. 1982.
[20] J. Max, "Quantizing for minimum distortion," *IRE Trans. Inform. Theory*, vol. IT-6, pp. 7–12, 1960.
[21] Y. Linde, A. Buzo, and R. M. Gray, "An algorithm for vector quantizer design," *IEEE Trans. Commun.*, vol. COM-28, pp. 84–95, Jan. 1980.
[22] R. M. Gray, J. C. Kieffer, and Y. Linde, "Locally optimal block quantizer design," *Inform. Contr.*, vol. 45, pp. 178–198, May 1980.
[23] A. Buzo, A. H. Gray, Jr., R. M. Gray, and J. D. Markel, "Speech coding based upon vector quantization," *IEEE Trans. Acoust., Speech, Signal Processing*, vol. ASSP-28, pp. 562–574, Oct. 1980.
[24] R. M. Gray and Y. Linde, "Vector quantizers and predictive quantizers for Gauss Markov sources," *IEEE Trans. Commun.*, vol. COM-30, pp. 381–389, Feb. 1982.
[25] P. C. Mahalanobis, "On the generalized distance in statistics," in *Proc. Indian Nat. Inst. Sci. (Calcutta)*, vol. 2, 1936, pp. 49–55.
[26] S. Kullback, *Information Theory and Statistics*. New York: Dover, 1968.
[27] T. W. Anderson, *An Introduction to Multivariate Statistical Analysis*. New York: Wiley, 1958.
[28] G. J. A. Hartigan, *Clustering Algorithms*. New York: Wiley, 1975.
[29] E. Diday and J. C. Simon, "Clustering analysis," in *Digital Pattern Recognition*, K. S. Fu, Ed. New York: Springer-Verlag, 1976, ch. 3.
[30] S. Tazaki, H. Ogawa, T. Gotoda, and Y. Yamada, "A quantitative estimate of the possibility of confusion of Japanese phonemes," *IEEE Trans. Acoust., Speech, Signal Processing*, vol. ASSP-27, pp. 296–298, June 1979.
[31] D. Wolf, "Statistical models of speech signals," *Inform. Syst. Theory in Digital Commun.*, NTG-Facherichte, vol. 65, VDE-Verlag, Berlin, 1978.
[32] R. J. Fontana, R. M. Gray, and J. C. Kieffer, "Asymptotically mean stationary channels," *IEEE Trans. Inform. Theory*, vol. IT-27, pp. 308–316, May 1981.
[33] L. Ya. Geronimus, *Orthogonal Polynomials*. New York: Consultants Bureau, Noble Offset Printers, 1961.
[34] J. D. Markel and A. H. Gray, Jr., "On autocorrelation with application to speech analysis," *IEEE Trans. Audio Electroacoust.*, vol. AU-21, pp. 69–79, Jan. 1973.
[35] R. J. McEliece, *The Theory of Information and Coding*. Reading, MA: Addison-Wesley, 1977.
[36] I. S. Gradshteyn and I. M. Ryzhik, *Table of Integral Series and Products*. New York: Academic, 1965.
[37] R. M. Gray, A. H. Gray, Jr., G. Rebolledo, and J. E. Shore, "Rate-distortion speech coding with a minimum discrimination information distortion measure," *IEEE Trans. Inform. Theory*, vol. IT-27, pp. 708–721, Nov. 1981.
[38] L. C. Stewart, R. M. Gray, and Y. Linde, "The design of trellis waveform coders," *IEEE Trans. Commun.*, Apr. 1982.

SPEECH AND SPEAKER INDEPENDENT CODEBOOK DESIGN IN VQ CODING SCHEMES

Herbert Reininger and Dietrich Wolf

Institut für Angewandte Physik der Johann Wolfgang Goethe-Universität
Frankfurt am Main, FRG

ABSTRACT

In vector quantization schemes usually speech and speaker dependent codebooks are applied in order to achieve good speech quality at medium bit rates. This paper deals with another approach: The speech waveforms are transformed into signals which ideally do no longer contain speech and speaker specific features. Thus these signals can be encoded by an universal vector quantizer. This concept is realized by a system called RELP-VQ. The performance of this RELP-VQ scheme was evaluated by SNR-measurements as well as by informal listening tests including female and male English and German speakers.

INTRODUCTION

Vector quantization (VQ) has been widely accepted as an efficient coding technique for various speech coding schemes at medium and low bit rates [1]. These schemes usually apply special codebooks based on a few speakers only. Therefore the schemes show considerable degradation of speech quality for speakers and/or for languages not included in the codebook design.

To avoid this drawback our system transforms a speech waveform into a signal which ideally does no longer contain any speaker and speech specific characteristics. The transformation is performed in two consecutive steps. At first, an adaptive linear predictor is applied to decorrelate the speech waveform since speech and speaker specific features are represented predominantly by the autocorrelation function. Secondly, the prediction error signal is normalized by adaptive gain control in order to provide a constant variance of the error signal. Thus, one obtains a speech and speaker independent quasi stationary signal which efficiently can be encoded by a vector quantizer with a non-adapted universal codebook. Since the normalized prediction error signal may be interpreted as a realization of a stationary random process, the vector quantizer also may be designed by sequences of this random process.

VQ CODING SCHEME

Coding schemes in which a prediction error signal is encoded and transmitted to the receiver generally may be referred to as RELP systems. Our system - depicted in Fig. 1 - uses a vector quantizer to encode the error signal and therefore is called RELP-VQ.

The prediction error signal d(n) is obtained by processing the speech waveform x(n) in segments of N samples with an adaptive linear predictor. The segment lengths range between 16 ms and 20 ms. The predictor parameters representing the side information are also encoded by a vector quantizer as described in detail in [2]. The samples of d(n)

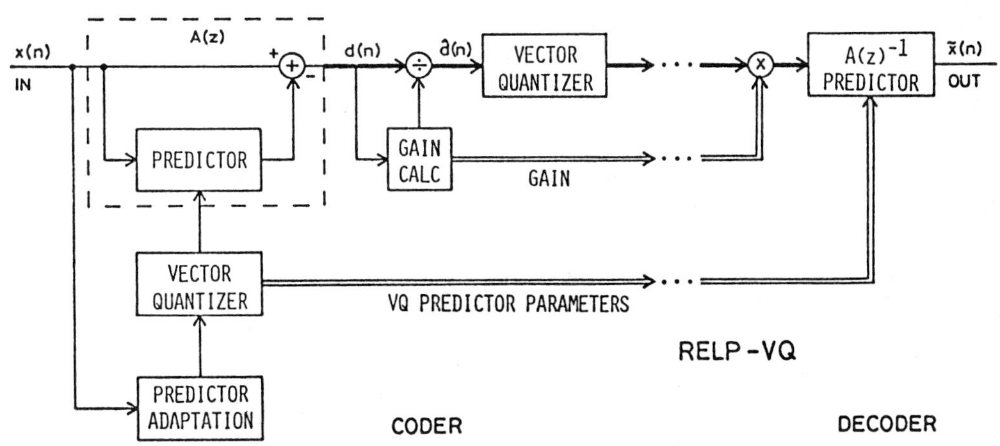

Fig. 1.

Block diagram of RELP-VQ

Reprinted from *Proc. IEEE Acoust., Speech, Signal Processing,* vol. 4, pp. 1700–1702, March 1985.

are divided by the RMS value of their corresponding segment to provide a nearly constant dynamic range of amplitudes. The normalized error $\hat{d}(n)$ is encoded by a vector quantizer. By the vector quantizer nonoverlapping blocks or vectors $\underline{x}$ of n successive samples of $\hat{d}(n)$ are encoded such that each vector $\underline{x}$ is mapped into that codebook vector $\underline{y}_i$ which has the minimum Euclidian distance. The vector $\underline{y}_i$ is element of a set of K vectors $\underline{y}_1, \ldots, \underline{y}_K$, the codebook Y, which represents $\hat{d}(n)$ with least MSE. The index i is transmitted to the receiver where it is decoded into the vector $\underline{y}_i$ using a copy of the codebook Y. In the receiver the quantized error signal is amplified according to the actual RMS value which is also quantized and transmitted to the receiver. Finally, the inverse prediction error filter $A(z)^{-1}$ is applied yielding a speech-like output signal $\tilde{x}(n)$.

EXPERIMENTS AND RESULTS

The RELP-VQ system with an 8th order adaptive linear predictor was simulated and analyzed on a HP-1000 minicomputer combined with a FPS AP-120B array processor. Speech signals of various female and male speakers of different languages of 5 minutes lengths each sampled at a rate of 8 kHz have been used in our investigations.

One- and two-dimensional amplitude densities of the speech signals, of the error signals, and of the normalized error signals have been measured. For all of the speakers the densities of the different signal types have been compared to each other. Typical results for the one-dimensional densities concerning four different female and male English and German speakers, respectively, are shown in Figs. 2a to 2c. In our measurements the gain factor has been adapted every 16 ms.

The results demonstrate that the densities of the original speech signals (Fig. 2a) differ from each other significantly - especially for large amplitude values. These differences are reduced by linear filtering (Fig. 2b) with the adaptive predictor. They are almost eliminated after normalization by adaptive gain control (Fig. 2c). The densities for the normalized error signals are then close to a Gaussian density. This observation is in accordance with the assumption of the well-known stochastic K_0-speech model which considers a speech signal to be a stationary Gaussian random process with a time-varying, also Gaussian distributed, variance [3]. This indicates that a vector quantizer designed for a statistically independent Gaussian random sequence which does not contain any speaker and language dependent features may be used also for optimum quantization of the normalized error signal.

For the normalized error signals of all of the speakers full and tree search codebooks were designed for vector quantizers with R = 1 bit per sample and R = 2 bits per sample and with dimensions n up to 8. In addition codebooks based on independent normally distributed random number sequences were designed.

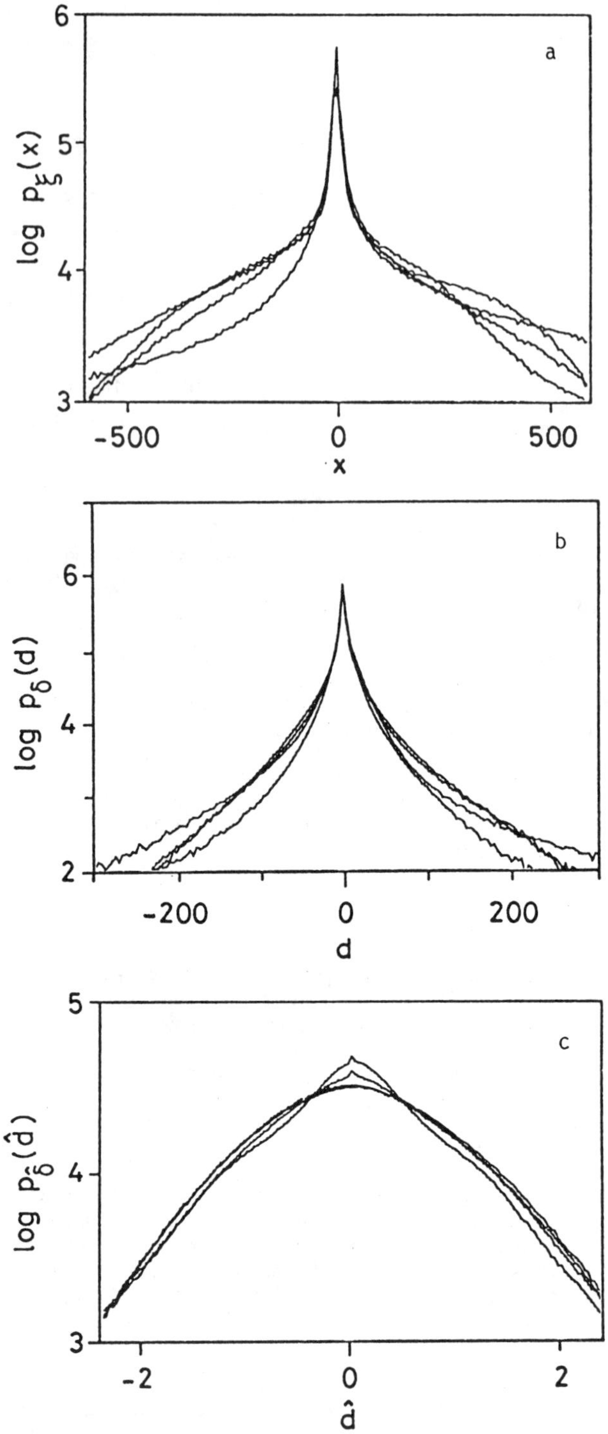

Fig. 2.
Amplitude densities $p_\xi(x)$ of speech signals (a), $p_\delta(d)$ of prediction error signals (b), and $p_{\hat{\delta}}(\hat{d})$ of normalized error signals (c) for various speakers.

The performance of the RELP-VQ scheme has been evaluated by signal-to-quantization noise ratio (SNR) measurements for the various configurations of n and R. Each speaker adapted codebook as well as a "Gaussian codebook" (G) has been used for encoding the waveforms of all of the other speakers. Table 1 summarizes the SNR values of the normalized error signals with n = 4, R = 2 bits per sample, and N = 128 samples (16 ms) obtained for the four speakers.

Table 1.

SNR values in dB of the normalized error signal in a RELP-VQ system with n = 4 and R = 2 bits per sample for a female English (fe), male English (me), female German (fg), male German (mg) speaker, and Gaussian noise (G). N = 128.

		Training				
		fe	me	fg	mg	G
Test	fe	8.78	8.51	8.39	8.09	8.23
	me	8.46	8.77	8.29	8.06	8.25
	fg	8.36	8.36	8.78	8.27	7.88
	mg	7.70	7.61	8.22	8.67	6.74

The results demonstrate that the performance of the system decreases only slightly if the test speaker is not identical with the training speaker employed in the codebook design. This statement also holds if the codebook design is based on the stochastic process. In the mean the differences between the SNR values of the normalized error signals remain far below 1 dB. It should be noted that these differences depend on the choice of N.

While the different RELP-VQ configurations can be compared objectively by means of the SNR measurements a final decision about their performances has to include subjective listening tests. Informal listening tests have shown a very good speech quality for all of the speakers although the SNR values of the complete RELP-VQ system ranging about 12 dB are rather low. Even with the "Gaussian codebook" a good speech quality is achieved.

CONCLUSION

The investigations of a RELP-VQ system indicate that speech and speaker independent VQ coding schemes can be realized with high speech quality at medium bit rates. Linear prediction analysis together with adaptive normalization of the error signal leads to nearly speech and speaker independent signals which are efficiently encoded by an universal vector quantizer. This vector quantizer can even be designed on the basis of a stochastic model of the normalized error signal.

REFERENCES

[1] R.M. Gray,
"Vector Quantization", IEEE ASSP Magazine 1, pp. 4-29; 1984.

V. Cuperman and A. Gersho,
"Adaptive Differential Vector Coding of Speech", Conference Record, Globe-Com 82, pp. 1092-1096; 1982.

H. Abut, R.M. Gray, and G. Rebolledo,
"Vector Quantization of Speech and Speech-Like Waveforms", IEEE Trans. ASSP 30, pp. 423-435; 1982.

[2] A. Buzo, A.H. Gray jr., R.M. Gray, and J.D. Markel,
"Speech Coding Based Upon Vector Quantization", IEEE Trans. ASSP 28, pp. 562-574; 1980.

[3] D. Wolf,
"Statistical Models of Speech Signals", Information and System Theory in Digital Communications, NTG-Fachberichte 65, Berlin 1978, pp. 1-9.

We thank the Deutsche Forschungsgemeinschaft for financial support.

Product Code Vector Quantizers for Waveform and Voice Coding

MICHAEL J. SABIN, STUDENT MEMBER, IEEE, AND ROBERT M. GRAY, FELLOW, IEEE

Abstract—**Memory and computation requirements imply fundamental limitations on the quality that can be achieved in vector quantization systems used for speech waveform coding and linear predictive voice coding (LPC). One approach to reducing storage and computation requirements is to organize the set of reproduction vectors as the Cartesian product of a vector codebook describing the shape of each reproduction vector and a scalar codebook describing the gain or energy. Such shape-gain vector quantizers can be applied both to waveform coding using a quadratic-error distortion measure and to voice coding using an Itakura-Saito distortion measure. In each case, the minimum distortion reproduction vector can be found by first selecting a shape codeword, and then, based on that choice, selecting a gain codeword.**

Several algorithms are presented for the design of shape-gain vector quantizers based on a training sequence of data or a probabilistic model. The algorithms are used to design shape-gain vector quantizers for both the waveform coding and voice coding application. The quantizers are simulated, and their performance is compared to that of previously reported vector quantization systems.

I. Introduction

Speech coding systems are generally classified as either waveform coders or voice coders. Waveform coders attempt to accurately reproduce the speech waveform itself, with typical data rates ranging from 8000 to 80 000 bits/s. Voice coders (vocoders) attempt to track the underlying process that generates the waveform and to encode parameters that describe that process. Of interest here are voice coders based on linear predictive coding (LPC) [2], [10], [13], [14], which have typical data rates ranging from 800 to 8000 bits/s.

Vector quantizers have been used in both waveform coding systems and voice coding systems. A vector quantizer maps input vectors drawn from the M-dimensional Euclidean space $\mathcal{R}^M$ into a finite set (codebook) of reproduction vectors (codewords) contained in the space $\mathcal{R}^K$. A vector quantizer (VQ) is described by M, the input vector dimension; K, the reproduction vector dimension; N, the number of reproduction vectors; $C = \{\hat{x}_i, i = 1, 2, \cdots, N\}$, the set of reproduction vectors; and $\hat{q}(x)$, the mapping of the input space into the set of reproduction vectors. In this paper, we restrict attention to the case of $M = K$, that is, the reproduction vector is of the same dimension as the input vector. We will refer to K as the block length.

Manuscript received February 2, 1983; revised September 6, 1983 and November 4, 1983. This research was supported by the Air Force Office of Scientific Research under Contract F49620-79-C-0058, by the Army Research Office under Contract DAAG29-80-K-0073, and by the Joint Services and Electronics Program at Stanford University.

The authors are with the Information Systems Laboratory, Department of Electrical Engineering, Stanford University, Stanford, CA, 94305.

A VQ can be decomposed into two blocks: an encoder, which views the input vector x and generates the index of the reproduction vector specified by $\hat{q}(x)$; and a decoder, which uses this index to generate the reproduction vector $\hat{x}_i$. A VQ can be used to communicate over a digital channel by placing the encoder at the transmitter and the decoder at the receiver and sending the index of the codeword across the channel. Operation of the decoder is simple: it stores the codebook and looks up the reproduction vector indexed by the encoder. Operation of the encoder is more complicated. In effect, it partitions the input space into a collection of bins according to $\hat{q}(x)$, one bin for each reproduction vector in the codebook, and determines in which bin an input vector is contained.

If we define a distortion measure $d(x; \hat{x})$ which represents the penalty or cost associated with reproducing a vector x by $\hat{x}$, then the best mapping $\hat{q}(x)$ is the one which selects as the reproduction vector for x the codeword $\hat{x}_i$ that minimizes $d(x; \hat{x}_i)$. With such a minimum distortion or nearest neighbor mapping, the encoder operates by computing $d(x; \hat{x}_i)$ for $i = 1, 2, \cdots, N$, and then selecting the value of i for which $d(x; \hat{x}_i)$ is minimized. This implies that the bin associated with a particular codeword $\hat{x}_i$ is the set of input vectors for which $\hat{x}_i$ is the minimum distortion codeword. This type of encoder is called a full-search VQ.

Vector quantizers used as waveform coders are described in [1] and [4]. The input vector to the VQ is taken to be K consecutive samples of the waveform. Each codeword represents a reproduction waveform. As in [1] and [4], we consider distortion measures of the form

$$d(x; \hat{x}) = (x - \hat{x})^T W (x - \hat{x}) \tag{1}$$

where W is some positive-definite matrix. Expanding this product gives

$$d(x; \hat{x}_i) = x^T W x - 2x^T W \hat{x}_i + \hat{x}_i^T W \hat{x}_i. \tag{2}$$

Since the first term does not depend on $\hat{x}_i$, it can be ignored when finding the nearest neighbor codeword. The encoder can then be designed to store $2W\hat{x}_i$ (a K vector) and $\hat{x}_i^T W \hat{x}_i$ (a scalar) and to compute $-2x^T W \hat{x}_i + \hat{x}_i^T W \hat{x}_i$ for each $i = 1, 2, \cdots, N$, choosing the codeword that minimizes this quantity.

Vector quantization applied to LPC voice coders is described in [3], [6], [16], [19], and [20]. In these schemes, a vector quantizer is used to encode and decode the all-pole model generated by an LPC analysis of a speech frame $S(z)$. (Coding of the excitation parameters is not considered here.) The model can be expressed as $\sigma/(1 + a_1 z^{-1} + a_2 z^{-2} \cdots + a_p z^{-p})$, de-

Reprinted from *IEEE Trans. Acoust., Speech, Signal Processing,* vol. ASSP-32, pp. 474–488, June 1984.

noted by $\sigma/A(z)$. The order p is typically 8-12. Once the model parameters $\{\sigma, a_1, a_2, \cdots, a_p\}$ have been obtained, they are coded by means of vector quantization. The input vector $\boldsymbol{x}$ to the VQ is the vector $[\sigma \quad a_1 \quad a_2 \cdots a_p]^T$ of model parameters. Each codeword is a vector $[\hat{\sigma}_i \quad \hat{a}_{i,1} \quad \hat{a}_{i,2} \cdots \hat{a}_{i,p}]^T$ that represents a reproduction model $\hat{\sigma}_i/(1 + \hat{a}_{i,1}z^{-1} + \hat{a}_{i,2}z^{-2} \cdots + \hat{a}_{i,p}z^{-p})$, denoted as $\hat{\sigma}_i/\hat{A}_i(z)$. The block length K is $p + 1$.

In [3], [6], and [16] and in this paper, the distortion measure chosen for LPC vocoding is the modified Itakura-Saito distortion. This measure is studied in detail in [5], [6], [10], and [15]. For our purposes, it will suffice to regard it as a measure of the dissimilarity between the power spectrum $|\sigma/A(e^{j\theta})|^2$ of the input model and the power spectrum $|\hat{\sigma}_i/\hat{A}_i(e^{j\theta})|^2$ of the reproduction model. For this case of autoregressive models, the distortion can be expressed as

$$d(\boldsymbol{x}; \hat{\boldsymbol{x}}_i) = \frac{\hat{\boldsymbol{a}}_i^T \boldsymbol{R}(\boldsymbol{x})\, \hat{\boldsymbol{a}}_i}{\hat{\sigma}_i^2} + \ln \hat{\sigma}_i^2 - \ln \sigma^2 - 1 \tag{3}$$

where $\hat{\boldsymbol{a}}_i$ is the vector $[1 \quad \hat{a}_{i,1} \quad \hat{a}_{i,2} \cdots \hat{a}_{i,p}]^T$ and $\boldsymbol{R}(\boldsymbol{x})$ is a $p+1$ by $p+1$ symmetric Toeplitz matrix $\{r_x(k-j),\ k = 0, 1, \cdots, p; j = 0, 1, \cdots, p\}$, with $r_x(k)$ given by

$$\int_{-\pi}^{\pi} |\sigma/A(e^{j\theta})|^2\, e^{jk\theta}\, (d\theta/2\pi).$$

Since the last two terms in (3) do not depend on $\hat{\boldsymbol{x}}_i$, they can be ignored when finding the nearest neighbor of an input vector. Thus, the encoding can be performed by computing $\hat{\boldsymbol{a}}_i^T \boldsymbol{R}(\boldsymbol{x})\ \hat{\boldsymbol{a}}_i/\hat{\sigma}_i^2 + \ln \hat{\sigma}_i^2$ for each $i = 1, 2, \cdots, N$ and choosing the codeword that minimizes this quantity. In [3] it is shown that this quantity can be efficiently computed as

$$\left[r_x(0)\, r_{\hat{a}_i}(0) + 2 \sum_{m=1}^{p} r_x(m)\, r_{\hat{a}_i}(m)\right] \hat{\sigma}_i^2 + \ln \hat{\sigma}_i^2$$

where $r_{\hat{a}_i}(k) = \sum_{m=0}^{p-k} \hat{a}_{i,m} \hat{a}_{i,m+k}$. For each codeword, the encoder stores $r_{\hat{a}_i}(0)/\hat{\sigma}_i^2, 2r_{\hat{a}_i}(1)/\hat{\sigma}_i^2, 2r_{\hat{a}_i}(2)/\hat{\sigma}_i^2, \cdots, 2r_{\hat{a}_i}(p)/\hat{\sigma}_i^2$, and $\ln \hat{\sigma}_i^2$, each a scalar quantity.

In a VQ, the rate r in bits per vector component is given by $r = K^{-1} \log_2 N$. Thus, for a fixed block length and an increasing rate or for a fixed rate and an increasing block length, the codebook size grows exponentially. This growth requires a corresponding increase in storage at both the encoder and decoder. In addition, the full-search encoder experiences a corresponding increase in computational effort. In a practical implementation, the storage and computational requirements limit the rate and blocklength that can be used.

Tree-search vector quantizers [3], [4], [9] reduce encoding complexity by searching a sequence of small codebooks to find the reproduction vector. However, their storage requirements are greater than the full-search VQ's, and the codewords they select are not, in general, optimum in the sense of minimizing $d(\boldsymbol{x}; \hat{\boldsymbol{x}})$. If we wish to use a full-search VQ with a large rate and block lengths, it is necessary to impose a structure on the codebook.

One such structure, called "gain separation," is described in [3] for the case of LPC vocoders. In this scheme, a reproduction vector $\hat{\boldsymbol{x}}_{ij}$ is formed as $[\hat{\sigma}_j \quad \hat{a}_{i,1} \quad \hat{a}_{i,2} \cdots \hat{a}_{i,p}]^T$ where $\hat{\sigma}_j$ is drawn from a set of scalars $\{\hat{\sigma}_j, j = 1, 2, \cdots, N_2\}$ called the gain codebook and $\{\hat{a}_{i,1}, \hat{a}_{i,2}, \cdots, \hat{a}_{i,p}\}$ represents one of a set of models $\{1/\hat{A}_i(z), i = 1, 2, \cdots, N_1\}$ called the model codebook. (As before, $\hat{A}_i(z) = 1 + \hat{a}_{i,1}z^{-1} + \hat{a}_{i,2}z^{-2} \cdots + \hat{a}_{i,p}z^{-p}$). Thus, the vector $\hat{\boldsymbol{x}}_{ij}$ corresponds to the reproduction model $\hat{\sigma}_j/\hat{A}_i(z)$. Since each pairwise combination of a gain and a model yields a reproduction vector, the total number of reproduction vectors is $(N_1)(N_2)$.

Referring to (3), the distortion between an input vector $\boldsymbol{x}$ and a reproduction vector $\hat{\boldsymbol{x}}_{ij}$ is

$$d(\boldsymbol{x}; \hat{\boldsymbol{x}}_{ij}) = \frac{\hat{\boldsymbol{a}}_i^T \boldsymbol{R}(\boldsymbol{x})\, \hat{\boldsymbol{a}}_i}{\hat{\sigma}_j^2} + \ln \hat{\sigma}_j^2 - \ln \sigma^2 - 1. \tag{4}$$

Finding the nearest neighbor to an input vector consists of finding the pair (i, j) that minimizes $\boldsymbol{a}_i^T \boldsymbol{R}(\boldsymbol{x}) \boldsymbol{a}i/\hat{\sigma}_j^2 + \ln \hat{\sigma}_j^2$. This can be done in two steps.

1) Select $\hat{\boldsymbol{a}}_i$ to minimize $\hat{\boldsymbol{a}}_i^T \boldsymbol{R}(\boldsymbol{x})\, \hat{\boldsymbol{a}}_i$.

2) Given $\alpha = \min_i \hat{\boldsymbol{a}}_i^T \boldsymbol{R}(\boldsymbol{x})\ \hat{\boldsymbol{a}}_i$, select $\hat{\sigma}_j$ to minimize $\alpha/\hat{\sigma}_j^2 + \ln \hat{\sigma}_j^2$.

Notice that since each $\hat{A}_i(z)$ is a monic polynomial, it is an "inverse filter" in the usual LPC vernacular. Then step 1) can be viewed as finding which inverse filter among the set of inverse filters represented by the model codebook produces the residual with the smallest energy when applied to the input speech frame. Step 2) scalar quantizes the value α of this minimum residual energy.

Computation of the residual energy $\hat{\boldsymbol{a}}_i^T \boldsymbol{R}(\boldsymbol{x})\, \hat{\boldsymbol{a}}_i$ can be performed in a manner identical to that described above for computing $\hat{\boldsymbol{x}}_i^T \boldsymbol{R}(\boldsymbol{x})\, \hat{\boldsymbol{x}}_i$. Each inverse filter is stored as $r_{\hat{a}_i}(0)$, $2r_{\hat{a}_i}(1)$, $2r_{\hat{a}_i}(2)$, $\cdots$, $2r_{\hat{a}_i}(p)$, and the residual energy is computed as $r_x(0)\, r_{\hat{a}_i}(0) + 2 \sum_{m=1}^{p} r_x(m)\, r_{\hat{a}_i}(m)$. Selecting $\hat{\sigma}_j$ to minimize $\alpha/\hat{\sigma}_j^2 + \ln \hat{\sigma}_j^2$ is performed by comparing α to a set of ordered thresholds representing the decision boundaries between consecutive gain values. With the gains indexed in ascending order, the threshold between $\hat{\sigma}_j$ and $\hat{\sigma}_{j+1}$ is given by

$$\left(\ln \frac{\hat{\sigma}_{j+1}^2}{\hat{\sigma}_j^2}\right)\left(\frac{1}{\hat{\sigma}_j^2} - \frac{1}{\hat{\sigma}_{j+1}^2}\right)^{-1}. \tag{5}$$

Fig. 1 illustrates the structure of the gain-separated VQ for LPC voice coding. It is shown in [3] that the structure results in a dramatic reduction in both storage and encoding complexity over a standard VQ with the same set of reproduction vectors.

It is natural to ask if a similar codebook structure could be used to reduce the storage and search effort in waveform coding VQ's. This question is addressed in Section II. First we introduce the "shape-gain" quantizer, a generalization of the gain-separated VQ of [3], and show its application to waveform coding. It is seen that by organizing the codebook of a waveform coding VQ as a pair of codebooks, one codebook of vectors (shapes) and one of scalars (gains), then the same sort of storage and computation advantages are obtained. The gain-separated VQ of [3] is seen to be an application of the shape-gain quantizer to LPC voice coding. The two cases (waveform and voice coding) are shown to have remarkably similar encoding structures despite the different natures of the quadratic-error and Itakura-Saito distortion measures. In

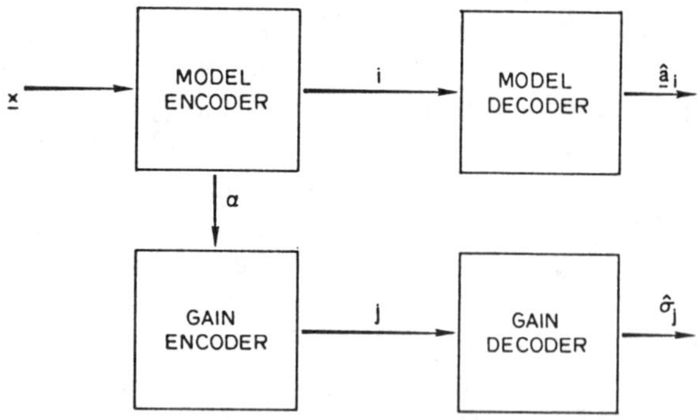

Fig. 1. Gain-separated VQ for LPC voice coding.

Section III, we discuss methods for designing shape-gain quantizers. The method of [3] is extended to the waveform coding case. Two new algorithms are presented: one jointly designs the two codebooks, while the other individually designs each codebook. Both algorithms are applicable to the waveform coding and LPC vocoding cases, and both overcome computational difficulties that occur in the method of [3]. In Section IV, we describe experimental simulations of waveform coding and LPC vocoding shape-gain quantizers designed by the algorithms.

II. Shape-Gain Quantizers

The gain-separated VQ of [3] for LPC vocoding can be abstracted as follows. Consider a VQ whose codebook is formed as the Cartesian product of a finite set of vectors and a finite set of scalars. That is, $C = \{(\hat{\boldsymbol{a}}_i, \hat{\sigma}_j), i = 1, 2, \cdots, N_1; j = 1, 2, \cdots, N_2\}$ where $\hat{A} = \{\hat{\boldsymbol{a}}_i, i = 1, 2, \cdots, N_1\}$ is a set of vectors drawn from $\mathcal{R}^K$ and $\hat{\Sigma} = \{\hat{\sigma}_j, j = 1, 2, \cdots, N_2\}$ is a set of scalars drawn from the nonnegative real numbers. (The use of the symbols $\hat{\boldsymbol{a}}_i$ and $\hat{\sigma}_j$ is in keeping with the notation of [3] for LPC voice coding. However, we are not yet attaching physical interpretation to the quantities.) We will call $\hat{A}$ the shape codebook and $\hat{\Sigma}$ the gain codebook, and will denote the formation of the product codebook by $C = \hat{A} \times \hat{\Sigma}$. Such a shape-gain vector quantizer (SGVQ) is described by M, the input vector dimension; K, the dimension of the shape codewords; N_1, the number of shape codewords; N_2, the number of gain codewords; $\hat{A}$, the shape codebook, $\hat{\Sigma}$, the gain codebook; and $\hat{\boldsymbol{q}}(\boldsymbol{x})$, the mapping of the input space into the product codebook $\hat{A} \times \hat{\Sigma}$. We restrict attention to the case of $M = K$ and refer to K as the block length.

Like the VQ, the SGVQ can be decomposed into an encoder-decoder structure. The encoder views the input vector $\boldsymbol{x}$ and generates the index pair (i, j) of the product codeword specified by $\hat{\boldsymbol{q}}(\boldsymbol{x})$. The decoder uses the index pair to look up the product codeword $(\hat{\boldsymbol{a}}_i, \hat{\sigma}_j)$. When used to communicate over a digital channel, the rate of an SGVQ is $\log_2 N_1 + \log_2 N_2$ bits/vector. Operation of the decoder is simple: it stores the shape codebook and the gain codebook and looks up the indexed shape codeword and gain codeword. Encoder operation is more complicated. It must partition the input space into bins according to $\hat{\boldsymbol{q}}(\boldsymbol{x})$, one bin for each product codeword in $\hat{A} \times \hat{\Sigma}$, and determine in which bin an input vector is contained.

If we define a distortion measure $d(\boldsymbol{x}; \hat{\boldsymbol{a}}, \hat{\sigma})$ which measures the penalty or cost associated with reproducing $\boldsymbol{x}$ by the pair $(\hat{\boldsymbol{a}}, \hat{\sigma})$, then the best mapping $\hat{\boldsymbol{q}}(\boldsymbol{x})$ is one which selects the pair $(\hat{\boldsymbol{a}}_i, \hat{\sigma}_j)$ that minimizes $d(\boldsymbol{x}; \hat{\boldsymbol{a}}_i, \hat{\sigma}_j)$. To find minimum distortion product codeword, the encoder can compute $d(\boldsymbol{x}; \hat{\boldsymbol{a}}_i, \hat{\sigma}_j)$ for $i = 1, 2, \cdots, N_1$ and $j = 1, 2, \cdots, N_2$, selecting the pair (i, j) that minimizes $d(\boldsymbol{x}; \hat{\boldsymbol{a}}_i, \hat{\sigma}_j)$.

To use a shape-gain vector quantizer for waveform coding, we associate a reproduction vector $\hat{\boldsymbol{x}}_{ij}$ with each product codeword $(\hat{\boldsymbol{a}}_i, \hat{\sigma}_j)$, with $\hat{\boldsymbol{x}}_{ij} = \hat{\sigma}_j \hat{\boldsymbol{a}}_i$. Since the gain codeword is nonnegative, the reproduction vector is in the same direction as the shape vector. We define the distortion between an input vector $\boldsymbol{x}$ and $(\hat{\boldsymbol{a}}_i, \hat{\sigma}_j)$ as the distortion between $\boldsymbol{x}$ and $\hat{\boldsymbol{x}}_{ij}$. Referring to (2), we have

$$d(\boldsymbol{x}; \hat{\boldsymbol{a}}_i, \hat{\sigma}_j) = \boldsymbol{x}^T \boldsymbol{W} \boldsymbol{x} - 2\hat{\sigma}_j \boldsymbol{x}^T \boldsymbol{W} \hat{\boldsymbol{a}}_i + \hat{\sigma}_j^2 \hat{\boldsymbol{a}}_i^T \boldsymbol{W} \hat{\boldsymbol{a}}_i. \tag{6}$$

To encode an input vector $\boldsymbol{x}$, the encoder finds the pair $(\hat{\boldsymbol{a}}_i, \hat{\sigma}_j)$ that minimizes (6). If each shape vector is normalized in the sense that $\hat{\boldsymbol{a}}_i^T \boldsymbol{W} \hat{\boldsymbol{a}}_i = 1$ for $i = 1, 2, \cdots, N_1$, then the pair $(\hat{\boldsymbol{a}}_i, \hat{\sigma}_j)$ that minimizes (6) can be found in two steps.

1) Select $\hat{\boldsymbol{a}}_i$ to maximize $\boldsymbol{x}^T \boldsymbol{W} \hat{\boldsymbol{a}}_i$.
2) Given $\alpha = \max_i \boldsymbol{x}^T \boldsymbol{W} \hat{\boldsymbol{a}}_i$, select $\hat{\sigma}_j$ to minimize $\hat{\sigma}_j^2 - 2\hat{\sigma}_j \alpha$.

The first step chooses the shape vector that has the highest correlation to the input vector. The second step scalar quantizes the value of this correlation in the least squares sense, since minimizing $\hat{\sigma}_j^2 - 2\hat{\sigma}_j \alpha$ is equivalent to minimizing $(\hat{\sigma}_j - \alpha)^2$. By indexing the gain codewords in ascending (or descending) order, the second step can be accomplished by comparing α to a set of ordered thresholds which represent the decision boundaries, i.e., the midpoints, between adjacent gain codewords.

Fig. 2 illustrates the structure of the waveform coding SGVQ. The shape encoder stores the vectors $\{\boldsymbol{W}\hat{\boldsymbol{a}}_i, i = 1, 2, \cdots, N_1\}$ and finds the index i of the most highly correlated vector. The correlation value α is passed to the gain encoder, which stores the decision thresholds between the gain values and outputs the index j of the nearest gain value. The shape decoder and gain decoder, respectively, look up the indexed shape and gain codeword, and the two outputs are multiplied to produce the reproduction vector.

To use the SGVQ for LPC vocoding, it is convenient to use a different notation from that used above. We replace the representation of the input model $\sigma/A(z)$ with the representation $1/X(z)$. Thus, $X(z)$ is $A(z)/\sigma$ and is not, in general, a monic polynomial. We define the input vector $\boldsymbol{x} = [x_0 \;\; x_1 \cdots x_p]^T$, with $X(z) = x_0 + x_1 z^{-1} + x_2 z^{-2} \cdots + x_p z^{-p}$. Similarly, we define $\hat{\boldsymbol{x}}_i = [\hat{x}_{i,0} \;\; \hat{x}_{i,1} \cdots \hat{x}_{i,p}]^T$, corresponding to the reproduction model $1/\hat{X}_i(z)$. We can write the modified Itakura-Saito distortion between the models represented by $\boldsymbol{x}$ and $\hat{\boldsymbol{x}}_i$ as

$$d(x; \hat{x}_i) = \hat{x}_i^T \boldsymbol{R}(x)\, \hat{x}_i - \ln \hat{x}_{i,0}^2 + \ln x_0^2 - 1 \tag{7}$$

with $\boldsymbol{R}(x)$ as before, using

$$r_x(k) = \int_{-\pi}^{\pi} |X(e^{j\theta})|^{-2}\, e^{jk\theta}\, (d\theta/2\pi).$$

We associate a reproduction vector $\hat{\boldsymbol{x}}_{ij}$ with each product codeword $(\hat{\boldsymbol{a}}_i, \hat{\sigma}_j)$ according to $\hat{\boldsymbol{x}}_{ij} = \hat{\boldsymbol{a}}_i / \hat{\sigma}_j$. This corresponds to a

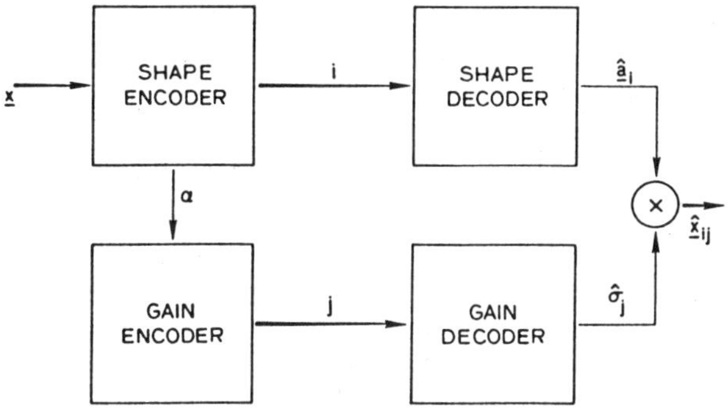

Fig. 2. SGVQ for waveform coding.

reproduction model of $\hat{\sigma}_j/\hat{A}_i(z)$ where $\hat{A}_i(z)$ is the polynomial $\hat{a}_{i,0} + \hat{a}_{i,1}z^{-1} + \hat{a}_{i,2}z^{-2} \cdots + \hat{a}_{i,p}z^{-p}$ corresponding to the shape vector $\hat{\boldsymbol{a}}_i$. We define the distortion measure between an input vector $\boldsymbol{x}$ and $(\hat{\boldsymbol{a}}_i, \hat{\sigma}_j)$ as the distortion between $\boldsymbol{x}$ and $\hat{\boldsymbol{x}}_{ij}$. Referring to (7), we have

$$d(\boldsymbol{x}; \hat{\boldsymbol{a}}_i, \hat{\sigma}_j) = \frac{\hat{\boldsymbol{a}}_i^T \boldsymbol{R}(\boldsymbol{x})\, \hat{\boldsymbol{a}}_i}{\hat{\sigma}_j^2} - \ln \frac{\hat{a}_{i,0}^2}{\hat{\sigma}_j^2} + \ln x_0^2 - 1. \tag{8}$$

If we constrain each shape vector so that $\hat{a}_{i,0} = 1$, then (8) reduces to (4) with $\sigma^2 = 1/x_0^2$. Then the two-step encoding method of [3] for finding the minimum distortion pair $(\hat{\boldsymbol{a}}_i, \hat{\sigma}_j)$ can be applied. The resulting structure is identical to that illustrated in Fig. 1.

It is interesting to note the similarities between the encoding structures of the waveform coding SGVQ and the LPC vocoding SGVQ. Figs. 1 and 2 illustrate the structures and Table I compares features. Observe that the vectors in each shape codebook have been normalized: for the waveform coder, $\hat{\boldsymbol{a}}_i$ has unit norm, i.e., $\hat{\boldsymbol{a}}_i^T \boldsymbol{W} \hat{\boldsymbol{a}}_i = 1$; for the LPC voice coder, $\hat{\boldsymbol{a}}_i$ has 1 as its first component, corresponding to monic $\hat{A}_i(z)$. This normalization is the key to the shape-gain encoding procedure, since it makes the choice of the optimum shape vector independent of the choice of gain. In both coders, each shape vector is stored as K real quantities. During encoding, the computation performed for each shape vector is a dot product between the input vector and the stored K tuple. For waveform coding, the maximum value of this dot product is sought; for LPC vocoding, the minimum value is sought. This maximum (minimum) value is compared to a table of thresholds representing the decision boundaries between consecutive gain codewords. Thus, in each case, the gain encoder stores the thresholds, not the gain values themselves.

Because of these similarities in the encoders of waveform coding and LPC voice coding SGVQ's, the amounts of storage and computation required by each are identical for equal block lengths and codebook sizes. Thus, the same storage and computation benefits described in [3] for the LPC vocoding case are obtained for the waveform coding case. Formulas for storage and computation are summarized in Table II. For reference, the corresponding formulas for standard VQ's are included. As with the SGVQ's, both types of VQ's have the same storage and computation requirements for equal block lengths and codebook sizes.

There is one other similarity worth noting. Referring to (6) with $\hat{\boldsymbol{a}}_i^T \boldsymbol{W} \hat{\boldsymbol{a}}_i = 1$, the quadratic-error distortion can be written as

$$[\boldsymbol{x}^T \boldsymbol{W} \boldsymbol{x} - ([\boldsymbol{x}^T \boldsymbol{W} \hat{\boldsymbol{a}}_i]^+)^2] + [\hat{\sigma}_j^2 - 2\hat{\sigma}_j \boldsymbol{x}^T \boldsymbol{W} \hat{\boldsymbol{a}}_i + ([\boldsymbol{x}^T \boldsymbol{W} \hat{\boldsymbol{a}}_i]^+)^2] \tag{9}$$

where $[\cdot]^+$ represents the larger of $(\cdot)$ and 0. The left term does not depend on $\hat{\sigma}_j$; the right term is nonnegative (for $\hat{\sigma}_j \geqslant 0$) and equals zero if and only if $\hat{\sigma}_j = [\boldsymbol{x}^T \boldsymbol{W} \hat{\boldsymbol{a}}_i]^+$. Thus, the left term is the "gain-optimized" distortion [5], [15] and is the value of the distortion when an input vector $\boldsymbol{x}$ is encoded into shape vector $\hat{\boldsymbol{a}}_i$ with gain value $[\boldsymbol{x}^T \boldsymbol{W} \hat{\boldsymbol{a}}_i]^+$. The right term indicates the contribution to the distortion when $\boldsymbol{x}^T \boldsymbol{W} \hat{\boldsymbol{a}}_i$ is quantized by $\hat{\sigma}_j$. Now, using (8) with $\hat{a}_{i,0} = 1$, we can write the Itakura-Saito distortion as

$$\left[\ln \frac{\hat{\boldsymbol{a}}_i^T \boldsymbol{R}(\boldsymbol{x})\, \hat{\boldsymbol{a}}_i}{1/x_0^2}\right] + \left[\frac{\hat{\boldsymbol{a}}_i^T \boldsymbol{R}(\boldsymbol{x})\, \hat{\boldsymbol{a}}_i}{\hat{\sigma}_j^2} - \ln \frac{\hat{\boldsymbol{a}}_i^T \boldsymbol{R}(\boldsymbol{x})\, \hat{\boldsymbol{a}}_i}{\hat{\sigma}_j^2} - 1\right]. \tag{10}$$

The left term does not depend on $\hat{\sigma}_j$; the right term is nonnegative and equals zero if and only if $\hat{\sigma}_j^2 = \hat{\boldsymbol{a}}_i^T \boldsymbol{R}(\boldsymbol{x})\, \hat{\boldsymbol{a}}_i$. Thus, the left term is the gain-optimized distortion (i.e., Itakura distortion) and is the value of the distortion when $\boldsymbol{x}$ is encoded with shape $\hat{\boldsymbol{a}}_i$ and gain $\sqrt{\hat{\boldsymbol{a}}_i^T \boldsymbol{R}(\boldsymbol{x})\, \hat{\boldsymbol{a}}_i}$. The right term is the contribution to the distortion when $\sqrt{\hat{\boldsymbol{a}}_i^T \boldsymbol{R}(\boldsymbol{x})\, \hat{\boldsymbol{a}}_i}$ is quantized by $\hat{\sigma}_j$. Thus, both distortion measures have the form

$$d(\boldsymbol{x}; \hat{\boldsymbol{a}}_i, \hat{\sigma}_j) = d'(\boldsymbol{x}; \hat{\boldsymbol{a}}_i) + d''(\hat{\sigma}_j, \sigma^*(\boldsymbol{x}, \hat{\boldsymbol{a}}_i)) \tag{11}$$

where d' is the gain-optimized distortion, $\sigma^*(\boldsymbol{x}, \hat{\boldsymbol{a}}_i)$ is either $\boldsymbol{x}^T \boldsymbol{W} \hat{\boldsymbol{a}}_i$ or $\sqrt{\hat{\boldsymbol{a}}_i^T \boldsymbol{R}(\boldsymbol{x})\, \hat{\boldsymbol{a}}_i}$, and d'' is the contribution due to quantizing $\sigma^*(\boldsymbol{x}, \hat{\boldsymbol{a}}_i)$ by $\hat{\sigma}_j$. We have the relation

$$d'(\boldsymbol{x}; \hat{\boldsymbol{a}}_i) = \inf_{\sigma \geqslant 0} d(\boldsymbol{x}; \hat{\boldsymbol{a}}_i, \sigma) = d(\boldsymbol{x}; \hat{\boldsymbol{a}}_i, [\sigma^*(\boldsymbol{x}, \hat{\boldsymbol{a}}_i)]^+). \tag{12}$$

Notice that in the SGVQ, the shape encoder minimizes d',

TABLE I
COMPARISON OF WAVEFORM AND LPC VOICE CODING SGVQ'S

	Waveform Coding	LPC Voice Coding
Normalization	$\hat{\mathbf{a}}_i^T \mathbf{W} \hat{\mathbf{a}}_i = 1$	$\hat{a}_{i,0} = 1$
Shape Vector Storage	$\mathbf{W}\hat{\mathbf{a}}_i$	$r_{\hat{\mathbf{a}}_i}(0), 2r_{\hat{\mathbf{a}}_i}(1), 2r_{\hat{\mathbf{a}}_i}(2), \cdots 2r_{\hat{\mathbf{a}}_i}(p)$
Shape Vector Encoding	max. $\mathbf{x}^T \mathbf{W} \hat{\mathbf{a}}_i$	min. $r_x(0) r_{\hat{\mathbf{a}}_i}(0) + 2 \sum_{m=1}^{p} r_x(m) r_{\hat{\mathbf{a}}_i}(m)$
Gain Thresholds	$\frac{1}{2}(\hat{\sigma}_j + \hat{\sigma}_{j+1})$	$\left(\ln \frac{\hat{\sigma}_{j+1}^2}{\hat{\sigma}_j^2}\right)\left(\frac{1}{\hat{\sigma}_j^2} - \frac{1}{\hat{\sigma}_{j+1}^2}\right)^{-1}$

TABLE II
ENCODER STORAGE AND COMPUTATION FORMULAS ($N = N_1 N_2$)

	SGVQ	VQ
Storage	$(K)(N_1) + (N_2)$	$(K+1)(N)$
Multiplications	$(K)(N_1)$	$(K)(N)$
Additions	$(K-1)(N_1)$	$(K)(N)$
Comparisons	$N_1 + N_2 - 2$	$N - 1$

while the gain encoder minimizes d''. We will refer to d' as the shape distortion and d'' as the gain distortion; d will be called the total distortion or simply distortion when the latter is unambiguous.

III. SGVQ Design

As is usually done, we adopt average distortion as the performance criterion for a quantizer. Let $\boldsymbol{X} = [X_1, X_2, \cdots, X_K]$ be a real random vector with cumulative distribution function $F(\boldsymbol{x}) = \Pr\{X_m \leqslant x_m;\ m = 1, 2, \cdots, K\}$. Then the performance criterion we adopt is $E\{d(\boldsymbol{X}; \hat{\boldsymbol{q}}(\boldsymbol{X}))\}$ where E is the expectation operator with respect to the distribution $F(\boldsymbol{x})$.[1] For the (full-search) SGVQ, the expected distortion can be written as $E\{\min_{i,j} d(\boldsymbol{X}; \hat{\boldsymbol{a}}_i, \hat{\sigma}_j)\}$. We would like to design a product codebook $\hat{A} \times \hat{\Sigma}$ that minimizes the quantity.

In both the waveform coding case and the LPC voice coding case, each pair $(\hat{\boldsymbol{a}}_i, \hat{\sigma}_j)$ corresponds to a specific reproduction vector $\hat{\boldsymbol{x}}_{ij}$. There are $(N_1)(N_2)$ reproduction vectors, but they cannot be chosen arbitrarily; they must be chosen according to some product codebook $\hat{A} \times \hat{\Sigma}$. Without the product code constraint, the problem reduces to the design of a standard VQ with $(N_1)(N_2)$ codewords. Thus, an SGVQ is a special case of a VQ so that, in general, a VQ can be designed to perform as well as, and presumably better than, an SGVQ of the same codebook size. The difference in performance could be large if the optimum VQ codebook for $F(\boldsymbol{x})$ does not resemble a product codebook. Nevertheless, the reduction in computation and storage offered by the SGVQ could make it an attractive alternative. (Recall that the only suboptimum aspect of the SGVQ is the codebook itself; the two-step encoding operation described above for the waveform coders and the LPC voice coders is optimum in that it finds the minimum distortion reproduction vector in the set.) These tradeoffs are investigated experimentally in Section IV. For now, we consider the problem of designing as good an SGVQ as possible for $F(\boldsymbol{x})$ without concern for its performance relative to an unconstrained VQ.

A design algorithm for ordinary unconstrained VQ's is described by Linde *et al.* [11] and Gray *et al.* [8]. The algorithm uses a distribution such as a sample histogram obtained from a training sequence of actual data to iteratively improve an initial quantizer. The algorithm has proved useful in designing VQ's for waveform coding [1], [4] and voice coding [3], [16] using the distortion measures considered here. We now use ideas from [3], [8], and [11] to extend this unconstrained VQ design algorithm to the design of good SGVQ's for waveform and voice coding. We shall refer to the basic design algorithm for unconstrained VQ's as the LBG algorithm or the generalized Lloyd algorithm. (The algorithm is a generalization of Lloyd's original algorithm for designing optimum PCM systems.)

A. Jointly Optimized Algorithm

Let $\hat{\boldsymbol{q}}(\boldsymbol{x})$ be an arbitrary mapping of the input space into a product codebook $\hat{A} \times \hat{\Sigma}$. Define the partition $\mathcal{S} = \{S_{ij}, i = 1, 2, \cdots, N_1; j = 1, 2, \cdots, N_2\}$ where $S_{ij} = \{\boldsymbol{x} \in \mathbb{R}^K : \hat{\boldsymbol{q}}(\boldsymbol{x}) = (\hat{\boldsymbol{a}}_i, \hat{\sigma}_j)\}$. We will denote the expected distortion of an SGVQ with product codebook $\hat{A} \times \hat{\Sigma}$ and partition $\mathcal{S}$ by $D(\hat{A}, \hat{\Sigma}, \mathcal{S})$. For convenience, we define $P_i = \cup_{j=1}^{N_2} S_{ij}$, the cell of the input space that maps into shape codeword $\hat{\boldsymbol{a}}_i$, and $Q_j = \cup_{i=1}^{N_1} S_{ij}$, the cell of the input space that maps into gain codeword $\hat{\sigma}_j$. We also define the shape mapping $\hat{\boldsymbol{a}}(\boldsymbol{x}) = \hat{\boldsymbol{a}}_i$ for $\boldsymbol{x} \in P_i$ and the gain mapping $\hat{\sigma}(\boldsymbol{x}) = \hat{\sigma}_j$ for $\boldsymbol{x} \in Q_j$.

Suppose we are given an SGVQ with product codebook $\hat{A} \times \hat{\Sigma}$ and partition $\mathcal{S}$. Then we can write

$$D(\hat{A}, \hat{\Sigma}, \mathcal{S}) = E\{d(\boldsymbol{X}; \hat{\boldsymbol{a}}(\boldsymbol{X}), \hat{\sigma}(\boldsymbol{X}))\} \geqslant E\{\min_{i,j} d(\boldsymbol{X}; \hat{\boldsymbol{a}}_i, \hat{\sigma}_j)\} \tag{13}$$

which is simply a restatement that the best mapping for an SGVQ is a minimum distortion mapping. Let $\mathcal{S}^+(\hat{A}, \hat{\Sigma})$ represent a minimum distortion partition of $\mathbb{R}^K$ for $\hat{A} \times \hat{\Sigma}$, i.e., $\mathcal{S}^+(\hat{A}, \hat{\Sigma}) = \{S_{ij}^+, i = 1, 2, \cdots, N_1; j = 1, 2, \cdots, N_2\}$ where $d(\boldsymbol{x}; \hat{\boldsymbol{a}}_i, \hat{\sigma}_j) \leqslant d(\boldsymbol{x}; \hat{\boldsymbol{a}}_k, \hat{\sigma}_l)$ for $\boldsymbol{x} \in S_{ij}^+$. Obviously, a minimum distortion partition need not be unique, but we can impose an arbitrary tie-breaking rule for resolving ties between several nearest neighbors. We can rewrite (13) as

$$D(\hat{A}, \hat{\Sigma}, \mathcal{S}) \geqslant D(\hat{A}, \hat{\Sigma}, \mathcal{S}^+(\hat{A}, \hat{\Sigma})). \tag{14}$$

For an SGVQ with a product codebook $\hat{A} \times \hat{\Sigma}$ and partition $\mathcal{S}$, we can write the expected distortion as

$$D(\hat{A}, \hat{\Sigma}, \mathcal{S}) = \sum_{i=1}^{N_1} E\{d(\boldsymbol{X}; \hat{\boldsymbol{a}}_i, \hat{\sigma}(\boldsymbol{X})) | \boldsymbol{X} \in P_i\} \Pr\{\boldsymbol{X} \in P_i\}. \tag{15}$$

Clearly,

$$D(\hat{A}, \hat{\Sigma}, \mathcal{S}) \geqslant \sum_{i=1}^{N_1} \inf_{\boldsymbol{u}_i \in \Lambda} E\{d(\boldsymbol{X}; \boldsymbol{u}_i, \hat{\sigma}(\boldsymbol{X})) | \boldsymbol{X} \in P_i\} \Pr\{\boldsymbol{X} \in P_i\} \tag{16}$$

where Λ is the set of allowable shape vectors. If each infimum in (16) is attained by some $\boldsymbol{u}_i^* \in \Lambda$ for each $i = 1, 2, \cdots, N_1$, then we can define $\hat{A}^+ = \{\boldsymbol{u}_i^*, i = 1, 2, \cdots, N_1\}$. Notice that $\hat{A}^+$ depends only on the partition $\mathcal{S}$ and the gain codebook $\hat{\Sigma}$; we denote this by writing $\hat{A}^+(\hat{\Sigma}, \mathcal{S})$. Inequality (16) can be expressed as

$$D(\hat{A}, \hat{\Sigma}, \mathcal{S}) \geqslant D(\hat{A}^+(\hat{\Sigma}, \mathcal{S}), \hat{\Sigma}, \mathcal{S}). \tag{17}$$

Similarly, we can write the expected distortion as

$$D(\hat{A}, \hat{\Sigma}, \mathcal{S}) = \sum_{j=1}^{N_2} E\{d(\boldsymbol{X}; \hat{\boldsymbol{a}}(\boldsymbol{X}), \hat{\sigma}_j) | \boldsymbol{X} \in Q_j\} \Pr\{\boldsymbol{X} \in Q_j\}. \tag{18}$$

Then

$$D(\hat{A}, \hat{\Sigma}, \mathcal{S}) \geqslant \sum_{j=1}^{N_2} \inf_{\lambda_j \geqslant 0} E\{d(\boldsymbol{X}; \hat{\boldsymbol{a}}(\boldsymbol{X}), \lambda_j) | \boldsymbol{X} \in Q_j\} \Pr\{\boldsymbol{X} \in Q_j\}. \tag{19}$$

[1] The usual convention of upper case for random vectors and lower case for dummy arguments is followed, e.g., $\boldsymbol{X}$ is a random vector, $\hat{\boldsymbol{q}}(\boldsymbol{x})$ is the function $\hat{\boldsymbol{q}}(\cdot)$ evaluated at $\boldsymbol{x}$, and $\hat{\boldsymbol{q}}(\boldsymbol{X})$ is the random vector resulting from $\hat{\boldsymbol{q}}(\cdot)$ operating on $\boldsymbol{X}$.

If each infimum in (19) is attained by some $\lambda_j^* \geqslant 0$ for $j = 1, 2, \cdots, N_2$, we define $\hat{\Sigma}^+ = \{\lambda_j^*, j = 1, 2, \cdots, N_2\}$. In this case, the dependence is on the partition $\mathcal{S}$ and the shape codebook $\hat{A}$, so we write $\hat{\Sigma}^+(\hat{A}, \mathcal{S})$. Then inequality (19) can be expressed

$$D(\hat{A}, \hat{\Sigma}, \mathcal{S}) \geqslant D(\hat{A}, \hat{\Sigma}^+(\hat{A}, \mathcal{S}), \mathcal{S}). \tag{20}$$

Inequalities (14), (17), and (20) suggest steps by which a given SGVQ can be improved. In (14), we see that one way to improve an SGVQ is to optimize the partition for the product codebook. This could be phrased as optimizing the encoder for the given decoder. Indeed, we have agreed that we will ultimately use an optimum encoder, i.e., we will have the encoder search the product codebook for the nearest neighbor to an input vector. In (17), we see that another way to improve an SGVQ is to optimize the shape codebook for the given partition and gain codebook. This could be described as optimizing the shape decoder for the given encoder and gain decoder. Similarly, (20) suggests optimizing the gain codebook for the given partition and shape codebook or optimizing the gain decoder for the given encoder and shape decoder. The following algorithm uses these steps to iteratively improve an SGVQ.

0) Initialization: Given $N_1, N_2, \epsilon \geqslant 0$, initial product codebook $(\hat{A}_0, \hat{\Sigma}_0)$, distribution $F(x)$. Set $m = 0, D_{-1} = \infty$.

1) Compute the optimum partition $\mathcal{S}^+(\hat{A}_m, \hat{\Sigma}_m)$.

2) Compute the average distortion $D_m = D(\hat{A}_m, \hat{\Sigma}_m, \mathcal{S}^+(\hat{A}_m, \hat{\Sigma}_m))$. If $(D_{m-1} - D_m)/D_m < \epsilon$, halt with $(\hat{A}_m, \hat{\Sigma}_m, \mathcal{S}^+(\hat{A}_m, \hat{\Sigma}_m))$ describing final quantizer. Else continue.

3) Compute the optimum shape codebook $\hat{A}_{m+1} = \hat{A}^+(\hat{\Sigma}_m, \mathcal{S}^+(\hat{A}_m, \hat{\Sigma}_m))$.

4) Compute the optimum partition $\mathcal{S}^+(\hat{A}_{m+1}, \hat{\Sigma}_m)$.

5) Compute the optimum gain codebook $\hat{\Sigma}_{m+1} = \hat{\Sigma}^+(\hat{A}_{m+1}, \mathcal{S}^+(\hat{A}_{m+1}, \hat{\Sigma}_m))$.

6) $m = m + 1$. Go to 1).

Each of steps 1), 3), 4), and 5) results in a quantizer whose average distortion is better (or at least as good) as the previous quantizer. Thus, the algorithm can only improve the initial quantizer. Characteristics of quantizers produced by the algorithm are discussed in [17]. A key result is that any quantizer that would not be changed by the algorithm (i.e., a "fixed point" quantizer) satisfies first-order differentiability requirements for a locally optimum SGVQ (i.e., is a "stationary point" quantizer) provided that the quantizer has zero probability of having more than one nearest neighbor to an input vector.

The main distinction between this algorithm and the LBG algorithm is that there is no step that corresponds to optimizing the decoder for the encoder. In the LBG algorithm, optimizing the decoder consists of finding the "centroid" of each encoder partition. Such a step does not, in general, yield a product code, even if, as in our case, the partition arises from a minimum distortion mapping into a product code. Thus, we turn instead to optimizing the shape decoder and gain decoder in separate steps so that the product code structure is preserved.

B. BGGM Algorithm

In [3], the SGVQ's for LPC vocoding are designed by separately optimizing the two codebooks. The method can also be used to design SGVQ's for the waveform coding case. The shape codebook is designed first, using the LBG algorithm with the shape distortion d' of (11). Then, with $\hat{a}(x)$ denoting the optimum mapping of the input space into the shape codebook, the gain codebook is designed by applying the LBG algorithm with gain distortion d'' to the quantity $\sigma^*(X, \hat{a}(X))$ where d'' and σ^* are as in (11). Since the shape codebook is designed independently of the gain codebook, it is reasonable to believe that the pair is not optimum in the sense of minimizing the average value of the total distortion d. Nevertheless, the method has given good results in the LPC vocoding case. In addition, in some applications, one may be interested in designing only the shape portion of the SGVQ, dealing with the gain in some other manner. In such a case, it is reasonable to design a shape codebook to minimize d'.

Design of the gain codebook is a straightforward application of the LBG algorithm. The LPC vocoding case is detailed in [3]. For the waveform coding case, substituting $\sigma^*(x, \hat{a}_i)$ for $x^T W \hat{a}_i$ in the second term of (9), rearranging, and using the definition of $\hat{a}(x)$ above yields

$$\begin{aligned} d''(\hat{\sigma}_j, \sigma^*(x, \hat{a}(x))) = {} & [\hat{\sigma}_j - \sigma^*(x, \hat{a}(x))]^2 \\ & + [([\sigma^*(x, \hat{a}(x))]^+)^2 \\ & - (\sigma^*(x, \hat{a}(x)))^2]. \end{aligned} \tag{21}$$

The second term does not depend on $\hat{\sigma}_j$, and thus does not enter into either the encoding of $\hat{\sigma}_j$ or the centroid calculations of the LBG algorithm. Thus, the problem reduces to the well-known case of (scalar) squared-error distortion [1], [4], [7], [9], [11], [12], [18] on the quantity $\sigma^*(X, \hat{a}(X))$ with the constraint that each codeword $\hat{\sigma}_j$ be nonnegative.

Design of the shape codebook runs into difficulty. In the LPC vocoding case, it is shown in [3] that the centroid calculation of the LBG algorithm is computationally difficult for the shape distortion d', so that an approximation must be used. It is shown in Appendix 2 that the waveform coding case suffers a similar problem, with an analogous approximation resolving the difficulty.

Henceforth we will refer to the method just described as the BGGM algorithm for both the waveform coding and LPC vocoding cases. We now show that the computational difficulties of the shape codebook design can be circumvented by using methods similar to those of the jointly optimized algorithm.

C. Individually Optimized Algorithm

Let $\hat{a}(x)$ be an arbitrary mapping of the input space into a shape codebook $\hat{A}$. Define the shape partition $\mathcal{P} = \{P_i, i = 1, 2, \cdots, N_1\}$ where $P_i = \{x \in \mathcal{R}^K : \hat{a}(x) = \hat{a}_i\}$. For the shape quantizer described by $(\hat{A}, \mathcal{P})$, denote the average shape distortion by $D'(\hat{A}, \mathcal{P})$. Observe, as in (13), that

$$D'(\hat{A}, \mathcal{P}) = Ed'(X; \hat{a}(x)) \geqslant E \min_i d'(X; \hat{a}_i). \tag{22}$$

As in Section III-A, we are led to define the optimum shape partition $\mathcal{P}^+(\hat{A}) = \{P_i^+, i = 1, 2, \cdots, N_1\}$ where $d'(x; \hat{a}_i) \leqslant d'(x; \hat{a}_k)$ for $x \in P_i^+$. Thus,

$$D'(\hat{A}, \mathcal{P}) \geqslant D'(\hat{A}, \mathcal{P}^+(\hat{A})). \tag{23}$$

With σ^* as in (11), we can write

$$D'(\hat{A}, \mathcal{P}) = \sum_{i=1}^{N_1} E\{d(X; \hat{a}_i, [\sigma^*(X, \hat{a}_i)]^+) | X \in P_i\} \Pr\{X \in P_i\}. \tag{24}$$

Then, as in (16),

$$D'(\hat{A}, \mathcal{P}) \geq \sum_{i=1}^{N_1} \inf_{u_i \in \Lambda} E\{d(X; u_i, [\sigma^*(X, \hat{a}_i)]^+) | X \in P_i\} \cdot \Pr\{X \in P_i\}. \tag{25}$$

Note that the right-hand side is identical to that in (16) except that $[\sigma^*(x, \hat{a}_i)]^+$ replaces $\hat{\sigma}(x)$. If each infimum is attained by some u_i^* in Λ, we can define $\hat{A}^+ = \{u_i^*, i = 1, 2, \cdots, N_1\}$. Since $\hat{A}^+$ depends on the partition $\mathcal{P}$ and on the mapping σ^*, which in turn depends on the codebook $\hat{A}$, we write $\hat{A}^+(\mathcal{P}, \hat{A})$. Then (25) can be expressed as

$$D'(\hat{A}, \mathcal{P}) \geq D'(\hat{A}^+(\mathcal{P}, \hat{A}), \mathcal{P}). \tag{26}$$

Inequalities (23) and (26) suggest steps by which the shape quantizer of an SGVQ can be improved with respect to average shape distortion. As before, (23) suggests optimizing the partition $\mathcal{P}$ for the codebook $\hat{A}$, i.e., optimizing the shape encoder for the shape decoder. Inequality (26) suggests the following step: replace $\hat{A}$ with the shape codebook that is optimum for the partition $\mathcal{P}$ and the mapping σ^* that is induced by $\mathcal{P}$ and $\hat{A}$. If one envisions $[\sigma^*]^+$ as the output of an infinitesimally spaced gain quantizer of an SGVQ, then this step consists of optimizing the shape decoder for the given encoder and decoder. The following algorithm uses these steps to iteratively improve the shape portion of an SGVQ.

0) Initialization: Given N_1, $\epsilon \geq 0$, initial shape codebook $\hat{A}_0$, distribution $F(x)$. Set $m = 0, D'_{-1} = \infty$.

1) Compute the optimum shape partition $\mathcal{P}^+(\hat{A}_m)$.

2) Compute the average shape distortion $D'_m = D'(\hat{A}_m, \mathcal{P}^+(\hat{A}_m))$. If $(D'_{m=1} - D'_m)/D'_m > \epsilon$, halt with $(\hat{A}_m, \mathcal{P}^+(\hat{A}_m))$ describing the final quantizer. Else continue.

3) Compute the optimum shape codebook

$$\hat{A}_{m+1} = \hat{A}^+(\mathcal{P}^+(\hat{A}_m), \hat{A}_m).$$

4) $m = m + 1$. Go to 1).

This algorithm is identical to the shape portion of the BGGM algorithm, except for step 3): here the codebook update depends on the current codebook as well as on the partition, rather than just on the partition. Unlike the corresponding step of the BGGM algorithm, step 3) results in a shape quantizer whose average shape distortion can be no worse than the previous shape quantizer. Thus, the algorithm can only improve the initial quantizer. It is shown in [17] that, as with the joint algorithm, a fixed point quantizer of this algorithm is a stationary point quantizer with respect to average shape distortion.

D. Centroid Calculations

To update the shape codebook as in step 3) of each algorithm, it is necessary to find u_i^* that minimizes the conditional distortion associated with the shape cell P_i so as to achieve the corresponding infimum in (16) and (25). In keeping with the terminology of the LBG algorithm, we will call u_i^* the centroid of cell P_i. Notice, however, that u_i^* depends on the gain mapping ($\hat{\sigma}$ or $[\sigma^*]^+$), as well as on P_i. To update the gain codebook in step 5) of the joint algorithm, it is necessary to find λ_j^* that minimizes the conditional distortion associated with gain cell Q_j so as to achieve each infimum in (19). We will call λ_j^* the centroid of cell Q_j, noting that it depends on the shape mapping $\hat{a}(x)$, as well as on Q_j.

The derivation of u_i^* and λ_j^* for the waveform and LPC voice coding cases is presented in Appendix A. The results are summarized in Table III. The shape centroid is in terms of an arbitrary gain mapping $\sigma(x)$. For the jointly optimized algorithm, $\sigma(x)$ is taken to be $\hat{\sigma}(x)$ of Section III-A. For the individually optimized algorithm, $\sigma(x)$ is taken to be $[\sigma^*(x, \hat{a}_i)]^+$ of Section III-B for shape cell P_i. For the BGGM algorithm in the waveform coding case, $\sigma(x)$ is taken to be $\sqrt{x^T W x}$ (Appendix B); for the LPC vocoding case, $\sigma(x)$ is the numerator of the model $\sigma/A(z)$ that corresponds to the input vector x [3].

In the waveform coding case, we see that u_i^* is the average vector (suitably normalized) of the vectors in cell P_i after each has been weighted by the gain mapping $\sigma(x)$. Thus, large vectors dominate the computation of u_i^*, due both to their large norm and to their tendency to map into large gain codewords. The expression for λ_j^* is simply the average correlation between each vector in cell Q_j and the shape vector into which it is mapped by $\hat{a}(x)$.

For the LPC voice coding case, u_i^* is found by solving a standard LPC autocorrelation-method problem. This can be done, for example, by Levinson's recursion [13], [14]. The autocorrelation matrix $\tilde{R}_i$ is the average of the autocorrelation matrices corresponding to each vector in cell P_i after each matrix has been normalized by $\sigma^2(x)$. Thus, low energy frames tend to influence the computation of u_i^* as much as do high energy frames, since the value of $\sigma^2(x)$ tends to be comparable to the energy of the frame. The expression for λ_j^* is simply the square root of the average residual energy when each vector in Q_j is filtered by the inverse filter corresponding to $\hat{a}(x)$.

Since no differentiability requirements are placed on the distribution $F(x)$, all three algorithms are useful for the very important case of designing quantizers that are well matched to a long training sequence $\{x_k, k = 1, 2, \cdots, L\}$ of actual data drawn from the source. In this case, we merely assign a probability of L^{-1} to each vector in the training sequence. Partitions become clusters of the training sequence and expectations become summations. The resulting calculations are summarized in Table IV, with the same interpretation of $\sigma(x)$ as in the general case.

In developing the centroid calculations, it was assumed that $\Pr\{X \in P_i\}$ and $\Pr\{X \in Q_j\}$ were nonzero. In practice, it is found that this is not always the case, that is, occasionally a

TABLE III
CENTROID CALCULATIONS

Waveform Coding	$\mathbf{u}_i^* = \frac{\mathbf{u}_i}{[\mathbf{u}_i^T \mathbf{W} \mathbf{u}_i]^{1/2}}$ where $\mathbf{u}_i = E\{\sigma(\mathbf{X})\mathbf{X} \mid \mathbf{X} \in P_i\}$
	$\lambda_j^* = E\{\mathbf{X}^T \mathbf{W} \dot{\mathbf{a}}(\mathbf{X}) \mid \mathbf{X} \in Q_j\}$
LPC Voice Coding	$\mathbf{u}_i^*$ solves minimize $\mathbf{u}_i^T \mathbf{R}_i \mathbf{u}_i$ subject to $\mathbf{u}_i^T = [1\ u_1\ u_2\ \cdots\ u_p]$ where $\mathbf{R}_i = E\left\{\mathbf{R}(\mathbf{X})/\sigma^2(\mathbf{X}) \mid \mathbf{X} \in P_i\right\}$
	$\lambda_j^* = \left\{E\{\dot{\mathbf{a}}^T(\mathbf{X})\mathbf{R}(\mathbf{X})\dot{\mathbf{a}}(\mathbf{X}) \mid \mathbf{X} \in Q_j\}\right\}^{1/2}$

TABLE IV
CENTROID CALCULATIONS FOR TRAINING SEQUENCE CASE

Waveform Coding	$\mathbf{u}_i^* = \frac{\dot{\mathbf{u}}_i}{[\dot{\mathbf{u}}_i^T \mathbf{W} \mathbf{u}_i]^{1/2}}$ where $\dot{\mathbf{u}}_i = \sum_{j=1}^{N_2} \sum_{k: \mathbf{x}_k \in S_{ij}} \sigma(\mathbf{x}_k)\mathbf{x}_k$
	$\lambda_j^* = \frac{1}{\lVert Q_j \rVert} \sum_{i=1}^{N_1} \sum_{k: \mathbf{x}_k \in S_{ij}} \dot{\mathbf{a}}_i^T \mathbf{W} \mathbf{x}_k$ where $\lVert Q_j \rVert$ is the number of training sequence vectors in cell Q_j
LPC Voice Coding	$\mathbf{u}_i^*$ solves minimize $\mathbf{u}_i^T \tilde{\mathbf{R}}_i \mathbf{u}_i$ subject to $\mathbf{u}_i^T = [1\ u_1\ u_2\ \cdots\ u_p]$ where $\tilde{\mathbf{R}}_i = \sum_{j=1}^{N_2} \sum_{k: \mathbf{x}_k \in S_{ij}} \frac{\mathbf{R}(\mathbf{x}_k)}{\sigma^2(\mathbf{x}_k)}$
	$\lambda_j^* = \left(\frac{1}{\lVert Q_j \rVert} \sum_{i=1}^{N_1} \sum_{k: \mathbf{x}_k \in S_{ij}} \dot{\mathbf{a}}_i^T \mathbf{R}(\mathbf{x}_k) \dot{\mathbf{a}}_i\right)^{1/2}$ where $\lVert Q_j \rVert$ is the number of training sequence vectors in cell Q_j

shape cell or gain cell is empty. When a cell is empty, it does not contribute to the expected distortion, as seen from (15), (18), and (24). Thus, an arbitrary assignment of the corresponding gain or shape codeword does not effect the average distortion. Several strategies for dealing with this situation are discussed in [11] for the LBG algorithm. These ideas can be applied here.

IV. EXPERIMENTAL SIMULATIONS

A. *Waveform Coding: SGVQ Versus VQ*

The jointly optimized algorithm was used to design SGVQ's to encode waveforms of natural, continuous speech sampled at 6.5 kHz. Squared-error distortion was used, i.e., W of (1) was chosen to be the identity matrix. The quantizers were designed on a training sequence of approximately 100 s of speech drawn from five male speakers. They were tested on a 12 s speech segment drawn from a sixth male speaker. Rates of 1 and 2 bits/sample were considered. Within a given rate and block length, various sizes of shape and gain codebooks were tried. The performance of the SGVQ's was compared to that of the standard VQ's described in [1], which were designed and tested using the same speech data used here.

To provide the design algorithm with an initial codebook pair $(\hat{A}_0, \hat{\Sigma}_0)$, we used a binary splitting procedure similar to that in [11] where quantizers of successively higher rates are designed. A higher rate quantizer is designed from an existing one by doubling the size of either the shape codebook or the gain codebook by "splitting" each shape or gain codeword. Splitting a codeword is performed by replacing it with two codewords, the original plus a slightly perturbed version. Since doubling the size of a codebook is performed by adding new codewords to the ones already present, it cannot increase average distortion in the full-search SGVQ. The order of the splitting procedure is as follows.

1) Design a zero-rate SGVQ of the desired dimension by running the algorithm for one iteration with $N_1 = 1$ and $N_2 = 1$ and with an arbitrary initial shape codeword and gain codeword. After the single iteration, each codeword is replaced with the appropriate centroid of the distribution or training sequence.

2) Split the shape vector. Using the resulting codebooks of two shapes and one gain as the initial codebook pair, run the design algorithm until it terminates.

3) Split the gain codeword. Using the resulting codebooks of two shapes and two gains as the initial codebook pair, run the design algorithm until it terminates.

4) If the gain codebook is at the desired size N_2, go to step 5). Otherwise, double the size of the gain codebook by splitting each gain codeword. Using the current shape codebook and the new gain codebook as the initial codebook pair, run the algorithm until it terminates. Repeat this step.

5) If the shape codebook is at the desired size N_1, halt with the desired SGVQ. Otherwise, double the size of the shape codebook by splitting each shape vector. Using the new shape codebook and the current gain codebook as the initial codebook pair, run the algorithm until it terminates. Repeat this step.

The gain codebook is built to full size first because gain encoding involves only scalar quantization, and hence is much faster than shape encoding. However, the two initial shape codewords generated in step 2) are used because they tend to arrange themselves antipodally. This provides a fuller range of gain codeword values than does a single shape codeword, since either $x^T W \hat{a}_1$ or $x^T W \hat{a}_2$ is usually nonnegative. In these experiments, a value of 0.1 was used for the termination threshold ϵ for intermediate steps where quantizers of less than the desired rate were designed. For the final full rate quantizers, a value of 0.001 was used.

Signal-to-quantization noise ratio (SQNR) is used to describe quantizer performance. It is chosen as the performance measure because of its practical interpretation and because it is the quantity that is optimized during design. SQNR is defined as the ratio of the average distortion of a zero rate vector quantizer of the appropriate dimension to the average distortion of the quantizer being described. The zero rate distortion is found in the initial step of the splitting procedure; in these experiments, it was found to be very nearly equal to the average energy in the unquantized sequence.

Table V(a) and (b) summarizes the results for the SGVQ's. The first three columns show the codebook sizes and the block length. The next three columns show the encoding computation required per sample of speech, obtained by dividing the formulas in Table II by the block length K to give per-sample rather than per-vector complexity. The next column gives the storage required by each of the encoder and the decoder, while the last two columns give the SQNR for the design sequence and for the test sequence.

Similarly, Table VI(a) and (b) summarizes the results for the VQ's in [1]. The first two columns show the codebook size and block length. The next three columns show encoding complexity per speech sample, again obtained from the formulas of Table II and adjusted to give per-sample complexity. The next column is encoder storage, and the last two columns are the design and test values of SQNR.

At block lengths 5, 6, 7, and 8 for rate one and at block lengths 2, 3, and 4 for rate two, there are several SGVQ's in Table V that can be compared to standard VQ's in Table VI. For the design sequence, the improvement in SQNR of the VQ over the best SGVQ for a particular block length and rate is $\frac{1}{2}$-1 dB. As discussed in Section III, this suggests that the optimum VQ codebook is not a product codebook. Fig. 3 illustrates the point. The horizontal axis is the index of each shape codeword of the SGVQ of Table V(a) that has $N_1 = 16$, $N_2 = 8$, and $K = 7$. The indexes have been arranged in ascending order according to the value of

$$E\{[X^T\hat{a}_i]^+ | X \in P_i\}, \tag{27}$$

i.e., the mean of the (unquantized) gain value assigned to each training sequence vector in shape cell P_i. This value is plotted in trace A of the figure. The standard deviation of the unquantized gain in P_i,

$$[E\{([X^T\hat{a}_i]^+)^2 | X \in P_i\} - E^2\{[X^T\hat{a}_i]^+ | X \in P_i\}]^{1/2}, \tag{28}$$

is plotted in trace B. The two traces show there is a definite dependence between shape index and gain value. This dependence is not exploited by the SGVQ, and hence the VQ shows better performance.

To compare the SGVQ's to the VQ's in terms of complexity, we adopt the number of multiplications per sample as the measure of encoding complexity. This is reasonable not only because multiplication is usually the dominant operation in a practical implementation, but also because inspection of Tables V and VI indicates that the number of additions and comparisons is comparable among equal rate quantizers of a given number of multiplications per sample. The SGVQ's of Table V(a) and (b) are organized into groups according to number of multiplications. In each group, there are three SGVQ's, with the second in the group having the highest SQNR for the design sequence. These SGVQ's were chosen as the best combination of N_1, N_2, and K for the particular rate and encoding complexity and are compared in Fig. 4(a) and (b) to the corre-

TABLE V(a)
SGVQ WAVEFORM CODING RESULTS—RATE ONE

N_1	N_2	K	Computation			Storage	SQNR (dB)	
			Mult.	Add.	Comp.		Design	Test
4	2	3	4	2.7	1.3	14	4.503	4.635
4	4	4	4	3.0	1.5	20	6.062	6.139
4	8	5	4	3.2	2.0	28	5.925	5.675
8	4	5	8	6.4	2.0	44	7.237	6.897
8	8	6	8	6.7	2.3	56	7.434	6.912
8	16	7	8	6.9	3.1	72	6.891	6.423
16	4	6	16	13.3	3.0	100	7.729	7.391
16	8	7	16	13.7	3.1	120	8.188	7.656
16	16	8	16	14.0	3.8	144	7.895	7.326
32	4	7	32	27.4	4.9	228	8.062	7.619
32	8	8	32	28.0	4.8	264	8.782	8.134
32	16	9	32	28.4	5.1	304	8.515	7.801
64	4	8	64	56.0	8.3	516	8.383	7.954
64	8	9	64	56.9	7.8	584	9.309	8.523
64	16	10	64	57.6	7.8	656	9.165	8.336
128	4	9	128	113.8	14.4	1156	8.677	8.107
128	8	10	128	115.2	13.4	1288	9.786	8.865
128	16	11	128	116.4	12.9	1424	9.738	8.816
256	8	11	256	232.7	23.8	2824	10.375	9.264
256	16	12	256	234.7	22.5	3088	10.406	9.035
256	32	13	256	236.3	22.0	3360	10.239	8.804

TABLE V(b)
SGVQ WAVEFORM CODING RESULTS—RATE TWO

N_1	N_2	K	Computation			Storage	SQNR (dB)	
			Mult.	Add.	Comp.		Design	Test
4	4	2	4	2.0	3.0	12	8.896	8.915
4	16	3	4	2.7	6.0	28	8.923	8.744
4	64	4	4	3.0	16.5	80	7.203	7.145
8	2	2	8	4.0	4.0	18	5.439	5.672
8	8	3	8	5.3	4.7	32	11.253	11.054
8	32	4	8	6.0	9.5	64	10.193	9.594
16	4	3	16	10.7	6.0	52	9.922	9.956
16	16	4	16	12.0	7.5	80	11.861	11.349
16	64	5	16	12.8	15.6	144	10.490	9.946
32	2	3	32	21.3	10.7	98	5.640	5.877
32	8	4	32	24.0	9.5	136	12.518	12.162
32	32	5	32	25.6	12.4	192	12.105	11.301
64	4	4	64	48.0	16.5	260	10.344	10.345
64	16	5	64	51.2	15.6	336	13.226	12.422
64	64	6	64	53.3	21.0	448	12.313	11.333
128	8	5	128	102.4	26.8	648	13.339	12.754
128	32	6	128	106.7	26.3	800	13.527	12.483
128	128	7	128	109.7	36.3	1024	12.609	11.547
256	4	5	256	204.8	51.6	1284	10.739	10.535
256	16	6	256	213.3	45.0	1552	14.460	13.428
256	64	7	256	219.4	45.4	1856	13.840	12.528

TABLE VI(a)
VQ WAVEFORM CODING RESULTS—RATE ONE

N	K	Computation Mult.	Computation Add.	Computation Comp.	Storage	SQNR (dB) Design	SQNR (dB) Test
4	2	4	4	1.5	12	5.24	5.25
8	3	8	8	2.3	32	6.09	6.01
16	4	16	16	3.8	80	7.09	6.99
32	5	32	32	6.2	192	7.88	7.57
64	6	64	64	10.5	448	8.53	8.05
128	7	128	128	18.1	1024	9.08	8.41
256	8	256	256	31.9	2304	9.74	8.82

TABLE VI(b)
VQ WAVEFORM CODING RESULTS—RATE TWO

N	K	Computation Mult.	Computation Add.	Computation Comp.	Storage	SQNR (dB) Design	SQNR (dB) Test
4	1	4	4	3.0	8	6.50	6.41
16	2	16	16	7.5	48	9.98	9.72
64	3	64	64	21.0	256	12.13	11.85
256	4	256	256	63.8	1280	13.48	12.70

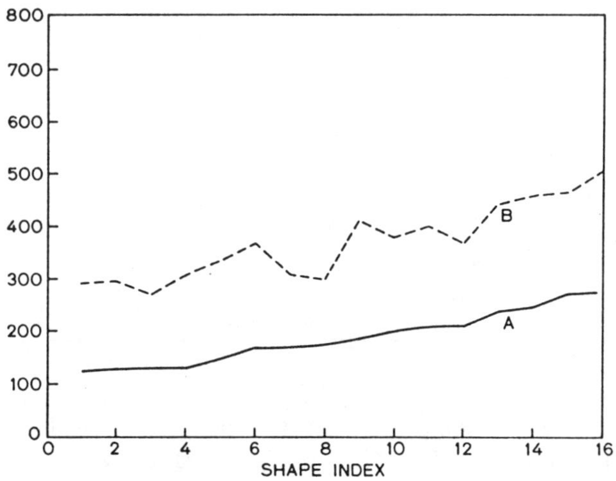

Fig. 3. Mean and standard deviation of gain versus shape (waveform coding, $N_1 = 16, N_2 = 8, K = 7$).

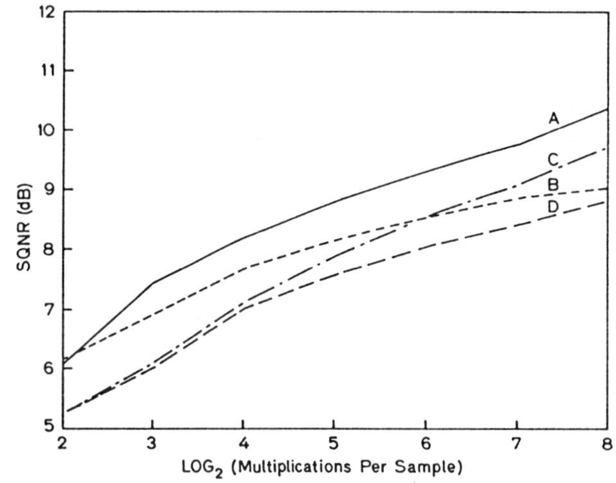

(a)

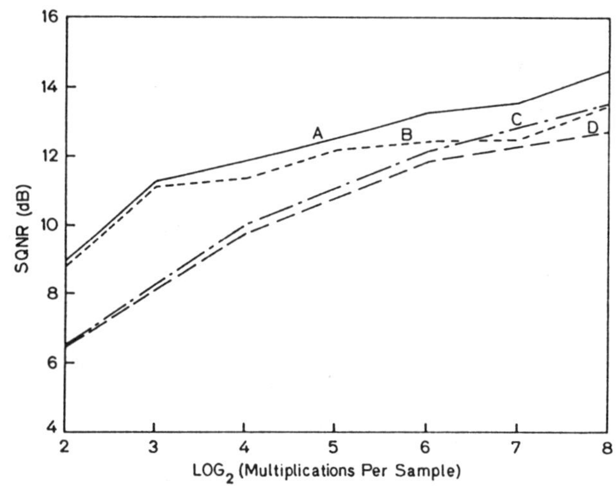

(b)

Fig. 4. (a) SQNR versus complexity for waveform coding SGVQ's–rate one. (b) SQNR versus complexity for waveform coding SGVQ's–rate two.

sponding VQ's. The horizontal axis is encoding complexity on a logarithmic scale, and the vertical axis is SQNR in decibels. The four traces are

A: SGVQ, design sequence;
B: SGVQ, test sequence;
C: VQ, design sequence;
D: VQ, test sequence.

The same SGVQ's chosen in Fig. 4(a) and (b) are compared in Fig. 5(a) and (b) to the corresponding VQ's in terms of encoder storage. A logarithmic scale is used on the horizontal axis for storage, with SQNR in decibels on the vertical axis. There are four traces, labeled as in Fig. 4(a) and (b).

Fig. 4(a) and (b) shows the superiority of the SGVQ's over VQ's of identical complexity for both the design sequence and the test sequence. At rate one, the improvement in SQNR is $\frac{1}{2}$-1 dB for the training sequence and $\frac{1}{4}$-1 dB for the test sequence. At rate two, the improvement is 1-3 dB for the training sequence and $\frac{1}{2}$-2 dB for the test sequence. Notice that the figures indicate increasing degradation in SQNR from the training sequence to the test sequence as the block length grows. Presumably, this effect could be reduced by using a longer training sequence.

Fig. 5(a) and (b) shows the superiority of the SGVQ's over VQ's of comparable storage requirements. At rate one, the improvement in SQNR is about $\frac{1}{2}$ dB for the training sequence and $\frac{1}{4}$ dB for the test sequence. At rate two, the improvement is 0-2 dB for both the training sequence and the test sequence.

B. Waveform Coding: Design Methods

The three design methods (BGGM, individually optimized, jointly optimized) were used to generate rate one and rate two SGVQ's using the same design and test data as in Section IV-A. Combinations of codebook sizes and block lengths were selected from those plotted in Figs. 4 and 5. For each size and

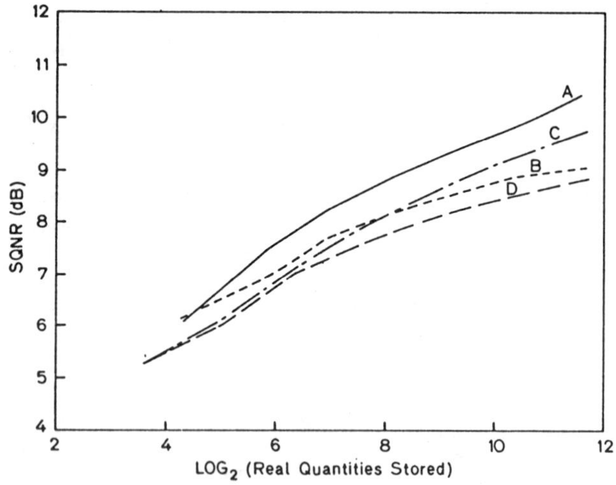

(a)

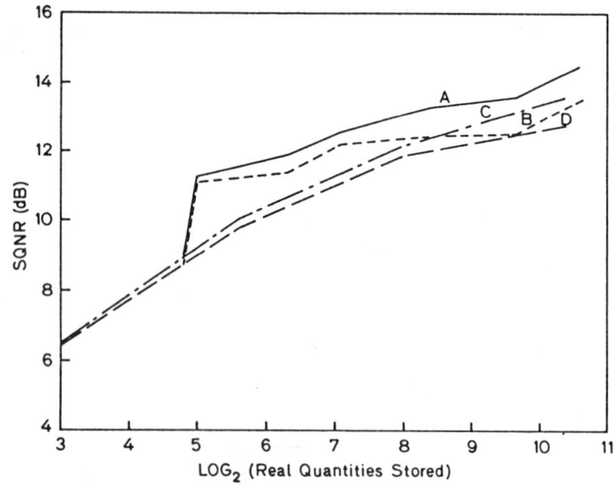

(b)

Fig. 5. (a) SQNR versus storage for waveform coding SGVQ's–rate one. (b) SQNR versus storage for waveform coding SGVQ's–rate two.

TABLE VII(a)
WAVEFORM SGVQ DESIGN METHOD RESULTS—RATE ONE

		Design			Test		
N_1 N_2 K	Design Method	Shape Dist.	Gain Dist.	Total Dist.	Shape Dist.	Gain Dist.	Total Dist.
	BGGM	.18941	.05781	.24721	.19359	.05151	.24510
4 4 4	I.O.	.18890	.05797	.24687	.19048	.05163	.24211
	J.O.	.18909	.05789	.24698	.19038	.05173	.24210
	BGGM	.13553	.01402	.14955	.15569	.01272	.16841
16 8 7	I.O.	.13551	.01402	.14953	.15609	.01269	.16879
	J.O.	.13552	.01401	.14953	.15593	.01273	.16866
	BGGM	.10156	.01419	.11575	.12635	.01213	.13848
64 8 9	I.O.	.10155	.01419	.11574	.12651	.01215	.13866
	J.O.	.10158	.01436	.11594	.12688	.01204	.13892
	BGGM	.08297	.00369	.08666	.12138	.00296	.12434
256 16 12	I.O.	.08287	.00369	.08656	.12135	.00295	.12430
	J.O.	.08286	.00366	.08653	.12139	.00295	.12434

TABLE VII(b)
WAVEFORM SGVQ DESIGN METHOD RESULTS—RATE TWO

		Design			Test		
N_1 N_2 K	Design Method	Shape Dist.	Gain Dist.	Total Dist.	Shape Dist.	Gain Dist.	Total Dist.
	BGGM	.12372	.00424	.12796	.12871	.00375	.13246
4 16 3	I.O.	.12307	.00411	.12719	.12844	.00378	.13221
	J.O.	.12305	.00414	.12719	.12822	.00374	.13195
	BGGM	.05930	.00427	.06357	.07038	.00411	.07450
16 16 4	I.O.	.05928	.00427	.06355	.07020	.00411	.07431
	J.O.	.05922	.00429	.06351	.06979	.00413	.07392
	BGGM	.04132	.00414	.04547	.05202	.00380	.05582
64 16 5	I.O.	.04135	.00415	.04550	.05203	.00382	.05585
	J.O.	.04133	.00415	.04548	.05201	.00382	.05584
	BGGM	.03016	.00408	.03424	.04118	.00376	.04493
256 16 6	I.O.	.03017	.00410	.03427	.04127	.00379	.04507
	J.O.	.03008	.00409	.03417	.04119	.00378	.04497

block length, the BGGM algorithm with the binary splitting procedure of [3] was run using a termination threshold of $\epsilon = 0.001$. The resulting codebook was used as the initial guess for each of the three design algorithms, with a termination threshold of $\epsilon = 0.0001$. In the jointly optimized algorithm, step 2) was modified to halt when the relative changes in both average shape distortion and average gain distortion dropped below ϵ, rather than halting based on the change in average total distortion. This provided a termination criterion comparable to that of the other two algorithms, which continue the design of the shape (or gain) codebook until the change in shape (or gain) distortion drops below ϵ.

The results of the simulation are summarized in Table VII(a) and (b). Columns 1–3 give the codebook sizes and block lengths and the fourth column indicates the design method. The next three columns are the average shape, gain, and total distortion for the design sequence where each has been normalized by the zero-rate distortion for the particular block length. The last three columns are the same distortion values for the test sequence.

In Table VIII(a) and (b), the reductions provided by one algorithm over another are tabulated for three of the distortion figures listed in Table VII. The fourth and fifth columns show the percent reduction in average shape distortion of the individually optimized (I.O.) algorithm over the BGGM algorithm for the design and test sequence. The next two columns show the reduction in average total distortion of the jointly optimized (J.O.) algorithm over the BGGM algorithm. The last two columns show the reduction in average total distortion of the J.O. algorithm over the I.O. algorithm. In each of the three comparisons, we would expect a nonnegative value in the design sequence column since the I.O. algorithm improves shape distortion at each step and the J.O. algorithm improves total distortion at each step. Such is the case in 18 of the 24 design sequence entries. In the test sequence case, 14 of 24 entries show improvement. Thus, the training sequence comparisons show a slight trend in improvement, while the test sequence comparisons do not really exhibit such a trend.

The most significant aspect of Table VIII is that in each comparison, the relative change in distortion values from one algo-

TABLE VIII(a)
WAVEFORM SGVQ DESIGN METHOD PERCENT REDUCTIONS—RATE ONE

			Shape Dist. I.O. vs. BGGM		Total Dist. J.O. vs. BGGM		Total Dist. J.O. vs. I.O.	
N_1	N_2	K	Design	Test	Design	Test	Design	Test
4	4	4	.269	1.606	.093	1.224	-.045	.004
16	8	7	.015	-.257	.013	-.148	.000	.077
64	8	9	.010	-.127	-.164	-.318	-.173	-.188
256	16	12	.121	.025	.150	.000	.035	-.032

TABLE VIII(b)
WAVEFORM SGVQ DESIGN METHOD PERCENT REDUCTIONS—RATE TWO

			Shape Dist. I.O. vs. BGGM		Total Dist. J.O. vs. BGGM		Total Dist. J.O. vs. I.O.	
N_1	N_2	K	Design	Test	Design	Test	Design	Test
4	16	3	.525	.210	.602	.385	.000	.197
16	16	4	.034	.256	.094	.779	.063	.525
64	16	5	-.073	-.019	-.022	-.036	.044	.018
256	16	6	-.033	-.219	.204	-.089	.292	.222

rithm to the next is very small. Furthermore, inspection of the final SGVQ's showed that the codebooks produced by the three algorithms were very similar. Thus, it would appear that for these applications, all three algorithms are of comparable practical value. (See Appendix C for an explanation of the similarity.) It was observed while performing the simulations that the I.O. and BGGM algorithms typically took half as much computation time as the J.O. algorithm to reach the same termination threshold. Since the I.O. algorithm has some optimal properties and since it is identical in implementation to the BGGM algorithm except for the choice of $\sigma(x)$ in the shape centroid, it would seem that it is the logical choice.

C. LPC Voice Coding: Design Methods

In [6], the performance of LPC vocoding SGVQ's designed with the BGGM algorithm is compared to the performance of standard VQ's. Our purpose here is to compare the performance of the BGGM algorithm to that of the two new algorithms. The three design methods were used to design SGVQ's to encode tenth-order ($p = 10$) autoregressive models generated by an LPC analysis system. The design sequence of models was extracted from the same speech data used in the waveform coding simulations. As described in [3], it was not necessary to compute the models themselves, but only the first $p + 1$ autocorrelation lags of each speech frame modeled. The speech frames were obtained using 128-point Hamming windows after preemphasis of the sequence by a filter with transfer function $1-0.9z^{-1}$. The frame rate was 50.8 Hz, with a total of 5000 vectors in the design sequence. Various sizes of shape and gain codebooks were tried. The SGVQ's were tested on the same speech data used to test the waveform coding SGVQ's. Using the same preprocessing scheme, a total of 600 vectors were in the test sequence. As in Section IV-B, the BGGM algorithm with binary splitting and termination threshold of $\epsilon = 0.001$ was used to generate initial codebooks. Then, each of the

TABLE IX
VOCODING SGVQ DESIGN METHOD RESULTS

			Design			Test		
N_1	N_2	Design Method	Shape Dist.	Gain Dist.	Total Dist.	Shape Dist.	Gain Dist.	Total Dist.
6	2	BGGM	.049885	.078779	.128664	.069769	.097558	.167324
		I.O.	.049574	.078878	.128450	.069899	.097421	.167320
		J.O.	.050527	.076081	.126608	.068821	.098370	.167192
6	3	BGGM		.020878	.070763		.023033	.092801
		I.O.		.020892	.070466		.022885	.092782
		J.O.	.049887	.020359	.070246	.069776	.023252	.093028
6	4	BGGM		.005404	.055288		.006917	.076685
		I.O.		.005415	.054987		.006945	.076843
		J.O.	.049681	.005311	.054993	.069922	.006793	.076715
6	5	BGGM		.001371	.051256		.001840	.071608
		I.O.		.001378	.050951		.001827	.071724
		J.O.	.049489	.001345	.050834	.069483	.001843	.071327
8	2	BGGM	.031916	.077032	.108948	.052618	.096337	.148957
		I.O.	.031680	.077024	.108704	.052588	.096000	.148588
		J.O.	.032655	.073505	.106160	.053735	.099777	.153512
8	3	BGGM		.020731	.052649		.022216	.074835
		I.O.		.020767	.052445		.022216	.074802
		J.O.	.031864	.019760	.051625	.053096	.022945	.076042
8	4	BGGM		.005467	.037383		.006392	.059010
		I.O.		.005477	.037157		.006307	.058894
		J.O.	.031645	.005055	.036699	.052780	.006489	.059270
8	5	BGGM		.001422	.033337		.001646	.054264
		I.O.		.001416	.033095		.001641	.054229
		J.O.	.031643	.001326	.032970	.052609	.001598	.054208
10	2	BGGM	.017287	.076081	.093368	.042720	.099086	.141808
		I.O.	.017125	.076144	.093267	.042644	.099178	.141822
		J.O.	.017613	.072465	.090077	.042934	.099998	.142934
10	3	BGGM		.020452	.037737		.022550	.065271
		I.O.		.020476	.037601		.022556	.065200
		J.O.	.017139	.019449	.036588	.042832	.022616	.065448
10	4	BGGM		.005203	.022491		.006576	.049295
		I.O.		.005202	.022326		.006577	.049221
		J.O.	.017020	.005076	.022096	.042565	.006293	.048857
10	5	BGGM		.001334	.018621		.001653	.044373
		I.O.		.001333	.018458		.001541	.044185
		J.O.	.017063	.001316	.018379	.042767	.001645	.044413

three algorithms was run with a termination threshold of $\epsilon = 0.0001$, with step 2) of the J.O. algorithm modified as above.

The results of the simulation are summarized in Table IX, which has the same format as Table VII described above, except that the column for block length is deleted (since all shapes have dimension 11). Also, some entries in the average shape distortion column have been deleted to emphasize the fact that a particular shape codebook is identical to the one above that has the same value of N_1 and the same design method.

Table X shows the percent reduction provided by one algorithm over another for several of the distortion figures of Table IX. The format is the same as Table VIII described above, with the block length column deleted. Inside the design sequence, a positive reduction in average shape distortion is exhibited by the I.O. algorithm over the BGGM algorithm. Also, for the design sequence, a positive reduction in average total distortion is exhibited by the J.O. algorithm over the BGGM and I.O. algorithms except for one entry. Thus, the I.O. and J.O. algorithms show the expected improvement in the appropriate distortion figures. As in the waveform coding case, the most significant observation is that the improvement exhibited by the new algorithms inside the design sequence is only slight and does not extend to the test sequence. Thus, all three algorithms appear of comparable practical value for these applications. With the same comments regarding implementation and computational effort, it would seem that the I.O. algorithm is the logical choice.

TABLE X
VOCODING SGVQ DESIGN METHOD PERCENT REDUCTIONS

		Shape Dist. I.O. vs. BGGM		Total Dist. J.O. vs. BGGM		Total Dist. J.O. vs. I.O.	
N_1	N_2	Design	Test	Design	Test	Design	Test
6	2	.623	-.186	1.598	.079	1.434	.077
6	3			.731	-.245	.312	-.265
6	4			.534	-.039	-.011	.167
6	5			.823	.392	.230	.554
8	2	.739	.057	2.559	-3.058	2.340	-3.314
8	3			1.945	-1.613	1.564	-1.658
8	4			1.830	-.441	1.233	-.638
8	5			1.101	.103	.378	.039
10	2	.937	.178	3.525	-.794	3.420	-.784
10	3			3.045	-.271	2.694	-.380
10	4			1.756	.889	1.030	.740
10	5			1.300	-.090	.428	-.516

It is observed in Table X that for a fixed gain codebook size (N_2), each of the three distortion comparisons increases as shape codebook size (N_1) increases. In the waveform coding case, such a trend could be explained by noting that the average number of training sequence vectors per shape cell decreases as N_1 increases. Appendix C reasons that, for waveform coding, the difference in the shape centroid calculations of the three algorithms will be most noticeable when the number of training vectors per cell is low. We conjecture that a similar reasoning holds for the vocoding shape centroid calculations, but have been unable to come up with such an analysis.

Another trend in Table X is that for a fixed shape codebook size, the reduction in total distortion by the J.O. algorithm over the other methods descreases as gain codebook size increases. In the case of J.O. versus I.O., this is expected: the shape centroid calculations differ by whether $[\sigma^*(x, \hat{a}_i)]^+$ or a quantized version is used for $\sigma(x)$, so that as N_2 increases and the quantization becomes more fine, the calculations become similar. For the case of J.O. versus BGGM, the same reasoning implies that, as N_2 increases, the amount of reduction approaches that of I.O. versus BGGM. Since the shape centroid computations of the I.O. and BGGM do not involve the gain codebook, it is expected that the reduction in total distortion for I.O. versus BGGM does not depend on N_2. The reduction in total distortion for I.O. versus BGGM is omitted from Table X for brevity, but inspection of Table IX verifies this behavior. Thus, the decreasing reduction in total distortion with increasing N_2 for J.O. versus BGGM is due to that for J.O. versus I.O.

V. CONCLUDING REMARKS

We introduced the shape-gain vector quantizer, a vector quantizer whose codebook is the Cartesian product of a codebook of vectors (shapes) and a codebook of scalars (gains). The SGVQ was applied to waveform coding with quadratic-error distortion and provided a performance/complexity tradeoff when compared to a standard VQ. Experimentally, Figs. 4 and 5 showed that the SGVQ requires roughly half as much storage and computation as does the VQ for equal SQNR. For LPC vocoding, it was seen that the SGVQ with Itakura-Saito distortion resulted in a structure identical to that of the gain-separated VQ of [3]. Both types of SGVQ's were seen to have very similar encoding structures. The design method of [3] was extended to waveform coding, and two new algorithms for both waveform and voice coding were presented that refine the technique. Experimentally, all three algorithms gave practically identical performance.

The jointly optimized algorithm as presented is not the only way in which the steps suggested by inequalities (14), (17), and (20) of Section III-A can be combined. For example, the steps suggested by (17) and (20) can be used to form a subiteration where the shape codebook and gain codebook are alternately updated while the partition is held fixed. This subiteration can be viewed as an attempt to optimize the product codebook for the given partition, mimicking the step of the LBG algorithm in which the decoder is optimized for the encoder. This subiteration can be combined with the step suggested by (14), i.e., optimizing the partition for the given product codebook, to yield an algorithm similar to the LBG algorithm. This algorithm is currently being studied.

It was noted in Section II that the normalization of the shape vectors is the key to selecting the optimum shape vector independently of the gain. Nevertheless, without the normalization, there is still some structure to exploit, resulting in a level of encoding complexity between that of the (normalized) SGVQ and the standard VQ of the same codebook size. In view of Fig. 3 and the related discussion in Section IV-A, relaxation of the normalization constraint should allow design of a better codebook. Modifications to the algorithms to allow the design of such codebooks are currently being studied, and experimental simulations to investigate the complexity/performance tradeoff are planned.

APPENDIX A
CENTROID CALCULATIONS FOR JOINTLY AND SEPARATELY OPTIMIZED ALGORITHMS

Shape Centroid: For an arbitrary gain mapping $\sigma(x)$, the conditional distortion associated with shape cell P_i times the probability of being in P_i is

$$\int_{P_i} d(x; \hat{a}_i, \sigma(x))\, dF(x). \tag{A.1}$$

Thus, we need to solve

$$u_i^* = \min_{u_i \in A}{}^{-1} \int_{P_i} d(x; u_i, \sigma(x))\, dF(x). \tag{A.2}$$

For the waveform coding case, A is the ellipsoid $\{a\colon a^T W a = 1\}$. Referring to (6), the quantity to minimize is

$$\int_{P_i} [x^T W x + \sigma^2(x)]\; dF(x) - 2u_i^T W \int_{P_i} \sigma(x)\, x\, dF(x). \tag{A.3}$$

The first term does not depend on u_i. By applying the Cauchy-

Schwarz inequality to the second term, (A.3) is minimized by taking u_i^* as the vector on Λ that is in the direction of $\int_{P_i} \sigma(x)\, x \, dF(x)$, i.e., in the direction of $E\{\sigma(X)\, X | X \in P_i\}$.

For the LPC vocoding case, Λ is the set of vectors of dimension $(p+1)$ whose first component is one. Referring to (8), the quantity to minimize is

$$u_i^T \left(\int_{P_i} \frac{R(x)}{\sigma^2(x)}\, dF(x) \right) u_i + \int_{P_i} [\ln \sigma^2(x) + \ln x_0^2 - 1]\, dF(x). \quad \text{(A.4)}$$

The second term does not depend on u_i. Minimizing the first term is a standard LPC autocorrelation method problem:

$$\text{minimize } u_i^T \tilde{R}_i u_i$$
$$\text{subject to } u_i^T = [1 \;\; u_1 \;\; u_2 \cdots u_p], \quad \text{(A.5)}$$

with $\tilde{R}_i = E\{R(X)/\sigma^2(X) | X \in P_i\}$. The value of u_i^* can be found, for example, by Levinson's recursion.

Gain Centroid (Jointly Optimized Algorithm): For an arbitrary shape mapping $\hat{a}(x)$, the conditional distortion associated with gain cell Q_j times the probability of being in Q_j is

$$\int_{Q_j} d(x; \hat{a}(x), \hat{\sigma}_j)\, dF(x). \quad \text{(A.6)}$$

Thus, we need to solve

$$\lambda_j^* = \min_{\lambda_j \geq 0}{}^{-1} \int_{Q_j} d(x; \hat{a}_i, \lambda_j)\, dF(x). \quad \text{(A.7)}$$

Referring to (6), the quantity to minimize in the waveform coding case is

$$\lambda_j^2 \int_{Q_j} dF(x) - 2\lambda_j \int_{Q_j} x^T W \hat{a}(x)\, dF(x) + \int_{Q_j} x^T W x\, dF(x). \quad \text{(A.8)}$$

For $\int_{Q_j} dF(x) > 0$, this expression is positive quadratic in λ_j and is minimized by

$$\lambda_j^* = \left(\int_{Q_j} x^T W \hat{a}(x)\, dF(x) \right) \left(\int_{Q_j} dF(x) \right)^{-1}, \quad \text{(A.9)}$$

which is $E\{X^T W \hat{a}(X) | X \in Q_j\}$. Of course, if this value is negative, then we set $\lambda_j^* = 0$.

For the LPC vocoding case, from (8), we have for the quantity to minimize

$$\frac{1}{\lambda_j^2} \left(\int_{Q_j} \hat{a}^T(x)\, R(x)\, \hat{a}(x)\, dF(x) \right) + \left(\int_{Q_j} dF(x) \right) \ln \lambda_j^2 + \int_{Q_j} [\ln x_0^2 - 1]\, dF(x). \quad \text{(A.10)}$$

By differentiating with respect to λ_j^2, it is seen that this quantity is minimized by

$$\lambda_j^* = \left(\int_{Q_j} \hat{a}^T(x)\, R(x)\, \hat{a}^T(x)\, dF(x) \right)^{1/2} \left(\int_{Q_j} dF(x) \right)^{-1/2} = (E\{\hat{a}^T(X)\, R(X)\, \hat{a}(X) | X \in Q_j\})^{1/2}. \quad \text{(A.11)}$$

Appendix B
Shape Centroid Calculation for BGGM Algorithm for Waveform Coding

The conditional shape distortion associated with cell P_i times the probability of being in P_i is

$$\int_{P_i} d'(x; \hat{a}_i)\, dF(x). \quad \text{(A.12)}$$

Thus, we need to solve

$$u_i^* = \min_{u_i \in \Lambda}{}^{-1} \int_{P_i} d'(x; u_i)\, dF(x). \quad \text{(A.13)}$$

In the waveform coding case, $\Lambda = \{a: a^T W a = 1\}$. From the left term in (9), the quantity to minimize is

$$\int_{P_i} x^T W x\, dF(x) - \int_{P_i} ([x^T W u_i]^+)^2\, dF(x). \quad \text{(A.14)}$$

The first term does not depend on u_i. Suppose for simplicity that $x^T W u_i$ is nonnegative for $x \in P_i$ so that $[x^T W u_i]^+ = x^T W u_i$. Rearranging the second term shows that minimizing (A.14) is equivalent to maximizing

$$[\sqrt{W} u_i]^T \left[\int_{P_i} (\sqrt{W} x)(\sqrt{W} x)^T\, dF(x) \right] [\sqrt{W} u_i] \quad \text{(A.15)}$$

where $\sqrt{W}$ is any square root of W. Subject to the constraint that $\tilde{u}_i^T W \tilde{u}_i = 1$, this requires finding the largest eigenvector of the middle bracketed quantity of (A.15). This is not an easy task. On the other hand, if we use the approximation and upper bound

$$x^T W u_i \approx \sqrt{x^T W x} \quad \text{for } \tilde{u}_i^T W \tilde{u}_i = 1, \quad \text{(A.16)}$$

then we can approximate the second term of (A.14) as

$$u_i^T W \int_{P_i} \sqrt{x^T W x}\, x\, dF(x). \quad \text{(A.17)}$$

By the Cauchy-Schwarz inequality, this is maximized by taking u_i as the vector on Λ in the direction of $\int_{P_i} \sqrt{x^T W x}\, x \cdot dF(x)$, i.e., in the direction of $E\{\sqrt{X^T W X}\, X | X \in P_i\}$.

Appendix C
Comparison of Waveform Coding Shape Centroid Calculations

For notational convenience, let E denote the conditional expectation operator in shape cell P_i. Assume that W is the identity matrix and that $x^T \hat{a}_i$ is nonnegative for $X \in P_i$. Then the three types of shape centroid calculations have the form $E[\sigma(X)\, X]$, suitably normalized, where $\sigma(x)$ is one of

1) $\sqrt{x^T x}$ (BGGM);
2) $x^T \hat{a}_i$ (I.O.);
3) $\min_{\sigma \in \Sigma}{}^{-1} (\sigma - x^T \hat{a}_i)^2$ (J.O.).

For any $\sigma(x)$, we can write

$$E[\sigma(X)\,X] = E[X^T \hat{a}_i X] + E[w(X)\,X^T \hat{a}_i]\,\hat{a}_i + E[w(X)\,(X^T - X^T \hat{a}_i \hat{a}_i)] \qquad (A.18)$$

where $w(X) = \sigma(x) - x^T \hat{a}_i$. Let $\vec{\lambda}$ be the ray with an endpoint at the origin and direction $\hat{a}_i$. For any of the above choices for $\sigma(x)$, $w(x)$ is symmetric about $\vec{\lambda}$, i.e.,

$$\omega(x) = \omega(x^T \hat{a}_i \hat{a}_i + (x - x^T \hat{a}_i \hat{a}_i)) = \omega(x^T \hat{a}_i \hat{a}_i - (x - x^T \hat{a}_i \hat{a}_i)). \qquad (A.19)$$

Now suppose that $\hat{a}_i$ is a fixed point of the I.O. algorithm, i.e., $\hat{a}_i$ is in the direction of $E[X^T \hat{a}_i X]$ and $\hat{a}_i^T \hat{a}_i = 1$. Then only the last term in (A.18) contributes to a direction other than $\hat{a}_i$. If it happens that X is symmetrically distributed (in P_i) about $\vec{\lambda}$, i.e., X and $[X^T \hat{a}_i \hat{a}_i - (X - X^T \hat{a}_i \hat{a}_i)]$ are identically distributed in P_i, then the last term in (A.18) is zero by the aforementioned symmetry of $\omega(x)$. This is because the contribution to the expectation due to a small region in P_i cancels the contribution due to the region's reflection through $\vec{\lambda}$. (This argument can be made rigorous.)

Of course, X does not, in general, possess such a symmetric distribution. However, since $\hat{a}_i$ is the centroid of P_i under the I.O. algorithm, some degree of symmetry about $\hat{a}_i$ is to be expected. Thus, it is reasonable that contributions to the expectation will tend to cancel out in the last term of (A.18) so that the term will be small in comparison to the others. This suggests that the three centroid calculations will yield similar results. In the case that the underlying distribution is the sample histogram of a training sequence, the symmetry will be least when the number of training vectors in P_i is low. Thus, for a short training sequence, it is likely that the three algorithms will give more differing results than for a large training sequence.

References

[1] H. Abut, R. M. Gray, and G. Rebolledo, "Vector quantization of speech and speech-like waveforms," *IEEE Trans. Acoust., Speech, Signal Processing*, vol. ASSP-30, pp. 423-435, June 1982.

[2] B. S. Atal and S. L. Hanauer, "Speech analysis and synthesis by linear prediction of the speech wave," *J. Acoust. Soc. Amer.*, vol. 50, pp. 637-655, Aug. 1971.

[3] A. Buzo, A. H. Gray, Jr., R. M. Gray, and J. D. Markel, "Speech coding based upon vector quantization," *IEEE Trans. Acoust., Speech, Signal Processing*, vol. ASSP-28, pp. 562-574, Oct. 1980.

[4] R. M. Gray and H. Abut, "Full searched and tree searched vector quantization of speech waveforms," in *Proc. 1982 IEEE Int. Conf. Acoust., Speech, Signal Processing*, May 1982, pp. 593-596.

[5] R. M. Gray, A. Buzo, A. H. Gray, Jr., and Y. Matsuyama, "Distortion measures for speech processing," *IEEE Trans. Acoust., Speech, Signal Processing*, vol. ASSP-28, pp. 367-376, Aug. 1980.

[6] R. M. Gray, A. H. Gray, Jr., G. Rebolledo, and J. E. Shore, "Rate-distortion speech coding with a minimum discrimination information distortion measure," *IEEE Trans. Inform. Theory*, vol. IT-27, pp. 708-721, Nov. 1981.

[7] R. M. Gray and E. D. Karnin, "Multiple local optima in vector quantizers," *IEEE Trans. Inform. Theory*, vol. IT-28, pp. 256-261, Mar. 1982.

[8] R. M. Gray, J. C. Kieffer, and Y. Linde, "Locally optimal block quantizer design," *Inform. Contr.*, vol. 45, pp. 178-198, May 1980.

[9] R. M. Gray and Y. Linde, "Vector quantizers and predictive quantizers for Gauss-Markov sources," *IEEE Trans. Commun.*, vol. COM-30, pp. 381-389, Feb. 1982.

[10] F. Itakura and S. Saito, "A statistical method for estimation of speech spectral density and formant frequencies," *Electron. Commun.*, vol. 53-A, pp. 36-43, 1970.

[11] Y. Linde, A. Buzo, and R. M. Gray, "An algorithm for vector quantizer design," *IEEE Trans. Commun.*, vol. COM-28, pp. 84-95, Jan. 1980.

[12] S. P. Lloyd, "Least squares quantization in PCM," Bell Lab. Memo., 1957; reprinted in *IEEE Trans. Inform. Theory*, vol. IT-28, pp. 127-135, Mar. 1982.

[13] J. Makhoul, "Linear prediction: A tutorial review," *Proc. IEEE*, vol. 63, pp. 561-580, Apr. 1975.

[14] J. D. Markel and A. H. Gray, Jr., *Linear Prediction of Speech*. New York: Springer-Verlag, 1976.

[15] Y. Matsuyama, A. Buzo, and R. M. Gray, "Spectral distortion measures for speech compression," Inform. Syst. Lab., Stanford Univ., Stanford, CA, Tech. Rep. 6504-3, Apr. 1978.

[16] G. Rebolledo, R. M. Gray, and J. P. Burg, "A multirate voice digitizer based upon vector quantization," *IEEE Trans. Commun.*, vol. COM-30, pp. 721-727, Apr. 1982.

[17] M. J. Sabin, "Product code vector quantization," Ph.D. dissertation, Stanford Univ., Stanford, CA, in preparation.

[18] M. J. Sabin and R. M. Gray, "Product code vector quantizers for speech waveform coding," in *Proc. IEEE Global Telecommun. Conf.*, Dec. 1982, pp. 1087-1091.

[19] D. Y. Wong, B. H. Juang, and A. H. Gray, Jr., "An 800 bps vector quantization LPC vocoder," *IEEE Trans. Acoust., Speech, Signal Processing*, vol. ASSP-30, pp. 770-780, Oct. 1982.

[20] D. Y. Wong, B. H. Juang, and A. H. Gray, Jr., "Recent developments in vector quantization for speech processing," in *Proc. 1981 IEEE Int. Conf. Acoust., Speech, Signal Processing*, Mar. 1981, pp. 1-4.

Efficient Bit Allocation for an Arbitrary Set of Quantizers

YAIR SHOHAM, MEMBER, IEEE, AND ALLEN GERSHO, FELLOW, IEEE

Abstract—This paper proposes a new bit allocation algorithm, capable of efficiently allocating a given quota of bits to an arbitrary set of different quantizers. This algorithm is useful in any coding scheme which employs bit allocation or, more generally, codebook allocation. It produces an optimal or very nearly optimal allocation, while allowing the set of admissible bit allocation values to be constrained to nonnegative integers. It is particularly useful in cases where the quantizer performance versus rate is irregular and changing in time, a situation that cannot be handled by conventional allocation algorithms.

I. Introduction

Bit allocation is a major concern in any coding scheme where a given quota of bits must be efficiently distributed among a number of different quantizers. In such schemes, each quantizer is capable of operating at (usually, a fixed number of) different rates by using different quantization tables. The rate is usually measured by the minimum number of bits required to address all the entries in the quantization table which is simply $\log_2 N$ where N is the size of the table. The performance of such a quantizer is characterized by its quantizer function (QF), defined by the average quantization distortion as a function of the rate. The set of QF's is used by the allocation algorithm to best determine the allocation strategy that minimizes the overall coding distortion. Most common of such coding schemes are the transform and the subband coders which will frequently be used as examples in various parts of this paper.

The problem of optimal bit allocation was first addressed (within the context of source coding) by Huang and Schultheiss [1] in their original paper on transform coding. They, however, suggested only an approximate solution to the problem. They assumed a set of identical unit variance quantizers each with a simple exponential QF. They solved the allocation problem for this set by allowing the bits to be any arbitrary real numbers, including negative values. Having such a solution at hand, they tried to find, by trial and error, a set of nonnegative integer allocations close to the continuous solution.

An improvement to this method was proposed by Segall [2]. He removed the requirement that the QF's be exponential, but required that they be strictly convex. As before, all the quantizers were assumed identical. The author provided an optimal solution for this case, allowing, again, the allocations to be real numbers which, however, were constrained to be nonnegative.

In most applications, the number of bits that can be assigned to a quantizer is bounded above by some maximum allowable allocation since there is a limit to how fine the quantizer resolution can be. This is particularly important for a vector quantizer since the codebook size grows exponentially with the number of bits. It is sometimes useful to put a lower bound on the number of bits, that is, to specify some minimum nonnegative allowable rate. This reduces the quantization table storage requirements. In subband coding, for example, limiting the minimum rate to 1 or 2 bits is known to produce better perceptual coding quality. The range of admissible rates, specified by the the lower and upper bounds, may, in general, be different for each quantizer. Also, the bit values within the admissible ranges are usually constrained to be integers for practical reasons. Neither of the above-mentioned allocation methods is applicable in the case of different (finite) ranges of admissible integer bit values.

An optimal allocation algorithm for nonnegative integer allocations was proposed by Fox [3]. Unlike Segall's solution, this algorithm (based on so-called marginal return analysis) is applicable to a set of different quantizers. However, the basic assumption is that all the QF's are convex and strictly decreasing with the number of bits. Also, as above, the case of upper and lower bounded allocations is not covered by this algorithm.

In many cases, the quantizers are not identical since the random sequences quantized by them do not have the same statistics, and each quantizer has been optimized for its own input sequence. Consider, for example, the subband coder where different spectral bands of speech are assigned different quantizers. It is known that the statistics of the high-frequency components of speech are significantly different from those of the lower frequency components.

Convexity of the QF's is essential for the above-mentioned algorithm to perform well. However, it may happen that the quantizer functions are not convex. Convexity is suggested by the rate-distortion (RD) theory in cases were the quantizer performance approaches the RD bound, which implies very long block coding. Practical quantizers operate on short data blocks and do not achieve the

Manuscript received September 19, 1986; revised February 15, 1988.
Y. Shoham is with the AT&T Bell Laboratories, Murray Hill, NJ 07974.
A. Gersho is with the Department of Electrical and Computer Engineering, University of California, Santa Barbara, CA 93016.
IEEE Log Number 8822383.

Reprinted from *IEEE Trans. Acoust., Speech, Signal Processing,* vol. ASSP-36, pp. 1445–1453, Sept. 1988.

RD performance, in which case convexity is not guaranteed. Nonconvex QF's may be generated by a set of quantization tables where not all of them perfectly match the source statistics. This usually happens when a quantization table (more generally, a vector codebook) is built of a multiple of subtables. A typical example to this case is the multistage vector quantizer [9] which is known to have an irregular QF. Nonconvexity of the QF may also happen if the set of quantization tables are not all equally well designed, or if the changing characteristics of a nonstationary source renders some of the tables inefficient. In extreme cases, the QF's may not even be monotone-nonincreasing with the number of bits. This strange anomaly may arise, for example, in multistage quantizers if adding one more bit to the rate by using one more 1-bit codebook is very inefficient and actually increases the distortion. The allocation algorithms mentioned above are not suitable for nonconvex or nonmonotone decreasing QF.

An optimal solution to the allocation problem can be obtained via dynamic programming [4]–[6]. However, the complexity of this technique is rather high. If R bits are to be allocated to a set of quantizers, dynamic programming requires on the order of R^2 arithmetic operations per quantizer. In coding schemes which use long blocks of data, the number of bits per block (R) is usually very high (hundreds, if not thousands), which makes dynamic programming unacceptable for this application.

In this paper we propose a new allocation algorithm, which efficiently solves the allocation problem for an arbitrary set of quantizers, under all the constraints mentioned above, with an acceptable degree of complexity. This algorithm can be applied to any coding scheme where bit allocation is performed. Moreover, since the algorithm performs efficiently for any set of quantizers, the allocation procedure can be dynamically updated by an on-line measurement of the distortion versus rate functions of all the quantizers. In particular, one can perform the "ultimate bit allocation" by measuring the distortion function for each single block. In other words, quantize each entry of the current block by its own quantizer, with all the admissible bit values, record the resulting distortions, and form the QF's. Then, apply the allocation algorithm and get the final bit allocation. This is, of course, a tedious operation, but may very well be worth it in terms of the increase in performance. Note that since, in this case, the measured distortions are not averaged, the QF's may be very erratic, that is, not convex and not even always decreasing with the rate. The main point is that such an ultimate bit allocation is fundamentally impossible with the traditional allocation algorithms.

II. The Allocation Problem

The allocation problem can be formulated as follows. We are given a set of M variable-rate quantizers and a set of M gains represented by the vector $G = (g_1, g_2, \cdots, g_M)$. The input to the kth quantizer is first normalized by the corresponding gain g_k and then quantized with i bits where i belongs to the set of nonnegative integer values $\{p_k, \cdots, q_k\}$. The resulting squared errors for all admissible bit values form the *normalized quantizer function* (NQF) of the kth quantizer, denoted by $\{\Theta_k(i), i = p_k, p_k + 1, \cdots, q_k\}$. Thus, if i bits are used, the actual squared error at the kth quantizer output is $(g_k)^2 \Theta_k(i)$.

Given a quota of bits R, the problem is to find the best way of distributing these bits among the quantizer so as to minimize the total squared error. In other words, find an allocation vector $B = (b_1, b_2, \cdots, b_M)$ which minimizes

$$\sum_{k=1}^{M} (g_k)^2 \Theta_k(b_k)$$

subject to

$$p_k \le b_k \le q_k, \qquad k = 1, \cdots, M$$

and

$$\sum_{k=1}^{M} b_k \le R.$$

Note that the domains $\{p_k, \cdots, q_k\}$ of admissible bit assignments are different for each quantizer (each k). In this work, we impose a practical constraint of nonnegative integer bit values. However, the algorithm can be generalized to allow for any finite set of real numbers as admissible bit assignments. Also, we obviously have $p_k \le q_k \le R$ for all k.

A typical example where such an allocation problem arises is subband coding. In this coding system, signals of several spectral bands are segmented into fixed-length blocks and quantized each by its own separate quantizer. The square root of the block energy is defined as the block gain. The data in the block are first normalized by the block gain and then fed to the quantizer. The (scalar or vector) quantizer is tuned to the characteristics of the normalized data and is able to operate at a predetermined range of bit rates. The problem, precisely formulated above, is how to split a total of R bits (the overall system rate) among all the bands so as to minimize the overall distortion, based on the measured gains and the quantizer characteristics.

The algorithmic solution to the problem, proposed here, is based on a little-known version of the Lagrange-multiplier method, as applied to allocation problems. In the next section, we state the main theorem associated with this method, which is the basis for our algorithm. Following this, the algorithm is developed and its efficiency is demonstrated and discussed.

III. Introduction of the Main Theorem

Let S be the finite set of all admissible allocation vectors. Let $H(B)$ be some real-valued function called the objective-function of B, defined for all allocation vectors B in S. Let $R(B)$ be some real-valued function called the constraint-function of B, defined for all B in S. Consider the following problem (called the constrained problem).

Given a constraint R_c, find

$$\min_{B \in S} H(B) \tag{1a}$$

subject to

$$R(B) \leq R_c. \tag{1b}$$

Theorem: For any $\lambda \geq 0$, the solution $B^*(\lambda)$ to the unconstrained problem

$$\min_{B \in S} \{H(B) + \lambda R(B)\} \tag{2}$$

is also the solution to the constrained problem (1) with the constraint $R_c = R(B^*(\lambda))$, that is, with $R(B) \leq R(B^*(\lambda))$. To simplify notation, we define $R^*(\lambda) = R(B^*(\lambda))$.

The proof of this theorem (which is actually quite simple) and some of its applications can be found in [7]. For completeness we repeat the proof here.

Proof: For the solution B^*, we have

$$H(B^*) + \lambda R(B^*) \leq H(B) + \lambda R(B)$$

for all B in S. Equivalently, we have

$$H(B^*) - H(B) \leq \lambda(R(B) - R(B^*))$$

for all B in S. Since this is true for all B in S, it is obviously true for all B in any subset of S. In particular, it is true for all B in a subset defined by

$$S^* = \{B: R(B) \leq R(B^*)\}.$$

Since $\lambda \geq 0$ we have

$$H(B^*) - H(B) \leq 0, \qquad B \in S^*$$

that is, B^* is the solution to the constrained problem with $R_c = R(B^*)$, and the theorem is proved.

It is important to properly appreciate what this theorem says. It does not guarantee any solution to the constrained problem (1). All it says is that to every nonnegative λ, there is a corresponding constrained problem whose solution is identical to that of the unconstrained problem. The main point is, however, that if $R^*(\lambda)$ happens to be equal to R_c, then $B^*(\lambda)$ is the desired solution to the constrained problem.

As we sweep λ over the range zero to infinity, sets of solutions $B^*(\lambda)$ and constraints $R^*(\lambda)$ are created. The key questions are whether or not one value of $R^*(\lambda)$ coincides with R_c and how to find the corresponding λ. These questions are discussed throughout the development of the algorithm which is based on the above theorem.

IV. Development of the Algorithm

The bit allocation problem formulated in the previous section is a constrained problem specified by

$$S = \{B: b_k \in \{p_k, \cdots, q_k\}, \qquad k = 1, \cdots, M\} \tag{3}$$

$$H(B) = \sum_{k=1}^{M} (g_k)^2 \Theta_k(b_k) \tag{4}$$

$$R(B) = \sum_{k=1}^{M} b_k. \tag{5}$$

For the purpose of this discussion, the squared gains and the normalized quantizer functions can be represented by

$$W_k(b_k) = (g_k)^2 \Theta_k(b_k), \qquad k = 1, \cdots, M. \tag{6}$$

Thus, the unconstrained problem can be written as

$$\min_{B \in S} \left\{ \sum_{k=1}^{M} W_k(b_k) + \lambda \sum_{k=1}^{M} b_k \right\}. \tag{7}$$

The crucial point is that the solution is obtained by minimizing each term of the sum in (7) separately. Denote by $b_k^*(\lambda)$ the kth component of $B^*(\lambda)$ and define

$$S_k = \{p_k, \cdots, q_k\} \tag{8}$$

then $b_k^*(\lambda)$ solves

$$\min_{i \in S_k} \{W_k(i) + \lambda i\}. \tag{9}$$

Given λ, one can solve for all b_k^*'s, sum them all up to get $R^*(\lambda)$, and then compare this value to the desired constraint R_c. If $R^*(\lambda) = R_c$, the desired solution has been achieved. $R^*(\lambda)$ is called the *solution rate* since it is the total number of bits associated with the solution $B^*(\lambda)$.

The solution $b_k^*(\lambda)$ may not be unique. The set of solutions to (9) for the kth quantizer and for a given λ are denoted by $\{b_k^*(\lambda)\}$. Therefore, the solution $B^*(\lambda)$ is also not necessarily unique. We denote the set of solutions for a given λ by $\{B^*(\lambda)\}$. The corresponding set of solution rates is denoted by $\{R^*(\lambda)\}$. The size of the solution set $\{B^*(\lambda)\}$ is at least the size of the set $\{R^*(\lambda)\}$, but may be larger since more than one solution may yield the same solution rate.

Let G be the (finite) set of all solution rates $\{R^*(\lambda)\}$ obtained by sweeping λ from 0 to ∞. A vector B, for which $R(B)$ is not in G, is called an inaccessible point in S since there is no λ for which $B = B^*(\lambda)$. It is desirable that the solution to our constrained problem not be an inaccessible point since an inaccessible solution cannot be found by solving the unconstrained problem for λ.

To proceed further in the development of the algorithm, the following lemmas are needed.

Lemma 1: Let $W(b)$ be a real-valued function over some bounded and closed domain Z on the real line. Let b_1 be a solution to

$$\min_{b \in Z} \{W(b) + \lambda_1 b\}$$

and let b_2 be a solution to

$$\min_{b \in Z} \{W(b) + \lambda_2 b\}$$

then,

$$(\lambda_1 - \lambda_2)(b_1 - b_2) \leq 0$$

for any function $W(b)$.

In other words, as λ increases, the minimizing value of b either decreases or remains unchanged.

Proof: By definition of b_1 and b_2, we have

$$W(b_1) + \lambda_1 b_1 \leq W(b_2) + \lambda_1 b_2$$

$$W(b_2) + \lambda_2 b_2 \leq W(b_1) + \lambda_2 b_1.$$

From this we get

$$W(b_1) - W(b_2) \leq \lambda_1(b_2 - b_1)$$

$$W(b_2) - W(b_1) \leq \lambda_2(b_1 - b_2).$$

Summing both sides of these two inequalities we get

$$0 \leq (b_1 - b_2)(\lambda_2 - \lambda_1)$$

and the lemma is proved.

Lemma 2: The solutions $b_k^*(\lambda)$ for all k and the corresponding solution rates $R^*(\lambda)$ are monotonically nonincreasing with λ, that is, if $\lambda_2 \geq \lambda_1 > 0$, then $b_k^*(\lambda_2) \leq b_k^*(\lambda_1)$ and $R^*(\lambda_2) \leq R^*(\lambda_1)$.

Proof: Apply Lemma 1 to (9) which proves Lemma 2 for $b_k^*(\lambda)$ for all k, and hence for the sum $R^*(\lambda)$.

Lemma 3: The solution sets $\{B^*(\lambda_1)\}$ and $\{B^*(\lambda_2)\}$ for two different values λ_1 and λ_2 can have at most one solution in common.

Proof: Suppose $\lambda_2 > \lambda_1$. By Lemma 2, the lowest value in $\{b_k^*(\lambda_1)\}$ is greater or equal to the greatest value in $\{b_k^*(\lambda_2)\}$. Therefore, at most one value (the lowest for λ_1 which could be equal to the greatest for λ_2) can be common to both sets. Since this is true for all k, the lemma is proved.

A corollary of this lemma is that two distinct values of λ cannot have the same set of solutions, unless this set contains a single solution. This implies that a set of multiple solutions $\{B^*(\lambda)\}$ occurs at an isolated value of λ since $\lambda + \epsilon$ has a different set of solutions $\{B^*(\lambda + \epsilon)\}$ (with at most one solution in common), regardless of how small $|\epsilon|$ is, as long as $\epsilon \neq 0$. The values of λ which correspond to multiple solutions are called *singular values* of λ.

Lemma 4: Let λ have a single solution $B^*(\lambda)$. If singular values less than λ exist, then the nearest one to λ, denoted by λ_1, has a solution $B^*(\lambda_1) = B^*(\lambda)$. Similarly, if singular values greater than λ exist, then the nearest one to λ, denoted by λ_2, has a solution $B^*(\lambda_2) = B^*(\lambda)$.

Proof: Since λ is nonsingular we have from (7)

$$W_k(b_k^*) - W_k(b_k) < \lambda(b_k - b_k^*); \qquad b_k \neq b_k^*;$$
$$k = 1, \cdots, M.$$

The multiple solutions associated with λ_2 imply that there are k's for which there are b_k's such that $b_k - b_k^* < 0$. Therefore, there exists $\epsilon > 0$ such that

$$W_k(b_k^*) - W_k(b_k) < (\lambda + \epsilon)(b_k - b_k^*)$$
$$k: b_k - b_k^* > 0$$

$$W_k(b_k^*) - W_k(b_k) = (\lambda + \epsilon)(b_k - b_k^*)$$
$$k: b_k - b_k^* < 0.$$

Let ϵ_m^+ be the minimum value of an ϵ over all k such that $b_k - b_k^* < 0$. Clearly, $\lambda + \epsilon_m^+$ is singular. It has at least two solutions, one of which is $B^*(\lambda)$. Also, $\lambda + \epsilon_m^+ = \lambda_2$ since it is the closest singular value to λ.

Using the same arguments, it is easily shown that there exists $\epsilon_m^- < 0$ such that $\lambda + \epsilon_m^- = \lambda_1$ and λ_1 has the solution $B^*(\lambda)$.

This lemma has two important corollaries. First, two adjacent singular values have one and only one solution in common. Second, all nonsingular values of λ between two adjacent singular values have the same (single) solution.

Fig. 1 illustrates a typical dependency of the solution rate $R^*(\lambda)$ on λ. It shows a decreasing staircase curve where the discontinuities correspond to singular values of λ. The black dots at the singular points represent multiple values of constraints which correspond to multiple solutions.

The main conclusion from the above lemmas is that it is not necessary to sweep λ over all possible values (all values on the nonnegative real line) in order to find all the solutions to the unconstrained problem. It is sufficient to check only the singular points, that is, only those values of λ for which more then one solution exists.

To find a constraint which satisfies $R^*(\lambda) = R_c$ (if it exists), it is not necessary to search over all the singular values. Because of the monotonicity of $R^*(\lambda)$, only a small subset of singular λ's needs to be searched. The search strategy which is the main issue in the proposed algorithm will soon be discussed.

Locating the desired solution requires the knowledge of the singular λ's. For this we have Lemma 5.

Lemma 5: Let λ_1 be singular, $\lambda > \lambda_1$, and $B^*(\lambda)$ be a common solution associated with λ and λ_1. Define the set

$$S_k^+ = \{b_k^* + 1, \cdots, q_k\}. \tag{10a}$$

This set is nonempty if $b_k^* + 1 \leq q_k$. Also, define the set

$$M^+ = \{k \in \{1, \cdots, M\} : S_k^+ \text{ is nonempty}\}. \tag{10b}$$

This defines the set of all the QF's for which an increase in bit assignment is possible. Then, the singular value λ_1, which is necessarily the closest to λ from below, is given in terms of λ by

$$\lambda_1 = \max_{k \in M^+} \max_{i \in S_k^+} \frac{W_k(b_k^*(\lambda)) - W_k(i)}{i - b_k^*(\lambda)}. \tag{10c}$$

Proof: Let $B^*(\lambda_1)$ be the common solution, that is,

$$b_k^*(\lambda_1) = b_k^*(\lambda), \qquad k = 1, \cdots, M.$$

Since λ_1 is singular, it has at least one more solution in addition to $B^*(\lambda_1)$. This means that there exists at least

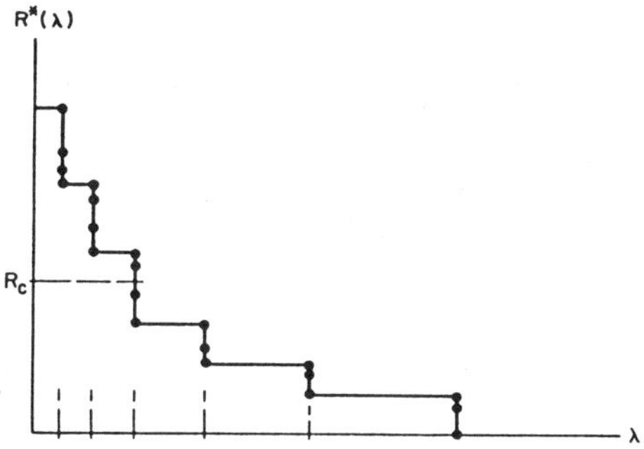

Fig. 1. Typical dependency of the solution rate $R^*(\lambda)$ on the Lagrange multiplier λ.

one value of k and at least one value of i such that

$$k \in M^+$$
$$i \in S_k^+$$

and

$$W_k(b_k^*(\lambda_1)) + \lambda_1 b_k^*(\lambda_1) = W_k(i) + \lambda_1 i.$$

For other values of k and i, we have

$$W_k(b_k^*(\lambda_1)) + \lambda_1 b_k^*(\lambda_1) \le W_k(i) + \lambda_1 i.$$

Now, replace $b_k^*(\lambda_1)$ by $b_k^*(\lambda)$ in the last two expressions, and express λ_1 as a function of the other terms. This gives

$$\lambda_1 \ge \frac{W_k(b_k^*(\lambda)) - W_k(i)}{i - b_k^*(\lambda)}, \qquad i \in S_k, \qquad k \in M^+.$$

The equality in the last expression holds if and only if i is a solution for some k in M^+. To find λ_1, we search for the maximum of the right term in the last inequality, that is, compute (10c). Thus, Lemma 5 is proved.

Note that if the sets S_k^+ are empty for all k, then there is no λ_1 satisfying $\lambda_1 < \lambda$ which means that there are no singular points below λ.

The pair k', i' which solves (10c) determines a solution to the unconstrained problem for λ_1. This solution is obtained by setting $b_{k'}^*(\lambda_1) = i'$ and $b_k^*(\lambda_1) = b_k^*(\lambda)$ for all other values of k.

Similarly, we have Lemma 6.

Lemma 6: Let λ_2 be singular, $\lambda_2 > \lambda$, and $B^*(\lambda)$ a common solution associated with λ and λ_2. Define the set

$$S_k^- = \{p_k, \cdots, b_k^*(\lambda) - 1\} \tag{11a}$$

which is nonempty only if $b_k^*(\lambda)) \ge 1$. Define the set

$$M^- = \{k : k \in \{1, \cdots, M\}; \quad S_k^- \text{ is nonempty}\}. \tag{11b}$$

Then, the singular value λ_2 is given in terms of λ by

$$\lambda_2 = \min_{k \in M^-} \min_{i \in S_k^-} \frac{W_k(i) - W_k(b_k^*(\lambda))}{b_k^*(\lambda) - i} \tag{11c}$$

which can be proved in the same way as for (10c). Note that if S_k^- are empty for all k, there is no λ_2 which satisfies $\lambda_2 > \lambda$.

As for (10c), the pair k', i' which solves (11c) determines a solution to the unconstrained problem for λ_2. This solution is obtained by setting $b_{k'}^*(\lambda_2) = i'$ and $b_k^*(\lambda_2) = b_k^*(\lambda)$ for all other values of k.

The last two lemmas say that by knowing a singular value, the next from below or from above can be found, and thus, all singular values and all the solutions $B^*(\lambda)$ can be located. This motivates the following basic search procedure which is the essence of the algorithm.

1) Obtain an initial value λ (in a way soon to be discussed).

2) Solve the unconstrained problem by using (9). If λ is not singular, there is only one such solution and one constraint $R^*(\lambda)$. If λ is singular, then there are at least two different solutions. Find the two solutions from $\{B^*(\lambda)\}$ with greatest and smallest constraints denoted by $R_l^*(\lambda)$ and $R_h^*(\lambda)$, respectively.

3) If the desired constraint R_c is such that

$$R_l^*(\lambda) \le R_c \le R_h^*(\lambda),$$

then obtain all solutions in $\{B^*(\lambda)\}$ [using (9)] and find the one for which the constraint, denoted by $R_a^*(\lambda)$, is the closest to R_c from below. If $R_c = R_a^*(\lambda)$, an optimal solution has been found. If not, an approximate solution, denoted by B_a^*, has been found. Stop the search.

4) If $R_c < R^*(\lambda)$ [or $R_c < R_l^*(\lambda)$], find the next singular λ using (11) [and the current solution $B^*(\lambda)$] which corresponds to $R^*(\lambda)$ [or $R_l^*(\lambda)$]. Then, go to Step 2.

5) If $R_c > R^*(\lambda)$ [or $R_c > R_h^*(\lambda)$], find the next singular λ using (10) and the current solution which corresponds to $R^*(\lambda)$ [or $R_h^*(\lambda)$]. Then, go to Step 2.

If the search terminates with only an approximate solution, which means that the optimal solution is an inaccessible point, then this solution can be adjusted by adding a few bits until the sum of the modified allocations becomes equal to R_c. The criterion for adding bits is to cause the greatest reduction in the distortion $H(B^*(\lambda))$. We add one bit at a time according to the following steps.

6) Find k' which solves

$$\min_{k \in M^+} \{W_k(b_k^* + 1) - W_k(b_k^*)\}. \tag{12}$$

7) Update the solution by adding 1 to $b_{k'}^*$.

8) Repeat Steps 6 and 7 until the sum of the allocations equals R_c.

Since B_a^* is not an optimal solution, it is natural to ask how far it is from the desired optimal solution. With the aid of the following lemma, a bound on the difference between the distortions of the optimal and approximate solutions can be found.

Lemma 7: Let B be the desired solution of the original constrained problem with the constraint R_c. Let $B^*(\lambda_1)$ and $B^*(\lambda_2)$ be two solutions such that

$$R^*(\lambda_1) \le R_c \le R^*(\lambda_2). \tag{13a}$$

Then,

$$H(B^*(\lambda_2)) \le H(B) \le H(B^*(\lambda_1)). \tag{13b}$$

Proof: The left side of (13a) directly implies the right side of (13b) by the main theorem. For the left side of (13b) we have

$$H(B^*(\lambda_2)) + \lambda_2 R^*(\lambda_2) \le H(B) + \lambda_2 R(B)$$
$$\le H(B) + \lambda_2 R_c.$$

Since $\lambda_2 \ge 0$ we have

$$H(B^*(\lambda_2)) - H(B) \le \lambda_2(R_c - R^*(\lambda_2)) \le 0$$

which means that the right side of (13b) implies the left side of (13b) and the lemma is proved.

Now, let $B^*(\lambda_1)$ be the solution at the end of the search (Step 3). By adding more bits to this solution, we reduce the distortion. However, upon reaching $R(B_a^*) = R_c$, the distortion $H(B_a^*)$ cannot be lower than $H(B)$ and cannot be higher than $B^*(\lambda_1)$. This means that (13b) still holds if we replace $H(B)$ by $H(B_a^*)$, which implies the following:

$$H(B_a^*) - H(B) \le H(B^*(\lambda_1)) - H(B^*(\lambda_2)). \quad (14a)$$

Thus, the difference in the distortions associated with the optimal and approximate solutions to the constrained problem is within a range determined by two solutions to the unconstrained problem which are the closest to the desired solution from above and below.

A more useful bound is expressed terms of the difference in decibels between the distortions of the optimal and approximate solutions. The Allocation Performance Bound (APB) is defined by

$$APB = -10 \log_{10} \frac{H(B_a^*) - H(B^*(\lambda_1)) + H(B^*(\lambda_2))}{H(B_a^*)} \quad (14b)$$

and from (14a) we have

$$10 \log \frac{H(B_a^*)}{H(B)} \le APB. \quad (14c)$$

This bound can be computed in an actual application of the algorithm to evaluate the approximation.

V. Variations of the Basic Algorithm

The computational effort associated with the basic algorithm described in the previous section can be significantly reduced, based on the following observation.

Let λ_1, λ_2, and λ_3 be three Lagrangian multipliers which satisfy

$$\lambda_1 \le \lambda_2 \le \lambda_3. \quad (15)$$

Then, by virtue of Lemma 2, the corresponding solutions satisfy

$$b_k^*(\lambda_3) \le b_k^*(\lambda_2) \le b_k^*(\lambda_1), \qquad k = 1, \cdots, M. \quad (16)$$

This means that, given $B^*(\lambda_1)$ and $B^*(\lambda_3)$, the solution for λ_2 can be searched for in a considerably smaller range than that specified in (9)–(11). The search regions for these expressions [see (8), (10a), and (11a)] are now redefined as

$$S_k = \{i : b_k^*(\lambda_3) \le i \le b_k^*(\lambda_1)\}, \quad k = 1, \cdots, M \quad (17)$$

$$S_k^+ = \{b_k^*(\lambda_3) + 1, \cdots, b_k^*(\lambda_1)\}, \quad k = 1, \cdots, M \quad (18)$$

$$S_k^- = \{b_k^*(\lambda_1), \cdots, b_k^*(\lambda_3) - 1\}. \quad (19)$$

With these modified search regions, we can describe the first variant of the basic algorithm.

Variant 1:

1) Find initial values λ_1 and λ_3 which satisfy

$$R^*(\lambda_1) \le R_c \le R^*(\lambda_3). \quad (20)$$

One way to do this is as follows.

1.1) Find an initial λ_1 (to be discussed in the next section).

1.2) Solve (9) with the old S_k as in (8) for all k and get $R^*(\lambda_1)$.

1.3) If $R^*(\lambda_1)$ is greater than R_c, set $\lambda_3 = \lambda_1$ and $B^*(\lambda_3) = B^*(\lambda_1)$. Increase λ_1 by adding to it some preselected update constant. Go to Step 1.2.

1.4) If $R^*(\lambda_1)$ is lower than R_c, decrease λ_1 by subtracting the update constant. Go to Step 1.2.

1.5) Repeat the above until the desired λ_1 and λ_3, for which (20) holds, have been obtained.

2) Continue as in the basic procedure but with the modified search regions as in (17)–(19). Note that as new solutions for λ's are found, the search regions can be dynamically modified, thereby, continuously decreasing the search intensity.

The value of the update constant of Step 1 determines the rate at which the procedure approaches the desired two initial solutions. If this constant is large, the two solutions can be obtained after one or two iterations. However, they may be far apart which means large search regions. If this constant is very small, many iterations are needed, but the resulting search regions are small since the two solutions will be close to each other. The best update constant should be experimentally found by actually measuring the overall search complexity while running the algorithm with various values of this constant.

It was observed during an experimental evaluation of the algorithm that more than two solutions for singular λ's were never detected. Therefore, we may assume that there cannot be more than two solutions for any singular λ with a very small risk, if at all, of skipping a solution which might have been the desired one.

Under this assumption, it is not necessary to search for all the solutions of a singular λ. The solution found by computing either (10c) or (11c) is the new solution of the new λ. This means that once λ_1 and λ_2 have been found, (9) need not be solved in order to find the next λ. This considerably reduces the search intensity.

We therefore have a new variant of the algorithm described as follows.

Variant 2:

1) Find two initial values λ_1 and λ_3 as in Variant 1.

2) Start from λ_3 and obtain successive singular values and solutions by successively solving (10c) for new λ_3 and new solution $B^*(\lambda_3)$. Update the search regions as described in Variant 1.

3) When $R^*(\lambda_3)$ becomes less than or equal to R_c, stop the search.

4) If necessary, adjust the solution $B^*(\lambda_3)$ to obtain a final approximate solution B_a^* as in Steps 6–8 of the basic algorithm.

VI. Initial Lagrange Multiplier

The computational complexity of the algorithm strongly depends on the initial λ since the closer this λ is to the final value, the fewer iterations are needed. The following method of obtaining an initial λ was empirically found to yield a low average computational intensity. The method is based on the fact that a necessary condition for a set $\{b_k\}$ to be a solution is

$$W_k(b_k) - W_k(b_k + 1) \leq \lambda \leq W_k(b_k - 1) - W_k(b_k)$$

$$p_k + 1 \leq b_k \leq q_k - 1$$

$$k = 1, 2, \cdots, M$$

which can very easily be verified. Since the above is true for all k, we can get an "average" initial λ by summing either one of the two inequalities in the last expression with respect to k. Using the inequality on the left, we obtain

$$\lambda \approx \frac{1}{M} \sum_{k=1}^{M} \{W_k(b_k) - W_k(b_k + 1)\}.$$

The b_k's are obviously unknown, therefore, we replace them by an "average" of their admissible values. The value $a_k = \text{INT}\{(q_k + p_k)/2\}$, where INT $\{\cdot\}$ means rounding to the nearest integer, was experimentally found to be a good replacement to the b_k's for this purpose. The initial λ is, therefore, computed by

$$\lambda = \frac{1}{M} \sum_{k=1}^{M} \{W_k(a_k) - W_k(a_k + 1)\}. \quad (21)$$

However, this initialization method is not necessarily the best one, and further study of this problem may yield a better solution.

VII. Evaluation of the Algorithm

In this section, we provide experimental results which demonstrate the efficiency of the proposed allocation algorithm. The algorithm was applied to a sequence of 1000 gain vectors $G_n = (g_{1,n}, \cdots, g_{64,n})$, $n = 1, \cdots, 1000$, with a given set of 64 NQF's $\Theta_k(b)$, $k = 1, \cdots, 64$. Three different sets of NQF's were considered: a set of 64 identical exponential NQF's over the bit range [0, 24] (Set 1); a set of 64 identical but nonconvex NQF's over [0, 24] (Set 2), and a set of different nonconvex NQF's each defined over a specific range $[0, q_k]$, $k = 1, \cdots, 64$ (Set 3).

The gain vectors were computed from a sequence of 1000 DCT blocks of length 256, extracted from a speech waveform. Each DCT block was partitioned into 64 4-dimensional subvectors and the RMS energy of the ith subvector was defined as g_i.

The experiment with Set 1 simulated a situation where all 64 4-D DCT subvectors had the same statistical characteristics and they were all quantized by the same *ideal* 4-D vector quantizer whose NQF was assumed to be $\Theta_k(i) = 2.2 \times 10^{0.5i}$. The quantizer factor 2.2 was actually obtained by averaging the performance of a 4-D vector quantizer over a long training set of 4-D normalized DCT subvectors.

The experiment with Set 2 simulated the case where all the subvectors in the DCT block were quantized by the same three-stage vector quantizer. Details on the structure of such a quantizer can be found in [8]. The main point is that the NQF of this quantizer is not convex, that is, the difference function $\Theta_k(i + 1) - \Theta_k(i)$ is not monotonically decreasing. This behavior is caused by the transitions from one stage to another in the quantization process. The NQF used in this experiment is depicted in Fig. 2. The figure shows the normalized distortion in decibels (to capture the whole range) versus number of bits.

The experiment with Set 3 simulated the case where each subvector in the DCT block was quantized by a separate vector quantizer, tuned to statistics of that subvector. Again, three-stage vector quantization was assumed. Furthermore, each quantizer had a different upper bound on the admissible bit assignment. The lowest bit assignment was zero for all the quantizers. The set of the 64 NQF's, which were found by actually designing 64 three-stage VQ's over a long training set of 4-D DCT subvectors, is shown in Fig. 3. The NQF's are shown in 6 separate groups of graphs for better clarity. A straight horizontal line at the end of a graph indicates a region of inadmissible bit assignments. In other words, an NQF is defined up to the beginning of such a line. Observe that these NQF's are not only nonconvex, they are sometimes not even monotonically decreasing.

The allocation algorithm was applied to the sequence of 1000 gain vectors with Set 1, Set 2, and Set 3. The total number of bits to allocate to each block was 128, corresponding to 2 bits per gain (0.5 bit per subvector component). In each case we measured the following.

1) The percentage of truly optimal allocations, out of the total 1000 allocations.

2) The signal-to-distortion ratio (SNR), defined as the ratio of the mean square average of the gain sequence to the average distortion, where the distortion was calculated for each block according to (4).

3) The average and maximum bounds (APB) on the deviation in performance from that of the optimal solution [see (14)].

4) The average of the arithmetic operations (mul. add.

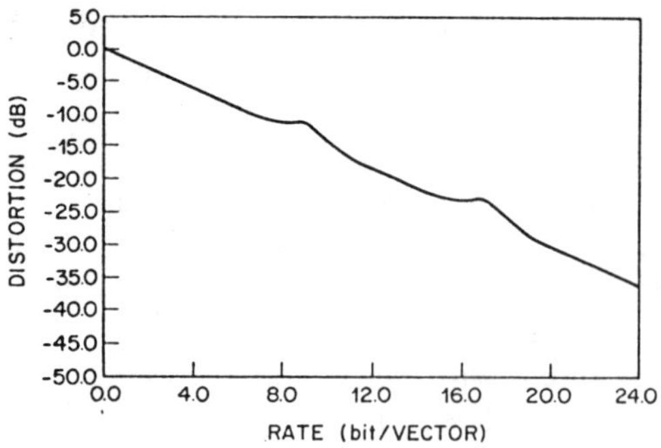

Fig. 2. Quantizer function of Set 2.

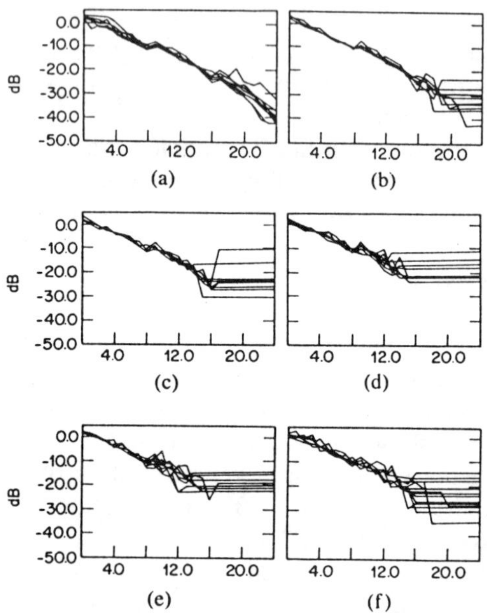

Fig. 3. The 64 quantizer functions of Set 3. (a) $k = 1, \cdots, 10$. (b) $k = 11, \cdots, 20$. (c) $k = 21, \cdots, 30$. (d) $k = 31, \cdots, 40$. (e) $k = 41, \cdots, 50$. (f) $k = 51, \cdots, 64$.

compare) performed by the algorithm per gain component.

The experimental results are summarized in Table I. As shown, the algorithm was fully optimal (100 percent optimal allocations) in the case of Set 1. It is not difficult to prove that this is always the case for convex NQF's. It simply means that in this case, there are no inaccessible points in the allocation space.

In the case of Set 2, 82 percent of the allocations were optimal. The status of the rest was unknown. However, for the 18 percent possibly nonoptimal allocations, the average bound on the increase in the distortion (compared to that of an optimal allocation) was only 0.21 dB and the maximum bound was 0.36 dB. This shows that although it is not strictly optimal in all cases, the algorithm is highly efficient.

We also applied the allocation algorithm proposed by Fox [3] to Set 2 and compared the resulting SNR to our result. As shown, the SNR produced by Fox's algorithm was 3.7 dB lower. This means that Fox's algorithm, which is known to be optimal for convex NQF's, breaks down when applied to a set of nonconvex NQF's, and may not even be used as an approximation to the optimal algorithm.

TABLE I
EXPERIMENTAL RESULTS

QUANTIZER SET	OPTIMAL ALLOCATIONS (%)	SNR (dB)	AVERAGE APB (dB)	MAXIMUM APB (dB)	AVERAGE OP./GAIN
SET 1	100	10.9	0	0	212
SET 2	82	14.3 (FOX: 10.6)	0.21	0.36	204 (FOX: 275)
SET 3	58	12.4	0.17	0.53	191

For Set 3, 58 percent of the allocations were optimal. The rest were unknown. The average bound, however, was 0.17 dB and the maximum one was 0.53 dB. This demonstrates that even in the case of nonconvex, nondecreasing, and nonidentical QF's, the algorithm still is essentially optimal.

VIII. Conclusion

In this paper, we propose an efficient bit allocation algorithm for an arbitrary set of quantizers. In contrast to earlier bit allocation algorithms, this algorithm is not based on prior assumptions about the nature of the given set of quantizers. Rather, bit allocation is optimized for the true performance of each individual quantizer, as measured from the data. The algorithm can be used in any coding scheme which requires bit allocation or, more generally, codebook allocation.

The algorithm is based on the Lagrangian method for integer allocation. Two variants of the basic algorithm are described. More work can be done to improve the algorithm and reduce its complexity, based on the same basic idea.

Experimental results, provided in this paper, demonstrate that the algorithm is essentially optimal and can be very useful in many applications.

References

[1] J. J. Huang and P. M. Schultheiss, "Block quantization of correlated Gaussian random variables," *IEEE Trans. Commun. Sys.*, vol. CS-11, pp. 289–296, Sept. 1963.

[2] A. Segall, "Bit allocation and encoding for vector sources," *IEEE Trans. Inform. Theory*, vol. IT-22, pp. 162–169, Mar. 1976.

[3] B. Fox, "Discrete optimization via marginal analysis," *Manage. Sci.*, vol. 13, no. 3, pp. 210–216, Nov. 1966.

[4] O. A. Zuniga, R. M. Haralick, and R. L. Klein, "Adaptive image data compression," in *Proc. IEEE ICASSP-79*, Feb. 1979, pp. 583–587.

[5] A. V. Trushkin, "Bit number distribution upon quantization of a multivariate random variable," *Problems of Inform. Transmission*, vol. 16, pp. 76–79, 1980 (Translated from Russian).

[6] ——, "Optimal bit allocation algorithm for quantizing a random vector," *Problems of Inform. Transmission*, vol. 17, pp. 156–161, 1981 (Translated from Russian).

[7] H. Everett, "Generalized Lagrange multiplier method for solving problems of optimum allocation of resources," *Operation Res.*, vol. 11, pp. 399–417, 1963.

[8] Y. Shoham, "Hierarchical vector quantization with application to speech waveform coding," Ph.D. dissertation, Univ. Calif. at Santa Barbara, Dep. Elec. Comput. Eng., Mar. 1985.

[9] B. H. Juang and A. H. Gray, "Multiple stage vector quantization for speech coding," in *Proc. Int. Conf. ASSP*, Paris, France, May 1982, pp. 597–600.

EMBEDDED CODING OF SPEECH : A VECTOR QUANTIZATION APPROACH.

Amine Haoui and David G. Messerschmitt.

Department of Electrical Engineering and Computer Science
University of California,
Berkeley, California 94720.

ABSTRACT

Embedded speech coders are characterized by the property that their output quality degrades gracefully as their bit rate is decreased. Design methods for vector quantizers with the embedded coding property are presented and their performance simulated for the medium-band of 32-8 Kbits/sec. Listening tests indicate that these coders can provide good quality speech at 32 and 24 Kbits/sec and intelligible speech down to 8 Kbits/sec.

1. Introduction

It is often the case in communication networks that the requests for transmission over a link of fixed capacity exceed that capacity. This usually results in some customers having to wait in a queue or be simply denied service. An alternative is to reduce the bit rate allocated to each customer by an amount which insures that all customers are accommodated at a lower grade of service. When the transmitted signal is the output of a speech encoder, this strategy is possible only if the reduction in the transmission rate results in an acceptable decrease in the quality of the decoder output.

An embedded coder (EC) is a source coder with the property that the fidelity of its reproduction of the input signal degrades gracefully as the bit rate is decreased in steps from a maximum rate of R_{max} to a minimum rate of R_{min}. The rule is that the encoder does not know the actual bit rate being transmitted, is told the order in which bits will be discarded (for example the transmitted bit stream might be byte oriented, with the bits discarded in order from least to most significant). This general and vague definition can fit any source coder as long as the term "gracefully" is not made more precise. Nevertheless, it does describe in simple terms the idea of embedded coding and more precise definitions can only be formulated for specific applications. In the context of speech signals, for example, one can elaborate on the definition by adding that the output speech must have toll quality at R_{max}, and must still be intelligible at R_{min} while the "graceful degradation" can be specified in terms of bounds on the rate of decrease of the signal-to-noise ratio with the bit rate.

This paper is primarily concerned with embedded coders for speech in the medium band ranging from an R_{max} of 32 Kbits/sec to an R_{min} of 8 Kbits/sec. The interest in this particular range is motivated by the fact that ADPCM provides toll quality speech at 32 Kbits/sec and that below 8 Kbits/sec the quality is unlikely to be acceptable. In this range of rates, we feel that vector quantizers are a natural choice for the implementation of EC's. It is the goal of this paper to develop design methods for EC's based on vector quantization, and to evaluate their performance by computer simulation.

2. Scalar Embedded Coders

The simplest example of an embedded coder is linear PCM with N bits per sample where, if the coding is appropriately chosen, losing the least significant bit (LSB) during transmission or otherwise will result in a resolution equivalent to $N-1$ bits, and loosing the second LSB to a $N-2$ bit resolution etc. . The degradation in performance is gradual from an N-bit resolution to a 1-bit resolution. Unfortunately the performance of PCM is not satisfactory for speech in the range of bit rates of interest.

ADPCM, which produces good quality speech at 32 Kbits/sec, does not perform well as an embedded coder. Dropping the least significant bit of the encoder output can result in a dramatic deterioration of the quality of the output speech because the predictors in the decoder and encoder would no longer have the same input signal and therefore won't be able to track each other. Goodman [1] proposes a modified ADPCM scheme, wherein only E of the D bits transmitted (the most significant ones) are used in the encoder and decoder prediction loops. The decoder can now track the encoder even if up to $E-D$ LSB's are dropped during transmission. For an adaptive quantizer at least two of the most significant bits must be transmitted since the sign bit cannot be used in a step size adaptation scheme. This means that D must be larger than two, which implies that embedded ADPCM can operate only at rates above 16 Kbits/sec for 8 kHz sampled speech.

We have simulated the performance of embedded ADPCM and the results are given in Section 5 where they are compared with those of the coders proposed in this paper.

3. Vector Quantizers

A K-dimensional vector quantizer (VQ) is a source coder which operates on K consecutive input samples at a time. A VQ is fully determined by a partition $\mathbf{S} = \{ S_1, \cdots, S_N \}$ of $\mathbb{R}^K$ called the encoder partition, a codebook $\mathbf{Y} = \{ \mathbf{y}_1, \cdots, \mathbf{y}_N \}$ of K-dimensional vectors called codewords and a set of N binary words $W = \{ w_1, \cdots, w_N \}$ of length $B = \log_2 N$ bits each. The encoder determines in which of the sets of the encoder partition the input vector lies and transmits to the decoder the binary word in W corresponding to that set. The decoder finds by table lookup the codeword corresponding to the binary word which is then the VQ output.

4. VQ Embedded Coders

Let Q_B be a K-dimensional vector quantizer of rate B bits/vector and let $\mathbf{Y}_0 = \{ \mathbf{y}_{0,1}, \cdots, \mathbf{y}_{0,N} \}$ be its codebook and $\mathbf{S}_0 = \{ S_{0,1}, \cdots, S_{0,N} \}$ its encoder partition. Let $W = \{ w_{0,1}, \cdots, w_{0,N} \}$ be the binary code of Q_B. If during transmission the least significant bit in every word is set to zero by the channel, the performance of the coder becomes equivalent to that of a VQ of rate $B-1$ bits/vector, which we designate by Q_{B-1}, whose codebook $\mathbf{Y}_1$ is a subset of $\mathbf{Y}_0$ of cardinality $N/2$ and whose encoder partition $\mathbf{S}_1$ satisfies the property that each one of its $N/2$ sets is the union of two sets in $\mathbf{S}_0$. More generally, if the channel replaces the i LSB's by zeros (for $i = 1, \cdots, L \leq B-1$) the per-

Reprinted from *Proc. IEEE Conf. Acoust., Speech, Signal Processing,* vol. 4, pp. 1703–1706, March 1985.

formance of the coder becomes equivalent to that of a vector quantizer Q_{B-i} of rate $B-i$ with codebook $\mathbf{Y}_i = \{ \mathbf{y}_{i,1}, \cdots, \mathbf{y}_{i,N_i} \}$ and encoder partition $\mathbf{S}_i = \{ S_{i,1}, \cdots, S_{i,N_i} \}$ which satisfy the following conditions

1) For each set $S_{i,j}$ in $\mathbf{S}_i$ there exists two sets $S_{i-1,j}^1$ and $S_{i-1,j}^2$ in $\mathbf{S}_{i-1}$ such that $S_{i,j} = S_{i-1,j}^1 \cup S_{i-1,j}^2$

2) $\mathbf{y}_{i,j}$ the codeword in $\mathbf{Y}_i$ corresponding to $S_{i,j}$ is identical to $\mathbf{y}_{i-1,j}^1$ the codeword in $\mathbf{Y}_{i-1}$ corresponding to $S_{i-1,j}^1$.

These two conditions will be referred to as the *embedded coder* conditions. We will call the set $Q = \{ Q_B, Q_{B-1}, \cdots, Q_{B-L} \}$ an embedded VQ coder with L rates and define its rate vector as $R = (B, B-1, \cdots, B-L)$ bits/vector. Although any VQ can be thought of as an embedded coder, the key consideration here is to determine ways to design embedded VQ coders whose performance at the various rates satisfies some given constraints. Let $\mathbf{x} \in \mathbf{R}^K$ be the input random vector with probability density $p_{\mathbf{x}}$ and let E denote the expectation with respect to $p_{\mathbf{x}}$. The performance of a K-dimensional quantizer will be measured by its mean-squared error $E(\|\mathbf{x} - Q(\mathbf{x})\|^2)$ where $\|\,\|$ denotes the Euclidean distance in R^K.

Consider first, for simplicity, the problem of designing an embedded VQ coder of rate vector $R = (B, B-1)$ bits/vector. We can start by designing the best possible rate B quantizer Q_B, using a generalized Lloyd algorithm [2]. The choice of the LSB's in the binary code W fully determines $\mathbf{Y}_1$ and $\mathbf{S}_1$ and thus Q_{B-1}. The idea is to assign two binary words which differ only in the LSB to two codewords which are close to each other in Euclidean distance so that replacing the LSB by zero will lead to a small increase in the error. The problem of determining the optimal assignment of binary words for which $E(\|\mathbf{x} - Q_{B-1}(\mathbf{x})\|^2)$ is minimum is solved in [3], where it is shown to be reducible to the *weighted perfect matching* problem in combinatorial optimization. Unfortunately the performance of the embedded coders obtained with this approach is not very good at the lower rates for the rate vectors of interest and this approach will not be considered further.

Alternatively, we can start by designing Q_{B-1} and then determine the best possible quantizer of rate B which satisfies the two embedded coding conditions. This second approach is called the *tree-expanded codebook* approach because of the tree structure of the rate B codebook thus obtained.

4.1. Tree-expanded Codebooks

Let Q_{B-1} be a K-dimensional VQ of rate $B-1$ with codebook $\mathbf{Y}_1 = \{ \mathbf{y}_{1,1}, \cdots, \mathbf{y}_{1,M} \}$ and encoder partition $\mathbf{S}_1 = \{ S_{1,1}, \cdots, S_{1,M}$ where $M = N/2 = 2^{B-1}$. We would like to determine the quantizer Q_B with codebook $\mathbf{Y}_0$ of size N and partition $\mathbf{S}_0$ which satisfies the embedded coding conditions and for which

$$E(\|\mathbf{x} - Q_B(\mathbf{x})\|)^2 \tag{4.1}$$

is minimum.

From the first condition we have that for each set $S_{1,i}$ in $\mathbf{S}_1$ there are two sets $S_{0,i}^1$ and $S_{0,i}^2$ in $\mathbf{S}_0$ such that

$$S_{1,i} = S_{0,i}^1 \cup S_{0,i}^2 \qquad i = 1, 2, \cdots, M. \tag{4.2}$$

With $\mathbf{y}_{0,i}^1$ and $\mathbf{y}_{0,i}^2$ denoting the two codewords in $\mathbf{Y}_0$ with decision regions $S_{0,i}^1$ and $S_{0,i}^2$ in $\mathbf{S}_0$ respectively, we have from the second condition that $\mathbf{y}_{0,i}^1 = \mathbf{y}_{1,i}$.

We can expand (4.1) in terms of the decision regions of Q_{B-1}

$$E(\|\mathbf{x} - Q_B(\mathbf{x})\|^2) = \sum_{i=1}^{M} E(\|\mathbf{x} - Q_B(\mathbf{x})\|^2 \mid \mathbf{x} \in S_{1,i}) Pr(\mathbf{x} \in S_{1,i}) \tag{4.3}$$

Each of the conditional mse terms in (4.3) can in turn be expanded as

$$E(\|\mathbf{x} - Q_B(\mathbf{x})\|^2 \mid \mathbf{x} \in S_{1,i}) E(\|\mathbf{x} - \mathbf{y}_{1,i}\|^2 \mid \mathbf{x} \in S_{0,i}^1) Pr(\mathbf{x} \in S_{0,i}^1 \mid \mathbf{x} \in S_{1,i})$$

$$+E(\|\mathbf{x} - \mathbf{y}_{0,i}^2\|^2 \mid \mathbf{x} \in S_{0,i}^2) Pr(\mathbf{x} \in S_{0,i}^2 \mid \mathbf{x} \in S_{1,i}) \tag{4.4}$$

The mse of Q_B is minimized if $S_{0,i}^1, S_{0,i}^2$ and $\mathbf{y}_{0,i}^2$ are chosen to minimize (4.4) independently for each $i \in \{ 1, \cdots, M \}$.

The problem is almost identical to that of designing an optimal memoryless VQ with a codebook size of two for the conditional distribution $p_{\mathbf{x} \mid \mathbf{x} \in S_{1,i}}$. The difference between the two problems is that in this case we have the additional constraint that one of the two codewords is fixed. As in the design of memoryless VQ's, the best we can do is derive an algorithm which converges to a fixed point which hopefully achieves a good value of the cost function. Algorithm A below, is one such algorithm and it is identical to the Lloyd-type algorithms with a provision for the additional constraint of the problem. This provision is in the codebook update step (step 3) where the fixed codeword $\mathbf{y}_0^1$ is left unchanged while the other codeword is replaced by the centroid of its decision region.

As is commonly the case in VQ design, the algorithm runs on a long training sequence TS of vectors from the source. Let TS_i be the subsequence of vectors from TS which lie in $S_{1,i}$.

Algorithm A

step 0: Initialize

Given a real $\epsilon > 0$, and an initial codeword $\mathbf{y}^0$. Let $m = 0$ and $d^0 = \infty$.

step 1: Partition TS_i

Let $U_1^m = \{ \mathbf{x} \in TS_i \mid \|\mathbf{x} - \mathbf{y}_{1,i}\| \leq \|\mathbf{x} - \mathbf{y}^m\| \}$

and $U_2^m = \{ \mathbf{x} \in TS_i \mid \|\mathbf{x} - \mathbf{y}_{1,i}\| > \|\mathbf{x} - \mathbf{y}^m\| \}$

step 2: Compute distortion and check for convergence

$$d^{m+1} = \frac{1}{|U_1^m|} \sum_{\mathbf{x} \in U_1^m} \|\mathbf{x} - \mathbf{y}_{1,i}\|^2 + \frac{1}{|U_2^m|} \sum_{\mathbf{x} \in U_2^m} \|\mathbf{x} - \mathbf{y}^m\|^2$$

If $\dfrac{d^m - d^{m+1}}{d^{m+1}} \leq \epsilon$ let $\mathbf{y}_{0,i}^2 = \mathbf{y}^m$, $S_{0,i}^1 = U_1^m$, $S_{0,i}^2 = U_2^m$ and Stop.

Else, go to step 3.

step 3: Update the codeword

$$\mathbf{y}^{m+1} = \frac{1}{|U_2^m|} \mathbf{x} \in U_2^m \mathbf{x}. \qquad \text{Go to step 1.}$$

The distortion d^m decreases with every iteration, and since it is bounded from below by zero, the algorithm always converges. Since it is essentially identical to the generalized Lloyd algorithm, the two should have the same convergence properties, like those derived in [4].

This design method is called the *tree-expanded codebook* method because the codebook $\mathbf{Y}_0$ thus obtained is a tree expansion of the given codebook $\mathbf{Y}_1$ in the sense that the encoder of Q_0 performs a tree search of $\mathbf{Y}_0$: First, $\mathbf{Y}_1$ is searched for the nearest neighbor, say $\mathbf{y}_{1,k}$, to the input vector $\mathbf{x}$. Then the nearest of $\mathbf{y}_{1,k}$ and $\mathbf{y}_{0,k}^2$ to $\mathbf{x}$ is chosen as the output of Q_0. Therefore, only $M+1$ Euclidean distances in $\mathbf{R}^K$ are required for the search, as opposed to $N = 2M$ for the full search. Note that the storage requirement of these tree-expanded codebooks is equal to that of full search codebooks of the same size. This is not the case for the standard tree-searched codebook which would have required $\dfrac{M+N}{M} = 1.5$ times more storage.

The method is easily generalized to any rate vector $R = (B-h, B)$ for arbitrary h. In that case, every region S_h^i in $\mathbf{S}_h$ must be partitioned into $p = 2^h$ regions $S_{0,i}^1, \cdots, S_{0,i}^p$. For each region $S_{0,i}^k$ a codeword $\mathbf{y}_{0,i}^k$ must be determined for $k = 2, \cdots, p$, while $\mathbf{y}_{0,i}^1 = \mathbf{y}_{i,i}$ as before. By repeated application of the algorithm we can design tree-expanded codebooks for any rate vector $(B_1, B_2, \cdots, B_L)$ starting from a given rate B_1 quantizer.

The low computational complexity without additional storage requirements together with the simplicity of the design make tree-expanded codebooks very attractive for the design of efficient embedded coders. The problem is that the codebook size for the larger rates (say 32 Kbits/sec) is excessive when the dimension exceeds three or four. At 32 kbits/sec, for example, and a sampling rate of 8 Khz we have 16 bits per vector for $K = 4$. This means that the codebook has $2^{16} = 65{,}536$ codewords. In order to resolve this problem, we need to look at *multistage coders* which allow us to span the range in a succession of stages, some of which can be tree-expanded codebook stages.

4.2. Multistage Coders

Multistage vector quantizers were first introduced by Juang and Gray [5], who were interested in their computational and storage advantages over ordinary and tree-searched VQ's. In the scalar case, a multistage VQ is simply a successive approximation analog to digital converter. In higher dimensions ($K>1$), the basic idea is to use the quantization error from the first stage as an input to the second stage etc ... . A two-stage VQ is shown in Fig. 1, and we will restrict our attention to this particular coder for the sake of clarity, the generalization to any number of stages being straightforward.

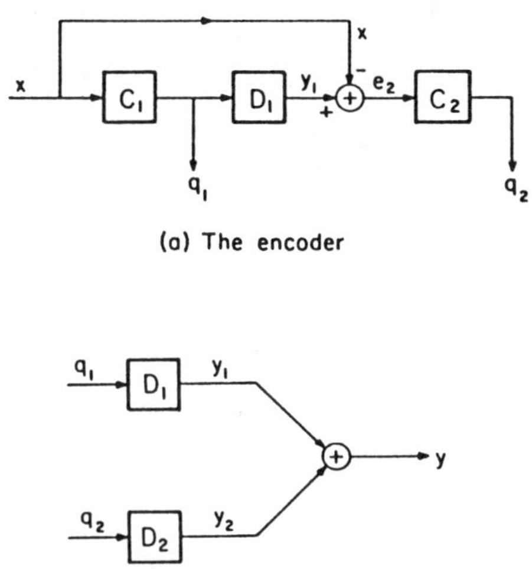

(a) The encoder

(b) The decoder

Figure 1. Two-stage VQ coder

In the block diagrams of Fig. 1, C_1 and C_2 are the first and second stage encoders, with corresponding decoders D_1 and D_2. Let Q_1 denote the first stage VQ, that is, the VQ with encoder C_1 and decoder D_1 and let Q_2 denote the second stage VQ. Let N_1 and N_2 be the codebook sizes of Q_1 and Q_2 respectively. The output q_1 of C_1 can take one of $M_1 = \log_2(N_1)$ values while q_2 can take one of $M_2 = \log_2(N_2)$ values.

The rate of the two-stage VQ is M_1+M_2, and the encoding requires the nearest neighbor search of a codebook of size N_1 and one of size N_2, that is, a total of $2^{M_1}+2^{M_2}$ Euclidean distance evaluations as opposed to $2^{M_1+M_2}$ for a full search single stage VQ of the bit rate. The reduction in computations can be substantial, and the storage is also reduced by a comparable factor. The price, of course, is a loss in performance which may be quite significant for small size codebooks, but is typically small for high rate coders

This is a natural structure for an embedded coder because if the bits representing q_2 are dropped during transmission, the multistage coder will be equivalent to its first stage VQ. If the memoryless VQ's are replaced by embedded coders designed using the tree-expanded method for example, we have a very flexible structure for embedded coding allowing the trade-off of complexity for performance. We now turn to the computer simulation results of the performance of such coders for the range 32-8 Kbits/sec and compare them to embedded ADPCM.

5. Simulation Results

We have simulated the performance of two VQ embedded coders, which we will refer to as EC1 and EC2, and that of the embedded ADPCM coder of [1]. The embedded ADPCM (EADPCM) coder has a fixed predictor of order one and an quantizer with four-bit precision and step-size adaptation with two step-size multipliers. Its rate vector is $R = (32,24,16)$ Kbits/sec. EC1 is a two stage VQ coder where each stage is a tree-expanded codebook VQ of dimension 5 and rate vector (8,16) Kbits/sec. EC2 has four stages, each an 8-dimensional full search VQ of rate 8 Kbits/sec. EC1 requires 128 multiplies/sample and a storage of 80 Kbits approximately, whereas EC2 has a much higher computational requirement (1,024 multiplies/sample) and comparable storage. EC1 and EC2 where both designed on a training sequence of 120,000 speech samples (8 kHz Sampling) and tested on a sequence of 72,000 samples from a speaker not in the training sequence. The signal-to-noise ratios of the three coders for the test sequence are compared in Table 1.

Rate (Kbits/sec)	32	24	16	8
EADPCM	17.92	15.83	11.89	-
EC1	22.23	17.75	14.14	7.73
EC2	22.62	18.98	14.54	9.49

Table 1.

Both VQ coders perform very well at 32 Kbits/sec and EC2 has almost 4 dB edge over EADPCM at that rate. The only real advantage in performance of EC2 over EC1 is at 8 Kbits/sec where the 1.8 dB difference is significant, but this is hardly sufficient to justify the order of magnitude difference in computations between the two coders.

We have listened to the output of the three coders and found the VQ coders to be almost as good as standard ADPCM at 32 Kbits/sec and much better than EADPCM. As the rate is dropped to 24 and 16 the output of the VQ coders becomes gradually noisier but the intelligibility is unaffected. At 16 Kbits/sec EADPCM is very scratchy and some segments are hardly intelligible. EC2 was still intelligible at 8 Kbits/sec but its quality at that rate is not acceptable.

6. Conclusion

VQ embedded coders can provide good quality speech at 32 and 24 Kbits/sec and acceptable quality at 16 Kbits/sec. Their performance at 8 Kbits/sec is poor which suggests that more powerful speech coders (e.g. subband coders) should be used as the first stage of the embedded coder followed the VQ stages

References

1. D.J. Goodman, "Embedded DPCM for Variable Bit Rate Transmission," *IEEE Transactions on Communications*, vol. COM-28, no. 7, pp. 1040-1046, July 1980.
2. Y. Linde, A. Buzo, and R.M. Gray, "An Algorithm for Vector Quantizer Design," *IEEE Transactions on Communications*, vol. COM-28, pp. 84-95, January 1980.
3. Amine Haoui, *Medium-Band Speech Coding: Vector Quantization with Memory*. Ph.D. Dissertation, Electrical Engineering Department, University of California at Berkeley.
4. M.J. Sabin and R.M. Gray, *Global Convergence and Empirical Consistency of the Generalized Lloyd Algorithm*. To be published
5. B.-H. Juang and A. H. Gray, "Multiple Stage Vector Quantization for Speech Coding," *Proceedings of the IEEE International Conference on Acoustics Speech and Signal Processing*, vol. 1, pp. 597-600, April 1982.

Vector Predictive Coding of Speech at 16 kbits/s

VLADIMIR CUPERMAN, MEMBER, IEEE, AND ALLEN GERSHO, FELLOW, IEEE

***Abstract*—Vector quantization, in its simplest form, may be regarded as a generalization of PCM (independent quantization of each sample of a waveform) to what might be called "vector PCM," where a block of consecutive samples, a vector, is simultaneously quantized as one unit. In theory, a performance arbitrarily close to the ultimate rate-distortion limit is achievable with waveform vector quantization if the dimension of the vector, k, is large enough.**

The main obstacle in effectively using vector quantization is complexity. A vector quantizer of dimension k operating at a rate of r bits/sample requires a number of computations on the order of $k2^{kr}$ and a memory of the same order. However, a low-dimensional vector quantizer (dimensions 4–8) achieves a remarkable improvement over scalar quantization (PCM). Consequently, using the vector quantizer as a building block and imbedding it with other waveform data compression techniques may lead to the development of a new and powerful class of waveform coding systems.

This paper proposes and analyzes a waveform coding system, adaptive vector predictive coding (AVPC), in which a low-dimensionality vector quantizer is used in an adaptive predictive coding scheme. In the encoding process, a locally generated prediction of the current input vector is subtracted from the current vector, and the resulting error vector is coded by a vector quantizer. Each frame consisting of many vectors is classified into one of m statistical types. This classification determines which one of m fixed predictors and of m vector quantizers will be used for encoding the current frame.

I. Introduction

THERE is currently a great need for a low-complexity speech coder at the rate of 16 kbits/s which achieves essentially "toll" quality, roughly equivalent to that of standard 64 kbit/s log PCM. Adaptive DPCM schemes are able to achieve this quality with low complexity for the new 32 kbit/s CCITT standard. At 16 kbits/s, the quality of ADPCM or adaptive delta modulation schemes is inadequate. More powerful methods such as subband coding or transform coding not only require a much higher level of complexity, but are also speech-specific, so that they are inadequate for voiceband data signals. In the last few years an important new approach to data compression, vector quantization, has emerged, which may provide the needed breakthrough for 16 kbit/s coding. For a comprehensive review of VQ, see Gray [26].

Vector quantization (VQ) has already made a significant impact on very low rate speech coding based on vocoder techniques [1]. Recently, vector quantization has been applied to waveform coding of speech [2]. The initial results show that direct use of VQ offers a significant improvement over scalar quantization, i.e., PCM, but it is not by itself competitive with other highly developed waveform coders.

In this paper, we describe a new approach to waveform coding by using VQ as a building block in a coding system rather than as a complete coder in itself. Specifically, the underlying idea of predictive coding schemes, such as DPCM, is generalized to vector signal processing so that the enhanced efficiency of VQ over scalar quantization can be exploited. In order to utilize the capability of predictors and vector quantizers to be "tuned" to specific waveform statistics, we introduce an adaptive scheme where a classifier is used to identify particular categories of local speech statistics. This scheme also has the potential capability of handling voiceband data signals. The basic ideas of vector predictive coding were first reported in [24]. The approach was also proposed in [27] for use in quantized control systems.

For waveform coding, VQ may be regarded as a generalization of PCM (independent quantization of each sample of a waveform) to what might be called "vector PCM," where a block of consecutive samples, regarded as a vector, is quantized as one unit. Rate-distortion theory guarantees a performance arbitrarily close to optimal for waveform vector quantizers if the dimension of the vector, k, is sufficiently large. The main obstacle in using the capability is implementation complexity. A vector quantizer of dimension k operating at a rate of r bits/component requires a number of computations of the order of $k2^{kr}$ and a memory of the same order.

The first approach we have considered to circumvent the complexity problem is based on suboptimal vector quantization. Various suboptimal vector quantization techniques, such as tree-structured codebooks or multistage quantizers, achieve an important reduction in computational complexity [1], [3], [4]. The resulting quantizers compare favorably with scalar quantization, but the acheived overall performance is not competitive with sophisticated scalar coding systems using compression techniques such as adaptive differential coding methods, transform coding, and subband coding.

A different approach for achieving a viable waveform coding system is to exploit the performance of a low-dimensionality vector quantizer by combining it with other waveform compression techniques. In this approach, the vector quantizer becomes a building block in the waveform coding system. Consequently, the entire system must be designed as a vector processing system, as a result of the presence of the vector quantizer. Thus, just as basic PCM is used as a building block in ADPCM and other more sophisticated coders, vector PCM can be used as a building block in a vector-based predictive coding system. This approach is based on the idea that the basic vector dimension is limited by complexity considerations to values that are much lower than the span of samples that contain significant correlation; hence, to further exploit the correlation between successive low-dimensional blocks (vectors) of speech samples, vector prediction is used.

This paper presents and analyzes a waveform coding system in which a low-dimensionality vector quantizer is used in an adaptive predictive coding scheme. In the encoding process, a locally generated prediction of the current input vector is subtracted from the current vector, and the error vector is coded by a vector quantizer. The vector quantizer observes a vector of error samples and identifies the closest matching error pattern from a codebook of stored patterns. A binary word, defining the address of that pattern in the codebook, is then transmitted to the receiver. The system is made adaptive by classifying each frame of speech, consisting of many

Paper approved by the Editor for Signal Processing and Communication Electronics of the IEEE Communications Society for publication after presentation at IEEE GLOBECOM, Miami, FL, December 1982. Manuscript received August 26, 1983; revised October 12, 1984. This work was supported in part by the Air Force Office of Scientific Research under Grant 82-0008.

V. Cuperman is with Calltalk, Ltd., Tel-Aviv 67 214, Israel.

A. Gersho is with the Department of Electrical and Computer Engineering, University of California, Santa Barbara, CA 93106.

Reprinted from *IEEE Trans. Commun.*, vol. COM-33, pp. 685–696, July 1985.

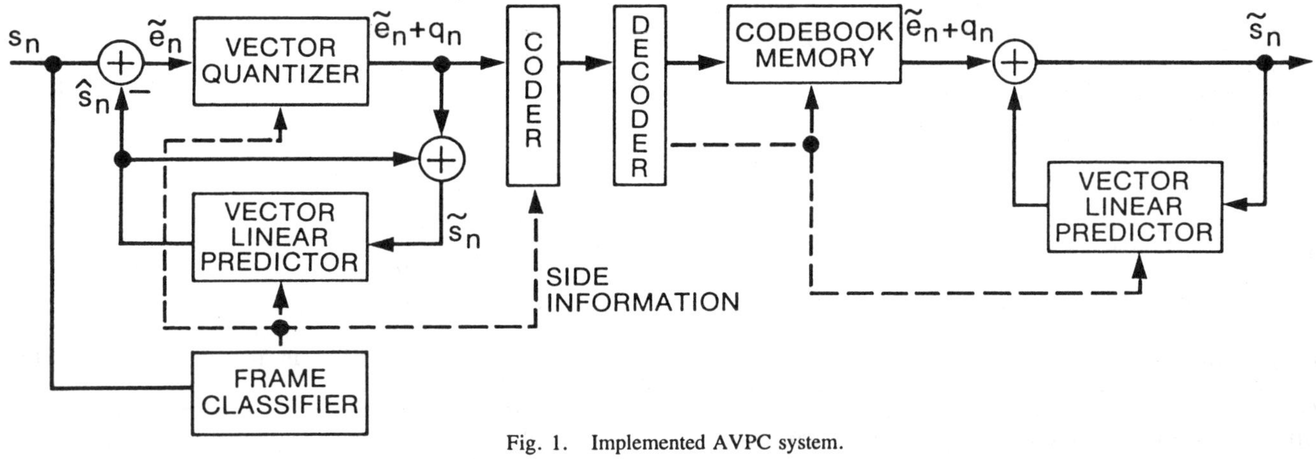

Fig. 1. Implemented AVPC system.

vectors, into one of m statistical types. This classification determines which one of m fixed predictors and of m vector quantizers will be used for coding the current frame. The class of each frame is transmitted as side information. The overall system is called adaptive vector predictive coding (AVPC).

The general configuration proposed for vector predictive coding is shown in Fig. 1. A vector linear predictor (VLP) locally generates a prediction $\hat{s}_n$ of the current vector s_n. The prediction is subtracted from the current input vector giving the prediction error vector $\tilde{e}_n$:

$$\tilde{e}_n = s_n - \hat{s}_n.$$

The prediction error vector is encoded by a vector quantizer, which selects from its codebook a codevector v_n to represent the error pattern. The difference between v_n and $\tilde{e}_n$ is the quantization error vector q_n. The feedback configuration is the same as that used in scalar DPCM to assure that the VLP in the transmitter and the identical VLP in the receiver have the same input $\tilde{s}_n$, which is also the reconstructed vector at the system output.

The coder output shown in Fig. 1 is actually the binary index of the codebook entry chosen to represent the prediction error $\tilde{e}_n$. Unless otherwise specified, the encoding process is assumed to perform a full search through the codebook for finding the closest match to $\tilde{e}_n$ under the chosen fidelity criterion.

The receiver has an "inverse" vector quantizer which simply uses the codebook as a table lookup, addressed by the received index, to retrieve the error pattern (codevector) v_n. The error pattern has the value $\tilde{e}_n + q_n$, which differs from the prediction error $\tilde{e}_n$ by the quantization error q_n. The receiver uses an identical VLP to reconstruct the output vector $\tilde{s}_n$ using the error pattern v_n.

Let ϵ_n be the overall error vector for the coding system, defined by

$$\epsilon_n = s_n - \tilde{s}_n.$$

Using the input-output relations for the two summation points in the transmitter structure, it is easy to see that

$$\epsilon_n = -q_n. \tag{1}$$

This relation proves that the variance of the overall system error, $\sigma_\epsilon{}^2$, is equal to the variance of $\sigma_q{}^2$ of the vector quantization error q_n, which is a fundamental property of the predictive coding configuration. This result is a direct generalization of the same well-known feature of scalar DPCM. (See [5] for a comprehensive theoretical review of DPCM with emphasis on adaptive prediction, and [6] for a broad tutorial description of quantization and DPCM.) Note that in the vector case, $\sigma_\epsilon{}^2$ and $\sigma_q{}^2$ are energies (variances) of the random vectors ϵ_n and q_n, respectively, defined as the expectations of the sum of squares of the components of the respective vectors.

The signal-to-noise ratio of the vector quantizer, $(\mathrm{SNR})_{\mathrm{VQ}}$, is a function of the statistics of the prediction error $\tilde{e}_n$. The determination of these statistics, given the statistics of the input vector s_n, is an analytically intractable problem because the feedback loop contains a nonlinear element (the vector quantizer). Nevertheless, the signal-to-noise ratio $(\mathrm{SNR})_{\mathrm{VQ}}$ is independent of the amplitude scaling of the prediction error vector, $\tilde{e}_n$, i.e., independent of its variance $\sigma_{\tilde{e}}{}^2$ [7].

On the basis of the above considerations, the variance of the quantization error may be written as the variance of the quantizer input signal divided by the signal-to-noise ratio:

$$\sigma_q{}^2 = \frac{\sigma_{\tilde{e}}{}^2}{(\mathrm{SNR})_{\mathrm{VQ}}}.$$

Then, the overall signal-to-noise ratio for the predictive vector coding system is given by

$$\mathrm{SNR} = \frac{\sigma_s{}^2}{\sigma_\epsilon{}^2} = (\mathrm{SNR})_{\mathrm{VQ}}\, G$$

where G, the prediction gain of the vector linear predictor in the given configuration, is defined by

$$G = \frac{\sigma_s{}^2}{\sigma_{\tilde{e}}{}^2}.$$

Thus, the signal-to-noise ratio for the predictive vector coding system is equal to the signal-to-noise ratio for the vector quantizer multiplied by the prediction gain. It should be noted that the prediction gain as defined above differs from the prediction gain associated with the unquantized prediction error. The relation between the actual prediction error and the prediction error in the absence of quantization will be discussed in the next section. The difference decreases as the quantization error decreases, i.e., as the number of output vectors (codebook size) in the vector quantizer increases.

II. Design of the Vector Predictive Coding System

The system design first requires the selection of a suitable quantitative measure of performance. Once a tractable measure has been selected, three major components to be designed are the vector predictor, the vector quantizer, and the speech classifier. This section of the paper examines these issues.

A. Vector Distortion Criteria for Coder Design

Vector signal processing conveniently allows more meaningful measures of fidelity to be used in optimizing system performance than does scalar processing. By regarding a vector of speech samples as a short segment of speech, it is possible to quantitatively measure the distance of distortion between the two vectors that to some degree reflects their perceptual dissimilarity.

A useful and general class of vector fidelity criteria can be defined as a weighted mean-square error:

$$e_{\mathrm{WMS}} = E\{(s_n - \hat{s}_n)' W (s_n - \hat{s}_n)\} \tag{2}$$

where s_n is the input vector, $\hat{s}_n$ is the corresponding output vector, and W is a positive definite weighting matrix. If W is the k-dimensional identity matrix I_k, the definition reduces to the sum of the mean-squared errors of the vector components, or in other words, the mean of the squared Euclidean distance between the vectors viewed as points in k-dimensional space.

More generally, the weighting matrix may be selected to reflect perceptually important differences between the vectors. Since speech is only a locally stationary process, the perceptually important features that need to be accurately reproduced must depend on the local or short-term statistics of the speech waveform. Hence, the desirable weighting matrix for the performance measure varies with time, and must be determined by the local (short-term) statistical character of the input signal itself.

Perceptually meaningful distortion criteria have been used in LPC coding by treating an entire frame of a hundred or so samples of speech as a single vector. The Itakura–Saito distortion measure, extensively used in linear predictive coding (LPC), may also be defined in the inner product form with a weighting matrix depending on the input vector s_n (see [8]). For low-dimensional vector coding, we do not have quite so meaningful distortion measures, but we still can do better than the standard mean-square error that is pervasive in studies of scalar coding systems.

For simplicity, in this paper we present only the special case where the weighting matrix is a scalar constant multiplied by the identity matrix. The most relevant local statistical characteristic that we can easily utilize is the short-term energy or power of the speech waveform. Suppose the speech waveform is partitioned into frames (for example, of length 10 ms, or 80 samples). Suppose also that each frame can be considered as a realization of a segment of a stationary process. Different frames may have different statistics, and are viewed as realizations of different random processes. The signal vectors are assumed to have dimensionality lower than the frame length, so that several vectors are contained in each frame. (In particular, $N_v = 10$ vectors per frame corresponds to $k = 8$ dimensions per vector and a frame length of 80 samples.) Then, if $s_n^{(l)}$ is a vector in the frame with the index l, the weighting matrix W_l is defined by

$$W_l = \frac{1}{E\|s_n^{(l)}\|^2} I_k.$$

This expectation is actually empirically estimated by time averaging the vector energies in the lth frame.

This gives a reasonable and simple performance measure for an individual frame of speech, within which we assume local stationarity of the speech waveform. However, we need a single performance measure for assessing the reconstructed quality of a record of speech of much longer duration than a single frame. The general approach, where a measure of performance is assigned to each frame or "segment" and these numbers are somehow combined to obtain an overall numerical indicator of performance, is what might be called a segmental approach. In this section we describe two particular segmental measures for assessing coder performance.

One way to average the frame performance measure over many frames is to use a geometric mean. The geometric mean of the weighted mean-square error measure, taken over K consecutive frames, we call the segmental mean square error, e_{SEG}:

$$e_{\mathrm{SEG}} = \left[\prod_{l=1}^{K} \frac{E\|s_n^{(l)} - \hat{s}_n^{(l)}\|^2}{E\|s_n^{(l)}\|^2}\right]^{\frac{1}{K}} \tag{3}$$

where $\hat{s}_n^{(l)}$ is the output vector corresponding to the input vector $s_n^{(l)}$. The expectation in the numerator can be empirically estimated by time averaging over all error vectors in the lth frame. This performance measure is essentially equivalent to the so-called segmental signal-to-noise ratio, or SEGSNR, previously introduced for evaluating conventional scalar speech coders [9]. This measure was found to give a better correspondence with the subjective quality of speech than the usual MSE measure, where the total mean-square error over the entire record is compared to the total signal energy over the entire record. The SEGSNR is obtained from

$$\mathrm{SEGSNR} = -10 \cdot \log(e_{\mathrm{SEG}})$$

so that an arithmetic average of the frame SNR values in decibels is actually being calculated [9]. In practice, a thresholding operation is also added to limit the maximum or minimum SNR value in each frame [6].

Unfortunately, the use of the SMSE measure for the design of waveform coders leads to intractable optimization problems. An alternate segmental criterion which we call the energy weighted mean square error, e_{EWMS}, is defined by a direct arithmetic (rather than geometric) average of the performance measure for each frame:

$$e_{\mathrm{EWMS}} = \frac{1}{K} \sum_{l=1}^{K} \frac{E\|s_n^{(l)} - \hat{s}_n^{(l)}\|^2}{E\|s_n^{(l)}\|^2}. \tag{4}$$

Since the arithmetic mean is always greater than the geometric mean ($e_{\mathrm{EWMS}} \geqslant e_{\mathrm{SEG}}$), by minimizing e_{EWMS}, an upper bound on e_{SEG} is minimized. The experimental results indicate that an improvement in both the segmental signal-to-noise ratio and the subjective quality of the reconstructed speech is obtained when using this criterion in coder design instead of the standard MSE.

B. Design of the Vector Linear Predictor

Although not so widely known as scalar linear prediction, the basic ideas of optimal linear prediction readily generalize to the case of a vector-valued time series. The optimal linear predictor for a stationary vector input process may be found by solving the Wiener–Hopf equation in matrix form [10], [11]. Recent work has also generalized these results to the degenerate case, where the current vector may be exactly predictable from a linear combination of previous vectors [12]. The prediction coefficients A_i are fixed matrices and are determined from the second-order statistics of the process using a generalization of the Levinson–Durbin algorithm. This is a precomputation that is performed once as part of the system design. In our application, the vector process is generated by "blocking" the original scalar speech process, that is, partitioning the waveform into consecutive blocks of k samples each and defining a vector composed of these k samples, consistently ordered. This particular vector process has some special statistical properties that simplify the computation of predictor coefficient matrices.

Because speech is a "locally stationary" process, better performance can be obtained by using predictor coefficients that are optimized for each frame. The disadvantage of this "frame-adaptive" prediction is that the coefficient matrices A_i must be recomputed in real time for each frame and transmitted to the receiver as side information, greatly increasing the complexity of the coding system.

A different compromise between performance, complexity, and rate is offered by using a small set of fixed predictors and selecting one according to the statistics of the input signal estimated for each speech frame. This method retains the adaptive character, while eliminating the need for computing predictor coefficient matrices "on line" for each frame and transmitting them to the receiver. The side information for this case is the index identifying which predictor from the finite set has been used for the current frame. It is only this index, rather than a full description of the predictor, which must be transmitted for each frame.

Suppose the speech waveform is partitioned into frames, so that each frame contains kN_v samples of N_v vectors, each of k contiguous samples. Each frame will be assigned to one particular statistical class by a classifier which uses some simple statistical characteristics of that frame to make the classification. The objective is to have some degree of statistical similarity for all frames belonging to a particular class. We then need to design a fixed predictor that can be used for all frames belonging to a particular class. Although this predictor will be optimal for the class, it will not be optimal for any specific frame in the class. For this reason, the quantization error will be considered a second-order effect and the design of the optimal predictor for a class will be presented for unquantized speech. Since the actual statistical characters of different frames belonging to the same class are not identical, we require a slight generalization of linear prediction theory to obtain the correct predictor design formulas.

Suppose that a given record of speech is partitioned into frames and all the frames that are identified as belonging to class c are extracted and numbered from 1 to L. Suppose that for each l, the lth frame is a realization of a particular stationary vector process P_l. The statistics of the process P_l will be estimated by time averaging within the speech frame with the index l. All the processes P_l for $l = 1, 2, \cdots, L$ belong to one class c, and a vector linear predictor is to be designed which is optimal for this class, by using as data the L observed frames belonging to class c. The following relations, leading to the needed predictor design for class c, will then apply in the same way for each other class.

Let $\{s_i^{(l)}\}$ be a sequence of random vectors chosen from the process P_l and define the covariance matrices:

$$C_{ij}^{(l)} = E\{s_{n-i}^{(l)} s_{n-j}^{(l)\prime}\} \tag{5}$$

where s' denotes the transpose of s. Actually, the matrix $C_{ij}^{(l)}$ depends on $i - j$ only, given the stationarity of P_l.

Consider the linear prediction $\hat{s}_n^{(l)}$ of $s_n^{(l)}$ given by

$$\hat{s}_n^{(l)} = \sum_{m=1}^{M} A_m s_{n-m}^{(l)} \tag{6}$$

where A_m are matrices of dimension $k \times k$ and k is the dimension of the vectors $s_i^{(l)}$. This estimate uses the same prediction matrices A_m for all the processes P_l, $l = 1, \cdots, L$.

We wish to find optimal predictor matrices A_m, using as fidelity criterion the energy weighted mean square error e_{EWMS} as given by (4), where now the averaging is taken over the L processes P_l belonging to the particular class c. Denote by σ_l^2 the variance of each vector in the process P_l:

$$\sigma_l^2 = E\|s_n^{(l)}\|^2 = \mathrm{tr}\,(C_{00}^{(l)}) \tag{7}$$

where tr denotes the trace of the matrix argument.

Then, it can readily be shown that the following generalized orthogonality principle gives the conditions for optimality of the predictor using the e_{EWMS} criterion:

$$\sum_{l=1}^{L} \frac{1}{\sigma_l^2} E\{(s_n^{(l)} - \hat{s}_n^{(l)}) s_{n-m}^{(l)\prime}\} = 0 \tag{8}$$

for $m = 1, 2, \cdots, M$. This form of the orthogonality principle may be used to derive a set of equations in A_j for finding the optimal prediction coefficients:

$$\sum_{j=1}^{M} A_j \tilde{C}_{ji} = \tilde{C}_{0i} \qquad \text{for } i = 1, 2, 3, \cdots, M \tag{9}$$

where the matrices $\tilde{C}_{ji}$ are defined by

$$\tilde{C}_{ji} = \frac{1}{L} \sum_{l=1}^{L} \frac{1}{\sigma_l^2} C_{ji}^{(l)}, \tag{10}$$

i.e., $\tilde{C}_{ji}$ are the weighted averages of the covariance matrices over the processes belonging to a given class.

Equation (9) is a generalized form of the multivariate Wiener–Hopf equation (the generalized error criterion e_{EWMS} was used and the optimization was performed over a class of processes rather than over a given process). This equation will be used to design the vector linear predictors used in adaptive vector predictive coding (AVPC). Note that the equation differs from the usual multivariate Wiener–Hopf equation for a stationary process by replacing the covariance matrices with a weighted average of the covariance matrices for each process P_l. The optimum solution under the standard (unweighted) MSE criterion for the nonstationary case may be obtained as a special case by setting $\sigma_l = \sigma =$ const. The relation (10) for the covariance matrices then reduces to

$$\tilde{C}_{ji} = \frac{1}{L} \sum_{l=1}^{L} C_{ji}^{(l)} \tag{11}$$

which is simply an unweighted average of the covariance matrices for the L processes.

To assess the performance achieved by vector linear prediction on typical speech waveforms, a series of experiments was executed. A speech database (see the Appendix) of about 128 000 samples (16 s of speech), containing six sentences spoken by four male talkers and two female talkers, was used for these experiments.

The segmental prediction gain G_{SEG}, used as a criterion for evaluating the predictor performance, is based on the segmental mean-square error and is defined by

$$G_{\mathrm{SEG}} = \left[\prod_{l=1}^{L} \frac{E\|s_n^{(l)}\|^2}{E\|s_n^{(l)} - \hat{s}_n^{(l)}\|^2} \right]^{\frac{1}{L}} \tag{12}$$

where a partition into frames of the speech waveform is assumed, and l is the frame index. For all the following results, a frame length of 7.5 ms was used unless otherwise indicated.

The vector linear prediction configuration used in this experiment is depicted in Fig. 2. The procedure may be regarded as a reversible "compression" by linear prediction: the transmitter compresses the input vector process s_n into the error process $s_n - \hat{s}_n$, and the receiver must reconstruct the waveform from the prediction error sequence. No quantization or other noise is added at this time.

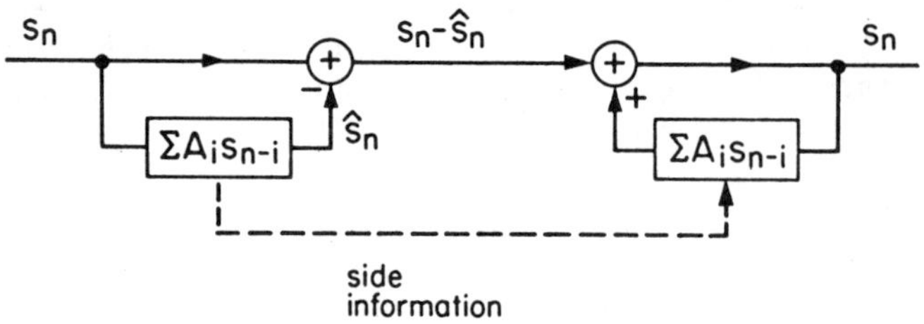

Fig. 2. Vector linear prediction configuration.

The experimental results which will be presented are based on first-order vector prediction ($M = 1$). The reason is that, for the vector dimensions k being used, there is a rapid decrease in the contribution of a prior "data" vector to the prediction gain achieved as the time lag between the data vector and the predicted vector increases. This is due to the exponential decrease of the autocorrelation function versus lag, which is characteristic of speech waveforms. On the other hand, the computational and memory complexity increase linearly with the predictor order.

Fig. 3 compares the segmental prediction gain in decibels versus dimension for two types of adaptive predictors. In the frame-adaptive prediction method, a separate predictor is designed for each frame using the statistical data obtained from that frame. This is a very high-complexity method, corresponding to the processing performed in LPC analysis, and gives us a point of reference for comparison with reduced complexity schemes. The second method uses switched prediction with one of three predictors selected for each frame by a simple three-way frame classification summarized in Table I. (The classifier will be discussed later.) The optimal predictors for each class were designed by separately solving the generalized Wiener–Hopf equations (9) for each class, using the energy-weighted mean-square error criterion. The classifier was used to select those speech frames belonging to each class, so that the weighted average covariance matrices could be computed for each class.

The results presented in Fig. 3 indicate that the predictor with switched coefficients achieves a good compromise between performance and complexity. The three-way "switched" predictors designed in this way will be used in the vector predictive coding system.

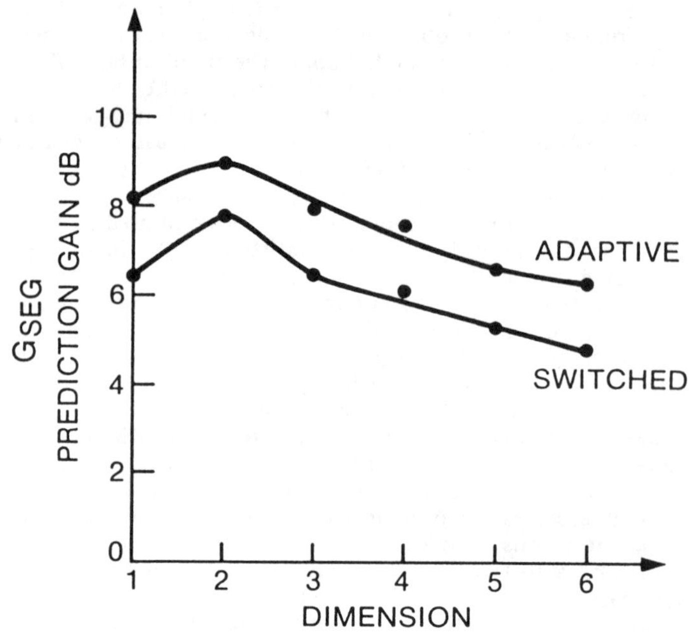

Fig. 3. Vector prediction gain versus vector dimension for frame-adaptive prediction and switched prediction.

TABLE I

Class 1	Class 2	Class 3
$r(1) \geq 0.94$ and $\sigma \geq 0.25\sigma_0$ Highly correlated and high energy	$0.94 > r(1) \geq 0.70$ and $\sigma \geq 0.25\sigma_0$ Moderately correlated and high energy	$r(1) < 0.70$ or $\sigma < 0.25\sigma_0$ Low correlated or low energy

C. Design of the Vector Quantizer

A vector quantizer is used to code the prediction error vector generated by the closed-loop vector predictive configuration shown in Fig. 1. Two design approaches were developed for optimizing the quantizer to the statistical character of the error vectors to be coded. Both approaches use a training sequence of speech data and a modified version of the Linde–Buzo–Gray (LBG) codebook design algorithm proposed in [13]. The first approach uses the prediction error vectors generated from the unquantized speech signal as the training sequence for codebook design, and will be called the open-loop design. The second approach uses actual prediction error vectors generated within the predictive coder loop, and will be called the closed-loop design. In this section, some aspects of the iterative codebook design algorithm are presented, followed by a detailed discussion of the two approaches studied.

1) Codebook Design Algorithms: In general, a vector quantizer is fully specified by the codebook of stored patterns or "codevectors" and the distance measure used by the encoding process to find the best match to an input vector. The codebook is designed on the basis of empirical data regarding the statistics of the vectors to be coded. These data consist of a large but finite set of "training" vectors that are chosen to be statistically representative of the signal source. Associated with each codebook is a partition of the training set into "cells" or clusters of training vectors, with each cluster corresponding to a particular codevector.

The codebook design begins with the selection of some initial codebook. Then, a simple but computationally demanding iterative algorithm leads to a final codebook design. The goal of the algorithm is to find a codebook whose member codevectors best represent the large set of training vectors used as input to the algorithm. The basic codebook design algorithm consists of iteratively performing the basic operation of generating the clusters for a given set of codevectors, then finding a new set of codevectors for these clusters. This algorithm always terminates in a finite number of steps and achieves a local optimum with respect to the training set for a given distortion measure.

The optimal codevectors in the LBG algorithm for the EWMSE criterion are generalized centroids given by

$$y_j = \left(\sum_{s \in S_j} \frac{1}{\sigma_s^2}\right)^{-1} \sum_{s \in S_j} \frac{s}{\sigma_s^2} \tag{13}$$

where σ_s^2 is the energy of the frame to which the current vector s belongs and S_j is the partition region for which the centroid is computed. This result is a special case of the centroid formula for the error criterion (2) and was first noted in [21]. If the distortion criterion is the MSE, the weighting matrix is the identity matrix and the previous relation becomes

$$y_j = \frac{1}{n_j} \sum_{s \in S_j} s \tag{14}$$

where n_j denotes the number of vectors in the partition S_j.

2) The Empty Cell Problem: One of the problems which frequently occurs when using this algorithm is that a codevector in the final codebook may be assigned to a cluster containing only a few, if any, input training vectors and, hence, may contribute very little to the reduction of overall distortion. The main reason for this "empty cell problem" is that the choice of the initial codebook determines to which of several local optima the algorithm converges. For some choices of the initial codebook, the final codebook, of size N, may contain a subset of N_1 codevectors which happen to be locally optimal for a quantizer of size N_1. The remaining $N - N_1$ codevectors may then be poorly utilized, having in some cases empty or nearly empty cells in spite of the overall codebook being locally optimal. A choice of the initial codebook which would avoid empty cells requires prior knowledge about the final codebook, which is generally unavailable.

One approach that partially alleviates this problem is to begin with a codebook of only two output points (one bit), and to build a codebook of $b + 1$ bits from a codebook of b bits by "splitting" every output point into two new output points. Although this technique gives good experimental results [13], [2], it does not guarantee the absence of empty cells from the resulting codebooks.

An alternate solution to the "empty cell problem" is to use some features of the ISODATA algorithm in pattern recognition [14]. This algorithm allows for splitting and lumping of clusters at every iteration, at the cost of a substantial increase in complexity.

As a compromise solution to the above considerations, a modified LBG type algorithm was used. In this modified version, after the clustering operation, the number of input vectors assigned to every cluster is checked. If a cell is empty, the corresponding codevector is deleted and a new codevector is created by "splitting" the codevector v_* representing the cluster with the highest distortion. The splitting operation retains the highest distortion codevector v_* but replaces the empty cell codevector with a perturbed version, $\tilde{z}_*$, of v_*. The highest distortion cluster is then split into two clusters. The perturbed codevector $\tilde{z}_*$ is obtained by multiplying all the components of the output vector by a constant factor. It can be shown that the result of such an operation can only result in a decrease in the overall distortion, so that the convergence of the modified algorithm is assured. In practice, we have found that convergence is achieved for speech data even when the cell is "nearly empty," containing one or two vectors.

3) Codebook Design for Predictive Coding: The codebook design was based on the training set of speech vectors residing in the speech database described in the Appendix. The classifier was used to select those speech frames belonging to one particular class. The design procedures described in this section were applied separately to obtain separate codebooks for each class.

The covariance matrices were computed by time averaging over each speech frame, using a vector generalization of the "autocorrelation" method known in scalar LPC [15]. Any element in a covariance matrix is equal to the scalar correlation coefficient of a given lag, which is computed by using the method used in scalar LPC analysis. Note that in this approach the covariance matrices are Toeplitz matrices, i.e., C_{ij} depends only on $|j - i|$.

The optimal prediction coefficients were then computed by solving the generalized Wiener–Hopf equations, (9), where the "weighted" covariance matrices $\tilde{C}_{ji}$ are given by (10) under the EWMSE criterion and by the relation (11) under the MSE criterion. These predictor matrices were used in the vector predictive coding systems.

Using the prediction matrices and the same training set of speech vectors, a training set of prediction error vectors for the unquantized input, e_n, was generated by applying the relation

$$e_n = s_n - \hat{s}_n = s_n - \sum_{j=1}^{M} A_j s_{n-j}. \tag{15}$$

In the open-loop approach, the vectors e_n were used as training data to design the codebook by applying the modified LBG iterative algorithm described previously.

The above vector quantizer design algorithm uses the prediction error e_n, generated by the unquantized input signal, rather than the prediction error $\tilde{e}_n$ that arises in the actual encoding process and is generated by the quantized signal $\tilde{s}_n$. To get an insight into this problem, a relation between the two prediction errors is needed.

In Section I, we noted that the overall system error is equal to the quantization error as given by (1). Thus, from (1), we have

$$\tilde{s}_n = s_n + q_n. \tag{16}$$

The actual prediction error $\tilde{e}_n$ is given by

$$\tilde{e}_n = s_n - \sum_{j=1}^{M} A_j \tilde{s}_{n-j}$$

so that, using (16), we get

$$\tilde{e}_n = s_n - \sum_{j=1}^{M} A_j s_{n-j} - \sum_{j=1}^{M} A_j q_{n-j}$$

or

$$\tilde{e}_n = e_n - \sum_{j=1}^{M} A_j q_{n-j}. \tag{17}$$

Hence, the difference between the two prediction errors is equal to the output of the prediction filter excited by the quantization error vectors.

To a first approximation, taking into account the "random" character of the quantization error q_n, for good quality quantization (large number of codewords), the averages computed over the quantities q_{n-j} in the relation (17) may be considered equal to zero. Now, consider that the computation of a centroid is based on the conditional expectation of a data vector, given a region of the quantizer partition. Note that the data vectors for the closed-loop design are the vectors $\tilde{e}_n$ and those for the open-loop design are e_n. Then, the relation (17) suggests that the centroids obtained by a closed-loop design and those computed by an open-loop design are essentially equal and the two design methods should give similar results.

The *closed-loop* design consists of the following main steps.

1) Using the training set of speech vectors, compute the covariance matrices and the predictor matrices.

2) Choose an initial codebook. A uniform codebook in the signal range may be used, an initial codebook may be designed by using the splitting procedure described in [13], or the result of an open-loop design can be used.

3) Keeping the codebook fixed, compute and store the set of prediction error vectors $\tilde{e}_n$, i.e., the relation (17), corresponding to the configuration of Fig. 1.

4) Using the set of vectors obtained in 3) as a training set, obtain a new codebook by a clustering and centroid computation. Replace the initial or the previous codebook with the new codebook.

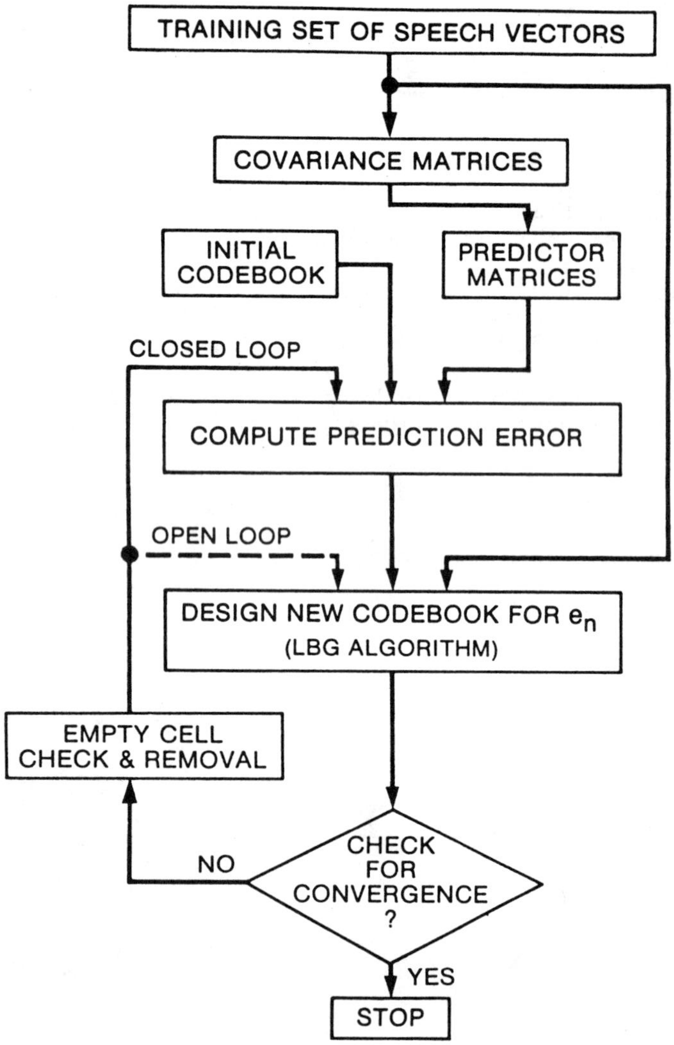

Fig. 4. Flowchart for closed-loop and open-loop codebook design.

5) Iterate steps 3) and 4) until no significant increase in the signal-to-noise ratio is observed.

The complete design algorithm for both approaches (open and closed loop) is depicted in the flowchart of Fig. 4.

The algorithm represented by steps 1)–5) was found experimentally to converge at least as fast as the open-loop algorithm, based on the fixed set of training vectors e_n. Better performance was always achieved by the closed loop design, with a signal-to-noise ratio improvement on the order of 1 or 2 dB. As expected, the closed-loop improvement decreased as the bit rate was increased. The experiments were executed at rates ranging from 1 to 2 bits/sample.

4) The Speech Classifier: An efficient coding system for nonstationary waveforms such as speech needs adaptive procedures, in order to match the coder parameters to the local statistical character of the waveform. Frame-adaptive vector linear prediction achieves good results in vector linear prediction of speech. The disadvantages are high complexity and the considerable overhead rate needed to transmit new prediction matrices for each frame. For example, if each matrix element requires an average of only 3 bits for transmission, 1.25 bits/sample are required only for transmitting prediction matrices for an AVPC system in dimension 5 with a frame length of 60 samples. As a result of complexity and rate considerations, the switched adaptive prediction is preferable to the AVPC system, in spite of a loss of more than 2 dB in performance compared to frame-adaptive prediction. (See the results presented in Fig. 3.)

An adaptive procedure for the vector quantizer is also needed. To track the nonstationarity of the speech waveform, the ideal solution would be to design a new codebook in real time for each frame of speech. This, however, is totally impractical because of the enormous computational complexity that would be needed and the excessive amount of side information needed to transmit a codebook in each frame. Good performance may be expected from a system which simply adapts the codebook to each talker. Such a system would avoid the performance degradation that results when a codebook designed on a given speech database is used to encode speech that is not in the database. (The results "inside" and "outside" the training sequence will be discussed in the next section.) The problem is that such a vector quantizer would require a delay of at least 5–10 s to design the optimal codebook for each talker, before transmission can begin.

The adaptive procedure which was chosen for the vector quantizer in the AVPC system is based on switching the codebooks in each frame according to the classification of the signal. Hence, in order to incorporate adaptive features into the vector predictive coding system, a device which will classify each speech frame into one of a few statistical classes is needed. Then, different codebooks and different prediction matrices will be used for each class. The information about the class will be transmitted to the receiver as side information.

The memory required for the system will increase in proportion to the number of classes (due to separate codebooks and predictors being used for each class). Also, as the number of classes increases, the complexity needed to perform the classification (feature extraction and selection) increases. The increase in the transmission rate due to the side information for m classes will be $\log_2 m$ bits per frame. Taking all these considerations into account, a classifier was designed for only three classes using two simple frame statistics, the variance and the first reflection coefficient. For application to digital telephony, a fourth class that could be added to the system would be used to identify voiceband data; however, the work reported here does not include this feature. This is certainly an "ad hoc" classifier, chosen for its simplicity. A classifier that is in some respects similar was used for switching predictors in scalar ADPCM [18]. Other studies are in progress to design classifiers in a more rigorous way, based on the same rationale that is used in codebook design.

There are also other ways to generalize and expand this classification approach to obtain better performance at the expense of increased complexity. One way would be to classify the frames using the basic acoustical categories of sounds (nasals, plosives, fricatives, etc.). Designing different codebooks and predictors for each such class will significantly improve the performance of the system, but the complexity of such a classifier will probably come close to the complexity of a speech recognition system.

The design of the classifier was based on an examination of the statistical characteristics of speech frames. Histograms of the first normalized correlation coefficient and of the frame variance for 1029 frames of speech sampled at 8 kHz were computed. On the basis of the analysis of these statistics, the classifier presented in Table I was chosen. The three classes were chosen to approximately divide the frames into three equally frequent groups, where each group covers a limited subrange of correlation and energy values.

Here $r(1)$ is the normalized autocorrelation coefficient for unit lag, σ is the frame variance, and σ_0 is the long-term variance computed over the entire speech database.

The first class contains highly correlated, high-energy frames. These frames generally contain voiced speech and are expected to give high prediction gain if an adequate predictor is designed for the class. The second class contains high-energy frames with lower correlation and, hence, lower prediction gain. Voiced speech with a high first formant may be an example of this class. The third class contains low energy and/or low correlation frames, as, for example, unvoiced sounds. It should be noted that this classification is not strictly related to the traditional voiced/unvoiced dichotomy. In fact, the voiced–unvoiced classification is not adequate in this context. Indeed, there are, for example, unvoiced sounds with high energy and correlation (plosives), which have very high predictability and should not be classified together with such unvoiced sounds as fricatives with very low predictability, etc. Typical examples of waveforms from the three classes are presented in Fig. 5.

Although these considerations are related to prediction gain or "predictability," they generally are also true for the vector quantizer. The performance of any given vector quantizer strongly depends on the energy level and autocorrelation of the input signal [7]. Since those frames which have energy and first autocorrelation coefficient in some given restricted range are grouped into the same class, an improvement in the performance achieved by each codebook can be expected.

5) Complexity of the AVPC System: Computational complexity is highly processor-dependent and is difficult to estimate without consideration of particular architectures. Nevertheless, we can make some preliminary assessment by assuming that multiplications are the most demanding basic operation needed. Although additions can often require equal execution times, the count of multiplications still gives a useful order-of-magnitude estimate of complexity. We shall see that the computational complexity of the AVPC system depends primarily on how the vector quantizer is implemented.

The classifier requires only two multiplications per sample to determine the frame energy and first autocorrelation value. Assuming that one-stage prediction is used, the vector linear predictor requires k multiplications per sample to perform the k by k matrix multiplication for input vectors with dimension k. A reduced computational complexity for the vector quantizer can be achieved by designing a codebook with a binary tree structure, allowing a rapid binary search implementation of the encoding operation [2]. This structural constraint does indeed degrade the performance of the quantizer, but the reduction in complexity may justify its use in some situations. In this case, the encoding operation requires only $2kr$ multiplications per sample, where r is the rate in bits per sample. On the other hand, a full-search implementation would require 2^{kr} multiplications per sample. The memory requirement for codebook storage is is $k2^{kr}$ words for a full-search codebook and double that figure for a binary-tree codebook. The predictor memory for a one-stage predictor is k^2 words.

In summary, the AVPC system with a full-search vector quantizer requires a number of multiplications per sample equal to

$$\mathrm{MUL}_{\mathrm{AVPC}} = 2^{kr} + k + 2.$$

If a binary tree is used for the vector quantizer implementation, the number of multiplications per sample becomes

$$\mathrm{MULB}_{\mathrm{AVPC}} = 2kr + k + 2.$$

Let c be the number of classes implemented in the AVPC system. Then, the memory required for the full-search implementation will be

$$\mathrm{MEM}_{\mathrm{AVPC}} = ck2^{kr} + ck^2$$

and for the binary-search implementation

$$\mathrm{MEMB}_{\mathrm{AVPC}} = 2ck2^{kr} + ck^2.$$

As an example, an AVPC system at a rate of 2 bits/sample, dimension $k = 4$, using binary search and three classes requires 22 multiplications per sample and a memory of about 6000 words and, hence, may be characterized as a medium complexity coder.

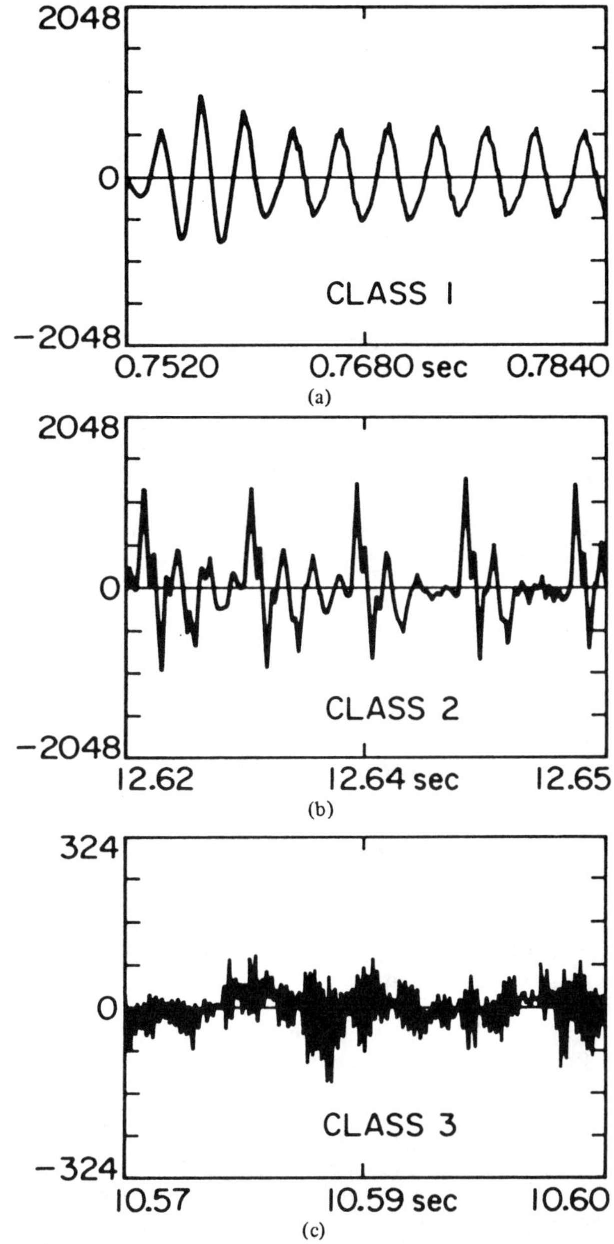

Fig. 5. Typical examples of speech waveforms for the three classes.

III. Experimental Results for the AVPC System

The AVPC system was simulated on both the VAX 11/780 and PDP 11/24 computers and tested using the speech database resident on these computers. Two main versions of the AVPC system were simulated, one in which the predictor and quantizer were optimized under the regular MSE criterion and the other optimized under the EWMSE criterion. In each version, the coding performance was evaluated by measuring the conventionally defined signal-to-noise ratios in both the MSE sense (SNR) and the segmental sense (SEGSNR), as frequently used by other speech coding researchers. Both versions of the AVPC system may be operated in a training mode or in a test mode. In the training mode, a locally optimal codebook is designed for the given input speech data, and the performance of the coder operating on these training data is measured. In the test mode, the coder performance is tested on a different set of speech data from the training set used to design the quantizer. The frame length for all experiments is 7.5 ms (60 samples) unless otherwise indicated.

One of the distinctive features of vector prediction of a blocked scalar process is that each component of the vector to be predicted has a different degree of predictability, depending on the time lag of the component from the observable data samples. This suggests that the AVPC system might introduce a coding noise that varies periodically with a period equal to the vector dimension, thereby introducing an undesirable periodic noise component. In fact, this does not happen, because the vector quantizer is designed to be matched to the statistics of the prediction error vectors.

To gain insight into the operation of the AVPC system, we first examine the variation of the signal-to-noise ratio with the component index inside a vector. Let s_n be an input vector, $\tilde{s}_n$ the reconstructed vector at the output of the system, and s_{ni}, $\tilde{s}_{ni}$ the ith components of these vectors. The vectors are defined from the scalar process, so that increasing the index i corresponds to moving forward in time. Then the MSE signal-to-noise ratio for the ith component may be defined by

$$(\mathrm{SNR})_i = \frac{E[s_{ni}^2]}{E[(\tilde{s}_{ni} - s_{ni})^2]}.$$

This definition is applied to each frame by computing the time average over the frame; then the geometric mean over all frames is found. The result obtained is the per-component segmental signal-to-noise ratio. The same expression defines the per-component prediction gain if the quantizer is eliminated, so that $\tilde{s}_{ni} - s_{ni}$ is simply the prediction error.

Fig. 6 presents the variation of both the per-component prediction gain and the per-component signal-to-noise ratio (SEGSNR) versus the component index i, for the AVPC system in dimension 5 at a rate of 2 bits/sample. Although the prediction gain significantly decreases as the vector component index increases, the input–output signal-to-noise ratio remains practically constant. This remarkable result is due to the fact that the vector quantizer compensates for the nonuniformity introduced by vector prediction. The compensating effect can be understood by examining the shape of the vectors in the codebook. Fig. 7 displays the superimposed codevectors for each of the three codebooks which are used to quantize the prediction error vectors. Each codevector consists of five samples ($k = 5$) which are interpolated in the plot to show a complete waveform segment. As can be seen, the shape of the vectors in each class reflects the statistical properties used to define that class.

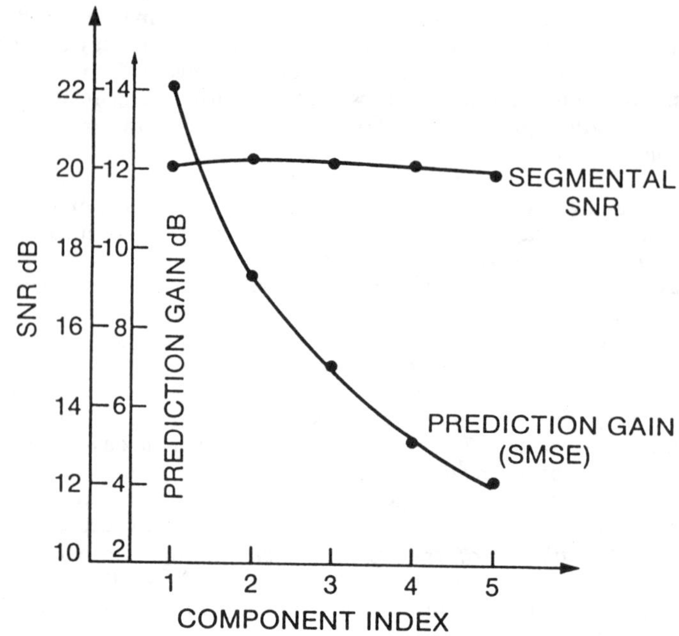

Fig. 6. Variation of segmental SNR with the component index for dimension 5 at 2 bits/sample.

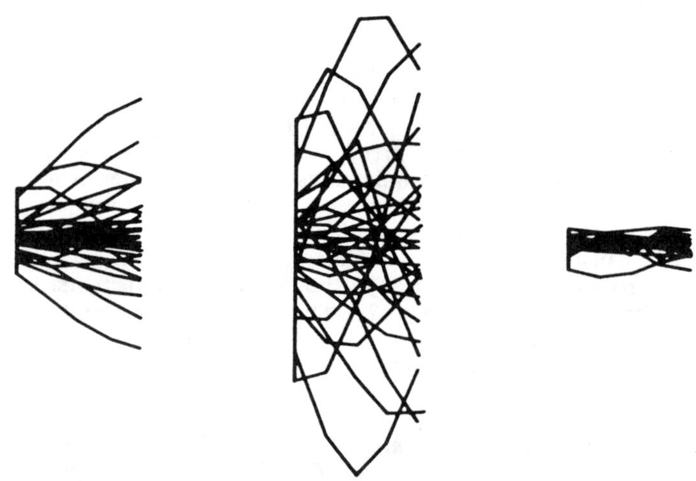

Fig. 7. Codebook for AVPC system for dimension 5 at 1 bit/sample.

Fig. 8 shows the overall performance of the AVPC system, evaluated by MSE and segmental signal-to-noise ratios versus the dimension of the input vectors. The MSE SNR was measured for the AVPC system optimized under the MSE criterion, and the SEGSNR for the system optimized under the EWMSE criterion. For comparison, the performances of some of the best waveform coding systems are noted in the same figure. The performance for standard ADPCM and pitch-adaptive DPCM are taken, respectively, from [16] and [17], the performance for sequential adaptive DPCM from [18], and the performance for tree and trellis coding from [19] and [20]. Fig. 8 shows that the AVPC system is competitive with such high-complexity systems as tree and trellis coding. The performance achieved by the AVPC system is significantly better than the performance of the known scalar versions of ADPCM. One should note that, generally, different researchers use different speech material, and this may produce slight variations in the SNR results (usually less than 1-2 dB). Hence, the comparison with other coding systems gives only an approximate idea of relative performance.

It is difficult to give a meaningful complexity comparison between the above-mentioned coding systems, since realistic measures of complexity are dependent on the architecture that is used. However, to give the reader at least a crude idea at this point, we estimate that the AVPC system (using binary search) would require about 22 multiplications/sample, compared to 16 for sequential scalar ADPCM and 110 multiplications/sample for tree-trellis coding. (At low rates, such as 1 bit/sample, tree-trellis coding may have a much lower complexity, using lookup tables for path selection.)

The AVPC system optimized using the EWMSE criterion achieves better segmental SNR (SEGSNR) than the same system designed using the usual MSE. Table II presents the SEGSNR results for the AVPC system designed using the EWMSE criterion (labeled EWMSE system in the table) and for the AVPC system designed using the MSE criterion (labeled MSE system). The EWMSE system achieves an improvement of 1.1–1.4 dB in SEGSNR. For both systems the rate was 2 bits/sample.

Informal listening tests of the processed speech indicate

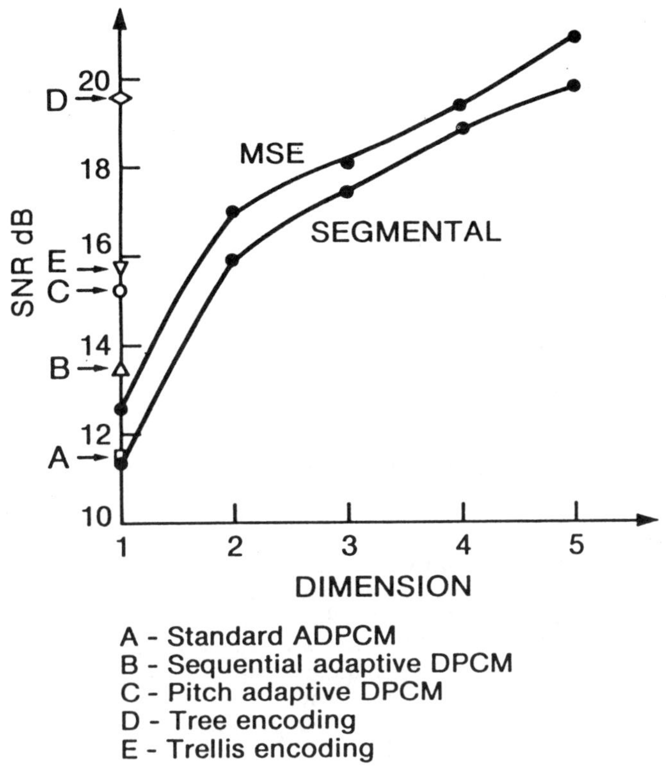

Fig. 8. MSE and segmental SNR versus dimension for the AVPC system.

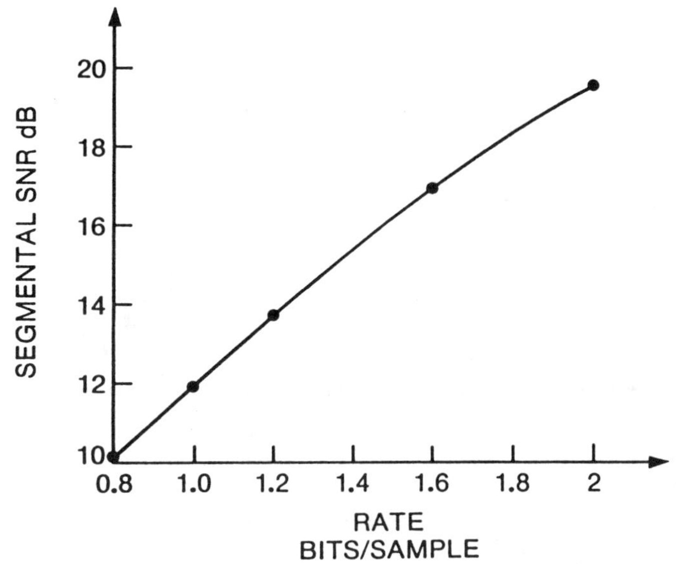

Fig. 9. Segmental SNR versus rate for dimension 5.

TABLE II
SEGSNR IN DECIBELS AT 2 BITS/SAMPLE

Dimension	1	2	3	4	5
EWMSE system	11.4	16.0	17.5	18.9	19.9
MSE system	10.3	14.6	16.4	17.8	18.6

good communications quality for rates close to 2 bits/sample and for vector dimensions 4 or 5. The subjective quality is better for the system optimized under the EWMSE criterion, verifying that this criterion is more suitable than the regular MSE for optimizing speech quality.

The results of Fig. 8 were obtained by encoding the same speech database as that used for designing the vector quantizer codebook and the predictor. This database (see the Appendix) contains 16 s of speech from six different talkers, four males and two females. Whenever the design of the codebook and the performance evaluation are done on the same data, the corresponding results are described as "inside the training sequence" or simply "training mode." The results for the tree and trellis coders noted in Fig. 8 were also inside the training sequence (optimum codebooks for the given data were used). When encoding speech which is not the design database, the results are described as "outside the training sequence" or "test mode." It is obvious that, to obtain good performance outside the training sequence, large databases containing a large variety of speech should be used for system design.

In the design procedure, each codevector in the codebook is defined by a centroid computation, averaging over a cluster of training vectors. Hence, the average number of training vectors per codevector, which we call the "training ratio" in vector quantizer design, should have a value acceptable from a statistical point of view. For example, at least 50–100 input vectors per codebook entry is desirable. Since the number of vectors in the codebook increases exponentially with the rate and the dimension, it follows that the size of the training sequence should also increase exponentially. In an example presented in [13], the same performance was obtained in both training and test modes by using a training sequence of 12 000 vectors for a codebook of size 16, i.e., a training ratio of 750. Such a high training ratio was not affordable with currently available resources for the codebook sizes of interest in this work. The limitation is primarily due to the considerable computer time needed for designing the codebook.

A first evaluation of the performance of the AVPC system outside of the training sequence was done by using the system designed on the database to encode a different talker (96 000 samples at 8 kHz sampling rate, male talker; see the Appendix). The SEGSNR in the test mode was found to be lower than the SEGSNR in the training mode by amounts ranging from 1 to 3.5 dB, the gap increasing with the dimension. The increase of this gap with the dimension is due to the exponential decrease of the training ratio with increasing dimension when the size of the training sequence is constant. (Recall that the number of codevectors grows exponentially with dimension for a fixed rate.)

The variation of performance of the AVPC system with the bit rate is presented in Fig. 9 for dimension 5. The signal-to-noise ratio (SEGSNR) in decibels increases smoothly with the rate, the curve being close to a straight line with a slope of about 9.5 dB/bit. This is different from the familiar scalar PCM slope of 6 dB/bit. In fact, the vector quantizer itself has a slope higher than 6 dB/bit in this range of rates. Note that the asymptotic theory of vector quantization gives 6 dB/bit in the high-rate region [7].

Fig. 10 shows the quantization error waveform of the AVPC system (dimension 5, rate 2 bits/sample) versus time for a typical speech waveform (15 frames of 60 samples each at 8 kHz sampling frequency). The original waveform is also shown in Fig. 10 with the same time scale. Fig. 11 presents the spectrum of quantization error compared to the spectrum of the original speech. The spectrum of the quantization error is generally flat, indicating a "white noise" type of distortion. This was confirmed by listening to the reconstructed speech.

It would be of interest to compare AVPC to ADPCM incorporating noise shaping [22], [23], which reportedly has good perceptual quality, but this was not done since SNR is an inadequate performance indicator for the latter technique, and adequate speech material for a subjective comparison was not available. A comprehensive subjective study would be a worthwhile future project.

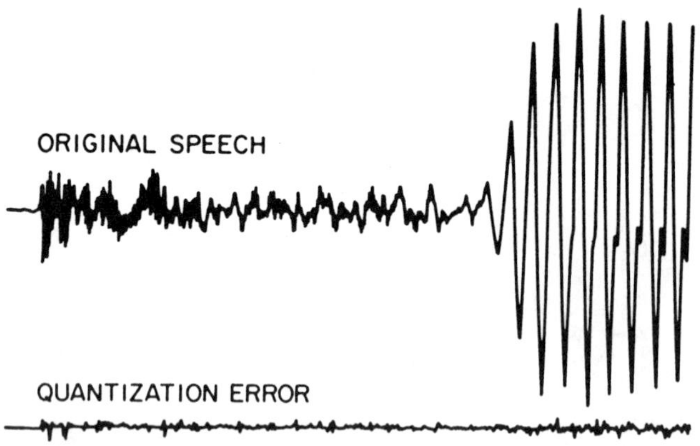

Fig. 10. Original speech and quantization error waveforms for dimension 5 at 2 bits/sample (16 kbits/s).

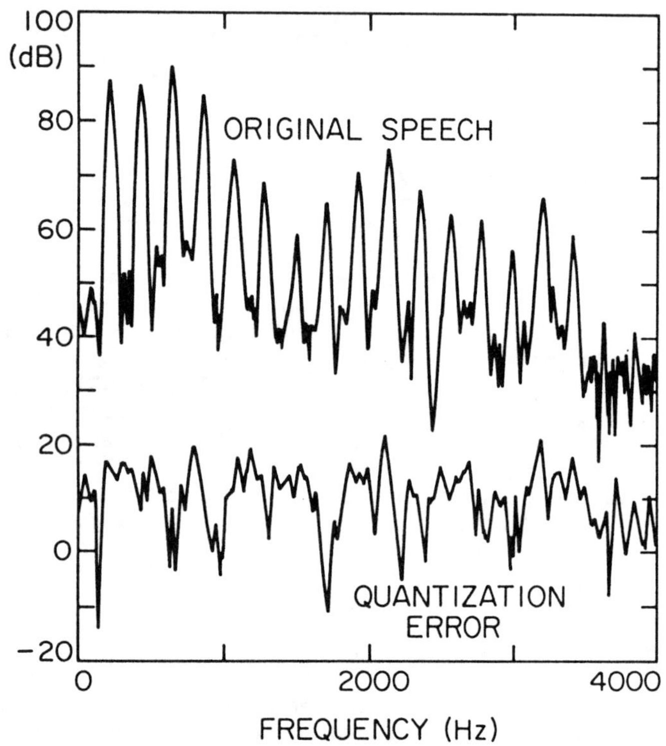

Fig. 11. Spectrum of the original speech and of the quantization error.

IV. Conclusions

Our results indicate that vector quantization may be used as a basis for a new and powerful class of waveform coding devices. These devices encode vectors, rather than discrete samples of the process, and use, together with vector quantization, vector linear prediction and adaptive techniques. An improvement of about 7 dB in SNR over the previous vector quantization studies was obtained using the adaptive vector predictive coder. The experimental results are significantly better than known results for ADPCM, the corresponding scalar coding method. At a rate of 2 bits/sample (16 kbits/s), a signal-to-noise ratio of about 20 dB was obtained, and the corresponding subjective speech quality may be described as good communications quality. These results compete favorably with such methods as tree and trellis coding at the same bit rate. We are optimistic that, with some further improvements and the use of larger training sets for the codebook design, the AVPC system may offer a viable option for a 16 kbit/s single-chip speech codec that is well within current VLSI capabilities.

TABLE III
AVERAGE PITCH VALUES

Talker	F1	F2	M1	M2	M3	M4
Pitch Hz	270	220	140	100	170	120

Appendix

The database contains 16 s of speech from six different talkers, four males and two females. The talkers will be denoted M1–M4 (male talkers) and F1 and F2 (female talkers), respectively. The database contains one utterance for each talker as follows.

F1) "The pipe began to rust while new."
F2) "Add the sum to the product of these three."
M1) "Open the crate but don't break the glass."
M2) "Oak is strong and also gives shade."
M3) "Thieves who rob friends deserve jail."
M4) "Cats and dogs each hate the other."

As can be seen, these utterances are well balanced, including voiced speech, plosives, fricatives, etc. The speech waveform was bandpass filtered at 200–3900 Hz and sampled at 8 kHz. A high-slope elliptical filter was used for low-pass filtering to avoid aliasing. The analog-to-digital conversion was performed using a 12 bit A/D converter. The main statistics of the sampled signal are: peak value 1792, mean value 1.165 (nominally zero), and standard deviation 279.23. Table III gives the average pitch for each talker.

The speech file used to test the performance outside of the training sequence contains 12 s of speech (96 000 samples at 8000 Hz sampling rate) from one male talker. The utterance used is:

M5) "When the sunlight strikes rain drops in the air they act like a prism and form a rainbow. The rainbow is a division of white light into many beautiful colors."

The average pitch for this utterance is about 150 Hz.

References

[1] A. Buzo, A. H. Gray, Jr., R. M. Gray, and J. D. Markel, "Speech coding based upon vector quantization," *IEEE Trans. Acoust., Speech, Signal Processing*, vol. ASSP-28, pp. 562–574, Oct. 1980.
[2] H. Abut, R. M. Gray, and G. Rebolledo, "Vector quantization of speech and speech-like waveforms," *IEEE Trans. Acoust., Speech, Signal Processing*, vol. ASSP-30, pp. 423–436, June 1982.
[3] B. H. Juang, "Multiple stage vector quantization for speech coding," in *Proc. IEEE ICASSP*, Paris, France, May 1982, pp. 597–600.
[4] A. Gersho and V. Cuperman, "Speech waveform coding using vector quantization," in *Proc. IEEE GLOBECOM*, San Diego, CA, 1983.
[5] J. D. Gibson, "Adaptive prediction in speech differential encoding systems," *Proc. IEEE*, vol. 68, pp. 488–525, Apr. 1980.
[6] N. S. Jayant and P. Noll, *Digital Coding of Waveforms*. Englewood Cliffs, NJ: Prentice-Hall, 1984.
[7] A. Gersho, "Asymptotically optimal block quantization," *IEEE Trans. Inform. Theory*, vol. IT-25, pp. 373–380, July 1979.
[8] R. M. Gray, J. Kieffer, and Y. Linde, "Locally optimal block quantization for sources without a statistical model," Inform. Syst. Lab., Stanford Univ., Stanford, CA, Tech. Rep. L-904-1, May 1979.
[9] P. Noll, "Adaptive quantizing in speech coding systems," in *Proc. Int. Zurich Sem. Digital Commun.*, Zurich, Switzerland, Mar. 1974, pp. B3.1–B3.6.
[10] P. Whittle, "On the fitting of multivariate autoregressions and approximate factorization of a spectral density matrix," *Biometrica*, vol. 50, pp. 129–134, 1963.
[11] R. A. Wiggins and E. A. Robinson, "Recursive solution to the multichannel filtering problem," *J. Geophys. Res.*, vol. 70, pp. 1885–1891, Apr. 1965.
[12] Y. Inouye, "Modeling of multichannel time series and extrapolation of matrix-valued autocorrelation sequences," *IEEE Trans. Acoust., Speech, Signal Processing*, vol. ASSP-31, pp. 45–55, Feb. 1983.

[13] Y. Linde, A. Buzo, and R. M. Gray, "An algorithm for vector quantizer design," *IEEE Trans. Commun.*, vol. COM-28, pp. 84–95, Jan. 1980.
[14] G. H. Ball and D. J. Hall, "A clustering technique for summarizing multivariate data," *Behavioral Sci.*, vol. 12, pp. 153–155, Mar. 1967.
[15] J. D. Markel and A. H. Gray, Jr., *Linear Prediction of Speech.* New York: Springer-Verlag, 1976.
[16] N. S. Jayant, "Digital coding of waveforms: PCM, DPCM and DM quantizers," *Proc. IEEE,* vol. 62, pp. 611–632, 1974.
[17] ——, "Pitch-adaptive DPCM coding of speech with two-bit quantization and fixed spectrum prediction," *Bell Syst. Tech. J.*, vol. 56, pp. 439–454, Mar. 1977.
[18] C. S. Xydeas, C. C. Evci, and R. Steele, "Sequential adaptive predictors for ADPCM speech encoders," *IEEE Trans. Commun.*, vol. COM-30, pp. 1942–1954, Aug. 1982.
[19] J. B. Anderson and J. B. Bodie, "Tree encoding of speech," *IEEE Trans. Inform Theory,* vol. IT-21, pp. 379–387, July 1975.
[20] L. C. Stewart, R. M. Gray, and Y. Linde, "The design of trellis waveform coders," *IEEE Trans. Commun.*, vol. COM-30, pp. 702–710, Apr. 1982.
[21] R. M. Gray and E. D. Karnin, "Multiple local optima in vector quantizers," *IEEE Trans. Inform. Theory,* vol. IT-28, pp. 256–261, Mar. 1982.
[22] M. R. Schroeder and B. S. Atal, "Predictive coding of speech signals and subjective error criteria," *IEEE Trans. Acoust., Speech, Signal Processing,* vol. ASSP-27, pp. 247–254, June 1979.
[23] J. Makhoul and M. Berouti, "Adaptive noise spectral shaping and entropy coding in predictive coding of speech," *IEEE Trans. Acoust., Speech, Signal Processing,* vol. ASSP-27, pp. 63–73, Feb. 1979.
[24] V. Cuperman and A. Gersho, "Adaptive differential vector coding of speech," in *Conf. Rec. IEEE GLOBECOM,* Dec. 1982, pp. 1092–1096.
[25] J. Adoul, J. Debray, and D. Dalle, "Spectral distance measure applied to the optimum design of DPCM coders with *L* predictors," in *Proc. IEEE ICASSP,* Denver, CO, Apr. 1980, pp. 512–515.
[26] R. M. Gray, "Vector quantization," *ASSP Mag.*, vol. 1, Apr. 1984.
[27] T. R. Fischer, "Quantized control with differential pulse code modulation," in *Proc. 21st Conf. Decision Contr.*, 1983.

Low-Rate Speech Encoding Using Vector Quantization and Subband Coding

by
Huseyin Abut, ECE Department, San Diego State University, San Diego, CA. 92182
and
Siegfried Ergezinger, Institut fur Allgemeine Nachrichtentechnik, Universitat Hannover, Federal Republic of Germany

ABSTRACT

Vector quantization (VQ), subband coding (SBC) and linear predictive coding (LPC) are three of the most effective data compression schemes used for medium-to-narrow band speech coding. In this study, we have attempted to improve the quality of encoded speech by using various combinations of these three coding methods. Waveform coders with rates 2400-9600 bits per second resulted in overall signal-to-distortion ratios of 6-12 dB. We have obtained somewhat lower values for segmented SNR's in the case of straight waveform encoding and higher values for the residually excited subband coded VQ quantizers as expected. However, informal listening tests yielded noticeable improvements over those of straight waveform VQ results. The best quality and the lowest transmission rates were achieved by a residually excited subband coded vector quantization system coupled with a three-way classifier with design parameters: LPC order P=14; 32-band complete binary tree QMF filter bank implementation of SBC and VQ waveform encoding with dimension K=32 at an overall bitrate of 3,100 bps.

INTRODUCTION

Applications of vector quantization to speech coding and waveform encoding problems have been one of the most studied topics during the past few years. It has been demonstrated that dramatic rate reduction could be achieved without any need for speech-specific heuristics. The quality of the reconstructed speech from simple waveform VQ's is comparable to --if not better than-- those of the more complex speech coding schemes.

As in the case of other schemes, the endoing rate, the quality of the reconstructed signal and the complexity are the most crucial design tradeoff issues. There are a number of simple through sophisticated modifications to improve the performance of VQ waveform encoders. First one is the notion of adaptation [1,6], where the codebooks are adapted to the signal shape or the spectra at regular intervals. At present there are no known completely adaptive real-time speech VQ systems. Only multiple codebook type adaptation schemes are available. Another avenue is the class of predictive waveform VQ's[1], in which the prediction gain and the error signal are encoded. To our knowledge at least, there has not been any quantum jump in performance from these two avenues.

We strongly believe and try to justify that quantum jumps in performance are possible from exploiting the quantization distortion decomposition properties of the speech signal by means of extensive subband coded and residually excited quantizers. Another important area for breakthrough in performance curves is the use of better distortion measures [10-11]. Atal and Schroeder have shown that it is possible to have a close-to-transparent quality at 4800 bps. However, the drawback is the mathematical complexity of their scheme. Finally, one can introduce a more improved source modeling --such as the composite source modeling of Meissner[11]-- and a universal coding scheme for VQ [3].

A larger number of people including the authors [1] have successfully designed waveform VQ's for speech. In these works, the research groups have obtained at least locally optimum quantizers operating in the range 6,500-16,000 bps with quantizer dimensions of 8 or less and the codebook size of 256 or less. Even though, on the order of 3.2 MOPS were needed for distortion computations two works have already reported successful real-time hardware implementations[3,7].

In this study, we attempt to improve the quality of the waveform VQ's by preprocessing the digitized speech samples with fixed and variable rate subband coders. In addition, we designed systems with complete 32-channel subband coded vector quantizers embedded in a residually excited LPC vocoding.

Virtually all waveform quantizers have used either a minimum mean square error distortion or a segmented SNR measure. However, the Itakura-Saito type measures were employed for encoding LPC parameters.

Reprinted from *Proc. IEEE Conf. Acoust., Speech, Signal Processing,* vol. 1, pp. 449–452, April 1986.

OCTAVE BAND SBC-VQ SYSTEMS

The subband coding technique as proposed by Crochiere [12] uses a bank of QMF filters and 2:1 decimators during encoding and 1:2 expanders during reconstruction. There are two types of SBC architectures, namely, octave band structures and the complete m-ary tree structures. In the case of octave band SBC, the tree has two branches at every node and high frequency branch is always a terminal node. Separate "good" quantizers are designed for each of these terminal nodes. The low frequency branch, however, is subdivided into two until the end of the splitting process is reached. The subsequent coder for this last level of subdivision is usually different from all the previous ones. A block diagram for this group of systems is shown in Figure 1. Since the bandwidths of bands 1,2,3 and 4-5 are not the same, different encoding rates could be effectively used for each band. This in turn leads to a variable-rate VQ encoding. In addition, a totally different type of quantizer could be designed for bands 4 and 5 since they are in the same octave level.

The order of QMF filter banks is another design consideration. It can be observed from Figure 2 that as the band index increases down the tree the energy content decreases. Hence, higher order filters are needed to control the transition regions and the spectral foldovers in the early stages of splitting. However, the subsequent filters have lower orders to achieve similar results.

In our experiments, we have used a database containing 640,000 samples with a sampling frequency of 8.0 kHz from five different male and female radio announcers. The QMF filter order for the first splitting level was 64 and then the orders were reduced to 48, 32, 16, and 8 for the subsequent stages of splitting.

For a target rate of 5,000 bps, we have designed an optimal VQ of dimension 32 and a rate 1/4 bit/sample for the first band. For the subsequent bands 2, 3, and 4 the quantizer rates were 1/2, 1, and 2 bits/sample and the corresponding dimensions were 16, 8, and 4, respectively. Finally, a single quantizer with a rate 2 and dimension 2 was designed for bands 5 and 6. In the first four stages one separate VQ was designed for strings of samples coming from a single channel; whereas the last VQ was designed from vectors of dimension two, which were formed by taking one sample from the lp branch and the other one from the hp branch. The resulting rate of 5,000 bps and the SNR and the segmented SNR values were 10.29 and 8.30 dB, respectively. The informal listening tests indicate that a reasonably good quality speech encoding is possible with some increased computational complexity when compared with the straight VQ waveform experiments. The reason for this good quality is most probably due to the large dimension of the first quantizer (K=32). A 32-component vector is equivalent to a 4.0 ms segment of speech and it is on the same order with pitch frequencies. Hence, the ever crucial role of pitch is taken into consideration without attempting to do any pitch extraction. Similar effects are also observable from the innovation vectors of Atal and Schroeder [10], in which they have adopted a vector size of 40.

RESIDUALLY EXCITED SUBBAND VQ SYSTEMS

As Gray pointed out [1], the residually excited linear predictive (RELP) VQ systems did not yield encouraging results at low bit rates. The possible reasons for that are (i) the small vector sizes used in encoding the residuals --typically 8 or less-- and (ii) the spectral and the statistical properties of the residuals are not taken into consideration in designing vector codebooks. In order to check the first point, we have experimented with dimensions 8, 16, and 32. We have noticed that the quality of the reconstructed speech was related to the vector size. Next, we have studied the power spectral densities of the original speech, the residual, the error resulting from the encoding+decoding operations and the distribution of the powers in various bands. They are shown in Figures 2 and 3, respectively.

The observations from these tests confirmed our conjectures and we have designed a residually excited subband coded waveform encoder with a block diagram as depicted in Figure 4. First, a 14-order LPC analysis was performed over frames of 20 msec speech segments. Next, while the normalized LPC coefficients and the gain were encoded by an LPC-VQ, the LPC filter using the same set of coefficients filtered out the residuals. Vectors of size 8 or 16 or 32 were formed from these residuals, which were then split into 8 or 16 or 32 bands using a complete binary tree QMF filter bank structure. For instance, the tree was terminated after three layers for an 8-band SBC. The terminal node filter outputs were simultaneously sampled to form vectors of dimension 8 or 16 or 32. These vectors were used in generating the waveform VQ codebooks in the design stage. Similarly obtained vectors were coded by a waveform VQ encoder operating at a rate 1.0 or 1/2 or 1/4 bit per sample for vector sizes of 8 or 16 or 32, respectively.

The simulation results from the residually excited and other VQ systems are summarized in Table 1. In test 6, we have encoded the residuals with an 8-dimensional rate one waveform VQ and 500 bps were needed for the LPC-VQ. Thus, the overall bit rate was 8500 bps and the SNR and the segmented SNR values were 6.48 and 6.72 dB, respectively. Similar results for a 3000 bps yielded 4.57

and 5.25 dB. The difference between the two SNR types increased to 0.68 dB from 0.24 dB as the bit rate decreased by a factor of almost three. These measurements and the quality of the reconstructed speech confirm our conjecture that better distortion measures would bring considerable improvement in performance, which was also the strength of Atal and Schroeder's work[10].

RESIDUAL SBC-VQ SYSTEMS WITH A CLASSIFIER

In order to attest our conjecture on the composite source modeling we have included a three-way classifier into the system of Figure 4. The block diagram for this modified system is depicted in Figure 5. The classification was based on energy and prediction gain measurements. Since these parameters are readily available from the LPC analysis the additional computational complexity is negligible. The only increase is in the memory requirements of the encoder+decoder pair. The test results yielded a 33% reduction in the mean-square distortion value corresponding to SNR and segmented SNR values of 5.39 and 5.80 dB, respectively. A similar improvement in quality was noticed.

DISCUSSIONS AND CONCLUSIONS

The results from Table 1 and the various listening tests show that there is a direct correlation between the SNR's --or the segmented SNR's-- and the quality of reconstruction if the VQ is a pure waveform quantizer.

Our objective of designing a fairly good quality waveform VQ system under 5000 bps with a manageable computational complexity has been achieved by the octave band SBC structure of Figure 1. The simulation results are given under entry "Test: 2" in Table 1. If the results are compared with that of the plain waveform VQ operating at a similar rate (Test:1) we see that there is an SNR gain of 2.65 dB. However, the quality difference in reconstructed speech is more significant than the SNR value reflects. The drawback of this system with respect to plain VQ and to systems proposed in Figures 4 and 5 is the need to design and keep track of form VQ's operating at four different rates.

In the case of residually excited subband coded VQ's, our objective was to decrease the bit rate further to the narrow band speech coding range without resorting to any pitch and voicing extraction mechanisms. The last five tests on Table 1 and the quality of the reconstructed speech segments indicate that this goal has been achieved. The trend in the segmented SNR values clearly justify the additional effort going into this route. Finally, the system of Figure 5 takes into consideration all of the improvement areas discussed in introductory paragraphs.

We believe that further improvements can be achieved if the system can incorporate the distortion measure and the speech production model recommended by Atal and Schroeder[10]. However, the final system should not require an untolerably high computational complexity to achieve that task. One possible solution is the use of fast search algorithms without impairing the quality, such as an m-ary tree search or a directed graph search instead of a full search VQ encoder.

ACKNOWLEDGMENTS

This work has been supported in part by the NSF Grant No: ECS-8302658 and grants from the Cubic Corporation of San Diego and the TRW Military Electronics Systems Division.

REFERENCES

[1] Gray, R.M., "Vector Quantization," IEEE ASSP Magazine, April 1984.

[2] Abut, H., R.M. Gray and G. Rebolledo, "Vector Quantization of Speech and Speech-Like Waveforms," IEEE Trans. on Acoust., Speech and Signal Processing, Vol. ASSP-30, pp. 423-435, June 1982.

[3] Tao, B.P.M., H. Abut and R.M. Gray, "Hardware Realization of Waveform Vector Quantizers," IEEE Journal on Selected Areas in Communication, Vol. SAC-2, pp. 343-352, March 1984.

[4] Steward, L.C., R.M. Gray and Y. Linde, "The Design of Trellis Waveform Coders," IEEE Trans. on Communications, Vol. COM-30, pp. 702-710, April 1982.

[5] Sabin, M.J. and R.M. Gray, "Product Code Vector Quantizers for Waveform and Voice Coding," IEEE Trans. on Acoust., Speech and Signal Processing, Vol. ASSP-34, pp. 474-488, June 1984.

[6] Gersho, A. and V. Cuperman, "Vector Quantization: A Pattern-matching Technique for Speech Coding," IEEE Communications Magazine, December 1983.

[7] Davidson, G., T. Stanhope, R. Aravind and A. Gerso, "Real-Time Speech Compression with a VLSI Vector Quantization Processor," Proc. IEEE Int. Conf. ASSP, pp. 1437-1440, March 1985.

[8] Abut, H. and S. A. Luse, "Vector Quantizers for Subband Coded Waveforms," Proc. IEEE Int. Conf. ASSP, pp. 10.5.1-10.5.9, March 1984.

[9] Rebolledo, G., R.M. Gray and J.P. Burg, "A Multi-rate Voice Digitizer Based upon Vector Quantization," IEEE Trans. on Communications, Vol. COM-30, pp. 721-727, April 1982.

[10] Schroeder, M. and B.S. Atal, "Code-Excited Linear Prediction (CELP): High-Quality Speech at Very Low Bit Rates," Proc. IEEE Int. Conf. ASSP, pp. 937-940, March 1985.

[11] Meissner, P. "Ein Mathematisches Modell fur Sprachsignale, Basierend auf einer mit Methoden der Mustererkennung

Durchgefuhrten Segmentierung," Phd. Dissertation, Universitat Hannover, 1983.
[12] Crochiere, R.E., "On the Design of Sub-Band Coders for Low Bit Rate Speech Communication," Bell System Tech. J., pp. 1411-1421, September 1981.

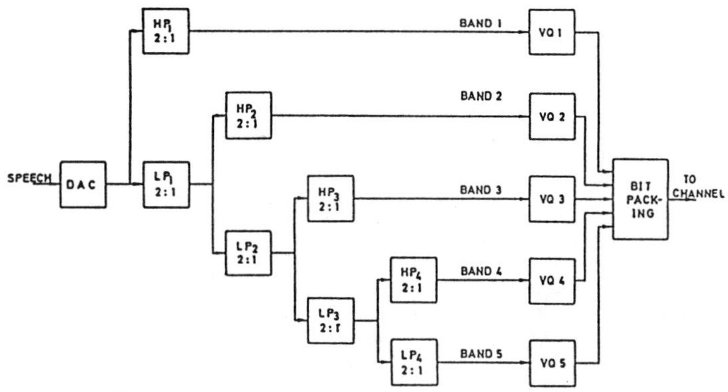

Figure 1. Octave Band Subband Coded VQ System.

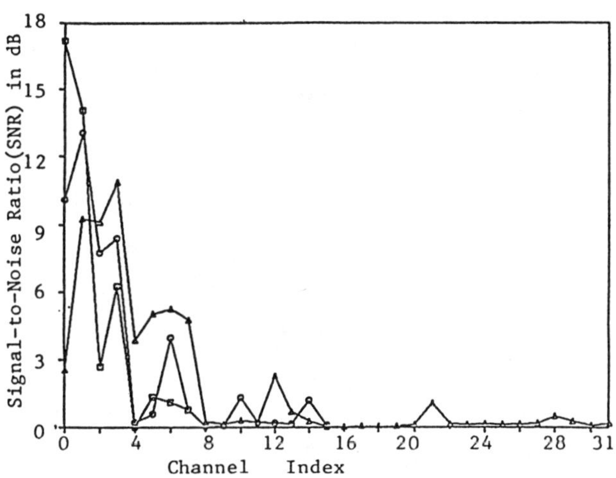

Figure 2. SNR Distribution for 8, 16, and 32-Channel SBC with a complete Tree.

Table 1. Comparison Between Speech Samples and Reconstructed Speech.

Test	LPC Order	Number of Bands	Rate for Wave-form	Overall Rate	SNR [dB]	SSNR128 [dB]
1	--	1	0.50	4000	7.64	--
2	--	--	0.63	5000	10.29	8.30
3	--	8	1.0	8000	11.11	10.93
4	--	16	0.5	4000	7.78	8.09
5	--	32	0.25	2000	5.83	6.38
6	14	8	1.0	8500	6.48	6.72
7	14	16	0.5	4500	4.46	5.35
8	14	32	0.25	2500	3.43	4.62
9	14	32	0.31	3000	4.57	5.25
10	14	32 k1	0.31	3100	5.39	5.80

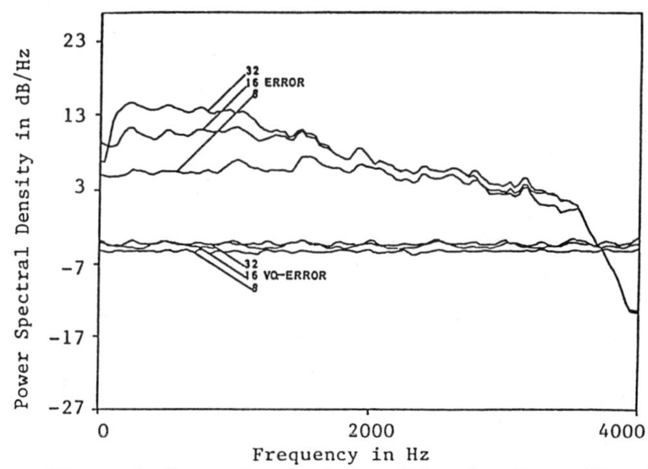

Figure 3. Power Spectral Densities for Error Signal and VQ-error for LPC order 14, Dimensions 8, 16, and 32.

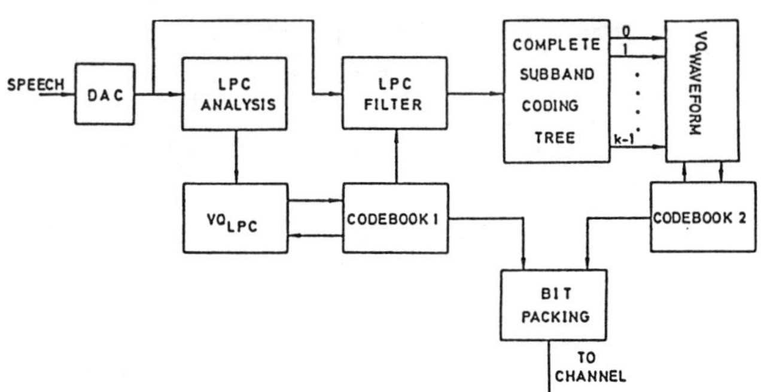

Figure 4. Residually Excited SBC-VQ

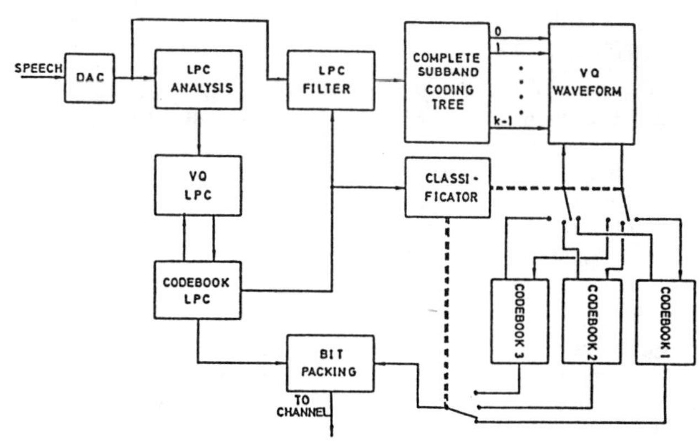

Figure 5. Residually Excited SBC-VQ with a Classifier

A High Quality Subband Speech Coder with Backward Adaptive Predictor and Optimal Time-Frequency Bit Assignment

F. K. Soong
R. V. Cox
N. S. Jayant
AT&T Bell Laboratories
Murray Hill, New Jersey 07974

Abstract. In this paper we propose a hybrid speech coder which utilizes properties of APC, ATC, and subband coding. Subband splitting is used to reduce the dynamic range of a full-band APC. As a result, the prediction or power gain is reduced and the instability problem associated with the APC coder is alleviated to a large extent. The optimal bit allocations used in ATC is extended in the new coder where the available coding bits are optimally allocated, both in the time and frequency domains. Frame boundary artifacts (such as those found in ATC) are not present due to the time-domain processing nature of SBC. In order to improve the efficiency in quantizing the side information (time and frequency domain signal power and the corresponding bit allocations), vector quantization is used. Adaptive backward predictors are used to further reduce the number of bits allocated to side information, leaving more bits to encode the prediction residual signals. At 16 kbps the coder achieves better than 20 dB segmental signal-to-noise ratio and sounds transparent for most speakers.

I. Introduction

At medium bit rates (9.6~16 kbps) several efficient speech coders, including adaptive predictive coding (APC) [1], adaptive transform coding (ATC) [2] and subband coding [3], are capable of coding speech signals with a performance ranging from communication quality to nearly toll quality.

In an APC coder, usually two predictors are used: a short-term predictor and a long-term predictor. The former predictor is used to remove short-term (i.e., within a pitch period) correlations of speech samples and the latter one is used to remove quasi-periodic long-term (i.e., across a pitch period) correlations of voiced speech. After removal of both short and long term correlations, the remaining prediction residuals can then be efficiently encoded by using techniques such as center clipping, entropy coding etc. However, despite the fact that APC encoded speech sounds good in general, some occasional "glitches" and "clicks" can still result from a temporary prediction loop instability [4]. At lower data rates, this prediction loop instability problem deteriorates very quickly.

In an ATC coder, the nonuniform power distribution of speech signals in a transform domain is exploited and bits are optimally allocated to different transform coefficients. As a result of this optimal bit allocation the signal to quantization noise ratio (SNR_q) in each transform block is maximized. However, due to the block processing nature in the transform domain, ATC coded speech signals are discontinuous at frame boundaries and some low frequency (i.e., frame-rate) rumbling noise can often be perceived even at a coding rate of 16 kbps.

In terms of coder complexity, a subband coder is much simpler than either APC or ATC. But a simple SBC coder can offer only communication output quality at 16 kbps. At an even lower bit rate such as 9.6 kbps, the output speech quality is rather poor and is not acceptable for many applications.

The aforementioned problems associated with APC, ATC and SBC can be alleviated to a large extent by using a hybrid subband coding scheme proposed by Honda and Itakura [5]. The coder called APC with adaptive bit allocation (APC-AB) is a sophisticated coder which is capable of alleviating or eliminating the instability of a full band APC, avoiding frame boundary discontinuities of ATC and, most importantly, improving significantly the quality of SBC.

In this paper we propose a new subband coder. In this new coder while many of the salient features of the APC-AB coder are used, the bit utilization is improved even further by adding two major features: (1) efficient vector quantization (VQ) is used to encode the side information for optimal bit allocations in time and frequency; (2) backward adaptive prediction (APB) and quantization (AQB) are adopted to replace forward adaptive prediction (APF) and quantization (AQF), respectively.

II. System Overview

A functional block diagram of the proposed speech coder is depicted in Fig. 1. Each functional block will be discussed in detail in the following sections.

Depicted in Fig. 2 are coding examples of subband signals, prediction errors, dynamic bit allocations in time and in frequency, reconstructed subband signals and reconstruction error signals for an unvoiced frame and in Fig. 3 for a voiced frame.

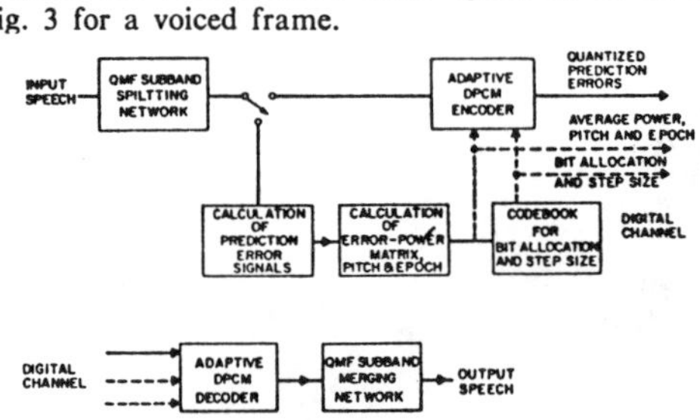

Fig. 1 A system block diagram of the proposed hybrid subband coder.

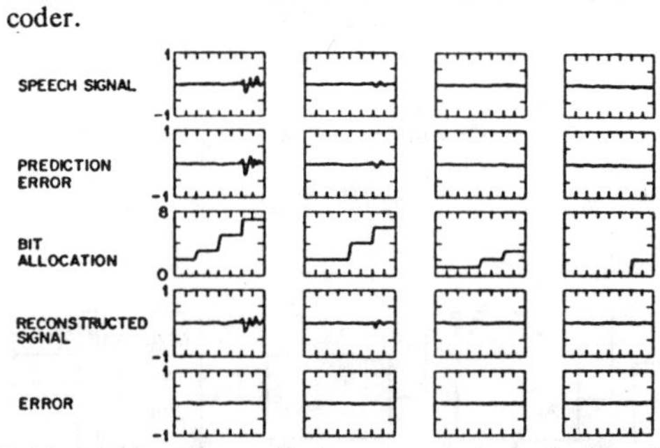

Fig. 2 Subband signals, prediction errors, dynamic bit allocations, reconstructed signals and reconstruction errors of 4 subbands in an unvoiced frame.

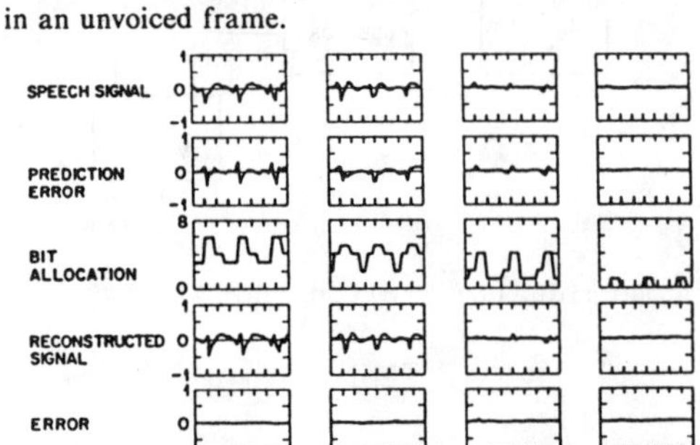

Fig. 3 Subband signals, prediction errors, dynamic bit allocations, reconstructed signals and reconstruction errors of 4 subbands in a voiced frame.

Reprinted from *Proc. IEEE Conf. Acoust., Speech, Signal Processing,* vol. 4, pp. 2387–2390, April 1986.

2.1 QMF Analysis and Synthesis Networks

As shown in Fig. 4, full band speech signals sampled at 8 kHz rate are first split into 5 consecutive, equally spaced subbands by using a quadrature minor filter (QMF) analysis network. Signals of each subband are then down-sampled at the corresponding Nyquist rate of 1.6 kHz. Also since the power of speech signals at the highest (the 5th) subband is, on the average, much lower than that of the other 4 lower bands, we can discard the 5th subband samples almost without affecting the quality of the synthesized speech. By using only 4 out of 5 subbands, the average bit rate per sample is increased by 25% (to 2.5 bits per sample at 16 kbps) and the flexibility of the dynamic bit allocation algorithm is therefore greatly enhanced.

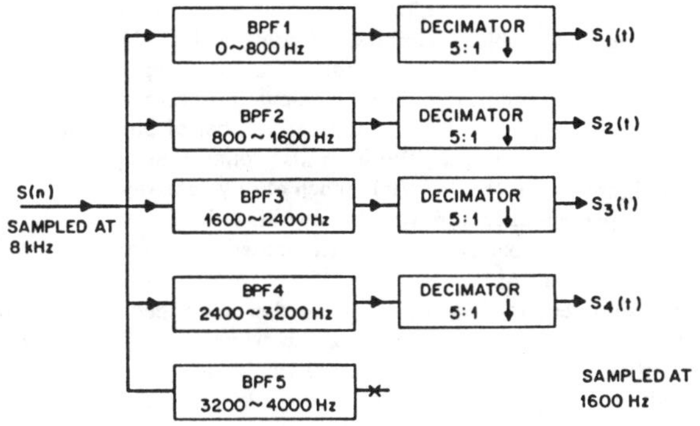

Fig. 4 The subband splitting QMF network of the proposed coder.

In our preliminary study [6], we have found that the interband aliasing effects due to quantization is almost imperceptible when optimally designed, 60 tap QMF filters are used for 5-band, equally spaced subband analysis and synthesis networks. Also we have found that there is no appreciable improvements of coding performance when a longer impulse response QMF filter say 90 or more taps is used. Based on the preliminary study results plus the fact that the inherent filter delay of 7.5 ms, the computational complexity and the storage requirement for this 60 tap QMF filter are acceptable for many applications, we choose this 60 tap QMF filter in our coder.

2.2 Backward Adaptive Prediction (APB)

At a bit rate of 32 kbps, the adaptive DPCM (ADPCM) speech coder with a backward adaptive pole-zero predictor recommended by CCITT [6] has been shown to give toll quality coding performance. At a lower bit rate of 16 kbps, however, the performance of the same ADPCM coder degrades dramatically due to two major problems. First, the fixed bit assignment per sample in ADPCM i.e., 2 bits/sample at 16 kbps, unavoidably yields rather crude quantizations. Consequently the short-time correlations of reconstructed (quantized) speech samples are very much different from the true (unquantized) speech samples and the backward prediction gain is therefore rather limited. Second, as in an APC coder, an occasional high power gain of the backward predictor can create a temporary unstable prediction loop and cause a large SNR_q loss. In this new coder, two important features, namely, a multi-band structure and a dynamic bit allocation in time and frequency domains, have made APB viable again. The first feature, a multi-band structure, reduces the spectral dynamic range in each subband and consequently improves the prediction loop stability appreciably. The second feature, a dynamic bit allocation in time and frequency domains, greatly improves the quantizer performance by assigning more bits to higher amplitude samples and less or even no bits to lower amplitude samples. The high quality reconstructed (quantized) speech samples are then used for adapting the backward adaptive predictors. In this study, we use two backward predictors, the standard gradient pole-zero predictor recommended by CCITT, and the fast recursive least squares (FRLS) lattice predictor. The former one is easier to implement due to its gradient-based adaptation and the latter one, though more complicated, delivers better performance than the CCITT predictor in our preliminary study. The adaptation algorithms which can be found elsewhere [6], are not repeated here.

2.3 Time and Frequency Adaptive Bit Allocation

The prediction error power pattern, which is used in the proposed coder for dynamic bit allocation in time and in frequency, is computed either pitch synchronously for voiced frames or pitch asynchronously for unvoiced frames using *quantized* prediction error samples.

The voiced and unvoiced decision is based on the detection of a quasi-periodicity in the baseband (1st-subband) signal. Using the correlation function of the 1st-subband samples, the present frame is determined to be voiced if a local maximum of the correlation function at time lag $\tau(\tau \neq 0)$ exceeds a pre-selected threshold. Once a frame is determined as voiced, the corresponding time lag τ is used as the pitch period estimate and the maximum amplitude of the baseband prediction error signals in the 1st full pitch period is also located. The position of the maximum relative to the beginning of the frame is denoted as the first pitch epoch. Each pitch period is then divided into 4 subblocks of approximately equal lengths and the prediction error power pattern is estimated pitch synchronously. For unvoiced frames, the whole frame (instead of the whole pitch period) is equally divided into four subblocks and the average prediction error power corresponding to each subband and subblock is similarly estimated. The division of a pitch period or a frame into subblocks is conceptually depicted in Fig. 5 for both voiced and unvoiced frames. For coding efficiency, the prediction error power matrix of every frame is further normalized by the corresponding average power in order to reduce the dynamic range. As depicted in Fig. 6, the prediction error power matrix is computed using the average power, the voiced and unvoiced (V/U) decision, plus the pitch period and the 1st pitch epoch position if the frame is voiced. The columns and rows of the output normalized error-power matrix correspond to the subblocks and subbands, respectively.

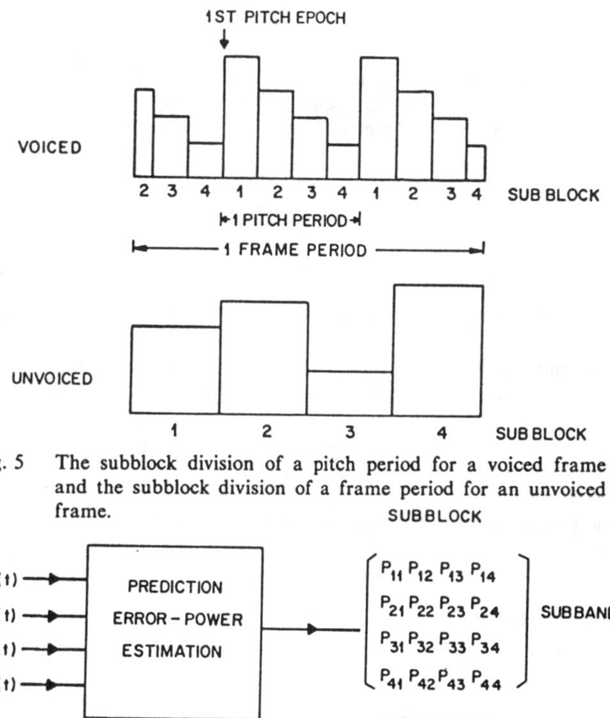

Fig. 5 The subblock division of a pitch period for a voiced frame and the subblock division of a frame period for an unvoiced frame.

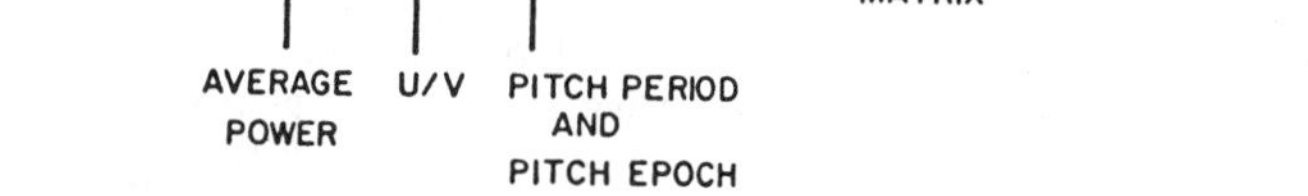

Fig. 6 A block diagram of the prediction error power matrix computation.

In a closed-loop predictive DPCM framework, the quantization error between the quantizer input and the output is the same as the error signal between the original signal and the reconstructed signal. Since the reconstruction error power equals the quantization noise power, the SNR_q can be maximized by minimizing the quantization noise power. For the specified structure of 4 subbands and 4 subblocks in the proposed coder, the quantization noise power is

$$\sigma^2 = \epsilon_*^2 \sum_{i=1}^{4} \sum_{j=1}^{4} 2^{-R_{ij}} b_i \ell_j P_{ij} \quad (1)$$

where R_{ij} and P_{ij} are the allocated bits and the residual error power of the ith subband and the jth subblock, respectively, b_i is the bandwidth of the ith subband normalized by the total effective bandwidth, in our case, $b_i = 0.25$, ℓ_j is the total number of samples of the jth subblock normalized by the frame length of each subband, and ϵ_*^2 is the quantizer performance factor assumed to be constant, i.e., independent of i and j.

The quantization noise power is minimized under the constraint of a given bit rate per sample

$$\bar{R} = \sum_{i=1}^{4} \sum_{j=1}^{4} b_i \ell_j R_{ij} \quad (2)$$

and the resultant bit allocation is obtained as

$$R_{mn} = \bar{R} + \frac{1}{2} \log_2 \left(\frac{P_{mn}}{\prod_{i=1}^{4} \prod_{j=1}^{4} P_{ij}^{b_i \ell_j}} \right) \quad (3)$$

for the mth subband and the nth subblock. The corresponding minimum distortion is

$$\sigma_{\min}^2 = \epsilon_*^2 \, 2^{-2\bar{R}} \prod_{i=1}^{4} \prod_{j=1}^{4} P_{ij}^{b_i \ell_j} \quad (4)$$

The SNR gain obtained by using this dynamic bit allocation in time and frequency can be decomposed into two components, frequency domain subband gain and temporal domain gain. The subband gain is

$$G_{sub} = \frac{P}{\prod_{i=1}^{4} P_i^{b_i}} \quad (5)$$

where $P_i \triangleq \sum_{j=1}^{4} P_{ij} \ell_j$ is the average power of the ith subband and $P \triangleq \sum_{i=1}^{4} P_i b_i$ is the average power of the 4 subbands.

The temporal domain gain is obtained using the nonuniform power distribution over subblocks and the gain obtained in the ith subband is

$$G_{i,temp} = \frac{P_i}{\prod_{j=1}^{4} P_{ij}^{\ell_j}} \quad (6)$$

The geometric mean of $G_{i,temp}$ over all 4 subbands is

$$G_{temp} = \prod_{i=1}^{4} [G_{i,temp}]^{b_i} \quad (7)$$

The total gain over the forward adaptive PCM (APCM) is the product of the subband gain and the temporal domain gain

$$G = G_{sub} \cdot G_{temp} \quad (8)$$

2.4 Vector Quantization of the Error Power Matrix

The prediction error power matrix is a 4×4 matrix. If the 16 components of the matrix are quantized separately as scalars, a moderate amount of bits, on the order of 20 ~ 25 bits per frame, is needed. In order to improve the coding efficiency, we propose to use vector quantization to quantize the whole matrix as an entity. The same idea of VQ was used by Heron et. al. [7] in quantizing LPC side information for frequency domain bit allocation. There is one more reason why VQ is an attractive quantization scheme for encoding the error power matrix; since adaptive backward quantization is used in the proposed coder and the step sizes are adapted in a backward manner, a precise knowledge of the prediction error power matrix or the corresponding quantizer step sizes is not as crucial as in adaptive forward quantization. Although the power matrix still needs to be quantized for dynamic bit allocations, a bit rate reduction of 2 ~ 3 times can be achieved by VQ over scalar quantization.

The prediction error power VQ codebook is generated using the algorithm proposed by Linde, Buzo and Gray [8] from a set of training vectors. A full search, eight-bit codebook which consists of 256 distinct entries is obtained by minimizing the average distortion (Euclidean distance) over all training vectors. In coding, the computed prediction error power matrix is vector quantized first and then used in conjunction with other side information to allocate bits dynamically to each subband and subblock samples according to Eq. (3). In practice, the bit assignment in Eq. (3) is modified to assign only non-negative integer bits. Also an iterative bit adjustment is used to subtract or add 1 bit to speech samples such that the total number of bits allocated per frame are exactly consumed. We have also found that by constraining the bit assignments to the 1st pitch epoch of any subband to be at least one bit, the perceptually annoying bandwidth reduction due to zero bit assignment to a whole subband is eliminated. Similar observations were made before [5,6].

The bit assignment procedure can be further simplified by deriving a companion VQ codebook of bit allocation using the VQ codebook of the prediction error power matrix. Once the bit allocation patterns are generated and prestored in the VQ codebook, given the index of the error power matrix VQ codebook entry, the corresponding bit allocation pattern can be retrieved directly without using Eq. (4). In our experiments, we have found no appreciable degradation of performance, either subjectively or objectively, by using the companion bit allocation VQ codebook for dynamic bit assignments.

III. Coder Performance

The parameters and bit allocations of the new coder are listed in Table I. The speech material used for deriving the VQ codebooks and for evaluating the coder performance consists of two different data bases. The first one which we shall call the training data base was used for both VQ codebook generation and coder evaluations. The training data base is a data base of 10 different short English sentences each spoken once by 7 (4 male and 3 female) speakers. The data base was recorded over high quality microphone in a sound booth. The second data base which we shall call the testing data base is a

Frame size	25 ms	
No. of subbands	4 (0~3.2 kHz)	
No. of subblocks	4	
Sampling rate	8 kHz	
Backward adaptive predictors	4th order LS lattice (all-pole)	
	CCITT gradient (pole-zero)	
Side information (bits/frame)	Error power matrix VQ codebook	8
	Pitch (voiced)	6
	1st pitch epoch (voiced)	5
	V/U	1
	Average error power	7
Main information (bits/frame)	Voiced	373 (2.33 bits/sample)
	Unvoiced	384 (2.4 bits/sample)

Table I. The system parameters and bit allocations of the new coder

miscellaneous data base which consists of several different sentences spoken by several speakers, both male and female. In addition to the fact that the speakers and the speech material of the testing data base are different from those of the training data base, the recording conditions of the testing data base are also less homogeneous than the training data base. Both data bases are sampled at 8 kHz sampling rate.

Two objective measures, segmental SNR (SNR_{seg}) in the time domain and the likelihood ratio spectral distortion in the frequency domain measure, are used to evaluate the coder performance. The likelihood ratio spectral distortion measure which is also called the gain normalized Itakura-Saito distortion measure is used here to measure the spectral distortion between the unquantized reconstructed speech and the quantized reconstructed speech signals. The distortion between a reference (unquantized) frame and a corresponding test (quantized) frame is of the following form

$$d_{LR} = \frac{\mathbf{a}^T \mathbf{R} \mathbf{a}}{\alpha} - 1 \qquad (9)$$

where $\mathbf{R}$ is the autocorrelation matrix of the reference signal, α is the corresponding LPC prediction residual and $\mathbf{a}$ is the LPC vector of the test signal. The likelihood ratio distortion measure, due to its sensitivity to spectral peaks, is a perceptually meaningful measure. In our evaluation, a standard 10th order LPC analysis is used. Also since d_{LR} is an asymmetrical measure, we measure the distortions between reference and test and vice versa.

The performance of the coder as measured by segmental SNR is given in Table II. For comparison purposes we also list the corresponding SNR_{seg} of the ADPCM coder with the CCITT predictor, the 4-band subband coder with fixed bit allocations and frequency domain dynamic bit allocations studied in our preliminary study [6]. All coders were operated at 16 kbps.

Parameters		ADPCM	4-Band Subband				New Coder			
Bit Allocation	Fixed	•	•		•					
	Dynamic in Frequency			•		•				
	Dynamic in Time and Frequency						•	•	•	•
Backward Predictor	Gradient Pole-Zero	•	•	•			•	•		
	FRLS				•	•			•	•
Data Base	Training	•	•	•	•	•	•		•	
	Testing							•		•
Segmental SNR		13.6	15.4	17.5	15.8	19.2	20.8	21.0	21.5	21.2

Table II. Segmental SNR's of an ADPCM coder, a 4-band subband coder and the new coder for different parameters at 16 kbps rate

Several observations can be made about the SNR_{seg} results.

(1) Subband splitting is helpful in reducing the high spectral dynamic range and in allocating the companion prediction loop instability of a full band ADPCM. Even for fixed bit allocations, the 4-band subband coder is able to improve the SNR_{seg} by 1 ~ 2 dB over ADPCM.

(2) When dynamic bit allocation is used in subband coding, even only in the frequency domain, an improvement of 2 ~ 3 dB in SNR_{seg} can be obtained over fixed bit allocations. When dynamic bit allocations are optimized both in time and in frequency, another 2 dB of SNR_{seg} is gained.

(3) The objective SNR_{seg} difference between the LS predictor and the pole-zero gradient predictor, is more appreciable in ADPCM and subband coders with fixed bit allocations and frequency domain dynamic bit allocations. The difference is not large in this new coder, i.e., being less than 1 dB. Furthermore, informal subjective listening indicates that the objective difference is almost imperceptible in this coder.

(4) The performance of the proposed coder is rather robust with respect to different testing material. The SNR_{seg} difference between the training data base and the testing base is negligible. Subjectively, we cannot detect any appreciable difference, either.

The likelihood ratio spectral distortions between quantized and unquantized reconstructed signals are given in Table III for the new subband coder. Due to the asymmetry of the distortion measure, in the parenthesis are also given the distortions when quantized and unquantized signals are reversed in Eq. (9). The decibel spectral distortion of d_{LR}, which is an approximation to the L_2 norm in the log spectral domain, is approximately 1.5 dB for a d_{LR} value of 0.07. For the low distortion level achieved by the new coder, the likelihood ratio distortion is virtually symmetrical. Also the distortion performance is rather robust to different data bases.

Parameters		New Coder			
Backward Adaptive Predictor	Gradient Pole-Zero	•	•		
	FRLS			•	•
Data Base	Training	•		•	
	Testing		•		•
d_{LR}		0.071 (0.072)	0.073 (0.070)	0.070 (0.072)	0.072 (0.070)

Table III. Likelihood ratio distortions of the new coder between quantized and unquantized speech signals. Values in parenthesis are obtained by reversing the order of quantized and unquantized

REFERENCES

[1] B. S. Atal, and M. R. Schroeder, "Adaptive Predictive Coding of Speech Signals," B.S.T.J., Vol. 49, pp. 1073-86, Oct. 1970.

[2] R. Zelinsky and P. Noll, "Adaptive Transform Coding of Speech Signals," IEEE Trans. on ASSP, Vol. ASSP-25, pp. 299-309, Aug. 1977.

[3] R. E. Crochiere, S. A. Webber and J. L. Flanagan, "Digital Coding of Speech in Sub-bands," B.S.T.J., Vol. 55, pp. 1069-1085, Oct. 1976.

[4] M. Kransner, M. Berouti and J. Makhoul, "Stability Analysis of APC Systems," Proc. ICASSP, pp. 627-630, 1981.

[5] M. Honda, and F. Itakura, "Bit Allocation in Time and Frequency Domain for Predictive Coding of Speech," IEEE Trans. on ASSP, Vol. ASSP-32, No. 3, pp. 465-473, June 1984.

[6] F. K. Soong, R. V. Cox and N. S. Jayant, "Subband Coding of Speech Using Backward Adaptive Prediction and Bit Allocation," Proc. ICASSP-85, pp. 43.1.1~43.1.4, Mar. 26, 1985.

[7] C. Heron, R. E. Crochiere and R. V. Cox, "A 32-band Sub-band/Transform Coder Incorporating Vector Quantization for the Transmission of Side Information," ICASSP-82, Apr. 1983.

[8] Y. Linde, A. Buzo and R. M. Gray, "An Algorithm for Vector Quantization," IEEE Trans. on Communication, Vol. COM-28, No. 1, pp. 84-95, Jan. 1980.

QUANTIZATION PROCEDURES FOR THE EXCITATION IN CELP CODERS

Peter Kroon and Bishnu S. Atal
Acoustics Research Department
AT&T Bell Laboratories, Murray Hill, NJ 07974

ABSTRACT

Past research on CELP (Code-Excited Linear Predictive) coders has mainly concentrated on the feasibility of the CELP concept and on the reduction of the computational complexity. In this paper we address the problem of finding and encoding the excitation parameters with a limited bit rate, such that high quality speech coding in the 4.8 - 7.2 kb/s range becomes feasible. First, we examine the effect of the various excitation parameters such as code book size, code book population, order of the long-term predictor and update rate on the quality of the reconstructed speech. Second, we investigate procedures for designing and incorporating quantizers for the parameters involved. Finally, using both scalar and vector quantization techniques for the LPC coefficients, we simulated 4.8 kb/s and 7.2 kb/s coders. We also report on the use of postfiltering to further improve the performance of the CELP coder.

INTRODUCTION

The Code-Excited Linear Predictive (CELP) coder [1,2] is expected to provide high quality speech at low bit rates (below 8 kb/s). The coder uses linear predictive techniques to remove the correlation in the speech signal. The remaining (residual) information is then modeled by samples from a stochastic code book, such that the resulting synthetic signal is "close" to the original speech signal. Figure 1 shows the basic structure of the coder.

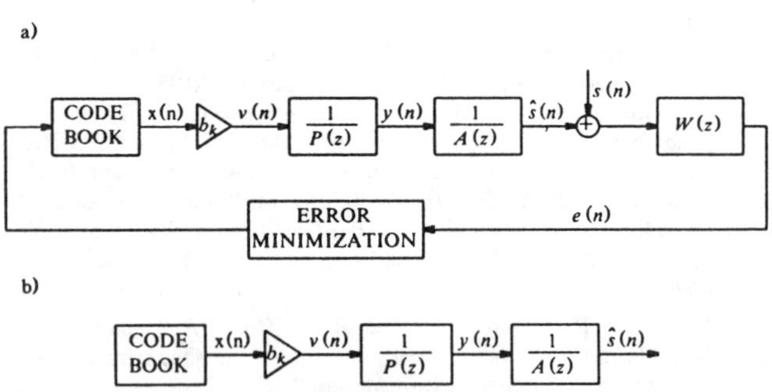

Figure 1. Basic structure CELP coder. (a) encoder, (b) decoder.

The code book contains N candidate sequences of length L. A selected code book sequence or vector is scaled by a factor b_k and is then fed into a cascade of two speech-adaptive all-pole filters. The filters $1/P(z)$ and $1/A(z)$ model the long-term correlations and the short-term correlation in the speech signal. The excitation for these time-varying linear filters is determined by a full search procedure in a code book using a weighted mean squared error criterion.

Prior research has mainly concentrated on the feasibility of the CELP concept [1,2] and on the reduction of the computational complexity [3]. A major unsolved problem, however, is to quantize the side information (predictor coefficients, gain etc.) without degrading performance. The side information can be divided into two more or less independent parts:

1. global spectral information (short-term predictor),
2. fine spectral information (excitation),

In this paper we mainly address the problem of determining and encoding the excitation parameters with a limited bit rate, such that when scalar or vector quantization for the short-term predictor parameters is used, high quality speech coding in the 4.8 - 7.2 kb/s range becomes feasible.

To evaluate the coder parameters and quantization procedures we used non-real-time computer simulations with real speech data. The speech data base consisted of approximately 120 s of speech spoken by 14 different speakers (6 male, 6 female and 2 children), and the sample input sentences had four different classes of phonemic characteristics [4]. All utterances were sampled with 16-bit resolution and an 8 kHz sampling rate, and were band-pass filtered (125-3400 Hz) prior to sampling.

The performance of the proposed coder structure is determined by many parameters but we will limit this discussion mainly to those related to the excitation. To examine the effect of a particular parameter, several analyses were performed with varying parameter settings. While varying the value of one parameter, all the other parameters were kept fixed at their default values. As default values we used those given in [2] since they were found to give a reasonable trade-off between speech quality and the number of bits required for their representation.

First we will discuss briefly some aspects of the short-term predictor and then we give a more detailed description of the long-term predictor. Code book issues, and procedures for determining and quantizing the excitation gain are discussed in the next section. We then combine the results to get a fully quantized stochastic coder. Finally, we describe the application of postfiltering to further enhance the quality of the reconstructed speech.

SHORT-TERM PREDICTOR

To get some insight into the effect of the short-term predictor parameters on the coder performance, we evaluated the CELP coder using different predictor parameter settings. The possible compensation effect of the excitation for small inaccuracies in the spectral envelope (as defined by the predictor) was neglected. The excitation analysis was performed with predictor parameters obtained with the *autocorrelation* method.

For three different *update intervals* of the predictor coefficients (10, 15 and 20 ms), we varied the order p between 8 and 16. As can be expected the best results were obtained with the highest update rate and order. As far as the order is concerned, we found that the performance drops for orders below 10. Also remarkable is the fact that the update rate of the predictor coefficients is not very critical. Using a 20 ms update interval is only slightly worse than a 10 ms update interval.

Reprinted from *Proc. IEEE Conf. Acoust., Speech, Signal Processing,* vol. 3, pp. 1649–1652, April 1987.

LONG-TERM PREDICTOR

The long-term correlation in the speech signal, due to the pitch, can be estimated by a long-term predictor, whose general form is given by

$$\frac{1}{P(z)} = \frac{1}{1 - \sum_{j=-I}^{J} \beta_j z^{-(M+j)}} . \qquad (1)$$

where M represents the correlation-lag and $\{\beta_j\}$ are long-term predictor coefficients (typical number of taps is between 1 and 3).

For the coder structure of figure 1, the pitch predictor parameters can be determined either in an open loop configuration or in a closed-loop configuration [5]. In the latter case, the parameters β and M are determined such that the output of the pitch generator due to its initial state is close (in the weighted sense) to the initial error signal (i.e. the speech signal with the effects of the filter memory subtracted). Once β and M have been determined, the excitation is determined as usual. The advantage of determining the pitch parameters within the analysis loop is that the pitch generator is then optimally contributing to the minimization of the weighted error. The difference in speech quality between closed- and open-loop computation of the pitch predictor parameters is significant.

For values of $M > L$, the output of the pitch-synthesizer is a function of the current input and the output in the previous frame only. This gives us the opportunity of simultaneously solving for β and b_k for given values of M and k. We found that this procedure gives a slightly better performance than the procedure that solves the equations sequentially.

The pitch predictor coefficients require many bits due to the short update intervals of the parameters (every 5 ms). We found that we could increase the interval to 10 ms, although, the degradation in performance was noticeable.

A problem that may arise in finding the pitch-predictor parameters is a large power gain (i.e. $\beta >> 1$), for transitions from silence to voiced speech. To prevent this, we limited the maximum value of β to a predefinded threshold (e.g. 1.8), and found that the performance of the coder was not affected by this constraint.

Obviously, the use of a pitch-predictor is important in the CELP coder. Without it, the random code book cannot effectively generate the periodic components in the excitation signal. On the other hand, during unvoiced regions the contribution of the pitch-predictor is much smaller. In these cases it might be an advantage not to use the pitch predictor at all and use the available extra bits for something else (e.g. larger code book).

Quantization

The correlation between the pitch predictor parameters M and β is small and to quantize these parameters, we will treat them separately. For a sampling rate of 8 kHz, a suitable range for M is between the bounds 16 and 143, which corresponds to a pitch range between 56 and 500 Hz and will cover the variation in pitch for a wide range of speakers. Reducing the bounds to, for example, 21 and 86 (i.e. 6 bits) gave a degradation of about 0.6 dB in the averaged segmental SNR. The value of M is directly quantized using 7 bits. We tried several differential schémes (e.g. encoding the difference between adjacent M) but found that each of them produced unacceptable distortions. The major reason is that the pitch predictor is not only representing the pitch in a speech signal, but any other long-term correlation. A differential coding scheme decreases the flexibility in adaptation, and as a result the performance drops.

The prediction coefficients $\{\beta_j\}$ are treated as vectors and encoded by vector quantization techniques. The dimension of the vector varies from 1 for a single-tap predictor to 3 for a three-tap predictor. In applying quantization we should keep in mind that our objective is to minimize the (weighted) squared error between the original and the reconstructed speech signal. This means that we have to search the available quantized values (code book) and select the one that minimizes the (weighted) mean squared error. However, to design the code book we used, for simplicity, the mean squared error between the predictor coefficients and the code book vectors as the distortion measure. The code books were designed with the LBG-algorithm [6]. using a large collection of predictor coefficients (obtained from CELP coder simulations). The value of M can be determined by a full search procedure, i.e. for each possible value of M, the whole β code book is searched for the best vector, until the best combination of β and M has been found. The resulting procedure is extremely complex. However, we found that equally good performance can be obtained by determining the value of M without quantization of β and then searching the code book for the best value of β. We designed code books of different sizes using different training sets and found that the code books were very robust, i.e. the performance for data in and outside the training set is not that different. The major reason for this behavior is that the pitch predictor coefficients have only a limited signal space, which does not vary too much for different speech data. Figure 2 shows the averaged SNR of the coders that was obtained with different code book sizes and different predictor orders. To compare the different curves we should look only at the number of bits per vector and not the number of bits per coefficient. We see from this figure that about 2 bits per coefficient are needed to match the performance to the unquantized case. We also see that if we only want to use 3 bits per vector, the performance of the one-tap is comparable to that of the multi-tap predictors.

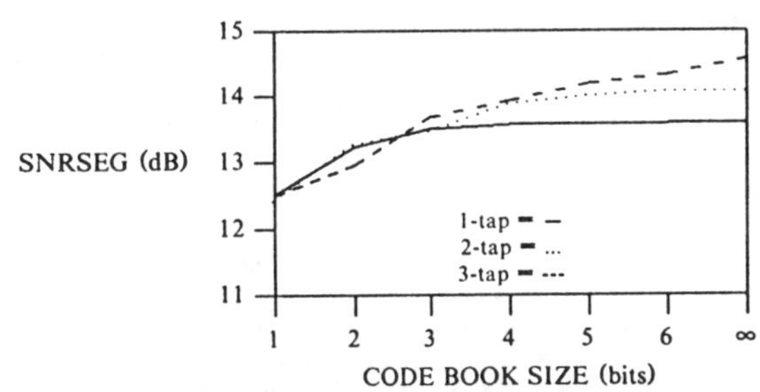

Figure 1. Segmental SNR as function of different code book sizes for the pitch predictor.

Using 7 bits for encoding the delay M and 3 bits for encoding the single-tap predictor coefficient we need a total of 10 bits for the pitch predictor, resulting in a 2 kb/s rate for 5 ms frames.

EXCITATION PARAMETERS

The excitation vector $\mathbf{v}$ can be seen as consisting of a gain b_k and a shape $\mathbf{x}$. The shape reflects the statistics of the signal to be encoded, and good results can be obtained with Gaussian samples [2]. However, the choice of the population is not that critical as long as the code book has enough variation. For example, we used code books populated with samples with respectively, a uniform, and Laplace distribution and obtained similar results as with the Gaussian code book. Also pulse-like code books where the code book vectors contain a small number (15%) of randomly placed pulses perform equally well.

The gain factor b_k plays an important role in the coder. Its sign effectively increases the code book size by one bit, without increasing the complexity. Its absolute value is adjusted such that the filtered excitation optimally matches the error signal. In other words the gain factor extends the effective code book size. For the final coder implementation the gain will be quantized and the effective code book size will be the sum of the number of bits used for encoding the index and the number of bits used for encoding the gain and its sign. To assess this effective increase in size, we evaluated the performance of the coder when the gain was quantized. A non-uniform quantizer was designed by using the LBG algorithm on a large training set of gain values. The gain was quantized by computing the weighted mean-squared error for each possible code book index and gain, and selecting the pair that minimized this error. This procedure is not computationally expensive since the various cross-correlation terms need to be computed only once for each index. We examined different bit allocations for the gain and found that a 5 bit representation does not introduce any significant reduction in the performance.

Assuming a 5 bit gain quantizer we see that the resulting bit rate, due to the gain, can be high. To reduce the bit rate it is desirable to reduce the update rate of the gain. One possibility is to extend the vector size L since there exists a direct coupling between the vector size L and the update rate of the gain. We found that for the same number of bits/sample larger vectors (at the expense of a higher complexity), at a lower bit rate (update rate of the gain is halved), gave a slight improvement in averaged SNR.

Multiple-gain

We have seen that the excitation vector can be described by a gain and a shape (index). Instead of using only one gain, we can define multiple gains for one particular shape. For example, we can divide an excitation vector **x** into n parts each of length L/n and compute the optimal gains $\{b_{k1},..,b_{kn}\}$ for each index k. The optimal values of the gain factors $\{b_{kn}\}$ can then be found by solving a set of n linear equations. Note that for $n = L$ this solution will give us a zero error, since we are then able to construct any residual signal. We found that the performance of a 7 bit code book with 2 gain values is about the same as that of a 10 bit code book with a single bit gain. In both cases the resulting bit rate (including gain quantization) will be about the same, but the multi-gain approach requires a search over fewer vectors. For encoding the gain vector we used again the code book approach and designed the code book by using the LBG algorithm on a large training set. Encoding the signs separately, we found that typically 6 or 7 bits are required for encoding a 2-dimensional gain vector.

In conclusion we can say that the most effective excitation description can be achieved by using large vectors (10 ms) with one single gain. The multiple-gain approach forms an interesting alternative for getting, at a lower complexity, about the same performance as that obtained with large code book vectors. Smaller vectors (5 ms) increase the bit rate due to the need for updating the gain at a higher rate.

BIT ALLOCATION

To obtain a fully quantized coder, we also need an efficient quantization scheme for the short-term predictor coefficients. In our experiments we found that a low rate/low distortion representation of these coefficients is crucial for a good coder performance, since the compensational effects of the excitation within the CELP model constraints are only marginal. We can either use high quality vector quantization schemes [7] or using efficient scalar quantization schemes based on Line-Spectral Pairs. Using the scalar differential scheme as described in [8], we are able to quantize a 10-th order filter with 36 bits per set, which requires 1800 b/s for a 50 Hz frame rate. The quantizer range was set between 0.0 and 0.10 and the bit allocation is {4,4,4,4,4,4,3,3,3,3}. Once the number of bits required for the short-term predictor has been established, we can determine our bit allocation for the excitation. Using our result from the previous sections we were able to construct a 7.2 kb/s coder whose performance is practically indistinguishable from the unquantized CELP coder as described in [2]. In this particular arrangement the excitation signal requires a total of 5.2 kb/s. For a 4.8 kb/s coder, this number has to be reduced to 3 kb/s (if 1.8 kb/s are needed for the short-term predictor parameters). A significant reduction can be obtained by reducing the update rate of the parameters. By reducing the update rate of the long-term predictor coefficients to 10 ms, we only need 1 kb/s instead of 2 kb/s, although the reduction in quality is noticeable. The remaining 2 kb/s can than be used for the excitation. Using 10 ms frames for the excitation we need only 500 b/s for the gain and the remaining 1500 b/s can be used for the code book index and sign. The resulting bit allocation is shown in Table 1.

TABLE 1. Example of a possible bit allocation for 4.8 kb/s.

PARAMETER	RATE (ms)	BITS	B/S
pitch predictor M	10	7	700
pitch predictor taps	10	3	300
code book index +sign	10	15	1500
gain	10	5	500
filter parameters	20	36	1800

Notice that many different allocations are possible, and it remains difficult to say which allocation is optimal. For example, rather than using the large 14 bit code book, we can use the multiple gain approach, and use a 12 bit code book and a 7 bit gain code book. Furthermore, in all our examples we used frames that where multiples of 5 ms, but many other combinations using different frame sizes are also possible. For example, when we applied vector quantization for the short-term predictor parameters [7], we used a 16 ms update interval for the predictor parameters and an 8 ms update interval for the code book and pitch predictor. The averaged SNRSEG values over the complete test data base (120 s) is around 11 dB for the 4.8 kb/s coder. The reduction in update rate and order, and the quantization of the LSP parameters, accounts for about 2 dB reduction in averaged segmental SNR. Furthermore, the degradation that is experienced is more an overall degradation, similar to that experienced with additive noise. No unexpected plops or clicks were perceived, and also there was none of the characteristic warble from which many VQ based systems suffer.

POSTFILTERING

Although the coder quality at 4.8 kb/s is acceptable, it is still perceived as rough or noisy. We can consider the coded speech to be the original signal corrupted by additive noise. Any speech enhancement technique that deals with this problem can be used to reduce the noise. We found that a simple postfilter as described in [9] improved the perceived quality of the speech. The postfilter emphasizes spectral peaks, thereby increasing the subjective signal-to-noise ratio, at the expense of a signal distortion. In general for a desired balance between noise suppression and speech distortion, the postfilter should have a response somewhere between an all-pass response and the signal spectrum. In the CELP coder the global spectral information is represented by the inverse of the short-term predictor, and the spectral fine information by that of the long-term predictor. Figure 3 shows a possible implementation of the postfilter in the CELP coder.

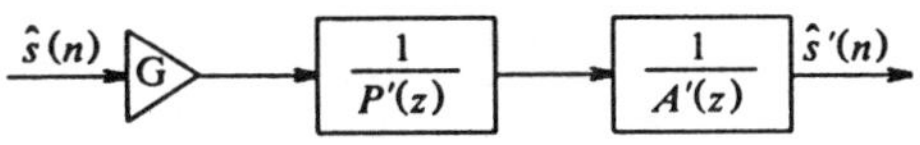

Figure 2. Possible implementation of postfiltering in a CELP coder.

The blocks $1/P'(z)$ and $1/A'(z)$ represent the modified long-term and short-term predictor, and G represents a scale factor which guarantees that the output signal $\hat{s}'(n)$ has about the same energy as the input $\hat{s}(n)$.

As mentioned before, one wants to control the effect of the postfiltering to find the optimum balance between noise suppression and distortion. One way of achieving this for the short-term filter is to use the postfilter

$$\frac{1}{A'(z)} = \frac{1}{1 - \sum_{j=1}^{p} \alpha^j a_j z^{-j}}, \quad 0 \leq \alpha \leq 1. \qquad (2)$$

The parameter α reduces the magnitude of the resonances and increases the bandwidth at the resonance frequencies The optimum value of α has to be determined by listening tests.

In general the postfilter introduces an amplification of the input signal, which is signal dependent. However, for a fixed value of α, some average value for G can be determined that will be appropriate for most inputs.

A commonly found implementation of the $1/P(z)$ filter uses only 1-tap (I=J=0).

$$\frac{1}{P'(z)} = \frac{1}{1 - \epsilon \beta_0 z^{-M}}, \quad 0 \leq \epsilon \leq 1. \qquad (3)$$

The frequency response is that of a comb-filter, whose amplitudes and bandwidths are controlled by the values of ϵ, β_0 and M. The parameter ϵ is used to change the response of $1/P'(z)$ to find the optimum balance between noise suppression and speech distortion.

Quality assessment was done using informal A-B comparison tests with 8 experienced listeners. For the global spectral postfiltering we found that a reasonable value of α lies somewhere between 0.3 and 0.4. For values larger than 0.4, the postfiltering introduces a significant distortion, and the speech sounds muffled. Values of α less than 0.2 make the effect of postfiltering almost non perceivable. Using the fine spectral postfilter, we found that the optimum value of ϵ also lies somewhere around 0.3. Generally, we concluded that postfiltering increases the subjective quality. Fine spectral postfiltering is about as effective as global spectral postfiltering, and the best results are obtained by using both forms of postfiltering.

CONCLUSION

In this paper we have described the effect of the different excitation parameters in the CELP coder on the quality of the reconstructed speech. We examined different procedures for computing the gain coefficients and the pitch predictor parameters. We found that the best results were obtained with a closed-loop computation of the parameters. A first order long-term predictor was found to be adequate in most cases.

A multiple gain approach was proposed to increase the effective code book size. By vector quantizing the gain vector, the shape code book can be made smaller without losing in performance.

Different bit allocations for the parameters involved have been investigated. Most of the degradation in the 4.8 kb/s schemes is due to the low update rate of the parameters. Other important issues are the fixed bit allocation and fixed frame sizes. We have already pointed out that for certain speech segments the long-term predictor does not reduce the error appreciably. Therefore in such cases some of the bits used for the long-term predictor should be allocated to some other parameters. We also demonstrated that postfiltering based on the spectral envelope and spectral fine structure increases the perceived quality of the CELP coder.

References

[1] B.S. Atal and M.R. Schroeder, "Stochastic coding of speech signals at very low bit rates," *Proc. IEEE Int. Conf. Communications*, p. 48.1 (May 1984).

[2] M.R. Schroeder and B.S. Atal, "Code-Excited Linear Prediction (CELP): high quality speech at very low bit rates," *Proc. IEEE Int. Conf. Acoust., Speech, Signal Processing*, pp. 937-940 (1985).

[3] I.M. Trancoso and B.S. Atal, "Efficient procedures for finding the optimum innovation in stochastic coders," *Proc. IEEE Int. Conf. Acoust., Speech, Signal Processing*, pp. 2375-2379 (1986).

[4] A.W.F. Huggins and R.S. Nickerson, "Speech quality evaluation using "phonemic-specific" sentences," *J. Acoust. Soc. Amer.*, (77(5)) pp. 1896-1906 (May 1985).

[5] P. Kroon, Ed.F. Deprettere, and R.J. Sluyter, "Regular-Pulse Excitation: a novel approach to effective and efficient multipulse coding of speech," *IEEE Trans. Acoust., Speech, Signal Processing* **ASSP-34**(5) pp. 1054-1063 (October 1986).

[6] Y. Linde, A. Buzo, and R.M. Gray, "An algorithm for vector quantizer design," *IEEE Trans. Communications* **COM-28** pp. 84-95 (Jan. 1980).

[7] Y. Shoham, "Vector predictive Quantization of the LPC spectral parameters for low-rate speech coding," *Proc. IEEE Int. Conf. Acoust., Speech, Signal Processing*, (1987).

[8] F.K. Soong and B.H. Juang, "Line spectrum pair (LSP) and speech data compression," *Proc. IEEE Int. Conf. Acoust., Speech, Signal Processing*, pp. 1.10.1-1.10.4 (March 1984).

[9] V. Ramamoorthy and N.S. Jayant, "Enhancement of ADPCM speech by adaptive postfiltering," *Bell Syst. Tech. J.* **63**(8) pp. 1465-1475 (October 1984).

CELP CODING FOR HIGH-QUALITY SPEECH AT 8 KBIT/S

Maurizio COPPERI and Daniele SERENO

CSELT - V. Reiss Romoli, 274 - 10148 Torino (ITALY)

ABSTRACT

A new speech coding technique at low bit-rate is presented in this paper. The coder is based upon a novel speech production model, independently developed by the authors [1,2] and by Atal and Schroeder [3,4], called CELP (Codebook Excited Linear Prediction). Differences exist between the two approaches, both in the strategy chosen to construct codebooks, and in the method to generate the innovation sequence. In this scheme, we split the incoming speech signal into two frequency bands in order to gain the benefits of the piecewise LP (Linear Prediction) approximation. Then, each residual signal is coded in blocks of 5-ms duration through an adaptive vector quantizer incorporating a noise shaping filter. Our results show that good quality speech can be obtained at 8 kbit/s.

1. INTRODUCTION

The advent of Vector Quantization (VQ) has disclosed new speech coding techniques at low and very low bit-rates. Triggered by the powerful codebook coding method, a new class of speech digitizers called CELP (Codebook Excited Linear Prediction) has proven to be efficient for encoding the speechwave at rates as low as 2.4 kbit/s [1,2], outperforming conventional pitch-excited vocoders. Impressive improvements in both naturalness and speaker recognizability have been obtained in previously reported experiments [1,2,3,4]. An extension of this technique (*) to a bit-rate of 8 kbit/s is presented herein. We exploit recent results of piecewise linear prediction analysis and noise spectral shaping to design a coder capable of providing good communication speech quality. Important issues are addressed, such as the structure that the excitation codebook should have and the fast search algorithm that could be used, in order to make the CELP coder real-time implementable. As this is a new area of research, further improvements are easily anticipated. Nevertheless, the preliminary results presented in this paper show that the CELP algorithm is a viable approach to overcome impairments of conventional coders even at bit-rates near to 8 kbit/s.

(*) Patent pending

2. CODER DESCRIPTION

The speech coder proposed in this study is depicted in Fig. 1. The input signal, sampled at 8 kHz, is split into two separate frequency bands by a quadrature mirror filter bank (32-tap FIR filters). Each subband signal, after 2:1 decimation, is encoded through our CELP algorithm. There are several benefits to be gained from this approach, as discussed in the following paragraphs.

2.1 Piecewise LP-VQ Method

According to the piecewise LP (PLP) approximation, each subband is analysed and coded in parallel. Since each half of the speech spectrum is approximated by an independent curve, without the constraint of continuity at the crossover frequency, the entire speech spectrum is fitted with better fidelity. Any degradation in one spectral region does not affect the entire spectrum. Moreover, every region can be handled in a different way, both in the analysis and in the quantization step, taking into account properties of speech production and perception. Using the PLP method, a substantial saving of computational load can be achieved due to the halved sampling frequency of each subband. On the other hand, this gain can be used to increase the order of prediction filters, so as to have an optimal matching between the LP model and the underlying speech spectrum and, consequently, a whiter residual signal. Finally, when VQ is used to quantize the LP parameters, the PLP approach permits a considerable reduction in memory requirements. Considering the reduced number of resonances to be removed in each subband, both the order of the LP filter and the codebook size can be kept sufficiently small.

2.2 VQ of LP Parameters

The LP parameters have been computed for each subband by means of the autocorrelation method, with a Hamming window length of 80 samples (20 ms at 4 kHz down-sampling rate) and a frame rate of 200 frames/s. This high adaptation speed allows the coder to follow fast transitions of the speech spectrum and to optimally shape the quantizing noise spectrum in the residual quantization stage, as we will see later on. Moreover, this overlapped windowing makes the parameter interpolation unnecessary. The codebooks of LP filter coefficients have been designed using the LBG algorithm [5],

Reprinted from *Proc. IEEE Conf. Acoust., Speech, Signal Processing,* vol. 3, pp. 1685–1688, April 1986.

which requires a long training sequence of vectors. The generation procedure was initialized with two initial codewords (splitting method) and a distortion threshold $\epsilon = 0.001$. Tree-structured codebooks have been chosen in order to get a significant reduction of the encoding complexity. In particular, a two-level configuration permits a good performance with reasonable memory/computation requirements. The optimized coder employs a 9-bit (5,4) codebook of 8-th order filters for the low frequency band, and a 6-bit (4,2) codebook of 4-th order filters for the high frequency band. The transmission rate of the spectral information is 3 kbit/s.

As distortion measure in the space of LP spectra, we use the likelihood ratio measure [2].

2.3 VQ of Residual Signals

The CELP coding technique outperforms other speech coders previously proposed at low bit-rates, by exploiting a rate-distortion motivated approach for the residual quantization. Furthermore, the quantizer incorporates a noise shaping filter, thus taking full advantage of the auditory masking effect. The split-band prediction is also useful to optimize individually the quantizers for the two frequency subbands. The basic concept of the CELP method is as follows. Each short block of residual signal, say 5 ms in duration corresponding to 20 samples, is normalized in amplitude with a 5-bit quantizer of gain terms. The scaled vector is then "compared" with the codevectors stored in the excitation codebook, to select that codevector which provides the subjectively closest reproduction of the speech input block. The index of the selected vector is sent to the receiver and used to synthesize a replica of the original signal. Three main problems arise in the coder design: i) construction of optimum excitation codebooks, ii) definition of a fast search algorithm, and iii) choice of the optimum perceptual weighting filter. These points are discussed next.

2.3.1 Codebook Design Algorithm

The excitation codebook must be generated in a manner consistent with both the distortion measure, adopted as nearest-neighbor rule, and the actual VQ configuration in the coding scheme.

Referring to Fig. 1, the error signal E is formed by taking the difference between the normalized residual vector R and a stored pattern S (the blocklength of these vectors is 20 samples). Then the error is frequency weighted by the filter W(z). The weighted error F is squared and averaged over the blocklength, and the best excitation pattern is selected in the codebook in order to minimize this squared error. The gain term of each input vector is fixed, and therefore the nearest-neighbor condition can be evaluated regardless its value. On the contrary, during the computation of the centroid of each cluster, the gain values σ_i must be associated to the corresponding input vectors R_i. Let L be the number of residual vectors in a cluster, then we have to minimize the squared error over the cluster:

$$\text{sq.err.} = \sum_{i=1}^{L} H_i \, \sigma_i^2 \sum_{j=1}^{20} [R_i(j) - S(j)]^2 \qquad (2)$$

where the weight H_i, related to the i-th vector R_i, is defined as:

$$H_i = [\sum_{j=1}^{20} F_j^2] \, / \, [\sum_{j=1}^{20} (R_i(j) - S(j))^2] \qquad (3)$$

The weighted centroid that minimizes (2) is given by:

$$S_c(j) = [\sum_{i=1}^{L} H_i \, \sigma_i^2 \, R_i(j)] \, / \, [\sum_{i=1}^{L} H_i \sigma_i^2] \qquad (4)$$

Therefore the codebook generation procedure can start from a given initial codebook, which is adapted to the training of residual vectors iteratively, by computing new sets of clusters and corresponding weighted centroids (optimal codevectors of the VQ codebook). The procedure terminates when the distortion drops below a fixed threshold. The weight H_i depends on the function W(z) used to frequency weighting the error E in the quantization stage. Since the synthetic speech signal is the output of an LP filter, which has been driven by the excitation signal S, we can model the output quantizing noise spectrum by using a proper filter W(z) in the loop for the selection of the optimal pattern S. The goal is to make the quantizing noise almost unperceptible, exploiting our knowledge of the speech production and perception processes. Different perceptual weighting functions can be used [6]. We have chosen a filter that de-emphasizes those errors which occur in the formant regions, thus W(z) is based on the entire detailed spectrum. Let 1-P(z) be the optimum LP inverse filter for the current frame, then a suitable choice for W(z) is given by [2]:

$$W(z) = [1 - P^*(z)]^{-1} = [1 - \sum_{i=1}^{M} a_i \gamma^i z^{-i}]^{-1} \qquad (5)$$

where γ ($0 < \gamma < 1$) controls the amount of noise power in the formant regions. The coefficients of the weighting filter can be computed on-line or stored in a codebook with the same ordering as in the LP codebook.

2.3.2 Codebook Structure and Fast Search Algorithm

So far we have not defined what structure the excitation codebook could assume. A basic idea is to use the LP model, selected in the inverse filtering stage, for choosing a corresponding excitation codebook designed to be optimal for the residual signals produced by that model. In this approach, there is a set of different codebooks (in general, one codebook for each LP model) which have a smaller dimension than that of a single codebook.

Thus, both the transmission rate and the search complexity are reduced. Fig. 2 shows the basic structure of this adaptive VQ driven by the LP information. The model classifier can also process very easily the LP codebook index in order to reduce to three or four the number of spectral classes and of the relative excitation codebooks. These excitation codebooks are generated individually but in the complete VQ structure, breaking the entire training sequence of residual vectors in subsequences according to the classifier decision. This method of waveform VQ adaptation based on spectral parameters is highly efficient here because does not require additional side-information. An alternative approach, described in [2], makes use of a single codebook generated on the whole training sequence. The codebook structure, as related to the LP models, is determined as soon as the final centroids are determined. It consists of the index maps of those N excitation vectors more frequently used by each spectral model. The search of the index set is done only once during the coder design, applying the full-search method to the excitation codebook. Fig. 3 depicts this latter codebook structure. The dimension required by the codebooks of innovation patterns to yield acceptable performance, the pattern length and the sophisticated nearest-neighbor condition do not permit a real-time operation of CELP coders using a standard full-search method. On the other hand, tree-search methods cannot be applied to the excitation codebooks, due to the particular type of their templates. Therefore we have devised a fast search algorithm, based on three steps as follows. First, an unweighted mean-squared error criterion is applied using the partial distance technique. This consists of computing the sum of the squares of the first k components of the difference between the actual innovation vector and a codevector. During the test of a new codevector (full-search), if the partial distance exceeds the previous minimum distance, the codevector is skipped without completing the distance computation. This technique reduces the number of mul/add operations to an average of 25% of that required by complete full search, while slightly increasing comparisons. The second step consists of keeping only those codevectors having lower partial distances, in order to perform the final choice in a reduced set of candidates. Our experiments have shown that the number of survivors can be limited to a 30% of the total number of codevectors. The final step consists of selecting the codevector that minimizes the weighted squared error, as described in par. 2.3.1, using a full-search method over all survivors. The last search is done at the cost in average of about 2M mul/add operations per second. Real-time implementation of this coder is not a difficult task, considering that present DSP chips can do 10 million instructions per second. However, the availability of faster chips or a parallel processing architecture could lead to a better coder performance, since the number of survivors for the last step can be increased.

3. RESULTS AND CONCLUSIONS

A problem inherent in this kind of coders is the huge amount of computation required in the generation phase of the excitation codebook. To circumvent this problem, the generation procedure has been implemented on the FPS-120B array processor, using a training sequence of 2 minutes in duration. In this preliminary experiment we have fixed the size of the excitation codebooks to 8 bits for the lower subband and to 7 bits for the higher one, thus the transmission rate for the excitation signals (shape and gain) is 5 kbit/s. For each subband, there are two excitation codebooks structured as in Fig. 2. Since no additional side information is required by the model classifier, one could design more codebooks, even one for each LP model, without altering the transmission rate. The better performance would be obtained at the expense of increased memory, while the computational load remains basically unchanged. This is a particularly interesting aspect of codebook-excited coders with adaptive VQ.

Listening tests have been conducted to compare the speech quality of the CELP coder at 8 kbit/s, with that of a Residual Excited Linear Prediction (RELP) coder at the same rate, described in [7].

This comparison is quite useful due to the different kind of noise shaping and processing of the residual signal. In our RELP scheme the residual signal is quantized with a small vector dimension and the output noise spectrum has the same general shape as the speech spectrum. The Signal-to-Noise Ratio (SNR) of the RELP becomes simply equal to that of the residual quantizer, since the LP inverse filtering of the input speech works as an adaptive preemphasis filter, and both the residual speech and the internal quantizing noise tend to have a flat spectrum. The RELP speech quality is judged to be perceptually superior to that of a simple adaptive quantizer, thus confirming the importance of noise shaping, but for optimal results one should adjust the envelope of the noise spectrum properly. This can be done, at low rates, by using a non-instantaneous quantizer that codes short blocks of samples (e.g. 5 ms in duration) minimizing a subjective error criterion, as the CELP coder does. By varying the parameter γ in eq. (5) it is possible to find the best tradeoff between SNR reduction and noise spectral shape, in order to achieve the optimal perceptual quality. As an example, Fig. 4 shows the CELP SNR performance evaluated on a sentence uttered by a male speaker. The corresponding subjective speech quality may be described as good communication quality. Fig. 5 shows a comparison of original and quantized waveform produced by the CELP coder. We see that, even if the error criterion is a subjective one, nevertheless the quantized version matches to a large extent the original signal .

These results confirm that the CELP system competes favorably at low bit-rates with conventional techniques such as RELP coders.

REFERENCES

[1] M. Copperi and D. Sereno, "Improved LPC excitation based on pattern classification and perceptual criteria", Proc. of 7th Int. Conf. on Pattern Recognition, pp. 860-862, Montreal, July 1984

[2] M. Copperi and D. Sereno, "Vector quantization and perceptual criteria for low-rate coding of speech" , Proc. of ICASSP, pp. 252-255, Tampa, March 1985

[3] B.S. Atal and M.R. Schroeder, "Stochastic coding of speech signals at very low bit rates", Proc. of ICC, pp. 1610-1613, Amsterdam, May 1984

[4] M.R. Schroeder and B.S. Atal, "Code-excited linear prediction (CELP): high-quality speech at very low bit rates", Proc. of ICASSP, pp. 937-940, Tampa, March 1985

[5] Y. Linde, A. Buzo and R.M. Gray: "An algorithm for vector quantizer design", IEEE Trans. on Commun., Vol. 28, pp. 84-95, 1980

[6] H.A. Hawkins et al., "Perceptual weightings and optimal pulse positioning in multipulse LPC speech coding", Proc. of ICASSP, pp. 477-480, Tampa, March 1985

[7] M. Copperi and D. Sereno, "Baseband and residual excited LPC coders based upon vector quantization", Proc. of Int. Conf. Digital Signal Processing, pp. 433-437, Florence, September 1984

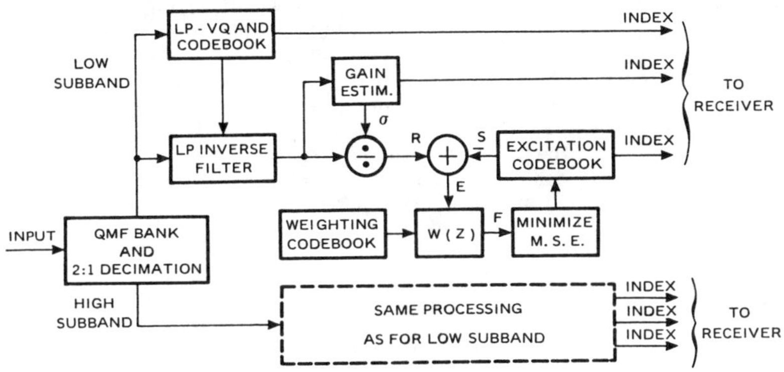

Fig. 1 - Block diagram of the split-band CELP coder (transmitter)

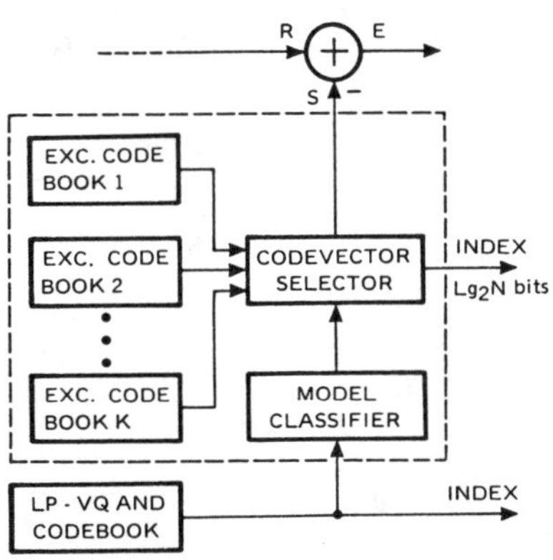

Fig. 2 - Basic structure of the multi-codebook excitation source. Each excitation codebook has N codevectors

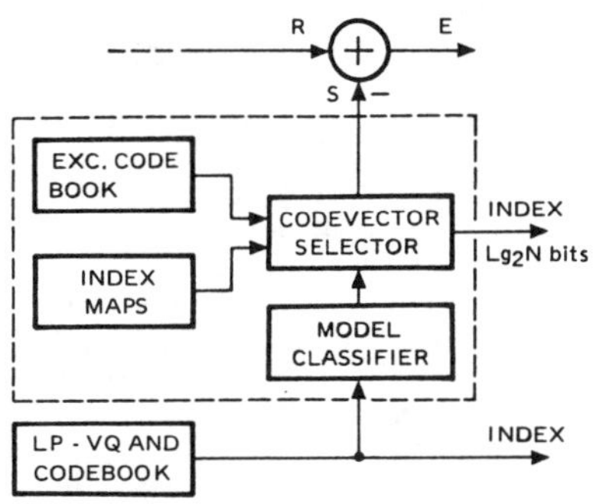

Fig. 3 - Basic structure of the single-codebook excitation source. The excitation codebook has T codevectors and each index map contains N indeces (N < T)

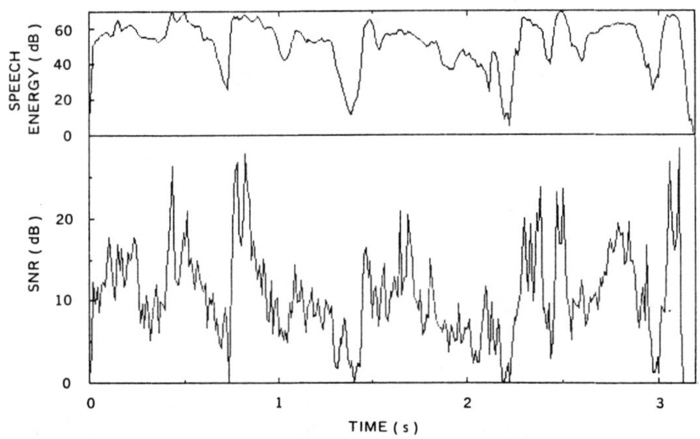

Fig. 4 - Segmental SNR for a sentence uttered by a male speaker

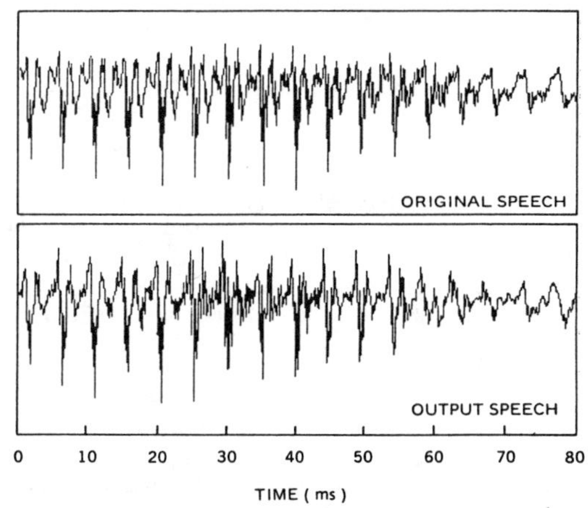

Fig. 5 - Comparison of speech waveforms

AN EXPANDABLE ERROR-PROTECTED 4800 BPS CELP CODER (U.S. FEDERAL STANDARD 4800 BPS VOICE CODER)

Joseph P. Campbell, Jr.
Vanoy C. Welch
Thomas E. Tremain

U.S. Government
Department of Defense, R5
Fort Meade, Maryland 20755-6000
USA

ABSTRACT

We describe a code excited linear predictive (CELP) coder, jointly developed by U.S. DoD and Bell Labs, that was recently selected as the U.S. Government standard 4800 bps voice coder [1]. The coder has the capability for future expansion and is robust to channel errors and noisy environments. Listening tests and DRT and DAM scores show the coder exceeding the performance of all Government standard speech coders operating at rates below 16 kbps and comparable to 32 kbps CVSD!

1. INTRODUCTION

The introduction of CELP in 1984 [2] sparked one of the speech coding community's greatest research efforts to achieve high quality speech at 4800 bps within reasonable computational complexity. We describe the salient features of our 4800 bps CELP coder including variable computational complexity that provides interoperability, frame rates, parameter coding and error protection. The frame rates and parameter coding were derived from subjective listening tests augmented with objective measures. The error protection strategy combines forward error correction (FEC), parameter coding and adaptive smoothing. FEC is applied to the bits that tests showed to be the most perceptually and objectively sensitive to errors given the parameter coding and adaptive smoothing of the algorithm. This technique resulted in selection of our integrated error protection system. The computational complexity reductions result in more than an order of magnitude improvement over traditional methods, while maintaining high quality speech. The coder can operate at different levels of computational complexity to provide interoperability between simple and powerful implementations. Common VQ stochastic code books can be used for synthesis, and subsets are permitted to reduce the computationally demanding analysis. In addition to subset VQ, parameters may be determined differently depending on processing power while maintaining compatibility with the standard (i.e., closed versus open loop quantization).

2. CELP ALGORITM DESCRIPTION

We use an 8 kHz sample rate and a 30 ms frame size with four 7.5 ms subframes [3]. CELP analysis consists of three basic functions: 1) short delay "spectrum" prediction, 2) long delay "pitch" search, and 3) residual "code book" search. CELP synthesis consists of the same three functions (performed in reverse order) with the addition of a fourth function, called a postfilter, to enhance the output speech. Our CELP coder's characteristics are summarized in Table 1 and are briefly described in the following.

Spectrum analysis is performed once per frame by open-loop, 10th order autocorrelation LPC analysis using no preemphasis and 15 Hz bandwidth expansion with a 30 ms Hamming window. The spectrum is coded using 34 bit independent nonuniform scalar quantization of line spectral pairs (LSP) [4]. The LSPs are linearly interpolated to form an intermediate set for each of the four subframes.

Pitch search is performed by closed-loop analysis using a modification of what is commonly called any one of the following: "self-excited" [5], "adaptive code book" [6] or "VQ" method. This method was found to be superior to the conventional "filtering approach" [3], especially for high pitched speakers. The pitch search is performed and coded four times per frame (every 7.5 ms). The pitch delay ranges from 20 to 147 lags every odd subframe, while even subframes are searched and coded within 32 lags relative to the previous subframe. This greatly reduces computational complexity and data rate while causing no perceivable loss in speech quality. The pitch gain is coded between -1 and +2 using absolute nonuniform scalar 5 bit quantization.

The minimum squared prediction error (MSPE) search criteria is modified to check the error at submultiples of the delay to determine if it is within 1 dB of the MSPE. The submultiple delay is selected if its error satisfies our modified criteria. This results in a smooth "pitch" delay contour which produces higher quality speech and is crucial to the synthesizer's smoother in the presence of bit errors.

Code book search is performed by closed-loop analysis using conventional MSPE criteria of the perceptually weighted error signal. We reserve 10 bits to allow a 1024 size code book. However, interoperable coders may use a subset of this code book (to reduce computational complexity). As with the pitch parameters, the code book index and gain are searched and coded four times per frame. Code book gain (plus and minus) is coded using 5 bit absolute nonuniform scalar quantization.

A special form of code book containing sparse, overlapping code words is used to compute fast convolutions and energies by exploiting recursive end-point correction algorithms [3,6]. The code book contains samples of a zero-mean, unit-variance, white Gaussian sequence center clipped at 1.2, resulting in approximately 75% sparsity (zero values). Each code word contains two new samples and all but two samples of the previous code word [7]. This form of a code book causes no degradation in speech quality relative to nonoverlapping, nonsparse code books and, as shown in Section 4, reduces computation by a factor of 20.

The postfilter is a short-delay pole-zero type with adaptive spectral tilt compensation as suggested by Chen [8]. Usually, the postfilter enhances the synthesized speech by 2 DRT and DAM points [1]. However, in some noisy environments, where LPC analysis models the noise, the noise is enhanced! In addition, postfiltering can be detrimental to tandem coding if not taken into consideration.

3. EXPANDABILITY

To provide 4800 bps voice coding today, a standard is required. To prevent this standard from be-

Reprinted from *Proc. IEEE Conf. Acoust., Speech, Signal Processing,* vol. 2, pp. 735–738, May 1989.

coming obsolete, we provide a bit for future expansion, as suggested by Atal [9]. As long as each expansion provides another expansion, the standard may continue to branch many times. For example, this bit could allow adaptive bit allocation [10,11] or adaptive postfiltering.

4. COMPUTATIONAL REQUIREMENTS

CELP's major computational requirements are the code book and pitch searches. As shown in [3], the remaining computations require only 1.3 MIPS.

Code book search computations are reduced by a factor of 20 relative to brute force techniques using recursive convolution and recursive energy calculations. Both recursions require an overlapped code book as described in Section 2. The approximate number of instructions (multiply, add, or multiply and add) per second (IPS) is:

$$IPS = [C + R + E] K / F$$

where:
$C = L [(L + 1) / 2 + A (N - 1)]$
$R = LN$
$E = L [1 + A (N - 1)]$
L = subframe size
A = nonzero code word samples percentage
N = code book size
K = code book searches per frame
F = frame rate in seconds

The convolution requires C instructions per subframe, correlation requires R instructions and energy E instructions. Therefore, a 1024 size code book search with L = 60, A = 0.25, K = 4 and F = 30 ms requires 12.5 MIPS. Slightly degraded speech (typically a loss of 3 DAM points [1]) can be obtained with a 256 size code book which requires only 3.3 MIPS.

Pitch search and code book search computations are very similar. The pitch search exploits recursive convolution. Depending on the amount of perceptual weighting, we use 0.8, the impulse response may be truncated. We exploit the pitch delay's natural smoothness by delta searching and coding even subframes. The combination of these techniques speeds the pitch search by a factor of 18 relative to brute force methods. The approximate number of instructions per second (IPS) is:

$$IPS = [(C + R + E) K + (C' + R' + E') K'] / F$$

where:
$C = L [(L+1) / 2 + N - 1]$
$C' = L [(L+1) / 2 + N' - 1]$
$R = E = LN$
$R' = E' = LN'$
L = subframe size
N = absolute pitch delays to search
N' = delta pitch delays to search
K = full pitch searches per frame
K' = delta pitch searches per frame
F = frame rate in seconds

The convolution requires C instructions per subframe, correlation requires R instructions and energy E instructions. The primed terms represent the delta pitch delay calculations. Therefore to search 128 pitch delays (N) and 32 pitch delta delays (N') with L = 60, A = 30, K = K' = 2 and F = 30 ms requires 2.2 MIPS.

5. ERROR PROTECTION

Parameter coding and continuity are the basis for the error protection strategy. We employ adaptive smoothers based on estimates of the channel error rate from a forward error correcting (15, 11) Hamming code. Therefore, the smoothers don't operate in error free conditions. We concentrated our efforts on preventing perceptually disturbing synthesis errors; i.e., squeaks and loud, clipped speech (blasts). Once these severe errors were under control, we then ranked bit sensitivity using the objective measures suggested in [12] to determine the protection required for each bit.

Spectrum parameters can be very sensitive. When the method of [13] is used to code LSPs, very low spectral distortion results, but differential LSPs are very sensitive to errors [14]. With only a slight increase in log spectral distortion (typically 0.6 dB) and only a slight decrease in DAM scores (typically less than the variance of the test [1]), LSPs can be coded independently and be very robust to errors. We guarantee LSP monotonicity (stable LPC filter) by repeating LSPs. If nonmonotonic LSPs are received, an attempt is made to correct them by repeating previous LSP pairs. If this correction fails (the LSPs are still nonmonotonic), the entire LSP frame is repeated. In addition, because we interpolate LSPs each subframe and each subframe has contributions from two LSP frames, the effects of errors are smoothed out.

The design of the LSP quantizer is crucial to the success of this approach. As suggested by Kang [4], we manually fit LSP distributions under the constraints that no adjacent LSPs could be coded to closer than 15 Hz (our LPC analysis bandwidth expansion factor) and to prohibit the first and last LSPs from being coded close to 0 and 0.5, respectively. These constraints limit the LPC filter's prediction gain. Therefore, undetected LSP transmission errors are unable to cause loud blasts and squeaks commonly associated with spectrum errors.

Code book gain magnitudes lend themselves to smoothing. The code book gain quantizer is nearly logarithmic, so we treat code book gains differently depending on the region of quantization. During quiet speech, small code book gains are generally transmitted and tight constraints are appropriate. In these regions, code book gain errors are very distracting and sound impulsive. In normal speech, larger code book gains are selected and looser constraints are appropriate. Large code book gain errors can cause blasting. Our smoother is able to eliminate the majority of these errors. The smoothing algorithm tests the average and variance of the neighboring code book gain magnitudes to determine suitable constraints to apply to the code book gain in question. If the variance and average exceed thresholds, the code book gain magnitude is replaced by its neighbor's average magnitude and its sign is preserved.

Code book indices exhibit chaotic behavior and, therefore, smoothing is inappropriate. Errors in the code book index result in noisy speech which is not perceptually disturbing. Therefore, the code book index is not protected.

Pitch delay exhibits smooth behavior and, to a lesser extent, so does pitch gain. Detailed listening tests revealed that these pitch parameters are the most sensitive because their corruption causes very rough speech. Therefore, they are smoothed and forward error corrected. The pitch delay is smooth because the search procedure, described in Section 2, is modified to favor smooth values and because even subframes are delta coded to a relatively small range. The pitch delay and code book gain smoothers operate similarly. The pitch delay smoother is simpler because the same thresholds are appropriate throughout the quantization range.

Pitch gain is smoothed by testing if the average of the change in magnitude of neighboring pitch gain values is within a threshold. If the magnitude of the pitch gain in question exceeds the average magnitude of its neighbors, it's magnitude is replaced by the average magnitude of its neighbors. As shown in Figure 1, pitch gain doesn't exhibit very smooth behavior and, therefore, this smoother is unable to provide performance equal to the other smoothers.

Forward error correction is done with a Hamming (15,11) single error correcting code that protects 10 pitch delay and pitch gain bits. Corruption of the pitch delay bits causes equally disturbing perceptual speech distortion, except for the least significant bit which is only slightly more robust than the others. Because the pitch delay smoother is able to compensate for large errors, we selected the 4th through 6th MSBs of the absolute pitch delays to protect. We chose to place all the FEC power on the absolute delay since correct absolute delays will aid the delta delay decoding. The delay smoother is able to resolve many delta delay problems because of their limited range. Since the pitch gain has many nonsmooth regions where smoothing fails, we protect it. This protection is for the MSB which is the most perceptually sensitive. For future expansion, we reserved 1 bit per frame. This is the 11th bit protected by the Hamming code.

6. PERFORMANCE

The CELP coder has none of the usual vocoder problems with background noise (in fact, it is faithfully reproduced) and even works well with multiple speakers. Informal listening tests indicated that the CELP's speech intelligibility and quality are comparable to 32 kbps CVSD!

We formally measure speech intelligibility and quality subjectively, using Dynastat's diagnostic rhyme test (DRT) and diagnostic acceptability measure (DAM). Typical test variances are 1 point for DRT and 2 points for DAM scores. Figure 2 shows subjective intelligibility and quality scores evaluated in four different environments for: input speech, plain old telephone service (POTS) and all the narrowband U.S. Government standard coders, including CELP. Coder performance in quiet, E-3A AWACS, 1 percent bit error rate and office environments is shown. CELP's performance is outstanding among other coders, especially in speech quality. CELP, at 4800 bps, breaks the performance barrier of most Government standards, providing "very good" intelligibility and "excellent" quality, comparable to 32 kbps CVSD. CELP will usher in a new era of narrowband speech coders capable of receiving wide user acceptance by providing very high quality speech.

Figure 1 shows CELP's performance in 5 percent bit errors. The figure compares an unprotected CELP (with monotonic LSPs) versus protected CELP. Note the regions of speech blasting, due to large excursions in the unprotected gains, that are not present in our protected version.

7. CONCLUSIONS

In this paper we describe our algorithm for a CELP coder that is practical to implement and is robust in real world conditions (noisy environments, nonspeech input and transmission errors). Any speech coder represents a set of compromises by its designers; our coder is only very slightly degraded in error free channels in order to provide robust performance. Most importantly, our algorithm sets a 4800 bps coding standard that will soon provide high quality speech in many applications, including secure voice, Mobile Satellite and Land Mobile Radio.

8. ACKNOWLEDGMENTS

We are grateful for our in-house support from: Ron Cohn, Dave Dalkowski, Jim Goble, Dave Kemp, Tina Kohler, Retha Sueda and Lynn Supplee. Many contributions were made by others, including: Dave Andre, Tom Barnwell, Shawn and Elizabeth Campbell, Allen Gersho, George Kang, Baastian Kleijn, Dan Lin and Craig Reese. We wish to give special recognition to the AT&T Bell Laboratories Acoustics Research Department, among them: Raymond Chen, Rich Cox, Frank Soong and especially to Bishnu Atal and Peter Kroon for their outstanding contributions.

9. REFERENCES

[1] Kemp, D., R. Sueda and T. Tremain, *An Evaluation of 4800 bps Voice Coders*, Proceedings of ICASSP, 1989.

[2] Atal, B. and M. Schroeder, *Stochastic Coding of Speech at Very Low Bit Rates*, Proceedings of ICC, 1984, pp. 1610-1613.

[3] Tremain, T., J. Campbell and V. Welch, *A 4.8 kbps Code Excited Linear Predictive Coder*, Proceedings of the Mobile Satellite Conference, 1988, pp. 491-496.

[4] Kang, G., Personal Communications, 1988.

[5] Rose, R. and T. Barnwell, *Quality Comparison of Low Complexity 4800 bps Self Excited and Code Excited Vocoders*, Proceedings of ICASSP, 1987, pp. 1637-1640.

[6] Lin, D., *Speech Coding Using Efficient Pseudo-Stochastic Block Codes*, Proceedings of ICASSP, 1987, pp. 1354-1357.

[7] Kleijn, W., D. Krasinski and R. Ketchum, *Improved Speech Quality and Efficient Vector Quantization in SELP*, Proceedings of ICASSP, 1988, pp. 155-158.

[8] Chen, J. and A. Gersho, *Real-Time Vector APC Speech Coding at 4800 bps with Adaptive Postfiltering*, Proceedings of ICASSP, 1987, pp. 2185-2188.

[9] Atal, B., Personal Communications, 1988.

[10] Kroon, P. and B. Atal, *Strategies for Improving the Performance of CELP Coders at Low Bit Rates*, Proceedings of ICASSP, 1988, pp. 151-154.

[11] Taniguchi, T. and S. Unagami, *Multimode Coding: A Novel Approach to Narrow and Medium Band Coding*, JASA, Vol. 84, 1988, Supplement No. 1.

[12] Cox, R., J. Hagenauer, N. Seshadri and C. Sundberg, *A Sub-Band Coder Designed for Combined Source and Channel Coding*, Proceedings of ICASSP, 1988, pp. 235-238.

[13] Soong, F. and B. Juang, *Optimal Quantization of LSP Parameters*, Proceedings of ICASSP, 1988, pp. 394-397.

[14] Cox, R, W. Kleijn and P. Kroon, *Robust CELP Coders for Noisy Background and Noisy Channels*, Proceedings of ICASSP, 1989.

	Spectrum	Pitch	Code Book
Update	30 ms	30/4 = 7.5 ms	30/4 = 7.5 ms
Order	10	1	1024
Analysis	Open loop; Correlation; 15 Hz exp.; 30 ms Hamming; no preemph.	Closed loop; Modified MSPE; VQ; weight=0.8; range=20-147	Closed loop; MSPE; weight=0.8; shift by 2; 75% sparsity
Bits/Frame	34 indep LSP; [3,4,4,4,4,3,3,3,3,3]	index: 7,5,7,5; gain(-1,2): 5*4	index: 10*4; gain(+/-): 5*4
Bit Rate	1133.3 bps	1466.7 bps	2000 bps

Note - The remaining 200 bps are used as follows: 1 bit per frame for synchronization, 4 bits per frame for forward error correction and 1 bit per frame to provide future expansion(s) of the coder.

Table 1. 4800 bps CELP Characteristics

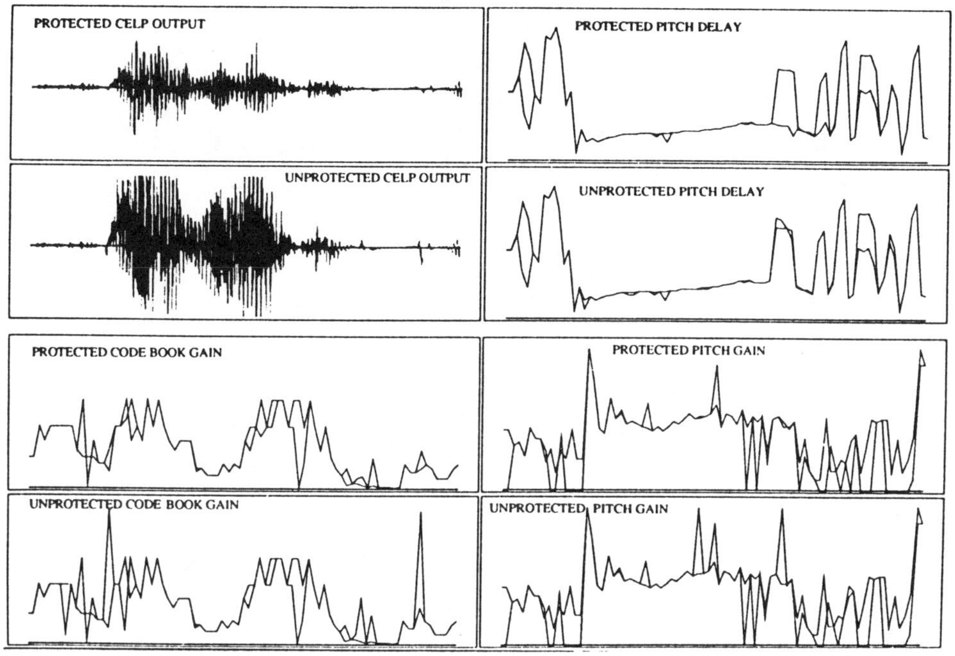

Figure 1. A demonstration of smoother effectiveness over a 5 percent random bit error rate channel.

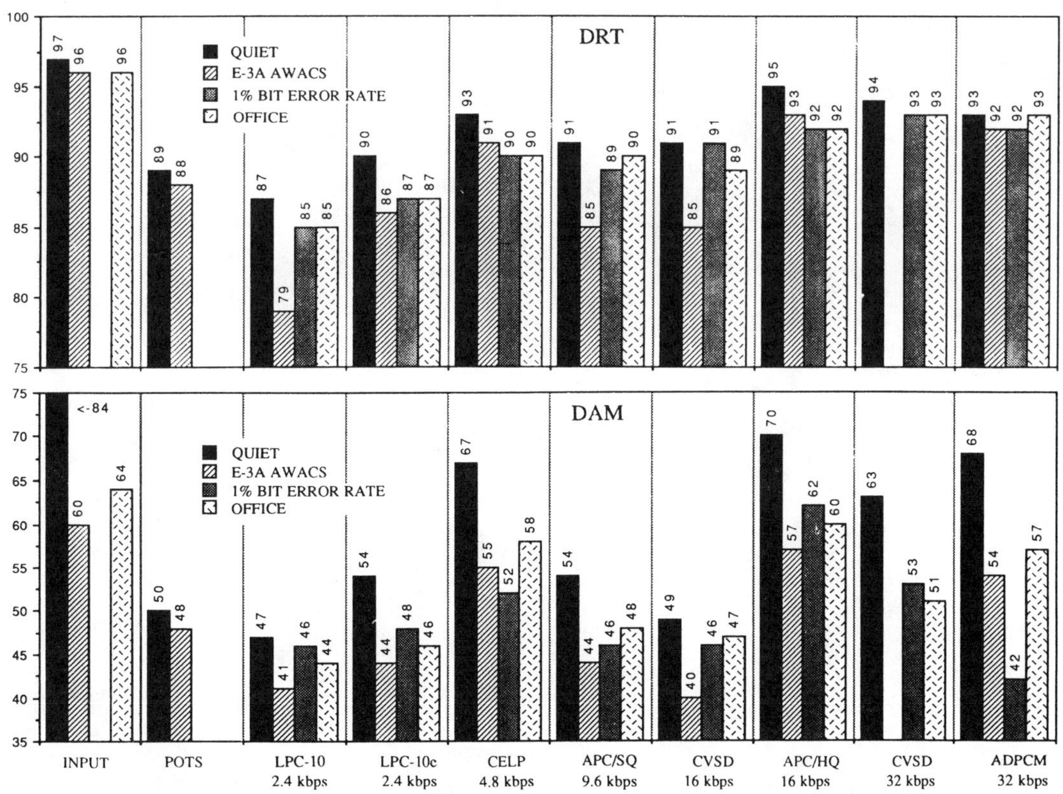

Figure 2. Speech coder intelligibility (DRT) and quality (DAM) comparison in quiet, E-3A AWACS (airborne command post), 1% bit error rate and office environments.

Part 4
Image Coding

IMAGE COMPRESSION USING NON-ADAPTIVE SPATIAL VECTOR QUANTIZATION

by R. L. Baker and R. M. Gray

Information Systems Laboratory, Stanford University
Stanford, California 94305

ABSTRACT

Vector quantization techniques have recently enjoyed much success compressing speech waveforms and speech spectral models. This paper presents results of applying two vector quantization methods to image compression. Both non-adaptively compress natural scene black-and-white digital images in the spatial domain by mapping an entire subimage block into the "closest" of a codebook of representative blocks via a squared-error distortion measure. No statistical image models are assumed. Performance of the rate-distortion coders is compared to that of a popular adaptive discrete cosine transform coder for 2.0, 1.0 and 0.5 bit-per-pixel rates.

INTRODUCTION

Vector quantization (VQ) has demonstrated pronounced success at producing low-rate compressed speech [1]-[7]. Application of this new compression technique to digitized images is just beginning. To our knowledge, the first report of using vector quantizers to encode blocks of intensity pixels was made by Gersho and Ramamurthi in a brief paper presented earlier this year [8]. Gersho's coder was designed by first segmenting a training image into edge cells and shade (non-edge) cells. One VQ codebook was designed for each cell type, minimizing squared error. The final codebook consisted of a concatenation of the two. The scheme worked remarkably well. Gersho also observed that this concatenated codebook was more robust to the type of image coded than designs using only a single unconcatenated codebook.

In this paper we discuss the advantages vector quantization techniques may enjoy in the area of image compression over conventional coding schemes. We also present results of two straightforward VQ techniques used to compress natural scene black-and-white images under a squared-error criterion. Performance of the non-adaptive VQs is compared with that of a popular adaptive discrete cosine transform coder (ADCTC) developed by Chen and Smith [9].

This work was supported in part by the Hughes Aircraft Company, Ground Systems Group, and by the Army Research Office under Contract DAAG-0073.

VECTOR QUANTIZATION

Image compression systems generally exploit two phenomena: interpixel statistical redundancy and psychophysical limitations of human observers. The former arises from high spatial correlations characteristic of photographs of natural scenes, while the latter emphasizes that information which cannot be perceived by the user need not be transmitted. Traditional image coders use predictor models or unitary transforms or combinations of both to identify and excise this wasted bandwidth. Each produce a stream of numbers which are individually quantized for transmission: one-dimensional source variables are mapped into one-dimensional representations according to some distortion criterion and statistical assumptions.

Vector quantization takes a different and powerful approach to the compression problem. Vector quantizers map an entire source vector into the least distorting or "nearest neighboring" representative vector belonging to a codebook according to a multi-dimensional distortion measure. For example, an 8-bit PCM encoded image can be compressed to 0.5 bits-per-pixel (bpp) by mapping each 4x4 pixel subblock into the best representation (its "nearest neighbor") found in a codebook of 2**8 4x4 subblocks. Eight bits unambiguously indicate the representation (codeword or codevector) selected, yielding a rate of 8/16 bpp. The trick is to develop a "good" codebook of 2**8 16-dimensional codewords from the set of all possible 4x4 subblocks, which numbers 2**128.

A computer-aided design algorithm developed by Linde, Buzo and Gray [10] is used to locate some of these "good" codebooks. There are many variations of the technique, but all follow one basic theme. We desire a codebook of vectors C = [y(i) ; i = 1, ... , N] which yields "good" re-

Reprinted from *Conf. Record of 16th Asilomar Conf. Circuits, Systems, Computers*, pp. 55-61, Oct. 1982.

presentations for each source vector x, where x is approximated by y(i), and d(x,y(i)) <= d(x,y(j)) for all j = 1, ... , N. Given a training sequence of source vectors, [x]; a distortion measure, d(• , •); and an initial guess for the codebook, C(0); we iterate two steps:

(i) Find the optimum representation (partition) of the training sequence by finding the nearest neighbor (minimum distortion) codevector for each training vector. That is, partition the training sequence into the sets

$$S(i) = [\ x : d(x,y(i)) <= d(x,y(j))\ ].$$

(ii) Compute the best reproduction vector (new codevector), y(i), for each partition S(i). y(i) is the minimum distortion vector, or centroid, of S(i) under d(• , •). For squared error, this is

$$y(i) = \frac{\sum_{x \in S(i)} x}{\|S(i)\|},$$

where "‖ ‖" denotes cardinality.

The distortion in step (ii) must decrease or not change. Should it decrease, the partition found in step (i) will no longer be optimum, and must be recomputed. This procedure repeats until the change in distortion becomes negligible. It has been shown that the codebook found will be a locally optimum quantizer for the training sequence [11].

This algorithm has several noteworthy features:

(1) No source model need be postulated to determine how to allocate bits within each block.

(2) Distortion measures need not be restricted to only those which are tractable. Any psychovisually meaningful measure can be used provided only that its centroid is computable.

We see that vector quantizers have the potential to simultaneously excise both the statistical redundancy within each input vector and any information which is psychovisually unimportant. Furthermore, the designer need not postulate underlying source statistics to obtain some approximation to the true rate-distortion quantizer for the source. A vector quantizer is at least locally optimum in the limit of an arbitrarily long training sequence.

THE CODERS

Two vector quantizer designs are investigated: a vector quantizer (VQ) of the PCM waveform; and a differential vector quantizer (DVQ). Only a single 256x256x8-bit per pixel black-and-white picture (USC's "woman with a hat" image) is used as a training sequence for the design of both coders' codebooks. We also restrict ourselves to minimizing the mean squared error (MSE) criterion. Although MSE does not correlate well with perceived image quality, it does permit numerical comparison of coder performance with other schemes minimizing MSE. In particular, we wish to compare simple vector quantizer coders to the Chen-Smith adaptive discrete cosine transform coder. Since it minimizes MSE, the vector quantizers should use this criterion, too. We first describe the VQ coder.

VQ Coder

The VQ coder represents each L x M pixel subimage block, x, of an image by an L x M pixel codevector, y(i). The codevector is selected from a codebook designed prior to the compression step by the LBG algorithm described above. Codebooks with 2**8 L x M codevectors were designed using the single image training sequence, where the subimage block dimensions were chosen to yield fixed coding rates of 2.0, 1.0 and 0.5 bpp.

It was found by experiment that square or nearly square blockings produced the most perceptually pleasing images. Blockings with high aspect ratios (e.g., 8x1 rather than 4x2) had objectionable edge artifacts in the direction orthogonal to the longer blocking dimension, and also suffered excessive busyness in the parallel direction. For this reason, blockings chosen for the three rates were made as nearly square as possible: 2x2, 4x2 and 4x4 pixels.

The vector quantizer described above charges a small number of codevectors (only 2**8) with the task of capturing both the average gray level of each subimage and its internal structure (edges, texture, etc.). A simple way to free the VQ to concentrate more intently on structure is to remove most of the DC energy from the subimage before the VQ step, transmitting it as side information. The vector quantizer is left with blocks representing only localized edge and texture information. This concept led to experimentation with a differential vector quantizer (DVQ).

DVQ Coder

The DVQ version we chose to implement approximates the sample mean of each subimage with some number of its most significant bits, leaving a vector quantizer of the type already described to quantize the pixel-by-pixel numerical difference between the original subimage and this approximated sample mean value. For example, a rate 1.0 bpp DVQ could allocate 4 bits to encode the sample mean of a 4x3 pixel subimage, leaving 8 bits to specify which 4x3 codevector best approximates the subimage less the 4-bit hard-limited mean. A codebook of representative difference subimages for the DVQ would then be designed from a training sequence of subimages minus their 4-bit hard-limited sample means.

We remark here that this version of a DVQ coder makes no attempt to efficiently transmit the sample mean of each subblock. Other schemes, such as single-element DPCM, could likely communicate a more accurate representation with fewer bits. Our purpose here is only to indicate that even simple coding structures using vector quantizers can acheive respectable performance.

The proper precision of the truncated 8-bit sample mean was determined experimentally. 1, 2, 3 and 4 of its most significant bits were saved after truncation and subtracted from the subimage, with blocking sizes selected to match the desired coding rates, given codebooks of size 2**8 codevectors. This unfortunately forced selections of some rectangular blockings with very high aspect ratios (e.g., 11x1 for the 3-bit mean, 1.0 bpp rate case). Nevertheless, all sample mean rates yielded nearly identical numerical distortions. The 4-bit case was slightly superior both numerically and perceptually for all tested images, and so it was used for subsequent comparison. Block sizes used for rates 2.0, 1.0 and 0.5 bpp were 3x2, 4x3 and 6x4, respectively.

COMPARATIVE PERFORMANCE

In order to gauge performance of the non-adaptive vector quantizers, we compared images compressed by the VQ and DVQ with those processed by a very good adaptive discrete cosine transform coder (ADCTC) developed by Chen and Smith [9]. This image coder has two levels of adaption:

(1) It segments subimage blocks into classes with similar a.c. energy.

(2) It derives the "optimum" bit allocation and quantizer for each segment of each image processed.

The first level exploits the fact that fewer bits are needed to encode subimage blocks which contain little energy, leaving more bits for highly energetic sections, while the second level customizes the quantizer for the statistics of each image. In our experiments subimage blocks are assigned to one of four equally populated segments, and each block is 16x16 pixels, ten to sixty-four times the size of those used by the vector quantizers.

Chen and Smith make the usual assumption (for the sake of tractability) in the bit allocation step that all a.c. transform coefficients are approximately zero-mean gaussian. They are not, but good performance is acheived nonetheless. Scalar quantizers for the transform coefficients are designed using the Lloyd-Max procedure [12], [13]. A subimage segment addressing map, bit allocation rules for each class, and a normalization factor all constitute overhead information. Symbols required to transmit them are included in the total bit rates cited. The coding procedure is entirely self-contained. As with the vector quantizers, it is a "one-shot" process, with no tweaking allowed.

Since all three coders minimize MSE,

$$MSE = \frac{1}{K} \sum_{k=1}^{K} [x_k - y_k(i)]^2 ,$$

signal to quantizing noise,

$$SNR = -10 * \log_{10} [MSE / K],$$

was selected as a comparative performance measure. Here d(x,y(i)) <= d(x,y(j)), j = 1, ... , N, and K = 256*256, the number of pixels in the image.

EXPERIMENTAL RESULTS

Figures 1 - 3 plot SNR versus bit rate for the three coders considered. Three images were encoded: the training image "woman", and two others, "peppers" and "baboon", at 2.0, 1.0 and 0.5 bpp rates. "Peppers" and "baboon" give an indication of VQ and DVQ performance on images outside the short training sequence.

The DVQ coder betters the adaptive transform coder at 1.0 and 0.5 bpp rates on the "woman" image; the VQ coder nearly ties it at 1.0 bpp and betters it at the 0.5 bpp rate. This good performance by both vector coders might be expected since this image was used as their training sequence. But even though the vector quantizers were trained on only this one image, are non-adaptive, and use small block sizes compared to the transform coder's 16x16 pixel blocking, the DVQ coder is still within two decibels of the adaptive transform coder for the "peppers" image at 1.0 and 0.5 bpp rates, and both coders surpass it for "baboon" image at 0.5 bpp. Experience with speech waveform vector quantizers suggests that codebooks trained on sequences of several to tens of representative images should show improved performance for images outside the training sequence, with slightly worse coding performance for those within.

Informal viewing of the compressed images judges the transform coder as excellent at 2.0 bpp. Images compressed by VQ and DVQ at 2.0 bpp also appear very good except for a slight staircasing of sharp edges. At 1.0 bpp the transform coder begins to acquire a "muddled" or "dirty" appearance over the entire image, which becomes quite pronounced at the 0.5 bpp rate. 16x16 subblocks belonging to the segment assigned the fewest bits also become noticeable as square, blurred regions. At these low rates the VQ and

DVQ coded images lack the "dirtyness," yet acquire a different, annoying artifact: "blockiness." Edges lack some of the definition present in the adaptive transformed images, and textures, such as the baboon's fur, are blurred. Observers generally express difficulty deciding which of the three coders produces the least objectionable images at 1.0 and 0.5 bpp, although the DVQ is often judged very slightly superior to the VQ.

ENHANCEMENTS

The previous section described experiments which show that even simply structured coders using a vector quantizer can acheive good coding efficiency. This performance can be improved by:

(1) Using distortion measures which correlate more directly with perceived fidelity.

(2) Building coding structures which throw at the VQ only those components of the image which it can code best with its limited blocksize, leaving what's left to other efficient quantizing schemes.

(3) Increasing the VQ's blocksize while maintaining a constant coding rate.

The first suggestion begs the proverbial problem in image coding: the need for a universally good fidelity criterion. But at least the application of candidate measures in VQ systems is potentially simpler since tractability is not an issue. (Gersho's segmentation technique is one way to perceptually improve performance of vector quantizers which minimize squared error.) The DVQ of this paper is an example of a step in the direction of suggestion two. Suggestion three warrants some further discussion.

For a given rate, vector quantizer codebooks grow exponentially with blocksize. If the entire codebook is searched to find the nearest neighbor of a source vector, computation also grows exponentially. These constraints severely restrict a VQ's maximum blocksize. The computation constraint will often be the more severe one in image coding applications. Search time must be minimized. One way to do just that has been suggested by Buzo, et al. [2]. Buzo's paper outlines one variation of the LBG algorithm which yields codebooks with a tree structure. These codebooks still grow exponentially with blocksize, but search time increases only linearly. Experience with tree coders for speech waveforms indicates that K-bit binary tree VQ coders approximate the performance of (K-1)-bit full searched VQ coders.

CONCLUSIONS

Two versions of vector quantizers were developed for compression of natural scene black-and-white digital images. Their performance was compared to that of a popular adaptive discrete cosine transform coder. The non-adaptive vector quantizers approached and in one case bettered the numerical performance of the adaptive transform coder at 1.0 and 0.5 bpp rates for images other than the one used to train the vector quantizer codebooks. Perceptually, the vector quantizers produce very good images at 2.0 bpp. At 1.0 and 0.5 bpp, VQ coded images are "cleaner" those of an adaptive transform coder, though more "blocked" and blurred. It is expected that vector quantizer coders using more perceptually accurate distortion measures, more carefully tailored coding structures, and larger, tree searched codebook constructions will yield still better results.

REFERENCES

[1] R. M. Gray, G. Rebolledo, and A. H. Gray, Jr., "'Optimal' Speech Compression," Proc. 13th Asilomar Conf. on Circuits, Systems & Computers, pp. 156-157, 1979

[2] A. Buzo, A. H. Gray, Jr., R. M. Gray, J. D. Markel, "Speech Coding Based Upon Vector Quantization," IEEE Trans. Acoust., Speech, Signal Processing, vol. ASSP-28, no. 5, pp. 562-574, Oct. 1980

[3] R. M. Gray, A. H. Gray, Jr., G. Rebolledo, and J. E. Shore, "Rate distortion speech coding with a minimum discrimination information distortion measure," IEEE Trans. Inform. Theory, vol. IT-27, no. 6, pp. 708-721, Nov. 1981

[4] D. Y. Wong, B. H. Juang, and A. H. Gray, Jr., "Recent Developments in Vector Quantization for Speech Processing" in Conf. Rec., 1981 IEEE Int. Conf. Acoust., Speech, Signal Processing, pp. 1-4, 1981

[5] G. Rebolledo, R. M. Gray, and J. P. Burg, "A Multirate Voice Digitizer Based Upon Vector Quantization," IEEE Trans. Commun., vol. COM-30, no. 4, pp. 721-727, Apr. 1982

[6] Y. Matsuyama and R. M. Gray, "Voice Coding and Tree Encoding Speech Compression Systems Based Upon Inverse Filter Matching," IEEE Trans. Commun., vol. COM-30, no. 4, p. 711-720, Apr. 1982

[7] D. Y. Wong, B. H. Juang, A. H. Gray, Jr., "An 800 bps Vector Quantized LPC Vocoder", to appear in IEEE Trans. on ASSP

[8] A. Gersho and B. Ramamurthi, "Image Coding Using Vector Quantization," Proc. Intl. Conf. Acoust., Speech, and Signal Proc., Paris, France, 1982

[9] W. H. Chen and C. H. Smith, "Adaptive coding of monochrome and color images," IEEE Trans. Commun., vol. COM-25, pp. 1285-1292, Nov. 1977

[10] Y. Linde, A. Buzo, and R. M. Gray, "An algorithm for vector quantizer design," IEEE Trans. Commun., vol. COM-28, pp. 84-95, Jan. 1980

[11] R. M. Gray, J. C. Kieffer, and Y. Linde, "Locally optimal block quantizer design," Inform. Contr., vol. 45, no. 2, pp. 178-198, May 1980

[12] J. Max, "Quantizing for Minimum Distortion," IRE Trans. on Inform. Theory, vol. IT-6, pp. 7-12, March 1960

[13] S. P. Lloyd, "Least Squares Quantization in PCM," IEEE Trans. Inform. Theory, vol. IT-28, pp. 129-137, Mar. 1982

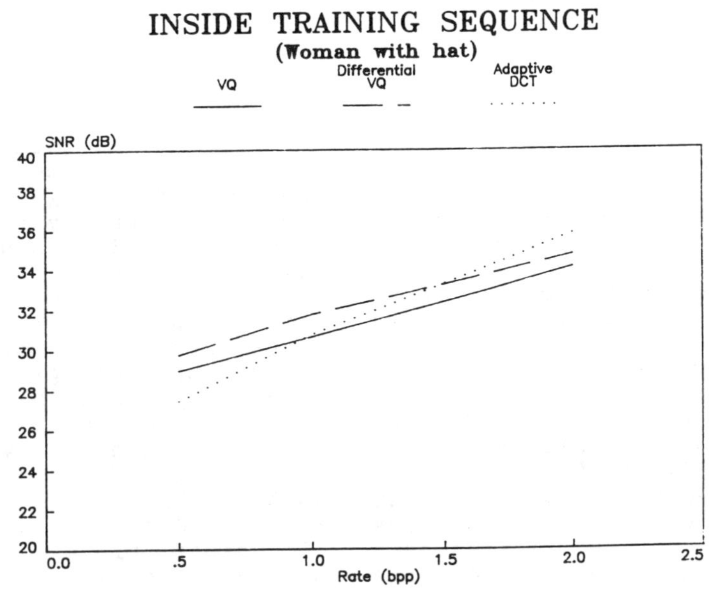

Figure 1. Vector quantizer (solid line), differential vector quantizer (dashes) and adaptive discrete cosine transform coder (dots) performance for "woman with hat" image. Vector quantizers used this image to design their codebooks.

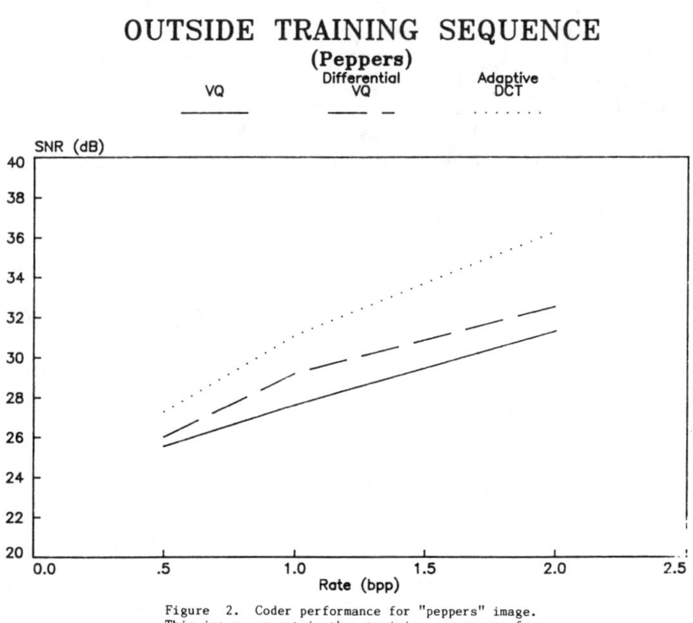

Figure 2. Coder performance for "peppers" image. This image was not in the training sequence for the vector quantizers.

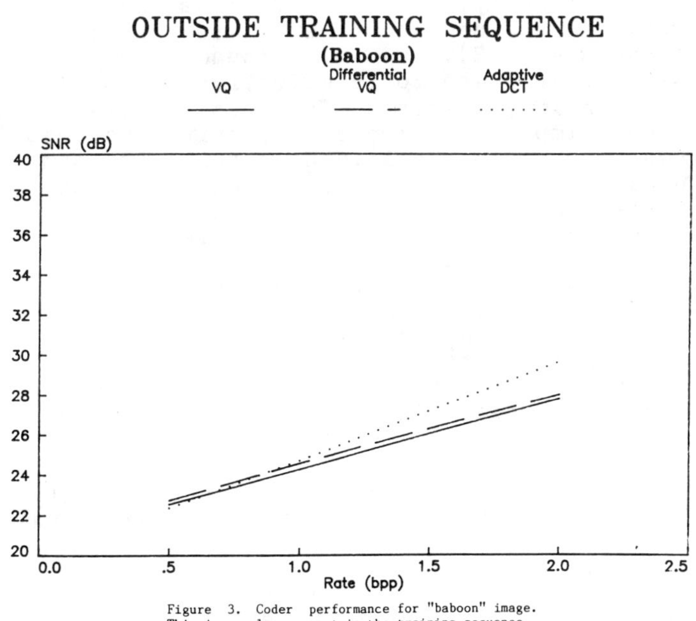

Figure 3. Coder performance for "baboon" image. This image also was not in the training sequence.

IMAGE CODING USING VECTOR QUANTIZATION

Allen Gersho and Bhaskar Ramamurthi

Dept. of Electrical & Computer Engineering
University of California,
Santa Barbara, CA 93106

Abstract

An image is partitioned into cells of of pxp pixels. Each cell is regarded as a vector of dimension p^2 and is encoded by searching through a codebook for a nearest matching representative vector. A binary word identifying the selected representative vector is assigned as the codeword to describe the original cell. The decoder uses this codeword to address a codebook. Each entry of the codebook contains a full precision digital representation of one of the N representative vectors. The codebook design is based on a clustering technique for vector quantizer design preceded by a classification of training cells into edge or shade cells. Results for coding rates from 0.5 to 1.5 bits/pixel are discussed. Vector quantization appears to be a powerful and promising technique for image coding.

Introduction

Recently, vector quantization has evolved as a new and powerful speech coding technique [1,2]. We have found the same general approach to be effective and powerful for image coding. The basic idea is very simple. An image is partitioned into two-dimensional cells. Each cell is compared with a set of standard cells (templates) and the codeword identifying the best match is transmitted. The receiver reconstructs the image using the corresponding templates in place of the original cells.

A key step, the design of the codebook containing the set of templates, is achieved using the vector quantizer design algorithm described by Linde, Buzo, and Gray [3]. This is in essence a clustering algorithm which uses a large training set of empirical data that is statistically representative of the data to be quantized (coded).

This work was supported in part by the California Space Institute and the Lawrence Livermore Laboratories.

The basic concept of block coding of images as described above is not new. In fact it is the most general way to do source coding given only a constraint on the dimensionality of the block. Shannon's rate distortion bound is based on this approach while allowing the dimension to approach infinity. Hilbert [4] used a similar approach to coding multispectral image data using a single pixel as a vector whose components give the intensity of different spectral bands at a given spatial location. Our work appears to be the first implementation of vector quantization to image coding where each vector describes a spatial block of pixel intensities. This allows the spatial correlation within a two dimensional block of pixels to be exploited in what is essentially an optimal manner.

To further improve performance, we have also used a simple adaptive technique for codebook design. Each training vector (cell) is first classified as an edge or shade (non-edge) cell and a separate codebook design is performed for each of the two cell types. The final codebook is a concatenation of the two codebooks. This type of segmentation has been used in other coding methods, for example Limb [5] used a high versus low detail segmentation for one dimensional differential coding. Tasto and Wintz [6] used a 3-way segmentation in transform coding and some authors have classified blocks using an "activity index" determined from transform coefficients [7].

Preliminary Experiments

Using the LBG codebook design algorithm, several codebooks were designed for different dimensionalities, k, (number of pixels per square cell), and different codebook sizes, N, (number of templates in the codebook). Each codebook was designed using at least 40N training vectors taken from one or more digital images in our data base. The distortion measure used was the mean-square-error. The vectors of dimension k (=pxp) were obtained by stacking pxp blocks of the picture.

The first study was intended to examine the reduction in bit rate achievable as the dimension is increased while attempting to maintain approximately constant subjective image quality. Three codebooks were designed with values of k,N

Reprinted from *Proc. IEEE Conf. Acoust., Speech, Signal Processing,* vol. 1, pp. 428–431, May 1982.

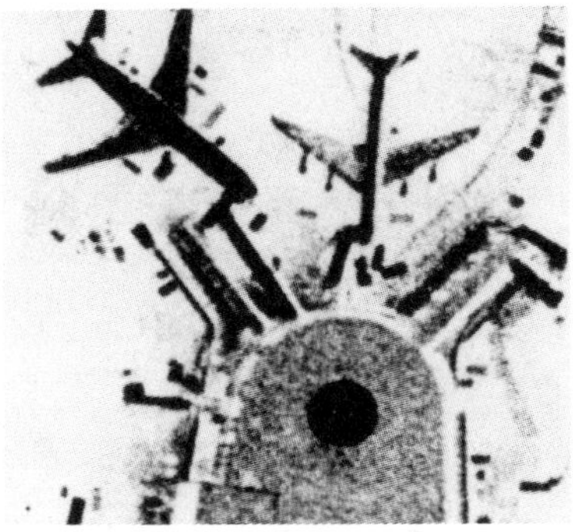

Fig. 1 LAX 8bpp

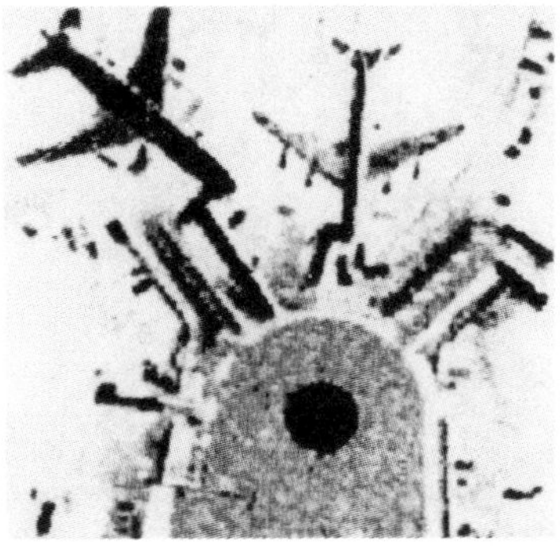

Fig. 2 4-dim. 1.5bpp

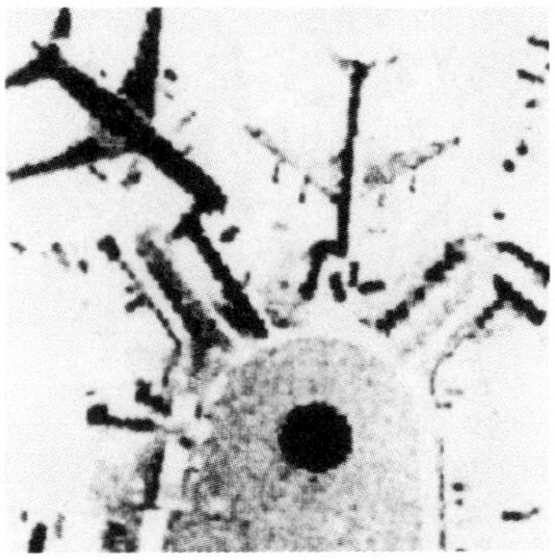

Fig. 3 9-dim. 0.8bpp

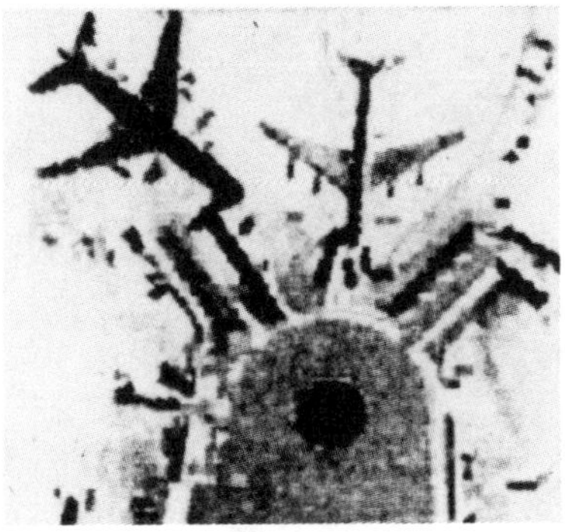

Fig. 4 16-dim. 0.5bpp

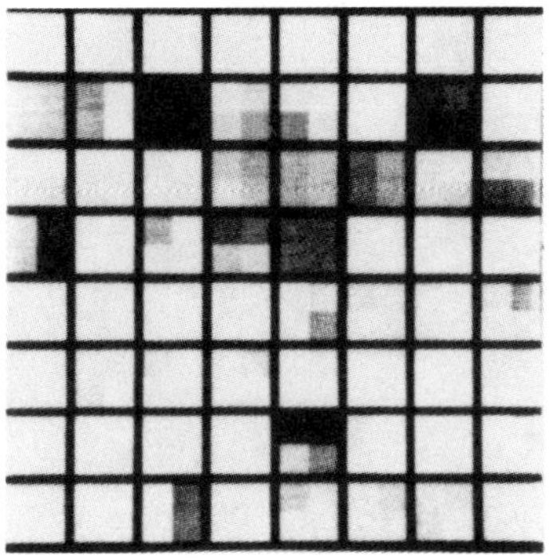

Fig. 5 4-d codebook (basic)

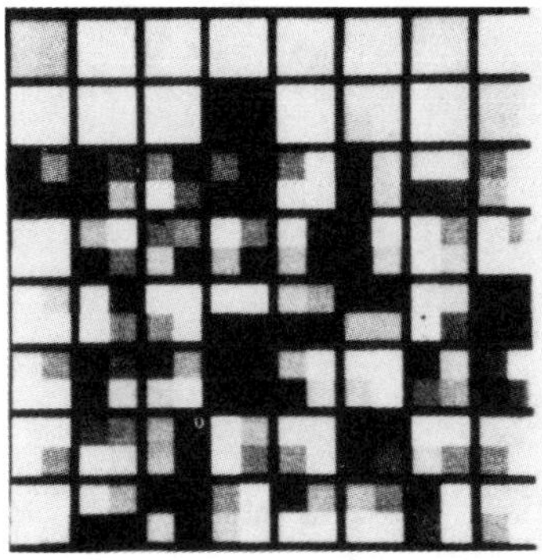

Fig. 6 4-d codebook (segmented)

as indicated: A: 4,64 ; B: 9,128; C: 16,256. The bit rate, given by $R = (\log_2 N)/k$, are 1.5, 0.78, 0.5 bpp (bits per pixel) respectively. Figure 1 shows a portion of the original image, LAX, used as the source of training vectors for all three cases. The original is composed of 1024 by 1024 pixels at 8 bits per pixel. Figures 2,3, and 4 show the reconstructed image after coding with quantizers A.B, and C respectively.

Considering the rather low bit rates and the conceptual simplicity of the method, the performance achieved is quite remarkable. The most noticeable artifact in all the reconstructed images is a 'staircase-effect', which causes the diagonal edges in the picture to be degraded to jagged ones. The contour of the 'staircase' follows that of the p x p templates that make up the edge, thus indicating that as we increase the dimensionality, this effect will be more noticable.

The staircase effect shows that there is a need for a larger set of templates containing portions of edges to allow greater flexibility to match cells that contain edges in the original image. For example, the 2 x 2 templates in the 4-dimensional codebook used to obtain Fig.3 are shown (enlarged) in Fig.5. Clearly, the number of patterns that can be used to represent edges are inadequate to handle the variety of edge cells that arise in the original image. We believe this effect is a consequence of using the mean square error criterion in the quantizer design algorithm. Rather than use a more complex distortion measure that is better subjectively, we adopt an alternate approach to improve the codebook design.

Edge/Shade Segmented Codebooks

After partitioning the image or images used for codebook design into cells. each cell is classified as either an edge or shade cell using an edge detection algorithm to determine whether or not the cell contains high contrast adjacent pixels. A training cell is classified as an edge cell if any two vertically or horizontally adjacent pixels in the cell differ by more than a pre-determined threshold. In particular, if M and m are the larger and smaller of two adjacent pixels in a cell. then the cell is classified as an edge when $(M-m)/M$ exceeds the threshold, t. Otherwise it is a shade cell. Once this splitting is done, we determine, in an ad-hoc fashion, the number, N_1, of the total of N templates that are to be edge cells.

We then use the LBG algorithm for designing the two codebooks separately (with different training-sets, and different codebook sizes). The final codebook is obtained by simply appending the two codebooks together. Encoding of pictures is now done in the usual way using this concatenated codebook. The resulting

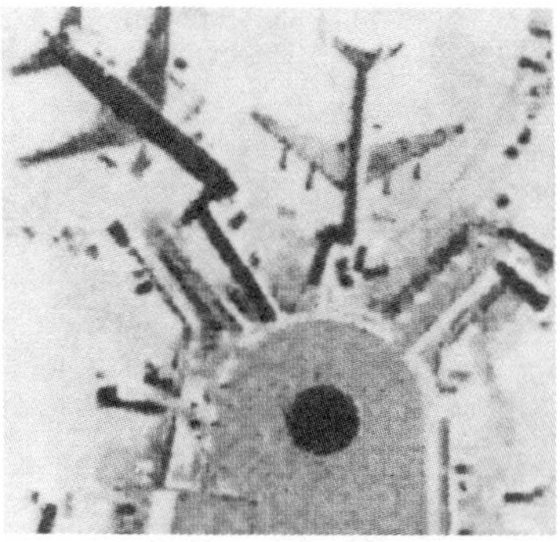

Fig. 7 9-dim. 0.8bpp

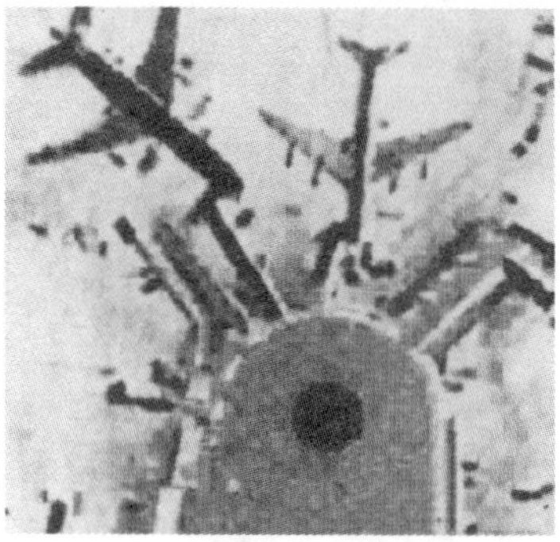

Fig. 8 16-dim. 0.5bpp

Fig. 9 Face 8bpp

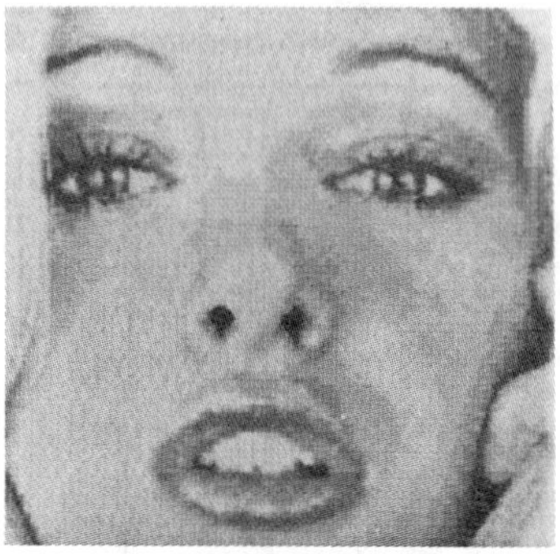

Fig. 10 9-dim..8bpp

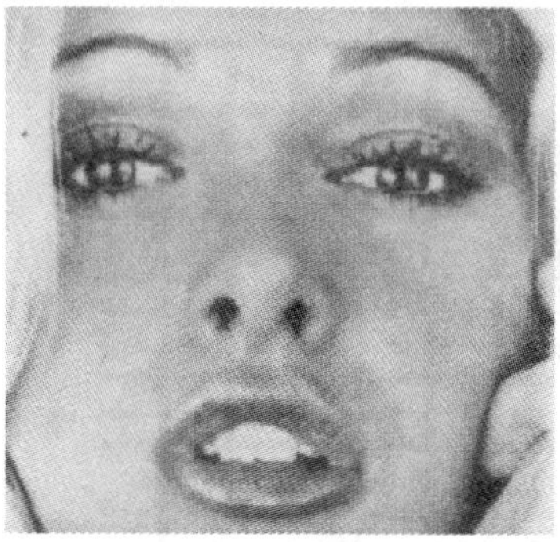

Fig. 11 9-dim. 0.8bpp

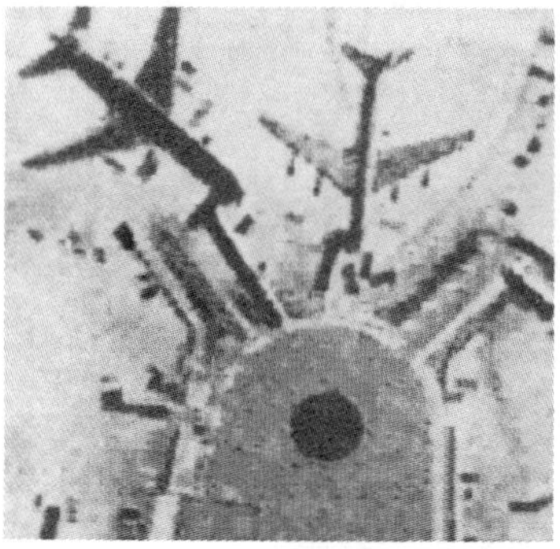

Fig. 12 9-dim. 0.8bpp

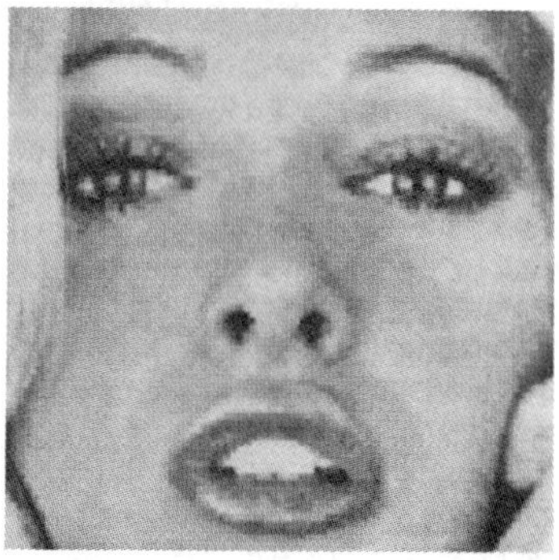

Fig. 13 9-dim. 0.8bpp

codebook is suboptimal with respect to the mean-square-error criterion. This was confirmed by the higher mean-square-error obtained for the picture encoded using the new codebook as compared to that achieved with the optimal codebook. Nevertheless. the segmented codebook gives distinctly better subjective quality for the same bit rate and the same encoding procedure.

Figures 7 and 8 show the results for 9 and 16 dimensions respectively using segmented codebooks B' and C' with the same dimensions and sizes as the previously considered codebooks B and C respectively. Each segmented codebook was chosen to have the number N_1 of edge templates equal to 75% of the codebook size. The threshold for edge-detection was set at 40%, a value found experimentally.

The improvement in the rendition of the edges, with no noticeable degradation otherwise, is significant. Note that the edges are better in the 9-dim. case (Fig.7), when compared with the 16-dim. case (Fig.8). The gross difference between the direct and segmented codebooks is apparent by comparing the corresponding codebooks for 4 dimensions shown in Figures 5 and 6.

Role of the Training Set in Codebook Design

For a codebook to be of practical value, it should perform well for a large class of images which were not part of the training data available for the codebook design. The quality of a codebook cannot be judged by its performance in coding images that were part of the training data used in the codebook design. To be effective, a codebook must be designed using a large enough set of training data to obtain a statistically adequate description of the class of images to be coded. The 'breadth' of this class will place a maximum limit on the amount of compression achievable at a given bit-rate, consistent with rate-distortion theory. Ultimately a codebook must be designed with a very large training set. We have not yet done such computationally expensive codebook designs, but we have obtained some preliminary results to examine the effect of coding images outside of the training set.

As a first step, a different image

obtained from the same scene as that shown in Fig.1 was encoded using the codebooks trained on Fig.1. No observable reduction in quality resulted, suggesting that the codebook designed using part of only one 1024 by 1024 pixel image is adequate to encode other images of a very similar nature. This was to be expected, as the training-image is quite representative of the class of images obtained from the same scene.

In order to study a worst-case situation, the woman's face shown in Fig.9, quite obviously from a completely different class of images than that of Fig.1, was encoded using the 9-dim. codebook trained on Fig.1. The result shown in Fig. 10 is seen to be considerably degraded. This is not surprising. as even the one-dimensional histograms of the test and training images are substantially different. On the other hand, if the image in Fig.9 is encoded using a 9-dim. codebook trained on itself (using the segmented design method with an edge-detection threshold of 10%. again fixed experimentally) the quality, at the same bit-rate of 0.78 bpp, is seen in Fig. 11 to be much better.

In order to study the effect of a proper choice of the training-set on encoding performance, a 9-dim. codebook of size 128 was designed with a training-set that contained cells from both Figs. 1 and 9. A segmented codebook was designed using 96 edge and 32 shade cells with edge threshold 40%. with $N_1=96$ and $N_2=32$, and the threshold for edge-detection being 0.4 as before. Both images were then encoded using this codebook. Figs. 12 and 13 show the encoded images. Since the bit-rate is again 0.78, a comparison can be made with Figs. 7 and 11. The images in Figs. 12 and 13 are poorer in quality when compared to those in Figs. 7 and 11, which were encoded using codebooks trained on each image individually. But the quality of Fig.13 is far superior to that of Fig. 10, which shows the same image encoded using a codebook trained on a completely different type of image (Fig.1). Some degradation in quality is to be expected since we have designed a more robust codebook for a wide class of images that includes features as disparate as human face texture and aircraft details. Thus, a proper and careful choice of the training set is essential to achieving a uniform encoding quality on a wide class of images.

Conclusions

This work is only a preliminary report of an ongoing investigation. A number of alternate techniques for improving the quality and reducing the encoding complexity of vector quantization are now being explored and will be reported in a later publication. In this paper, we have established the viability of vector quantization as a novel and effective technique for the coding of images. Its strength lies in its ability to fully exploit the redundancy present in a given spatial block of pixels. We have shown that a reduction in the bit-rate by an order of magnitude is possible with only limited degradation, and with low dimensionality.

Acknowledgement

We wish to thank A. Meyyappan for his assistance with the preparation of the photographs.

References

[1] A. Buzo, A.H. Gray, Jr., R.M. Gray, and J. D. Markel, "Speech Coding based Upon Vector Quantization," IEEE Trans. Acoust., Speech and Signal Proc., vol. ASSP-28, no. 5, pp. 562-574, Oct. 1980.

[2] B. Juang, D.Y. Wong, and A. H. Gray, Jr., "Recent Developments in Vector Quantization for Speech Processing," Proc. Intl Conf. Acoust., Speech, and Signal Proc., April 1981.

[3] Y. Linde, A. Buzo, and R. M. Gray, "An Algorithm for Vector Quantizer Design," IEEE Trans. on Communications, vol. COM-28, pp. 84-95, Jan. 1980.

[4] E. E. Hilbert, "Cluster Compression Algorithm A Joint Clustering/Data Compression Concept" Jet Propulsion Laboratory, Publication 77-43, Dec. 1, 1977.

[5] J.O. Limb, "Adaptive Encoding of Picture Signals," presented at the Symposium on Picture Bandwidth Compression MIT, April 1969.

[6] M. Tasto and P.A. Wintz, "Image coding by adaptive block quantization," IEEE Trans. on Commun. Tech., vol. COM-19, no. 6, December 1971, pp. 957-971.

[7] A. Habibi, "Survey of Adaptive Image Coding Techniques," IEEE Trans. on Commun., Vol. COM-25, no. 11, November 1977, pp. 1275-1284.

VECTOR QUANTISER OF VIDEO SIGNALS

T. MURAKAMI
K. ASAI
E. YAMAZAKI

Information Systems & Electronics Development Laboratory
Mitsubishi Electric Corporation
325 Kamimachiya, Kamakura City
Kanagawa Prefecture 247, Japan

Indexing terms: Signal processing, Video signals

An efficient design method is presented for a vector quantiser of 16-dimensional vectors based on a long training sequence of normalised video signals. The basic properties are demonstrated by examples.

Introduction: A vector quantiser is a mapping of k-dimensional Euclidean space R^k into a finite set of Y in R^k, where an input vector $X = [X_1, X_2, \ldots, X_k] \in R^k$. $Y = [y_1, y_2, \ldots, y_N]$, and $y_i = [y_{i1}, y_{i2}, \ldots, y_{ik}]$ is a representative or output vector in R_i for each i. $I = [1, 2, \ldots, N]$ is an index set of Y.

Then, a vector quantiser is defined as the composite of a coder C and a decoder D as follows:[1]

$$C: R^k \rightarrow 1 \qquad D: I \rightarrow Y$$

and $Q = D \,.\, C$.

That is, the vector quantiser converts an input vector into one of a set of output vectors Y for minimum distortion in R^k. Using the Euclid norm, the distortion $d(X, y_i)$ is defined by the following expression:

$$d(X, y_i) = \left[k^{-1} \sum_{j=1}^{k} (X_j - y_{ij})^2 \right]^{1/2}$$

Design of vector quantiser for video signals: The performance of a vector quantiser depends on the extraction of a compact set of output vectors corresponding to the rudimentary structural components of video signals.

Here, a block of input video samples $S_l = [S_1, S_2, \ldots, S_{16}]_l$ is composed of 4×4 lattice sampled data with an 8-bit codeword. The input vector X_l is formed by the following conversion:

$$m_l = (1/16) \sum_{K=1}^{16} S_k$$

$$\sigma_l = \left[(1/16) \sum_{k=1}^{16} (S_k - m_l)^2 \right]^{1/2}$$

$$X_k = (S_k - m_l)/\sigma_l$$

$$X_l = [X_1, X_2, \ldots, X_{16}]_l$$

where l is the block sequence number of video signals.

Then, the probability function $P(X)$ of input vectors X is approximately similar for video signals in the different scene.

We have generated a set of 256 16-dimensional output vectors using a clustering approach developed by Linde, Buzo and Gray[2] based on a long training sequence of these input vectors X_l. The clustering flow shown in Fig. 1 converges to minimise the total distortion D_c and to optimise locally a set of output vectors. Hence, we adopt the mini-max distortion measure to fasten and simplify the vector quantising operation

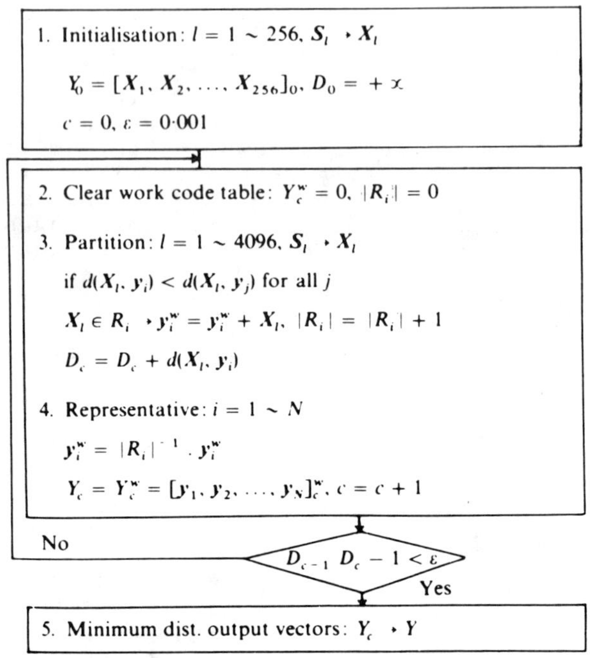

Fig. 1 *Generation of output vectors*

instead of the Euclid norm as follows:

$$D_c = \sum_l \min_i d(X_l, y_i)$$

$$\min_i d(X, y_i) = \min_i \left[\max_k |X_k - y_{ik}| \right]$$

In Fig. 1,

$\|R_i\|$ = number of input vectors included in cell R_i

ε = distortion threshold

w = suffix of work code table

c = count of clustering loop

N = number of output vectors (= 256)

The schematics of the vector quantiser of video signals designed using a set of output vectors obtained is shown in Fig. 2.

In a coder, after a block of input video samples S_l has been normalised by the average value m_l and the standard deviation σ_l, the input vector X_l is converted into the index i of the output vector y_i for the minimum distortion. m_l, σ_l and i are transmitted as coding data.

In a decoder, the index i is converted into the corresponding output vector y_i, and a block of output video samples $\hat{S}_l$ is reproduced by the following process:

$$\hat{S}_k = \sigma_l \,.\, y_{ik} + m_l \qquad k = 1, 2, \ldots, 16$$

$$\hat{S}_l = [\hat{S}_1, \hat{S}_2, \ldots, \hat{S}_{16}]_l$$

Reprinted with permission of *Electronics Letters,* vol. 7, pp. 1005–1006, Nov. 1982.

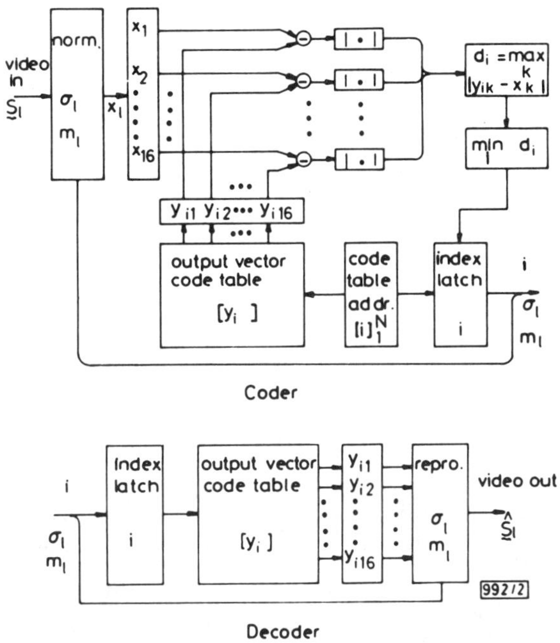

Fig. 2 *Schematic diagram of vector quantiser for video signals*

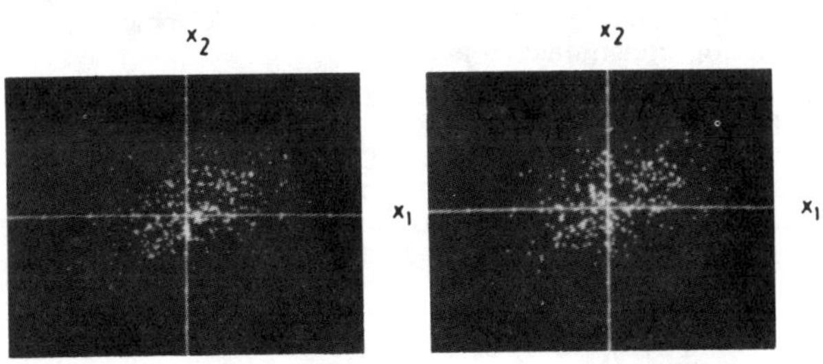

a b

Fig. 3 *Projection of output vectors in the X_1–X_2 plane*
a 'Girl' train
b 'Aerial' train

We have simulated such a 16-dimensional 256-level vector quantiser using a set of output vectors based on a training sequence of 4096 blocks in 'aerial scene' or 'girl' composed of 512 × 512 sampled data. In Fig. 3 is shown the similar projections of output vectors $[y_i]_1^{256}$ on the X_1–X_2 plane obtained by 'aerial scene' or 'girl' training. One division of each X axis is 1.

Fig. 4 is a reproduced image 'girl' using 256 output vectors obtained by 'aerial scene' training. Fig. 5 is a reproduced image 'aerial scene' using 256 output vectors obtained by 'girl' training. In Table 1, the Sp-p/N RMS ratio and the effective entropy of coder output data m_l, σ_l and i is shown.

Now, if the average value of a block of video signals composed of 3 × 2 lattice sampled data is simply quantised at the same data rate into an 8-bit codeword, the Sp-p/N RMS in such a scalar quantiser is 32·1 dB in 'girl' and 22·9 dB in 'aerial scene'. Further, the coding performance of the vector quantiser is far superior on the subjective evaluation owing to a multidimensional quantisation rather than on the traditional distortion measure corresponding to the Sp-p N RMS ratio.[3]

Conclusion: We have ensured the efficiency of vector quantisation for video signals by means of normalising each block of video signals with regard to the intensity and the amplitude.

Fig. 4 *Reproduced image 'girl' ('aerial scene' train)*

Fig. 5 *Reproduced image 'aerial scene' ('girl' train)*

Table 1 Sp-p/N RMS AND EFFECTIVE ENTROPY OF REPRODUCED IMAGES

Items / Reproduced image	Sp-p/N RMS	Entropy
	dB	bit/pel
1. 'Girl'		
1.1 'Girl' training	35·0	1·19
1.2 'Aerial' training	34·8	1·19
2. 'Aerial scene'		
2.1 'Girl' training	24·6	1·27
2.2 'Aerial' training	25·7	1·28

Further, we have proposed an implementation of the fas vector quantiser due to mini-max distortion measure.

References

1 GERSHO, A.: 'On the structure of vector quantizers', *IEEE Trans.*, 1982, **IT-28**, pp. 157–166
2 LINDE, A., BUZO, A., and GRAY, R. M.: 'An algorithm for vector quantizer design', *ibid.*, 1980, **COM-28**, pp. 84–95
3 MURAKAMI, T., ASAI, K., and OHTA, T.: 'Vector quantizer for video signals'. IECE Jpn., Tech. Rep., 1982, IE82-39, pp. 25–32 (in Japanese)

IMAGE VECTOR QUANTIZATION WITH A PERCEPTUALLY-BASED CELL CLASSIFIER

Bhaskar Ramamurthi and Allen Gersho

Department of Electrical and Computer Engineering
University of California
Santa Barbara, CA 93106

ABSTRACT

Vector quantization (VQ) has made it possible to utilize perceptually meaningful techniques for direct space-domain image coding. A simple 2 or 3 way classified codebook approach [2,3] allocates the perceptually important edges with more resolution than the easily encoded monotone regions of an image. In this paper, we introduce a major extension of the classification approach to include edge orientation and location, thereby exploiting an important feature of the human visual mechanism. In particular, each 4 x 4 block of pixels is classified into one of 31 classes for the case of 16 dimensional VQ. The encoding and codebook design complexity is significantly reduced, allowing us to use large codebooks designed from a large database of training images. We present images encoded at 0.7 and 0.8 bits per pixel using this scheme with 16-dimensional vectors. Only a small fraction of one bit per pixel is needed to code the monotone regions of an image; the rest of the bitrate is used to achieve a high level of edge integrity.

INTRODUCTION

Over the years, many researchers have attempted to include psychovisual features into a performance measure in order to minimize the perceived distortion in image coding. A major hurdle to perceptually-based coding is that with any subjectively meaningful measure, we must simultaneously code a spatially contiguous region of pixels rather than quantize a pixel at a time. The eye does not add up distortions on a pixel-by-pixel basis to decide on the total distortion. Very recently, multidimensional quantization, or vector quantization, has been recognized as a viable technique, as a result of the development of effective codebook design algorithms and advances in digital processing and storage technology. Vector quantization paves the way for making substantial progress in the use of meaningful distortion measures. For a tutorial-review of vector quantization, see [1].

Vector quantization (VQ) falls in the generic group of block coding techniques, such as transform coding (TC) and block truncation coding (BTC), all of which code a block of source samples as one entity. VQ is, in fact, the optimal encoding method for a given distortion measure and block size. Even though TC and BTC perform block coding, the quantization is done a sample at a time (in the frequency or spatial domain, as the case may be) with a distortion measure that depends only on the sample being quantized. This precludes the use of perceptually-based distortion measures, although some perceptual control is possible by bit allocation in the transform domain. Further, TC is by far the most widely used block coding scheme where quantization is done in the frequency domain, making it difficult to exploit space-domain psychovisual effects such as masking. On the other hand, VQ intrinsically quantizes a block of samples as a single entity. Any distortion measure involving all the samples in the block can be used for VQ, provided the codebook design procedure for that measure is known.

Although perceptually meaningful distortion measures are preferable, much of the early work on image VQ has been based on the mean-squared error (MSE) criterion [2]-[5]. Some results with the minimax criterion have also been reported [6,7]. Still,the MSE, or at most a weighted-MSE, is the most practical measure since optimal codebook design algorithms are not known for other measures, and any more complex measure adds to the already high computational complexity of VQ.

This work was supported in part by the State of California Microelectronics Program and Santa Barbara Research Center.

EDGE/SHADE CLASSIFICATION IN IMAGE VQ

Rather than attempting to use a complex perceptually meaningful distortion measure, the classification approach separates image vectors into perceptually distinct classes and uses the MSE criterion and a separate codebook for each class. The MSE is in fact a good distortion measure that selects the better match out of two candidate codevectors, ***provided all perceptually important features are similar for the two vectors.***

It was observed in [2] that high-detail image cells (vectors) that determine the edges in the image contribute most significantly to the perceived distortion, and must be given greater emphasis. A simple classification of the training vectors into "edge" (high detail) and "shade" (low detail) types, followed by a separate codebook design for each class was proposed for this purpose. The encoding was done in the usual VQ way, by searching through the entire codebook for the best (MSE sense) codevector. In this manner, the codebook was perceptually improved and the MSE criterion was retained to choose the best codevector from the entire codebook. The classification was subsequently extended to include a third "mid-range" class of cells that lie in between edge and shade types [3]. Instead of searching through the entire codebook in the encoding process, the input vector is classified and only the appropriate subclass of the codebook is then searched, using the usual MSE. Thus, the perceptually important feature of coding a vector of one class with codevector of its own type is achieved, while avoiding the complexity of a perceptually-based distortion measure. It was shown in [3] that this codebook design and encoding procedure was approximately equivalent to using a weighted-MSE distortion measure, with the weighting depending on the perceptual category of the input vector. Although this was a step

Reprinted from *Proc. IEEE Conf. Acoust., Speech, Signal Processing,* vol. 2, pp. 32.10.1–32.10.4, March 1984.

towards a perceptually-based coder, the main effect was to generate codebooks with a larger proportion of high-detail codevectors than would be obtained with a standard MSE-based codebook design.

EDGE-ORIENTED CLASSIFICATION

It is of great perceptual importance to preserve the fidelity of edges in location and angular orientation. We therefore extend the classification idea to include a large number of perceptual classes that identify a variety of edge cells in addition to the shade and mid-range classes. The categories of edge cells depend on both the orientation (direction) and the location of the edge within the cell. This extended classification also permits the incorporation of both the spatial frequency response of the eye and the visual masking effect while still using only the weighted-MSE distortion measure. For sixteen dimensional vectors (4×4 blocks), four edge orientations are distinguished : horizontal, vertical, and the two diagonals. Three edge locations are distinguished for the horizontal and vertical orientations, and four for the diagonal cases. For higher dimensional cells, a larger number of directions and locations can be used. Each orientation-location pair defines a particular class. We further subdivide each edge class into two classes, one with dark-to-light transitions across the edge, and the other with light-to-dark. This subclassification is not essential since the MSE will, in any case, select the best codevector from the edge class that contains both subclasses. However, use of this classification reduces the complexity of encoding and codebook design by half with almost no degradation of the codebook or of the encoding performance. This gives 28 edge classes for 16 dimensional vectors. Vectors that are not easily identified with any of the 28 edge classes or with the shade or mid-range classes are assigned to the "mixed" class.

Subcodebooks are designed separately for each class as before, and after classifying the input vector, a partial search through the appropriate subcodebook (using the MSE) completes the encoding process. The total codebook size, N, determines the rate, $\log_2 N / k$ bits per pixel (bpp). The sizes of the subcodebooks are parameters in our control; the larger the size is for one subcodebook, the more weight is given to distortions of that class.

The classified-codebook approach is approximately equivalent to the use of a weighted-MSE distortion measure, where the weighting function depends on both the input vector *and* the codevector. The weight is infinite for coding an input vector of one class with a codevector from another class, and has some preselected fixed value for codevectors from the same class. The infinite weighting forces the appropriate class to be selected. This can be modified for very low rate applications to allow "neighboring" classes to be considered under certain conditions. The fixed weighting within a class is relevant only for the codebook design step. We can argue as in [3] that, for large N, the size of each subcodebook is proportional to the fixed weight for that class. We directly choose the subcodebook sizes and experiment with various combinations to determine a good mix.

Our edge classification is intimately related to the well-known edge detectors in the human visual system [8]. The visual system detects edges in an image with each particular detector "tuned" to a particular edge orientation. Different portions of the visual field are "classified" before being further processed and mapped into a mental image. By mimicking one of the early processing stages of the human visual system, the critical issue of edge integrity is effectively handled without using a complex distortion measure.

Encoding complexity, a critical consideration for VQ, is greatly reduced by this approach. Instead of having to compute the MSE with each of the N codevectors, we now have to do so only for the codevectors in the appropriate subcodebook. Since the size of each subcodebook is of the same order, the encoding complexity is significantly reduced by a factor of thirty or more. Note also that as the dimensionality increases, the number of classes increases (more edge orientations and locations), and this compensates partly for the increased complexity due to the increase in the codebook size.

The complexity of codebook design is also greatly reduced, since both the subcodebook size *and* the size of the training-set for each class are smaller. This is particularly significant, since it is now computationally feasible to employ huge training-sets (30-40 images) to design highly robust VQ codebooks that are not scene-dependent. Further, all the complexity-reducing schemes described in the literature [1] are still separately applicable for *any or all* of the subcodebooks. This means that we can now utilize complexity reducing methods that are specifically tailored to particular subclasses of image cells. We will illustrate the use of such techniques for the design of a 13-bit codebook ($N = 2^{13}$) in 16 dimensions with the complexity of a 7-bit codebook.

THE CLASSIFICATION ALGORITHM

We now describe the essential features of the classification rule. The input vector is defined as the two-dimensional block of pixels from which it was obtained, $a(i,j)$. Two sets of gradient tables are formed, one each for the horizontal (x) and vertical (y) directions. Let $h_{ij} = |a(i,j+1)-a(i,j)|$, $s_{ij} = sgn[a(i,j+1)-a(i,j)]$, and $m_{ij} = [a(i,j+1)+a(i,j)]/2$. The y-table contains s_{ij} in the $(i,j)^{th}$ position if $h_{ij} > T_e m_{ij}$, and zero otherwise. T_e is a pre-selected threshold (0.2, in our experiments) for detecting edges. The x-table is similarly computed for the pair of pixels $a(i+1,j)$ and $a(i,j)$. At the same time, two counters are maintained for the x and y directions. The x-counter is incremented if $h_{ij} > T_s m_{ij}$, and the y-counter is incremented in a similar fashion. T_s is another pre-selected threshold for detecting shade vectors (0.025 in our case, a value indicated by Weber's Law).

A vector is deemed to be a shade vector if both the counters are separately less than a threshold (2, in our 16-dimensional case). A vector is an edge vector if there are more than a minimum number, again chosen to be 2, of +1's or −1's in either table. Otherwise it is a mid-range vector. If it is an edge vector, further processing is done on the gradient-tables to determine the orientation and location of the edge. For brevity, we omit this portion of the algorithm.

For a typical image, about 70-80% of the vectors are mid-range or shade, 2-3% are "mixed", and the rest are approximately equally divided among the edge classes. The division between shade and mid-range varies widely and depends on the quality of the original image. Only those vectors containing distinctly random gradient changes were classified as mixed. In fact, the algorithm classifies vectors with two edges running through them as belonging to one of the edge classes, since this was found to be perceptually acceptable for 16-dimensional vectors. Fig. 1 magnifies typical vectors from a horizontal edge class, a diagonal edge class, the mixed class, and the mid-range class respectively from top to bottom. It is clear from this figure that a vector belonging to one of the edge classes is perceptually distinct from a vector belonging to another edge class. The figure also helps us to visualize the vectors in the mixed and mid-range classes.

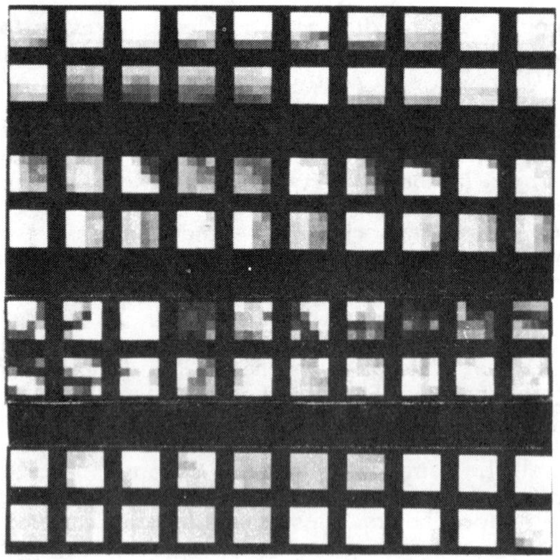

Fig. 1 : Typical Classes

CODEBOOK DESIGN AND RESULTS

Classified codebooks were designed for two rates, 0.7 and 0.8 bpp. These rates are ideal for illustrating the flexibility this approach provides for making various bit-assignments for the various classes and for controlling the complexity. Different complexity reducing techniques that are suited to the different classes are used. The training-set was obtained from eight widely different images. Two other images not belonging to the training set were encoded to study the performance of the codebooks. One of them is shown in Fig. 2.

The mid-range class has nearly half the total number of vectors in a typical image. The subcodebook for this class, though small in comparison to N, is quite large in comparison with the other subcodebooks, in order to obtain good quality for mid-range input vectors. It is important to reduce the search complexity for this class. For a vector in the mid-range class the discrete cosine transform has significant energy only for a few coefficients. Since the DCT is a distance-preserving transformation, it is immaterial whether we design the codebooks and perform the encoding in the spatial or frequency domain. However, the frequency domain has the advantage that many of the low energy coefficients can be set to zero with negligible effect on the distance between two vectors. Thus, by performing the search operation in the frequency domain, we can significantly reduce the effective dimensionality of the vectors for computational purposes. For the mid-range codebook, the DCT of each codevector is stored. For encoding, an input vector classified as mid-range, is transformed in real-time and a reduced complexity search yields the optimum codevector. The decoder of course uses only the original space-domain codebook for reconstructing the image. The only additional complexity of this technique is that of transforming the input vector. We conducted experiments to measure the loss in performance due to ignoring high frequency components of the mid-range vectors. We found that even if only the 6 lowest frequency (out of 16) DCT coefficients are retained, the increase in the MSE due to the truncation is negligible. Although this method appears similar to transform coding with zonal sampling, it should be noted that optimal VQ pattern matching is taking place and not transform coding. The DCT is used only as a more efficient way of describing the mid-range image vectors.

For both the 0.7 and 0.8 bpp cases, the shade and mid-range codebooks had 64 and 256 codevectors respectively. For the 0.7 bpp case, the horizontal (6 in number) and vertical (6) edge sub-codebooks were each of size 40, while the diagonal sub-codebooks (16) had 78 codevectors each. The total codebook size, N, adds up to 2048 ($64 + 256 + 40 \times 12 + 78 \times 16$). This combination gave approximately equal SNR for each of the classes. The mixed class was encoded with the mid-range subcodebook. This class was not assigned any additional bits since there are too few total bits available, and the other classes are perceptually more important than the mixed class. The image in Fig. 3 shows the encoding result for the original image of Fig. 2. This image shows significantly less edge distortion than all previous results obtained with VQ. Further, the encoding of the mid-range vectors is of very high quality, which is particularly significant considering the robustness of a scene-independent codebook designed with a multiple image training-set. Since the quality of encoding for the shade and mid-range classes is already much higher than that for the edge classes, any additional bits may be devoted to improve the coding of the edges and the mixed class vectors.

To illustrate the flexibility of the classified-codebook approach for controlling edge quality, we present results for 0.8 bpp. For this case, the edge subcodebook sizes were doubled. The mixed class was allocated 4096 codevectors, but to keep the search complexity low, the product codebook approach of [4] was adopted. The mean was quantized with 5 bits, and the difference vector with 128 codevectors. In addition to reducing the complexity, this separation guarantees that, at the very least, the mean value is reproduced with small error. This feature is important since the mixed class consists of vectors of high activity that the training-set may not typify very well. The result is shown in Fig. 4. The improvement over Fig. 3 is, of course, only at the edges. The staircase effect is reduced even further than in Fig. 3. This is clearly visible when the images are viewed on a monitor, but can also be seen in some parts of the photograph (edges of the hat, for example). The mixed class is more effectively encoded in Fig. 4. This can be seen on close scrutiny of the portion of the image which shows the fur on the hat. The reader should be aware that the reproduction of the images in this publication may be inadequate for a critical evaluation of the results.

CONCLUSIONS

A fundamentally new approach to image VQ with perceptually-oriented coding based on classification of the blocks has been developed. Practical codebooks based on a large training-set have been designed and tested outside the training-set. The subjective quality of the coded images is better than that produced by existing VQ-based coders, and comparable to that of any other known coding scheme at 0.75 bpp. The complexity is low, considering that we are doing VQ at these rates in 16 dimensions. Most important of all, various complexity reducing schemes, each suited to a particular class, can be employed at the same time. Some of these schemes are also desirable from other considerations, as in the case of the mixed class at 0.8 bpp.

It is possible to easily incorporate many perceptual features into the distortion measure used for a particular class. For example, by taking the logarithm of each element of the vector before any processing is done, we can incorporate the effect of the well-established first processing stage in the human visual system. Note that the classification remains unchanged with or without the logarithmic operation (for equivalent T_e and T_s). One can use a weighted-MSE distortion measure in the frequency domain for the mid-range class. The weighting function

Fig. 2 : Original : 8 bpp

Fig. 3 : 16-dim VQ : 0.7 bpp

Fig. 4 : 16-dim VQ : 0.8 bpp

would be the spatial frequency response of the eye - a well-known function that is very meaningful for the mid-range class. Since the location and orientation of the edge in each edge class is precisely known, the masking effect could be exploited by weighting errors in the pixels on either side of the edge less than the errors in the rest of the pixels. This implies a different weighted-MSE measure with a constant weighting matrix for each class, which is not very complex to implement.

The flexibility of using the spatial or frequency domain allows us to exploit space-domain and frequency-domain perceptual features simultaneously. The classification approach, equivalent to a weighted-MSE measure with a diagonal weighting matrix, achieves the effect of using a very complex input-dependent weighted-MSE measure that is perceptually meaningful. To our knowledge, no other existing coder is able to exploit so many perceptual features simultaneously. We are currently evaluating the refinements described above. A more comprehensive presentation of this work, with the refinements also included, will be given elsewhere.

REFERENCES

[1] R. M. Gray, "Vector Quantization," IEEE ASSP Magazine, to appear, April 1983.

[2] A. Gersho and B. Ramamurthi, "Image Coding Using Vector Quantization," Proc. Int'l Conf. on Acoustics, Speech and Signal Processing, April 1982.

[3] B. Ramamurthi and A. Gersho, "Low-Rate Image Coding Using Vector Quantization," IEEE Global Communications Conf. Record, pp. 184-187, Nov. 1983.

[4] R. L. Baker and R. M. Gray, "Image Compression Using Non-adaptive Spatial Vector Quantization," Proc. 16th Asilomar Conf. on Circuits, Systems and Computers, 1982.

[5] R. L. Baker and R. M. Gray, "Differential Vector Quantization of Achromatic Imagery," Picture Coding Symposium, March 1983.

[6] T. Murakami, K. Asai and E. Yamazaki, "A Design of Vector Quantizers for Video Signals," Picture Coding Symposium, March 1983.

[7] Y. Yamada, and S. Tazaki, "A Fundamental Study of Vector Quantization of Images," (in Japanese), Japanese Instt. of Elect. and Commn. Engrs., IT82-21, 1982.

[8] D. H. Hubel and T. Wiesel, "Brain Mechanisms of Vison" Scientific American, September 1979, pp. 130-146.

A Matrix Quantizer Incorporating

The Human Visual Model

K.S.Thyagarajan, S.ParthasarathY, and H.Abut

Department of Electrical and Computer Engineering
San Diego State University
San Diego, CA 92182

Abstract

This paper discusses the design of a matrix quantizer incorporating the human visual model for image encoding. It is well known that in image processing square error distortion measure is not the most suitable yardstick for the evaluation of subjective quality of reconstructed images. Since in almost all image data compression systems the human observer is the ultimate destination, one may take into account some parameters of the human visual system in computing the distance measure used in the design of an encoding algorithm. The matrix quantizer design and encoding algorithms have been modified to include a reasonably accurate model to reflect the human visual perception process. It is found that the reconstructed images in this case are sharper and exhibit much less staircase effect usually found in matrix quantizers that do not include such a model. The results were compared for rate one and Ø.56 bits per pixel (bpp) with those of straight matrix quantizers of similar rate and size.

Introduction

Encoding of speech using vector quantization (VQ) resulted in higher data compression at acceptable-to-good quality [1,2]. Since in an image, pixels lying in an area are highly correlated, the concept of VQ was extended to encoding of images [3,4,5]. Since image data is two-dimensional a submatrix of dimension k*l of pixels are blocked at a time for encoding and classification. Also an image block of size k*l can be transformed into another matrix of size k*l using orthogonal transformations such as Hadamard and then quantized in the transform domain.

In order to accomodate the human visual perception model the k*l matrices are preprocessed by the model discussed below. The encoder contains a codebook of N = 2**(k*l*r) matrices of dimension k*l elements. The preprocessed image to be encoded is partitioned into submatrices of dimension k*l, the encoder then compares the input matrix with matrices stored in the previously optimized codebook to find the best match with respect to a distance measure, and transmits its index.

The decoder has a replica of the same codebook, and decoding is simply a table look-up process from the same codebook.

The codebook design algorithm needs a partitioning procedure and a distance measure. The distortion measure should, as usual, be mathematicaly tractable and easily computable. The distortion measure should also be subjectively meaningful so that small or large quantitative values for the distortion measure correspond to good or bad subjective quality. Though the squared error distortion is widely used, it should be emphasized that in the context of image preocessing, this measure does not yield a subjectively good quality.

Mannos and Sakrison [6] have shown that the inclusion of the human visual model in the encoder results in crisper pictures at the same bit rate as compared to the encoder that uses a simple square error distortion as the optimality criterion. Mannos and Sakrison [6] have considered encoding of images wherein the square-error distortion is used in the image transform domain rather than in the pixel domain itself. The quantization rule is to transmit the Fourier transform coefficients of the image under consideration which exceed a certain threshold value and the rest are not transmitted. At the receiver the image is reconstructed from the received transform coefficients which are nonzero while the rest is padded with zeros. This paper discusses the matrix quantization of images incorporating the human visual model.

Reprinted from *Proc. IEEE Conf. Acoust., Speech, Signal Processing,* vol. 1, pp. 141-144, March 1985.

Human Visual Model

It is natural to take into account some or all of the parameters governing the human visual phenomenon in the design of efficient image transmission systems. It is a known fact that we are more sensitive to some spatial frequencies than others and that we are more sensitive to intensity errors in grey areas than in white. The psychophysical experiments on the visibility of gratings reveal that the contrast sensitivity ratio is a nonlinear function of the spatial frequencies and is of the form $k*(f/f_0)*\exp\{-f/f_0\}$ [6], where k is a constant, f is the the frequency in Hz per degree (Hz/deg) of the sinusoidal grating and f_0 is the frequency at which the contrast sensitivity ratio is a maximum. f_0 is experimentally observed to vary between 3-5 Hz/deg. It is further known that the retinal response to light intensity is not linear [7]. However, after an initial nonlinear point transformation, it is reasonable to assume that the remainder of the system to be linear[8].

Thus, the human visual model consists of a nonlinear mapping of intensity followed by a linear processor. The linear processor has a system function which is the contrast sensitivity ratio as given above. Mannos and Sakrison [6] have found that the best fit for the contrast sensitivity function is of the form

$$H(f)=(a+bf)\cdot \exp\{-(c\cdot f)**1.1\} \qquad (1)$$

where a = 0.050, b = 0.2964, c = 0.114 and $f = f_x + f_y$; f_x and f_y are the horizontal and vertical spatial frequencies respectively. H(f)) in (1) has a peak response of unity at f = 8 Hz/deg and a dc response of 0.05. Circular symmetry was chosen since we usually do not want any preferred data orientation and also we want to keep the parameters of the visual model at a minimum.

Distance Measure

The design of the matrix quantizer calls for a distortion measure. For ease of computation we still use the square error distance measure of the form

$$d(\underline{X},\hat{\underline{X}})=tr(\underline{X}-\hat{\underline{X}})^t\cdot(\underline{X}-\hat{\underline{X}}) \qquad (2)$$

where $\underline{X}$ and $\hat{\underline{X}}$ are the original and the reconstructed image submatrices respectively. However, the square error is computed in the filtered image plane rather than the pixel plane itself. Specifically, let x(m,n) be the original image to be encoded, and let w(m,n) be the nonlinearly mapped image such that

$$w(m,n)=x(m,n)**1/3 \qquad (3)$$

Of different possible nonlinear mapping, the one in (2) was found to give good subjective results [6]. The intensity compressed image w(m,n) is then linearly processed by a 2-D filter whose transfer function has the form given by (1). Let y(m,n) be the filtered image. This image signal y(m,n) is then used as the training sequence to design optimal codebook, the optimality criterion being the minimization of the average square-error distortion of the form

$$D=E\{d(\underline{Y},\hat{\underline{Y}})\} \qquad (4)$$

where E refers to the expectation operation. The design algorithm can be found in [1-3].

Experimental Procedure

Images of size 512 * 512 pixels are considered as the training database with 8 bpp resolution both in the design of codebooks and in the testing of quantizers. First an image is divided into groups of 64*64 matrices and they are compressed by point operation according to (3). It is then filtered by the linear system given in (1). Due to memory size limitation of the computing system, the filtering is performed using an overlap and save technique.

The filter function in (1) is a real continuous function of the spatial frequencies with circular symmetry. It is discretized to a 64x64 point 2-D function and a linear phase term is added. Then its point-spread function is determined by taking the inverse discrete Fourier transform. To minimize aliasing distortion in the spatial domain, a 2-D Kaiser window is used on the point-spread function. The window size is chosen to be 17x17.

Next, the filtered image is used as the training sequence in the design of a codebook using the procedure given in [1-3]. The square error distortion measure is applied to the filtered image pixels to design the codebook. Once a good codebook is designed, the filtered image is then matrix quantized. Before decoding, the visual model must be removed by filtering the quantized image using the inverse of the system function given in (1).

Inverse filtering is an ill-conditioned operation and calls for a cautious approach. Since the inverse filter characteristics have very high response at high frequencies, noise tends to predominate in this region. A possible cure is a Wiener filtering type approach with a proper model for the noise and image power spectral densities. In that procedure, the quantization noise is treated as a white noise source with a flat spectrum while the image is considered to exhibit an exponential type of autocorrelation. This gives rise to an inverse filter characteristic that is significantly less spiky at high frequencies. Finally the inverse filtered image is expanded by cubing each pixel.

A block diagram of the encoder and decoder are shown in figures 1 and 2 respectively.

Conclusions

In this paper we have discussed the results of incorporating the human visual model in the encoding process. A 1/3 power law was used for the intensity mapping of the input image and followed by a linear filter that best fits the contrast sensitivity function obtained from psychophysical experiments. A wiener filter was used for inverse filtering to remove the visual model from the quantized image. It was found that the reconstructed image was sharper when compressed when compared to images using straight MQ at the same bitrate (0.56 bpp).

Acknowledgements

This work has been supported in part by grants from the Cubic Corporation of San Diego and TRW Military Electronic Systems Division.

References

[1] Y.Linde, A.Buzo and R.M. Gray, "An Algorithm for Vector Quantizer Design", IEEE Transactions on Communications; Vol. Com-28, pp 84-95, Jan 1980

[2] H.Abut, R.M. Gray and G.Rebolledo, "Vector Quantization of Speech and Speech-like Waveforms", IEEE Transactions on Acoust., Speech and Digital Signal Processing Vol. ASSP-30, pp 423-435, June 1982

[3] K.S.Thyagarajan, H.Abut and H. Bheda, "A Matrix Quantizer for Image Processing", Proc. ICASSP-84

[4] Ramamurthy, B., Gersho, A., and Sekey, A., "Low-Rate Image Coding Using Vector Quantization," Proc. of the Conf. GLOBECOM, Oct. '83

[5] Gray, R.M., "Vector Quantization," IEEE ASSP Magazine, April '84

[6] Mannos J.L. and Sakrison D.J., " The Effects of a Visual Fidelity Criterion on the Encoding of Images," IEEE Transactions on Information Theory, Vol. IT-20, No. 4, July '74

[7] Marimont, R.B., "Linearity and the Mach Phenomenon," J. Opt. Soc. Amer., vol. 55, Mar. '63

[8] Whiteside, J.A., and Davidson, M.L., "Symmetrical Appearance of Bright and Dark Mach Bands from an Exponential Illumination Gradient," J. Opt. Soc. Amer., vol. 61, July '71

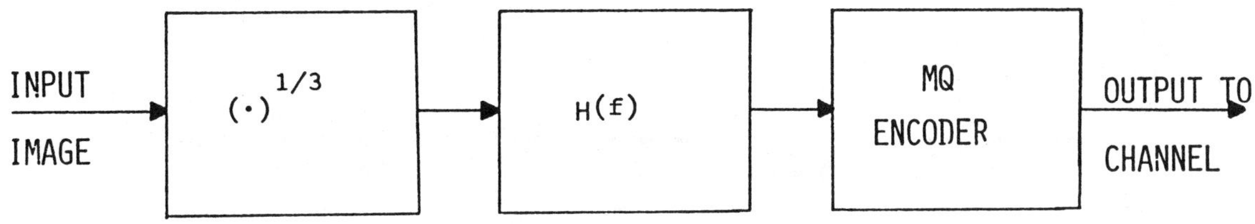

Figure 1. Matrix Quantizer Encoder Block Diagram with the Human Visual Model.

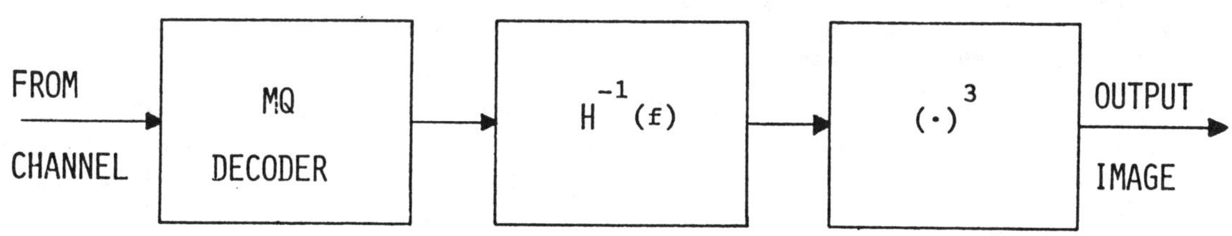

Figure 2. Matrix Quantizer Decoder Block Diagram with the Human Visual Model.

HIGH QUALITY IMAGE CODING WITH A MODEL-TESTING VECTOR QUANTIZER AND A HUMAN VISUAL SYSTEM MODEL

V. Ramamoorthy
N. S. Jayant

AT&T Bell Laboratories
Murray Hill, New Jersey 07974

ABSTRACT

A new class of image quantizers called the model-testing vector quantizer (MTVQ) is introduced. This is a finite-state quantizer with the additional ability of making a decision about whether the current input does indeed conform to the assumed finite-state model or not. By a careful design, a model-testing quantizer can be made to decompose the input image into a conforming component that can be coded with a low bit rate VQ coder and a nonconforming component that traditionally needs a high bit rate for transparent coding. In this paper, the nonconforming component is coded at relatively lower bit rates by utilizing well-known linear and nonlinear properties of the human visual system. We employ an adaptive spatial and intensity quantization procedure that removes the perceptually irrelevant information from the nonconforming component. The differential brightness discrimination of the eye is also utilized. Resulting coded images are indistinguishable from the originals at bit rates that range from 1.0 to 3.5 bits/pixel for a wide variety of 512×512 images. These bit rates can be regarded as upper bounds for the perceptual entropy of the corresponding images.

INTRODUCTION

It is well known that a vector quantizer (VQ) is an efficient coder offering a good trade-off between distortion and bit-rate. With typical image inputs, successive VQ outputs are highly correlated and this high correlation can be exploited to further reduce the bit rate. *Reduced search vector quantizers, predictive vector quantizers, finite-state vector quantizers* — all attempt to cut down the transmission rate by considering the block-to-block correlations existing in the image [1-10]. In particular, finite-state vector quantizers use a single composite model to realize this bit-rate reduction capability. However, at bit rates of interest, the resulting coded images contain visible quality degradations such as "blockiness", "staircase effect", and "edge discontinuities". The reasons for such degradations can be traced back to the design of the codebook, the distortion measure and the finite-state machine structure which tends to become too restrictive at the edge boundaries.

In this paper we take a different approach: we propose a two-component image model that is more appropriate for high quality image coding. We identify one of the components as the part that conforms to the image model derived from a training set. The other component is the portion that does not conform to the model. By decomposing the input source into such conforming and nonconforming components, we are able to use two different encoding techniques and this helps provide very high levels of image quality.

A two-component composite model for images is not a new idea. What is new in this work is the idea of a model-testing finite-state vector quantizer (MTVQ) which is capable of measuring the "distance" between the source input and the assumed model as a part of the encoding process. Such a "distance" measurement naturally leads to a decision whether the input is behaving according to the assumed model or not.

Since our aim is to encode the image inputs with the highest possible subjective quality, we attempt to make use of the established psychophysical properties of human vision in encoding the nonconforming source inputs. In particular, we propose the use of a class of zero perceptual distortion codes for encoding these components. The requirement of zero perceptual distortion naturally leads to high rate codes in contrast to the low rate codes typically associated with the finite-state vector quantizer. The average bit rate of the complete encoding system and the overall subjective quality depend on the ratio of the conforming and nonconforming source inputs. This ratio can be controlled by a proper threshold setting.

MODEL-TESTING FINITE-STATE VECTOR QUANTIZER

Consider a memoryless full-search K-dimensional VQ with a codebook $\mathcal{Y} = \{\mathbf{y}_i;\, i \in \mathcal{N}\}$, $\mathcal{N} = \{1, 2, 3, 4 \cdots N\}$ with a distortion measure $d(\mathbf{x}_n, \hat{\mathbf{x}}_n) = ||\mathbf{x}_n - \hat{\mathbf{x}}_n||^2$ between the input $\mathbf{x}_n$ and output $\hat{\mathbf{x}}_n \in \mathcal{Y}$. The channel symbol u_n is defined by

$$u_n = \min_{i \in \mathcal{N}}{}^{-1}\, d[\mathbf{x}_n, \mathbf{y}_i]$$

where the $\min^{-1}$ notation means that u_n is the index ℓ for which the minimum distortion is achieved over all code words contained by $\mathcal{Y}$. We use 512×512 images and consider 4×4 blocks in the image as the input sequence $\{\mathbf{x}_n\}$. Since the blocks considered are small in size, a set of blocks in a given neighborhood is expected to show high correlation. Now consider a specific neighborhood of blocks to the left and the top of current block $\mathbf{x}_n$. Based on this specific neighborhood, it is possible to define a 2-D Markov chain that portrays the dependencies between the transmitted vector indices. From a memoryless VQ, and from the above definition of neighborhood, it is possible to obtain a $N^2 \times N$ conditional probability matrix

$$\mathbf{P} = (P_{(i,j)}(k))$$
$$P_{(i,j)}(k) = \text{Prob}\,[u_n = k \mid u_{n-1} = j;\; u_{n-128} = i]$$
$$P_{(i,j)}(k) \geqslant 0 \text{ and } \sum_{k=1}^{N} P_{(i,j)}(k) = 1; \text{ for all } i, j, k \in \mathcal{N}$$

The N^2 rows of the matrix P are uniquely addressed by the ordered pair (i, j) with $i, j \in \mathcal{N}$. On the other hand, we can also denote the rows by $s \in S = \{1, 2, 3 \cdots N^2\}$ with $s = i + (j-1)N$. The matrix P is estimated from the training set of images used in the LBG design algorithm [4,5]. Consider an $N^2 \times N$ ordered predictor matrix $\mathbf{T}$ derived from $\mathbf{P}$:

$$\mathbf{T} = (T_s(k)) \quad s \in S,\; k \in \mathcal{N}$$

such that

Reprinted from *IEEE Conf. Acoust., Speech, Signal Processing,* vol. 2, pp. 1164–1167, April 1988.

$$T_s(1) = \max_{k \in \mathcal{N}}{}^{-1} P_s(k); \quad T_s(N) = \min_{k \in \mathcal{N}}{}^{-1} P_s(k)$$

$$P_s(T_s(\ell)) \geqslant P_s(T_s(\ell+1)) \quad \text{for} \quad \ell = 1, 2, \ldots, N-1$$

The $\max^{-1}$ and $\min^{-1}$ notations indicate the value of the argument k for which the maximum and the minimum values are respectively achieved. The matrix **T** thus contains an ordered set of code vector indices for each value of s such that the most probable code vector index is placed in $T_s(1)$ and the next to the most probable code vector index is placed in $T_s(2)$,... and so on. For a given value of $s \in S$, the row vector $T_s(\ell)$, $\ell = 1, 2, \ldots, N$ provides the code vector indices in the decreasing order of their probability of occurrences.

A *model-testing finite-state* VQ is described by a state space $S = \mathcal{N} \times \mathcal{N} = \{1, 2, 3, \ldots, \mathcal{N}^2\}$, each state s defined by the ordered pair (u_{n-128}, u_{n-1}) and each state codebook identical to $\mathcal{Y}$. We call this coder as 'model-testing', because it gives the information about when a finite-state VQ is to be employed for encoding and when it is not appropriate for encoding. The encoding operations are as follows: Given an input $\mathbf{x}_n \in \mathbb{R}^K$ and a state s_n defined by the ordered pair $(u_{n-128} = i, \ u_{n-1} = j)$, the vector $\mathbf{x}_n$ is encoded into the reconstruction vectors pointed by the s_n-th row of T i.e. $T_{s_n}(k)$, $k \in \mathcal{N}$. The corresponding distortions are given by a distortion vector

$$\mathbf{D}_n = \left[d_n^{(1)}, d_n^{(2)}, d_n^{(3)}, \ldots, d_n^{(N)}\right]; \quad d_n^{(\ell)} = d[\mathbf{x}_n, \mathbf{y}_{T_{s_n}(\ell)}] \quad \ell \in \mathcal{N}$$

Since $\{T_{s_n}(\ell), \ \ell = 1, 2 \cdots N\}$ contains the ordered set of code vector indices, the first component of $\mathbf{D}_n$, namely $d_n^{(1)}$, just contains the distortion between the input $\mathbf{x}_n$ and the most probable code vector for the given state s_n. The component $d_n^{(2)}$ contains the distortion between the input $\mathbf{x}_n$ and the second most probable code vector, etc. Observe that the vector $\mathbf{D}_n$ contains valuable information about the current image activity. If $d_n^{(1)}$ happens to be the minimum value of $\{d_n^{(\ell)}, \ \ell = 1, 2 \cdots N\}$, then the input $\mathbf{x}_n$ is indeed closest to the most probable code vector predicted and indexed by $T_{s_n}(1)$. The inference that can be drawn from this occurrence is that our Markov model is indeed a good fit for the input $\mathbf{x}_n$. Since the finite-state quantizer is built from the model, it is appropriate to use this quantizer for encoding $\mathbf{x}_n$. On the other hand, if $d_n^{(p)}$, $p >> 1$ happens to be the minimum of $\{d_n^{(\ell)}, \ \ell = 1, 2 \cdots N\}$, then we can infer that the input $\mathbf{x}_n$ is not behaving according to the assumed model and a decision to switch to a different coder may be made.

In this paper, we just consider a specific channel symbol release strategy defined by $i_n = \min^{-1} \{d_n^{(\ell)}\}$ where the minimization is over $\ell \in \mathcal{N}$. The channel symbol i_n is the super index of the component of $\mathbf{D}_n$ having the minimum distortion. Since $\mathbf{D}_n$ is generated using $\{T_{s_n}(\ell), \ \ell = 1, 2, \ldots, N\}$, $T_{s_n}(i_n)$ is the index of the code vector closest to the input $\mathbf{x}_n$ in the Euclidean sense. Since the matrix T is assumed to be available at the decoder, the decoder knowing s_n and i_n can reconstruct the decoded output of $\hat{\mathbf{x}}_n = \mathbf{y}_{T_{s_n}(i_n)}$. Observe that the state information s_n is present both at the encoder and the decoder as s_n is defined by the ordered pair (u_{n-1}, u_{n-128}).

Since we have a probabilistic description of the image, we can check for the image activity by the following threshold comparisons: The encoder issues a binary sequence $\{b_n\}$ which describes the conformity to the assumed model:

$$b_n = 0 \quad \text{(Input conforms to the model)} \qquad \text{if } i_n \leqslant I_{\text{threshold}}$$
$$b_n = 1 \quad \text{(Input does not conform to the model)} \qquad \text{if } i_n > I_{\text{threshold}}$$

The input block with $b_n = 0$ is called a conforming block and the input block with $b_n = 1$ is called a nonconforming block.

In summary the model-testing finite-state VQ transmitter is completely described by the predictor matrix **T**, the instantaneous hard-decision quantizer used for emitting the channel symbols, the activity detector and the state transition rule f. The basic finite-state model employed by the encoder is defined by the codebook $\mathcal{Y}$ and the predictor matrix **T**. The model-testing finite-state VQ is defined by a state transition rule f: $S \times N \rightarrow S$. Since the state at time n, s_n, is defined by the ordered pair $(u_{n-128} = i, \ u_{n-1} = j)$, the state at time $n+1$ is given by $(u_{n+1-128} = i', \ u_n = j')$ where $u_n = j' = T_{(i,j)}(i_n) = T_{s_n}(i_n)$. The decoder, on receiving the channel symbol i_n, emits the reconstruction vector

$$\hat{\mathbf{x}}_n = \mathbf{y}_{T_{s_n}(i_n)}$$

Since the matrix **T**, the codebook $\mathcal{Y}$, and the state-transition rule f are known to the decoder, the encoder and the decoder work synchronously and the decoder tracks the encoder. Our design consisted of the following parameters: $K = 16$, $(4 \times 4$ pixel block), $N = 256$ and the matrix **T** had $256 \times 256 \times 256$ elements. We used five images as a training set to obtain the codebook $\mathcal{Y}$, the matrices **P** and **T**. A test image called 'WOMANHAT' is used to verify our design. Seven other test images are also used. The threshold $I_{\text{threshold}}$ is found by careful experimentation.

CODING STRATEGY

When the source is found to conform with the finite-state model, the coding is done with a finite-state vector quantizer. When the source does not conform to the finite-state model, the input source block is coded as given below. Entropy coding is employed to compress the bit stream transmitted. We only employ very simple two level or three level noiseless codes that are not 'tuned' to the statistics of $\{i_n\}$.

A. Viewing Conditions and Quality Criterion

An important consideration in the perceptually based coding schemes is the viewing condition. All our experiments were conducted with a Sun 3/110 workstation with a 19" screen having 1152×900 pixel resolution to display the 512×512 images on an area of 6.25" × 6.25". The viewing is carried out at an approximate distance of 25" perpendicular to the display screen. The background room illumination can be varied to the convenience of the viewer. We found that a dark room with minimal illumination is helpful in differentiating images displayed on the screen. In the experiments dealing with the subjective quality, the original uncoded image and the coded image were displayed side by side and the viewer was asked to examine the images and find the coding distortion in the coded image. We used the following criterion to judge the quality of the coded image: If the viewer took more than 20 seconds to find a coding distortion, we declared the coded image to be of high quality. Our '20 second' criterion is indeed a stringent requirement from the coding point of view; many simple VQ schemes did not pass this quality criterion.

B. First Level of Coding: Adaptive Amplitude and Spatial Quantization of the Nonconforming Blocks

A nonconforming block is divided into four 2×2 subblocks. Each of these four subblocks is quantized to have only two amplitude levels. The subblocks are spatially divided in the following manner: Let x_1, x_2, x_3, and x_4 be the pixel intensities in a subblock. Let $y_1(i), y_2(i), y_3(i)$ and $y_4(i)$ be the encoded pixel intensities for $i = 1, 2, \ldots, 8$. The relationship between these two quantities are:

$$y_1(1) = x_1$$
$$y_2(1) = y_3(1) = y_4(1) = (x_2 + x_3 + x_4)/3$$
$$y_2(2) = x_2$$
$$y_1(2) = y_3(2) = y_4(2) = (x_1 + x_3 + x_4)/3$$
$$y_4(3) = x_4$$
$$y_1(3) = y_2(3) = y_3(3) = (x_1 + x_2 + x_3)/3$$
$$y_3(4) = x_3$$
$$y_1(4) = y_2(4) = y_4(4) = (x_1 + x_2 + x_4)/3$$
$$y_1(5) = y_2(5) = (x_1 + x_2)/2$$
$$y_3(5) = y_4(5) = (x_3 + x_4)/2$$
$$y_1(6) = y_3(6) = (x_1 + x_3)/2$$
$$y_2(6) = y_4(6) = (x_2 + x_4)/2$$
$$y_1(7) = y_4(7) = (x_1 + x_4)/2$$

$$y_2(7) = y_3(7) = (x_2 + x_3)/2$$
$$y_1(8) = y_2(8) = y_3(8) = y_4(8) = (x_1 + x_2 + x_3 + x_4)/4$$

What we are attempting is eight distinct spatial quantization levels on a 2×2 block. Such a quantization leaves at most two separate regions of constant luminance within this subblock. Since there are four such subblocks, we have a total of $8^4 = 4096$ spatial quantization levels on a 4×4 block. We call each of these 4096 levels as a "binary pattern." Each pattern contains eight distinct light intensity levels, two for each of the four subblocks. The intensity levels are computed for each 2×2 block. The encoding is done by minimizing the squared error sum $\sum (x_i - y_i(k))^2$, $i = 1$ to 4, with respect to k $(1, 2, \ldots, 8)$. The value of k that minimizes the sum is transmitted as 3 bits of pattern information. Since $y_i(k)$'s (for a given value of k) have at most two different values, these two values are encoded by two 8 bit numbers. The basic assumption in the above encoding procedure is that in a 2×2 subblock, one can have a maximum of two separate intensity regions. Because of the high spatial correlation existing in the image intensity, such an assumption is valid.

It is well known that the relative sensitivity of the human visual system has a bandpass characteristic with respect to the spatial frequencies. The spatial frequency masking studies by Stromeyer and Julesz [11] suggest that the visual system is selectively tuned to different spatial frequencies. The experimental results of Hall and Hall [12], Campbell [13], Davidson [14], clearly support such a model and the idea of a Modulation Transfer Function (MTF) of the human visual system is widely accepted. Cornsweet [15] and Granrath [16] give an excellent summary of the basics of this model. For white light and pupil diameter of 3 mm, Hall and Hall [12], found that −3 dB cross-over point for the lowpass filter of the visual system occurs at 6.6 cycles/degree and the response is less than −26 dB for frequencies above 30 cycles/degree. The visual model developed in [12] is essentially a nonlinear model portraying the system bandwidth variations as a function of the input contrast. As the input contrast increases, the bandwidth of the system decreases. From the overall spatial response of this model, (Figure 11 of [12]), we find that the −3 dB cross-over frequency changes from about 4 cycles/degree to about 8 cycles/degree as the contrast is decreased. We assume a simple static linear model with a fixed cross-over frequency of about 6 cycles/degree, which corresponds to the midpoint of the range of variations described above. For a "normal" viewer at an approximate distance of 25" from the display screen, the displayed images in our experiments subtend a solid angle of $14.25^\circ \times 14.25^\circ$. Operating on the 2×2 pixel blocks can thus correspond to a spatial frequency of $256/14.25 = 17.96$ cycles/degree which is well above the cross-over point of the visual system model. That is, the spatial frequency at which adjacent pixels in 2×2 coded blocks can be distinguished by the difference in their average intensity (due to Mach effect) occurs at 17.96 cycles/degree. For this spatial frequency, the attenuation of the visual system, from [12], is around 20 dB. Small changes in intensity at this frequency is practically invisible. Since our quantization operations are just averaging over neighboring pixels, the changes caused are expected to be very small. Hence we conclude that the differences created in the image intensities due to the quantization operations performed on the 2×2 blocks will be invisible to the viewer and the viewer would not be able to distinguish between the uncoded original block and the coded one.

C. Second Level of Coding: Coding of Nonconforming Blocks Using Nonlinear Properties of the Human Visual System

It is well known that the perceived brightness of a region in an image is a nonlinear function of the light intensity of the region and its surroundings. In general, it is very difficult to make use of the nonlinear properties of vision mechanisms in image coding as these properties are very sensitive to the context information and the a priori knowledge of the human viewer. We employ the visual thresholds of intensity discrimination to achieve bit rate reduction. If the mean intensity value of a nonconforming block is y_m and the maximum deviation from the mean within the 4×4 block is y_r, then bit rate reduction without any reduction in quality can be obtained if y_m and y_r satisfy certain visual threshold conditions. If y_r is small and y_m falls below the lower saturation threshold $I_{\ell s}$, then y_m can be quantized to the lower saturation threshold and only this value is transmitted to reconstruct the entire 4×4 block. Similarly if y_r is small and y_m is greater than the upper saturation threshold I_{us}, then only I_{us} is transmitted as the quantized value of the entire block. If y_m is the linear region in between I_{us} and $I_{\ell s}$ and if $|y_r|$ is less than or equal to the dynamic threshold I_d, then the mean value y_m can be used to reconstruct the entire block. By transmitting only the mean or the threshold values, bit rate compression is attained without any perceptual degradation. On the other hand, if the threshold conditions are not satisfied, then the coding is done as explained in Section B. We performed carefully designed experiments to arrive at these thresholds.

Table 1 shows the results for the test image 'WOMANHAT' when the contrast discrimination properties of the eye are taken into account. The measured values of mean and maximum deviation over a 4×4 block are used to select the quantization as described above. Table 1a describes the decision thresholds used in our experiments. Resulting bit rates and SNR values are shown in the Table 1b. We see that the setting $I_{\text{threshold}} = 2$ gives rise to a bit rate of 1.688 bits/sample and a SNR value of 28.941 dB. Even if the SNR is low, the overall quality is very good. As the threshold $I_{\text{threshold}}$ is increased, the SNR values increases as less and less blocks get affected by the replacement strategies with increases in $I_{\text{threshold}}$. For the setting $I_{\text{threshold}} = 8$, the bit rate is 1.436 bits/pixel and the SNR is 29.755 dB. The overall quality of the coded image is very good and almost as good as the uncoded original.

The methods described above — in particular, the coding method using the Table 1a, were tested further with seven new images not used in the training. The effect of our limited training procedure with only five images appears as an increased percentage of nonconforming blocks. The bit rate is a function of the percentage of the nonconforming blocks and it varies from image to image in the range of 1-3.5 bits/pixel. Photo 1 shows the uncoded original 'WOMANHAT' on the left and the coded image on the right. The bit rate for the coded image is 1.688 bit/pixel and the threshold setting is $I_{\text{threshold}} = 2$, and $I_d = 8$. For this setting, 53.8% of the blocks are nonconforming blocks. Photo 2 shows the uncoded original 'BABOON' on the left and the coded image on the right. For this coded image, the bit rate is 3.406 bit/pixel, $I_{\text{threshold}} = 16$ and $I_d = 8$. The percentage of nonconforming blocks is 69.68%. The image "BABOON" has a large number of active regions and requires a high bit rate. The other six images require a considerably lower bit rate, in the range of 1 to 2 bits per pixel. All of the above bit rates can be regarded as upper bounds on the perceptual entropies of the respective inputs.

REFERENCES

[1] A. Gersho and V. Cuperman, "Vector Quantization: A pattern-matching technique for speech coding," *IEEE Commun. Mag.*, Vol. 21, pp. 15-21, Dec. 1983.

[2] R. M. Gray, "Vector quantization," *IEEE ASSP Mag.*, Vol. 1, pp. 4-29, April 1984.

[3] J. Makhoul, S. Roucos and H. Gish, "Vector quantization in speech coding," *Proc. IEEE*, Vol. 73, pp. 1551-1558, Nov. 1985.

[4] Y. Linde, A. Buzo and R. M. Gray, "An algorithm for vector quantizer design," *IEEE Trans. Commun.*, Vol. COM-28, pp. 84-95, Jan. 1980.

[5] R. M. Gray and Y. Linde, "Vector quantizers and predictive quantizers for Gauss-Markov sources," *IEEE Trans. Commun.*, Vol. COM-30, pp. 381-389, Feb. 1982.

[6] A. Haoui and D. G. Messerschmitt, "A fixed delay speech coder with variable length binary coding," *Proc. of IEEE Int. Conf. on Comm.*, ICC85, Chicago, pp. 45.7.1-45.7.4, June 1985.

[7] M. O. Dunham and R. M. Gray, "An algorithm for the design of labeled-transition finite-state vector quantizers," *IEEE Trans. Commun.*, Vol. COM-33, pp. 83-89, Jan. 1985.

[8] J. Foster, R. M. Gray and M. O. Dunham, "Finite-state vector quantization for waveform coding," *IEEE Trans. Info. Theory*, Vol. IT-31, pp. 348-359, May 1985.

[9] V. Ramamoorthy and Shakir Abdul-Jabbar, "A reduced search vector quantizer for speech," *Proc. of IEEE-IECEJ-ASJ Int. Conf. on Acoustics, Speech and Signal Processing*, ICASSP-86, Tokyo, Japan, pp. 56.3.1-56.3.4, April 1986.

[10] H. Tseng, M. J. Sabin and E. A. Lee, "Fuzzy vector quantization applied to hidden Markov modelling," *Proc. of IEEE Int. Conf. on Acoustics, Speech and Signal Processing*, ICASSP-87, Dallas, Texas, pp. 15.5.1-15.5.4, April 1987.

[11] C. F. Stromeyer III and Bela Julesz, "Spatial-Frequency Masking in Vision: Critical Bands and Spread of Marking," *Journal of the Opt. Soc. of America*, Vol. 62, No. 10, pp. 1221-1232, October 1972.

[12] C. F. Hall and E. L. Hall, "A nonlinear model for the spatial characteristics of the human visual system," *IEEE Trans. on Systems, Man and Cybernetics*, Vol. SMC-7, No. 3, pp. 161-170, March 1977.

[13] F. W. Campbell, "Visual acuity via linear analysis," in *Proc. of the Symposium on Information Processing in Sight Sensory Systems*, Pasadena, CA, Nov. 1965.

[14] M. Davidson, "Perturbation approach to spatial brightness interaction in human vision," *Journal of Opt. Soc. Amer.*, Vol. 58, pp. 1300-1309, 1968.

[15] Tom N. Cornsweet, "Visual Perception," Academic Press, New York, 1970. Ch. XI to XIV.

[16] D. J. Granrath, "The Role of Human Visual Models in Image Processing," *Proc. IEEE*, Vol. 69, No. 5, pp. 552-561, May 1981.

Table 1

Coding results for the test image 'WOMANHAT'. The conforming blocks are coded with the finite-state VQ. $I_d = 8$. The nonconforming blocks are divided into three categories: a) blocks that can be replaced by a constant value of zero, b) blocks that can be replaced by their mean value and c) blocks that need transmission of both the mean and the deviations from the mean value. The categories are made using the mean value y_m and the maximum deviation of the mean value y_r as shown in the table below:

a)

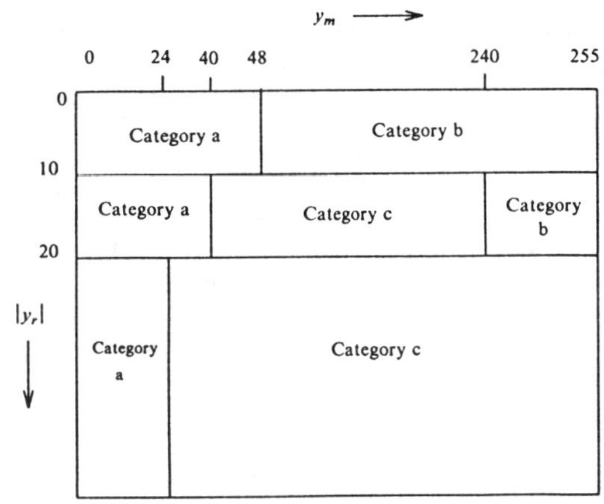

b)

$I_{threshold}$	Percentage of Nonconforming Blocks With Only Mean Value Transmission	Percentage of Nonconforming Blocks P_{nc}	Rate, r bits/pixel	SNR dB
2	21.2%	53.8%	1.688	28.941
4	11.4%	40.8%	1.536	29.426
8	7.8%	34.1%	1.436	29.755
16	6.6%	31.3%	1.407	30.137
32	5.8%	29.0%	1.381	30.313
64	4.5%	24.4%	1.293	30.963
128	3.5%	16.9%	1.064	30.447
256	–	–	0.562	30.542

Transparent Quality

Figure 1

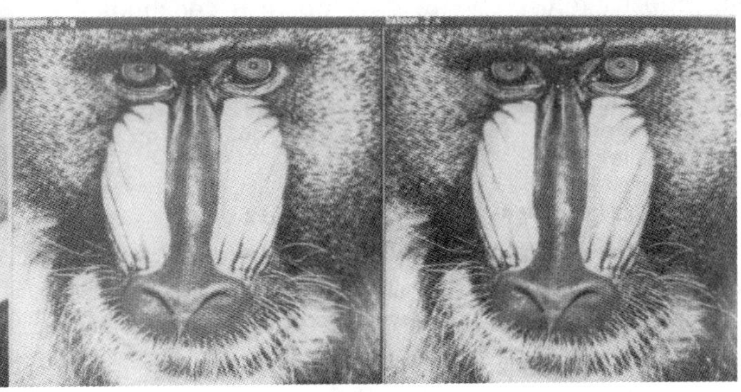

Figure 2

A New Vector Quantization Clustering Algorithm

WILLIAM H. EQUITZ, MEMBER, IEEE

Abstract—The Pairwise Nearest Neighbor (PNN) algorithm is presented as an alternative to the Linde–Buzo–Gray (generalized Lloyd) algorithm for vector quantization clustering. The PNN algorithm derives a vector quantization codebook in a diminishingly small fraction of the time previously required, without sacrificing performance. In addition, the time needed to generate a codebook grows only like $O(N \log N)$ in training set size, and is independent of the number of code words desired. Using this new method, one can either minimize the number of code words needed subject to a maximum allowable distortion or minimize the distortion subject to a maximum rate. The PNN algorithm can be used with squared error and weighted squared error distortion measures. Simulations on a variety of images encoded at 1/2 bit per pixel indicate that PNN codebooks can be developed in roughly 5 percent of the time required by the LBG algorithm.

I. Vector Quantization

VECTOR quantization [1], [2] is a process in which data to be encoded are broken into small "blocks," or vectors, which are then sequentially encoded vector by vector. The idea is to identify a set, or "codebook," of possible vectors which are representative of the information to be encoded. The vector quantization encoder pairs up each source vector with the closest matching vector from the codebook, thus "quantizing" it. The actual encoding is then simply a process of sequentially listing the identity of the code words which were deemed to most closely match the vectors making up the original data. The decoder has a codebook identical to the encoder, and decoding is a trivial matter of piecing together the vectors whose identity has been specified. The key to this method, of course, is to have a good codebook of representative vectors, typical of the data to be sent. To date, the method used almost exclusively for developing a codebook has been the algorithm known as the Linde–Buzo–Gray (LBG) algorithm [3]. This algorithm is also sometimes referred to as the generalized Lloyd algorithm (GLA), since it is a vector generalization of a clustering algorithm due to Lloyd [4].

A. The Generalized Lloyd, or "LBG," Algorithm

The LBG algorithm for deriving a codebook based on a set of training vectors is iterative, and can be described as follows. The initialization step involves choosing the starting codebook of vectors. This could be a codebook used previously, or something arbitrary, such as evenly spaced points in the vector space. The iteration begins by assigning each training vector to its "best fit" code word, based on some distortion measure and an exhaustive search. Next, given the set of vectors assigned to a particular code, that code is modified to minimize its error relative to the training vectors currently assigned to it. This two step process continues iteratively. The process is terminated when the overall error between the training vectors and the codes they are assigned to changes by a small enough fraction between one iteration and the next. The codebook is then considered to be determined and, given arbitrary error tolerances, this codebook reduces the coding error to at least a local minimum. The execution time of this algorithm is uncertain because the required number of iterations cannot be predicted ahead of time. Experience indicates that execution time grows quickly as the training set gets larger, as the number of code words increases, and as the vector dimension increases.

In this algorithm, by far the most complicated task is to come up with an acceptable initial codebook. Since the LBG algorithm can only find *local* minima, it is important that one start out in the general neighborhood of the correct solution, lest one become stranded at a local minimum distant from the global minimum. As stated previously, one possible way to initialize the LBG algorithm is to use a codebook previously developed for some other purpose. Alternatively, one might initialize with evenly spaced code words in the vector space or use a so-called "splitting" technique as described in [1]. However, perhaps the best simple initialization is to randomly choose a sampling from the training set for use as the initial codes. In practice, the "random initialization" is typically implemented by choosing evenly spaced elements in the training sequence (e.g., $\vec{x}$, $\vec{x}_{k+1}$, $\vec{x}_{2k+1}$, $\cdots$, $\vec{x}_{(C-1)k+1}$ where $\vec{x}_i$ is the ith training vector, N = the number of training vectors, C = the number of code words desired, and $k = N/C$).

B. Other Clustering Algorithms

Other authors have mentioned that one could develop vector quantization codebooks based on pattern recognition "clustering" techniques, and at least one has been implemented [5]. However, the efficacy of these techniques has not been demonstrated, and these techniques often suffer from the defect that they cannot specify ahead of time how many clusters will result. Additionally, many typical clustering algorithms are no less complicated than the LBG algorithm and suffer from the defect that they

Manuscript received April 15, 1987; revised December 17, 1988. This paper is based in part on work supported by a National Science Foundation Graduate Fellowship, and was presented in part at 1987 ICASSP.

The author is with the IBM Almaden Research Center, 650 Harry Rd., Department K52/802, San Jose, CA 95120-6099.

IEEE Log Number 8929999.

Reprinted from *IEEE Trans. Acoust., Speech, Signal Processing*, vol. 37, pp. 1568–1575, Oct. 1989.

cluster points with the aim of classifying the points rather than minimizing reconstruction error.

II. Pairwise Nearest Neighbor Clustering

In this section, a new algorithm, the Pairwise Nearest Neighbor (PNN) algorithm, is presented as a substitute for the LBG algorithm. This new algorithm significantly reduces needed computation without sacrificing performance. This algorithm can also be considered as an initializer for the LBG algorithm, providing better performance than either algorithm can achieve separately. The PNN algorithm is designed for use with squared error and weighted squared error distortion measures. The full search version of this algorithm will first be presented, and then an efficient approximation will be described. A preliminary version of this work first appeared in [6].

A. Full Search Pairwise Nearest Neighbor Clustering

The process of generating vector quantization code words from a training set is equivalent to the process of grouping the training set into "clusters," where each cluster is to be represented by a single code word. The Pairwise Nearest Neighbor (PNN) algorithm begins with a separate cluster for each vector in the training set and merges together two clusters at a time until the desired codebook size is achieved. At the start of the clustering process, one converts N clusters, each containing one vector, into the optimal $(N - 1)$ cluster codebook by merging together into a single cluster the two closest training vectors. The code word for this new cluster is chosen to minimize the error incurred by replacing these two vectors with a single code word. In other words, it is the centroid of the two vectors now in the new cluster.

For instance, in Fig. 1 we start with six training vectors of two-dimensional data. We consider each training vector to be a separate cluster. The two components of each vector are represented as x and y coordinates on the graph and each cluster centroid is represented in the diagram with an X. Before any merges, each cluster centroid has the number "1" next to it, signifying that it is made up of just one training vector. After one merging step there are five clusters, with the two closest cluster centroids deleted and replaced by a single cluster centroid half way between the two deleted X's. This new centroid has a "2" next to it, signifying that it now represents two training vectors.

Unfortunately, once clusters have more than one member, things become more complicated. However, given K clusters, we can always optimally move to $K - 1$ by merging the two clusters which result in the best tradeoff between merging close clusters and affecting few training vectors. If the members of a cluster can be approximated by their centroid, then this step-by-step optimality will lead us to a good overall clustering.

Consider the example in Fig. 2. This diagram represents a typical merge in the PNN algorithm. We start with five clusters and merge together the clusters consisting of

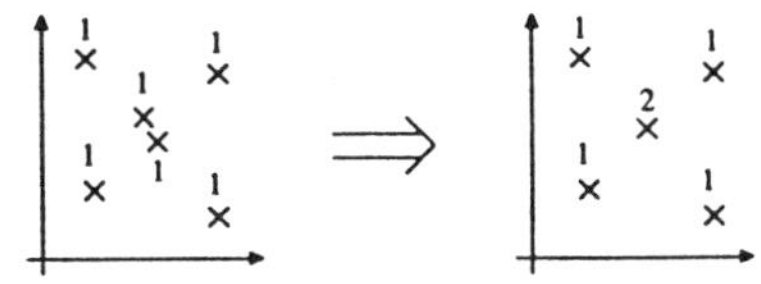

Fig. 1. First merge in PNN algorithm.

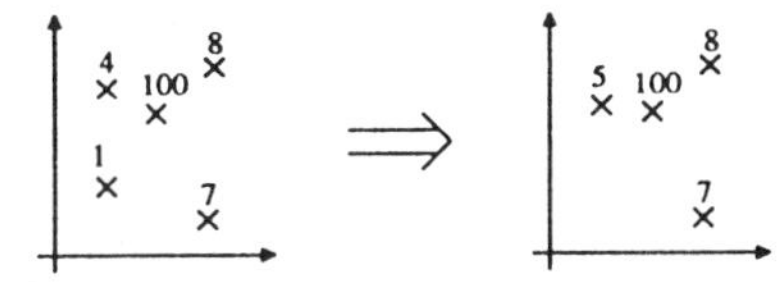

Fig. 2. Typical merge in PNN algorithm.

four and one training vectors, respectively. We merge together these two clusters rather than the clusters with four and one-hundred training vectors, which appear "closer," because the larger error introduced to each training vector in our chosen two clusters is outweighed by the fact that our choice of clusters affects many fewer training vectors.

The clustering process is halted when we are satisfied with our clusters, and then the centroid of each cluster is used as a code word. Stopping criteria are described in more detail below. Of course, it is *not* true that the "optimal" size C codebook will necessarily be achievable by sequentially developing the optimal codebooks of decreasing size, but this is the approximation on which the algorithm is built.

The pair of clusters which will introduce the least error when merged can be calculated as follows. If our distortion measure is squared error (the calculations are similar for weighted squared error), and we use the following notation:

C_i = ith cluster of training vectors
C_{ij} = cluster formed by merging ith and jth clusters
n_i = number of training vectors in C_i
n_{ij} = number of training vectors in C_{ij}
$\bar{x}_i$ = centroid (mean) of the training vectors in C_i
$\bar{x}_{ij}$ = centroid (mean) of the training vectors in C_{ij}
S_i^2 = average squared error between $\bar{x}_i$ and the training vectors in C_i
S_{ij}^2 = average squared error between $\bar{x}_{ij}$ and the training vectors in C_{ij}
$\langle x, y \rangle$ = inner product of x and y,

then

$$n_{ij} = n_i + n_j \tag{1}$$

$$\bar{x}_{ij} = \frac{n_i\bar{x}_i + n_j\bar{x}_j}{n_i + n_j} \tag{2}$$

$$n_{ij}S_{ij}^2 = \sum_{x \in C_{ij}} |x - \bar{x}_{ij}|^2 \tag{3}$$

$$= \sum_{x \in C_i} |x - \bar{x}_{ij}|^2 + \sum_{x \in C_j} |x - \bar{x}_{ij}|^2, \tag{4}$$

where

$$\sum_{x \in C_i} |x - \bar{x}_{ij}|^2 = \sum_{x \in C_i} (|x|^2 - 2\langle x, \bar{x}_{ij}\rangle + |\bar{x}_{ij}|^2) \quad (5)$$

$$= n_i(S_i^2 + |\bar{x}_i|^2)$$

$$- 2n_i\langle \bar{x}_i, \bar{x}_{ij}\rangle + n_i|\bar{x}_{ij}|^2 \quad (6)$$

$$= n_i S_i^2 + n_i|\bar{x}_i - \bar{x}_{ij}|^2 \quad (7)$$

$$= n_i S_i^2 + n_i\left|\frac{n_i\bar{x}_i + n_j\bar{x}_i - n_i\bar{x}_i - n_j\bar{x}_j}{n_i + n_j}\right|^2 \quad (8)$$

$$= n_i S_i^2 + n_i\left|\frac{n_j\bar{x}_i - n_j\bar{x}_j}{n_i + n_j}\right|^2 \quad (9)$$

$$= n_i S_i^2 + \frac{n_i n_j^2}{(n_i + n_j)^2} \cdot |\bar{x}_i - \bar{x}_j|^2. \quad (10)$$

Consequently,

$$n_{ij}S_{ij}^2 = n_i S_i^2 + n_j S_j^2 + \frac{n_i n_j^2}{(n_i + n_j)^2}|\bar{x}_i - \bar{x}_j|^2$$

$$+ \frac{n_i^2 n_j}{(n_i + n_j)^2}|\bar{x}_j - \bar{x}_i|^2 \quad (11)$$

$$= n_i S_i^2 + n_j S_j^2 + \frac{n_i n_j(n_i + n_j)}{(n_i + n_j)^2}|\bar{x}_i - \bar{x}_j|^2 \quad (12)$$

$$= n_i S_i^2 + n_j S_j^2 + \frac{n_i n_j}{n_i + n_j}|\bar{x}_i - \bar{x}_j|^2. \quad (13)$$

We interpret the last term in (13) as the squared error introduced by merging clusters C_i and C_j, and the idea is to choose the clusters C_i and C_j which minimize this quantity. Notice that the only statistics one need keep track of for each cluster are $\bar{x}_i$ and n_i. In fact, C_i can be considered to be a vector of "weight" n_i located at the centroid of the cluster $\bar{x}_i$. In this way, the distortion introduced by merging two clusters can be considered a "weighted distance" between the two centroids.

If one is interested in tracking the error introduced as the clustering proceeds, one might also keep track of S_i^2 for each cluster. However, one should recognize that this value is just an upper bound on the distortion the real encoder will introduce. This is because the calculation above assumes that all the training vectors in a cluster are closer to the centroid of their own cluster than to the centroid of a different cluster. In practice, this may not be the case, and the training vectors may thus be encoded with strictly less distortion. Using this method allows one to terminate the clustering process when a certain distortion relative to the training set is reached, rather than when a certain number of clusters are obtained.

We can now see that there are two possible termination criteria. First, we may choose to terminate when we have reduced our training set to a predetermined number of clusters. Alternately, we may choose to continue merging clusters as long as the average error introduced by representing the training data by the cluster centroids stays below some predetermined threshold. Since the number of clusters determines the coding rate, we see that these termination criteria correspond respectively to minimizing distortion, subject to a rate constraint, and to minimizing rate, subject to a distortion constraint.

B. Fast Search PNN Clustering

The PNN codebook development algorithm is a matter of progressively merging together pairs of clusters with minimal weighted distance between their centroids. The key to quick execution of this algorithm is quickly finding the closest pairs of centroids among an essentially randomly distributed set. The obvious way is to explicitly find each point's nearest neighbor, but this is very slow (a log N search at best) and leads to closest pair computation cost on the order of N.log N for each merge. However, it is possible to efficiently find good, if suboptimal, pairs of clusters to merge as described below. We do not care if we merge the absolute closest pair of clusters at each step, as long as close clusters get merged eventually. We call this approximated PNN algorithm the "fast" PNN algorithm.

The way we accomplish the computational saving is to search for a vector's near neighbors only within a small region, with neighborhoods being defined by a *k-d* tree partitioning of the multidimensional space. In the past [6]–[8], *k-d* trees have been used in vector quantization coding as a means of performing the nearest neighbor searches required by the LBG algorithm, but in our case we will merely be using the partition of *k*-dimensional space induced by organizing the training vectors into a *k-d* tree structure.

1) K-d Trees:

Structure: K-d trees (short for *k*-dimensional trees) were developed by Bentley [9], [10] and provide a data structure which allows for log(N) multidimensional "nearest-neighbor" searches to be accomplished. *K-d* trees (see Fig. 3) consist of a set of interconnected nodes and a set of terminal nodes, or "buckets," located at the lowest level of the tree. The nodes serve to organize the data, the buckets hold the data. Similar to binary search trees, each node in a *k-d* tree partitions data into two sets based on some scalar threshold, but unlike simple binary trees, Bentley's *k-d* trees partition *vector* data at each node by performing a threshold test on a single coordinate of each vector. The coordinate being tested is the same for all vectors being "partitioned" at a given node, but can be different at each node of the *k-d* tree.

A node in a *k-d* tree is defined by four elements. The first element is the index i of the coordinate used for partitioning the data at this node. The second element is the threshold t for partitioning data. Any vector in the subtree "below" this node whose ith coordinate has a value less than t is located in the left branch of this subtree and,

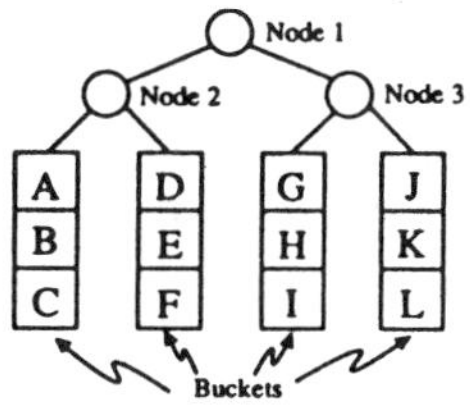

Fig. 3. Example of a *K-d* tree.

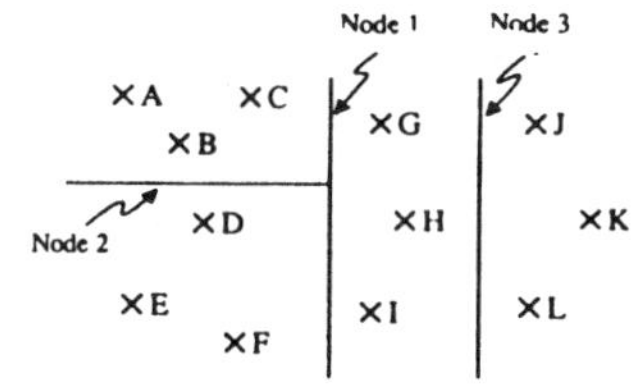

Fig. 4. *K-d* partition corresponding to example.

conversely, any vector whose ith coordinate is greater than t is located in the right branch. It is a matter of convention in which branch a vector belongs if its ith coordinate happens to equal t. The last two elements of a node are pointers to the left and right branches, or children. These children are *k-d* trees in their own right, organizing their "half" of the data.

Buckets are special nodes which point to, or "contain," data rather than other nodes. In a completely organized tree, each bucket would contain only one vector, but usually it is more efficient to have a bucket contain a small number of vectors which are all similar. This way fewer nodes (and levels in the tree) are needed. The vectors in a given bucket can be stored in a finite array if it is guaranteed that there will never be more than a certain number of vectors per bucket or they can be stored as a linked list.

One can consider each particular *k-d* tree to be a partitioning of k-dimensional with each node corresponding to a hyper-plane parallel to all but one coordinate axis. For example, consider the two-dimensional case in Fig. 4 which corresponds to the tree in Fig. 3. Here there are 12 vectors, represented by the letters A–L. The top node acts as the first partition, dividing the two-dimensional region into two half-planes. The next level of the tree divides these half-planes once again, and so on. The lines (hyperplanes in general) dividing up the "space" correspond to nodes, and the regions defined by the nodes correspond to buckets. Each partition could, in theory, be done with respect to any coordinate, but to get the most effective partitions one would normally try to use different coordinates at different nodes.

Building a K-d Tree: K-d trees, like binary trees, are constructed recursively from the top down, and are designed specifically to partition a particular set of data. The first parameter one fixes is the number of vectors one will allow in each bucket. Then, if one wants to build a tree with this many or fewer vectors, they are all assigned to a single bucket. If there are more vectors than will fit into a single bucket, a single coordinate i and a threshold t are chosen and the set of vectors is divided into two halves based on the value of their ith coordinate. As described above, if a vector's ith coordinate is less than t it is assigned to the left child, and if it is greater than t it goes to the right child. If the ith coordinate *equals* t it goes to one child or the other based on convention. Each of these two sets of vectors are then recursively formed into a new *k-d* tree, and so on, until all the vectors are assigned into buckets.

At each node, one is faced with the difficult question of which coordinate to use for partitioning the data. One simple approach is to choose cyclically among the coordinates. On the top level of the tree partition on the basis of the first coordinate; on the second level, use the second, etc. When all the coordinates have been used once, start over again with the first one. A better method is to choose for each node the coordinate which best "spreads out" the data. That is to say, one might choose to partition the data on the basis of the coordinate which has the largest variance associated with it. The reason for choosing this as the "split coordinate" is that if the data are very spread out along a particular dimension, then presumably differences in that coordinate are more "significant" in some sense than differences in another, more densely grouped coordinate. Choosing in this second manner has the effect of splitting the data on the basis of uncorrelated coordinates.

To achieve a maximally balanced tree, one should use the median coordinate value from among the vectors involved as the split threshold. Friedman *et al.* [10] report that a bucket size averaging around eight entries seems to be optimal for a wide range of problems.

2) Using K-d Trees to Perform Fast PNN Clustering: We use *k-d* trees in the following way (see Fig. 5). When looking for pairs of clusters to merge, we only consider pairs where both cluster centroids are assigned to the same region of k-dimensional space as defined by the buckets of a *k-d* tree.

We start by organizing the training set in a *k-d* tree. Then we repeatedly perform the following steps until we are satisfied with our clustering. As usual, we consider the training vectors to be the first set of cluster centroids. After the tree is created, candidate pairs for merging are generated by doing local comparisons within each *k-d* bucket. Within each bucket, the pair of clusters which will introduce the least distortion when merged is called the bucket's "candidate." Next, a fixed fraction of these candidate pairs (such as 50 percent) are merged based on the distortion their merge would introduce. The candidate pair which would introduce the least distortion is always merged, and then the pair which would introduce the second least amount, and so on, until the desired fraction of candidates have been merged. The reason for merging only a fraction of the candidate pairs at each pass is that some partitions (buckets) will not have *any* close pairs. At this point, we can stop if we want, either because we have the correct number of clusters or because we have reached the maximum desired distortion between the

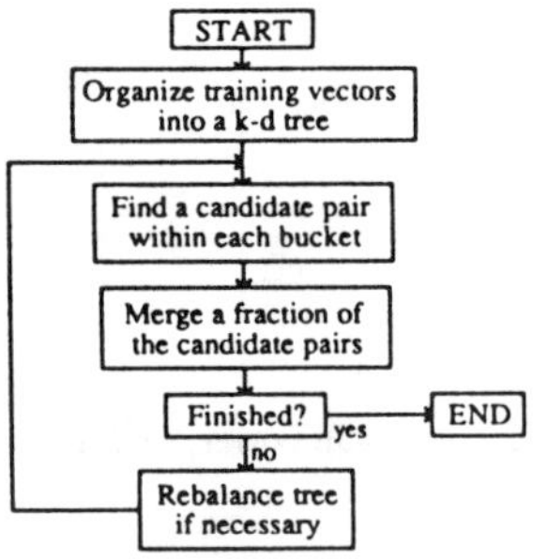

Fig. 5. The PNN algorithm.

training set and our cluster centroids. If we decide to continue, the *k-d* tree is then rebalanced to account for the loss of the merged cluster centroids and the addition of the results of these merges. Buckets which now contain too many cluster centroids are split, and buckets which now contain too few cluster centroids are combined with neighboring buckets. This tree readjustment has the effect of keeping the number of vectors in each bucket roughly constant. With this newly adjusted tree, new candidate pairs are generated and the process continues.

3) Complexity of the Fast PNN Algorithm: Bentley shows that the amount of computation needed to build the *k-d* tree is $O(kN \log N)$ where there are N k-dimensional vectors to be organized, so if T equals the number of vectors in the training set, the complexity of the first step in the fast PNN algorithm is $O(T \log T)$. This computation is performed once.

If there are initially T clusters (one for each training vector) and if at the end there are C clusters, there must be a total of $(T - C)$ merges performed. Since T is typically much larger than C, we lose little by upper bonding the number of merges performed by T. The cost of an individual merge is the cost of deleting two vectors from the tree and adding one. This is of complexity $O(\log T)$ so the complexity of all the merges combined is $O(T \log T)$.

At each pass in the fast PNN algorithm, a fixed fraction of the candidate pairs of cluster centroids are actually merged, so the total number of buckets searched during the entire clustering process is $(T - C)/\gamma$, where γ is the fraction of candidates merged at any given step. Since buckets are kept at a small constant size, searching within a bucket incurs a constant computational cost, and the overall complexity attributable to searching for candidates is $O(T)$. In addition, the operation of determining which candidates to merge is linear in the number of candidates produced at any stage, since it involves only finding a percentile (median, for example) and comparing all candidates to this percentile. Since finding a percentile is linear is the size of the data set, at any stage the complexity of determining which candidates to merge is proportional to the number of buckets searched. As above, the total number of buckets searched for a candidate is $O(T)$, so the overall cost of generating candidate pairs is $O(T)$.

Consequently, the complexity of the PNN clustering process is $O(T \log T)$ in the size of the training set and is essentially independent of the size of codebook generated (even though larger codebooks take slightly less time than small ones).

III. Application to Image Coding

In this section we describe an application of the PNN algorithm to vector quantization coding of still images. We chose image coding because we felt this was the application in which the computational difficulties were most severe. However, in theory the PNN algorithm could serve as an effective substitute for the LBG algorithm in any vector quantization application.

A. Vector Quantization Picture Coding

We used the typical [11]–[14] approach of coding digitized images by dividing them into blocks of pixels and then using these blocks as vectors. Our blocks were 4 × 4 pixels in size and the images we used were 512 × 512 pixels. Our pictures were digitized to 8 bits per pixel (256 possible gray levels) and we chose to generate a codebook with 256 codewords, resulting in a coding rate of 1/2 bit per pixel.

While organizing the training vectors in *k-d* trees, the coordinate used to partition the data at each node in the tree was the coordinate with the greatest variance. The split threshold was the median value of this coordinate. When a child was given eight or fewer vectors they were placed in a bucket.

All tests were performed on either a Vax 11/780 or a Vax 11/750 with programs written in the C programming language. Execution time was measured in Berkeley UNIX accounting units. These "units" are measured in seconds and are similar to CPU seconds, except that they are adjusted for system load. All performance tests shown in a single table were run on the same computer. All tests involving the LBG algorithm were run until there was less than a 0.1 percent change in the distortion introduced by representing the training set with the codebook being developed. Most tests were run using vectors from a single image as the training set, although larger training sets were also tested with similar results.

For LBG coding, a "trick" which increased performance significantly was to initialize not with evenly spaced vectors from the training set beginning with the first vector, but rather with evenly spaced vectors from the training set offset by a small number. The reason for the performance enhancement was that the first training vector corresponded to the block of pixels in the top left corner of the image, and evenly spaced vectors beginning with this vector included a large fraction of vectors lying along the left boundary of the picture. This was a problem because digitized images often have artifacts along the edges corresponding to windowing or other effects, and initializing the LBG algorithm with unrepresentative vectors degraded its performance significantly. We also always ran the LBG algorithm as enhanced by the use of *k-d* tree nearest neighbor searching. This alone reduced computation by a factor of two as reported in [6] and [7]. We also used other computational enhancements such as those cited in [15] and [16].

B. Full Search PNN versus "Fast" PNN

The "full search" PNN algorithm of Section II-A is a slow and costly process. Since it requires an unacceptable amount of computation, the pertinent question is how much the "fast" PNN algorithm degrades performance as compared to the full search algorithm. It was found that the fast PNN algorithm increased coding error (squared error in this case) by approximately 0.4–0.6 dB for single image encoding. This figure increased as the overall coding error increased.

C. Execution Time of PNN versus LBG

The fast implementation of the PNN algorithm was compared to the LBG algorithm (enhanced to use *k-d* tree searching) and was found in this application to require less than 5 percent of the amount of time needed by LBG algorithm (see Table I). Recall that for these examples, the training set is a single picture ($T = 16{,}384$ for 512×512 pictures by 4×4 codewords), although in practice the training set might be composed of several pictures which belong to a certain "class of pictures."

D. Numerical PNN Quality versus LBG Quality

The quality of images generated by any VQ design algorithm is of great importance because even an extremely fast algorithm is useless if it produces bad pictures. What is desired is a fast alternative with performance equivalent to LBG performance, and as Table II shows, PNN distortion is comparable in all cases. It should be noted that the codebooks generated by the PNN algorithm are suboptimal since the LBG algorithm can always improve on them by running a few iterations. On the other hand, the LBG algorithm is not necessarily globally optimal. A typical picture is shown in Fig. 6 along with the images produced by each algorithm blown up to show detail (see Figs. 6–8). These images were coded at 1/2 bit per pixel.

TABLE I
EXECUTION TIME OF LBG VERSUS PNN

Execution time (in UNIX accounting units)			
	Codebook Development Algorithm		
Picture	LBG with random initialization		fast PNN
	iterations	time (secs)	time (secs)
baboon	17	10020	361
lake	25	8443	363
airport	27	16089	373
lena	25	6930	366
peppers	31	8602	364
plane	33	9577	400

TABLE II
CODED PICTURE ERROR OF LBG VERSUS PNN

Total Squared Pixel Error for Decoded Pictures ($\times 10^7$)			
	Codebook Development Algorithm		
Picture	LBG initialized with random training vectors	fast PNN	LBG initialized with fast PNN codes
baboon	6.55	7.21	6.54
lake	2.60	2.65	2.38
airport	1.75	1.54	1.38
lena	1.17	1.26	1.11
peppers	1.27	1.33	1.20
plane	1.42	1.44	1.25

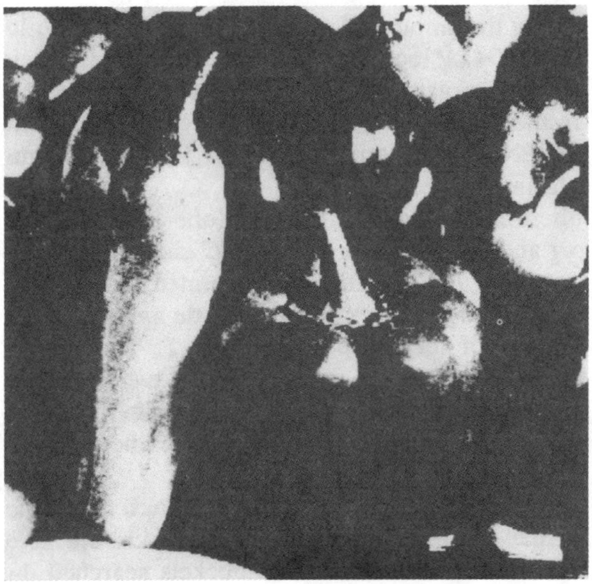

Fig. 6. "Peppers" original.

E. PNN as LBG Initializer

As can be seen by Table II, when the output from the PNN clustering algorithm was used as the initializer for the LBG algorithm, total coding error was lower in all cases than for the LBG codebook with random initialization. In addition, with the PNN initializer the LBG algorithm always converged in fewer iterations (often half as many), so the computation time overall was lower also, as one or two iterations alone take more time than the entire execution of the fast PNN algorithm. Consequently, it is clear that even if the PNN algorithm were to be considered unacceptable because it usually does not generate an "optimal" codebook, it appears to be an excellent alternative to random initialization for the LBG algorithm, and used this way results in excellent performance as well as computational savings.

F. Performance Outside Training Set

Tests were performed to determine if the PNN algorithm would continue to perform comparably with the LBG algorithm when coding pictures outside the training set. In these cases, picture reconstruction was understandably worse, but picture quality with PNN codebooks remained comparable to that achieved using LBG codebooks. The computational advantage of the PNN

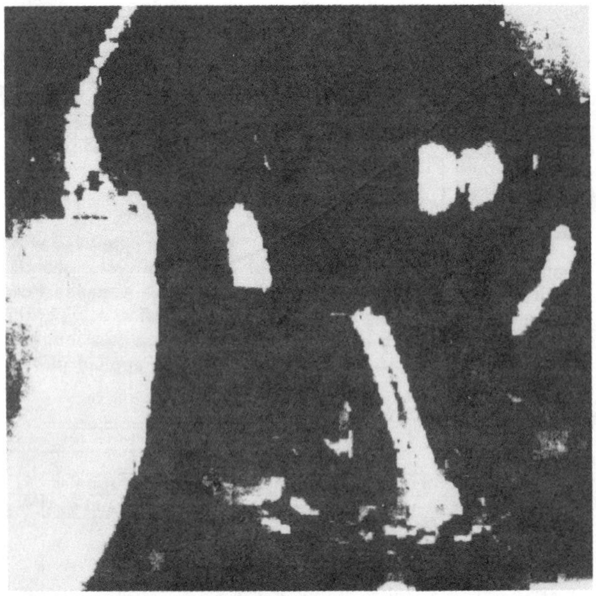

Fig. 7. Blowup (250 × 250 pixels) of "Peppers" coded with LBG codebook.

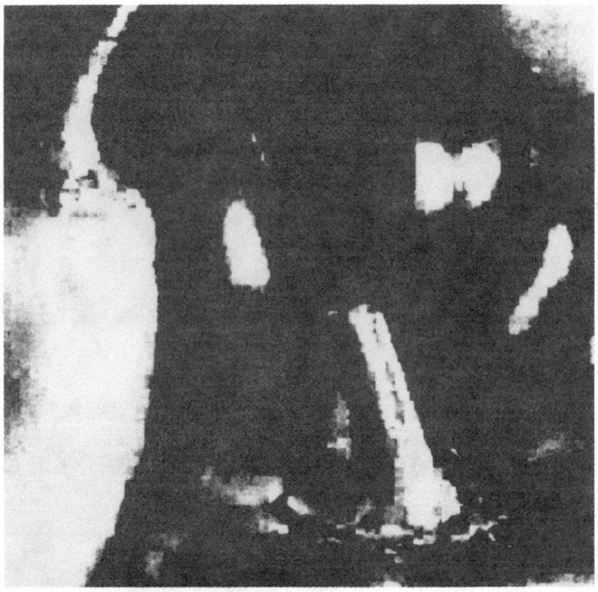

Fig. 8. Blowup (250 × 250 pixels) of "Peppers" coded with fast PNN codebook.

algorithm over the LBG algorithm became more pronounced as the size of the training set increased.

IV. Conclusions

The Pairwise Nearest Neighbor (PNN) algorithm was presented as a noniterative way to generate vector quantization codebooks comparable to those generated by the LBG algorithm. This algorithm is useful for applications involving squared error and weighted squared error distortion measures. The PNN algorithm's main "feature" is that it develops codebooks in a diminishingly small fraction of the time previously required. The PNN algorithm is not sensitive to initializations and is guaranteed to terminate in a finite amount of time. The PNN algorithm is shown to take slightly *less* time the larger the codebook desired and allows one the option of either minimizing coding distortion subject to a rate constraint, or minimizing rate subject to a distortion constraint. In addition, since practical implementations of the PNN algorithm use local nearest neighbor searching, processing can be done in parallel to take advantage of the added speed that extra hardware can bring.

When applied to image coding with small training sets, the time required to execute the PNN algorithm is demonstrated to be just 5 percent of that typically required by the LBG algorithm. Reconstructed pictures generated with this new algorithm were shown to be as good as those generated with the standard algorithm. It is interesting to note that the PNN algorithm makes it computationally feasible to develop a different codebook for each image. Sending the codebook to the receiver before the encoded image would take an additional 1/8 bit per pixel with parameters as stated.

It was also shown that using PNN code words as an LBG initialization for image coding results in much better codebooks (with fewer iterations) than does "random initialization." This provides for a way to generate better codebooks and indicates that the LBG algorithm using random training vectors as an initializer typically converges to suboptimal codebooks, a fact often not given adequate attention. It is the opinion of the author, however, that the main usefulness of the PNN algorithm is as a fast alternative to the LBG algorithm which allows vector quantization to be used in situations where it had previously been computationally prohibitive, such as in repetitive experimental work, or in situations with large training sets or codebooks.

References

[1] R. M. Gray, "Vector quantization," *IEEE ASSP Magazine*, pp. 4–29, Apr. 1984.

[2] A. Gersho, "On the structure of vector quantizers," *IEEE Trans. Inform. Theory*, vol. IT-28, pp. 157–166, Mar. 1982.

[3] Y. Linde, A. Buzo, and R. M. Gray, "An algorithm for vector quantizer design," *IEEE Trans. Commun.*, vol. COM-28, pp. 84–95, Jan. 1980.

[4] S. P. Lloyd, "Least squares quantization in pcm," *IEEE Trans. Inform. Theory*, vol. IT-28, pp. 129–137, Mar. 1982 (reprint of 1957 paper).

[5] L. Miclet and M. Dabouz, "Low bit rate transmission of speech by vector quantization of the spectrum," Dep. Commun. Ecole Nationale Superieure des Telecommunications, Paris, 1985, unpublished.

[6] W. H. Equitz, "Fast algorithms for vector quantization picture coding," Master's thesis, Mass. Inst. Technol., June 1984.

[7] ——, "Fast algorithms for vector quantization picture coding," in *Proc. ICASSP*, Dallas, TX, Apr. 1987, pp. 18.1.1–18.1.4.

[8] A. Lowry, S. Hossain, and W. Millar, "Binary search trees for vector quantization," in *Proc. ICASSP*, Dallas, TX, Apr. 1987, pp. 51.8.1–51.8.4.

[9] J. L. Bentley, "Multidimensional binary search trees used for associative searching," *Commun. ACM*, vol. 18, no. 9, pp. 509–517, Sept. 1975.

[10] J. H. Friedman, J. L. Bentley, and R. A. Finkel, "An algorithm for finding best matches in logarithmic expected time," *ACM Trans. Math. Software*, vol. 3, no. 3, pp. 209-226, Sept. 1977.
[11] R. L. Baker and R. M. Gray, "Image compression using non-adaptive spatial vector quantization," in *Proc. 16th Asilomar Conf. Circuits, Syst., Comput.*, 1982, pp. 55-61.
[12] T. Murakami, K. Asai, and E. Yamazaki, "Vector quantizer of video signals," *Electron. Lett.*, vol. 18, no. 23, pp. 1005-1006, Nov. 1982.
[13] A. Gersho and B. Ramamurthi, "Image coding using vector quantization," in *Proc. ICASSP*, Paris, France, 1982, pp. 428-431.
[14] Y. Yamada, K. Fujita, and S. Tazaki, "Vector quantizer of video signals," in *Proc. Annu. Conf. IECE*, 1980, p. 1031.
[15] D.-Y. Cheng, A. Gersho, B. Ramamurthi, and Y. Shoham, "Fast search algorithms for vector quantization and pattern matching," in *Proc. ICASSP*, San Diego, CA, Mar. 1984, pp. 9.11.1-9.11.4.
[16] C.-D. Bei and R. M. Gray, "An improvement of the minimum distortion encoding algorithm for vector quantization," *IEEE Trans. Commun.*, vol. COM-33, pp. 1132-1133, Oct. 1985.

VECTOR QUANTIZATION OF COLOR IMAGES

Tokumichi Murakami   Kohtaro Asai   Atsushi Itoh

Mitsubishi Electric Corporation

Information Systems and Electronics Development Laboratory

Kamakura City, Kanagawa Pref., JAPAN

ABSTRACT

The new vector quantization methods are presented for high efficiency coding and transmission of color images. In order to encode three color components Y, C^1, C^2 in individual or combined form effectively, techniques are developed which resulted in vector quantizer using only three factors ; mean, deviation and normalized vector. Individual component VQ is studied expanding to dynamic multi-stage VQ for motion color picture transmission at lower bit rates. Another component VQ can encode high resolution color images sufficient at 1.0 bit/color pixel. The color video codec using dynamic multi-stage VQ can be presented which transmits color motion pictures at the rate of 4-8 frames per second at 64 kbit/s.

MEAN SEPARATED AND NORMALIZED VQ

Vector Quantization is a mapping of a k-dimensional input vector into one of the representative vectors in the signal space R^k. Expressing an input vector by $\underline{X}$, quantizing operation is expressed by the following function.

$$Q(\underline{X}) = \underline{y}_i$$

$$\text{if } d(\underline{X},\underline{y}_i) < d(\underline{X},\underline{y}_j) \text{ for all } j \neq i$$

$\underline{y}_i$ gives the optimum approximation of $\underline{X}$ among the set of output vectors $[\underline{y}]=[\ \underline{y}_1,\underline{y}_2,..,\underline{y}_n\ ]$ and $d(\cdot)$ denotes the distortion measure [1].

Direct mapping of a source input vector into an output one requires a large size codebook, a set of output vectors. Therefore, several applications of VQ to encode signals with a small codebook effectively have been studied. The mean separated and normalized VQ represents an input vector by mean, deviation and normalized vector [2].

Let a source input vector and normalized input vector be $\underline{S}=[\ s_1,s_2\ ..,s_k\ ]$ and $\underline{X}=[\ x_1,x_2,..,x_k\ ]$.

$$\mu = \Sigma s_j\ /\ k \qquad \text{(mean)}$$

$$\sigma = [\ \Sigma (s_j-\mu)^2\ /\ k\ ]^{1/2} \qquad \text{(deviation)}$$

$$x_j = (\ s_j-\mu\ )\ /\ \sigma \qquad \text{(normalized element)}$$

In the case of forming an input vector from a block of pixels, $\underline{X}$ corresponds to the information of rudimental waveform pattern. Then, the probability distribution $P(\underline{X})$ of normalized input vectors $\underline{X}$ is approximately similar for various images. So based on a long training sequence of these input vectors $\underline{X}$, the universal codebook of output vectors $[\underline{y}]$ can be generated using the clustering methods by Linde et al [3].

The reconstructed source vector is expressed as

$$\hat{\underline{S}} = \sigma\cdot\underline{y}_i + \mu\cdot\underline{U}$$

$$(\ \underline{U} = [\ 1,1,...,1\ ]\ )$$

In the case that the Euclid distance is used as the distortion measure, the above VQ method can be approximated using the inner product as below.

$$Q(\underline{S}) = \sigma^*\cdot\underline{y}_i + \mu\cdot\underline{U}$$

$$\text{if } (\underline{S}-\mu\cdot\underline{U},\underline{y}_i) > (\underline{S}-\mu\cdot\underline{U},\underline{y}_j) \text{ for all } j \neq i$$

$$\sigma^* = (\underline{S}-\mu\cdot\underline{U},\underline{y}_i)$$

$$= |\underline{S}-\mu\cdot\underline{U}|\ |\underline{y}_i|\ \cos\theta_i$$

$$= \sigma\cdot\cos\theta_i$$

$$\text{when } \sigma = |\underline{S}-\mu\cdot\underline{U}| \text{ and } |\underline{y}_i| = 1$$

Then, the corrected deviation σ^* is obtained from the inner product so that $Q(\underline{S})$ might be the optimum approximation of $\underline{S}$ [4].

Thus, an input vector is converted into the three factors ; mean, deviation and normalized vector and they are respectively encoded.

The deviation and the distortion measure can be calculated as follows to fasten and simplify the operation.

$$\sigma = \Sigma|s_j-\mu|/k$$

$$d(\underline{X},\underline{y}_i) = \Sigma|x_j-y_{ij}| \qquad \text{(Chebychef distance)}$$

μ and σ are able to be coded with any coding scheme for scalar data.

VECTOR QUANTIZATION FOR CCLOR STILL PICTURES

Vector Quantization for Color Images

When the mean separated and normalized VQ method is being applied to color images, how the previous three factors are defined must be investigated.

At first, luminance and two chrominance components are used in color image coding, since the chrominance components can be sub-sampled without degrading picture quality.

Here, two approaches are studied. One is the individual component VQ, which encodes each spec-

Reprinted from *Proc. IEEE Conf. Acoust., Speech, Signal Processing,* vol. 1, pp. 133–136, April 1986.

tral component through the same process as for monochrome pictures. The other is the combined component VQ, which encodes the input vector defined in the component-combined signal space.

By the former method, the codebook can be designed more easily than the latter. The latter can realize sufficient compression ratio without greatful sub-sampling of the chrominance components.

It seems to be necessary to classify these methods for the system requirements in picture quality or transmission rate.

Individual Component VQ

Let luminance and chrominance components be Y, C^1 and C^2. Supposing three blocks of 4X4 samples of Y, C^1 and C^2 at a 4:1:1 sample rate in horizontal and a 2:1:1 in vertical as shown in Fig.1, one block of Y occupies the same area as two pixels of C^1 and C^2.

Defining all input vectors to be uniform in the dimension offers good matching for hardware implementation. It enables to process all vectors in the time division multiplex. Each vector is respectively encoded after converting into the three factors.

The result of the computer simulation is shown in Table 1. The source pictures are (1) a girl with carnation, (2) seashore. Both pictures are consisted of 512X480 pixels with 8 bit per each component (R,G,B). C^1 and C^2 are given by B-Y and R-Y. These vectors are 16-dimensional as shown in Fig.1. The distortion measure is defined as Chebychef distance between input and output vectors. The number of output vectors is 512 for Y and 512 for C^1 and C^2. Both codebooks are generated

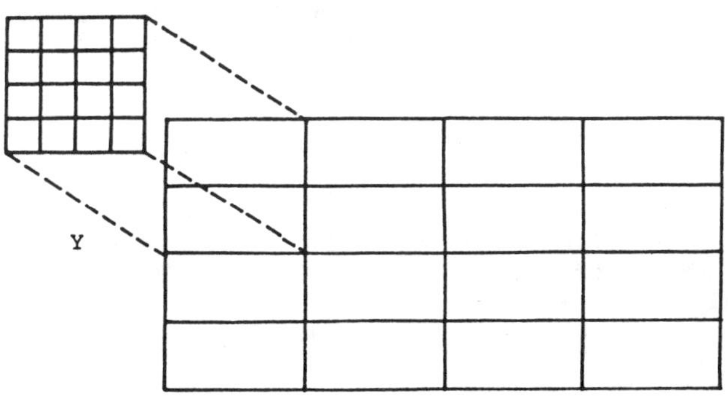

Fig.1 Blocking Structure for Individual Component VQ

	Sp-p/Nrms (dB)			Entropy (bit/color pixel)
	Y	C^1	C^2	
Pic.1	38.2	37.3	38.6	1.08
Pic.2	30.2	31.8	32.6	1.30

Table 1 Performance of Individual Component VQ

through the clustering of the input vectors derived from the above two pictures. The mean and deviation are DPCM-coded in TDM format with 33-level quantizer for mean and 17-level quantizer for deviation. Two dimensional prediction is adopted.

A good picture quality is kept. However, the sub-sampling pattern is seen at the sharp chrominance edges of the reconstructed pictures. These noises can be supressed through the low-pass filter.

This method can be easily expanding to the coding of motion video signals using DVQ which is described later.

Combined Component VQ

In this section, an input vector is defined in the signal space that both spatial and spectral components are combined. A block diagram of this coding scheme is shown in Fig.2 [5].

Consider a block of 4X4 Y sample and two blocks of 2X2 C^1, C^2 samples. These three blocks are positioned on the same spatial area as C^1 and C^2 are sub-sampled at 1/2 in both horizontal and vertical directions as shown in Fig.3.

Now, let a source input vector be $\underline{S}$ as follows.

$$\underline{S} = [\ \underline{Y}, \underline{C}^1, \underline{C}^2\ ]$$
$$= [\ Y_1, Y_2, .., Y_{16}, C^1{}_1, .., C^1{}_4, C^2{}_1, .., C^2{}_4\ ]$$

Removing each block mean from corresponding elements of the three blocks, the 24-dimensional input vector obtained is zero-mean. The amplitude deviation is defined as

$$\sigma = [\ \Sigma|Y_j-\mu_Y| + \Sigma|C^1{}_j-\mu_{C^1}| + \Sigma|C^2{}_j-\mu_{C^2}|\ ]/24.$$

μ_Y, μ_{C^1} and μ_{C^2} denote each block mean. And the normalized input vector $\underline{X}$ is

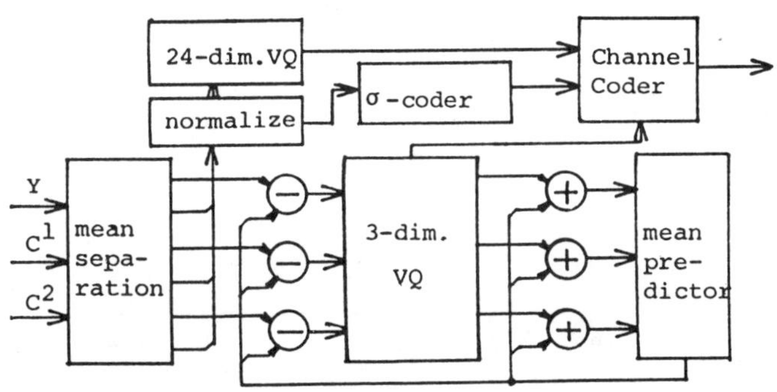

Fig.2 Block diagram of Combined Component VQ

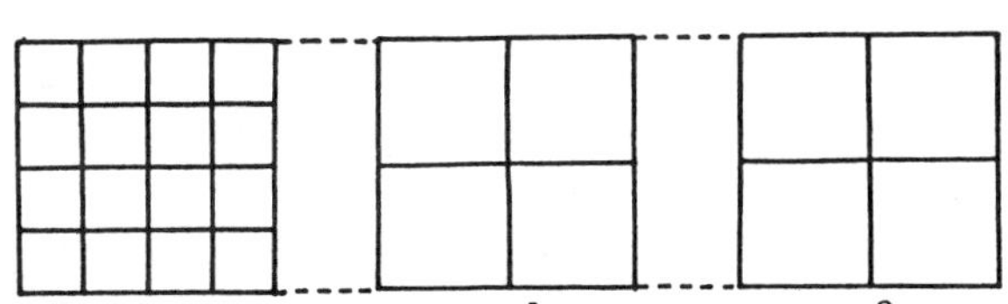

Fig.3 Blocking Structure for Combined Component VQ

	Sp-p/Nrms (dB)			Entropy (bit/color pixel)
	Y	C^1	C^2	
Pic.1	38.4	37.7	37.7	1.01
Pic.2	30.2	31.2	31.8	1.30

Table 2 Performance of Combined Component VQ

$$\underline{X} = [\ \underline{Y}-\mu_Y\cdot\underline{U}_Y, \underline{C}^1-\mu_{C^1}\cdot\underline{U}_{C^1}, \underline{C}^2-\mu_{C^2}\cdot\underline{U}_{C^2}\ ]\ /\ \sigma$$

$$\underline{U}_Y = [\ 1,1,1,\ldots,1,0,0,0,0,0,0,0,0\ ]$$

$$\underline{U}_{C^1} = [\ 0,0,0,\ldots,0,1,1,1,1,0,0,0,0\ ]$$

$$\underline{U}_{C^2} = [\ 0,0,0,\ldots,0,0,0,0,0,1,1,1,1\ ]$$

Further, regarding the three block means, it is possible to derive 3-dimensional combined vector. It gives a good solution that the predictive errors of three means are combined to design a universal codebook for 3-dimensional vector. Let the 3-dimensional predictive error vector be

$$\underline{\varepsilon} = [\ \varepsilon\mu_Y, \varepsilon\mu_{C^1}, \varepsilon\mu_{C^2}\ ].$$

Thus, a block of color pixels are compacted into three factors ; mean(3-dimensional vector), deviation and normalized vector(24-dimensional). Supposing that $\hat{X}$ is the mapped output vector and $\hat{\mu}_Y, \hat{\mu}_{C^1}, \hat{\mu}_{C^2}$ are the DPCM-decoded means, the reconstructed source vector $\hat{S}$ is obtained as follow.

$$\hat{\underline{S}} = \sigma\cdot\hat{\underline{X}} + \hat{\mu}_Y\cdot\underline{U}_Y + \hat{\mu}_{C^1}\cdot\underline{U}_{C^1} + \hat{\mu}_{C^2}\cdot\underline{U}_{C^2}$$

The result of the computer simulation is shown in Table 2. The number of output vectors is 512 for 24-dimensional vector and 256 for 3-dimensional vector. Both codebooks are designed through the clustering algorithm. It is necessary to design the 3-dimensional output vectors symmentrically in the signal space. The deviation is DPCM-coded with 33-level quantizer.Compared with the individual component VQ, the resolution of chrominance is improved at lower bit rates.

VECTOR QUANTIZATION FOR COLOR MOTION PICTURES

A new vector coding technique and its coding control method is described in this chapter. They are proposed for the purpose of a design for color video codec which enables to transmit color video signals at lower bit rates such as 64 kbit/s. In case of video signals, it is necessary to take into consideration of real-time operation, robust coding performance and less complexity. Therefore, we have presented an interframe adaptive vector quantization (AVQ) technique in the previous work.

Aiming at higher performance coding of video signals, a dynamic vector quantization (DVQ) technique is introduced and a feedback multi-stage connection of DVQ and AVQ, which is called " Dynamic Multi-stage VQ (DMVQ) ", is proposed [6]. Further, the coding control method for a very low-bit-rate transmission of color video signals are studied. The scheme of the above coding system is shown in Fig.4. In Fig.4, the input vector is $\underline{S}_l$, the output vector at the first stage is $\hat{\underline{S}}_l$, the residual differential vector between $\underline{S}_l$ and $\hat{\underline{S}}_l$ is $\underline{\varepsilon}_l$, the reconstructed differential vector at the subsequent stage is $\hat{\underline{\varepsilon}}_l$ and the reproduced vector is $\hat{\hat{\underline{S}}}_l$.

Pre-processing for Color Video Signals

The input color video signals are converted into luminance (Y) and two chrominance (C^1,C^2) signals. The sampling ratio of them is 4;1:1 in the horizontal direction and 2:1:1 in the vertical direction, respectively. The spatial resolution of Y signal is 352(H)X240(V) pixels per frame. The C^1, C^2 signals are time interleaved with the horizontal blanking period of the Y signal alternately for the purpose of time division operation block by block.

Dynamic Vector Quantizer

DVQ with a dynamic codebook at the first stage performs coarse quantization for Y signals block by block which are composed of 16X8 pixels. A dynamic output vector in the dynamic codebook is sequentialy generated from position-shifted blocks within

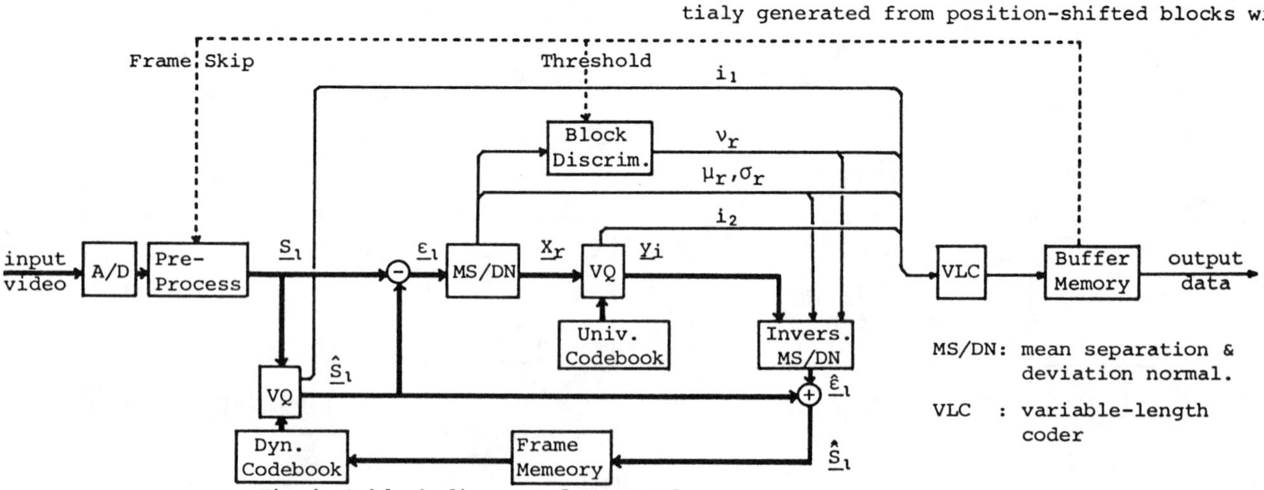

Fig.4 A block diagram of DMVQ coder

privious frames, the frequent generating blocks for a long period and fixed mean value blocks. If the indicator of the position-shifted block is denoted by $\underline{u_i}$, and the 16X8 dimensional input vector and output vector is $\underline{S}, \underline{\hat{S}_{ui}}$ respectively, DVQ can be expressed as follows.

$$Q(\underline{S}) = \underline{\hat{S}_{ui}} \quad (\underline{u_i} = [\, u_1, u_2 \,]_i \quad i = 1,2,\ldots,)$$
$$\text{if } d(\underline{S}, \underline{\hat{S}_{ui}}) < d(\underline{S}, \underline{\hat{S}_{uj}}) \text{ for all } j \neq i$$

Where $\underline{u_i}$ can be regarded and locally optimized as the motion vector according to frame-to-frame displacement of objects. In this expanding case, the other output vectors at the first stage are effective for a scene change or a new background after movement. DVQ can reduce the power of interframe differential signals as to improve the coding performance of AVQ for video signals. Since an algorithm for the dynamic codebook generation in a decoder is the same as in a coder, the index i of $\underline{u_i}$ has only to be transmitted.

Adaptive Vector Quantizer

The residual vectors $\underline{\varepsilon_1}$ are devided into eight residual input vectors $\underline{\varepsilon_r}$ consisted of 16 = 4X4 lattice pixels. After the mean separation and deviation normalization, AVQ carries out 16-dimensional vector quantization of the effective blocks and the conditional replenishment. Denoting the block discrimination parameter ν_r, the conditional replenishment according to the threshold decision is executed as follows.

if $|\mu_r|, \sigma_r$ < Threshold (T_1),

then $\nu_r = 0$ (ineffective ; replenishment)

else $\nu_r = 1$ (effective $\underline{\varepsilon_r}$; quantized).

The normalized vector $\underline{X_r}$ on the effective block is vector quantized with an universal codebook which has designed by a clustering training procedure of the LGB algorithm.

Multi-stage Connection of DVQ and AVQ

Consequently, the transmitted data, that is output parameters of the dynamic multi-stage VQ, are the index i_1 of output vector within the dynamic codebook, the block discrimination parameter for conditional replenishment ν_r, the index i_2 of output vector within the universal codebook, the mean μ_r and the deviation σ_r. They are converted into the suitable codewords through variable-length coder, respectively. The ν_r is encoded by block run-length coder and, the μ_r and the σ_r are encoded by DPCM coder, respectively. The average codeword length of each parameter is that, i_1=3.0 bit/16X8 pixel, ν_r=0.5 bit/4X4 pixel, i_2=9.0 bit/4X4 pixel, μ_r=3.5 bit/4X4 pixel and σ_r=3.5 bit/4X4 pixel.

The amount of information transmitted by above encoding procedure fluctuates according to the size of moving area. In order to transmit coding data over the constant rate channel, the threshold T_1 for block discrimination in AVQ is feedback controlled according to the transmission buffer memory occupancy and the amount of information can be smoothed out. Further, aiming to prevent the system from inoperativeness caused by channel error, a forward double error correcting code or (255,239) BCH code and a CRC check code are added to the transmitted data.

Coding Performance

In Fig.5 is shown the coding performance of the above DMVQ with coding control by computer simulation at a transmission rate of 64 kbit/s. Actual color video signals on one person speaking scene is used. The block size and the number of output vectors are (16X8,32) for DVQ and (4X4,512) for AVQ, respectively. Except for the amount of information that occured through the channel coding process, the effective transmission rate for encoded data is limited to 59 kbit/s. Then, color video signals have been reproduced with average SN ratio 37.4 dB at 0.1-0.2 bit/pixel in case of moving areas 10-20 % of the frame-to-frame. Consequently, it is confirmed that color motion video signals consisted of 352X240 pixels can be transmitted at the rate of 4-8 frames per second in case of 64 kbit/s.

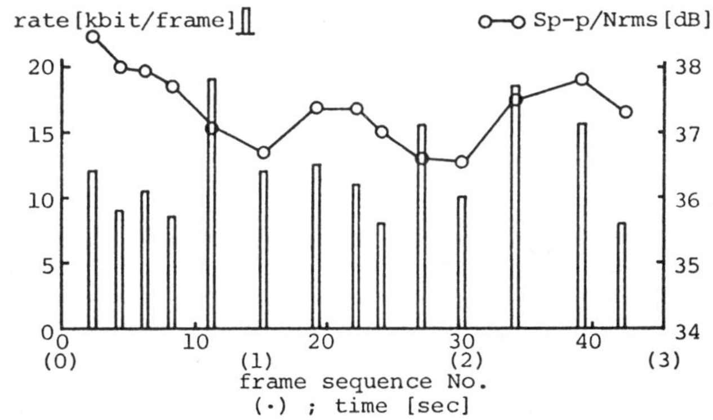

Fig.5 Coding performance of DMVQ

CONCLUSIONS

The mean separated and normalized VQ results in the superior coding performance of color images. Using these presented VQ technique, the color video codec at a low-bit-rate has been implemented.

REFFERENCES

[1]A.Gersho and V.Cuperman,"Vector Quantization : A pattern matching technique for speech coding", IEEE Commun.Mgn.21,9(1984).

[2]T.Murakami, et al,"Vector Quantiser of Video Signals",Electronics Letters,18,23(1982).

[3]Y.Linde, A.Buzo and R.M.Gray,"An Algorithm for Vector Quantizer Design",IEEE Trans.,Commun., COM-28,1(1980).

[4]T.Murakami,"High Efficiency Picture Coding Using Vector Quantization",Tech.Rep. IECE Japan, IT85-61(1985).

[5]T.Murakami, et al,"Vector Quantization for Color Images",J.Inst.TV Engrs. Japan,39,10(1985).

[6]T.Murakami, et al,"Dynamic Multi-stage Vector Quantization of Images",Trans. of IECE Japan,J68-B, 1(1985).

Image Compression Using Adaptive Vector Quantization

MORRIS GOLDBERG, MEMBER, IEEE, PAUL R. BOUCHER, AND SEYMOUR SHLIEN

Abstract—**The paper demonstrates the application of adaptive vector quantization techniques to the coding of monochromatic and color pictures. Codebooks of representative vectors are generated for different portions of the image. By means of extensive simulations, the performance of the coder for monochrome images is estimated as a function of the various coding parameters—vector dimensionality, number of representative vectors, and degree of adaptivity. It is shown that for a vector size of dimension 4, there is a zone of operation, ranging from 1.0 to 1.5 bits/pixel, where adaptive vector quantization is advantageous. Despite the simplicity of the decoder, the performance of the adaptive vector quantizer is found to be comparable with block transform coding schemes for both monochrome and color pictures.**

I. Introduction

Vector quantization is an effective scheme for performing speech compression [1], [2]. It has also been applied in image compression yielding acceptable quality images at about 1.0 bit/pixel for monochrome, and 2.25 bits/pixel for color [2]–[5].

The first step in image vector quantization is the decomposition of the image into a set of vectors. These vectors can be defined in a number of ways, for example, as the color components of a pixel [3], as the intensity values of a spatially contiguous group of pixels [4], or as transformed components of these groups [6]. In the next step, a codebook of representative vectors is generated using an iterative clustering algorithm such as K-means [7], [8], which is then used to quantize the vectors to the closest representative vector of the codebook. Compression is achieved by replacing the representative vector by a label. Reconstruction of the image is then performed by table lookup using the label as an entry to a table containing the representative vectors of the codebook. Thus, the display memory need only store the compressed image, which is then reconstructed during refresh time by table lookup. Two other attractive features of vector quantization are that no assumptions need be made about the underlying statistics, and that different distortion measures can be used [2].

In vector quantization, a fixed codebook is generally generated from a training set of images. To maintain a low distortion over all possible images, a large codebook is therefore required. In addition, to obtain coding efficiency, large dimensional vectors must be used. This entails practical difficulties not only in generating the codebook, but also in encoding the input images and storing the codebook at the decoder.

The basic assumption made in this paper is that individual images, and more specifically, subparts of the images would only need to address a small subset of the large fixed codebook. The adaptive techniques presented in this paper involve generating and transmitting separate codebooks for either individual images or subparts of the image. The savings in bit rate achieved by using smaller individual codebooks compensate for the additional overhead for transmitting all the codebooks.

If the codebooks are to be generated adaptively on an image-to-image basis, then the computational load required for generating the different codebooks can become significant. It is shown below that these requirements decrease when intraframe adaptive codebook generation is used.

The paper is organized as follows. In the following section, a fairly extensive review of image vector quantization is provided. The third section begins with a description of the intraframe adaptive codebook generation technique. There then follows a presentation of the experimental results on using a spatially adaptive codebook design procedure, which are compared to those from block transform methods. The fourth section describes experimental results on two procedures for vector quantizing color imagery, which are again compared to those from cosine transform coding. The concluding section suggests some areas for future research.

II. Review

Vector quantization of images was carried out as early as 1975 by Hilbert [9] primarily for the coding of multispectral Landsat imagery. The vectors were formed from the intensity values associated with individual pixels in the four Landsat bands, thus exploiting the spectral correlation. In 1980, Yamada *et al.* [10] reported results on using vectors formed from spatially contiguous pixels. Recent experiments on combining both spectral and spatial information were described in [5]. Applications of vector quantization to image sequences were also reported [13], [14]. The technique has also been used for texture generation [15] and for the design of refresh memories for image display [16], [17].

Hilbert's work [3], [9] was performed in the context of clustering for classification of Landsat imagery, not that of vector quantization. The premise taken was that the compressed images would be used primarily for classification purposes. Thus, compression schemes based upon clustering appeared to be ideal. This feature of the scheme was later verified experimentally [18]–[20].

An important contribution made by Hilbert was the use of locally adaptive codebooks. The image is divided into a number of subimages and different codebooks are then generated for each subimage. This technique reduces codebook generation time and, by adjusting to the spatially nonstationary behavior of images, yields lower distortions.

Paper approved by the Editor for Signal Processing and Communication Electronics of the IEEE Communications Society. Manuscript received July 11, 1984; revised June 14, 1985. This paper was presented at the International Conference on Acoustics, Speech, and Signal Processing, San Diego, CA, March 1984. This work was supported by a NATO Fellowship, by the Communications Research Centre, Ottawa, and by a grant from the Natural Sciences and Engineering Research Council of Canada, and was performed at the University of Ottawa and at the Ecole Nationale Supérieure des Télécommunications, Paris, France.

M. Goldberg is with the Department of Electrical Engineering, University of Ottawa, Ottawa, Ont., Canada K1N 6N5.

P. R. Boucher is with Bell Northern Research, Ile-des-Soeurs, Verdun, P.Q., Canada.

S. Shlien is with the Communications Research Centre, Ottawa, Ont., Canada.

IEEE Log Number 8406876.

Reprinted from *IEEE Trans. Commun.*, vol. COM-34, pp. 180–187, Feb. 1986.

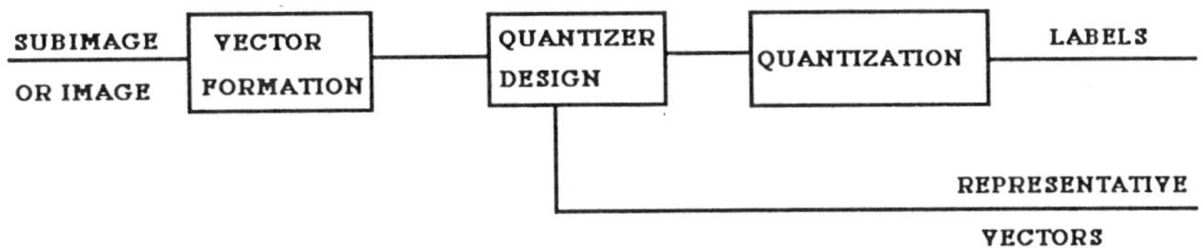

Fig. 1. Block diagram of the operations involved in adaptive image vector quantization.

Similar experiments were performed by Lowitz [21], [22] using eight-band multispectral data. The Karhunen–Loève expansion was the first calculated and only the first two principal components were used for codebook generation.

As discussed in [4], using spatially contiguous pixels introduces a problem in the reproduction of edges. One proposed solution [4], [23]–[25] is to first identify potential edge vectors and then to cluster these separately. Although the mean-squared error sense may increase, a definite visual improvement results. It is noted that certain textured images may consist predominantly of microedges; vector quantization has shown good results for these images [15]. The spatial and spectral information can also be combined into a vector [5].

Various methods for preprocessing the vectors before clustering have also been proposed. These include removal of mean [27], [28], predictive coding [26], Hadamard transform [6] and S-transform [29]. The techniques of [6] have also been extended to sequences of images [13]. Postprocessing of the coded images, in the form of dithering and spatial filtering, has also been suggested [17], [30], [31].

In high-quality color image display systems, a refresh memory with 24 bits/pixel is normally required. Vector quantization has been proposed for reducing the memory requirements [16], [17]. The main refresh memory would contain the class labels and a fast lookup table, the appropriate intensity values for each class. In [16] the vector quantization is effectively performed on the spectral information using a three-dimensional Peano scan to group the vectors. In [17], on the other hand, the 3-D color space is divided so that each color contains approximately the same number of pixels. It is clear that many of the other image vector quantization methods could also be of value for reducing the refresh memory.

III. Adaptive Vector Quantization of Monochrome Image

Nonadaptive image vector quantization employs a *single* codebook generated from a large set of training images. To maintain coding fidelity the codebook must be large, which, in turn, entails more bits for coding the labels and increased coder/decoder complexity. Adaptive methods which track the local behavior of the image statistics, not only from image to image, but within an image, can improve overall performance [32], [33].

For any given image, the adaptive scheme proposed consists of computing a separate codebook (or set of codebooks) from the vectors of the image. Thus, smaller codebooks than the nonadaptive case are required to maintain a given level of distortion. Both the codebook and the coded image must be transmitted to the decoder. The effective bit rate is thus determined from the amount of data required to transmit both the codebook and the coded elements (labels) of the image.

A block diagram of the operations involved in image adaptive vector quantization is shown in Fig. 1. The image is first decomposed into a sequence of vectors. The codebook is then computed using Lloyd and McQueen's clustering algorithm [7] on the given vectors (details are provided in the Appendix); that is, the training set is drawn from the image being coded. In spatially adaptive vector quantization, the image is first partitioned into separate rectangular regions or subimages and, for each, separate codebooks are computed independently. For each subimage, the locally generated representative vectors of the codebook are transmitted, followed by the codewords or labels for the vectors of the image.

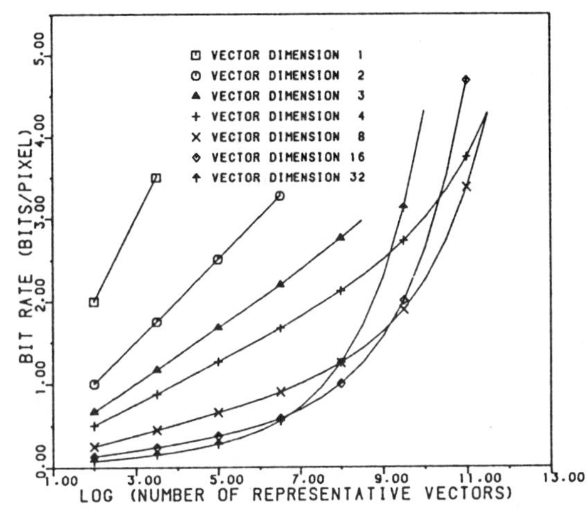

Fig. 2. This graph shows the bit rate for vector quantization as a function of the log {number of representative vectors} for different vector dimensions. Here the image is of size 256 by 256 pixels, with 8 bit resolution.

The average bit rate for adaptive image vector quantization contains two components: r_c, associated with the transmission of the individual codeword, and r_0, the overhead, related to the transmission of the representative vectors of the codebook. The bit rate is related to the following parameters:

- M number of pixels in the subimage
- K number of representative vectors, where K is normally a power of 2
- n dimension of vector
- b number of bits per representative vector component

by the following relationship:

$$r = (\log_2 K)/n + (K \cdot n \cdot b/M) \text{ bits/pixel} \qquad (1)$$

$$r = r_c + r_0. \qquad (2)$$

The rate r, as a function of K, is shown in graphical form in Fig. 2. Here $b = 8$, $M = 256 \times 256$, as n varies from 1 to 32. The exponential increase in rate for the higher dimensional vectors is due to the overhead component r_0.

Figs. 3–5 show the rate distortion curves corresponding to, respectively, r_c, r_0, and r, for the different block sizes indicated. These values were calculated from the 256 by 256 pixel image shown in Fig. 9, where the number of bits per pixel $b = 8$.

$$\text{NMSE} = \frac{\text{MSE of approximation}}{\text{variance of image}} \times 100. \qquad (3)$$

The distortion has been measured by the percentage normalized mean squared error. It is readily apparent that for a given distortion, r_c decreases with increasing vector dimension, whereas r_0 increases. This confirms experimentally the results of Fig. 2. A vector formed from blocks of 3 by 3 pixels is seen, from Fig. 5, to be optimum over a wide range of values.

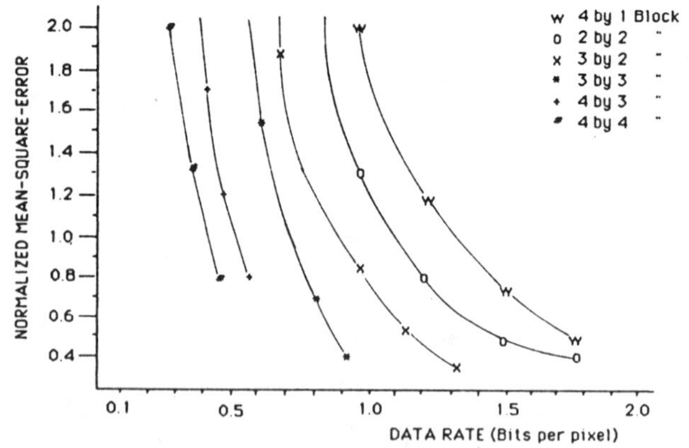

Fig. 3. Rate distortion curves for the image of Fig. 9 for different vector sizes and configurations. Only the component required for transmission of the codewords is included in the rate. The larger the vector size, the more efficient the coding becomes.

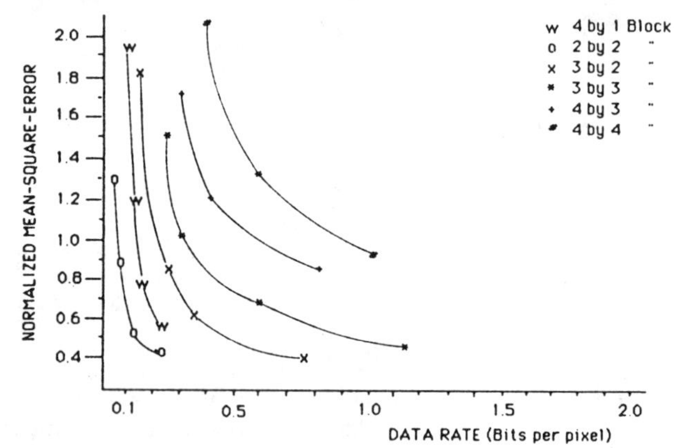

Fig. 4. Rate distortion curves for the image of Fig. 9 for different vector sizes and configurations. Here only the component of the rate required for transmission of the codebook of representative vectors, the overhead, is included. In this case the smaller the block size, the smaller the contribution to the overall rate becomes.

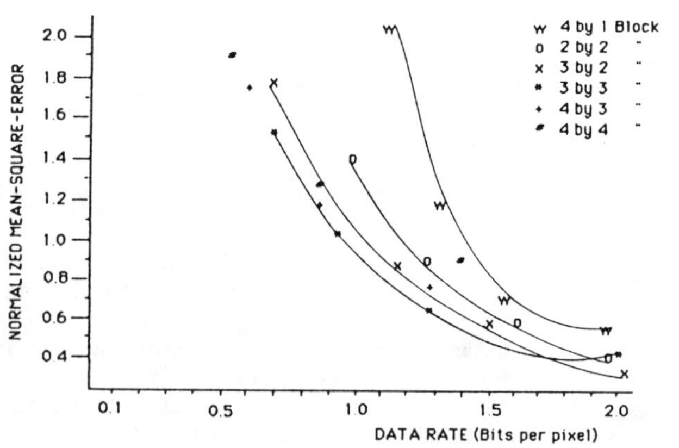

Fig. 5. Rate distortion curves for the image of Fig. 9 for different vector sizes and configurations. Here both components of the rate, overhead and codeword transmission, are included. Vectors formed from blocks of 3 by 3 pixels are seen to be optimum over a large range of bit rates.

As a first approximation, the computing time varies with

$$CT = PKM_s n$$

where

P is a constant
K is the number of representative vectors
M_s is the number of training samples
n is the vector dimension.

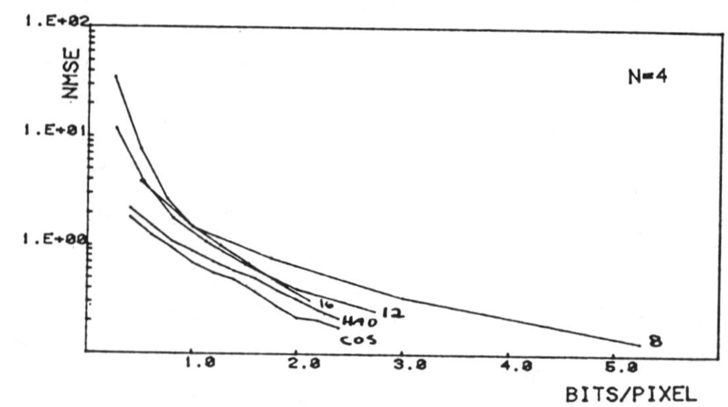

Fig. 6. Rate distortion curves for the image of Fig. 9 for a vector formed from blocks of 2 by 2 pixels, illustrating the performance of adaptive codebook generation. The five curves correspond to, respectively: 8—256 pixels; 12—4096 pixels; 16—the complete image; Had—Hadamard transforms on 16 by 16 pixel blocks; and Cos—cosine transforms on 16 by 16 pixel blocks.

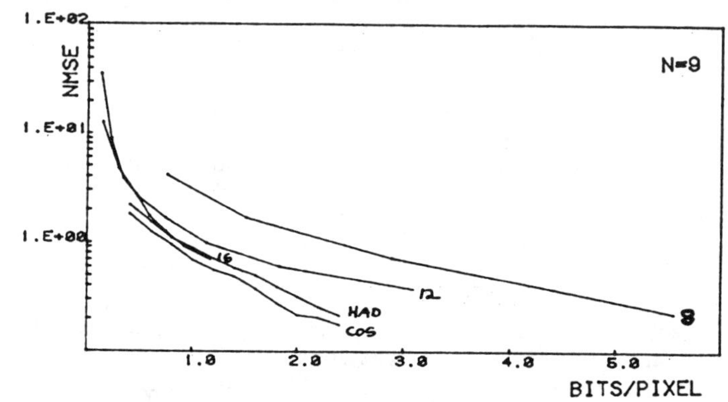

Fig. 7. Rate distortion curves for the image of Fig. 9 for a vector formed from blocks of 3 by 3 pixels, illustrating the performance of adaptive codebook generation. The five curves correspond to, respectively: 8—256 pixels; 12—4096 pixels; 16—the complete image; Had—Hadamard transforms on 16 by 16 pixel blocks; and Cos—cosine transforms on 16 by 16 pixel blocks.

Decreasing K, M_s, or n, therefore, reduces the computing time but, of course, affects bit rate and NMSE. An effective manner of decreasing K and M_s, without serious degradation, is to divide the image into smaller sections and generate separate codebooks for each section, that is, spatially adaptive codebook generation. This approach is based upon the observation that image statistics are not spatially stationary. Thus, smaller codebooks are required to maintain the same distortion, which, in turn, implies that both K and M_s can be reduced. Computing times were decreased by up to a factor of 4 when adaptive codebook generation was used.

The results depicted in Figs. 3–5 were obtained by using a single codebook adapted to the image at hand. The next set of results are obtained by spatially adaptive vector quantization, where a different codebook is calculated for subimages.

Figs. 6–8 correspond, respectively, to the use of vectors of dimension 4, 9, and 16, where the vectors are formed from square blocks of pixels. The results are presented in the form of rate distortion curves for different subpicture size. The curves are labeled according to the $\log_2$ {subpicture size}; for example, the label 12 corresponds to a subpicture of size 64 × 64 pixels. For vectors of dimension 9 and 16 (Figs. 7 and 8), spatially adaptive codebook generation results in a degraded performance. For example, it is necessary to double the bit rate to achieve the same distortion on a subpicture size of 1024 as on the full image. On the other hand, for vectors of dimension 4 (Fig. 6), up to 1.5 bits/pixel, adaptive codebook

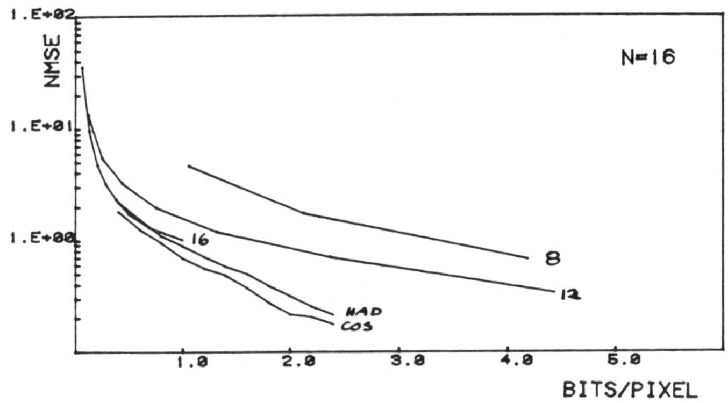

Fig. 8. Rate distortion curves for the image of Fig. 9 for a vector formed from blocks of 4 by 4 pixels, illustrating the performance of adaptive codebook generation. The five curves correspond to, respectively: 8—256 pixels; 12—4096 pixels; 16—the complete image; Had—Hadamard transforms on 16 by 16 pixel blocks; and Cos—cosine transforms on 16 by 16 pixel blocks.

Fig. 9. Extract from monochrome CCITT test image "Frank" of size 256 by 256 pixels, with 8 bits grey level resolution. The lower right-hand portion of this image has been repeated at a different density to show the details of the shadowed regions.

generation has comparable performance characteristics. This behavior is due to the overhead r_0 for codebook transmission. The higher the bit rate or the dimension, the larger the codebook size, and thus the larger the contribution of the overhead r_0 to the overall bit rate. The smaller the subpicture size, the more significant this contribution becomes. Techniques such as varying the number of representative vectors on a subpicture basis, and combining or cascading the codebook, have been successfully used to decrease the overhead components [4], [21] and could yield similar savings in the present situation.

The measured results, as seen from the rate distortion curves, are generally confirmed by a visual inspection of the corresponding images. Adaptive codebook generation gives better results than indicated for the 2 by 2 pixel case as illustrated in Figs. 10 and 11. Fig. 10 shows the result of using 16 representative vectors for the entire image (nonadaptive codebook generation); the bit rate is 1.01 and the normalized mean-squared error is 1.31. The image is of good quality but some false contouring appears in the shadowed regions. Fig. 11 shows the coded image using adaptive codebook generation; the image of Fig. 9 is divided into 64 nonoverlapping 32 by 32 pixel local sources, and a separate codebook is generated for each local source. With eight representative vectors for each local source, the bit rate is 1.00 and the normalized mean-squared error decreases to 1.22; furthermore, the false contouring effects are attenuated.

There appear to be distinct operating regions which depend upon the particular block size and technique. For example, using blocks of 2 by 2 pixels and adaptive codebook generation, below 1.0 bit/pixel the resultant images are of poor quality, whereas by 1.5 bits/pixel the resultant images are almost indistinguishable from the original. An example is shown in Fig. 12, where 16 representative vectors are used for each local source, for a bit rate of 1.5, and a normalized mean-squared error of 0.76. For the nonadaptive 2 by 2 pixel case, the corresponding operating region ranges from approximately 1.25 to 1.75 bits/pixel, while for the 3 by 3 pixel case the range is 0.9–1.25 bits/pixel.

The performance of the vector quantization coding scheme was compared to the block cosine and Hadamard transform coding schemes [34]. The picture was partitioned into 16 by 16 pixel nonoverlapping blocks and each block was transformed using either the two-dimensional discrete cosine transform or the Hadamard transform. The transform coefficients were scalar quantized using a varying number of bits chosen according to the average coefficient energy [35]. The dc coefficient was simply linearly quantized to 8 bits, while the ac

Fig. 10. Vector quantization of test image. The vectors are formed from blocks of 2 by 2 pixels; the codebook contains 16 representative vectors; the bit rate is 1.01, and the normalized mean-squared error is 1.31.

Fig. 11. Adaptive codebook generation. The vectors are formed from blocks of 2 by 2 pixels; separate codebooks with eight representative vectors are generated for each of 64 nonoverlapping squares of 32 by 32 pixels from the image of Fig. 9; the bit rate is 1.0, and the normalized mean-squared error is 1.22.

Fig. 12. Adaptive codebook generation. The vectors are formed from blocks of 2 by 2 pixels; separate codebooks with 16 representative vectors are generated for each of 64 nonoverlapping squares of 32 by 32 pixels from the image of Fig. 9; the bit rate is 1.5, and the normalized mean-squared error is 0.76.

components were quantized using the Max-quantizer [34, p. 143] and a Laplacian distribution model [36], [37]. Results for both the cosine and Hadamard transforms at various bit rates are superimposed in Figs. 6–8. The adaptive vector quantization coding scheme never outperforms the block cosine transform, but it becomes comparable at certain bit rates depending upon the dimensionality of the vector, n.

Visual comparison of the vector quantized and cosine transformed reconstructed monochromatic images did not suggest that either technique is superior at 1.5 bits/pixel, although a blocking effect is noticeable in Fig. 13(a) near the left eye. At 1.0 bit/pixel, contouring and poor reproduction of edges were the principal shortcomings of the adaptive vector quantization scheme, while the cosine transformed image had a "snowy" and blurred appearance, as well as a minor blocking effect in the eyes [see Fig. 13(b)].

IV. Vector Quantization of Color Images

The adaptive vector quantization techniques above were extended to color images represented by the three tristimulus signals R, G, B. Before processing, these values were transformed to the NTSC YIQ coordinate system [34, p. 596]. This transformation tends to decorrelate the information and to compact most of the energy in the luminance component Y. In the NTSC system, both chrominance components I and Q are severely band limited; in the results described below, the spatial resolution of the I and Q are thus decreased by a factor of 2 to yield I' and Q'. Both the vector quantization and the cosine transform techniques are now applied to the Y, I', and Q' signals. Results are presented for two different adaptive vector quantization schemes and compared to cosine transform coding.

The first scheme simply treats each component or plane Y, I', and Q' as if it were a monochrome image. Each plane is decomposed into vectors formed from blocks of 2 by 2 pixels and a separate codebook generated for each group of 32 by 32 vectors. Half of the available bits were allocated to the Y plane and a quarter each to the I' and Q' planes.

The second scheme, called combined coding, involves first decomposing the YIQ image into blocks of 2 by 2 pixels. The spatial resolution of the chrominance planes I and Q is reduced by 2; thus, two chrominance coefficients, one from each of the I' and Q' planes, can be associated with each 2 by 2 pixel block of Y. Vector quantization is then applied to the six-dimensional vector thus formed, using the adaptive codebook generation scheme on 32 by 32 pixels. Combined coding

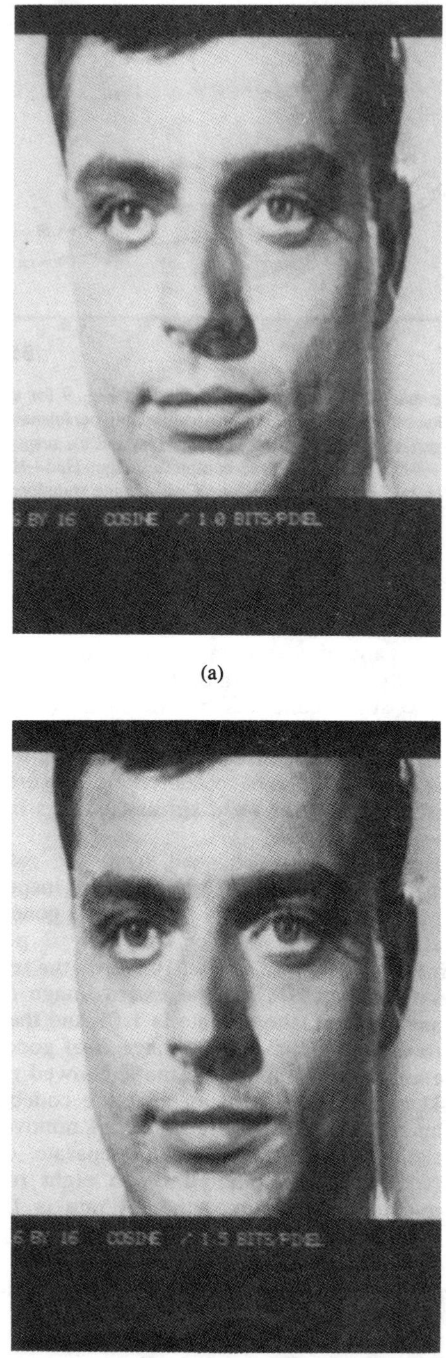

(a)

(b)

Fig. 13. Cosine block transform coding. The images shown have been cosine transformed in blocks of 16 by 16 pixels. (a) This image has been coded with 1.0 bits per pixel and the normalized mean-squared error is 0.67. (b) This image has been coded at 1.5 bits per pixel and the normalized mean-squared error is 0.37.

attempts to exploit the residual correlation remaining between the Y, I', and Q' planes.

The results are presented in Table I using the distorting measures of (3) for each of the components R, G, and B. These values are measured from a 256 by 256 pixel portion of the CCITT "Boat" image shown in Fig. 14, displayed in RGB with 24 bits/pixel. The results for the cosine transform are obtained by separately coding the Y, I', and Q' planes. Results for three different bit rates are shown in Table I; approximate rates of 1.0, 1.5, and 2.3 bits/pixel.

TABLE I
EXPERIMENTAL RESULTS FOR THE "BOAT" IMAGE OF FIG. 14

Coding scheme	Bit rate	Normalized MSE *R*	*G*	*B*
Cosine	2.25	4.03	1.88	5.98
VQ Plane	2.45	5.06	2.17	9.78
VQ Combined	2.31	4.57	2.22	6.39
Cosine	1.50	4.54	2.64	7.01
VQ Plane	1.39	5.94	3.74	12.03
VQ Combined	1.53	5.42	2.90	7.37
Cosine	1.00	5.51	3.73	8.48
VQ Plane	1.08	8.10	4.10	14.04
VQ Combined	1.00	6.81	3.97	8.79

Fig. 14. Extract from CCITT "Boat" *RGB* image, of size 256 by 256 pixels, with 24 bits/pixel.

Vector quantization of the combined spectral and spatial vectors is seen to outperform individually vector quantizing the planes, as is to be expected. Although cosine transform coding performs vector quantization of the combined spectral and spatial vectors, the visual appearance of the latter is more appealing. This is shown in Figs. 15 and 16, where the results on the "Boat" image are given for both the methods. At high bit/pixel rates, 2.25 for cosine and 2.31 for vector quantization of combined spatial and spectral vectors, both images are of general good quality, but the cosine coded image exhibits some speckled noise in the shadowed region. At low bit rates, 1.0 bit/pixel, the cosine coded image has a very snowy appearance, while the vector quantized image suffers from poorer reproduction of the edges.

V. CONCLUSIONS

This paper has presented experimental results on the use of adaptive codebook generation for the vector quantization of images. In this technique, the image is divided into a number of nonoverlapping subimages, for each of which a different codebook is generated. The first set of experiments was

(a)

(b)

Fig. 15. Coded "Boat" image. (a) Cosine block transform for 16 by 16 pixel blocks and a bit rate of 2.25. (b) Codebook adaptive vector quantization on combined spatial and spectral vectors at a bit rate of 2.31. Separate codebooks are generated for each of 64 nonoverlapping squares of 32 by 32 pixels.

(a)

(b)

Fig. 16. Coded "Boat" image. (a) Cosine block transform for 16 by 16 pixel blocks and a bit rate of 1.00. (b) Codebook adaptive vector quantization on combined spatial and spectral vectors at a bit rate of 1.00. Separate codebooks are generated for each of 64 nonoverlapping squares of 32 by 32 pixels.

performed on monochrome images. The results demonstrate that, for vectors formed from blocks of 2 by 2 pixels, there is a potential operating zone from 1.0 to 1.5 bits/pixel, where adaptive codebook generation is advantageous. At these rates, adaptive vector quantization outperforms nonadaptive methods; furthermore, the resultant images were seen to be comparable with cosine transform coding. As regards implementation, adaptive codebook generation requires fewer computations than nonadaptive techniques and holds out the potential for real-time compression. The experiments on color imagery demonstrate that adaptive vector quantization yields results comparable to cosine transform coding. Here the vectors used correspond to both the spatial and spectral information.

The images coded by vector quantization show two types of defects: poor edge reproduction and false contouring. Techniques are needed for preprocessing the vectors so that the vectors corresponding to these errors are isolated and given preference in the codebook generation stage. Further research is also needed in removing the redundancy between the different codebooks used in the different subimages. The potential for improved performance had already been demonstrated [3], [21].

Appendix

The experimental results described in this paper were obtained by using the Lloyd–McQueen algorithm as described in [7], where the unweighted mean-squared error (MSE) was minimized. The initial seeds were found by the diagonal seeding method described in [3]. The centroids were computed after each complete pass of the training set, where the set is formed from the entire picture or subpicture. The following stopping criterion, with a value of 0.0005, was used to test for convergence:

$$\frac{\text{MSE } (t-1) - \text{MSE } (t)}{\text{MSE } (t)}$$

where MSE (t) is the mean-squared error after the tth iteration.

References

[1] Y. Linde, A. Buzo, and R. M. Gray, "An algorithm for vector quantizer design," *IEEE Trans. Commun.*, vol. COM-28, pp. 84–95, Jan. 1980.

[2] R. M. Gray, "Vector quantization," *IEEE ASSP Mag.*, pp. 4–29, Apr. 1984.

[3] E. E. Hilbert, "Cluster compression algorithm: A joint clustering/data compression concept," Jet Propulsion Lab., Pasadena, CA, Publ. 77-43.

[4] A. Gersho and B. Ramamurthi, "Image coding using vector quantization," in *Proc. IEEE Int. Conf. Acoust., Speech, Signal Processing,* May 1982, pp. 428–431.

[5] P. Boucher and M. Goldberg, "Colour image compression by vector quantization," in *Proc. IEEE Int. Conf. Acoust., Speech, Signal Processing,* San Diego, CA, Mar. 1984, pp. 29.6.1–29.6.4.

[6] R. A. King and N. M. Nasrabadi, "Image coding using vector quantization in the transform domain," *Pattern Recogn. Lett.*, vol. 1, pp. 323–329, 1983.

[7] S. P. Lloyd, "Least squares optimization in PCM," Bell Lab. Tech. Note, 1957; also in *IEEE Trans. Inform. Theory,* vol. IT-28, pp. 129–137, Mar. 1982.

[8] J. T. Tou and R. C. Gonzalez, *Pattern Recognition Principles.* Reading, MA: Addison-Wesley, 1974, pp. 94–109.

[9] E. E. Hilbert, "Joint pattern recognition/data compression concept for ERTS multispectral data," *SPIE,* vol. 66, "Efficient transmission of pictorial information," Aug. 1975.

[10] Y. Yamada, K. Fujita, and S. Tazaki, "Vector quantization of video signals," in *Proc. Annu. Conf. IECE,* 1980, p. 1031.

[11] T. Murakami, K. Asai, and E. Yamazaki, "Vector quantizer of video signals," *Electron. Lett.*, pp. 1005–1006, Nov. 1982.

[12] Y. Yamada and S. Tazaki, "Vector quantizer design for video signals" (in Japanese), *Trans. IECE Japan*, vol. J66-B, pp. 965–972, 1983.

[13] N. M. Nasrabadi and R. A. King, "A new image coding technique using transform vector quantization," in *Proc. IEEE Int. Conf. Acoust., Speech, Signal Processing,* San Diego, CA, Mar. 1984, pp. 29.9.1–29.9.4.

[14] H. F. Sun and M. Goldberg, "Image sequence coding using vector quantization," in *Proc. Int. Commun. Energy Conf.,* Montreal, P.Q., Canada, Oct. 1984, pp. 266–269.

[15] F. Schmitt, M. Goldberg, N. Nge-Ndifor, and P. Boucher, "Texture representation and synthesis by vector quantization," in *Proc. Int. Conf. Pattern Recogn.,* Montreal, P.Q., Canada, Aug. 1984, pp. 1222–1225.

[16] R. J. Stevens, A. F. Lehar, and F. H. Preston, "Manipulation and presentation of multidimensional image data using the Peano scan," *IEEE Trans. Pattern Anal. Mach. Intell.,* vol. PAMI-5, pp. 520–526, Sept. 1983.

[17] P. Heckbert, "Color image quantization for frame buffer display," *ACM Comput. Graphics,* vol. 16, pp. 297–307, July 1982.

[18] G. E. Lowitz and J. M. Vivier, "Classification of raw and compressed multispectral data," in *Data Reduction,* vol. 1 (Euro. Space Agency CR(P)-1387), Matra, France, July 1980.

[19] ——, "Cascaded clustering of compressed multispectral image data," *Data Rate Reduction,* vol. 3 (Euro. Space Agency CR(P)-1387), Matra, France, July 1980.

[20] D. J. Stanley, "Interpretation of remotely sensed image data and impact of cluster compression," Euro. Space Agency CR(P)-1450, Feb. 1981.

[21] G. E. Lowitz, "Compression des données images par reconnaissance des formes et 'clustering'," in *Proc. Int. Conf. Earth Observ. from Space, Management of Planet. Resources,* Toulouse, France, Mar. 1978, pp. 243–250.

[22] ——, "Image data reduction frame," in *Proc. Int. Conf. Spacecraft On-Board Data Management,* Nice, France, Oct. 1978, pp. 291–294.

[23] B. Ramamurthi, A. Gersho, and A. Sekey, "Low-rate image coding using vector quantization," *IEEE Conf. Rec. GLOBECOM,* San Diego, CA, Dec. 1983, pp. 5.6.1–5.6.4.

[24] B. Ramamurthi and A. Gersho, "Image vector quantization with a perceptually-based call classifier," in *Proc. IEEE Int. Conf. Acoust., Speech, Signal Processing,* San Diego, CA, Mar. 1984, pp. 32.10.1–32.10.4.

[25] ——, "Image coding using segmented codebooks," in *Proc. Int. Picture Coding Symp.,* Mar. 1983.

[26] S. Cabrera and I. D. Landau, "Codage différentiel adaptatif d'images avec quantification vectorielle," in *Proc. 9ème Colloque Traitement du Signal et Appl., GRETSI Conf.,* Nice, France, 1983, pp. 571–576.

[27] R. L. Baker and R. M. Gray, "Differential vector quantization of achromatic imagery," in *Proc. Int. Picture Coding Symp.,* 1983, pp. 105–106.

[28] ——, "Image compression using non-adaptive spatial vector quantization," in *Proc. IEEE Conf. Circuits, Syst., Comput.,* 1982, pp. 55–61.

[29] P. Boucher and M. Goldberg, "Transform image coding by vector quantization," in *Proc. GRETSI, 9th Symp. Signal Processing Appl.,* Nice, France, May 1983, pp. 629–633.

[30] M. Goldberg *et al.,* "Amélioration de la qualité d'images fortement compressées par quantification vectorielle," in *Proc. GRETSI: 1er Colloque Image,* Biarritz, France, May 1984.

[31] B. Ramamurthi and A. Gersho, "Edge oriented spatial filtering of images with application to post-processing of vector quantized images," in *Proc. IEEE Int. Conf. Acoust., Speech, Signal Processing,* San Diego, CA, May 1984, pp. 48.10.1–48.10.4.

[32] A. Habibi, "Survey of adaptive image coding techniques," *IEEE Trans. Commun.,* vol. COM-25, pp. 1275–1284, Nov. 1977.

[33] B. R. Hunt, "Nonstationary statistical image models (and their application to image data compression)," *Comput. Graphics, Image Processing,* vol. 12, pp. 173–186, 1980.

[34] W. K. Pratt, *Digital Image Processing.* New York: Wiley-Interscience, 1978.

[35] W. H. Chen and C. H. Smith, "Adaptive coding of monochrome and color images," *IEEE Trans. Commun.,* vol. COM-25, pp. 1285–1292, 1977.

[36] R. C. Reininger and J. D. Gibson, "Distributions of the two-dimensional DCT coefficients for images," *IEEE Trans. Commun.,* vol. COM-31, pp. 835–839, June 1983.

[37] W. C. Adams, Jr. and C. E. Giesler, "Quantizing characteristics for signals having Laplacian amplitude probability density function," *IEEE Trans. Commun.,* vol. COM-26, pp. 1295–1297, 1978.

Variable Block Vector DPCM: Vector Predictive Coding of Color Images

Charles W. Rutledge

AT&T Bell Laboratories
Holmdel, New Jersey 07733

Abstract

Vector DPCM is a method of image coding that combines vector quantization coding with DPCM coding. This paper describes Variable Block Vector DPCM that is a Vector DPCM method where the block size is changed depending upon the error of the prediction in each segment of the picture. This method initially assumes an 8 pixel x 8 pixel block size and then will reduce to four 4 pixel x 4 pixel blocks if excessive prediction errors result in detailed areas. Any of these 4x4 blocks can again be reduced to four 2x2 blocks for very detailed areas. The bit rate for color images ranges from approximately 1 bit/pixel for a excellent results to .46 for good results. The variable block method also provides a more consistent image quality throughout all areas of the pictures processed. Like Vector DPCM, this method requires only one universal codebook.

Introduction

The Vector DPCM coding method is a cross between DPCM coding and Vector Quantization (VQ) coding. The Vector DPCM process is a method in which an image is broken up into blocks and then these blocks are processed in a raster format. Each block of a picture is predicted from bordering pixels found in the previously processed blocks above and to the left of the block being processed. After each block is predicted, it is then compared to the corresponding block in the original picture. A differential block is thus created from the original and predicted block and this differential block is then vector quantized. The vector quantization codes are transmitted, decoded into coded differential blocks, and combined with a prediction at the receiver in a similar manner as is typical with a DPCM process.

Variable Block Vector DPCM is a modification of Vector DPCM in that the block size is not constant. The block size may be as large as 8x8 pixels when the original prediction is very good or as small as 2x2 pixels in very detailed areas. The block size is determined by the magnitude of prediction errors between the pixels of an original block and the prediction of that block. The process starts by assuming an 8x8 pixel block size and if the total prediction errors throughout the block are within a given limit, a special code is sent that indicates that only the prediction will be used. If the sum of excessive prediction errors is outside the given limit then the 8x8 block is divided into four 4x4 blocks and a prediction is then performed on each of these. Each of these four blocks is treated separately. If the error resulting from a 4x4 block prediction is within the given limits, the error is vector quantized and a vector code is sent representing the prediction error found in that 4x4 block. Again, if the sum of excessive prediction errors is greater that a specified limit, then the 4x4 block is divided into four 2x2 blocks and the Vector DPCM process is performed on each of these, resulting in transmission of a 2x2 vector code. Though this method is slightly more complex and will produce varying bit rates dependent upon the complexity of the picture, a more consistent quality is provided across the entire image and only a very small amount of information relating to non-detailed sections of the image will need to be sent. This coding procedure is only slightly more complex than straight Vector DPCM and can be implemented inexpensively in hardware with just a few VLSI chips and some support.

Vector Quantization has a long history, but has just recently been picking up momentum as a viable method of coding images. The best general reference for basic vector quantization, written by Gray[1] , outlines the basics for vector quantization and describes various applications and details of more advanced vector quantization systems. A good theoretical reference on vector quantization was done by Gersho [2]. He discusses various mathematical properties of vector quantization. The more recent developments using vector quantization are predictive schemes where by some type of prediction is made to remove signal redundancy followed by vector quantization. Baker and Gray[3][4] introduced a method where the mean of each

Reprinted from *Proc. IEEE,* vol. 1, pp. 130–135, June 1987.

image block was removed and the residual signals were then vector quantized.

Variable Block Vector DPCM is a new method of predictive vector quantization that provides a more efficient structure than previous methods and provides image quality better than the methods mentioned above. The background, method, and experimental results, of Variable Block Vector DPCM are to follow in this paper. Section 2 discusses background issues concerning Vector DPCM [5]. The method and some attractive features of Variable Block Vector DPCM are described in Section 3. Section 4 documents Vector DPCM results on monochrome images and Section 5 presents the results on color images.

Vector DPCM

The Vector DPCM method is actually a combination of DPCM and Vector Quantization coding. This method processes a picture block by block as opposed to pixel by pixel as is typically done in DPCM coding. Each block is predicted and compared to the original picture block. The difference between the predicted and original blocks is then vector quantized and vector codes transmitted. At the decoder the blocks are again predicted and the difference vectors as represented by the transmitted vector codes are added to the prediction to form a reconstructed block.

The prediction of each block is made using the edge border pixels of the block above it and to the left of it. These blocks are used because they are previously processed blocks, since the picture is processed from left to right and from top to bottom. Each pixel of a block is predicted by using the average of the pixel above and to the left of the pixel to be predicted. Thus each pixel of a block is predicted in a raster scan type format. If the pixel above or to the left of the pixel to be predicted is a member of a previously processed block the actual coded value is used in the averaging, otherwise the pixel will be a member of the block being presently processed and only the predicted value is used. The result is that the entire block is predicted using some combination of the values of the border pixels of the above and behind previously processed blocks (Figure 1).

This prediction is then subtracted from the corresponding block taken from the original

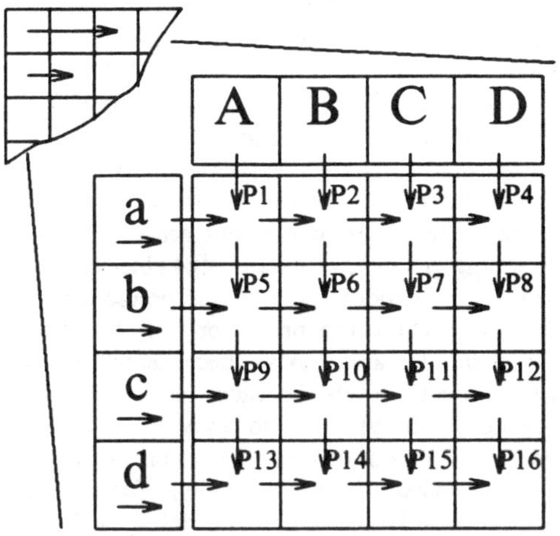

Figure 1. Method of Vector DPCM Prediction

picture, and a difference block is created that represents the error between the two blocks. This difference block is then vector quantized, and a vector code is transmitted that is representative of a vector whose values are the most similar to the values of the difference block. Simultaneously, the value of each segment of the vector is added to the value of the prediction of each corresponding pixel to form the coded block, the same block as will be reconstructed at the decoder in the absence of transmission errors. This coded block is then input into the predictor for use in the prediction of the blocks in the picture that have not yet been processed.

The decoder is very simple, can be constructed inexpensively, and will operate very fast. The decoder also operates as part of the encoder as with typical DPCM coding. If designed in VLSI, both could be designed on the same chip simply sharing similar parts. At the decoder the vector code is received and each element of that vector is added to the prediction of the corresponding pixel created in the same manner as in the encoder. The result is a reconstructed block that is displayed and also input back into the predictor as is done in the encoder.

Variable Block Vector DPCM

Variable Block Vector DPCM is basically the original Vector DPCM process with the added ability to change the block size in accordance to the detail found in each area of an image.

It is ideal to transmit only as much information as is required to describe an image, and this is what is accomplished by varying the block size. In flat areas where the actual information contained is very small, 8x8 blocks based only on the prediction are used, thus providing an appropriate low bit rate. As the detail increases in gradient areas or areas containing moderate detail, 4x4 blocks are used. This provides for a better prediction to be made because the prediction will be made over a smaller area. Also, because there are 1/4 as many pixels in each block a small codebook will be able to provide a better general representation of the error found in such predicted areas. Finally, for areas containing a large amount of detail such as sharp edges, numbers, text, etc, 2x2 blocks are used providing an even greater advantage in prediction and codebook representation. Of course when transmitting 2x2 blocks the bit rate is rather high (one code for every 4 pixels), which is appropriate because much of the image information important for human understanding is contained in these areas. But these areas also occupy only a small percentage of the total image.

The decision to change block sizes is determined by two conditions. The first condition is the detection of a small number of large errors and the other condition is the detection of a large number of small errors . The values of pixel error and pixel limits may be varied to allow the ability to tradeoff between bit rate and picture quality.

The image is scanned in a raster format first assuming an 8x8 block size. An 8x8 block of the image will be predicted and compared to the corresponding block found in the original image as would be done in straight Vector DPCM. If the computed differential errors between the predicted and original blocks are within the limits earlier described a special code (0) is transmitted. In this case, the prediction is close enough to the original block that no correction for error is required and only the prediction will be used. If the differential errors exceed the limits, then that 8x8 block is divided into four 4x4 blocks. These four 4x4 blocks are then scanned in raster format so as to complete the processing of the original 8x8 block before starting another. Again each 4x4 block is predicted as in Vector DPCM and if the differential errors are within the limits set for 4x4 blocks then the differential error block is vector quantized. Because the 4x4 blocks only describe moderate detail, a small codebook of only 63 codes is required. Thus codes 1 through 63 are 4x4 vector codes. After the vector quantization, one of these codes is transmitted, the vector quantized error will be added back to the prediction, and this sum will be simultaneously entered back into the predictor.

If the differential errors exceed these limits, the 4x4 block will be divided up into four 2x2 blocks. These 2x2 blocks are operated on in a similar manner processing the four blocks in a raster format to complete processing of the original 4x4 block. But in processing the 2x2 blocks there will be no checking of error limits because the blocks are at the smallest size. Because of the highly detailed areas that are being described by the 2x2 blocks, the errors found in these areas are the highest and require the most codes. I have used 192 codes to represent 2x2 blocks. Thus code numbers 64 to 255 are 2x2 code blocks. This creates a total of 256 codes or 1 byte per code. After the prediction of these blocks the differential error is vector quantized as with typical Vector DPCM. After the four 2x2 blocks are processed then the coder goes back to processing 4x4 blocks and similarly after completion of the four 4x4 blocks the raster processing of the 8x8 blocks continues.

The attractive features of Vector DPCM have been described in detail before [5]. Vector DPCM provides bit rates equivalent to that of Vector Quantization, but with higher quality. Thus this method provides a bit rate advantage over DPCM and a quality advantage over typical Vector Quantization. Because the blocks formed from the difference between the predicted and actual block contain much more general features than the picture itself, a codebook can be created that is much more robust and will be useful for a large variety of pictures.

Variable Block Vector DPCM adds features that make this process even more attractive. The main improvement is that a different number of pixels per code are transmitted depending upon the detail of the picture. Thus, in effect, information is transmitted only where the relevant information lies in the picture. By using different size blocks this method concentrates on very detailed areas and edges where the prediction will be the poorest. This method will also provide a more consistent quality throughout the picture. With the Variable Block process an unnoticeable or only slightly noticeable error will be allowed throughout the image to achieve lower bit rates. This method will also provide generally better predictions throughout the image because, as the detail becomes greater, the blocks become smaller. Thus, the area and distance over which the prediction is made will become much less.

A large asset of the Vector DPCM method is that it is simple, and thus should result in small, inexpensive hardware. This process is relatively simple as compared to many other coding techniques, especially those that operate in these bit rates. Also there will be reduced vector quantization encoding times because not every block will have to be compared to 256 codes. The 8x8 blocks that make up a large area of a typical image require no vector quantization, the 4x4 blocks that make up the majority of a typical image requires a codebook of only 63 codes, and finally the 2x2 blocks that make up only a small percentage of a typical picture requires vector quantization with a codebook of 192 codes. Because the vector quantization processing is the most time consuming function of the coder, the time savings discussed above will far outweigh the added processing of possible multiple predictions over the same area created by block size changes. No additional information will need to be transmitted to describe the processing block size. This information will be inherent in the code number. Also the decoder will operate just as fast, and only require the slightly more complex predictor.

Monochrome Results

The test processing was done on monochrome images to make a direct comparison between Vector DPCM and Variable Block Vector DPCM so as to evaluate the increased quality obtained by implementing the variable block sizes. The three images used were all 8 bit/pixel, 512x512 images of various subjects, specifically: 1) a womans portrait (Lena), 2) a F16 fighter jet, and 3) various colored peppers. The codebooks were created using the LBG algorithm suggested by Linde, Buzo, and Gray [6]. The codebooks were created using 4 different images none of which were the three images to be used in the comparison. Three quality levels were created so that a user could trade off quality for bit rate. These quality levels were created by simply changing the error levels at which the algorithm switches from 4x4 to 2x2 blocks. The error levels for switching from 8x8 to 4x4 blocks always remain the same because it was discovered that raising the error levels to decrease the bit rate produced unacceptable errors in what was expected to be a generally smooth background. In Quality 0 and 1, 63 codes were used for 4x4 blocks and 192 codes where used for 2x2 blocks. When using the Quality 2 partitions, no 2x2 blocks are used. Thus, all 255 codes are dedicated to 4x4 blocks.

The first comparison made was using the image of Lena processed at all three quality levels. The results of the comparison are found in Table 1. As can be observed the use of variable block sizes provides less error and/or lower bit rates. It should also be noted that the improvement is not simply less error but also the distribution of error. The variable block method will permit slight and unnoticeable error in the background but will create sharper more accurate edges. Figure 2 displays the image of Lena processed using the 0 Quality level. The following tables show similar results for the F16 and Peppers: It can also be noticed that even though the bit rate is variable the bit rate for each of these three images is very similar in each quality group. But it is also realized that the bit rate could change significantly for a very simple or very complex image.

In Figure 3 is a picture of the break down into block sizes for the image of Lena processed using Quality 0 limits. It is clearly seen that the 8x8 blocks are found in larger smooth areas, while the 2x2 blocks are found along edges and in detailed areas.

Color Results

The color images used in this investigation were all 512 x 512 24-bit RGB images. The images were converted to the YUV components [7] before they were processed. The Y component represents the luminance of a color pixel. The U and V components are the two chrominance components and have a

Table 1

Lena 512x512 image

Method	Block size	Codes	Bit/pel	Sq Error/pixel	Quality
Vector DPCM	3x3	256	0.89	25.2	
Vector DPCM	4x4	256	0.50	46.5	
VB Vector DPCM	Variable	256	.79	18.8	0
VB Vector DPCM	Variable	256	.54	32.4	1
VB Vector DPCM	Variable	256	.41	49.8	2

Table 2

F16 512x512 image

Method	Block size	Codes	Bit/pel	Sq Error/pixel	Quality
Vector DPCM	3x3	256	0.89	32.1	
Vector DPCM	4x4	256	0.50	59.8	
VB Vector DPCM	Variable	256	.77	19.5	0
VB Vector DPCM	Variable	256	.56	32.6	1

Table 3

Peppers 512x512 image

Method	Block size	Codes	Bit/pel	Sq Error/pixel	Quality
Vector DPCM	3x3	256	0.89	33.6	
Vector DPCM	4x4	256	0.50	53.2	
VB Vector DPCM	Variable	256	.74	27.1	0
VB Vector DPCM	Variable	256	.55	35.9	1

spatial bandwidth of about 1/5 of the Y component. Each of these three components were coded separately, then brought back together and converted back into RGB. Because the U and V components have a lower spatial bandwidth [7], they can be spatially subsampled without deteriorating overall coding performance, thus achieving lower bit rates than coding each chrominance component at the highest spatial resolution. Both the U and V components are sub-sampled, and the images are then coded in the usual Variable Block Vector DPCM manner. After the codes are transmitted and decoded, the U and V components must be enlarged to their original size using bi-linear interpolation.

The images used to demonstrate the results of the Variable Block Vector DPCM process on color images are color versions of the images used to provide the monochrome results. The Y, U, and V codebooks were all created separately using the LBG algorithm as described in the monochrome section. Thus, the images processed were not images used in the codebook training set. Table 4 shows the results of a picture of a woman (Lena) coded using various quality levels and compared to images coded using straight Vector DPCM.

Table 5 is a comparison of a two different pictures all coded with the same codebooks as Lena using a separate codebook for each component, thus demonstrating the independence of the codebook and high quality at a reasonably low bit rate. The results were all very good using the same set of codebooks. It should also be noted that much of the error found in these images is due to the sub-sampling that doesn't have a great affect upon visual image quality.

Discussion

This compression technique can probably be slightly improved in a couple of ways. The codebooks were all created using four pictures as described earlier. The codebooks may be improved by using a different set of pictures or a larger variety of pictures. Thus, the proper selection of pictures used to generate the codebook is not clear. Also, various methods of selecting vector codes (hierarchical trees, etc) could be explored in order to decrease the amount on encoding time.

Acknowledgements

I would like to thank Hsueh-Ming Hang and Joe Othmer for numerous valuable discussions and Bill Ninke for his support of this work.

Table 4

Lena 256 Vector Codes Independent Codebooks for Y, U , and V.

Method	Sub-Sampled	Sq error/pel	Bits/pel	Quality
Vector DPCM	3-1	52	1.08	
Vector DPCM	2-1	63	.75	
Vector DPCM	4-1	87	.56	
VB Vector DPCM	2-1	48	1.00	0
VB Vector DPCM	4-1	59	.86	0
VB Vector DPCM	4-1	73	.61	1
VB Vector DPCM	4-1	94	.46	2

Table 5

Various pictures 256 Vector Codes 4-1 Sub-Sampled Chrominance.

Picture	Sq error/pel	Bits/pel	Quality
F16	78	.83	0
F16	90	.62	1
Peppers	114	.85	0

References

[1] Gray R. M. "Vector Quantization", *IEEE ASSP Magazine*, pp. 4-29, April 1984.

[2] Gersho A. "On the Structure of Vector Quantizers", *IEEE Trans. on Information Theory*, Vol. IT-28, No. 2, pp. 157-166, March 1982.

[3] Baker R. L. and Gray R. M. "Image Compression Using Non-Adaptive Spatial Vector Quantization", in *Proc. 16th Asilmar Conf. Circuits, Systems & Computers*, pp. 55-61, October 1982.

[4] Baker R. L. and Gray R. M. "Differential Vector Quantization of Achromatic Imagery", *Proc. International Picture Symposium*, pp 105-106, March 1983.

[5] Rutledge, C. W. "Vector DPCM: Vector Predictive Coding of Color Images", *Con. Rec. Globecom'86*, pp. 1158 - 1164, 1986.

[6] Linde Y., Buzo A., and Gray R. M. "An Algorithm for Vector Quantizer Design", *IEEE Trans. on Communications*, Vol. COM-28, No. 1, pp. 84-95, January 1980.

[7] Limb J. O., Rubinstein C. B., and Thompson J. E. "Digital Coding of Color Video Signals -- a Review", *IEEE Trans. on Communications*, Vol. COM-25, pp. 1349-1385, November 1977.

Figure 2. Lena

Figure 3. Lena - Division of Blocks

Linear Block Prediction With Source Classification For Image Encoding Applications

K.S.Thyagarajan, and Sanjiv Bhatt
Electrical & Computer Eng. Dept., SDSU, San Diego, CA 92182
and Dept. of EE, Rutger°s University, NJ

Abstract

In this paper we will discuss the results of linear block prediction of images. An image can be considered to be a zero-mean real vector process by removing the mean vector. Vectors of dimension k=rxr are formed from subblocks of size rxr, and a subblock is predicted as a linear combination of p, q previous blocks along the row and column respectively. The coefficient matrices are chosen so as to minimize the mean square error over a given prediction frame of N x N pixels. In this paper we deal only with prediction along the rows. A Levinson type recursive algorithm can be used to obtain the coefficient matrices. As an application to image coding, the residual vectors are encoded using a vector quantizer (VQ) in a closed-loop fashion forming a vector DPCM (VDPCM). The prediction frames are classified according to their mean and variances using K-means algorithm and codebooks are generated for each individual class. This source classification will utilize the codebook more efficiently and result in lower data rate or better quality.

1. INTRODUCTION

It is well known that linear prediction techniques are used extensively in speech processing for different purposes such as for modeling, source coding, voice recognition, etc. In a similar manner images have been modeled as autoregressive (AR) processes, and the results of linear prediction for speech have been obtained for images as well. The simplest form of prediction, though, is the pel prediction in one dimension. Adding another dimension not only increases the computational complexity but also introduces some stability problems [1].

The concept of sample prediciton of speech signals has been advanced to encompass vector prediction by Cuperman and Gersho [2], and a gradient algorithm for the vector prediction by Chang and Gray [3], where speech vectors formed by grouping consecutive k samples are predicted based on past vectors. Since image pixels have, in general, block correlation, i.e., correlation over a finite spatial extent, it is natural to be able to predict a subblock based on previous blocks. In most practical systems, images are obtained in frames. Hence one can use either a causal or noncausal prediction model. This paper discusses the autoregressive (AR) modeling of images using causal linear block prediction. Further image subblocks are classified acoording to their mean and variance so that they can be represented more accurately. As an application of block prediction and classification, a vector quantizer is designed to encode the residual image to achieve data rate reduction.

2. Linear Block Prediction

Let us consider an image as a zero mean real vector random process. Eventhough an image has a nonzero mean intensity value, one can subtract the mean vector and consider the resultant process as a zero mean vector process. The mean vector can then be added at the output. This procedure has been adopted for pixel prediction in 2-D [1]. The vectors X_i are formed from rxr subblocks of the image where the first row of r pixels are the first r entries of X_i, the second row of r pixels are the second r entries of X_i, etc. Thus the vectors are k-dimensional where k=rxr. Implying causality, the present vector $X_{m,n}$ can be predicted as a linear combination of row and column blocks as given by

$$X_{m,n} = -\sum_{i}^{p}\sum_{j=0}^{q} A_{ij}\, X_{m-i,n-j} + \varepsilon_{m,n} \quad (1)$$

where $X_{m,n}$ and $\varepsilon_{m,n}$ are respectively the k-dimensional original vectors and prediction error vectors, and A_{ij} are the kxk coefficient matrices, p and q are the prediction orders along the row and column respectively. Due to the complexity of computation, we confine our discussion to prediction along one dimension. Then, (1) can be written as

$$X_n = -\sum_{i=\emptyset\,1}^{p} A_i\, X_{n-i} + \varepsilon_n \quad (2)$$

The coefficient matrices are chosen so as to minimize the mean square error between the actual and the predicted image blocks.

Reprinted from *Proc. IEEE,* vol. 2, pp. 1308–1311, April 1988.

The mean square error is

$$E(\|\epsilon_n\|^2) = tr(\sum_i \sum_{j=0}^{p} A_i R_{i-j} A_j^t) \quad (3)$$

where $A_0 = I$ and R_k is the kxk autocorrelation matrix given by

$$R_k = E(X_n X_{n+k}^t) \quad (4)$$

and t denotes the matrix transpose. It should be observed that $R_{-k} = R_k^t$. Define the block matrix R as

$$R(p) = \begin{bmatrix} R_0 & R_1 & \cdots & R_p \\ R_{-1} & R_0 & \cdots & R_{p-1} \\ \vdots & & & \\ R_{-p} & \cdots & \cdots & R_0 \end{bmatrix} \quad (5)$$

In the case of multichannel real zero-mean and wide-sense stationary process, It is shown in [4] that if R is positive semi-definite, then the minimization of (3) with respect to the coefficient matrices (A_i) with $A_0 = I$ exists. In the case of images, vectors formed from pixel blocks can be considered as a multichannel process and therefore the minimization of (3) is achieved by solving the set of equations

$$\sum_{i=0}^{p} A_i R_{-i} = \Gamma_p$$
$$\sum_{i=0}^{p} A_i R_{k-i} = \underline{0}, \quad k=1,2,..,p \quad (6)$$

In (6), Γ_p is the covariance of the minimum error with $tr(\Gamma_p)$ equal to the minimum mean square error. Also Γ_p is a k x k symmetric matrix and is uniquely determined from the autocorrelation matrices (R_0, R_1, ..., R_p). When dealing with a sample image data, the autocorrelation matrices are determined as follows [5] :

$$R_k = (r_{ij}^k) \quad (7)$$

where

$$r_{ij}^k = (1/N^2) \sum_{m=0}^{N-1} \sum_{n=0}^{N-k} x_i^{m,n} \, x_j^{m,n+k} \quad (8)$$

and N x N is the prediction frame size and p, the prediction order, $x_i^{m,n}$ is the ith element of the vector at location (m,n) in the image. Since the direct evaluation of (6) is inefficient especially when the order is large, we use the recursive algorithm to compute the A matrices and the corresponding minimum error covariance matrix Γ_p [4]. This is an extention of the multichannel Levinson recursive algorithm to the general degenerate case[4]. The transfer function H(z) of the AR model obtained above can be written as

$$H(z) = (A_0 + A_1 z^{-1} + \ldots + A_p z^{-p}) \sqrt{\Gamma_p} \quad (9)$$

For stabilty the poles of H(z) must lie inside the unit circle in the z-plane. As long as R(p) is positive semidefinite, H(z) is guaranteed to be stable [4].

3. Vector DPCM

In the context of linear block prediction, one can encode the residuals separately using a VQ. For a discussion on VQ, the reader is referred to [6]. The drawback in this appraoch is that the predictor input at the encoder is not matched to that at the decoder, with the result that the reconstructed image has streaks similar to the scalar DPCM. Instead one can use a vector DPCM where the residuals are quantized and used in the prediction loop. Since the input to the predictor at the encoder is identical to that at the decoder, the reconstructed image does not suffer from the drawback that the open-loop suffers. In closed loop operation, first a codebook is generated in an openloop fashion. Using this as the initial codebook a new error image is generated while the quantized error is used in the feedback. Next, the codebook is updated using the newly generated residuals. This process is continued until negligilble change in mimimum distortion is achieved [2]. However, it has been found that the codebook generated in openloop is sufficient to be used in the VDPCM without further optimization [7]. The degradation is minimal.

4. Source Classification

It is well known that the human visual system is more sensitive to edge details than to shade areas. Therefore, the codebook should reflect this information. However, a single codebook is not adequate to contain all of the information about the image since usually it is generated based on pixel intensity. One possibility is to use a composite source where each subsource characterizes a particular feature of the group of pixels under consideration. One can choose a number of features to classify an image block. When the classification is done in the pixel domain, one naturally uses edge/nonedge detectors to determine whether a small block (usually 4x4) has an edge or shade. Such classification, as it depends upon geometric apporaches, may not be visually compatible. Instead, one can use the mean and variance of a prediction frame to determine the class. Since these two parameters represent a much larger area of an image, they contain much more information than that obtained by the geometric method on a smaller block. The variance is a measure of edge activity while the mean represents the background. These two parameters may be sufficient for many applications.

Another possible parameter set is the Fourier transform coefficients or DCT coefficients. The Fourier transform coefficient at the top left-hand corner of an image block contains the dc or the mean value of that block, while the coefficients along the rows and columns portray the edge and perhaps texture features of the block. Similar argument holds good for the DCT. Since the phase is not a concern in image coding, it may be advantageous to use the DCT rather than the Fourier transform. Better yet, instead of using merely the Fourier coefficients, a computationally more effiecient method would be to use the mean, and sum of the squares of the absolute values of the Fourier or DCT coefficients along the first row and column of the transform block as feature vectors in the classifier. Yet another parameter set could be the autocorrelation coefficients obtained from the image block under consideration. In any case, a classification based on the spectral measure could be visually more compatible than that based on the geometric approach. An image encoder using source classification and VDPCM is shown in figure 6. First the original image to be encoded is divided into blocks of size NxN called the prediction frames, and the mean and variance of each prediction frame is computed. The individual mean and variance are fed to the classifier to determine the class of the source. The class index obtained from the classifier is used to access the corresponding codebook in the encoder, and prediction matrix, and is also sent as side information. Now the mean of the current prediction frame is removed and the mean-removed frame is input to the VDPCM. The VDPCM outputs the index of the codeword selected by the encoder corresponding to the current residual vector of dimension k. The decoder operation is fairly straightforward.

5. Experimental results

The block prediction scheme described above was used to obtain the AR model parameters of the UC girl image of size 512 x 512 with 8 bpp. Different prediction order values were used for a fixed prediction frame size, and the corresponding minimum error was obtained. Fig.1 shows the variation of the prediction error versus the prediction order for a prediction frame size of 32 x 32 pixels. The ordinate represents the error per pixel normalized with respect to the energy of the original image. In fig.2 is shown the variation of the prediction error versus the prediction frame size for a fixed order of 3. As per our assumption, the residual process is zero-mean. This is evident from fig.3, where a profile of the residual image corresponding to row 100 is shown. The corresponding original profile is shown in fig.4. The histogram of the residual image is shown in fig.5 where it can be seen that the distribution is confined to a much smaller range as compared to the original intensity image.

Next, a source classifier is designed using only the mean and variance parameters of blocks of size 64x64. The prediction matrices of the classes are determined as the centroids of the respective clusters. Codebooks of different lengths are generated for each class using the LBG algorithm [6]. Finally, the UC girl image is encoded via the classified VDPCM scheme as described above. The image quality is found to be better than the one without the source classification. In the experiment, the loop was not iterated for codebook optimization and the codebooks were generated in open loop. But the quantizer was used in the loop to predict pixel blocks.

6. Conclusions

The linear block prediction technique has been applied to obtain the AR model of an image. As expected, the prediction error decreases as the order is increased for a fixed frame size. Further, it is found that the error decreases rapidly as the prediction frame size is increased with the fixed order. Beyond a prediction frame size of 64x64, It is seen that the rate of decrease of error is relatively small. As an application to image coding, a vector DPCM has been designed to encode the residuals. It has been found that quantizing the residuals independently after prediction causes problems in the reconstruction. Hence it is necessary to include the matrix quantizer in the prediction loop so that the input to the decoder is the same as the error transmitted. This is what is called D*PCM by Jayant and Noll [8]. The visual quality of the reconstructed image is good with an SNR of about 18 dB. The compressed image with a source classifier has a better visual quality compared to the one without the classifier.

This work was supported in part by Naval Ocean Systems Center (NOSC) under grant N660001-85-D-0203.

7. References

[1] Maragos, P.A., Schafer, R.W., and Mersereau, R.M., "Two-Dimensional Linear Prediction and Its Application to Adaptive Predictive Coding of Images," IEEE Tran. ASSP, vol.ASSP-32, pp1213-1229, December 1984

[2] Cuperman, V., and Gersho A., "Vector Predictive Coding of Speech at 16 kbits/s," IEEE Tran. Com., vol. COM-33, pp 685-696, July 1985

[3] Chang, P.C., and Gray, R., "Gradient Algorithms for Designing Predictive Vector Quantizers," IEEE Trans. ASSSP, vol.34, pp.679-690, August, 1986

[4] Inouye Y., "Modeling of Multichannel Time Series and Extrapolation of Matrix-Valued Autocorrelation Sequences," IEEE Tran. ASSP, vol. ASSP-31, pp 45-55, February 1983

[5] Jenkins, G.M., and Watts, D.G., Spectral Analysis and its applications, Holden-Day, 1968

[6] Gray, R.M., "Vector Quantization," IEEE ASSP Magazine, April 1984

[7] Thyagarajan, K.S., and Bhatt Sanjiv, "Linear Block Prediction of Images For Encoding Applications," ICC-87, June 1987

[8] Jayant, N.S., and Peter Noll, Digital Coding of Waveforms, Prentice-Hall, 1985

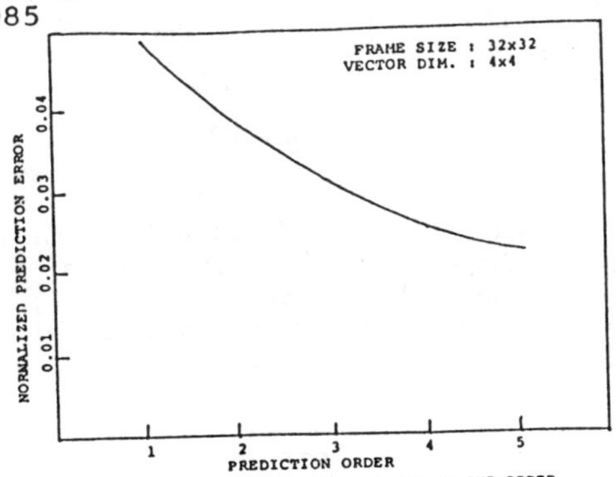

FIG. 1. VARIATION OF ERROR VERSUS ORDER

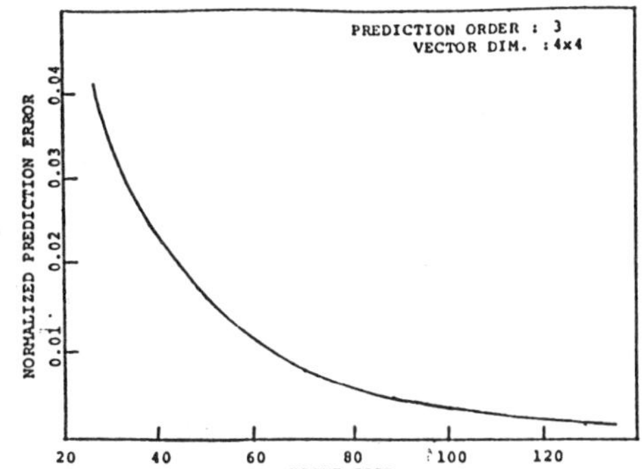

FIG. 2. VARIATION OF ERROR VERSUS FRAME SIZE

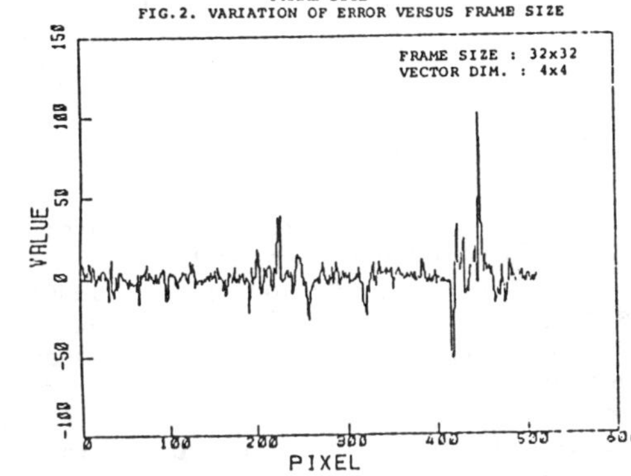

FIG. 3. PREDICTION ERROR PROFILE OF ROW 100

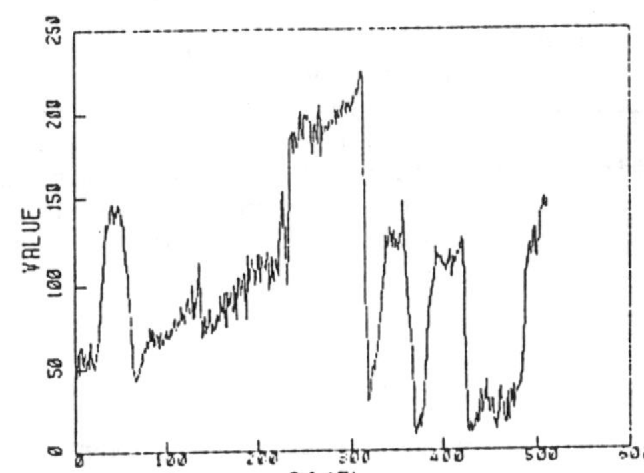

FIG. 4. ORIGINAL IMAGE PROFILE ROW 100

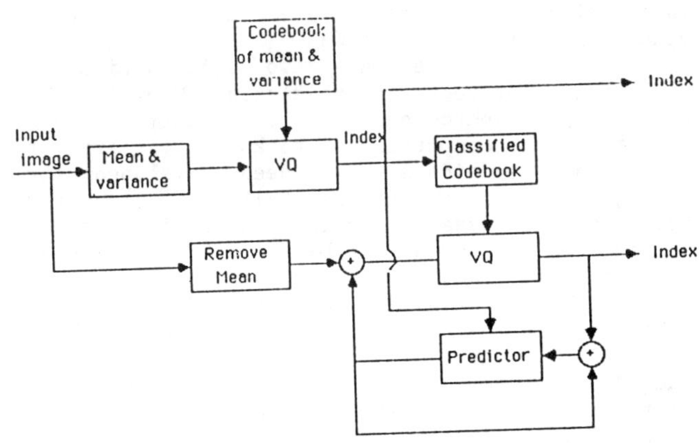

Figure 7. Block diagram of a VDPCM with source classifier

ADAPTIVE DISCRETE COSINE TRANSFORM CODING WITH VECTOR QUANTIZATION FOR COLOR IMAGES

Kiyoharu Aizawa , Hiroshi Harashima and Hiroshi Miyakawa

Department of Electrical Engineering,
The University of Tokyo
7-3-1 Hongo Bunkyo-ku , Tokyo , 113 Japan

ABSTRACT

A new vector quantization scheme in discrete cosine transform domain, named DCT-VQ and its application to color image coding are described. In this scheme, DCT-domain is partitioned into vectors which are normalized and vector-quantized using universal vector quantizers designed with multidimensional Laplacian distribution. Adaptive coding scheme is also introduced to obtain better reconstruction of images. The color image coder employs the above scheme and encodes separately three components converted from R,G,B signals.

The simulations have shown that adaptive DCT-VQ exhibits better performance than a conventional adaptive cosine transform coding with scalar quantization. The decomposition of DCT-block into vectors results in much less complex coder than a vector quantizer in original space domain.

1 INTRODUCTION

Vector quantization has raised considerable interest since it can nearly achieves optimal rate-distortion performance in principle, and its applications to low rate image coding have been widely investigated.[3]-[10]

However, there are two difficulties in application of vector quantization. One is the reduction of vector quantizer complexity because of its large computational task. A variety of suboptimal techniques have been developed for it [1]. The other is as follows. A vector quantizer is usually designed in original space domain by a cluster algorithm named LBG algorithm developed by Linde, Buzo and Gray [2]. Since this procedure uses training data set derived from specific images, a vector quantizer shows considerable degradation of quality for images not included in the codebook design [3]. Then it is difficult to use a universal vector quantizer in original space domain.

One approach to avoid this drawback is to vector-quantize transformed signals which ideally does not contain any statistical variations. For example, separating-mean-normalized vector quantization [4] may be considered one of this approach. If a probability distribution of transformed signals can be modeled as a well-defined standard distribution, a universal vector quantizer can be designed from the model.

From this point of view, in this paper, a vector quantization scheme in discrete cosine transform domain, named DCT-VQ [5][6] is presented. In this scheme DCT-domain is partitioned into vectors which are normalized and vector-quantized. Since normalized ac-components are well modeled as Laplacian distribution with zero mean and unit variance, they can be efficiently quantized using universal vector quantizers designed with multidimensional Laplacian distribution with zero mean and unit variance. Then DCT-VQ provides a coding scheme using an universal vector quantizer independent of specific images. The bit allocation which corresponds to the size of the codebook of each vector quantizer can be optimized based on an asymptotic theory. In order to improve the quality of reconstructed images, adaptive coding scheme based on activity index is incorporated in the above encoding procedure ,which is named adaptive DCT-VQ.

The section 2 describes the adaptive DCT-VQ in more detail. In section 3 its application to color image coding is presented. In section 4 the performance of adaptive DCT-VQ and conventional adaptive cosine transform coding with scalar quantization are compared.

2 ADAPTIVE DCT-VQ [7]

Coding

A block diagram of adaptive DCT-VQ for monochrome images is shown in Figure 1. Each operation is described below.

(1) An input image is divided into 8x8 pixel sub-blocks and a two dimensional DCT is performed on each sub-block.

(2) Transform sub-blocks are classified into four classes according to their level of activity (ac-energy) such that each class contains the same number of sub-blocks. Adaptivity is provided by this operation since the coding parameters ,i.e., bit allocation and normalization coefficients are determined in each class[14].

(3) Each transform sub-block is partitioned into vectors V_i(i=0,..,14) according to Figure 2. V_0 is a dc transform coefficient and the other vectors are composed of ac transform coefficients.

(4) The variance of each component of vector V_i (i=1,..,14) is computed within each class. These variances are used to determine the normalization coefficient and the bit assignment for each vector of that class.

Reprinted from *Proc. IEEE Int. Conf. Acoust., Speech, Signal Processing*, vol. 2, pp. 985-988, April 1986.

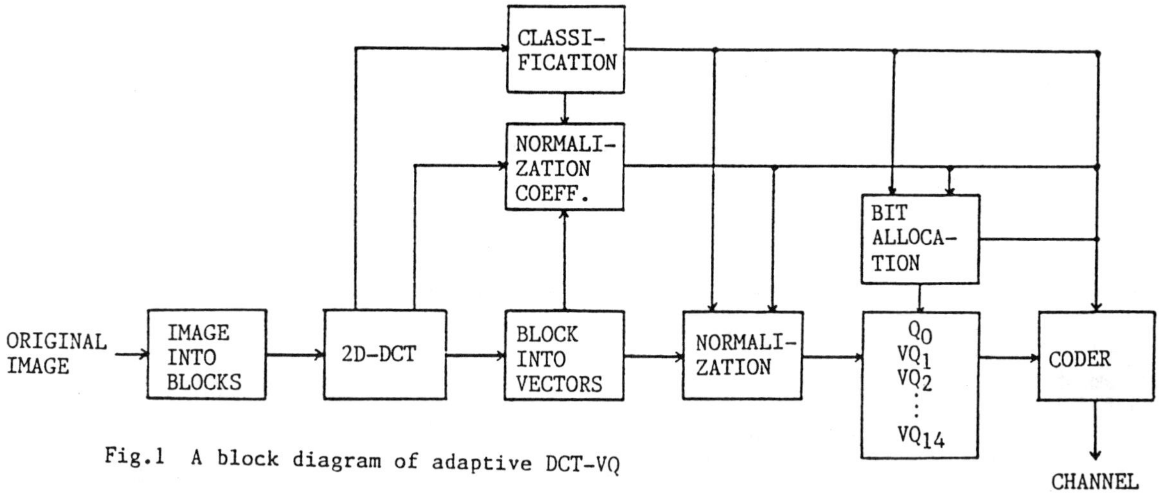

Fig.1 A block diagram of adaptive DCT-VQ

(5) Each vector V_i(i=1,..,14) is normalized by the normalization coefficient $\hat{\sigma}_i$. Here $\hat{\sigma}_i$ is defined as

$$\hat{\sigma}_i = \sqrt{(\sigma_{i1}^2 \cdot \sigma_{i2}^2 \cdot \cdot \cdot \cdot \sigma_{ik_i}^2)^{1/k_i}} \tag{1}$$

where σ_{ij}^2 represents the variance of jth component of vector V_i and k_i is the dimension of V_i.

(6) The number of bits assigned to vector V_i corresponding to the size of the codebook of each vector quantizer VQ_i is given by

$$b_i = \frac{k_i}{2} \log_2\left(\frac{\frac{C_{ki}}{k_i} \hat{\sigma}_i^2}{\bar{D}} \right) \tag{2}$$

$$C_{ki} = \frac{2}{\pi} \left(\frac{k_i+2}{k_i} \right)^{k_i+1} \Gamma\left(\frac{k_i}{2}+1\right)^{2/k_i} \tag{3}$$

$\Gamma(\cdot)$: Gamma Function

where $\bar{D}$ is a parameter which determines bit rate. The derivation of the bit allocation is described later.

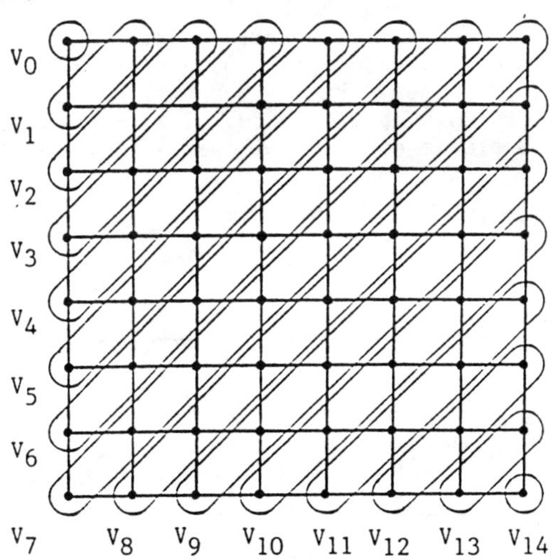

Fig.2 Decomposition of transform sub-block into vectors

(7) The dc transform coefficient, V_0, is quantized with a uniform quantizer with 8bits. V_i(i=1,..,14) is vector-quantized by VQ_i using the codebook which size corresponds to the bit allocation. The design method of vector quantizer is described later.

(8) Output indices of vector quantizer, and overhead information ,i.e., the bit allocation ,the normalization coefficients and the classification information are also transmitted.

<u>The Design Method and The Structure of Vector Quantizer</u>

Ac transform coefficients can be characterized as follows: (1)the variance of the transform coefficients of each vector are almost the same [11]; (2)transform coefficients are almost decorrelated; and (3)ac transform coefficients are well approximated by Laplacian distribution with zero mean [12]. Therefore, components of each normalized vector are considered to be well modeled as Laplacian distribution with zero mean and unit variance.

Consequently each vector quantizer VQ_i can be designed with multidimensional Laplacian distribution with zero mean and unit variance using LBG algorithm.

Two stage structure [1] is introduced into each VQ_i ,as shown in Figure 3, to deal with large bit assignment which appears in the bit allocation of a class of high activity. The second stage vector quantizer is generated from error vectors formed by taking difference between the input vector and the quantized vector by first stage.

Here, both the first stage and second stage vector quantizer are set to have codebooks up to 8bits each. If the bit assignment is less than 8bits, input vector is quantized only by the first stage vector quantizer. If the bit assignment is beyond 8bits, input vector is quantized by first stage quantizer with 8bits and error vector is quantized by second stage quantizer with remaining bits.

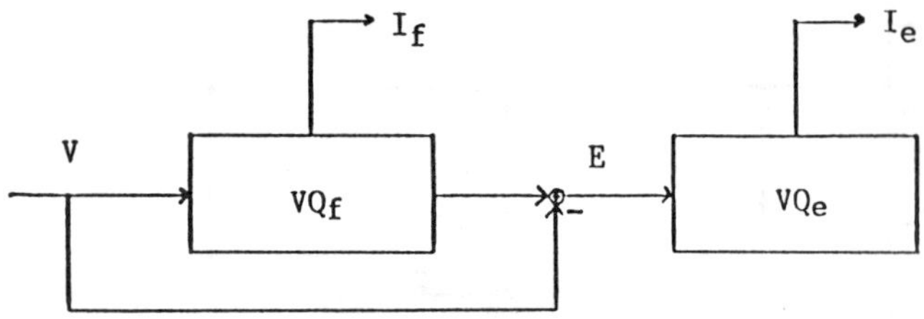

Fig.3 Two stage structure of VQ_i
VQ_f:first stage vector quantizer
VQ_e:second stage vector quantizer

Bit Assignment

Zador's lower bound [13] is used to evaluate vector quantization error. The coding distortion in each vector can be measured by equation(4) when ac transform coefficients are modeled as Laplacian distribution.

$$D_i = C_{ki}\,\hat{\sigma}_i^2\,2^{-2b_i/k_i} \qquad (4)$$

where D_i denotes the coding distortion in vector V_i and b_i is the number of assigned bits to V_i. $\hat{\sigma}_i$, C_{ki} are given in equation (1),(3) respectively.

To minimize total distortion in the sub-block, D_i must be constant ,i.e.,

$$D_i = \bar{D} \quad ; \text{ for all } i \qquad (5)$$

which results the optimum bit allocation equation(2).

Complexity

Large amount of complexity is a problem of vector quantization. In this coding scheme transform sub-block is decomposed into vectors V_i(i=0,..,14) which are separately encoded. If there are W_i codewords in each codebook, the product codebook has a size of $W=\prod W_i$. Since each codebook are searched independently, the total search complexity is much less than that of the equivalent size codebook in space domain.

For example, multiplications of vector quantizer in space domain and that of DCT-VQ at rate of 0.5bit per pixel are compared in table 1.

Encoding task could be more reducible ,since each vector quantizer can operate in parallel in hardware realization of DCT-VQ.

table 1. Comparison of the number of multiplications at rate of 0.5 bit/pel

coding method	multipications/8x8sub-block
8x8sub-block VQ in space domain	2.56×10^9
4x4sub-block VQ in space domain	1.64×10^4
8x8sub-block DCT-VQ	about 1.0×10^3 in average

3 COLOR IMAGE CODING

Figure 4 shows a block diagram of a color image coding system using adaptive DCT-VQ. To take advantage of the spectral redundancy three source tristimulus signals R,G,B are first converted into new three component signals which are then separately encoded by adaptive DCT-VQ.

Which color coordinate conversion is more suitable for data compression ? Usually YIQ or YUV coordinate are used. In this paper ,however, we decide to use a DCT for the color coordinate conversion. If the mean squared error in R,G,B signals is adopted for a distortion measure, a DCT is considered to be more preferable. Since a DCT is a orthogonal transform , it is suitable for the bit allocation ,i.e., parameters determining bit assignment can be set to the same value among new three color components. Besides a DCT has a superior energy compaction property.

New three components converted from R,G,B by a DCT are named C_0,C_1,C_2. The (0,0)coefficient of C_0 which corresponds to the average of R,G,B is uniformly quantized with 8 bits, while those of C_1,C_2 are nonuniformly quantized with assigned bits.

4 EXPERIMENTAL RESULTS

The color image "COUPLE" of SIDBA (a picture data base) is used as a test image. "COUPLE" is composed of 256x256 pels coded with 8bits per each R,G,B planes. Figure 5 shows the simulation results of coding performance. SNR is calculated for R,G,B planes each as follows:

$$SNR = 20\log_{10}\left(\frac{255}{\sqrt{M.S.E.}}\right) \qquad (6)$$

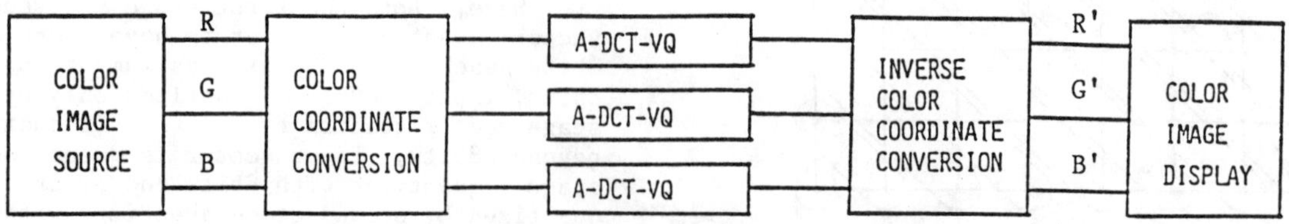

Fig.4 Color image coding system
A-DCT-VQ:adaptive DCT-VQ

In figure 5 A-DCT-VQ indicates the proposed scheme. A-DCT-SQ denotes the color image coder employing the conventional adaptive discrete cosine transform coding with scalar quantization on 8x8 sub-block according to the method described by Chen and Smith [14]. This coder also uses a DCT for the color coordinate conversion.

The simulation results shows that A-DCT-VQ coder outperforms A-DCT-SQ coder more than 1dB in each R,G,B signals.

A computer simulation shows that the A-DCT-VQ coder requires a computation time of only 1.5 times as much as the A-DCT-SQ coder.

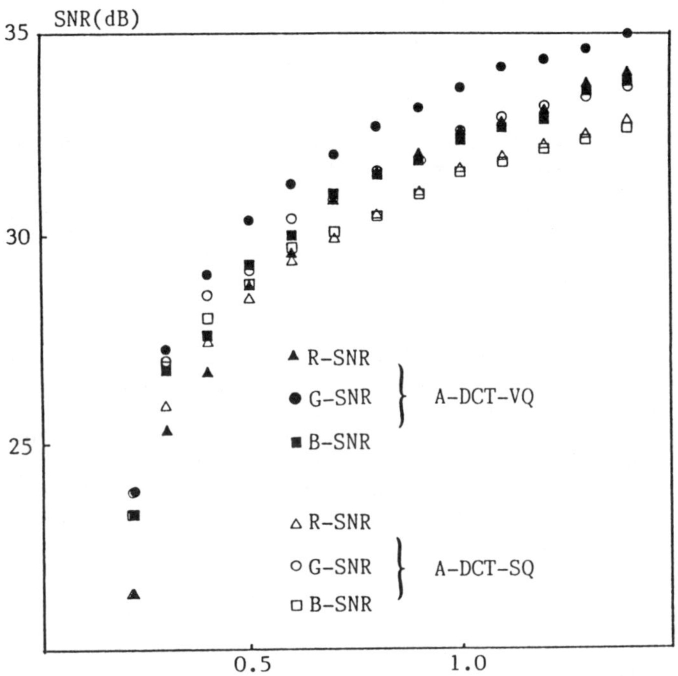

Fig.5 Comparison of coding performance

5 CONCLUSION

An adaptive vector quantization scheme in discrete cosine transform domain named adaptive DCT-VQ is presented. This scheme uses universal vector quantizers designed independently of specific images. The coding complexity of DCT-VQ is much less than a vector quantizer in space domain because of the decomposition of the transform sub-block into vectors. An Application of this scheme to color image coder using the DCT color coordinate conversion is also described.

The simulation results have shown that the coder employing adaptive DCT-VQ outperforms the conventional adaptive cosine transform coding.

ACKNOWLEDGMENT

We are grateful to Prof.Takahiro Saito of Kanagawa University for his helpful discussions.

REFERENCES

[1]R.M.Gray, "Vector Quantization" ,IEEE ASSP Magagine ,pp4-29 (April,1984)

[2]Y.Linde,A.Buzo and R.M.Gray, "An algorithm for vector quantizer design" ,IEEE Trans. on Commu.,COM-28,1,pp.84-95 (Jan. 1980)

[3]H.Takeo,T.Saito,H.Harashima and H.Miyakawa, "A study of mismatch in vector quantization of picture signals", IECE Japan, Tech. Rep.,IE83-103, (Dec.1983) (in Japanese)

[4]T.Murakami and K.Asai, "Vector quantization for video signals", J. Inst. TV Engrs. of Japan, vol.38,5,pp.452-457 (May 1984) (in Japanese)

[5]K.Aizawa,H.Harashima and H.Miyakawa, "Vector quantization of picture signals in DCT domain", IECE Japan, Tech. Rep.,IE84-86,(Dec.1984) (in Japanese)

[6]K.Aizawa,H.Harashima and H.Miyakawa, "Vector quantization of picture using discrete cosine transform(DCT-VQ)" , J. Inst. TV Engrs of Japan, vol.39,10,pp.920-925 (Oct. 1985) (in Japanese)

[7]K.Aizawa,H.Harashima and H.Miyakawa, "Adaptive vector quantization of picture signals in discrete cosine transform domain", IECE Japan, Tech. Rep.,IE85-5(Apr. 1985) (in Japanese)

[8]T.Saito, H.Takeo, K.Aizawa, H.Harashima and H.Miyakawa, "Multistage gain/shape vector quantization of picture signals" ,IECE Japan Tech. Rep.,IE85-16 (May 1985) (in Japanese)

[9]T.Saito,T.Komatsu,H.Harashima and H.Miyakawa, "Singular value decomposition(SVD) image coding scheme using vector quantization technique-SVD vector quantization" , J. Inst. TV Engrs of Japan, vol.39,10,pp.963-971(Oct.1985) (in Japanese)

[10]N.M.Nasrabadi, "Use of vector quantizers in image coding" , Proc. Inter. Conf. Acoustics, Speech and Signal Processing, Florida, March,1985

[11]A.Habibi,"Survey of adaptive image coding techniques", IEEE Trans. on Commun. COM-25,11,pp.1275-1284 (Nov. 1977)

[12]R.C.Reininger and J.D.Gibson, "Distributions of the two dimensional DCT coefficients for images", IEEE Trans. on Commun., COM-31, 11, pp.835-839 (June 1984)

[13]P.L.Zador, "Asymptotic quantization error of continuous signals and the quantization dimension", IEEE Trans. on IT, IT-28, 2, pp.139-149 (Mar. 1982)

[14]W.H.Chen and H.Smith, "Adaptive coding of monochrome and color images", IEEE Trans. on Commun., COM-25,11,pp.1285-1292 (Nov.1977)

Image Sequence Coding Using Vector Quantization

MORRIS GOLDBERG, MEMBER, IEEE, AND HUIFANG SUN

Abstract—**In this paper, a new interframe coding technique based upon vector quantization is presented. This algorithm has as its basis two-dimensional block vector quantization, at the frame level, onto which is grafted the concept of adaptive codebook replenishment and frame replenishment. This algorithm includes three strategies: *label replenishment*, *label replenishment with mean shift*, and *label replenishment with mean shift and selective codeword replacement*. The last two strategies efficiently update the codebook to track the changes in local statistics on a frame basis. Compared with three-dimensional block vector quantization [16], for bit rates between 0.5 and 0.65 the normalized mean squared error (NMSE) is reduced by a factor ranging between 2 and 2.5 by the use of replenishment.**

I. Introduction

HISTORICALLY, the techniques used to reduce the bandwidth required for transmitting digitized pictures have generally followed two paths: transform domain encoding or spatial domain encoding. The goal is to process the correlated data (image picture elements) in such a way that an uncorrelated set of data results. The processed data are then quantized and coded, so that data reduction is achieved [1]. Schemes for image sequence coding (interframe coding) are usually based upon frame replenishment [2], transform [3], predictive [4], and hybrid coding techniques [5].

In this paper, we propose a new algorithm for image sequence coding based upon vector quantization [8]–[21]. In vector quantization, the image data to be coded are first processed to yield a set of vectors. Then, a codebook of representative vectors (codewords) is generated using an iterative clustering algorithm such as K-means [6], or the closely related generalized Lloyd clustering algorithm proposed by Linde, Buzo, and Gray [7]. The vectors in the set are then individually quantized to the closest codeword of the codebook. Compression is achieved by replacing the representative vector (codeword) with a label. Reconstruction of the images can be implemented by table lookup techniques; the label is simply used as an address for a table containing the representative vectors (codewords). The algorithm of this paper has as its basis two-dimensional block vector quantization, at the frame level, onto which is grafted the concept of adaptive codebook replenishment and label replenishment. The underlying rationale is that a separate codebook, tailored to each frame, may be more efficient than one designed for the entire sequence. A smaller codebook, more representative of the local statistics, could conceivably give the same distortion at lower bit rate. However, the need for updating the codebook on a frame-to-frame basis increases the overhead and so could nullify these savings. In vector quantization, each vector is encoded by two distinct pieces of information: the label and the corresponding codeword. As both these pieces can be permitted to change independently, the concept of adaptivity can be applied to each. Comparing each input vector to the spatially corresponding one in the previous frame, there are three possibilities: a) the labels are the same; b) a new label is required; c) a new codeword is required. In the first case, the vector is in a nonchanging area, and so no new information need be transmitted. In the second case, an acceptable new codeword in the existing codebook can be found, so that only the new label need be transmitted. In the last case, the closest codeword in the existing codebook yields too large an error, so that a new codeword must be determined. In other words, the codebook must be updated to track the changes of the local statistics, by dropping certain codewords and adding new ones.

In the experiments described below, an extract of the CCETT image sequence of a woman talking on a telephone, consisting of 20 frames of size 112 × 96 pixels and quantized with 8 bits/pixel, has been used to perform the computer simulations. Significantly improved performance is obtained when label replenishment is combined with codebook replenishment. Furthermore, it is also demonstrated from the experimental results that adaptive codebook replenishment is better able to track the changes in the local frame statistics.

The rest of the paper is organized as follows. Section II reviews the main features of vector quantization. Section III describes an algorithm which uses vector quantization combined with frame replenishment, and an adaptive codebook replenishment scheme to track the local statistics. Finally, Section IV presents some discussions and conclusions.

II. Main Features of Vector Quantization

A. Basic Principle of Vector Quantization

An N-level vector quantizer Q is a mapping from a K-dimensional vector set $\{V\}$ into a finite codebook $W = \{w_1, w_2, \cdots, w_N\}$.

$$Q : V \rightarrow W. \tag{1}$$

In other words, it assigns to an input vector v a representative vector (codeword) w. The vector quantizer Q is completely described by the codebook $W = \{w_1, w_2, \cdots, w_N\}$, together with the disjoint partition $R = \{R_1, R_2, \cdots, R_N\}$, where

$$R_i = \{v : Q(v) = w_i\} \tag{2}$$

and w and v are K-dimensional vectors. The partition should ideally minimize the quantization error [22].

A block diagram of the various steps involved in vector quantization, as applied to image coding, is depicted in Fig. 1. The first step in image vector quantization is the decomposition of the image into a set of vectors $\{V\}$. In the second step, a subset of $\{V\}$, $\{T\}$, is chosen as a training set. In the third step, a codebook is generated from the training set $\{T\}$, normally with the use of an iterative clustering algorithm. The quantizer or coding step involves searching, for each input vector, the closest codeword w in the codebook W; the

Paper approved by the Editor for Image Processing of the IEEE Communications Society. Manuscript received September 9, 1985. This paper was presented in part at the International Conference on Acoustics, Speech, and Signal Processing, Tampa, FL, March 1985. This work was supported in part by an operating grant from the Natural Sciences and Engineering Research Council of Canada.

M. Goldberg is with the Department of Electrical Engineering, University of Ottawa, Ottawa, Ont., Canada K1N 6N5.

H. Sun is with the Department of Electrical Engineering, University of Ottawa, Ottawa, Ont., Canada K1N 6N5, on leave from the Shanghai Aeronautical Radio Research Institute, Shanghai, China.

IEEE Log Number 8609091.

Reprinted from IEEE *Trans. Commun.*, vol. COM-34, pp. 703–710, July 1986.

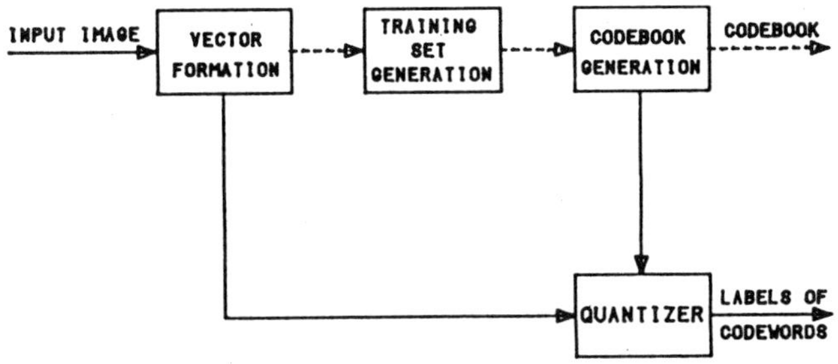

Fig. 1. Principle of vector quantization. The dashed lines correspond to training set generation, codebook generation, and transmission.

corresponding label of this codeword is then transmitted. Thus, the design decisions in implementing image vector quantization are as follows:

a) vector formation,
b) training set generation,
c) codebook generation,
d) quantization.

In this paper, the first three aspects are of particular concern.

B. Vector Formation

The first step in vector quantization is vector formation; that is, the decomposition of the images into a set of vectors. Many different decompositions have been proposed; examples include the color components of a pixel [11], the intensity values of a spatially contiguous block of pixels [8], [9], these same intensity values but now normalized by the mean and variance [17], the transformed coefficients of the block of pixels [14], [16], and the adaptive linear predictive coding (LPC) coefficients for a block of pixels [15].

C. Training Set Generation

An optimal vector quantizer should ideally match the statistics of the input vector source [7]. However, if the statistics of an input vector source are unknown, a training set representative of the expected input vector source is used to design the vector quantizer. If the expected input vector source has a large variance, then a large training set (and codebook) is needed. To alleviate this problem the input vector source can be subdivided. Gersho and Ramamurthi [8] divide the single input vector source into two more homogeneous sources corresponding to "edge" and "shade" vectors, respectively. Separate training sets are then used for each source and the resultant codebooks are then concatenated. In Boucher and Goldberg [18], small local input sources, corresponding to portions of the image, are used as the training sets; thus, the codebook can better match the local statistics.

D. Codebook Generation

The key step in vector quantization is the development of a good codebook. The optimal codebook, using the mean squared error (MSE) criterion (for the lack of a better measure), must simultaneously satisfy two necessary conditions [22].

1) The input vector source V is partitioned into N closed sets or Voronoi regions $\{R_i : 1 \le i \le N\}$ determined by the minimum distance rule:

$$R_i = \{v : |v - w_i| \le |v - w_j| \text{ for all } i \neq j\}. \quad (3)$$

2) The corresponding codewords are defined by

$$w_i = E\{v | v \in R_i\}, \qquad i = 1, 2, \cdots, N. \quad (4)$$

In other words, the codewords of the optimal codebook are the mean value of the corresponding Voronoi regions under the MSE measurement criterion. The K-means [6] or the closely related generalized Lloyd clustering algorithm proposed by Linde, Buzo, and Gray [7] are typically used to generate the codebook. Both these clustering algorithnms are an iterative process, minimizing a performance index calculated from the distances between the sample vectors and their cluster centers. These algorithms have the advantage of not requiring any knowledge of the underlying statistics of the input source and additionally minimizing a criterion related to the quantization error. However, these clustering algorithms can only assure a local optimum, which depends upon the initial codebook or cluster seeds [6].

Two basic approaches have been developed to obtain the initial codebook. In the first [26], the starting point involves finding a small codebook with only two codewords, and then recursively splitting the codebook until the required number of codewords are generated. This approach is referred to as binary splitting. The second starts with initial seeds for the required number of codewords, these seeds being generated by preprocessing the training sequence. Two examples of the latter approach are parametric or diagonal seeding [23] and mode-seeking seeding [24]. In the first, the seeds are located along the diagonals of the Euclidean hyperspace [23]. In the second, the seeds lie at the modes of the histogram [24].

E. Quantization

In vector quantization, the process of quantization involves finding the closest codeword for each input vector. In this paper, the straightforward expedient of searching the entire codebook is feasible, as the codebooks are relatively small.

III. Adaptive Vector Quantization by Replenishment

In an image sequence, successive frames are usually very similar. A simple interframe coding technique which exploits this similarity is frame replenishment coding [2]. In this method, only the information necessary to describe the new grey level values of the pixels that change between successive frames is transmitted. The implementation uses two frame memories, one at the receiver and another at the transmitter to store the required reference frames. The incoming signal is compared pixel by pixel with the reference frame. When the magnitude of the frame-to-frame difference is greater than a given threshold, it is considered as significant, and the new signal value is replaced in both frame memories. Thus, the receiver memory is made to track the transmitter memory. Studies [4], [25] indicate that the pixels of the changing area need not be transmitted independently of one another. Various techniques such as predictive coding, motion-compensation, and others have been used to exploit the correlation between these pixels [4].

In this section, a new algorithm which combines the concept of frame replenishment with two-dimensional block vector quantization is proposed. Each frame is first divided into two-dimensional blocks, and vectors are then extracted from the resultant blocks, either directly in the spatial domain or indirectly through the transform domain. The vectors corresponding to the first frame are taken as the training set for generating the first codebook. In encoding subsequent frames, the operation of replenishment can be applied to both labels (label replenishment) and codewords (codebook replenishment). A more detailed description of each operation follows.

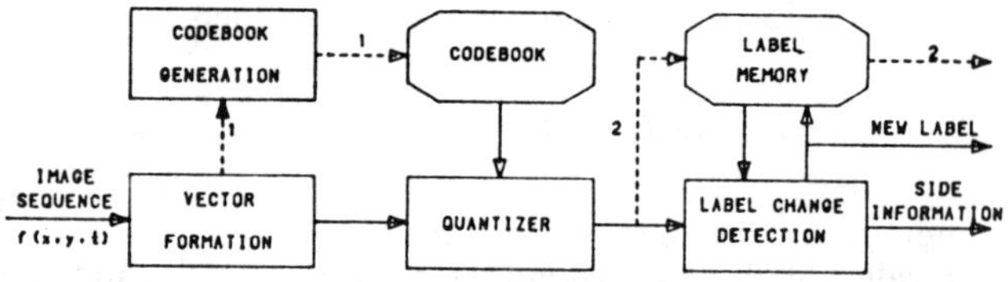

Fig. 2. Schematic diagram for label replenishment on the direct two-dimensional vector quantization. Dashed line 1 corresponds to the codebook generation step using the first frame. Dashed lines 2 correspond to storage and transmission of the labels of the first frame. The remaining solid lines correspond to coding the subsequent frames.

A. Label Replenishment

The basic principle of operation of label replenishment using direct two-dimensional block vector quantization, illustrated in Fig. 2, consists of two phases. The first phase is codebook generation represented by the dashed line 1, where the first frame is used as a training set. The first frame is then quantized, and the labels are stored in the label memory and transmitted as indicated by dashed line 2. The next phase, represented by the solid line, corresponds to coding subsequent frames. Each vector is first quantized, the label is then compared to the one in the label memory, and if it differs, the memory is replenished. An extra bit is used as side information to indicate whether the label needs to be replenished. The bit rate for this coding scheme can be calculated as follows:

$$\begin{aligned} \text{BR} &= \text{overhead} + \text{side information} + \text{new label} \\ &= \text{BR}_o + \text{BR}_s + \text{BR}_{\text{nl}} \\ &= N \times B_c / N_p + 1/N_b + N_c \times (\log_2 N)/N_f \qquad (5) \end{aligned}$$

where

- BR_o the bit rate for the codebook overhead
- BR_s the bit rate for the side information
- BR_{nl} the bit rate for transmitting the new labels
- B_c the average number of bits per codeword
- N the number of codewords in the codebook
- N_p the number of pixels in the input source
- N_b the number of pixels per block
- N_c the number of new labels
- N_f the number of pixels per frame.

Note that the first two terms, the overhead required for transmitting the codebook and the side information, are fixed. The third term depends upon the number of labels changing between successive frames, and is therefore variable.

Returning to the vector formation step in Fig. 2, each frame is first divided into two-dimensional blocks. The two-dimensional Hadamard transform is first applied to the block (4 × 4 are used below), thus decorrelating the components. The block mean, containing the most information, is scalar quantized; the remaining coefficients are first scalar quantized, where the number of bits are allocated according to the energy distribution of the coefficients [27], and then vector quantized [see Fig. 3(a)]. The encoding scheme requires, in addition to a quantizer, two separate frame memories, and two associated bits of side information: one for the label as before and one for the block mean [see Fig. 3(b)]. For this scheme, the bit rate can be calculated as follows:

$$\begin{aligned} \text{BR} &= \text{overhead} + \text{side information} + \text{new label} + \text{new mean} \\ &= \text{BR}_o + 2 \times \text{BR}_s + \text{BR}_{\text{nl}} + \text{BR}_{\text{nm}} \qquad (6) \end{aligned}$$

where the term $\text{BR}_{\text{nm}} = B_m \times N_c/N_f$ is the bit rate for the new

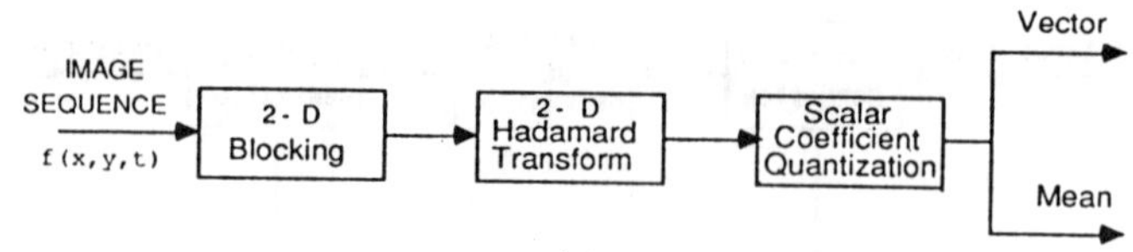

(a)

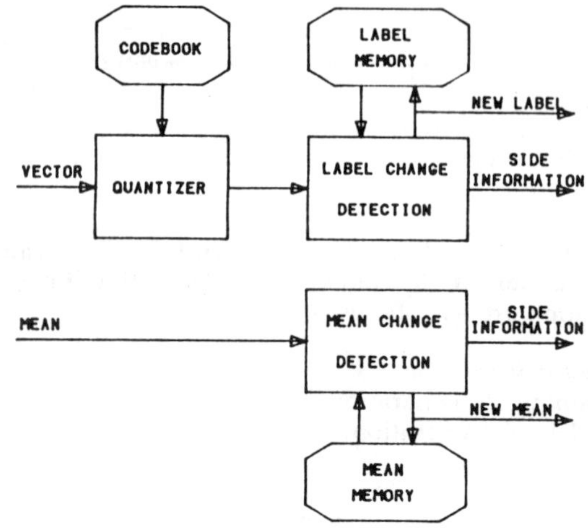

(b)

Fig. 3. (a) Vector formation and block mean separation step. The images are first divided into two-dimensional blocks and Hadamard transformed. The coefficients are scalar quantized with a variable number of bits. The block mean (zero-order) component is separated, and the remaining coefficients are used to form the vector. (b) Encoder of two-dimensional Hadamard transformation with vector quantization involves a quantizer and two separate frame memories, and therefore, two bits of side information are needed: one for the label and one for the block mean.

cluster means (codeword). The remaining terms are the same as in (5).

Comparing with (5) above, this scheme requires the same overhead, twice as much side information, and an extra term for the new mean. The rate distortion curves shown in Fig. 4 are obtained by using four different threshold values, 2, 4, 6, and 8, for replenishing the block means. For these values, four different codebook sizes, 64, 128, 256, and 512, are used as well. The distortion is measured as the percentage normalized mean squared error (%NMSE), and is defined as follows:

$$\text{NMSE} = \frac{\Sigma\,(X_i - \hat{X}_i)^2}{\Sigma\,X_i^2}\ 100 \text{ percent}$$

where X_i and $\hat{X}_i$ are the original and quantized grey levels, respectively. From the rate distortion curves shown in Fig. 4, it can be seen that at low bit rates (lower than 0.6 bits/pixel) large threshold values give lower distortion, whereas at higher bit rates (higher than 0.6 bits/pixel) small threshold values give better performance. If the threshold value is too low, insignificant changes are replenished, resulting in only small additional reduction in distortion for a disproportionate increase in bit rate. A large threshold value would result in many significant changes not being replenished. A threshold value of 4 is chosen as a compromise in the subsequent experiments.

B. Codebook Replenishment

In vector quantization, each input vector is encoded by two pieces of information: the label and the corresponding codeword or representative vector. In the algorithm presented in the last section, changes in the input source were tracked by replenishing only the label component. The underlying assumption made is that vectors from the changing areas can be acceptably encoded with codewords from the existing code-

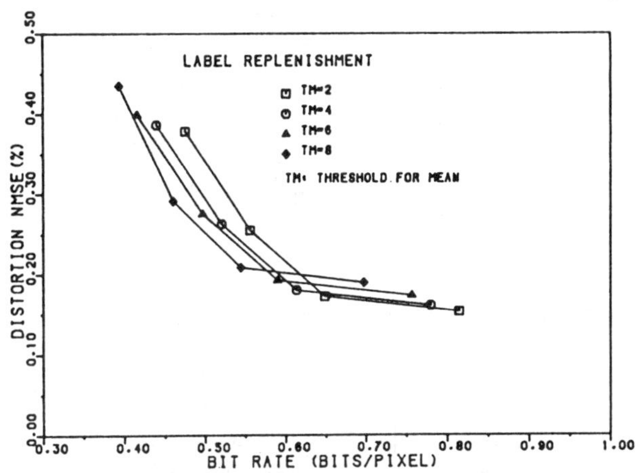

Fig. 4. Rate distortion curves for different threshold values used to replenish the block mean term. Here, a block size of 4 × 4 is used, and after Hadamard transform, the mean (zero-order) component is separated and the remaining coefficients are scalar quantized and used to form a vector. The codebook size is 256. At low bit rates (lower than 0.6 bits/pixel), large threshold values give lower distortion, whereas, for higher bit rates, small threshold values are better.

book; thus, only a new label need be transmitted. The main drawback to this approach is that a codebook which is representative of the entire sequence must be found. From the point of view of implementation, this implies that the statistics of the entire sequence be known beforehand. Furthermore, as demonstrated above, if the individual frames have different statistics, then a comparatively large codebook is required, so increasing the bit rate. An alternative strategy would be, therefore, to design a codebook which more closely matches the local frame statistics. One such algorithm for intraframe coding is described by Boucher and Goldberg [18], and uses a small codebook designed for different portions of the image. However, the new codebooks increase both the overhead and computation.

Instead of generating an entirely new codebook, it may be more advantageous to only update or replenish the codebook, especially if the changes are not large. The problem then becomes how to detect the changes and replenish the codebook. At same time, however, if the codebook is replenished in too dramatic a fashion, the amount of label replenishment increases correspondingly. Thus, it may be more advantageous, with respect to the overall performance, to restrict the type of replenishment for the codebook. With this observation in mind, two strategies for codebook replenishment are presented: *mean shift* and *mean shift combined with selective codeword replacement.*

1) Codebook Replenishment by Mean Shift: The main idea behind the mean shift algorithm is that the codebook is replenished by simply updating the individual representative vectors. Recall that from (4), these representative vectors are chosen to be the expected mean value over the Voronoi regions. If the local (frame) input sources are used to estimate the means, then changes in the local statistics on a frame-to-frame basis can be tracked. It should be noted that the mean shift algorithm has an important merit; it preserves as much as possible the original codebook structure, thus reducing the need for label replenishment. This scheme was first proposed in [20]; a detailed presentation follows.

The steps involved in the mean shift codebook replenishment are shown in schematic form in Fig. 5. The first frame is normally used to generate, in the standard manner, the initial codebook which is then transmitted:

$$w_i^1 = E\{v^1 | v^1 \in R_i^1\}, \qquad i=1, 2, \cdots, N \tag{8}$$

where v^1 is the training set drawn from the first frame. For

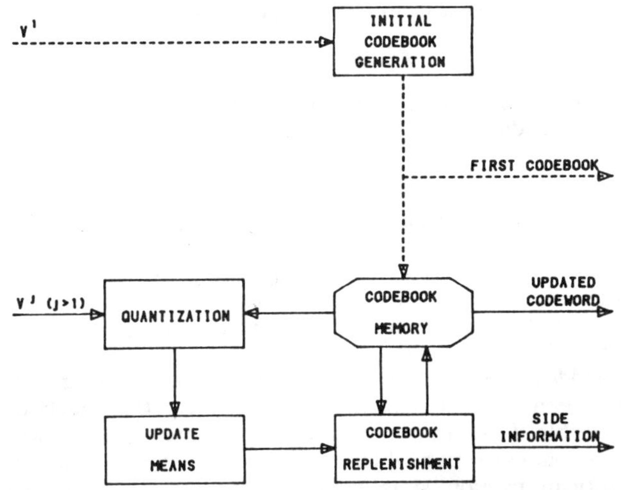

Fig. 5. Codebook replenishment by mean shift. The dashed lines correspond to the initial codebook generation step. The solid lines correspond to coding the subsequent frames. The input vectors are first quantized by the previous codebook stored in the codebook memory, and then used to update the codewords and the corresponding Voronoi regions. The current codewords are compared with the previous ones, and replenishment carried out if necessary.

subsequent frames, in particular the jth frame, there are three distinct steps.

a) Quantization: Each input vector v^j is assigned to the nearest Voronoi region R_k^{j-1}, using the nearest distance rule:

$$|v^j - w_k^{j-1}| = \operatorname*{Min}_i \ \{|v^j - w_i^{j-1}|\}. \tag{9}$$

Note that the codebook from the previous frame is used for the quantization. The label k is then compared to the one in the label memory and sent if it differs.

b) Updating Representative Vectors (Means): The vectors of the current frame, v^j, are used to update the representative vectors (cluster means) and corresponding Voronoi region:

$$w_i^j = E\{v^j | v^j \in R_i^{j-1}\}, \qquad i=1, 2, \cdots, N \tag{10}$$

$$R_i^j = \{v^j : |v^j - w_i^j| \le |v^j - w_m^j|, \text{ for all } i \neq m\}. \tag{11}$$

Note that v^j are the vectors formed from the jth frame in the sequence.

c) Codebook Replenishment: The current representative vectors $\{w^j\}$ are compared with the previous set $\{w^{j-1}\}$. If the difference is greater than a preset threshold T_h, the codebook is replenished at both the transmitter and receiver:

$$w_i^j = w_i^{j-1}, \qquad \text{if } |w_i^j - w_i^{j-1}| \le T_h \tag{12}$$

and

$$w_i^j = E\{v^j | v^j \in R_i^{j-1}\}, \quad \text{if } |w_i^j - w_i^{j-1}| > T_h. \tag{13}$$

It is easily demonstrated that by tracking the shift in mean, the distortion is reduced. Let D_0 and D_s be the distortion before and after the shift in mean; then

$$D_0 = E\{(v^j - w^{j-1})^2\}$$
$$= E\{(v^j - w^j + w^j - w^{j-1})^2\}.$$

As the cross terms are zero,

$$E\{(v^j - w^j)(w^j - w^{j-1})\} = 0.$$

Thus, D_0 can be rewritten as

$$D_0 = E\{(v^j - w^j)^2\} + E\{(w^j - w^{j-1})^2\}. \tag{15}$$

However, the first term corresponds to D_s,

$$D_s = E\{(v^j - w^j)^2\} \tag{16}$$

so that we can conclude that the distortion is reduced by using the new codebook,

$$D_s \leq D_0. \tag{17}$$

We note in passing that the mean shift algorithm is equivalent to using the previous codebook $\{w^{j-1}\}$ as seeds for the clustering algorithm used for codebook generation. The new codebook $\{w^j\}$, however, is formed by applying only one iteration of this algorithm to the input vectors, $\{v^j\}$, of the jth frame.

2) Mean Shift and Selective Codeword Replacement: The mean shift algorithm for replenishing the codebook is applicable if the frame-to-frame changes are small. However, when changes occur which alter the local statistics in a more significant fashion, simply shifting the mean may no longer be sufficient. The codebook must, therefore, be replenished in a more complex fashion. One possible strategy would be to find a completely new codebook by applying an iterative clustering algorithm with the present means as seeds. However, this could change the correspondence at the label level between the frames. An alternative strategy is to create new codewords that can represent the changes, while at the same time preserving some of the previous codewords. A measure based upon the maximum expected distortion for each Voronoi region is introduced below and used to decide whether a new codeword should be created.

Recall that the vector quantizer Q is completely described by the codebook, $W^1 = \{w_1^1, w_2^1, \cdots, w_N^1\}$, together with the disjoint partition, $R^1 = \{R_1^1, R_2^1, \cdots, R_N^1\}$ for the first frame, where $R_i^1 = \{v^1 : Q(v^1) = w_i^1\}$, and v^1 is in the training set. We now define the maximum distortion of the region R_i^1:

$$d_{i\,\max} = \lim_{v \in R_i} |v^1 - w_i^1|. \tag{18}$$

For the jth frame with respect to $d_{i\,\max}$, a new set of regions, $R'^j = \{R_1'^j, R_2'^j, \cdots, R_N'^j\}$ is defined as follows:

$$R_i'^j = \{v^j : |v^j - w_i^{j-1}| \leq d_{i\,\max} \text{ and } |v^j - w_i^{j-1}| \leq |v^j - w_h^{j-1}| \text{ for } i \neq h\}. \tag{19}$$

The first constraint defines a hypersphere in the Euclidean space centered at w_i, while the second constraint removes those subregions of the hypersphere which intersect with the other Voronoi regions (R_h, $h \neq i$). If the complement of $R_i'^j$ with respect to R_i^j is taken, then a new region O_i^j is defined such that

$$R_i = R_i'^j \cup O_i^j. \tag{20}$$

The test for creating new codewords will now depend upon the number of vectors lying in the region O_i^j, that is, quantized as the ith codeword, but with error greater than $d_{i\,\max}$. An adaptive codebook replenishment strategy which combines the mean shift with selective codeword replacement is illustrated in Fig. 6. The following three phases can be distinguished.

a) Initial Codebook Generation: As before, the first frame is used as a training set to generate a codebook $\{w^1\}$. In addition, the maximum distortions $\{d_{i\,\max}\}$ with respect to the training set are evaluated and stored.

b) Quantization and Mean Shift: For subsequent frames, in particular the jth frame, each vector is assigned to the nearest codeword in the codebook $\{w^{j-1}\}$. After quantization, the same algorithm as in Section III-B-1, the mean shift, is used, that is,

$$w_i^j = w_i^{j-1}, \qquad \text{if } |w_i^j - w_i^{j-1}| \leq T_h \tag{21}$$

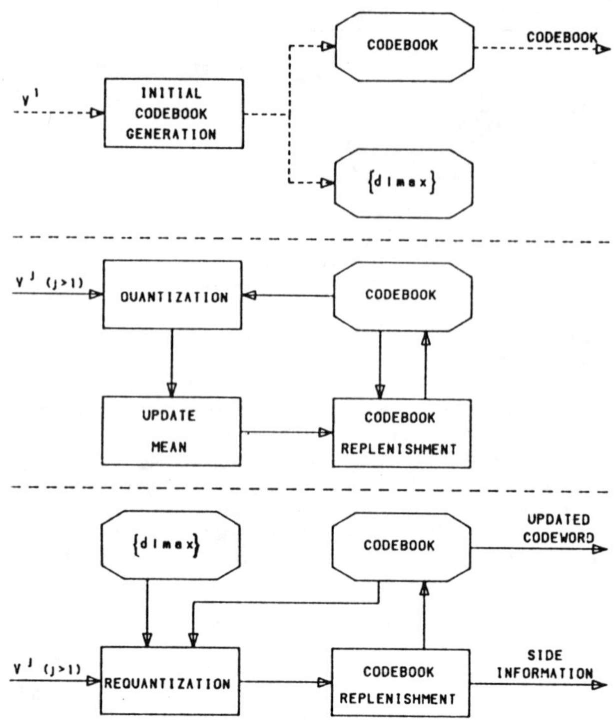

Fig. 6. The schematic diagram of mean shift with selective codeword replacement. The procedure includes: i) *initial codebook generation:* the first frame is taken as a training set to generate the codebook and the maximum distortions $\{d_{i\,\max}\}$; ii) *quantization and mean shift:* for the subsequent frames, each vector is assigned to the nearest codeword, and used to update the region mean, the codebook is then updated by shifting the mean; iii) *requantization and replacement:* the vectors of each Voronoi region, R_i^j, are assigned to two regions, one where the distortion is less than $d_{i\,\max}$, and the second where it is greater than $d_{i\,\max}$. The strategy is now to replace those w_i^j which correspond to the empty or near empty regions, $\{R_i^j\}$, by the mean of those $\{O_k^j\}$ which are more populated.

and

$$w_i^j = E\{v^j | v^j \in R_i^{j-1}\}, \qquad \text{if } |w_i^j - w_i^{j-1}| > T_h. \tag{22}$$

However, the updated codewords and side information are not as yet transmitted.

c) Requantization and Replenishment: During this phase the vectors of the jth frame, v^j, are assigned either to $\{O_i^j\}$ or $\{R_i'^j\}$, by using (19). In other words, the vectors of each Voronoi region, R_i^j, are assigned to two regions, one where the distortion is less than $d_{i\,\max}$, and the second where it is greater than $d_{i\,\max}$. The strategy is now to replace those w_i^j which correspond to the empty or near empty regions. The new regions are chosen among $\{O_k^j, k = 1, \cdots, N\}$ and the corresponding new codewords, which are transmitted, are calculated as follows:

$$w_i^j = E\{v^j | v^j \in O_k^j\}. \tag{23}$$

Computer simulations have been performed for both codebook replenishment algorithms; the results are provided in the next section. The bit rate can also be calculated with (6), but the overhead portion, BR_o, now includes the new codewords.

C. Experimental Results

The three replenishment-based coding schemes described above—label replenishment, label replenishment combined with mean shift, and label replenishment combined with mean shift and selective codeword replacement—were applied to the CCETT Picturephone sequence. The corresponding rate distortion curves are presented in Fig. 7, where the bit rate and normalized mean-squared error (NMSE) have been averaged

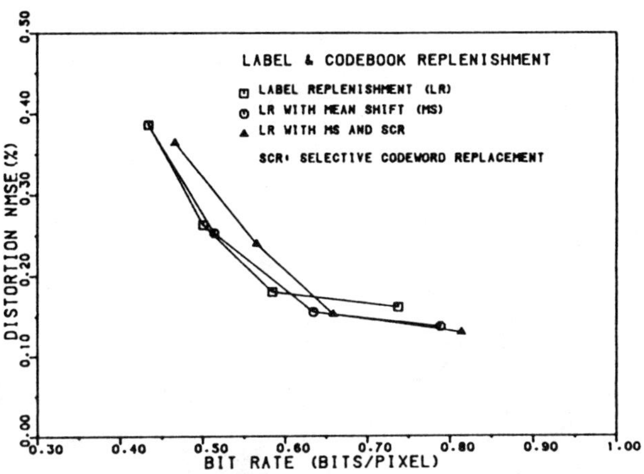

Fig. 7. Rate distortion curves for label and codebook replenishment, where the bit rates and the NMSE values are the averages over 20 frames. Three coding schemes are shown, label replenishment, label replenishment with mean shift, and label replenishment with mean shift and selective codeword replacement. At a bit rate above 0.6, label replenishment combined with codebook replenishment gives a 15 percent improvement in NMSE over straight label replenishment.

over the 20 frames; the four values given correspond to using codebooks of size 64, 128, 256, and 512, respectively. Fig. 8 shows the rate and distortion values on a frame basis. The first three entries correspond to the use of a codebook with 256 codewords; for the fourth entry, 512 codewords are used.

In terms of the average rate distortion, at low bit rates, label replenishment and label replenishment with mean shift are very similar. Both are better than selective codeword replacement. At bit rates above 0.6, label replenishment combined with codebook replenishment (both mean shift and mean shift with selective codeword replacement) gives about a 15 percent improvement in NMSE over straight label replenishment. A possible explanation is that at the lower bit rate, there are fewer codewords, forcing each to represent broad classes of vectors. Hence, in the bit rate calculation formula (6), the component for new labels, BR_{nl}, tends to be small, and the overhead (BR_o) becomes an important factor in the bit rate, whereas at higher bit rates, the situation is reversed: there are more codewords giving a finer representation of the classes of vectors. The finer representation is, therefore, more sensitive to small interframe variations and in (6), the component for new labels (BR_{nl}) becomes dominant. Of further note is the only marginal increase in performance achieved by selective codeword replacement compared to the mean shift. This may be due to the increase of the overhead with only marginal reduction in distortion.

An important feature of codebook replenishment, illustrated in Fig. 8(a), is the ability to track interframe variations and, by suitably modifying the codebook, to maintain a more constant distortion over the entire sequence. In Fig. 8(a), it is seen that the three replenishment coding schemes behave identically for the first nine frames, where there is little motion. At frame 10, an abrupt change begins and the distortion for label replenishment increases by about 40 percent and remains high for the rest of the sequence. However, label replenishment combined with codebook replenishment is able to adapt to the change and maintain almost the same level of distortion throughout the sequence. In terms of bit rate, codebook replenishment requires more bits, but as illustrated in Fig. 8(b), they are allocated selectively to the latter frames which contain more changes.

An important observation is the significant improvement of two-dimensional vector quantization with replenishment as compared to using vector quantization on the three-dimensional blocks (see [16]). For bit rates between about 0.5 and 0.65, the NMSE is reduced by a factor of at least 2. This can be seen from Table I, where the data for 3-D vector quantization are reproduced from [16].

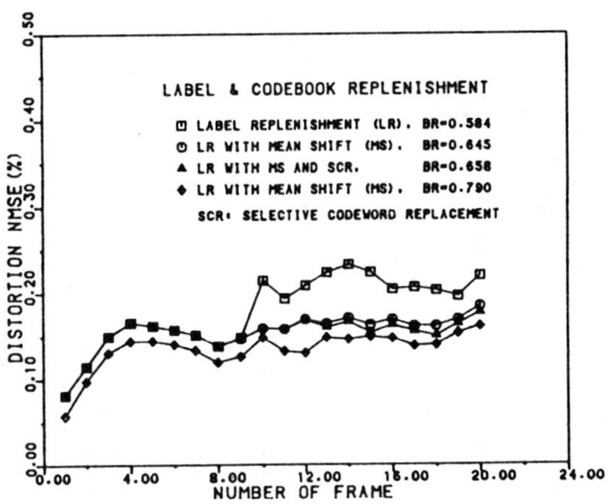

(a)

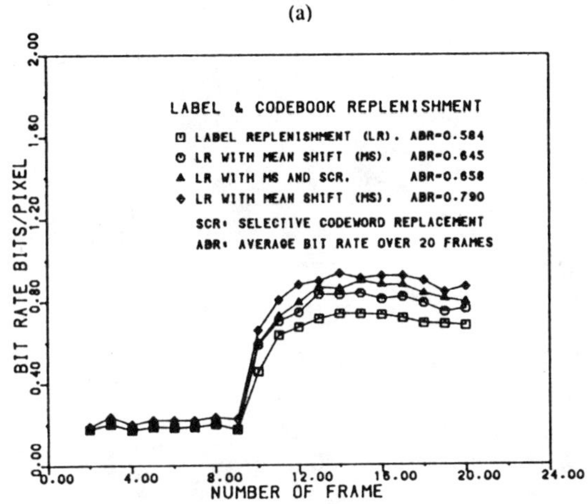

(b)

Fig. 8. (a) The distortion (NMSE) as a function of the frame number for the three replenishment schemes. Note that label replenishment combined with codebook replenishment maintains a relatively constant distortion even in the frames (10–20) containing changes. The first three schemes use 256 codewords and the last 512 codewords. (b) The bit rate as a function of the frame number for the three replenishment schemes. Codebook replenishment allocates more bits to the frames (10–20) containing changes, thus resulting in the more constant distortion shown in Fig. 8(a).

TABLE I
COMPARATIVE RATE DISTORTION VALUES FOR THREE-DIMENSIONAL VECTOR QUANTIZATION AND TWO-DIMENSIONAL VECTOR QUANTIZATION WITH REPLENISHMENT. THE HT-VQ DATA ARE FROM [16].

	Bit Rate	%NMSE	Codebook size
Label Replenishment (LR)	0.584	0.180	256
LR with Mean Shift (MS)	0.645	0.154	256
LR With MS and Selective Codeword Replacement	0.658	0.152	256
LR with MS	0.790	0.135	512
HT-VQ (12 coefficients)	0.494	0.461	512
HT-VQ (18 coefficients)	0.609	0.422	512

The visual results presented in Fig. 9 corroborate the rate distortion curves. In Fig. 9(a) is shown the coded images of the first frame using, respectively, the three-dimensional block Hadamard transform with vector quantization on the 12 lowest nonzero-order coefficients (from [16]); label replenishment with 256 codewords; and label replenishment with 512

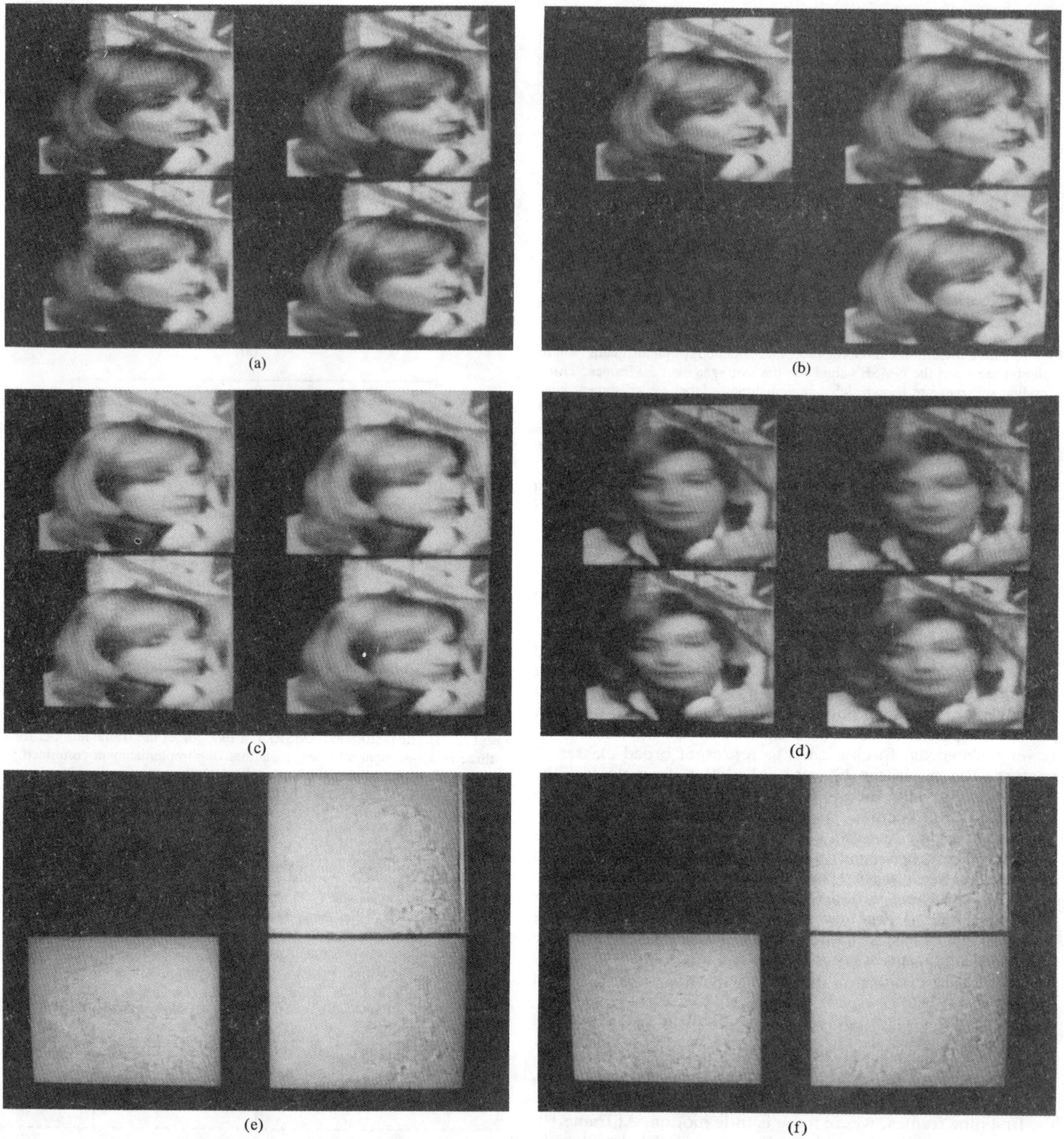

Fig. 9. Image sequence of a woman talking on a telephone, consisting of 20 frames of size 112 × 96 pixels and quantized with 8 bits/pixel. (a) The pictorial results for the first frame (the top left is the original). The three coded images correspond to, respectively, the three-dimensional block Hadamard transform with vector quantization on the 12 lowest nonzero order coefficients (bottom left), label replenishment with 256 codewords (top right), and label replenishment with 512 codewords (bottom right). Note that the images coded with label replenishment are clearly superior. (b) The pictorial results for the fourth frame. The two coded images correspond to, respectively, label replenishment with 256 codewords (top right) and label replenishment with 512 codewords (bottom right). A slight degradation is noticeable in the area of the lips, especially for the smaller codebook. (c) The pictorial results for the 10th frame. The three coded images correspond to, respectively, label replenishment with 256 codewords (top right), label replenishment with 256 codewords combined with mean shift (bottom left), and label replenishment with 512 codewords combined with mean shift (bottom right). The improvements are especially noticeable for the mean shift cases over the right eye and the highlights of the hair. (d) The pictorial results for the 14th frame, which is in a zone of rapid change. The three coded images correspond to, respectively, label replenishment with 256 codewords (top right), label replenishment with 256 codewords combined with mean shift (bottom left), and label replenishment with 512 codewords combined with mean shift (bottom right). Note that as the speaker turns her head a new portion of the background has been exposed, and that better reproduction is achieved with the mean shift. (e) The error images corresponding to the 10th frame. Note that the error is more uniformly distributed for the schemes with mean shift, which results in better reproduction. (f) The error images corresponding to the 14th frame. Note that the error is more uniformly distributed for the schemes with mean shift, which results in better reproduction.

codewords. The images coded with label replenishment are clearly superior. Fig. 9(b) shows the images of the fourth frame coded by label replenishment. A slight degradation is noticeable in the area of the lips, especially for the small codebook. Fig. 9(c) presents the results for frame 10, which begins to exhibit significant interframe changes. The three coded images correspond to, respectively, label replenishment with 256 codewords, label replenishment with 256 codewords combined with mean shift, and label replenishment with 512 codewords combined with mean shift. In both cases where the mean shift is used for codebook replenishment, improved reproductions are visible over the right eye and in the highlights of the hair. Finally, Fig. 9(d) presents the results for the 14th frame, which is in a zone of rapid change. Note that as the speaker turns her head, a new portion of the background has been exposed and better reproduction is achieved with the mean shift. To contrast the differences between the three coded images, the three corresponding error images are shown in Fig. 9(e) and (f), respectively, for the 10th frame and the 14th frame, where middle grey corresponds to no error. The improvement with codebook replenishment is confirmed by the error images. The error is more uniformly distributed through the image and, in particular, shows that the edges have in general been well reproduced.

IV. Discussion and Conclusions

In this paper, a new interframe coding technique based upon vector quantization is presented. This algorithm has as its basis two-dimensional block vector quantization, at the label level, onto which is grafted the concept of adaptive codebook replenishment and frame replenishment. For label replenishment, the underlying assumption made is that vectors from changing areas can be acceptably encoded with codewords from the existing codebook. However, if the individual frames have different statistics, the codebook, which is obtained from the first frame, must be updated to track these changes. This is implemented by label replenishment with mean shift and label replenishment with mean shift and selective codeword replacement. These strategies are very efficient and exploit the advantages of both frame replenishment and vector quantization. Compared to three-dimensional block vector quantization [16], at bit rates between 0.5 and 0.65, the NMSE is reduced by a factor of at least 2 by the use of the replenishment strategies. However, a practical consideration is the need to use a variable bit rate. In exchange for this variable rate, however, the scheme is highly adaptive and allocates more bits to those frames with large changes, thus maintaining a more constant level of distortion throughout the sequence.

Acknowledgment

This research was supported in part by an NSERC operating grant. We also thank Dr. R. Baker for his comments and suggestions on an earlier version of this paper.

References

[1] W. K. Pratt, Ed., *Image Transmission Techniques.* New York: Academic, 1979.

[2] W. F. Mount, "A video encoding system using conditional picture elements replenishment," *Bell Syst. Tech. J.*, vol. 48, pp. 2545–2554, Sept. 1969.

[3] J. A. Roese, W. K. Pratt, and G. S. Robinson, "Interframe cosine transform image coding," *IEEE Trans. Commun.*, vol. COM-25, pp. 1329–1339, Nov. 1977.

[4] B. G. Haskell, F. W. Mounts, and J. C. Candy, "Interframe coding of videotelephone pictures," *Proc. IEEE*, vol. 60, pp. 792–800, July 1972.

[5] A. Habibi, "Hybrid coding of pictorial data," *IEEE Trans. Commun.*, vol. COM-22, pp. 614–624, May 1979.

[6] J. Tou and R. C. Gonzalez, *Pattern Recognition Principles.* Reading, MA: Addison-Wesley, 1974.

[7] Y. Linde, A. Buzo, and R. M. Gray, "An algorithm for vector quantizer design," *IEEE Trans. Commun.*, vol. COM-28, pp. 84–89, Jan. 1980.

[8] A. Gersho and B. Ramamurthi, "Image coding using vector quantization," in *Proc. IEEE ICASSP*, May 1982, pp. 428–431.

[9] R. L. Baker and R. M. Gray, "Image compression using nonadaptive spatial vector quantization," in *Proc. IEEE Conf. Circuits Syst. Comput.*, 1983, pp. 55–61.

[10] P. Boucher and M. Goldberg, "Transform image coding by vector quantization," in *Proc. 9ème Colloq. Traitement du Signal et ses Appl.*, GRETSI, Nice, France, May 1983, pp. 629–632.

[11] E. E. Hilbert, "Joint pattern recognition/data compression concept for ERTS multispectral data," *SPIE*, vol. 66, "Efficient transmission of pictorial information," Aug. 1975.

[12] G. E. Lowitz, "Image data reduction," *Proc. Int. Conf. Spacecraft On-Board Data Management*, Nice, France, Oct. 1978, pp. 291–294.

[13] R. M. Gray, "Vector quantization," *IEEE ASSP Mag.*, pp. 4–29, Apr. 1984.

[14] R. A. King and N. M. Nasrabadi, "Image coding using vector quantization in the transform domain," *Pattern Recogn. Lett.*, vol. 1, pp. 323–329, July 1983.

[15] H. F. Sun and M. Goldberg, "Image coding using LPC with vector quantization," in *Proc. IEEE Int. Conf. Digital Signal Processing*, Florence, Italy, Sept. 1984, pp. 508–512.

[16] ——, "Image sequence coding using vector quantization," in *Proc. IEEE Int. Commun. Energy Conf.*, Montreal, P.Q., Canada, Oct. 1984, pp. 266–269.

[17] T. Murakami, K. Asai, and E. Yamazaki, "Vector quantizer of video signals," *Electron. Lett.*, vol. 7, pp. 1005–1006, Nov. 1982.

[18] P. Boucher and M. Goldberg, "Color image compression by adaptive vector quantization," in *Proc. IEEE ICASSP*, San Diego, CA, Mar. 1984, pp. 29.6.1–29.6.4.

[19] N. M. Nasrabadi and A. K. King, "A new image coding technique using transform vector quantization," in *Proc. IEEE ICASSP*, San Diego, CA, Mar. 1984, pp. 29.9.1–29.9.4.

[20] H. F. Sun and M. Goldberg, "Adaptive vector quantization for image sequence coding," in *Proc. IEEE ICASSP*, Tampa, FL, Mar. 1985, pp. 339–342.

[21] A. Gersho and M. Yano, "Adaptive vector quantization by progressive codevector replacement," in *Proc. IEEE ICASSP*, Tampa, FL, Mar. 1985, pp. 133–136.

[22] A. Gersho, "On the structure of vector quantizer," *IEEE Trans. Inform. Theory*, vol. IT-28, pp. 157–166, Mar. 1982.

[23] E. E. Hilbert, "Cluster compression algorithm: A joint clustering/data compression concept," Jet Propulsion Lab. Pub. 73-43.

[24] J. R. MacCalla and M. V. Chang, "Multispectral data compression using hybrid cluster coding," in *Proc. AIAA Commun. Satellite Syst. Conf.*, 1980, pp. 404–407.

[25] A. N. Netravali and J. D. Robins, "Motion-compensated television coding: Part I," *Bell Syst. Tech. J.*, vol. 58, pp. 631–670, Mar. 1979.

[26] A. Buzo, A. H. Gray, R. M. Gray, and J. D. Markel, "Speech coding based upon vector quantization," *IEEE Trans. Acoust., Speech, Signal Processing*, vol. ASSP-28, pp. 562–574, Oct. 1980.

[27] A. Habibi and P. A. Wintz, "Image coding by linear transformation and block quantization," *IEEE Trans. Commun.*, vol. COM-19, pp. 50–62, Feb. 1971.

A FINITE STATE/FRAME DIFFERENCE INTERPOLATIVE VECTOR QUANTIZER FOR LOW RATE IMAGE SEQUENCE CODING

Hsiao-hui Shen and Richard L. Baker

Integrated Circuits and Systems Laboratory
Department of Electrical Engineering
University of California
Los Angeles, California 90024
(213) 825-2327

ABSTRACT

Vector quantization (VQ) is an effective spatial domain image coding technique at sub-1.0 bpp rates, mapping discrete k-dimensional vectors into a digital sequence suitable for communication or storage. This paper presents a Finite State Vector Quantizer suitable for both intraframe and interframe coding. Interblock and interframe memory is expressed through a state variable driven by an improved classifier which keys on a block's neighbor's intensity gradients and its own interframe motion. A bilinear interpolator reduces the energy of foreground residuals encoded by the VQ, improving edge rendition and limiting blocking artifacts. Background blocks are vector predictive coded using simple frame differencing and codebooks built with a gradient design algorithm. The codec at 0.4 bpp outperforms a similar, but non-finite state intraframe VQ at 1.0 bpp by 4 dB.

I. INTRODUCTION

Vector quantizers are receiving wide attention as a promising spatial domain image compression technique at sub-1.0 bit per pixel (bpp) rates. A simple VQ operates on a block (or *vector*) of k pixels $\mathbf{x}$, locating its least distorting representation $\hat{\mathbf{x}}_i$ in a codebook $C = \{\hat{\mathbf{x}}_i, i = 1, \ldots, N_c\}$. Only the index i is sent to the receiver, which has a copy of C, yielding a rate of $k^{-1}\log_2 N_c$ bits per pixel. An excellent introduction to VQ may be found in [1]. Early applications to image coding include [2, 10].

Since the number of codevectors in a fixed rate VQ codebook grows exponentially with block size, VQ systems use relatively small blocks (typically less than 6×6 pixels) compared to conventional transform codecs. These blocks are often too small to exploit much of the correlation found in typical images. Edges become "staircased" at low rates and blocking patterns appear in smooth regions. Coping strategies include systems which classify blocks prior to the VQ (Classified VQ) [3-4], make VQ codebooks adaptive over small regions of an image [5], generalize DPCM to the vector case (Predictive VQ) [8-9], use product VQ codebooks (Product VQ) [6, 10-12], send a low resolution bilinear interpolated image across a side channel while VQ'ing the residual (Interpolative VQ) [13], or post-filter at the receiver to remove coding artifacts [7].

We have found that a blend of the Classified VQ, PVQ, Product VQ and IVQ strategies, cast into a Finite State VQ coding structure, can be very effective for both intraframe and interframe coding. We discuss salient design issues for the interframe coding application in this paper. Section II briefly reviews FSVQ and introduces the basic components of an interframe FSVQ, described in greater detail in Section III. Section IV presents experimental results.

II. INTERFRAME FINITE STATE VECTOR QUANTIZERS

A Finite State VQ uses a *state* variable $s \in \{s_i, i = 1, \ldots, N_s\}$ to identify which *sub-codebook* $C_s = \{\hat{\mathbf{x}}_i, i = 1, \ldots, N_c\}$ in the codec's *super codebook* $C = \bigcup_{s=1,\ldots,N_s} C_s$ to use for encoding a source vector $\mathbf{x}$. Ideally, the *next state function* f (which defines s) identifies the C_s which contains the best reproduction $\hat{\mathbf{x}}$ in all of C. If f uses only information available at the decoder, the system mimics a VQ having rate $k^{-1}\log_2 N_s N_c$ bpp, while sending only $k^{-1}\log_2 N_c$ bpp. The key to success is finding a good f.

Early FSVQs were used to encode speech [15-16]. Aravind and Gersho presented the first FSVQ for still images, using a novel perceptually based classifier to drive the codec's two-dimensional state [17]. Its performance at 0.375 bpp rivaled that of a memoryless Classified VQ operating at 0.7 bpp. We extended their FSVQ to image sequence sources at last year's ICASSP [18]. This paper presents an improved classifier (next state function), adds frame differencing for background blocks, uses an IVQ for the foreground (moving) areas, and includes a simple adaptive rate scheme to improve edge rendition. The result is nearly 7 dB better SNR at less than two-thirds the rate of the system in [18]. None of the designs presented here use motion compensation nor conditional replenishment techniques, although these could be incorporated in obvious ways.

This work was supported by Bell Communications Research and the Eastman Kodak Research Laboratories, through the University of California MICRO program.

A. Intraframe and Interframe Decomposition

Image sequences exhibit their greatest correlation from frame to frame in background (non-moving) areas. Perhaps the simplest vector predictor to exploit this memory uses the jth block from the previous frame $\hat{\mathbf{x}}_j^{k-1}$ to predict the current block $\mathbf{x}_j^k$ (Figure 1a). (Reconstructed blocks at the receiver are denoted with a hat.) A Frame Difference VQ (FDVQ) vector quantizes the residual,

$$\mathbf{r}_j^k = \mathbf{x}_j^k - \hat{\mathbf{x}}_j^{k-1}, \tag{1}$$

so the effective bit rate increases with each successive frame. Motion Compensation VQ (MCVQ) uses a more general version of the predictor, searching over a neighborhood surrounding $\hat{\mathbf{x}}_j^{k-1}$. Typically its decision is made using $\mathbf{x}_j^k$, which forces the predictor to transmit a motion vector as side information.

We have found that FDVQ can attain nearly perfect rendition of background areas just a few frames after a scene change. However, the performance of both FSVQ and the more powerful MCVQ is less than satisfactory when faced with rapidly moving foreground areas, which become peppered with quantization noise. Adding filtering within the prediction loop helps alleviate the problem (see, e.g., [22], elsewhere these proceedings). Intraframe algorithms, which cannot iteratively improve the rendition of a background area, tend to do better on these foreground areas as their quantization noise does not accumulate from frame to frame.

The interframe FSVQ presented here is a hybrid which capitalizes on the best of both techniques: interframe prediction using frame differencing (or, potentially, motion compensation) is used for background regions, while an intraframe predictor is applied to moving foreground regions. During the codebook design phase, a motion detector determines which scheme to use on a given block. After the codebooks are built, the encoder may either use the same motion detector or simply try both schemes on every block, choosing the one which yields least distortion. This is, in fact, what is done in the simulations in Section IV.

B. Basic Components

We see that an interframe FSVQ needs at least three sub-systems.

(1) *Vector predictor.* The predictor attempts to whiten and reduce the energy of vectors presented to the VQ. As mentioned, the type of predictor used depends upon whether a block is in the moving foreground or the stationary background. Intraframe prediction, such as that used by Vector Predictive Coding [8], Mean-Residual VQ [11-12], Predicted Mean VQ [18], or Interpolative VQ [13] may be applied in the former case; interframe prediction, such as frame differencing or motion compensation, may be used for the latter.

(2) *Block motion detector.* The motion detector directs the vector predictor to operate either in the interframe or intraframe mode. Its influence is greatest during the design phase, as it determines which vectors are used to train interframe versus intraframe codebooks.

(3) *Block classifier and next state function.* This is perhaps the most important element in an FSVQ codec. We have found that FSVQ performance is strongly influenced by the function used. Of particular interest are structures which drive the next state function with a classifier, as first introduced by Aravind and Gersho [17, 23]. The classifier serves as a front-end processor to the VQ. While the VQ typically minimizes squared error (MSE), we note that the classifier can operate on some criterion other than MSE, enabling the encoder's overall decision making process to optimize some hybrid performance measure.

The distinction is important. Since MSE is a per-letter, rather than a per-word measure, it is ignorant of the structure of perceptually important two-dimensional features within a block, such as edges and textures. A classifier which keys on such structures can parse source blocks into perceptually distinct sets. For each set, a VQ codebook can then be designed to minimize the squared reproduction error of all blocks *within the set.*

The union of all such sub-codebooks forms a super codebook. A sub-optimal encoder could search the super codebook by first classifying the source block. The indicated sub-codebook would then be searched under the MSE cri-

Reprinted from *Proc. IEEE Int. Conf. Acoust., Speech, Signal Processing*, vol. 2, pp. 1188–1191, April 1988.

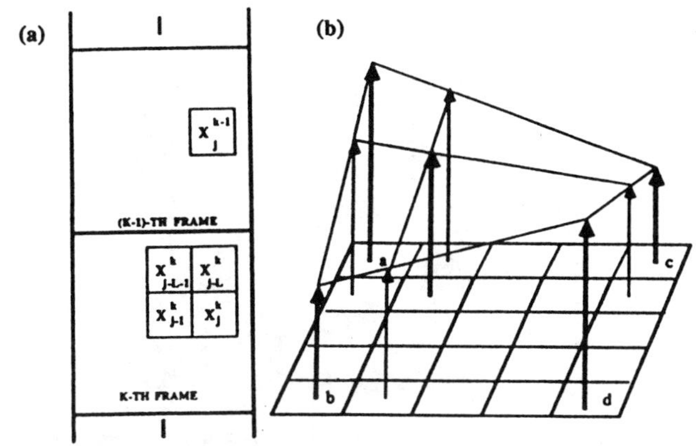

Figure 1. (a) Geometric relationship of source blocks. (b) Bilinear interpolation.

terion for the best match. We see that codevectors comprising the super codebook represent those which minimize a hybrid distortion criterion composed of the classifier's parsing decision criteria and the VQ encoder's Euclidean (MSE) measure. Such codebooks should yield perceptually better performance than a MSE-optimized tree codebook of the same size and structure. The classifier used in this paper is quite similar to part of the one described in [23], although it was arrived at independently.

III. SOME SPECIFIC DESIGNS

This section presents some specific FSVQ component designs with which we have experimented. Details of the FSVQ codebook design procedures essentially follow those in [18], and are omitted here for brevity.

A. Vector Predictors

Vector predictors prior to the VQ whiten the source and remove energy. If the predictor accurately estimates average block intensity, most FSVQ codevector patterns can be used at nearly any background intensity. This improves edge rendition and reduces blocking artifacts for a given VQ codebook size. Near zero codevectors represent shade blocks, while the majority of the codebook consists of edge and texture codevectors.

The FSVQ in Figure 2 forms vector predictions from bilinear interpolations of foreground blocks and from pixels in the previous frame for background blocks. The residual is quantized by the FSVQ. We discuss these predictors next.

Bilinear Interpolation (Foreground only)

In [18] we used linear predictors to form a scalar estimate of each current block's mean intensity, be it from the foreground or background, using neighboring reconstructed block mean values (Predicted Mean, or PM). That predictor improves MSE performance by a couple of decibels, but simulations show that excessive blocking occurs near sharp intensity gradients.

We have since experimented with a simple yet very effective scheme recently proposed by Hang and Haskell, called an Interpolative VQ [13]. The IVQ is reminiscent of Yan and Sakrison's two-component image coding [24]. The original image is sub-sampled on a two-dimensional grid, and the samples sent to the receiver. Next, an approximation to a low pass version of the image is formed by bilinearly interpolating (BI) between these points (Figure 1b, [14]) and the residual (approximately high pass image) is vector quantized. The reconstructed image exhibits much less blocking at a given rate than does one encoded by the Mean/Residual VQ [10-12] and Predicted Mean VQ [18], largely because a good *vector* valued prediction is made, rather than a scalar one.

The sub-sampled pixels may be DPCM coded to reduce side information rate. Also, the higher the decimation, the lower this rate becomes, but performance suffers. We have found that an 8x8 decimation and conventional PCM coding gives good results, at a cost of 0.125 bpp. We use the predictor only on foreground blocks, so the actual rate is somewhat less. Perceptually, the improvement over PM is striking. (See Figures 3a and 3b.)

Frame Differencing (Background only)

As mentioned earlier, the vector prediction used for each background block in the current frame x_j^k is simply the corresponding reconstructed block in the previous frame $\hat{x}_j^{k-1}$. The VQ encodes the residual, increasing the block's effective rate with each successive frame. Once accurately reconstructed, the block could be ignored until the next scene change or until it is occluded by foreground objects.

Frame Differencing VQ is a special case of a Predictive VQ having a non-adaptive predictor, so the steepest descent and stochastic gradient PVQ design algorithms recently reported by Chang and Gray [9] can be used to iteratively

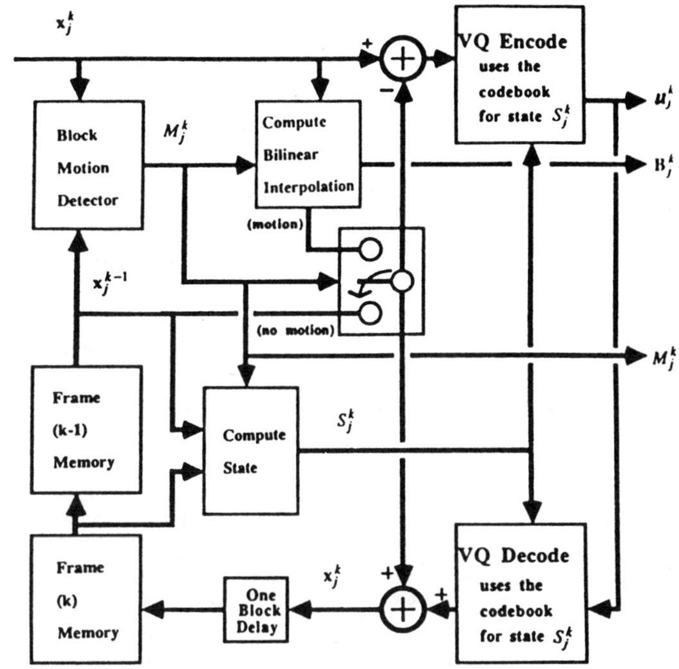

Figure 2. Finite State/Frame Difference Interpolative VQ block diagram.

improve an FDVQ codebook. In our interframe FSVQ, only one state, hence only one sub-codebook, is assigned to background block residuals. We use their steepest descent algorithm to design this sub-codebook. The codebook is first built using residuals formed without a quantizer in the prediction loop. This codebook is then embedded in the loop and a new set of residuals generated. The old codebook is then trained on these residuals, which is used to generate another set of residuals, etc.

We have found that frame differencing improves background fidelity by about 2 dB, although perceptually the improvement is quite noticeable. About 1 dB of the improvement comes from the iterative design algorithm.

B. Motion Detector

The vector predictor used for x_j^k and its associated state variable s_j^k depends upon whether x_j^k appears in the foreground or background. In Figure 2, this decision is controlled by a motion detector. Backward or forward detection may be used, but the latter performs best for fast motion [20]. Our experiments have used only forward motion detection.

The detector computes, for example, a correlation coefficient, a squared error distortion, or an absolute error distortion between x_j^k and its reconstructed ancestor in the previous frame $\hat{x}_j^{k-1}$ and compares the result to a threshold. These computations must be made prior to quantizing the current block, so the one bit outcome M_j^k is unavailable to the decoder without side information. In our experiments, we include one bit of overhead per block for M_j^k. Noiseless coding schemes, such as run length coding, could reduce this rate.

C. Block Classifier and Next-State Function

Two objectives of the classifier and next state function are to:

(1) Parse incoming vectors into perceptually distinct sets, to help combat the shortcomings of the VQ's MSE criterion.

(2) Consistently select the sub-codebook which contains the best reproduction vector, even when driven by noisy reconstructed data.

We have found that both objectives can be satisfactorily met by having the FSVQ classifier key on edges, insuring that blocks are encoded by a sub-codebook dominated by similarly oriented edge patterns.

In [18], we classified reconstructed blocks according to their intensity gradient in one of four directions. There were only 24 states in all, sixteen for foreground blocks, four for background blocks, and four for blocks along the image border. Results presented in the next section use an improved classifier with many more states. It is quite similar to the edge classifying section of Arvind and Gersho's intraframe FSVQ classifier [23].

The classifier operates as follows. Assume x_j^k is the 4x4 block we wish to encode. As noted in [23], any edge present in the pixels bordering x_j^k will, with high probability, extend into the block. Let W denote the pixels in a neighboring 2x2 window,

$$\mathbf{W} = \begin{bmatrix} e\,f \\ g\,h \end{bmatrix}.$$

We want to identify the direction of dominant edges in $\mathbf{x}_j^k$ using information available at the receiver, so the classifier convolves two 2×2 kernels

$$\begin{bmatrix} 1 & 1 \\ -1 & -1 \end{bmatrix}, \quad \begin{bmatrix} 1 & -1 \\ 1 & -1 \end{bmatrix}$$

over the last two rows of pixels in the upper reconstructed block $\hat{\mathbf{x}}_{j-L}^k$ and along the right two columns of pixels in the left block $\hat{\mathbf{x}}_{j-1}^k$ (Figure 1a). This yields two numbers per 2×2 window position, say F_1 and F_2. If $F = F_1^2 + F_2^2$ exceeds a threshold,

$$T = \begin{cases} k, & \text{if } W_{max} < b_1 \\ a_1\,(W_{max} - b_1) + k, & \text{if } b_1 < W_{max} < b_2 \\ a_2\,(W_{max} - b_2) + a_1\,(b_2 - b_1) + k, & \text{if } b_2 < W_{max} \end{cases} \qquad (2)$$

where $W_{max} = max\{e, f, g, h\}$, then we declare an edge to be present in the window with orientation

$$\phi = \tan^{-1} \frac{F_2}{F_1}. \qquad (3)$$

By comparing the magnitudes and signs of F_1 and F_2, (3) quantizes into four orientations, each with a white-to-black or black-to-white transition, or eight possibilities. The window in the upper block containing the strongest edge measured by F, and the window in the left block containing the strongest edge are identified. If the F for both windows exceeds T, there are 8·8=64 possible combinations of edge orientations, for 64 states. If only one window exceeds the threshold, we obtain 2·8=16 more states. If neither exceeds the threshold, we still are interested in the intensity gradient directions that are present, and so we designate 8·8=64 more states, for a total of 144 states. Our 145th state is used for all background (frame difference encoded) blocks. We selected values for the constants k, a_1, a_2, b_1, b_2 in equation (2) that maximize the Euclidean distance between the centroids of classified training data, in an effort to meet objective (1) above.

This classifier can be generalized to arbitrary block sizes in an obvious way, leaving the number of classes (hence, N_s) unchanged. We note that the classifier could be still more precise about edges without expending side information. For example, the class could also depend upon the *positions* of the dominant 2x2 windows, and edge content in the upper left block $\hat{x}_{j-L-1}^k$ could also be included. This would greatly increase the number of states, but it is likely that the size N_c of their corresponding sub-codebooks could be decreased without affecting performance. This is presently under investigation.

IV. EXPERIMENTAL RESULTS

Six implementations are compared in Table 1. Cases 1 and 2 are from [18]. In Cases 3-6, the FSVQ codebooks trained on data from a set of 36 128x120 B&W sequences of four frames each and one 192 frame sequence, all sampled at 15 frames per second. This is fewer training vectors than desired, but a consequence of having modest computing resources. The test sequence is 64 B&W 128x120 frames, sampled at 15 frames per second, which does not appear in the training data. About 60% of its pixels are in the foreground. Sub-codebook sizes include 64, 256, and 1024 codevectors. Figure 4 plots the peak-to-peak SNR for each case on a frame-by-frame basis.

Rates are computed as follows. One bit per block is sent by the motion detector to specify motion. Its block sizes are 4x2, 4x4 or 8x8. The bilinear interpolator uses PCM to encode the 8 bit corner pixels. About 60% of the test sequence blocks are in the foreground, so an 8x8 BI block size adds about (8/64)(0.60) = 0.075 bpp. The Predicted Mean cases require no side information, except to PCM encode the mean values for top and left border blocks (about 0.05 bpp for 4x2 blocks, 0.03 bpp for 4x4 blocks). Finally, states are computed from classes using reconstructed data; hence, no side information is needed to send them.

A. Performance

Cases 1-3 illustrate the effects of the finite state and frame differencing structures. Case 1 has no classifier, nor next state function, and is entirely intraframe with one 8 bit codebook. Case 2 adds a finite state structure with 8 bit sub-codebooks that uses the old classifier [18]. 0.125 bpp overhead is used for the motion bit. Case 3 uses the new 145 state classifier and frame differencing with much smaller 6 bit sub-codebooks. We see that its SNR performance at under 0.4 bpp betters that of the intraframe, non-finite state PMVQ at 1.0 bpp by 4 dB.

Case 4 replaces Case 3's Predicted Mean scalar predictor with the bilinear interpolation vector predictor for foreground blocks. One pixel is PCMed for each 8x8 block. SNR increases by only 1.3 dB, but the photos show that blocking is markedly reduced. (Error images are magnified in intensity by four.) Case 5 uses larger, 8 bit sub-codebooks.

Case 6 illustrates a simple variable rate FS/FD IVQ. 10-bit sub-codebooks are used to represent the edge areas (about 25% of our test image sequence and 80 states), 6 bits for the background (about 40% of the sequence and one state), and 8 bits for the rest (64 states). The total rate is about 0.62 bpp, where the last two terms are for motion and bilinear interpolation. From Figure 3d, we see the edges are much sharper, making the encoded sequence difficult to distinguish from the original. Other variable rate schemes, such as conditional replenishment and variable codebook tree depth search, are presently under investigation.

(a)

(b)

(c)

(d)

Figure 3. (a) Original 128x120 frame numbers 36 and 37.
(b) Encoded frames for FS/FD PMVQ, 0.42 bpp (Case 3).
(c) FS/FD IVQ, 0.51 bpp (Case 4).
(d) Variable rate FS/FD IVQ, 0.62 bpp (Case 6).

Case	Structure	Vector Predictor	Block Classifier	Frame Difference	Motion Detector	Number States	Block Size	Subcodebook Size (bits)	Rate (bpp)	SNR (dB)	Curves (Fig.4)	Photos
1	PMVQ	4x2 PM	No	No	No	1	4x2	8	1.05	26.9	1	see [18]
2	FS/PMVQ	4x2 PM	see [18]	No	4x2 MSE	24	4x2	8	1.17	29.2	2	see [18]
3	FS/FD PMVQ	4x4 PM	Improved	Yes	8x8 MSE	145	4x4	6	0.42	30.9	3	Fig.3b
4	FS/FD IVQ	8x8 BI	Improved	Yes	4x4 MSE	145	4x4	6	0.51	32.2	4	Fig.3c
5	FS/FD IVQ	8x8 BI	Improved	Yes	4x4 MSE	145	4x4	8	0.64	34.7	5	No
6	FS/FD IVQ	8x8 BI	Improved	Yes	4x4 MSE	145	4x4	6-10 (var.)	0.62	36.0	6	Fig.3d

Table 1. Implementations.

B. A Custom ASIC Implementation

The FS/FD IVQ design may appear complex but its encoder is not difficult to implement. A simplified classifier would contain only adders, subtractors and shift registers (no multipliers). An entire codebook (about 1.5 megabytes for Case 6) is small enough to store in RAM chips. The VQ portion of the codec could be implemented with a custom VLSI chip set designed by our research team last year for real time video VQ coding [21]. The block classifier and bilinear interpolator simulated here would require a couple of additional ASICs. Motion compensation can also be added to improve performance for sequences which include camera panning and slow motion. We have sent a basic cell for such a chip out for fabrication and testing. The entire codec could fit on a single printed circuit board.

V. CONCLUSION

We have presented an improved interframe FSVQ. It uses a bilinear interpolator for foreground blocks, frame differencing for background areas, and an improved edge classifier. Background codebooks are iteratively improved using a steepest descent design algorithm. At 0.4 bpp the codec outperforms a very similar, non-finite state VQ having only intraframe memory which operates at 1.0 bpp. A few custom ASIC chips would permit the codec to be implemented on a single board.

REFERENCES

[1] R. M. Gray, "Vector Quantization," *ASSP Magazine*, pp. 4-29, Apr. 1984.

[2] A. Gersho and B. Ramamurthi, "Image Coding using Vector Quantization," *Proc. ICASSP*, May 1982, pp. 428-431.

[3] B. Ramamurthi, A. Gersho, and A. Sekey, "Low-rate Image Coding Using Vector Quantization," *Globecom '83*, pp. 5.6.1-5.6.4, Nov. 1983.

[4] B. Ramamurthi and A. Gersho, "Classified Vector Quantization of Images," *IEEE Trans.*, vol. COM-34, no. 11, pp. 1105-1115, Nov. 1986.

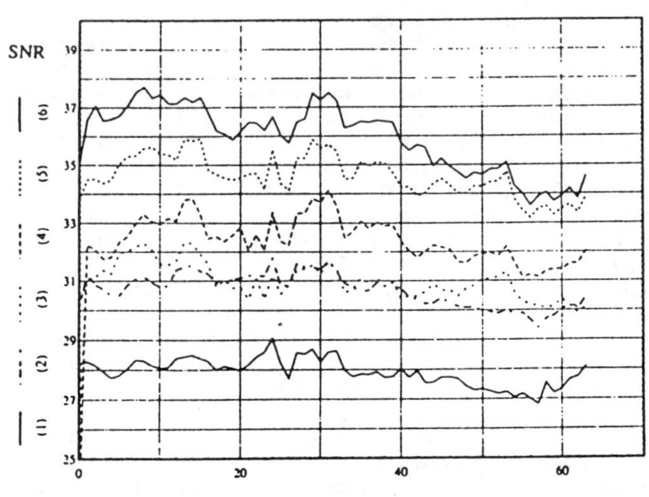

Figure 4. Peak-to-peak SNR versus frame number of the test sequence.

[5] M. Goldberg, Paul R. Boucher, Seymour Shlien, "Image Compression using Adaptive Vector Quantization," *IEEE Trans.*, vol. COM-34, no. 2, pp. 180-187, Feb. 1986.

[6] T. Murakami, K. Asai, E. Yamazaki, "Vector quantizer of video signals," *Electronic Letters*, vol. 7, pp. 1005-1006, Nov. 1982.

[7] B. Ramamurthi and A. Gersho, "Nonlinear Space-Variant Postprocessing of Block Coded Images," *IEEE Trans.*, vol. ASSP-34, no. 5, pp. 1258-1268, Oct. 1986.

[8] V. Cuperman and A. Gersho, "Vector Predictive Coding of Speech at 16 kb/s," *IEEE Trans.*, vol. COM-33, no. 7, pp. 685-696, July. 1985.

[9] Pao-Chi Chang and R. M. Gray, "Gradient Algorithms for Designing Predictive Vector Quantizers," *IEEE Trans.*, vol. ASSP-34, no. 4, pp. 679-690, Aug. 1986.

[10] R. L. Baker and R. M. Gray, "Image Compression Using Non-adaptive Spatial Vector Quantization," *Proc. 16th Asilomar Conf. on Circuits, Systems & Computers*, pp. 55-61, 1982

[11] R. L. Baker, "Vector Quantization of Digital Images," Ph.D. dissertation, Stanford U., June 1984.

[12] S. E. Budge and R. L. Baker, "Compression of Color Digital Images Using Vector Quantization in Product Codes," *Proc. ICASSP*, pp. 4.5.1-4.5.4, 1985.

[13] H. Hang and B. Haskell, "Interpolative Vector Quantization of Color Images," submitted for publication.

[14] W. K. Pratt, "Digital Image Processing," New York: Wiley, 1978.

[15] J. Foster, R. M. Gray, and M. O. Dunham, "Finite-State Vector Quantization for Waveform Coding," *IEEE Trans.*, vol. IT-31, no. 3, pp. 348-359, May 1985.

[16] M. O. Dunham and R. M. Gray, "An algorithm for the Design of Labeled-Transition Finite-State Vector Quantizers," *IEEE Trans.*, vol. COM-33, no. 1, pp. 83-89, Jan. 1985.

[17] R. Aravind and A. Gersho, "Low-Rate Image Coding with Finite-State Vector Quantization," *Proc. ICASSP*, pp. 137-140, Apr. 1986.

[18] R. L. Baker and H. H. Shen, "A Finite-State Vector Quantizer for Low-Rate Image Sequence Coding," *Proc. ICASSP*, pp.760-763 , Apr. 1987.

[19] I. Abdou and W. Pratt, "Quantitative Design and Evaluation of Enhancement /Thresholding Edge Detectors," *Proc. IEEE*, vol. 67, no. 5, pp. 753-763, May. 1979.

[20] S. Elnahas and K. Tzou, "Hybrid Interframe Coding of Video Signals with Backward-Acting Motion Detection," *Proc. ICASSP*, pp. 165-167, Apr. 1986.

[21] R. Dianysian and R. L. Baker, "A VLSI Chip Set for Real Time Vector Quantization of Image Sequences," *Proc. ISCAS*, pp. 221-224, May 1987.

[22] J. Salinas and R. L. Baker, "A Motion Compensated Vector Quantizer with Filtered Prediction," *Proc. ICASSP*, Apr. 1988.

[23] R. Aravind and A. Gersho, "Image Compression based on Vector Quantization with Finite Memory," *Optical Engineering*, vol. 26, no. 7, pp. 570-580, July 1987.

[24] J. K. Yan and D. J. Sakrison, "Encoding of Images Based on a Two-Component Source Model," *IEEE Trans.*, vol. COM-25, no. 11, Nov. 1977.

Image Coding Based on Segmentation Using Region Growing

K.S.Thyagarajan, ECE Dept., SDSU, San Diego, CA 92182
Helge Bohlmann, ECE Dept., SDSU, San Diego & Institut
fur Allgemeine Nachrichtentechnik, Universitat Hannover, Germany
Huseyin Abut, ECE Dept., SDSU, San Diego, CA 92182

Abstract

This paper describes a composite source model for images. An image is segmented into uniform and homogeneous regions using centroid linkage region growing algorithms. The region homogenity is determined by the Student T-statistics. Excessive regions resulting from region growing are merged according to region merging rules. The initially segmented image is then clustered into classes with the help of the K-means algorithm. The image classes are shown in this paper as visual aids in judging the classification procedure. The class numbers and the corresponding pixel counts are also included. Finally, as an application of composite source models, an image is encoded using matrix quantizers with separate codebook for each class of the image.

1. Introduction

Source models for images play an important role in coding and recognition systems as it does in speech processing. In coding applications source modeling is of paramount importance because it offers a theoretically sound basis for the measurement of the coder performance. An image source can be considered to be a random process whose statistical properties can be determined. This information will be useful in exploiting the statistical dependencies among pixels so as to achieve low bit rates. However, a single source model of an image is often inadequate to fully represent the image details. It is, therefore, meaningful to view an image as being made up of a composite of subsources and a switch process. The overall output is simply the output of a particular subsource being selected currently by the swtich, The advantage of such a composite source model is that it adequately represents the complete image details, for example, the individual subsources might represent edges, shades, and textures.

The main task in obtaining the source model parameters is the segmentation of the image into homogeneous regions. There are various segmentation techniques available, and the choice depends on the intended application. If one classifies an image based on textural features, then the composite source structure will reflect the different textures inherent in the given image. On the other hand, if edge-shade information is used to classify an image, then some of the subsources will correspond to edges with different orientations while others will correspond to homogeneous regions with different textures. The study of a composite source model for images is described in this paper. An image is first segmented into uniform and homogeneous regions using region growing algorithms. Then some of the regions are merged with the help of region merging techniques. The merging step is necessary since the region growing usually results in an excessive number of regions. After merging the regions, the subsources are obtained by clustering the image pixels in the feature domain by means of a clustering algorithm, such as the K-means algorithm. As an application of composite source modeling, a classified image was encoded using a matrix quantizer. Optimal codebooks were designed for the subsources. For more details on matrix or vector quantizer design, the reader is referred to [1].

1.1. Segmentation

In order to decompose an image into its composite sources, one has to segment it into appropriate regions. Segmentation is a means of partitioning an image into uniform and homogeneous regions with respect to some characteristics, such as gray tone or texture. Formally one can define segmentation as a partition of an image field X into disjoint nonempty subsets X1, X2,..., Xn such that (1) the union of the subsets is the whole image, (2) each subset is directly connected or connected, (3) a uniformity predicate P(Xi) is TRUE for each region, and (4) the uniformity predicate for the union of any two adjacent regions is FALSE. The uniformity predicate used in this work is the Student T statistics. The segmentation can be achieved by a number of techniques, each with its own merits and demerits [4]. Here we used the centroid linkage region growing technique which is briefly described in the next section.

1.2. Region Growing

A common method to generate a complete segmentation of an image is the region

Reprinted from *IEEE Conf. Acoust., Speech, Signal Processing,* vol. 2, pp. 752–755, April 1987.

growing method. Starting from one pixel (atomic region), adjacent pixels are merged with this region as long as P(y) = TRUE, where Y is a subset of the image X, and P(.) is an appropriate predicate. If no pixel can be found showing a similarity with the neighboring region a new pixel is chosen to form a new atomic region. The next region growing process begins. This is continued until all pixels are labeled. The regions can be grown either by parallel processing [5] or by sequential processing [4]. In growing regions sequentially, an image is scanned left to right and top to bottom. Let R be a segment of M pixels in the neighborhood of a pixel with an intensity y as shown in figure 1. Let X and S2 be the mean and scatter for R as defined by

$$\bar{X} = (1/M) \sum_{r,c \in R} x_{r,c} \quad (1)$$

$$S^2 = \sum_{r,c \in R} (x_{r,c} - \bar{X})^2 \quad (2)$$

where r and c denote the row and column index of each pixel. Compute the T staistic given by

$$T = ((M(M-1)/(M+1))(y-\bar{X})^2/S^2)^{1/2} \quad (3)$$

Under the assumption that all pixels in R as well as y under consideration are i.i.d's, then T has a Student distribution with M-1 degrees of freedom. If T is significantly small, y is added to R and the mean and scatter are updated using

$$\bar{X}_{new} = (M\bar{X}_{old}+y)/(M+1) \quad (4a)$$

$$S^2_{new} = S^2_{old}+(y-\bar{X}_{new})+M(\bar{X}_{new}-\bar{X}_{old})^2 \quad (4b)$$

In order to be statistically consistent, we use an α level statistical significance test. The fraction α represents the probability that a T statistic with M-1 degrees of freedom will exceed the value $t_{M-1}(\alpha)$. If the observed T from (3) is larger than $t_{M-1}(\alpha)$, then we infer the difference to be significant. If the pixel and the segment come from the same population, then α is the probability that the test provides an incorrect answer. Here, α is a user provided parameter. For a small M, $t_{M-1}(\alpha)$ is high and it is low for large M. The T test implies that as the region gets larger, y has to be closer to the region's mean in order to merge it with R. In order to keep M high it is possible to initialize the region scatter and the region size with some reasonable values, so that division by zero is avoided. Note that S2 is zero for M = 1.

The above segmentation algorithm produces a large number of regions. To avoid regions opened on edge zones, region growing is allowed only if the local gradient of the actual pixel is smaller than a predefined threshold. Otherwise the pixel is marked as an edge pixel and not taken into account in the region growing process.

1.3. Region Merging

The segmentation described above usually results in a large number of regions. In order to reduce this number, one has to merge the neighboring regions which can be accomplished either in the pixel domain or feature domain. In the pixel domain, one looks for similarities in the neighboring regions to merge. Again a T-test can check for the significance of the difference between the region means. This can be done in a second pass over an image using a sequential tracking algorithm. During subsequent merging steps the threshold of the merging criterion can also be changed until an appropriate result is achieved.

In the case of measurement-space guided spatial clustering, first some features are extracted for every region. Possible features are average gray level, variance, and difference between the maximum and minimum values of the pixels. After extracting the features, an ISODATA algorithm [6] can be used to find data clusters in the feature space which represent regions with similarities in the pixel domain.

1.4. Edge Classification

As mentioned earlier, an image can be classified based on edge-shade information or textural details. In this study, we have attempted to use only the edge and shade information of an image. The edge-shade classification can be carried out directly on a given image using templates without segmenting the image. In this paper the edge-shade information is derived in the following manner. After region growing and merging steps are done, each pixel has a label. This is then replaced by a new label corresponding to the class number generated from the clustering algorithm. This array of new labels has a much simpler structure than tha original image itself, so that boundary directions are easier to find with simple pattern matching methods. A square block is said to belong to a homogeneous region if more than m pixels of this block have the same label. The direction of the boundary can be found using template matching techniques[2].

2. Experimental Procedure

An original image of size 512x512 was segmented into homogeneous regions using the T-statistic. The significance level was initially set at 0.5. A single pass generated about 20,700 regions. In order to reduce the number of regions a multiple pass over the entire image was performed using region merging in the pixel domain and the T-statistic. This reduced the number of regions to 17,156.

After this merging, since more reliable region statistics could be obtained, the significance level was decremented stepwise to 0.0001 while at every level the same multiple pass merging was carried out. The final number of regions was 2,695. Figure 2 shows the segmented image of the girl's picture prior to merging. It has 2,695 uniform regions. The classified images of the same picture are shown in figures 3 through 6 .

Next, using the original and the segmented images, region merging was done in the feature domain using the combinations of (i) average region gray level, F1, (ii) variance of region gray levels, F2, (iii) 1/(1+F2), F3, (iv) difference between the maximum and minimum gray levels, F4, (v) average sobel magnitude, F5, (vi) average sobel phase, F6, (vii) horizontal edge, F7, (viii) high frequency spot, F8, (ix) V-shape, F9, (x) vertical line, F10. Of all the possible and meaningful combinations, the mean gray level of the regions, F1, was chosen as it resulted in better visual appearance. Table 1 shows the number of elements per class and the feature chosen for clustering. Instabilities were encountered in clustering when other feature vectors were used.

Table 1

Feature selected for clustering = F1				
# classes	# elements in each class			
2	1290	1405		
3	1127	973	595	
5	732	712	701	380

As an application to image coding, matrix quantizers were designed for each class. In order to encode the image under consideration, it was divided into say, 4x4 pixel blocks. Each block was classified as edge or nonedge (shade) block using the templates in [2] on the segmented image. To decide whether a block is edge or shade, and if edge its direction, the following rule was used: Decide shade if more than "m" pixels of the block have the same label. Otherwise the highest value of the convolution with the template and the segmented image block indicates the orientation. Table 2 shows the SNR values for different edge-shade codebooks.

3. Conclusion

A composite source modeling of images has been described in this paper. The individual sources are obtained by segmenting the image. A complete segmentation is accomplished by centroid-linkage region growing algorithm using the T-statistics. Excessive regions resulting from initial segmentation are merged by a merging process in the pixel domain. Finally, the segmented image is decomposed into classes or subsources by clustering the pixels in an appropriate feature space using K-means algorithm. Though there are various features that can be used to form the cluster space, it was found that the region mean gray level yielded good visual appearance. Since the classification was based on edge-shade details, naturally the subsources reflect these properties more than the textural details. Further work in this type of source modeling will contribute to better encoding schemes. The authors wish to acknowledge the grant support from NOSC which made the study possible.

Table 2

Edge-Shade Class & SNR in dB			
# shade	# edge	# codewords	SNR
4	4	512	21.6
4	2	512	20.14
4	1	512	20.14
2	4	512	20.99
2	2	512	19.66
2	1	512	18.88
1	4	512	20.47
1	2	512	19.27
1	1	512	18.56

4. References

[1] Robert,M.Gray, "Vector Quantizer," ASSP Magazine, 1984

[2] Thyagarajan,K.S., Abut,H., and Bheda,H.,"A Fast Matrix Quantizer For Image Encoding," ICASSP 1985

[3] Ramamurthy,B. and Gersho,A., "Edge-Oriented Spatial Filtering of Images with Application to Post Processing VQ Images," ICASSP 1984

[4] Haralick,R.M., and Shapiro,L.G., "Image Segmentation Techniques," Computer Vision, Graphics and Image Processing, vol.29, 1985.

[5] Ernst,D., et. al., "Processing of Remote Sensing Data by a Region Growing Algorithm," Proc. 3rd Int.Joint Conf. on Pattern Recognition , 1976.

[6] Duda,R.D., and Hart,P.E., "Pattern Classification and Scene Analysis," Wiley, New York, 1973.

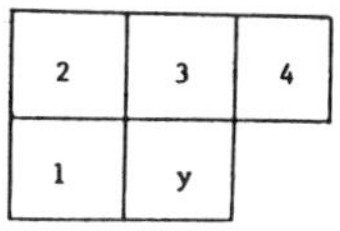

Fig. 1

Fig.2 Segmented image prior to merging,α=0.0001,2695 regions

Fig.3 Image class #0

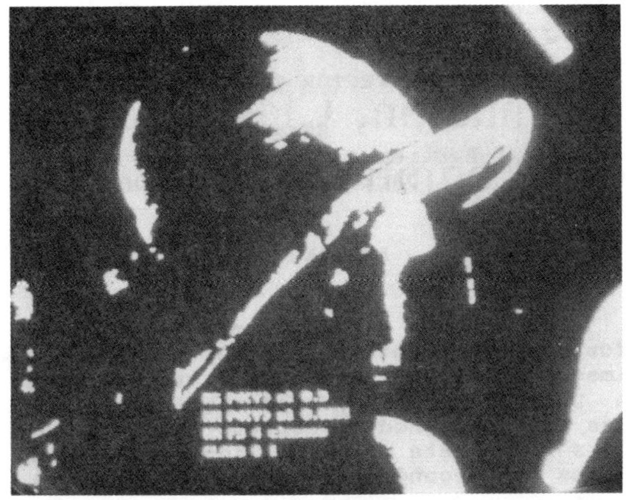

Fig.4 Image class #1

Fig.5 Image class #2

fig.6 Image class #3

Part 5
Segmentation, Classification, and Recognition

VECTOR QUANTIZATION AND MARKOV SOURCE MODELS APPLIED TO SPEECH RECOGNITION

Roberto BILLI

CSELT - Centro Studi e Laboratori Telecomunicazioni
Via G. Reiss Romoli, 274 - 10148 TORINO (ITALY)

ABSTRACT

An isolated word recognizer based on vector quantization at the acoustic level and on stochastic modeling at the phonetic level is described.

The power of this approach lies in its best utilization of the training data. The first experimental results obtained are encouraging and suggest that further optimization is possible.

I. INTRODUCTION

Many recognition systems take advantage of the structural information present in speech and base their description at each linguistic level on lower level categories. Necessarily at the acoustic level a finite number of primitives must be used to model the continuous variation of speech.

Several attempts were made to spectrally describe speech in terms of a manually or automatically extracted set of prototypes. In the Harpy system [1] 98 talker-specific linear prediction spectral templates were used. The template selection required initially a careful manual choice, but could automatically adapt to a new talker. In a more recent work Klatt [2] proposed to identify spectral sequences for each phonetic transition and to encode this knowledge in a network using only a finite set of static spectra. Extensive work was made at IBM [3] on the design of centisecond labeling acoustic processors using from 33 to 200 hand-picked prototypes.

Unfortunately the experience employable and the accuracy achievable in the manual choice of acoustic prototypes is at the expense of a hard and long human work. It is therefore interesting and promising the attempt recently made at IBM to use prototypes automatically extracted from a few minutes of speech by means of a cluster analysis algorithm. A similar procedure called vector quantization is used in some kind of vocoders and was proved to be practically feasible and theoretically superior to other coding techniques. In fact Buzo et al. [4] proved that, for a given distortion, vector quantization can reduce of 15-20 bits the average number of bits per frame transmitted with respect to scalar quantization.

A complex acoustic event such as a consonant to vowel transition cannot be described as a unique sequence of spectral states, rather it is better to have alternative descriptions which can be easily encoded in a graph structure, as in the Harpy system. A lot of observation and experience is required to compile this knowledge source by hand. Also in this case, as in the choice of spectral templates, an automatic method, even if less accurate and refined, can process a much larger amount of data definitely obtaining more general results.

The first attempt to automatically model a speech process as a sequence of spectral prototypes is due to Baker [5]. The method is based on stochastic modeling and, at the phonetic level, can be seen in some sense as a generalization of the network used in Harpy. The most appealing advantage is that stochastic models can be automatically and extensively trained and can effectively incorporate much of the information supplied in this process.

After the implementation of these ideas in the Dragon system [5], other experiments were carried out at IBM following the same approach and achieving good results [6]. The main application of the above mentioned concepts has been in the field of continuous speech recognition, because the difficulty of the task justifies the complexity of this approach.

We think that stochastic modeling may be a powerful and adequate technique also in the field of isolated word recognition, particularly when the vocabulary is large and/or speaker independent operation is desired. In this paper we discuss the choices adopted and the first results achieved in the initial development of an isolated word recognizer based on vector quantization and stochastic modeling at the phonetic level for Italian.

II. SELECTION OF SPECTRAL TEMPLATES

In the first stage of our speech recognition system a spectral parametric representation is obtained every 15 msec using the energies out of a 13 filter bank designed according to the ear's articulatory bands and covering the 300 ÷ 3400 Hz frequency range, clearly intended for use with telephone speech. Subsequently we process each frame of spectral parameters to reduce redundancy, contemporarily obtaining a representation based on a finite number of elements. This is the main but not the only advantage of vector quantization. In fact it can provide optimal quantization of a parameter vector taking into account its statistical properties as well as its relationship to some distortion measure. The relationship among the components of a reference and a coded parameter vector, subject to a maximum distortion value, can be very complex as was clearly pointed out by Wong [7] in the case of LPC parameters and Itakura's distortion measure. This suggests the inefficiency of coding each parameter separately.

Furtherly the source statistics can be very useful. We analyzed 6000 frames of speech taken from isolated words uttered by one talker and corresponding to 1 minute and a half of speech. In figure 1A a scatter plot of the first two spectral components F1, F2 is shown. Even though it is only a two-dimensional cross section of the 13-dimensional space, it points out the non-uniform distribution of points. Consequently it is useless to use bits of information to represent space portions which are never or very rarely occupied by speech frames.

The best utilization of the above mentioned properties is obtained through a cluster process. We employed a simple distortion measure, commonly used for spectral analysis, that is the sum of the absolute differences between the energies out of corresponding filters taken in decibels. The cluster algorithm used was the automatic iteration scheme "K-means" [8]. In our implementation the number of clusters is not predetermined but depends on the choice of two parameters: the greatest cluster radius and the smallest distance between centers.

While in the design of cluster vocoders a codebook of more than 10 bits is used, for speech recognition the number of spectral templates is generally comprised from 100 to 200. In fig. 1B the two dimensional scatter

Reprinted from *Proc. IEEE Conf. Acoust., Speech, Signal Processing,* vol. 1, pp. 574–577, May 1982.

plot of a 132 points codebook is shown. Clearly this representation gives only a rough idea of the position of each center with respect to clusters.

In vector quantization for each frame of incoming speech parameters a prototype is chosen from the codebook to minimize the distortion measure. Although the computational aspects are not object of this paper, we observe that the processing requirements can be reduced in a practical implementation by choosing a suboptimal tree search procedure at this stage [4].

III. STOCHASTIC MODELING OF SPEECH BASED ON PHONETIC ELEMENTS

Following the Baker's idea [5] a probabilistic model is first formulated to describe a speech process and then to make inferences in a speech recognition system. More specifically, after the vector quantization process, the sequences of spectral states observed are represented as a probabilistic function of a Markov process. This is completely defined specifying the number of internal states, the set of transition probabilities $\{p_{ij}\}$ the set of output symbols $\{a_k\}$, which correspond to the observed spectral states, and the set of output probabilities $\{q_{ij}(a_k)\}$. The internal states have no physical meaning but serve only to represent the memory of the process and the statistical dependency among subsequent symbols.

Such a model can describe speech segments of arbitrary length, provided the number of states is sufficiently high. In fact it was used at the work, syllable or phoneme level [6]. Yet the choice of the basic speech unit may be crucial for a speech recognition system. We based our choice on the requirement that any word had to be described and recognized by the simple concatenation of stochastic models belonging to a unique and restricted set. This introduces a segmentation which is clearly in contrast with the continuous nature of speech. However, by carefully defining the models, most of the coarticulation effects can be included inside them, reducing therefore the approximation effects. We took advantage from the fact that in the Italian language the word pronunciation is often related to the written form, moreover syllables are uttered quite independently from each other.

A tentative set of elements was chosen, including all the vowels (V), the consonant to vowel transitions (CV), plus some diphtongs and a group of particular sounds (CCV). A set of transcription rules was specified to convert each word from its lexical form to a unique sequence of phonetic elements. Fig.2 is an example of how the stochastic models of these elements can be connected together to represent a word of the kind (VCVCV). The final state of each graph has no outgoing branch and so it allows to concatenate two models without recomputing the probababilities of the merging branches.

The recognition strategy follows a Top-Down approach: a word is first hypothesized at the lexical level; the transcription rules are then applied to identify which phonetic models should be connected to obtain the composite word model; finally the hypothesis is verified at the acoustic level by computing the probability that the model generates the observed spectral state sequence. This probability can be computed in an efficient way by using the Viterbi algorithm [10]. This is analogous to dynamic programming used in pattern matching strategies. However instead of computing string similarities, here the likelihood that the input strings belongs to each of the word models is computed.

To estimate all the probabilities needed for each phonetic model, a training process is carried out on a set of strings out of the vector quantizer, obtained using a speech segment in which several occurrences of each phonetic element are present. Baum [9] proved that an iterative a posteriori reestimation of the probabilities can give an approximation to maximum likelihood estimates. To do this it is only necessary to specify initial probabilities for the Markov process. A particular form of this estimation method, called the Forward-Backward algorithm [6], was used. This name derives from the particular way in which it organizes the computations operating with a trellis scheme.

The algorithm is not easy to describe and the interested reader is addressed to the exhaurient formulation in [6], [9]. Here we want only to point out one of its interesting properties. The method has an intrinsic capability of segmenting speech, which only in part relies on the initial probability estimate, but mainly derives from the presence of the same phonetic element in different words and positions. The segmentation accuracy improves therefore during the training, as the probability estimates are refined. The main advantage of this property is that whole words, uttered in a natural way, can be used for training.

Given a probabilistic finite state machine with M states, M^2 output producing transitions and M^2 null transitions are possible in principle. However too many transitions require to estimate an excessive number of parameters. Consequently the training data have to be increased and if the initial probability estimates are not sufficiently reliable there is the possibility of obtaining inaccurate results from training. Therefore we chose to limit the allowed transitions and set to zero the remaining initial probabilities. These will remain zero also after the training process.

If the states are numbered from 1 to M we considered only the recursion probability $p_{i,j}$, which permits to obtain variable length sequences of the same state, and the probabilities of skipping one or two states $p_{i,i+1}$, $p_{i,i+2}$. For each of these transitions there are the output probabilities for the codebook spectral states. In the experiments described in sec. IV uniform values were assigned to all of these probabilities for each phonetic model.

Better choices are possible, as that of computing initial estimates based on the comparison of hand labeled with machine labeled strings [6], but this procedure is quite tedious and unreliable.

Another method, not yet tested, may be suitable for most of the phonetic models corresponding to speech segments that can exist and be pronounced separately: it consists of using the probabilities obtained from training on isolate models, as initial estimates for training on words.

IV. EXPERIMENTAL RESULTS

A vocabulary of 100 words including digits, city names and command words was repeated 5 times by one speaker in 5 different sessions using a telephone microphone. The first two repetitions, giving a total of 6000 frames, were used to derive a 132 element codebook by cluster analysis. To train the models of the phonetic inventory described in Sec. III a smaller vocabulary of 25 out of the 100 words was used and the first two repetitions were employed.

The Forward-Backward algorithm was iterated three times, having verified that further training did not significantly modify the probabilities.

All the models used in the experimental evaluation consisted of the same number of states, although a more efficient choice should be that of using a smaller number of states for almost stationary sounds, like vowels, and a higher number of states for the more transitional elements. We tested two values of M, that is M = 3 and M = 5.

The recognition results, reported in Tab, 1 refer to the test strings obtained from the last three repetitions of each word, as well as to the training strings. In fact, as each phonetic model is trained on different realizations, moreover belonging to different phonetic contexts, it is not assured a priori that even the training strings are always recognized. From Tab. 1, however, it can be seen that errors were confined to test strings. Most of the errors occurred for words including less frequent phonetic elements, suggesting that a more extensive training is required, using several occurrences of each sound placed in different positions inside the word.

It can be seen that better results are obtained with M = 3, probably as a smaller number of parameters can be better estimated even with limited training.

In the second experiment we carried out only a recognition test on a new vocabulary of ten words using the phonetic models trained in the first experiment. Some of the observed errors occurred for very similar words (TREDICI-SEDICI) (TRENO-MENO) and some for a particular word (NO) which was pronounced quite differently from the phonetic element "NO" appearing in the training words. The results are shown in Tab. 2.

It can be noticed that about the same percentage of errors as the previous experiment is obtained, although now the phonetic elements appear often in a different position inside the word.

V. CONCLUSIONS

Several factors affecting the system's recognition performance have again to be optimized.

1) The dimension of the codebook used in vector quantization
2) How many and which phonetic elements are employed
3) The number of states in each stochastic model
4) The initial probability estimate
5) The size and the kind of training data.

Probably the only valid criterion for this optimization is an experimental evaluation of the recognition accuracy in many different choices of the above mentioned factors.

The experiments carried out, based on tentative choices, give only an approximate idea of the system's performance, but suggest that further improvement is possible.

REFERENCES

[1] D.H.Klatt, "SCRIBER and LAFS: Two New Approaches to Speech Analysis" in W.A.Lea, "Trends in Speech Recognition" Prentice Hall 1980 - pp. 543-552.

[2] B.Lowerre, "The Harpy Speech Recognition Systems" Ph. D. dissertation Computer Science Dept., CMU 1977.

[3] L.R.Bahl, R.Bakis, P.S.Cohen, A.G.Cole, F.Jelinek, B.L.Lewis, R. L.Mercer, "Recognition Results with Several Experimental Acoustic Processors" - Proc. ICASSP 1979.

[4] A.Buzo, A.H.Gray, R.M.Gray, J.D.Markel, "Speech Coding Based Upon Vector Quantization" - IEEE Trans. on ASSP, Vol. ASSP-28, No. 5, Oct. 1980.

[5] J.K.Baker, "The DRAGON System: an Overview" - Proc. IEEE Symposium on Speech Recognition, Pittsburgh, Pa., 1974.

[6] F.Jelinek, "Continuous Speech Recognition by Statistical Methods"- Proc. IEEE, Vol. 64, No. 4, April 1976.

[7] D.Y.Wong, B.H.Juang, A.H.Gray, Jr., "Recent Developments in Vector Quantization for Speech Processing" - ICASSP 1981.

[8] J.A. Hartigan, "Clustering Algorithms", J. Wiley & Sons., N.Y. 1975.

[9] L.E. Baum, "An inequality and Associated Maximization Technique in Statistical Estimation for Probabilistic Functions of a Markov Process" - Inequalities, Vol. III, 1972, pp. 1-8.

[10] G.D. Forney, Jr., "The Viterbi algorithm" - Proc. IEEE, Vol. 61, pp. 268-278, Mar. 1973.

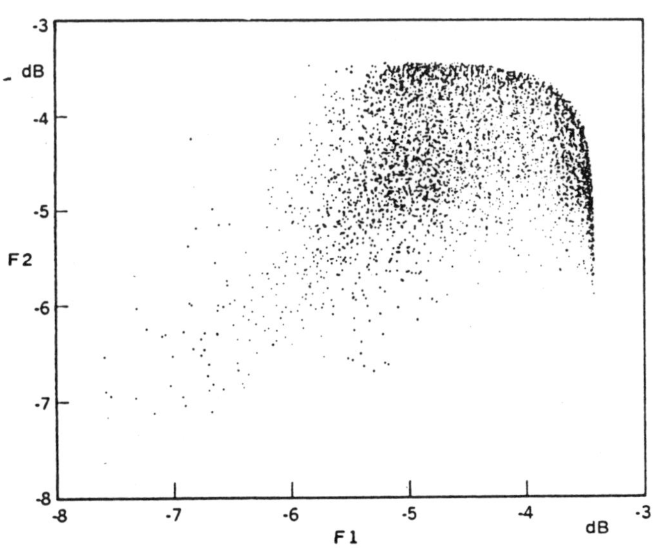

Fig. 1A - Two dimensional scatter plot of source vectors

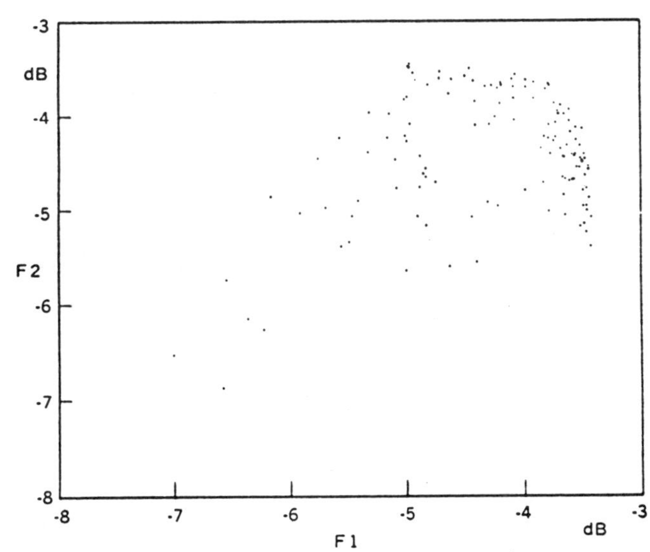

Fig. 1B - Two dimensional scatter plot of codebook vectors

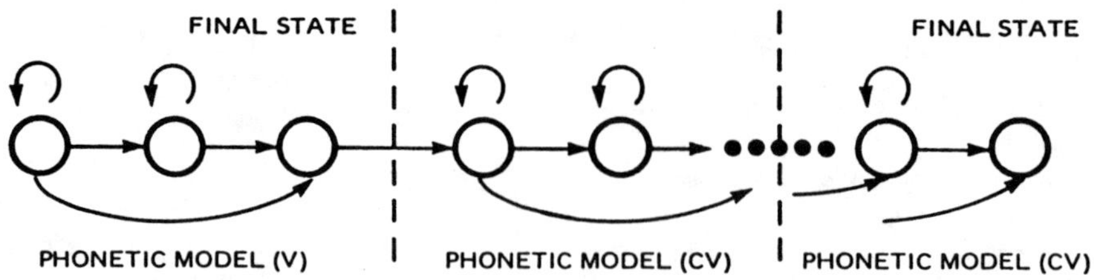

Fig. 2 - Composite stochastic model of a word

	DATA SIZE (Strings)	ERRORS M = 5	ERRORS M = 3
Training	50	0	0
Test	75	20	6

Tab.1 - Experiment with training and testing on the same vocabulary

	DATA SIZE (Strings)	ERRORS M = 5	ERRORS M = 3
Test	50	11	4

Tab.2 - Experiment with different vocabularies used for training and testing

DISCRETE UTTERANCE RECOGNITION BASED UPON SOURCE CODING TECHNIQUES

A. BUZO*. H.MARTINEZ**, C.RIVERA*

* Facultad de Ingeniería Univ. Nac. Aut. de México
** Instituto de Investigaciones Eléctricas

ABSTRACT

A speaker-independent isolated word recognition system is described which is based on some techniques and results from rate-distortion speech coders . The recognition system can be viewed as a minimum distortion or nearest-neighbor system where the distortion measure is defined between an observed sequence of frames of speech and a reference pattern. The patterns are sequences of sets of LPC models. Every one of the sets of each pattern consist of a collection of LPC models that "best" reproduces a given frame of a word from a training sequence. The Itakura Saito distortion measure is used to design the system (or selection of the patterns) and for the decision step.

INTRODUCTION

Recently theoretical development of the Itakura-Saito distortion measure [1,2,3] and applications of classical techniques of information theory have been applied to design very low rate-compression systems [1,4]. These compression systems as the standard LPC system [5,6] map a frame of sampled speech into a speech reproduction model. With these systems, an encoded version of an isolated word (a sequence of speech frames) is a sequence of models. The fidelity criterion used, to design these rate-compression systems, has been the average amount of the Itakura-Saito distortion measure, between the frames of sampled speech and the selected reproduction models, over all frames in a given training sequence.

In the isolated word recognition system that we present, a sampled version of an isolated word is encoded into a code word from a dictionary using a nearest-neighbor mapping, with a distortion measure or cost for identifing a given word, defined over all frames of speech samples and a pattern which is selected using the same distortion measure.

The emphasis here is in the selection of the reference patterns and the distortion measure used to encode the speech samples. The reader is referred to [7], an expanded version of this paper, for further details as well as the implementation of the system.

COMPRESSION SYSTEM

Rate-distortion speech coders with the Itakura-Saito distortion measure as a fidelity criterion [1] and optimum (in the Itakura-Saito distortion measure sense) LPC speech compression systems [4] map in one and two steps, respectively a frame of speech samples $\bar{X}_i=(X_{ni},\ldots,X_{ni+n-1})$ into a filter $\sigma/A(Z)\varepsilon C$ (where $A(Z)=\Sigma\, a_k Z^{-k}$, $a_0=1$), or simply $\sigma/A\varepsilon C$, where C is a finite set of Mth-order stable autoregressive filters, plus some information of an excitation process, basically a pitch parameter [5,6], that is used to generate the driving process of the filters. The receiver then uses the filters and the pitch information to synthesize the speech with a locally generated driving process. It has been observed [1,4], that when the speech is synthesized, without using the pitch information, by driving the filters with white noise to produce "whispered" speech, many listeners found the whispered speech (even using a small code book C of 128 code words, resulting in a rate of about 400 bits per second) to be understandable. Following this observation and previous work [8,9,10], we focus on an identification system based on the filter models σ/A.

The map rule used in the Rate-distortion speech coder [1] is defined by

$$\bar{X}_i \rightarrow (\sigma/A)_i \text{ if } \rho(\bar{X}_i,(\sigma/A)_i)=\min_{\sigma/A\varepsilon C}\rho(X_i,\sigma/A) \quad (1)$$

where

$$\rho(\bar{X}_i,\sigma/A)=\bar{r}_i\bar{b}^t-1-\ln\sigma^2_{Mi} \quad (2)$$

is the modified Itakura-Saito distortion measure, and

Reprinted from *Proc. IEEE Conf. Acoust., Speech, Signal Processing,* vol. 1, pp. 539–542, May 1982.

$$\bar{r}_i = (1, r_{xi}(0), 2r_{xi}(1), \ldots, 2r_{xi}(M))$$

$$\bar{b} = (\ln\sigma^2, r_a(0)/\sigma^2, \ldots, r_a(M)/\sigma^2)$$

$$\sigma_{Mi}^2 = \frac{\det R_i^{(M)}}{\det R_i^{(M-1)}}$$

$$r_{xi}(k) = \frac{1}{n} \sum_{j=i}^{i+n-k-1} X_j X_{j+k}$$

$$r_a(k) = \sum_{i=1}^{M-k} a_i a_{i+k}$$

where $R_i^{(M)}$ is the matrix defined by

$$R_i^{(M)} = \left\{ r_{xi}(k-J),\ k=0,\ldots,M,\ J=0,\ldots,M \right\}$$

If in (1) C is the continuum of all Mth-order stable autoregressive filters, the map defined in (1) is the identification step of a standard LPC system using the autorregressive method, which is followed by a digital approximation of the model, to complete a compression system. If the digital approximation is made properly in the Itakura-Saito sense [4], the system becomes equivalent in performance to the rate distortion speech encoder.

When a sequence of speech frames $\bar{X}^L = (\bar{X}_0, \ldots, \bar{X}_{L-1})$ is encoded, using (1), into a sequence of all pole models $\omega = ((\sigma/A)_0, \ldots, (\sigma/A)_{L-1}) \in {}^{L}\bar{X}_0^{1}$ C, a distortion or cost for reproduction of the speech, by driving, in a proper manner [5,6], the all pole models by a zero mean unit power noise, could be given by

$$d(\bar{X}^L, \omega) = \frac{1}{L} \sum_{i=0}^{L-1} \rho(\bar{X}_i, (\sigma/A)_i) \qquad (3)$$

This distortion measure has been used with success to measure the performance and design (i.e. selection of the code book C, to obtain minimun distortion) of the rate-distortion speech coder [1] and the optimum LPC compression system [4].

The distortion measure that we use to select the pattern of each word ω in the dictionary W is based on (3). In the next section we present the criterion used to obtain these reference patterns.

PATTERN SELECTION

To select the patterns a training sequence for each word from a given dictionary W is used. Each training sequence consist of several repetitions of the same word from a large variety of speakers. Throughout this paper we assume that the beginning and ending frame of all isolated words has been accurately located. We also assume, at this moment, that all words in the training sequences have a fixed number of frames, say L, assumption that will be dropped later on. Hence the problem of finding the reference patterns, one for each word in the dictionary, from the training sequences can be formulated as a selection of models $((\sigma/A)_0, \ldots, (\sigma/A)_{L-1})$ for each word $\omega \in W$, to be used to compute the cost for encoding a sequence of L frames of sampled speech from an unknown isolated word $\bar{X}^L$, into a specific word from the dictionary, using a distortion like (3) and a nearest-neighbor rule given by

$$\bar{X}^L \to w_i \text{ if } d(\bar{X}^L, w_i) = \min_{w \in W} d(\bar{X}^L, w) \qquad (4)$$

Several alternate sets of parameters that represent uniquely the all pole models σ/A [11] are candidates, to make an average over a training sequence, in order to select the patterns. Some of these possible sets are the reflection coefficients and gain, poles and gain, predictor coefficients and gain, autocorrelation coefficients of $\{X_i\}$, autocorrelation coefficients of $\{a_i/\sigma\}$ and gain, cepstral coefficients, etc.

Although the above sets provide equivalent information about the models σ/A, taking the average of each one of these sets over a given training sequence, will result in a diferent vector of models $((\sigma/A)_0, \ldots (\sigma/A)_{L-1})$. Henceforth the selection must be based on a given criterion and a given distortion measure for encoding $\bar{X}^L$ into a vector of all pole models.

One of these criterions is to measure the average distortion for encoding the speech samples $\bar{X}^L$ from an unknown word into the encoded version of a word $\omega_i \in W$, where the average is taken over all repetitions of the word ω_i that are in the training sequence.

Another criterion is to measure the distortion for encoding the unknown word $\bar{X}^L$ into the vector of reproduction models that, with respect to a given distortion measure, best reproduces all the repetitions of the ω_i word in the training sequence of ω_i.

The set of parameters over which the average should be taken, using these two criterions, will depend of the selection of the distortion measure.

Using [3] as distortion measure, we obtain for the first criterion, that the average must be taken over the vector b, given in (2), which we call the autocorrelation vector of the model σ/A, this follows after taking the average value of (3) over all encoded versions of the word ω in its training sequence, which results after some algebra, in

$$<d(\bar{X}^L,\omega)> = \frac{1}{L} \sum_{0}^{L-1} (\bar{r}_{x_i} <\bar{b}_i|\omega>^t - 1 - \ell n\, \sigma_{Mi}^{t\,2})$$

where

$$<\bar{b}_i|\omega> = \frac{1}{N} \sum_{k=0}^{N} b_i^{(k)}, i = 0, \ldots, L-1$$

N is the number of repetitions, of the word ω, in the training sequence, and $\bar{b}_i^{(k)}$ are the autocorrelation vectors of the models $(\sigma/A)_i^{(k)}$, which are the encoded version, using (1), of the ith frame of the word ω in the training sequence. Calling $\overline{(\sigma/A)}_i$ to the model with autocorrelation vector $<\bar{b}_i|\omega>$ we can write

$$<d(\bar{X}^L,\omega)> = \frac{1}{L} \sum_{i=0}^{L-1} \rho(\bar{X}_i,\overline{(\sigma/A)}_i)$$

The second criterion with the same distortion measure, is based upon the rate distortion speech compression system [1]. The goal using this technique is to select a sequence of filters $((\sigma/A)_0,\ldots,(\sigma/A_{L-1}),$ for each word $\omega\varepsilon W$, which minimize the average distortion for encoding all repetitions of the word w in the training sequence. From [1,7] the selection of this pattern reduces to choose $(\sigma/A)_i$ that minimizes the average distortion.

$$\frac{1}{N} \sum_{k=0}^{N-1} \rho(\bar{X}_i^{(k)}, (\sigma/A)_i)$$

which is simply [1] the standard LPC model σ/A for the average sample autocorrelation of the ith-frame of the word ω in the training sequence, therefore the pattern is chosen from the average autocorrelation of $\{X_i\}$ for each frame.

If instead of using the Itakura-Saito distortion measure in (3), we use a diferent distortion, another set of parameters from which the patterns must be computed could result. For example using the Log Spectral deviation [2], the average should be taken over the cepstral coefficients for both criterions.

In the next section, we drop the assumption of having a fixed number of frames, and we present a recognition scheme, based on the minimum distortion criterion with the Itakura Saito distortion measure.

RECOGNITION SYSTEM

Since there are large variations among speakers pronouncing the same word, in general, in a training sequence for a given word there are large variations in the time scale of the phonetic contents for each word, which could be compressed or expanded, resulting in a variable number of frames for each phonetic content, and a variable number of frames for the word. This feature of the speech creates a problem for obtaining a reference pattern which produces the minimun distortion between a test isolated word and the pattern. This problem has been solved [9] by obtaining not just one pattern for each word, but a multiplicity of patterns which characterize these variations. The same problem affects the recognition procedure, where the time scales of the unknown word and the reference pattern are generally not aligned. A solution for the recognition procedure problem, asuming a reference pattern (or a set of reference patterns for each word), has been done [8,10] by using nonlinear dynamic time warping algorithms.

Here we present an alternative to solve these problems. To present the scheme, suppose that we want only to recognize words from the collection of training sequences (one training sequence, with N repetitions of the word, for each word in the dictionary). Therefore the reference patterns could be defined by the sets of code books $C_\omega = \{C_{i|\omega}\}_{i=0}^{N_\omega - 1}$, where N_ω is the frame number of the most expanded word in the training sequence of ω, and $C_{i|\omega} = \{(\sigma/A)^{(k)}\}_{k=0}^{N-1}$ is a code book with the LPC models, as code words, of the ith-frame for the training sequence of ω, with number of frames greater or equal than i. According to these patterns we encode a sampled isolated word $\bar{X}^t$ from the training sequences into the ω_i word using (4) and

$$d(\bar{X}^L,\omega) = \frac{1}{L} \sum_{i=0}^{L-1} \min_{\sigma/A\varepsilon C_{i|\omega}} \rho(\bar{X}_i,\sigma/A) \qquad (6)$$

which is equal to zero, when $\omega=\omega_i$.

The basic assumption of our scheme, assumption that is enforced by our results, is that if the training sequences, for each word in the dictionary, are large enough (i.e. they contain several repetitions of each word by a large population of speakers), then when a test isolated word, that is in the dictionary but not in the training sequence, is encoded using the above rule (where we replace L in the right hand side of (5) by max $\{L, N\omega\}$), a correct decision is taken, even if the number of computations is large, This assumption suggests to reduce the number of code words or models at least in one of the $C_{i|\omega}$ code books in a proper manner, which can be done using the algorithm described in [1,12]. This algorithm allows to obtain the best code book $C_{i|\omega}$ of size J that minimize.

$$\sum_k \min_{\sigma/A \in C_{i|\omega}} \rho(X_i^{(k)}, \sigma/A)$$

where the summation is done over the ith frame of the training sequence of the ω word.

This reduction in the number of code words of each of the $C_{i|\omega}$ code books should be done until the scheme starts to fail.

For practical purposes the code books $C_{i|\omega}$ can be created starting with one code word, and adding new codewords, using the same algorithm of [1,12], until the scheme does not fail any more. We have designed a recognition system based in this criterion for a dictionary with 20 words, having as an average three codewords or models in each code book $C_{i|\omega}$.

The system was designed using a training sequence for each word, consisting of forty repetitions of the word from four male speakers (ten repetitions for each speaker). The speech was preemphasized, Hamming windowed and bandpass filtered, the sampling rate was 6500 Hz and we use 10th-order autoregressive filters as elements of the code books $C_{i|\omega}$. To test the system we use four male speakers not used in the training sequences, which pronounced one time each word in the dictionary with no errors. At the time of this writting we are doing more test of the system and we expect to have more convincing results of the presentation of this paper.

COMMENTS

A speaker-independent isolated word recognition system based on applications of some results of rate distortion speech coders has been given. The system uses the Itakura Saito distortion measure for the design and the recognition procedure. Although the size of the dictionary, that we use, has been small, we consider that our results hold considerable promise for obtaining and automated word recognition system. From our tests of the system, it seems that there is no need for a dynamic time warping technique, although this technique could be used to make a more robust decision.

REFERENCES

1. R.M. Gray, A.H. Gray Jr., G. Rebolledo and J.E. Shore. "Rate-Distortion Speech Coding with a Minimum Discrimination Information Distortion Measure" IEEE Trans I.T., Vol. 27 pp. 708-721 November 1981.
2. Gray, R.M., Buzo, A., Gray Jr., A.H., and Matsuyama, V. "Distortion Measures for Speech Processing" IEEE, Trans. ASSP. Vol. 28 pp. 367-376. August 1980
3. Itakura, F. and Saito, S. "Analysis Synthesis Telephony based on the Maximum Likelihood Method" in Proc. 6th Int. Congr. Acoust. Tokyo, Japan, pp. C-17-20, 1968.
4. A. Buzo, A. H. Gray Jr., R.M. Gray and J.D. Markel "Speech Coding Based Upon, Vector Quantization" IEEE trans. Vol. 28, pp. 562-574, October 1980.
5. Markel, J.D. and Gray Jr., A.H., Linear Prediction of speech, Springer-Verlang, N.Y., 1976.
6. Rabiner, L.R. and Schafer, R.W. Digital Processing of Speech signals. Prentice-Hall, N.J., 1978.
7. Buzo, A. Rivera C. and Martínez H. Paper in progress.
8. Itakura, F. "Minimum Prediction Residual Applied to Speech Recognition". IEEE trans. ASSP, Vol. 23, pp. 67-72, Feb. 1975.
9. Rabiner, L.R., Levinso, S.E., Rosenberg, A.E., Wilpon J.C. "Speaker-Independent Recognition of Isolated Words Using Clustering Techniques" IEEE trans. ASSP Vol. 27, pp. 336-349, Aug. 1979.
10. Myers, C., Rabiner, L.R., Rosenberg A.E., "Performance Trade offs in Dynamic Time Warping Algorithms for Isolated Word Recognition" IEEE, trans. ASSP, Vol. 28, pp. 622-636, Dec. 1980.
11. Viswanathan R., Makhoul J., "Quantization Properties of Transmission Parameters in Linear Predictiva Systems". IEEE trans. ASSP, Vol. 3, Jun 1975.
12. Linde Y., Buzo A., and Gray R.M., "An algorithm for vector quantizer design", IEEE trans. Commun. Vol. 28, pp. 89-95, Jan 1980.

Discrete Utterance Speech Recognition Without Time Alignment

JOHN E. SHORE, SENIOR MEMBER, IEEE, AND DAVID K. BURTON, MEMBER, IEEE

Abstract—**The results of a new method are presented for discrete utterance speech recognition. The method is based on rate-distortion speech coding (speech coding by vector quantization), minimum cross-entropy pattern classification, and information-theoretic spectral distortion measures. Separate vector quantization code books are designed from training sequences for each word in the recognition vocabulary. Inputs from outside the training sequence are classified by performing vector quantization and finding the code book that achieves the lowest average distortion per speech frame. The new method obviates time alignment. It achieves 99 percent accuracy for speaker-dependent recognition of a 20-word vocabulary that includes the ten digits, with higher accuracy for recognition of the digit subset. For speaker-independent recognition, the method achieves 88 percent accuracy for the 20-word vocabulary and 95 percent for the digit subset. Background of the method, detailed empirical results, and an analysis of computational requirements are presented.**

I. Introduction

CURRENTLY successful approaches to discrete utterance speech recognition involve time alignment [1], [2]. From an unknown input utterance, a feature vector is obtained every 10–30 ms by making a set of measurements. The resulting sequence of feature vectors is classified by comparing it to a set of prestored reference sequences derived from training data and finding the best match. An important step in these comparisons is the alignment of the input sequence in time with each reference sequence. In the simplest form of alignment, the endpoints of the input sequence are aligned with the endpoints of the reference sequence and the intervening input data is stretched or compressed linearly in time. Because variations in speaking rates are nonlinear, the best current systems use more sophisticated methods in which the time axis of the input data is transformed by a constrained, nonlinear warping function that is chosen by dynamic programming to provide the best possible match between the input sequence and the reference sequence [3], [4], [5]. Such methods generally are called dynamic time warping (DTW).

The time-alignment approach is successful—a recent test of commercial speech recognizers yielded accuracies in the 90 percent to 99 plus percent range for a 20-word vocabulary [2]. The approach is also intuitively compelling—it seems obvious that accurate recognition requires the exploitation of time sequence information.

It appears, however, that time sequence information is less critical than is commonly assumed. In this paper we present recent results of a new method for discrete utterance speech recognition. The new method does not use time alignment—indeed, no time sequence information is used at all. Nevertheless, the method achieves 99 percent accuracy for the same 20-word vocabulary used in [2]. The method is based on a variety of ideas and methods from information theory and related fields, namely, rate-distortion speech coding (speech coding by vector quantization) [6], [7], minimum cross-entropy pattern classification [8], and information-theoretic spectral distortion measures [9]. We reported initial results in [10], [11]. Similar work has been reported by authors from Japan [12], Mexico [13], and the United States (Bell Laboratories) [14].

II. Background

Speech coding by vector quantization [6], [7] is a narrow-bandwidth speech coding technique based on linear predictive coding (LPC). Input speech is divided into a sequence of fixed-length segments called frames—typical frame lengths are 20–30 ms. Using estimates of the sample autocorrelation function that are measured in each frame, the shape of the speech spectrum in each frame is coded in terms of the identity of a prestored set of LPC parameters that defines an autoregressive model and is called a *codeword*. The parameters used are the inverse filter gain squared σ^2 and sample coefficients a_i, $i = 1, \cdots, M$, with $a_0 = 1$. The collection of possible codewords is called a *code book*. Let $C = \{C_1, C_2, \cdots, C_N\}$ be a code book of N codewords C_i, each defining an autoregressive model. Let S_j be the autocorrelation estimates from the jth frame of the speech to be coded. Then the shape of the spectrum for the jth frame is coded by identifying the codeword C_b that "best represents" S_j according to the "nearest-neighbor

Editor's Note: One of the duties of the Editor is to serve as Associate Editor for Applications and Miscellaneous Topics. The applications area has always been a difficult one for the TRANSACTIONS: many readers would like to see more applications of information theory in the TRANSACTIONS, but it is not always clear when a paper fits such a description. In addition, many applications oriented papers often tend to drift to other, more applications oriented TRANSACTIONS. On seeing an early draft of this paper, I found it to meet the qualifications of an applications paper admirably, and I invited the authors to publish it in this TRANSACTIONS. It is an empirical study of an information theoretic classification technique applied to the important area of speech recognition. It combines information measures, pattern recognition, and source coding techniques that have appeared among our pages to develop a novel approach to isolated word recognition.

Manuscript received December 22, 1982; revised March 31, 1983.

The authors are with the Computer Science and Systems Branch, Information Technology Division, Naval Research Laboratory, Washington, DC 20375.

Reprinted from *IEEE Trans. Inform. Theory,* vol. IT-29, pp. 473–491, July 1983.

rule"

$$d(S_j, C_b) = \min_i d(S_j, C_i), \tag{1}$$

for some distortion measure d. The distortion measure used in [6], [7] is the Itakura–Saito distortion [15], [9].

Speech coding by vector quantization using the Itakura-Saito distortion has strong connections with information theory. In particular, under suitable assumptions the Itakura–Saito distortion can be shown to be a special case of the asymptotic cross-entropy rate between two stochastic processes [9], [7], [16]. Cross-entropy—also called discrimination information, directed divergence, Kullback-Leibler number, etc.—is a measure of information dissimilarity [17], [18]. Cross-entropy minimization, which can be viewed as a general method of inference about probability distributions [19], is useful in a variety of applications [8], [16], [20]–[23]. Speech coding by vector quantization is a particular example [6]. Not only can it be derived directly by cross-entropy minimization [7], it can be derived as a special case of a more general minimum cross-entropy classification method [8]. Specifically, vector quantization using the Itakura–Saito distortion measure is equivalent to choosing the codeword C_b such that the set of input speech parameters S_j provides the least additional information beyond what C_b provides [8].

Vector quantization code books are designed to minimize the average distortion that results from encoding a long training sequence of speech frames. In particular, if $T_j, j = 1,\cdots, L$ is such a training sequence, the code book C is designed so that

$$\frac{1}{L}\sum_{j=1}^{L} d(T_j, C_b^{(j)}) \tag{2}$$

achieves at least a local minimum, where $C_b^{(j)}$ is the codeword resulting from encoding T_j. If the training sequence comprises a representative sample of the speech to be coded, then C should also encode speech from outside the training sequence with a similarly small distortion. In practice, code books are designed by an iterative, clustering technique. The original ideas were published in [24]; for details about the vector quantizer design algorithm, see [6], [7], [25]. Put simply, the algorithm divides the L frames of the training sequence into N clusters of frames such that all of the frames in any particular cluster have similar spectrum shapes. The N codewords are the centroids of these clusters. Usually, the size of the code book C is a power of 2—i.e., $N = 2^R$. The code book is then known as a *rate R* code book because R bits must be transmitted to identify the best codeword for each speech frame.

III. Description of Approach

In speech coding by vector quantization, a single code book is designed from a training sequence that is as long as computational constraints permit and that is chosen to be representative of all speech to be encoded by the system. In our approach to discrete utterance recognition, we use multiple code books—one for each word in the recognition vocabulary—and we design each code book using a relatively short training sequence containing repetitions of one word in the recognition vocabulary. For example, a code book for the word SEVEN is designed by running the vector quantizer design algorithm on a training sequence of several repetitions of the word SEVEN. When an unknown word is to be classified, every frame of the word is encoded as in (1) using each code book. The average distortion over all frames of the input is computed for each code book, and the input is classified as the word corresponding to the code book yielding the lowest average distortion. A threshold on the average distortion can be used to reject input words that are not in the recognition vocabulary. Note that the basic approach can be used for both single-speaker and multiple-speaker training sequences.

To be more precise, let V be the number of words in the recognition vocabulary. Then there are V code books C_k, $k = 1,\cdots, V$—one for each word—which together comprise a *code book set*. Let C_{ki}, $i = 1,\cdots, N_k$, be codewords in C_k, where N_k is the size of C_k. Let there be L frames of speech in the utterance to be classified, and let S_j be the set of autocorrelation estimates from the jth frame ($j = 1,\cdots, L$). Finally, let D_k be the *average distortion* resulting from coding the utterance with the kth code book,

$$D_k = \frac{1}{L}\sum_{j=1}^{L} d(S_j, C_{kb}^{(j)}), \tag{3}$$

where $C_{kb}^{(j)}$ is the codeword resulting from encoding S_j with code book C_k as in (1). Then the utterance is classified as the rth word in the recognition vocabulary, where

$$D_r = \min_k D_k. \tag{4}$$

Note that this procedure classifies the unknown utterance without performing any kind of time alignment or normalization.

A problem can occur if one word in the recognition vocabulary is contained within another word. This would happen, for example, if the words SEVEN and SEVENTEEN were both in the recognition vocabulary. In this case, it is likely that an utterance of SEVEN would result in low average distortions from two code books. One way to deal with this is to include in the classification decision the extent to which each code book is "spanned" by the input utterance—if the correct code book is C_m, then we expect that b in $C_{kb}^{(j)}$ from (3) will vary over more of C_m than over C_j, $j \neq m$. To test this idea, we tried classifying by minimizing D_k/A_k instead of D_k in (4), where A_k is the fraction of the codewords in code book C_k that were selected during the classification of the input utterance. We refer to A_k as the *span fraction*.

If desired, one can select a set of threshold values $D_r^\dagger$ and require $D_r < D_r^\dagger$ in (4) for a valid classification. This can improve classification reliability and can also be useful in rejecting words outside of the recognition vocabulary.

A. Distortion Measures

We used several alternatives for the distortion measure d in (1)–(3). Let $f(\vartheta)$ and $\hat{f}(\vartheta)$ be two power spectra. Then the *Itakura–Saito distortion* between them is

$$d_{IS}(f,\hat{f}) = \int_{-\pi}^{\pi} \frac{d\vartheta}{2\pi}\left[\frac{f}{\hat{f}} - \ln\frac{f}{\hat{f}} - 1\right]. \tag{5}$$

As we mentioned earlier, the theoretical significance of d_{IS} arises from maximum-likelihood classification [15] and also from d_{IS} being the asymptotic cross-entropy between the stochastic processes underlying f and $\hat{f}$ [9], [7], [16].

We are concerned with power spectrum estimates f and $\hat{f}$ that have the autoregressive (LPC) form

$$f(\vartheta) = \frac{\sigma^2}{|A(z)|^2}, \tag{6}$$

where

$$A(z) = \sum_{k=0}^{M} a_k z^{-k}$$

and $z = \exp(i\vartheta)$. In this case, (5) can be expressed as [9]

$$d_{IS}(f,\hat{f}) = \frac{\alpha}{\hat{\sigma}^2} + \ln(\hat{\sigma}^2) - \ln(\sigma^2) - 1, \tag{7}$$

where

$$\alpha = r(0)\hat{r}_a(0) + 2\sum_{n=1}^{M} r(n)\hat{r}_a(n),$$

$$\hat{r}_a(n) = \sum_{i=0}^{M-n} \hat{a}_i\hat{a}_{i+n},$$

and where $r(n)$ are the time-domain autocorrelations of $f(\vartheta)$.

The Itakura–Saito distortion between a power spectrum and a scaled version of itself is

$$d_{IS}(f,\lambda f) = \frac{1}{\lambda} + \ln\lambda - 1,$$

which shows that the use of d_{IS} in (1)–(3) could lead to problems if an overall amplifier gain can vary during training and classification. To avoid such problems, two gain-insensitive versions of d_{IS} have been introduced. The first is called the *gain-optimized* Itakura-Saito distortion [9],

$$d_{GO}(f,\hat{f}) \equiv \min_{\lambda>0} d_{IS}(f,\lambda\hat{f})$$
$$= \ln\int_{-\pi}^{\pi}\frac{d\vartheta}{2\pi}\left[\frac{f}{\hat{f}}\right] - \int_{-\pi}^{\pi}\frac{d\vartheta}{2\pi}\ln\left[\frac{f}{\hat{f}}\right]. \tag{8}$$

In the case of LPC forms (6), d_{GO} can be expressed as

$$d_{GO}(f,\hat{f}) = \ln(\alpha) - \ln(\sigma^2). \tag{9}$$

The second gain-insensitive measure, the *gain-normalized* Itakura-Saito distortion [9] is defined for spectra of the LPC form (6):

$$d_{GN}(f,\hat{f}) \equiv d_{IS}\left(\frac{f}{\sigma^2},\frac{\hat{f}}{\hat{\sigma}^2}\right)$$
$$= \frac{\alpha}{\sigma^2} - 1. \tag{10}$$

For power spectra that have the LPC form (6), the following relationships hold among the three foregoing distortion measures:

$$d_{GO} = \ln(1 + d_{GN})$$

$$d_{GO} = \ln\left[\frac{\hat{\sigma}^2}{\sigma^2}\left(d_{IS} + \ln\frac{\sigma^2}{\hat{\sigma}^2} + 1\right)\right].$$

During classification, the input speech frames provide the argument f in (7), (9), or (10).

B. Code Book Generation

Each classification code book C_k is designed from a separate training sequence containing repetitions of the kth word in the recognition vocabulary. For speaker-dependent experiments, the training sequence for each code book is spoken by one person and the code books are used to classify additional utterances from the same person. For multiple-speaker and speaker-independent experiments, the training sequence for each code book is spoken by several people and the code books are used to classify additional utterances from these people or utterances from different people.

We used two different types of classification code books. The first, called *clustered* code books, are full-search, optimal, vector quantization code books mentioned in Section III and fully described in [6], [7]. Our clustered code book sets were generated either to fixed-size or to fixed-distortion criteria. As the name implies, in a *fixed-size* code book, the size N_k is specified ahead of time and the design algorithm chooses N_k codewords that minimize the average distortion resulting from encoding the training sequence. All code books in a fixed-size code book set have the same size. For a *fixed-distortion* code book, the design algorithm increases the code book size until it can design a code book that encodes the training sequence with an average distortion that is less than or equal to a prespecified value. All code books in a fixed-distortion code book set are generated with the same average distortion threshold and can therefore have different sizes. The use of fixed-distortion code books for discrete utterance classification was suggested by Rabiner [14]. Because each fixed-distortion code book is only as large as necessary to satisfy the distortion criterion, it follows that fixed-distortion code books might lead to the same classification performance as fixed-size code books but with fewer total codewords. This in turn would lead to smaller memory requirements and faster classification performance. Furthermore, fixed-distortion code books have the intuitive advantage that they are approximately equal in terms of a measure that should be connected closely to classification performance—average distortion in classifying the known training sequence. This measure may not, however, predict how well the code book set can discriminate among the different words in the recognition vocabulary. For both fixed-size and fixed-distortion code books, code book sizes are limited for convenience to powers of 2, i.e., $N_k = 2^{r_k}$, where r_k is the *rate* of C_k—typical classification code books contain on the order of 32 codewords ($rk = 5$).

The second basic type of code book we used is the *unclustered code book*. These are generated simply by making a codeword out of each frame in the training sequence. Training sequences for unclustered code books usually are much shorter than training sequences for clustered code books. Our motivation for considering unclustered code books was computational efficiency and convenience—generating them obviously is much easier than generating clustered code books. Although we did not expect them to perform as well as clustered code books, we thought that their performance might be sufficiently good for some applications.

In generating clustered code books, we used both d_{IS} and d_{GN} as the distortion measure in (2). We didn't use d_{GO} because it led to difficulties in computing cluster centroids—in the case of d_{IS} and d_{GN}, the centroid of a set of spectra can be computed simply by averaging the autocorrelations of the members (see [6], [8]). We refer to clustered code books generated with d_{IS} and d_{GN} respectively as IS and GN code books. In performing classification of unknown utterances, we used all three distortion measures (d_{IS}, d_{GN}, and d_{GO}) as the distortion measure in (3), except that d_{IS} was used only in the case of IS code books. (GN code books are generated by setting the gain to 1 before computing centroids, so it doesn't make sense to use the gain-sensitive measure d_{IS}.)

Spectrum shapes that result from analyzing nearly silent frames can be quite arbitrary. In order to avoid cluttering up code books with codewords that that would result from including such frames, we used an energy threshold during code book generation—frames with energy below the threshold were ignored. Similarly, another threshold was used to ignore the low-energy frames of an input utterance during classification.

C. Figures of Merit

We used two figures of merit in evaluating the experiments. The first is simply the number of classification errors made. The second attempts to quantify the extent to which the classifications are correct or incorrect. In particular, suppose that the input utterance is the mth word in the recognition vocabulary. For correct classification, D_m should be the smallest of the average distortions (3), i.e., $D_r = D_m$ (see (4)). Define

$$D^* = \min_{k \neq m} D_k \tag{11}$$

as the smallest average distortion of all code books except the correct one, and define

$$F = \frac{D^* - D_m}{D_m}. \tag{12}$$

If the classification is correct, $F > 0$; if the classification is incorrect, $F < 0$. For correct classifications, F is the fractional difference between the distortion of the correct code book, and the distortion of the next best choice—a large value of F means that the correct code book stands out clearly from the other choices. For each experiment, we computed the number of errors, the average value of F (F_{av}), and the standard deviation of F (F_σ).

D. Experiments

In this subsection we list the various experiments reported in the remainder of the paper. Section IV contains a brief summary of some previously reported experiments. The rest of the paper reports on experiments with the data base used in the test of commercial recognizers that we mentioned earlier [2]. The data base is described in Section V-A, and our experimental parameters are defined in Section V-B. The rest of Section V contains the results of speaker-dependent classification tests on all 16 speakers in the data base using fixed parameters that we selected after studying a single speaker. In particular, we performed the following experiments, which are listed according to the corresponding subsection of Section V:

C. Single-speaker (WMF) study of classification performance versus code book rate for fixed-size IS and GN code books and various classification distortion measures (Fig. 1);
Single-speaker (WMF) study of classification performance versus autoregressive model order for fixed-size code books (Fig. 2);
Single-speaker (WMF) study of classification performance versus average distortion for fixed-distortion IS and GN code books (Figs. 3–4);

D. Full-vocabulary classification performance for all 16 speakers using fixed parameters based on the WMF study (Tables I–III);
Digit-subset classification performance for all 16 speakers using fixed parameters based on the WMF study (Tables IV–VI).

Section VI contains the results of various speaker-dependent performance studies. The following experiments are listed according to the corresponding subsection of Section VI:

A. Classification performance versus code book rate for fixed-size IS and GN code books (Table VII);
Classification performance versus average distortion for fixed-distortion IS and GN code books (Figs. 5–6);

B. Classification performance versus fixed-size code book rate for classification by average distortion divided by span fraction (Table VIII);

C. Results for data downsampled to 8000 samples per second from 12 500 samples per second (Table IX);

D. Full-vocabulary results for classification with unclustered code books (Tables X and XII);
Digit-subset results for classification with unclustered code books (Table XIII);

E. Results for classification for power-normalized data (Table XIV).

Section VII contains results of the following multiple-speaker and speaker-independent experiments.

A. Results for classification of 4 speakers using a fixed-size code book set designed from utterances of all four speakers (Table XVI);

B. Single-speaker (RLD) study of speaker-independent classification performance versus fixed-size code book rate, versus size of training sequence, and versus method of estimating autoregressive model (autocorrelation or Burg technique) (Table XVII);
C. Full-vocabulary, speaker-independent classification performance for 8 male speakers using parameters based on the RLD study and two training sequence lengths (Table XVIII–XX);
Digit-subset classification performance for 8 male speakers using parameters based on the RLD study and two training sequence lengths (Tables XXI–XXIII).

IV. Summary of Previous Experiments

In this section we summarize some previous experiments in speaker-dependent classification; utterances from individual speakers were classified using code books designed from training sequences spoken by the same speaker. These experiments, conducted using a small data base, were reported in [10]. The experiments used a data base of the digits ZERO through NINE, each spoken eleven times by one male speaker—110 utterances in all. The beginning and end of each utterance were marked by hand. The speech was passed through an anti-aliasing filter, sampled at 6500 samples/s, and divided into frames of 128 samples (approximately 20 ms). Tenth order LPC analysis was performed on each frame by means of Levinson recursion on autocorrelations that were estimated with Hamming windowing and 90 percent preemphasis. Eight utterances were used as a training sequence in designing a fixed-size, rate-4 (size 16), IS code book for each word in the vocabulary. This left three utterances of each word outside of the training sequences. The thirty total utterances outside of the training sequences were classified using the rate-4 code books and d_{GO} for d in (3). No errors were made, with $F_{av} = 1.20$. The thirty utterances were classified again, this time using d_{IS} in (3). The results were similar: no errors were made, with $F_{av} = 1.05$.

We also performed limited studies with one- and four-utterance unclustered code books. Classification with d_{GO} was much better than with d_{IS}, and increasing the size of the unclustered code books improved performance. The unclustered code books, which were all larger than the clustered code books, generally performed worse than the clustered code books. For details, see [10].

V. Speaker-Dependent Performance on Large Data Base

All of our subsequent experiments were conducted using a much larger data base that was prepared by Texas Instruments, Inc. (TI), during a systematic test of existing discrete utterance recognition systems [2]. Some preliminary results were reported in [11].

Such a data base can be useful in both tuning and testing a recognition algorithm. In order to balance the conflict between tuning and unbiased testing, we chose the following procedure: we tuned the algorithm based on prior experience and on a preliminary study using one of the speakers in the TI data base. We then tested the results (without changing parameters) on the other speakers in the data base. The results of that study are reported in this section. Subsequently, we varied some of the parameters and performed studies on the entire data base in order to gain additional insights. The results of those studies are reported in subsequent sections.

A. TI Data Base

The TI data base [2] consists of twenty words: the digits ZERO through NINE and the ten control words YES, NO, ERASE, RUBOUT, REPEAT, GO, ENTER, HELP, STOP, and START. Eight male and eight female speakers each recorded twenty-six repetitions of each word in the vocabulary, for a total of 8320 utterances. The data was recorded on analog tape under tightly controlled conditions: the noise level was low, the speech level was restricted to a ± 3 dB range, the acoustic environment was unvarying and all errors in the input words were eliminated. After collection, the data was low-pass filtered and sampled at 12 500 samples per second. We received the data in digital form on magnetic tape. Each utterance, preceded and followed by short segments of ambient noise, was contained in a separate file.

We used automatic endpoint detection for both training-sequence and classification utterances in all of our experiments with the TI data base. Our endpoint-detection algorithm is based upon ideas described in [26], [27]. The algorithm first analyzes the background noise to determine its average magnitude A, and then uses the results to set three thresholds for the average adjusted magnitude—the average of $|S(t) - A|$ over 10 ms, where $S(t)$ is the speech waveform. The three thresholds—a start threshold, a high threshold, and an end threshold—are used to find significant "energy clumps" in the data. The end threshold is required to be lower than the start threshold. Briefly, a word is detected when the average adjusted magnitude satisfies the following sequence of criteria:

1) it rises above the start threshold;
2) it remains above the start threshold until it rises above the high threshold;
3) it drops below the end threshold; and
4) it doesn't exceed the start threshold for another 150 ms.

When all four criteria are satisfied, the endpoints are defined as the points at which (1) and (3) were satisfied. If (1)–(3) are satisfied, but (4) is not, the algorithm applies (3) and (4) again starting at the point where (4) failed. The purpose of this correction is to avoid being confused by short periods of low energy that occur in the middle of certain words (e.g., REPEAT).

Like the algorithm in [27], ours does not use zero-crossing information to try and distinguish between noise and sibilant speech. For this reason, the algorithm should also work well on lowpass-filtered speech such as exists in telephone channels or military communication channels.

B. Experimental Parameters

In this subsection we describe the various experimental parameters associated with code book generation and utterance classification. The parameters associated with code book generation are as follows:

a) code book type (clustered or unclustered);
b) number of utterances in the training sequence;
c) energy threshold E_{min}, where E is computed by

$$E = \sum_{i=1}^{W} x_i^2$$

(Here, W is the analysis window width (see below), and x_i are the time-domain samples of the TI data after optional preemphasis and Hamming windowing.); and
d) LPC analysis parameters for determining inverse filter gain and sample coefficients of autoregressive models (see (6)).

For clustered code books, we also have

e) distortion measures used in clustering (d_{IS} or d_{GN});
f) method of determining code book size (fixed-rate or fixed-distortion); and
g) code book rate (for fixed-rate code books) or maximum average distortion (for fixed-distortion code books).

The parameters associated with utterance classification are as follows:

a) LPC analysis parameters for determining inverse filter gain and sample coefficients;
b) energy threshold; and
c) distortion measure (d_{IS}, d_{GO}, or d_{GN}).

The LPC analysis parameters for autoregressive modeling, relevant both to code book generation and to utterance classification, are

a) analysis method (autocorrelation or Burg);
b) number of points to shift between successive speech frames (N);
c) analysis window width (number of points within a speech frame that contribute to the analysis of the frame) (W);
d) data window (rectangular or Hamming);
e) preemphasis factor (0–98 percent); and
f) filter order (M).

For consistency, the LPC analysis parameters used in classifications were always chosen to match those used in generating the code books.

C. Parameter Selection

For the initial study, we selected clustered code books with the first 10 utterances of each word as the training sequences. (This choice for the training sequences is the same as that in [2].) The TI data has little background noise, and we chose an energy threshold of $E_{min} = 250$ after examining a few utterances. We selected the autocorrelation method of LPC analysis using Levinson recursion, along with $N = W = 250$ (20 ms), Hamming windowing, and 90 percent preemphasis.

Because we believed that classification performance would depend strongly on the remaining parameters, we conducted a preliminary study on a single speaker (WMF) to determine them. In particular, we investigated 1) code book size and method of determining size; 2) LPC filter order (M); and 3) distortion measures for clustering and classification.

We began by designing rate-2 through rate-6, fixed-size code book sets with both d_{IS} and d_{GN}. In each case, the first ten utterances of each word were used as training sequences. The remaining 320 (total) words were classified by all five code book sets using d_{IS}, d_{GO}, and d_{GN} with the IS code books and using d_{GO} and d_{GN} with the GN code books. The LPC filter order was fixed at $M = 16$. The results are shown in Fig. 1. Three trends are apparant:

1) independent of distortion measure types, the error rate tends to decrease with increasing code book rate;
2) classification using d_{GO} yields the best classification performance; and
3) increasing the code book rate decreases the performance differences resulting from the various distortion measures.

Next, we studied the effect of analysis filter order (M). We generated fixed-size, rate-5, IS and GN code book sets for $M = 8$, 10, 12, 14, 18, and 20. Only d_{GO} was used as a classification distortion measure. The results are shown in Fig. 2. For a particular value of M the results for IS code books were usually better than those for GN code books, but both IS and GN code books exhibited the same general trend: the error rates initially decreased smoothly with increasing M and became constant for large M.

Next, we evaluated the choice between fixed-size and fixed-distortion code books. We designed fixed-distortion IS and GN code book sets with five different distortion thresholds and with parameter settings otherwise the same as those for the fixed-size study that was summarized in Fig. 1, except that only d_{GO} was used as a classification distortion measure. The performance of the IS code book sets is plotted in Fig. 3 as a function of the actual average distortion of the code book set. (The actual average is different from the design threshold because the code book sizes are limited to powers of 2.) Also plotted in Fig. 3 is the performance of the fixed-size, IS code book sets from Fig. 1 for the case of classification with d_{GO}. For these fixed-size code book sets, the performance is plotted as a function of the average distortion with which the code books in the set encoded their training sequences. For both the fixed-distortion and fixed-size code book sets, the average code book size appears in parentheses next to the plotted point. Note that all code books in a fixed-size code book set have the size indicated. Analogous results for GN code book sets are shown in Fig. 4. Comparing fixed-size code book sets to fixed-distortion code book sets with

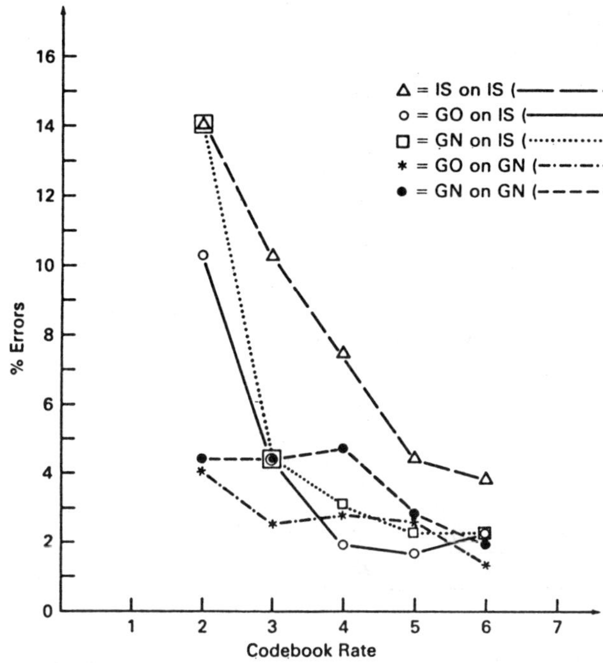

Fig. 1. Single-speaker study of error rate versus code book rate. LPC filter order is 16. "IS on IS" means classifications were performed using d_{IS} with IS code books; "GO on IS" means classifications were performed using d_{GO} with IS code books; "GN on IS" means classifications were performed using d_{GN} with IS code books; "GO on GN" means classifications were performed using d_{GO} with GN code books; and "GN on GN" means classifications were performed using d_{GN} with GN code books.

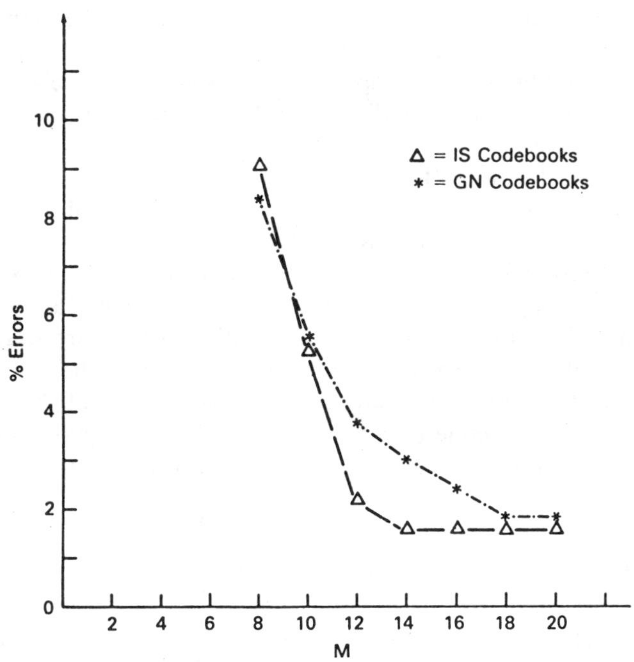

Fig. 2. Single-speaker study of error rate versus LPC filter order (M). Code book rate is 5. Classification distortion measure is d_{GO}.

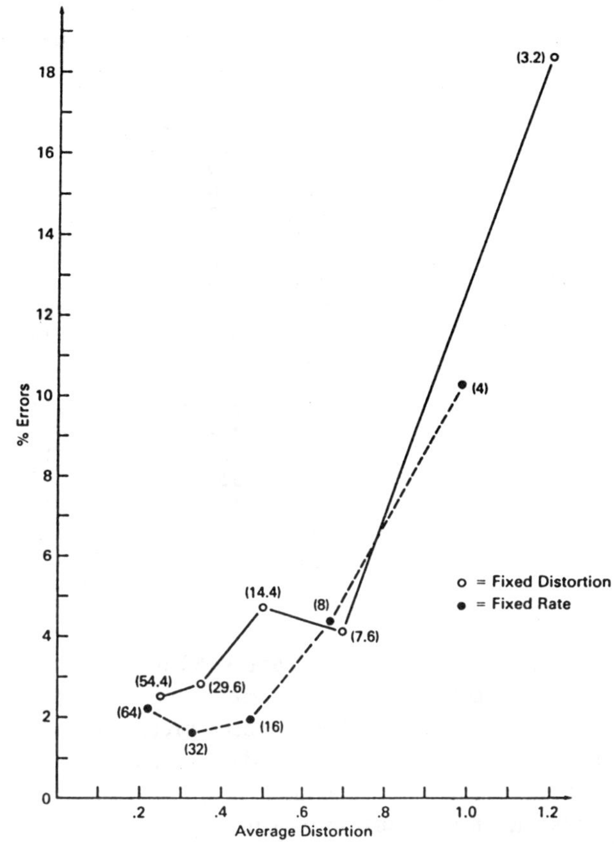

Fig. 3. Single-speaker study of error rate versus average code book distortion using IS code books. Classification distortion measure is d_{GO}. LPC filter order is 16. For each code book set, the average number of codewords per code book is shown in parenthesis.

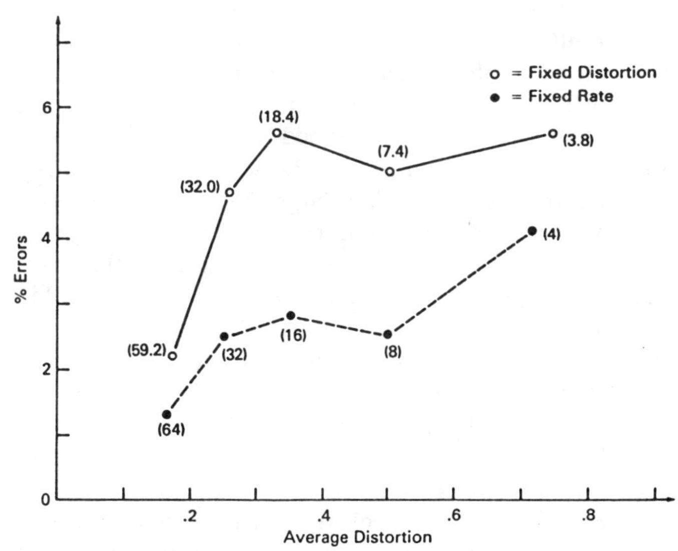

Fig. 4. Single-speaker study of error rate versus average code book distortion using GN code books. Classification distortion measure is d_{GO}. LPC filter order is 16. For each code book set, the average number of codewords per code book is shown in parenthesis.

approximately the same average code book size, it is apparent that the fixed-size code books performed better. We therefore selected fixed-size code books for the full data base experiment reported in the next subsection. Since the results in Figs. 3–4 are somewhat counterintuitive, however, we explored the issue again in studies on the full data base reported in Section VI.

D. Results for the Full Data Base

For the initial study on the entire data base, we used the same parameter settings that we fixed for the preliminary study on WMF: clustered code books with 10-utterance

TABLE I
INITIAL RESULTS FOR FULL TI DATA BASE

Speaker	No. Class	IS Code Books Errors	IS % Right	IS F_{av}	IS F_σ	GN Code Books Errors	GN % Right	GN F_{av}	GN F_σ
WMF	320	5	98.4	0.499	0.328	8	97.5	0.475	0.302
RLD	320	1	99.7	0.653	0.466	2	99.4	0.664	0.437
RGL	320	0	100.0	0.909	0.431	0	100.0	0.894	0.423
MSW	320	6	98.1	0.618	0.372	0	100.0	0.592	0.337
GRD	320	4	98.8	0.592	0.384	4	98.8	0.590	0.378
TBS	320	13	95.9	0.606	0.388	7	97.8	0.591	0.346
KAB	320	4	98.8	0.615	0.369	8	97.5	0.618	0.392
REH	320	2	99.4	0.825	0.503	1	99.7	0.852	0.511
CJP	320	4	98.4	0.629	0.323	1	99.7	0.648	0.325
DFG	320	5	98.4	0.612	0.338	1	99.7	0.621	0.329
GNL	320	2	99.4	0.883	0.554	2	99.4	0.897	0.542
JWS	320	2	99.4	0.941	0.570	1	99.7	0.947	0.568
HNJ	320	4	98.8	0.830	0.494	3	99.1	0.836	0.469
SAS	320	3	99.1	0.798	0.423	2	99.4	0.804	0.428
SJN	320	0	100.0	0.909	0.502	0	100.0	0.931	0.501
ALK	320	1	99.7	0.723	0.409	0	100.0	0.693	0.410
all	5120	56	98.9	0.728	0.456	40	99.2	0.728	0.450

training sequences, $E_{\min} = 250$, and LPC analyses by means of Levinson recursion on autocorrelations estimated with Hamming windowing and 90 percent preemphasis. Based on the results of the WMF study, we chose fixed-size, rate-5 (size 32), IS and GN code books, LPC order $M = 16$, and classification distortion measure d_{GO}.

Shown in Table I are the results for all 16 speakers for both the IS and GN code books. The first 8 speakers are male, the rest are female. The performance for both code book types was approximately the same—error rates of about 1 percent. Although we are testing an algorithm as opposed to a commercial device, it is worth noting that these results are better than six of the seven commercial devices tested with the same data base [2]. The results show that much more can be done without time-sequence information than is commonly assumed.

The results in Table I exhibit the so-called "goat-sheep" phenomenon [2], in which a large fraction of the errors occur within a small segment, the "goats," of the population. For the GN code books, over half of the errors occurred for just three speakers: KAB, TBS, and WMF. WMF and TBS were also hard speakers for the IS code books. Contrary to what is usually found [2], our results were better for female speakers than they were for male speakers, and we have no explanation for this trend.

In Tables II and III, we present summaries of the specific errors in the form of confusion matrices for the total IS and GN code book results. Each row comprises the results for one word in the recognition vocabulary; the columns correspond to the different classification decisions. Each row contains the results of classifying all utterances of one word in the vocabulary.

Some of the error classes shown in the confusion matrices are easily understandable. For example, the NO ↔ ONE confusions are not surprising—if a recording of NO is played backwards, it sounds like ONE and vice versa. It follows that the two words have many similar spectra and could be confused by a method that ignores time sequence information. This is an example of a general class of potential confusions. The GO ↔ NO confusion is another example of this general class. Because there is a time-sequence similarity as well, it is a confusion that occurs with many other methods of word recognition. The unilateral confusions like SIX → YES are probably examples of the same phenomenon—more of the spectra in SIX are similar to spectra in YES than vice versa. As we mentioned in Section III, one possible way to reduce confusions between words that comprise similar spectra is to classify by minimizing average distortion divided by span fraction. This possibility was the basis for one of the studies reported in Section VI.

Since many word-recognition applications involve only the digits ZERO through NINE, we also obtained results for this restricted case. The results for each speaker are shown in Table IV, with summary confusion matrices in Tables V and VI. As one would expect, the digit results are much better than those for the full vocabulary. For the digits, there was an extreme example of a "goat"—nine of the ten errors with IS code books and five of the eight errors with GN code books occurred for one male speaker (TBS).

VI. Speaker-Dependent Studies

In order to gain additional insights into the method in general as well as into the effects of the various parameters, we performed a variety of additional studies. In particular, we studied performance as a function of code book size, IS versus GN code books, fixed-size versus fixed-distortion code books, average distortion classification versus average distortion divided by span fraction, clustered versus unclustered code books, performance on 4 kHz. bandwidth data, and performance on power-normalized data.

A. Clustered Code Book Type and Classification Measure

We repeated the experiment described in Section V-D using fixed-size, IS and GN code book sets with rates 2, 3,

TABLE II
FULL DATA BASE CONFUSION MATRIX FOR IS CODE BOOKS

	0	1	2	3	4	5	6	7	8	9	ENTER	ERASE	GO	HELP	NO	RUBOUT	REPEAT	STOP	START	YES
0	256	·	·	·	·	·	·	·	·	·	·	·	·	·	·	·	·	·	·	·
1	·	253	·	·	·	1	·	·	·	·	·	·	·	·	2	·	·	·	·	·
2	·	·	255	·	·	·	·	·	·	·	1	·	·	·	·	·	·	·	·	·
3	·	·	·	253	·	·	·	·	·	·	·	·	·	·	·	·	3	·	·	·
4	·	·	·	·	256	·	·	·	·	·	·	·	·	·	·	·	·	·	·	·
5	·	·	·	·	·	251	·	1	·	·	·	·	·	1	·	3	·	·	·	·
6	·	·	·	·	·	·	248	3	·	·	·	2	·	·	·	·	·	·	·	3
7	·	·	·	·	·	·	·	256	·	·	·	·	·	·	·	·	·	·	·	·
8	·	·	·	·	·	·	·	·	252	·	·	·	·	·	·	·	4	·	·	·
9	·	·	2	·	·	·	2	·	1	251	·	·	·	·	·	·	·	·	·	·
ENTER	·	·	·	·	·	·	·	·	·	·	256	·	·	·	·	·	·	·	·	·
ERASE	·	·	·	·	·	·	·	·	·	·	·	256	·	·	·	·	·	·	·	·
GO	·	·	·	·	·	·	·	·	·	·	·	·	242	·	13	1	·	·	·	·
HELP	·	·	·	·	·	·	1	·	·	·	·	·	·	255	·	·	·	·	·	·
NO	·	·	4	·	·	·	·	·	·	·	·	·	3	·	249	·	·	·	·	·
RUBOUT	·	·	·	·	·	·	·	·	·	·	·	·	·	·	·	256	·	·	·	·
REPEAT	·	·	·	·	·	·	·	·	·	·	·	·	·	·	·	·	256	·	·	·
STOP	·	1	·	·	·	·	·	·	·	·	·	·	·	·	·	·	·	252	3	·
START	·	·	·	·	·	·	·	·	·	·	·	·	·	·	·	·	·	·	256	·
YES	·	·	·	·	·	·	·	·	·	·	·	·	1	·	·	·	·	·	·	255

TABLE III
FULL DATA BASE CONFUSION MATRIX FOR GN CODE BOOKS

	0	1	2	3	4	5	6	7	8	9	ENTER	ERASE	GO	HELP	NO	RUBOUT	REPEAT	STOP	START	YES
0	256	·	·	·	·	·	·	·	·	·	·	·	·	·	·	·	·	·	·	·
1	·	254	·	·	·	·	·	·	·	·	·	·	·	·	2	·	·	·	·	·
2	·	·	255	1	·	·	·	·	·	·	·	·	·	·	·	·	·	·	·	·
3	·	·	·	254	·	·	·	·	·	·	·	·	·	·	·	·	2	·	·	·
4	·	·	·	·	256	·	·	·	·	·	·	·	·	·	·	·	·	·	·	·
5	·	·	·	·	·	247	·	·	·	·	·	·	·	5	·	4	·	·	·	·
6	·	·	·	·	·	·	252	1	·	·	·	1	·	·	·	·	·	·	·	2
7	·	·	·	·	·	·	·	256	·	·	·	·	·	·	·	·	·	·	·	·
8	·	·	·	·	·	·	·	·	255	·	·	·	·	·	·	·	·	·	·	1
9	·	·	2	·	·	·	1	·	·	253	·	·	·	·	·	·	·	·	·	·
ENTER	·	·	·	·	·	·	·	·	·	·	256	·	·	·	·	·	·	·	·	·
ERASE	·	·	·	·	·	·	·	·	·	·	·	256	·	·	·	·	·	·	·	·
GO	1	·	·	·	·	·	·	·	·	·	·	·	247	1	7	·	·	·	·	·
HELP	·	·	·	·	·	·	·	·	·	·	·	·	·	256	·	·	·	·	·	·
NO	·	5	·	·	·	·	·	·	·	·	·	·	·	2	249	·	·	·	·	·
RUBOUT	·	·	·	·	·	·	·	·	·	·	·	·	·	·	·	256	·	·	·	·
REPEAT	·	·	·	·	·	·	·	·	·	·	·	·	·	·	·	·	256	·	·	·
STOP	·	·	·	·	·	·	·	·	·	·	·	·	·	·	·	·	·	255	1	·
START	·	·	·	·	·	·	·	·	·	·	·	·	·	·	·	·	·	·	256	·
YES	·	·	·	·	·	·	·	·	·	·	·	1	·	·	·	·	·	·	·	255

TABLE IV
INITIAL RESULTS FOR DIGIT SUBSET

Speaker	No. Class	IS Code Books				GN Code Books			
		Errors	% Right	F_{av}	F_σ	Errors	% Right	F_{av}	F_σ
WMF	160	1	99.4	0.714	0.371	2	98.8	0.670	0.364
RLD	160	0	100.0	1.043	0.543	0	100.0	1.059	0.550
RGL	160	0	100.0	1.144	0.507	0	100.0	1.117	0.503
MSW	160	0	100.0	0.923	0.458	0	100.0	0.901	0.432
GRD	160	0	100.0	0.915	0.459	0	100.0	0.943	0.466
TBS	160	9	94.4	0.850	0.510	5	96.9	0.810	0.444
KAB	160	0	100.0	0.970	0.511	1	99.4	0.970	0.520
REH	160	0	100.0	1.255	0.686	0	100.0	1.239	0.710
CJP	160	0	100.0	0.983	0.462	0	100.0	0.996	0.459
DFG	160	0	100.0	0.923	0.378	0	100.0	0.934	0.412
GNL	160	0	100.0	1.428	0.702	0	100.0	1.424	0.677
JWS	160	0	100.0	1.540	0.845	0	100.0	1.512	0.816
HNJ	160	0	100.0	1.215	0.614	0	100.0	1.188	0.576
SAS	160	0	100.0	1.138	0.507	0	100.0	1.146	0.549
SJN	160	0	100.0	1.441	0.698	0	100.0	1.412	0.690
ALK	160	0	100.0	1.200	0.607	0	100.0	1.124	0.542
all	2560	10	99.6	1.105	0.611	8	99.7	1.090	0.600

TABLE V
DIGITS-ONLY CONFUSION MATRIX FOR IS CODE BOOKS

	0	1	2	3	4	5	6	7	8	9
0	256	·	·	·	·	·	·	·	·	·
1	·	255	·	·	·	1	·	·	·	·
2	·	·	256	·	·	·	·	·	·	·
3	·	·	·	256	·	·	·	·	·	·
4	·	·	·	·	256	·	·	·	·	·
5	·	·	·	·	·	255	·	1	·	·
6	·	·	·	·	·	·	253	3	·	·
7	·	·	·	·	·	·	·	256	·	·
8	·	·	·	·	·	·	·	·	256	·
9	·	·	2	·	·	·	2	·	1	251

TABLE VI
DIGITS-ONLY CONFUSION MATRIX FOR GN CODE BOOKS

	0	1	2	3	4	5	6	7	8	9
0	256	·	·	·	·	·	·	·	·	·
1	·	256	·	·	·	·	·	·	·	·
2	·	·	255	1	·	·	·	·	·	·
3	·	·	·	256	·	·	·	·	·	·
4	·	·	·	·	256	·	·	·	·	·
5	·	·	·	·	·	253	·	2	·	1
6	·	·	·	·	·	·	255	1	·	·
7	·	·	·	·	·	·	·	256	·	·
8	·	·	·	·	·	·	·	·	256	·
9	·	2	·	·	·	1	·	·	·	253

4, and 6. The overall results and those from the previous rate-5 experiment are shown in Table VII. The GN code books generally performed better, but the difference was substantial only for code book rates 2 and 3. The computational complexity of generating IS and GN code books is essentially the same, but the GN code books require no gain term so they require slightly less storage. Perhaps the most surprising result in Table VII is the remarkably good performance of the rate-2 code book sets. With only four codewords per code book, the GN code books had an error rate of only 2.3 percent.

Next, we repeated the experiment using fixed-distortion IS and GN code book sets that were designed to five different average distortion thresholds. The results and those for the fixed-size code book sets (Table VII) are plotted in Figs. 5–6—overall performance is plotted as a function of the average distortion with which the code books in the set encoded their training sequences. Each point in Figs. 5–6 is based on the classification of 5120 utterances outside the training sequences. The results confirm the conclusion of the WMF study (Figs. 3–4)—fixed-size code books perform better with fewer total codewords. For example, for almost every fixed-distortion point in Fig. 6, one can find a fixed-size point with higher average distortion but with fewer average codewords per code book and lower error rate. The higher average distortion of the fixed-size code books is reasonable since they have fewer average codewords per code book than the fixed-distortion code books. The lower error rate of the fixed-size code books is, however, puzzling.

An examination of the errors showed that the fixed-distortion code books performed slightly better on some mul-

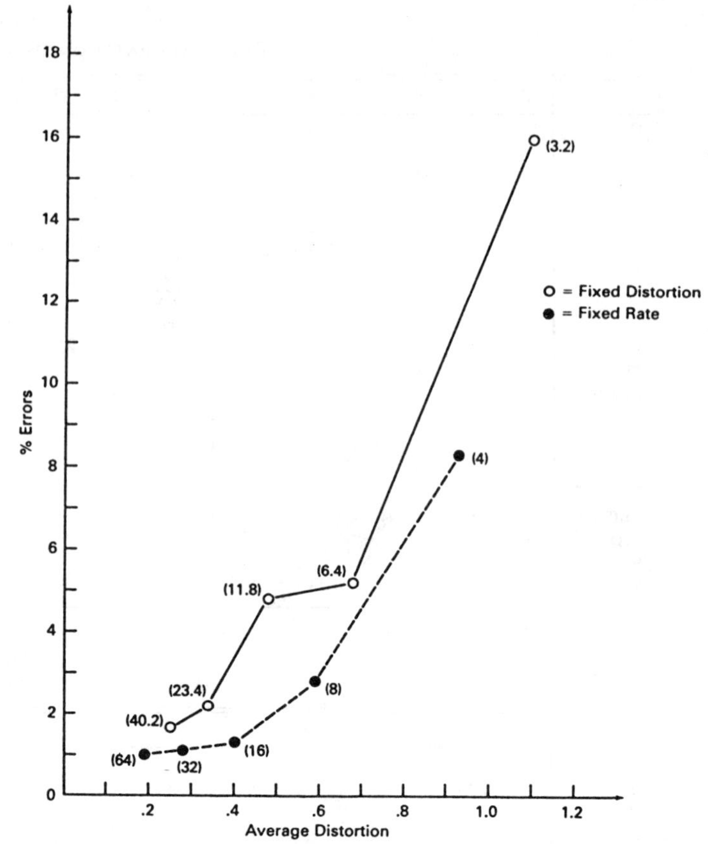

Fig. 5. Full data base study of error rate versus average code book distortion using IS code books. Classification distortion measure is d_{GO}. LPC filter order is 16. For each code book set, the average number of codewords per code book is shown in parenthesis.

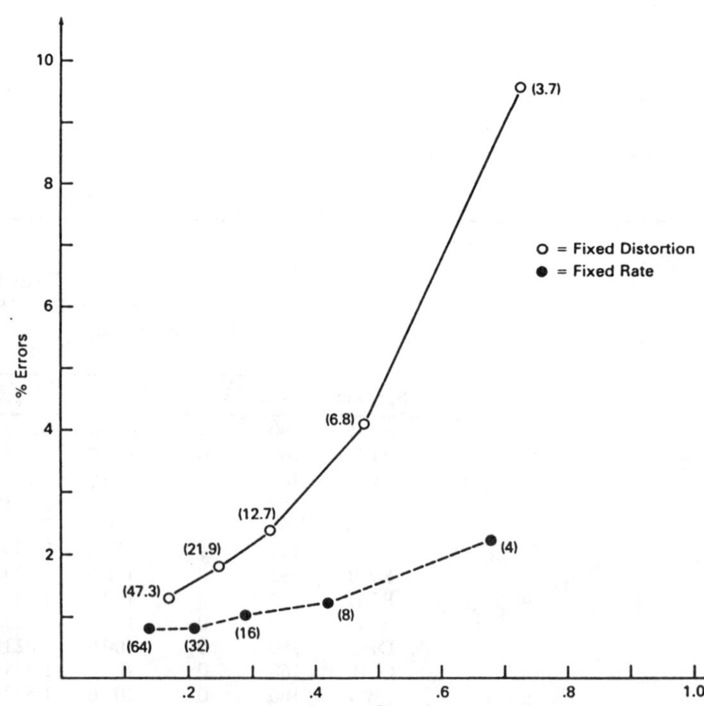

Fig. 6. Full data base study of error rate versus average code book distortion using GN code books. Classification distortion measure is d_{GO}. LPC filter order is 16. For each code book set, the average number of codewords per code book is shown in parenthesis.

TABLE VII
RESULTS OF RATE STUDY FOR FIXED-SIZE CODE BOOKS

Code Book Rate	No. Class	IS Code Books				GN Code Books			
		Errors	% Right	F_{av}	F_σ	Errors	% Right	F_{av}	F_σ
2	5120	423	91.7	0.342	0.301	116	97.7	0.447	0.294
3	5120	144	97.2	0.519	0.373	62	98.8	0.597	0.373
4	5120	69	98.7	0.647	0.416	52	99.0	0.684	0.422
5	5120	56	98.9	0.728	0.456	40	99.2	0.728	0.450
6	5120	52	99.0	0.765	0.472	41	99.2	0.745	0.455

TABLE VIII
RESULTS OF SPAN FRACTION STUDY

Code Book Rate	No. Class	IS Code Books				GN Code Books			
		Errors	% Right	F_{av}	F_σ	Errors	% Right	F_{av}	F_σ
2	5120	460	91.0	0.382	0.340	138	97.3	0.504	0.348
3	5120	224	95.6	0.709	0.530	123	97.6	0.810	0.537
4	5120	99	98.1	1.018	0.706	78	98.5	1.054	0.684
5	5120	82	98.4	1.155	0.780	73	98.6	1.124	0.756
6	5120	76	98.5	1.161	0.799	64	98.8	1.100	0.749

tisyllabic words, but considerably worse on some of the monosyllabic words. The confusions that the fixed-distortion code books made on monosyllabic words were generally a superset of those made by the fixed-size code books. The additional confusions made by the fixed-distortion code books were usually into multisyllabic words that corresponded to code books containing about twice as many codewords as those of the monosyllabic words, the most frequent examples being FIVE → RUBOUT, SIX → ERASE, THREE → REPEAT, and EIGHT → REPEAT. It seems that the spectra in these larger code books were sufficiently varied so that a low average distortion resulted from encoding some of the shorter words in the recognition vocabulary. As pointed out by an anonymous referee, this explanation suggests that classification by minimizing average distortion divided by span fraction might work better for fixed-distortion code books than for fixed-size code books.

Our results concerning fixed-size versus fixed-distortion code books points out that the key to high recognition accuracy is how well the code books discriminate among the vocabulary words, not simply how well the training sequence is characterized. Apparently, fixed-size code books discriminate better than fixed-distortion code books.

B. Average Distortion Versus Span Fraction

We repeated the fixed-size code book set experiments (Table VII) with identical parameters except that classification was performed by minimizing average d_{GO} divided by span fraction instead of just average d_{GO}. The results, shown in Table VIII, are always worse than the corresponding results in Table VII.

In response to the referee's comment mentioned in the previous subsection, we repeated the fixed-distortion code book experiments described in that subsection, except that classification was performed by minimizing average distortion divided by span fraction. Indeed, some improvement resulted. In terms of the results plotted in Figs. 5–6, the performance improved slightly for the fixed-distortion code books designed to distortion thresholds greater than about 0.4—below this point the performance was worse. In no case, however, did performance improve to the level of the equivalent fixed-rate code book.

C. Performance on 4 kHz Bandwidth Data

The excellent results shown in Tables I–VII were for high-fidelity data sampled at 12 500 samples/s. This represents a larger bandwidth than available over most commercial and military telephone and radio telephone networks. Our original experiments, reported in [10] and summarized in Section IV, were performed with data that was sampled at 6500 samples/s. The performance of d_{GO} classification using fixed-size, rate-4 code books was good, and it implied that our technique could be useful with telephone-bandwidth speech, but the data base was extremely small (a total of 110 utterances). To obtain more data on the classification of telephone-bandwidth speech, we performed some classification experiments on the TI data base after down sampling the data to 4000 Hz. The sampling rate conversion was carried out by [28]:

1) padding the signal with zeros to create a signal sampled at 200 000 samples/s;
2) lowpass filtering the signal at 3900 Hz; and
3) down sampling the signal to 8000 samples/s.

The interpolation/decimation filter consisted of a fourth-order elliptical filter cascaded with a third-order Chebyshev filter. After converting the sampling rate, we ran the endpoint detection algorithm described in Section V-A.

The 4000 Hz bandwidth and the following LPC analysis conditions were chosen for compatibility with the Navy's 2.4-kbs LPC-10 [29]: $N = 180$, $W = 128$, $M = 10$, and Hamming windowing. We continued to use Levinson recursion, although LPC-10 uses the covariance method. As

TABLE IX
RESULTS FOR 4 kHz BANDWIDTH DATA

Speaker	No. Class	IS Code Books				GN Code Books			
		Errors	% Right	F_{av}	F_{σ}	Errors	% Right	F_{av}	F_{σ}
WMF	320	7	97.8	0.738	0.491	5	98.4	0.720	0.464
GRD	320	5	98.4	0.659	0.481	6	98.1	0.687	0.499
CJP	320	2	99.4	0.758	0.436	2	99.4	0.773	0.429
JWS	320	5	98.4	1.114	0.745	2	99.4	1.167	0.743
all	1280	19	98.5	0.817	0.579	15	98.8	0.837	0.581

TABLE X
RESULTS FOR 10-UTTERANCE UNCLUSTERED CODE BOOKS

Speaker	No. Class	Avg. Size	d_{IS} Classification				d_{GO} Classification			
			Errors	% Right	F_{av}	F_{σ}	Errors	% Right	F_{av}	F_{σ}
WMF	320	225.8	21	93.4	0.532	0.389	3	99.1	0.518	0.312
TBS	320	248.3	18	94.4	0.585	0.464	7	97.8	0.618	0.336
RLD	320	225.0	6	98.1	0.658	0.457	1	99.7	0.663	0.449
CJP	320	321.0	1	99.7	0.812	0.388	2	99.4	0.669	0.336
all	1280	255.0	46	96.4	0.647	0.439	13	99.0	0.617	0.367

TABLE XI
RESULTS FOR FIXED-SIZE, RATE-5, CLUSTERED CODE BOOKS (FROM TABLE I)

Speaker	No. Class	IS Code Books				GN Code Books			
		Errors	% Right	F_{av}	F_{σ}	Errors	% Right	F_{av}	F_{σ}
WMF	320	5	98.4	0.499	0.328	8	97.5	0.475	0.302
TBS	320	13	95.9	0.606	0.388	7	97.8	0.591	0.346
RLD	320	1	99.7	0.653	0.466	2	99.4	0.664	0.437
CJP	320	4	98.8	0.629	0.323	1	99.7	0.648	0.325
all	1280	23	98.2	0.597	0.385	18	98.6	0.595	0.364

before, we generated fixed-size, rate-5 clustered IS and GN code books using 10-utterance training sequences, and an energy threshold of $E_{min} = 250$. We tested two male and two female speakers using average d_{GO} as a classification criterion. The results, which should be compared with rows in Table I that correspond to the same speakers, are shown in Table IX. For the 1280 classification trials, the error rates for the narrow-bandwidth tests are approximately equal to those of the full bandwidth tests. Although there was some overlap, different errors tended to be made in the tests at different bandwidths.

Based upon these results, we believe that our approach is capable of achieving a high level of performance on telephone or military-channel bandwidth data.

D. Results for Unclustered Code Books

Unclustered code books are generated simply by making a codeword out of every frame in the training sequence. Unclustered code books therefore can be viewed as a limiting case of clustered code books, namely when one requires that the resulting code book encode the training sequence with zero distortion. Since the clustering procedure attempts to find codewords that are representative of a training sequence, this suggests that the effectiveness of clustering can be evaluated by comparing the performance of clustered and unclustered code books designed from the same training sequence.

Accordingly, for four speakers we generated unclustered code books from the 10-utterance training sequences that were the basis of the fixed-size, rate-5 clustered code book experiments discussed in Section V and summarized in Table I. All other parameters were the same, except that we classified the utterances outside of the training sequences using both d_{IS} and d_{GO} instead of just d_{GO}. The results are shown in Table X. For ease of comparison, the appropriate four lines from Table I are collected in Table XI. Comparing Tables X and XI shows that classification using d_{GO} worked better with the unclustered code books than either the clustered IS or GN code books. The differences were small, however, which attests to the effectiveness of the clustering procedure—the clustered code books were about one eighth the size of the unclustered code books.

If one views the use of clustered code books as a method of condensing unclustered code books, the foregoing results suggest the potential of using other methods for condensing the training sequences, as has been done in some related work by Miclet and Nehame [30]. Note that unclustered code books are used as an initial guess in the k-means clustering technique [31]. One general class of methods for condensing the training sequence involves picking a subset that satisfies an optimality criterion. Possible methods include that of Hart [32]. Another class of methods involves picking an arbitrary subset, for example, by using fewer utterances in unclustered code book train-

TABLE XII
RESULTS FOR 1-UTTERANCE UNCLUSTERED CODE BOOKS

Speaker	No. Class	Avg. Size	d_{IS} Classification Errors	% Right	F_{av}	F_σ	d_{GO} Classification Errors	% Right	F_{av}	F_σ
WMF	320	21.9	132	58.8	0.100	0.529	31	90.3	0.339	0.280
TBS	320	23.9	142	55.6	0.081	0.500	41	87.2	0.481	0.462
RLD	320	24.4	83	74.1	0.235	0.397	30	90.6	0.366	0.326
CJP	320	30.4	41	87.2	0.439	0.415	14	95.6	0.408	0.264
all	1280	25.2	398	68.9	0.214	0.485	116	90.9	0.354	0.292

TABLE XIII
RESULTS FOR CLASSIFICATION OF DIGIT SUBSET WITH 1-UTTERANCE UNCLUSTERED CODE BOOKS

Speaker	No. Class	Avg. Size	d_{IS} Classification Errors	% Right	F_{av}	F_σ	d_{GO} Classification Errors	% Right	F_{av}	F_σ
WMF	160	22.4	77	51.9	0.303	0.975	14	91.3	0.462	0.316
TBS	160	23.3	64	60.0	0.276	0.768	11	93.1	0.496	0.338
RLD	160	24.4	30	81.3	0.494	0.535	3	98.1	0.677	0.398
CJP	160	30.9	4	97.5	0.602	0.362	0	100.0	0.657	0.278
all	640	25.3	175	72.7	0.419	0.713	28	95.6	0.573	0.348

TABLE XIV
RESULTS OF d_{IS} CLASSIFICATION USING FIXED-SIZE, RATE-5, CLUSTERED IS CODE BOOKS FOR POWER-NORMALIZED DATA

Speaker	No. Class	Normalized Data Errors	% Right	F_{av}	F_σ	Unnormalized Data Errors	% Right	F_{av}	F_σ
WMF	320	9	97.2	0.494	0.293	14	95.6	0.518	0.356
TBS	320	21	93.4	0.587	0.459	20	93.8	0.575	0.459
RLD	320	5	98.4	0.669	0.461	6	98.1	0.679	0.479
KAB	320	8	97.5	0.699	0.452	8	97.5	0.640	0.375
all	1280	43	96.6	0.612	0.430	48	96.3	0.603	0.425

ing sequences. The "random quantizers" in [33] are unclustered code books comprising a random selection of frames from the training sequence. Some limited results with 1-utterance unclustered code books in our original experiments suggested that surprisingly good performance could be obtained [10]. To obtain more data, we repeated the experiments reported in Table X, except that we used only one utterance from the training sequences for the unclustered code books. In each of the four experiments, the same 320 utterances as before were classified. The results are shown in Table XII. The results for classification with d_{GO} show that about 90 percent accuracy can be expected using single-utterance, unclustered code books. For the digit subset, accuracy increases to about 95 percent (Table XIII). Since unclustered code books are so easy to generate, these results are quite good. They suggest that the method could be used for an easy-to-program, easy-to-train, "poor-man's" discrete utterance speech recognizer.

E. Power-Normalized Itakura–Saito Classification

Classification using d_{IS} in (3) can only be used for code books that include gain terms, i.e., for IS clustered code books and unclustered code books. Results from Fig. 1, Table X, and Table XII show consistently that classification with d_{IS} in these cases is always inferior to classification with d_{GO}. These results are somewhat disturbing, since d_{IS} is a special case of asymptotic cross entropy [9], [7], [16], and since classification with d_{IS} is optimal in a well-defined information-theoretic sense [8].

The problem appears to be that small code books are not capable of reflecting the large gain variations that occur, a conjecture supported by the large difference in performance for d_{IS} classification between 1- and 10-utterance unclustered code books (see Tables X and XII). It follows that decreasing the gain variations in the training and classification data might improve the performance of d_{IS}. In an attempt to remove large gain variations for similar spectra, but still to allow different spectra to have characteristic gain terms, we normalized each utterance so that they all had the same average power. This power-normalized data was then used to build code books and used as classification data. We didn't expect dramatic differences since TI restricted the level of the speech data to within a ± 3 dB range when they recorded it.

We generated fixed-size, rate-5 clustered IS code books for four male speakers using the power-normalized utterances and parameter settings otherwise equivalent to the experiments reported in Table I. For each speaker, we classified the 320 utterances outside of the training sequence using d_{IS}. We then repeated the four experiments, except that we used the original, unnormalized data. The results are shown in Table XIV. For ease in comparing with the previous results for classification with d_{GO}, the

TABLE XV
RESULTS OF d_{GO} CLASSIFICATION USING FIXED-SIZE, RATE-5, CODE BOOKS (FROM TABLE I)

		IS Code Books				GN Code Books			
Speaker	No. Class	Errors	% Right	F_{av}	F_σ	Errors	% Right	F_{av}	F_σ
WMF	320	5	98.4	0.499	0.328	8	97.5	0.475	0.302
TBS	320	13	95.9	0.606	0.388	7	97.8	0.591	0.346
RLD	320	1	99.7	0.653	0.466	2	99.4	0.664	0.437
KAB	320	4	98.8	0.615	0.369	8	97.5	0.618	0.392
all	1280	23	98.2	0.593	0.395	25	98.1	0.587	0.379

TABLE XVI
RESULTS OF MULTIPLE SPEAKER EXPERIMENT

		IS Code Books				GN Code Books			
Speaker	No. Class	Errors	% Right	F_{av}	F_σ	Errors	% Right	F_{av}	F_σ
WMF	320	18	94.4	0.302	0.210	17	94.7	0.303	0.214
TBS	320	15	95.3	0.318	0.244	15	95.3	0.317	0.245
REH	320	8	97.5	0.478	0.332	7	97.8	0.509	0.341
GRD	320	22	93.1	0.365	0.283	19	94.1	0.353	0.257
all	1280	63	95.1	0.366	0.280	58	95.5	0.317	0.245

appropriate four lines from Table I are collected in Table XV. The results for d_{IS} classification with normalized data are still significantly worse than the results for d_{GO} classification with unnormalized data. The results for d_{IS} classification with normalized data may be slightly better than those for unnormalized data, but clearly there is no significant improvement.

VII. MULTIPLE-SPEAKER AND SPEAKER-INDEPENDENT EXPERIMENTS

We use the terms *multiple-speaker* and *speaker-independent* recognition when a set of speakers contribute to the training sequence of a code book. Multiple-speaker recognition is the case when the resulting code book is used for classifying utterances from these same speakers. Speaker-independent recognition is the case when the code book is used for classifying utterances from a speaker who is not in the training set.

In this section we present the results of one multiple-speaker classification test, several preliminary speaker-independent parameter studies, and a male speaker-independent experiment. The data and the LPC analysis conditions in these experiments were the same as in the speaker-dependent experiment discussed in Section V (Table I). The number of speakers in the training sequence, the number of utterances per speaker in the training sequence, and the code book rate were varied in the preliminary experiments.

A. Multiple-Speaker Test

To gain an appreciation for the degradation introduced by interspeaker variations in the training sequence, a four-speaker training sequence was used to generate IS and GN code books. In particular, fixed-size, rate-5 clustered code books were designed from two utterances each from REH, GRD, WMF, and TBS (all males). Additional utterances from the same four speakers were then classified using d_{GO}. The results are shown in Table XVI. As expected, classification accuracy is less than that of the speaker-dependent case (Table I), but the degradation is surprisingly small.

B. Speaker-Independent Parameter Studies

Using the same four speakers as in the multiple-speaker experiment, we generated four additional code book sets: fixed-size, rate-5 and rate-6 clustered IS and GN code books using six utterances per speaker in the training sequences. We then classified all 520 utterances from RLD, a male speaker not included in the training set, using d_{GO}. The classification accuracies for all six code book sets are shown in the first row of Table XVII. The results do not strongly favor any of the code book sets.

The second row in Table XVII was motivated by results in [34]. There, the Burg technique for estimating LPC parameters was used in speech coding by vector quantization. The Burg technique led to code books that encoded the training sequence with lower average distortion than the code books designed from LPC parameters estimated by the autocorrelation method. To see if the Burg technique would lead to better classification code books, we estimated LPC parameters and equivalent autocorrelations with the Burg technique, using 90 percent preemphasis, order $M = 16$, and frame size $N = 250$. We then generated new multiple-speaker code books using the results together with the usual energy threshold of $E_{min} = 250$, and we repeated the classification of RLD except that we again used Burg estimation for the autocorrelations. The results were about the same as those for the autocorrelation method.

For additional information, we compared the spectra in the code books designed using the two methods of estimating LPC parameters—they were similar but hardly identical. Next we compared how well the code books repre-

TABLE XVII
CLASSIFICATION OF RLD USING d_{GO} ON VARIOUS SPEAKER-INDEPENDENT CODE BOOKS
(r = FIXED-SIZE CODE BOOK RATE)
(u = NO. OF UTTERANCES PER SPEAKER IN TRAINING SEQUENCE)

		IS Code Books			GN Code Books		
Method	No. Class	$r = 5$ $u = 2$	$r = 5$ $u = 6$	$r = 6$ $u = 6$	$r = 5$ $u = 2$	$r = 5$ $u = 6$	$r = 6$ $u = 6$
Auto	520	77.5	77.3	78.1	74.2	77.1	78.5
Burg	520	77.9	81.9	78.6	72.5	80.4	78.7

TABLE XVIII
SPEAKER-INDEPENDENT CLASSIFICATION USING d_{GO} ON RATE-5, CLUSTERED GN-CODE BOOKS

		2-Utterance Training				9-Utterance Training			
Speaker	No. Class	Errors	% Right	F_{av}	F_σ	Errors	% Right	F_{av}	F_σ
WMF	520	61	88.3	0.183	0.162	63	87.9	0.188	0.164
RLD	520	107	79.4	0.139	0.178	91	82.5	0.173	0.198
RGL	520	61	88.3	0.229	0.196	41	92.1	0.256	0.201
MSW	520	65	87.5	0.215	0.194	45	91.4	0.237	0.210
GRD	520	119	77.1	0.134	0.158	106	79.6	0.121	0.164
TBS	520	85	83.7	0.159	0.184	77	85.2	0.162	0.170
KAB	520	83	84.0	0.178	0.187	57	89.0	0.221	0.192
REH	520	41	92.1	0.270	0.217	18	96.5	0.321	0.254
all	4160	622	85.1	0.189	0.191	498	88.0	0.210	0.205

sented the training sequences by examining the average distortion for encoding them. We computed the code book set distortion AD by averaging the average distortion with which each code book encoded its training sequence. As in [34], we computed the RMS log spectrum error for each code book set, $SE = 6.142(e^{AD} - 1)^{1/2}$. The average difference in SE between the autocorrelation method and the Burg technique code books was 0.05 dB for IS code books and 0.06 dB for GN code books. Although the code books designed using the Burg technique had smaller average distortion, the difference appears to be too small to have a significant effect on classification accuracy.

As an additional test, we classified the 520 utterances of one female speaker (HNJ) using d_{GO} on the 4-male, rate-5 code books designed from two utterances per speaker. The classification accuracy was only 58.5 percent for the IS code books and 37.9 percent for the GN code books. These are significantly worse than the results for RLD (a male).

C. Speaker-Independent Results

We performed speaker-independent experiments using the 8 males in the TI data base. To increase the total number of classifications, we classified all 520 utterances of each male in turn using code books designed from the other seven. Except for the training sequences, parameters were the same as those in the speaker-dependent experiment discussed in Section V (Table I)—rate-5 clustered GN code books, $E_{min} = 250$, autocorrelation method of LPC analysis with Hamming windowing and 90 percent pre-emphasis, $N = W = 250$, and $M = 16$. For each speaker, we generated two GN code book sets—one from training sequences of two utterances per remaining speaker and one from nine utterances per remaining speaker. The two-utterance training sequences were subsets of the nine-utterance training sequences. We are not suggesting that multiple repetitions by the same speaker is a good way to train a speaker-independent word recognizer; we did it only to evaluate the adequacy of using only seven speakers in the training sequence. Only d_{GO} was used as a distortion measure for classification.

Table XVIII contains the results for the 8 male speakers in the TI data base. Tables XIX and XX are the confusion matrices for the two different code book sets. Accuracy is higher for the code books designed from the longer training sequences. This suggests that higher accuracy would result from a training sequence with more speakers. Comparing Tables III, XIX, and XX shows that the same types of errors occur in both the speaker-dependent and speaker-independent cases.

As in the speaker-dependent case, we obtained speaker-independent results for the subset of the vocabulary consisting only of the digits. Table XXI shows the classification results and Tables XXII–XXIII are the confusion matrices. Again, the higher accuracy of the nine-utterance per speaker code books suggests that better performance could be achieved by increasing the number of speakers in the training sequence. The results show that speaker-independent digit classification can be performed with about 95 percent accuracy, which is quite good.

VIII. COMPUTATIONAL CONSIDERATIONS

Most of the software for these experiments was written in FORTRAN-77 and run on a DEC VAX11/750 with a floating point accelerator. Generating the fixed-size, rate-5 clustered code books required 1–1.5 minutes of execution time each. Classification of a single utterance with these

TABLE XIX
SPEAKER-INDEPENDENT CONFUSION MATRIX FOR 2 - UTTERANCE PER SPEAKER TRAINING SEQUENCES

	0	1	2	3	4	5	6	7	8	9	ENTER	ERASE	GO	HELP	NO	RUBOUT	REPEAT	STOP	START	YES
0	201	·	3	1	·	·	·	·	·	·	·	·	1	·	·	1	1	·	·	·
1	1	156	·	·	1	·	·	1	·	14	·	·	1	4	26	4	·	·	·	·
2	6	·	195	·	1	·	1	·	·	·	1	·	1	·	·	1	2	·	·	·
3	1	·	·	166	·	·	·	·	·	·	3	5	·	·	·	·	33	·	·	·
4	·	2	·	·	196	4	·	·	·	·	·	·	·	·	·	4	·	·	2	·
5	·	·	1	·	·	162	·	·	·	3	·	·	·	9	1	13	·	18	2	·
6	·	·	1	·	·	·	187	1	·	·	·	2	·	·	·	·	·	3	·	14
7	·	·	·	·	·	·	3	189	·	·	·	·	·	·	·	·	·	16	·	·
8	·	·	·	12	·	·	2	·	130	·	1	5	·	·	·	·	39	·	·	19
9	·	5	·	·	·	11		5	·	166	3	·	·	6	3	9	·	·	·	·
ENTER	·	·	1	1	·	·	·	·	·	·	204	·	·	·	·	·	2	·	·	·
ERASE	·	·	1	2	·	·	·	·	·	·	·	203	·	·	·	·	·	·	1	1
GO	2	12	4	·	·	·	·	·	·	·	·	·	121	41	19	9	·	·	·	·
HELP	·	3	·	·	·	6	·	·	·	·	·	·	5	187	2	4	·	1	·	·
NO	·	26	·	·	1	·	·	1	·	1	·	·	14	28	134	2	·	1	·	·
RUBOUT	2	·	·	·	·	2	·	·	·	3	·	·	·	·	·	200	·	1	·	·
REPEAT	2	·	·	21	·	·	·	·	·	·	·	11	·	·	·	·	174	·	·	·
STOP	·	·	·	·	·	2	·	1	·	·	·	·	·	·	·	1	·	179	25	·
START	9	·	·	·	·	·	2	·	·	·	·	·	1	·	·	5	·	2	188	1
YES	·	·	·	·	·	·	1	·	·	·	·	6	·	·	·	·	·	1	·	200

TABLE XX
SPEAKER-INDEPENDENT CONFUSION MATRIX FOR 9-UTTERANCE PER SPEAKER TRAINING SEQUENCES

	0	1	2	3	4	5	6	7	8	9	ENTER	ERASE	GO	HELP	NO	RUBOUT	REPEAT	STOP	START	YES
0	198	·	1	·	·	·	·	1	·	·	·	3	3	·	·	1	1	·	·	·
1	·	145	·	·	1	3	·	1	·	9	·	·	·	5	41	2	·	·	1	·
2	4	·	196	·	1	·	1	·	·	·	·	·	3	·	·	1	2	·	·	·
3	1	·	·	168	·	·	·	·	·	·	1	8	·	·	·	·	30	·	·	·
4	·	·	·	·	204	·	·	·	·	·	·	·	·	1	·	1	·	·	2	·
5	·	·	·	·	·	180	·	·	·	2	·	·	·	14	·	5	·	4	3	·
6	·	·	·	·	·	·	191	1	·	·	·	6	·	·	·	·	·	·	·	10
7	·	·	·	·	·	·	·	196	·	·	·	·	·	·	·	·	·	2	·	10
8	·	·	·	·	4	·	·	·	161	·	1	7	·	·	·	·	18	·	·	17
9	·	6	·	·	·	14	·	2	·	170	7	·	·	3	4	2	·	·	·	·
ENTER	·	·	·	·	3	·	·	·	·	·	203	1	·	·	·	·	1	·	·	·
ERASE	·	·	·	·	1	·	·	·	·	·	·	207	·	·	·	·	·	·	·	·
GO	3	·	1	·	·	·	·	·	·	·	·	·	134	45	16	9	·	·	·	·
HELP	·	1	·	·	·	10	·	·	·	·	·	·	1	192	·	2	·	1	1	·
NO	2	24	·	·	1	·	·	1	·	·	·	·	8	26	143	1	·	2	·	·
RUBOUT	·	·	·	·	·	2	·	1	·	3	·	·	·	·	·	199	·	2	1	·
REPEAT	·	·	·	7	·	·	·	·	1	·	1	5	·	·	·	·	194	·	·	·
STOP	·	·	·	·	·	2	·	·	·	·	·	·	·	·	·	2	·	176	28	·
START	·	·	·	·	·	1	·	1	·	·	·	1	·	·	·	1	·	·	202	2
YES	·	·	·	·	·	·	1	·	·	·	·	4	·	·	·	·	·	·	·	203

TABLE XXI
SPEAKER-INDEPENDENT CLASSIFICATION OF DIGITS USING d_{GO} ON RATE-5, CLUSTERED GN-CODE BOOKS

		2-Utterance Training				9-Utterance Training			
Speaker	No. Class	Errors	% Right	F_{av}	F_σ	Errors	% Right	F_{av}	F_σ
WMF	260	20	92.3	0.317	0.223	18	93.1	0.320	0.221
RLD	260	7	97.3	0.283	0.197	6	97.7	0.339	0.224
RGL	260	9	96.5	0.398	0.250	6	97.7	0.408	0.274
MSW	260	20	92.3	0.360	0.259	12	95.4	0.391	0.279
GRD	260	31	88.1	0.252	0.194	22	91.5	0.252	0.215
TBS	260	10	96.2	0.330	0.229	9	96.4	0.317	0.200
KAB	260	14	94.6	0.309	0.197	11	95.8	0.378	0.228
REH	260	15	94.2	0.409	0.282	2	99.2	0.499	0.327
all	2080	126	93.9	0.332	0.236	86	95.9	0.363	0.259

TABLE XXII
SPEAKER-INDEPENDENT, DIGIT-SUBSET CONFUSION MATRIX FOR 2-UTTERANCE PER SPEAKER TRAINING SEQUENCES

	0	1	2	3	4	5	6	7	8	9
0	203	1	3	1	·	·	·	·	·	·
1	1	188	·	·	1	1	·	1	·	16
2	6	1	199	·	1	·	1	·	·	·
3	3	·	·	204	·	·	·	·	1	·
4	·	3	·	·	200	5	·	·	·	·
5	·	6	·	·	·	192	·	2	·	8
6	·	·	1	·	·	·	202	5	·	·
7	·	·	·	·	·	·	3	205	·	·
8	1	·	·	21	·	·	5	·	181	·
9	·	5	·	·	·	16	·	7	·	180

TABLE XXIII
SPEAKER-INDEPENDENT, DIGIT-SUBSET CONFUSION MATRIX FOR 9-UTTERANCE PER SPEAKER TRAINING SEQUENCES

	0	1	2	3	4	5	6	7	8	9
0	203	·	1	2	·	·	·	2	·	·
1	·	191	·	·	2	4	·	1	·	10
2	6	·	199	·	2	·	1	·	·	·
3	2	·	·	205	·	·	·	·	1	·
4	·	1	·	·	206	1	·	·	·	·
5	·	1	·	·	·	198	·	1	·	8
6	·	·	·	1	·	·	205	2	·	·
7	·	·	·	·	·	·	2	206	·	·
8	2	·	·	5	·	·	3	·	198	·
9	·	8	·	·	·	15	·	2	·	183

code books took about 1 s/code book. All of the software was designed for research purposes; we are confident that specially designed programs would run considerably faster.

In general, let N = code book size, M = LPC model order, and L = length of an input utterance in frames. Then GN code books each require $N(M+1)$ storage locations, and the classification of an input utterance requires NL distortion calculations per code book. Several approaches could be used to reduce these requirements. For example, some codewords might contribute more to incorrect classifications than to correct classifications. If they can be found and removed, code book storage would decrease and classification speed would increase. Also, one could track the accumulating average distortions during classification, and reject some of the hypotheses without having to compute the average distortion of every code book over the entire input utterance.

Even without such improvements, it is instructive to compare the computational requirements of the vector quantization (VQ) approach with those of DTW. Both requirements are dominated by the number of distortion calculations. In typical DTW approaches, the reference template and the input utterance are linearly normalized to the same length L before performing DTW, and appropriate constraints are applied to the search path [5]. High recognition accuracies can then be achieved with αL^2 distortion computations per reference template, where α is in the approximate range 0.2 to 0.3 [5]. Thus the ratio P of the number of distortion calculations required by the VQ approach to the number required by the DTW approach is about $P \approx N/\alpha L$. For fixed-rate code books with $N = 2^R$, and for a nominal value of $\alpha \approx 0.25$, the ratio becomes $P \approx 2^{R+2}/L$. We shall assume that a typical input utterance is $L \approx 32$ frames long (640 ms at 20 ms per frame)—this is perhaps too large, but it is conveniently a power of two. It follows that the ratio of distortion calculations becomes

$$P \approx 2^{R-3}. \tag{13}$$

For our best results—achieved with rate-5 code books—(13) shows that DTW requires fewer distortion calculations. But for rate-2 and rate-3 code books, which still achieve excellent recognition accuracies of about 98 percent (see Table VII), the VQ approach requires fewer or about the same number of distortion calculations.

A DTW approach using ordered, graph-searching techniques can reduce the number of distortion computations by an additional factor of about 2.5 [35], which changes the exponent in (13) from $R-3$ to about $R-2$. It does so, however, at the expense of a more complicated control structure. Whether the overall computational requirement is reduced depends strongly on the hardware available for distortion calculations [35].

The foregoing comparison does not apply to the speaker-independent case, since conventional speaker-independent recognition systems usually have several templates for each word in the recognition vocabulary. Since the VQ approach still requires only one code book, considerably fewer distortion calculations are required for the VQ approach than for DTW. The moderately good performance of the VQ approach in this case suggests its use as a preprocessor for DTW systems [14].

During classification, the input speech frames provide the argument f in (7), (9), or (10). It follows that both the time domain autocorrelations $r(n)$ and the LPC gain squared σ^2 must be known for each input frame, which in turn means that an LPC analysis must be performed. For d_{IS} and d_{GO}, however, the gain enters as a constant term $(\ln(\sigma^2))$ that contributes a constant term in the computation of the average code book distortions (3). The classification can therefore be performed without this term, which means than no LPC analysis of the input utterance is required. However, because only the relative values of the classification distortion will be known in this case, an overall rejection threshold cannot be used.

IX. CONCLUSION

The strongest conclusion from our study is that, for speaker-dependent recognition of a 20-word vocabulary, 99 percent accuracy results from classification using the average distortion of specially designed vector quantization code books. For the digit subset, performance is considerably better. This performance is about the same as that achieved by DTW, which shows that much more can be done without time-sequence information than is commonly assumed. Other strong conclusions are as follows:

a) The method achieves almost 98 percent accuracy with rate-2 code books—only four codewords per code book.

b) The method achieves 88 percent accuracy for speaker-independent recognition of the 20-word vocabulary and 95 percent accuracy for the digit subset.
c) Clustered code books should be designed to fixed-rate rather than fixed-distortion criteria.
d) With 1-utterance unclustered code books, the method achieves about 90 percent accuracy for speaker-dependent recognition of the 20-word vocabulary and 95 percent accuracy for the digit subset. This approach has the advantages of fast training and no requirement for code book design software.

For suitably chosen vocabularies, our results show that characteristic spectra contain enough information for recognition, and that information-theoretic clustering does a good job of extracting that information from training data.

Given the performance obtained without exploiting time sequence information, it seems worthwhile to consider ways of increasing performance by incorporating some time sequence information. We see two basic approaches. One is to observe the sequence of code words that are closest to frames of an input utterance when the utterance is classified with the correct code book. If typical code book "trajectories" can be defined, it might be possible to incorporate them into the classification algorithm. Another approach is to divide utterances into sections and to use a separate code book for each section, an approach suggested by Buzo [13]. Preliminary experiments with this approach show substantial improvements in both classification accuracy and computational efficiency [36].

Acknowledgment

We thank J. Buck, A. Buzo, R. Johnson, and L. Rabiner for helpful discussions. We thank J. Buck for writing some of the software and for providing various additional assistance. We thank R. M. Gray for providing some of the vector quantization software and for reviewing an earlier draft of this paper, and we thank T. Schalk for his help in obtaining the data base. We also benefited from reviews by two anonymous referees.

References

[1] L. R. Rabiner and S. E. Levinson, "Isolated and connected word recognition-theory and selected applications," *IEEE Trans. Commun.*, vol. COM-29, pp. 621–659, 1981.
[2] G. R. Doddington and T. B. Schalk, "Speech recognition: Turning the y to practice," *IEEE Spectrum*, pp. 26–32, Sept. 1981.
[3] F. Itakura, "Minimum prediction residual principle applied to speech recognition," *IEEE Trans. Acoust., Speech, Signal Processing*, vol. ASSP-23, pp. 67–72, Feb. 1975.
[4] H. Sakoe and S. Chiba, "Dynamic programming algorithm optimization for spoken word recognition," *IEEE Trans. Acoust., Speech, Signal Processing*, vol. ASSP-26, pp. 43–49, Feb. 1978.
[5] C. Myers, L. R. Rabiner, and A. E. Rosenberg, "Performance tradeoffs in dynamic time warping algorithms for isolated word recognition," *IEEE Trans. Acoustics, Speech, Signal Processing*, vol. ASSP-28, pp. 623–635, Dec. 1980.
[6] A. Buzo, A. H. Gray Jr., R. M. Gray, and J. D. Markel, "Speech coding based upon vector quantization," *IEEE Trans. Acoust., Speech, Signal Processing*, vol. ASSP-28, pp. 562–574, Oct. 1980.
[7] R. M. Gray, A. H. Gray, Jr., G. Rebolledo, and J. E. Shore, "Rate-distortion speech coding with a minimum discrimination information distortion measure," *IEEE Trans. Inform. Theory*, vol. IT-27, pp. 708–721, Nov. 1981.
[8] J. E. Shore and R. M. Gray, "Minimum-cross-entropy pattern classification and cluster analysis," *IEEE Trans. Pattern Anal. and Machine Intell.*, vol. PAMI-4, pp. 11–17, Jan. 1982.
[9] R. M. Gray, A. Buzo, A. H. Gray, Jr., and Y. Matsuyama, "Distortion measures for speech processing," *IEEE Trans. Acoust., Speech, Signal Processing*, vol. ASSP-28, pp. 367–376, Aug. 1980.
[10] J. E. Shore and D. Burton, "Discrete utterance speech recognition without time normalization," in *Proc. ICASSP82*, Rep. IEEE 82CH1746-7, pp. 907–910, May 1982.
[11] J. E. Shore and D. Burton, "Discrete utterance speech recognition without time normalization—Recent results," in *Proc. 1982 Int. Conf. Pattern Recognition*, Rep. IEEE 82CH1801-0, pp. 582–584, Oct. 1982.
[12] R. Hamabe, Y. Yamada, M. Murata, and T. Namekawa, "A speech recognition system using inverse filter matching technique," in *Proc. Ann. Conf. Inst. of Television Engineers*, Kyushu University, June 1981, (in Japanese).
[13] H. G. Martinez, C. Riviera, and A. Buzo, "Discrete utterance recognition based upon source coding techniques," in *Proc. ICASSP82*, Rep. IEEE 82CH1746-7, pp. 539–542, May 1982.
[14] L. R. Rabiner, Bell Laboratories, private communication.
[15] F. Itakura and S. Saito, "Analysis synthesis telephony based on the maximum likelihood method," in *Reports of the 6th Int. Cong. Acoustics*, 1968.
[16] J. E. Shore, "Minimum cross-entropy spectral analysis," *IEEE Trans. Acoust., Speech, Signal Processing*, vol. ASSP-29, pp. 230–237, Apr. 1981.
[17] S. Kullback, *Information Theory and Statistics*. New York: Dover, 1969.
[18] J. E. Shore and R. W. Johnson, "Properties of cross-entropy minimization," *IEEE Trans. Inform. Theory*, vol. IT-27, pp. 472–482, July 1981.
[19] ——, "Axiomatic derivation of the principle of maximum entropy and the principle of minimum cross-entropy," *IEEE Trans. Inform. Theory*, vol. IT-26, pp. 26–37, Jan. 1980.
[20] R. W. Johnson and J. E. Shore, "Minimum-cross-entropy spectral analysis of multiple signals," *IEEE Trans. Acoust., Speech, Signal Processing*, vol. ASSP-31, June 1983.
[21] ——, "Multi-signal minimum-cross-entropy spectrum analysis with weighted priors," NRL Report in publication, Naval Research Laboratory, Washington, D.C., 1983. (Submitted to *IEEE Trans. Acoustics, Speech, Signal Processing.*)
[22] R. W. Johnson, J. E. Shore, D. Burton, and J. Buck, "Speech noise reduction by means of multi-signal minimum cross-entropy spectral analysis," in *Proc. ICASSP83*, Rep. IEEE 83CH1841-6, pp. 1129–1132, April 1983.
[23] J. E. Shore, "Information theoretic approximations, for M/G/1 and G/G/1 queuing systems," *Acta Informatica*, vol. 17, pp. 43–61, 1982.
[24] S. P. Lloyd, "Least squares quantization in PCM," Bell Laboratories Tech. Rep., Murray-Hill, NJ, 1957.
[25] Y. Linde, A. Buzo, and R. M. Gray, "An algorithm for vector quantizer design," *IEEE Trans. Commun.*, vol. COM-28, pp. 84–95, Jan. 1980.
[26] L. R. Rabiner and M. Sambur, "An algorithm for determining the endpoints of isolated utterances," *Bell. Syst. Tech. J.*, vol. 54, pp. 297–315, Feb. 1975.
[27] L. Lamel *et al.*, "An improved endpoint detector for isolated word recognition," *IEEE Trans. Acoustics, Speech, Signal Processing*, vol. ASSP-29, pp. 777–785, Aug. 1981.
[28] R. E. Crochiere and L. R. Rabiner, "Interpolation and decimation of digital signals—A tutorial review," in *Proc. IEEE*, vol. 69, pp. 300–331, Mar. 1981.
[29] T. E. Tremain, "The government standard linear predictive coding algorithm: LPC-10," *Speech Tech.*, vol. 1, pp. 40–49, April 1982.
[30] L. Miclet and A. Nehame, "Experience a reconnaissance de la parole par prediction linear," Dept. Systeme et Communication Rep. ENST-C-79020, ENST, Sept. 1979.
[31] J. MacQueen, "Some methods for classification and analysis of multivariate observations," in *Proc. 5th Berkeley Symp. Math., Stat.,*

and Prob., vol. 1, Univ. California Press, pp. 281–286, 1967.

[32] P. E. Hart, "The condensed nearest neighbor rule," *IEEE Trans. Inform. Theory*, vol. IT-14, pp. 515–516, 1968.

[33] S. Roucos, R. Schwartz, and J. Makhoul, "Segment quantization for very-low-rate speech coding," in *Proc. ICASSP82*, Rep. IEEE 82CH1746-7, pp. 1565–1568, May 1982.

[34] G. Rebolledo, R. M. Gray, and J. P. Burg, "A multirate voice digitizer based upon vector quantization," *IEEE Trans. Commun.*, vol. COM-30, pp. 721–727, April 1982.

[35] M. K. Brown and L. R. Rabiner, "An adaptive, ordered, graph search technique for dynamic time warping for isolated word recognition," *IEEE Trans. Acoustics, Speech, Signal Processing*, vol. ASSP-30, pp. 535–544, Aug. 1982.

[36] J. E. Shore, D. Burton, and J. Buck, "A generalization of isolated word recognition using vector quantization," in *Proc. ICASSP83*, Rep. 83CH1841-6, pp. 1021–1024, April 1983.

Network-Based Connected Digit Recognition Using Vector Quantization

Marcia A. Bush
Gary E. Kopec

Schlumberger Palo Alto Research
3340 Hillview Ave.
Palo Alto, CA 94304

Abstract

This paper describes a network-based approach to speaker-independent connected digit recognition. The digits are modeled by a pronunciation network whose arcs represent classes of acoustic-phonetic segments. Each arc is associated with a *matcher* for rating an input speech interval as an example of the corresponding segment class. The matchers are based on vector quantization of LPC spectra and the use of gross acoustic features for pruning. Recognition involves finding a minimum quantization distortion path through the network by dynamic programming. The system has been evaluated using a portion of a large multi-dialect database developed by Texas Instruments (TI). Using a baseline network of concatenated independent digit models, string and digit accuracies of 86% and 97% respectively have been obtained.

1 Introduction

During the past several years, a variety of approaches to speaker-independent isolated and connected digit recognition have been explored. Of the proposed methods, dynamic time warping (DTW) of whole-word propotypes[1] and probabilistic methods based on finite-state hidden Markov models (HMM)[2] have been the most successful and popular. For example, DTW using a level-building algorithm has achieved string recognition accuracies exceeding 90% in small-scale connected-digit experiments[1]. Although the results thus far have been encouraging, achieving very high speaker-independent recognition accuracy across a wide range of talkers is likely to be difficult with DTW or HMM because these techniques are based on rigid formulations of the speech recognition problem which make it difficult to exploit important sources of acoustic-phonetic constraint. For example, a finite-state HMM is ill-suited for representing inter-segment duration dependencies.

This paper describes a network-based connected digit recognizer developed within a general framework for phonetically-based speech recognition. The recognizer is an extension of a previously-developed isolated digit recognition system[3]. A primary long-term motivation for the system is to investigate the embedding of various forms of acoustic-phonetic knowledge within a recognizer founded on well-formulated optimization principles. In common with other network-oriented systems such as Harpy[4] the proposed method is based on finding a minimum-cost path through a pronunciation network. It differs from other network-based recognizers in that local acoustic pattern matching is performed at the level of segments rather than individual spectral analysis frames. As a result, the system can exploit a wider variety of acoustic-phonetic constraints.

The recognition system is organized in terms of three components— a finite-state pronunciation network, a set of acoustic pattern matchers, and a dynamic programming search algorithm. A preliminary version of the system has been evaluated using a subset of a large multi-dialect database developed by Texas Instruments. Using a baseline concatenated digit pronunciation network and simple LPC-based acoustic matchers, digit and string accuracies of 86% and 97% have been obtained when the length of the input string is known apriori. The corresponding scores for unknown string length are 82% and 96%.

The rest of this paper is organized as follows. Sections 2–4 discuss the pronunciation networks, acoustic pattern matchers, and dynamic programming algorithm. Section 5 describes the system evaluation experiments. Finally, section 6 identifies directions for future work.

2 Pronunciation Networks

Information about the acoustic-phonetic structure of the digit vocabulary is represented using a finite-state pronunciation network similar to those in Harpy[4]. A pronunciation network is a collection of *nodes* interconnected by labeled, directed *branches*. Branches represent classes of acoustic-phonetic segments and a path through the network corresponds to a realization (i.e., a pronunciation) of a digit string as a sequence of such segments.

Pronunciation networks are described using a Lisp-based language which supports the hierarchical specification of networks in terms of two principal types of components— *primitive branches* and *composite networks*. Primitive branches are the "terminal symbols" of a network specification and correspond to user-defined acoustic-phonetic segment classes. A composite network is a collection of nodes interconnected by directed branches. The branches of a composite network may be either primitive branches or instances of other composite networks.

The primitive segments used to model the 11 digits (*one–nine*, *oh* and *zero*) are listed in Table 1. As suggested by their labels, the segments correspond roughly to phonemic or subphonemic units. Each of the eleven isolated digits is modeled as a composite network constructed from the primitive segments identified in Table 1. A graphical representation of the composite network for *zero* is shown in Fig. 1a. A baseline network for connected digit recognition is formed by concatenating indepen-

Reprinted from *Proc. IEEE Conf. Acoust., Speech, Signal Processing,* vol. 3, pp. 1197–1200, March 1985.

dent isolated digit models. Fig. 1b, for example, is a network for recognition without apriori knowledge of string length. Additional branches may be added to a baseline network to account for coarticulatory phenomena at digit boundaries. The experiments discussed in this paper however, deal solely with baseline networks and are intended to provide a point of reference for evaluating more sophisticated representations.

3 Acoustic Pattern Matchers

Each of the primitive branches in a pronunciation network is associated with an acoustic pattern *matcher*. A matcher is a (Lisp) procedure for rating an interval of an unknown utterance as an instance of the corresponding acoustic-phonetic class. A matcher is invoked with a specification of the interval to be rated and returns a numerical "distance" between the interval and the acoustic-phonetic category. Few constraints are placed on the form of a matcher, with the result that arbitrary features, patterns and distance metrics may be used.

Two types of segment matchers have been investigated. The first type (the baseline matcher) is based on vector quantization (VQ) of linear prediction (LPC) spectra and is similar to VQ-based whole-word models recently proposed for isolated word recognition[5]. Each segment class is represented as an ordered sequence of VQ codebooks. The codebooks for a matcher are created by clustering LPC spectra from labeled training examples of the corresponding segment class. For example, in the current system, which uses two size-8 codebooks per matcher, the first codebook for the **UW** in *two* is generated using the LPC spectra from the first half of each labeled **UW** segment in the training corpus, and the second codebook using spectra from the second half.

The matcher score for a candidate interval of input speech is

$$D(i,j) = \sum_{k=i}^{\frac{i+j}{2}-1} d_{GO}(a^k, \mathbf{A}_1) + \sum_{k=\frac{i+j}{2}}^{j-1} d_{GO}(a^k, \mathbf{A}_2),$$

where

$$d_{GO}(a, \mathbf{A}) = \min_{\hat{a}\in\mathbf{A}} d_{GO}(a, \hat{a}),$$

where $\{a^k \mid k = 0 \ldots K-1\}$ is the sequence of LPC coefficient vectors for the input utterance, $\mathbf{A}_1$ and $\mathbf{A}_2$ are the matcher codebooks, $d_{GO}(a, \hat{a})$ is the *gain-optimized* Itakura-Saito distortion [5] between input vector a and codebook entry $\hat{a}$, and i and j (where $0 \le i < j \le K$) are the bounds of the candidate interval in terms of analysis frame number.

The second type of matcher is a modification of the baseline matcher in which gross acoustic measurements are used to rule out certain candidate intervals regardless of their codebook quantization distortions. Preliminary experiments with the baseline matchers revealed numerous recognition errors in which model branches were matched to short acoustically inappropriate input segments. The large local distortions of the short mismatched segments were masked by long intervals of relatively low distortion. One approach to solving this problem is for a matcher to reject a candidate segment completely (i.e., give it a score of $+\infty$) if the gross acoustic characteristics of the candidate are inappropriate for the segment category represented by the matcher. The use of gross acoustic features to rule out recognition hypotheses is motivated by recent proposals for pruning large lexicons on the basis of broad phonetic segmentations[6].

The segment rejection criteria took the form of constraints on the minimum and maximum values of two short-time spectral features:

1. Normalized low frequency energy (NLFE) — energy in the band 0–750 Hz (LFE) relative to the maximum value of LFE over the entire utterance;
2. Spectrum tilt (ST) — energy at 4kHz relative to energy at 0 Hz.

Three sets of constraints were defined in terms of bounds on the allowed values of NLFE or ST within certain classes of segments:

1. The maximum value of NLFE within the vowel nucleus of a stressed syllable must be at least −20 dB.
2. The minimum value of NLFE within a voiceless fricative (i.e. **S**, **F** or **TH**) must be less than −30 dB. Similarly, the minimum value within the aspirated stop release **TR** must be less than −40 dB.
3. The maximum value of ST in an **F**, **TH** or **TR** segment must exceed 0 dB. Similarly, ST must exceed +10 dB within **S**.

The constraint on NLFE within a stressed syllable nucleus is motivated by the occurance of recognition errors in which unstressed model segments span most of an input digit while the stressed segments are reduced to minimum duration and placed within lower-amplitude regions. For example, when *oh* is misclassified as *seven*, the stressed vowel **EH** is often assigned minimum duration and located within the initial portion of the *oh*, well before the point of maximum amplitude. A similar phenomena frequently accompanies digit insertion errors. The NLFE and spectrum tilt constraints on **S**, **F**, **TH**, and **TR** are motivated by errors in which voiceless model segments are matched with silent or sonorant regions. This occurs, for example, when *eight* is misclassified as *six* or *oh* is misclassified as *seven*. The numerical thresholds in the constraints were based on informal visual examination of short-time parameter tracks. To avoid excessive tuning they were restricted to be multiples of 10 dB.

4 Pronunciation Network Search

Recognizing an unknown utterance involves finding a path through the pronunciation network, and a corresponding partition of the utterance into a sequence of contiguous intervals, which together minimize the sum of the matcher scores for the intervals in the partition. The unknown is recognized as that string of digits for which the best path is a possible pronunciation. Computing the best path proceeds in two steps[7].

First, a dynamic programming algorithm is used to recursively compute, for each node in the pronunciation network and for each analysis frame of the input utterance, the score of the best partial path which terminates on that node at that time. For each node n, and time $t = 1 \ldots K$, the node score

$$S(n,t) = \min_{(p,m,l_{min},l_{max})\in J(n)} \; \min_{l=l_{min}\ldots l_{max}} S(p,t-l) + D_m(t-l,t)$$

is computed and stored, where $J(n)$ is the set of branches terminating on n and each branch is represented by a tuple (p, m, l_{min}, l_{max}) identifying the node from which the branch originates (p), the segment matcher (m), the minimum allowed segment duration (l_{min}), and the maximum duration (l_{max}). The score returned by matcher m for a candidate segment between $t-l$ and t is denoted $D_m(t-l,t)$.

In addition to the value of $S(n,t)$, the branch in $J(n)$ which minimizes the node score is also recorded. During the second step of best path determination, these "backpointers" are used to trace the best path backwards from the final node to the start node. The segment duration limits (l_{min} and l_{max}) were proportional to the minimum and maximum segment durations observed in the training data.

5 Recognition Experiments

Two recognition experiments have been performed using the baseline pronunciation networks and the matchers described above. The utterances for the experiments were taken from the adult-talker portion of a large isolated and connected digits database developed by Texas Instruments[8]. The TI database contains 77 utterances for each of 111 men and 114 women from 22 geographical regions of the United States. The utterances for each talker include two repetitions of each of the eleven isolated digits (*one–nine*, *oh*, and *zero*) and 11 strings each of length 2, 3, 4, 5 and 7. The 112 TI-designated "training" talkers were divided into two groups of 56 talkers each, designated TI-TRNA and TI-TRNB. The corpora used in the recognition experiments were the 1- and 5-digit subsets of TI-TRNA and TI-TRNB and will be refered to by names of the forms TI-TRNA-x and TI-TRNB-x, where x identifies the string length. The TI data was received in digitized form at a sample rate of 20kHz and digitally downsampled to an 8Khz rate (4kHz bandwidth). LPC analysis was performed every 10 msecs using a 25 msec Hamming window and 11 poles. The waveforms were first-differenced prior to analysis.

Table 2 summarizes the results of the first experiment, in which the system was trained on the 1232 isolated digit tokens in TI-TRNA-1 and tested on the 616 utterances in each of TI-TRNA-5 and TI-TRNB-5. Training segments from TI-TRNA-1 were created automatically using a bootstrap procedure in which the system was first trained on a small set of handmarked isolated-digit utterances and then run over TI-TRNA-1. The segmentations generated during the recognition of TI-TRNA-1 were used to define the training segments. Both types of matchers were used and the system was run both with and without apriori knowledge of the input string length.

As indicated in the second column of Table 2a, the system with baseline matchers achieved string recognition accuracies of about 66% and 78% respectively, for unknown and known string lengths. Columns 3–6 report the number of individual digit matches, substitutions, insertions and deletions. These statistics were obtained by using a dynamic programming algorithm to find the Levenshtein distance [9] (i.e., minimum total number of substitutions, insertions and deletions for alignment) between each input digit string and the corresponding recognition hypothesis. Column 7 reports the number of matches as a percentage of the total number of digits. The alignment statistics suggest that knowledge of string length served primarily to reduce the number of insertions and did not affect the digit match rate.

Table 2b contains results for the modified matchers which incorporate gross acoustic features. As indicated, the string recognition accuracies increased to about 77% and 82% for unknown and known string lengths. The alignment statistics indicate a large reduction in insertions, a small increase in deletions and a small improvement in match rate.

Table 3 summarizes the results of the second experiment, in which training segments were obtained from TI-TRNA-5 using the segmentations generated during the first experiment. This training condition is analogous to the use of embedded training data in DTW connected digit recognition[1]. Only the modified matchers were used. As expected, recognition performance on TI-TRNA-5 (the training data) increased significantly, by 7% and 11% respectively for known and unknown length strings. Performance on TI-TRNB-5 improved slighly, to 82% for unknown and 86% for known string length.

6 Summary and Future Directions

The system achieved speaker-independent string recognition accuracies of 82% and 86%, for unknown and known string lengths. This performance is somewhat below that reported for DTW[1], although the results are encouraging for an initial system. Furthermore, a wider range of talkers were used in the current experiments and the utterances included the troublesome digit *oh*.

There are three main aspects of the current system which could be significantly improved. First, as noted above, the baseline pronunciation network simply concatenates independent digit models and thus ignores constraints associated with word juncture phenomena. Second, matchers based on VQ distortion are ill-suited for explicitly detecting important acoustic cues such as formant trajectories and glottalization. Finally, the relatively unsophisticated treatment of segment durations contributes to problems such as the short mismatched segments noted previously. Approaches to correcting these deficiencies are currently under investigation.

References

[1] L. Rabiner, J. Wilpon, A. Quinn, and S. Terrace, "On the application of embedded digit training to speaker-independent connected digit recognition", *IEEE Trans. on Acoustics, Speech and Signal Processing*, **ASSP-32**, no. 2, Apr., 1984.

[2] L. Rabiner, S. Levinson and M. Sondhi, "On the application of vector quantization and hidden Markov models to speaker-independent, isolated word recognition" BSTJ **62**(4) (Apr. 1983).

[3] G. Kopec and M. Bush, "Network-based isolated digit recognition using vector quantization", submitted to IEEE Transactions on Acoustics, Speech and Signal Processing, (June, 1984).

[4] B. Lowerre, *The HARPY Speech Recognition System*, PhD thesis, Carnegie-Mellon Univ., Pittsburgh, PA (1976).

[5] D. Burton, J. Shore and J. Buck, "Isolated-word speech recognition using multi-section vector quantization codebooks", to appear in *IEEE Trans. on Acoustics, Speech and Signal Processing*.

[6] D. Shipman and V. Zue "Properties of large lexicons: implications for advanced word recognition systems", in *Proceedings, 1982 IEEE International Conference on Acoustics, Speech and Signal Processing*, Paris, France (Apr 1982).

[7] H. Ney, "The use of a one-stage dynamic programming algorithm for connected word recognition", *IEEE Trans. on Acoustics, Speech and Signal Processing*, vol. **ASSP-32**, no. 2 (Apr. 1984).

[8] G. Leonard, "A database for speaker-independent digit recognition", in *Proceedings, 1984 IEEE International Conference on Acoustics, Speech and Signal Processing*, San Diego, CA (Mar 1984).

[9] D. Sankoff and J. Kruskal (eds.), *Time Warps, String Edits, and Macromolecules: The Theory and Practice of Sequence Comparison*, Addison-Wesley, Reading, MA, 1983.

Digit	# Segments	Segment Labels
oh	1	OW
1	3	W AH N
2	2	TR UW
3	3	TH R_i IY
4	3	F AO R_f
5	4	F AY_1 AY_2 V
6	4	S IH KS S
7	5	S EH V AX N
8	3	EY TS TR
9	4	N AY_1 AY_2 N
zero	5	Z IY R_{m1} R_{m2} OW
	37	25

Table 1: Acoustic-phonetic segment classes used in digit pronunciation network. The labels are similar to ARPABET.

Testing Corpus	% correct string	no. match	no. sub.	no. ins.	no. del.	% correct digit	% correct length
TI-TRNA-5	65.7	2908	145	100	27	94.4	82.6
TI-TRNB-5	66.7	2922	129	119	29	94.9	80.0
TI-TRNA-5	77.3	2909	153	18	18	94.4	(known)
TI-TRNB-5	80.0	2933	129	18	18	95.2	(known)

(a)

Testing Corpus	% correct string	no. match	no. sub.	no. ins.	no. del.	% correct digit	% correct length
TI-TRNA-5	77.4	2923	114	19	43	94.9	90.6
TI-TRNB-5	77.8	2928	114	15	38	95.1	92.2
TI-TRNA-5	83.4	2957	122	1	1	96.0	(known)
TI-TRNB-5	82.3	2955	119	6	6	95.9	(known)

(b)

Table 2: Connected digit recognition performance of concatenated isolated digit models trained on TI-TRNA-1. (a) Baseline matchers. (b) Matchers using gross acoustic features.

Testing Corpus	% correct string	no. match	no. sub.	no. ins.	no. del.	% correct digit	% correct length
TI-TRNA-5	88.1	3001	64	13	15	97.4	95.8
TI-TRNB-5	81.5	2958	98	23	24	96.0	92.9
TI-TRNA-5	90.3	3010	68	2	2	97.7	(known)
TI-TRNB-5	85.6	2977	101	2	2	96.7	(known)

Table 3: Connected digit recognition performance of concatenated isolated digit models trained on TI-TRNA-5. Matchers using gross acoustic features.

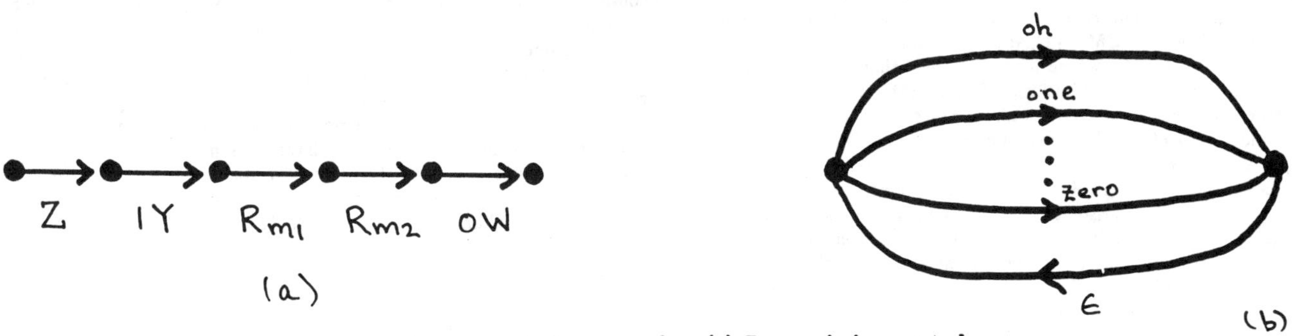

Figure 1: Pronunciation networks. (a) Pronunciation network for *zero*. (b) Concatenated isolated digit network. The branch labeled ϵ is a null transition.

Recent Developments in the Application of Hidden Markov Models to Speaker-Independent Isolated Word Recognition

B. H. Juang
L. R. Rabiner
S. E. Levinson
M. M. Sondhi

AT&T Bell Laboratories
Murray Hill, New Jersey 07974

ABSTRACT. In this paper we extend previous work on isolated word recognition based on hidden Markov models by replacing the discrete symbol representation of the speech signal by a continuous Gaussian *mixture* density. In this manner the inherent quantization error introduced by the discrete representation is essentially eliminated. The resulting recognizer was tested on a vocabulary of the 10 digits across a wide range of talkers and test conditions, and shown to have an error rate at least comparable to that of the best template recognizers and significantly lower than that of the discrete symbol hidden Markov model system. Several issues involved in the training of the continuous density models and in the implementation of the recognizer are discussed.

I. Introduction

Template based approaches with dynamic programming (DP) have been demonstrated to be one of the most effective methods for isolated word recognition. Several alternative approaches, however, have been proposed, due to:

1. the high computational cost of the DP approach;
2. the difficulties in extending the DP recognition paradigm to more difficult problems — e.g., continuous speech;
3. the desire to use a parametric model to represent the speech, rather than the non-parametric template;
4. the desire to use speech units smaller than words — e.g., syllables, demisyllables, phonemes.

These alternative approaches include using vector quantization (VQ) in the DP computation, using word-based vector quantization to eliminate the DP processing, using VQ as a front end preprocessor, and using hidden Markov models (HMM's) to represent the speech signal. Among these, the HMM recognizer is of great interest because of its potential low cost and its capability of modelling various events (phonemes, syllables, etc.) in the speech signal with efficient parametric representations. In our previous work [1], we studied how to apply discrete observation (i.e., vector quantized LPC vectors from a fixed size codebook) HMM's in isolated word, speaker independent speech recognition applications over dialed-up telephone lines. Work performed at IBM [2], CMU [3], and more recently at Phillips [4] has used continuous HMM's where it was assumed that all parameters of interest had Gaussian distributions. The HMM's to be discussed in this paper are based on continuous, *mixture* density models of the distribution of LPC derived parameter vectors. We have devised training procedures for obtaining maximum likelihood estimates of the parameters of the mixture distribution and applied the models to the problem of recognizing isolated digits. Our results show that the average error rate of such HMM recognizers are essentially identical to if not better than that of the best template approaches using DP methods, and considerably lower than that of an HMM recognizer with a discrete symbol VQ front end.

II. The Continuous Mixture Density HMM

Figure 1 shows the type of HMM we are considering here. It is based upon a left-to-right Markov chain which starts in state 1 and ends in state N. The observed signal is assumed to be a stochastic function of the state sequence of the Markov chain. The state sequence itself is unobservable (hidden). The goal is to choose the parameters of the hidden Markov model to optimally match the observed characteristics of a given signal.

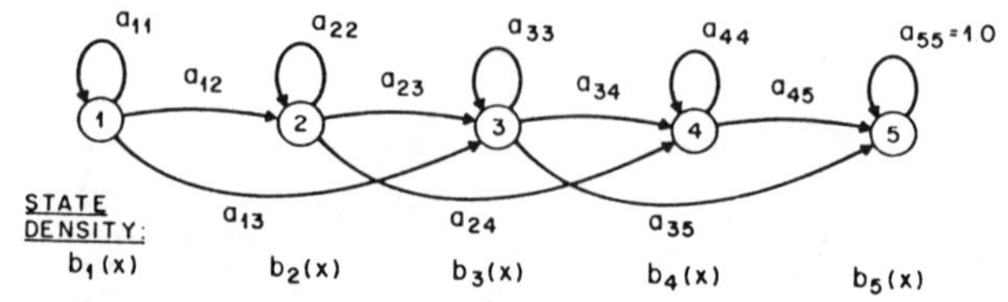

Fig. 1 Representation of left-to-right hidden Markov model.

The parameters which characterize the HMM of Figure 1 are the following: 1) N, the number of states in the model; 2) $A = [a_{ij}]$, $1 \leq i, j \leq N$, the state transition matrix where a_{ij} is the probability of making a transition from state i to state j; and 3) B, the observation probability function. If we assume that the signal to be represented by the HMM consists of a sequence of observation vectors $\mathbf{O} = \{O_1, O_2, \ldots, O_T\}$, where each O_t is a vector which characterizes the signal at time t, then we can consider two types of observation probability functions, namely discrete and continuous. Discrete type HMM's have been extensively discussed in [1]. In the continuous case we have the probability density function $B = \{b_j(\mathbf{x})\}$, $1 \leq j \leq N$, where $b_j(\mathbf{x})d\mathbf{x}$ is the probability that the vector O_t lies between $\mathbf{x}$ and $\mathbf{x} + d\mathbf{x}$. The types of density functions allowed for $b_j(\mathbf{x})$, for which a reestimation algorithm exists, include strictly log concave densities [5], elliptically symmetric densities [6], and more recently mixtures of strictly log concave or elliptically symmetric densities [7]. In this paper we will consider Gaussian mixture densities of the form

$$b_j(\mathbf{x}) = \sum_{k=1}^{M} C_{jk} \mathcal{N}(\mathbf{x}, \boldsymbol{\mu}_{jk}, \mathbf{U}_{jk}) \tag{1}$$

where $\mathcal{N}(\mathbf{x}, \boldsymbol{\mu}, \mathbf{U})$ denotes a D-dimensional normal density function of mean vector $\boldsymbol{\mu}$ and covariance matrix $\mathbf{U}$.

To summarize the discussion above, a complete specification of a continuous mixture density HMM requires choosing values (parameter estimates) for the following: 1) N — number of states in the model; 2) M — number of mixtures; 3) D — number of dimensions in each vector; 4) $A = [a_{ij}]$ — state transition matrix; 5) $C = [c_{jk}]$ — mixture gain matrix; 6) $\boldsymbol{\mu} = [\mu_{jkd}]$ — means of the mixture components; and 7) $\mathbf{U} = [U_{jkde}]$ — covariance matrices of the mixture components. For the work to be presented here, we have chosen $N = 5$ states on the basis of previous studies with discrete symbol models [1]. Also our signal observation vectors are cepstral coefficient vectors derived from an 8th order LPC analysis of the speech signal.

2.1 Training the HMM

For each word, v, in a vocabulary of V words ($V = 10$ for the digits), an HMM is designed; i.e., the set of parameters above is estimated from a training set of data representing multiple occurrences of the vocabulary word by multiple talkers. The procedure for obtaining model parameter estimates is shown in Figure 2. We assume a training set of data consisting of Q sequences of observations, where each sequence, $\mathbf{O}^i = \{O_1^i, O_2^i, \ldots, O_{T_i}^i\}$, $1 \leq i \leq Q$ is the set of

Reprinted from *Proc. IEEE Conf. Acoust., Speech, Signal Processing,* vol. 1, pp. 9–12, March 1985.

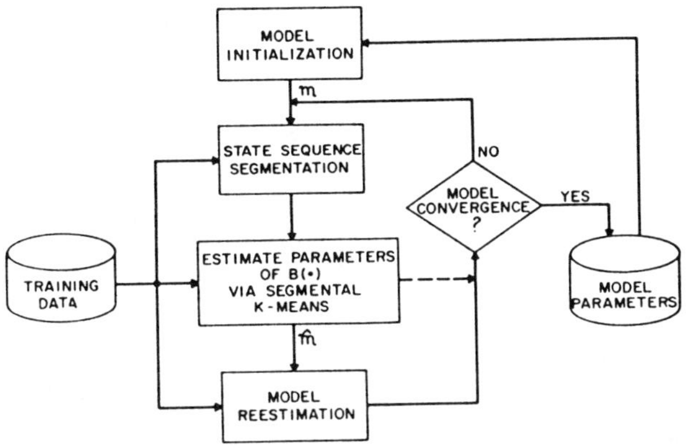

Fig. 2 The training procedure used to estimate parameter values.

vectors (observations) constituting a single occurrence of the word. An initial model estimate is assumed. This initial estimate (unlike the one required for reestimation) can be chosen randomly, or based upon any prior knowledge.

The second step in the training procedure is to segment each word occurrence, $\mathbf{O}^i$, into states based on the current model. This segmentation is achieved by finding the optimum state sequence via the Viterbi algorithm. The result of segmenting each of the Q training sequences is, for each of the N states, a set of the observations that occur within each state according to the current model. This step, in effect, isolates the observation stochastic processes from the underlying Markov chain so that the initial estimate of the observation statistics is not interfered by the Markov chain estimate. To illustrate this and the need for mixture densities, we compare the marginal distribution $b_j(\mathbf{x})|_{\mathbf{x}=(\cdots x_n \cdots)}$ of a 5 term Gaussian mixture distribution with diagonal covariances against a histogram of the actual observations in the corresponding state in Figure 3 for a 9 dimensional representation. The need for values of $M > 1$ is seen in the histogram of the 1st, 2nd, 4th, and the 8th parameters.

Following the above segmentation, a segmental K-means procedure is used to cluster the vectors in each state into a set of M clusters (using a Euclidean distortion metric and a VQ design algorithm). From the clustering, an updated set of model parameters is derived as follows: 1) $\hat{c}_{jk}$ = number of vectors classified in cluster k of the jth state/number of vectors in state j; 2) $\hat{\mu}_{jkd}$ = dth component of mean

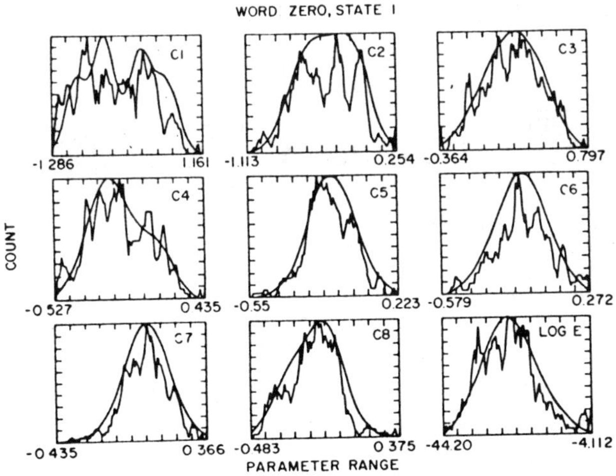

Fig. 3 Comparison of estimated density (jagged contour) and model density (smooth contour) for each of the 9 components of the representation vector.

of vectors classified in cluster k of state j; and 3) $\hat{U}_{jkrs}$ = (r, s)th component of covariance matrix of vectors classified in cluster k of state j. The transition matrix coefficients, a_{ij}, are not changed according to this procedure since no information about state transitions is retained. At this point the formal reestimation procedure is used to reestimate optimal values (in a maximum likelihood sense) of *all* model parameters. The resulting model is then compared to the previous model by computing a distance score which reflects the statistical similarity of the HMM's. If the model distance score exceeds a threshold, the model is updated and the overall training loop is repeated. The iteration stops when the distance falls below a prechosen threshold.

Since the steps of segmenting the training sequences into states, and clustering the vectors via a VQ clustering procedure are relatively inexpensive (in a computational sense), whereas reestimation is an exceedingly costly procedure, a practical implementation of the training procedure of Figure 2 is to bypass the step of model reestimation until local model convergence is obtained, and then apply the reestimation procedure at the final step.

Aside from using the reestimation algorithm to achieve maximum likelihood, we also developed a parameter estimation procedure based upon histogram fitting (for each observation state). The results obtained from this histogram fitting algorithm are comparable to those from the maximum likelihood estimate.

2.2 The HMM Recognizer

Once the HMM's have been trained on each vocabulary word, the recognition strategy is straightforward. Figure 4 shows a block diagram of the recognizer. The speech signal, $s(n)$, for the unknown word is first analyzed using an 8th order LPC analysis. The speech sampling rate is 6.67 kHz, and overlapping sections of 45 msec of speech are analyzed every 15 msec to give a set of 8 LPC coefficients. An LPC transformation algorithm is used to convert the LPC representation to an LPC derived cepstrum. Then, for each vocabulary word model, the optimum state sequence is found via the Viterbi algorithm and the log likelihood score for the optimal path is computed. The decision rule assigns the unknown word to the vocabulary word whose model has the highest log likelihood score.

2.3 Incorporation of Duration into the Recognizer

Inherently, each state in the HMM has an exponential duration probability. A state j, with a probability a_{jj} of returning to itself, has a state duration probability of

$$p_j(\ell) = (1-a_{jj})a_{jj}^{\ell-1}$$

where ℓ is the number of frames occurring in state j. To better exploit the state sequence information in the recognizer, we considered two alternatives, namely modification of the scoring procedure to include an internal durational model, and application of a post-processing durational model on the state sequence as determined by the Viterbi algorithm. In either case, in the training phase, we estimate a state duration probability of the form

$p_j(\ell/T)$ = probability of being in state j for (ℓ/T) of the word, where T is the number of frames in the word and ℓ is the number of frames in state j.

For each word and each state, the quantity $p_j(\ell/T)$ is estimated for 25 values of ℓ/T from 0 to 1.

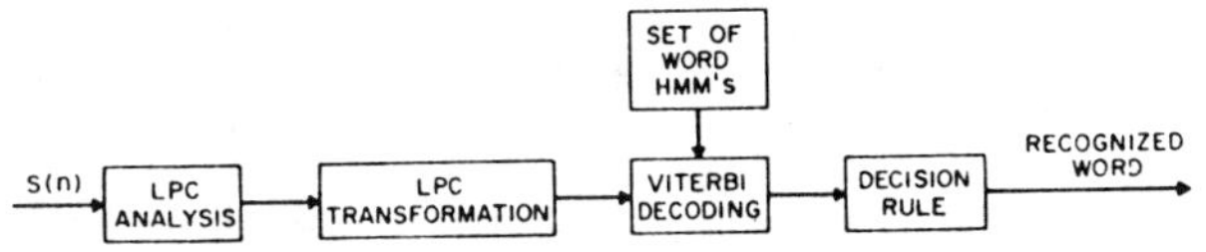

Fig. 4 Block diagram of the HMM recognizer.

For scoring a given observation sequence using the internal durational model, a recursion of the Viterbi procedure is required. The recursion is considerably more costly than the implementation of the standard Viterbi scheme.

The post-processor durational model, on the other hand, uses the original Viterbi alignment procedure. Then for each word, the optimal state sequence is determined, and the duration of each state is obtained via a backtracking procedure. The log likelihood is then augmented by the log duration probabilities (suitably weighted) to give the final score for the recognition decision, as shown in Eq. (2):

$$\log \hat{f} = \log f + \alpha \sum_{j=1}^{N} \log (p_j(\ell_j/T)) \ . \qquad (2)$$

III. Experimental Evaluations

Four sets of spoken digits recorded over standard dial-up telephone lines, were used to evaluate the performance of the HMM recognizer with the mixture density representation. These consisted of the following:

DIG 1 — 100 talkers (50 male, 50 female), 1 replication of each digit by each talker.

DIG 2 — Same 100 talkers and recording conditions as DIG 1; recordings made several weeks later than those of DIG 1.

DIG 3 — 100 new talkers (50 male, 50 female), 1 averaged occurrence of each digit by each talker obtained from averaging a pair of robust tokens of the digit. The transmission conditions (i.e., analog front end, filter cutoff frequencies, etc.) differed slightly from those used in recording the DIG 1 and DIG 2 databases.

DIG 4 — A second group of 100 new talkers (50 male, 50 female), 20 recordings of each digit by each talker. A random sampling of 1 of the recordings of each digit by each talker was used. The transmission conditions again differed somewhat from those used in recording the other sets.

For model training, only one digit set (either DIGI or DIG4) was used; for testing and performance evaluation, each of the 4 sets was used.

3.1 Diagonal Versus Full Covariance Matrices

Two forms for the U matrices of Eq. (1) were considered, namely diagonal matrices (with assumed zero correlation between components of the representation), and full covariance matrices. Recognition tests with diagonal covariance matrices using $M = 1$, 3 and 5, and full covariance matrices using $M = 1$ only were performed. The results showed that performance with the full covariance matrix with $M = 1$ was better than that obtained using only the diagonal covariance matrix with $M = 1$ or 3; however for $M = 5$ the performance with the diagonal matrix was comparable to that of the full covariance matrix with $M = 1$. Both full covariance and diagonal covariance matrices were used in subsequent recognition tests.

3.2 Applicability of Word Clustering to Model Generation

We also considered combining the word clustering procedure with model estimation, to give more than one HMM per word. We tested this idea in the following way. First a single HMM per word was created on training set DIG 1; next the 2 cluster per word template set was used to segment the 100 token training set into two groups. For each group a single HMM was created; hence a total of 2 HMM's per word was used in the performance evaluation. The potential disadvantage of this procedure should be clear, i.e., the training data per model available for estimating HMM parameters is half that used for the single model case and hence the estimates may be less reliable. Results of experiments with each of the 4 test sets and with 1 and 2 models per word showed that in the diagonal covariance case, a performance improvement of 0.65% was realized using 2 models per digit, and in the full covariance case the improvement was 0.25%.

3.3 Effects on Different Number of Mixtures

Using models trained on DIG1, the number of mixtures, M, was varied from 1 to 7, in steps of 2, for the diagonal covariance case, and from 1 to 2 for the full covariance case to see the effects on recognition performance. The results of these tests on the 4 digit data bases are given in Table I. The results show an improvement in performance from an average digit error rate of 2.95% for $M = 1$ down to an average digit error rate of 1.95% for $M = 5$. Results for $M = 7$ show a slight increase in average digit error rate to 2.23%. This result seems to indicate that the improvement in modelling the statistics of the "limited" observations from using more mixture terms is offset by the accompanying effect of broadened fitting range that helps incorrect words during recognition.

For the full covariance case the effect of increasing M from 1 to 2 is an increase in digit error rate by 0.95%. Hence we again see the evidence of sparser training data that makes model estimates unreliable and not robust for open recognition tests over a wide range of signal conditions.

3.4 Effects of Energy and Duration

To study the effects of including energy in the signal representation, and of including the durational model in the testing, a series of recognition runs were made using the diagonal covariance matrix models with $M = 5$, using 2 models per word. The results of these recognition tests are given in Table II. The durational model was implemented as a post-processor computation in all cases. The results show clearly that the addition of either energy or duration uniformly improves the performance of the HMM recognizer. Furthermore the combination of both energy and durational model yields better performance than either factor individually. The biggest improvements in performance were obtained for test sets DIG 3 and DIG 4 where the transmission characteristics of the speech were different from those of DIG 1 and DIG 2. In these cases the addition of energy and duration model make the system more robust because these features are, for the most part, insensitive to differences in transmission conditions.

Covariance Matrix		Average Digit Error Rate (%)				
	M	DIG 1	DIG 2	DIG 3	DIG 4	Overall
Diagonal	1	1.1	1.3	3.2	6.2	2.95
	3	0.2	1.1	4.1	5.2	2.65
	5	0.1	0.7	3.0	4.2	2.0
	7	0.0	0.8	3.1	5.0	2.23
Full	1	0.2	0.9	2.9	4.7	2.18
	2	0.0	1.2	6.0	5.3	3.13

Table I. Comparison of Performance of HMM Recognizer with Different Values of M

	Average Digit Error Rate (%)				
Condition	DIG 1	DIG 2	DIG 3	DIG 4	Overall
No Energy No Duration	0.3	2.5	4.3	8.0	3.78
Energy but No Duration	0.3	0.9	2.5	5.5	2.3
Duration but No Energy	0.1	1.3	3.3	5.4	2.53
Energy and Duration	0.1	0.7	2.8	4.2	1.95

Table II. Comparison of Performance of HMM Recognizer

3.5 Comparison of Internal and Post-Processor Durational Models

The two different implementations of the durational model, namely the internal durational model and the post-processor durational model were also compared. In both cases the same state-duration probability density function was used, with a multiplier of $\alpha = 3.0$. (This factor was locally optimum based on preliminary experimentation.) The results show that the performance of the HMM recognizer with the post-processor duration model was uniformly slightly better than for the recognizer with the internal durational model. Across the 4 data sets the improvement in performance was almost 0.7%.

3.6 Effects of Different Training Sets

As noted earlier, the recognizer used in the above experiments was trained on data set DIG1. To study the effects of training data, we also trained a new series of HMM's on DIG4, which is less homogeneous in recording conditions than the other 3 data sets. These HMM's use single multivariate Gaussian density with a full covariance matrix. Recognition results based on this new set of HMM's are shown in the first row of Table III. Results of the recognizer trained on DIG1 are shown in the second and third rows of Table III. It is seen that this new recognizer achieves a performance improvement of about 0.2%. What is more significant is that the error rate resulting from this has much less variation across the different test data sets than that obtained from the earlier training set.

IV. Comparison with Previous Recognizers

The above results characterize what can be achieved by the continuous mixture density HMM recognizer. Another way of measuring the effectiveness of the approach is to compare the current performance results with those of alternative recognition systems based on discrete densities (i.e., VQ symbols) [1], and based on templates [8]. Such a comparison is given in the last 2 rows of Table III. For the discrete density HMM recognizer, results are given only for the DIG 2 data set where the performance is significantly worse than that of the HMM recognizer with a continuous mixture density. For the template-based DTW recognizer, the results, based on the latest clustering procedure are comparable to those of the continuous density HMM recognizer. Since the template-based DTW recognizer has been studied for about 10 years and has been highly optimized in its performance, the equality between the HMM recognizer and the DTW recognizer, at least for the digits vocabulary, is highly significant.

Training Set	Cov. Type	Type of Recognizer	Average Digit Error Rate (%)				
			DIG 1	DIG 2	DIG 3	DIG 4	Overall
DIG4	Full	HMM-Continuous	2.5	1.7	2.1	0.8	1.78
DIG1	Full	$M = 1$	0.2	0.0	2.2	4.7	1.93
DIG1	Diagonal	HMM-Continuous $M = 5$	0.1	0.7	2.8	4.2	1.95
DIG1	NA	HMM-Discrete	–	2.9	–	–	–
DIG1	NA	DTW-Templates	0.0	0.6	2.7	3.9	1.8

Table III. Comparison of Performance of Several Recognizers

V. Summary

In the previous sections we have developed and tested an HMM isolated word recognition technique which uses a continuous mixture density model for the probability densities of the feature vector. Based on experimentation with the recognizer, in a speaker independent mode, using a vocabulary of 10 digits, the following general results were obtained:

1. The proposed model training procedure, with an iterative K-means loop for estimating initial values for the means and covariances of the components of the mixture model, works extremely well in practice and was able to converge to a local maximum of the likelihood function in a small number of iterations (typically 2-4 in most cases).
2. Mixture models with diagonal covariance matrices need a larger number of mixtures than mixture models with full covariance matrices in order to give the same performance.
3. Combining the techniques of clustering and HMM models can lead to small improvements in the performance of the HMM recognizer.
4. The addition of a word normalized energy contour (as an extra dimension to the feature vector) as well as durational information uniformly improves performance of the HMM recognizer and makes it more robust to differences in talker populations and transmission conditions.
5. The combination of normalized energy and durational information works better than either factor alone in the HMM recognizer.
6. The durational model of the HMM recognizer can be conveniently implemented as a post-processor to the Viterbi decoding procedure.

REFERENCES

[1] L. R. Rabiner, S. E. Levinson, and M. M. Sondhi, "On the Application of Vector Quantization and Hidden Markov Models to Speaker-Independent, Isolated Word Recognition," *Bell System Tech. J.*, Vol. 62, No. 4, pp. 1075-1105, April 1983.

[2] F. Jelinek, "Continuous Speech Recognition by Statistical Methods," *Proc. IEEE*, Vol. 64, pp. 532-556, April 1976.

[3] J. K. Baker, "The Dragon System — An Overview," *IEEE Trans. on Acoustics, Speech, and Signal Proc.*, Vol. ASSP-23, No. 1, pp. 24-9, February 1975.

[4] H. Bourlard, C. J. Wellekins and H. Ney, "Connected Digit Recognition Using Vector Quantization," *Proc. IEEE ICASSP '84*, pp. 26.10.1-26.10.4, San Diego, CA, March 1984.

[5] L. E. Baum, T. Petrie, G. Soules, and N. Weiss, "A Maximization Technique Occurring in the Statistical Analysis of Probabilistic Functions of Markov Chains," *Ann. Math. Stat.*, Vol. 41, pp. 164-171, 1970.

[6] L. R. Liporace, "Maximum Likelihood Estimation for Multivariate Observations of Markov Sources," *IEEE Trans. on Information Theory*, Vol. IT-28, pp. 729-734, September 1982.

[7] B. H. Juang, S. E. Levinson, and M. M. Sondhi, "Maximum Likelihood Estimation for Multivariate Normal Mixture Observations of Markov Chains," submitted for publication.

[8] L. R. Rabiner, S. E. Levinson, A. E. Rosenberg, and J. G. Wilpon, "Speaker Independent Recognition of Isolated Words Using Clustering Techniques," *IEEE Trans. on Acoustics, Speech, and Signal Proc.*, Vol. ASSP-27, No. 4, pp. 336-349, August 1979.

A Vector-Quantization-Based Preprocessor for Speaker-Independent Isolated Word Recognition

KUK-CHIN PAN, MEMBER, IEEE, FRANK K. SOONG, MEMBER, IEEE, AND LAWRENCE R. RABINER, FELLOW, IEEE

***Abstract*—In this paper, we propose a speaker-independent isolated word recognition system whose performance is comparable to that of a conventional isolated word recognizer, but whose computation is greatly reduced. The structure of the proposed recognizer consists of a word-based vector quantization (VQ) preprocessor, followed by a conventional DTW postprocessor. The purpose of the preprocessor is essentially to eliminate from further consideration all words in the vocabulary which are unlikely recognition candidates. In some cases, the preprocessor will be able to eliminate all word candidates except one; for such cases, there is no further processing required for word recognition. In all other cases (i.e., when more than one word candidate is passed on), a dynamic time warping (DTW) processor is used to resolve finer acoustical distinctions among the remaining word candidates. The performance of this type of recognizer (i.e., using a word-based preprocessor and a standard DTW comparison to make finer distinctions) is affected by a number of factors involved with the details of exactly how the system is implemented—e.g., the distortion measure used in the preprocessor and in the DTW comparison, the size of the VQ codebook for each vocabulary word, the decision thresholds of the preprocessor, etc. Several of these factors were studied experimentally using testing databases consisting of isolated digits and words from a vocabulary of 129 airline terms. The results show that the proposed preprocessor has the capability of reducing computation for recognition by up to an order of magnitude, while maintaining the same performance as that obtained using a DTW comparison without the preprocessor. A somewhat smaller reduction in memory over the straight DTW implementation is also obtained in the proposed approach.**

I. Introduction

SEVERAL different approaches have been proposed for recognizing isolated words; the most popular and the most successful systems are still the ones based on a conventional statistical pattern recognition model [1]–[5]. Such a model is depicted in Fig. 1. The speech signal is first analyzed and a time series of feature vectors that characterize the spectral content of the speech sounds is extracted. Typically, the feature vectors are the short-time spectra of the speech signal obtained from an analysis via a bank of bandpass filters or through a linear predictive coding (LPC) analysis. The resulting test pattern (i.e., the temporal sequence of feature vectors) is then compared to a set of prestored reference patterns, and the corresponding similarity (distance) score for each reference pattern is determined. The most successful pattern comparison method for speech recognition has been the so-called dynamic programming or dynamic time warping (DTW) algorithm in which reference and test word patterns are dynamically time aligned and the resultant alignment path of maximum similarity is obtained. Associated with the pattern matching algorithm is a distance measure. This distance measure can be as simple as summing the difference of magnitudes between frames of the time-aligned reference and test patterns [1] or as complex as that generated using LPC distances [2]. The final stage of the pattern recognition model is the decision block which makes a recognition decision (or decisions) based on the similarity or distance scores provided by the pattern comparison block. The most commonly used decision rule for recognition is the nearest neighbor rule which chooses the recognized word as the reference word with the smallest distance score. A generalized version of this rule, called the K-nearest neighbor (KNN) rule is often used in a speaker-independent word recognition environment to achieve more robust, and usually better, recognition performance [3].

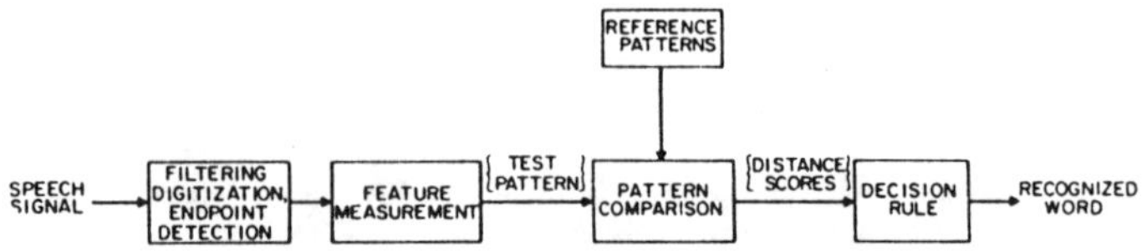

Fig. 1. Block diagram of pattern recognition model for isolated word recognition.

The DTW-based isolated word recognition approach described above has been shown to yield very high recognition accuracies compared to other approaches [1]–[5]. However, it does so at a high computational cost. The high number of computations required by the DTW-based algorithm has been a major obstacle to its widespread use in many applications. Recently, several different approaches have been proposed where the amount of computation is highly reduced over that required for the DTW method. These approaches include using hidden Markov models (HMM's) [6], vector-quantization-based DTW recognition [7], [8], and a recognizer without time alignment [9]. The first two alternatives are still under investigation and show great potential, but at the current time, they yield degraded performance with most reasonable size vector quantizers. The third low computation alternative

Manuscript received July 17, 1984; revised October 10, 1984. This work is based on the M.S. thesis of K. C. Pan, submitted to the Massachusetts Institute of Technology, Cambridge, May 1984.

K.-C. Pan was with the AT&T Bell Laboratories, Murray Hill, NJ 07974. He is now the Hewlett-Packard Corporation,

F. K. Soong and L. R. Rabiner are with AT&T Bell Laboratories, Murray Hill, NJ 07974.

Reprinted from *IEEE Trans. Acoust., Speech, Signal Processing,* vol. ASSP-33, pp. 546–560, June 1985.

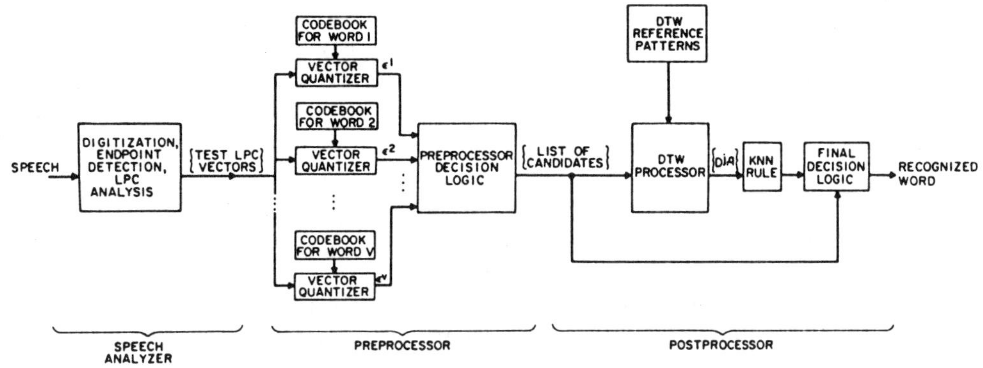

Fig. 2. Block diagram of isolated word recognizer incorporating a word-based VQ preprocessor and a DTW-based postprocessor.

used a word-based vector quantizer without relying on any temporal information. For speaker-trained isolated word recognition and with a highly nonconfusable vocabulary of 20 words, Shore and Burton reported 99 percent word recognition accuracy [9]. The results were significantly worse for speaker-independent word recognition, where an accuracy of 88 percent was achieved with the same vocabulary. This approach required relatively low computation, but its performance in a speaker-independent mode was not adequate for many applications. Recently, this method has been altered to add temporal information directly into the codebook design, with greatly improved recognition performance [10].

In this paper, we propose a complete speaker-independent, isolated word recognition system using a word-based VQ preprocessor to screen out all word candidates which are unlikely to match the unknown word. Based upon the average frame distortions of an unknown utterance with respect to a word-based VQ codebook for each vocabulary word, the preprocessor eliminates as many unlikely word candidates as possible. (In many cases, all word candidates but one are eliminated and no further processing is required.) Consequently, the preprocessor eliminates or greatly alleviates the computational load of the DTW processor. In addition, the distances required for the DTW processor are the same vector quantization distances computed in the preprocessor in our system; hence, the amount of computation of the entire recognizer is reduced even further.

The organization of this paper is as follows: in Section II we describe in detail the overall recognition system which incorporates a word-based VQ as a preprocessor. In this section, we explain the use of a hybrid distortion (distance) measure which combines the LPC likelihood ratio distance with an added energy distance. We also discuss the method used to generate the word-based VQ codebooks. In Section III, we describe the three databases used in our experiments and explain the quantitative performance measures used. In Section IV, we present the results of several pilot experiments in which we chose parameter and threshold values for the overall recognition system. In Section V, we present overall recognition results obtained by testing the recognizer on three different databases. In Section VI, we summarize our findings and discuss some directions for future research.

II. Word-Based VQ as a Preprocessor in an Isolated Word Recognition System

A. Overall Recognition System

A block diagram of the proposed isolated word speech recognizer is given in Fig. 2. The system consists of three major blocks, namely, an LPC spectral analyzer, a word-based VQ preprocessor, and a DTW postprocessor. The input speech signal is first digitized, endpointed into words, and spectrally analyzed. As a result, a time series of LPC vectors is thus formed. The preprocessor includes a set of word-based VQ codebooks. Each word in the vocabulary is characterized by its own codebook. For an unknown test word, the series of LPC vectors extracted from the analyzer is vector quantized by each individual codebook, and the corresponding frame distortions are accumulated over the duration of the word. The average frame distortion of each codebook is calculated, and the preprocessor decision logic chooses a list of possible word candidates for further processing. In the case when only a single word candidate is selected for postprocessing, the preprocessor makes a final recognition decision, and no further processing is required. Otherwise, no final decision will be made by the preprocessor and the set of "good" candidates are passed to the DTW postprocessor (i.e., unlikely candidates which are of large average distortion are eliminated by the preprocessor). By incorporating this preprocessor into the standard recognizer structure of Fig. 1, we can dramatically alleviate the computational load of the DTW postprocessor or even eliminate it completely if a final recognition decision can be made by the preprocessor. In addition, the DTW postprocessor can be even further simplified by vector quantizing the reference word templates with the VQ codebooks stored in the preprocessor. In this way, in the recognition phase, the distances (or the frame distortions) needed in the DTW process are readily available through a simple table-lookup procedure since all distances have been computed in the preprocessor stage. Hence, the computational complexity of the DTW processor is further reduced.

1) The VQ Preprocessor: The recognition procedures of the VQ preprocessor can be formally stated as follows. Assume that there are N frames in a test utterance (word to be recognized) which can be represented spectrally by N LPC vectors $a_1, \cdots, a_N$. Also assume that there are V words in the recognition vocabulary and the LPC codebook for the ith word is $b_1, b_2, \cdots, b_L$ where L is the codebook size. (We assume a fixed codebook size for each word.)

The average distortion by encoding (vector quantizing) the test utterance with the ith codebook is given by

$$\epsilon^i = \frac{1}{N} \sum_{n=1}^{N} \min_{1 \le l \le L} [d(a_n, b_l^i)] \tag{1}$$

where $d(a_n, b_l^i)$ is the distance between the two LPC vectors a_n and b_l^i.

The decision rule logic of the preprocessor consists of two possible decisions based on the average distortions $\{\epsilon^i | 1 \le i \le V\}$. Define the word i^* as the word whose codebook yields the minimum averaged distortion, i.e.,

$$i^* = \operatorname*{argmin}_{1 \le i \le V} \epsilon^i \tag{2}$$

and the word k^* as the word whose codebook yields the second smallest distortion, i.e.,

$$k^* = \operatorname*{argmin}_{\substack{1 \le i \le V \\ i \ne i^*}} \epsilon^i \tag{3}$$

The decision rules of the preprocessor are as follows.

Rule 1 (Final Word Candidate Rule):

$$\text{Choose word } v_{i^*} \text{ iff } \epsilon_{i^*} \le d_1 \text{ and } \epsilon^{k^*} - \epsilon^{i^*} \ge d_2 \tag{4}$$

where d_1 and d_2 are two empirically chosen thresholds for the best absolute average distance and the separation between the best and the second best candidates. Rule 1 basically chooses word v_{i*} when its average frame distortion is sufficiently small that a good confidence match is achieved, and when no other good matches occur among all other vocabulary words. If Rule 1 fails, then Rule 2 is used to select all likely candidates and pass them to the DTW postprocessor.

Rule 2 (Valid Candidates Rule):

Pass word v_i to DTW postprocessor iff $\epsilon^i - \epsilon^{i^*}$

$$\le d_3 \quad i = 1, 2, \cdots, V. \tag{5}$$

This rule says that any word whose average VQ distortion is within some threshold of the minimum distortion is considered a valid candidate for further DTW processing.

2) DTW Processor: In the DTW processor, the test utterance is compared to reference patterns of the candidates that were not eliminated by the preprocessor. We use $Q = 12$ reference patterns for each word in the recognition vocabulary to account for speaker differences [3]. Hence, Q time warps are performed for each word candidate passed on to the DTW processor, with each time warp resulting in a distance score between the test and reference patterns along the resulting time alignment path.

The KNN rule is used to select the recognized word using the distance scores for all the reference patterns of the DTW candidates. If we assume that there are J word candidates (passed from the preprocessor), and we define $D^{j,q}$ as the DTW distance of the qth reference pattern ($q = 1, 2, \cdots, Q$) of the jth candidate word ($j = 1, 2, \cdots, J$), then by reordering the Q DTW distances for each word, we have

$$D^{j,[1]} \le D^{j,[2]} \le \cdots \le D^{j,[Q]} \tag{6}$$

and the average distance of the best K patterns for each word (i.e., the K nearest neighbor (KNN) rule) is then given by

$$r^j = \frac{1}{K} \sum_{k=1}^{K} D^{j,[k]}. \tag{7}$$

The final word recognition decision is then given by j^* where

$$j^* = \operatorname*{argmin}_{1 \le j \le J} r^j. \tag{8}$$

With $Q = 12$, K is typically chosen as 2.

There are three different ways to prepare the reference and test word patterns for the DTW process, namely, 1) no vector quantization of either reference or test vectors, 2) vector quantizing both the reference and the test word patterns, and 3) vector quantizing the reference patterns only. The first method is expected to achieve the best recognition performance, in general, because no quantization error is introduced. By vector quantizing both the reference and test patterns (in the second method), we can replace the distance computations in the DTW processing by simple table-lookup operations, but at a price of performance degradation. However, it has been shown that the performance degradation due to vector quantizing the test word patterns is absolutely unnecessary [7], [8]. This is because if we want to vector quantize the test word patterns, we have to compute the distances between each VQ codebook vector and each frame of the test utterance. These distances are exactly the same distances needed in the DTW process if we vector quantize only the reference word patterns. Therefore, no computation overhead is required if we choose not to vector quantize the test pattern and the resulting performance degradation is eliminated. The implementation of this procedure is straightforward. The frame distortions, computed in the preprocessor stage, are stored in a table as illustrated in Fig. 3. In the DTW process, the distance between an input frame, e.g., a_i, and the vector quantized reference frame, e.g., b_j^k, can be easily retrieved from the distance table of Fig. 3. There are two possible implementations in vector quantizing the reference patterns. We can vector quantize a reference word pattern either by its own word-based codebook or by all codebooks. Except for the one time effort in quantizing the reference word patterns, these two implementations require exactly the same amount of computation in the recognition process.

In summary, three different ways can be used to prepare word templates, namely,

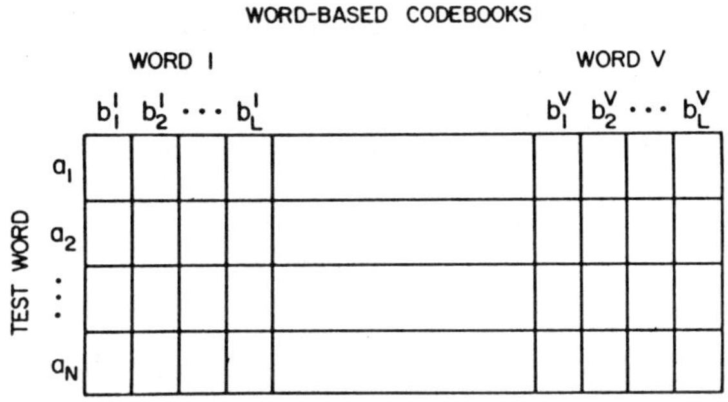

Fig. 3. Distance computation in the word-based VQ preprocessor. By saving each individual distance computation in a table, a reduction in quantization error in the DTW stage is achieved.

1) no VQ, use unquantized test and reference patterns
2) VQ/REF, SELF VQ, vector quantize each reference pattern with its own codebook only
3) VQ/REF, ALL VQ, vector quantize each reference pattern with all codebooks.

B. Log Likelihood Ratio LPC Distance and Additional Energy Information

The LPC spectral analysis can be interpreted as a spectral matching method which minimizes the difference between the input signal spectrum and all-pole model spectrum. The spectral difference used is commonly known as the Itakura–Saito (I–S) distortion measure, and has the form

$$D_{IS} = \frac{1}{2\pi} \int_0^{2\pi} \left[\log \left[\frac{S(\omega)}{S_{in}(\omega)} \right] + \frac{S_{in}(\omega)}{S(\omega)} - 1 \right] d\omega \quad (9)$$

where $S_{in}(\omega)$ is the input signal spectrum and $S(\omega)$ is the resultant LPC all-pole model spectrum given by

$$S(\omega) = \frac{\sigma^2}{|1 + a_1 e^{-j\omega} + a_2 e^{-2j\omega} + \cdots + a_p e^{-jp\omega}|^2}. \quad (10)$$

The same distortion measure can be used to measure the spectral difference between two all-pole spectra $S_R(\omega)$ and $S_T(\omega)$. The corresponding Itakura–Saito distortion measure can be computed in the autocorrelation domain as

$$D_{IS}(S_R, S_T) = \frac{a_R' V_T a_R}{\sigma_T^2} + \log \left(\frac{\sigma_T^2}{\sigma_R^2} \right) - 1 \quad (11)$$

where $a_R = [1, a_1, a_2, \cdots, a_p]'$, V_T is the Toeplitz matrix of the autocorrelation of $S_T(\omega)$, and σ_T^2 and σ_R^2 are the LPC gain terms for the two frames. Without modification, the original I–S distortion measure is not appropriate for speech recognition applications because it is sensitive to the absolute gain. For example, the I–S distortion between a frame of speech and an amplified version of itself is

$$D_{IS} = \log \left[\frac{\sigma_T^2}{\sigma_R^2} \right] \neq 0, \quad (12)$$

which is undesirable. To accommodate this gain sensitivity problem, there exist two modifications of the I–S distortion, namely, a gain-normalized I–S distortion measure and a gain-optimized measure. The first one sets the gains σ_T^2 and σ_R^2 to be equal, and the resultant distortion measure (likelihood ratio) is

$$D_{LR} = D_{IS} \left(\frac{S_R(\omega)}{\sigma_R^2}, \frac{S_T(\omega)}{\sigma_T^2} \right) = \frac{a_R' V_T a_R}{\sigma_T^2} - 1. \quad (13)$$

The second one, proposed by Itakura [2], is of the form

$$D_I = \log \left(\frac{a_R' V_T a_R}{\sigma_T^2} \right). \quad (14)$$

Both modifications yield roughly the same recognition performance. Since the likelihood ratio measure needs no logarithm operations and it is more mathematically tractable in VQ codebook generation, it has been widely used [11], [12].

Incorporating naively the absolute energy (or gain) information in the distortion measure such as the original I–S distortion is inappropriate for speech recognition since gain variations, between recordings, can lead to large distances between inherently similar spectra. However, since the energy contour of an utterance carries some phonetically relevant prosodic information, by adding energy (or gain) information to the spectral distortion measure *properly*, we should be able to improve the recognizer performance. The method we use is the one proposed by Brown and Rabiner [13]. The combined distortion measure between two frames of speech, i.e., frame n of a test pattern $T(n)$ with an LPC vector a_T and frame m of a reference pattern $R(m)$ with an LPC vector a_R, can be written as

$$D(T(n),R(m)) = D_{CLR}(T(n),R(m)) + \alpha f(D_E(T(n),R(m))). \quad (15)$$

The likelihood ratio distance has been modified to include a clip threshold d_{CLIP} and the resultant clipped likelihood ratio distance D_{CLR} is given as

$$D_{CLR}(T(n),R(m)) = \min \left\{ \frac{a_R' V_T a_R}{a_T' V_T a_T} - 1, d_{CLIP} \right\} \quad (16)$$

where V_T is the Toeplitz autocorrelation matrix of the test frame and d_{CLIP} is a suitably chosen maximum allowable value for D_{LR} (in practice, we use a value of 2.5) which prevents any single distance from greatly altering the average frame distortions in the preprocessor or greatly interfering with the DTW optimal path search. The energy distance $D_E(T(n), R(m))$ is given by

$$D_E(T(n),R(m)) = |(E^R(m) - E^R_{\max}) - (E^T(n) - E^T_{\max})| \quad (17)$$

where $E^R(m)$ is the log energy (in decibels) of the frame $R(m)$ and $E^R_{\max}$ is the maximum energy (in decibels) over all frames of the whole utterance, and similarly for $E^T(n)$ and $E^T_{\max}$. In this manner, each frame energy is normalized with respect to the maximum point in the energy contour of the entire utterance and the resultant $E^R(m) - E^R_{\max}$ (or $E^T(n) - E^T_{\max}$) is the relative energy measured in decibels down from $E^R_{\max}$ (or $E^T_{\max}$).

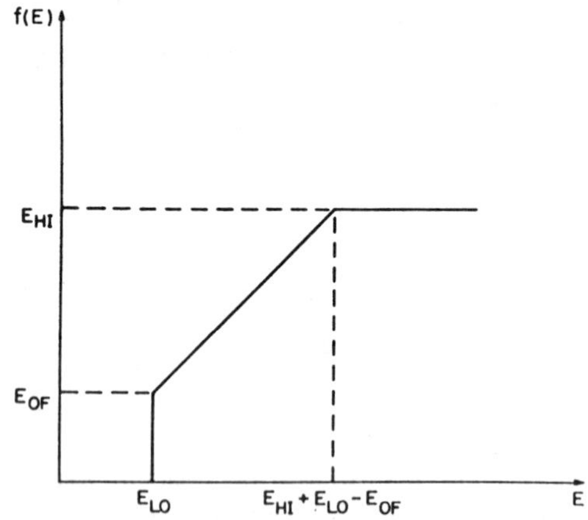

Fig. 4. Nonlinearity applied to log energy difference between frames for the energy distance calculation.

The nonlinear function f on the energy distance is defined as

$$f(E) = \begin{cases} 0 & |E| \le E_{LO} \\ |E| - E_{LO} + E_{OF} & E_{LO} < |E| \le E_{HI} + E_{LO} - E_{OF} \\ E_{HI} & E_{HI} + E_{LO} - E_{OF} < |E| \end{cases} \quad (18)$$

where E_{LO}, E_{OF}, and E_{HI} are three appropriately chosen thresholds. The function $f(E)$ is shown in Fig. 4. It assigns no penalty for any $|E|$ less than E_{LO} (i.e., small energy difference) to account for insignificant variations of energy. For large energy differences, $f(E)$ is clipped at an appropriate level E_{HI} to prevent it from becoming unbounded. Energy differences between these two extremes are linearly weighted by a factor of α.

C. Generation of Word-Based LPC Vector Codebooks

In the preprocessor, each word is represented by the VQ codebook. Each codebook is designed from a training sequence that comprises the LPC vectors from a large number of utterances of the particular word spoken by many different talkers. These same training sequences are used in template creation; hence, no additional recording is required for codebook generation over that required for template generation.

Assume a set of training LPC vectors c_i, $i = 1, \cdots, I$. We design a codebook of an optimal set of LPC vectors b_l, $l = 1, \cdots, L$ such that for this given size L, the average distortion in replacing each vector in the training set c_i by the closest vector b_l is minimized. This average distortion is given as

$$D_L = \frac{1}{I} \sum_{i=1}^{I} \min_{1 \le l \le L} [d(c_i, b_l)]. \quad (19)$$

The iterative algorithm we used to solve (19) is essentially the one proposed by Linde *et al.* [11] or commonly called the binary-split algorithm. There are several variations to the above-mentioned algorithms in terms of different splitting procedures and the way to repopulate any empty cell in the training procedure. However, Rabiner *et al.* [14] have found that several different variations in the training procedure gave rise to insignificant differences in the resulting codebooks. Therefore, the binary-split procedure is chosen because computationally it is the most efficient.

III. Descriptions of Databases and Performance Measure

A. Databases

Three different databases were used to test the proposed overall isolated word recognizer. They consisted of two different vocabularies, a ten-word digits vocabulary (zero through nine) and a 129-word airline vocabulary [15], [16]. Two different databases were used for the digit vocabulary and one for the airline vocabulary.

The first database for the ten-word digits vocabulary [3] was generated by 100 talkers, equally divided between male and female. Each talker gave two repetitions of all the ten digits with a period of four weeks separating the recording of the two repetitions. Hence, for each repetition, there were 100 utterances of each word in the vocabulary, giving a total of 1000 utterances for the entire repetition. We shall call this database the REFERENCE database, denoting the first repetition as the training set and the second repetition as the test set.

The second database for the digits vocabulary, which we shall call the KLS database [17], was generated by a set of ten talkers, also equally divided between male and female. These ten talkers were different from the 100 talkers who generated the REFERENCE database. Each talker spoke each digit 20 times, giving 200 total utterances per talker, and a total of 2000 utterances for the entire database. These ten talkers were chosen from a larger population of 100 talkers on the basis of their observed higher error rates in previous recognition experiments.

The airline vocabulary comprised 129 words that are commonly used in airline information and reservation systems, such as city names, days of the week, digits, words used to combine the terms into sentences, etc. The database for the airline vocabulary consisted of a training set and a test set. The training set was generated by a set of 100 talkers (50 males and 50 females), each of them speaking all the words in the recognition vocabulary once. This gave 100 utterances for each of the 129 words in the recognition vocabulary. The test set for the airlines database was generated by a set of 20 talkers who were different from those who generated the training set. This set of 20 talkers was also evenly divided between male and female; every talker spoke each word in the vocabulary once. All three databases described above have been used in previous recognition experiments [3], [16], [17]. All the isolated utterances were recorded off a standard dialed-up telephone line, and the endpoints of words were carefully monitored for accuracy. The two sets of data for the digits vocabulary were used to give an indication of the performance of the overall recognizer for a small size, standard vocabulary. The airline database was intended to show the feasibility of using the proposed recognizer on a much bigger and harder vocabulary that had many more acoustically similar words than the simple digits vocabulary.

B. Performance Measures

Recognition accuracy, computational complexity, and storage requirements are the three main criteria we use to judge the performance of the overall speech recognizer. To quantitatively characterize the errors (as well as the correct decisions) made by the recognizer at different stages of the overall system, we define the following parameters.

C1: Average fraction of correct final decisions made by the preprocessor, i.e., when a single candidate is chosen by the preprocessor.

E1: Average fraction of errors made by the preprocessor when a final recognition decision is made.

E2: Average fraction of errors made by the preprocessor when it fails to pass the true word candidate to the DTW postprocessor.

C3: Average fraction of correct decisions made by the DTW postprocessor.

E3: Average fraction of errors made by the DTW postprocessor.

Two useful performance parameters can be derived from the above definitions, namely, the following.

γ: Average fraction of all recognition trials in which the preprocessor makes a final decision, correct or wrong. Obviously, $\gamma = C_1 + E_1$.

β: Average fraction of candidates passed to the DTW postprocessor when no final decision is made by the preprocessor.

The breakdown of all the sources of errors and correct decisions and the corresponding stages where the errors and correct decisions are made is depicted in Fig. 5. The total error rate of the overall system is the sum of the preprocessor errors $E1 + E2$ and the DTW postprocessor errors $E3$.

C. Computational Complexity

A major consideration in evaluating the performance of the overall recognizer of Fig. 1 is its computational complexity. To understand this issue, we have to examine the required computation in the preprocessor C_{PRE}, the DTW postprocessor C_{DTW}, and the overall system. To quantify these concepts, we first define C_{PRE} as

$$C_{\mathrm{PRE}} = V \cdot N \cdot L \qquad \text{(distance calculations)} \qquad (20)$$

where a distance computation requires approximately $(p+1)$ additions and $(p+1)$ multiplications for a pth-order LPC analysis using the likelihood ratio distance. The energy distance adds negligible computation if the logarithms are computed before the distance calculation.

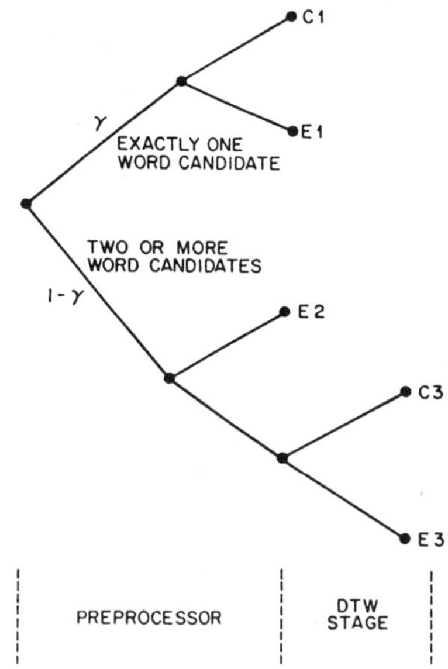

Fig. 5 A breakdown of the potential sources of error in the preprocessor and in the DTW stage of the recognizer.

The computational complexity of the DTW processor C_{DTW} is readily shown to be a function of the vocabulary size V, the number of reference templates for each word Q, and the type of DTW implementation (i.e., whether or not VQ is used). There are two components to C_{DTW}, namely, distance calculations, and combinatorics. When VQ is used, the cost of distance calculations becomes negligible (they become table lookups), and when no VQ is used, the cost of distance calculations dominates the computation. Using the approximation that the cost of a combinatorics calculation is about 1/5 that of a full distance calculation (as is the case on most general-purpose computers), we get the following results:

$$C_{\mathrm{DTW}} = \tfrac{6}{5}\, VQ\, N^2/3 \qquad \text{(distance calculations)} \qquad (21)$$

and

$$C_{\mathrm{DTW/VQ}} = \tfrac{1}{5}\, VQ\, N^2/3 \qquad \text{(distance calculations).} \qquad (22)$$

The overall computation of the recognizer of Fig. 2, C_{ALL}, can now be expressed in terms of C_{PRE}, C_{DTW}, γ, and β. The resulting expression is of the form

$$C_{\mathrm{ALL}} = C_{\mathrm{PRE}} + (1-\gamma)\beta C_{\mathrm{DTW}} \qquad (23)$$

when no VQ is used in the DTW processor, and

$$C_{\mathrm{ALL/VQ}} = C_{\mathrm{PRE}} + (1-\gamma)\beta\, C_{\mathrm{DTW/VQ}} \qquad (24)$$

when a VQ is used in the DTW processor.

From (20)–(24), we get

$$C_{\mathrm{ALL}} = NV[L + (1-\gamma)\beta\, \tfrac{6}{15}\, QN] \qquad (25a)$$

$$C_{\mathrm{ALL/VQ}} = NV[L + (1-\gamma)\beta\, \tfrac{1}{15}\, QN]. \qquad (25b)$$

We can readily define computational gain ratios as

$$R_{ALL} = \frac{C_{DTW}}{C_{ALL}} \tag{26a}$$

and

$$R_{ALL/VQ} = \frac{C_{DTW}}{C_{ALL/VQ}} \tag{26b}$$

which express the reduction in computation of the overall recognizer as compared to a conventional DTW system without VQ. Consider "typical" values of the system parameters for a digits vocabulary. For this case, we have $\gamma = 0.9$, $\beta = 0.2$, $V = 10$, $Q = 12$, $N = 40$, $L = 16$, giving

$$R_{ALL} = 10$$

$$R_{ALL/VQ} = 11.5.$$

Hence, a reduction in computation of more than 10 to 1 is entirely feasible with the proposed approach.

D. Storage Considerations

In our storage analysis, we will only be concerned with the amount of memory or storage required by the VQ codebooks, the test utterance, the reference patterns for the DTW processor, and the local distance storage required when DTW processing is done using vector-quantized reference patterns. We will neglect the differences in program storage requirements between the overall recognizer and that of the DTW processor for the reasons that the program storage requirements are negligible compared to the factors mentioned above. Recall that there are V words in the recognition vocabulary, L LPC vectors in each VQ codebook, Q DTW reference templates per word, and we assume that there is an average of N frames of speech per reference pattern, and each reference or test template is of $(p+1)$ dimensions. Under these assumptions, the amount of memory required by the preprocessor and the DTW processor is given by

$$M_{PRE} = V \cdot L \cdot (p+1) \quad \text{(floating-point numbers)} \tag{27a}$$

$$M_{DTW} = V \cdot Q \cdot N \cdot (p+1) \quad \text{(floating-point numbers).} \tag{27b}$$

The memory required to store the test utterance is

$$M_{TEST} = N \cdot (p+1) \quad \text{(floating-point numbers).} \tag{28}$$

The memory required to store the local distances computed in the preprocessor state is given by

$$M_{DIST} = N \cdot V \cdot L \quad \text{(floating-point numbers).} \tag{29}$$

When using vector-quantized reference patterns for the DTW processor, the LPC vectors of the text utterance do not have to be retained since the DTW algorithm will only require the table of local distances computed by the preprocessor. The memory required for storing the distance table then becomes $M_{DIST} = N \cdot L$. In addition, each vector of the reference patterns need only be stored as an integer index of the corresponding codebook vector, as opposed to the $(p+1)$ floating-point numbers required for each LPC vector. The resultant memory requirement is then

$$\begin{aligned} M_{DTW/VQ} &= V \cdot Q \cdot N \quad &\text{(fixed-point numbers)} \\ &= \tfrac{1}{2} V \cdot Q \cdot N \quad &\text{(floating-point numbers)} \end{aligned} \tag{30}$$

where we have assumed the storage requirement for a fixed-point number to be half the storage requirements for a floating-point number.

With the above assumptions, the total amount of memory M_{ALL} is

$$\begin{aligned} M_{ALL} &= M_{PRE} + M_{TEST} + M_{DTW} \quad &\text{(DTW/no VQ)} \\ M_{ALL/VQ} &= M_{PRE} + M_{DIST} + M_{DTW/VQ} \quad &\text{(DTW/VQ).} \end{aligned} \tag{31a}$$

The storage requirement of a straight DTW recognizer (without a VQ preprocessor) is

$$M_{ALL} = M_{DTW} + M_{TEST}. \tag{32}$$

IV. Initial Pilot Experiments

Before evaluating the proposed recognizer, we first carried out some pilot experiments to study the effects of various system parameters on the recognizer performance. In particular, we studied the following.

1) Performance of the preprocessor when energy information is incorporated into the distortion measure.

2) Performance of the preprocessor when the decision thresholds d_1, d_2, d_3 are varied.

3) Effects of adding energy information to the distortion measure on the performance of the DTW processor.

4) Effects of using word-based VQ codebooks for quantizing the reference templates in the DTW processor.

All pilot experiments were conducted using the REFERENCE database unless specified otherwise.

A. The Effects of Adding Energy Information to the Distortion Measure in the Preprocessor

Three different sets of parameters were used to find the best way to incorporate energy information in the distortion measure. We used several different sets of parameters, α, E_{LO}, E_{HI}, and E_{OF}, including the following.

1) $\alpha = 0.0$ (LPC likelihood ratio only, no energy); this condition is labeled LR in the text.

2) $\alpha = 0.1$ $E_{LO} = 6.0$, $E_{HI} = 20.0$, $E_{OF} = 0.0$ (LR with energy information, smooth offset); this condition is labeled LRSO in the text.

3) $\alpha = 0.1$ $E_{LO} = 6.0$, $E_{HI} = 20.0$, $E_{OF} = 6.0$ (LR with energy information, discontinuous offset); this condition is labeled LRDO in the text.

The first case with $\alpha = 0.0$ is the original likelihood

TABLE I(a)
AVERAGE DISTORTIONS FOR LPC AND ENERGY (E) DISTANCES AS A FUNCTION OF CODEBOOK SIZE AND TYPE OF DISTORTION MEASURE NOTE THE ENERGY DISTANCE SHOWN HAS NOT BEEN SCALED DOWN BY THE WEIGHTING FACTOR α

Distortion Measure	Codebook-Size 4		8		16		32	
	LPC	E	LPC	E	LPC	E	LPC	E
LR	.55	—	.41	—	.31	—	.24	—
LRSO	.65	1.3	.50	.53	.37	.30	.28	.20
LRDO	.70	2.47	.55	.79	.42	.27	.32	.09

TABLE I(b)
AVERAGE DIGIT ERROR RATES OF THE PREPROCESSOR AS A FUNCTION OF CODEBOOK SIZE AND DISTORTION MEASURE

Distortion Measure	Codebook-Size 4	8	16	32
LR	17.3	11.8	9.9	7.2
LRSO	8.4	6.7	4.7	5.0
LRDO	9.5	6.8	4.8	4.8

ratio distortion measure with no additional energy information. The second and third distortion measures are a linear combination of the likelihood ratio LPC spectral distortion measure and a linearly weighted ($\alpha=0.1$) energy distance. The LRSO distortion adds a gradually increasing energy distance to the LPC distance, whereas the LRDO distortion has a discontinuous energy distance step at $E_{LO} = 6$ dB; hence, a heavy penalty is paid when the energy difference between frames exceeds 6 dB.

Using the training set of the REFERENCE database (which comprised 100 utterances for each of the ten digits), different size word-based VQ codebooks consisting of 4, 8, 16, and 32 LPC vectors were designed for each digit. Each of the three distortion measures LR, LRSO, and LRDO were used. A breakdown of the resultant average distortions for different codebook sizes and distortion measures is given in Table I(a). As expected, the average distortions decrease with increasing codebook sizes. It is also interesting to see that, for a given codebook size, the LPC spectral distortions are higher for LRSO and LRDO than for LR. This is a natural consequence of incorporating energy into the distortion measure since, implicitly, some of the codebook vectors are used to encode the energy information. Comparing the breakdown of the LPC spectral distortions and the energy distortions, we also note that by using LRDO as the overall distortion measure, a greater emphasis is placed on energy distortion. Therefore, for larger codebook sizes, LRDO achieves lower energy distortions at the expense of higher spectral distortions than the LRSO. With the codebooks generated through these three different distortion measures, we used the preprocessor as a complete recognizer (i.e., no DTW processing was used) and measured its performance. The test set of the REFERENCE database was used in these tests. The recognition error rates (average fraction times 100) of the preprocessors are shown as a function of codebook size and distortion measures in Table I(b).

The results of Table I(b) show that codebooks designed on the basis of the VQ distortion and energy distortion (LRSO and LRDO) performed significantly better than codebooks designed with LPC spectral distance alone (LR). The performance difference between LRSO and LRDO is not significant for codebook sizes larger than 4. These results strongly indicate that by using energy information, we can improve the performance of the preprocessor. We also see that the performance saturates at codebook size 16 for both the LRSO and LRDO distortion measures. For this digits database, a codebook of size 16 seems to be adequate to characterize the vector space spanned by most of the utterances. Thus, although having more than 16 vectors might allow more outlying frames in the training sequence to be represented in the word-based codebook, in the recognition phase, this tends to reduce both the average distortion of incorrect words and the average distortion of the correct word; hence, no recognition performance improvement is obtained.

B. Effects of Preprocessor Decision Thresholds on Performance

We recall that the word-based preprocessor had three decision thresholds, namely,

- d_1 = threshold on average word distance for best candidate to be considered as a valid recognition candidate
- d_2 = threshold on difference in average word distance between best and second best word candidates for only a single candidate to be chosen by the preprocessor
- d_3 = threshold on difference in average word distance between best and any word candidate for passing on the word candidate to the DTW postprocessor.

For convenience, and since it is reasonable, we set $d_3 = d_2$ in all simulations. We also used the LRSO distortion measure in all experiments to be described in this section.

The threshold values d_1 and $d_2 = d_3$ were systematically varied, and we measured their effects on the system error rate parameters $C1$, $E1$, $(1 - \gamma)$, $E2$, and β. Figs. 6 and 7 show typical plots of the variations of these parameters as a function of d_1 (for the fixed value $d_2 = d_3 = 0.05$) and for codebook sizes varying from 4 to 32. From Fig. 6, we see that for small values of d_1, values of $(1 - \gamma)$ are very close to 1, indicating that very few recognition decisions were made by the preprocessor. As d_1 increases, $(1 - \gamma)$ quickly approaches a steady-state value of about 0.1, indicating that about 90 percent of the words were uniquely recognized by the preprocessor. The resulting steady-state values of $C1$ and $E1$ are about 0.9 and 0.0, indicating that at these thresholds, essentially the preprocessor made final decisions about 90 percent of the time, and all the preprocessor decisions were correct. Similar behavior of $E2$ and β are noted in Fig. 7 for this set of $E2$ and β are noted in Fig. 7 for this set of decision thresh-

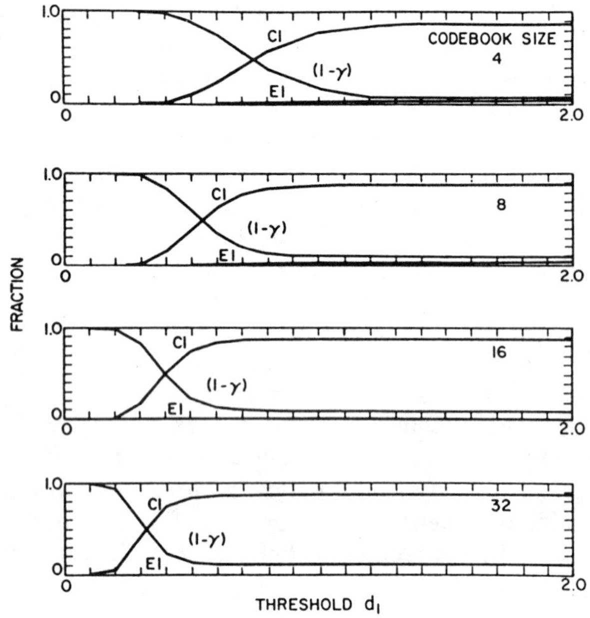

Fig. 6. Plots of the variation of C_1, E_1, and $(1 - \gamma)$ as a function of the preprocessor threshold d_1 for several codebook sizes for the digits vocabulary.

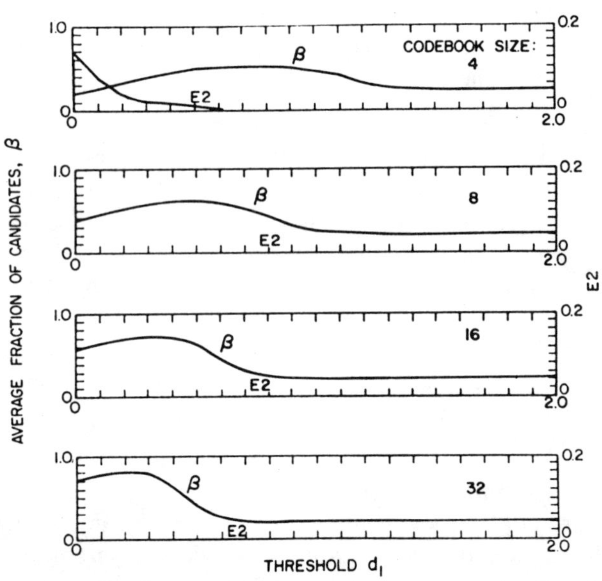

Fig. 7. Plots of the variation of β and E_2 as a function of the preprocessor threshold d_1 for several codebook sizes for the digits vocabulary.

olds. The steady-state values of β and $E2$ for large values of d_1 are about 0.22 and 0, indicating that an average of 2.2 words get passed to the DTW postprocessor and the correct word is essentially always one of the words passed on.

From Figs. 6 and 7, we also see that for smaller size codebooks, in order for the preprocessor to make the same percentage of recognition decisions [i.e., fixed values of $(1-\gamma)$], d_1 had to be set to a higher value since, intrinsically, a smaller size codebook yielded larger average distortions.

Based on curves of the type shown in Figs. 6 and 7, it is clear that a good strategy is to choose the smallest value of d_1 for which the parameters $C1$, $E1$, $E2$, β, and $(1 - \gamma)$

TABLE II
AVERAGE DIGIT RECOGNITION ERROR RATES FOR SEVERAL DISTANCE METRICS AND METHODS OF QUANTIZATION

Distance Variant	Data Base	
	REFERENCE	KLS
LR	2.7	3.6
LRSO	2.1	2.8
LRDO	2.8	3.3

(a)

Quantization in DTW	Data Base		
	REFERENCE	KLS	AIRLINE
NO VQ	2.1	2.8	10.2
VQ/REF	2.8	4.2	14.0
VQ/REF VQ/TEST	3.8	4.0	—

(b)

Codebook Size	Quantization Method	
	VQ/REF, SELF VQ	VQ/REF, ALL VQ
4	4.7	3.2
8	2.3	2.8
16	2.1	1.8
32	1.9	2.1

(c)

are all at steady-state values. We will defer a choice of $d_2 = d_3$ until later in this section since this threshold parameter has the most influence on the performance tradeoffs of interest.

C. Effects of VQ and Energy Metrics on DTW Performance

To provide benchmarks on the performance of the overall recognizer structure of Fig. 2, we ran several experiments using only the DTW processor. The first such experiment used each of the three distance measures, namely, LR (LPC likelihood distance, no energy), LRSO (LPC likelihood distance combined with energy with no offset energy distance), and LRDO (LPC likelihood distance combined with energy with a 6 dB offset). Each of these distance metrics was used in recognition tests using the REFERENCE and the KLS digit sets. Results of these experiments are given in Table II(a), which shows average digit error rates for the three types of distance measures. The results show that the LRSO measure achieved the best performance for both databases. This shows that energy information, when used properly, can be used to improve recognition performance, even for a simple vocabulary such as the digits. The results also show that both the energy discontinuity measure (LRDO) and the measure without energy (LR) performed essentially the same, namely, about 0.5 percent worse than the LRSO measure. On the basis of these results, the LRSO measure was used exclusively in all the remaining DTW processing.

The second experiment considered the effect of using a

word-independent VQ on the performance of the DTW word recognizer (i.e., again without the preprocessor). In this case, a VQ with 128 codebook entries was used, and the testing sets included all three databases, namely, the digit sets of the REFERENCE and KLS databases, and the AIRLINE words. The results of this experiment are given in Table II(b), which shows average digit error rates as a function of the type of quantization of the reference patterns in DTW. The results show a substantial deterioration in performance, for all three sets, when the VQ was used on just the reference templates alone. An even greater performance degradation occurred when the VQ was used on *both* the reference and test patterns. Thus, as mentioned earlier, word-independent VQ leads to performance degradation in return for reduced computation. For the AIRLINE vocabulary, it can be seen that a larger performance degradation occurs than for the digits. Sufficient evidence exists that for large vocabularies (e.g., the AIRLINE set), the required size of the VQ codebook must be on the order of 512–1024 entries [7], [8].

The last experiment considered the performance of the DTW processor when the reference patterns were quantized using either the individual word codebooks (SELF VQ) or using all codebooks for the vocabulary (ALL VQ). For this test, only the REFERENCE digits were used. Results of this test are given in Table II(c), which shows average digit error rates (average fraction times 100) as a function of the individual word codebook size. It can be seen that with SELF VQ, codebook size 4 gave degraded performance (there was too much quantization error); however, for codebook sizes 8, 16, and 32, the performance was essentially identical to that of the DTW system *without* VQ! For the ALL VQ case, a small degradation in performance occurred for the small codebooks (sizes 4 and 8); however, for codebook sizes 16 and 32, we again achieved the same performance as the DTW processor without VQ.

D. Summary of Pilot Experiment Results

The results of the pilot experiments show the following.

1) Incorporating energy information into the distance metric improves the recognizer performance.

2) The preprocessor decision thresholds can be set to allow the preprocessor to make a final decision on close to 90 percent ($\gamma = 0.9$) of all trials reliably for the REFERENCE database. When the preprocessor was unable to make a final decision, it eliminated close to 80 percent ($\beta = 0.2$) of all candidates.

3) Vector quantization can be incorporated into the DTW process to significantly reduce computation at a cost of a small increase in recognition error rate.

4) Word-based VQ codebooks perform better than word-independent VQ codebooks, and are comparable to the case when no VQ is used at all in the DTW stage

V. Performance of the Overall Recognition System

To evaluate the performance of the complete speaker-independent, isolated word recognizer of Fig. 2, three evaluations were performed. The first evaluation used the REFERENCE digits testing database and studied the behavior of the overall system as a function of the codebook size, the preprocessor decision threshold d_2, and as a function of the method for quantization of the reference patterns used in the DTW processing. The second evaluation used the KLS digits database and made similar measurements. The third evaluation used the AIRLINE vocabulary database and, using the best values of d_2 and DTW implementation from the digit runs, measured the resulting error rate and the reduction in computation.

A. Results on the REFERENCE Digits Database

For all experiments to be described in this section, a value of $d_1 = 2.4$ was used. This value guaranteed the minimum preprocessor sensitivity to codebook size, vocabulary, etc. The first tests performed used a DTW implementation with no VQ (Method 1), one with VQ from individual word codebooks (SELF VQ or Method 2), and one with VQ from all word codebooks (ALL VQ or Method 3), and measured the various recognizer error rates ($C1$, $E1$, $E2$, $C2$, $E2$) as a function of L, codebook size, and d_2, the preprocessor decision threshold.

The results of these recognition tests are given in Table III (in which the average fractions are converted to error rates by multiplying by 100) and Fig. 8-11, which show plots of overall error rate (Fig. 8), average fraction of candidates passed to the DTW processor (Fig. 9), average fraction of decisions made by the preprocessor (Fig. 10), and computational complexity (Fig. 11) as a function of d_2 and codebook size. (The results in Figs. 8–11 are for Method 1; similar results were obtained for Methods 2 and 3.)

The results of Table III and Figs. 8–11 show the following.

1) As d_2 increases (especially for codebook sizes 4 and 8), the overall system error rate decreases dramatically, the average number of DTW candidates increases slightly, the percentage of decisions made by the preprocessor falls, and the computational complexity rises. This is because by increasing d_2, the preprocessor is deferring decisions (and hence potential errors) to the DTW processor at an increasing rate. For larger codebook sizes (16 and 32 vectors per word), the error rate is much less sensitive to d_2; hence, for smaller values of d_2, we can achieve very high performance with a very modest amount of computation.

2) At small values of d_2, for each codebook size, the preprocessor made most of the errors; at larger values of d_2, the preprocessor error rate was almost negligible and the DTW processor made most of the errors. For larger codebook sizes (e.g., 16 and 32 vectors), optimum performance was attained with reasonably small values of d_2.

3) The introduction of VQ (via either Methods 2 or 3) led to essentially no increased error rate for all codebook sizes and for all values of d_2. Hence, the proposed quantization schemes are very good ones for use with this system.

To see how the performance of the resulting recognizer

TABLE III
AVERAGE RATES (FRACTION * 100) OF CORRECT AND INCORRECT DECISIONS FOR BOTH THE VQ-BASED PREPROCESSOR AND FOR THE THREE METHODS OF QUANTIZATIONN IN THE DTW PROCESSOR AS A FUNCTION OF CODEBOOK SIZE AND VALUE OF THE DECISION THRESHOLD d_2

					DTW IMPLEMENTATION								
					METHOD 1			METHOD 2			METHOD 3		
Codebook Size	d_2	$C1$ *100	$E1$ *100	$E2$ *100	$C3$ *100	$E3$ *100	Overall Error Rate	$C3$ *100	$E3$ *100	Overall Error Rate	$C3$ *100	$E3$ *100	Overall Error Rate
4	0.05	87.7	4.7	0.3	7.1	0.2	5.2	6.6	0.7	5.7	7.2	0.1	5.1
4	0.075	84.5	4.0	0.5	10.7	0.3	4.8	10.0	1.0	5.5	10.7	0.3	4.8
4	0.1	81.0	2.5	0.8	15.0	0.7	4.0	13.9	1.8	5.1	14.8	0.9	4.2
4	0.125	77.0	1.9	0.6	19.6	0.9	3.4	18.0	2.5	5.0	18.9	1.6	4.1
4	0.15	73.1	1.1	0.8	24.1	0.9	2.8	22.2	2.8	4.7	23.4	1.6	3.5
8	0.05	87.9	3.0	0.9	7.9	0.3	4.2	7.7	0.5	4.4	7.8	0.4	4.3
8	0.075	83.8	2.4	0.7	12.6	0.5	3.6	12.3	0.8	3.9	12.4	0.7	3.8
8	0.1	79.4	1.3	0.8	17.9	0.6	2.7	17.3	1.2	3.3	17.5	1.0	3.1
8	0.125	73.7	0.5	0.9	24.1	0.8	2.2	23.4	1.5	2.9	23.6	1.3	2.6
8	0.15	67.8	0.4	0.3	30.7	0.8	1.5	30.0	1.5	2.2	30.1	1.4	2.1
16	0.05	88.8	1.6	0.7	8.4	0.5	2.8	8.2	0.7	3.0	8.2	0.7	3.0
16	0.075	83.9	1.1	0.3	14.0	0.7	2.1	13.9	0.8	2.2	13.9	0.8	2.2
16	0.1	78.1	0.6	0.3	20.1	0.9	1.8	20.0	1.0	1.9	25.6	1.3	1.8
16	0.125	72.6	0.3	0.2	25.5	1.4	1.9	25.5	1.4	1.9	25.6	1.3	1.8
16	0.15	65.2	0.2	0.2	32.8	1.6	2.0	32.9	1.5	1.9	33.0	1.4	1.8
32	0.05	87.6	1.0	0.1	10.4	0.9	2.0	10.3	1.0	2.1	10.4	0.9	2.0
32	0.075	79.4	0.7	0.1	18.6	1.2	2.0	18.8	1.0	1.8	18.7	1.1	1.9
32	0.1	71.7	0.2	0.2	26.5	1.4	1.8	26.7	1.2	1.6	26.5	1.4	1.8
32	0.125	62.5	0.2	0.0	35.8	1.5	1.7	36.0	1.3	1.5	35.8	1.5	1.7
32	0.15	54.7	0.0	0.0	43.7	1.6	1.6	43.9	1.4	1.4	43.7	1.6	1.6

compares to that of the conventional DTW recognizer (with no preprocessor), and the preprocessor alone, we compared error rate, memory, and computation for these systems. Using a value of $d_2 = 0.075$ and $L = 16$ for the codebook size, we get the following comparisons.

	Overall Error Rate (%)	Memory (k)	Computation (k)
DTW Alone	2.1	44	76.8
Method 1	2.1	45.3	8.5
Method 2	2.2	45.3	6.4
Preprocessor Alone	5.2	—	6.4

The results show a 9-to-1 reduction in computation with *no loss* in accuracy or gain in memory for Method 1 over DTW alone; similarly, we get a 12 -to -1 reduction in computation with no loss in accuracy or gain in memory for Method 2 over DTW alone. When we use the preprocessor alone as a complete recognizer (i.e., by setting $d_1 = \infty$, $d_2 = 0$ so a single candidate is selected each time), the error rate increases by 3.1 percent over DTW alone. Thus, without the temporal information provided by the DTW processor, the performance of the word recognizer is greatly degraded.

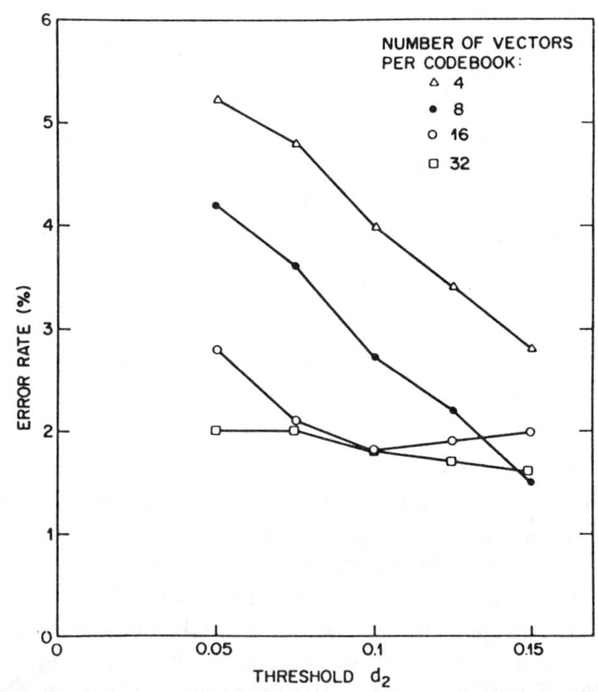

Fig. 8. Plots of total error rate versus preprocessor decision threshold d_2 for several codebook sizes for the digits vocabulary.

B. Results on the KLS Digits Database

The KLS digits database was selected as a somewhat more difficult test of the recognizer since the error rate using the DTW recognizer alone was somewhat higher than for the REFERENCE digits set. Based on varying codebook sizes and values of d_2, the following results were obtained with this database.

Recognition Method	Overall Error Rate (%)	Computation (k)
DTW Alone	2.8	76.8
Method 1 ($L = 16$)	3.1	18
Method 2 ($L = 16$)	3.5	8
Preprocessor Alone ($L = 32$)	10.9	12.8

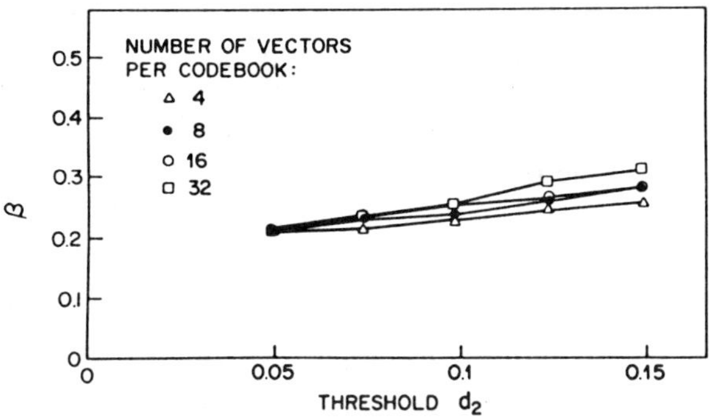

Fig. 9. Plots of average fraction of candidates sent to the postprocessor β versus preprocessor decision threshold d_2 for several codebook sizes for the digits vocabulary.

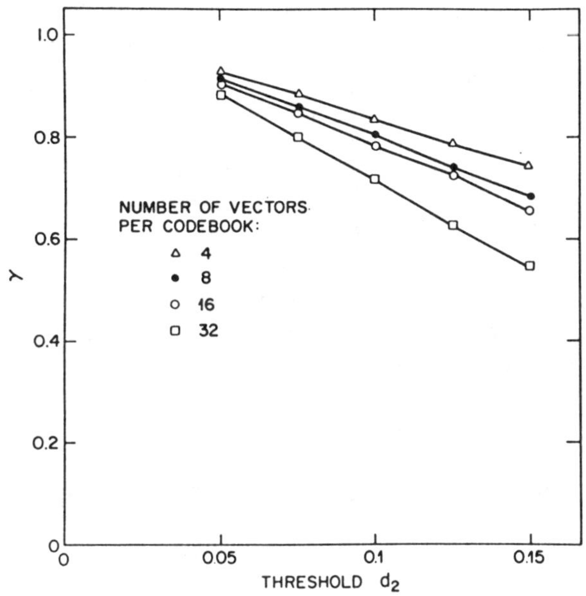

Fig. 10. Plots of average fraction of decisions made by the preprocessor γ versus preprocessor decision threshold d_2 for several codebook sizes for the digits vocabulary.

These results again show that the preprocessor alone gave very bad performance, whereas the overall system of Fig. 2 gave performance that was only slightly degraded from that of the DTW processor alone, with reductions in computation of 4.3 to 1 for Method 1 and 9.6 to 1 for Method 2, as compared to the conventional DTW processor.

C. Results on the AIRLINE Database

The tests using the AIRLINE database were intended to check the extensibility of the overall recognizer to larger, more complex vocabularies than the digits. This vocabulary, with 129 words, is rich in similarly sounding words (e.g., MAY, MAKE, MANY, etc.) and would have a tendency to make the preprocessor less useful as a total recognizer. For this test, there was no experimentation with system variables; a codebook size of 16 vectors was used for each word and only two values of d_2 were considered. A comparison on the best results on the various recognizers for this vocabulary is as follows.

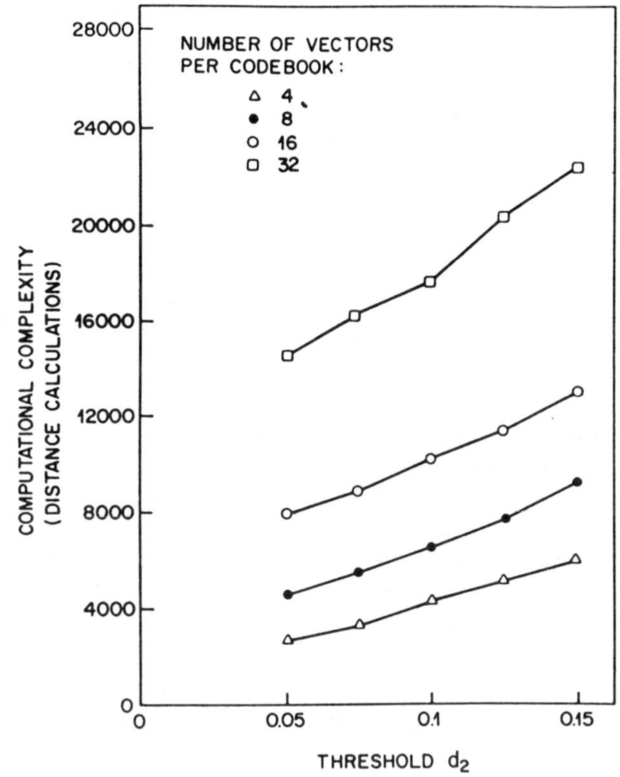

Fig. 11. Plots of computational complexity of the overall recognizer versus preprocessor decision threshold d_2 for several codebook sizes for the digits vocabulary.

Recognition Model	Overall Error Rate (%)	Computation (k)
DTW Alone	10.2	990
Method 2	12.8	107.3
Preprocessor Alone	35.8	82.5

The results given above again show that, when forced to make a unique decision, the preprocessor alone performs very poorly; however, when used in conjunction with the DTW processor, the error rate increases only by 2.6 percent over that of the DTW processor alone with a reduction in computation of close to 9.2 to 1. This tradeoff in accuracy for computation is one which is probably quite acceptable in many proposed applications of this vocabulary, especially ones with tasks which can be used to correct recognition errors via syntax and/or semantics of the proposed application [15].

For the AIRLINE vocabulary, one other comparison was made, namely, we measured the performance of the DTW system alone with a word-independent VQ used on all reference patterns. The VQ codebook that we used had 128 vectors. The resulting error rate of the system was 14.5 percent and the computation was about $\frac{1}{6}$ that required for the DTW system without VQ. Thus, the combined recognizer (Method 2 above) outperformed the VQ based DTW implementation by almost 2 percent in error rate and required about half the computational effort.

VI. Discussion

The results presented in the previous two sections have shown the following.

1) The proposed speaker-independent, isolated word recognition structure consisting of a word-based VQ preprocessor and a DTW postprocessor is indeed a viable structure for recognition. Its performance, measured in terms of word error rate, was comparable to or better than that of a conventional DTW recognizer. However, it required significantly less computation than the DTW recognizer.

2) VQ methods can be incorporated into the DTW processor with little or no loss in performance (word error rate) and with greatly reduced computation.

3) Temporal information is indeed necessary for a high-performance word recognizer, even for a vocabulary as simple as the ten digits.

The philosophy of the proposed recognizer is simple, namely, that since the DTW processor performs so well, it is not acceptable to use any preprocessor which seriously degrades the overall performance of the system. Hence, the job of the preprocessor is to reliably eliminate word candidates which are highly unlikely matches to the spoken word. It is not required, or desirable, for the preprocessor to make fine recognition distinctions; the DTW does this job quite well. The remarkable result is that for simple vocabularies, such as the digits, the preprocessor can make a final decision (i.e., it eliminates all word candidates but one) about 80 percent of the time, and even when it cannot make a unique choice, it usually only passes about two word candidates to the DTW processor. For the more complex vocabulary of airline words, a final decision was made much less often; however, again only about 20 percent of the words were passed on to the DTW processor. Thus, the gain in performance of the proposed system is mainly in computation effort, which is reduced somewhere between 2–20 times, at the expense of little or no increase in word error rate.

Before giving some general comments, it is worthwhile reviewing the major results presented in the previous section on each of the three datasets that were studied.

A. Summary of Results on REFERENCE Digits Set

The results on the REFERENCE digits set showed the following.

1) The best error rate achieved was 1.4 percent (versus 2.1 percent for the DTW processing alone), using Method 2 for VQ of the reference patterns, with an $L = 32$ codebook for each word and with $d_2 = 0.15$. For this system, the values of $(1 - \gamma)$ and β were 0.453 and 0.28, respectively, and a 5.3-to-1 reduction in computation and a 1.5-to-1 reduction in storage (over the conventional DTW processor) was obtained.

2) For maintaining the same error rate as the DTW processor alone (i.e., 2.1 percent), the system variables were $L = 8$ codebooks, Method 2 reference patterns, $d_2 = 0.15$. This implementation gave values of $(1 - \gamma)$ and β of 0.32 and 0.25, respectively, and led to an 18-to-1 reduction in computation and a 4.8-to-1 reduction in memory.

It is interesting to note that for system implementations with error rates lower than that of the DTW processor, the preprocessor actually eliminated some DTW errors by not passing these candidates to the postprocessor.

B. Summary of Results on KLS Digits Set

The results on the KLS digit set showed the following.

1) The DTW stage with either form of VQ (e.g., Methods 2 or 3) had a somewhat higher error rate than the DTW stage without VQ (Method 1).

2) For maintaining the same error rate as the conventional DTW recognizer, the system implementation required $L = 32$ codebooks, $d_2 = 0.2$, with Method 1 DTW processing. This system yielded a 2.8 percent error rate, and required 2.4 times less computation than the DTW implementation. Values of $(1 - \gamma)$ and β of 0.76 and 0.34 resulted from this choice of system variables.

3) If a slight degradation in performance were allowed (i.e., an error rate of 3.4 percent), then Method 2 VQ representations of the reference templates could be used and a 4.5-to-1 reduction in computation and a 1.7-to-1 reduction in memory could be achieved.

When the performance of the overall system is compared to that of a conventional DTW system with a word-independent VQ (with 128 codebook vectors), the performance of the proposed system was always (for all codebook sizes, methods of implementation of the reference quantization, etc.) comparable to or better than the DTW/VQ system, and the computation was less by a factor of two.

C. Summary of Results on AIRLINE Test Set

The results on the AIRLINE test set showed the following.

1) Compared to the standard DTW implementation, the proposed system has a 2.4 percent increase in error rate (from 10.2 percent for DTW to 12.6 percent for the proposed system), but achieved about a 9-to-1 reduction in computation and almost a 3-to-1 reduction in memory.

2) Compared to the DTW/VQ, using a 128-word codebook for the reference patterns, the error rate of the proposed system was 2 percent lower than that of the DTW/VQ system, and the computation was about half the amount.

D. General Comments

The experimental results have indicated that the overall recognizer proposed in this paper is certainly a very attractive method for speaker-independent isolated word recognition. We have seen that the strength of the overall recognizer lies not only in its reduced computational requirement, but also in its ability to achieve a high recognition accuracy. For the digits vocabulary, the overall recognizer was always able to achieve a reduction in computational complexity over the conventional DTW processor and still maintain a comparable (or even lower) error rate. For a recognition system with a large vocabulary, it is clear that some form of preprocessing is essential to eliminate unlikely candidates so that the DTW phase of the recognition system would be used only on reasonably close word candidates. In this manner, with a given amount of computational power, a larger vocabulary can be handled than without such a preprocessor.

There are several areas in which the overall recognizer can still be improved further. In the implementation of the overall recognizer, the computational cost of the preprocessor, although significantly lower than the computational cost of the DTW stage, increases linearly with the number of words in the vocabulary. With V words in the recognition vocabulary and L vectors in each word-based VQ codebook, there are a total of LV codebook vectors in the system. One would expect a certain number of these LV codebook vectors to overlap each other since every codebook was designed separately. In research on vector quantization for speech coding, it has been found that 1024 or 512 LPC vectors are sufficient to encode all speech [11], [12]. Hence, one method to reduce the computational requirement of the overall recognizer, when a large vocabulary is used, is to start with a general word-independent codebook with 512 or 1024 vectors. From this word-independent codebook, 16 or 32 vectors can then be selected to form the word-based codebook for each word. In this way, regardless of the number of words in the vocabulary, the maximum number of codebook vectors in the recognition system is 512 or 1024, as opposed to LV. Although the word-based codebooks generated in this way are expected to be suboptimal, it is not clear if the accuracy of the overall recognizer will degrade.

Alternatively, a slightly different approach to reduce the computational cost of the preprocessor is to generate the optimal word-based VQ codebooks individually and then remove some of the vectors that are very close to each other in the LPC space. In other words, all vectors from the optimal word-based codebooks are reclustered and the centroid of each cluster is used to replace the vectors within that cluster. The total number of codebook vectors can thus be reduced. One problem that might be caused by the above approaches is that the reduced number of codebook vectors in the system may not provide a fine enough spectral resolution for dynamic time warping with vector quantization. A possible solution to this problem is to use a slightly different structure for the preprocessor in which the optimal word-based codebooks for the likely candidates selected by the preprocessor are used to requantize the test word. In this way, the DTW/VQ processor will still have the same fine spectral resolution provided by the optimal word-based codebooks.

Another method to reduce the computational cost in the vector quantizer is to use a different search algorithm. In this research, we have used a full search algorithm in which each test vector is compared to every codebook vector in order to find the best matching codebook vector. A binary search algorithm can be used instead to reduce the computational cost of the preprocessor. The binary search algorithm is suboptimal and has been found to give slightly degraded performance in speech recognition [7]. Shikano [7] has proposed an alternative search method that can lead to almost optimal (full search) performance, but that only requires, on the average, the computational cost of the binary search algorithm. It should be noted that both the binary search algorithm and the Shikano method are more effective for large VQ codebooks and can only be used in the overall recognizer using Method 1 (unquantized test and reference patterns for the DTW algorithm) or using vector quantized patterns for both test and reference.

It should also be pointed out that the savings in computational complexity which we have found in the overall recognizer is an average overall recognition trials for each of the databases used. The amount of computation required to recognize each utterance is not fixed. However, if this is a problem for any application, it can easily be solved by requiring the preprocessor to pass a fixed percentage of candidates to the DTW stage for all test utterances. In this way, the response time of the overall recognizer can be set to a fixed value.

One of the main weaknesses in the preprocessor is the loss of temporal information in the recognition process. It appears that some form of time alignment of the test and reference patterns is essential to eliminate this class of errors. If some temporal information can be brought back, even in very crude forms as suggested by Buzo [18] and investigated by Burton *et al.* [10], the performance of the word-based VQ approach can be significantly improved when it is used either as a complete word recognizer or as a preprocessor of an overall recognizer.

VII. Summary

We have proposed a structure for isolated word recognition in which a word-based VQ preprocessor is used to screen out bad matches to an unknown test word and eliminate them from further consideration. In many cases, especially with a small, noncomplex vocabulary such as the digits, the screening process eliminates all words except one. In such cases, no further processing is needed for recognition. Otherwise, a conventional DTW processor is used to resolve any fine acoustical distinctions between words which were passed by the preprocessor. An evaluation of this new recognition system showed that implementations could be achieved for which the performance degradation (i.e., word error rate) was small or negligible, and for which the reduction in computation was as large as 20 to 1 in some cases. The major weakness in the preprocessor was the lack of temporal information for doing the screening. We are currently considering several alternatives for including such temporal information in the preprocessor.

References

[1] G. M. White and R. B. Neely, "Speech recognition experiments with linear prediction, bandpass filtering, and dynamic programming,"

IEEE Trans. Acoust., Speech, Signal Processing, vol. ASSP-24, pp. 183–188, Apr. 1976.

[2] F. I. Itakura, "Minimum prediction residual principle applied to speech recognition," *IEEE Trans. Acoust., Speech, Signal Processing*, vol. ASSP-23, pp. 67–72, Feb. 1975.

[3] L. R. Rabiner, S. E. Levinson, A. E. Rosenberg, and J. G. Wilpon, "Speaker-independent recognition of isolated words using clustering techniques," *IEEE Trans. Acoust., Speech, Signal Processing*, vol. ASSP-27, pp. 336–349, Aug. 1979.

[4] G. R. Doddington and T. B. Schalk, "Speech recognition: Turning theory to practice," *IEEE Spectrum*, vol. 18, pp. 26–32, Sept. 1981.

[5] L. R. Rabiner and S. E. Levinson, "Isolated word connected word recognition—Theory and selected applications," *IEEE Trans. Commun.*, vol. COM-29, pp. 621–659, May 1981.

[6] L. R. Rabiner, S. E. Levinson, and M. M. Sondhi, "On the application of vector quantization and hidden Markov models to speaker-independent, isolated word recognition," *Bell Syst. Tech. J.*, vol. 62, pp. 1075–1105, Apr. 1983.

[7] K. Shikano, "Spoken word recognition based upon vector quantization of input speech," *Trans. Committee Speech Res.*, pp. 473–480, Dec. 1982.

[8] L. R. Rabiner, K. C. Pan, and F. K. Soong, "On the performance of isolated word speech recognizers using vector quantization and temporal energy contours," *AT&T Bell Lab. Tech. J.*, vol. 63, pp. 1245–1260, Sept. 1984.

[9] J. E. Shore and D. K. Burton, "Discrete utterance speech recognition without time alignment," *IEEE Trans. Inform. Theory*, vol. IT-29, pp. 473–491, July 1983.

[10] D. K. Burton, J. T. Buck, and J. E. Shore, "Parameter selection for isolated word recognition using vector quantization," in *Proc. IEEE Int. Conf. Acoust., Speech, Signal Processing*, Mar. 1984, pp. 9.4.1–9.4.4.

[11] Y. Linde, A. Buzo, and R. M. Gray, "An algorithm for vector quantization," *IEEE Trans. Commun.*, vol. COM-28, pp. 84–95, Jan. 1980.

[12] B. Juang, D. Wong, and A. H. Gray, Jr., "Distortion performance of vector quantization for LPC voice coding," *IEEE Trans. Acoust., Speech, Signal Processing*, vol. ASSP-30, pp. 294–303, Apr. 1982.

[13] M. K. Brown and L. R. Rabiner, "On the use of energy in LPC-based recognition of isolated words," *Bell Syst. Tech. J.*, vol. 61, pp. 2971–2987, Dec. 1982.

[14] L. R. Rabiner, M. M. Sondhi, and S. E. Levinson, "Note on the properties of a vector quantizer for LPC coefficients," *Bell Syst. Tech. J.*, vol. 62, pp. 2603–2616, Oct. 1983.

[15] S. E. Levinson, A. E. Rosenberg, and J. L. Flanagan, "Evaluation of a word recognition system using syntax analysis," *Bell Syst. Tech. J.*, vol. 57, pp. 1619–1626, May 1978.

[16] J. G. Wilpon, L. R. Rabiner, and A. Bergh, "Speaker-independent isolated word recognition using a 129-word airline vocabulary," *J. Acoust. Soc. Amer.*, vol. 72, pp. 390–396, Aug. 1982.

[17] A. E. Rosenberg, K. L. Shipley, and D. E. Bock, "A speech data base facility using a computer controlled cassette tape deck," *J. Acoust. Soc. Amer.*, suppl. 1, vol. 72, p. 580, Fall 1982.

[18] A. Buzo, C. Riviera, and H. G. Martinez, "Discrete utterance recognition based upon source coding techniques," in *Proc. IEEE Int. Conf. Acoust., Speech, Signal Processing*, May 1982, pp. 539–542.

CONNECTED DIGIT RECOGNITION USING VECTOR QUANTIZATION

H. Bourlard, Chr. J. Wellekens

Philips Research Laboratory, Brussels, Belgium

H. Ney

Philips GmbH Forschungslaboratorium, Hamburg, West-Germany

ABSTRACT

The principles of classification applied to the representation of the words in a vocabulary lead to the clustering of the acoustic vectors into prototype vectors. For a small number of prototypes, recognition scores comparable to those observed with unclustered vocabularies are obtained with a highly reduced computation time. Two different forms (deterministic and stochastic) of the single-level recognition method for concatenated words are described and the improvements obtained by vector quantization are put into evidence. The use of prototypes in the training phase of the finite stochastic automata representing a vocabulary word is also described.

1. THE RECOGNITION PRINCIPLE

Speech recognition is essentially based on the comparison between an utterance and a representation of the vocabulary words obtained by a training phase. Having defined some form of distance between words (dissimilarity measure), the nearest word in the vocabulary to the utterance can be identified. Two main distortions are generally observed between an utterance and its representation. The first one is a non-linear warping of the time scale : the word has been pronounced with a speed different from that of its model. It would be sufficient to find the shape of this distortion to transform the word into its model and to consider the distance as zero. The second difference concerns the pronunciation itself so that a dissimilarity remains in spite of an optimal restoration of the time scale.

To be able to estimate the dissimilarity and to find the optimal warping of the time scale, the speech is sampled and divided into frames of 10 msec. During such a time interval, the human vocal apparatus may be assumed as time independent. Data corresponding to each frame are then compressed according to different methods (DFT, cepstrum coefficients, LPC parameters ...) and are gathered in an "acoustic vector" the dimension of which is a measure of the compression rate. Both the reference words and the word or sequence of words to be recognized are characterized by acoustic vectors.

If the hereabove described distortions are taken into account, it is impossible to set up a one-to-one correspondence between the utterance vectors and the reference vectors. Let us consider the (generally rectangular) matrix of the "local" distances i.e. the distances between the utterance vectors and the reference vectors. A path in this matrix starting from the first vectors to the last vectors of the words to be compared is computed such that it minimizes the sum of the local distances (accumulated distance). The distance between two words is then defined as this minimal accumulated distance. The path in constrained by some conditions of continuity, monotonicity and slope. These conditions can be summarized as a definition of the predecessors of an entry. The dynamic programming principle (DTW : dynamic time warping) is best suited to solve this problem of isolated word recognition and its use can be generalized to connected speech recognition [1] as will be recalled in section 3. A straightforward computation provides the dimensions of the local distance matrix. If each local distance is Euclidean, the computation of the matrix requires too much time even for a small vocabulary. Initially, the clustering has been used to reduce the amount of computations. Another advantage is an increase of the discrimination between similar words (section 4).

2. CLUSTERING

The number of acoustic vectors required to describe the words of a vocabulary is very high. The distance of each vector of an utterance to be recognized must be computed versus each vector of this set. The definition of prototype vectors in the space of the acoustic vectors allows the reduction of the number of distances to be computed. Indeed, each acoustic vector is replaced by the "nearest" prototype and each vocabulary word will be a sequence of prototype vectors with repetitions so that the distances will be computed only versus the prototypes. Exhaustive experiments on a data base have put into evidence that 50 to 100 prototypes are sufficient to characterize the acoustic vector space and allows the same recognition scores to be reached as without clustering in the case of speaker dependent recognition and German digits. The clustering problem is stated as follows : find a partition of the set of the N vectors $v(j)$ into n disjoint classes for which the mean vectors $u(i)$ are known as cluster centers or prototype vectors and such that the sum over the class of the mean distances to the center is minimized.

Several iterative algorithms exist but none is able to guarantee a global optimality of the result which depends on the initial guesses. The classical K-means algorithm [2] , which has been used for our experiments, is described as an example. Its choice has been dictated more by its simplicity than by a performance analysis.

- Among the N vectors v(j), a subset of n vectors u(k) is chosen,
- for each v(j), the n distances |v(j)-u(k)| are computed,
- each vector v(j) is associated to a class i according to the rule : $\min_k |v(j)-u(k)|$ is reached for k=i,
- in each class, any center u is computed as mean vector of the class,
- one iterates by going back to point 2 as long as the content of the classes or the centers of the classes vary from one iteration to the other.

The experiments have been performed on sets of N vectors of dimension 15 with N up to 1000 and for a number of classes n ranging from 10 to 200. The time required for the search is high but not prohibitive as the training phase is realized off-line once for all. The quality control of the results can be checked either analytically or experimentally. In the first case, the distances between the centers is computed and are compared with the mean distance of the vectors of each class to their center. One easily realizes that each class forms an isolated cloud in the vector space (true clustering) or that the algorithm provides a simple quantization of the space in which the vectors are uniformly distributed. The experimental verification is related to the representation of the vocal signals themselves. The following results can be observed :

- multiple representations of the same word provide very similar representations in terms of prototype vectors,
- by comparing the quantified representations of two words containing the same phonemes, it is possible to identify these phonemes in the words,
- already for a small number of classes, recognitions performed on a clustered vocabulary give scores similar to those obtained without clustering.

3. DETERMINISTIC SINGLE LEVEL RECOGNITION METHOD

Dynamic programming for time alignment is well known for isolated words recognition. Its use in connected word recognition is described in [1] and shortly recalled here. Let us agree that each column of the accumulated distances matrix corresponds to an acoustic vector of the utterance and that the rows correspond to the acoustic vectors of the reference words connected end to end. Let us choose Itakura's predecessors [3] (fig.1) with the following modifications :

- the predecessors of an entry belonging to a line representing the first vector of a reference word are the preceding entry on the same row and the entries of the preceding column corresponding to the last vector of each reference word,
- the jump over a vector of a reference word is forbidden if, by doing so, one crosses the boundary between two reference words ; this crossing is allowed only though a simple diagonal move (fig. 2).

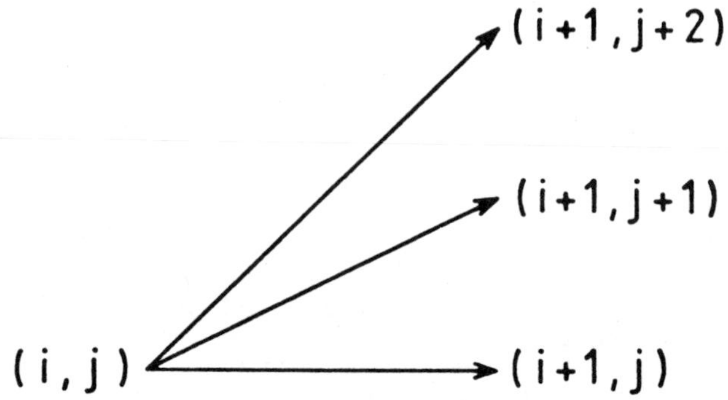

Fig. 1. Itakura's predecessors

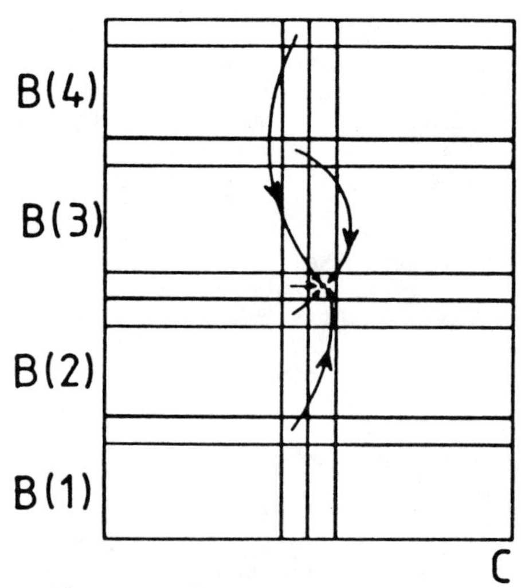

Fig.2. Modified predecessors

The minimal accumulated distance corresponding to a given entry is recursively computed from the predecessors by taking into account possible weight factors applied to the local distance according to the considered predecessor. Once the matrix has been filled in, it is sufficient to compare in the last column the accumulated distances in the rows corresponding to the terminations of the reference words : the lowest distance indicates the last word of the utterance. From this entry, the optimal path traces back through the matrix, from each entry to the predecessor selected in the forward matrix filling. This sequence of transition forces jumps from one reference word to another and the path identifies the sequence of utterances (fig. 3.).

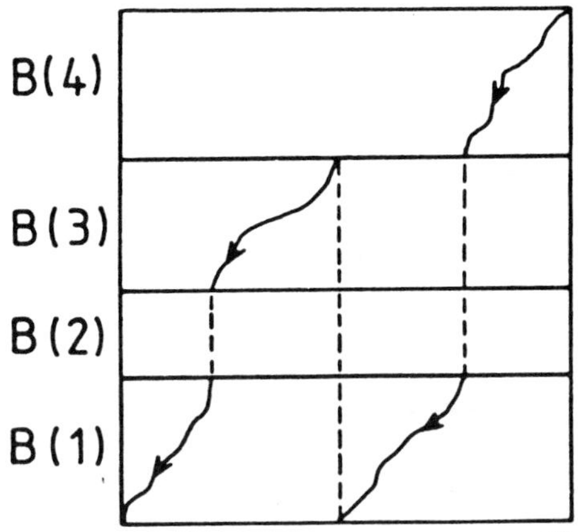

Fig. 3. Optimal path

The recognition experiments on a clustered vocabulary have been performed with this single-level algorithm but with a slightly modified definition of the predecessors.
The reference word vectors have been replaced by the prototype of their own class. It is trivial to estimate the computation economy realized by using a clustered vocabulary. Another advantage of the clustering is the improvement in the discrimination between words containing identical syllables. After clustering, the accmulated distance between these words will come essentially from the phonetically different parts and will allow a better decision than if this distance would be merged with unsignificant local differences (f.i. noise) coming from the phonetically identical parts.

4. RESULTS USING THE DETERMINISTIC METHOD

In the reported experiment, for each 10 msec time-frame, the 16 first cepstral coefficients are computed. The first coefficient is proportional to the gain of the acquisition system and, by neglecting it, an automatic normalization of the signal amplitude is realized. The acoustic vectors are thus constituted by 15 cepstral coefficients. Data are provided from 4 different speakers (males and females). During the training session, each speaker utters each digit twice in an isolated manner. The sentences to be recognized are constituted by 100 sequences of 7 digits uttered by each speaker without artificial pauses between digits. Recognition and clustering are performed independently for each speaker. The sequences to be recognized are unclusted. The predecessors of an entry (i,j) are the entries (i-1, j), (i-1, j-1) and (i,j-1) with the corresponding weight factors 2, 1, 1/2. These kinds of errors appear : the deletion of a digit, the insertion of an extra digit or the substitution of a digit by another one. The number of errors as a function of the number of clustering classes (10 to 100) is shown in Table I.

NMB of classes / SPEAKER		10	20	40	50	60	80	100	200
	D	6	5	5	4	4	4	5	4
1	I	2	1	1	2	1	2	4	2
	S	17	7	7	16	6	4	8	9
	T	25	13	13	22	11	10	17	15
	D	0	0	0	0	0	0	0	0
2	I	1	0	0	0	0	0	0	0
	S	5	1	2	0	1	1	1	1
	T	6	1	2	0	1	1	1	1
	D	0	0	0	0	0	0	0	0
3	I	0	0	0	0	0	0	0	0
	S	16	15	9	2	2	1	4	2
	T	16	15	9	2	2	1	4	2
	D	9	4	5	4	4	5	6	10
4	I	0	0	2	0	1	0	0	0
	S	91	32	34	20	19	24	23	19
	T	100	36	41	24	24	29	29	29

Table I : number of deletions (D), insertions (I), substitutions (S) and total number of errors (T).

The experiments show that a minimum error rate is obtained already for a rather low number of classes (about 70) and that it increases again for a higher number of classes or without any clustering, although this increase is not necessarily statistically significant. This minimum of error rate as a function of the number of classes could be due to the fact that phonetically similar parts in the different words of the vocabulary are represented by the same prototypes and the fluctuations of the local distances without statistical meaning are thus partially avoided. The reduction of the computation load is very important : the distances versus 70 prototype vectors are computed instead of 1000 local distances (about 50 vectors for each of both versions of each digit). The number of local distance computations is decreased by a factor 14 while the global recognition time reduction including the combinatorics required for each matrix entry is about a factor of 10.
A further improvement of the recognition time may be obtained by reducing combinatorics as a result of decreasing the number of vectors required for a word representation. Indeed, after clustering, a high number of repetitions of prototypes are observed which may be suppressed much as the predecessor weights are adequately adjusted. This weight adaptation is easily implemented in a hidden Markov word model (section 5).

5. STOCHASTIC MODELS

In the deterministic methods as described in section 3, the accumulated distance corresponding to an entry is the sum of the accumulated distance corresponding to the chosen predecessor and of the local distance weighted by a factor depending on the chosen transition. An alternative solution is to accept an additive weighting rather than a multiplicative one and then assimilate the distances and the additive weights as the negative of logarithms of probabilities. The meaning of these probabilities is evident if one accepts that the acoustic vectors of an uttered word are produced when reaching a state of a hidden Markov chain(fig. 4). A path from an initial state to a final one in the Markov chain generates an utterance of the word: the chain is thus a production model of all possible pronunciations of a word for a given speaker. The states correspond to the vectors of the reference words.

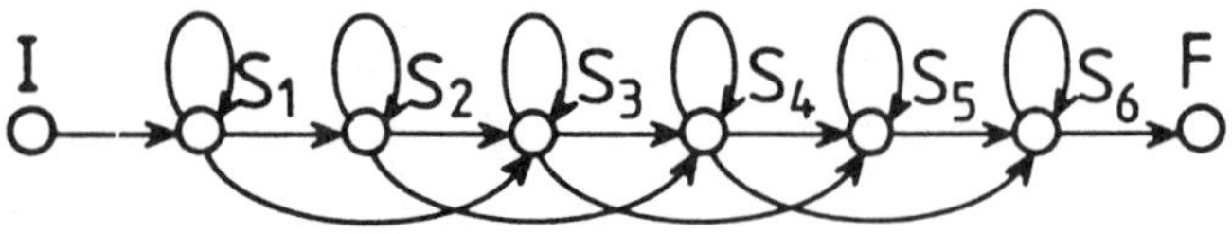

Fig. 4. Bakis model

The accumulated distances correspond to the probabilities to reach a given state while having pronounced the preceding vectors ; the transition probabilities between two states correspond to the weighting terms; the probabilities corresponding to the local distances are related to those of emitting a given vector in a given state. These emission probabilities generally obey some probability density function (a Gaussian law in our experiments) the parameters of which (mean vector and covariance matrix) are determined by a training of the model on several utterances of the word. Isolated word recognition from these stochastic models also uses a dynamic programming algorithm where, instead of searching in a matrix a path minimizing the accumulated distance, one maximizes in the Markov chains the accumulated probability of utterance : the Markov chain providing the highest probability indicates the uttered word. This particular form of the algorithm is known as the Viterbi algorithm.
The commonly used Markov chain is the Bakis model [4]; three transitions are allowed from each state: to itself (insertion of an acoustic vector), to the next state (substitution of an acoustic vector) , to the second following state by jumping over a state (omission of an acoustic vector) fig. (4). The path from the initial state to the final one using only substitutions is the basis of the model; it corresponds to the standard pronunciation.
The first phase in the training of the Bakis model is the search of the prototypes by clustering of the set of the vectors produced by several utterances of the same word. It can be observed that each utterance contains repeated sequences of the same prototype. The utterance which, after elimination of the repetitions, has the greatest length is selected. Then each prototype which was repeated, is replaced by a pair of these prototypes. The obtained sequence of vectors forms the basis of a Bakis model. The doubling of the initially repeated prototypes has been suggested by experimental results : in the Bakis model, a prototype cannot be jumped over if it appears as a pair of states.
In the second phase of the training, for each utterance the optimal path is determined in the model with guessed values of the parameters : the mean vectors and the covariance matrices associated to each state are respectively the prototypes and arbitrarily chosen diagonal matrices while the transition probabilities are all equal to 1/3. Then, estimated values for the parameters are computed by a simple count procedure and the process is iterated with the updated parameters up to stabilization of their values. As a consequence, clustering is used to determine the length of the chain and the tied states (different states having the same mean vector and covariance matrix) and also to provide initial mean vectors. Statistical concatenated words recognition is performed by adapting the single-level method described in section 3 to adaptive additive weights.

6. RESULTS USING STATISTICAL MODELS

The same data base is used as in section 4. A training is performed on 20 strings of 7 German digits. It consists of a Viterbi algorithm constrained to recognize known strings. A clustering with 60 prototypes is performed on isolated utterances of the vocabulary words and provides the topology of the word models.One hundred strings of 7 German digits uttered by the same four speakers are then recognized. Table II summarizes the results. The use of adaptative weights appears to be essential for a high score.

SPEAKER	1	2	3	4
	D I S T	D I S T	D I S T	D I S T
NMB of CLASSES 60	1 1 0 2	0 0 0 0	0 0 3 3	1 0 6 7

Table II : number of deletions (D), insertions (I), substitutions (S) and total number of errors (T).

7. REFERENCES

[1] H. Ney, "Experiments in connected word recognition", Proc. Intern. Conf. on Acoust., Speech, Signal Processing 1983, pp. 288-291, Boston.

[2] J. T. Tou and R. C. Gonzales, Pattern recognition principles, Addison-Wesley Reading, 1974

[3] F. Itakura, "Minimum prediction residual principle applied to speech recognition", IEEE Trans. Acoust., Speech, Signal Processing, vol. ASSP-23, pp. 67-72, Feb. 1975.

[4] F. Jelinek, "Continuous speech recognition by statistical methods", Proc IEEE, vol. 64, pp. 532-556, April 1976.

BYBLOS: The BBN Continuous Speech Recognition System

Y.L. Chow, M.O. Dunham, O.A. Kimball, M.A. Krasner,
G.F. Kubala, J. Makhoul, P.J. Price, S. Roucos,
and R.M. Schwartz

BBN Laboratories Incorporated
10 Moulton Street
Cambridge, MA 02239

Abstract

In this paper, we describe BYBLOS, the BBN continuous speech recognition system. The system, designed for large vocabulary applications, integrates acoustic, phonetic, lexical, and linguistic knowledge sources to achieve high recognition performance. The basic approach, as described in previous papers [1, 2], makes extensive use of robust context-dependent models of phonetic coarticulation using Hidden Markov Models (HMM). We describe the components of the BYBLOS system, including: signal processing frontend, dictionary, phonetic model training system, word model generator, grammar and decoder. In recognition experiments, we demonstrate consistently high word recognition performance on continuous speech across: speakers, task domains, and grammars of varying complexity. In speaker-dependent mode, where 15 minutes of speech is required for training to a speaker, 98.5% word accuracy has been achieved in continuous speech for a 350-word task, using grammars with perplexity ranging from 30 to 60. With only 15 seconds of training speech we demonstrate performance of 97% using a grammar.

1. Introduction

Speech is a natural and convenient form of communication between man and machine. The speech signal, however, is inherently variable and highly encoded. Vast differences occur in the realizations of speech units related to context, style of speech, dialect, talker. This makes the task of large vocabulary continuous speech recognition (CSR) by machine a very difficult one. Fortunately, speech is also structured and redundant: information about the linguistic content in the speech signal is often present at the various linguistic levels. To achieve acceptable performance, the recognition system must be able to exploit the redundancy inherent in the speech signal by bringing multiple sources of knowledge to bear. In general, these can include: acoustic-phonetic, phonological, lexical, syntactic, semantic and pragmatic knowledge sources (KS). In addition to designing representations for these KSs, methodologies must be developed for interfacing them and combining them into a uniform structure. An effective and coherent search strategy can then be applied based on global decision criteria. Practical issues that need to be resolved include computation and memory requirements, and how they could be traded off to obtain the desired combination of speed and performance.

In BYBLOS, we have explored many issues that arise in designing a large and complex system for continuous speech recognition. This paper is organized as follows. Section 2 gives an overview of the BYBLOS system. Section 3 describes our signal processing frontend. Section 4 describes the trainer system used for phonetic model knowledge acquisition. Section 5 describes the word model generator module that compiles word HMMs for each lexical item. Section 6 describes the syntactic/grammatical knowledge source that operates on a set of context-free rules describing the task domain to produce an equivalent finite state automaton used in the recognizer. Section 7 describes the BYBLOS recognition decoder using combined multiple sources of knowledge. Finally, Section 8 presents some figures and discussions on BYBLOS recognition performance.

2. Byblos System Overview

Figure 1 is a block diagram of the BYBLOS continuous speech recognition system. We show the different modules and knowledge sources that comprise the complete system, the arrows indicating the flow of module/KS interactions. The modules are represented by rectangular boxes. They are, starting from the top: Trainer, Word Model Generator, and Decoder. Also shown are the knowledge sources, which are represented by the ellipses. They include: Acoustic-Phonetic, Lexical, and Grammatic knowledge sources. We will describe briefly the various modules and how they interact with the various KSs.

Acoustic-Phonetic KS

The Trainer module is used for the acquisition of the acoustic-phonetic knowledge source. It takes as input a dictionary and training speech and text, and produces a database of context-dependent HMMs of phonemes.

Lexical KS

The Word Model Generator module takes as input the phonetic models database, and compiles word models phonetic models. It uses the dictionary - the lexical KS, in which phonological rules of English are used to represent each lexical item in terms of their most likely phonetic spellings. The lexical KS imposes phonotactic contraints by allowing only legal sequences of phonemes to be hypothesized in the recognizer, reducing the search space and improves performance. The output of the Word Model Generator is a database of word models used in the recognizer.

Reprinted from *Proc. IEEE Conf. Acoust., Speech, Signal Processing,* vol. 1, pp. 89–92, April 1987.

Grammatical KS

More recently, we have been working on representation and integration of higher levels of knowledge sources into BYBLOS, including both syntactic and semantic KSs. By incorporating both of these KSs into BYBLOS in the form of a grammar into our recognizer, we demonstrate improved recognition performance. In Section 6, we describe the Grammatical KS in more detail.

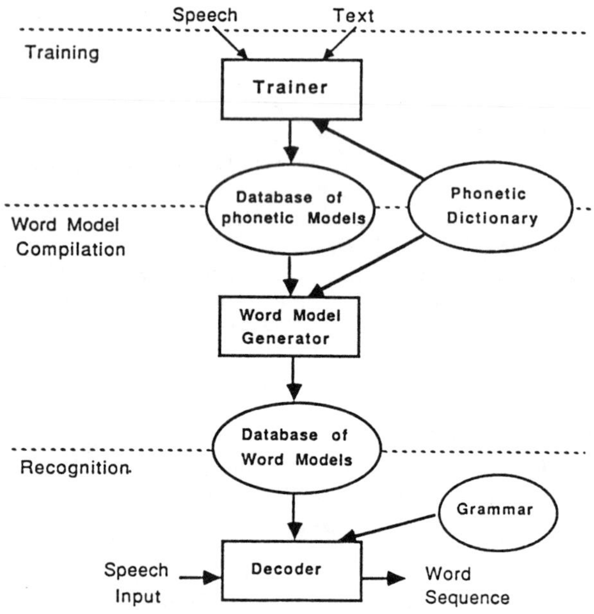

Figure 1: BYBLOS System Diagram.

3. Signal Processing and Analysis Component

The BYBLOS signal processing frontend performs feature extraction for the acoustic models used in recognition. Sentences are read directly into a close talking microphone in a natural but deliberate style in a normal office environment. The input speech is lowpass filtered at 10 kHz and sampled at 20 kHz. Fourteen Mel-frequency cepstral coefficients (MFCC) are computed from short-term spectra every 10 ms using a 20 ms analysis window. This MFCC feature vector is then vector quantized to an 8-bit (256 bins) representation. The vector quantization (VQ) codebook is computed using the k-means clustering algorithm with about 5 minutes of speech. We perform a variable-frame-rate (VFR) compression in which strings of up to 3 identical vector codes are compressed to a single observation code. We found this VFR procedure speeds up computation with no loss in performance.

4. Training/Acquisition Of Phonetic Coarticulation Models

The training system in BYBLOS acquires and estimates the phonetic coarticulation models used in recognition. Given that we model speech parameters as probabilistic functions of a hidden Markov chain, we make use of the Baum-Welch (also known as the Forward-Backward) algorithm [3] to estimate the parameters of the HMMs automatically from spoken speech and corresponding text transcription. For each training utterance, the training system takes speech and text, and builds a network of phonemes using the dictionary. It first builds the phonetic network for the word by using the phonetic transcription provided by the dictionary. The phonetic network is expanded into a triphone network so that each arc completely defines a phonetic context up to the triphone. These triphone networks of the word are then concatenated to form a single network for the sentence, which in general can take into account within word as well as across-word phonological effects. The training system then compiles a set of phonetic context models for each triphone arc in the network. It then runs the forward-backward algorithm to estimate the parameters of the phonetic context models. The Trainer operates in two modes: speaker-dependent and speaker-adapted. Associated with these two modes are two distinct methods for training the parameters of the hidden Markov models described below.

Speaker-Dependent

This is the algorithm used to find the parameters of the HMMs that maximizes the probability of the observed data given the model. This method produces HMMs that are finely tuned to a particular speaker, therefore in general would work well only for this speaker. Typically about 15 minutes of speech from a speaker is required for speaker-dependent training.

Speaker-Adapted

This is a new method of training that transforms HMM models of one speaker to model the speech of a second speaker [4]. This procedure estimates a probabilistic spectral mapping from a well-trained prototype speaker to a new speaker. Using this method it is possible for a new speaker to used the system with as little as 15 seconds of speech.

5. Word Model Generator

Prior to recognition, word HMMs are computed for each word in the vocabulary. The word model generator takes as input two objects: a database of phonetic HMMs as obtained in training, and a dictionary that contains phonetic spellings for each word. For each phoneme in each word of the lexicon, it first finds in the phonetic HMM database all the context models that are relevant to this phoneme in its particular phonetic environment. It then combines this set of phonetic models with appropriate weights to produce a single HMM for each phoneme in the word. This combination process saves computation by precompiling the many levels of phonetic context models that can occur for a given phonetic context into a single representation. The output of the word model generator is a database of word HMMs serving as the input to the decoder.

6. Grammatical Knowledge Source

To solve the CSR problem requires major advances in two areas: acoustic modeling and language modeling. A good acoustic model is essential in making fine phonetic distinctions when needed. However, it is not sufficient by itself to solve the CSR problem. In a complex task with large vocabulary where the number of hypothesized word candidates is large, the probability for acoustic confusability can be high, and the recognizer could make errors. A conceptually simple yet effective way to restrict the number of words that are allowed to be hypothesized, and therefore decrease probability of acoustic similarity, is to incorporate a grammar into the recognizer. It is well known that recognition performance improves as vocabulary size decreases. Similarly, when syntactical information is used to reduce the number of words that can legally follow a given sequence of words, a recognizer is expected to make fewer errors. The purpose for using a grammar then, is to improve recognition performance, with an added benefit of reducedcomputation.

Grammar Design and Implementation

We approach the implementation of a grammar in BYBLOS in two stages. First, we create a description of the task domain language using a modified context-free notation. Typically this description is based on a representative set of sentences that characterizes the task domain, and is designed to capture generalizations of the linguistic phenomena found in them. Second, we use a tool that transforms this description into structures in our recognizer that provide the corresponding grammatical constraints. This tool provides us with a general facility for capturing in BYBLOS an approximation of any language expressible in context-free grammars (CFG) expressed as context-free rules. We elected to implement the grammatical constraints in the form of a finite state automaton (FA) similar to those described in [5].

At the first stage in generating a grammar, we use a context-free notation augumented with variables in order to simplify the process of describing a language. For example, this notation would allow a rule that says a noun phrase of any number can be replaced by an article and a noun of the same number; ordinary context-free notation would require two rules that are identical except that one would be for singular number and the other for plural.

Our system first translates the augmented notation into ordinary CFGs and then constructs a FA based on these rules. Because context-free grammars can accept recursive languages and a FA cannot, recursion is approximated in the FA by limiting the number of levels of recursion. Such an approximation is reasonable for most task languages, since spoken sentences do not ordinarily use more than a few levels of recursion.

7. Recognition Search Strategy

Once the FA is compiled from the context-free description of the task domain, it is ready to be used in the decoder. An important characteristic of a recognizer is the search strategy that is used to find the word sequence that best matches the input speech. We believe that an optimum search strategy avoids making local decisions; the search decision should be made globally, based on scores from all the KSs. One such search paradigm is the one used in BYBLOS: the search is made top down, linguistically driven, with tightly coupled KSs.

The FA is convenient for deploying such a search strategy. It is used as follows in our recognizer. We associate with each transition in the FA a hidden Markov model for the word. This model is used to compute the probability of the acoustic event (sequence of VQ spectra) given the occurrence of the word at that place in the grammar. Before the start of recognition, the initial state of the FA where a legal sequence of words can begin is initialized to unity, and all the other states are initialized to zero. For each 10 ms frame of the input speech, the scores for the states in all the words in the FA network are updated using modified Baum-Welch algorithm [2]. In addition to state updates within a word, a word can have a score propagated to its initial state from its best scoring predecessor word. This simple state update operation is repeated every 10 ms for each FA transition until the end of the utterance is reached. The decoder output is then computed by tracing back through the FA network to find the highest scoring sequence of words that end in the terminal state of the FA.

One potential problem associated with using a FA grammar for recognition is that computation is expected to be proportional to the number of transitions in the FA. This number can be quite large for complex languages. However, in our experience with different grammars in our recognizer, we find that a beam search effectively reduces the computation to a very manageable level while maintaining the same performance as that of an exhaustive search.

8. Byblos Recognition Performance

In [2], we presented word recognition results for a 334-word electronic mail task. In speaker-dependent mode, we demonstrated performance of 90% across several speakers without the use of a grammar (i.e., branching factor of 334). Since then, we have tested the system along many dimensions: two task domains, FA grammars with varying perplexities, varying amounts of adaptation speech, and different speaker types. The results are tabulated in Figure 2. Below we describe the different conditions in more detail.

Task Domains

The two task domains tested are: Electronic Mail (EMAIL) and Naval Database Retrieval (FCCBMP). Both tasks have vocabulary sizes of approximately 350 word (334 for EMAIL, 354 for FCCBMP). A description of the task domain language was created using CFG. The CFGs were designed to capture generalizations of linguistic phenomena found in example task domain sentences.

Grammars

Two finite state grammars were generated for each task domain: Command and Sentence. The Command Grammar in each case was designed to cover only the command subset of

Grammar/ Perplexity \ Training Time	EMAIL		FCCBMP	
	Command (20)	Sentence (30)	Command (22)	Sentence (30)
15 minute	98.4	98.8	99.6	99.5
2 minute	97.9	94.9	96.6	96.2

Figure 2: BYBLOS Recognition Results. Two task domains (EMAIL and FCCBMP), two grammars for each task (Command and Sentence), and varying amounts of training speech (2 minutes and 15 minutes). Also shown are maximum perplexity measures for the grammars.

the language; the Sentence Grammar was designed to cover all of the language, which included both command and question type constructs. The maximum perplexity measures of the grammars, as proposed in [6], are shown in Figure 2. In both tasks, the sentence grammars have a higher perplexity than their command counterparts.

Adaptation Time

As described in Section 2, The BYBLOS operate in two modes, speaker-dependent and speaker-adapted. In speaker-dependent mode, 15 minutes of training speech is required for a speaker. This mode in general will give word accuracy in the 98.5+ range. In the speaker-adaptive mode, anywhere from 2 minutes down to 15 seconds of speech from a new speaker is needed to "adapt" the HMM parameters to the new speaker. The performance in this case is 97%.

Speaker Type

We have tested BYBLOS on several speakers with different dialects, including a female speaker, a non-native speaker, and 3 naive (uncoached) speakers. The recognition results for these speakers showed little deviation typical male speakers of standard American dialects.

9. Summary

We have presented BYBLOS, a system for large vocabulary continuous speech recognition. We showed how we integrate multiple sources of knowledge to achieve high recognition performance. In recognition experiments, we demonstrated consistent performances across task domains, grammars, adaptation time, and speaker type.

We are currently working to improve various aspects of the system, including: a real time implementation of the recognizer, search strategy, acoustic modeling, and language modeling. In the future, we plan to work on integration of speech and natural language for speech understanding applications.

Acknowledgement

This work was supported by the Defense Advanced Research Projects Agency and was monitored by the Space and Naval Warfare Systems Command under Contract No. N00039-85-C-0423.

References

1. R.M. Schwartz, Y.L. Chow, O.A. Kimball, S. Roucos, M. Krasner, and J. Makhoul, "Context-Dependent Modeling for Acoustic-Phonetic Recognition of Continuous Speech", *IEEE Int. Conf. Acoust., Speech, Signal Processing*, Tampa, FL, March 1985, pp. 1205-1208, Paper No. 31.3.

2. Y.L. Chow, R.M. Schwartz, S. Roucos, O.A. Kimball, P.J. Price, G.F. Kubala, M.O. Dunham, M.A. Krasner, and J. Makhoul, "The Role of Word-Dependent Coarticulatory Effects in a Phoneme-Based Speech Recognition System", *IEEE Int. Conf. Acoust., Speech, Signal Processing*, Tokyo, Japan, April 1986, pp. 1593-1596, Paper No. 30.9.1.

3. L.R. Bahl, F. Jelinek, and R.L. Mercer, "A Maximum Likelihood Approach to Continuous Speech Recognition", *IEEE Trans. Pattern Analysis and Machine Intelligence*, Vol. PAMI-5, No. 2, March 1983, pp. 179-190.

4. R.M. Schwartz, Y.L. Chow, G.F. Kubala, "Rapid Speaker Adaptation using a Probabilistic Spectral Mapping", *IEEE Int. Conf. Acoust., Speech, Signal Processing*, Dallas, TX, April 1987, Elsewhere in these proceedings

5. R.G. Goodman, *Analysis of Languages for Man-Machine Communication*, PhD dissertation, Carnegie-Mellon University, May 1976.

6. M.M. Sondhi and S.E. Levinson, "Computing Relative Redundancy to Measure Grammatical Constraint in Speech Recognition Tasks", *IEEE Int. Conf. Acoust., Speech, Signal Processing*, Tulsa, OK, April 1978, pp. 409-412.

A VQ-BASED PREPROCESSOR USING CEPSTRAL DYNAMIC FEATURES FOR LARGE VOCABULARY WORD RECOGNITION

SADAOKI FURUI

NTT Electrical Communications Laboratories

Musashino-shi, Tokyo, 180 Japan

ABSTRACT

This paper proposes a new VQ (Vector Quantization)-based preprocessor for use in a method which reduces the amount of computation necessary in speaker-independent large vocabulary isolated word recognition. A speech wave is analyzed by time functions of instantaneous cepstrum coefficients and short-time regression coefficients for both cepstrum coefficients and logarithmic energy. A universal VQ codebook for these time functions is constructed based on a multi-speaker, multi-word database. Next, a separate codebook is designed as a subset of the universal codebook for each word in the vocabulary. These word-specific codebooks are used for front-end preprocessing to eliminate word candidates whose distance scores are large. A dynamic time-warping processor based on a word dictionary, in which each word is represented as a time-sequence of the universal codebook elements (SPLIT method), then resolves the choice among the remaining word candidates. Effectiveness of this method has been ascertained by recognition experiments using a database consisting of words from a vocabulary of 100 Japanese city names uttered by 20 male speakers.

I. INTRODUCTION

Recently, a VQ-based approach has been proposed to reduce the amount of computation in large vocabulary isolated word recognition [1]. In this method, word-based VQs are used for preprocessing to eliminate word candidates whose distortion scores are large, with a DTW processor then resolving the choice among the remaining word candidates. The VQ codebooks are constructed by clustering the spectral envelopes of the utterances of each word taking the speaker-variability into consideration. Since each codebook simply represents a set of instantaneous spectra, that is, since there is a lack of transitional spectral information, the ability of the VQ-based processor decreases as the vocabulary grows in size.

A technique for incorporating a temporal structure into the preprocessor has been proposed to alleviate this difficulty [2]. In this method, the probability density function of the time of occurrence for each vector in the codebook is estimated from a study of the training sequences. The spectral distance score of the VQ is combined with a temporal distance score for each frame in the word. Although this method improves performance, it is problematic for two reasons: processing can start only after endpoint detection, and the performance depends on the accuracy of endpoint detection. Furthermore, the method using word-based codebooks has the disadvantage that the amount of distance calculation increases in proportion to the vocabulary size.

This paper proposes a new method for improving the performance of preprocessing. The method is based on the incorporation of dynamic and instantaneous spectral features at every short interval, and on the introduction of a universal codebook which is commonly used as a basic codebook for all words. The usefulness of dynamic features in word recognition and speaker recognition has been ascertained in previous research [3]-[6].

II. FEATURE EXTRACTION AND DATABASE

The speech wave is passed through a low-pass filter having a cutoff frequency of 4 kHz and digitized at an 8-kHz rate. Word endpoints (beginning and ending positions) are detected, and linear predictive coding (LPC) analysis is performed on all frames within the word. The LPC analysis is a tenth-order analysis of 32-ms frames, spaced every 8 ms along the word. Each overlapping 32-ms section of speech is windowed using a Hamming window. The results of the LPC analysis are the set of LPC cepstrum coefficients and logarithmic energy (these will be called simply cepstrum coefficients and energy hereafter).

Time functions of the cepstrum coefficients and the energy over the N frame intervals are extracted every 8 ms, and their dynamic characteristics indicated over the interval are represented by regression coefficients. The speech wave is then represented by a set of cepstrum coefficients and regression coefficients of both cepstrum and energy every 8 ms. The energy value itself is not used since it is insignificant for phoneme perception.

A vocabulary of 100 Japanese city names uttered by 20 male speakers (2000 samples) was used for the test utterances, and the same vocabulary uttered by a different set of four male speakers was used for the training utterances. These four speakers, considered to represent the entire range of male voices, were selected from among 30 male speakers.

III. PREPROCESSING EXPERIMENT USING WORD-BASED CODEBOOKS

3.1. Method

As the first step in the experiments, the effectiveness of feature parameters for preprocessing was evaluated using word-based codebooks constructed independently for each word. As shown in Fig. 1, feature parameter sets (cepstrum coefficients and regression coefficients for both cepstrum and energy) of all speech frames uttered by the four training speakers are clustered for each word to produce a codebook representing the parameter sets of the word. The clustering is done using the method proposed in [7]. Since the parameter sets consist of different kinds of features, each feature parameter is multiplied by appropriate weighting factor obtained in [4] during the distance calculation for clustering. Each input utterance is vector-quantized by codebooks of all vocabulary words, and word candidates having relatively small distortion scores averaged over all input frames are selected.

Two methods were investigated for word candidate selection: Method A, in which a fixed number of word candidates having relatively small distortions are selected, and Method B, in which word candidates having distortions smaller than a previously determined threshold are selected. In the latter case, the threshold, θ, is determined based on either of the equations,

Reprinted from *Proc. IEEE Conf. Acoust., Speech, Signal Processing,* vol. 2, pp. 1127–1130, April 1987.

$$\theta = \begin{cases} \theta_0 \quad \text{(fixed)} & \text{[Method B-1]} \\ \theta_0 + \min_n\{d_n\} & \text{[Method B-2],} \end{cases}$$

where d_n is the distortion using the codebook of the n-th word.

3.2. Effectiveness of Feature Parameters in Preprocessing

The first experiment was performed using Method A, under the conditions that the cepstrum (C) and regression coefficients of both cepstrum (C') and energy (E') were all used as feature parameters, and that the codebook size for each word was set at 64. The symbol ' indicates the regression coefficient. Figure 2 shows the preprocessing error rate, that is, the ratio indicating that the true word is excluded from the selected candidates, as a function of the number of frames for regression analysis. The number of frames was varied between 7 and 15 (between 56 and 120 ms). Although the error rate variation as a function of the number of frames is small when the number of frames is less than 13 and the number of candidates is between 2 and 10,

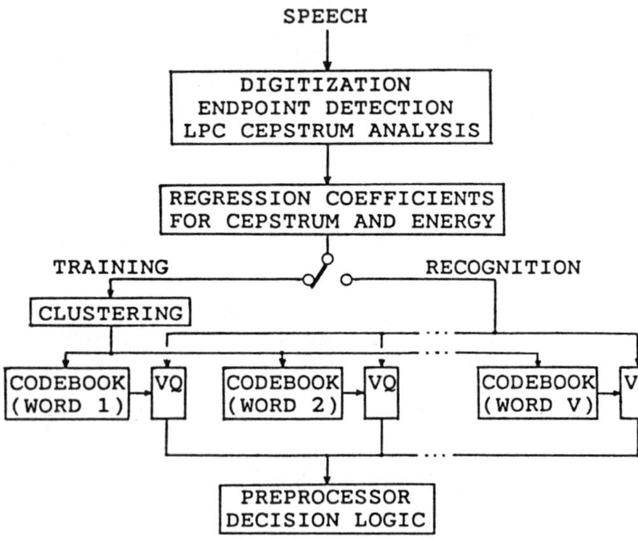

Fig. 1 - Block diagram of a VQ preprocessor based on word-based codebooks.

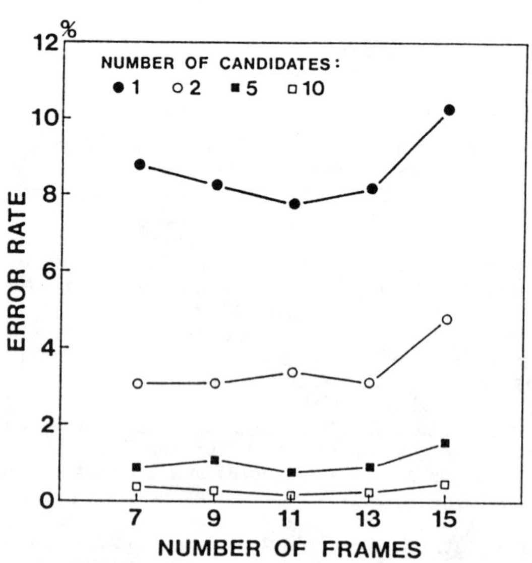

Fig. 2 - Relation between number of frames used for regression analysis and preprocessor error rate.

the error rate has a minimum at 11 frames when only one candidate is selected.

Preprocessing performances for various parameter combinations were compared in the second experiment under the condition that the number of frames was temporally fixed at 11 based on the results of the first experiment. Figure 3 shows the results for five parameter conditions: (C) and (C') only, and (C,E'), (C',E'), and (C,C',E') combinations. The resulting error rate is shown as a function of the number of selected candidates. The error rate derived using the regression coefficients of the cepstrum is 1/3 or less than that using the cepstrum itself. As is clearly indicated, the error rate is largely reduced by combining the regression coefficients of both cepstrum and energy with the cepstrum. When the (C,C',E') combination is used and 10 candidates are selected, the probability that the true word is included in the candidates is 99.8 %.

The experimental results for Method B-2 are plotted in Fig. 4. The results for three parameter conditions, (C), (C,E'), and (C,C',E'), are indicated by the preprocessing error rate and the ratio of the number of selected words to vocabulary size as a function of the threshold, θ_0. This shows that the error rate and the number of candidates can be greatly reduced by combining the regression coefficients with the cepstrum. When $\theta_0 = 0.2$, the probability that the true word is included in the candidates is 99.8 %, and the number of selected words averages 4.5 out of 100. These results are better than those obtained by Method A. Method B-1 was also examined for comparison. The results indicated that the number of selected words is 9.9, while the percentage of true word selection is only 92.4 % under the optimum threshold condition. This means that Method B-2 is far more effective than Method B-1. Accordingly, Methods A and B-1 were removed from the investigation.

The results for Method B-2, when the codebook size is varied between 8, 16, 32, and 64, are shown in Fig. 5. The difference between the results for 32 and 64 is clearly very small. Therefore, all subsequent experiments used codebooks having a size of 32.

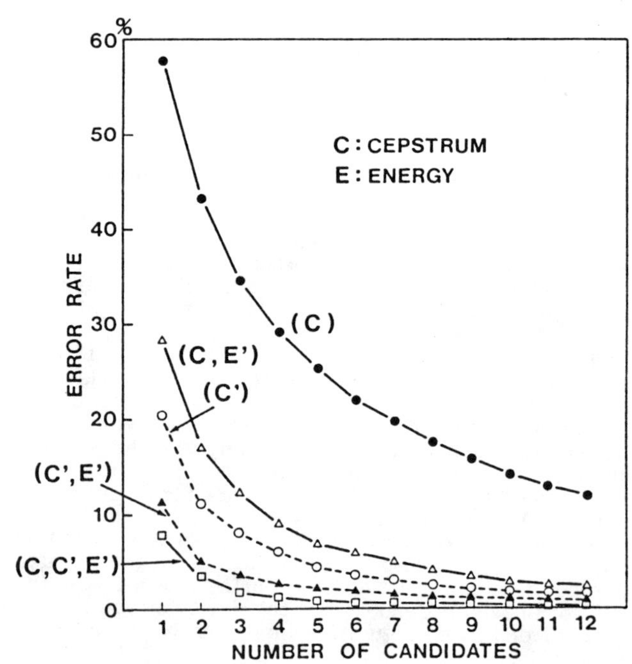

Fig. 3 - Comparison of preprocessor performances for five feature parameter conditions (Method A: Preselection by distance rank order).

IV. PREPROCESSING EXPERIMENT USING UNIVERSAL CODEBOOK

4.1. Method

In the above experiments, codebooks for preprocessing were constructed independently for each word. Although this method can efficiently reduce the number of candidate words, the number of calculations for the total system including preprocessing and postprocessing can be reduced only very slightly. This is because the distance between each input speech frame and each codebook element vector must be calculated for every vocabulary word; hence, the number of distance calculations increases in proportion to the vocabulary size.

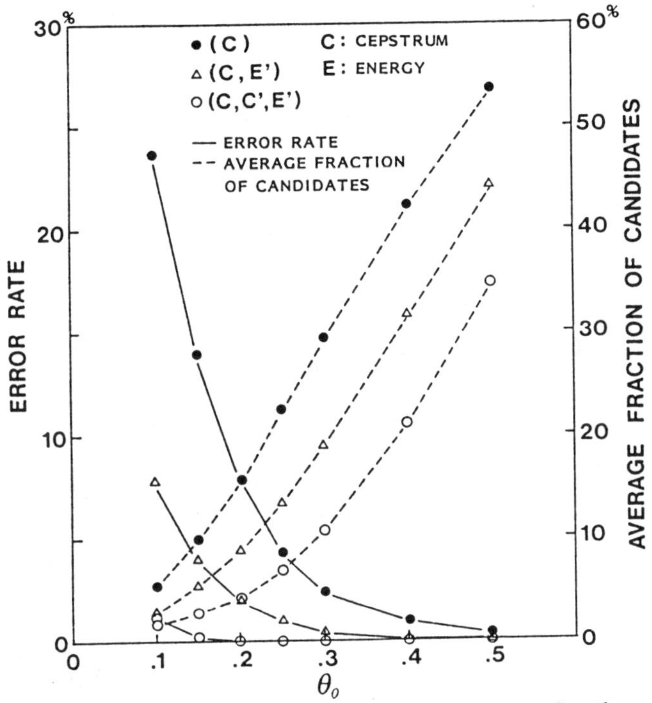

Fig. 4 - Comparison of preprocessor performances for three feature parameter conditions (Method B-2: Preselection by biased distance threshold).

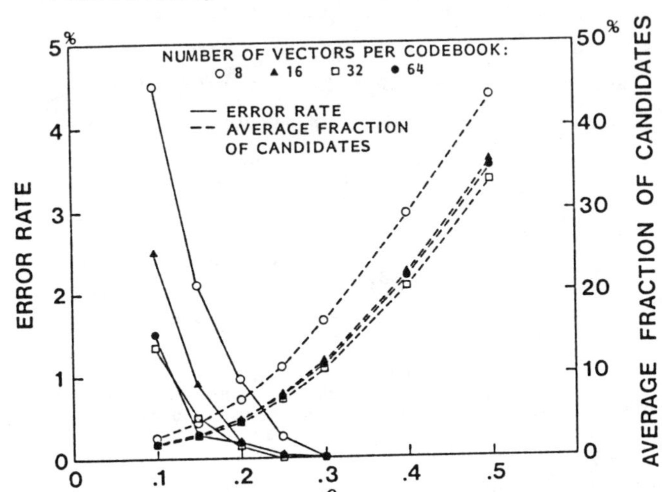

Fig. 5 - Comparison of preprocessor performances for four codebook-size conditions.

In this section, a new method diagrammed in Fig. 6 is investigated, in which a universal codebook is used for all vocabulary words. The universal codebook is constructed by clustering all element vectors of the word-based codebooks, which are in turn constructed for each word as described in the previous section. The codebook of each word for preprocessing is then built by selecting universal codebook elements which closely match the word-based codebook elements. With this method, distances are calculated between each input speech frame and each universal codebook element to construct a distance matrix. Since the codebook for each word is a subset of the universal codebook, the VQ distortion using the codebook of each word can be easily obtained by referring to the distance matrix in the same way as the SPLIT method [8].

The ratio of the number of distance calculations for the present method to that for the previous method is $M / (L \times 100)$, where L is the codebook size for each word, and M is the universal codebook size. For example, when L=32 and M=1024, the ratio is $1024/3200 \simeq 1/3$. Notably, the ratio decreases in inverse proportion to the vocabulary size.

The experimental results given in the previous section indicate that the preprocessing performance variation as a function of the number of frames for regression analysis is small in a range of less than 13 frames. On the other hand, recognition experiments based on the DTW matching method using the combination of cepstrum and regression coefficients for both cepstrum and energy without preprocessing [4] indicate that the optimum length is 7 frames. Based on both sets of results, the number of frames for polynomial expansion was set at 7 in all subsequent experiments.

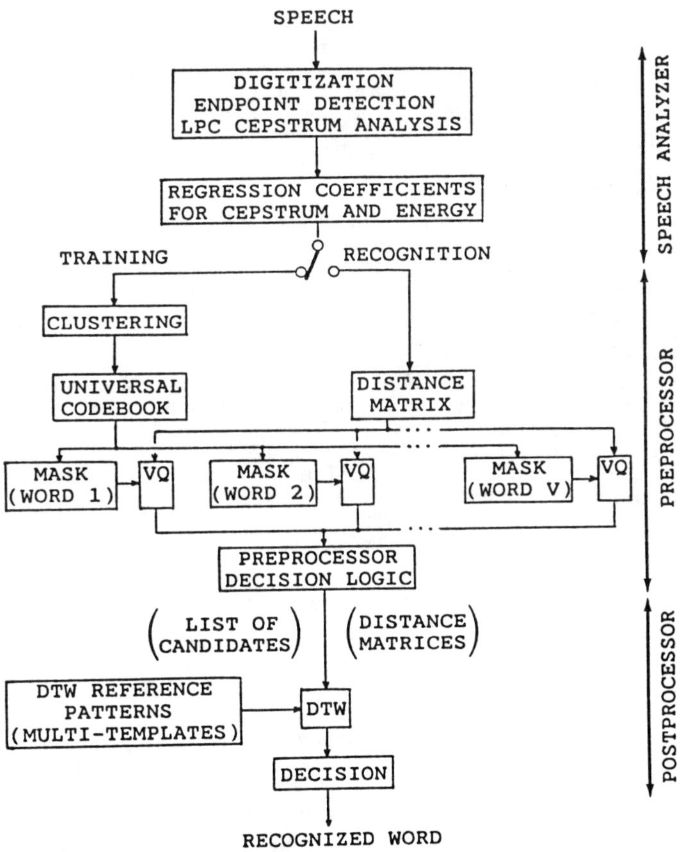

Fig. 6 - Block diagram of isolated word recognizer incorporating a VQ preprocessor based on a universal codebook and a SPLIT postprocessor.

4.2. Experimental Results

Figure 7 shows the results of preprocessing using the universal codebook as well as those using the word-based codebooks. The effect of the universal codebook size was investigated by varying it between 512 and 1024. When the universal codebook size is 1024, a preprocessing performance almost the same as that using word-based codebooks can be obtained. When $\theta_0 = 0.2$, the number of candidate words is 6.8 on an average, and the true word inclusion percentage is 99.9 %. A precise analysis of the experimental results indicates that for a quarter of the input utterances (496 samples), the preprocessor eliminates all word candidates except one. For such cases, there is clearly no further processing required for word recognition.

V. COMBINATION OF PREPROCESSING AND DTW

5.1. Experimental Results

Overall recognition performance was compared between the method based on the combination of preprocessing and DTW processing, and the DTW-only method without preprocessing. DTW was performed based on the SPLIT method using a 1024-element universal codebook. Four reference templates which correspond to four training speakers were represented by the time sequences of universal codebook elements for each word. A distance matrix between input speech and universal codebook elements was used for both preprocessing and DTW. The word whose DTW distance was smaller than any other word was selected as the final decision.

Table 1 shows the recognition results for various experimental conditions. When a universal codebook and Method B-2 are used and $\theta_0=0.2$, DTW is performed for only 6.8 % of the vocabulary words, as described in the previous section. Then, a recognition error rate of 2.0 %, which is the same as that obtained without preprocessing, can be finally achieved. The total amount of calculation necessary in this condition is almost 1/10 of that without preprocessing.

5.2. Discussion of VQ Method

In the above-mentioned experiments, three kinds of feature parameters (C,C',E') were clustered at once as elements of a single vector. On the other hand, in paper [6], cepstrum (C) and regression coefficients (C') were clustered separately and the VQ distortions associated with each parameter set were combined by a weighted summation during speaker verification experiments. An additional experiment was performed to compare the performances achieved by these two methods. The results indicated that the error rate for the separate clustering method is twice as large as the method proposed in this paper. This means that instantaneous and dynamic features convey phonetic information through mutual interaction.

VI. CONCLUSION

A new VQ-based preprocessor for selecting candidate words has been proposed for large vocabulary speaker-independent word recognition. Each word-specific codebook is built as a subset of a universal codebook. A feature parameter vector consists of instantaneous and transitional parameters for both cepstrum coefficients and logarithmic energy. The input utterance is passed through VQ processors using the codebook of each word, and the words whose VQ distortion scores are relatively small are selected as candidate words. A DTW processor based on the SPLIT method then resolves the choice among the word candidates.

Experiments confirmed two principal results. First, the number of candidate words for DTW processing is reduced to 1/15 of the vocabulary while maintaining the same recognition accuracy (98.0 %) as that obtained without the preprocessor. Second, the number of calculations is reduced to almost 1/10 of that of the method without preprocessing. Further investigations will include evaluation experiments using a large vocabulary database.

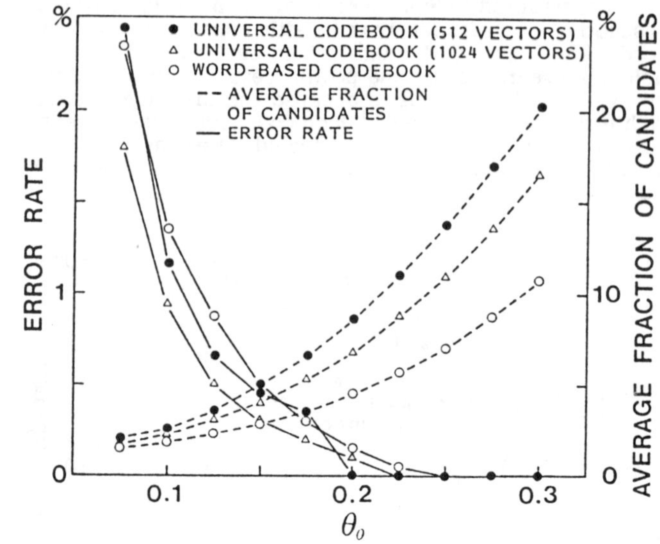

Fig. 7 - Comparison of preprocessor performances for three codebook conditions.

Table 1 - Experimental results for the word recognizer incorporating a VQ preprocessor and a DTW-based postprocessor.

		ERROR RATE (%) PRE-SELECTION ALONE	ERROR RATE (%) PRE-SELECTION & DTW	NUMBER OF CANDI-DATES FOR DTW
DTW ALONE		-	2.0	100
WORD-BASED CODEBOOK	PRESELECTION BY BEST ONE	7.9	-	-
	TOP TEN	0.2	2.1	10
	<θ (θ_0=.2)	0.2	2.1	4.5
UNIVER-SAL CODEBOOK	PRESELECTION BY BEST ONE	10.9	-	-
	TOP TEN	0.4	2.3	10
	<θ (θ_0=.2)	0.1	2.0	6.8

REFERENCES

[1] K-C.Pan,F.K.Soong and L.R.Rabiner, "A vector-quantization-based preprocessor for speaker-independent isolated word recognition," IEEE Trans. Acoust.,Speech,Signal Processing, ASSP-33,3, pp.546-560 (1985)

[2] A.F.Bergh,F.K.Soong and L.R.Rabiner, "Incorporation of temporal structure into a vector-quantization-based preprocessor for speaker-independent,isolated-word recognition," AT&T Tech.J., 64,5,pp.1047-1063 (1985)

[3] S.Furui, "Cepstral analysis technique for automatic speaker verification," IEEE Trans. Acoust.,Speech,Signal Processing, ASSP-29,2,pp.254-272 (1981)

[4] S.Furui, "Speaker-independent isolated word recognition using dynamic features of speech spectrum," IEEE Trans. Acoust., Speech,Signal Processing,ASSP-34,1,pp.52-59 (1986)

[5] S.Furui, "Speaker-independent isolated word recognition based on emphasized spectral dynamics," Proc.IEEE Int.Conf. Acoust., Speech,Signal Processing,Tokyo,37.10,pp.1991-1994 (1986)

[6] F.K.Soong and A.E.Rosenberg, "On the use of instantaneous and transitional spectral information in speaker recognition," Proc. IEEE Int.Conf.Acoust.,Speech,Signal Processing,Tokyo,17.5, pp.877-880 (1986)

[7] Y.Linde,A.Buzo and R.Gray, "An algorithm for vector quantizer design," IEEE Trans.Commun.,COM-28,1,pp.84-95 (1980)

[8] N.Sugamura,K.Shikano and S.Furui, "Isolated word recognition using phoneme-like templates," Proc.IEEE Int.Conf. Acoust.,Speech, Signal Processing,Boston,16.3,pp.723-726 (1983)

A Vector Quantization Approach to Speaker Recognition

F. K. Soong
A. E. Rosenberg
L. R. Rabiner
B. H. Juang

AT&T Bell Laboratories
Murray Hill, New Jersey 07974

ABSTRACT. In this study a vector quantization (VQ) codebook was used as an efficient means of characterizing the short-time spectral features of a speaker. A set of such codebooks were then used to recognize the identity of an unknown speaker from his/her unlabelled spoken utterances based on a minimum distance (distortion) classification rule. A series of speaker recognition experiments was performed using a 100-talker (50 male and 50 female) telephone recording database consisting of isolated digit utterances. For ten random but different isolated digits, over 98% speaker identification accuracy was achieved. The effects, on performance, of different system parameters such as codebook sizes, the number of test digits, phonetic richness of the text, and difference in recording sessions were also studied in detail.

I. Introduction

Automatic speaker recognition has long been an interesting and challenging problem to speech researchers [1-10]. The problem, depending upon the nature of the final task, can be classified into two different categories: speaker verification and speaker identification. In a speaker verification task, the recognizer is asked to verify an identity claim made by an unknown speaker and a decision to reject or accept the identity claim is made. In a speaker identification task the recognizer is asked to decide which out of a population of N speakers is best classified as the unknown speaker. The decision may include a choice of "none of the above" (i.e., a choice that the specific speaker is not in a given closed set of speakers). The input speech material used for speaker recognition can be either text-dependent (text-constrained) or text-independent (text-free). In the text-dependent mode the speaker is asked to utter a prescribed text. The utterance is then compared with some prestored pattern(s) of the same text. In the text-independent mode the speaker can, in theory, speak any speech material with no constraint. The unconstrained speech input is compared with model(s) which characterizes the speaker's features. In general, due to higher acoustic-phonetic variability of the text-independent input, longer training material is required to characterize (model) a speaker more reliably than in the text dependent mode. On the other hand, a fairly short, usually sentence long, utterance is adequate for text-dependent speaker recognition. In general, recognition procedures are quite different for text dependent and text independent systems. In the text dependent mode, samples of the same speech events in reference and test utterances are compared, usually by establishing a non-linear time alignment between the utterances. There is no possibility of time alignment in true text-independent systems. Talker characterization is carried out either by statistical averaging over selected acoustic features, or by locating comparable speech events in test and reference utterances.

In this paper we propose a new approach to the speaker recognition problem. The approach is partially motivated by the success of two word-based VQ speech recognition systems. In one, Pan, Soong and Rabiner [11] used word-based VQ codebooks in an isolated word recognition preprocessor to substantially alleviate the computational complexity of a dynamic programming based speech recognition system. In the other, Shore and Burton [12] used word-based VQ codebooks and reported good performance in speaker-trained isolated-word recognition experiments. Here, instead of using word-based VQ codebooks to characterize the phonetic contents of isolated words, we propose to use speaker-based VQ codebooks to characterize the variability of short-time acoustic features of speakers.

II. Speaker-based VQ Codebook Approach to Speaker Characterization and Recognition

A set of short-time raw feature vectors of a speaker can be used directly to represent the essential acoustical, phonological or physiological characteristics of that speaker if the training set includes sufficient variations. However such a direct representation is not practical when the number of training vectors is large. The memory requirements for storage and computational complexity in the recognition phase eventually become prohibitively high. Therefore an efficient way of compressing the training data had to be found. In order to compress the original data to a small set of representative points, we used a VQ codebook with a small number of codebook entries.

The speaker-based VQ codebook generation can be summarized as follows: Given a set of I training feature vectors, $\{\mathbf{a}_1, \mathbf{a}_2, \ldots, \mathbf{a}_I\}$ characterizing the variability of a speaker, we want to find a partitioning of the feature vector space, $\{S_1, S_2, \ldots, S_M\}$, for that particular speaker where, S, the whole feature space is represented as $S = S_1 \cup S_2 \cup \cdots \cup S_M$. Each partition, S_i, forms a convex, nonoverlapping region and every vector inside S_i is represented by the corresponding centroid vector, $\mathbf{b}_i$, of S_i. The partitioning is done in such a way that the average distortion

$$D = \frac{1}{I} \sum_{i=1}^{I} \min_{1 \leq j \leq M} d(\mathbf{a}_i, \mathbf{b}_j) \tag{3}$$

is minimized over the whole training set. The distortion (distance) between the vectors $\mathbf{a}_i$ and $\mathbf{b}_j$ is denoted as $d(\mathbf{a}_i, \mathbf{b}_j)$.

In this study we use short-time LPC vectors as feature vectors. The corresponding distortion measure to measure the similarity between any two feature vectors is the LPC likelihood ratio distortion measure. The likelihood ratio distortion between two LPC vectors $\mathbf{a}$ and $\mathbf{b}$ is defined as

$$d_{LR}(\mathbf{a}, \mathbf{b}) = \frac{\mathbf{b}^T \mathbf{R}_a \mathbf{b}}{\mathbf{a}^T \mathbf{R}_a \mathbf{a}} - 1 \tag{4}$$

where $\mathbf{R}_a$ is the autocorrelation matrix of speech input data associated with the vector $\mathbf{a}$. Using this distortion measure, and the VQ codebook training algorithm proposed by Linde, Buzo and Gray [13], we generated speaker-based VQ codebooks of different sizes.

The speaker identification system based on this VQ codebook approach is depicted in Fig. 1. The input speech signal is sampled, endpointed and LPC analyzed giving the sequence of vectors $\mathbf{a}_1, \mathbf{a}_2, \ldots, \mathbf{a}_L$. The resultant LPC vectors are vector quantized (encoded) using the N codebooks corresponding to the N different speakers. The quantization

Reprinted from *Proc. IEEE Conf. Acoust., Speech, Signal Processing,* vol. 1, pp. 387–390, March 1985.

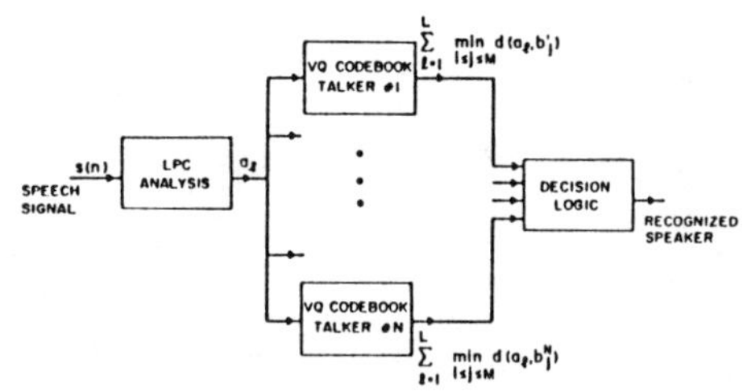

Fig. 1 Block diagram of the speaker-based VQ speaker recognition system.

errors (or distortions) with respect to each codebook are individually accumulated across the whole test token. The average distortion with respect to the ith codebook (speaker) is

$$D^i = \frac{1}{L} \sum_{\ell=1}^{L} \min_{1 \le j \le M} d(\mathbf{a}_\ell, \mathbf{b}_j^i) \tag{5}$$

The N resultant average distortions are compared to find the mimimum. The final speaker recognition decision is given by

$$i^* = \underset{1 \le i \le N}{\operatorname{argmin}} D^i \tag{6}$$

A speaker-verification system has a similar structure except only the codebook of the claimed identity is used and the resultant average distortion is compared with a preset threshold to reject or to accept the identity claim made by the unknown talker.

III. Data Base, LPC Analysis and Experiments

The proposed speaker-based VQ approach to speaker characterization is applicable to both text-dependent and text-independent speaker recognition. Since collecting a truly linguistically unconstrained database of many speakers is not a trivial task, we decided to use a much more constrained database as a first step to test the idea. Over a period of 2 months, a 100-talker (50 male and 50 female), digit vocabulary database was collected. Each of 100 talkers spoke 200 isolated digits (20 utterances per digit) over standard, local, dialed-up telephone lines in 5 different recording sessions. In each recording session the talker was asked to speak 4 sets of digit strings. Each string consisted of 10 randomly ordered, different isolated digits. The 200 isolated digits were split into two parts. The first one hundred digits were used for training (codebook generation) while the second one hundred digits were used for testing (recognition). The recognition experiments were performed on a digit-independent basis. In other words the speaker was allowed to say any random, isolated digit strings.

The analog speech input samples were first bandlimited from 200 Hz to 3200 Hz and then sampled at a 6.67 kHz sampling rate. The speech samples were pre-emphasized by a first order filter whose transfer function was $H(z) = 1 - 0.95z^{-1}$. A 45 ms Hamming window was used to window the preemphasized speech data and an 8th order autocorrelation analysis was performed. The resultant 9 autocorrelation coefficients were then used to find the LPC feature vector which represented the short-time LPC spectral information of the corresponding frame. The LPC analysis was performed every 15 ms (i.e., 30 ms overlapping between adjacent frames was used).

The following experiments were performed to evaluate the effects of different speaker recognition parameters on the recognition performance. In particular we varied:

(1) The number of digits in the test utterance.
Test tokens of one, two, four and ten random but different digits were used in the 100 talker identification experiments.

(2) The size of the codebook.
Six codebooks with 2, 4, 8, 16, 32 and 64 codebook vectors were used.

(3) The digits used in the test utterances.
Sets of 10 repeated digits were used as test tokens. The results were compared with the results obtained by using 10 random but different digits as test tokens.

(4) The frames used in the recognizer.
Voiced frames were extracted from the test utterance by using a simple but effective voiced/unvoiced detector. The voiced/unvoiced decision was made based on the frame energy and prediction gain. The results were compared with the results obtained when all speech frames were used.

(5) The time span between the training and testing material.
The 100 test utterances were divided into 3 different groups according to the recording session sequence. This experiment was designed to study whether the intra-speaker variations and possible channel variations of different recording sessions affected the recognizer performance.

IV. Results and Discussion

4.1 Effects of Codebook Size and Number of Digits in the Test Token

The speaker identification error percentage is plotted as a function of codebook size in Fig. 2. The codebook size is represented by the

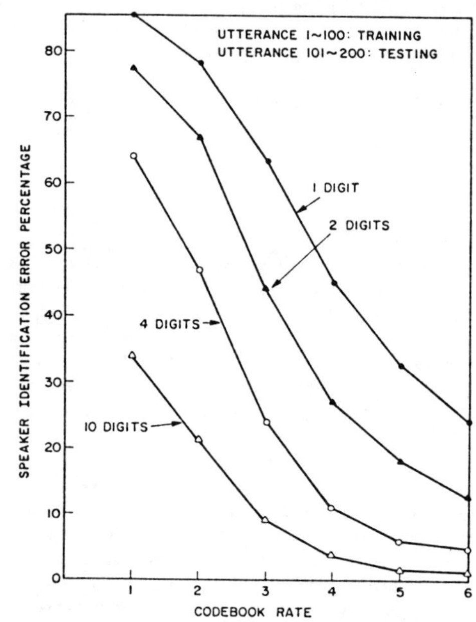

Fig. 2 Identification error percentage versus codebook rate for test sequences of 1, 2, 4, and 10 digits.

number of vector entries, M or by the corresponding rate, $M = 2^R$. Four curves corresponding to 4 different test token lengths (i.e., 1 digit, 2 digits, 4 digits and 10 digits) are shown in the figure. The identification error rate decreases when either the codebook size or the test token length increases. For a test token of ten isolated digits, the error rate dropped from 34% to 1.5% when the codebook size was increased from 2 to 64 vectors. With a codebook of 64 vectors the error rate dropped from 24% to 1.5% when the test token length was increased from 1 to 10 digits.

To further illustrate the effects of codebook size on the recognition error rate Fig. 3 shows the averages, standard deviations and histograms, respectively, of inter-speaker and intra-speaker distortions of the 100 speaker identifications. Results using two different codebook (i.e. 8 vectors (rate 3) and 32 vectors (rate 5)) are given in this figure. The average distortions of both codebooks are fairly

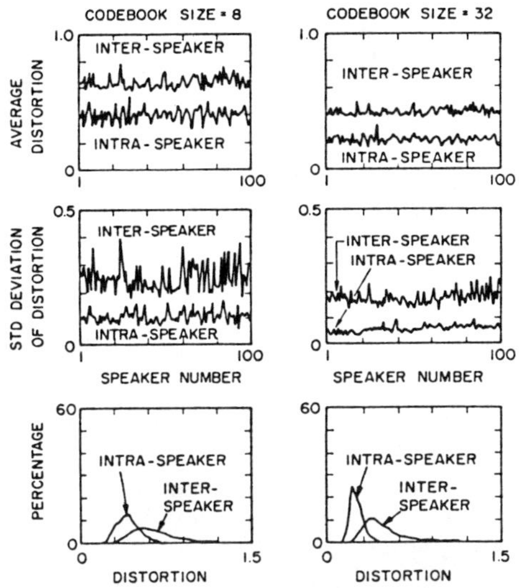

Fig. 3 The effects of different codebook sizes.

constant across all 100 talkers and only a small amount of random fluctuations are observed. It is seen that the separations between the inter- and the intra-speaker average distortions are roughly the same for the two different size codebooks, or, equivalently when the codebook sizes increase, both the inter- and intra-speaker average distortions decrease by more or less the same amount. Also we observe that the standard deviations of both the inter- and intra-speaker distortions are reduced when the codebook size is increased from 8 to 32. The inter- and the intra-speaker distortion histograms clearly show that when the separation between the inter- and intra-speaker average distortion are held relatively fixed, reduction of the standard deviation decreases the overlapping area between the two histograms. As a result of this reduction of overlapping area, the recognition error rate is reduced.

In Fig. 4 the averages, standard deviations and histograms of inter- and intra-speaker distortions of three different token lengths (1 digit, 4 digit and 10 digits) are shown. A codebook of 32 vectors was used for all cases. The average distortions are identical for all three token lengths, but the standard deviations of the distortions are significantly different. The intra-speaker standard deviation decreases much faster than the inter-speaker standard deviation as the number of digits in the test increases. The overlapping region between the two histograms in Fig. 4(c) is smaller for longer test tokens.

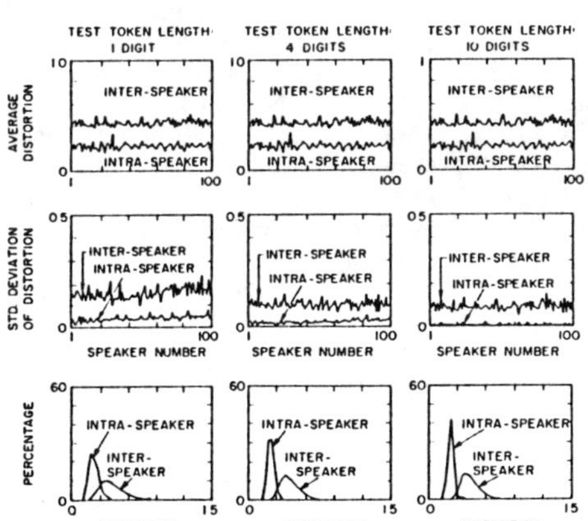

Fig. 4 The effects of different token lengths.

4.2 Effects of Repeated Digits and Different Digits

The speaker identification results obtained using 10 repeated digits are shown in Fig. 5. The digit "9" achieved the best results while the digits "3", "4" and "6", provided the worst scores. The good result obtained for digit "9" may be explained by the fact that it has a relatively long duration and it has a strong nasal-vowel coarticulation. The nasal-vowel coarticulation has long been known to be effective for speaker recognition purposes [14]. Since it is a relatively rapid event, the nasal-vowel coarticulation is unlikely to be modified consciously by the speaker. Therefore it serves as a reliable talker identification feature with inherently low intra-speaker variability.

The digits "6" and "8", each having a stop gap with silence frames, do not achieve good recognition scores. This should not be surprising because the short-time spectra in the stop gap carry no speaker related information. The short-time spectrum of the silence (or the background noise) is matched by any codebook with a large variance. As a consequence the average distortion of the true talker is corrupted by a random distortion making a misrecognition more likely.

It is interesting to note that the performance of 10 different digits, shown as the dashed line in Fig. 5, is better than the performance of

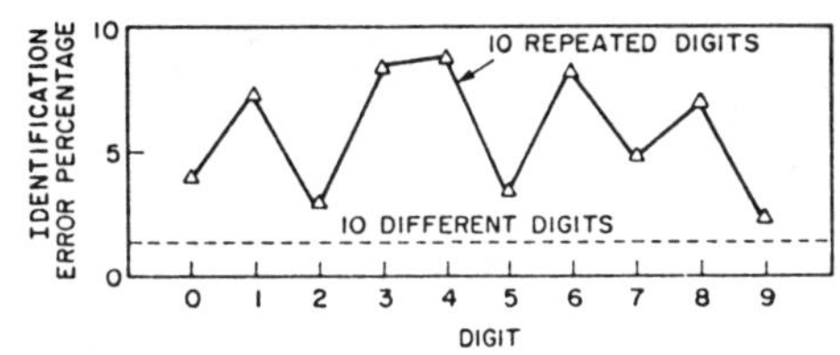

Fig. 5 Identification error percentage versus digit for repeated digits and different digits.

any 10 repeated digits. This result indicates that statistically uncorrelated information could, and did, improve the speaker recognition performance.

4.3 Effects of Using Only Voiced Speech

The results obtained using only frames of voiced speech are presented along with the results obtained using all speech frames in Fig. 6. In this figure, the speaker identification error rate obtained using only voiced frames is consistently worse than the error rate obtained using

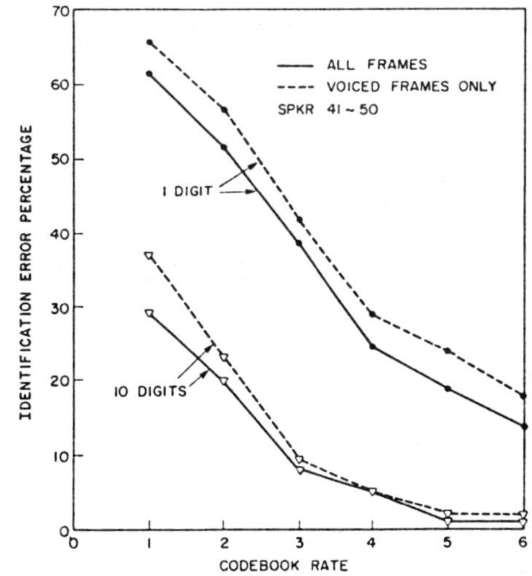

Fig. 6 Identification error percentage versus codebook rate for 1 and 10 digit test utterances using either voiced frames or all frames.

all speech frames for different codebook sizes and test token lengths, although for 10 digit test utterances the difference is small. These results are somewhat surprising because it is commonly believed that the voiced frames are more effective than the unvoiced frames in characterizing the acoustical features of a speaker. Also, the likelihood ratio distortion used in our experiments is most effective in measuring the spectral distortion (similarity) between voiced speech frames. The results can be explained as follows. In the nonparametric VQ codebook approach, all of the feature vectors (both voiced and unvoiced) are well represented. Since, in the training phase, we do not deliberately remove the unvoiced frames, we should use them in the recognition phase, too. In this new nonparametric VQ approach we not only eliminate the need to separate voiced frames from the input data, but also we improve the speaker recognition performance by using all the speech data.

4.4 Effects of Different Recording Sessions

The identification error rate plotted as a function of the recording session number is shown in Fig. 7. The codebook was generated from the first 100 utterances, or equivalently, utterances recorded in the 1st, the 2nd and the first half of the 3rd session. The remaining 100 utterances were grouped according to their corresponding recording session numbers to form three different test sets. The first test set gave a significantly better result than the other two sets. This can be

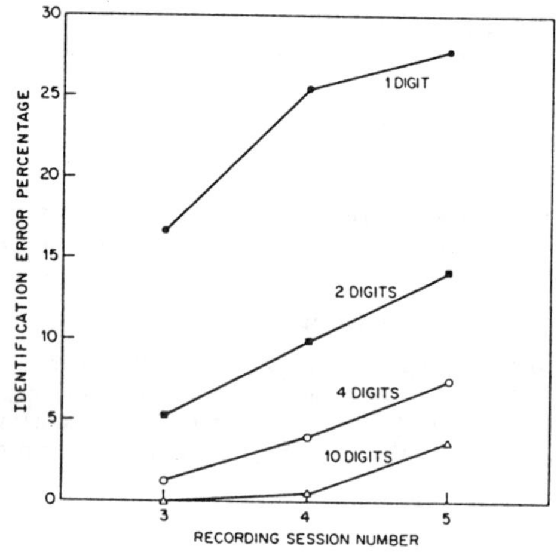

Fig. 7 Identification error percentage versus recording session number.

explained as follows. The forty utterances in the third recording session formed a rather homogeneous data set. Since the first twenty digits of this homogeneous set were used as part of the VQ codebook training set, the remaining twenty digits should be well represented by the resultant codebook. In the recognition phase when the remaining 3rd session utterances were used, the effects of intra-speaker variation and channel differences were negligibly small and the distortion (quantization error) was as low as the training data. On the other hand, the digit utterances recorded in the 4th and the 5th sessions were less correlated with the training utterances than the 3rd session, and the channel difference and the intra-speaker variations degraded the recognizer performance. Since the performance of the 5th recording utterances was even worse than the 4th recording utterances, we observe that the longer the separation between the training and the test recordings, the worse the performance. This result clearly indicates the need to update the VQ codebook from time to time.

V. Summary and Discussion

We have proposed a speaker-based VQ codebook approach to speaker recognition application. The VQ codebook was used as an efficient means to characterize a speaker's feature space and was employed as a minimum distance classifier in the proposed speaker recognition system. The results of the 100 talker speaker recognition experiments are good; given a 10 digit long test token and a codebook of 64 vectors a recognition rate of over 98% was achieved. We summarize our results as follows:

(1) Both larger codebook size and longer test token length (more digits in the test utterance) can be used to improve the recognition performance. Ten different digits, when used as test tokens, outperformed any of the repeated digit test tokens. The digit "9" when used as a test token outperformed other digits.

(2) Phonetically rich test tokens give better performance than phonetically poor test tokens.

(3) It is recommended that all speech frames be used for both training and recognition in our approach.

(4) It is recommended that the VQ codebook be updated from time to time to alleviate the performance degradation due to different recording conditions and intra-speaker variations.

The proposed approach, although tested only on a highly linguistically constrained database (i.e., digits), is believed to be extensible to truly text-independent speaker recognition environment.

REFERENCES

[1] Pruzansky, S., "Pattern-Matching Procedure for Automatic Talker Recognition," *JASA*, Vol. 35, pp. 354-358, March 1963.

[2] Li, K. P., Dammann, J. E. and Chapman, W. D., "Experimental Studies in Speaker Verification Using an Adaptive System," *JASA*, Vol. 40, pp. 966-978, 1966.

[3] Pruzansky, S. and Mathews, M. V., "Talker-Recognition Procedure Based on Analysis of Variance," *JASA*, Vol. 36, pp. 2041-2047, 1964.

[4] Atal, B., "Automatic Recognition of Speakers from their Voices," *Proc. IEEE*, Vol. 64, pp. 460-475, April 1976.

[5] Markel, J. D., Oshika, B. and Gray, A. H., "Long-Term Feature Averaging for Speaker Recognition," *IEEE Trans. Acoust., Speech and Signal Processing*, Vol. ASSP-25, pp. 330-337, August 1977.

[6] Markel, J. D. and Davis, S. B., "Text-Independent Speaker Recognition from a Large Linguistically Unconstrained Time-spaced Data Base," *IEEE Trans. Acoust., Speech and Signal Processing*, ASSP-27, No. 1, pp. 74-82, February 1979.

[7] Rosenberg, A. E. and Sambur, M. R., "New Techniques for Automatic Speaker-Verification," *IEEE Trans. Acoust., Speech and Signal Processing*, Vol. ASSP-23, pp. 169-176, 1975.

[8] Furui, S., "Cepstrum Analysis Technique for Automatic Speaker Verification," *IEEE Trans. on Acoust., Speech and Signal Processing*, ASSP-29, No. 2, pp. 254-272, April 1981.

[9] Li, K. P. and Wrench, E. H., "An Approach to Text-Independent Speaker Recognition with Short Utterances," *Proc. ICASSP*, pp. 555-558, 1983.

[10] Schwartz, R., Roucos, R. and Berouti, M., "The Application of Probability Density Estimation to Text-Independent Speaker Identification," *Proc. ICASSP*, pp. 1649-1652, 1982.

[11] Pan, K. C., Soong, F. K. and Rabiner, L. R., "A Vector Quantization Based Preprocessor for Speaker-Independent Isolated Word Recognition," submitted for publication.

[12] Shore, J. E. and Burton, D. K., "Discrete Utterance Speech Recognition Without Time Alignment," *IEEE Trans. on Inform. Theory*, Vol. IT-24, No. 4, pp. 473-491, July 1983.

[13] Linde, Y., Buzo, A. and Gray, R. M., "An Algorithm for Vector Quantization," *IEEE Trans. on Communications*, Vol. COM-28, No. 1, pp. 84-95, January 1980.

[14] Su, L-S, Li, K. P. and Fu, K. S., "Identification of Speakers by Use of Nasal Coarticulation," *JASA*, Vol. 56, pp. 1876-1882, December 1974.

Text-Dependent Speaker Verification Using Vector Quantization Source Coding

DAVID K. BURTON, MEMBER, IEEE

Abstract—Several vector quantization approaches to the problem of text-dependent speaker verification are described. In each of these approaches, a source codebook is designed to represent a particular speaker saying a particular utterance. Later, this same utterance is spoken by a speaker to be verified and is encoded in the source codebook representing the speaker whose identity was claimed. The speaker is accepted if the verification utterance's quantization distortion is less than a prespecified speaker-specific threshold. The best approach achieved a 0.7 percent false acceptance rate and a 0.6 percent false rejection rate on a speaker population comprising 16 admissible speakers and 111 casual imposters. The approaches are described, and detailed experimental results are presented and discussed.

I. Introduction

Speaker verification by machine consists of automatically authenticating the identity claimed by a speaker, given only samples of the speaker's voice. It has been an area of active study for more than twenty years resulting in two categories of approaches. In one, verification decisions are based on speech that is selected by the speaker and not known ahead of time by the verification system; this is called *text-independent* verification. In the other category, the verification system is trained on *a particular utterance.* This *same utterance* is later spoken by the individual in question and a verification decision is made based on this utterance; this is called *text-dependent* speaker verification. In this paper, we describe and evaluate several new approaches to the *text-dependent* speaker verification problem.

A typical approach to text-dependent speaker verification consists of selecting parameters that can be derived from the speech waveform and then representing each speaker by a time-series of these parameters (called a reference template) obtained from a particular utterance. The parameters are normally chosen with the hope that they reflect speaker-specific, organic differences in the structure of the vocal apparatus or, perhaps, that the time series of parameters will reflect learned differences in the use of the vocal apparatus to produce a particular utterance. After obtaining a reference for each speaker to be verified, an unknown speaker claims an identity and speaks the appropriate utterance. This utterance is analyzed, and a time-series of parameters is obtained. The unknown speaker's parameters are then aligned in time with the reference stored for the speaker whose identity was claimed and the decision to accept or reject the speaker is based on a measure of the similarity between the two time-series of parameters. Examples of parameters that have been used in this way are pitch [1], short-time energy [1], short-time spectra [2], and linear predictive coding (LPC) coefficients or parameters derived from these coefficients [3].

In addition to the template matching approach described above, statistical methods are sometimes used [4], [5]. These methods use large amounts of training data to estimate the underlying probability densities of parameters that are chosen to represent a speaker. Once the probability densities are specified, statistical decision theory methods [6] are used to verify a speaker.

We approach the text-dependent speaker verification problem from a different viewpoint. We consider a speaker of a particular utterance as an information source, and we model this information source by using a standard information-theoretic source coding method called *vector quantization* (VQ). VQ is a source coding technique [7] that has been used successfully in both speech coding [8] and speech recognition [9]–[11]. In VQ, each source vector is coded as one of a prestored *set of codewords*, called a *codebook*, by finding the codeword that minimizes the distortion between itself and the source vector. For speech, a VQ codebook is designed from a training sequence containing typical speech [12]. The training sequence is divided into frames (typically 20 ms), linear predictive analysis is done on each frame, and a clustering algorithm is used on this set of LPC coefficient vectors to obtain a codebook of representative vectors or codewords. The codebook is designed to minimize the average quantization distortion between itself and the training sequence.

To use VQ source coding in speaker verification, we represent each speaker by a VQ codebook designed from a training sequence composed of repetitions of a particular utterance. Later, this same utterance is spoken by an unknown speaker with a claimed identity. This test utterance is coded in the codebook representing the speaker whose identity was claimed, and the resulting quantiza-

Manuscript received May 16, 1985; revised June 24, 1986. This work was done while the author was employed at the Naval Research Laboratory, Washington, DC.

The author is with Entropic Speech Incorporated, Washington Research Laboratory, Washington, DC 20003, a division of Entropic Processing, Inc., Cupertino, CA 95014.

IEEE Log Number 8610885.

Reprinted from *IEEE Trans. Acoust., Speech, Signal Processing,* vol. ASSP-35, pp. 133–143, Feb. 1987.

tion distortion is compared to a threshold. If the distortion is less than the threshold, the speaker is accepted.

In addition to our source coding point of view, our speaker verification approach is quite different from other approaches in several ways. No attempt is made to align-in-time a test sequence with a stored reference sequence (indeed, no reference sequence exists), and no explicit estimate is made of the underlying probability density function of the parameters chosen to represent the speaker. Our verification procedures are, however, closely related to optimal information-theoretic methods of classification that use the *information dissimilarity* between two vectors as a discrimination measure [13], [14]. Preliminary results were reported in [15].

We previously used VQ source coding in isolated word recognition [9], [16], [17]. The methods used in those approaches to represent a word (design a codebook) and to compare an unknown input word to the stored codebooks (classify an input utterance) are the same as the ones described in this paper to represent and verify a speaker. The differences are in the application of the ideas and in the use of thresholds to make decisions.

In related work, Li and Wrench represented speakers with vectors from a general-speech VQ codebook in a text-independent speaker recognition scheme [18], and much more recently and akin in spirit to our approaches, Soong *et al.* [19] and Shikano [20] reported on text-independent speaker recognition approaches. In Soong's approach, each speaker is represented by a VQ codebook designed to represent that speaker saying the names of the 10 decimal digits. An unknown speaker then says any of the 10 digits, and the average quantization distortion resulting from encoding the spoken digits is used as a discrimination measure to identify the speaker. In Shikano's approach, each speaker is represented by a VQ codebook designed from unconstrained continuous speech, and different speech is used for recognition. Reported results for both methods are quite good.

The rest of this paper is organized as follows. Section II describes three ways to represent a speaker by using VQ source coding. Section III explains our speaker verification approach. Section IV presents experimental results, and Section V concludes with a summary and general discussion.

II. Background

In this section we briefly describe three ways to design a source model of a speaker; for detailed descriptions of the methods, see [9], [16], and [17]. Following these descriptions is a list of the distortion measures and LPC parameters used in the speaker verification experiments.

First we establish some notation and define some terms. Upper and lower case Roman and italic letters (e.g., n, N, q, Q) denote scalars; lower case italic letters with bars (e.g., $\bar{c}$) denote vectors; upper case italic letters with bars denote sets of vectors (e.g., $\bar{C} = \{\bar{c}_1, \bar{c}_2, \cdots, \bar{c}_N\}$); bold lower case Roman letters denote sequences of vectors (e.g., $\mathbf{c} = [\bar{c}_i; i = 1, \cdots, K]$); and bold upper case Roman letters denote sets of vector sequences (e.g., $\mathbf{C} = \{\mathbf{c}_1, \mathbf{c}_2, \cdots, \mathbf{c}_N\}$). In what follows, a set means an unordered collection of elements; a sequence means a time-ordered collection of elements.

Throughout, all vectors consist of LPC coefficients and a gain term. $\mathbf{t} = [\bar{t}_1, \bar{t}_2, \cdots, \bar{t}_P]$ is a P-vector training sequence obtained from M catenated repetitions of an utterance by a speaker. $\mathbf{v} = [\bar{v}_1, \bar{v}_2, \cdots, \bar{v}_L]$ is an L-vector test sequence corresponding to an utterance obtained from a speaker for verification purposes. $\bar{C}$ represents a VQ codebook, and finally, $\mathbf{C}$ represents a matrix quantization codebook.

A. Single Section Vector Quantization

For speaker verification, a single-section VQ codebook $\bar{C}$ is designed to minimize the average distortion that results from encoding a training sequence $\mathbf{t}$

$$\sum_{p=1}^{P} d(\bar{t}_p, \bar{c}_B), \tag{1}$$

where $\bar{c}_B$ is the codeword that results from encoding speech segment $\bar{t}_p$,

$$d(\bar{t}_p, \bar{c}_B) = \min_i d(\bar{t}_p, \bar{c}_i),$$

and d is an appropriate vector distortion measure. This codebook represents a speaker saying a particular word.

The average quantization distortion D_{avg} that results from coding a verification utterance $\mathbf{v}$ in codebook $\bar{C}$ is

$$D_{avg} = \frac{1}{L} \sum_{l=1}^{L} d(\bar{v}_l, \bar{c}_B). \tag{2}$$

We use this average quantization distortion in making the verification decision.

This approach is called *single section* to distinguish it from the approach described in the next section, in which each speaker is represented by a sequence of single section codebooks.

B. Multisection Vector Quantization

In multisection VQ, we represent each speaker by a time-ordered sequence of single section codebooks, which we call a multisection codebook. A speaker is verified by dividing a verification utterance $\mathbf{v}$ into sections corresponding to the sections of the multisection codebooks, doing VQ section by section with the appropriate multisection codebook, and computing the average distortion.

To be more specific, let F_q be the number of frames in the qth utterance in the training sequence $\mathbf{t}$ for multisection codebook $\bar{C}$, where $q = 1, \cdots, M$; and let U_{fq} be the fth frame in the qth training utterance, where $f = 1, \cdots, F_q$ and $P = \Sigma_{q=1}^{M} F_q$ is the total number of training vectors in $\mathbf{t}$. The multisection codebook $\bar{C}$ consists of a sequence of VQ *section codebooks* $\bar{C}_j$, where the section codebook $\bar{C}_j$ is designed using (1) and n frames from each training utterance. That is, $\bar{C}_j$ is designed from the frames U_{fq}, where $f = (j - 1)n + 1, \cdots, jn$, and $q = 1, \cdots, M$. For example, $\bar{C}_1$ is designed from the first n frames of each training utterance, $\overline{C}_2$ from the second n frames, and so on. We call n the *section length*—it is the number of frames that are spanned per section. Finally, let $\{\overline{c}_{j1}, \overline{c}_{j2}, \cdots, \overline{c}_{jN_j}\}$ be codewords in section codebook $\overline{C}_j$.

D_{avg} is the average distortion that results from coding the verification utterance $\mathbf{v}$ with the codebook $\overline{C}$,

$$D_{avg} = \frac{1}{L} \sum_{j=1}^{S} d_j, \tag{3}$$

where S is the number of section codebooks in $\overline{C}$,

$$d_j = \sum_{l=(j-1)n+1}^{\min[jn, L]} \min_i d(\overline{v}_l, \overline{c}_{ji})$$

is the total distortion from coding the jth section of the utterance $\mathbf{v}$ with the jth section codebook $\overline{C}_j$ of $\overline{C}$, L is the number of vectors in $\mathbf{v}$, and n is the section length. The verification decision is made using this average distortion.

C. Matrix Quantization

In matrix quantization, instead of coding a *single* source vector in a codebook containing characteristic vectors, we code a *time-ordered sequence* of source vectors in a codebook containing characteristic vector sequences. Given $\mathbf{t}$, we find the matrix quantization codebook $\mathbf{C}$ containing codework matrices (or vector sequences) $\mathbf{c}_j = [\overline{c}_{j1}, \overline{c}_{j2}, \cdots, \overline{c}_{jK}]$ that minimizes

$$\sum_{p=1}^{P-K+1} D(\mathbf{t}_p, \mathbf{c}_B),$$

where $\mathbf{c}_B$ is the codeword matrix that results from coding the sequence of training vectors

$$\mathbf{t}_p = [\overline{t}_p, \overline{t}_{p+1}, \cdots, \overline{t}_{p+K-1}],$$

by using the nearest neighbor rule

$$D(\mathbf{t}, \mathbf{c}_B) = \min_j D(\mathbf{t}, \mathbf{c}_j),$$

and where the distortion between a speech segment $\mathbf{t}$ and the jth codeword matrix is

$$D(\mathbf{t}, \mathbf{c}_j) = \sum_{l=1}^{K} d(\overline{t}_l, \overline{c}_{jl}). \tag{4}$$

We call K the *codeword matrix size.* The MQ codebook design algorithm [21] we used is a generalized version of the VQ design algorithm developed by Linde *et al.* [12].

To use MQ in speaker verification, we represent each speaker by a codebook $\mathbf{C}$, just as in the VQ approaches above. A verification utterance is processed by dividing it into overlapping sequences of K frames, coding each K frame sequence in the speaker-codebook $\mathbf{C}$, and computing the average quantization distortion between the utterance and the codebook. To be specific, for a verification utterance $\mathbf{v}$, the average distortion resulting from coding it with codebook $\mathbf{C}$ is

$$D_{avg} = \frac{1}{L-K+1} \sum_{l=1}^{L-K+1} D(\mathbf{v}_l, \mathbf{c}_B) \tag{5}$$

where $\mathbf{v}_l = [\overline{v}_l, \overline{v}_{l+1}, \cdots, \overline{v}_{l+K-1}]$.

D. Distortion Measures

Based on results from previous work on isolated word recognition [9], we used the *gain normalized Itakura–Saito* distortion measure (d_{GN}) in (1) and (4) to generate codebooks. For power spectrum estimates f and $\hat{f}$ having the autoregressive (LPC) form

$$f(\theta) = \frac{\sigma^2}{|A(z)|^2},$$

where

$$A(z) = \sum_{k=0}^{M} a_k z^{-k}$$

and $z = \exp(i\theta)$, the d_{GN} distortion is given by

$$d_{GN}(f, \hat{f}) = \frac{\alpha}{\sigma^2} - 1,$$

where

$$\alpha = r(0)\, \hat{r}_a(0) + 2 \sum_{n=1}^{M} r(n)\, \hat{r}_a(n),$$

$$\hat{r}_a(n) = \sum_{i=0}^{M-n} \hat{a}_i \hat{a}_{i+n},$$

and where $r(n)$ are the time-domain autocorrelations of $f(\theta)$.

We chose the *gain optimized Itakura–Saito* distortion measure (d_{GO})

$$d_{GO}(f, \hat{f}) = \ln(\alpha) - \ln(\sigma^2),$$

for the verification distortion measure in (2), (3), and (5), again based on results from previous work [9]. d_{GN} would also be a good choice [9]. Properties of these distortion measures are discussed in [22].

E. LPC Parameters

LPC parameters for both codebook generation and speaker verification were generated using the autocorrelation method of linear predictive analysis with Hamming windowing. We chose analysis conditions for compatibility with the Navy's 2.4-kbits/s LPC-10 system [23]: analysis window width = 128 points, filter order = 10, and preemphasis = 94 percent.

III. Speaker Verification Approach

Usually, no information is available for the characteristics of specific unacceptable speakers, and the main problem in applying these source coding approaches to speaker verification is in formulating a criterion for rejecting a speaker. To decide whether to reject a speaker (given an utterance), we associate a threshold with each speaker codebook. An unknown speaker (utterance) is rejected if its distortion exceeds the threshold. To design thresholds for a speaker, we estimate parameters for two Gaussian distributions: the *in-class* distribution of distortions obtained by encoding utterances from a speaker in his or her codebook, and the *out-of-class* distribution of distortions resulting from encoding utterances spoken by other speakers. We chose the threshold to equalize the overlap area of the two distributions, thus equalizing the expected numbers of imposter acceptances (false acceptances) and rejections of admissible speakers (false rejections).

In more detail, the threshold computation is as follows. For each speaker, encode that speaker's training data with his or her codebook. Compute the mean distortion μ_i^{in} resulting from encoding the training data from speaker i in speaker i's codebook, and compute the corresponding standard deviation σ_i^{in}. Also compute μ_i^{out}, the mean distortion resulting from encoding utterances *not* spoken by speaker i using the codebook for speaker i, and the corresponding standard deviation σ_i^{out}. To equalize the number of false acceptances and false rejections, the threshold T_i is chosen to be an equal number of standard deviations away from each mean, giving

$$T_i = \frac{\mu_i^{\text{in}} \sigma_i^{\text{out}} + \mu_i^{\text{out}} \sigma_i^{\text{in}}}{\sigma_i^{\text{out}} + \sigma_i^{\text{in}}}$$

This method of threshold determination assumes Gaussian distributions. Some previous studies by Buck [24], however, showed that the logarithms of average distortions are more nearly Gaussian than the distortions themselves; so the thresholds T_i were based on the statistics of the logarithms of distortions, instead of simply the distortion as shown in (2), (3), and (5).

(It was pointed out by a referee that using the Gaussian assumption is not necessary in determining the thresholds, and it may be part of the cause of a bias in the error rates toward false rejections that is reported in Sections III-D–F. The referee suggested using the empirical distributions obtained from the histograms of the true talker's distortions and of the imposter training data distortions to find the thresholds. This, we believe, would be a good way of determining the thresholds.)

To verify a speaker, the verification utterance **v** is coded in the appropriate codebook and the average log distortion is computed. This distortion value is compared to the threshold associated with that codebook. If the distortion value exceeds the threshold, the speaker is rejected; otherwise the speaker is accepted.

Preliminary experiments indicated that verification accuracy using a single verification utterance is poor [15]. To improve the verification accuracy, we based the verification decision on the results for several words. The next section describes our approach to extending this method to multiple words.

A. Multiple Word Approach

In previous work [15], we examined three ways of extending our method to more than one word. All three methods achieved about the same verification accuracy, and based on those results, we used the simplest of the three methods in this work.

A separate codebook is designed to represent each word that a speaker is required to say; if W words are spoken, there are W codebooks designed for each speaker. Next, for each speaker, separate thresholds are computed for each word, just as described in Section III. Now, an unknown speaker claims an identity and says the W words in the appropriate order. Each word is coded in the appropriate codebook, the distortion is compared to the threshold, and a verification decision is made for each word. For example, if a speaker is requested to say *zero* and *nine*, first the *zero* utterance is encoded with the *zero* codebook for that speaker and the resulting distortion is compared to the threshold; next the *nine* utterance is encoded with the *nine* codebook. To make a final verification decision, we use a majority rule; that is, the decision made in a majority of the individual word verification tests is used as the overall decision. In the case of ties, the speaker is rejected.

IV. Experiments

We first describe the speech databases and how the databases were partitioned for use in separate parameter studies and evaluation tests. We then list the parameters varied in the experiments. This is followed by three subsections; each subsection describes the verification results using one of the source models described in Section II.

A. Databases

We combined two databases to do these experiments, both collected by Texas Instruments Inc. (TI). The main difference in the databases is the resolution of the A/D converters. One database was digitized with a 12-bit converter; the second was digitized at a later time with a 16-bit converter. The differences in the two databases may bias the results slightly, but we do not believe the bias is significant.

Data for designing the codebooks to represent the speakers, determining the parameters for the in-class distributions, and testing verification accuracy, came from the database described in [25]. It contains 26 utterances of each digit (*zero* through *nine*) by 16 speakers (8 male and 8 female). We call this database TI-1. The database used for determining the parameters for the out-of-class distributions and for testing the imposter rejection capabilities of the methods contains two utterances of each of the 10 digits from 108 adult male and 112 adult female speakers [26]. (This database also contains 5 additional speakers that are in TI-1; these speaker's data were not used in our work.) This database is divided into two parts: a *training* part containing 54 male and 55 female speakers, and a *testing* part containing 54 male and 57 female speakers. We call this second database TI-2.

Automatic endpoint detection for both training and test utterances was used in our experiments. Our endpoint-detection algorithm is based on ideas presented in [27]

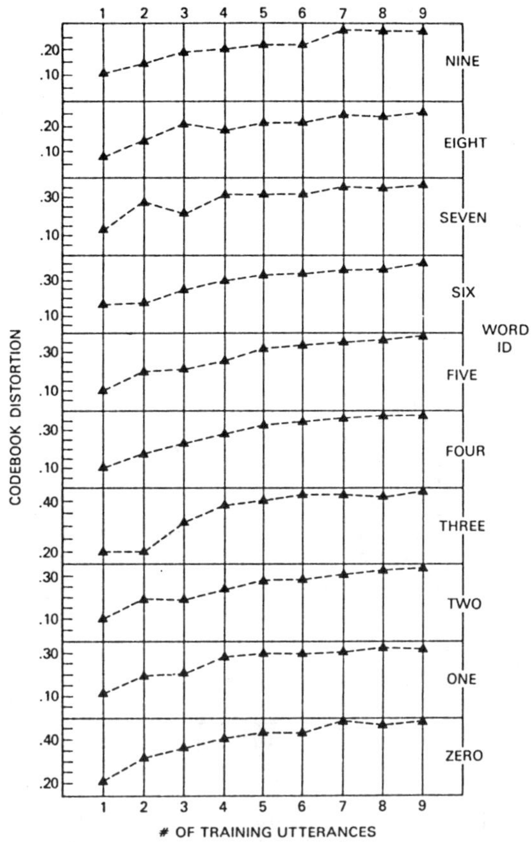

Fig. 1. The codebook design distortion for each of the 10 digits as a function of the number of codebook training utterances—one speaker's results.

and [28]. Briefly, the algorithm first analyzes the background noise to determine its average magnitude and then uses the result to set various thresholds that are used to find significant "energy clumps" in the data. See [9] for more details.

B. Database Partition

We first determined the number of training utterances required to model a speaker saying a digit. To do this, the average codebook-design distortion as a function of the number of training utterances was examined. For each speaker–digit combination, we designed a sequence of single-section codebooks, each containing 8 codewords. The first codebook was designed using a one-utterance training sequence, and each new codebook in the sequence was designed by increasing the number of training utterances by one. The average codebook-design distortion was recorded for each codebook. (See Fig. 1 for the results from a typical speaker.) We expected the average distortion to increase with each additional training utterance until the training data contained all the normal variations in a speaker's pronunciation of a word, and we then expected the distortion to stop increasing significantly with additional training data. We felt that a flattening of the distortion curve indicated that all the speaker's typical pronunciations were present. Fig. 1 shows that generally the distortion curve is flat by the 7th utterance. In addition, considering all 10 digits and the 16 speakers in TI-1, the average number of training utterances needed to reach 90 percent of the maximum codebook-design distortion was 8.

Based on the results of these experiments, in all three parameter studies described below, we designed digit codebooks for each speaker in the TI-1 database using the first 8 utterances of each digit. These 8 training utterances plus the next 4 utterances were used to estimate the parameters for the in-class distribution for each speaker–digit model. The next 7 digits in TI-1 were the speaker supplied verification data. For the parameter studies, the TI-2 *training* data were divided into two parts. One part containing the first 29 male and 30 female speakers was used to estimate the out-of-class distribution parameters; the second part containing the rest of the training portion of TI-2 (50 speakers) was used as imposter data.

Based on the results of the parameter studies, we chose several sets of words and codebook parameters to use in the full database tests of the three source models. In these tests, the training data for each speaker codebook again consisted of the first 8 utterances of a digit. The in-class parameter estimation data, however, consisted of the first 16 utterances. The remaining 10 utterances of each digit in TI-1 were the verification data. We used all 109 speakers in the training portion of TI-2 to estimate the out-of-class distribution parameters and the 111 speakers in the test portion as the imposters.

C. Experimental Parameters

We varied the codebook size—the number of codewords in a codebook—in the experiments. For single section codebooks, the codebook size is always a power of 2—i.e., $N = 2^R$, and we call R the codebook rate. For multisection codebooks, the size of each constituent section codebook is also a power of 2, and we call the section codebook rate R_S. Similarly, the size of matrix quantization codebooks is a power of 2, and we call the matrix codebook rate R_M.

In addition to the codebook rates, we varied the section length n for multisection codebooks and the matrix size K for matrix quantization codebooks during the parameters studies. The parameters used during verification always matched those used in designing the codebooks.

There are several parameters affecting the design of codebooks and thus the verification results that we did not vary. For one, after endpoint detection, we preprocessed the training and verification data by dividing each utterance into 24 equal length frames. This was done to provide a rough form of time normalization to all the utterances; previously, this was shown to help in isolated word recognition [16]. Also, since the spectra corresponding to nearly silent frames can be quite arbitrary and unspeechlike, to avoid cluttering codebooks with codewords that would result from including such frames, we used an energy threshold to ignore them. For single-section and multisection codebooks, we used an energy (sum-of-squares of data points) threshold of 250; this threshold was used both in codebook generation and speaker verification. For matrix quantization, we handled low-energy

TABLE I
SPEAKER VERIFICATION STUDY: SINGLE SECTION CODEBOOKS, MAJORITY RULE, AND ALL 10 DIGITS

Codebook Rate (R)	Number Of Imposter Attempts	False Acceptances	Number Of Admissible Attempts	False Rejections
1	800	7	112	15
2	800	1	112	7
3	800	0	112	7
4	800	0	112	6

TABLE II
SPEAKER VERIFICATION STUDY: A SINGLE CODEBOOK PER SPEAKER AND $R = 3$

Digit Spoken	Number Of Imposter Attempts	False Acceptances	Number Of Admissible Attempts	False Rejections	$\sqrt{FA*FR}$ %
ZERO	800	24 (3.0%)	112	10 (8.9%)	5.2
ONE	800	9 (1.1%)	112	19 (17.0%)	4.3
TWO	800	14 (1.8%)	112	10 (8.9%)	4.0
THREE	800	17 (2.1%)	112	17 (15.2%)	5.7
FOUR	800	12 (1.5%)	112	13 (11.6%)	4.2
FIVE	800	19 (2.4%)	112	26 (23.2%)	7.5
SIX	800	14 (1.8%)	112	17 (15.2%)	5.2
SEVEN	800	9 (1.1%)	112	16 (14.3%)	4.0
EIGHT	800	32 (4.0%)	112	15 (13.4%)	7.3
NINE	800	20 (2.5%)	112	20 (17.9%)	6.7

TABLE III
SPEAKER VERIFICATION RESULTS: $R = 3$, MAJORITY RULE, AND ALL 10 DIGITS

Speaker ID	Number Of Imposter Attempts	False Acceptances	Number Of Admissible Attempts	False Rejections	$\sqrt{FA*FR}$ %
TBS	222	0	10	0	0.0
WMF	222	0	10	0	0.0
RLD	222	1	10	0	0.0
GRD	222	1	10	1	2.1
KAB	222	5	10	2	6.7
MSW	222	0	10	0	0.0
REH	222	0	10	0	0.0
RGL	222	0	10	0	0.0
CJP	222	0	10	0	0.0
DFG	222	0	10	0	0.0
ALK	222	0	10	0	0.0
HNJ	222	0	10	0	0.0
GNL	222	0	10	1	0.0
JWS	222	0	10	0	0.0
SJN	222	0	10	1	0.0
SAS	222	0	10	1	0.0
Totals	3552	7 (0.2%)	160	6 (3.8%)	0.9

frames in the following manner. The first K-1 low-energy frames in a sequence were replaced with flat-spectrum frames with energy equal to 250; if more than K-1 frames occurred in a sequence, we ignored all but the first K-1. The reason for keeping the K-1 silent frames was to preserve the information in the transitions from silence-to-speech and vice versa, while eliminating any long all-silent training and verification segments.

D. Single Section Results

Parameter Studies: We varied the codebook rate R in these experiments. Verification decisions were made using all 10 digits and the majority-rule classifier, which was described in Section III-A. The results are listed in Table I for R ranging from 1 to 4. Most of the verification errors were false rejections. This implies that the thresholds T are too small. For R equal to 3 and 4, all errors were caused by just 2 of the 16 speakers.

For R equal to 3, we measured the verification accuracy of each digit individually; the results are in Table II. No single digit reliably verifies the speakers, and the individual digit results are also biased toward false rejections. The last column in Table II contains the square root of the product of the false-acceptance rate and the false-rejection rate ($\sqrt{FA * FR}$); this is considered a good overall performance measure [29].

Verification Tests: Because of the bias toward false rejections in the parameter study, we added more utterances to the data used in estimating the in-class distribution parameters; the new training set contained the 8 utterances used to design the codebooks and 8 additional utterances. (This was also done in the verification tests reported in Sections IV-E and F.) We felt that by including more utterances that were not in the codebook design set, the in-class distribution for a speaker would better represent new utterances from that speaker. Rate-3 codebooks were used in the verification tests because they did best in the parameter study and also because rate-3 codebooks yielded good speaker-trained isolated word recognition results [9].

The results, using all 10 digits in the verification decision, are listed by individual speaker in Table III. The majority of errors were caused by KAB and GRD; they also were the two difficult speakers in the parameter study. The results are still biased toward false rejections, how-

TABLE IV
SPEAKER VERIFICATION RESULTS: $R = 3$, MAJORITY RULE, AND 5 DIGITS

Digit Subset	Number Of Imposter Attempts	False Acceptances	Number Of Admissible Attempts	False Rejections	$\sqrt{FA*FR}$ %
01247	3552	19 (0.5%)	160	6 (3.8%)	1.4
35689	3552	27 (0.8%)	160	4 (2.5%)	1.4
25678	3552	25 (0.7%)	160	8 (5.0%)	1.9

TABLE V
SPEAKER VERIFICATION STUDY: MULTISECTION CODEBOOKS, MAJORITY RULE, ALL 10 DIGITS, 800 IMPOSTER ATTEMPTS, AND 112 ADMISSIBLE ATTEMPTS

Codebook Rate (R_S)	$n=12$ # FA	$n=12$ # FR	$n=8$ # FA	$n=8$ # FR	$n=4$ # FA	$n=4$ # FR	$n=2$ # FA	$n=2$ # FR	$n=1$ # FA	$n=1$ # FR
0	18	19	6	9	0	10	0	10	0	10
1	1	6	1	6	0	8	0	5	0	5
2	0	7	0	6	0	5	–	–	–	–

TABLE VI
SPEAKER VERIFICATION STUDY: A MULTISECTION CODEBOOK PER SPEAKER, $R_S = 2$, AND $n = 4$

Digit Spoken	Number Of Imposter Attempts	False Acceptances	Number Of Admissible Attempts	False Rejections	$\sqrt{FA*FR}$ %
ZERO	800	19 (2.4%)	112	12 (10.7%)	5.1
ONE	800	11 (1.4%)	112	13 (11.6%)	4.0
TWO	800	8 (1.0%)	112	16 (14.3%)	3.8
THREE	800	21 (2.6%)	112	18 (16.1%)	6.5
FOUR	800	13 (1.6%)	112	13 (11.6%)	4.3
FIVE	800	16 (2.0%)	112	31 (27.7%)	7.4
SIX	800	9 (1.1%)	112	15 (13.4%)	3.8
SEVEN	800	6 (0.8%)	112	24 (21.4%)	4.1
EIGHT	800	33 (4.1%)	112	9 (8.0%)	5.7
NINE	800	16 (2.0%)	112	19 (17.0%)	5.8

ever, the bias is smaller. $\sqrt{FA * FR}$ for this test was 0.9 percent.

Next several subsets of the digits were tested, each consisting of 5 digits. Based on the single digit parameter study, we used the best (0, 1, 2, 4, and 7), the worst (3, 5, 6, 8, and 9), and an arbitrary (2, 5, 6, 7, and 8) set of five digits. Results are listed in Table IV. Again, KAB and GRD were difficult speakers, but many other speakers contributed to the errors. Based on the $\sqrt{FA * FR}$ values for these tests, the verification accuracies obtained by representing each speaker by five words were significantly worse than those obtained using all 10 words. Generally, the degradation in performance was caused by additional false acceptances, and the overall performance was closer to the design goal of equal error rates for the two types of errors.

E. Multisection Results

Parameter Studies: We varied both the section length n and the section codebook rate R_S in these experiments. Table V shows the results. Generally for a fixed R_S value, better results are achieved using smaller n values. For $R_S = 2$, tests using n equal to 1 and 2 were not done because of insufficient codebook training data. Using $n = 4$ and $R_S = 2$, we tested the verification performance of the individual digits; the results are in Table VI. Again, as in the single section approach, no single digit gives good overall results and the errors are biased toward false rejections.

Verification Tests: We used $n = 4$ and $R_S = 2$ for the verification tests. These conditions were chosen because they both did well in the parameter study and in previous isolated word recognition work [16]. Table VII contains the results using all 10 digits to make the verification decision. No speaker was particularly difficult, as KAB and GRD were when using the single section approach, and, in general, the results are closer to the design goal of equal false-rejection and false-acceptance error rates than were the single section results. $\sqrt{FA * FR}$ was 0.6 percent.

Again, we did verification tests using the best (1, 2, 4, 6, and 7), the worst (0, 3, 5, 8, and 9) and an arbitrary (0, 1, 2, 3, and 4) set of five digits; the results are in Table VIII. The verification performance of the various five-digit subsets corresponded well with the expected performance based on the single digit study—i.e., the best five-digit set had the smallest $\sqrt{FA * FR}$, the worst set had the largest $\sqrt{FA * FR}$, and the arbitrary set had a $\sqrt{FA * FR}$ between the other two. The only consistently difficult speaker in these tests was KAB; averaged over the 3 five-digit tests, he had a false acceptance rate of 3.3 percent.

F. Matrix Quantization

Parameter Study: We varied the codebook rate R_m and the matrix size K in these experiments. For each K value, the maximum R_M was limited by the amount of codebook training data (poor codebooks often result if insufficient training data are used). The results are listed in Table IX. No obvious relationship between K and R_M is shown in these results. Using $R_M = 3$ and $K = 8$ (these conditions are also good for isolated word recognition [17]), we measured the verification performance of the individual

TABLE VII
SPEAKER VERIFICATION RESULTS: $R_S = 2$, $n = 4$, MAJORITY RULE, AND ALL 10 DIGITS

Speaker ID	Number Of Imposter Attempts	False Acceptances	Number Of Admissible Attempts	False Rejections	$\sqrt{FA*FR}$ %
TBS	222	1	10	0	0.0
WMF	222	0	10	0	0.0
RLD	222	0	10	0	0.0
GRD	222	1	10	0	0.0
KAB	222	2	10	0	0.0
MSW	222	0	10	0	0.0
REH	222	0	10	0	0.0
RGL	222	0	10	0	0.0
CJP	222	2	10	0	0.0
DFG	222	0	10	0	0.0
ALK	222	3	10	0	0.0
HNJ	222	0	10	0	0.0
GNL	222	0	10	2	0.0
JWS	222	0	10	0	0.0
SJN	222	0	10	0	0.0
SAS	222	0	10	0	0.0
Totals	3552	9 (0.3%)	160	2 (1.3%)	0.6

TABLE VIII
SPEAKER VERIFICATION RESULTS: $R_S = 2$, $n = 4$, MAJORITY RULE, AND 5 DIGITS

Digit Subset	Number Of Imposter Attempts	False Acceptances	Number Of Admissible Attempts	False Rejections	$\sqrt{FA*FR}$ %
12467	3552	26 (0.7%)	160	1 (0.6%)	0.7
03589	3552	34 (1.0%)	160	5 (3.1%)	1.7
01234	3552	17 (0.5%)	160	3 (1.9%)	0.9

TABLE IX
SPEAKER VERIFICATION STUDY: MATRIX QUANTIZATION CODEBOOKS, MAJORITY RULE, ALL 10 DIGITS, 800 IMPOSTER ATTEMPTS, AND 112 ADMISSIBLE ATTEMPTS

Codebook Rate (R_M)	$K=4$		$K=8$		$K=12$		$K=24$	
	# FA	# FR	# FA	# FR	# FA	# FR	# FA	# FR
2	6	11	5	16	4	14	2	8
3	0	10	0	8	0	9	–	–
4	0	8	–	–	–	–	–	–

TABLE X
SPEAKER VERIFICATION STUDY: A MATRIX QUANTIZATION CODEBOOK PER SPEAKER, $R_M = 3$, AND $K = 8$

Digit Spoken	Number Of Imposter Attempts	False Acceptances	Number Of Admissible Attempts	False Rejections	$\sqrt{FA*FR}$ %
ZERO	800	24 (3.0%)	112	12 (10.7%)	5.7
ONE	800	6 (0.8%)	112	14 (12.5%)	3.2
TWO	800	12 (1.5%)	112	16 (14.3%)	4.6
THREE	800	29 (3.6%)	112	21 (18.8%)	8.2
FOUR	800	12 (1.5%)	112	13 (11.6%)	4.2
FIVE	800	45 (5.6%)	112	33 (29.5%)	12.9
SIX	800	11 (1.4%)	112	16 (14.3%)	4.5
SEVEN	800	11 (1.4%)	112	22 (19.6%)	5.2
EIGHT	800	36 (4.5%)	112	14 (12.5%)	7.5
NINE	800	17 (2.1%)	112	22 (19.6%)	6.4

digits; these results are in Table X. As in the single section and multisection approaches, the errors are biased toward false rejections.

Verification Tests: The full database results using $R_M = 3$, $K = 8$, and all 10 digits are listed in Table XI. Once again, KAB was a difficult speaker. We tested the best (1, 2, 4, 6, and 7), the worst (0, 3, 5, 8, and 9), and an arbitrary (0, 1, 2, 3, and 4) five digits; the verification results are in Table XII. The relative performance of the five-digit sets did not correspond exactly with the expected results based on the individual digit performances, but the worst digit set did produce the poorest results.

V. SUMMARY AND DISCUSSION

The verification performances ($\sqrt{FA * FR}$) of the three source models when using only a single digit per speaker were similar—roughly varying from 4 to 8 percent depending on the digit, and consistently, the digits 1, 2, 4, and 7 individually did best in the speaker verification tests. When the individual digits were joined with the majority rule classifier, however, the verification performances of the three approaches were no longer equivalent. The multisection VQ source model did best when using the 10- and 5-digit sets of verification words (for the 10 digits, $\sqrt{FA * FR} = 0.6$ percent; for the best 5-digit set, $\sqrt{FA * FR} = 0.7$ percent). In addition, the multisection VQ approach came closer to satisfying the design goal of equal error rates, and the results on the 5-digit subsets corresponded more closely to the expected results (based on $\sqrt{FA * FR}$ for the individual digits). The next best source model was the single section approach, although

TABLE XI
SPEAKER VERIFICATION RESULTS: $R_M = 3$, $K = 8$, MAJORITY RULE, AND ALL 10 DIGITS

Speaker ID	Number Of Imposter Attempts	False Acceptances	Number Of Admissible Attempts	False Rejections	$\sqrt{FA*FR}$ %
TBS	222	1	10	0	0.0
WMF	222	0	10	0	0.0
RLD	222	0	10	1	0.0
GRD	222	1	10	1	2.1
KAB	222	3	10	3	6.4
MSW	222	0	10	0	0.0
REH	222	0	10	0	0.0
RGL	222	2	10	0	0.0
CJP	222	0	10	0	0.0
DFG	222	0	10	0	0.0
ALK	222	0	10	0	0.0
HNJ	222	0	10	0	0.0
GNL	222	0	10	2	0.0
JWS	222	0	10	0	0.0
SJN	222	0	10	1	0.0
SAS	222	0	10	1	0.0
Totals	3552	7 (0.2%)	160	9 (5.6%)	1.1

TABLE XII
SPEAKER VERIFICATION RESULTS: $R_M = 3$, $K = 8$, MAJORITY RULE, AND 5 DIGITS

Digit Subset	Number Of Imposter Attempts	False Acceptances	Number Of Admissible Attempts	False Rejections	$\sqrt{FA*FR}$ %
12467	3552	17 (0.5%)	160	8 (5.0%)	1.5
03589	3552	44 (1.2%)	160	9 (5.6%)	2.6
01234	3552	15 (0.4%)	160	6 (3.8%)	1.3

the differences in $\sqrt{FA * FR}$ values between the single section VQ (for the 10 digits, $\sqrt{FA * FR} = 0.9$ percent) and the MQ approach (for the 10 digits, $\sqrt{FA * FR} = 1.1$ percent) were small.

All three source models did well in the speaker verification tests. In retrospect, because all three use the same training data and contain essentially the same short-time spectrum information, simply stored differently, the similarity of their performance should not be surprising. At the heart of the three approaches is the clustering algorithm [12]; it designs all three codebook types. Both the multisection VQ and the MQ approach are straightforward generalizations of single section VQ that contain information about the time ordering of the speech spectra. One can see this by considering the multisection VQ approach with a section length n equal to the normalization length (24 in this study) and considering the MQ approach with the matrix size K equal to 1. Each approach reduces to single section VQ under the appropriate condition.

The single section VQ source model captures only the short-time spectrum shape information. This spectrum shape information is useful in speaker verification because it contains estimates of formant frequencies, relative formant amplitudes, and formant bandwidths, and these are correlated with the locations and physical sizes of the speech articulators. As such, the single section results are a measure of how well the short-time spectrum alone can characterize a speaker. In addition, because the codebook spectra are unordered, the single section VQ source model is directly applicable to text-independent speaker recognition, as pointed out by Soong [19] and Shikano [20].

It is generally believed, however, that examining parameters as a function of time is valuable in speaker verification for two reasons: 1) many of the speaker-characteristic properties of speech are the result of idiosyncrasies in the speaking habits of people, and 2) by considering the time sequence of parameters, the emphasis is on how the parameters vary rather than the exact value of a parameter. The multisection VQ and the MQ approaches are two different ways of incorporating some phonetic duration and coarticulation information into the verification process while maintaining the information-theoretic source model approach. Multisection VQ models a source by dividing it into several independent, time-ordered subsources. When used for speaker verification, a multisection VQ codebook provides an accurate representation of the speech spectra in an utterance, and multisection VQ coding enforces limits on the position of phonemes, the duration of phonemes, and to a lesser degree, the coarticulated phonemes in an utterance. Based on our results, adding this information improves the verification performance. As noted earlier, MQ models an utterance with a single codebook that contains an unordered set of time-ordered speech spectrum sequences. (Thus, MQ codebooks could be used in text-independent speaker recognition, just as single section codebooks are used.) These spectrum sequences correspond to stable continuant sounds or transitions from one sound to another. Thus, for large K values, MQ coding includes coarticulation and phonetic duration information. Our results show that MQ's use of this extra information does not improve the verification performance relative to the single section VQ results. This, of course, is contrary to our intuition, and it is unclear why the additional information inherent in the MQ approach does not improve the verification performance.

In addition to short-time spectra, a speaker will say an utterance with characteristic pitch [30] and stresses [1]. Because these are roughly independent of the spectrum shape information, improvements in the verification accuracy could be achieved by adding pitch and short-time energy information to the verification process.

As an aside, it is interesting to consider how VQ speech coding could defeat a speaker verification or identification system. Our single section VQ results show that the source model found by using the Linde, Buzo, and Gray clustering algorithm [12] is an accurate representation of the short-time spectra produced by a speaker. Thus, to impersonate a speaker, one needs only to obtain training data spoken by that speaker and to design a VQ codebook with these data. Anyone could talk through this codebook (via VQ speech coding), and the resulting speech would be characteristic of the speaker who provided the training data. It seems this procedure would defeat any speaker recognition system that relies solely on short-time spectrum representations.

Finally, the connection between these speaker verification approaches and our previous isolated word recognition approaches needs emphasis. The parameters (codebook size, section length, and matrix size) and the source model (codebook) design procedure used in each of the speaker verification tests are exactly those used in our previous work on isolated word recognition [9], [16], [17]. In those studies, accuracies for speaker-trained recognition of the digits exceeded 99 percent, and the computational and memory storage requirements were shown to be quite small. The very good speaker verification and isolated word recognition results achieved using these approaches point toward a combined speaker–speech recognition system, and Shikano describes one example of how this could be done [20]. He proposes using VQ speaker recognition as a preprocessor for speaker-independent isolated word recognition. In his system, the speaker first is recognized using VQ codebooks, and then previously stored speaker-trained references are used to recognize the words. Thus, speaker-dependent recognition accuracies are obtained in a speaker-independent speech recognition system. In addition, however, adding VQ speech recognition to a speaker verification system should improve its performance. For example, suppose each speaker is represented by a set of speaker-trained VQ codebooks, as we described above. After a speaker claims an identity and wants to be verified, the verification system tells the speaker what words to say. The system first recognizes the words using the set of codebooks that were designed for the speaker whose identity was claimed and then compares the individual word distortions to the prestored verification thresholds. If the words are incorrectly recognized (which is much more likely to happen for an imposter) or the speaker fails the distortion tests, the speaker is rejected. This procedure would certainly reduce the number of false acceptances, although it may increase the number of false rejections.

Acknowledgment

I thank J. Buck for writing most of the software, R. Johnson for pointing out the VQ speaker impersonation scheme, and J. Shore for comments on this paper. In addition, I thank T. Schalk and G. Leonard for their help in obtaining the databases.

References

[1] R. C. Lummis, "Speaker verification by computer using speech intensity for temporal registration," *IEEE Trans. Audio Electroacoust.*, vol. AU-21, pp. 80–89, Apr. 1973.
[2] S. Pruzansky, "Pattern-matching procedure for automatic talker recognition," *J. Acoust. Soc. Amer.*, vol. 35, pp. 354–358, Mar. 1963.
[3] B. S. Atal, "Effectiveness of linear predictive characteristics of the speech wave for automatic speaker identification and verification," *J. Acoustic. Soc. Amer.*, vol. 55, pp. 1034–1312, 1974.
[4] W. S. Mohn, Jr., "Two statistical feature evaluation techniques applied to speaker identification," *IEEE Trans. Comput.*, vol. C-20, pp. 979–987, Sept. 1971.
[5] P. D. Bricker *et al.*, "Statistical techniques for talker identification," *Bell Syst. Tech. J.*, vol. 50, pp. 1427–1454, Apr. 1971.
[6] H. L. Van Trees, *Detection, Estimation, and Modulation Theory—Part I.* New York: 1968.
[7] R. M. Gray, "Vector quantization," *IEEE ASSP Mag.*, pp. 4–29, Apr. 1984.
[8] A. Buzo, A. H. Gray, Jr., R. M. Gray, and J. D. Markel, "Speech coding based upon vector quantization," *IEEE Trans. Acoust., Speech, Signal Processing*, vol. ASSP-28, pp. 562–574, Oct. 1980.
[9] J. E. Shore and D. K. Burton, "Discrete utterance speech recognition without time alignment," *IEEE Trans. Inform. Theory*, vol. IT-29, pp. 473–491, July 1983.
[10] N. Sugamura, K. Shikano, and S. Furiu, "Isolated word recognition using phoneme-like templates," in *Proc. ICASSP 1983, IEEE Int. Conf. Acoust., Speech, Signal Processing*, Boston, MA, Apr. 1983, pp. 723–726.
[11] L. R. Rabiner, S. E. Levinson, and M. M. Sondhi, "On the application of vector quantization and hidden Markov models to speaker-independent, isolated word recognition," *Bell Syst. Tech. J.*, vol. 62, pp. 1075–1105, Apr. 1983.
[12] Y. Linde, A. Buzo, and R. M. Gray, "An algorithm for vector quantizer design," *IEEE Trans. Commun.*, vol. COM-28, pp. 84–95, Jan. 1980.
[13] S. Kullback, *Information Theory and Statistics.* New York: Dover, 1968; New York: Wiley, 1959.
[14] J. E. Shore and R. M. Gray, "Minimum-cross-entropy pattern classification and cluster analysis," *IEEE Trans. Pattern Anal. Machine Intell.*, vol. PAMI-4, pp. 11–17, Jan. 1982.
[15] J. T. Buck, D. K. Burton, and J. E. Shore, "Text-dependent speaker recognition using vector quantization," in *Proc. 1985 ICASSP, IEEE Int. Conf. Acoust., Speech, Signal Processing*, Tampa, FL, Mar. 1985, pp. 11.5.1–11.5.4.
[16] D. K. Burton, J. E. Shore, and J. T. Buck, "Isolated-word speech recognition using multi-section vector quantization codebooks," *IEEE Trans. Acoust., Speech, Signal Processing*, vol. ASSP-33, pp. 837–849, Aug. 1985.
[17] D. K. Burton, "Applying matrix quantization to isolated word recognition," in *Proc. ICASSP 1985, IEEE Int. Conf. Acoust., Speech, Signal Processing*, Tampa, FL, Mar. 1985, pp. 1.8.1–1.8.4.
[18] K. P. Li and E. H. Wrench, Jr., "An approach to text-independent speaker recognition with short utterances," in *Proc. ICASSP 1983, IEEE Int. Conf. Acoust., Speech, Signal Processing*, Boston, MA, Apr. 1983, pp. 555–558.
[19] F. Soong *et al.*, "A vector quantization approach to speaker recognition," in *Proc. 1985 ICASSP, IEEE Int. Conf. Acoust., Speech, Signal Processing*, Tampa, FL, Mar. 1985, pp. 11.7.1–11.7.4.
[20] K. Shikano, "Text-independent speaker recognition experiments using codebooks in vector quantization," Carnegie-Mellon Univ., Apr. 1985, private communication (Abstract in program: 109th Meet. Acoust. Soc. Amer.).
[21] C. Tsao and R. M. Gray, "Matrix quantizer design for LPC speech using the generalized Lloyd algorithm," *IEEE Trans. Acoust., Speech, Signal Processing*, vol. ASSP-33, pp. 537–545, June 1985.
[22] R. M. Gray, A. Buzo, A. H. Gray, Jr., and Y. Matsuyama, "Distortion measures for speech processing," *IEEE Trans. Acoust., Speech, Signal Processing*, vol. ASSP-28, pp. 367–376, Aug. 1980.
[23] T. E. Tremain, "The government standard linear predictive coding algorithm: LPC-10," *Speech Technol.*, vol. 1, pp. 40–49, Apr. 1982.
[24] J. T. Buck, "Vector quantization code book distortions as features

for maximum likelihood classification of isolated words,'' in *Proc. 1984 IEEE Global Telecommun. Conf. (GLOBECOM)*, Atlanta, GA, Nov. 1984, pp. 9.3.1-9.3.5.

[25] G. R. Doddington and T. B. Schalk, ''Speech recognition: Turning theory to practice,'' *IEEE Spectrum*, vol. 18, pp. 26-32, Sept. 1981.

[26] R. G. Leonard, ''A database for speaker-independent digit recognition,'' in *Proc. 1984 ICASSP, IEEE Int. Conf. Acoust., Speech, Signal Processing*, San Diego, CA, Mar. 1984, pp. 42.11.1-42.11.4.

[27] L. R. Rabiner and M. Sambur, ''An algorithm for determining the endpoints of isolated utterances,'' *Bell Syst. Tech. J.*, vol. 54, pp. 297-315, Feb. 1975.

[28] L. Lamel *et al.*, ''An improved endpoint detector for isolated word recognition,'' *IEEE Trans. Acoust., Speech, Signal Processing*, vol. ASSP-29, pp. 777-785, Aug. 1981.

[29] G. R. Doddington, ''Voice authentication gets the go-ahead for security systems,'' *Speech Technol.*, vol. 2, pp. 14-23, Sept./Oct. 1983.

[30] B. S. Atal, ''Automatic speaker recognition based on pitch contours,'' *J. Acoust. Soc. Amer.*, vol. 52, pp. 1687-1697, 1972.

Part 6
Real-Time Implementation

Hardware Realization of Waveform Vector Quantizers

BERTRAM P. M. TAO, MEMBER, IEEE, HÜSEYIN ABUT, SENIOR MEMBER, IEEE, AND ROBERT M. GRAY, FELLOW, IEEE

***Abstract*—A real-time full search vector quantization system for speech waveform coding is implemented using LSTTL and CMOS devices. The system consists of low-pass filters, A/D and D/A converters, an algorithm for discriminating voiced and unvoiced speed, a full search vector quantizer encoder and decoder, and a microprocessor-based controller.**

The system is designed to operate at two possible rates: one bit/sample using a dimension 8 vector quantizer (6500 bits/s) or 2 bits/sample using a dimension 4 vector quantizer (13 000 bits/s). In both cases the codebooks have rate 8 bits/vector. Separate codebooks were designed for voiced and unvoiced speech based on a training sequence of 640 000 samples containing five different speakers.

The subjective and quantitative results are compared to both simulations and with a real-time array processor based implementation.

I. Introduction

WITH the increasing importance of digital speech in military, industrial, and consumer applications, a variety of techniques have been developed. The low rate systems operating under 4800 bits/s are mostly voice coders such as LPC systems which transmit a parametric representation of a frame of speech. Higher rate systems are mostly waveform coders that transmit information sufficient to produce a reproduction that "looks like" the original. The low rate systems are useful in systems with severe memory or communication capacity constraints, but they are complex and are sensitive to background noise, communication channel errors, and multiple speakers. Waveform coders require more rate, but they are usually simpler and more robust against all kinds of errors.

Numerous sophisticated waveform coders have been developed and implemented during the past few years. An example of such a sophisticated and implementable coding system is a subband coder using adaptive transform coding or adaptive DPCM in the subbands. Such techniques are implementable using special purpose chips or general purpose DSP chips. Most such schemes eventually use some form of scalar quantization following a transformation on input blocks. For example, the block may be Fourier transformed and then the Fourier coefficients separately quantized with varying bit allocations. Such techniques can be further fine tuned by adaptation.

From an information theoretic point of view, however, scalar quantization, even if preceded by optimal transformations and incorporating adaptation, is not the optimal means of quantizing a sequence of random vectors if by optimal we mean in the sense of minimizing an average distortion. The basic theorems of Shannon theory state that even for memoryless processes, one can always do better by quantizing an entire vector rather than separately handling components. Intuitive ad hoc schemes such as transform coding and DPCM are popular, however, because they tend to be simple, they work quite well for a variety of sources, they are easy to implement, and, in addition, no better techniques have been available. In particular, Shannon theory promises the existence of optimal vector quantizers, but provides no constructive techniques for designing them.

During recent years many simulation studies have been conducted on a family of design algorithms for vector quantizers that use a generalization of Lloyd's original PCM design technique, a technique that is currently commonly used for minimum average cost cluster analysis and pattern recognition. (See, for example, the papers and references in the March 1982 Special Issue on Quantization of the IEEE Transactions on Information Theory for both the classical papers and modern surveys of these algorithms.) The design technique is known to produce at least a locally optimal vector quantizer for a given distortion measure and for either a probabilistic source model or a very long training sequence of data produced by the source to be compressed.

Simulations have been conducted for a variety of information sources, including speech waveforms [1]. The transmitter or encoder is conceptually simple since it finds a minimum distortion or nearest neighbor codeword for an observed source vector by searching a codebook of reproduction vectors or templates in a ROM where the codebook is designed off-line by the design algorithm. This operation requires no transforming, variable bit allocation, adap-

Manuscript received September 14, 1982; revised June 15, 1983. This work was supported in part by the Joint Services Electronics Program at Stanford University, Stanford, CA 94305. Parts of this paper were presented at the IEEE International Conference on Acoustics, Speech, and Signal Processing, Paris, France, April 1982, and the IEEE International Symposium on Information Theory, Les Arcs, France, June 1982.

B. P. M. Tao is with Hycom, Inc., Irvine, CA 92714.

H. Abut is with the Department of Electrical and Computer Engineering, San Diego State University, San Diego, CA 92182.

R. M. Gray is with the Information Systems Laboratory, Stanford University, Stanford, CA 94305.

Reprinted from *IEEE J. Selected Areas in Commun.*, vol. SAC-2, pp. 343–352, March 1984.

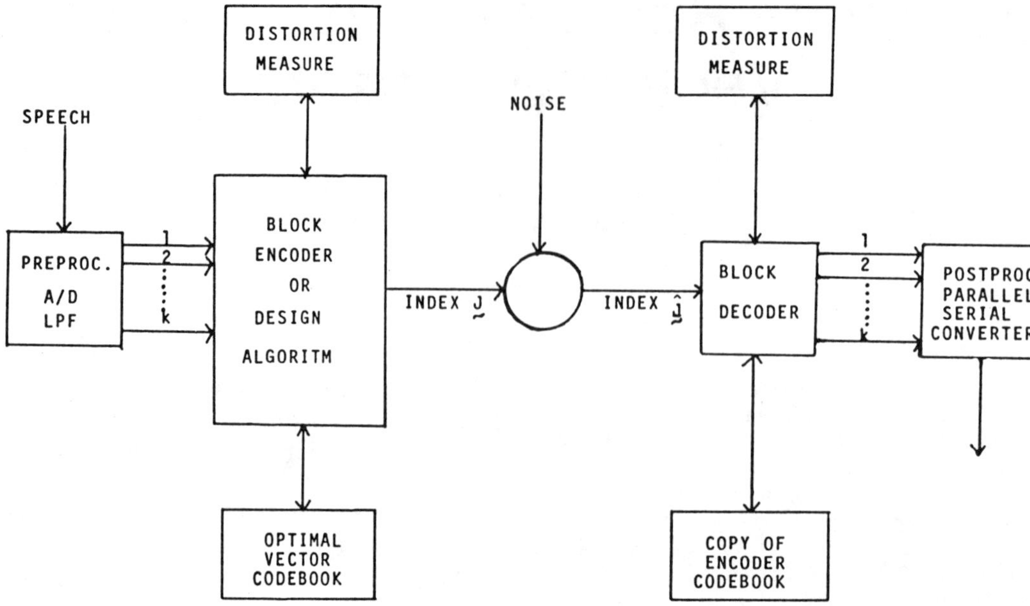

Fig. 1. General block diagram of a vector quantization system applied to speech waveforms.

tation, prediction, or other operations. It does, however, require very large memories and a large search effort. Both requirements precluded hardware implementation during the early days of algorithm development. With faster microprocessors and cheaper and larger memories these problems are no longer significant. The research reported in this paper was aimed at developing a real-time hardware vector quantizer using mostly "off-the-shelf" economical and industry standard IC devices coupled with two special purpose processors—one to perform the vector quantization and another to extract voicing information. A similar project was accomplished simultaneously but independently using a similar approach by Y. Yamada of Osaka University [4].

The quantizer was designed using such components rather than a DSP chip for two reasons. Several of the more famous DSP chips were not available to us when this project was begun, and almost all of the available chips were not well matched in structure or in timing to our particular application. While capable of doing signal processing of high complexity, the chips were not efficient for simple nearest neighbor searches of large codebooks. For example, the Texas Instruments TMS320 chip has the smallest basic cycle time of 200 ns and it is optimized for particular DSP applications. The internal architecture of this chip, however, is such that it would require a minimum of 4 cycles to perform distortion computations, which is 300 ns slower than needed here.

The paper is organized as follows. Section II contains a description of vector quantizers for waveform coding. Section III summarizes some old and new simulation results, including some real-time simulations using an array processor. Sections IV–VI describe the implementation in detail. In the final section the results of the various experiments are compared and possible improvements and modifications are described.

II. Waveform Vector Quantizers

A. Full Search Quantizers

A block diagram of a k-dimensional vector quantizer (VQ) applied to speech waveforms is shown in Fig. 1. This encoding scheme is described by a vector dimension k, a rate of R bits/vector (or $r = R/k$ bits/sample or $R_B = r \cdot F_S$ bits/s if the sampling rate is F_S), a codebook $\hat{A} = \{y_0, \cdots, y_{N-1}\}$ of $N = 2^R$ real k-dimensional vectors, an encoder mapping α that maps an input source vector $x = (x_0, \cdots, x_{R-1})$ into a binary R-tuple or a binary channel vector $u = (u_0, \cdots, u_{R-1}) = \alpha(x)$ and a decoder mapping which transforms the received binary vector into a reproduction vector $y = \beta(u) \in \hat{A}$.

The decoder is a simple table lookup procedure. The optimal vector codebook $\hat{A}$ is stored into a sufficiently large memory to hold an accurate digital representation of $N = 2^R$ k-dimensional vectors; the binary channel vector $\hat{j}$ —possibly different from the encoder output j due to channel bit errors—is interpreted as the binary index of the reproduction vector to be output.

As in the Shannon theory of information, we define a nonnegative distortion measure $d(x, y)$ that measures the distortion or cost of reproducing an input vector x as an output vector y and then attempt to minimize the average distortion $Ed(x, \beta(\alpha(x)))$. Having average distortion as a performance criterion, the optimum encoder for a given codebook $\hat{A}$ follows the minimum distortion rule:

$$\alpha(x) = u \quad \text{if} \quad d(x, \beta(u)) \leqslant d(x, \beta(v)) \quad v \neq u \quad (1)$$

that is, send the binary vector which minimizes the distortion between the input vector and the reproduction. This is called a full search encoder since in general the encoder must search all $N = 2^R$ codewords to find the minimum distortion vector. On the other hand, for a given encoder α the optimum codebook $\hat{A}$ is formed as $\alpha(\boldsymbol{u}) = \text{cent}(\boldsymbol{u})$ where $\text{cent}(\boldsymbol{u})$ denotes the generalized centroid or center of gravity of $\boldsymbol{u}$ defined as the vector y minimizing the conditioned expectation $E(d(x, y)|\alpha(x) = \boldsymbol{u})$, if such a centroid exists. Using a long training sequence to estimate averages, these two facts yield the basic iterative VQ design technique detailed in [1] and used here.

A major problem with such full search VQ's is the large search effort required to select a minimum distortion codeword. For example, for a 1 bit/sample 8-dimensional code with a squared error distortion measure, the encoder must compute 256 error energies of 8-dimensional vectors. In order to reduce the search effort of a VQ, some form of a tree searching technique can be employed in the encoder/decoder structure leading to a tree search vector quantization (TSVQ) system. Such systems trade off performance for speed and are considered for waveform coding applications in Gray and Linde [2] and in Gray and Abut [3]. Our focus here, however, is to demonstrate the implementability of a full search VQ of single and multiple codebook systems.

B. Multiple Codebook Systems

In vector quantization and similar systems multiple codebooks have often proved a useful means of providing codes that are better suited to the local stationary behavior of sources such as speech. Single codebooks are designed for the long run average behavior and this can often result in a lack of reproduction words that match the short term behavior well. For example, words that match the mixture of a very noisy source (unvoiced speech) and an almost periodic source (voiced speech) may not well match source words from one or the other. Multiple codebooks may be designed for these separate modes of behavior and thereby better match the local variations. Such multiple codebooks often provide better performance for a given block length for sources such as speech whose short term locally stationary behavior can vary widely from the long term globally stationary behavior. For example, the tree encoding systems of Stewart *et al.* [9] match waveform coders to one of several possible LPC models for the locally stationary behavior. More simply, Wong *et al.* [8] used a simple voiced/unvoiced/silence decision to design separate codebooks for voiced and unvoiced sounds in an LPC vector quantization system. The decision mechanism is used to preclassify the incoming data and the subsets of the training sequence are then used to design the separate codes. Such codes can be viewed as a simple form of adaptive coding where only a relatively small number of codes are permitted.

We considered a simple two-codebook vector quantizer using the voiced/unvoiced/silence decision approach. The particular classifier was chosen because it was relatively simple, because it seemed a natural means of classifying speech for a two-codebook system, and because it had proven useful in an LPC vector quantization system. No claim is made that this is the best means of classification; in fact the results indicate that this is not a good choice for a waveform coding system and that the performance gains do not justify the added complexity. The system does, however, demonstrate the feasibility of simple multiple codebook systems and other classification schemes, e.g., the vector quantized LPC classifiers of Stewart *et al.* [9] for tree encoding systems, may yield more significant improvement.

Fig. 2 is a block diagram of the two-codebook vector quantizer encoder using voiced/unvoiced classification where a decision of silence results in a single codeword rather than a codebook. Analog speech is bandlimited and sampled at a sampling rate of $F_S = 6400$ samples/s. Every 10 ms a block of 64 samples is fed to the classifier and a vector quantizer buffer. The classifier issues a voiced/unvoiced/silence decision SUV every 10 ms based on a 20 ms data window. A 10 ms buffer is required to maintain synchronization since the quantizer needs a class decision before doing anything. The VQ takes k samples at a time and one SUV information and then searches through the voiced or unvoiced codebooks to find the minimum distortion codeword. In the case of a silence input from the classifier the quantization stage can be bypassed totally. The binary coder is a 1:1 mapping of SUV information and the quantizer outputs onto $\{0,1\}^R$. Every 10 ms, $J = 64/k$ indexes and one SUV side information is transmitted. The classifier information of only 2 bits is required for each frame of 10 ms duration. If we assume that the source changes slowly compared to the vector lengths studied here, the data compression achieved by this technique can be significant as shown in the following numerical example. Let a particular speech segment of 100 frames —almost one second of speech and 64 samples/frame—be classified as 40 percent voiced, 20 percent unvoiced, and 40 percent silence. If all of these 100 frames were encoded by a rate $r = 2$ and dimension $k = 4$ quantizer, 12 800 bits would be needed. (There is no need for a side information channel.) If the scheme to be defined in (2) is used, then the number of bits needed for voiced, unvoiced, and silent portions would be 5200, 1320, and 80 bits, respectively. The total number of bits reduces to 6600 of which 200 bits are side information. This corresponds to a data rate reduction of nearly 50 percent.

Due to the computation requirements of the different blocks shown in Fig. 2, the system requires two parallel processors working in pipeline structure—one for the classifier and another one for the vector quantizer—both of which have sample-level fast operations. These two processors, the sample-data system, buffers, memories, the binary coder, and other units are integrated and controlled by a medium-speed general-purpose microprocessor, μP. Details of design and the performance results of this two-codebook VQ system will be discussed in Sections V and VI.

III. Computer Simulation Results

Vector quantizers were designed and tested in six experi-

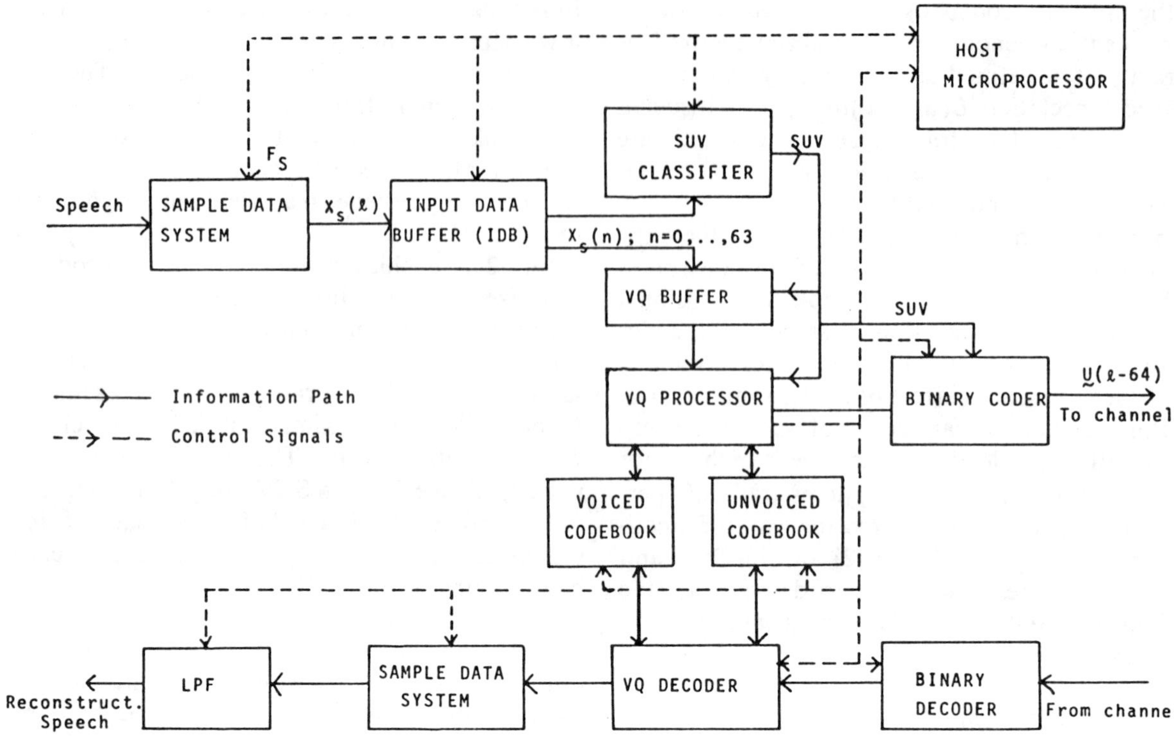

Fig. 2. Block diagram of the two-codebook VQ system.

ments for a variety of rates, dimensions, and data sets. Except for the two-codebook variable-rate cases, duplicate runs were made on separate computers using different programs. The first three experiments compare full search and tree search designs. The last three experiments investigate the performance of multiple codebook systems. Results from the first three and the last three experiments are summarized in Tables I and II, respectively.

Experiment 1: In Abut, Gray, and Rebolledo [1], codes were designed for rate $r=1$ bit/sample and dimensions $k=1,2,\cdots,8$ and rate $r=2$ bits/sample and dimensions $k=1,\cdots,4$. Full search VQ was used on a training sequence of 640 000 samples from five male speakers. The designed codebooks have been tested on a separate sequence of 78 000 samples recorded by a different speaker.

Experiment 2: In Gray and Abut [3], all of the above cases have been duplicated for binary tree searched vector quantizers and are presented here for comparison.

Experiment 3: The first experiment was repeated on training and test sequences in German. The design was accomplished off-line as usual, but the tests were conducted in real time using an array processor. There are a number of differences between the English and the German databases. The German data contained 20 min of recording from a German radio newscast spoken by several professional speakers. The data set was digitized at a rate of 8000 samples/s, whereas the English one had a rate of 6500 samples/s. Due to the limited computer storage capacity, the German data were then compressed into 8 bits using a 15 segment logarithmic quantizer. Full search VQ's were designed using a subset of 256 000 samples from the decompanded version of this database and a different set of 120 000 samples were used in testing the codebooks.

Experiment 4: The first database used in experiments 1 and 2 were decomposed into classes of voiced, unvoiced, and silent segments using a classifier similar to that in [5]. An optimal full search voiced codebook of rate one and dimension 8 was designed based on voiced segments only. Similarly, an unvoiced codebook of the same rate and dimension was obtained from unvoiced and silent segments. Class SNR's as well as the overall SNR were computed.

Experiment 5: The previous experiment was repeated for rate 2 and dimension 4.

Experiment 6: Variable-Rate Universal Coding: The voiced data subset was employed in designing a rate $r=2$ dimension 4 codebook. A rate $r=1$ dimension 4 codebook was obtained for the unvoiced segments. Finally, only 2 bits of SUV classifier side information are transmitted for every silence interval of 10 ms. The overall transmission rate including the side information channel is given by

$$R_B = (\text{SUV} + 64 \cdot B) \cdot (\text{frame rate}) \text{ bits/s} \tag{2}$$

where

$$B = \begin{cases} 0 & \text{silence} \\ 1 & \text{unvoiced} \\ 2 & \text{voiced} \end{cases}$$

the frame rate is 100 frames/s and SUV is a 2 bit side information.

Observation of Table I indicates that vector quantization provides a 6–8 dB performance improvement over optimal scalar quantization at the rates considered.

The quantitative results from the German database conform to those in English, even though the testing conditions were significantly different and, in particular, the

TABLE I
SIMULATION RESULTS FROM SINGLE CODEBOOK EXPERIMENTS

		Experiment 1		Experiment 2		Experiment 3	
Rate	Dimension	Full Search English SNR (db)		TSVQ English SNR (db)		Full Search German SNR(db)	
Bits per sample	k	Design	Test	Design	Test	Design	Real-Time Test
1	1	2.01	2.05	2.01	2.05	2.16	-
1	2	5.24	5.25	5.12	5.14	3.76	-
1	4	7.09	6.99	6.37	6.42	5.29	-
1	8	9.74	8.82	8.85	8.02	8.89	8.25
2	1	6.50	6.41	6.27	6.21	6.46	-
2	2	9.98	9.72	9.42	9.17	8.13	-
2	4	13.46	12.70	12.60	11.88	12.87	11.90

TABLE II
SIMULATION RESULTS FROM MULTIPLE CODEBOOK EXPERIMENTS

	Experiment 4		Experiment 5		Experiment 6	
	Design	Test	Design	Test	Design	Test
Encoder Rates (bits/sample)						
Voiced Frames	1.0	1.0	2.0	2.0	2.0	2.0
Unvoiced Frames	1.0	1.0	2.0	2.0	1.0	1.0
Dimensions						
Voiced Frames	8	8	4	4	4	4
Unvoiced Frames	8	8	4	4	4	4
Class SNR's (db)						
Voiced Frames	10.73	9.05	14.84	13.24	14.84	13.24
Unvoiced Frames	9.06	6.43	10.42	8.52	5.97	5.21
Overall SNR (db)	10.32	8.74	13.47	12.53	11.12	11.59
Side Information Rate (bits/s)	100	100	100	100	200	200
Overall Transmission Rate (bits/s)	6,600	6,600	13,100	13,100	6,852	7,655

sampling rate was different. A dozen or more subjects, bilingual in both languages, concluded uniformly that the German reconstructed speech sounded better than the English ones. This is attributed to the faster sampling rate in the German database.

The SNR values for TSVQ's are within 1 dB of the full search results. Moreover, both systems were rated very close in numerous informal listening tests.

It is clear from Table II that the two-codebook system of Fig. 2 did not yield significantly superior results over that of the single codebook case. The improvemènt was always less than 1 dB. The variable rate cases of experiment 6, however, resulted in SNR's of 11.12 dB and 11.59 dB for the training and the test data, respectively. These numbers are within 1–2 dB of the results in experiment 5, but the observed rate reduction is very significant (41.6–47.7 percent). In addition, the recordings from these two experiments were presented to subjects and they have indicated that the differences were negligibly small.

IV. Implementation of a Single Codebook System

Here we describe the system structure for a single codebook waveform vector quantizer with an encoding rate of either 1 bit/sample at dimension 8 or 2 bits/sample at dimension 4. Thus, both systems use 8 bit codebooks. The hardware architecture and the flowchart of the system are depicted in Fig. 3(a) and (b), respectively. It consists of the following components:

1) sample-data system consisting of a low-pass filter and an analog-to-digital converter (ADC).

2) 64-sample input data buffer (IDB).

3) quantization processor $Q(\cdot)$.

4) microprocessor system (μP), including a Z80A CPU, RAM, EPROM, and I/O buffer logic.

5) codebook ROM.

6) output driver including a digital-to-analog converter (DAC), and a reconstruction LPF.

The overall number of devices is 82, comprising common "off-the-shelf" LSTTL, CMOS, and NMOS devices. The following is a description of the major functional modules and their respective chip count.

A. Preprocessor

The analog speech $x(t)$ is filtered by a seventh-order Cauer low-pass filter with a cutoff frequency of 3150 Hz

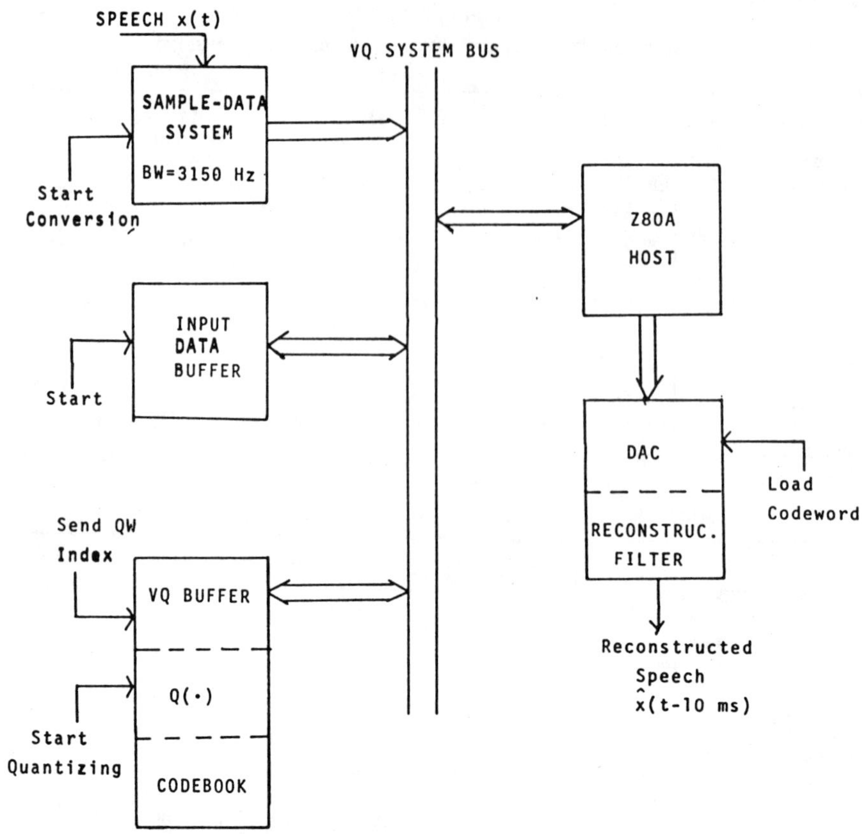

(a)

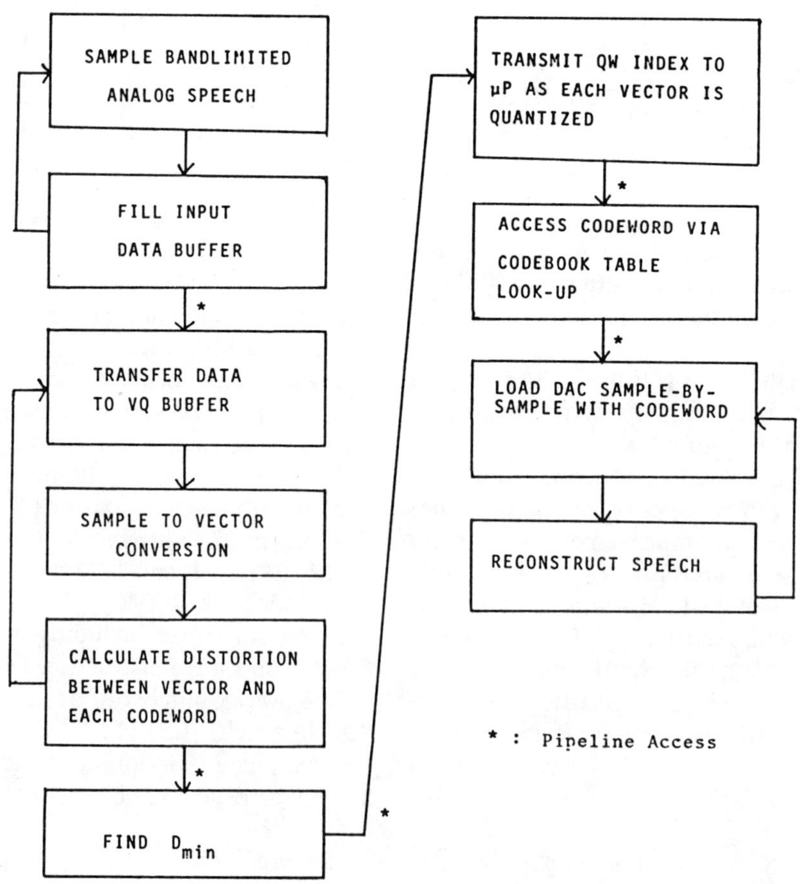

Fig. 3. (a) The hardware architecture of the single codebook VQ system. (b) Flow diagram of the single codebook VQ system.

and the gain is manually adjusted to match the dynamic range of the ADC, ± 5 V. The filtered signal is digitized at a sampling rate of $F_S = 6400$ Hz, and the ADC accuracy is 8 bits± 1 LSB. The sampled 8 bit data are then sequentially and byte-wise loaded into the 64-sample IDB. A new frame is started every 64 samples, and after the last sample of the current frame but before the first sample of the next frame, all 64 samples are transferred byte-wise to the quantization processor buffer (VQ buffer). This transfer is completed in 57 μs, at a clock rate of $F_C = 800$ ns/sample, and is initiated by the μP by forcing $Q(\cdot)$ and IDB into a special direct-memory-access (DMA) mode. The preprocessor module requires 8 devices including the VQ buffer, consisting of mostly analog chips, a RAM, and a FIFO.

B. Quantization

Following the DMA, $Q(\cdot)$ processor starts the first of $J = 64/k$ quantization cycles independent of the μP. The samples are accessed from the VQ buffer in modulo-k base to simulate a serial-to-parallel converter where k is the vector size or the dimension of the system. The sequence of vectors and samples can be expressed in the following manner:

$$x^l = \{x^l(n) - ((k\cdot l + n))_m;\ n = 0,\cdots,k-1,\quad m = 0,\cdots,N-1,\ l = 0,\cdots,J-1\} \quad (3)$$

where x^l is the lth of $J = 64/k$ vectors in which $x^l(n)$ is the nth sample, and $N = 256$ is the number of codewords in the codebook ROM $\hat{A}$. The operation $((\cdot))_m$ denotes the modulo addressing of the VQ buffer. Each vector x^l is quantized independently of adjacent vectors in full search cycles and each cycle is initiated by the μP.

The input vectors x^l are compared to 8 bit accuracy in a mean-square distortion sense to every codeword y_i for $i = 0,\cdots,255$ in the codebook ROM, after which a minimum distortion codeword is selected. The codebook entries in ROM are extracted from the final codebooks of experiment 4. Since the system has an 8 bit architecture, high precision numbers from computer simulations are scaled down to 8 bit numbers before burning the PROM's.

The quantization process takes place in two pipelined sections of hardware. The first one computes the squared-error distortion between x^l and y_i at a particular codeword i, and the second module keeps track of a running minimum distortion.

Due to design requirements, one full search cycle time (FSC) has been kept to 500 ns per distortion computation. This speed is accomplished by the use of high-speed memories, low-power Schottky logic, and a table lookup for the squaring operation. In addition, any accumulation overflow is monitored in parallel with the distortion computation by employing a look-ahead logic. In particular, if an overflow occurs, it is flagged without waiting until the end of the calculation, therefore no extra time is required for overflow detection.

Since one distortion calculation is carried out every 500 ns, the time needed for one full pass through the codebook is

$$\begin{aligned} T_{FSC} &= (8\ \text{samples/vector})\ (500\ \text{ns/sample}) \\ &\quad \cdot (256\ \text{vectors/cycle}) \\ &= 1.024\ \text{ms/cycle}. \end{aligned}$$

Actually, T_{FSC} is a few microseconds more than 1.024 ms in order to account for the delay resulting from the pipeline synchronization. After each T_{FSC}, the final codebook index is loaded into the quantizer word mail box register (QW). Every k sample clocks or 1.248 ms the host μP reads the value of QW, but only after a full search cycle is completed. The quantizer module requires 45 chips including SSI and MSI circuits, and a high-density bipolar RAM.

C. Postprocessing

If the system is in the encoding mode then the host μP sends the index of the minimum distortion codebook to the binary coder so that the digital signal is conditioned for the channel. If it is in the decoding mode, however, then the μP points to the codeword j in the ROM with the first byte in row $j\cdot k + k - 1$, the second byte in row $j\cdot k + k - 2$, etc. The bytes are accessed synchronously with the sampling rate F_S and are sent to the output driver section for digital-to-analog conversion. This process is completely controlled by the μP. After the DAC, the reconstructed signal is filtered by an LPF similar to that in the preprocessing stage.

After sample reconstruction, $Q(\cdot)$ sends a flag to the μP to indicate completion. Although the μP software can predict the end of the cycle, this flag is monitored as a safety measure in case of synchronization problems. At this point, a new frame is loaded into the IDB, and the μP starts another 64-sample DMA cycle from the IDB to VQ buffer, which, in turn, begins another quantization cycle. This stage needs nine chips including analog devices.

D. Host Microprocessor μP

It is observable from the system architecture of Fig. 4 that the primary functions of the μP are timing, control, and the interfacing of various modules. There is very little numerical processing undertaken by the host. In addition to the Z80A system, there are six modules on the μP bus: one general purpose working space RAM as a data buffer, a ROM containing the software, DAC system, an I/O buffer for interfacing with the VQ system bus, an external multiplier chip for high-speed multiplications (needed in the clipping stage of the classifier module and it is used only in the design of the two-codebook encoding scheme), and a ROM for the codebook $\hat{A}$. The total number of chip count here is 20.

E. Testing

The system has been tested both quantitatively by monitoring the resulting minimum distortion d_{min} and subjec-

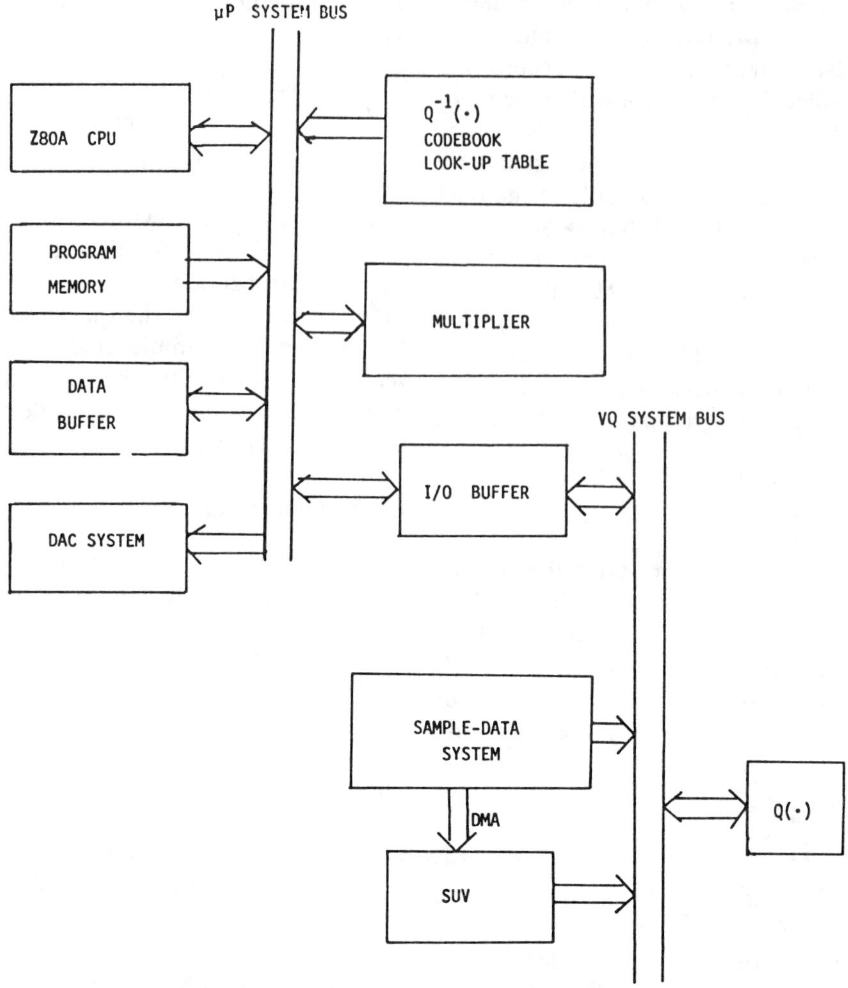

Fig. 4. Bus architecture for the VQ system and the host microprocessor.

tively by informal listening tests. The hardware was designed for a dynamic range of 8 bits with a resolution of one LSB. Therefore, the accuracy of quantization is 1 part in 256 and the squared error distance between any two vectors can be a minimum of 1 bit of full scale 8 bits. In the quantitative measurements, vectors from a test data set were used; and hence, $d_{\min}$ for each measurement was exactly known and the minimum distortion codebook index j could be predicted. For this particular test, all 256 codewords in the ROM were used as the input data sequence and the resulting $d_{\min}$ was zero. The minimum distortion codeword indices were found without a single error.

Next the system was tested with real speech prerecorded and played back on a cassette tape deck containing the English alphabet, digits, words, and phrases. The distortion was very low during the silent intervals because the codebook ROM contained a null vector and vectors with magnitudes close to zero. The test yielded fairly intelligible but noisy output for voiced speech segments. There were, however, severe errors in encoding fricative sounds because the employed codebook was not optimal for unvoiced sounds. The minimum distortion measurements were recorded for each vector and the accumulated measurements resulted in an overall SNR of 6.25 dB. This figure is about 2 dB below the simulation results of Section III. We believe that the major cause of degradation is due to the dynamic range problem (i.e., the incompatibility between the codebook and input signal variances) and the absence of a reasonable AGC mechanism to facilitate an 8 bit A/D conversion.

V. Design of a Two-Codebook VQ System

We present the design structure of a two-codebook waveform quantizer with a rate of 2 bits/sample and dimension 4 for voiced data, 1 bit/sample and dimension 4 for unvoiced sounds, and a null vector for silent periods. The basic block diagram of the system is given in Fig. 2. Here we discuss the actual design of various modules. The overall system of Fig. 2 can be broken into the following building blocks:

a) preprocessor comprising a preamplifier, an LPF, the sample-data system (ADC), and an input buffer (IDB).

b) SUV classifier.

c) quantization processor $Q(\cdot)$.

d) voiced and unvoiced codebooks $\hat{A}_v$, $\hat{A}_u$.
e) host microprocessor system μP.
f) transmitter coder (XMTR) and receiver decoder (RCVR).
g) postprocessor.

We will not be discussing items a), c), and g) since they are identical to the corresponding blocks in the case of the single codebook design.

SUV Classifier

It is basically a pitch detector working as a silence/unvoiced/voiced (SUV) classifier, issuing one decision every 10 ms based on 20 ms analysis windows with 50 percent overlapping. The reason why we build a full pitch detector for this restricted task has been to avoid duplication of work in our present effort on designing pitch synchronous speech compression systems based on VQ encoding techniques. The SUV classifier module is a simplified version of the trilevel clipping and autocorrelation method by Dubnowski *et al.* [6]. The simplification entails the use of 20 ms windows, instead of 30 ms, in which the clipping levels are determined every 10 ms. Every 10 ms a new segment is loaded, thus new levels are computed 100 times/s. While the new segment is loaded, the trilevel clipping process and subsequent autocorrelation process take place. The output of this module is a 2 bit SUV information. The design details of this module can be found in [7].

Voiced and Unvoiced Codebooks, $\hat{A}_v$, $\hat{A}_u$

Once a SUV decision is made, the quantization processor $Q(\cdot)$ switches to the appropriate codebook and it freezes there until a different SUV value shows up. This switching and freezing action is done by the host microprocessor.

Microprocessor System μP

The Z80A CPU is the heart of this part. The μP monitors the system as in Section IV using the Z80A software. Specific tasks are:

1) initialize the system.
2) start each processing cycle.
3) load and control the SUV queing register, Q_n.
4) perform clipping level computations.
5) SUV decisions.
6) coder and decoder actions.
7) I/O control.

Transmitter Coder (XMTR) and Receiver Decoder (RCVR)

The transmitter coder packs the data and the side information of each 10 ms frame, standardizes the package in some known format, and sends the final bit stream through the channel. In the receiver side the above functions are undone by a binary decoder.

The data packets are 66 bits long, consisting of a 2 bits SUV header and 64 bits codebook indexes. The indexes are 8 bits long for each voiced vector of dimension 4. In the case of unvoiced data, however, the indexes are only 4 bits. Finally, for silent periods a string of 64 zeros are attached to the appropriate header. Experiments have shown that more than 30 percent data compression could be achieved if the zeros in silent periods are suppressed.[1]

As previously mentioned, the postprocessor reconstructs a replica of the original speech with a 10 ms delay. All the design details of the functional units in this section are explained by Tao [7].

VI. Concluding Remarks

Hardware realizations of both single codebook full search vector quantizers and multiple codebook full search vector quantizers were successfully completed and tested.

Comparison of the resulting performance with that of several old and new simulations indicates that the performance of the hardware VQ's were about 2 dB inferior to the simulated VQ's. In addition, this degradation was clearly noticeable in listening tests as well. We believe that this is primarily due to the absence of any AGC mechanism in A/D and D/A conversions. The reason for not including such a block was to follow the simulations exactly.[2] To improve the performance, an AGC mechanism is recommended to match the dynamic range of the input speech to that of the codebook. In this way, absolute accuracy to 8 bits can be maintained at the cost of a few bits of side information for the gain factor. However, the hardware developed in this study proved that waveform coding data compression systems can be implemented using full search 8 bit codebooks with readily available devices. It should also be noted that the architecture of the VQ processor designed here is within VLSI capabilities [7] and it can be developed into a single-chip codec. Parallel use of new 16 or 32 bit devices should also improve the quality of such coders by permitting higher accuracy codebooks and the resolution in distortion computations. The architecture of such a system would be essentially that outlined here modified to take advantage of the better microprocessor. The implementability of full search VQ of up to 8 bits was further attested by the real-time simulations conducted with the aid of an array processor. The superior quality of these simulations supports our belief that improved codebook and arithmetic accuracy will yield improved quality.

[1] This percentage is a very conservative figure and the thresholds in the classifier have been adjusted to make errors, if any, in favor of classifying a silent period as voiced or unvoiced depending upon what comes before or after the questionable segment. Depending on the data set and the classifier thresholds, close to 50 percent data rate reduction can be achieved as in TASI systems.

[2] Training database and the test sequence were used in designing the codebooks without any signal shaping or normalization.

Acknowledgment

The authors would like to thank National Semiconductor Speech Processing Laboratory for permission to use their facilities and for supplying the necessary electronic parts. The authors would also like to acknowledge Prof. D. Wolf and his research team at Institut für Angewandte Physik der Johann-Wolfgang-Goethe Universität, Frankfurt, Germany, for their help in realizing the real-time simulations.

References

[1] H. Abut, R. M. Gray, and G. Rebolledo, "Vector quantization of speech and speech-like waveforms," *IEEE Trans. Acoust., Speech, Signal Processing*, vol. ASSP-30, pp. 423–435, June 1982.

[2] R. M. Gray and Y. Linde, "Vector quantizers and predictive quantizers for Gauss Markov sources," *IEEE Trans. Commun.*, vol. COM-30, pp. 381–389, Feb. 1982.

[3] R. M. Gray and H. Abut, "Full search and tree searched vector quantization of speech waveforms," *Proc. IEEE ICASSP'82*, May 1982, Paris, France.

[4] Y. Yamada, private communication.

[5] B. S. Atal and R. L. Rabiner, "A pattern recognition approach to voiced-unvoiced-silence classification with applications to speech recognition," *IEEE Trans. Acoust., Speech, Signal Processing*, vol. ASSP-24, pp. 201–212, June 1976.

[6] J. J. Dubnowski, R. W. Schafer, and L. R. Rabiner, "Real-time digital hardware pitch detector," *IEEE Trans. Acoust., Speech, Signal Processing*, vol. ASSP-24, pp. 2–8, Feb. 1976.

[7] B. Tao, "A real-time speech data compression system using vector quantization," M.Sc. thesis, Dep. Elec. Comput. Eng., California State Univ., Long Beach, 1982.

[8] D. Y. Wong, B.-H. Juang, and A. H. Gray, Jr., "An 800 bit/s vector quantization LPC vocoder," *IEEE Trans. Acoust., Speech, Signal Processing*, vol. ASSP-30, pp. 770–780, Oct. 1982.

[9] L. C. Stewart, R. M. Gray, and Y. Linde, "The design of trellis waveform coders," *IEEE Trans. Commun.*, vol. COM-30, pp. 702–710, Apr. 1982.

REAL-TIME SPEECH COMPRESSION WITH A
VLSI VECTOR QUANTIZATION PROCESSOR

Grant Davidson, Terry Stanhope, R. Aravind, and Allen Gersho

Department of Electrical and Computer Engineering
University of California
Santa Barbara, CA 93106

ABSTRACT

Vector Quantization is an attractive block coding scheme which allows more efficient use of a given channel capacity. The computational complexity of the codebook search, however, limits the practical applications of Vector Quantization -- especially in real-time situations. This paper presents a special purpose processing architecture which supports real-time search of large codebooks. The core of the computation required for a squared-error distortion measure is accomplished using a VLSI pattern matching chip. We describe a pipeline architecture which reduces the effective computation time for a single vector component to one period. The chip supports a selectable vector dimension of 2, 4, 6, or 8 and can be used with any codebook size consistent with the input vector rate and the chip's throughput. We have built and tested a 4-micron NMOS chip which supports up to four million squared-error distance calculations per second. We outline three approaches to real-time Vector Quantization which use this chip to do the pattern matching: two waveform coding processors using Vector Pulse Code Modulation and Adaptive Vector Predictive Coding, and a novel approach to rapid codebook design.

1. MOTIVATION

Vector Quantization (VQ) has recently emerged as a promising new direction for speech waveform coding [1]-[3]. In its most rudimentary form, VQ may be used as a generalization of PCM, called VPCM for Vector PCM, where a block or *vector*, of consecutive samples of the input waveform is treated as one entity. The vector, of dimension k, is the input to the VQ-encoder. which compares this vector with a *finite* set of stored reference vectors known as *patterns* or *codevectors* of the same dimension. Using a suitable measure of fidelity, the best matching pattern in the *codebook*, the set of stored patterns, is selected by the encoder to represent the input vector. The address or *index* of this codevector is transmitted over the channel to the decoder. This index identifies the chosen codevector to the decoder, which does a table look-up on a copy of the codebook to fetch the codevector. This codevector forms the encoded version of the original input vector. The procedure is repeated for successive input vectors, allowing the decoder to reconstruct an approximation to the original waveform. The size of the codebook, N, is the number of stored vectors in the codebook and determines the bit-rate, $r = \log_2 N / k$ in bits/sample where $\log_2 N$ bits are used to represent an index.

In more sophisticated VQ-based encoding systems the input vector to the pattern-matching operation rather than directly containing the waveform samples is extracted from the original waveform in a variety of ways. See for example [1]-[3]. Imbedded in all such schemes is the basic VQ building block, which we call the *codebook-search processor*. This processor receives a vector and with access to a codebook, performs the pattern-matching operation. The fidelity measure used by the processor is in practice a dissimilarity or *distortion* measure (d.m.), which quantifies the distortion in approximating the input by a particular codevector. The most widely used d.m. is the squared-error d.m., where the distortion is given by the sum of the squared differences between corresponding components of the input and codevector.

The attractiveness of VQ is its theoretical optimality in minimizing average distortion for a given rate as the dimension approaches infinity. Unfortunately, the price paid for this optimality of VQ is an exponential growth in computational complexity with increasing dimension k. For example, a VPCM coder with exhaustive search computes the distortion between the input and *every* codevector before deciding on the best match. This computation requires 2^{kr} squaring operations per sample. Hence the encoding complexity is the critical limitation impeding the practical use of VQ-based waveform coders. Low-dimensional vector quantizers (of moderate complexity) will not adequately exploit the power of VQ. Although VQ has been studied for several years by computer simulation, this paper appears to be the first report of a real-time pipeline architecture VQ implementation. One prior publication has reported a real-time VQ implementation using off-the-shelf components, but with an inevitably higher device count [4].

We now consider in more detail the essential but intensive search operation which is central to all VQ-based coders employing the squared-error d.m.. Let $\mathbf{x}$ be a vector input to the codebook-search processor, $\{\mathbf{a}_i\}_{i=1}^{N}$ be the codebook, and let $x(j)$ denote the jth component of $\mathbf{x}$. For k-dimensions and codebook size N, the following sequence of operations are required for every vector $\mathbf{x}$:

Codebook-Search Algorithm:

```
set E = ∞
for i = 1 to N
    set e_i = 0
    for j = 1 to k
        set d_i(j) = x(j) - a_i(j)
        set e_i = e_i + [d_i(j)]^2
    if e_i < E  set I = i and E = e_i
return I
```

On examining the above sequence we can make these observations:

a) The input data is processed sequentially in several stages. Each stage performs a specialized task and passes the results to the next stage. In fact, as we shall see later, the squaring operation can itself be

Reprinted from *Proc. IEEE Conf. Acoust., Speech, Signal Processing,* vol. 4, pp. 1437–1440, March 1985.

decomposed into three sequential steps.

b) The vector components are processed independently and in an identical manner until the e_i are calculated.

c) The entire process is highly repetitive: the top-level procedure calls the same routines for every codevector and the same procedure repeats for every input vector.

The process is eminently suited to a pipeline implementation. A dedicated pipelined architecture would perform the various stages simultaneously, allowing significantly greater throughput than would be possible with a general-purpose signal processor (such as the TMS320) which performs one operation at a time. In Section 2, we describe such an implementation of the above codebook-search algorithm centered around a semi-custom VLSI chip which performs the pattern-matching arithmetic. In Section 3, we examine several applications of this implementation to real-time VQ-based encoding systems.

2. THE CODEBOOK SEARCH PROCESSOR

In this section we describe a particular implementation of the codebook-search algorithm, referred to as the *Codebook-Search Processor (CSP)*. This project extends the work that was reported in [4] where real-time VQ was implemented without the use of custom VLSI. Our CSP performs the basic operations required of all VQ coders which use the squared-error distortion measure, and efficiently finds the address or index of the codevector which is nearest to the input vector. The CSP is organized around a chip, called the *Pattern Matching Chip* (PMC), that we custom-designed to perform the pattern matching computations. The addressing and overall control is left to off-chip support circuitry.

The way that a codebook is searched and the size of the codebook determines the throughput of a real-time vector quantizer. Given T_{vc}, the time required to do a squared difference calculation for a single vector component in the codebook-search algorithm, the throughput TP for exhaustive full-search using sequential processing of the vector components is given by:

$$TP = 1/(T_{vc} \times k \times N) \textit{ input vectors / sec.}$$

Our approach was to design VLSI hardware which minimizes T_{vc} in the sequential processing of vector components for a given clock rate.

The CSP architecture is shown in Figure 1. It receives input vector components in a bit-parallel, word-serial fashion and produces one index (having $\log_2 N$ bits) at the end of each search period. The input words are received serially and grouped into a k-dimension vector using the FIFO. When a new codebook search begins, k words from the FIFO are transferred in parallel to the *Vector Register* (VR) which holds that input vector for the ensuing nearest-neighbor search. The VR is addressed sequentially so that one component per clock cycle is transferred to the PMC.

In this architecture, each codebook is stored in one memory bank, and each bank is composed of either ROM or RAM. We have shown several memory banks in Figure 1 to allow for VQ encoders with multiple codebooks. Codevectors are stored sequentially using k contiguous memory locations per vector. The Codevector Data Bus provides access to the memory by devices external to the CSP. The Address Generator supplies addressing information for the codebook. The Address Generator selects the memory bank and accesses one component of one codevector at a time. In the simplest case, the Address Generator is a binary counter which is incremented once each machine cycle. The Address Bus allows random access of the codebook by an external device.

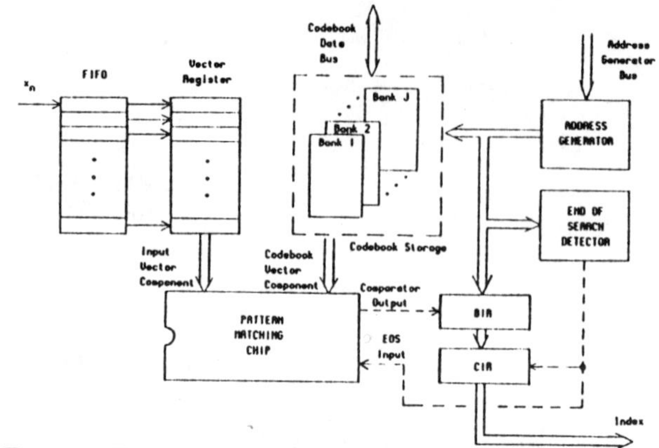

Figure 1. The Codebook Search Processor Architecture

Whenever the PMC finds a codevector which approximates the input vector better than the best found so far, it toggles the Comparator Output line, which in turn latches the index for this codevector into the *Best Index Register* (BIR). The *End-of-Search* (EOS) Detector signals the PMC when the current search is complete and simultaneously transfers the contents of the BIR to the *Channel Index Register* (CIR). The number stored in the CIR is then the index corresponding to the codevector which approximates the input vector best.

The Pattern Matching Chip (PMC)

As suggested in the first section, we selected a pipelined architecture for the CSP. There are seven pipeline stages in the PMC which do all the arithmetic for the pattern matching. Each stage operates in a single clock cycle and all the stage outputs are latched so that they can be run in parallel. Thus, the effective T_{vc} is a single clock period.

Figure 2 shows a register transfer diagram of the chip. We selected a vector component bit width of twelve bits to match the available 12-bit A/D converters. Rather than using a general multiplier to accomplish the squaring portion of the calculation, we employ six table lookup ROMs to generate partial products which are then combined to obtain the squared result using three stages. By efficiently partitioning the partial products, only 2944 bits of storage are required to generate the 24-bit result. Since this technique utilizes unsigned input data, an absolute value stage was added to the processing flow.

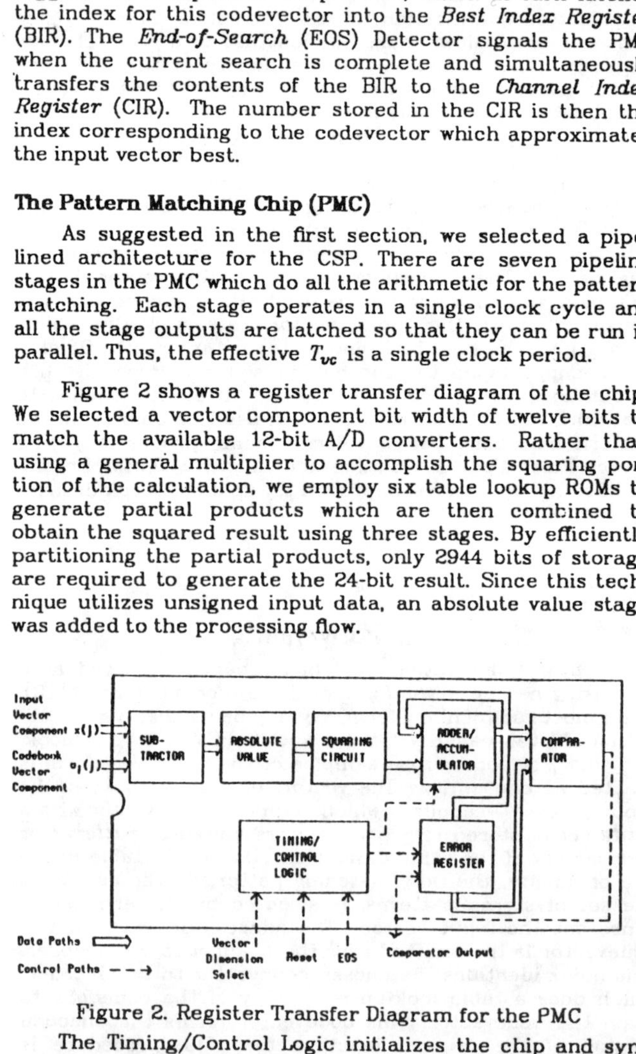

Figure 2. Register Transfer Diagram for the PMC

The Timing/Control Logic initializes the chip and synchronizes the data transfers through the various pipeline stages. The EOS signal initializes the Error Register to its maximum value. A Reset signal clears the Adder/Accumulator and sets up the timing to accumulate the squared differences for the k components of each codevector. This accumulated value is then compared to the value in the Error Register. If the new value is less than the error value, the Comparator Output signal is generated and this new value is latched into the Error Register. One codevector component is fed to the pipeline each clock period. However, because of the delays in the pipeline the Comparator Output is not generated at the same time as the last component of a codevector is applied to the input. Therefore, the off-chip control logic must compensate for this fixed seven clock-period delay.

We adopted the 4-micron NMOS technology using the Mead-Conway approach [5] because good design tools were available for this technology, many established designs for elementary cells could be used [6], and future improvements in NMOS technology through scaling will allow improved T_{vc} performance for the design.

The PMC is designed to run with a clock frequency up to 4MHz, so the minimum T_{vc} is 250 ns. It supports a selectable vector dimension, k, of size 2, 4, 6, or 8. The codebook size, N, is limited by k, T_{vc}, and the input vector data rate. The design requires 12,000 transistors. The chip is packaged in 40 pin DIP and dissipates 0.9 watts of power. It works well and is being incorporated into a waveform coder which is discussed in the next section.

3. APPLICATIONS OF THE CODEBOOK SEARCH PROCESSOR

The Codebook Search Processor is quite general in nature and can be employed in many types of real-time VQ coders. In particular, it is applicable to virtually any VQ system that is based on the squared-error distortion measure. We discuss in this section how the CSP module can be used to implement two speech compression systems, VPCM and AVPC, and then describe the application of the CSP to a laboratory hardware unit for rapid design of VQ codebooks.

3.1. Vector Pulse Code Modulation (VPCM)

VPCM, described in Section 1, is the simplest form of a VQ-based speech coder. By exploiting the high correlation between neighboring samples in a speech waveform, it outperforms a scalar quantizer with the same bit-rate. The disadvantage of VPCM is that for feasible dimensions, up to $k = 8$ or so, the subjective performance is still inferior to the simulated results of more sophisticated VQ coders such as AVPC [3] and HVQ [7] for the same bit-rates. Nevertheless, the inherent simplicity of this coding scheme makes it ideal for implementing in hardware.

VPCM is in essence a special case of the CSP shown in Figure 1. The basic architecture of a VPCM encoder is identical to the CSP architecture, except for the following simplifications:

a) Only a single bank of codebook ROM (with kN storage locations) is required.

b) The Codebook Data Bus and Address Generator Bus are not needed.

Furthermore, the Address Generator for a VPCM coder consists of a simple modulo-kN counter. The EOS Detector can be an address decoder which flags the PMC once every kN machine cycles.

The block diagram of a half-duplex communication system based on VPCM is presented in Figure 3. Here we have supplemented the CSP with preprocessing and postprocessing elements along with a table look-up unit. The Preprocessor bandlimits the analog signal $x(t)$ and then converts the filtered signal to a discrete-time, discrete-amplitude source x_n. As described in Section 2, the CSP forms vectors from x_n and transmits an index every k sample periods. The receiver uses these indices to perform a simple table look-up procedure (using a copy of the transmitter codebook) to generate the discrete-time, discrete-amplitude signal $\hat{x}_n$ which approximates x_n. Finally, the Postprocessor generates the approximate analog signal $\hat{x}(t)$ using a D/A converter followed by a reconstruction lowpass filter.

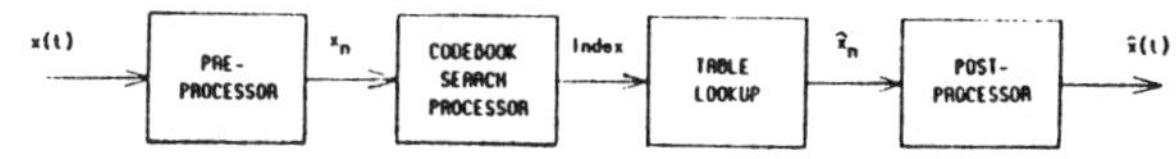

Figure 3. VPCM Speech Communication System

We have constructed a single-board VPCM speech communication system having the form described above. This coder is designed to operate at rates up to 2.25 bits per sample using a codebook containing 512 vectors. Various rates can be used by appropriately selecting the codebook size and dimension parameters (N and k) from the following sets: $N \in \{128, 256, 512\}$, $k \in \{4, 8\}$. The input speech is sampled at 8 kHz and bandlimited to 3.6 kHz. Standard 12-bit A/D and D/A converters are used in this implementation. The circuitry for the coding system shown in Figure 3 occupies an area of $18 \times 20\ cm^2$ and contains a total of 44 analog and digital IC's. Of these, 36 devices are used in the transmitter and the remaining 8 comprise the receiver. We are currently testing this single-board coder.

3.2. Adaptive Vector Predictive Coding (AVPC)

An alternate approach to designing a waveform coding system is to combine a low-dimensionality vector quantizer with other waveform compression schemes. This idea is embodied in Adaptive Vector Predictive Coding (AVPC), in which a vector quantizer is imbedded within a vector-based adaptive predictive coding system. AVPC extends DPCM to vector processing where a prediction $\hat{\mathbf{s}}_n$ of the current input vector is subtracted from the current vector $\mathbf{s}_n$ and the error $\tilde{\mathbf{e}}_n$ is coded using a codebook containing typical error vectors.

Adaptation is performed on a frame-by-frame basis Each speech frame consisting of many vectors is classified into one of m types. Associated with these m classes are m fixed predictors and m codebooks, all of which are optimized for the particular class to which they belong. The classification determines which predictor-codebook pair will be used for encoding the current frame.

For an Mth-order vector linear predictor the backward prediction $\hat{\mathbf{s}}_n$ of $\mathbf{s}_n$ is given by:

$$\hat{\mathbf{s}}_n = \sum_{j=1}^{M} A_j\, \tilde{\mathbf{s}}_{n-j},$$

where A_j is a kth-order predictor matrix and $\tilde{\mathbf{s}}_n$ is the reconstructed version of $\mathbf{s}_n$. As shown in [3], a simple adaptive scheme is *switched prediction*, in which the A_i matrices are predesigned and stored for use in the transmitter and receiver. The side information is an index identifying which predictor-codebook pair has been chosen for the current frame.

A real-time AVPC system as described above can be designed using the CSP and a separate processing element called a *Vector Prediction Unit* (VPU) which classifies each input vector and calculates the error vector.

General-purpose DSP chips such as the TMS320 or μPD7720 have architectures that are well-suited for the tasks mentioned above. One such processor can perform most or all of the VPU operations required for switched adaptive prediction.

A more powerful version of AVPC also discussed in [3] uses *frame-adaptive* predictors rather than switched predictors. Certain forms of frame-adaptive prediction can also be implemented with a single DSP chip. We can therefore implement very powerful AVPC coders by combining the CSP with one or more DSP chips.

The basic architecture of a general AVPC system is shown in Figure 4. The general CSP unit depicted in Figure 1 is used here without the Address Generator Bus which is not needed for AVPC. The CSP in this case has one bank of codebook ROM (of size kN storage locations) for each of the m classes. The speech vector $\mathbf{s_n}$ is processed by the VPU, which then transfers the error vector $\tilde{\mathbf{e}}_n$ to the CSP. When the CSP finishes the nearest-neighbor calculation, the codebook index is transferred back to the VPU and transmitted together with appropriate side information. The VPU also uses this index to perform a table look-up procedure (via the Codebook Data Bus) to obtain the quantized error vector $Q(\tilde{\mathbf{e}}_n)$ for use in calculating the next prediction vector $\hat{\mathbf{s}}_n$.

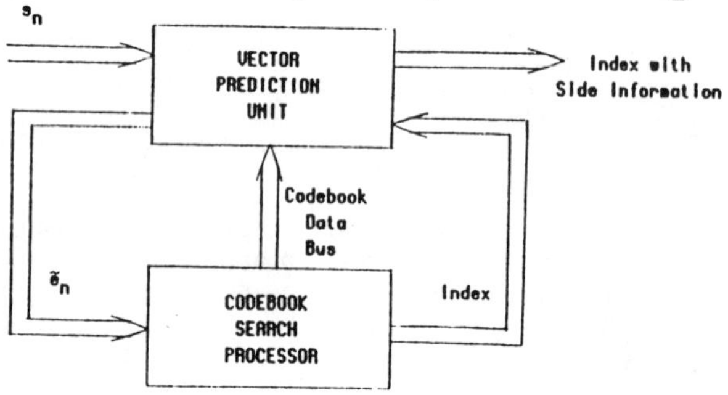

Figure 4. A General AVPC Encoder Architecture

Note that this system will perform considerably better than VPCM, and yet by adding a DSP chip, we have only modestly increased hardware complexity.

3.3. The Rapid Codebook Design Processor (RCDP)

Effective codebook design is a critical factor in the performance of a VQ-based speech coder. A locally optimal codebook can be designed using a large training sequence of typical source vectors taken from a speech data base. One algorithm of this type is the Linde-Buzo-Gray (LBG) algorithm [8]. In the first step, the training vectors are encoded using an initial codebook and the appropriate distortion measure. This operation partitions the training set into *cells* or *clusters* so that each cell is associated with one codevector. In the second step, a new set of codevectors is derived by finding the centroid of each cell and replacing the associated codevector with that centroid vector. These two steps are repeated iteratively until the algorithm converges.

If the final codebook is to effectively represent the range of speech data that the coder will encounter, the speech data base must contain a variety of male and female speakers. The *training ratio*, defined as the ratio of the number of training vectors M to the size of the codebook N, must also be very large. As a result, this design technique is very computationally demanding.

We have developed a special-purpose architecture for a *Rapid Codebook Design Processor* (RCDP) based on the PMC which allows the use of much larger training ratios than are usually practical with general-purpose computers. The key feature is that one iteration of the LBG design procedure can be performed in *real-time* using the RCDP. By real-time we mean that the RCDP can process incoming training vectors at least as fast as the rate at which the vectors arrive in real-time from a speech source.

The implications of performing an LBG iteration in real-time are highly significant. Suppose that a very long speech training set is created by recording analog voice signals onto magnetic tape. Then we can design codebooks by playing this recording into the RCDP once for each iterative step in the LBG algorithm. The time required to complete one iteration is therefore equal to the length in seconds of the training set.

The basic architecture of the RCDP is shown in Figure 5. Preprocessor and postprocessor blocks have been omitted for clarity. This architecture is directly suitable for designing codebooks for VPCM coders using the squared-error distortion measure.

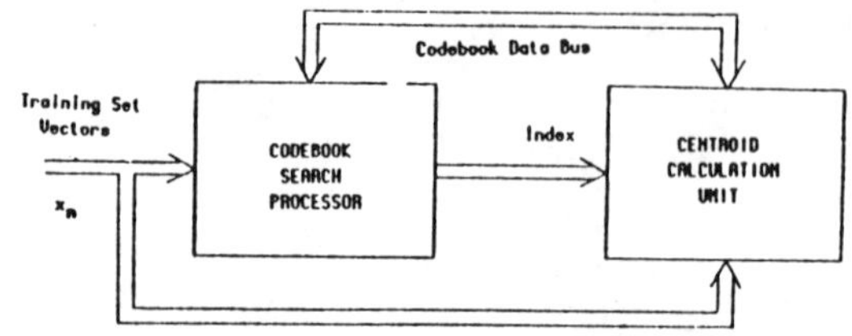

Figure 5. A Basic RCDP Architecture

Computational tasks in the RCDP are performed by two separate processing units. The CSP finds the best matching codevector to each input training vector, and the Centroid Calculator Unit (CCU) computes cell centroids and transfers the improved codebook to the CSP at the end of each iteration. The architecture shown in Figure 5 is capable of executing all of these functions. One general-purpose DSP chip may be used to realize the CCU.

The architecture presented in Figure 5 may easily be expanded to accommodate applications where the codebook contains parameter vectors extracted from a speech waveform by adding an appropriate real-time parameter-extractor to the front-end of the RCDP.

4. CONCLUSIONS

We have identified a general processor architecture, called the CSP, which can be incorporated into virtually any type of VQ speech coder using the squared-error distortion measure. The primary component in this architecture is a pattern matching processor which we have implemented in the form of a single VLSI chip. This chip utilizes a pipelined architecture and efficiently finds the codevector which is nearest, in a mean-square sense, to a given input vector. Using the current or enhanced versions of the Pattern Matching Chip, many state-of-the-art VQ coders may be realized with a minimal amount of hardware by combining one or more CSP modules with general-purpose DSP chips.

REFERENCES

[1] R. M. Gray, "Vector Quantization," IEEE ASSP Magazine, April 1984.

[2] A. Gersho and V. Cuperman, "Vector Quantization: A pattern-matching technique for speech coding," IEEE Communications Magazine, December 1983.

[3] V. Cuperman and A. Gersho, "Vector Predictive Coding of Speech at 16 Kb/s," IEEE Trans. on Commun., (to appear)

[4] B. P. M. Tao et al, "Hardware Realization of Waveform Vector Quantizers, " IEEE Journal on selected areas in Communications, March 1984.

[5] C. Mead and L. Conway, Introduction to VLSI Systems, Addison-Wesley, 1979.

[6] J. Newkirk and R. Mathews, The VLSI Designer's Library, Addison-Wesley, 1983.

[7] Y. Shoham and A. Gersho, "Pitch-synchronous Transform Coding of Speech at 9.6 Kb/s based on Vector Quantization," ICC 1984, pp. 1179-1182.

[8] Y. Linde, A. Buzo and R. M. Gray, "An Algorithm for Vector Quantizer Design," IEEE Trans. on Commun., January 1980.

Systolic Architectures for Vector Quantization

GRANT A. DAVIDSON, MEMBER, IEEE, PETER R. CAPPELLO, MEMBER, IEEE, AND ALLEN GERSHO, FELLOW, IEEE

Abstract—**The intensive computational demands of vector quantization (VQ) for important applications in speech and image compression and speech recognition have motivated the need for dedicated processors with very high throughput capabilities. Systolic architectures offer one of the most promising approaches for fulfilling the demanding VQ speed requirements in many applications. We propose a novel family of architectural techniques which offer efficient computation of weighted Euclidean distance measures for nearest neighbor codebook searching. The general approach uses a single metric comparator chip in conjunction with a linear array of inner product processor chips. Very high VQ throughput can be achieved for many speech and image processing applications. Several alternative configurations allow reasonable tradeoffs between speed and VLSI chip area required.**

I. Introduction

In the last few years, Vector Quantization (VQ) has emerged as a valuable technique for speech coding, speech recognition, and image and video coding [1]–[3]. In essence, VQ is an efficient general-purpose method for the digital representation of a set of parameters. Although the concept is not new to the literature of pattern recognition, it is quite new to the signal processing community where the intensive number crunching requirements of data compression have led to a rapid accumulation of a body of techniques and tools for effectively employing VQ. In some respects, the expertise with VQ for data compression has outstripped the established knowledge from the area of pattern recognition where VQ would be described as a pattern recognition technique. While the effectiveness and importance of VQ in data compression and in speech recognition is undisputed, a major obstacle to its use is the computational complexity of real-time implementations. The complexity barrier has motivated many reduced complexity techniques that enforce structural constraints on the VQ codebook with the usual price of degraded performance. On the other hand, it has been recognized that dedicated processor architectures for VQ have the potential of vastly improving the computational power achievable with current and emerging technologies. This paper focuses on a very promising architectural approach for such processors—the use of systolic VLSI arrays.

Manuscript received April 7, 1987; revised November 4, 1987. This work was supported in part by the Jet Propulsion Laboratory under a contract with NASA, the University of California MICRO Program, the General Electric Company, the National Science Foundation, and by the Office of Naval Research under Contracts N00014-84-K-0664 and N00014-85-K-0553.

G. A. Davidson was with the Department of Electrical and Computer Engineering, University of California, Santa Barbara, CA 93106. He is now with Dolby Laboratories, San Francisco, CA 94103.

P. R. Cappello is with the Department of Computer Science, University of California, Santa Barbara, CA 93106.

A. Gersho is with the Department of Electrical and Computer Engineering, University of California, Santa Barbara, CA 93106.

IEEE Log Number 8822838.

A vector quantizer is a mapping of an input or test vector to an output or *reproduction* vector that is selected from a finite *codebook* of stored vectors. The codebook contains N vectors and these are usually predesigned off line and stored in a permanent memory. The development of an effective codebook design algorithm [4] has provided the impetus for the proliferation of VQ-based algorithms in recent years. In this paper, a *vector* always refers to a k-dimensional ordered set of real numbers. The components may represent signal samples or numerical values of certain parameters or features that have been extracted from a signal. The vector may be regarded as analog although the components are ordinarily described with finite precision digital arithmetic. In the most direct application of VQ to waveform or image compression, a group of contiguous signal samples is blocked into a vector so that each vector simply describes a small segment of the original signal. This leads to efficient exploitation of the correlation between samples within an individual vector.

The efficiency of VQ comes from its role as a *pattern matching* technique. In essence, the vector of input samples is a *pattern* that must be approximated by one of the finite set of prototype patterns or templates that is stored in the codebook. To describe this pattern, we simply identify the *address* or *index* of that pattern in the codebook of standard patterns that "best" approximates the input pattern. This implies that the encoding process requires some form of search through the codebook and a measure of dissimilarity or *distortion* between an input vector and a stored prototype. The most widely used distortion measure and the most tractable analytically is the squared Euclidean distance between two vectors, often referred to simply as the mean-square error (MSE). The patterns in the codebook are called *codevectors*. Efficient coding or representation is achieved by replacing the original vector (described with kw bits where w is the wordlength per component of a vector) with the address of the selected codevector, which requires only $\log_2 N$ bits. (Usually N is chosen to be a power of 2.)

If the codebook has N codevectors, the quantity $R = \log_2 N$ is simply the number of bits needed to represent (i.e., to uniquely address) an input vector by specifying a

Reprinted from *IEEE Trans. Acoust., Speech, Signal Processing,* vol. ASSP-36, pp. 1651–1664, Oct. 1988.

particular codevector. The *resolution*, $r = R/k$, is the average number of bits per component needed to represent an input vector. In compression applications where the input vector represents a block of samples, the performance improves as the dimension is increased for a fixed resolution. In other applications, the dimension is fixed but the resolution must be increased to improve performance. In both cases, the codebook size, $N = 2^{rk}$, grows exponentially as the quality is improved. The critical encoding task of VQ is to find the best matching codevector from the codebook for a given input vector. The search process requires the computation of a dissimilarity measure between the input vector and each of the N codevectors in order to select the most similar codevector for each input vector. Typically this search must be performed for each of a sequence of input vectors presented to the VQ processor at a very high rate. The complexity of this computation is $O(N)$ per input vector, and the very demanding throughput requirement is the key obstacle to real-time implementation of VQ for many applications of practical interest.

A. Systolic Arrays

Systolic arrays and algorithms were first discussed by Kung and Leiserson, who characterize systolic algorithms [5] as follows:

> ''The important feature common to all of our algorithms is that their data flows are very *simple* and *regular*, and they are *pipeline algorithms*.''

Some of the ideas that underlie systolic arrays predate VLSI technology (e.g., Hennie's [6] iterative array). Recent attention, however, is due precisely to their suitability for VLSI implementation. Kung gives a tutorial introduction to systolic arrays [7], and maintains a list of systolic papers [8].

B. Review of Previous Work

Until recently, there has been little attention devoted to processor architecture design for VQ-based speech or image coding implementations. Even today, a vast portion of research effort in vector quantization is focused on algorithm development by computer simulation. This situation is caused by the complexity mismatch between algorithms which can be simulated (off line) on a general-purpose computer and those which lie within the lesser capabilities of real-time processors. Nevertheless, significant strides have been taken in the design of real-time VQ-based speech and image coders. Rapid growth in VLSI technology is beginning to close the gap between simulated and real-time algorithms. DSP chip transistor counts have been tripling every 4 years since 1978. We can now implement very sophisticated algorithms in a seemingly ever-diminishing area on VLSI chips. Furthermore, the design and VLSI implementation of dedicated, high performance processor architectures is becoming more attractive as chip layout and fabrication procedures are streamlined.

Hardware vector quantizer implementations have begun to appear in the literature. The first reported realization was a speech waveform vector quantizer built from off-the-shelf devices [9]. It was based on a moderately pipelined architecture, necessary for achieving the high-throughput rates for VQ. The first reported application of VLSI to VQ was reported in [10], in which a special-purpose architecture was described for efficiently performing codebook searches in real time. The architecture is based on a heavily pipelined VLSI pattern matching chip which performs distortion computations and comparisons. The chip requires Nk clock cycles to complete one codebook search, where N and k are the codebook size and dimension, respectively. A two-stage speech waveform vector quantizer [11] has also been reported.

Recently, systolic architecture concepts have been applied to hardware VQ realizations, enabling even higher vector throughput rates. A cascadable NMOS VLSI chip set implementing a bit-level systolic array architecture for both speech and image coding has been designed, fabricated, and tested [12]. This design is sufficiently general to accommodate any distortion measure which can be expressed as a Euclidean inner product. At a clock rate of 10 MHz, a linear array of $k + 1$ chips can compute and compare 10 million distortion terms per second (for any vector dimension between 1 and 16) while maintaining 24-bit precision in the distortion accumulations. A similar architecture was subsequently described in [13] and [14] together with a hardware realization for image coding using LSI technology—although this design is not purely systolic due to the presence of ''long wires.'' Additionally, a full-search architecture based on associative memories is presented in [14]. Dianysian and Baker [15] described a VLSI systolic architecture based on the linear array architecture in [12] but more suitable for use with tree-structured codebooks.

Other systolic processor architectures have been proposed. A word-level systolic architecture for data set clustering (codebook design) was presented in [16]. Based on the mean-squared error distortion measure, this design achieves a period for one search of $P = O(k)$ cycles; however, it also contains long wires. Other work includes a bit-serial architecture and cascadable VLSI chip developed specifically for image vector quantization [17], and a linear systolic array for speech recognition using dynamic time warping based on the Minkowsky (L_1) norm [18].

In this paper, we review the basic architecture described in [12] and present three alternative design configurations suitable for VLSI implementation. These new architectural directions, also constructed with bit-level systolic arrays, offer a wider range of throughput rates than the original design. The period for one codebook search ranges from $O(Nk)$ cycles for the design lowest in area to $O(1)$ cycles for the most demanding VQ applications (such as video compression). Furthermore, several of the designs contain only locally connected cells and input data ''broadcasting'' is never required. Single large memory elements for codevector storage can be avoided by distributing component bits throughout the array.

The remaining sections of this paper are organized as follows. We motivate the use of special-purpose architectures for real-time VQ in Section II by briefly outlining some applications and discussing the throughput requirements for each. In Section III, we show how the inner product plays an important role in a variety of distortion measures. A review of the bit-level linear array systolic architecture for VQ is given in Section IV. In Section V, we present three new extensions of the basic linear array. Section VI briefly describes a VLSI systolic chip set implementation based upon the architectural concepts presented in this paper. A performance summary of all the architectures is given in the conclusions.

II. Applications

In order to motivate the need for high throughput VQ capabilities, we briefly describe three specific VQ applications in speech and image compression to illustrate the computational complexity that must be handled for practical applications. While there are many applications to speech recognition and other areas of pattern recognition, we focus here on data compression examples for speech and image applications.

A. Waveform Coding of Speech

By grouping samples of a speech waveform into blocks of k samples each, VQ can be used to directly represent each block by a selected matching codevector from a codebook of predesigned waveform patterns. This rudimentary approach corresponds to Shannon's original concept *block source coding* and assuming a satisfactory distortion measure is asymptotically optimal for encoding the waveform as the dimension k approaches infinity. If the dimension is not large enough, correlation in the waveform cannot be adequately exploited and inadequate performance results. Waveform VQ applications have been considered where the dimension has ranged from 4 to as high as 20 [19], [1], [20]. In all of these cases, the largest codebook size considered was 1024 to avoid excessive complexity. For reasonable dimensions, prediction techniques can remove long-term correlation and VQ can handle the correlation in the *residual* waveform rather than the speech waveform itself. Experience with various coding algorithms and techniques indicates that a 5 ms block ($k = 40$ samples at 8 kHz sampling rate) and a resolution in the range 0.5–1 should provide a transparent quality speech coder (some side information is also needed for coding prediction parameters). This gives a codebook size of at least 2^{20}, requiring one-million distortion computations every 5 ms for a computation rate of 8 Gigaflops assuming a simple MSE distortion measure is used. This complexity is well beyond the limits of the best programmable processor technology available today (such as the TMS320C30).

A major conceptual advance in VQ speech compression that has emerged recently is *Code-Excited Linear Prediction*, or *Vector Excitation Coding* [21]–[24]. (An earlier version of this technique [25] used trellis coding instead of VQ.) Each segment of a speech waveform is reconstructed in a decoder from transmitted information specifying a) a speech production filter model, and b) an excitation signal segment. The reconstructed speech segment is the output of the filter whose input is the excitation. The encoder performs a search for the excitation codevector that gives best performance but the distortion measure is now a time varying one that is very demanding to compute. In effect, the distortion computation depends on the parameters of the production model and a time-varying perceptual error weighting filter.

Clearly, much higher complexities arise in this application compared to the use of an ordinary fixed distortion measure. Nevertheless, we have found that systolic architectures offer the potential for real-time implementation of a vector excitation coder with codebooks of size 1024 or higher.

B. Waveform Coding of Images

Digital transmission of images offers another area of application for VQ that is complexity bound. In raw form, a typical 512 by 512 pixel image is stored at 8 bits per pixel or 256 Kbytes per image. VQ is particularly efficient in this application. Consider the encoding task of the following typical situation: to transmit one picture per second, 2.1 Mbits with no compression would require a transmission rate of 2.1 Mbits/s. If VQ at 0.5 bits/pixel is used, the channel bit rate of 131 kbits/s would suffice, or 65 kbits/s for a 0.5 picture per second transmission rate. In spatial domain image VQ, typically blocks of pixels in the range of $k = 16$ up to $k = 36$ have been studied. At a rate of 0.5 bits/pixel with $k = 36$, then $N = 2^{18}$ which requires a search through 262 144 codevectors each of dimension 36. Various complexity reducing techniques are available which force a special structure in the codebook or use product codebooks at the price of compromising quality. Clearly, superior processor architectures for VQ offer an important potential solution for digital image transmission.

C. Real-Time VQ Codebook Design

The performance of any vector quantizer strongly depends upon how well the codebook represents the statistical character of the source to be coded. The Generalized Lloyd (GL) algorithm described in [4] has become the prevalent technique for designing codebooks for VQ. In each iteration, a new codebook is obtained and each training vector in a large training set must be encoded by a search through the codebook. The training ratio—ratio of number of training vectors to number of codevectors—should be as large as possible to have a statistically adequate representation of the signal. A typical value often viewed as adequate is 100. However, complexity has limited the training ratio that can be used to empirically characterize the source statistics. In many applications, suitably large training sets would tax the resources of most mainframe computers used for codebook design.

Recently, a custom hardware processor was proposed to perform fast codebook design with large training sets [26]. The processor uses the pipeline IC chip for VQ reported in [10], yet it still suffers substantial performance limitations due to the inadequate throughput for many codebook design applications. With the rapid growth in the use of VQ for data compression, it is becoming evident that more advanced processor architectures are needed to achieve effective hardware subsystems for VQ codebook design.

III. VQ Distortion Measures and Computational Issues

A *distortion measure* between an input vector x and a codevector y is a nonnegative real-valued function $d(x, y)$ which measures the degree of dissimilarity between x and y. The distortion can be regarded as the cost of reproducing the input x with the reproduction vector y. The usual VQ encoding process requires N distortion computations for an exhaustive search through the codebook to find the codevector that achieves the least distortion. The most direct encoding algorithm for VQ is simply as follows:

Nearest Neighbor Encoding Rule

1) Set $d \leftarrow d_{-1}$ and $j \leftarrow 0$.
2) Compute $d_j = d(x, y_j)$.
3) If $d_j < d$, set $d \leftarrow d_j$. Set $i \leftarrow j$.
4) If $j < N - 1$, set $j \leftarrow j + 1$ and go to step 2.
5) Stop.

The resulting value of i gives the encoder output $C(x)$, and the final value of d is the distortion between x and y_i. The initial value, d_{-1}, must be larger than any expected distortion value and is usually set to the largest positive number that can be represented with the processor's arithmetic.

The encoding algorithm can be viewed as an *exhaustive search* algorithm where the computation of distortion is performed sequentially on every codevector in the codebook, keeping track of the "best so far" and continuing until every codevector has been tested. Furthermore, the *vector rate*, f_v, in vectors per second, is also rather high in many applications so that the number of distortion calculations that must be performed per unit time, given by Nf_v, implies a very demanding computational complexity, typically involving many millions of arithmetic operations per second.

The most common and important distortion measure is the *squared error distortion measure* defined by the squared Euclidean distance between the two vectors

$$d(x, y) = \|x - y\|^2 = \sum_{i=0}^{k-1} (x_i - y_i)^2. \qquad (1)$$

In many cases, this measure is not adequately tied to subjective quality assessments, but generalizations permitting input-dependent weightings often achieve improved correspondence with perceptual degradation and are often only slightly more complicated.

Although (1) can be computed directly by performing squaring operations, an alternate and often preferable expression for the distortion is based on computing inner products. By expanding (1), we get

$$d(x, y_j) = \|x\|^2 - 2(x \cdot y_j + e_j), \qquad (2)$$

where

$$e_j = -\tfrac{1}{2}\|y_j\|^2 = -\tfrac{1}{2}\sum_{i=0}^{k-1} y_{ji}^2, \qquad (3)$$

and the inner product is given by

$$x \cdot y_j = \sum_{i=0}^{k-1} x_i y_{ji}. \qquad (4)$$

Equation (2) shows that minimization of the distortion for a fixed input vector x is equivalent to maximizing an inner product of two $k+1$-dimensional vectors. (The constant e_j can be treated as an additional component of the vector y_j, and the input vector x can be given the corresponding additional component value of unity.) Thus, the codebook search process can be regarded as a search for that (augmented) codevector which has the largest inner product with the given (augmented) input vector.

In some applications, the distortion measure of interest is the *weighted squared error* given by

$$d_W(x, y_j) = (x - y_j)^T W(x - y_j) \qquad (5)$$

where W is a fixed $k \times k$ symmetric positive definite matrix, and the vectors x and y_j are regarded as column matrices of order k. Equation (5) can be expanded and simplified to yield

$$d_W(x, y_j) = d_W(x, 0) - 2x \cdot \tilde{y}_j + y_j \cdot \tilde{y}_j, \qquad (6)$$

where $\tilde{y}_j = Wy_j$ and 0 is the zero vector. It is evident that the minimization of this distortion measure over the codevectors y_j is equivalent to maximizing the expression

$$H_j = x \cdot \tilde{y}_j + \tilde{e}_j \qquad (7)$$

where $\tilde{e}_j = -\tfrac{1}{2} y_j \cdot \tilde{y}_j$. Here, we see that the problem reduces to maximizing the inner product of an augmented input vector with a transformed and augmented codevector.

As a final illustration of the ubiquitous inner product maximization for VQ, we consider the important issue of coding or representing Linear Predictive Coding (LPC) parameters. A common metric which is useful for LPC is the well-known Itakura–Saito distortion measure [27], [28]. This measure is a spectral-error matching function which is defined in terms of the power spectral density of windowed speech samples and the magnitude squared of a conventional LPC all-pole filter model. In much the same way that the weighted mean-square error of (5) is minimized by maximizing the inner product of (7), the Itakura–Saito measure is minimized by minimizing an inner product of the form [29]

$$\alpha = R_{aa}(0)\, R_{xx}(0) + 2 \sum_{i=1}^{p} R_{aa}(i)\, R_{xx}(i), \qquad (8)$$

where $R_{aa}(i)$ and $R_{xx}(i)$ are autocorrelations of the predictor coefficients for the LPC model of interest, and the input speech samples, respectively. In the codebook search procedure, an optimal codevector is selected by minimizing (8), the energy of the residual signal obtained by inverse filtering the windowed speech samples. Note that (8) is again a simple inner product between a fixed input vector [whose components are $R_{xx}(i)$] and a stored "codevector" whose first component is $R_{aa}(0)$ and whose remaining components are $2R_{aa}(i)$.

We have thus illustrated how the inner product plays a central role in a variety of distortion measures and its computation for the VQ encoding task is of central interest for the implementation of many coding and recognition systems.

IV. A Linear-Array Systolic VQ Processor

We have shown that several useful distortion measures for VQ coding of speech and image waveforms can be expressed in terms of an inner product. Fortunately, inner products are ideally suited for computation with systolic arrays. We now describe a systolic array architecture which efficiently performs distortion and comparison computations in VQ. A word-level architecture is given first, followed by a functionally equivalent design which uses bit-level systolic multiplier arrays.

A. Word-Level Systolic VQ Processor

A word-level systolic processor for pattern matching is presented in Fig. 1. This linear array is composed of k *Inner Product Processor* (IPP) blocks and a *Comparator Processor* (CP). The computation performed by the array is given by

$$i_n = \min_{0 \le j \le N-1}^{-1} d_j = \min_{0 \le j \le N-1}^{-1} \left(\sum_{i=0}^{k-1} x_i^n y_{j,i} + e_j \right) \qquad (9)$$

where the inverse means "output the index i_n which achieves the minimum." Maximization of the expression can be achieved if desired by negating $y_{j,i}$ and e_j before storing them in ROM. The input vector components at time n (x_i^n) remain stationary in the array for the duration of one codebook search, while the components of y_j ($y_{j,i}$) enter the array skewed in time by one clock cycle, with $y_{0,0}$ entering first. The first distortion term d_0 is produced k clock cycles later by the last IPP in the array. The remaining d_j are produced on subsequent clock cycles, and distortion comparisons occur at the same rate. The e_j quantities are precomputed constants associated with the jth codevector, and are input to the first IPP in the array. Since one index is output from the CP every N clock cycles, the period $\boldsymbol{P}$ of this architecture is N cycles.

Functional descriptions of the two processor types are presented in Figs. 2 and 3. IPP's operate on their three input words to produce the output $d_{out} = xy + d_{in}$. The CP compares its input word d_j to the contents of the *Minimum Distortion Register* (MDR) and saves the most negative of these words in the MDR on the subsequent clock cycle. The index of the codevector which yields the smaller distortion is also saved by the CP after each comparison operation.

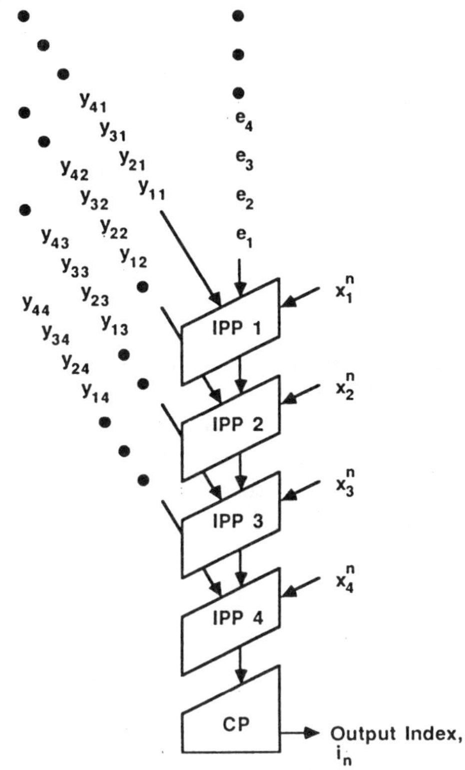

Fig. 1. Linear systolic array architecture for vector quantization.

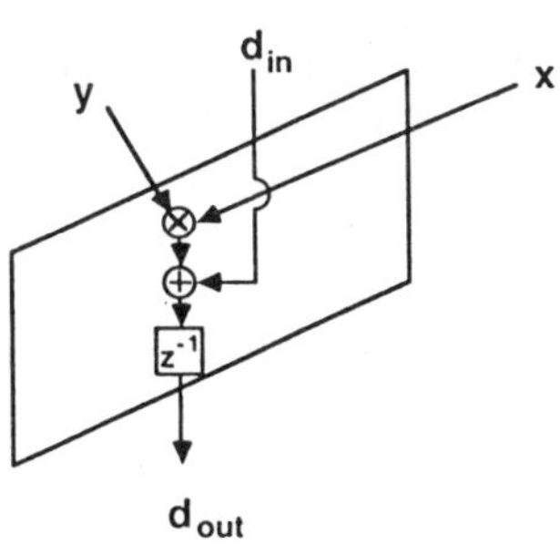

Fig. 2. Inner Product Processor word-level block diagram.

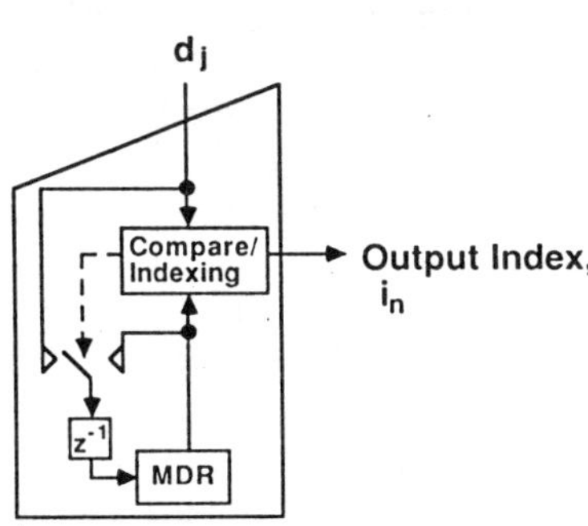

Fig. 3. Comparator Processor word-level block diagram.

In performing a sequence of exhaustive search procedures, the pipeline can be kept continuously full when changing from one input vector to the next by changing the input to the ith IPP from x_i^n to x_i^{n+1} synchronously as the last codevector component $y_{N-1,i}$ changes to $y_{0,i}$. The *latency* of this architecture, defined here as the elapsed time from when the first component of a given codevector enters the array until the associated distortion comparison is finished, is $k + 1$ cycles.

In general, the number of processors required to implement this word-level design is $k + 1$. Considerable savings in the number of processors (and hence the area) can be realized by reducing the number of IPP's from k to k/M ($1 < M \leq k$) and by having each IPP compute M partial distortion terms for each input vector instead of only one. The number of processors and the area are reduced by a factor of M in this case, and the period P also drops proportionally. This architecture will be described in Section V-B.

B. Bit-Level Systolic VQ Processor

The architecture presented in Fig. 1 offers a rather high-level view of a pipelined pattern matching processor. As such, it is conceptually useful but somewhat cumbersome to implement on a VLSI chip. An alternate representation which describes the architecture on a lower level, i.e., the bit level, is desired.

There are several advantages to a bit-level design. Since the maximum clock frequency for such a chip is equivalent to the inverse of the propagation time for a single cell, a much shorter cycle time is achievable compared to processors which operate on the word level. Furthermore, the regularity of systolic designs makes them very easy to lay out in VLSI. Even with 4 μm NMOS, a relatively primitive technology by current standards, it is feasible to place at least one IPP with an input wordlength of 12 bits on a single chip [12].

We use the word-level model in Fig. 1 as a starting point to construct a pipelined processor with the same period as the word-level design. This is accomplished by substituting a bit-level systolic two's-complement multiplier/adder for every IPP in Fig. 1, and a bit-level comparator for the CP. This transformation from word to bit level necessitates an increase in latency for one distortion computation and comparison from $k + 1$ to $2wk + 5w + \log_2 N$ clock cycles, where w is the wordlength of the incoming x_i^n and $y_{j,i}$ vector components. For medium to large N, this extra latency is negligible compared to the overall processing delay of N cycles introduced by one codebook search. The CP also can be implemented entirely with bit-level systolic processors, as shown in Section IV-B-2. The time complexity for quantizing a vector with this bit-level systolic architecture is $O(kw + N)$.

1) Inner Product Processor: McCanny and McWhirter have introduced an iterative array for multiplying two w-bit two's-complement numbers [30]. This modular building block can be incorporated into a wide variety of functional units for high speed signal processing (FIR filters, convolution arrays, matrix/vector processors, etc.). In this paper, the array serves as the basic computational element in a family of systolic architectures for vector quantization.

The basic arrangement and structure of the cells in this multiplier array are presented for the case of $w = 3$ in Fig. 4. The array is composed of $2w^2$ cells which are latched full adders. Each cell has four inputs and outputs, and performs the bit-level logic operations shown in Fig. 4. Here $X(i)$ and $Y(i)$ represent individual bits of the ith pair of numbers to be multiplied, $D_{in}(i)$ is one bit of accumulating partial product, and $C_{in}(i)$ is the carry-in from an adjoining cell. Outputs of each cell are latched and passed to adjacent cells on successive clock periods.

The numbers to be multiplied, x_i and y_i, are input along the upper right and left edges of the array, respectively. The value e_i, which is added to the product $x_i y_i$, is also input along the upper array edges. Skewing latches (not shown in Fig. 4) are necessary along the perimeter of the array to ensure that the individual bits of x_i, y_i, and e_i arrive at the right cell at the right time. Bits of x_i and y_i propagate in a southwesterly and southeasterly direction through the array, respectively. Multiplication occurs as individual bits of x_i meet those of y_i and one-bit partial product terms are generated. The one-bit sum output terms $D_{out}(i)$ are accumulated as they move downward along the columns of the array. Carry-out bits $C_{out}(i)$ propagate in the same direction as the bits of x_i.

Successive bits of the result emerge along the bottom edge of the array skewed in time (LSB first). Although the latency of a single array of this type is $3w$ clock cycles, it is possible to start a new multiply/add operation, and complete a previous one, on every cycle.

The cells along the upper left array edge (containing dots) use the ones-complement of y_i. Moreover, the MSB of x_i is tied to the carry-in as well as the "$X(i)$" input of the uppermost cell. The $w - 1$ remaining carry inputs are tied to logic "0." The cells which contain a "0" serve as simple delay elements for x_i and e_i. In practice, they can be constructed using the same full-adder cell shown in Fig. 4 with $Y(i)$ and $C_{in}(i)$ tied to logic "0."

Now we turn our attention to the case when a sum-of-products computation is desired

$$d_n = \sum_{i=0}^{k-1} x_i^n y_i + e. \tag{10}$$

To implement this with bit-level arrays, we create a vertical stack of k multiplier array sections to carry out the inner product calculation, as indicated by Fig. 1. The k values of x_i^n are input time skewed on the word level, as before, along the right side of the array. Furthermore, the bits of x_i^n, y_i, and e must be skewed in time as described above. Individual product terms accumulate as they propagate downward along columns of the array. The latency of this array is $O((2k + 2)w)$ cycles. To accommodate wordlength growth as the terms accumulate, the array must have $2w + \lceil \log_2 k \rceil$ columns.

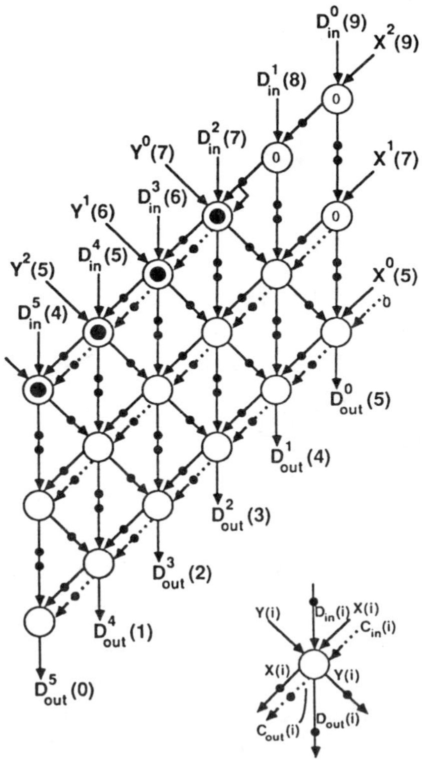

$$D_{out}(i) = D_{in}(i) \oplus (X(i) \bullet Y(i)) \oplus C_{in}(i)$$

$$C_{out}(i) = (D_{in}(i) \bullet X(i) \bullet Y(i)) + (C_{in}(i) \bullet X(i) \bullet Y(i)) + (D_{in}(i) \bullet C_{in}(i))$$

Fig. 4. 3 × 3 bit systolic multiplier array. Cells containing dots use the complement of b_n. Detail of a single cell is shown in the lower part of the figure. Heavy dots denote unit delay elements. Superscripts denote bit number within the word.

The Inner Product Processor (IPP) consists of one such multiplier array with the addition of some synchronization logic. In a complete VQ processor configuration, the first IPP receives a synchronization signal from external logic which has a period of N clock cycles (going high whenever a new input vector is presented to the IPP). The IPP does not explicitly use the synchronization signal, but passes it to the next IPP in the array with an appropriate delay. This signal is eventually passed by the last IPP in the stack to the CP coincidently with the LSB of the first distortion term of a new codebook search. This synchronization signal triggers various initialization operations in the CP which are necessary to complete the current codebook search and start a new one, as described in Section IV-B-2.

IPP arrays can also be used to calculate the transformation

$$d_l^n = \sum_{i=0}^{k-1} (x_i^n y_{l,i}) + e_l, \quad 0 \leq l \leq L - 1. \quad (11)$$

Equation (11) can be computed with a period $P = L$ cycles using a single linear inner product array. An alternative, higher performance scheme can be implemented with a two-dimensional array consisting of L inner product sections stacked side by side. Each section contains k vertically stacked IPP modules. The x_i^n values pass directly from one inner product section to the next in a southwesterly direction. As before, the bits of $y_{l,i}$ move in a southeasterly direction, and the entire structure computes a matrix–vector product with period $P = 1$. In Section V-A, we show how this structure can be augmented in a straightforward manner to implement a *completely pipelined* [31] systolic vector quantization processor with $P = 1$.

2) *Comparator Processor:* The linear IPP array described in the previous section produces one distortion term per clock cycle. To complete a VQ processor design, some mechanism must be provided to find the smallest distortion and identify the index of the associated codevector. These functions are performed by the Comparator Processor (CP).

Fig. 5 shows a bit-level systolic CP design appropriate for use with one or more IPP modules. The figure depicts the case when 4-bit distortion terms are being compared using a 3-bit codebook. The four main components are a triangular *inverse skewing array*, a comparator string, a *counter*, and an *index register* (IR). The purpose of the inverse skewing array is to delay the mth distortion bit in a $2w$-bit distortion word by $2(2w - m - 1)$ cycles so that the MSB of the distortion value enters the comparator string first (recall that the distortion values leave the IPP array LSB first). In an actual VLSI implementation, high uniformity of the VQ processor can be achieved by constructing these delay cells from the full adders of Fig. 4, wiring $X(i)$, $Y(i)$, and $C_{in}(i)$ to logic "0."

The comparator string contains the MDR, which at any point in time holds the minimum distortion obtained during the codebook search. Comparisons between the distortion and MDR words occur systolically [12]. Starting with the MSB, corresponding bits from the two words are compared to find the first occurrence of a difference in the bit patterns. As soon as a difference is detected, the smaller of the two words can be identified, and remaining bit comparisons in the comparator string are inhibited. A two-bit decision status (DS) signal carries information regarding the current state of the distortion-MDR comparison.

A synchronization signal, SY (provided by the adjacent IPP module), is used to reset the MDR to its largest permissible value at the beginning of every codebook search. When the result of the first comparison in a new codebook search is produced, the systolic counter is reset to 0 by SY and increments by one on each subsequent clock period. In this way, the counter identifies the index associated with the distortion which has just finished passing through the comparator string. The counter is a sum-save carry-transfer systolic counter. A 1-bit decision signal D (derived from DS) determines whether or not the current counter output is saved in the IR. The IR is updated if and only if the current distortion is less than the previous best ($D = 1$). At the end of the last comparison, the optimal index i_n will be in the IR. The SY signal unloads the IR systolically prior to the start of a new codebook search.

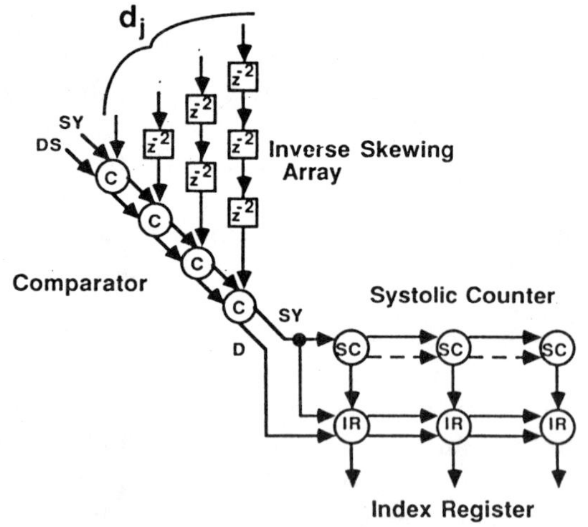

Fig. 5. 4-bit systolic Comparator Processor using 3-bit codebook indexes.

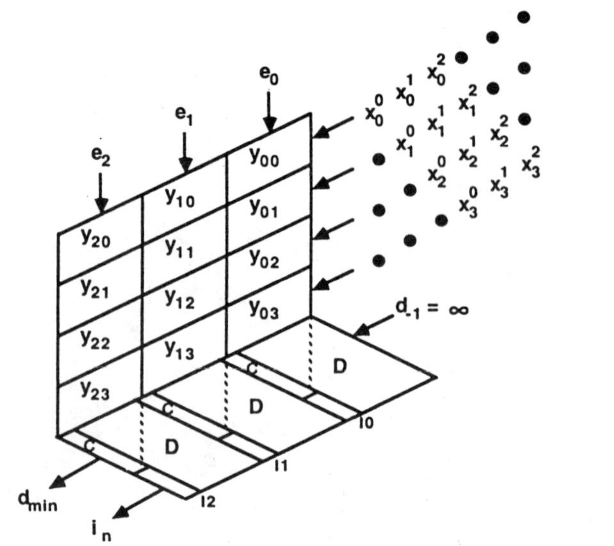

Fig. 6. High-performance systolic architecture for vector quantization ($N = 3$, $k = 4$).

V. Extensions of the Linear Systolic Array Architecture

In this section, we describe three new bit-level systolic processor designs for VQ. All three of the designs are based on the linear array architecture described in the previous section. We begin by extending the linear array to a two-dimensional array with very high throughput.

A. Two-Dimensional Array Architecture

For applications where the $P = N$ performance of the linear array VQ architecture is not sufficient, it is desirable to employ a higher throughput processor. This section discusses a two-dimensional systolic architecture designed for the most demanding vector quantization applications. The period for complete quantization of one source vector is only one cycle. This implies that a VLSI realization running at 10 MHz can fully quantize a stream of 10 million input vectors per second (f_v = 10 MHz) *regardless of codebook size or dimension*. For example, the input vector rate f_v for 16-dimensional vectors from 512 × 512 pixel images at 30 frames per second is 491 520 vectors per second—well within the capability of this architecture. A second advantage of the design is that the codebook can be distributed throughout the array, eliminating the need for external codebook memory and addressing logic.

Fig. 6 depicts a VQ processor design for $N = 3$ codevectors of dimension $k = 4$. The main computational structure is a $k \times N$ array of modified IPP modules. The lettering shown within each IPP designates the particular codebook component stored there—one codevector per array column. The bits of $y_{j,i}$ are stored within the individual cells of the IPP, as described in [32], eliminating the usual codevector propagation paths shown in Fig. 4. The e_j terms are input to the array by appropriately "hardwiring" the uppermost IPP's $D_{in}(i)$ input lines to logic "0" or "1." The stream of input vectors to be quantized, x^n, is input along the right edge of the array as before.

To complete the design, the $k \times N$ processor array is augmented with bit-level arrays of distortion comparison (C), index selection (I), and delay (D) blocks. One block of each type is added to the bottom of every inner product column. After distortions are generated by the inner product columns and flow downward to the C block, they are compared to distortions moving southwest from the adjacent D block. The smallest of the two words is passed to the next inner product column for subsequent comparison, and the associated index travels in parallel with it (skewed in time). The optimal index i_n and the associated minimum distortion $d_{\min}$ are available (one per clock period) along the lower left array edge.

Details of the C, I, and D block interconnections are shown in Fig. 7. The distortion comparison strings are identical to those described in Section IV-B-2 except that the MDR register is not present. Distortion bits arriving from the top and right are compared, and the "winning" bit is immediately passed to the adjacent D block cell. The I block is a 2-to-1 systolic multiplexer which outputs the index it receives from the northeast if the distortion associated with that index is greater than or equal to the distortion emanating from the IPP column directly above. Otherwise, it outputs the codevector index j associated with that column. The multiplexer select input is driven by the decision status output of the comparator string.

In this design, we trade latency and chip area for an extremely short period. Compared to the linear array, a reduction in the period from N cycles to 1 cycle is traded for an increase in the area from $O(kw^2)$ to $O(Nkw^2)$. The latency from codebook search start to finish also increases from $O((kw + N))$ cycles to $O((k + N)w)$. Since in [12] $k + 1$ chips were required for the linear array, an upper bound on the number of VLSI chips required for an implementation of the two-dimensional architecture is $N(k + 1)$ chips. For large N and/or k, this chip count is rather prohibitive; however, multiple IPP's can be implemented on one chip if 1 μm VLSI technology is used in-

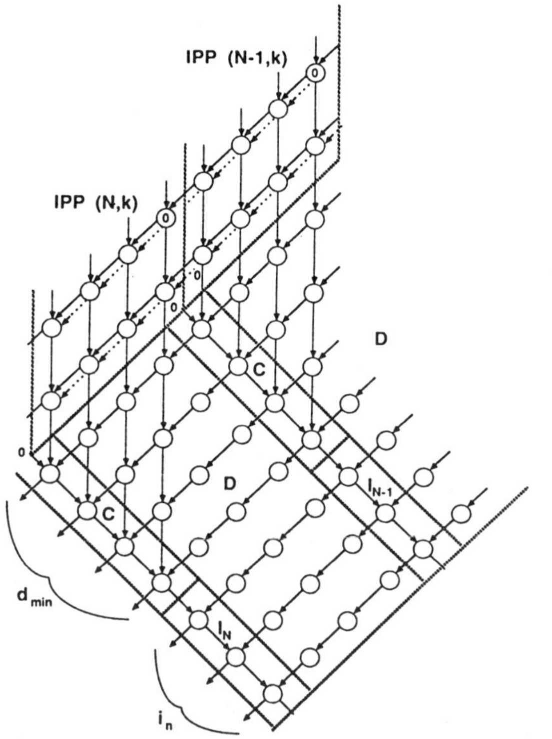

Fig. 7. Detail showing interconnections of comparison (C), delay (D), and index multiplexer (I) blocks in the high-performance architecture. The SY signal associated with the conventional comparison string is not necessary here. Unit delay elements between cells have been omitted from the figure for clarity.

stead of 4 μm as in [12] (see Section VI for VLSI implementation details).

B. Sublinear Array Architecture

One may want a throughput rate other than that which results from using k IPP's and a CP. This section discusses the case when one wants less hardware, and can tolerate a commensurate reduction in throughput. In the discussion that follows, we take both the codebook size and the dimension of the vector space to be powers of 2: $N = 2^n$ and $k = 2^{d=m+c}$. Our architecture accommodates codebook sizes and vector dimensions that are not powers of two. We assume powers of two only to simplify the discussion.

It is well known that a cascade of IPP's can compute an inner product in a pipelined fashion [33], [32]. The idea pursued here is to 1) break up the inner product into pieces (of length 2^c), 2) compute each piece on a cascade of 2^c IPP's, and 3) sum these ($k/2^c = 2^m$) partial inner products:

$$d_j = \sum_{i=0}^{k-1} x_i^n y_{ji} + e_j = \sum_{i=0}^{2^m-1} \sum_{l=0}^{2^c-1} x_{i2^c+l}^n y_{j,i2^c+l} + e_j. \tag{12}$$

We obtain a full k-dimensional inner product with one cascade of $k/2^m = 2^c$ IPP's. This is done by reusing the cascade $k/2^c = 2^m$ times for each codevector. Several architectural requirements arise as follows.

1) For each inner product, each of the 2^c IPP's computes 2^m products. We need to deliver 2^m different x and y vector components to each IPP.

2) We need a separate accumulator at the bottom of the cascade to sum the $k/2^c = 2^m$ partial inner products.

3) To efficiently use the cascade, it needs to be pipelined. This complicates our component delivery system.

Before we present an architecture that fulfills these requirements, another building block is introduced: the *recirculating* queue (*requeue*). Imagine a queue of fixed length that has its head feeding into its tail: items that are dequeued at the head are enqueued automatically on the tail. This is depicted in Fig. 8. Let Q be a requeue. The operation "*enqueue*(Q)" recirculates an item on the requeue. The operation "$x \leftarrow$ *enqueue*(Q)" does two things. It 1) dequeues an item from Q, assigning it to variable x; and 2) recirculates it onto Q. We need to allow for the insertion of new items (and deletion of old items, since the queue is of fixed length). To accomplish this, the enqueue operation has an optional argument, a new queue item. When this argument is present, the new item is inserted on the tail of the queue, instead of recirculating the dequeued item (see Fig. 9). The operation "*enqueue*(Q, *new*)" 1) dequeues an item from Q, and 2) inserts the item whose value is *new* onto the tail of the requeue. The operation $x \leftarrow$ *enqueue*(Q, *new*) similarly 1) dequeues an item from Q, assigning the value to x; and 2) inserts the item whose value is *new* onto the tail of the requeue. The requeue data structure can be used in a variety of ways. One simple use has the structure of the nested iterations in Fig. 10.

This is the use that we have in mind. We use a recirculating queue to deliver x and y components. A recirculating queue can be implemented in many ways, among them are the following:

1) "directly" with custom circuitry,

2) with a dual-ported RAM,

3) with a standard RAM, if dequeueing and enqueueing do not need to be overlapped,

4) with a ROM, if new items are never inserted (i.e., only recirculation is used on a hard-coded recirculating queue).

One more component must be modified before we present the architecture: the array multiplier. The modified array multiplier is better suited for our inner product cascade. To see what must be modified, consider the array multiplier in Fig. 4. The y values flow southeast through what we refer to as the y-columns of the array. Notice that the high order $w + 1$ y-columns all receive the same y-bit: the most significant bit (MSB). These high-order y-columns thus can be tied to a common input, that propagates the MSB systolically, as in Fig. 11. If we vertically juxtapose two such array multipliers, then delivering the y operand to the bottom array multiplier becomes a problem. To alleviate this, we augment the array with an extra set of y-bit lines. Each such line's sole purpose is to deliver a y-bit, through one array multiplier, to the array multiplier below. This is depicted in Fig. 12. Fig. 13

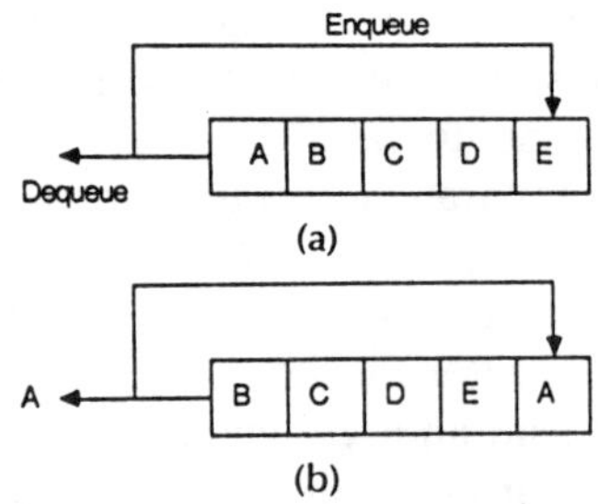

Fig. 8. Schematic of fixed length, automatically recirculating queue. Part (b) shows the queue just after recirculation operation. Item A is output as well as recirculated.

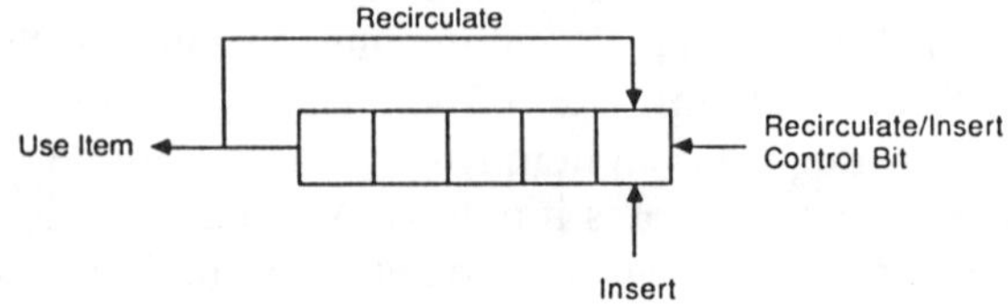

Fig. 9. Schematic of a requeue. The control bit indicates whether or not a new item is to be inserted.

```
Q: requeue;
while (there exists a vector to be quantized) do
    for i := 1 to N do      /*for each codevector */
    for j := 1 to 2^m do
    for l := 1 to 2^c do
        {
        if i ≠ N
        then x ← enqueue(Q) /* use and recirculate component */
        else x ← enqueue(Q, read(new_x)) /* use component and insert
            new one */
        }
```

Fig. 10. The *iteration* structure for quantizing vectors. Only operations on the x component recirculating queue are shown. (The initial insertion of x components is not shown.)

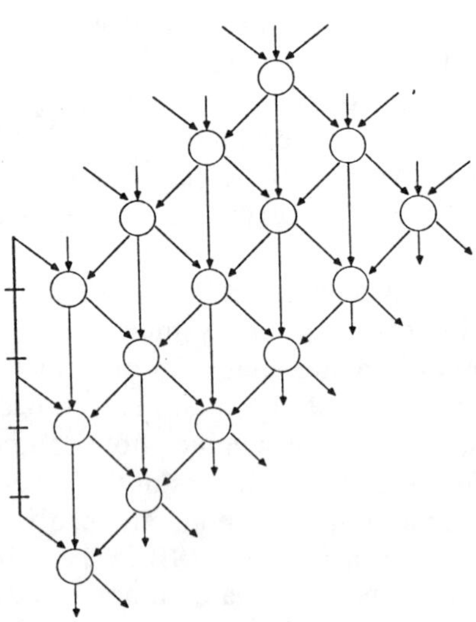

Fig. 11. Schematic of array multiplier with high-order b-columns attached to a systolically delivered MSB input.

shows two juxtaposed array multipliers where the product produced by the top array multiplier is delivered to the bottom array multiplier. This is the form of the array multiplier used in our partial inner product computation.

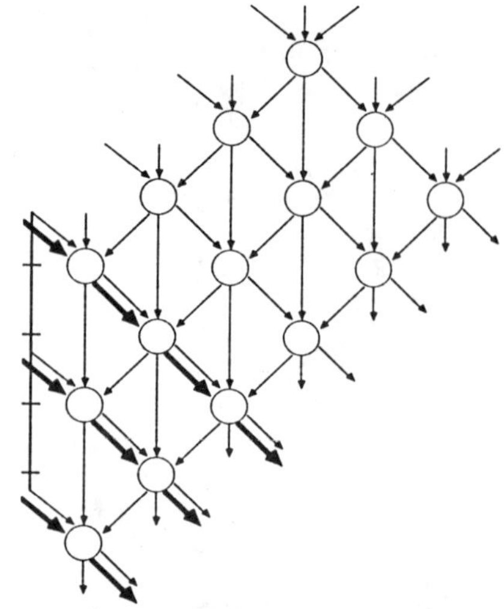

Fig. 12. Schematic of modified array multiplier. Emboldened arrows represent b-bit distribution lines.

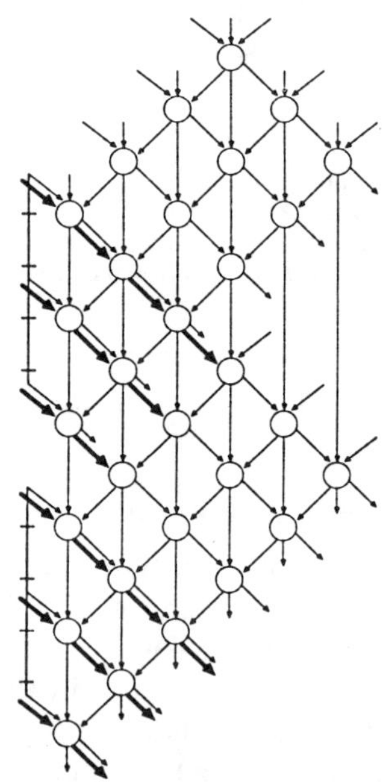

Fig. 13. Schematic of two vertically juxtaposed modified array multipliers.

We now are ready to present the vector quantizer that uses fewer IPP's. Its diagram is given in Fig. 14. It is comprised of 2^c IPP's, each of which is flanked by two recirculating queues, one on the left and one on the right. The recirculating queue on the right is used to deliver x components. It has a capacity of 2^m x components. They are recirculated N times, one for each codevector. The recirculating queue on the left contains the codevectors (recoded). It is recirculated one time for each x input, and therefore has a capacity of $N \cdot 2^m$ components. This codevector component recirculating queue has no need for insert; it is initialized with the recoded codevectors "once and for all." This recirculating queue therefore may be implemented with ROM.

The operation of the assemblage is simple: at each cycle a new partial inner product is initiated at the top of the

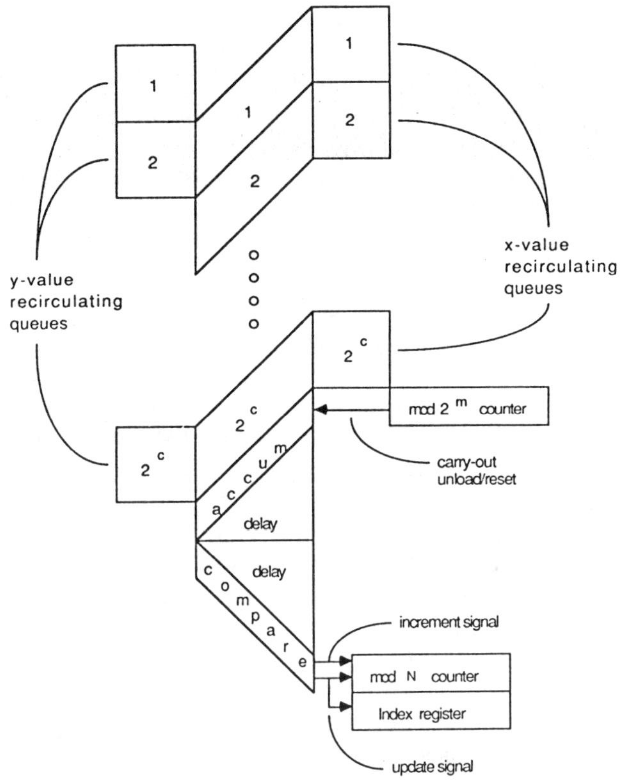

Fig. 14. Architectural schematic of vector quantizer.

IPP cascade, and another is completed at the bottom. Partial inner products are delivered to the accumulator at the bottom where they are summed. Every 2^m cycles, the accumulator has summed one complete inner product, and needs to unload and reinitialize. This is the purpose of the *mod* 2^m counter (which operates systolically). Its carry-out serves as an unload and reinitialize signal for the accumulator. The subsequent delay and comparison are performed as previously described. Every time a comparison is finished, the comparator sends an increment signal to the *mod N* counter. Another signal generated by the comparator is the *update* signal. The *update* signal is set to "1" when, and only when, a new minimum distortion is found. When the update signal is "1," the contents of the *mod N* counter are loaded systolically into the Index register. Among the codevectors that have been examined, the index of the codevector that best approximates the input vector thus is loaded into the Index register. After all codevectors have been examined, this register contains the index of that codevector which produces the minimum distortion.

The time complexity for quantizing a vector with this architecture is $O((2^c w + 2^m N))$. This complexity reduces to that of the standard design when $m = 0$ implying $c = d$. Instead of using $k = 2^d$ IPP's, only 2^c are used: a 2^m factor of IPP's is traded for a 2^m factor of time. This design also requires the recirculating queues, however.

C. Superlinear Array Architecture

This section discusses the case when one wants a larger throughput, and is willing to accept a commensurate increase in hardware. Again, to simplify the discussion, we take the codebook size to be a power of 2: $N = 2^{n=m+c}$.

We partition the codebook. An input vector is quantized with respect to each part of the codebook, producing for each part a 1) minimum distortion and 2) partial codevector index. Of these codevectors, the one with the minimum distortion is chosen, and its full index is constructed. We are interested in increasing the throughput of the vector quantizer, but are not particularly interested in reducing the latency of quantizing an individual vector. The architecture consists of 1) a collection of partial quantizers, 2) a *distortion* bus (*d*-bus), 3) an *index* bus (*i*-bus), and 4) a chip that completes the quantization.

The partial quantizers are described first. Fig. 15 gives a schematic of a partial quantizer. It is essentially a cascade of IPP's followed by a CP. Since this component group operates in (what is now) a familiar way, most detail is omitted from this discussion. The codebook is partitioned into 2^m parts, each containing $N/2^m = 2^c$ codevectors. Each partial quantizer is responsible for quantizing with respect to one part of the codebook: a partial quantizer's comparator is responsible for finding the minimum distortion among $N/2^m = 2^c$ codevectors. The *mod* 2^c counter on the left generates an *unload/reset* signal for the comparator. This signal propagates systolically through the comparator. The comparator unloads (and resets) in a bit-serial manner, communicating its distortion bits off-chip one at a time. The unload is from high-order bits to low-order bits, which is important for subsequent comparisons. When the *unload/reset* signal has propagated completely through the comparator (causing all its bits to be unloaded), it feeds into the Partial Index register, causing it to unload onto its *i*-bus in a similar manner.

Fig. 16 gives the scheme by which the partial quantizers are attached to the busses, and the busses are attached to the completion chip. There are 2^m partial quantizers. If the *i*th partial quantizer completes its partial quantization cycle j, then the $i + 1$st partial quantizer is synchronized to complete its partial quantization on cycle $j + w_d$, where w_d is the number of bits in the comparator (distortion register). Each partial quantizer therefore has time to feed bit-serially its result to the completion chip. The completion chip is given schematically in Fig. 17. It determines, of these partial results, which is the overall minimum, and constructs the full codevector index. We assume here that the number of bits in the distortion register is greater than or equal to c, the size of the Partial Index register.[1]

The completion chip now is described. Its tasks are to

1) determine which partial quantizer has the minimum distortion value,

2) save its partial index, which is the low-order part of the codevector index, and

3) construct the high-order portion of the index. This is just the m-bit index of the "winning" partial quantizer. These $m + c$ bits form the complete codevector index.

[1]If this assumption is incorrect, then adjacent partial quantizers can be synchronized to complete c cycles apart.

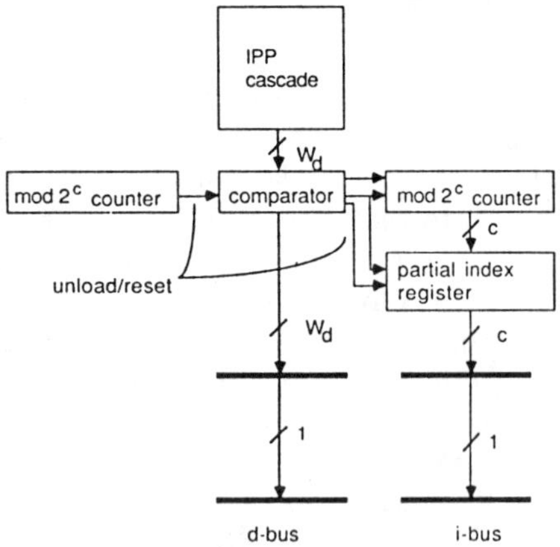

Fig. 15. Schematic of a partial quantizer.

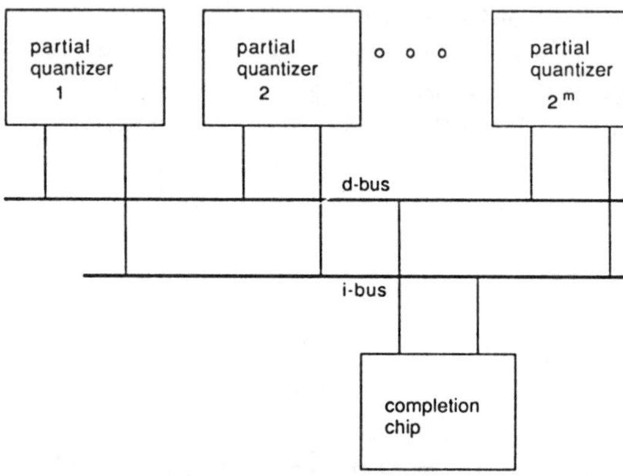

Fig. 16. Schematic of bus interconnection between partial quantizers and completion chip. All lines are 1 bit wide.

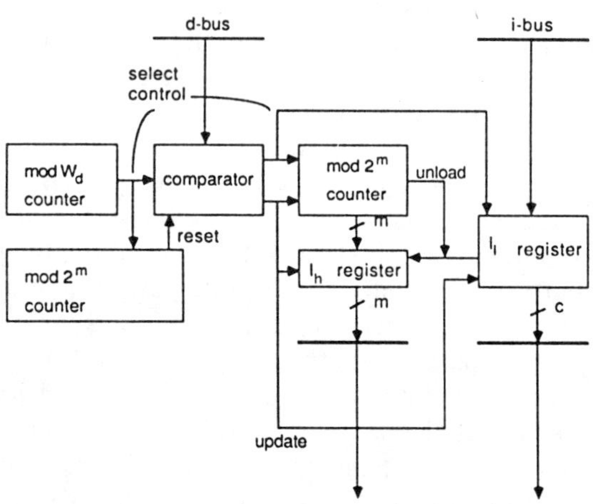

Fig. 17. Schematic of completion chip.

We first describe that part of the completion chip that determines the globally minimum distortion. Distortion register bits enter the chip serially. Unloading a partial quantizer's distortion register must be synchronized with the comparison operation taking place in the completion chip. Every w_d cycles, the comparison of a new distortion value begins. This w_d-cycle is controlled by the *mod* w_d counter on the left. Its high-order carry-out serves as a *select* signal. Every w_d cycles, this signal goes high. It propagates systolically across the comparator. The bit position of the comparator that has the *select* signal of high value compares its bit to that on the *d*-bus. In other ways, the completion chip comparator operates as does a partial quantizer comparator. The completion chip has 2^m comparisons to do for each input vector. Therefore, after every 2^m comparisons, the comparator is reset. This is controlled by the *mod* 2^m counter on the left, which itself is incremented by the carry-out signal of the *mod* w_d counter.

We now discuss the way in which the high-order part of the codevector index is constructed. The high-order part of the index is $[0, \cdots, 2^m - 1]$, depending on which partial quantizer held the globally minimum distortion. The *mod* 2^m counter and I_H register are used to construct the high-order part of the codevector index. Once the *select* signal propagates through the comparator, it is used to increment the *mod* 2^m counter. This happens, therefore, once every w_d cycles, once for each distortion value. If the comparator finds a new minimum distortion, then the *update* signal causes the *mod* 2^m counter to systolically unload its contents into the I_H register. The high-order carry-out of the *mod* 2^m counter is used to signal the complete quantization of one input vector. Its value serves as a signal to systolically unload the *I* registers, both high and low; the concatenation of their values is the codevector index we seek.

We finally discuss the way that the low-order part of the codevector is maintained. At the same time that the *select* signal is used to increment the *mod* 2^m counter, it propagates into the I_L register. It propagates systolically across this register where it is used as a *select* signal the same way that it was used as a *select* signal for the comparator. That is, it is used to synchronize the unloading of a partial index from a partial quantizer into the I_L register. But we do not want the partial index from a partial quantizer loaded into the I_L register unless the corresponding distortion is the current minimum. The *update* signal emanating from the comparator, used to control the load of the I_H register, also is used to determine whether the bits entering on the *i*-bus will be loaded or discarded.

The period to quantize one vector using this architecture is max $\{2^m w_d, 2^c\}$.

There is a variation of this architecture that has a higher throughput. If 1) w_d-bit busses are used instead of 1-bit busses, and 2) partial quantizers are synchronized to complete one cycle after another, instead of w_d cycles, then the period can be reduced to $O(2^m)$. In this design we must have $2^c \geq 2^m$. That is, it can be used to reduce the period between quantized vectors to $O(\sqrt{N})$. Although busses are wider, in this design the completion chip is somewhat less complex; the select control signal and *mod* w_d counter are eliminated.

Further reduction is possible if we use more than one completion chip. If we augment the high bandwidth completion chip to unload its distortion, as well as its index, then we can "tree" the completion chips so that a "tree" of height h can quantize vectors with a period of $O(N^{1/h})$. For example, with a codebook of 4096 vectors, and a completion chip "tree" of height 4, as indicated in Fig. 18, each of the 512 partial quantizers will quantize the input vector to 1 of 8 possibilities. With this architecture, the period for complete quantization is 8 cycles/vector. Bus speeds may limit chip cycle time.

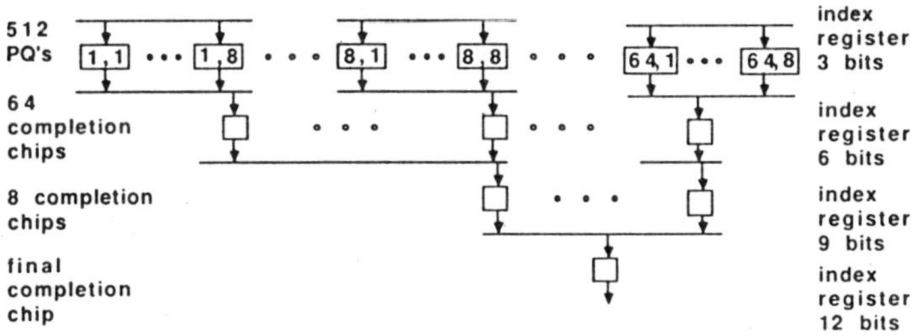

Fig. 18. Schematic of "tree" completion architecture. In the figure, N = 4096 and the height of the "tree" is 4.

VI. VLSI Realizations

To demonstrate the feasibility of implementing real-time systolic VQ architectures, we have realized the IPP and CP systolic architectures in VLSI. The two basic computational elements (IPP and CP) of all the architectures presented in this paper were implemented on separate cascadable chips. This approach facilitated the implementation of VQ processors with a wide range of vector dimensions. The chips are primarily intended for use in a linear array architecture. With slight modification, they may serve in the two-dimensional array as well. The chips are implemented in 4 μm NMOS technology, and hence consume significantly more silicon area than is possible with current VLSI technology. For a 1 μm implementation, it is possible to fit at least 16 IPP's on a single chip; however, Pin-Grid-Array (PGA) package pin counts will limit packing density when moderate or large wordlengths are required.

We adopted the Mead–Conway approach [34] to VLSI design. Both chips are essentially full-custom (only the bonding pads and some unit delay elements were taken from a cell library). The Berkeley–Caesar and Magic CAD tools were used for chip layout, design simulation, and prefabrication testing. The layout time required was approximately 1 man-month for each chip. Samples were produced by the MOSIS facility through the support of the National Science Foundation. Chip testing was conducted using facilities at UCSB. We currently have a fully functional version of the CP and, based on preliminary test results, of the IPP as well. The maximum clock rate for the chip set was found to be 9 MHz based on the measured speed of the more critical IPP chip.

The layout of the IPP was considerably simplified by designing one full-adder cell, and then duplicating it in two dimensions to obtain the entire array. The multiplier input word width is 12 bits, and accumulations are performed with 24-bit precision. This chip contains 234 full-adder cells realized using a function-block approach. The logic functions for generating $D_{out}(i)$ and $C_{out}(i)$ (see Fig. 4) were realized by putting their truth tables down in silicon. The function-block approach was taken to simplify full-adder cell design; other more efficient designs are available which would reduce cell area significantly. The IPP contains about 66 000 transistors, and occupies 8450 $\times$ 7250 μm^2 of silicon. The total pin count is 80, and the IPP is packaged in an 84-pin PGA package.

The CP was implemented by laying out three basic bit-level cells for the inverse skewing array, comparator string, and counter/index register. These fundamental cells were then duplicated and appropriately arranged to build a complete processor. The width of the CP d_j input word is consistent with the IPP output word (24 bits). The counter/index register width is 16 bits, accommodating codebooks with size up to 65 536 vectors. The CP is contained in an 84-pin PGA package (46 of these are used), has 6736 transistors, and is 5866 $\times$ 8216 μm^2 in area. Our test results indicate that 12 of the 14 CP chips we received from the foundry operate correctly.

VII. Conclusions

We have presented a family of powerful and flexible systolic processor architectures for vector quantization of speech and image waveforms. The designs are suitable for a very wide variety of real-time coding and recognition applications ranging from direct waveform VQ to more sophisticated techniques for LPC parameter vector quantization and speech recognition. An appropriate processor architecture can be selected to match the computational demands of the application while minimizing the hardware complexity (chip area). Codebook search period for these architectures ranges from Nk cycles for the design lowest in area to only 1 cycle for the highest performance design.

In the past, powerful VQ-based compression algorithms have been simulated on general-purpose computers, but their complexity has often far outstripped the capability of today's conventional real-time processors. As a result, hardware implementations typically use suboptimal codebook search schemes to reduce complexity to a manageable level. The new systolic architectures for VQ make exhaustive codebook search routines much more attractive since processor performance (codebook search period) can be matched to virtually any application without compromising optimality. For applications which are computationally less demanding but still require the use of a special-purpose processor (instead of a general-purpose DSP chip), one of the systolic designs with a smaller area and higher codebook search period can be used. Furthermore, the architectures are suitable for implementing many suboptimal schemes such as tree-structured codebooks and multistage VQ.

A performance summary of the systolic architecture family is given in Table I. The table shows the asymptotic area and time complexities as well as the exact period for each of the four processor types described. The period is given in units of clock cycles, and both the time and period figures apply to one complete codebook search.

The architectures are arranged in increasing order of area A and performance (as measured by their period). For typical values of N, k, and w, the AP product for all architectures is approximately the same, implying that performance and hardware complexity are directly proportional. In almost all source compression applications, time complexity is of secondary importance. However, time complexity does play a role in applications incorporating feedback (such as finite-state VQ) since each input vector cannot be processed until the immediately pre-

TABLE I
SUMMARY OF SYSTOLIC VQ ARCHITECTURE PERFORMANCE CHARACTERISTICS

Architecture	Area	Period	Time
Sub-Linear Array†	$O(2^c w^2 + \log N)$	$2^m N$	$O(2^c w + 2^m N)$
Linear Array	$O(kw^2)$	N	$O(kw + N)$
Super-Linear Array‡	$O(2^m kw^2)$	$\max\{2^m w_d, 2^c\}$	$O((k + 2^m)w + 2^c)$
Two-Dimensional Array	$O(Nkw^2)$	1	$O((k + N)w)$

ceding input vector has been quantized. In general, time complexity increases with processor area.

All of the usual advantages of systolic architectures apply to the designs presented in this paper as well, including ease of VLSI implementation and high clock rate capability due to short signal propagation paths. Furthermore, the feasibility of implementing these architectures using VLSI technology has already been demonstrated. As VLSI technology continues to improve, the silicon area required to implement these architectures will continue to diminish, making them even more attractive.

REFERENCES

[1] A. Gersho and V. Cuperman, "Vector quantization: A pattern-matching technique for speech coding," *IEEE Commun. Mag.*, vol. 21, pp. 15-21, Dec. 1983.

[2] R. M. Gray, "Vector quantization," *IEEE ASSP Mag.*, vol. 1, pp. 4-29, Apr. 1984.

[3] J. Makhoul, S. Roucos, and H. Gish, "Vector quantization in speech coding," *Proc. IEEE*, vol. 73, pp. 1551-1588, Nov. 1985.

[4] Y. Linde, A. Buzo, and R. M. Gray, "An algorithm for vector quantizer design," *IEEE Trans. Commun.*, vol. COM-28, pp. 84-95, Jan. 1980.

[5] H. T. Kung and C. E. Leiserson, "Algorithms for VLSI processor arrays," in *Introduction to VLSI Systems.* Menlo Park, CA: Addison-Wesley, 1980.

[6] F. C. Hennie, *Finite-State Models for Logical Machines.* New York: Wiley, 1968.

[7] H. T. Kung, "Why systolic architectures," *Computer*, vol. 15, no. 1, pp. 37-45, Jan. 1982.

[8] ——, "A listing of systolic papers," Dep. Computer Sci., Carnegie-Mellon Univ., Pittsburgh, PA, Dec. 1986.

[9] B. P. M. Tao, H. Abut, and R. M. Gray, "Hardware realization of waveform vector quantizers," *IEEE J. Select. Areas Commun.*, vol. SAC-2, pp. 343-352, Mar. 1984.

[10] G. Davidson and A. Gersho, "Application of a VLSI vector quantization processor to real-time speech coding," *IEEE J. Select. Areas Commun.*, vol. SAC-4, Jan. 1986.

[11] T. M. Liu and Z. Hu, "Hardware realization of multistage speech waveform vector quantizer," *Proc. IEEE Int. Conf. Acoust., Speech, Signal Processing*, Tokyo, Japan, Apr. 1986.

[12] P. Cappello, G. Davidson, A. Gersho, C. Koc, and V. Somayazulu, "A systolic vector quantization processor for real-time speech coding," in *Proc. IEEE Int. Conf. Acoust., Speech, Signal Processing*, Tokyo, Japan, Apr. 1986.

[13] B. P. Tao, H. Abut, and F. Mehran, "LSI architecture for VQ systolic array systems," in *Proc. IEEE Military Commun. Conf.*, Oct. 1986, pp. 49.4.1-49.4.5.

[14] H. Abut, B. P. Tao, and J. Smith, "Vector quantizer architectures for speech and image coding," in *Proc. IEEE Int. Conf. Acoust., Speech, Signal Processing*, Dallas, TX, Apr. 1987.

[15] R. Dianysian and R. Baker, "A VLSI chip set for real time vector quantization of image sequences," in *Proc. IEEE Int. Symp. Circuits Syst.*, May 1987, pp. 221-224.

[16] L. M. Ni and A. K. Jain, "A VLSI systolic architecture for pattern clustering," *IEEE Trans. Pattern Anal. Machine Intell.*, vol. PAMI-7, pp. 80-89, Jan. 1985.

[17] B. E. Nelson and C. J. Read, "A bit-serial VLSI vector quantizer," in *Proc. IEEE Int. Conf. Acoust., Speech, Signal Processing*, Tokyo, Japan, Apr. 1986.

[18] F. Jutland, N. Demassieux, and D. Vicard, "VDP: A high-performance vector distance processor," in *Proc. IEEE Int. Conf. Acoust., Speech, Signal Processing*, Tokyo, Japan, Apr. 1986.

[19] H. Abut, R. M. Gray, and G. Rebolledo, "Vector quantization of speech and speech-like waveforms," *IEEE Trans. Acoust., Speech, Signal Processing*, vol. ASSP-30, pp. 423-435, June 1982.

[20] J. H. Chen and A. Gersho, "Vector adaptive predictive coding of speech at 9.6 kb/s," in *Proc. IEEE Int. Conf. Acoust., Speech, Signal Processing*, Tokyo, Japan, Apr. 1986, pp. 1693-1696.

[21] M. R. Schroeder and B. S. Atal, "Code-excited linear prediction (CELP): High-quality speech at very low bit rates," in *Proc. IEEE Int. Conf. Acoust., Speech, Signal Processing*, Tampa, FL, Mar. 1985, pp. 937-940.

[22] G. Davidson and A. Gersho, "Complexity reduction methods for vector excitation coding," in *Proc. IEEE Int. Conf. Acoust., Speech, Signal Processing*, Tokyo, Japan, Apr. 1986, pp. 3055-3058.

[23] M. Copperi and D. Sereno, "CELP coding for high-quality speech at 8 kbit/s," in *Proc. IEEE Int. Conf. Acoust., Speech, Signal Processing*, Tokyo, Japan, Apr. 1986, pp. 1685-1688.

[24] G. Davidson and A. Gersho, "Multiple-stage vector excitation coding of speech waveforms," in *Proc. IEEE Int. Conf. Acoust., Speech, Signal Processing*, vol. I, New York, Apr. 1988, pp. 163-166.

[25] L. C. Stewart, "The design of trellis waveform coders," *IEEE Trans. Commun.*, vol. COM-30, pp. 702-210, Apr. 1982.

[26] D. Chase and A. Gersho, "Real-time VQ codebook generation hardware for speech processing," in *Proc. IEEE Int. Conf. Acoust., Speech, Signal Processing*, vol. III, New York, Apr. 1988, pp. 1730-1733.

[27] F. Itakura and S. Saito, "Analysis synthesis telephone based upon the maximum likelihood method," in *Conf. Rec. 6th Int. Congr. Acoust.*, Tokyo, Japan, 1968.

[28] R. M. Gray, A. Buzo, A. H. Gray, Jr., and Y. Matsuyama, "Distortion measures for speech processing," *IEEE Trans. Acoust., Speech, Signal Processing*, vol. ASSP-28, pp. 367-376, Aug. 1980.

[29] A. Buzo, A. H. Gray, R. M. Gray, and J. D. Markel, "Speech coding based upon vector quantization," *IEEE Trans. Acoust., Speech, Signal Processing*, vol. ASSP-28, pp. 562-574, Oct. 1980.

[30] J. V. McCanny and J. G. McWhirter, "Completely iterative, pipelined multiplier array suitable for VLSI," *IEE Proc.*, vol. 129, no. 2, pp. 40-46, Apr. 1982.

[31] P. R. Cappello and K. Steiglitz, "Completely pipelined architectures for digital signal processing," *IEEE Trans. Acoust., Speech, Signal Processing*, vol. ASSP-31, pp. 1016-1023, Aug. 1983.

[32] ——, "A note on 'free accumulation' in VLSI filter architectures," *IEEE Trans. Circuits Syst.*, vol. CAS-32, pp. 291-296, Mar. 1985.

[33] P. B. Denyer and D. J. Myers, "Carry-save arrays for VLSI signal processing," in *Proc. Int. Conf. VLSI*, Edinburgh, Scotland, Aug. 1981, pp. 151-160.

[34] C. Mead and L. Conway, *Introduction to VLSI Systems.* Reading, MA: Addison-Wesley, 1980.

VECTOR QUANTIZER ARCHITECTURES FOR SPEECH AND IMAGE CODING

by
Huseyin Abut, ECE Dept., SDSU, San Diego, CA 92182
Bertram P.M. Tao, Voice Innovations Inc, 90 Elksford, Irvine, CA 92714
Jack L. Smith, TRW-MED, One Rancho Carmel, San Diego, CA 92128

ABSTRACT

We present a number of architectures for vector quantization (VQ) of speech and images using VLSI and VHSIC technologies. A Dual Distortion Processor Module (DPM) has been designed to compute the error vectors at a rate of 10 million vector operations per second in a systolic configuration. An array processor controller (APC) administers the system and determines the nearest neighbor matching codeword in either a full-search or a tree-search manner. A real-time system was built and tested in 0.5 bit per pixel (bpp) image coding application. We also present an architecture using VHSIC technology based on fuzzy associative memory (FAM) chips. In this case, the system has been configured in a VME bus environment and the overall number crunching task is handled by VHSIC chips.

1. INTRODUCTION

The fundamental number crunching operation in a VQ encoding system is a minimum distortion codeword selection process for a given distortion measure, which involves a full or tree search through a previously optimized codebook. Speech and image encoders using vector quantizer algorithms require 2**(K.r) distance computations for each vector of size K or for each matrix of size K=l*k, where r is the encoding rate in bits per symbol. For instance, for an image frame of size 256 x 256 with an encoding rate of 0.5 bit per pixel, codebook size of 256, dimensionality of K=4x4 and a frame rate of 7.5 frames per second, the number of computations is on the order of 8 million vector/matrix operations per second. In order to address the real-time implementation of such a problem, one has to resort to either very fast hierarchical searching techniques or to parallel processing [1,2].

In this paper, we present three architectures based on VLSI and VHSIC technologies to tackle the class of problems with the aforementioned magnitude; in particular, two different versions of systolic array structures using the Distortion Processor Module (DPM) [3] and a wide-word architecture using Fuzzy Associative Memory (FAM) techniques [4].

2. SYSTEM FUNCTION

Figure 1 illustrates the data flow model of our systolic array. The movement is two-dimensional, proceeding both horizontally and vertically across the component processing elements (PE). The action represents the desired parallel computation, whereby valid kth-order outputs y(k) are available for every input sample x(k) and codebook component c(k) at the kth PE. In such an arrangement, a distortion value is available at the last stage after every cycle.

Since each error component y(k,n) is fed forward to the next stage at each successive clock cycle, the output of PE(k) represents the kth-order error vector. The total accumulated error Y(n) = y(K-1,n) is available only at the last stage. The values of y's at any given time n are related by the following system function:

$$y(k,n) = y(k-1,n-1)+d[x(k),c(j,k,n)] \quad \text{for } j=0,..,J-1 \text{ and } k=K-1,..,0 \qquad (1)$$

where

$$d(k) = d[x,c] = [x(k)-c(k)]^{2} \qquad (2)$$

Equation (1) completely describes the process at any stage in the system. Hence, a total accumulated distortion value is available at every time interval n. To realize the PE function, a distortion processor module (DPM) has been designed. A dual DPM has subsequently been designed for VLSI implementation for expediting the full-search process and also in facilitating the binary tree-search method.

3. DPM AND APC MODULES

Figure 2 shows the architecture of the DPM. There are three distinct elemental operations (EO): 1) Absolute Magnitude Difference Function (AMDF); 2) Squaring; and 3) Accumulation. The composite of the EO's gives the partial accumulated squared-error distortion of (2). The pipelined architecture of the DPM enables a 3-fold increase in throughput compared to systems with no pipes, without requiring a similar speed-up in EO's. The effective processing speed is 100 ns per EO, which facilitates an equivalent 10 million vector operations per second.

This particular DPM architecture causes 4 units of delay (one for each pipeline register) in the array system timing,

Reprinted from *Proc. IEEE Conf. Acoust., Speech, Signal Processing,* vol. 2, pp. 756–759, April 1987.

and the net delay for the overall system is K+4 (K=dimension). Input data to the DPM must be synchronized to the overall system in order to keep all pipes full. This is especially crucial in a binary tree configuration. Synchronous operation is maintained by the Array Processor Controller (APC).

The APC performs the following tasks: 1) system synchronization; 2) distortion comparisons; 3) codeword selection; and 4) overhead control. Array synchronization is done by delaying the DPM inputs by the appropriate number of cycles. In a full-search task, the distortion comparison determines the minimum of the current nearest-neighbor mse and the mse between the input and the current codeword. In a tree-search task, however, the minimum distortion between the input and upper and lower branch codes is determined.

Codeword selection in the full-search mode is completely predictable; since the index starts from row 1 and proceeds to row N (N=number of codewords). For a tree-search situation, however, the next codeword is a function of the current level and the branch selection of codes at that level.

Finally, data adminstration, timing, and control operations constitute the overhead control section. The APC's reconfigurable architecture is shown in figures 3 and 4, and elucidated in the next two sections, respectively.

4. A FULL-SEARCH SYSTEM ARCHITECTURE

Figure 3 illustrates the configuration of a full-search system. For K-dimensional vectors, K DPM's (or K/2 dual DPM's) and one APC can be used. In the example of figure 3, K=16, and the number of levels is 256. The inputs to the kth DPM are the kth components of the input and code vectors. Note that the APC uses a FIFO to buffer a block of input data. The FIFO facilitates the circular action needed to access the input vector multiple times throughout the quantization process.

Once the FIFO is loaded, the input vector and codeword are presented to the DPM's for distortion computation. An Index Counter is used to access the codewords from the codebook, which is divided into 16 components; one for each vector dimension. The Counter action is such that it is incremented after every cycle (100 ns) from codeword #1 to #256.

The output of the kth DPM is the kth distortion component, which is passed to the next stage. The final vector distortion is available at the 16-th stage to the APC, which performs a comparison between the magnitudes of the running minimum mse and the current vector distortion. The smaller of the two is stored in a holding register, and the Index Counter is also stored (in another holding register) if a new codeword is selected. Due to the systolic action of the system, a complete vector distortion is computed every cycle, with the exception of the first 20 due to the initial overhead delay.

The best match is determined at the final level. Overall, there are exactly 256 vector computations per input vector, hence it requires 25.6 μs for quantization plus an additional 2 μs for overhead.

5. A TREE-SEARCH SYSTEM ARCHITECTURE

Figure 4 illustrates the configuration of a 9-level complete binary tree-search system with K=16 dimensions. For our real-time applications, only one dual DPM and one APC are needed. Enough memory for 512 16-dimensional codewords is required which are allocated to two codebook memory devices (one for each DPM). The codebook organization is such that one codebook memory contains the first half of each vector (dimensions 1 through 8), while the other comprises dimensions 9-16.

In a binary tree-search, one of two codewords is selected at each level of the tree, which represents the current best match. The selection is determined by computing the distortion between the input vector and both the upper and lower-branch codewords (performed by the DPM's). The code which produces the minimum of the two is chosen. The next set of test codewords branch from the resulting node and the final code resides at the root (bottom) of the tree.

An Index Shift Register (ISR) stores the accumulated path through the tree. Since the outcome of each decision is binary, one bit is sufficient to represent the path at each node level. The bit is shifted into the ISR, whose contents give the upper bits of the next codeword address. To complete the address, an independent 4-bit counter accesses the individual vector components. This process continues through the root level, at which time the channel symbol can be transmitted.

Since there are only two DPM's, then a feedback path from the second DPM to the first must be connected. This path facilitates a continuous accumulation of distortion terms as the vector components are processed through each DPM. As a result, a distortion value is computed every 8 cycles and stored in a temporary holding register.

For efficiency, two input vectors are quantized at a time and in an interleaved fashion. The reason is to keep the DPM pipelines full at all times. Without interleaving, the overhead incurred in determining the codebook addresses would significantly delay the supply of data to the DPM's. However, with this design it is possible to achieve nearly 100% efficiency, therefore, a 9-level binary-tree searched VQ requires 144 cycles plus an initial overhead delay of 8 cycles, for a total time of 15.2 μs at 100 ns per cycle.

6. A VHSIC FULL SEARCH ARCHITECTURE

Unlike the hardware structures discussed above, the FAM does not utilize systolic processing techniques on individual components of vectors or matrices, but instead employs VHSIC technology in a wide-word architecture. Throughput is obtained via the use of parallel processing on two fronts: (1) processing all vector components simultaneously and (2) processing multiple input vectors at a given time. The key to this technique is a VSHIC implementation of very high speed comparison circuitry which is configured for dedicated processing of assigned vector components. Thus, the distortion computations of the nearest neighbor selection procedure are performed inherently within the VSHIC circuitry and only the final value is passed out to the next stage of processing.

Depending upon the specifics of the application and the overall bit rate the codebook and data is either entered and stored in the FAM local memory or is accessed on demand basis from an external ROM. Input vectors to be quantized, described as rectangular regions in a codebook vector space, are loaded into a comparison item memory, also in the FAM. These elements are depicted in figure 5. If the encoding rate is sufficient enough multiple input test vectors or matrices can be accomodated by the FAM. Searching is then initiated against a designated block of codebook vectors. During the search process, all component fields of all input vectors are compared against successive codewords through the use of the bank(s) of the VHSIC comparators. The codewords or the input vectors may be up to 96 bits in width --or even 192 bits-- with component field widths externally programmable. This last feature is extremely important in block coding techniques containing vectors of varying bit resolution. One such example is the class of LPC-based VQ and MQ vocoding schemes, in which the coefficients with high indices do not require as much resolution as the lower order ones. The FAM structure discussed in this paper permits a varying word lengths with minimal constraints both for the incoming test vectors and the codebook entries.

Thus, the FAM is applicable to a wide variety of VQ and MQ tasks. Applications flexibility is obtained though the use of an industry-standard VME bus interface, programmable, configurable vector formats, very large codebook size and the ability to partition the codebook memory.

7. AN IMAGE CODING APPLICATION

A full search VQ image coder with a rate of 0.5 bpp, a block size of 4x4 pixels, 8-bits per pixel, 256x256 pixels per frame and a frame refreshing rate of 7.5 frames per second has been implemented. The overall rate of quantization was 30,720 matrices per second, which led to an actual processing speed of 10 million vector operations per second or equivalently, 480 million elemental operations per second.

Initial software simulations were performed using a database of 6 pictures of varying scenes. All testing was done outside the training database. The system prototype was implemented from simple CMOS and TTL devices, which resulted in no significant difference relative to the simulations, and the SNR values were almost identical: 17.68 dB (simulation) versus 17.69 dB (hardware).

Figure 6 displays an original picture (upper left and lower right), the simulated logic emulation results (upper right) and the hardware output (lower left).

8. ACKNOWLEDGEMENTS This project has been partially supported by Naval Oceans System Center, TRW-MED, Cubic Corporation, and Voice Innovations.

9. REFERENCES

[1] B.P.M. Tao, H. Abut, and R.M. Gray, "Hardware Realization of Waveform Vector Quantizers," IEEE Journal on Selected Areas in Communications, Vol. SAC-2, pp. 343-352, March 1984.

[2] G. Davidson and A. Gersho, "Application of a VLSI Vector Quantization Processor to Real-Time Speech Coding," IEEE Journal on Selected Areas in Communications, Vol. SAC-4, January 1986.

[3] B.P.M. Tao, H. Abut, and Farokh Mehran, "An LSI Architecture for VQ/MQ Systolic Array Systems", Proc. IEEE MILCOM'86, Vol. 3, pp. 49.4.1, Monterey, CA, Oct 1986.

[4] T.A. Nodes, J.L. Smith, and R. Hecht-Nielsen, "A Fuzzy Associative Memory Module and its Application to Signal Processing," Proc. IEEE ICASSP'85, Vol. 4, pp. 1511-1514, Tampa, Fl., March 1985.

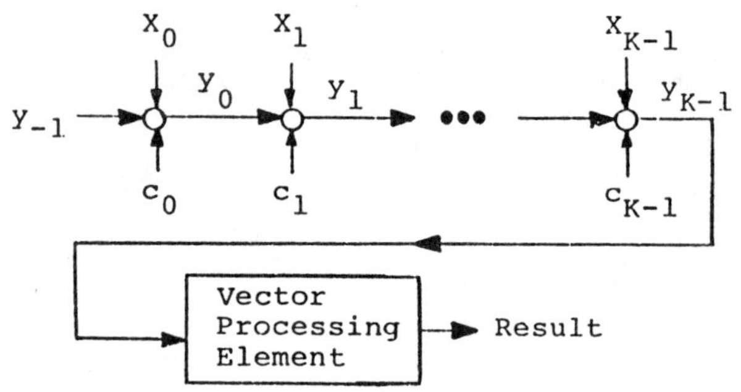

Figure 1. Systolic Array Architecture for VQ Systems. o represents PE.

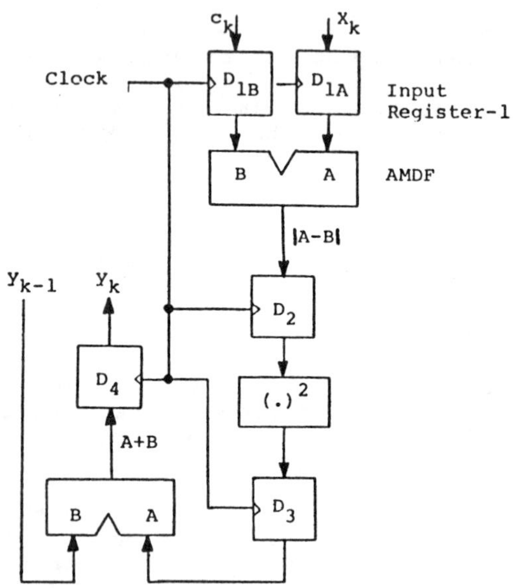

Figure 2. k-th Distortion Processor where $y_k = y_{k-1} + |x_k - c_k|^2$

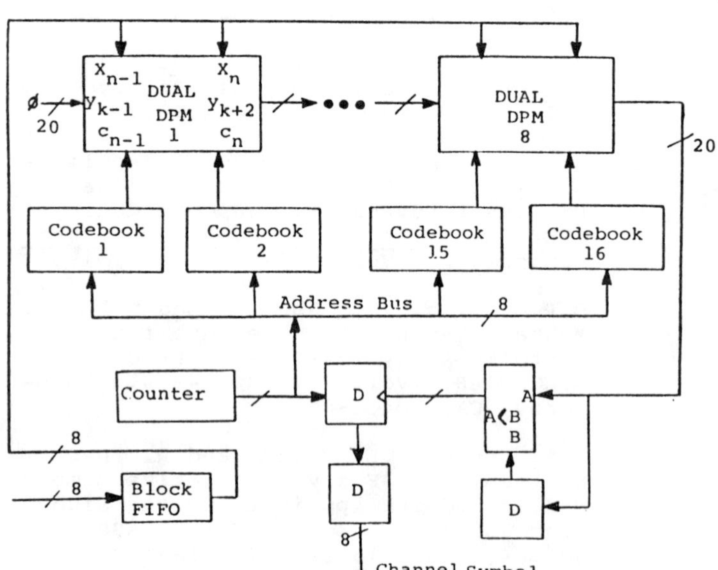

Figure 3. Configuration Using 8 Dual DPM's and one APC for 256-level, 16-dimensional Full-Search VQ.

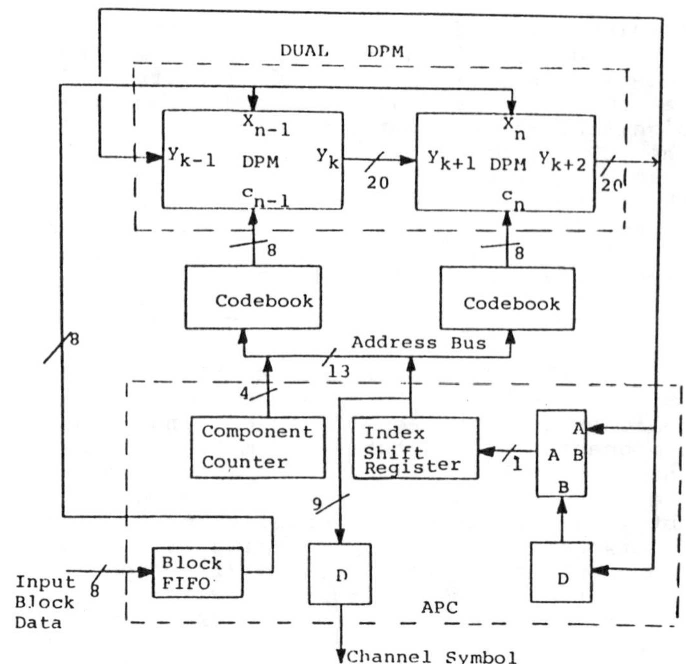

Figure 4. Configuration Using one Dual DPM and one APC for 9-level, 16-dimensional, Binary Tree-Searched VQ.

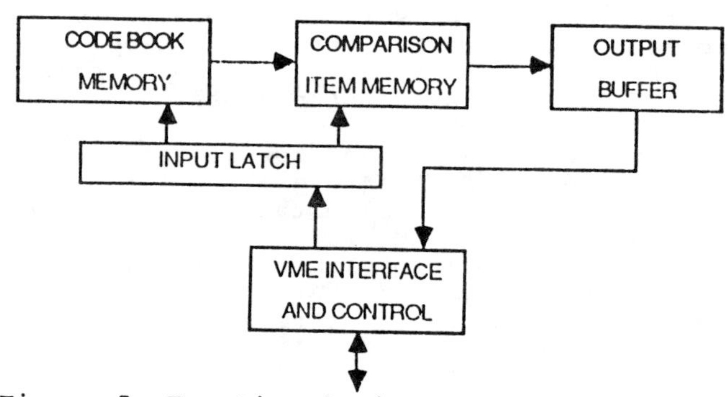

Figure 5. Functional Diagram of a VSHIC Full-Search Architecture

Figure 6. Results of 0.5 bpp VQ Image Encoding System. Upper left and lower right = original, upper right = prototype output, lower left = software.

REAL-TIME VECTOR APC SPEECH CODING AT 4800 BPS WITH ADAPTIVE POSTFILTERING[1]

Juin-Hwey Chen [2] *and Allen Gersho*

Communications Research Laboratory
Department of Electrical and Computer Engineering
University of California, Santa Barbara, CA 93106

ABSTRACT

An improved Vector APC (VAPC) speech coder at 4800 bps produces speech with very good communications quality while maintaining a complexity low enough to allow a real-time implementation with at most two commercially available DSP chips. The VAPC algorithm combines APC with vector quantization and incorporates analysis-by-synthesis, perceptual noise weighting, and adaptive postfiltering. A novel adaptive postfiltering technique helps to achieve an essentially inaudible level of coding noise. Real-time software has been developed for an implementation using the AT&T DSP32 floating-point processor chip. The overall complexity of the implemented VAPC system is about 3 million multiply-adds/second of computation and 6 kwords of memory.

1. Introduction

Vector Adaptive Predictive Coding, or VAPC [1] , is a new speech coding algorithm which greatly improves the performance of conventional Adaptive Predictive Coding (APC) [2] by using Vector Quantization (VQ) techniques. In VAPC, we first use a long-delay predictor, or *pitch predictor*, to remove the long-term redundancy due to pitch quasi-periodicity. We then use a short-delay *LPC predictor* to remove the redundancy remaining in the pitch-prediction residual. The final residual is then quantized by a particular type of *gain-adaptive vector quantizer* [3] , [4] that performs the codebook search in an analysis-by-synthesis manner. Noise spectral shaping is achieved by using a perceptual weighting filter in the codebook search. The decoder obtains the coded speech by exciting the LPC synthesis filter and the pitch synthesis filter with the quantized residual.

The work described in this paper is an extension of our previous work on 9600 bps VAPC [1] . We have made several improvements to our previous VAPC system, including better VQ schemes for the two predictors and an optimized residual VQ codebook. The speech quality has been noticeably improved by these schemes; however, at 4800 bps the coding noise was still audible despite the use of perceptually-based noise spectral shaping.

To further improve the perceptual quality of coded speech, we have developed a novel adaptive postfiltering technique that reduces the perceived level of coding noise without introducing significant distortion in the filtered speech. For high-quality input speech, the coding noise is reduced to an essentially inaudible level after the postfiltering. At 4800 bps, VAPC with postfiltering has produced coded speech that is natural-sounding, clean, intelligible, and quite close to the original.

The VAPC algorithm and the postfiltering scheme were developed for real-time hardware implementability. In particular, the postfilter only requires 0.3 million multiply-adds/second of computation to implement. To reduce the overall complexity of the coder, we have developed a simplified version of VAPC for hardware implementation. This version achieves almost the same speech quality as obtained by a more complex version, and the overall complexity of VAPC with postfiltering is only about 3 million multiply-adds/second and 6 kwords of data storage. This is within the capacity of commercially available programmable DSP chips. We have chosen to implement this low-rate, low-complexity, and high-quality speech coder using the AT&T DSP32 floating-point signal processor chip.

2. Coder Structure and Encoding Operation

We have shown in [1] the step-by-step evolution from a direct VAPC structure to an equivalent but computationally more efficient one, shown in Fig. 1. Here we briefly describe this efficient structure and its encoding operation. More detail can be found in [1] .

For each speech frame, the transmitter first performs the required analysis to determine the pitch period, the pitch predictor, the LPC predictor, the weighting filter, and the gain. These parameters, except for the weighting filter, are then quantized and sent to the receiver as side information (not shown in Fig. 1).

The *zero-state-response codebook*, or *ZSR codebook*, is the key to complexity reduction. It is obtained in the following way. Let G be the quantized residual gain, $A(z)$ be the quantized LPC inverse filter, and $W(z)$ be the weighting filter with the form

$$W(z) = \frac{1 - P(z/\lambda)}{1 - P(z/\gamma)} \ , \quad 0 < \gamma < \lambda \leq 1 \ , \qquad (1)$$

where $P(z)$ is the *unquantized* LPC predictor that is obtained by performing LPC analysis on the *original* speech. At 4800 bps, we use a vector dimension of 20 and a codebook size of 128 for residual VQ. At the beginning of each frame, we first determine G, $W(z)$, and $A(z)$. Then, we pick one codevector at a time from the gain-normalized codebook of gain-adaptive VQ, convolve it with the impulse response of the "weighted LPC synthesis filter" $GW(z)/A(z)$, and retain only the first 20 samples. The resulting vector is a codevector in the ZSR codebook, and this procedure is repeated until all the 128 codevectors are exhausted.

After the speech analysis and the ZSR codebook computation are finished, the transmitter in Fig. 1 operates as follows. In encod-

[1] This work was performed for the Jet Propulsion Laboratory, California Institute of Technology, sponsored by the National Aeronautics and Space Administration.

[2] Juin-Hwey Chen is now with Codex Corporation, 20 Cabot Blvd., Mansfield, MA 02048.

Reprinted from *Proc. IEEE Conf. Acoust., Speech, Signal Processing,* vol. 4, pp. 2185–2188, April 1987.

ing each input vector, the prediction $\hat{s}(n)$ generated by the pitch predictor is first subtracted from the original speech vector $s(n)$. The pitch-prediction residual vector $\mathbf{d}(n)$ is filtered by the weighting filter $W(z)$. Then, the zero-input-response vector of the filter $GW(z)/A(z)$, which is the ringing of the filter from previously selected gain-normalized codevectors, is computed and is subtracted from the weighted pitch-prediction residual vector $g(n)$. The resulting vector $v(n)$ is compared with the 128 codevectors in the ZSR codebook. The index of the nearest-neighbor codevector is sent to the receiver and is also used to extract the corresponding codevector in the gain-normalized codebook. The extracted codevector $y(n)$ is used to reset the memory of the filter $GW(z)/A(z)$, and a scaled version of it is also used to excite the LPC synthesis filter, whose output is then used in pitch prediction of the following speech vectors. The process continues until all input vectors in a frame are encoded.

At the receiver, the side information and the VQ index are decoded. The decoded codevector is scaled up by the gain G, and the resulting vector is used to excite the LPC synthesis filter and the pitch synthesis filter to obtain the reconstructed speech. An adaptive postfilter is then used to enhance the reconstructed speech.

3. Vector Quantization of Pitch and LPC Predictors

Our previous VAPC scheme quantizes the coefficients of the three-tap pitch predictor using a three-dimensional vector quantizer with the Mean-Squared Error (MSE) of the coefficients as the distortion criterion. Although such a quantizer gives reasonable results, we have found that better prediction gain can be achieved by using the energy of the open-loop pitch-prediction residual as the distortion criterion. A direct approach for the codebook search would be to use each candidate pitch predictor stored in the codebook to inverse filter the input speech and pick the predictor that minimizes the residual energy. However, such an approach requires 2 million multiply-adds/second for a codebook size of 64 — the size we use in VAPC. We have developed an efficient search method that is at least 20 times faster for this codebook size [5].

The quantization of LPC parameters plays a crucial role in VAPC. The VAPC coder reported in [1] uses a 10-bit vector quantizer to quantize the log-area ratio (LAR) parameters. Ten-bit VQ of the LARs is inadequate, as the coded speech has quite audible spectral distortion. On the other hand, we have found that 10-bit VQ of the Line-Spectrum Pair (LSP) parameters [6] gives an improved speech quality. To achieve even better quality, we use a two-stage vector quantizer [7] for the LSP parameters, where each stage uses a 10-bit codebook. To maintain the same overall bit-rate, we decrease the residual encoding rate by increasing the vector dimension from 16 to 20. Although the residual encoding rate is decreased, the speech quality is still improved noticeably by using 20-bit LSP VQ instead of 10-bit LAR VQ.

4. Vector Quantization of the Prediction Residual

In [1], we designed the residual VQ codebook based on the open-loop residual using the codebook design algorithm for forward gain-adaptive VQ [3]. Such a codebook gives better speech quality than a codebook containing Gaussian random numbers as used in stochastic coders. We later developed an even better closed-loop codebook design method for generating a codebook that minimizes the perceptually weighted error between the coded and the original speech waveforms. When compared with the Gaussian codebook, the optimized codebook improves the coder SNR by nearly 1 dB, and it also significantly improves the speech quality.

5. Adaptive Postfiltering

In addition to the above improvements, we use adaptive postfiltering to further improve the perceptual quality of coded speech. The idea of our adaptive postfiltering scheme is explained below.

The noise masking effect of human auditory perception [8] is exploited in VAPC by using noise spectral shaping. However, in noise spectral shaping, lowering noise components at certain frequencies can only be achieved at the price of increased noise components at other frequencies [9]. Therefore, at bit-rates as low as 4800 bps where the average noise level is quite high, it is very difficult, if not impossible, to force noise below the masking threshold at all frequencies. Since speech formants are much more important to perception than spectral valleys, our philosophy is to preserve the formant information by keeping the noise in the formant regions as low as is practical during encoding. Of course, in this case, the noise components in spectral valleys may exceed the threshold; however, these noise components can be attenuated later by a postfilter. In performing such postfiltering, the speech components in spectral valleys will also be attenuated. Fortunately, the *limen*, or *just noticeable difference (JND)*, for the intensity of spectral valleys can be quite large [10]. Therefore, by attenuating the components in spectral valleys, the postfilter only introduces minimal distortion in the speech signal, but it achieves a substantial noise reduction.

Adaptive postfiltering has been used successfully in enhancing ADPCM-coded speech [11], [12]. The postfilter used in [11] is derived from the two-pole six-zero ADPCM synthesis filter by moving the poles and zeros radially toward the original. If we extend this idea directly to an all-pole LPC synthesis filter $1/[1-\hat{P}(z)]$, we will get $1/[1-\hat{P}(z/\alpha)]$ as the corresponding postfilter, where $0 < \alpha < 1$. Such an all-pole postfilter indeed reduces the perceived noise level; however, sufficient noise reduction can only be achieved with severe muffling in the filtered speech. This is due to the fact that the frequency response of this all-pole postfilter generally has a lowpass spectral tilt for voiced speech. As an example, we show in Fig. 2 an LPC spectrum and the corresponding frequency responses of the all-pole postfilter $1/[1-\hat{P}(z/\alpha)]$ for $\alpha = 0.9, 0.8, 0.5$, and 0. The lowpass spectral tilt for nonzero α is clearly seen in Fig. 2.

The spectral tilt of the all-pole postfilter $1/[\hat{1}-P(z/\alpha)]$ can be easily reduced by adding zeros having the same phase angles as the poles but with smaller radii. The transfer function of the resulting pole-zero postfilter has the form

$$H(z) = \frac{1-\hat{P}(z/\beta)}{1-\hat{P}(z/\alpha)} \ , \ 0 < \beta < \alpha < 1 \ . \tag{2}$$

The frequency response of $H(z)$ can be expressed as

$$20 \log \left| H(e^{j\omega}) \right| = 20 \log \frac{1}{\left| 1-\hat{P}(e^{j\omega}/\alpha) \right|} - 20 \log \frac{1}{\left| 1-\hat{P}(e^{j\omega}/\beta) \right|} . \tag{3}$$

Therefore, in logarithmic scale, the frequency response of the pole-zero postfilter $H(z)$ is simply the difference between the frequency responses of two all-pole postfilters. Typical values of α and β are 0.8 and 0.5, respectively. From Fig. 2, we see that the response for $\alpha = 0.8$ has both formant peaks and spectral tilt, while the response for $\alpha = 0.5$ has spectral tilt only. Thus, with $\alpha = 0.8$ and $\beta = 0.5$ in Eq. (2), we can at least partially removed the spectral tilt by subtracting the response for $\alpha = 0.5$ from the response for $\alpha = 0.8$. The resulting frequency response of $H(z)$ is shown in the upper plot of Fig. 3.

In informal listening tests, we have found that the muffling effect was significantly reduced after the numerator term $[1-\hat{P}(z/\beta)]$ was included in the transfer function $H(z)$. However, the filtered speech remained slightly muffled even with the spectral-tilt compensating term $[1-\hat{P}(z/\beta)]$. To further reduce the muffling effect, we add a first-order filter which has a transfer function of $[1-\mu z^{-1}]$, where μ is typically 0.5. Such a filter provides a slightly highpassed spectral tilt and thus helps to reduce muffling. This first-order filter is used in cascade with $H(z)$, and a combined frequency response with $\mu = 0.5$ is shown in the lower plot of Fig. 3.

The first-order filter $[1-\mu z^{-1}]$ can also be made adaptive to better track the change in the spectral tilt of $H(z)$. In computer simulations, however, we have found that even a fixed filter with μ = 0.5 gives quite satisfactory results. Therefore, we use a fixed value of μ for simplicity.

To avoid occasional large gain excursions, we add an automatic gain control (AGC) at the output of the adaptive postfilter. The purpose of AGC is to scale the enhanced speech such that it has roughly the same power as the unfiltered noisy speech. This is achieved by estimating the power of the unfiltered and filtered speech separately and then use the ratio of the two values as the scaling factor. Let $\{s(n)\}$ be the sequence of either unfiltered or filtered speech samples; then, the speech power $\sigma^2(n)$ is estimated by using

$$\sigma^2(n) = \zeta\sigma^2(n-1) + (1-\zeta)s^2(n)\,, \quad 0 < \zeta < 1. \tag{4}$$

A suitable value of ζ is 0.99.

The complexity of the postfilter described in this section is only a small fraction of the overall complexity of VAPC. In simulations, this postfilter achieves significant noise reduction with almost negligible distortion in speech. To test for possible distorting effects, we applied the adaptive postfiltering operation to clean, uncoded speech and found that the unfiltered original and its filtered version sound essentially the same, indicating that the distortion introduced by this postfilter is negligible.

6. Coder Performance

For clean speech, the adaptive postfilter has reduced the coding noise of 4800 bps VAPC to an essentially inaudible level. The coded speech sounds quite close to the unquantized original. The main difference is that the original sounds somewhat "richer".

We have tested the 4800 bps VAPC system using speech corrupted by either white noise or vehicle noise (where the speech was recorded in a moving truck). For input SNR as low as 6 dB, the lowest SNR tested, VAPC still performs quite well. Such robustness to background noise is not surprising since VAPC is basically a waveform-coding approach. In fact, we have found that the power level of the background noise is even reduced after VAPC coding and postfiltering.

We have also studied the robustness of VAPC to transmission errors when there is no error protection at all. Our simulation is based on bit errors produced in a simulated mobile satellite channel provided by the Jet Propulsion Laboratory. The simulated channel errors have an average bit-error rate of 0.11% (error probability $\approx$ 0.001); however, the bit errors are rather bursty.

Surprisingly, even without error protection, VAPC still works reasonably well. Out of 12 to 15 error bursts that occurred in 30 seconds of continuous speech, usually only two or three bursts caused brief audible glitches in the simulation, while all other error bursts did not have any audible effect. We think this is because the prediction residual in VAPC has a much smaller amplitude than the speech; hence, if the errors only affect the residual VQ index, the effect in the decoded speech waveform is only a relatively small amplitude disturbance. On the other hand, the speech waveform is expected to be altered more drastically if the bit errors cause incorrect decoding of side information. Such degradation can be reduced by parameter smoothing, Gray coding, or explicit error protection of side information. A pseudo-Gray coding technique for assigning binary words to codevectors has been developed in our laboratory and offers useful error control without adding redundant bits

7. Real-Time Implementation

We have developed real-time software for a hardware prototype 4800 bps VAPC coder which we are currently implementing using the commercially available AT&T DSP32 chip. In its newest version, this chip has a processing power of 6 million instructions per second (MIPS). To facilitate the real-time implementation, we have used a simplified version of the 4800 bps VAPC coder in our real-time software. This version has a much lower complexity but gives nearly the same speech quality.

Pitch detection and LPC parameter quantization are computationally intensive in the full-complexity version of VAPC. To reduce the complexity, we first developed an efficient pitch-detection algorithm. This algorithm uses 4:1 decimation to reduce the computation required in computing correlations; after the time delay corresponding to the maximum correlation of the decimated samples is quickly identified, we perform a full search in the neighborhood for the maximum correlation of the original speech samples. The second simplification is to replace the 20-bit LSP vector quantizer by a simpler 24-bit scalar reflection-coefficient quantizer [13].

The residual VQ codebook search is also computationally intensive. In the real-time implementation we use an inner-product approach which is more efficient than the conventional difference-square approach of computing the MSE distortion. Given a test vector $\mathbf{v}$ and N ZSR codevectors $\mathbf{z}_j$, $j = 1, 2, \ldots, N$, the j-th MSE distortion can be computed as

$$\|\mathbf{v}-\mathbf{z}_j\|^2 = \|\mathbf{v}\|^2 - 2\left[\mathbf{v}^T\mathbf{z}_j - \frac{1}{2}\|\mathbf{z}_j\|^2\right] \tag{5}$$

At the beginning of each frame, we can compute and store $\frac{1}{2}\|\mathbf{z}_j\|^2$, $j = 1, 2, \ldots, N$; then, in the actual codebook search, we find the codevector $\mathbf{z}_j$ that maximizes the difference between the inner product $\mathbf{v}^T\mathbf{z}_j$ and the stored half norm square $\frac{1}{2}\|\mathbf{z}_j\|^2$. With the DSP32 processor and for the dimension and codebook size we used, the difference-square approach of the codebook search requires about 2.5 MIPS to implement, while the inner-product approach only requires about 1.5 MIPS.

The complexity of the simplified VAPC coder is about 3 million multiply-adds/second and 6 kwords of data memory. However, due to the overhead in implementation, a single 16 MHz DSP32 chip was not sufficient for implementing the coder. Therefore, we required two DSP32 chips to implement the coder. With the faster DSP32 chip now available, which has an instruction cycle time of 160 ns rather than 250 ns, we expect that the coder can be implemented using only one DSP32 chip.

8. Conclusions

We have improved our previous VAPC coder so that it now gives very good communication-quality speech at 4800 bps. Better VQ schemes for the predictors, the optimized residual VQ codebook, and the postfilter all contribute to quality improvement. The adaptive postfilter developed in this work introduces negligible distortion while achieving a notable reduction of perceived noise level. VAPC gives natural-sounding, clean, and intelligible speech at 4800 bps; in addition, it is reasonably robust both to transmission errors and to background noise in the input speech, and its complexity is low enough to allow real-time implementation using only one or at most two DSP chips.

References

[1] J. H. Chen and A. Gersho, "Vector Adaptive Predictive Coding of Speech at 9.6 kb/s," *Proc. ICASSP*, pp. 1693-1696, Tokyo, Japan, Apr. 1986.

[2] B. S. Atal and M. R. Schroeder, "Adaptive Predictive Coding of Speech Signals," *Bell Syst. Tech. J.*, vol. 49, pp. 1973-1986, Oct. 1970.

[3] J. H. Chen and A. Gersho, "Gain-Adaptive Vector Quantization for Medium-Rate Speech Coding," *Conf. Rec., IEEE Int'l Conf. Commun.*, pp. 1456-1460, Chicago, June 1985.

[4] J. H. Chen and A. Gersho, "Gain-Adaptive Vector Quantization with Application to Speech Coding," to appear in *IEEE Trans. Commun.*

[5] J. H. Chen, *Low-Bit-Rate Predictive Coding of Speech Waveforms Based on Vector Quantization* , Ph. D. dissertation, Univ. of Calif., Santa Barbara, Jan. 1986.

[6] F. K. Soong and B. H. Juang, "Line-Spectrum Pair (LSP) and Speech Data Compression," *Proc. ICASSP*, pp. 1.10.1-1.10.4, San Diego, Mar., 1984.

[7] B. H. Juang and A. H. Gray, "Multiple Stage Vector Quantization for Speech Coding," *Proc. ICASSP*, pp. 597-600, Paris, France, Apr. 1982.

[8] M. R. Schroeder, B. S. Atal, and J. L. Hall, "Optimizing Digital Speech Coders by Exploiting Masking Properties of the Human Ear," *J. Acoust. Soc. Am.*, vol. 66, no. 6, pp. 1647-1652, Dec. 1979.

[9] B. S. Atal and M. R. Schroeder, "Predictive Coding of Speech Signals and Subjective Error Criteria," *IEEE Trans. Acoust., Speech, and Signal Processing*, vol. ASSP-27, no. 3, pp. 247-254, June 1979.

[10] J. L. Flanagan, *Speech Analysis, Synthesis, and Perception*, Academic Press, New York, 1972.

[11] V. Ramamoorthy and N. S. Jayant, "Enhancement of ADPCM Speech by Adaptive Postfiltering," *AT&T Bell Labs Tech. J.*, pp. 1465-1475, Oct. 1984.

[12] N. S. Jayant and V. Ramamoorthy, "Adaptive Postfiltering of 16 kb/s-ADPCM Speech," *Proc. ICASSP*, pp. 829-832, Tokyo, Japan, Apr. 1986.

[13] O. Ghitza and J. L. Goldstein, "Scalar LPC Quantization Based on Formant JND's," *IEEE Trans. Acoust., Speech, and Signal Processing*, vol. ASSP-34, no. 4, pp. 697-708, Aug. 1986.

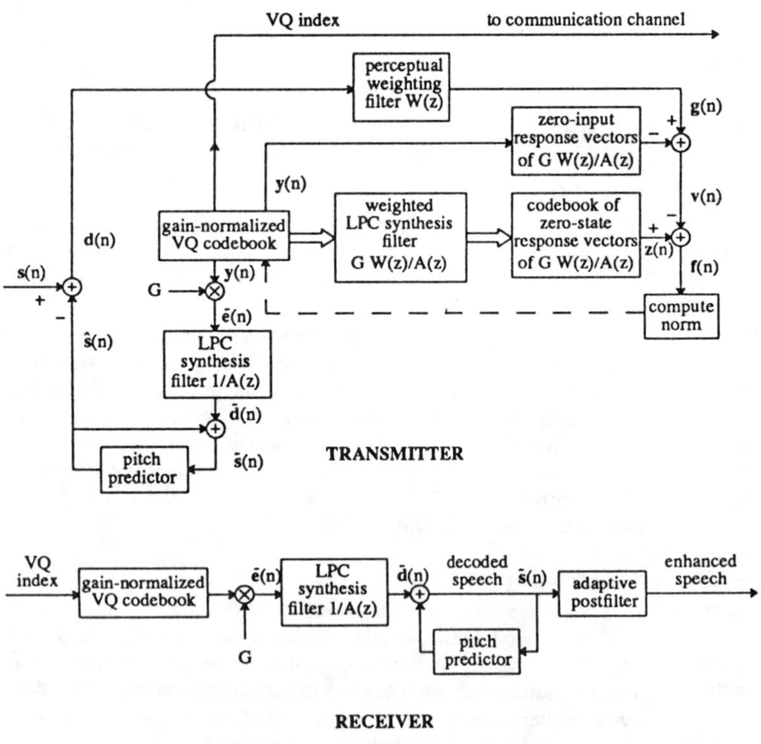

Fig. 1 Structure of the VAPC speech coder

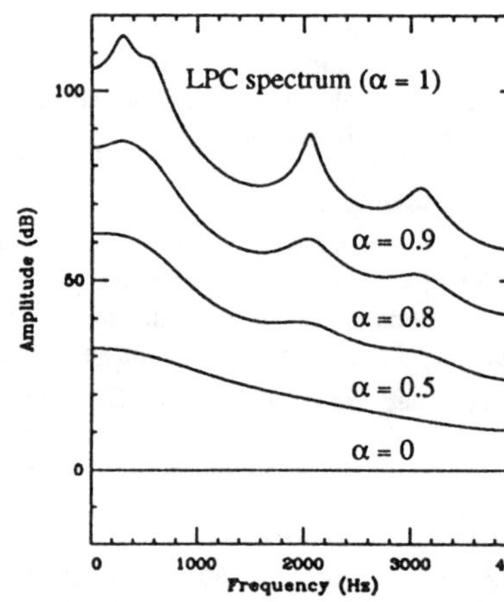

Fig. 2 Example of the LPC spectrum and the corresponding frequency response of the all-pole postfilter $1/[1-\hat{P}(z/\alpha)]$ for different values of α. The offset between adjacent plots is 20 dB.

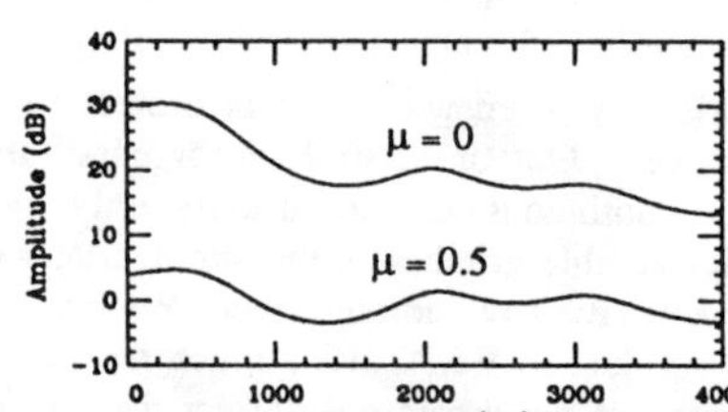

Fig. 3 Frequency responses of the postfilter $[1-\mu z^{-1}][1-\hat{P}(z/\beta)]/[1-\hat{P}(z/\alpha)]$ corresponding to the LPC spectrum shown in Fig. 2. In both plots, $\alpha = 0.8$ and $\beta = 0.5$. The offset between the two plots is 20 dB.

A MULTIPLE RATE LOW RATE VOICE CODEC

Joseph Rothweiler and Jack Carmody

ITT Defense Communications Division
492 River Road
Nutley, NJ 07110

ABSTRACT

Algorithms for a multiple rate (2400, 800, and 400 bps) CODEC are described. The U.S. Government standard LPC-10 algorithm is used for the 2400 bps rate. Lower rates are obtained by applying the source coding techniques of vector quantization and trellis coding to the LPC analysis output.

In this paper, the algorithms used are described, along with data rate control and formatting techniques for use with a fixed rate modem. Simulation test results for different vector quantizer codebook sizes and structures are presented, along with DRT test results showing the voice quality achieved at each rate.

Fortran simulations of the algorithms have been performed. Real time implementation of the CODEC is in progress, with completion expected by the end of 1985.

1. INTRODUCTION

Current tactical secure voice systems use a standard 2400 bps Linear Predictive Coding algorithm for speech coding [1]. Lower data rates are desirable due to the greater channel capacity and higher jamming resistance that can be achieved. However, lower rate systems typically have a number of limitations, including reduced intelligibility, speaker dependence, vocabulary restrictions, and sensitivity to noise and channel errors. The purpose of this investigation is to develop a multirate real time coder which uses a compatible set of data compression algorithms so that higher rates can be used when feasible to obtain the highest possible quality, and progressively lower rates can be used as the transmission environment becomes more severe.

Other goals are to develop robust bit formatting techniques to allow operation in a high error environment and to investigate design techniques to allow a high degree of speaker independence.

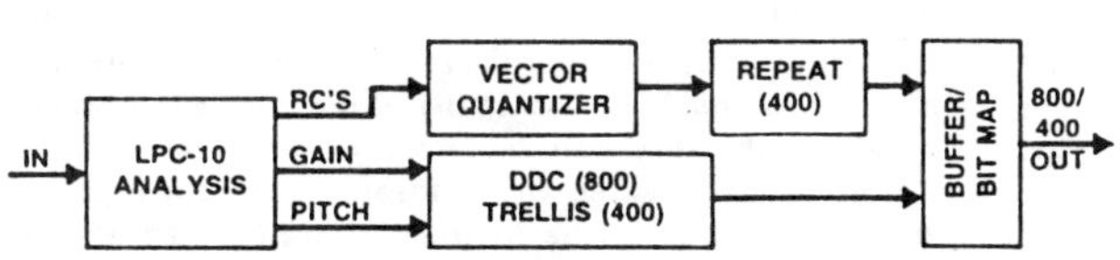

Fig. 1. Low rate CODEC approach.

2. APPROACH

The algorithm structure used as a basis for the CODEC is shown in Fig. 1. As indicated, the standard LPC-10 analysis is used to compute the speech reflection coefficients, amplitude, pitch, and voicing. For lower rates, vector quantization and trellis coding are used to encode the parameters in a more efficient form. Finally, buffering and formatting is performed to interface with a fixed rate modem and to provide synchronization bits and error protection. The algorithms form an integrated set in that the modules of each rate are used as part of the lower rate coders. Bit allocations for the various rates are indicated in table 1.

3. DEVELOPMENT OF CODEBOOKS FOR VECTOR QUANTIZATION

TABLE 1. Bit allocations for low rate coding.

Component	2400	800	400
spectrum	41	10	6
gain	5	3.7	1
pitch	7	4	1
framing and sync	1	0.3	1

Due to the importance of the codebook structure to the performance of a vector quantization algorithm, a series of experiments was performed to determine the optimum configuration for the codebooks. Specific questions of interest included the effect of codebook size on intelligibility, as well as the effects of different distortion measures, training population size, and varying codebook structures. The data generated in these experiments is presented in Fig. 2. Comments on specific issues follow.

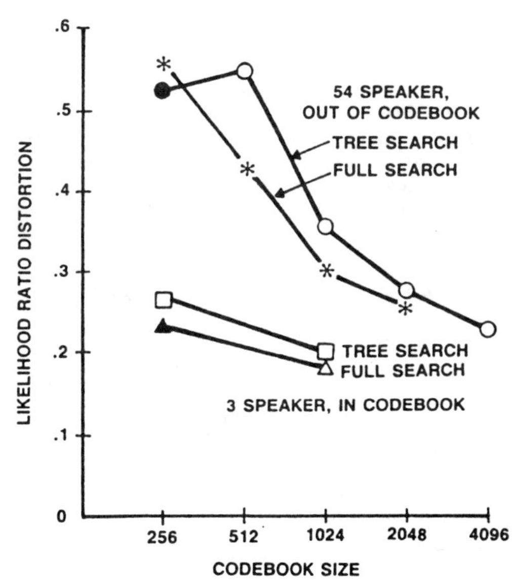

Fig. 2. Codebook size and training population comparisons.

3.1. Distortion measures.

The three distortion measures which were of interest are the likelihood ratio and the log area ratio using the L1 and L2 norms. These measures are defined as:

Reprinted from *Proc. IEEE Conf. Acoust., Speech, Signal Processing,* vol. 1, pp. 248–251, March 1985.

$$d(LR)=\frac{1}{2\pi}\int_{-\pi}^{\pi}\left|\frac{A(\omega)}{A_M(\omega)}\right|^2 d\omega - 1$$

$$d(LAR_{L1})=\sum_{i=1}^{P}\left|LAR_i - LAR_{M,i}\right|$$

$$d(LAR_{L2})=\sum_{i=1}^{P}\left|LAR_i - LAR_{M,i}\right|^2$$

The likelihood ratio is widely used because of its analytical tractability and ease of implementation in autocorrelation LPC analyzers [2]. However, the log area ratio is more suitable for integration with LPC-10, which uses the covariance method and computes reflection coefficients. To compare the distance measures, a 1024 element codebook was designed using the Linde, Buzo, Gray [3] algorithm, and DRT testing was performed. Measured DRT scores are 76.7, 74.9, and 73.6 for LR, LAR_{L1}, and LAR_{L2}, respectively. Both the DRT scores and informal listening tests showed no significant differences between the measures. Therefore, for this application, the L1 log area ratio is used.

3.2. Codebook size and training population.

To test the effects of training population size, two training sets were used. The first set consists of 3 speakers including the test speaker; it is representative of a codebook optimized for a given speaker. The second set contains 54 male and female speakers and does not include the test speaker. Codebooks were designed for each set with codebooks from 256 to 4096 elements. Conclusions from these tests are:

(1) For equivalent distortion levels, optimizing a codebook for a specific speaker is equivalent to a factor of 4 increase in codebook size.

(2) Speech quality increases with codebook size. Quality is acceptable with a 1024 element codebook, although better performance is achieved at 4096 elements.

Given these results, a 1024 element large population codebook was chosen as the best compromise between intelligibility and bit rate.

3.3. Codebook structure

Three codebook structures were tested: (1) a 1024 element codebook using an exhaustive search, (2) a 32 by 32 tree structure, and (3) a split codebook containing a 31 by 32 tree for voiced vectors and a 1 by 32 section for unvoiced vectors (Fig. 3).

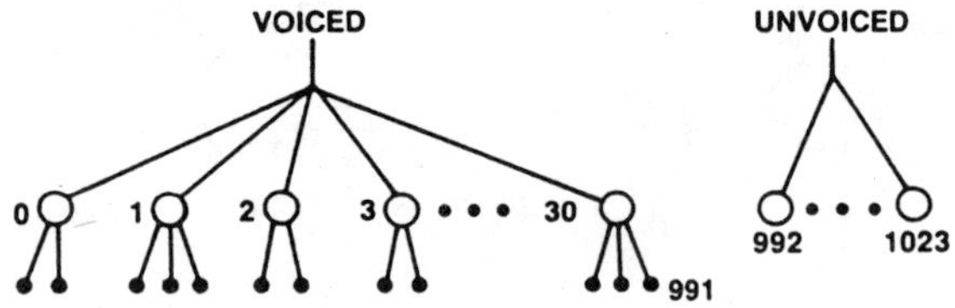

Fig. 3. Split tree codebook structure.

As in previous results [4], the performance of the 1024 by 1 and 32 by 32 structures was found to be virtually identical. The split codebook gives slightly better performance than the tree structure, since each section can be independently optimized for the different types of data, and it also simplifies the coding, since the voicing decision is implicit in the vector code, and does not need to be integrated into the pitch code.

Two comments: although measurements of the distortion due to the unvoiced segments (in the split codebook) were up to a factor of seven greater than those distortions for the voiced segments (using the split codebook), the overall speech quality remained good. This validates the size of the unvoiced codebook. Secondly, for the second level of the tree search the likelihood ratio and log area ratio measures were found to produce equivalent results; that is, when the log area ratio vector quantizer was forced to choose the same root as the likelihood ratio quantizer, the same final vector was selected. The converse is also true.

4. VARIABLE RATE VECTOR CODING

It has been shown by a number of authors [11] that additional data compression of an LPC speech signal can be achieved by repeating frames which contain little, if any, new information. This temporal redundancy removal is exploited in these algorithms by repeating frames where the distortion measured for the newly received vector and that for the previously received vector are within a set threshold. This requires that an additional (frame repeat) bit be sent each frame. The maximum number of bits to be transmitted at 400 bps is then thirteen (13). To reach 400 bps, the average bit rate must be nine (9) bits per frame. (Frame refers to a DoD LPC frame [22.5 milliseconds in duration].) This requires a forty percent repeat rate.

However, such a frame repeat system becomes variable rate. For many applications a fixed rate system is preferable. Our solution to the variable rate-to-fixed rate conversion employs an adaptive threshold control (ATC), illustrated in Figure 4.

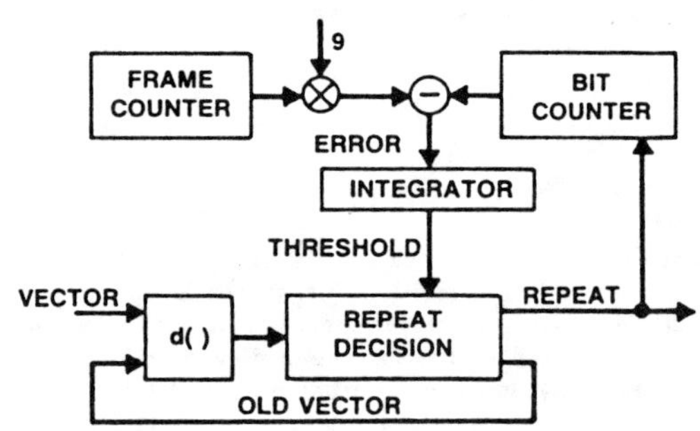

Fig. 4. Bit rate control loop.

The ATC is a feedback control loop. This control loop compares the number of bits that have been transmitted by the modem (at exactly 9 bits per frame) with the total number of bits that have been generated by the frame repeat coder, which produces either 3 or 13 bits per frame. The difference between these values is the number of bits currently available for transmission. If this value is above a preset threshold, too many 13 bit frames are being sent. In this case the threshold for repeat is increased, forcing the number of repeats up and the bit rate down. Conversely, if the number of bits being generated is too low, too many three bit segments are being sent. Therefore, the threshold is decreased which reduces the number of repeats and increases the bit rate. This control loop smooths out the fluctuations of the bit rate to produce an average data rate of exactly 400 bps, and it also limits the short term excursions above and below the average bit rate. Behavior of the control loop with a typical speech segment is illustrated in Figure 5.

In order to avoid synchronization problems in the presence of bit errors, a blocked bit format is used. This format requires exactly 8 repeats in each frame of 180 bits, for a repeat rate of 40%. Since each parameter is in a fixed location within the frame, a repeat bit error does not garble the data, but instead causes a timing error between the vector codes and the excitation parameters. Correction bits within the block allow correct timing to be maintained in the presence of moderate bit error rates (up to about 0.5%).

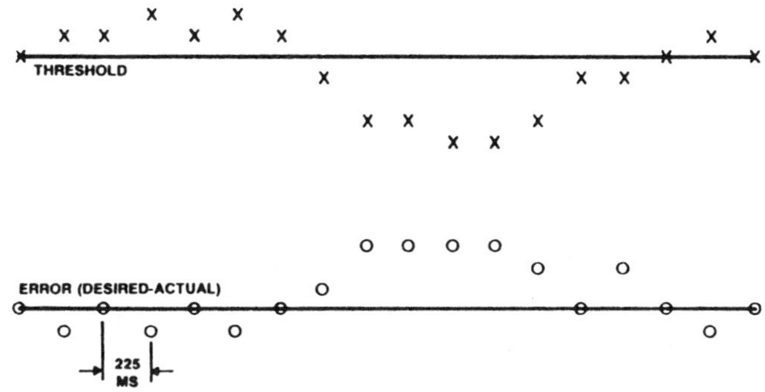

Fig. 5. Example of threshold adjustment.

5. EXCITATION CODING

At 800 bps, the Direct-Differential coding scheme is used [5]. This approach is simple to implement, adds minimal buffering delay (2 frames), and gives voice quality which is virtually identical to the coding used in LPC-10.

At 400 bps, the data rate must be minimized at the expense of higher complexity and delay. Two techniques which have been shown to allow coding at rates near the rate distortion bound are waveform vector quantization [6] and tree or trellis coding[7]. Although comparable accuracy can be achieved with either approach, vector quantization generally adds several frames of delay since each encoded vector is formed by buffering several data symbols. For this reason, the fake process trellis coder (Fig. 6) was chosen as the preferred approach.

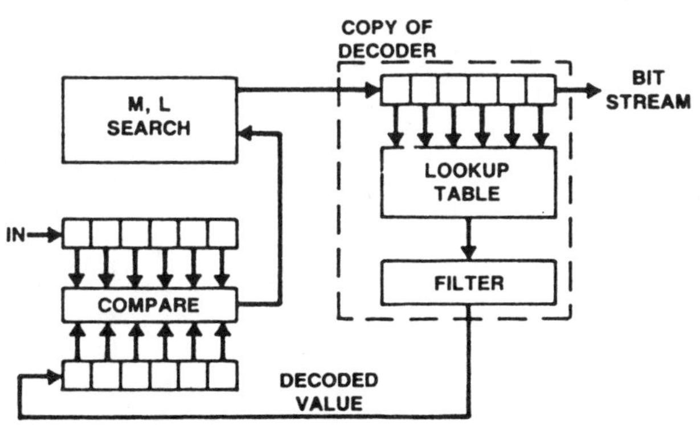

Fig. 6. Fake process trellis encoder for pitch or gain.

The fake process trellis coder achieves its coding efficiency because it is matched to both the spectrum and the statistical distribution of the signal being coded. For the 400 bps coder, these parameters were estimated from an actual speech sample consisting of 30 seconds of male speech. By plotting the signal autocorrelation function and the observed probability density function, it was found that a first order Gauss-Markov model with $\alpha = 0.9$ gave an excellent match to the gain values. A similar model with $\alpha = 0.97$ was used for the pitch.

A first order Gauss-Markov model has the form [8]

$$y_i = \alpha y_{i-1} + x_i$$

where y is the information symbol, x is a gaussian random process, and alpha is a model dependent coefficient. The rate distortion bound for this process is

$$D(R) = \frac{1-\alpha^2}{2^{2R}}$$

where R is the rate in bits per symbol.

Although this model is only an approximation to the true signal statistics, it may still be of interest to compare the computed rate distortion bound for this model with the values obtained from real speech. Table 2 gives the computed rate distortion bound and the measured performance for the trellis coder, as well as the Direct-Differential coder and the standard LPC-10 coder. The trellis coder uses the M-L [7] search algorithm with L = 6, and values of M as indicated. Test results are measured using a file of 50 seconds of male speech, which does not include the design data.

TABLE 2 Gain coder Signal to Noise Ratio in DB as a function of search depth, compared with scalar coders.

Rate Bits/sample	D(R)	Trellis search depth M=6	M=4	M=2	DDC	LPC
1	13.23	12.14	12.08	10.34		
2	19.25	18.8	17.3			
3	25.27				17.4	
5	37.3					25.3

Conclusions from this analysis are:

(1) A search depth of 4 is adequate for the gain trellis. Larger search depths produce no significant improvement in intelligibility.

(2) The trellis coder is very nearly optimum, operating only about 1 db away from the approximate rate distortion bound (as computed from the estimated signal statistics).

6. SIMULATION TEST RESULTS

Overall performance of the low rate coding algorithms was tested using the Diagnostic Rhyme Test [10]. Tests were made under conditions of clean speech, clean speech with 0.1% transmission errors, and speech with background acoustic noise (F-15 cockpit environment). Results of these tests are depicted graphically in Figure 7. In all cases, the low rate algorithms use a 3 speaker codebook designed using the test speakers.

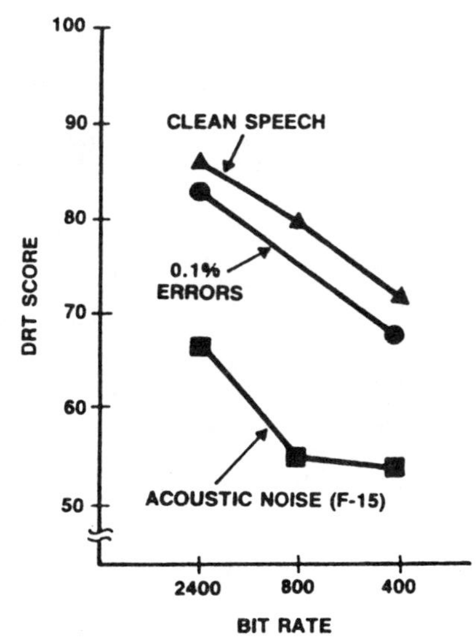

Fig. 7. Diagnostic Rhyme Test scores.

Conclusions of the tests are:

(1) The decrease in intelligibility with reductions in the bit rate is quite modest. There is a drop of about 5 DRT points from 2400 bps to 800 bps, and a similar drop from 800 to 400 bps.

(2) The low rate algorithms are relatively insensitive to transmission errors. The test results show that 400 bps suffers a degradation comparable to the degradation exhibited by a standard 2400 bps coder at the same error rate. Listening tests confirm that bit errors tend to distort only 1 or a few frames, and that both the trellis coder and the frame repeat coder recover quickly.

(3) Fair performance is achieved in the presence of background noise. A large decrease occurs at 800 bps, and almost no decrease from 800 to 400 bps. It may be noted here that, since LPC is used as a front end to the low rate algorithms, any of the noise compensation methods that have been developed for LPC (Spectral subtraction, noise cancelling microphones [9], etc) should also produce corresponding improvements in the quality of the low rate algorithms.

A demonstration tape will be played, showing performance of the algorithms for speakers in and outside the codebook, as well as performance in noisy environments and with bit errors.

7. SUMMARY

An approach to a multirate low rate voice coding system has been presented. A compatible family of low rate coding techniques are used to allow the efficient implementation of multiple rates in a single processor. Test results have been presented showing that good voice quality is achieved at all rates and that the system is capable of operating with transmission errors and background noise.

Currently, work is currently being done on the real time implementation of this CODEC by modifying an existing low rate voice tactical terminal. Completion of this effort is expected by the end of 1985.

8. ACKNOWLEDGEMENTS

The authors would like to thank Dr. David Wong for helpful discussions and comments in the course of this development effort, and Mr. L. Fransen of Naval Research Laboratories for providing several of the speech segments used in this project.

9. REFERENCES

1. T. E. Tremain, "The Government Standard Linear Predictive Coding Algorithm: LPC-10," *Speech Technology* Vol. 1, No. 2, pp. 40-49, Apr. 1982.

2. Robert M. Gray, Andres Buzo, Augustine H. Gray, Jr., Yasuo Matsuyama, "Distortion Measures for Speech Processing," *IEEE Trans. Acoust., Speech, and Signal Proc.*, Vol. ASSP-28, No. 4, Aug. 1980, pp. 367-376.

3. Yoseph Linde, Andres Buzo, and Robert M. Gray, "An algorithm for Vector Quantizer Design," *IEEE Trans. Comm.*, Vol. COM-28, No. 1, Jan. 1980, pp. 84-95.

4. Yasuo Matsuyama and Robert M. Gray, "Universal Tree Encoding of Speech," *IEEE Trans. Info. Theory*, Vol. IT-27, No. 1, Jan. 1981, pp. 31-40.

5. David Y. Wong, Biing-Hwang Juang, and Augustine H. Gray, Jr., "An 800 bit/s Vector Quantization LPC Vocoder," *IEEE Trans. Acoust., Speech, Signal Proc.*, Vol. ASSP-30, No. 5, Oct. 1982, pp. 770-780.

6. Allen Gersho and Vladimir Cuperman, "Vector Quantization: A Pattern- Matching Technique for Speech Coding," *IEEE Communications Magazine*, pp. 15-21, Dec. 1983.

7. Yoseph Linde and Robert M. Gray, "A Fake Process Approach to Data Compression", *IEEE Trans. Comm.*, Vol. COM-26, No. 6, June 1975, pp. 840-847.

8. Lawrence C. Stewart, "Trellis Data Compression," *Stanford University Report No. L905-1*, July 1981.

9. G. S. Kang and S. S. Everett, *Improvement of the Narrowband Linear Predictive Coder Part 1 — Analysis Improvements*, NRL Report 8645, Dec. 27, 1982.

10. William D. Voiers, "Evaluating Processed Speech using the Diagnostic Rhyme Test," *Speech Technology*, Vol. 2, No. 1, pp. 30-39, Jan. 1983.

11. D. Y. Wong, B. H. Juang, and D. Y. Cheng, "Very Low Data Rate Speech Compression with LPC Vector and Matrix Quantization," *ICASSP-83* pp. 65-68.

A VLSI Chip Set for Real Time Vector Quantization of Image Sequences

Raffi Dianysian and Richard L. Baker

Integrated Circuits and Systems Laboratory, Department of Electrical Engineering
University of California, Los Angeles, CA 90024
(213) 825-2327

ABSTRACT

This paper describes the architecture and implementation of a VLSI chip set that vector quantizes (VQ) image sequences in real time. The chip set forms a programmable Single-Instruction, Multiple-Data (SIMD) machine which can implement various vector quantization encoding structures. Its VQ codebook may contain an unlimited number of codevectors, N, having dimension up to $K = 64$.

Under a weighted least squared error criterion, the engine locates at video rates the best codevector in full-searched or large tree searched VQ codebooks. The ability to manipulate tree structured codebooks, coupled with parallelism and pipelining, permits searches in as short as $O(\log N)$ cycles. A full codebook search results in $O(N)$ performance, compared to $O(KN)$ for a Single-Instruction, Single-Data (SISD) machine. With this VLSI chip set, an entire video codec can be built on a single board that permits real-time experimentation with very large codebooks.

I. INTRODUCTION

Vector Quantization (VQ) is a new and promising method of data compression. The encoder in a VQ system steps through a simple, but computationally intensive vector matching procedure. An image is broken into blocks of K pixels (*source vectors*), $\mathbf{x}$, and the encoder compares each with N potential replacements (*codevectors*), $\{\hat{\mathbf{x}}^i, i=1,..., N\}$, prestored in a *codebook*. It then transmits the index i of the best matching (least distorting) codevector to the receiver. The receiver uses the index to access $\hat{\mathbf{x}}^i$ from an identical codebook and places it in the corresponding block location. An excellent survey paper by Gray [1] discusses a variety of VQ structures and codebook building techniques.

Most conventional low rate video compression techniques perform a unitary transform (such as the discrete cosine transform) on each block of pixels prior to the quantizing step and then inverse transform the quantized data at the decoder [2]. Implementing the transform in real time can be expensive. By operating directly on a block of pixels, VQ avoids the complexity associated with transforms. Complexity in VQ systems concentrates at the encoder. The decoder is a essentially a table look up which is easy to implement, making VQ systems attractive for broadcast applications.

Reconstructed image fidelity is best when the VQ uses large codebooks, but large codebooks may require too much time to fully search. By searching them non-exhaustively, a highly parallel, pipelined VQ encoder can accommodate real time video data.

Several VLSI chips have been designed for real time VQ encoding of speech. In 1985, Davidson and Gersho successfully fabricated a chip that encodes 12-bit speech with codebooks having as many as 375 entries [3]. The following year, Cappello, *et. al*, described a highly parallel, systolic version which operates on an entire vector of samples and has a projected throughput of 10^7 distortion computations per second [4]. Its 12-bit resolution, necessary for high quality speech encoding, precluded placement of more than one distortion computing element per chip. The design cannot easily search tree structured codebooks.

This is the first VLSI VQ encoding engine we are aware of that is built for real time image coding. Designed to search tree structured codebooks of unlimited size, it computes one weighted squared error distortion per clock for dimensions as large as $K = 64$. The engine can also be used to construct codebooks using real time video training data. The chips are being laid out in 3 μm CMOS. Projected clock rate exceeds 10 MHz. Key components were sent in February to the MOSIS facility for fabrication and complete designs will be submitted this summer.

In Section II, we describe the weighted squared error distortion computed by the VQ engine. We identify the constraints a real time video source imposes on VQ and show how tree structured codebooks can meet them. Section III describes how the need to search tree structures shaped our design. Sections IV and V discuss implementation of the chip set's key processing modules.

II. VQ ENCODING

Distortion Computation

VQ compression systems based on the squared error distortion measure perform well on imagery provided they embed the VQ in a suitable preprocessing structure. For example, *Mean/Residual VQ* (M/RVQ) [5, 6], removes each block's sample mean prior to the VQ. *Classified VQ* [7] parses incoming blocks according to their internal structure and uses a different codebook for each class. *Weighted* squared error measures can further improve fidelity. In all cases, the optimum VQ identifies which K-dimensional codevector, $\hat{\mathbf{x}}^i$, in a codebook is the least distorting representation of a source vector of preprocessed pixels, $\mathbf{x}$. Mathematically, it computes

$$i = \min_{i=1,\ldots,N}{}^{-1} \left\{ \sum_{k=1}^{K} w_k^i \mid x_k - \hat{x}_k^i \mid^2 \right\}, \tag{1}$$

where the inverse means "return the index i which achieves the minimum." i is sent to the receiver which has a copy of the codebook and reproduces $\mathbf{x}$ with $\hat{\mathbf{x}}^i$. The weight vector $\mathbf{w}^i$ could be a constant or a function of the input vector $\mathbf{x}$, its neighbors, or $\hat{\mathbf{x}}^i$.

Direct evaluation of (1) requires subtract, square, multiply and accumulate operations. Expanding and dividing it by two, we obtain

$$i = \min_{i=1,\ldots,N}{}^{-1} \left\{ \sum_{k=1}^{K} \frac{w_k^i (x_k)^2}{2} - \sum_{k=1}^{K} w_k^i \hat{x}_k^i x_k + \sum_{k=1}^{K} \frac{w_k^i (\hat{x}_k^i)^2}{2} \right\}. \tag{2}$$

We see that the first summation in (2) does not affect the minimization. As pointed out by Cappello, *et. al*, in [4], if the weight w_k^i is restricted to be a function of the codevector $\hat{\mathbf{x}}^i$, the last summation,

This work was supported by Advanced Micro Devices, Inc., through the University of California MICRO Program, and NASA, under grant number NAG2-304, Supplement No. 2.

Reprinted from *Proc. IEEE Int. Symp. Circuits and Systems,* vol. 1, pp. 221–224, May 1987.

$$c^i \triangleq \sum_{k=1}^{K} \frac{w_k^i (\hat{x}_k^i)^2}{2},$$

is a constant which can be pre-computed and stored for each codevector. Also, the quantity

$$z_k^i \triangleq -w_k^i \hat{x}_k^i$$

can be stored in the codebook instead of $\hat{x}_k^i$. (2) simplifies to

$$i = \min_{i=1,\ldots,N}^{-1} \left\{ \sum_{k=1}^{K} z_k^i x_k + c^i \right\}, \tag{3}$$

which just involves a multiply and accumulate.

Real Time Encoding Constraints

An optimum VQ encoder must identify the least distorting codevector for each source vector. If a source vector contains K pixels, and if f_f frames h pixels high and w pixels wide arrive each second, then an encoder capable of computing f_d distortions per second has time to compare at most M codevectors to each source vector, where

$$M = \frac{K f_d}{h w f_f}. \tag{4}$$

M places an upper bound on the number of codevectors permitted in a fully searched codebook. Typically, it is too small to insure good VQ performance. As an example, assume the engine must encode a 256 x 256 pixel, $f_f = 15$ frames per second image source at 0.5 *bits per pixel* (bpp). Since a codebook containing N K-dimensional codevectors has *rate* $R = (\log_2 N)/K$ bpp, an $N = 256$, $K = 16$ (e.g., 4 x 4 block size) codebook would yield the desired rate. If the encoder computes $f_d = 10^7$ distortions per second, then $M = 162$, which is not enough to fully search the codebook. The largest 0.5 bpp codebook this engine could fully search is $N = 128$, $K = 14$. ($M = 142$ in this case.)

Unfortunately, an $N = 128$ codebook is too small to reconstruct good quality images [6]. A larger 0.5 bpp codebook, such as $N = 2^{14}$, $K = 28$ (e.g., 7 x 4 block size), would perform much better, but has too many entries to be searched in real time. ($M = 284$ in this case.)

Tree-Structured Codebooks

One way to accommodate large codebooks is to tree search them, which drops search time from $O(N)$ to $O(\log N)$. It has been found experimentally that suitably designed tree structured codebooks of size N perform about as well on images as fully searched codebooks having size between $N/2$ and N [6].

To continue our example, the $N = 2^{14}$, $K = 28$ codebook could be designed as a two-level tree having 128 branches per level. (4) yields $M = 284$. To search the codebook, we compute 128 distortions at each level, or 256 in all, which is less than M.

III. VQ ENCODING ENGINE

Engine Description

Fig. 1 depicts the system's overall architecture, a cascade of J *Multiple Multiply-ACcumulate* (MMAC) chips terminated by one *Next Address Selector* (NAS) chip. The MMAC cascade performs the dot product in (3); the NAS chip does the minimization. By indicating which incoming distortions become new minima, NAS selects the appropriate codevector address for the next level.

Source vectors are fetched one pixel at a time by the appropriate MMAC chip and buffered internally in a FIFO. Multiplexing the source vectors reduces MMAC pin count and simplifies external logic. The engine's high clock rate assures that at most one pixel will arrive per clock cycle from a typical video source.

Pipeline Latency

We see from the example in the previous section that an f_d on the order of 10 million would permit real time video compression. We can achieve this throughput rate by breaking the distortion computation (3) into many small, fast stages, forming a *pipeline*.

Pipelining creates *latency* between the *State Machine* which feeds codevectors to the cascade and the NAS module which selects the best codevector. This latency does not affect full codebook searches. A tree search, however, has *data dependency*, and the affects of latency on its search process must be considered. The encoder cannot select the best branch to the next level until all distortions for the current level have exited the pipe. Rather than wait for them, we buffer and operate on several source vectors at a time, cycling all of them through one level of the tree before proceeding to the next level. The number of source vectors that have to be buffered on chip increases with latency.

To reduce pipeline latency, we use a *two-level systolic array* (Fig. 2). First, a fine grain systolic array of five *Multiply-ACcumulate* (MAC) blocks accumulates products. Each set of five MAC blocks forms a MMAC chip. Second, a coarse grain systolic array of MMAC chips tallys the sums. For J MMAC chips, (3) becomes

$$i = \min_{i=1,\ldots,N}^{-1} \left\{ \sum_{j=0}^{J-1} \left[\sum_{k=5j}^{5j+4} z_k^i x_k \right] + c^i \right\}. \tag{5}$$

The tree-like summation adds only one delay per chip to the pipe. Total delay for a $K = 64$ system ($J = 13$ MMAC chips) is only 26 clocks. By requiring codebook trees to have at least eight branches per level, at most five source vectors must share the pipe. Chip area used for buffering is not excessive and the constraint on codebook structure is minor.

IV. MULTIPLE MULTIPLY-ACCUMULATE (MMAC)

Fig. 3 illustrates the internal architecture of the MMAC chip. Each of its five MAC blocks multiplies one element of the source vector, x_k, with the corresponding weighted codevector element, z_k^i, and adds the result to the accumulated sum output by the preceding MAC block (see (5)). The *Tally Section* adds these partial sums to partial sums accumulated from previous MMAC modules. c^i is input to the Tally Section of the first MMAC module. Delays skew codevector elements and data as needed (Fig. 2).

MAC Block

A circulating shift register at the kth MAC block holds the kth element of five interleaved source vectors. A FIFO grabs the next set as they arrive on the input bus. The multiplier and accumulator both use

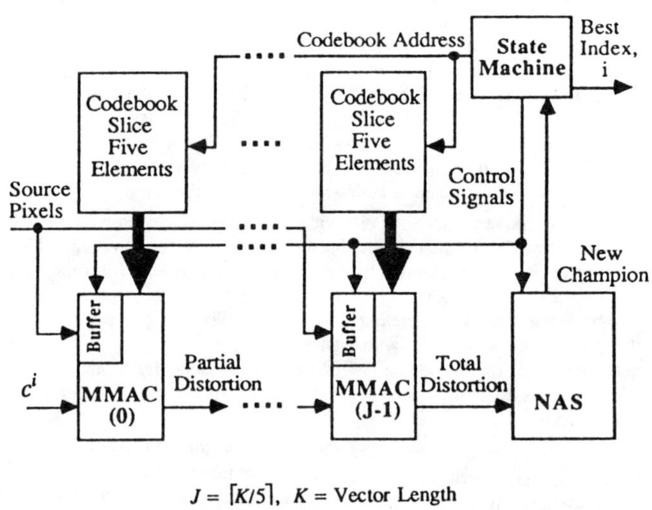

Fig. 1: Board-level description of the VQ encoder. The Multiple Multiply ACcumulate (MMAC) and Next Address Selector (NAS) chips tree search the codebook to identify a least distorting representation of incoming pixels.

Carry Save Adders (CSAs) [5] to circumvent a carry propagate problem. A CSA has the same combinational logic as a conventional carry ripple adder, but its interconnections are modified to output the carry vector along with the pseudo sum instead of feeding it back to the adder's input. The freed input connection is then used to input a third number, making the CSA ideal for accumulation.

Tally Section

Limited pin count precludes using a CSA for tallying, as both the pseudo sum and the saved carry would have to be brought in to and sent out of the chip. 18-bit carry propagation adders would create a speed bottleneck. Output pad delay further worsens the situation. To combat the problem, we halve the length of the carry propagation, performing the addition in two stages. We further boost speed by using a dynamic Manchester carry adder [10].

As the distortion accumulates through the MMAC chip cascade, the delay between it and each MMAC chip's internal sum increases. Depending on the position of the MMAC chip in the cascade, a programmable delay skews the internally accumulated distortion (Fig. 2). The amount of delay in each chip is set at boot time.

Pin-Bandwidth Constraint

The level of integration in the MMAC chip is dominated more by pin count than by area. Pins are needed to propagate the accumulated distortion from chip to chip and for a high bandwidth bus which fetches an entire codevector every clock cycle from codebook storage RAM into the MMAC cascade. To reduce pin count, we tailor the size of data buses to the application.

The human eye distinguishes about six bits of gray [2]. An additional bit is needed to handle preprocessing structures that output signed data (as in the M/RVQ, [6]), so 7-bit wide data buses are used to transfer x_k and z_k^i into the MMAC chips. Simulations show that the two least significant bits at the input and output of the MMAC Tally Section can be discarded with no noticeable degradation in performance. This saves four pins. The pin requirements allow five blocks to be fabricated on a MOSIS standard 6900x6800 sq. µm 84-pin chip, a size which gives acceptable yields.

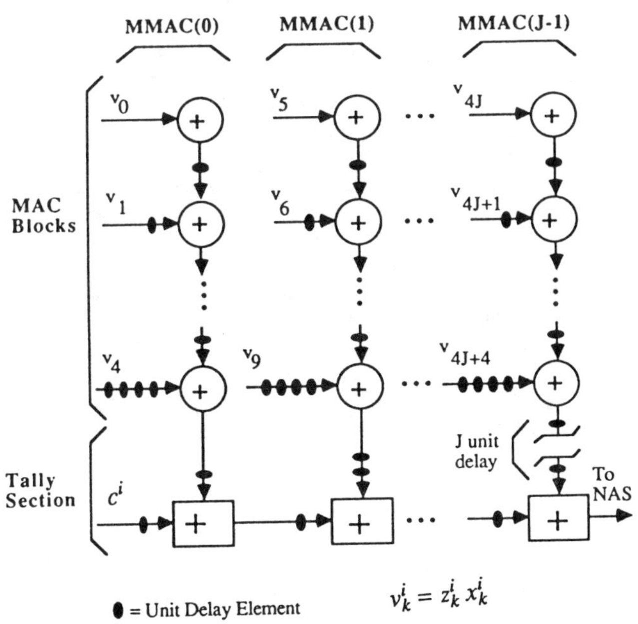

Fig. 2: The cascade of MMAC chips forms a two-level systolic array. This reduces pipeline latency, which means fewer source vectors have to be interleaved.

V. NEXT ADDRESS SELECTOR (NAS)

The simpler of the two chips, the NAS module identifies the smallest distortion computed at each level of the tree (Fig. 4). It consists of a *Comparator* and a *Minimum Distortion Register* (MDR).

Each time a new level of the tree structure is entered, the *Control Store* overrides the comparator to load the first distortion into the MDR. Whenever the comparator identifies a subsequent distortion that is smaller than the one in the MDR, it updates the MDR and flags the State Machine. A register within the State Machine then latches a pointer that addresses the corresponding group of codevectors in the next level of the codebook. The Control Store loads the pointer into the *State Counter* after all branches in the level are evaluated. To account for pipeline latency, all signals from the State Machine destined for the NAS module are delayed by an amount equivalent to the pipelength. Data traveling in the opposite direction is delayed by the number of interleaved source vectors.

To keep up with the flux of distortions produced by the MMAC cascade, we use fast dynamic carry generation circuitry similar to carry lookahead networks [9]. Moreover, we pipeline the comparison.

Pipelining

The pipelined NAS must compare a continuous stream of 18-bit two's complement distortions (challengers) against the current minimum

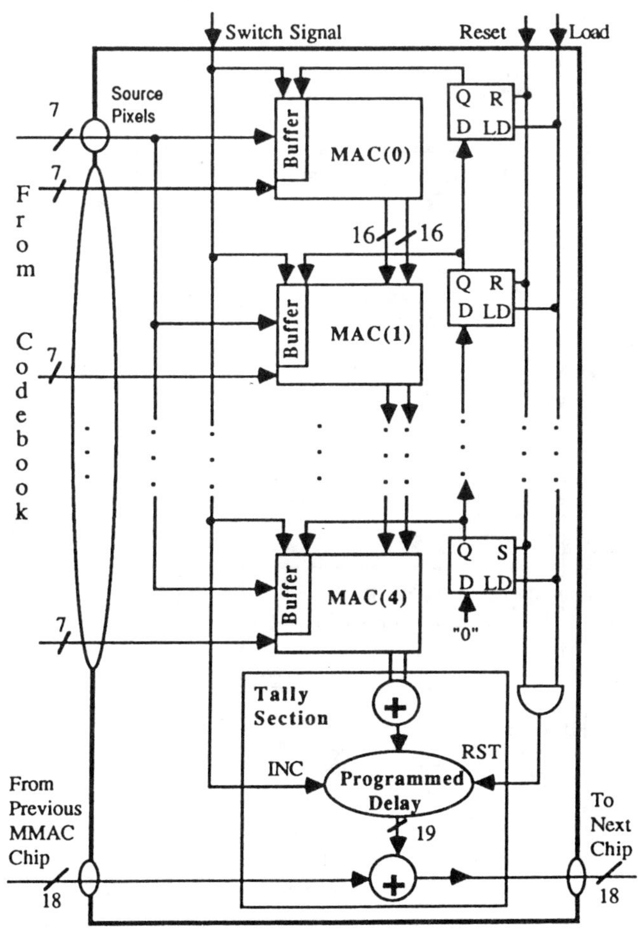

Fig. 3: The Multiple Multiply ACcumulate (MMAC) chip buffers elements of the interleaved source vectors and computes the dot product between source and code vectors.

distortion (champion). This poses a problem. We need to replace the champion with a victorious challenger without disturbing data flow in the pipe. The replacement should not affect the results of partial comparisons of succeeding challengers.

We tackle the problem by using an architecture similar to the Comparator Processor (CP) chip in [4]. We pipeline NAS in three 6-bit stages, with the most significant 6-bit digit of the skewed challenger entering the pipe first. If the challenger wins at any stage of the comparison, then it was equivalent to the champion at all of the comparator's earlier (more significant) stages. Those digits need not be altered. The challenger replaces only the lower digits of the old champion, one digit per clock, as it rolls down the pipe. Partial comparisons of succeeding challengers already in the pipe remain valid since their least significant digits have yet to be compared. By operating on 6-bit digits, the NAS adds only 3 delays to overall latency. The CP in [4], which uses 1-bit digits, has a much more regular structure than the NAS. However, it would lengthen the pipe by 18 cycles.

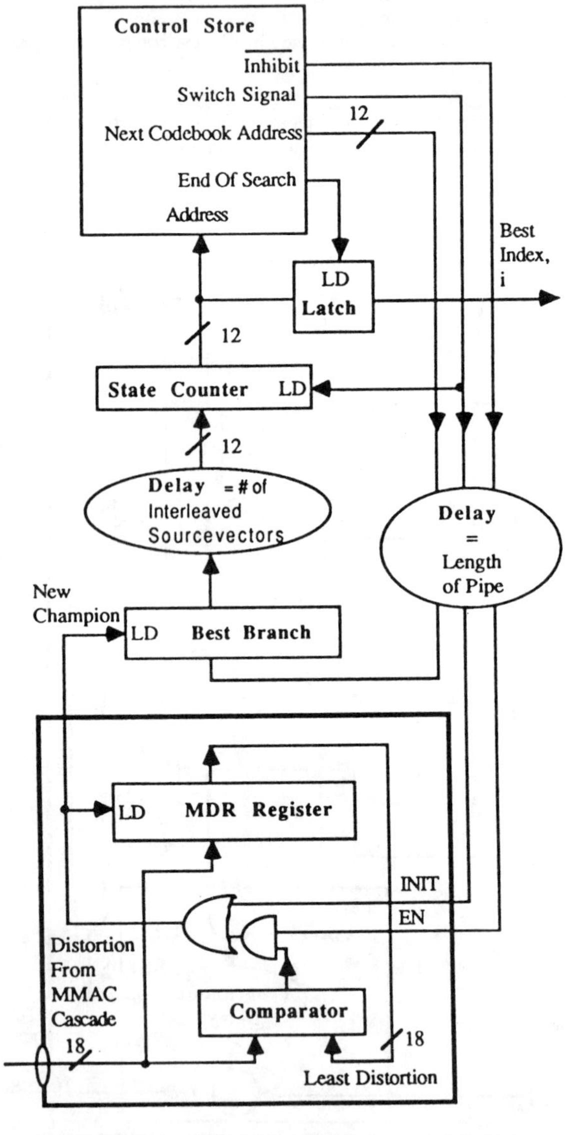

Fig. 4: Functional description of the Next Address Selector (NAS) chip and State Machine. The NAS keeps track of the lowest distortion, while the engine's State Machine generates the address of the next codevector.

VI. CONCLUSIONS

A pair of highly pipelined VQ distortion computing chips has been described. The chips will form the heart of a real time, single board VQ-based video coding system that can both design and use large tree structured codebooks. Depending on the tree's structure, searches occur in $O(\log N)$ to $O(N)$ time and use a weighted squared error distortion.

The problem of data dependency, which occurs when descending a tree, is nullified by interleaving the distortion computations of several source vectors. A two-level systolic array shortens pipeline latency, reducing the number of source vectors which must be interleaved and buffered on chip.

The chip set is currently being implemented in 3 μm CMOS, with a projected clock rate of 10 MHz. A future 2 μm CMOS implementation ($\lambda = 1$ μm) would shrink die size by 30% and increase speed by about 30%.

ACKNOWLEDGEMENT

Ley Chai Lau and Hsiao-hui Shen assisted in the layout of the chips. Prof. V. C. Tyree provided instruction on the use of Berkeley-Magic CAD tools and patiently described pertinent VLSI concepts. Prof. M. D. Ercegovac and Prof. T. Lang helped analyze different arithmetic structures.

REFERENCES

[1] R. M. Gray, "Vector Quantization," *IEEE ASSP Magazine*, April 1984, pp. 4-29.

[2] W. K. Pratt, *Image Transmission Techniques*, Academic Press, New York, 1979.

[3] G. Davidson and A. Gersho, "Application of a VLSI Vector Quantization Processor to Real-Time Speech Coding," *IEEE Trans.*, Vol. SAC-4, No. 1, Jan 1986, pp. 112-124.

[4] P. Cappello, G. Davidson, A. Gersho, C. Koc, V. Somayazulu, "A Systolic Vector Quantization Processor for Real Time Speech Coding," *ICASSP '86*, pp. 2143-2146.

[5] R. L. Baker and R. M. Gray, "Image Compression Using Non-adaptive Spatial Vector Quantization," *Proc. 16th Asilomar Conf. on Circuits, Systems & Computers*, pp. 55-61, 1982.

[6] R. L. Baker, "Vector Quantization of Digital Images," Ph.D. dissertation, Stanford U., June 1984.

[7] B. Ramamurthi and A. Gersho, "Image Vector Quantization with a Perceptually-Based Cell Classifier," *ICASSP*, San Diego, March, 1984.

[8] A. Buzo, A. H. Gray, Jr., R. M. Gray, J. D. Markel, "Speech Coding Based Upon Vector Quantization," *IEEE Trans. on ASSP*, Vol. ASSP-28, No. 5, pp. 562-574, Oct. 1980.

[9] M. D. Ercegovac and T. Lang, *Digital Systems and Hardware/Firmware Algorithms*, New York, John Wiley, 1985.

[10] N. H. E. Weste and K. Eshragian, *Principles of CMOS VLSI Design*, Massachusetts, Addison-Wesley, 1985.

Bibliography

[1] A. A. Abdelwahab and S. C. Kwatra, "Image data compression with vector in the transform domain," *Proc. Int. Commun. Conf.*, pp. 1285–1288, 1986.

[2] A. A. Abdelwahab and S. C. Kwatra, "Component coding of color video signal with VQ in the transform domain," *Proc. IEEE Region 10 Conf.*, vol. 2, pp. 433–437, 1987.

[3] M. Abe, S. Nakamura, K. Shikano, and H. Kuwabara, "Voice conversation through vector quantization," *Proc. ICASSP'88,* vol. 1, pp. 655–658, April 1988.

[4] H. Abut, R. M. Gray, and G. Rebolledo, "Vector quantization of speech and speech-like waveforms," *Proc. ICASSP'81,* vol. 1, pp. 12–15, March–April 1981.

[5] H. Abut, R. M. Gray, and G. Rebolledo, "Vector quantization of speech and speech-like waveforms," *IEEE Trans. Acoust., Speech, and Signal Processing,* vol. ASSP-30, pp. 423–435, June 1982.

[6] H. Abut and S. A. Luse, "Vector quantizers for subband coded waveforms," *Proc. ICASSP'84,* vol. 1, pp. 10.6.1–10.6.4, March 1984.

[7] H. Abut and S. Ergezinger, "Low rate speech coding using VQ and SBC," *Proc. ICASSP'86,* vol. 1, pp. 449–452, April 1986.

[8] H. Abut, B. P. M. Tao, and F. Mehran, "LSI architecture for VQ/MQ systolic array systems," *Proc. Milcom,* vol. 3, pp. 49.4.1–49.4.5, Oct. 1986.

[9] H. Abut, B. P. M. Tao, and J. L. Smith, "Vector quantizer architectures for speech and image encoding," *Proc. ICASSP'87,* vol. 2, pp. 756–759, April 1987.

[10] S. Adlersberg and V. Cuperman, "Transform domain vector quantization for speech signals," *Proc. ICASSP'87,* vol. 4, p. 1938–1941, April 1987.

[11] J.-P. Adoul, C. Collin, and D. Dalle, "Block encoding and its applications to data compression of PCM speech, *Proc. Canadian Commun. and Power Conf.,* pp. 145–148, Oct. 1978.

[12] J.-P. Adoul, S. Morrisette, and M. Rudko, "Bit-rate halving algorithm for PCM encoded speech using a new bidimensional data compression scheme," *Proc. ICASSP'79,* pp. 432–435, April 1979.

[13] J.-P. Adoul, J.-P. Debray, and D. Dalle, "Spectral distance measure applied to the optimum design of DPCM coders with L predictors," *Proc. ICASSP'80,* pp. 512–515, April 1980.

[14] J.-P. Adoul and P. Mabilleau, "4800 bps RELP vocoder using vector quantization for both filter and residual representations," *Proc. ICASSP'81,* vol. 1, pp. 601–604, March–April 1981.

[15] J.-P. Adoul, C. Lamblin, and A. LeGuyader, "Baseband speech coding at 2400 bps using spherical vector quantization," *Proc. ICASSP'84,* vol. 1, pp. 1.12.1–1.12.4, March 1984.

[16] J.-P. Adoul, F. Didelot, P. Mabilleau, and S. Morissette, "Generalization of the multipulse coding for low bit rate speech coding purposes: the generalized decimation," *Proc. ICASSP'85,* vol. 1, pp. 256–259, March 1985.

[17] J.-P. Adoul, "Quantification vectorielle des Signaux: approche Algebrique," *Annales des Telecommunicationes,* tome 41, no. 3–4, pp. 158–177, 1986 (in French).

[18] J.-P. Adoul and M. Delprat, "Design algorithm for variable-length vector quantizers," *Proc. 24th Annu. Allerton Conf. Commun., Control and Computing,* pp. 1004–1011, Oct. 1986.

[19] J.-P. Adoul, "Speech coding algorithms and vector quantization," in *Advanced Digital Communications and Signal Processing,* K. Feher, Ed. Englewood Cliffs: NJ, Prentice Hall, pp. 133–181, 1987.

[20] J.-P. Adoul and C. Lamblin, "A comparison of some algebraic structures for CELP coding of speech," *Proc. ICASSP'87,* vol. 4, pp. 1953–1956, April 1987.

[21] J.-P. Adoul, P. Mabilleau, M. Delprat, and S. Morissette, "Fast CELP coding based on algebraic structures," *Proc. ICASSP'87,* vol. 4, pp. 1957–1960, April 1987.

[22] K. Aikawa and K. Shikano, "Spoken word recognition using vector quantization in power-spectrum vector space," *Trans. Inst. Electron. Commun. Eng. Japan,* vol. J68-D, no. 3, pp. 316–322, 1985.

[23] K. Aizawa, H. Harashima, amd H. Miyakawa, "Vector quantization of picture signals in DCT domain," *Tech. Rep., IECE Japan,* (IE84-86), Dec. 1984 (in Japanese).

[24] K. Aizawa, H. Harashima, and H. Miyakawa, "Adaptive vector quantization of picture signals in discrete cosine transform domain," *Trans. IECE Japan,* vol. J69-B, no. 3, pp. 228–236, March 1986 (in Japanese). Also, in English, *Electronics and Communications in Japan Part I,* vol. 70, no. 5, Scripta Technica, Inc., New York: John Wiley & Sons, 1987.

[25] K. Aizawa, H. Harashima, and H. Miyakawa, "Adaptive discrete cosine transform coding with vector quantization for color images," *Proc. ICASSP'86,* vol. 2, pp. 985–988, April 1986.

[26] K. Aizawa, H. Harashima, and H. Miyakawa, "Adaptive discrete cosine transform coding with vector quantization," *Proc. of Int. Picture Coding Symp.,* pp. 225–226, April 1986.

[27] K. Aizawa, H. Harashima, and H. Miyakawa, "Vector quantization of picture using discrete cosine transform (DCT-VQ)," *J. Inst. Telev. Engrs. Japan,* vol. 39, pp. 920–925, Oct. 1986 (in Japanese).

[28] K. Aizawa, T. Saito, and H. Harashima, "Adaptive discrete cosine transform coding with vector quantization for color images," *Trans. IECE Japan,* vol. J69-B, no. 10, pp. 1156–1159, Oct. 1986 (in Japanese).

[29] M. Akamine and K. Miseki, "ARMA model based speech coding at 8 kb/s," *Proc. ICASSP'89,* vol. 1, pp. 148–151, May 1989.

[30] A. Akansu and M. S. Kadur, "Discrete Walsh-Hadamard transform vector quantization for motion-

compensated frame difference signal coding,'' *Proc. ICASSP'89,* vol. 3, pp. 1862–1865, May 1989.

[31] A. Aktas and H. Hoge, ''Real-time recognition of subword units on a hybrid multi-DSP/ASIC based acoustic front end,'' *Proc. ICASSP'89,* vol. 1, pp. 148–151, May 1989.

[32] L. Alvarez and N. Garcia, ''Some results on vector quantization,'' in *Signal Processing III: Theories and Applications,* I. T. Young, J. Biemond, R. P. W. Duin, and J. J. Gerbrands, Eds. North Holland, Amsterdam: Elsevier Science Publishers B. V., 1986, pp. 797–800.

[33] D. O. Anderton and C. K. Rushforth, ''Waveform coding of voiceband data signals at 16 kb/s,'' *Proc. ICASSP'87,* vol. 3, pp. 1328–31, April 1987.

[34] A. H. Ang and T. S. Durrani, ''Adaptive transform coding using vector quantization,'' *Conf. Record Second Int. Conf. on Image Processing and Its Applications,''* pp. 11–16, June 1986.

[35] T. H. Applebaum, B. A. Hanson, and H. Wakita, ''Weighted cepstral distance measures in vector quantization based speech recognizers,'' *Proc. ICASSP'87* vol. 2, pp. 1155–1158, April 1987.

[36] R. Aravind and A. Gersho, ''Low-rate image coding with finite state vector quantization,'' *Proc. ICASSP'86* vol. 1, pp. 137–140, April 1986.

[37] R. Aravind and A. Gersho, ''Image coding below 0.5 bits per pixel using vector quantization,'' in *Signal Processing III: Theories and Applications,* I. T. Young, J. Biemond, R. P. W. Duin, and J. J. Gerbrands, Eds. North-Holland, Amsterdam: Elsevier Science Publishers B. V., 1986, pp. 757–760.

[38] R. Aravind and A. Gersho, ''Image compression based on vector quantization with finite memory,'' *Optical Engineering,* vol. 26, no. 7, pp. 570–580, July 1987.

[39] Y. Ariki, F. R. McInnes, and M. A. Jack, ''Hierarchical Phoneme discrimination by hidden Markov modeling using cepstrum and formant information,'' *Proc. ICASSP'89,* vol. 1, pp. 663–666, May 1989.

[40] Y. Asakawa, A. Ichigawa, S. Yajima, and K. Yamasaki, ''Speech coding method using fuzzy vector quantization,'' *Proc. ICASSP'89* vol. 2, pp. 755–758, May 1989.

[41] B. S. Atal, ''Stochastic gaussian model for low-bit rate coding of LPC area parameters,'' in *Proc. ICASSP'87,* vol. 4, pp. 2404–2407, April 1987.

[42] B. S. Atal, R. V. Cox, and P. Kroon, ''Spectral quantization and interpolation for CELP coders,'' *Proc. ICASSP'89,* vol. 1, pp. 69–72, May 1989.

[43] A. Averbuch, L. Bahl, R. Bakis, P. Brown, A. Cole, G. Daggett, S. K. Das, K. Davies, S. De Gennaro, P. de Souza, E. Epstein, D. Fraleigh, F. Jelinek, S. Katz, B. Lewis, R. Mercer, A. Nadas, D. Nahamoo, M. Picheny, G. Shichman, and P. Spinelli, ''An IBM PC based large-vocabulary isolated-utterance speech recognizer,'' *Proc. ICASSP'86,* vol. 1, pp. 53–56, April 1986.

[44] E. Ayanoglu and R. M. Gray, ''The design of predictive trellis waveform coders using the generalized Lloyd algorithm,'' *IEEE Trans. Commun.,* vol. COM-34, pp. 1073–1080, Nov. 1986.

[45] E. Ayanoglu, ''Optimal quantization of noisy sources,'' *Proc. ICASSP'88,* vol. 1, pp. 569–572, April 1988.

[46] M. J. Bage, ''Interframe predictive coding of images using hybrid vector quantization,'' *IEEE Trans. Commun.,* vol. COM-34, pp. 411–415, April 1986.

[47] L. R. Bahl, S. K. Das, P. V. deSouza, F. Jelinek, S. Katz, L. R. Mercer, and M. A. Picheny, ''Some experiments with large-vocabulary isolated-word sentence recognition,'' *Proc. ICASSP'84* vol. 2, pp. 25.5.1–26.5.2, March 1984.

[48] L. R. Bahl, P. F. Brown, P. V. deSouza, and R. L. Mercer, ''Speech recognition with continuous-parameter hidden Markov models,'' *Proc. ICASSP'88,* vol. 1, pp. 40–43, April 1988.

[49] L. R. Bahl, R. Bakis, P. V. deSouza, and R. L. Mercer, ''Obtaining candidate words by polling in a large vocabulary speech recognition system,'' *Proc. ICASSP'88,* vol. 1, pp. 489–492, April 1988.

[50] L. R. Bahl, P. F. Brown, P. V. deSouza, R. L. Mercer, and M. A. Picheny, ''Acoustic Markov models used in the Tangora speech recognition system,'' *Proc. ICASSP'88,* vol. 1, pp. 497–500, April 1988.

[51] L. R. Bahl, R. Bakis, J. Bellagarda, P. F. Brown, D. Burshtein, S. K. Das, P. V. deSouza, P. S. Gopalakrishnan, F. Jelinek, D. Kanevski, R. L. Mercer, A. J. Nadas, D. Nahamoo, and M. A. Picheny, ''Large vocabulary natural language continuous speech recognition,'' *Proc. ICASSP'89,* vol. 1, pp. 465–467, May 1989.

[52] R. L. Baker and R. M. Gray, ''Image compression using non-adaptive spatial vector quantization,'' *Record of the Sixteenth Asilomar Conf. on Circuits, Systems and Computers,* pp. 55–61, Oct. 1982.

[53] R. L. Baker and R. M. Gray, ''Differential vector quantization of achromatic imagery,'' *Proc. of the Int. Picture Coding Symp.,* March 1983.

[54] R. L. Baker and H.-H. Shen, ''A finite-state vector quantizer for image sequence coding,'' *Proc. ICASSP'87,* vol. 2, pp. 760–763, April 1987.

[55] R. L. Baker and J. L. Salinas, ''A motion compensated vector quantizer with filtered prediction,'' *Proc,. ICASSP'88,* vol. 2, pp. 1324–1327, April 1988.

[56] E. S. Barnes and N. J. A. Sloane, ''The optimal lattice quantizer in three dimensions,'' *SIAM J. of Alg. Disc. Math.,* vol. 4, pp. 30–41, March 1983.

[57] E. S. Barnes and N. J. A. Sloane, ''New lattice packings of spheres,'' *Can. J. Math.,* vol. 35, pp. 117–130, 1983.

[58] J. Barrilleaux, R. Hinkle, and S. Wells, ''Efficient vector quantization for color image encoding,'' *Proc. ICASSP'87,* vol. 2, pp. 740–743, April 1987.

[59] C.-D. Bei and R. M. Gray, ''Simulation of vector trellis encoding systems,'' *IEEE Trans. Commun.,* vol. COM-34, pp. 214–218, March 1986.

[60] A. F. Bergh, F. K. Soong, and L. R. Rabiner, ''Incorporation of temporal structure into a vector-quantization-based preprocessor for speaker-independent, isolated-word recognition, *AT&T Tech. J.,* vol. 64, pp. 1047–1063, 1985.

[61] A. Bergstrom and P. Hedelin, ''Code-book driven glottal pulse analysis,'' *Proc. ICASSP'89,* vol. 1, pp. 53–56, May 1989.

[62] L. Bertorello and M. Copperi, ''Design of a 4.8/9.6 kbps baseband LPC coder using splitband and vector quantization,'' *Proc. ICASSP'83,* vol. 3, pp. 1312–1315, April 1983.

[63] H. Bheda, K. S. Thyagarajan, and H. Abut, ''Fast matrix quantizers for image encoding,'' *Proc. ICASSP'85,* vol. 1, pp. 137–140, March 1985.

[64] R. Billi, ''Vector quantization and Markov models applied to speech recognition,'' *Proc. ICASSP'82,* pp. 574–577, May 1982.

[65] S. Bingol, B. Sankur, and W. Steenaart, ''Joint source channel coding in vector quantization,'' in *Digital Signal Processing,* A. Luque, A.R. Figueireas-Vidal, and V. Cappellini, Eds. North-Holland, Amsterdam: Elsevier Science Publishers, B.V., pp. 137–139, 1985.

[66] M. E. Blain and T. R. Fischer, ''Optimum rate allocation in pyramid vector quantizer transform coding of imagery,'' *Proc. ICASSP'87,* vol. 2, pp. 729–732, April 1987.

[67] H. Bonneau and J. L. Gauvain, ''Vector quantization for speaker adaptation,'' *Proc. ICASSP'87,* vol. 3, pp. 1434–1437, April 1987.

[68] A. Bos, J. H. Conway, and N. J. A. Sloane, ''Further lattice packings in high dimensions,'' in *Mathematika, A Journal of Pure and Applied Mathematics,* vol. 29, part 2, no. 58, pp. 171–180, Dec. 1982.

[69] P. Boucher and M. Goldberg, ''Transform image coding by vector quantization,'' ''Neuvieme collogue sur le Traitement du Signal et ses Applications,'' GRETSI, pp. 629–632, May 1983, Nice, France.

[70] P. Boucher and M. Goldberg, ''Color image compression by adaptive vector quantization,'' *Proc. ICASSP'84,* vol. 3, pp. 29.6.1–29.6.4, March 1984.

[71] H. Bourlard, C. J. Wellekens, and H. Ney, ''Connected digit recognition using vector quantization,'' *Proc. ICASSP'84,* vol. 1, pp. 26.10.1–26.10.4, March 1984.

[72] H. Bourlard, Y. Kamp, and C. J. Wellekens, ''Speaker dependent connected speech recognition via phonemic Markov models,'' *Proc. ICASSP'85,* vol. 3, pp. 1213–1216, March 1985.

[73] M. Breeuwer, ''Transform coding of images using directionally adaptive vector quantization,'' *Proc. ICASSP'88,* vol. 2, pp. 788–791, April 1988.

[74] H. Brehm and K. Trottel, ''Vector quantization of spherically invariant random processes,'' in *Signal Processing II: Theories and Applications,* H. Schussler, Ed. North-Holland, Amsterdam: Elsevier Science Publishers B. V., pp. 383–386, 1983.

[75] H. Brehm and M. Herbert, ''Lattice quantizer in speech coding,'' *Proc. ICASSP'89,* vol. 1, pp. 140–143, May 1989.

[76] J. T. Buck, D. K. Burton, and J. E. Shore, ''Text-dependent speaker recognition using vector quantization,'' *Proc. ICASSP'85,* vol. 1, pp. 391–394, March 1985.

[77] J. T. Buck, ''Vector quantization codebook distortions as features for maximum likelihood classification of isolated words,'' Record of Conf. Globecom, pp. 9.3.1–9.3.4, Nov. 1984.

[78] J. A. Bucklew, ''Two results on the asymptotic performance of quantizers,'' *IEEE Trans. Inf. Theory,* vol. IT-30, pp. 341–348, March 1984.

[79] J. A. Bucklew, ''Companding and random quantization in several dimensions,'' *IEEE Trans. Inform. Theory,* vol. IT-31, pp. 106–109, Jan. 1985.

[80] J. A. Bucklew and N. C. Gallagher, Jr., ''Quantization schemes for bivariate gaussian random variables, in *Quantization,* P. F. Swaszek, Ed. New York: Van Nostrand Reinhold, pp. 245–251, 1985.

[81] S. Budge and R. L. Baker, ''Compression of color digital images using vector quantization in product codes,'' *Proc. ICASSP'85,* vol. 1, pp. 129–132, March 1985.

[82] S. E. Budge, T. G. Stockham, Jr., D. M. Chabries, and R. W. Christiansen, ''Vector quantization of color digital images within a human visual model,'' *Proc. ICASSP'88,* vol. 2, pp. 816–819, April 1988.

[83] D. K. Burton, J. E. Shore, and J. T. Buck, ''A generalization of isolated word recognition using vector quantization,'' *Proc. ICASSP'83,* pp. 1021–1024, April 1983.

[84] D. K. Burton, J. T. Buck, and J. E. Shore, ''Parameter selection for isolated word recognition using vector quantization,'' *Proc. ICASSP'84,* vol. 1, pp. 9.4.1–9.4.4, March 1984.

[85] D. K. Burton, ''Applying matrix quantization to isolated word recognition,'' *Proc. ICASSP'85,* vol. 1, pp. 29–32, March 1985.

[86] D. K. Burton and J. E. Shore, ''Speaker-dependent isolated word recognition using speaker-independent vector quantization codebooks augmented with speaker-specific data,'' *IEEE Trans. Acoust., Speech, and Signal Processing,* vol. ASSP-33, pp. 440–443, April 1985.

[87] D. K. Burton, J. E. Shore, and J. T. Buck, ''Isolated-word speech recognition using multisection vector quantization codebooks,'' *IEEE Trans. Acoust., Speech, and Signal Processing,* vol. ASSP-33, pp. 837–849, Aug. 1985.

[88] D. K. Burton, ''Text-dependent speaker verification using vector quantization source coding,'' *IEEE Trans. Acoust., Speech, and Signal Processing,* vol. ASSP-35, pp. 133–143, February 1987.

[89] M. Bush, G. E. Kopec, and N. Lauritzen, ''Segmentation in isolated word recognition using vector quantization,'' *Proc. ICASSP'84,* vol. 2, pp. 17.11.1–17.11.4, March 1984.

[90] M. Bush and G. E. Kopec, "Network-based connected digit recognition using vector quantization," *Proc. ICASSP'85,* vol. 3, pp. 1197–1199, March 1985.

[91] A. Buzo, R. M. Gray, A. H. Gray, Jr., and J. D. Markel, "Optimal quantizations of coefficient vectors in LPC speech," *1978 Joint Meeting of the Acoustical Soc. of America and the Acoustical Soc. of Japan,* December 1978, Honolulu, HI.

[92] A. Buzo, A. H. Gray, Jr., R. M. Gray, and J. D. Markel, "A two-step compression system with vector quantizing," *Proc. ICASSP'79,* pp. 52–55, April 1979.

[93] A. Buzo, A. H. Gray, Jr., R. M. Gray, and J. D. Markel, "Speech coding based upon vector quantization," *Proc. ICASSP'80,* vol. 1, pp. 15–18, April 1980.

[94] A. Buzo, A. H. Gray, Jr., R. M. Gray, and J. D. Markel, "Speech coding based upon vector quantization," *IEEE Trans. Acoust., Speech, and Signal Processing,* vol. ASSP-28, pp. 562–574, Oct. 1980.

[95] A. Buzo, H. G. Martinez, and C. Riviera, "Discrete utterance recognition based upon source coding techniques," *Proc. ICASSP'82,* pp. 539–542, May. 1982.

[96] J. P. Campbell, Jr., V. C. Welch, and T. E. Tremain, "An expandable error-protected 4800 BPS CELP coder (U.S. federal standard 4800 BPS voice coder)," *Proc. ICASSP'89,* vol. 2, pp. 735–738, May 1989.

[97] P. Cappello, G. Davidson, A. Gersho, C. Kok, and V. Somayazolu, "A systolic vector quantization processor for real time speech coding," *Proc. ICASSP'86,* vol. 3, pp. 2143–2146, April 1986.

[98] M. Caravero, R. Pieraccini, and F. Ranieri, "Definition and evaluation of phonetic units for speech recognition by hidden Markov models," *Proc. ICASSP'86,* vol. 3, pp. 2235–2238, April 1986.

[99] L. Cellario, G. Ferraris, and D. Sereno, "A 2-ms delay CELP coder," *Proc. ICASSP'89,* vol. 1, pp. 73–76, May 1989.

[100] A. E. Cetin and V. Weerackody, "Design of vector quantizers using simulated annealing," *IEEE Trans. Circuits and Syst.,* vol. CAS-35, Dec. 1988.

[101] W. Y. Chan, J.-H. Chen, G. Davidson, and A. Gersho, "A prototype 4800 bps voice terminal for NASA's mobile satellite experiment," *Proc. Globecom,* pp. 1135–1139, Nov. 1987, Tokyo, Japan.

[102] P.-C. Chang, J. May, and R. M. Gray, "Hierarchical vector quantizers with table-lookup encoders," *Record of Int. Commun. Conf.,* pp. 1452–1455, June 1985.

[103] P.-C. Chang and R. M. Gray, "Gradient algorithms for designing predictive vector quantizers," *IEEE Trans. Acoust., Speech, and Signal Processing,* vol. ASSP-34, pp. 679–690, Aug. 1986.

[104] P.-C. Chang, R. M. Gray, and J. May, "Fourier transform vector quantization for speech coding," *IEEE Trans. Commun.,* vol. COM-35, pp. 1059–1068, Oct. 1987.

[105] D. Chase and A. Gersho, "Real-time VQ codebook generation hardware for speech processing," *Proc. ICASSP'88,* vol. 3, pp. 1730–1733, April 1988.

[106] C. H. Chen, "Laplacian pyramid image data compression," *Proc. ICASSP'87,* vol. 2, pp. 737–739, April 1987.

[107] J.-H. Chen and A. Gersho, "Vector adaptive predictive coding of speech at 9.6 kbits/s," *Proc. ICASSP'86,* vol. 3, pp. 1693–1696, April 1986.

[108] J.-H. Chen and A. Gersho, "Gain adaptive vector quantization for medium-rate speech coding," *Record of Int. Commun. Conf.,* pp. 1456–1460, June 1985.

[109] J.-H. Chen and A. Gersho, "Covariance and autocorrelation methods for linear prediction," *Proc. ICASSP'87,* vol. 3, pp. 1545–1548, April 1987.

[110] J.-H. Chen and A. Gersho, "Real-time vector APC speech coding at 4800 bps with adaptive postfiltering," *Proc. ICASSP'87,* vol. 4, pp. 2185–2188, April 1987.

[111] J.-H. Chen, G. Davidson, A. Gersho, and K. Zeger, "Speech coding for the mobile satellite experiment," *Record of Int. Commun. Conf.,* vol. 2, pp. 756–763, June 1987.

[112] J.-H. Chen and A. Gersho, "Gain-adaptive vector quantization application to speech coding," *IEEE Trans. Commun.,* vol. COM-35, pp. 918–930, Sept. 1987.

[113] T. C. Chen, "A fast algorithm for uniform vector quantization," *Proc. ICASSP'87,* vol. 3, pp. 1344–1347, April 1987.

[114] D. Y. Cheng, A. Gersho, B. Ramamurthy, and Y. Shoham, "Fast search algorithms for vector quantization and pattern matching," *Proc. ICASSP'84,* vol. 1, pp. 9.11.1–9.11.4, March 1984.

[115] D. Y. Cheng and A. Gersho, "A fast codebook search algorithm for nearest-neighbour pattern matching," *Proc. ICASSP'86,* vol. 1, pp. 265–268, April 1986.

[116] Y. S. Cheung and S. T. Leung, "Speaker-independent isolated word recognition, using word-based vector quatization and hidden Markov models," *Proc. ICASSP'87,* vol. 2, pp. 1135–1138, April 1987.

[117] P. A. Chou, T. Lookabough, and R. M. Gray, "Entropy-constrained vector quantization," *IEEE Trans. Acoust., Speech, and Signal Processing,* vol. ASSP-37, pp. 31–42, Jan. 1989.

[118] P. A. Chou, T. Lookabough, and R. M. Gray, "Optimal pruning with applications to tree-structured source coding and modeling," *IEEE Trans. Inform. Theory,* vol. IT-35, pp. 299–315, March 1989.

[119] Y.-L. Chow, R. Schwartz, S. Roucos, O. Kimball, P. Price, F. Kubala, M. O. Dunham, M. Krasner, and J. Makhoul, "The role of word-dependent coarticulatory effects in a phoneme based speech recognition system," *Proc. ICASSP'86,* vol. 3, pp. 1593–1596, April 1986.

[120] Y.-L. Chow, M. O. Dunham, O. A. Kimball, M. A. Krasner, F. Kubala, J. Makhoul, P. J. Price, S. Roucos, and R. M. Schwartz, "BYBLOS: the BBN continuous speech recognition system," *Proc. ICASSP'87,* vol. 1, pp. 89–92, April 1987.

[121] J. H. Chung and R. W. Schafer, "A 4.8 Kbps

homomorphic vocoder using analysis-by-synthesis excitation analysis,'' *Proc. ICASSP'89,* vol. 1, pp. 144–147, May 1989.

[122] A. Ciaramella and G. Venuti, ''Vector quantization firmware for an acoustical front-end using TMS32020,'' *Proc. ICASSP'87,* vol. 4, pp. 1895–1898, April 1987.

[123] F. Class, A. Kaltenmeier and H. Katterfeldt, ''Comparison of two continuous connected word recognition principles: hidden Markov modeling and dynamic time warping,'' in *Signal Processing III: Theories and Applications,* I.T. Young, J. Biemond, R. P. W. Duin, and J. J. Gerbrands, Eds. North-Holland, Amsterdam, Elsevier Science Publishers B. V., 1986.

[124] R. A. Cohen and J. W. Woods, ''Sliding block entropy coding of images,'' *Proc. ICASSP'89,* vol. 3, pp. 1731–1734, May 1989.

[125] J. H. Conway and N. J. A. Sloane, '' Voronoi regions of lattices, second moments of polytopes and quantization,'' *IEEE Trans. Inform. Theory,* vol. IT-28, pp. 211–226, March 1982.

[126] J. H. Conway and N. J. A. Sloane, ''Fast quantizing and decoding algorithms for lattice quantizers and codes,'' *IEEE Trans. on Inform. Theory,* vol. IT-28, pp. 227–232, May 1982.

[127] J. H. Conway and N. J. A. Sloane, ''On the Voronoi regions of certain lattices,'' *SIAM J. Alg. Discrete Math.,* vol. 5, pp. 294–305, 1983.

[128] J. H. Conway and N. J. A. Sloane, ''A fast encoding method for lattice, codes and quantizers, *IEEE Trans. Inform. Theory,* vol. IT-29, pp. 820–824, 1983.

[129] J. H. Conway, and N. J. A. Sloane, ''A lower bound on the average error of vector quantizers,'' *IEEE Trans. Inform. Theory,* vol. IT-31, pp. 106–109, Jan. 1985.

[130] J. H. Conway, and N. J. A. Sloane, *Sphere Packings, Lattices and Groups,* New York: Springer-Verlag, 1988.

[131] M. Copperi, ''A variable-rate embedded-code search waveform coding,'' *Proc. ICASSP'82,* vol. 1, pp. 212–215, May 1982.

[132] M. Copperi and D. Sereno, ''9.6 kbits/s piecewise LPC residual excited coder using multiple-stage vector quantization,'' *Proc. ICASSP'84,* vol. 1, pp. 10.5.1–10.5.4, March 1984.

[133] M. Copperi and D. Sereno, ''Improved LPC excitation based on pattern classification and perceptual criteria,'' *Proc. of Seventh Int. Conf. on Pattern Recognition,* vol. 2, pp. 860–862, July–August 1984.

[134] M. Copperi and D. Sereno, ''Baseband and residual excited LPC coders based upon vector quantization,'' *Proc. Int. Conf. on Digital Signal Processing,* pp. 433–437, Sept. 1984.

[135] M. Copperi and D. Sereno, ''Vector quantization and perceptual criteria for low-rate coding of speech,'' *Proc. ICASSP'85,* vol. 1, pp. 252–255, March 1985.

[136] M. Copperi, D. Sereno, and L. Bertorello, ''16 kbits/s split-band APC coder using vector quantization and dynamic bit allocation,'' *Proc. ICASSP'86,* vol. 2, pp. 845–848, April 1986.

[137] M. Copperi and D. Sereno, ''CELP coding for high quality speech at 8 kbits/s,'' *Proc. ICASSP'86,* vol. 3, pp. 1685–1688, April 1986.

[138] M. Copperi, L. Bertorello, F. Perosino, and D. Sereno, ''Predictive coding of speech using vector quantization,'' in *Digital Communications,* E. Biglieri and G. Prati, Eds. North-Holland, Amsterdam: Elsevier Science Publishers B.V., pp. 283–291, 1986.

[139] M. Copperi and D. Sereno, ''A 16 Kbit/s split-band APC coder for digital mobile radio systems,'' *Proc. Second Nordic Sem. on Digital Land Mobile Radio Commun.,* pp. 243–247, Oct. 1986.

[140] M. Copperi and D. Sereno, ''Feature extraction and product codes in vector excited coders,'' *Proc. ICASSP'87,* vol. 4, pp. 1942–1945, April 1987.

[141] M. Copperi, ''Rule-based speech analysis and application to CELP coding,'' *Proc. ICASSP'88,* vol. 1, pp. 143–146, April 1988.

[142] R. V. Cox, S. L. Gay, Y. Shoham, S. R. Quackenbush, N. Seshadri, and N. S. Jayant, ''New directions in subband coding,'' *IEEE J. Selected Areas Commun.,* vol. JSAC-6, pp. 391–409, February 1988.

[143] V. Cuperman and A. Gersho, ''Adaptive differential vector coding of speech,'' *Record of Globecom Conf.,* pp. 1092–1096, Dec. 1982.

[144] V. Cuperman and A. Gersho, ''Vector predictive coding of speech at 16 Kb/s,'' *IEEE Trans. Commun.,* vol. COM-33, pp. 685–696, July 1985.

[145] V. Cuperman, ''Vector transform quantization for speech coding, *Record of Globecom Conf.,* Dec. 1986.

[146] V. Cuperman, ''On adaptive vector transform quantization for speech coding,'' *IEEE Trans. Commun.,* vol. COM-37, pp. 261–267, March 1989.

[147] E. Daly and T. R. Hsing, ''Variable bit rate vector quantization of video images for packet-switched networks,'' *Proc. ICASSP'88,* vol. 2, pp. 1160–1163, April 1988.

[148] G. Davidson, T. Stanhope, R. Aravind, and A. Gersho, ''Real-time speech compression with a VLSI vector quantization preprocessor,'' *Proc. ICASSP'85,* vol. 4, pp. 1437–1440, March 1985.

[149] G. Davidson and A. Gersho, ''Application of a VLSI vector quantization processor to real-time speech coding,'' *IEEE J. Selected Areas Commun.,* vol. SAC-4, pp. 112–124, Jan. 1986.

[150] G. Davidson and A. Gersho, ''Complexity reduction method for vector excitation coding,'' *Proc. ICASSP'86,* vol. 4, pp. 3055–3058, April 1986.

[151] G. Davidson, M. Yong, and A. Gersho, ''Real-time vector excitation coding of speech at 4800 bps,'' *Proc. ICASSP'87,* vol. 4, pp. 2189–2192, April 1987.

[152] G. Davidson and A. Gersho, ''Multiple-stage vector coding of speech waveforms,'' *Proc. ICASSP'88,* vol. 1, pp. 163–166, April 1988.

[153] G. Davidson, P. Cappello, and A. Gersho, ''Systolic

architectures for vector quantization," *IEEE Trans. Acoust., Speech, and Signal Processing,* vol. ASSP-36, pp. 1651–1664, Oct. 1988.

[154] A. Dembo and D. Malah, "Statistical design of analysis/synthesis systems with quantization," *IEEE Trans. on Acoust., Speech, and Signal Processing,* vol. ASSP-36, pp. 328–341, March 1988.

[155] T. Denker, and J. Dziumbla, P. List, and D. Wolf, "Vector quantization of speech waveforms," in *Signal Processing II: Theories and Applications,* H. Schussler, Ed. North-Holland, Amsterdam: Elsevier Science Publishers B.V., pp. 387–389, 1983.

[156] K. Dezhgosha, M. Jamali, and S. C. Kwatra, "Real-time VSLI architecture for VQ-based high-quality image coding algorithm," *Proc. ICASSP'89,* vol. 3, pp. 2425–2429, May 1989.

[157] R. Dianysian and R. L. Baker, "A VLSI chip set for real-time vector quantization of image sequences," *Proc. Int. Symposium on Circuits and Systems,* May 1987.

[158] S. S. Dixit and Y. Feng, "Adaptive vector quantization of video for packet switched networks," *Proc. ICASSP'89,* vol. 3, pp. 1870–1873, May 1989.

[159] I. Dologlou and A. Gersho, "A vector companding technique for fast codebook search in waveform coding of speech," *Proc. 1987 European Conf. Circ. Theory and Design,* pp. 133–138, Sept. 1987.

[160] L. S. Dooley, W. A. Evans, W. A. Mahmoud, and L. A. M. Bennett, "The performance of a fast vector quantization algorithm upon digital marine sound synthesis," in *Signal Processing III: Theories and Applications,* I. T. Young, J. Biemond, R. P. W. Duin, and J. J. Gerbrands, Eds. North-Holland, Amsterdam: Elsevier Science Publishers B.V., pp. 461–464, 1986.

[161] R. Drogo de Jacovo, R. Montagna, F. Perrosino, and D. Sereno, "Some Experiments of 7 kHz Audio coding at 16 kbit/s," *Proc. ICASSP'89,* vol. 1, pp. 192–195, May 1989.

[162] M. O. Dunham and R. M. Gray, "An algorithm for the design of labeled-transition finite-state vector quantizers," *IEEE Trans. Commun.,* vol. COM-33, no. 1, Jan. 1985.

[163] J. Dziumbla, H. Reininger, and D. Wolf, "Reduktion der Ubertragunstrate eines RELP Vocoders mit Hilfe der Vektorquantisierung," *Conf. Record, URSI-CONF. KLEINHEUBACH,* 1989 (in German).

[164] A. A. El Gamal, L. A. Hemachandra, I. Shperling, and V. K. Wei, "Using simulated annealing to design good codes," *IEEE Trans. Inform. Theory,* vol. IT-33, pp. 116–123, Jan. 1987.

[165] Y. Ephraim, D. Malah, and B.-H. Juang, "On the application of hidden Markov models for enhancing noisy speech," *Proc. ICASSP'88,* vol. 1, pp. 533–536, April 1988.

[166] Y. Ephraim and R. M. Gray, "A unified approach for encoding clean and noisy sources by means of waveform and autoregressive model vector quantization," *IEEE Trans. on Inform. Theory,* vol. IT-34, pp. 826–834, July 1988.

[167] W. Equitz, "Fast algorithms for vector quantization picture coding," *Proc. ICASSP'87,* vol. 2, pp. 725–728, April 1987.

[168] W. H. Equitz, "A new vector quantization clustering algorithm," *IEEE Trans. Acoust., Speech, and Signal Processing,* vol. ASSP-37, pp. 1568–1575, Oct. 1983.

[169] E. P. Farges and M. A. Clements, "Hidden Markov models applied to very low bit rate speech coding," *Proc. ICASSP'86,* vol. 1, pp. 433–436, April 1986.

[170] E. P. Farges and M. A. Clements, "An analysis-synthesis hidden Markov model of speech," *Proc. ICASSP'88,* vol. 1, pp. 323–326, April 1988.

[171] H. G. Fehn and P. Noll, "Multipath search coding of stationary signals with applications to speech," *IEEE Trans. Commun.,* vol. COM-30, pp. 687–701, April 1982.

[172] M.-W. Feng, F. Kubala, R. Schwartz, and J. Makhoul, "Improved speaker adaptation using text dependent spectral mappings," *Proc. ICASSP'88,* vol. 1, pp. 131–134, April 1988.

[173] M.-W. Feng, R. Schwartz, F. Kubala, and J. Makhoul, "Iterative normalization for speaker-adaptive training in continuous speech recognition," *Proc. ICASSP'89,* vol. 1, pp. 612–615, May 1989.

[174] Y. Feng and N. M. Nasrabadi, "A dynamic address-vector quantization algorithm based on inter-block and inter-color correlation for color image coding," *Proc. ICASSP'89,* vol. 3, pp. 1755–1758, May 1989.

[175] B. Fette, W. Clark, C. Jaskie, M. Tugenberg, and W. Yip, "Experiments with a high quality low complexity 4800 bps residual excited LPC (RELP) vocoder," *Proc. ICASSP'88,* vol. 1, pp. 263–266, April 1988.

[176] W. A. Finnamore, "Entropy coded vector quantizer," in *Digital Signal Processing,* V. Capellini and A. G. Constantinides, Eds. North-Holland, Amsterdam: Elsevier Science Publishers B.V., pp. 438–441, 1987.

[177] T. R. Fischer and R. M. Dicharry, "Vector quantization design for Gaussian, gamma and Laplacian sources," *Proc. Int. Commun. Conf.,* pp. 1085–1089, 1983.

[178] T. R. Fischer and R. M. Dicharry, "Vector quantization of memoryless Gaussian, gamma and Laplacian sources," *IEEE Trans. Commun.,* vol. COM-32, pp. 1065–1069, Sept. 1984.

[179] T. R. Fischer, "A Laplacian vector quantizer," *Proc. 22nd Annu. Allerton Conf.,* Oct. 1984, University of Illinois.

[180] T. R. Fischer and K. T. Malone, "Contour vector quantization and waveform coding," *Proc. ICASSP'85,* vol. 4, pp. 1707–1710, March 1985.

[181] T. R. Fischer and K. T. Malone, "Transform coding of speech with pyramid vector quantization," *Proc. Milcom,* pp. 620–623, Oct. 1985.

[182] T. R. Fischer, "A pyramid vector quantizer," *IEEE Trans. Inform. Theory,* vol. IT-32, pp. 568–583, July 1986.

[183] T. R. Fischer, M. E. Blain, and K. T. Malone, "Pyramid vector quantization of color imagery," *Record of Int. Commun. Conf.,* vol. 2, pp. 1133–1137, June 1987.

[184] T. R. Fischer, "Geometric source coding and vector quantization," *IEEE Trans. Inform. Theory,* vol. IT-35, pp. 137–145, Jan. 1989.

[185] L. Fissore, P. Laface, G. Micca, and R. Pieraccini, "Very large vocabulary isolated utterance recognition: a comparison between one pass and two pass strategies," *Proc. ICASSP'88,* vol. 1, pp. 203–206, April 1988.

[186] T. Fjallbrant, W. Kou, and F. Mekuria, "ATCVQ systems for speech coding (one DSP implementation)," in *Digital Signal Processing,* V. Capellini and A. G. Constantinides, Eds. North-Holland, Amsterdam: Elsevier Science Publishers B.V., pp. 306–309, 1987.

[187] T. Fjallbrant, "Discrete cosine transform magnitude and phase coefficients used for vector quantization in a speech coder," *Proc. ISCAS'88,* vol. 2, pp. 1087–1089, June 1988.

[188] T. Fjallbrant, F. Mekuira, and W. Kou, "A voiced/unvoiced classified vector quantized speech transform coder implementation on a TMS 32020 signal processor," *Proc. ISCAS'88,* vol. 2, pp. 1333–1336, June 1988.

[189] J. K. Flanagan, D. R. Morrel, R. Frost, C. J. Read, and B. E. Nelson, "On the global quality of vector quantization codebooks generated by simulated annealing," *Proc. ICASSP'89,* vol. 3, pp. 1759–1762, May 1989.

[190] C. Fong and M. Baraniecki, "Adaptive transform coding with multi-band vector quantization for video signals," *Proc. ICASSP'88,* vol. 2, pp. 1184–1187, April 1988.

[191] J. Foster, R. M. Gray, and M. O. Dunham, "Finite-state vector quantization for waveform coding," *IEEE Trans. Inform. Theory,* vol. IT-31, pp. 348–359, March 1985.

[192] E. Frangoulis, "Vector quantization of the continuous distributions of an HMM speech recognizer based on mixture of continuous distributions," *Proc. ICASSP'89,* vol. 1, pp. 9–12, May 1989.

[193] W. Frank and A. Lacroix, "Vector quantization of side information based on generalized lattice techniques," in *Signal Processing III: Theories and Applications,* I. T. Young, J. Biemond, R. P. W. Duin, and J. J. Gerbrands. Eds. North-Holland, Amsterdam: Elsevier Science Publishers B.V., pp. 489–491, 1986.

[194] T. Fujii and N. Ohta, "A load balancing technique for video signal processing on a multicomputer type DSP," *Proc. ICASSP'88,* vol. 4, pp, 1981–1984, April 1988.

[195] R. R. Furner, R. W. Christiansen, and D. M. Chabries, "Motion compensated vector quantization," *Proc. ICASSP'86,* vol. 2, pp. 989–992, April 1986.

[196] S. Furui, "A VQ-based preprocessor using cepstral dynamic features for large vocabulary word recognition," *Proc. ICASSP'87,* vol. 2, pp. 1127–1130, April 1987.

[197] S. Furui, "A VQ-based preprocessor using cepstral dynamic features for speaker-independent large vocabulary word recognition," *IEEE Trans. Acoust., Speech, Signal Processing,* vol. ASSP-36, pp. 980–987, July 1987.

[198] S. Furui, "Unsupervised speaker adaptation method based on hierarchical spectral clustering," *Proc. ICASSP'89,* vol. 1, pp. 286–289, May 1989.

[199] N. C. Gallagher, Jr., and J. A. Bucklew, "Properties of minimum mean squared error block quantizers," in *Quantization,* P. F. Swaszek, Ed. New York: Van Nostrand Reinhold, pp. 121–123, 1985.

[200] K. Ganesan, M. Marlot, and P. Mehta, "An efficient algorithm for combining vector quantization and stochastic modeling for speaker-independent speech recognition," *Proc. ICASSP'86,* vol. 2, pp. 1069–1071, April 1986.

[201] R. Garcia-Gomez, F. J. Casajus-Quiros, and L. Hernandez-Gomez, "Vector quantized multipulse-LPC," *Proc. ICASSP'87,* vol. 4, pp. 2181–2184, April 1987.

[202] A. Gersho, "Asymptotically optimal block quantization." *IEEE Trans. on Inform. Theory,* vol. IT-25, pp. 373–380, July 1979. Also in *Quantization,* P. Swaszek, Ed. New York: Van Nostrand Reinhold Co., 1985.

[203] A. Gersho, "A structural approach to vector quantization," *Proc. Nat. Telecomm. Conf.,* pp. F4.1–F4.5, Nov. 1981.

[204] A. Gersho, "On the structure of vector quantizers," *IEEE Trans. Inform. Theory,* vol. IT-28, pp. 157–166, March 1982.

[205] A. Gersho and B. Ramamurthi, "Image coding using vector quantization," *Proc. Calspace Investigators Conf.,* pp. 41–43, March 1982.

[206] A. Gersho and B. Ramamurthi, "Image coding using vector quantization," *Proc. ICASSP'82,* vol. 1, pp. 428–431, May 1982.

[207] A. Gersho, "Quantization," in *Tutorials in Modern Communications,* V. B. Lawrence, J. L. LoCicero, and L. B. Milstein, Eds. Rockville MD: Computer Science Press, pp. 5–14, 1983.

[208] A. Gersho and V. Cuperman, "Speech waveform coding using vector quantization," *Conf. Record, Globecom'83,* pp. 193–197, Dec. 1983.

[209] A. Gersho and V. Cuperman, "Vector quantization: a pattern matching technique for speech coding," *IEEE Commun. Mag.,* pp. 15–21, Dec. 1983.

[210] A. Gersho and Y. Shoham, "Hierarchical vector quantization of speech with dynamic codebook allocation," *Proc. ICASSP'84,* vol. 1, pp. 10.9.1–10.9.4, March 1984.

[211] A. Gersho, T. Ramstad, and I. Versvik, "Fully vector

quantized sub-band coding with adaptive codebook allocation,'' *Proc. ICASSP'84,* vol. 1, pp. 10.7.1–10.7.4, March 1984.

[212] A. Gersho, ''Adaptive vector quantization,'' Annales des Telecommunications, vol. 41, no. 9–10, pp. 470–480, Sept.–Oct. 1986.

[213] A. Gersho, ''Vector quantization: a direction in source coding,'' in *Digital Communications,* E. Biglieri and G. Prati, Eds. North-Holland, Amsterdam: Elsevier Science Publishers B.V., pp. 267–281, 1986.

[214] A. Gersho and M. Yano, ''Adaptive vector quantization by progressive codevector replacement,'' *Proc. ICASSP'85,* vol. 1, pp. 133–136, March 1985.

[215] A. Gersho, and R. M. Gray, *Vector Quantization and Signal Compression,* Norwell, MA: Kluwer Academic Press Publishers, 1989.

[216] J. D. Gibson, ''Lattice vector quantization,'' *Proc. 29th Midwest Symposium on Circuits and Systems,* Aug. 1986, Lincoln, NE.

[217] J. D. Gibson and K. Sayood, ''Lattice quantization,'' in *Advances in Electronics and Electron Physics,* New York; Academic, vol. 72, pp. 259–330, 1988.

[218] J. D. Gibson, T. R. Fischer, and B. Koo, ''Estimation and vector quantization of noisy speech,'' *Proc. ICASSP'88,* vol. 1, pp. 541–544, April 1988.

[219] B. Gold, R. P. Lippmann, and M. L. Malpass, ''Some neural net recognition results on isolated-words,'' *Proc. IEEE 1st Int. Conf. Neural Networks,* June 1987.

[220] B. Gold and R. P. Lippmann, ''A neural network for isolated word recognition,'' *Proc. ICASSP'88,* vol. 1, pp. 44–47, April 1988.

[221] M. Goldberg, ''Amelioration de la qualited images fortement compressees par quantification vectorielle,'' CESTA Colloque Image, pp. 263–268, May 1984, (in French).

[222] M. Goldberg, P. Boucher, and S. Shlien, ''Image compression using adaptive vector quantization,'' *IEEE Trans. Commun.,* vol. COM-34, pp. 180–187, Feb. 1986.

[223] M. Goldberg and H. Sun, ''Image sequence coding using vector quantization,'' *IEEE Trans. Commun.,* vol. COM-34, pp. 703–710, July 1986.

[224] M. Goldberg and H.-F. Sun, ''Image sequence coding by 3-D vector quantization,'' *IEEE Proc. Commun., Radar, and Signal Processing,* vol. 133, pp. 482–487, Aug. 1986.

[225] M. Goldberg and F. Schmitt, ''Texture representation by adaptive vector quantization,'' *8th Int. Conf. on Pattern Recognition,* pp. 672–674, Oct. 1986.

[226] M. Goldberg and H.-F. Sun, ''Frame adaptive vector quantization for image sequence coding,'' *IEEE Trans. Commun.,* vol. COM-36, pp. 629–635, May 1988.

[227] M. Gotze and Y. Du, ''Adaptive transform coding of images using vector quantization, in *Signal Processing III: Theories and Applications,* I. T. Young, J. Biemond, R. P. W. Duin, and J. J. Gerbrands, Eds. North-Holland, Amsterdam: Elsevier Science Publishers B.V., pp. 789–792, 1986.

[228] W. Granzon, P. Noll, and C. Volmary, ''Channel error resistant adaptive transform coding of speech signals,'' in *Signal Processing III: Theories and Applications,* I. T. Young, J. Biemond, R. P. W. Duin, and J. J. Gerbrands, Eds. North-Holland, Amsterdam: Elsevier Science Publishers B.V., pp. 457–460, 1986.

[229] R. M. Gray, A. Buzo, Y. Matsuyama, A. H. Gray, Jr., and J. D. Markel, ''Source coding and speech compression,'' *Proc. 1978 Int. Telemetry Conf.,* 24, pp. 871–878, Dec. 1978.

[230] R. M. Gray, A. H. Gray, Jr., and G. Rebolledo, ''Optimal speech compression,'' *Proc. 13th Asilomar Conf. Circuits, Systems, and Computers,* Nov. 1979.

[231] R. M. Gray, J. C. Kieffer, and Y. Linde, ''Locally optimal block quantizer design,'' *Inform. Contr.,* vol. 45, pp. 178–198, May 1980.

[232] R. M. Gray, A. Buzo, A. H. Gray, Jr., and Y. Matsuyama, ''Distortion measures for speech processing,'' *IEEE Trans. Acoust., Speech, and Signal Processing,* vol. ASSP-28, pp. 367–376, Aug. 1980.

[233] R. M. Gray, A. H. Gray, Jr., G. Rebolledo, and J. E. Shore, ''Rate distortion speech coding with a minimum discrimination information distortion measure,'' *IEEE Trans. Inform. Theory,* vol. IT-27, pp. 708–721, Nov. 1981.

[234] R. M. Gray, and Y. Linde, ''Vector quantizers and predictive quantizers for Gauss-Markov sources,'' *IEEE Trans. Commun.,* vol. COM-30, pp. 381–389, Feb. 1982.

[235] R. M. Gray and E. Karnin, ''Multiple location optima in vector quantizers,'' *IEEE Trans. Inform. Theory,* vol. IT-28, pp. 256–261, March 1982.

[236] R. M. Gray and H. Abut, ''Full search and tree-searched vector quantizers,'' *Proc. ICASSP'82,* May 1982.

[237] R. M. Gray, ''Vector quantization,'' *IEEE ASSP Mag.,* vol. 1, pp. 4–29, April 1984.

[238] R. M. Gray, ''Multivariate quantization,'' in *Multivariate Analysis VI,* P. R. Krishnaiah, Ed. North-Holland, Amsterdam: Elsevier Science Publishers B.V., 1985.

[239] V. N. Gupta, M. Lennig, and P. Mermelstein, ''Integration of acoustic information in a large vocabulary word recognizer,'' *Proc. ICASSP'87,* vol. 2, pp. 697–700, April 1987.

[240] R. Hamabe, Y. Yamada, Y. Murata, and T. Namekawa, ''A speech recognition system using inverse filter matching techniques,'' *Proc. of the Annu. Conf. of Inst. Telev. Eng.,* June 1981 (in Japanese).

[241] B. Hammer and M. Schielein, ''Image coding by vector quantization of M-Hadamard transform coefficients,'' *SPIE,* vol. 594, Image Coding, pp. 72–78, 1985.

[242] B. Hammer, A. v. Brandt, and M. Schielein, ''Hierarchical encoding of image sequences using multistage

vector quantization," *Proc. ICASSP'87,* vol. 2, pp. 1055–1058, April 1987.
[243] H.-M. Hang and J. W. Woods, "Predictive vector quantization of images," *IEEE Trans. Commun.,* vol. COM-33, pp. 1208–1219, Nov. 1985.
[244] H.-M. Hang and B. G. Haskell, "Interpolative vector quantization of color images," *IEEE Trans. Commun.,* vol. COM-36, pp. 465–470, April 1988.
[245] A. Haoui and D. G. Messerschmitt, "Predictive vector quantization," *Proc. ICASSP'84,* vol. 1, pp. 10.10.1–10.10.4, March 1984.
[246] A. Haoui and D. G. Messerschmitt, "Embedded coding of speech: a vector quantization approach," *Proc. ICASSP'85,* vol. 4, pp. 1702–1706, March 1985.
[247] f. j. harris, K. S. Thyagarajan, and A. Onal, "Using overlapped windows to improve reconstruction fidelity of vector quantized LPC image compression algorithms," *Record 21st Asilomar Conf. Signals, Systems & Computers,* vol. 2. pp. 789–792, Nov. 1987.
[248] N. He, A. Buzo, and F. Kuhlmann, "A frequency domain waveform speech compression system based on product vector quantizers," *Proc. ICASSP'86,* vol. 4, pp. 3031–3034, April 1986.
[249] L. A. Hernandez-Gomez, F. J. Casajus-Quiros, and R. Garcia-Gomez, "Vector quantization for multiple pulse speech coding," in *Digital Signal Processing,* V. Capellini and A. G. Constantinides, Eds. North-Holland, Amsterdam: Elsevier Science Publishers B.V., pp. 310–314, 1987.
[250] L. A. Herandez-Gomez, F. J. Casajus-Quiros, and R. Garcia-Gomez, "High quality vector adaptive transform coding at 4.8 Kb/s," *Proc. ICASSP'88,* vol. 1, pp. 167–170, April 1988.
[251] L. A. Hernadez-Gomez, C. Garcia-Mateo, and F. Casajus-Quiros, "Short-time synthesis procedures in vector adaptive transform coding of speech," *Proc. ICASSP'89,* vol. 2, pp. 762–765, May 1989.
[252] C. D. Heron, R. E. Crochiere, and R. V. Cox, "A 32-band subband/transform coder incorporating vector quantization for dynamic bit allocation," *Proc. ICASSP'83,* vol. 3, pp. 1276–1279, April 1983.
[253] A. Higgins and R. E. Wohlford, "A new method for text-independent speaker recognition," *Proc. ICASSP'86,* vol. 2. pp. 869–872, April 1986.
[254] Y. S. Ho and A. Gersho, "Variable-rate multi-stage vector quantization for image coding," *Proc. ICASSP'88,* vol. 2, pp. 1156–1159, April 1988.
[255] Y. S. Ho and A. Gersho, "Classified transform coding of images using vector quantization," *Proc. ICASSP'89,* vol. 3, pp. 1890–1893, May 1989.
[256] H. Holzlwimmer, A. v.Brandt, and W. Tengler, "A 64 Kbit/s motion compensated transform coder using vector quantization with scene adaptive codebook," *Conf. Record, ICC'87,* vol. 1, pp. 151–156, June 1987.
[257] T. R. Hsing, "Video compression techniques: networking aspects," *Proc. ISCAS'88,* vol. 1, pp. 223–226, June 1988.
[258] J. J. Y. Huang and P. M. Schultheiss, "Block quantization of correlated Gaussian random variables," in *Quantization,* P. F. Swaszek, Ed. New York: Van Nostrand Reinhold, pp. 237–244, 1985.
[259] S.-S. Huang and R. M. Gray, "Conditional histogram vector quantization for spellmode recognizer," *Proc. ICASSP'87,* vol. 4, pp. 1930–1933, April 1987.
[260] X. D. Huang and M. A. Jack, "Unified techniques for vector quantisation and hidden Markov modeling using semi-continuous models," *Proc. ICASSP'89,* vol. 1, pp. 639–642, May 1989.
[261] L. Hui and K. N. Ngan, "Predictive classified vector quantizer," *Proc. IEEE Region 10 Conf.,* vol. 2, pp. 408–412, 1987.
[262] S. Iai and K. Irie, "8 kbits/s speech coder with pitch adaptive vector quantizer," *Proc. ICASSP'86,* vol. 3, pp. 1697–1700, April 1986.
[263] M. A. Ireton and C. S. Xydeas, "On improving vector excitation coders through the use of spherical lattice codebooks (SLCs)," *Proc. ICASSP'89,* vol. 1, pp. 57–60, May 1989.
[264] N. S. Jayant and P. Noll, *Digitial Coding of Waveforms, Principles, and Applications to Speech and Video,* Englewood Cliffs, NJ: Prentice Hall, 1985.
[265] N. S. Jayant and J.-H. Chen, "Speech coding based on time-varying bit allocations to excitation and LPC parameters," *Proc. ICASSP'89,* vol. 1, pp. 65–68, May 1989.
[266] D. G. Jeong and J. D. Gibson, "Lattice vector quantization for image coding," *Proc. ICASSP'89,* vol. 3, pp. 1743–1746, May 1989.
[267] R. Johnson, J. E. Shore, J. T. Buck, and D. K. Burton, "Speech noise reduction by means of multi-signal minimum-cross-entropy spectral analysis," *Proc. ICASSP'83,* vol. 3, pp. 1129–1132, April 1983.
[268] B.-H. Juang, D. Y. Wong, and A. H. Gray, Jr., "Distortion performance of vector quantization for LPC voice coding," *IEEE Trans. Acoust., Speech, and Signal Processing,* vol. ASSP-30, pp. 294–304, April 1982.
[269] B.-H. Juang and A. H. Gray, Jr., "Multiple stage vector quantization for speech coding," *Proc. ICASSP'82,* vol. 1, pp. 597–600, May 1982.
[270] B.-H. Juang, L. R. Rabiner, S. E. Levinson, and M. M. Sondhi, "Recent developments in the application of hidden Markov models to speaker-independent isolated word recognition," *Proc. ICASSP'85,* vol. 1, pp. 9–12, March 1985.
[271] B.-H. Juang, "Design and performance of trellis vector quantizers for speech signals," *Proc. ICASSP'86,* vol. 1, pp. 437–440, April 1986.
[272] B.-H. Juang, L. R. Rabiner, and J. G. Wilpon, "On the

use of bandpass liftering in speech recognition,'' *Proc. ICASSP'86,* vol. 1, pp. 765–768, April 1986.

[273] B.-H. Juang and. L. R. Rabiner, ''Signal restoration by spectral mapping,'' *Proc. ICASSP'87,* vol. 4, pp. 2368–2371, April 1987.

[274] B.-H. Juang, ''Design and performance of trellis vector quantizers for speech signals,'' *IEEE Trans. Acoust., Speech, and Signal Processing,* vol. ASSP-36, pp. 1423–1431, Sept. 1988.

[275] S. Katagiri, E. McDermott, and M. Yokota, ''A new algorithm for representing acoustic feature dynamics,'' *Proc. ICASSP'89,* vol. 1, pp. 322–325, May 1989.

[276] Y. Kato, N. Mukawa, and J. Watanabe, ''Hybrid vector and scalar quantization for image coding,'' *Conf. Record, Picture Coding Symp.* pp. 78–79, 1986.

[277] S. Kaul and M. Shridhar, ''Tree structures for implementation of a vector quantized speech coding system,'' *Proc. ICASSP'85,* vol. 4, pp. 1855–1858, March 1985.

[278] T. Kawabata and K. Shikano, ''Island-driven continuous speech recognizer using phoneme-based HMM word spotting,'' *Proc. ICASSP'89,* vol. 1, pp. 461–464, May 1989.

[279] D. P. Kemp, R. A. Sueda, and T. E. Tremain, ''An evaluation of 4.8 kbit/s voice coders,'' *Proc. ICASSP'89,* vol. 1, pp. 200–203, May 1989.

[280] J. C. Kieffer, ''Exponential rate of convergence for Lloyd's method I,'' *IEEE Trans. Inform. Theory,* vol. IT-28, pp. 205–210, March 1982.

[281] C.-S. Kim, J. Bruder, M. T. J. Simith, and R. M. Mersereau, ''Subband coding of color images using finite state vector quantization,'' *Proc. ICASSP'88,* vol. 2, pp. 753–756, April 1988.

[282] C.-S. Kim, M. T. J. Smith, and R. M. Mersereau, ''An improved SBC/VQ scheme for color image coding,'' *Proc. ICASSP'89,* vol. 3, pp. 1941–1944, May 1989.

[283] J. W. Kim and S. U. Lee, ''Discrete cosine transform-classified VQ technique for image coding,'' *Proc. ICASSP'89,* vol. 3, pp. 1831–1834, May 1989.

[284] T. Kim, ''New finite state vector quantizers for images,'' *Proc. ICASSP'88,* vol. 2, pp. 1180–1183, April 1988.

[285] R. A. King and N. M. Nasrabadi, ''Image coding using vector quantization in the transform domain,'' in *Pattern Recognition Letters,* 1, North-Holland, Amsterdam: Elsevier Science Publishers, B.V., pp. 323–329, 1983.

[286] W. B. Kleijn, D. J. Krasinski, and R. H. Ketchum, ''Improved Speech Quality and Efficient Vector Quantization in CELP,'' *Proc. ICASSP'88,* vol. 1, pp. 155–158, April 1988.

[287] T. Kohonen, *Self-Organization and Associative Memory,* Berlin: Springer-Verlag Publishers, 1984.

[288] T. Kohonen, K. Masisara, and T. Saramaki, ''Phonotopic Maps–Insightful Representation of Phonological Features for Speech Representation,'' *Proc. IEEE 7th Int. Conf. Pattern Recognition,* 1984, Montreal Canada.

[289] T. Kohonen, G. Barna and R. Chrisley, ''Statistical Pattern Recognition with Neural Networks: Benchmarking Studies,'' *Proc. IEEE Inter. Conf. Neural Networks,* vol. 1, pp. 61–68, July 1988.

[290] A. M. Kondoz and B. G. Evans, ''Vector Quantized Transform Coder,'' *Electronics Letters,* vol. 23, no. 24, pp. 1286–1288, 1987.

[291] A. M. Kondoz and B. G. Evans, ''CELP Base-Band Coder for High Quality Speech Coding at 9.6 to 2.4 KBPS,'' *Proc. ICASSP'88,* vol. 1, pp. 159–162, April 1988.

[292] A. M. Kondoz and B. Evans, ''Improved quality CELP based-band coding of speech at low bit-rate,'' *Proc. ICASSP'89,* vol. 1, pp. 128–131, May 1989.

[293] J. M. Koo and C. K. Un, ''Isolated word recognition based on finite state vector quantization and hidden Markov modeling,'' *Proc. IEEE Region 10 Conf.,* vol. 3, pp. 1282–1286, 1987.

[294] G. E. Kopec, ''Voiceless stop consonant identification using LPC spectra,'' *Proc. ICASSP'84,* vol. 3, pp. 42.1.1–42.1.4, March 1984.

[295] G. E. Kopec, ''Formant tracking using hidden Markov models,'' *Proc. ICASSP'85,* vol. 3, pp. 1113–1116, March 1985.

[296] G. E. Kopec, ''A family of formant trackers based on hidden Markov models,'' *Proc. ICASSP'86,* vol. 2, pp. 1225–1228, April 1986.

[297] G. E. Kopec, ''Formant tracking using hidden Markov models and vector quantization,'' *IEEE Trans. Acoust., Speech, and Signal Processing,* vol. ASSP-34, pp. 709–729, Aug. 1986.

[298] W. Kou, ''A transform product code vector quantizer for digital image coding,'' *Proc. Globecom,* Dec. 1986.

[299] W. Kou and Z. Hu, ''A new fast search algorithm for VQ,'' *Proc. Int. Phoenix Conf. Commun. and Computers,* pp. 87–89, March 1986.

[300] W. Kou and Z. Hu, ''Fast search algorithms for VQ,'' *Proc. Int. Conf. Pattern Recognition,* pp. 1007–1009, 1986.

[301] H. Koyama and A. Gersho, ''A new technique for multipulse excited LPC speech coding using vector quantization,'' *Proc. Phoenix Conf. Commun. and Computers,* March 1986.

[302] H. Koyama and A. Gersho, ''Fully-vector quantized multipulse LPC at 4800 bps,'' *Proc. ICASSP'86,* vol. 1, pp. 445–448, April 1986.

[303] P. Kroon and E. F. Deprettere, ''Quantization procedures for regular-pulse excitation coders,'' in *Signal Processing III: Theories and Applications,* I. T. Young, J. Biemond, R. P. W. Duin, and J. J. Gerbrands, Eds. North-Holland, Amsterdam: Elsevier Science Publishers B.V., pp. 449–451, 1986.

[304] P. Kroon, R. J. Sluyter, and E. F. Deprettere, ''A low

complexity regular pulse coding scheme with a reduced transmission delay,'' *Proc. ICASSP'86,* vol. 4, pp. 3085–3088, April 1986.

[305] P. Kroon and B. S. Atal, ''Quantization procedures for the excitation in CELP coders,'' *Proc. ICASSP'87,* vol. 3, pp. 1649–1652, April 1987.

[306] P. Kroon and B. S Atal, ''Strategies for improving the performance of CELP coders at low bit rates,'' *Proc. ICASSP'88,* vol. 1, pp. 151–154, April 1988.

[307] F. Kubala, Y. Chow, A. Derr, M. Feng, O. Kimball. J. Makhoul, P. Price, J. Rohlicek, S. Roucos, R. Schwartz, and J. Vandegrift, ''Continuous speech recognition results of the BYBLOS system on the DARPA 1000-Word resource management database,'' *Proc. ICASSP'88,* vol. 1, pp. 291–294, April 1988.

[308] F. Kuhlmann and J. A. Bucklew, ''Piecewise uniform vector-quantizers,'' *IEEE Trans. Inform. Theory,* vol. IT-34, pp. 1259–1263, Sept. 1988.

[309] C. Lamblin, J.-P. Adoul, D. Massaloux, and S. Morrisette, ''Fast CELP coding based on the Barnes-Wall lattice in 16 dimensions,'' *Proc. ICASSP'89,* vol. 1, pp. 61–64, May 1989.

[310] B. Landell, J. Taylor, and R. E. Wohlford, ''Effects of vector quantization on a continuous speech recognition system,'' *Proc. ICASSP'84,* vol. 2, pp. 26.11.1–26.11.4, March 1984.

[311] T. Langlais, J. Mason, and Z. Picel, ''Real time implementation of 16 Kbit/s sub-band coder with vector quantization,'' in *Signal Processing III: Theories and Applications,* I. T. Young, J. Biemond, R. P. W. Duin, and J. J. Gerbrands, Eds. North-Holland, Amsterdam: Elsevier Science Publishers B.V., pp. 419–422, 1986.

[312] I. Lecomte, M. Lever, L. Lelievre, M. Delprat, and A. Tassy, ''Medium band speech coding for mobile radio communications,'' *Proc. ICASSP'88,* vol. 1, pp. 231–234, April 1988.

[313] C.-H. Lee, F. K. Soong, and B.-H. Juang, ''A segment model based approach to speech recognition,'' *Proc. ICASSP'88,* vol. 1, pp. 501–504, April 1988.

[314] H. J. Lee and D. T. L. Lee, ''A gain-shape vector quantizer for image coding,'' *Proc. ICASSP'86,* vol. 1, pp. 141–144, April 1986.

[315] K.-F. Lee and H.-W. Hon, ''Large-vocabulary speaker-independent continuous speech recognition using HMM,'' *Proc. ICASSP'88,* vol. 1, pp. 123–126, April 1988.

[316] K.-F. Lee, H.-W. Hon, M.-Y. Hwang, S. Mahajan, and R. Reddy, ''The SPHINX speech recognition system,'' *Proc. ICASSP'89,* vol. 1, pp. 445–448, May 1989.

[317] K.-F. Lee and H.-W. Hon, ''Speaker independent phone recognition using hidden Markov models,'' *IEEE Trans. Acoust., Speech, and Signal Processing,* vol. ASSP-37, pp. 1641–1648, Nov. 1983.

[318] S. U. Lee and D. S. Kim, ''Vector quantizer based on DCT-domain classification,'' *Proc. IEEE Region 10 Conf.,* vol. 2, pp. 413–417, 1987.

[319] A. LeGuyader, P. Combescure, C. Lamblin, M. Mouly, and J. F. Zurcher, ''A robust 16 kbits/s vector adaptive predictive coder for mobile communications,'' *Proc. ICASSP'86,* vol. 2, pp. 857–860, April 1986.

[320] A. Leguyader, D. Massaloux, and J. P. Petit, ''Robust and fast CELP coding of speech signals,'' *Proc. ICASSP'89,* vol. 1, pp. 120–123, May 1989.

[321] S. E. Levinson, L. R. Rabiner, and M. M. Sondhi, ''Speaker independent isolated digit recognition using hidden Markov models,'' *Proc. ICASSP'83,* vol. 3, pp. 1049–1052, April 1983.

[322] S. E. Levinson, L. R. Rabiner, and M. M. Sondhi, ''An introduction to the application of the theory of probabilistic functions of a Markov process to automatic speech recognition,'' *Bell Syst. Tech. J.,* vol. 62, no. 4, pp. 1035–1074, April 1983.

[323] J. Li, W. A. Perlman, and C. N. Manikopoulos, ''A fast search algorithm for vector quantization,'' *Proc. ICASSP'89,* vol. 2, pp. 759–761, May 1989.

[324] K. P. Li and E. H. Wrench, Jr., ''Text-independent speaker recognition with short utterances,'' *Proc. ICASSP'83,* vol. 2, pp. 555–558, April 1983.

[325] Y. Linde, A. Buzo, and R. M. Gray, ''An algorithm for vector quantizer design,'' *IEEE Trans. Commun.,* vol. COM-28, pp. 84–95, Jan. 1980.

[326] R. P. Lippmann, ''Neural nets for computing,'' *Proc. ICASSP'88,* vol. 1, pp. 1–6, April 1988.

[327] P. List, H. Reininger, and D. Wolf, ''Ein effizienter Algorithmus zur Bestimmung Optimaler Codebuch-vectoren,'' DFG-Kolloquium Digitale Signalverarbeitung, Tagungsband 9, pp. 87–90, 1986, Stutgart, FRG (in German).

[328] T.-M. Liu and Z. Hu, ''Hardware realization of a multistage speech waveform vector quantizer,'' *Proc. ICASSP'86,* vol. 4, pp. 3095–3098, April 1986.

[329] T.-M. Liu, ''A new hardware realization of speech waveform vector quantizer,'' *Proc. IEEE-Academia Sinica Workshop ASSP,* April 1986.

[330] Y. J. Liu and J. Rothweiler, ''A high quality speech coder at 400 bit/s,'' *Proc. ICASSP'89,* vol. 1, pp. 204–206, May 1989.

[331] S. P. Lloyd, ''Least squares quantization in PCM,'' *Bell Lab. Tech. Note* (1957). Published in *IEEE Trans. Inform. Theory,* vol. IT-28, part II, pp. 127–135, March 1982.

[332] T. Lookabaugh and R. M. Gray, ''High-resolution quantization theory and vector quantizer advantage,'' *IEEE Trans. Inform. Theory,* vol. IT-35, pp. 1020–1033, Sept. 1989.

[333] A. Lowry, H. Sqama, and W. Millar, ''Binary search trees for vector quantization,'' *Proc. ICASSP'87,* vol. 4, pp. 2205–2208, April 1987.

[334] L. Lu, G. Cohen, and P. Godlewski, ''Composite permutation coding of speech waveforms,'' *Proc. ICASSP'86,* vol. 4, pp. 2359–2362, April 1986.

[335] L. Lu. G, Cohen, and P. Godlewski, "Permutation codes for speech waveforms," *Proc. IEEE Academia Sinica Workshop on ASSP,* April 1986.

[336] P. Mabilleau and J.-P. Adoul, "Medium band speech coding using a dictionary of waveforms," *Proc. ICASSP'81,* vol. 2, pp. 804–807, March–April 1981.

[337] J. MacQueen, "Some Methods for classification and analysis of multivariate observations," *Proc. 5th Berkeley Symp. on Math. Stat. and Prob.,* University of California Press, pp. 281–297, 1967.

[338] A. Madiseti, R. Subramanian, and R. V. Algazi, "A radius-bucketing approach to fast vector quantization encoding," *Proc. ICASSP'89,* vol. 3, pp. 1767–1770, May 1989.

[339] W. A. Mahmoud and L. A. M. Bennett, "Speech recognition based upon VQ reference codebooks generated from voiced, unvoiced and silence classified data," *Int. Symp. Circuits and Syst.,* vol. 2, pp. 567–569, 1986.

[340] J. Makhoul, S. Roucos, and H. Gish, "Vector quantization in speech coding," *Proc. IEEE,* vol. 73, pp. 1551–1588, Nov. 1985.

[341] K. T. Malone and T. R. Fischer, "Contour-gain vector quantization," *IEEE Trans. Acoust., Speech, and Signal Processing,* vol. ASSP-36, pp. 862–870, June 1988.

[342] C. N. Manikopoulos and H. F. Sun, "Finite state vector quantization with an interleaved image source designed for reconstruction of lost packets," *Proc. ICASSP'89,* vol. 3, pp. 1894–1897, May 1989.

[343] K. Mano, S. Ishige, and K. Shirai, "Phoneme recognition in connected speech using both static and dynamic properties of spectrum described by vector quantization," *Proc. ICASSP'86,* vol. 3, pp. 2243–2246, April 1986.

[344] J. R. B. de Marca and N. S. Jayant, "An algorithm for assigning binary indices to the codevectors of a multidimensional quantizer," *Conf. Record, ICC'87,* vol. 2, pp. 1128–1132, June 1987.

[345] J. P. Maresq and C. Labit, "Vector quantization in transformed image coding," *Proc. ICASSP'86,* vol. 1, pp. 145–148, April 1986.

[346] J. Mariani, "Recent advances in speech processing," *Proc. ICASSP'89,* vol. 1, pp. 429–440, May 1989.

[347] S. Marshall, R. Chapman, T. S. Durrani, and L. L. Scharf, "Contour coding of images," *Proc. ICASSP'86,* vol. 2, pp. 1009–1012, April 1986.

[348] E. Masgrau-Gomez, J. B. Marino-Acebel, and F. Vallerdu, "A continuously adaptive vector predictive coder (AVPC) for speech encoding," *Proc. ICASSP'86,* vol. 4, pp. 3027–3030, April 1986.

[349] E. Masgrau-Gomez and J. B. Marino-Acebel, "Adaptive vector predictive speech coding with sample-by-sample update at 16 KBPS," in *Signal Processing III: Theories and Applications,* I. T. Young, J. Biemond, R. P. W. Duin, and J. J. Gerbrands, Eds. North-Holland, Amsterdam: Elsevier Science Publishers B.V., pp. 431–434, 1986.

[350] V. J. Mathews, R. W. Waite, and T. D. Tran, "Image compression using vector quantization for linear (one-step) prediction errors," *Proc. ICASSP'87,* vol. 2, pp. 733–736, April 1987.

[351] V. J. Mathews and M. Khorchidian, "Multiplication-free vector quantization using the L_1 distance measure," *Proc. ICASSP'89,* vol. 3, pp. 1747–1750, May 1989.

[352] S. Matsumura, "Upper bit separation coding of images using vector quantization," *Proc. ICASSP'89,* vol. 3, pp. 1751–1754, May 1989.

[353] Y. Matsuyama and R. M. Gray, "Voice coding and tree encoding speech compression system based upon inverse filter matching," *IEEE Trans. Commun.,* vol. COM-30, pp. 711–720, April 1982.

[354] Y. Matsuyama, "Joint time-spectral vector quantization and inverse filter set," *Proc. ICASSP'86,* vol. 1, pp. 441–444, April 1986.

[355] Y. Matsuyama, "Variable region vector quantization, space warping and speech/image compression," *Proc. ICASSP'87,* vol. 4, pp. 2201–2204, April 1987.

[356] Y. Matsuyama, "Vector quantization with variable dimensions," *Proc. IEEE Region 10 Conf.,* vol. 2, pp. 423–427, 1987.

[357] Y. Matsuyama, "Variable length vector quantization and time warping," *The XI Symposium on Information Theory and its Applications,* pp. 227–230, 1986.

[358] B. Mazor, C. Hudson, and D. Borkowski, "Transform subbands coding with channel error control, *Proc. ICASSP'89,* vol. 1, pp. 172–175, May 1989.

[359] B. Mazor, D. Veeneman, E. Mintz, "Adaptive subbands excited transform (ASET) coding," *Proc. ICASSP'86,* vol. 4, pp. 3075–3078, April 1986.

[360] E. McDermott and S. Katagiri, "Shift-invariant, multi-category phoneme recognition using Kohonen's LVQ2," *Proc. ICASSP'89,* vol. 1, pp. 81–84, May 1989.

[361] G. Mensa, R. Montagna, and F. Rusina, "Comparison between vector and scalar quantization in variable rate subband coders," *Proc. ICASSP'86,* vol. 4, pp. 2383–2386, April 1986.

[362] J. Menez, C. Galand, M. Rosso, and F. Bottau, "Adaptive code excited linear predictive coder (ACELP)," *Proc. ICASSP'89,* vol. 1, pp. 132–135, May 1989.

[363] N. Moayeri, D.L. Neuhoff, and W. E. Stark, "Fast vector quantizers," *Proc. 23 Annu. Allerton Conf.,* pp. 347–353, Oct. 1985, Monticello, IL.

[364] N. Moayeri and D. L. Neuhoff, "Tree based fast quantization," *Proc. 21 Annu. Conf. Inform. Sci. Syst.,* pp. 250–255, March 1987, Baltimore, MD.

[365] N. Moayeri and D. L. Neuhoff, "Decision trees for vector quantizer codebook searching," *Proc. ICASSP'88,* vol. 1, pp. 255–258, April 1988.

[366] S. Morishima, K. Aizawa, and H. Harashima, "An

intelligent facial image coding driven by speech and phoneme,'' *Proc. ICASSP'89,* vol. 3, pp. 1795–1798, May 1989.

[367] T. Moriya and M. Honda, ''Speech coder using phase equalization and vector quantization,'' *Proc. ICASSP'86,* vol. 4, pp. 1701–1704, April 1986.

[368] T. Moriya and M. Honda, ''Transform coding of speech with weighted vector quantization,'' *Proc. ICASSP'87,* vol. 3, pp. 1629–1632, April 1987.

[369] T. Moriya and H. Suda, ''An 8 Kbit/s transform coder for noisy channels,'' *Proc. ICASSP'89,* vol. 1, pp. 196–199, May 1989.

[370] T. Murakami, K. Asai, and E. Yamazaki, ''Vector quantizer for video signals,'' *Electronic Letters,* vol. 7, pp. 1005–1006, Nov. 1982.

[371] T. Murakami, K. Asai, and E. Yamazaki, ''A design of vector quantizers for video signals,'' *Proc. Int. Picture Coding Symp,* pp. 25–26, March 1983.

[372] T Murakami and K. Asai, ''Vector quantization for video signals,'' *J. Inst. Telev. Eng. Jap.,* vol. 38, pp. 452–457, May 1984 (in Japanese).

[373] T. Murakami, K. Asai, A. Itoh, and E. Yamazaki, ''Interframe vector coding of color video signals, *Proc. Int. Picture Coding Symp.,* July 1984.

[374] T. Murakami, K. Asai, and A. Itoh, ''Vector quantization for color images,'' *J. Inst. Telev. Eng. Jap.,* vol. 39, p. 10, 1985 (in Japanese).

[375] T. Murakami, K. Asai, and A. Itoh, ''Dynamic multistage vector quantization of images,'' *Trans. IECE Japan,* vol. J68-B, no. 1, 1985 (in Japanese). Also translated into English in *Electronics and Communications in Japan,* Scripta Technica, Inc., Part 1, vol. 69, no. 3, New York: John Wiley & Sons, Inc., 1986.

[376] T. Murakami, K. Asai, A. Itoh, ''Vector quantization of color images, *Proc. ICASSP'86,* vol. 1, pp. 133–136, April 1986.

[377] T. Murakami, K. Asai, A. Itoh, ''Design of component vector quantiser for video signals,'' *Proc. Int. Picture Coding Symp.,* pp. 249–250, April 1986.

[378] T. Murakami, K. Asai, A. Itoh, ''Design method of symmetric vector quantiser,'' *Electronics Letters,* vol. 22, no. 16, pp. 848–849, July 1986.

[379] T. Murakami, K. Asai, A. Itoh, ''64 Kbps motion video codec using vector quantizer,'' *Proc. Int. Picture Coding Symp.,* pp. 1–2, June 1987.

[380] T. Murakami, K. Asai, A. Itoh, ''Progressive vector quantization of color images,'' *Proc. Globecom,* pp. 1392–1396, Nov. 1987.

[381] T. Murakami, K. Kamizawa, M. Kameyama, and S. Nakegawa, ''A DSP architecture for 64 KBPS motion video codec,'' *Proc. ICAS'88,* vol. 1, pp. 227–230, June 1988.

[382] T. Murakami, N. Kinjo, K. Kamizawa, and T. Shimada, ''A 24 bit DSP for motion video codec and software development support system,'' *Proc. ICASPP'89,* vol. 3, pp. 1929–1932, May 1989.

[383] H. Murveit and m. Weintraub, ''1000-word speaker-indendent continuous-speech recognition using hidden Markov models,'' *Proc. ICASSP'88,* vol. 1, pp. 115–118, April 1988.

[384] E. Mwangi and C. S. Xydeas, ''The use of phoneme-like templates in isolated word recognition without time alignment,'' in *Signal Processing III: Theories and Applications,* I. T. Young, J. Biemond, R. P. W. Duin, and J. J. Gerbrands, Eds. North-Holland, Amsterdam: Elsevier Science Publishers B.V., pp. 561–564, 1986.

[385] A. Nadas, R. L. Mercer, L. R. Bahl, R. Bakis, P. S. Cohen, A. G. Cole, F. Jelinek, and B. L. Lewis, ''Continuous speech recognition with automatically selected acoustic prototypes obtained by either bootstrapping or clustering,'' *Proc. ICASSP'81,* pp. 1153–1155, March–April 1981.

[386] A. Nadas, D. Nahamoo, and M. A. Picheny, ''Speech recognition using noise-adaptive prototypes,'' *Proc. IEEE ICASSP'88,* vol. 1, pp. 517–520, April 1988.

[387] A. Nadas, D. Nahamoo, and M. A. Picheny, ''Normalization of speech by adaptive transformations based on vector quantization,'' *Proc. IEEE ICASSP'88,* vol. 1, pp. 521–524, April 1988.

[388] D. Nahamoo, ''Clustering acoustic prototypes with self organizing distortion measures,'' *Proc. IEEE ICASSP'86,* vol. 1, pp. 753–756, April 1986.

[389] H. Nakaide, K. Aizawa, and H. Harashima, ''Fast search algorithms for vector quantization in orthogonal transform domain,'' *Trans. IECE Japan,* vol. J69-B, no. 10, pp. 1102–1110, Oct. 1986.

[390] S. Nakamura and K. Shikano, ''Speaker adaptation applied to HMM and neural networks,'' *Proc. ICASSP'89,* vol. 1, pp. 89–92, May 1989.

[391] N. M. Nasrabadi and R. A. King, ''A new image coding technique using transforms and vector quantization,'' *Proc. ICASSP'84,* vol. 2, pp. 29.9.1–29.9.4, March 1984.

[392] N. M. Nasrabadi, ''Use of vector quantizers in image coding,'' *Proc. ICASSP'85,* vol. 1, pp. 125–128, March 1985.

[393] N. M. Nasrabadi and R. A. King, ''Image coding using vector quantization,'' *IEEE Trans. Commun.,* vol. COM-36, pp. 957–971, Aug. 1988.

[394] N. M. Nasrabadi, S. E. Lin, and Y. Feng, ''Interframe hierarchical vector quantization,'' *Proc. ICASSP'89,* vol. 3, pp. 1739–1742, May 1989.

[395] B. E. Nelson and C. J. Read, ''A bit-serial VLSI vector quantizer,'' *Proc. IEEE ICASSP'86,* vol. 3, pp. 2211–2214, April 1986.

[396] M. Nishimura, ''HMM-based speech recognition using dynamic spectral feature,'' *Proc. ICASSP'89,* vol. 1, pp. 298–301, May 1989.

[397] M. J. Noah, ''Optimal Lloyd-Max quantization of LPC speech parameters,'' *Proc. ICASSP'84,* vol. 1, pp. 1.8.1–1.8.4, March 1984.

[398] F. Oliveri, G. Conte, and M. Guglielmo, ''A technique

using a one-dimensional mapping for vector quantisation of images,'' *Proc. ICASSP'86,* vol. 1, pp. 149–152, April 1986.

[399] S. Ono and K. Ozawa, ''2.4 KBPS pitch prediction multi-pulse speech coding,'' *Proc. ICASSP'88,* vol. 1, pp. 175–178, April 1988.

[400] S. O'Shaughnessy, ''Speech enhancement using vector quantization and a formant distance measure,'' *Proc. ICASSP'88,* vol. 1, pp. 549–552, April 1988.

[401] S. Othman and S. Wilson, ''Image sequence coding by classified vector quantization and block matching,'' *Proc. ICASSP'89,* vol. 3, pp. 1913–1916, May 1989.

[402] Y. Ozturk and H. Abut, ''Multichannel linear prediction and applications to image coding,'' *Archive für Electronik und Ubertragunstechnik,* vol. 43, pp. 312–320, Sept./Oct. 1983.

[403] K. K. Paliwal, ''Low bit-rate image coding using 2-D linear prediction and 2-D stochastic excitation,'' *Proc. IEEE ICASSP'88,* vol. 2, pp. 800–803, April 1988.

[404] K. K. Paliwal and V. Ramasubramanian, ''Effect of ordering the codebook on the efficiency of the partial distance speech algorithm for vector quantization, *IEEE Trans. Commun.,* vol. 37, pp. 538–540, May 1989.

[405] K.-C. Pan, F. K. Soong, L. R. Rabiner, and A. F. Bergh, ''An efficient vector-quantization preprocessor for speaker-independent isolated word recognition,'' *Proc. IEEE ICASSP'85,* vol. 2, pp. 875–878, March 1985.

[406] K.-C. Pan, F. K. Soong, and L. R. Rabiner, ''A vector quantization based preprocessor for speaker-independent isolated word recognition,'' *IEEE Trans. Acoust., Speech, and Signal Processing,* vol. ASSP-33, pp. 546–560, June 1985.

[407] D. B. Paul and P. E. Blakenship, ''Two distance measure-based vocoder quantization algorithms for very-low rate applications: frame-fill and spatial vector quantization,'' *Proc. of ICC,* pp. 3G.5.1–3G.5.4, June 1982.

[408] D. B. Paul, ''A 500-800 bps adaptive vector quantization vocoder using a perceptually motivated distance measure,'' *Record of Globecom, Conf.,* pp. 1079–1082, Dec. 1982.

[409] D. B. Paul, ''An 800 bps adaptive vector quantization vocoder using perceptual distance measure,'' *Proc. IEEE ICASSP'83,* vol. 1, pp. 73–76, April 1983.

[410] W. A. Pearlman, ''Polar quantization of a complex Gaussian random variable,'' in *Quantization,* P. F. Swaszek, Ed. New York: Van Nostrand Reinhold, pp. 252–259, 1985.

[411] R. Pieraccini and R. Billi, ''Experimental comparison among data compression techniques in isolated word recognition,'' *Proc. ICASSP'83,* vol. 2, pp. 1025–1028, April 1983.

[412] D. Pollard, ''Quantization and the method of K-means,'' *IEEE Trans. Inform. Theory,* vol. IT-28, pp. 199–204, March 1982.

[413] S. Quackenbush, ''Hardware implementation of a 16 KBPS subband coder using vector quantization,'' *Proc. ICASSP'88,* vol. 1, pp. 386–389, April 1988.

[414] Q. Qureshi and T. R. Fischer, ''A hardware pyramid vector quanitzer,'' *Proc. ICASSP'87,* vol. 3, pp. 1402–1405, April 1987.

[415] L. R. Rabiner, S. E. Levinson, and M. M. Sondhi, ''On the application of vector quantization and hidden Markov models to speaker-independent isolated word recognition,'' *Bell Syst. Tech. J.,* vol. 62, pp. 1075–1106, April 1983.

[416] L. R. Rabiner, M. M. Sondhi, and S. E. Levinson, ''Note on the properties of a vector quantizer for LPC coefficients,'' *Bell Syst. Tech. J.,* vol. 62, no. 8, pp. 2603–2616, Oct. 1983.

[417] L. R. Rabiner, M. M. Sondhi, and S. E. Levinson, ''A vector quantizer incorporating both LPC shape and energy,'' *Proc. ICASSP'84,* vol. 2, pp. 17.1.1–17.1.4, March 1984.

[418] L. R. Rabiner, S. E. Levinson, and M. M. Sondhi, ''On the use of hidden Markov models for speaker-independent recognition of isolated words from a medium-size vocabulary,'' *AT&T Bell Lab. Tech. J.,* vol. 63, no. 4, pp. 627–642, April 1984.

[419] L. R. Rabiner, M. M. Sondhi, and S. E. Levinson, ''A vector quantizer combining LPC parameters and its application to isolated word recognition'' *AT&T Tech. J.,* vol. 63, pp. 721–735, May–June 1984.

[420] L. R. Rabiner, K. C. Pan, and F. K. Soong, ''On the performance of isolated word speech recognizers using vector quantization and temporal energy contours,'' *AT&T Bell Lab. Tech. J.,* vol. 63, no. 7, pp. 1245–1260, Sept. 1984.

[421] L. R. Rabiner, B.-H. Juang, S. E. Levinson, and M. M. Sondhi, ''Recognition of isolated digits using hidden Markov models with continuous mixture densities,'' *Bell Syst. Tech. J.,* vol. 64, pp. 1211–1234, 1985.

[422] L. R. Rabiner, and S. E. Levinson, ''A speaker-independent, syntax directed connected word recognition system based on hidden Markov models and level building,'' *IEEE Trans. Acoust., Speech, Signal Processing,* vol. ASSP-33, no. 3, pp. 561–573, June 1985.

[423] L. R. Rabiner, and F. K. Soong, ''Single frame vowel recognition using vector quantization with several distance measures,'' *AT&T Tech. J.,* vol. 64, no. 10, Dec. 1985.

[424] L. R. Rabiner, C. H. Lee, B.-H. Juang, and J. G. Wilpon, ''HMM clustering for connected word recognition,'' *Proc. ICASSP'89,* vol. 1, pp. 405–408, May 1989.

[425] P. A. Ramamoorthy and T. Tran, ''A hybrid coding involving ADM and vector quantization for digital video image compression,'' *Proc. ICASSP'86,* vol. 1, pp. 153–156, April 1986.

[426] P. A. Ramamoorthy and B. Potu, "Bit serial systolic chip set for real-time image coding," *Proc. ICASSP'87,* vol. 2, pp. 787–790, April 1987.

[427] V. Ramamoorthy, "An improved modulo-PCM speech encoder with an imbedded vector quantizer," *Proc. ICASSP'86,* vol. 4, pp. 2395–2398, April 1986.

[428] V. Ramamoorthy and A. Abdul-Jabbar, "A reduced search vector quantizer for speech," *Proc. ICASSP'86,* vol. 4, pp. 3035–3038, April 1986.

[429] V. Ramamoorthy and N. S. Jayant, "High quality image coding with a model-testing vector quantizer and a human visual system model," *Proc. ICASSP'88,* vol. 2, pp. 1164–1167, April 1988.

[430] B. Ramamurthi and A Gersho, "Image coding using segmented codebooks," *Proc. Int. Picture Coding Symp.,* pp. 44–45, March 1983.

[431] B. Ramamurthi, A. Gersho, and A. Sekey, "Low-rate image coding using vector quantization," *Record of Globecom Conf.,* pp. 184–187, Nov. 1983.

[432] B. Ramamurthi and A. Gersho, "Image vector quantization with a perceptually-based cell classifier," *Proc. ICASSP'84,* vol. 2, pp. 32.10.1–32.10.4, March 1984.

[433] B. Ramamurthi and A Gersho, "Edge-oriented spatial filtering of images with application to post processing VQ images," *Proc. ICASSP'84,* vol. 3, pp. 48.10.1–48.10.4, March 1984.

[434] B. Ramamurthi and A. Gersho, "Nonlinear space-variant postprocessing of block coded images," *IEEE Trans. Acoust., Speech, Signal Processing,* vol. ASSP-34, pp. 1258–1268, Oct. 1986.

[435] B. Ramamurthi and A. Gersho, "Classified vector quantization of images," *IEEE Trans. Commun.,* vol. COM-34, pp. 1105–1115, Nov. 1986.

[436] C. J. Read, D. M. Chabries, R. W. Christiansen, and J. K. Flanagan, "A method for computing the DFT of vector quantized data," *Proc. ICASSP'89,* vol. 2, pp. 1015–1018, May 1989.

[437] G. Rebolledo, R. M. Gray, and J. P. Burg, 'A multirate voice digitizer based upon vector quantization," *IEEE Trans. Commun.,* vol. COM-30, pp. 721–727, April 1982.

[438] H. Reininger and D. Wolf, "Adaptive filtering of binary quantized speech," in *Signal Processing II: Theories and Applications,* H. Schussler, Ed. North-Holland, Amsterdam: Elsevier Science Publishers B.V., pp. 379–381, 1983.

[439] H. Reininger and D. Wolf, "Speech and speaker-independent codebook design in VQ coding schemes," *Proc. ICASSP'85,* vol. 4, pp. 1700–1702, March 1985.

[440] H. Reininger and D. Wolf, "Fast search algorithms for speech coding schemes using vector quantization," in *Signal Processing III: Theories and Applications,* I. T. Young, J. Biemond, R. P. W. Duin, and J. J. Gerbrands, Eds. North-Holland, Amsterdam: Elsevier Science Publishers B.V., pp. 453–456, 1986.

[441] H. Reininger and D. Wolf, "Verfahren der Sprachcodierung fur Bitraten unter 16 kb/s-Aktueller Stand und Entwicklungs-aspekte," *NTG-Fachberichte,* vol. 94, pp. 156–161, 1986 (in German).

[442] H. Reininger and D. Wolf, "A robust and efficient high quality speech coding scheme for transmission rates below 16 Kbits/s," in *Digital Signal Processing,* V. Capellini and A. G. Costantinides, Eds. North-Holland, Amsterdam: Elsevier Science Publishers B.V., pp. 323–327, 1987.

[443] K. D. Rines and N. C. Gallager, Jr., "The design of two-dimensional quantizers using prequantization," *IEEE Trans. Inform. Theory,* vol. IT-28, pp. 232–239, March 1982.

[444] K. D. Rines and N. C. Gallager. Jr., "The design of multidimensional quantizers using prequantization," in *Quantization,* P. F. Swaszek, Ed. New York: Van Nostrand Reinhold, pp. 140–147, 1985.

[445] E. A. Riskin, T. Lookabaugh, P. A. Chou, and R. M. Gray, "Variable rate vector quantization for medical image compression with applications to progressive transmission,' *Record of Medical Imaging III Conf.,* Jan. 1989.

[446] E. A. Riskin, E. M. Daly, and R. M. Gray, "Pruned tree-structured vector quantization in image coding," *Proc. ICASSP'89,* vol. 3, pp. 1735–1737, May 1989.

[447] A. Rosenberg and F. K. Soong, "Evaluation of a vector quantization talker recognition system in text independent and text dependent modes," *Proc. ICASSP'86,* vol. 2, pp. 873–876, April 1986.

[448] M. C. Rost and K. Sayood, "Investigation of lattice vector quantization," *Proc. 27th Midwest Symp. Circuits Syst.,* pp. 149–152, June 1984.

[449] M. C. Rost and K. Sayood, "The root lattices as low bit rate vector quantizers, *IEEE Trans. Inform. Theory,* vol. IT-34, pp. 1053–1058, Sept. 1988.

[450] J. Rothweiler and J. Carmody, "A multiple rate low rate voice codec," *Proc. ICASSP'85,* vol. 1, pp. 248–251, March 1985.

[451] J. Rothweiler, "Performance of a real time low rate voice codec," *Proc. ICASSP'86,* vol. 4, pp. 3039–3042, April 1986.

[452] S. Roucos, R. M. Schwartz, and J. Makhoul, "Vector quantization for very-low-rate coding of speech," *Record of Globecom Conf.,* pp. 1074–1078, Dec. 1982.

[453] S. Roucos, R. M. Schwartz, and J. Makhoul, "A segment vocoder at 150 B/S," *Proc. ICASSP'83,* vol. 1, pp. 61–64, April 1983.

[454] S. Roucos and R. M. Schwartz, "A comparison of methods for 300-400 B/S vocoders," *Proc. ICASSP'83,* vol. 1, pp. 69–72, April 1983.

[455] S. Roucos, V. Viswanathan, C. Henry, and R. M. Schwartz, "Word recognition using multisensor speech input in high ambient noise," *Proc. ICASSP'86,* vol. 1, pp. 737–740, April 1986.

[456] S. Roucos, A. Wilgus, and W. Russel, "A segment

vocoder algorithm for real-time implementation," *Proc. ICASSP'87,* vol. 4, pp. 1949–1952, April 1987.
[457] F. Rusina, G. Mensa, and R. Montagna, "A subband coder for digital mobile radio application," in *Signal Processing III: Theories and Applications,* I. T. Young, J. Biemond, R. P. W. Duin, and J. J. Gerbrands, Eds. North-Holland, Amsterdam: Elsevier Science Publishers B.V., pp. 423–426, 1986.
[458] C. W. Rutledge, "Vector DPCM: vector predictive coding of color images," *Record of Globecom Conf.,* 1986.
[459] C. W. Rutledge, "Variable block vector DPCM: vector predictive coding of color images," *Proc. ICC'87,* vol. 1, pp. 130–135, June 1987.
[460] M. J. Sabin and R. M. Gray, "Product code vector quantizers for speech waveform coding," *Record of Globecom Conf.,* pp. 1087–1091, Dec. 1982.
[461] M. J. Sabin and R. M. Gray, "Product code vector quantizers for waveform and voice coding," *IEEE Trans. Acoust., Speech, Signal Processing,* vol. ASSP-32, pp. 474–488, June 1984.
[462] M. J. Sabin and R. M. Gray, "Global convergence and empirical consistency of the generalized Lloyd algorithm," *IEEE Trans. Inform. Theory,* vol. IT-32, pp. 148–155, March 1986.
[463] R. J. Safranek, K. Mackay, N. S. Jayant, and T. Kim, "Image coding based on selective quantization of the reconstruction noise in the dominant sub-band," *Proc. ICASSP'88,* vol. 2, pp. 765–768, April 1988.
[464] T. Saito, H. Takeo, H. Harashima, and H. Miyakawa, "Gain/shape vector quantizer for multidimensional spherically symmetric random source," *Trans. IECE Japan,* vol. J68-B, no. 8, pp. 904–911, Aug. 1985 (in Japanese). Also translated into English in *Electronics and Communication in Japan,* Scripta Technica, Inc., New York: John Wiley & Sons, 1986.
[465] T. Saito, T Komatsu, H. Harashima, and H. Miyakawa, "Singular value decomposition (SVD) image coding scheme using vector quantization," *J. Inst. Telev. Eng. Jap.,* vol. 39, pp. 876–884, Oct. 1985 (in Japanese).
[466] T. Saito, H. Takeo, K. Aizawa, H. Harashima, and H. Miyakawa, "Adaptive discrete cosine transform image coding using gain/shape vector quantizers," *Proc. IEEE ICASSP'86,* vol. 1, pp. 129–132, April 1986.
[467] T. Saito, H. Takeo, K. Aizawa, H. Harashima, and H. Miyakawa, "Discrete cosine transform coding using gain/shape vector quantizers and its application to color image coding," *Proc. Int. Picture Coding Symp.,* pp. 225–226, April 1986.
[468] T. Saito, H. Takeo, A. Moriya, and H. Harashima, "Method for designing gain/shape vector quantizer for spherically symmetric random source and its applications," *Trans. IECE Japan,* vol. J69-B, no. 10, pp. 1162–1165, Oct. 1986 (in Japanese).
[469] T. Saito, H. Takeo, K. Aizawa, and H. Harashima, "Adaptive discrete cosine transform coding with gain/shape vector quantizers for picture signals," *Trans. IECE Japan,* vol. J69-B, no. 10, pp. 1731–1741, Oct. 1986 (in Japanese).
[470] T. Saito, Y. Miyakawa, T. Komatsu, and H. Harashima, "Vector quantization using interpolation for picture signals," *Trans. IECE Japan,* vol. J69-B, no. 12, pp. 1849–1852, Dec. 1986 (in Japanese).
[471] T. Saito, H. Takeo, K. Aizawa, H. Harashima, and H. Miyakawa, "Discrete cosine transform coding with gain/shape vector quantizers for picture signals," *Trans. IECE Japan,* vol. J69-B, no. 3, pp. 161–170 (in Japanese). Also translated into English in *Electronics and Communication in Japan,* Scripta Technica, Inc., New York: John Wiley & Sons, 1987.
[472] T. Saito, I. Akiyoshi, and H. Harashima, "Laplacian pyramid vector quantization," *Proc. Int. Picture Coding Symp.,* pp. 124–125, June 1987.
[473] T. Saito, T. Komatsu and H. Harashima, "A vector quantizer with short-time memory for digital image coding," *Proc. ISCAS'88,* vol. 1, pp. 217–221, June 1988.
[474] T. Saito, H. Eguchi, R. Abe, T. Komatsu, and H. Harashima, "Self organizing pattern-matching coding for picture signals," *Proc. ICASSP'89,* vol. 3, pp. 1671–1674, May 1989.
[475] R. A. Salami, L. Hanzo, and D. G. Appleby, "A fully vector quantized self-excited vocoder," *Proc. ICASSP'89,* vol. 1, pp. 124–127, May 1989.
[476] K. Sayood, J. D. Gibson, and M. C. Rost, "An algorithm for uniform vector quantizer design," *IEEE Trans. Inform. Theory,* vol. IT-30, pp. 805–814, Nov. 1984.
[477] K. Sayood, and S. J. Blankenau, "A fast quantization algorithm for lattice quantizer design," *Proc. ICASSP'88,* vol. 2, pp. 1168–1171, April 1988.
[478] F. Schmitt, M. Goldberg, and M. Montagner, "Hierarchical method of vector quantization: an application to texture coding," *CESTA Image Coding Symp.,* pp. 544–550, April 1986.
[479] M. R. Schroeder and B. S. Atal, "Code-excited linear prediction (CELP): high-quality speech at very low bit rates," *Proc. IEEE ICASSP'85* vol. 3, pp. 937–940, March 1985.
[480] M. Schultheiss and A. Lacroix, "On the performance of CELP algorithms for low rate speech coding," *Proc. ICASSP'89,* vol. 1, pp. 152–155, May 1989.
[481] R. Schwartz, Y. Chow, O. Kimball. S. Roucos, M. Krasner, and J. Makhoul, "Context-dependent modeling for acoustic-phonetic recognition of continuous speech," *Proc. ICASSP'85,* vol. 3, pp. 1205–1208, March 1985.
[482] R. Schwartz, O. Kimball, F. Kubala, M.-W. Feng, Y.-L. Chow, C. Barry, and J. Makhoul, "Robust smoothing methods for discrete hidden Markov models," *Proc. ICASSP'89,* vol. 1, pp. 548–551, May 1989.
[483] H.-H. Shen and R. L. Baker, "A finite state/frame difference interpolative vector quantizer for low rate image sequence coding," *Proc. ICASSP'88,* vol. 2,

pp. 1188–1191, April 1988.
[484] K. Shikano, "Spoken word recognition based upon vector quantization of input speech," *Trans. Comm. Speech Res., The Acoust. Soc. Japan,* vol. S82-15, pp. 113–120, Dec. 1982 (in Japanese).
[485] K. Shikano, K.-F. Lee, and R. Reddy, "Speaker adaptation through vector quantization," *Proc. ICASSP'86,* vol. 4, pp. 2643–2646, April 1986.
[486] K. Shikano, "Improvement of word recognition results by trigram model," *Proc. IEEE ICASSP'87,* vol. 3, pp. 1261–1264, April 1987.
[487] K. Shirai, T. Kobayashi, and J. Yazawa, "Estimation of articulatory parameters by table look-up method and its application for speaker-independent phoneme recognition," *Proc. IEEE ICASSP'86,* vol. 3, pp. 2247–2250, April 1986.
[488] K. Shirai, N. Aoki, and N. Hosaka, "Multi-level clustering of acoustic features for phoneme recognition based on mutual information," *Proc. ICASSP'89,* vol. 1, pp. 604–607, May 1989.
[489] Y. Shiraki and M. Honda. "LPC speech coding based on variable-length segment quantization," *IEEE Trans. Acoust., Speech, Signal Processing,* vol. ASSP-36, pp. 1437–1444, Sept. 1988.
[490] Y. Shoham and A. Gersho, "Pitch-Synchronuous Transform Coding of Speech at 9.6 Kb/s Based on Vector Quantization," *Record of Int., Comm. Conf.,* pp. 1179–1182, May 1984.
[491] Y. Shoham and A. Gersho, "Efficient codebook allocation for an arbitrary set of vector quantizers," *Proc. ICASSP'85,* vol. 4, pp. 1696–1699, March 1985.
[492] Y. Shoham, "Vector predictive quantization of the spectral parameters for low rate speech coding," *Proc. ICASSP'87,* vol. 4, pp. 2181–2184, April 1987.
[493] Y. Shoham and A. Gersho, "Efficient bit allocation for an arbitrary set of quantizers," *IEEE Trans. Acoust., Speech, Signal Processing,* vol. ASSP-36, pp. 1445–1453, Sept. 1988.
[494] Y. Shoham, "Cascaded likelihood vector coding of the LPC information," *Proc. ICASSP'89,* vol. 1, pp. 160–163, May 1989.
[495] J. E. Shore, "Minimum cross-entropy spectral analysis," *IEEE Trans. Acoust., Speech, Signal Processing,* vol. ASSP-29, pp. 230–237, April 1981.
[496] J. E. Shore and R. M. Gray, "Minimum-cross-entropy pattern classification and cluster analysis," *IEEE Trans. Pattern Anal. Machine Intell.,* vol. PAMI-4, pp. 11–17, January 1982.
[497] J. E. Shore and D. K. Burton, "Discrete utterance speech recognition without time alignment," *Proc. ICASSP'82,* pp. 907–910, May 1982.
[498] J. E. Shore and D. K. Burton, "Discrete utterance speech recognition without time alignment," *IEEE Trans. Inform. Theory,* vol. IT-29, pp. 473–491, July 1983.
[499] J. E. Shore, D. K. Burton, and J. T. Buck, "A generalization of isolated word recognition using vector quantization," *Proc. ICASSP'83,* pp. 1021–1024, April 1983.
[500] S. Singhal, "On encoding filter parameters for stochastic coders," *Proc. ICASSP'87,* vol. 4, pp. 1633–1636, April 1987.
[501] N. J. A. Sloane, "Recent bounds for codes, sphere packings and related problems obtained by linear programming and other methods," in *Papers in Analysis, Algebra and Statistics,* R. Lidl, Ed. Providence RI: Amer. Math. Soc., pp. 153–185, 1982.
[502] N. J. A. Sloane, "The packing of spheres," Scientific American, pp. 116–125, Jan. 1984.
[503] N. J. A. Sloane, "Les Empilements de Spheres," Pour la Science, pp. 44–55, March, 1984 (in French).
[504] A. R. Smith, J. N. Deneberg, T. B. Slack, C. C. Tan, and R. E. Wohlford, "Application of sequential pattern learning system to connected speeh recognition," *Proc. ICASSP'85,* vol. 3, pp. 1201–1204, March 1985.
[505] M. R. Soleymani and S. D. Morgera, "A high-speed search algorithm for vector quantization," *Proc. ICASSP'87,* vol. 4, pp. 1946–1948, April 1987.
[506] M. R. Soleymani and S. D. Morgera, "An efficient nearest neighbor search method," *IEEE Trans. Commun.,* vol. COM-35, pp. 667–679, June 1987.
[507] M. R. Soleymani and S. D. Morgera, "A fast MMSE encoding technique for vector quantization," *IEEE Trans. Commun.,* vol. COM-37, pp. 656–659, June 1989.
[508] F. K. Soong, A. E. Rosenberg, L. R. Rabiner, and B.-H. Juang, "A vector quantization approach to speaker recognition," *Proc. ICASSP'85,* vol. 1, pp. 387–390, March 1985.
[509] F. K. Soong, A. E. Rosenberg, L. R. Rabiner, and B.-H. Juang, "Application of vector quantization to speaker verification," *Proc. ICC'85,* pp. 1465–1469, June 1985.
[510] F. K. Soong and A. E. Rosenberg, "On the use of instantaneous and transitional spectral information in speaker recognition, *Proc. ICASSP'86,* vol. 2, pp. 877–880, April 1986.
[511] F. K. Soong, R. V. Cox, and N. S. Jayant, "A high quality subband speech coder with backward adaptive predictor and optimal time-frequency bit assignment," *Proc. ICASSP'86,* vol. 4, pp. 2387–2390, April 1986.
[512] F. K. Soong, R. V. Cox, and N. S. Jayant, "On the backward adaptive predictor and optimal time-frequency bit assignment of high quality subband speech coder," *Proc. ISCAS'86,* pp. 1343–1346, May 1986.
[513] F. K. Soong, A. E. Rosenberg, B.-H. Juang, and L R. Rabiner, "A vector quantization approach to speaker recognition," AT & T *Tech. J.,* vol. 66, pp. 14–26, March–April 1987.
[514] F. K. Soong and A. E. Rosenberg, "On the use of instantaneous and transitional spectral information in speaker recognition," *IEEE Trans. Acoust., Speech, Signal Processing,* vol. ASSP-36, pp. 871–879, June 1988.
[515] T. V. Sreenivas, "Modelling LPC-residue by compo-

nents for good quality speech coding,'' *Proc. ICASSP'88,* vol. 1, pp. 171–174, April 1988.

[516] L. C. Stewart, R. M. Gray, and Y. Linde, ''The design of trellis waveform coders,'' *IEEE Trans. Commun.,* vol. COM-30, pp. 702–710, April 1982.

[517] T. G. Stockham and Z. Xie, ''Optimal previsualized image vector quantization,'' *Proc. ICASSP'89,* vol. 3, pp. 1763–1766.

[518] N. Sugamura and S. Furui, ''Large vocabulary word recognition using psuedo-phoneme templates,'' *Trans. IECE Japan,* vol. J65-D, no. 8, pp. 1041–1048, Aug. 1982 (in Japanese).

[519] N. Sugamura, K. Shikano, and S. Furui, ''Isolated word recognition using phoneme-like templates,'' *Proc. ICASSP'83,* vol. 2, pp. 723–726, April 1983.

[520] N. Sugamura and S. Furui, ''Isolated word recognition using strings of phoneme-like templates (SPLIT),'' *J. Acoust. Soc. Japan,* vol. E-5, no. 4, pp. 243–252, 1984.

[521] H.-F. Sun and M. Goldberg, ''An adaptive block coding scheme,'' *IEEE Int. Elec. Electron. Conf.,* pp. 226–229, Sept. 1983.

[522] H.-F. Sun and M. Goldberg, ''Image coding using LPC with vector quantization,'' *Proc. Int. Conf. Digital Signal Processing,* pp. 508–512, Sept. 1984.

[523] H. F. Sun and M. Goldberg, ''Image sequence coding using vector quantization,'' *Proc. IEEE Commun. Energy Conf.,* Oct. 1984.

[524] H. F. Sun and M. Goldberg, ''Adaptive vector quantization for image sequence coding,'' *Proc. ICASSP'85,* vol. 1, pp. 339–342, March 1985.

[525] H. F. Sun and M. Goldberg, ''Frame adaptive vector quantization, *Proc. ICC'86,* pp. 1878–1882, June 1986.

[526] H. F. Sun and M. Goldberg, ''Medical image sequence coding using two stage adaptive vector quantization,'' *Record of SPIE Med. Imag. Conf. and Int. Symp.,* vol. 767, pp. 281–285, Feb. 1987.

[527] H.-F. Sun and C. N. Manikopoulos, ''Intra-frame image coding by adaptive vector quantization with variable block sizes,'' *Proc. SPIE,* vol. 845, pp. 32–35, Oct. 1987.

[528] H.-F. Sun and M. Goldberg, ''Radiographic image sequence coding using two-stage adaptive vector quantization,'' *IEEE Trans. Med. Imag.,* vol. 5, pp. 118–126, June 1988.

[529] T. Svendsen, ''Multi-dimensional quantization applied to predictive coding of speech,'' *Proc. IEEE ICASSP'86,* vol. 4, pp. 3063–3066, April 1986.

[530] P. F. Swaszek and J. B. Thomas, ''Optimal circularly symmetric quantizers,'' *J. Franklin Inst.,* vol. 313, pp. 373–384, June 1982; also in *Quantization,* P. F. Swaszek, Ed. New York: Van Nostrand Reinhold, Co., 1985.

[531] P. F. Swaszek and J. B. Thomas, ''Multidimensional spherical coordinates quantization,'' *IEEE Trans. Inform. Theory,* vol. IT-29, pp. 570–576, July 1983; also in *Quantization,* P. F. Swaszek, Ed. New York: Van Nostrand Reinhold, Co., 1985.

[532] P. F. Swaszek, ''Further notes on circularly symmetric quantizers,'' *Proc. John Hopkins CISS,* pp. 794–800, March 1983.

[533] P. F. Swaszek, ''Uniform spherical coordinate quantization of spherically symmetric sources,'' *IEEE Trans. on Commun.,* vol. COM-33, pp. 518–521, June 1985.

[534] P. F Swaszek, ''Asymptotic performance of dirichlet rotated polar quantizers,'' *IEEE Trans. Inform. Theory,* vol. IT-31, pp. 537–540, July 1985.

[535] P. F. Swaszek Ed. *Quantization,* New York: Van Nostrand Reinhold Co., 1985.

[536] P. F. Swaszek, ''Vector quantization,'' in *Communications and Networks: A Survey of Recent Advances,* I. F. Blake and H. V. Poor, Eds. New York: Springer-Verlag, pp. 362–389, 1986.

[537] P. F. Swaszek and T. Ku, ''Asymptotic performance of unrestricted polar quantizers,'' in *Quantization,* P. F. Swaszek, Ed. New York: Van Nostrand Reinhold, pp. 266–271, 1985; also in *IEEE Trans. Inform. Theory,* vol. IT-32, pp. 330–333, March 1986.

[538] P. Swaszek, ''Vector quantization for image compression,'' *Proc. Conf. Inform. Sci. Syst.,* pp. 254–259, 1986, Princeton, N.J.

[539] P. F. Swaszek and A. Narasimhan, ''Quantization of the correlated Gaussian source,'' *Proc. Princeton CISS,* pp. 784–789, March 1988.

[540] T. Taniguchi and R. M. Gray, ''Multi mode coding: application to CELP,'' *Proc. ICASSP'89,* vol. 1, pp. 156–159, May 1989.

[541] B. P. M. Tao, H. Abut, and R. M Gray, ''Hardware realization of waveform vector quantizers,'' *IEEE J. Selected Areas Commun.,* vol. SAC-2, pp. 343–352, March 1984.

[542] A. Tassy and L. Miclet, ''Speaker-independent French digit recognition using word-based vector quantization and hidden Markov models,'' *Proc. IEEE ICASSP'86,* vol. 4, pp. 2683–2686, April 1986.

[543] S. Tazaki, Y. Yamada, and R. M. Gray, ''Evaluation of two dimensional quantization,'' in *Digital Signal Processing,* Cappelini and Constantinides, Eds. New York: Academic, 1980, pp. 117–122.

[544] L. D. Thede and S. C. Kwatra, ''Image sequence coding by classified vector quantization using quaternary decomposition with simulated annealing optimization of codebooks,'' *Proc. ICASSP'89,* vol. 3, pp. 1909–1912, May 1989.

[545] D. L. Thomson, ''Parametric models of the magnitude/phase spectrum for harmonic speech coding,'' *Proc. IEEE ICASSP'88,* vol. 1, pp. 378–381, April 1988.

[546] K. S. Thyagarajan, H. Bheda, and H. Abut, ''Matrix quantizers for image processing,'' *Proc. IEEE ICASSP'84,* vol. 3, pp. 40.2.1–40.2.3, March 1984.

[547] K. S. Thyagarajan, S. Parthasarathy, and H. Abut, ''Design of a matrix quantizer incorporating human visual model,'' *Proc. IEEE ICASSP'85,* vol. 1, pp. 141–144, March 1985.

[548] K. S. Thyagarajan, H. Abut, and H. Bohlmann,

"Image coding based on segmentation using region growing," *Proc. ICASSP'87,* vol. 2, pp. 752–755, April 1987.

[549] K. S. Thyagarajan and M. Viswanathan, "Low bit-rate image coding techniques," *Proc. IEEE ICASSP'87,* vol. 2, pp. 748–751, April 1987.

[550] K. S. Thyagarajan and S. Bhatt, "Linear block prediction of images for encoding applications," *Record of ICC'87 Conf.,* vol. 2, pp. 992–996, June 1987.

[551] K. S. Thyagarajan and H. Gadjali, "Subband coding and matrix quantization of images," *Record of IASTED Conf. Signal Processing and Applied Filtering,* Aug. 1987.

[552] K. S. Thyagarajan and S. Bhatt, "Linear block prediction with source classification for image encoding applications," *Proc. IEEE ICASSP'88,* vol. 2, pp. 1308–1311, April 1988.

[553] K. S. Thyagarajan and H. Sanchez, "Image sequence coding using interframe VDPCM and motion compensation," *Proc. ICASSP'89,* vol. 3, pp. 1858–1861, May 1989.

[554] P.-Y. Ting, C.-Y. Tseng, and L.-S. Lee, "New speech recognition approaches based upon finite state vector quantization with structural constraints," *Proc. IEEE ICASSP'88,* vol. 1, pp. 187–190, April 1988.

[555] N. Tishby, "Information theoretic factorization of speaker and language in hidden Markov models with application to speaker recognition," *Proc. IEEE ICASSP'88,* vol. 1, pp. 87–90, April 1988.

[556] K. Toshioka and O. Watanuki, "Vector quantization for isolated word recognition using hidden Markov models," *Proc. 1985 Fall Conf., Acoust. Soc. Japan* (in Japanese).

[557] S. A. Townes, A. Gersho, T. P. Barnwell III, R. C. Rose, G. D. Davidson, and J.-H. Chen, "4800 b/s speech compression techniques for mobile satellite systems, *Proc. IEEE Milcom,* pp. 32.6.1–32.6.5, Oct. 1986.

[558] A. Tran, T.-M. Liu, K.-H. Tzou, and E. B. Vogel, "An efficient pyramid image coding system," *Proc. IEEE ICASSP'87,* vol. 2, pp. 744–747, April 1987.

[559] I. M. Trancoso, J. S. Rodrigues, L. B. Almeida, J. S. Marques, A. Serralheiro, D.Santos, and J. M. Tribolet, "Quantization issues in harmonic coders," *Proc. IEEE ICASSP'88,* vol. 1, pp. 382–385, April 1988.

[560] C. Tsao and R. M. Gray, "An approach to speaker-dependent vowel recognition using vector quantization," in *Towards Robustness in Speech Recognition,* W. Lea, Ed. Santa Barbara, CA: Speech Science Publications, 1983.

[561] C. Tsao and R. M. Gray, "An endpoint detector for LPC speech using residual error look-ahead for vector quantization applications," *Proc. IEEE ICASSP'84,* vol. 2, pp. 18B.7.1–18B.7.4, March 1984.

[562] C. Tsao and R. M. Gray, "Matrix quantizer design for LPC speech using the generalized Lloyd algorithm," *IEEE Trans. Acoust., Speech, Signal Processing,* vol. ASSP-33, pp. 537–545, June 1985.

[563] C. Tsao and R. M. Gray, "Shape-gain matrix quantizers for LPC speech," *IEEE Trans. Acoust., Speech, Signal Processing,* vol. ASSP-34, pp. 1427–1439, Dec. 1986.

[564] H.-C. Tseng and T. R. Fischer, "Transform and hybrid transform/DPCM coding of images using pyramid vector quantization," *IEEE Trans. Commun.,* vol. COM-35, pp. 79–86, Jan. 1987.

[565] H.-P. Tseng, M. J. Sabin, and E. A. Lee, "Fuzzy vector quantization applied to hidden Markov modeling," *Proc. IEEE ICASSP'87,* vol. 2, pp. 641–644, April 1987.

[566] G. Tu, G. Verbeek, J. Rommelaere, and A. Osterlinck, "Adaptive coding in transform domain using VQ and scalar correlation," *Proc. IEEE Region 10 Conf.,* vol. 2, pp. 418–422, 1987.

[567] V. R. Updikar and J. P. Raina, "BTC image coding using vector quantization," *IEEE Trans. Commun.,* vol. COM-35, pp. 352–356, March 1987.

[568] D. J. Vaisey and A. Gersho, "Variable block-size image coding *Proc. ICASSP'87,* vol. 2, pp. 1051–1054, April 1987.

[569] D. J. Vaisey and A. Gersho, "Simulated annealing and codebook design," *Proc. ICASSP'88,* vol. 2, pp. 1176–1179, April 1988.

[570] H. Van Helden, F. Booman, and J. Biemond, "A generalized tree structured algorithm for vector quantization of images," *Proc. Int. Picture Coding Symp.,* July 1984.

[571] I. Versvik and H. C. Guren, "Subband coding with vector quantization," *Proc. ICASSP'86,* vol. 4, pp. 3099–3102, April 1986.

[572] D. Vicard and L. Miclet, "Steady part recognition of continuous speech for acoustic-phonetic decoding," *Proc. ICASSP'86,* vol. 3, pp. 2263–2266, April 1986.

[573] A. Waibel, T Hanazawa, G. Hinton, K. Shikano, and K. Lang, "Phoneme recognition: neural networks vs. hidden Markov models," *Proc. IEEE ICASSP'88,* vol. 1, pp. 107–110, April 1988.

[574] E. Walach, D. Chevion, and E. Karnin, "On modification of block truncation coding approach to image compression," in *Signal Processing III: Theories and Applications,* I. T. Young, J. Biemond, R. P. W. Duin, and J. J. Gerbrands, Eds. North-Holland, Amsterdam: Elsevier Science Publishers B.V., pp. 727–730, 1986.

[575] L. Wang and M. Goldberg, "Lossless progressive image transmission by residual vector quantization," *13th Biennial Symp. Commun.,* pp. c.1.9–c.1.12, June 1986.

[576] L. Wang and M. Goldberg, "Progressive image transmission by multistage transform coefficient quantization," *Proc. IEEE ICC'86,* pp. 419–423, June 1986.

[577] L. Wang and M. Goldberg, "Progressive image transmission using vector quantization in transform domain," *Proc. IEEE Phoenix Conf. Comput. Commun.,* pp. 178–182, Feb. 1987.

[578] L. Wang and M. Goldberg, "Progressive image transmission using vector quantization in images in quadtree form," *Proc. Int. Conf. Digital Signal Processing,* Sept. 1987, Florence, Italy.
[579] L. Wang and M. Goldberg, "Progressive image transmission by transform coefficient residual error quantization," *IEEE Trans. Commun.,* vol. COM-36, pp. 75–87, Jan. 1988.
[580] L. Wang and M. Goldberg, "Pyramid transform coding using vector quantization," *Proc. IEEE ICASSP'88,* vol. 2, pp. 812–815, April 1988.
[581] S. Wang and A. Gersho, "Phonetically-based vector excitation coding of speech at 3.6 kbit/s," *Proc. ICASSP'89,* vol. 1, pp. 49–52, May 1989.
[582] Y. Wang and S. Mitra, "Edge-preserving image coding based on local modelling of images," *Proc. IEEE ICASSP'88,* pp. 1320–1323, April 1988.
[583] O. Watanuki and T. Kaneko, "Speaker-independent isolated word recognition using hidden Markov models," *Proc. IEEE ICASSP'86,* vol. 4, pp. 2679–2682, April 1986.
[584] P. H. Westerink, J. Biemond, and D. E. Boekee, "Subband coding of images using predictive vector quantization," *Proc. IEEE ICASSP'87,* vol. 3, pp. 1378–1381, April 1987.
[585] P. H. Westerink, J. Biemond, D. E. Boekee, and J. W. Woods, "Sub-band coding of images using vector quantization," *IEEE Trans. Commun.,* vol. COM-36, pp. 713–719, June 1988.
[586] S. G. Wilson, "Magnitude/phase quantization of independent Gaussian variates," in _Quantization,_ P. F. Swaszek, Ed. New York: Van Nostrand Reinhold, pp. 260–265, 1985.
[587] D. Wolf, H. Reininger, and J. Dzuimbla, "A comparative study of various speech encoding schemes using vector quantization," *Proc. ICASSP'84,* vol. 1, pp. 10.8.1–10.8.4, March 1984.
[588] D. Wolf, P. List, and H. Reininger, "Vector quantization of random model sources," *Proc. Dixieme Colloque sur le Traitement du Signal et ses Applications,* pp. 1123–1127, May 1985.
[589] D. Y. Wong, B.-H. Juang, and A. H. Gray, Jr., "Recent developments in vector quantization for speech processing," *Proc. IEEE ICASSP'81,* vol. 1, pp. 1–4, March–April 1981.
[590] D. Y. Wong, B.-H. Juang, and A. H. Gray, Jr., "An 800 bit/s vector quantization LPC vocoder," *IEEE Trans. Acoust., Speech, Signal Processing,* vol. ASSP-30, pp. 770–779, Oct. 1982.
[591] D. Y. Wong, B-H. Juang, and D. Y. Cheng, "Very low data rate speech compression with LPC vector and matrix quantization," *Proc. ICASSP'83,* vol. 1, pp. 65–68, April 1983.
[592] D. Y. Wong, and B.-H. Juang, "Voice coding at 800 BPS and lower data rates with LPC vector quantization," *Proc. IEEE ICASSP'82,* vol. 2, pp. 606–609, May 1982.
[593] H. C. Woo and J. D. Gibson, "Multipulse-based codebooks for CELP coding at 7 kbits/s," *Proc. ICASSP'89,* vol. 1, pp. 136–139, May 1989.
[594] F. H. Wu and K. Ganesan, "Comparative study of algorithms for VQ design using conventional (LBG) and neural-net based approaches," *Proc. ICASSP'89,* vol. 2, pp. 751–754, May 1989.
[595] L. Xu, J. Oglesby, and J. S. Mason, "The optimization of perceptually-based features for speaker identification," *Proc. ICASSP'89,* vol. 1, pp. 520–523, May 1989.
[596] Y. Yamada, K. Fujita and S. Tazaki, "Vector quantization of video signals," *Proc. Annu. Conf. IECE,* p. 1031, 1980 (in Japanese).
[597] Y. Yamada and S. Tazaki, "A fundamental study of vector quantization of images," *Trans. IECE,* Japan, vol. IT82-21, 1982 (in Japanese).
[598] Y. Yamada and S. Tazaki, "Vector quantizer design for video signals," *Trans. IECE,* Japan, vol. J66-B, pp. 965–972, 1983 (in Japanese).
[599] Y. Yamada and S. Tazaki, "A method for constructing successive approximation vector quantizers for video signals," *Proc. Annu. Conf. Telev. Eng. Japan,* pp. 6–12, 1983 (in Japanese).
[600] C.-L. Yeh, "Color-image sequence compression using adaptive binary-tree vector quantization with codebook replenishment," *Proc. IEEE ICASSP'87,* vol. 2, pp. 1059–1062, April 1987.
[601] M. Yong, G. D. Davidson, and A. Gersho, "Encoding of LPC spectral parameters using switched-adaptive interframe vector prediction," *Proc. IEEE ICASSP'88,* vol. 1, pp. 402–405, April 1988.
[602] M. Yong and A. Gersho, "Subband vector excitation coding with adaptive bit allocation," *Proc. ICASSP'89,* vol. 2, pp. 743–746, May 1989.
[603] W. S. Youn and C. K. Un, "Isolated word recognition based on finite-state vector quantization." *Proc. IEEE ICASSP'86,* vol. 1, pp. 717–720, April 1986.
[604] X. Yuan, V. K. Ingle, and D. G. Manolakis, "An adaptive image coding algorithm based on noncausal LPC and vector quantization," *Proc. ICASSP'88,* vol. 2, pp. 1172–1175, April 1988.
[605] P. L. Zador, "Asymptotic quantization error of continuous signals and the quantization dimension," *IEEE Trans. Inform. Theory,* vol. IT-28, part II, pp. 139–149, March 1982.
[606] K. Zeger and A. Gersho, "Real-time vector predictive coding of speech," *Record of ICC'87,* vol. 2, pp. 1147–1152, June 1987.
[607] K. Zeger and A. Gersho, "Zero redundancy channel coding in vector quantization," *Electronics Letters,* vol. 23, no. 12, pp. 654–656, June 1987.
[608] P. Zheng and H.-J. Zhang, "A new idea of codebook design in vector quantization of speech," *Proc. IEEE ICASSP'87,* vol. 3, pp. 1352–1353, April 1987.

Addendum to Bibliography

The following additions to the bibliography have been compiled during the publication process.

[609] M. H. Ahmad-Fadzil and T. J. Dennis, "Sample selection for subband VQ coder with an improved motion estimation technique," *Proc. IEEE ICASSP'90,* paper 30.M6.1, Apr. 1990.

[610] A. N. Akansu and M. Kadur, "Subband coding of video with adaptive vector quantization," *Proc. IEEE ICASSP'90,* paper 30.M6.8, Apr. 1990.

[611] M. Akamine and K. Miseki, "CELP coding with and adaptive density pulse excitation model," *Proc. IEEE ICASSP'90,* paper 5.S1.8, April 1990.

[612] M. Antonini, M. Barlaud, and P. Mathieu, "Image coding using vector quantization in the wavelet transform domain," *Proc. IEEE ICASSP'90,* paper 44.M9.10, April 1990.

[613] B. S. Atal, V. Cuperman, and A. Gersho, Eds. *Advances in Speech Coding,* Norwell, MA: Kluwer Academic Publishers, 1990.

[614] S. A. Atungsiri, A. M. Kondoz, and B. G. Evans, "Error detection and control for the parametric information in CELP coders," *Proc. IEEE ICASSP'90,* paper 14.S4b.4, April 1990.

[615] J. L. Boxerman and H. J. Lee, "Variable block-sized vector quantization of grayscale images with unconstrained tiling," *Proc. ICASSP'90,* paper 44.M9.5, April 1990.

[616] F. Casajus-Quiros, L. A. Hernandez-Gomez, and C. Garcia-Mateo, "Analysis and quantization procedures for a real-time implementation of a 4.8 Kb/s CELP coder," *Proc. IEEE ICASSP'90,* paper 14.S4a.2, April 1990.

[617] W. Y. Chan and A. Gersho, "High fidelity audio transform coding with vector quantization," *Proc. IEEE ICASSP'90,* paper 6A1.5, April 1990.

[618] J. H. Chen, "A robust low delay CELP speech coder at 16 kb/s," *Record of Globecom Conf.,* Nov. 1989. Also in "AT&T low delay code-excited linear predictive coder," contribution to CCITT Working Party T1Y1.2, doc. no. T1Y1.2/89-35, Jan. 1989.

[619] J. H. Chen, M. J. Melchner, and R. V. Cox, "Real-time implementation of a 16 kb/s low-delay CELP speech coder," *Proc. IEEE ICASSP'90,* paper 14.S4a.4, April 1990.

[620] P. A. Chou, "Conditional entropy-constrained vector quantization of linear predictive coefficients," *Proc. IEEE ICASSP'90,* paper 14.S4a.8, April 1990.

[621] R. A. Cohen and J. W. Woods, "Entropy constrained SBPVQ for image coding," *Proc. IEEE ICASSP'90,* paper 44.M9.3, April 1990.

[622] R. V. Cox, S. L. Gay, Y. Shoham, S. R. Quackenbush, N. Seshadri, and N. S. Jayant, "New directions in subband coding," *IEEE J. Selected Areas Commun.,* vol. SAC-6, pp. 391–409, Feb. 1988.

[623] V. Cuperman, R. Pettigrew, J.-H. Yao, A. Gersho, and J. Shynk, "Low delay vector excitation coding of speech at 16 Kbit/s," *Canadian Conf. Commun.,* pp. 396–400, Sept. 1989.

[624] V. Cuperman, A. Gersho, R. Pettigrew, J. Shynk, and J.-H. Yao, "Backward adaptation techniques for low delay vector excitation coding of speech," *Proc. Globecom,* pp. 1242–1246, Nov. 1989.

[625] V. Cuperman, A. Gersho, R. Pettigrew, J. Shynk, and J.-H. Yao, "Backward adaptive and analysis-by-synthesis configurations for low-delay speech coding," in *Advances in Speech Coding,* B. S. Atal, V. Cuperman, and A. Gersho, Eds., Norwell, MA: Kluwer Academic Publishers, 1990.

[626] G. Davidson and A. Gersho, "Vector excitation or audio coder for transmission or storage," U.S. Patent No. 4,868,867, Sept. 1989.

[627] R. Drogo De Jacovo, R. Montagna, and D. Sereno, "Vector quantization and perceptual criteria in SVD CELP coders," *Proc. IEEE ICASSP'90,* paper 5.S1.9, April 1990.

[628] Y. Ephrahim, D. Malah, and B.-H. Juang, "On the application of hidden Markov models for enhancing noisy speech," *IEEE Trans. Acoust., Speech, Signal Processing,* vol. ASSP-37, pp. 1846–1856, Dec. 1989.

[629] Y. Feng and N. Nasrabadi, "A dynamic finite-state vector quantizer scheme," *Proc. IEEE ICASSP'90,* paper 44.M9.1, April 1990.

[630] T. R. Fischer, J. D. Gibson, and K. Koo, "Estimation and noisy source coding," *IEEE Trans. Acoust., Speech, Signal Processing,* vol. ASSP-38, pp. 23–34, Jan. 1990.

[631] N. Fournier and Y. Grenier, "A real-time vector excited adaptive predictive coder at 9.6 kbits/s," *Proc. IEEE ICASSP'90,* paper 14.S4a.1, April 1990.

[632] E. Frangoulis, "Vector quantization of HMM distributions: further results," *Proc. IEEE ICASSP'90,* paper 34.S10.18, April 1990.

[633] S. Furui, "Unsupervised speaker adaptation based on hierarchical spectral clustering," *IEEE Trans. Acoust., Speech, Signal Processing,* vol. ASSP-37, pp. 1923–1930, Dec. 1989.

[634] C. Galand and J. Menez, "A 2 ms delay adaptive code excited linear predictive coder," *Proc. IEEE ICASSP'90,* paper 33.S9.2, April 1990.

[635] C. Garcia-Mateo, F. Casajus-Quiros, and L. A. Hernandez-Gomez, "Multi-band vector excitation coding of speech at 4.8 Kbps," *Proc. IEEE ICASSP'90,* paper 5.S1.4, April 1990.

[636] A. Gersho, "Adaptive quantization," *Proc. Seminar on Quantization and Adaptivity,* pp. 41–49, May 1985, Nice, France.

[637] A. Gersho, "Principles of quantization," in *Quantization,* P. F. Swaszek, Ed. New York: Van Nostrand Reinhold Co., pp. 7–13, 1985.

[638] A. Gersho and W. Y. Chan, "A compact hardware speech codec for NASA's mobile satellite experiment," *Proc. Speech Tech'88,* New York, April 1988.

[639] A. Gersho, W. Y. Chan, G. Davidson, J. H. Chen, and

M. Yong, Speech coding at 4800 bps for mobile satellite commun.,'' *Proc. Mobile Satellite Conf.*, pp. 485–490, May 1988, Pasadena, CA.

[640] A. Gersho and V. Cuperman, ''A high level description of the consortium for speech coding's low delay vector excitation coder (LD-VXC),'' *Contribution to CCITT Study Group XV,* March 1989.

[641] I. A. Gerson and M. A. Jasiuk, ''Vector sum excite linear prediction (VSELP),'' *Proc. IEEE Workshop on Speech Coding for Telecom.*, Sept. 1989.

[642] I. Gerson and M. A. Jasiuk, ''Vector sum excited linear prediction (VSELP) speech coding at 8 KBS,'' *Proc. IEEE ICASSP'90,* paper 33.S9.3, April 1990.

[643] R. M. Gray, ''Three-searched block source codes,'' *Proc. 18th Annu. Allerton Conf.,* 1980.

[644] R. M. Gray and E. A. Riskin, ''Tree-structured vector quantization for progressive transmission image coding,'' in *Advances in Commun. and Signal Processing, Lecture Notes in Control and Inform. Sciences,* vol. 129, pp. 142–150, Berlin, Heidelberg: Springer-Verlag, 1989.

[645] R. M. Gray, *Source Coding Theory,* Norwell, MA: Kluwer Academic Publishers, 1990.

[646] D. D. Guisto and G. Vernazza, ''Color image coding by an advanced vector quantizer,'' *Proc. IEEE ICASSP'90,* paper 44.M9.2, April 1990.

[647] R. Hagen and P. Hedelin, ''Low bit-rate spectral coding in CELP, a new method,'' *Proc. IEEE ICASSP'90,* paper 14.S4a.6, April 1990.

[648] Y. Hirata and S. Nakegawa, ''Comparison among time-delay neural networks, LVQ2, discrete observation HMM and continuous observation HMM,'' *Proc. IEEE ICASSP'90,* paper 35.S10.6, April 1990.

[649] Y.-S. Ho and A. Gersho, ''Variable-rate contour-based interpolative vector quantization for image coding,'' *Proc. Globecom,* pp. 1890–1893, Nov. 1988.

[650] Y.-S. Ho and A. Gersho, ''A pyramidal image coder using contour-based interpolative vector quantization,'' *Proc. Int. SPIE Conf. Visual Commun. and Image Processing,* vol. 1199, pp. 733–740, Nov. 1989.

[651] Y.-S. Ho and A. Gersho, ''Contour-based postprocessing of encoded images,'' *Proc. Int. SPIE Conf. Visual Commun. and Image Processing,* vol. 1199, pp. 1440–1449, Nov. 1989.

[652] Y.-S. Ho and A. Gersho, ''A variable rate image coding scheme using vector quantization and clustering interpolation,'' *Proc. Globecom,* pp. 898–902, Nov. 1989.

[653] S.-S. Huang and R. M. Gray, ''Spellmode recognition based on vector quantization,'' *Speech Commun.,* vol. 7, pp. 41–53, March 1988.

[654] H. Iwamida, S. Katagiri, E. McDermott, and Y. Tohkura, ''A hybrid speech recognition system using HMMs with an LVQ-trained codebook,'' *Proc. IEEE ICASSP'90,* paper 35.S10.1, April 1990.

[655] N. S. Jayant, ''High-quality coding of telephone speech and wideband audio,'' *IEEE Commun. Mag.,* Jan. 1990.

[656] D. G. Kimber, M. A. Bush, and G. N. Tajchman, ''Speaker-independent vowel classification using hidden Markov models and LVQ2,'' *Proc. IEEE ICASSP'90*, paper 35.S10.3, April 1990.

[657] W. B. Kleijn, D. J. Krasinski, and R. H. Ketchum, ''Analysis and improvement of the vector quantization in CELP,'' *Proc. MSAT Conf.,* Jet Propulsion Labs, May 1988, Pasadena, CA.

[658] W. B. Kleijn, ''Source-dependent channel coding for CELP,'' *Proc. IEEE ICASSP'90,* paper 5.S1.1, April 1990.

[659] K.-F. Lee, H.-W. Hon, and R. Reddy, ''An overview of the SPHINX speech recognition system,'' *IEEE Trans. Acoust., Speech, Signal Processing,* vol. ASSP-38, pp. 35–45, Jan. 1990.

[660] T. Lookabaugh, E. A. Riskin, P. A. Chou, and R. M. Gray, ''Variable rate vector quantization for speech, image, and video compression,'' *IEEE Trans. Commun.* (in press).

[661] K. Mano and T. Moriya, ''4.8 kbit/s delayed decision CELP coder using tree coding,'' *Proc. IEEE ICASSP'90,* paper 5.S1.6, April 1990.

[662] P. Maragos and K. L. Young, ''Fractal excitation signals for CELP speech coders,'' *Proc. IEEE ICASSP'90,* paper 46.S12.8, April 1990.

[663] M. W. Marcellin, T. R. Fischer, and J. D. Gibson, ''Predictive trellis coded quantization of speech,'' *IEEE Trans. Acoust., Speech, Signal Processing,* vol. ASSP-38, pp. 46–55, Jan. 1990.

[664] M. W. Marcellin and T. R. Fischer, ''Trellis coded quantization of memoryless and Gauss-Markov Sources,'' *IEEE Trans. Commun.,* vol. COM-38, January 1990.

[665] J. McAuliffe, L. Atlas, and C. Rivera, ''A comparison of the LBG algorithm and Kohonen neural network paradigm for image vector quantization,'' *Proc. IEEE ICASSP'90*, paper 44.M9.9, April 1990.

[666] X. Michel, R. Leonardi, and A. Gersho, ''Unsupervised segmentation of texture images,'' *Proc. Int. SPIE Conf. Visual Commun. and Image Processing,* vol. 1001, Nov. 1988.

[667] P. Monet and C. Labit, ''Codebook replenishment in classified pruned tree-structured vector quantization of image sequences,'' *Proc. IEEE ICASSP'90,* paper 44.M9.7, April 1990.

[668] T. Moriya and M. Honda, ''Transform coding of speech using a weighted vector quantizer,'' *IEEE J. Selected Areas Commun.,* vol. SAC-6, pp. 425–431, Feb. 1988.

[669] M. Niranjan, ''CELP coding with adaptive output-error model identification,'' *Proc. IEEE ICASSP'90,* paper 14.S4a.3, April 1990.

[670] M. Nishiguchi, S. Gee, W.-Y. Chan, and A. Gersho, ''High fidelity audio coating using vector quantization,'' *1989 Workshop Applications Signal Processing to Audio and Acoust.,* Oct. 1989.

[671] N. Phamdo and N. Farvardin, ''Coding of speech LSP parameters using TSVQ with interblock noiseless cod-

ing,'' *Proc. IEEE ICASSP'90,* paper 14.S4a.7, April 1990.

[672] J. R. Ohm, ''Still image coding using predictive tree-VQ with subband decomposition,'' *Proc. IEEE ICASSP'90,* paper 44.M9.4, April 1990.

[673] M. Ostendorf and J. R. Rohlicek, ''Joint quantizer design and parameter estimation for hidden Markov models,'' *Proc. IEEE ICASSP'90,* paper 47S13.7, April 1990.

[674] K. Ozawa and I. Miyazaki, ''A hybrid speech coding based on multi-pulse and CELP at 3.2 kb/s,'' *Proc. IEEE ICASSP'90,* paper 46.S12.9, April 1990.

[675] C. I. Podilchuk and N. S. Jayant, ''Sparse-vector codebooks for the quantization of non-dominant subband in image coding,'' *Proc. IEEE ICASSP'90,* paper 30.M6.6, April 1990.

[676] T. F. Quateri and R. J. McAulay, ''Noise reduction using a soft-decision sine-wave vector quantizer,'' *Proc. IEEE ICASSP'90,* paper 55.S.15b.4, April 1990.

[677] Q. A. Qureshi and T. R. Fischer, ''A hardware processor for implementing the pyramid vector quantizer,'' *IEEE Trans. Acoust., Speech, Signal Processing,* vol. ASSP-37, pp. 1135–1142, April 1989.

[678] L. R. Rabiner, J. G. Wilpon, and B.-H. Juang, ''A segmental k-means training procedure for connected word recognition,'' *AT&T Tech. J.,* pp. 21–40, May–June 1986.

[679] X. Ran and N. Farvardin, ''Combined VQ/DCT coding of images using interblock noiseless coding,'' *Proc. IEEE ICASSP'90,* paper 44.M9.6, April 1990.

[680] E. A. Riskin, T. Lookabaugh, P. A. Chou, and R. M. Gray, ''Variable rate vector quantization for medical image compression,'' *IEEE Trans. Med. Imag.* (in press).

[681] Y. Shiraki and M. Honda, ''Speaker adaptation algorithms based on the piece-wise moving segment quantization method,'' *Proc. IEEE ICASSP'90,* paper 46.S12.5, April 1990.

[682] H. Y. Su, J.-P. Adoul, and C. Laflamme, ''On reducing computational complexity of codebook search in CELP coder through the use of algebraic codes,'' *Proc. IEEE ICASSP'90,* paper 14.S4a.3, April 1990.

[683] P. F. Swaszek, ''The effect of mismatch on polar coordinate quantizer,'' *Proc. Johns Hopkins CISS,* pp. 700–705, March 1985.

[684] P. F. Swaszek, ''Modified block vector quantizers,'' *Proc. Johns Hopkins CISS,* pp. 495–500, March 1989.

[685] T. Taniguchi, Y. Tanaka, A. Sasama, and Y. Ohta, ''Principal axis extracting vector excitation coding: High quality speech at 4.8 kb/s.'' *Proc. IEEE ICASSP'90,* paper 14.S4b.7, April 1990.

[686] T. D. Tran, V. J. Mathews, and C. K. Rushforth, ''A baseband residual vector quantization algorithm for voiceband data signals,'' *IEEE Trans. Commun.,* vol. COM-37, pp. 949–955, September 1989.

[687] I. M. Trancoso and C. Ribeiro, ''CELP: a candidate for GSM half-rate coding,'' *Proc. IEEE ICASSP'90,* paper 33.S9.5, April 1990.

[688] T. E. Tremain, D. J. Rahikka, V. C. Welch, and J. P. Campbell, Jr., ''CELP coding for land mobile radio applications,'' *Proc. IEEE ICASSP'90,* paper 33.S9.4, April 1990.

[689] K. Truong and R. M. Merserau, ''Hypercube structured codebooks for image vector quantization based upon the Kohonen self-organizing feature map algorithm,'' *Proc. IEEE ICASSP'90,* paper 44.M9.8, April 1990.

[690] J. Vaisey and A. Gersho, ''Variable rate image coding using quad-trees and vector quantization,'' *Proc. European Signal Processing Conf.,* pp. 1133–1136, Sept. 1988, Grenoble, France.

[691] S. Vaseghi, ''Finite state CELP for variable rate coding of speech,'' *Proc. IEEE ICASSP'90,* paper 5.S1.10, April 1990.

[692] L. Wang and M. Goldberg, ''Progressive image transmission using vector quantization in pyramid form,'' *IEEE Trans. Commun.,* vol. COM-37, pp. 1339–1349, Dec. 1989.

[693] S. Wang and A. Gersho, ''Phonetic segmentation for low rate speech coding,'' in *Advances in Speech Coding,* B. S. Atal, V. Cuperman, and A. Gersho, Eds., Norwell, MA: Kluwer Academic Publishers, 1990.

[694] P. Wilhelm, D. R. Haynor, Y. Kim, A. C. Nelson, and E. A. Riskin, ''Lossy image compression for digital medical imaging systems,'' *Proc. Int. SPIE Conf. Med. Imag. IV,* vol. 1232, Feb. 1990.

[695] D. Wolf, M. Falkhausen, and S. A. Euler, ''Improved training and recognition algorithms with VQ-based hidden Markov models,'' *Proc. IEEE ICASSP'90,* paper 35.S10.16, April 1990.

[696] E. Yair, K. Zeger, and A. Gersho, ''Conjugate gradient methods for designing vector quantizers,'' *Proc. IEEE ICASSP'90,* paper 14.S4b.8, April 1990.

[697] Y. Yamada, S. Tazaki, and R. M. Gray, ''Asymptotic performance of block quantizers with difference distortion measures,'' *IEEE Trans. Inform. Theory,* vol. IT-26, pp. 6–14, Jan. 1980.

[698] Y. Yamada, S. Tazaki, M. Kasahara, and T. Namekawa, ''Variance mismatch of vector quantizers, *IEEE Trans. Inform. Theory,* vol. IT-30, pp. 104–107, Jan. 1984.

[699] K. Yamasaki, Y. Asakawa, and A. Ichikawa, ''A 2.4 kbps speech coding based on fuzzy vector quantization,'' *Proc. IEEE ICASSP'90,* paper 46.S12.9, April 1990.

[700] M. Yong and A. Gersho, ''Vector excitation coding with dynamic bit allocation,'' *Proc. IEEE Globecom'88,* pp. 290–294, Dec. 1988.

[701] M. Yong and A. Gersho, ''Efficient encoding of the long-term predictor in vector excitation coders,'' in *Advances in Speech Coding,* B. S. Atal, V. Cuperman, and A. Gersho, Eds., Norwell, MA: Kluwer Academic Publishers, 1990.

[702] G. Yu, W. Russel, and R. Schwartz, "Discriminant analysis and supervised vector quantization," *Proc. IEEE ICASSP'90,* paper 47.S13.2, April 1990.

[703] K. Zeger and A. Gersho, "Vector quantizer design for noisy channels," *Proc. Int. Commun. Conf.*, pp. 1593–1597, June 1988.

[704] K. Zeger and A. Gersho, "A stochastic relaxation algorithm for improved codebook design," *Electronics Letters,* vol. 25, no. 14, pp. 896–898, July 1989.

[705] K. Zeger and A. Gersho, "Pseudo-Gray coding," *IEEE Trans. Commun.* (in press).

Authors' Index to Bibliography

A

Abdelwahab, A. A., 1–2
Abdul-Jabbar, A., 428
Abe, M., 3, 474
Abut, H., 4–9, 63, 237, 402, 541, 546–548
Adlersberg, S., 10
Adoul, J.-P., 11–21, 309, 336, 682
Ahmad-Fadzil, M. H., 609
Aikawa, K., 22
Aizawa, K., 23–28, 366, 389, 466–467, 469, 471
Akamine, M., 29, 611
Akansu, A. N., 30, 610
Akiyoshi, I., 472
Aktas, A., 31
Algazi, R. V., 338
Almeida, L. B., 560
Alvarez, L., 32
Anderton, D. O., 33
Ang, A. H., 34
Antonini, M., 612
Aoki, N., 488
Applebaum, T. H., 35
Appleby, D. G., 475
Aravind, R., 36–38, 149
Ariki, Y., 39
Asai, K., 370–380
Asakawa, Y., 40, 699
Atal, B. S., 41–42, 305–306, 479, 613
Atlas, L., 665
Atungsiri, S. A., 614
Averbuch, A., 43
Ayanoglu, E., 44–45

B

Bage, M. J., 46
Bahl, L. R., 43, 47–51, 385
Baker, R. L., 52–55, 81, 157, 483
Bakis, R., 43, 49, 51, 386
Baraniecki, M., 190
Barlaud, M., 612
Barna, G., 289
Barnes, E. S., 56–57
Barnwell III, T. P., 557
Barrilleaux, J., 58
Barry, C., 482
Bei, C.-D., 59
Bellagarda, J., 51
Bennett, L. A. M., 160, 339
Bergh, A. F., 60, 405
Bergstrom, A., 61
Bertorello, L., 62, 136, 138
Bhatt, S., 550–552
Bheda, H., 63, 546
Biemond, J., 570, 584–585
Billi, R., 64, 411
Bingol, S., 65
Blain, M. E., 66, 183
Blankenship, P. E., 407
Blankenau, S. J., 477
Boekee, D. E., 584–585
Bohlmann, H., 548
Bonneau, H., 67
Booman, F., 570
Borkowski, D., 358
Bos, A., 68
Bottau, F., 362
Boucher, P., 69–70, 222
Bourlard, H., 71–72
Boxeman, J. L., 615
Brandt, A. v., 242, 256
Breeuwer, M., 73
Brehm, H., 74–75
Brown, P. F., 43, 48, 50–51
Bruder, J., 281
Buck, J. T., 76–77, 83–84, 87, 267, 499
Bucklew, J. A., 78–80, 199, 308
Budge, S., 81–82
Burg, J. P., 437
Burshtein, D., 51
Burton, D. K., 76, 83–88, 267, 497–499
Bush, M., 89–90, 656
Buzo, A., 91–95, 229, 232, 248, 325

C

Campbell, Jr., J. P., 96, 688
Cappello, P., 97, 153
Caravero, M., 98
Carmody, J., 450
Casajus-Quiros, F. J., 201, 249–251, 616, 635
Cellario, L., 99
Cetin, A. E., 100
Chabries, D. M., 82, 195, 436
Chan, W. Y., 101, 617, 638, 639, 670
Chang, P.-C., 102–104
Chapman, R., 347
Chase, D., 105
Chen, C. H., 106
Chen, J.-H., 101, 107–112, 266, 557, 618–619, 639
Chen, T. C., 113
Cheng, D. Y., 114–115, 591
Cheung, Y. S., 116
Chevion, D., 574
Chow, P. A., 117–118, 445, 620, 660, 680
Chow, Y. L., 119–120, 307, 481–482
Chrisley, R., 289
Christiansen, R. W., 82, 195, 436
Chung, J. H., 121
Ciaramella, A., 122
Clark, W., 175
Class, F., 123
Clements, M. A., 169–170
Cohen, G., 334–335
Cohen, P. S., 385
Cohen, R. A., 124, 621
Cole, A. G., 43, 385
Collin, C., 11
Combescure, P., 319
Conte, G., 398
Conway, J. H., 68, 125–130

Copperi, M., 62, 131–141
Cox, R. V., 42, 142, 252, 511–512, 619, 622
Crochiere, R. E., 252
Cuperman, V., 10, 143–146, 208–209, 613, 623–625, 640

D

Daggett, G., 43
Dalle, D., 11, 13
Daly, E. M., 147, 446
Das, S., 43, 47, 51
Davidson, G., 97, 101, 111, 148–153, 557, 601, 626
Davies, K., 43
Debray, J.-P., 13
Delprat, M., 18, 21, 312
DeGennaro, S., 43
Delprat, M., 18, 21, 313
de Marca, J. R. B., 344
Dembo, A., 154
Deneberg, J. N., 504
Denker, T., 155
Deprettere, E. F., 303–304
Derr, A., 307
de Souza, P., 43, 47–51
Dezhgosha, K., 156
Dianysian, R., 157
Dicharry, R. M., 177–178
Didelot, F., 16
Dixit, S. S., 158
Dologlou, I., 159
Dooley, L. S., 160
Drogo de Jacovo, R., 161, 627
Du, Y., 227
Dunham, M. O., (*See* Ostendorf, M.)
Durrani, T. S., 34, 347
Dzuimbla, J., 155, 163, 587

E

Eguchi, H., 474
El Gamal, A. A., 164
Ephraim, Y., 165–166, 628
Epstein, E., 43
Equitz, W., 167–168
Ergezinger, S., 7
Evans, B. G., 290–292, 614
Evans, W. A., 160

F

Falkhausen, M., 695
Farges, E. P., 169–170
Farwrdin, N., 671, 679
Fehn, H. G., 171
Feng, M.-W., 172, 173, 307, 482
Feng, Y., 158, 174, 394, 629
Ferraris, G., 99
Fette, B., 175
Finnamore, W. A., 176
Fischer, T. R., 66, 177–184, 218, 341, 414, 564, 630, 663–664, 677
Fissore, L., 185
Fjallbrant, T., 186–188
Flanagan, J. K., 189–436
Fong, C., 190
Foster, J., 191
Fournier, N., 631
Fraleigh, D., 43
Frangoulis, E., 192, 632
Frank, W., 193
Frost, R., 189
Fujii, T., 194
Fujita, K., 596
Furner, R. R., 195
Furui, S., 196–198, 518–520, 633

G

Gadjali, H., 551
Galand, C., 362, 634
Gallagher, Jr., N. C., 80, 199, 443–444
Ganesan, K., 200, 594
Garcia, N., 32
Garcia-Gomez, R., 201, 249, 250
Garcia-Mateo, C., 251, 616, 635
Gauvain, J. L., 67
Gay, S. L., 142, 662
Gee, S., 670
Gersho, A., 36–38, 97, 101, 105, 107–112, 114–115, 143–144, 148–153, 159, 202–215, 254–255, 301–302, 430–435, 490–491, 493, 557, 568–569, 581, 601–602, 606–607, 613, 617, 623–626, 636–640, 649–652, 666, 670, 690, 693, 696, 700–701, 703–705
Gibson, J. D., 216–218, 266, 476, 593, 630, 663
Gish, H., 340
Godlewski, P., 334–335
Gold, B., 219–220
Goldberg, M., 69, 70, 221–226, 478, 521–526, 528, 575–580, 692
Gopalakrishnan, P. S., 51
Gotze, M., 227
Granzon, W., 228
Gray, Jr., A. H., 91–94, 229–230, 232–233, 268–269, 589–590
Gray, R. M., 4–5, 44, 52–53, 59, 91–94, 102–104, 117–118, 162, 166, 191, 215, 229–238, 259, 325, 332, 353, 437, 445, 446, 460–461, 496, 516, 540–541, 543, 560–563, 643–645, 653, 660, 680, 697
Grenier, R., 631
Guglielmo, M., 398
Guisto, D. D., 646
Gupta, V. N., 239
Guren, H. C., 512

H

Hagen, R., 647
Hamabe, R., 240
Hammer, B., 241–242
Hanazawa, T., 573
Hang, H.-M., 243–244
Hanson, B. A., 35
Hanzo, L., 475
Haoui, A., 245–246
Harashima, H., 23–28, 366, 389, 464–474
harris, f. j., 247
Haskell, B. G., 244
Haynor, D. R., 694
He, N., 248
Hedelin, P., 61, 647
Hemachandra, L. A., 165
Henry, C., 455
Herbert, M., 75
Hernandez-Gomez, L., 201, 249–251, 616, 635
Heron, C. D., 252
Higgins, A., 253
Hinkle, R., 58
Hinton, G., 573
Hirata, I., 648
Ho, Y.-S., 254–255, 649–652
Hoge, H., 31
Holzlwimmer, H., 256
Hon, H.-W., 315–317, 359
Honda, M., 367–368, 489, 668, 681

Hosaka, N., 488
Hsing, T. R., 147, 257
Hu, Z., 299–300, 328
Huang, J. J. Y., 258
Huang, S.-S., 259, 653
Huang, X. D., 260
Hudson, C., 358
Hui, L., 261
Hwang, M.-Y., 316

I

Iai, S., 262
Ichigawa, A., 40, 699
Ingle, V. K., 604
Ireton, M. A., 263
Irie, K., 262
Ishige, S., 343
Itoh, A., 373–380
Iwamida, H., 654

J

Jack, M. A., 39, 260
Jamali, M., 156
Jasiuk, M. A., 641–642
Jaskie, C., 175
Jayant, N. S., 142, 264–265, 344, 429, 463, 511–512, 622, 655, 675
Jelinek, F., 43, 47, 51, 385
Jeong, D. G., 266
Johnson, R., 267
Juang, B.-H., 165, 268–274, 313, 421, 424, 508–509, 513, 589–592, 628, 678

K

Kadur, M. S., 30, 610
Kaltenmeier, A., 123
Kameyama, M., 381
Kamizawa, K., 381–382
Kamp, Y., 72
Kaneko, T., 583
Kanevski, D., 51
Karnin, E., 235, 574
Kasahara, M., 698
Katagiri, S., 275, 360, 654
Kato, Y., 276
Katterfeldt, H., 123
Katz, S., 43, 47
Kaul, S., 277
Kawabata, T., 278
Kemp, D. P., 279
Ketchum, R. H., 286, 657
Khorchidian, M., 351
Kieffer, J. C., 231, 280
Kim, C.-S., 281–282
Kim, D. S., 318
Kim, J. W., 283
Kim, T., 284, 463
Kim, Y., 694
Kimball, O., 119–120, 307, 481–482
Kimber, D. G., 656
King, R. A., 285, 391, 393
Kinjo, N., 382
Kleijn, W. B., 286, 657, 658
Kobayashi, T., 487
Kohonen, T., 287–289
Kok, C., 97
Komatsu, T., 465, 470, 473–474
Kondoz, A. M., 290–292, 614
Koo, B., 218
Koo, J. M., 293
Koo, K., 630
Kopec, G. E., 89–90, 294–297
Kou, W., 186, 188, 298–300
Koyama, H., 301–302
Krasinski, D. J., 286, 657
Krasner, M., 119–120, 481
Kroon, P., 42, 303–306
Ku, T., 537
Kubala, F., 119–120, 172–173, 308, 482
Kuhlmann, F., 248, 308
Kuwabara, H., 3
Kwatra, S. C., 1–2, 156, 544

L

Labit, C., 345, 667
Lacroix, A., 193, 480
Laface, P., 185
Laflamme, C., 682
Lamblin, C., 15, 21, 309, 319
Landell, B., 310
Lang, K., 573
Langlais, T., 311
Lauritzen, N., 89
Lecomte, I., 312
Lee, C.-H., 313, 424
Lee, D. T. L., 314
Lee, E. A., 565
Lee, H. J., 314, 615
Lee, K.-F., 315–317, 359, 485
Lee, L.-S., 554
Lee, S. U., 283, 318
Leguyader, A., 15, 319–320
Lelievre, L., 312
Lenning, M., 239
Leonardi, R., 666
Leung, S. T., 116
Lever, M., 312
Levinson, S. E., 270, 321–322, 415–419, 421–422
Lewis, B. L., 43, 385
Li, J., 323
Li, K. P., 324
Lin, S. E., 394
Linde, Y., 231, 234, 325, 516
Lippmann, R. P., 219–220, 326
List, P., 155, 327, 588
Liu, T.-M., 328–329, 558
Liu, Y. J., 330
Lloyd, S. P., 331
Lookabough, T., 117–118, 332, 445, 660, 680
Lowry, A., 332
Lu, L., 334–325
Luse, S. A., 6

M

Mabilleau, P., 14, 16, 21, 336
Mackay, K., 463
MacQueen, J., 337
Madiseti, A., 338
Mahajan, S., 316
Mahmoud, W. A., 160, 339
Makhoul, J., 119–120, 172–173, 307, 340, 452–453, 481–482
Malah, D., 154, 165, 628
Malone, K. T., 180–181, 183, 341
Malpas, M. L., 219

Manikopoulos, C. N., 323, 342, 527
Mano, K., 343, 661
Manolakis, D. G., 604
Maragos, P., 662
de Marca, J. R. B., 344
Marcellin, M. W., 663–664
Maresq, J. P., 345
Mariani, J., 346
Marino-Acebal, J. B., 348–349
Markel, J. D., 91–94, 229
Marlot, M., 200
Marques, J. S., 559
Marshall, S., 347
Martinez, H. G., 95
Masgrau-Gomez, E., 348–349
Masisara, K., 288
Mason, J., 311, 595
Massaloux, D., 309, 320
Mathews, V. J., 350–351, 686
Mathieu, P., 612
Matsumura, S., 352
Matsuyama, Y., 229–230, 353–357
May, J., 102, 104
Mazor, B., 358–359
McAulay, R. J., 676
McAuliffe, J., 665
McDermott, E., 275, 360, 654
McInnes, F. R., 39
Mehran, F., 8
Mehta, P., 200
Mekuira, F., 186–188
Melchner, M. J., 619
Menez, J., 362, 634
Mensa, G., 361, 457
Mercer, L. R., 43, 47–51, 385
Mermelstein, P., 239
Mersereau, R. M., 281–282, 689
Messerschmitt, D. G., 245–246
Micca, G., 185
Michel, X., 666
Miclet, L., 542, 572
Millar, W., 333
Mintz, E., 359
Miseki, K., 29, 611
Mitra, S., 582
Miyakawa, H., 23–27, 464–467, 470–471
Miyazaki, I., 674
Moayeri, N., 363–365
Monet, P., 667
Montagna, R., 161, 361, 457, 627
Montagner, M., 478
Morgera, S. D., 505–507
Morishima, S., 366
Moriya, T., 367–369, 468, 661, 668
Morrel, D. R., 189
Morrisette, S., 12, 16, 21, 309
Mouly, M., 319
Mukawa, N., 276
Murakami, T., 370–392
Murata, Y., 240
Murveit, H., 383
Mwangi, E., 384

N

Nadas, A., 43, 51, 385–387
Nahamoo, D., 43, 51, 386–388
Nakaide, H., 389
Nakamura, S., 3, 390
Nakegawa, S., 381, 648
Namekawa, T., 240, 698
Narasimhan, A., 539
Nasrabadi, N. M., 174, 285, 391–394, 629
Nelson, B. E., 189, 395
Neuhoff, D. L., 363–365
Ney, H., 71
Ngan, K. N., 261
Niranjan, M., 669
Nisiguchi, M., 670
Nishimura, M., 397
Noah, M. J., 397
Noll, P., 171, 228, 264

O

Oglesby, J., 595
Ohm, J. R., 672
Ohta, N., 194, 685
Oliveri, F., 399
Onal, A., 247
Ono, S., 399
O'Shaughnessy, D., 401
Ostendorf, M., (M. O. Dunham, also) 119–120, 162, 191, 673
Osterlinck, A., 566
Othman, S., 400
Ozawa, K., 399, 674
Ozturk, Y., 402

P

Paliwal, K. K., 403–404
Pan, K.-C., 405–406, 420
Parthasarathy, S., 547
Paul, D. B., 407–409
Pearlman, W. A., 324, 410
Perosino, F., 138, 161
Petit, J. P., 320
Picel, Z., 311
Picheny, M., 43, 47, 50–51, 386–387
Pieraccini, R., 98, 185, 411
Podilchuk, C. I., 675
Pollard, D., 412
Potu, B., 426
Price, P., 119–120, 307

Q

Quackenbush, S., 142, 413, 622
Quateri, T. F., 676
Qureshi, Q., 414, 677

R

Rabiner, L. R., 60, 270, 273–273, 321–322, 405–406, 415–424, 508–509, 513, 678
Rahikka, D. J., 688
Raina, J. P., 567
Ramamoorthy, P. A., 425–426
Ramamoorthy, V., 427–429
Ramamurthi, B., 114, 205–206, 430–435
Ramasubramanian, V., 404
Ramstad, T., 211
Ranieri, F., 98
Read, C. J., 189, 395, 436
Rebolledo, G., 4, 5, 230, 233, 437
Reddy, R., 316, 485
Reininger, H., 163, 327, 438–442, 587–588
Ribeiro, C., 687
Rines, K. D., 443–444

Riskin, E. A., 446–447, 644, 660, 680, 694
Riviera, C., 95, 665
Rodrigues, J. S., 559
Rohlicek, J., 307, 673
Rommelaere, J., 566
Rose, R. C., 557
Rosenberg, A., 447, 508–510, 513–514
Rosso, M., 362
Rost, M. C., 448–449, 476
Rothweiler, J., 330, 450–451
Roucos, S., 119–120, 301, 340, 452–456, 481
Rudko, M., 12
Rushforth, C. K., 33, 686
Rusina, F., 361, 457
Russel, W., 456, 702
Rutledge, C. W., 458–459

S

Sabin, M. J., 460–461, 565
Safranek, R. J., 463
Saito, T., 28, 464–474
Salami, R. A., 475
Salinas, J. L., 55
Sanchez, H., 553
Sankur, B., 65
Santos, D., 559
Saramaki, T., 288
Sasama, A., 685
Sayood, K., 217, 448–449, 476–477
Schafer, R. W., 121
Scharf, L. L., 347
Schielein, M., 241–242
Schmitt, F., 225, 478
Schroeder, M. R., 479
Schultheiss, M., 480
Schultheiss, P. M., 258
Schwartz, R., 119–120, 172–173, 307, 452–455, 481–482, 702
Sekey, A., 431
Sereno, D., 99, 132–140, 161, 627
Serralheiro, A., 559
Seshadri, N., 142, 622
Shen, H.-H., 54, 483
Shichman, G., 43
Shikano, K., 3, 22, 278, 390, 484–486, 519, 573
Shimada, T., 382
Shirai, K., 343, 487–488
Shiraki, Y., 489, 681
Shlien, S., 222
Shoham, Y., 114, 142, 210, 490–494, 622
Shore, J. E., 76, 83–84, 86–87, 233, 267, 495–599
Shperling, I., 164
Shridhar, M., 277
Shynk, J., 623–625
Singhal, S., 500
Slack, T. B., 504
Sloane, N. J. A., 56–57, 68, 125–130, 501–503
Sluyter, R. J., 304
Smith, A. R., 504
Smith, J. L., 9
Smith, M. T. J., 281–282
Soleymani, M. R., 505–507
Somayazolu, Y., 97
Sondhi, M. M., 270, 321–322, 415–419, 421
Soong, F. K., 60, 313, 405–406, 420, 423, 447, 508–514
Spinelli, P., 43
Sqama, H., 333
Sreenivas, T. V., 515
Stanhope, T., 148
Stark, W. E., 363
Steenaart, W., 65
Stewart, L. C., 516
Stockham, Jr., T. G., 82, 517
Su, H. Y., 682
Subramanian, R., 338
Suda, H., 369
Sueda, R. A., 279
Sugamura, N., 518–520
Sun, H.-F., 223–224, 226, 342, 521–528
Svendsen, T., 529
Swaszek, P., 530–539, 683–684

T

Tajchman, G. N., 656
Takeo, H., 464, 466–469, 471
Tan, C. C., 504
Tanaka, Y., 685
Taniguchi, T., 540, 685
Tao, B. P. M., 8–9, 541
Tassy, A., 312, 542
Taylor, J., 310
Tazaki, S., 543, 596–599, 697–698
Tengler, W., 256
Thede, L. D., 544
Thomas, J. B., 530–531
Thomson, D. L., 545
Thyagarajan, K. S., 63, 247, 546–553
Ting, P.-Y., 554
Tishby, N., 555
Tohkura, Y., 654
Toshioka, K., 556
Townes, S. A., 557
Tran, A., 558
Tran, T. D., 350, 425, 686
Trancoso, I. M., 559, 687
Tremain, T. E., 96, 279, 688
Tribolet, J. M., 559
Trottel, K., 74
Truong, K., 689
Tsao, C., 560–563
Tseng, C.-Y., 554
Tseng, H.-C., 564
Tseng, H.-P., 565
Tu, G., 566
Tugenberg, M., 175
Tzou, K.-H., 558

U

Un, C. K., 293, 603
Updikar, V. R., 567

V

Vaisey, D. J., 568–569, 690
Vallerdu, F., 348
Vandegrift, J., 307
Van Helden, H., 570
Vaseghi, S., 691
Veeneman, D., 359
Venuti, G., 122
Verbeek, G., 566
Vernazza, G., 646
Versvik, I., 211, 571
Vicard, D., 572
Viswanathan, M., 549
Viswanathan, V., 455

Vogel, E. B., 558
Volmary, C., 228

W

Waibel, A., 573
Waite, R. W., 350
Wakita, H., 35
Walach, E., 574
Wang, L., 575–580, 692
Wang, S., 581, 693
Wang, Y., 582
Watanabe, H., 276
Watanuki, O., 556, 583
Wei, V. K., 164
Weintraub, M., 383
Welch, V. C., 96, 688
Wellekens, C. J., 71–72
Wells, S., 58
Westerink, P. H., 584–585
Wilgus, A., 456
Wilhelm, P., 694
Wilpon, J. G., 272, 424, 678
Wilson, S. G., 401, 586
Wohlford, R. E., 253, 310, 504
Wolf, D., 155, 163, 327, 438–442, 587–588, 695
Wong, D. Y., 268, 589–592
Woo, H. C., 593
Woods, J. W., 124, 243, 585, 621
Wrench, Jr., E. H., 324
Wu, F. H., 594

X

Xie, Z., 517
Xu, L., 595
Xydeas, C. S., 263, 384

Y

Yair, E., 696
Yajima, S., 40
Yamada, Y., 240, 543, 596–599, 697–698
Yamasaki, K., 40, 699
Yamazaki, E., 370–371, 373
Yano, M., 214
Yao, J.-H., 623–625
Yazawa, J., 487
Yeh, C.-L., 600
Yip, W., 175
Yokota, M., 275
Yong, M., 151, 601–602, 639, 700–701
Youn, W. S., 603
Young, K. L., 662
Yu, G., 702
Yuan, X., 604

Z

Zador, P. L., 605
Zeger, K., 111, 607–608, 696, 703–705
Zhang, H.-J., 608
Zheng, P., 608
Zurcher, J. F., 319

Author Index

A

Abut, H., 258, 312, 348, 402, 485, 513
Aravind, R., 495
Atal, B. S., 320
Asai, K., 342, 364
Aizawa, K., 386

B

Baker, R. L., 333, 398, 525
Billi, R., 407
Bhatt, S., 381
Bohlmann, H., 402
Boucher, P. R., 368
Bourland, H., 457
Burton, D. K., 415, 473
Bush, M. A., 434
Buzo, A., 88, 411

C

Campbell, J. P., Jr., 328
Cappello, P. R., 499
Carmody, J., 521
Chen, J. H., 517
Cheng, D. Y., 233
Chou, P. A., 195
Chow, Y.-L., 461
Conway, J. H., 118, 134
Copperi, M., 324
Cox, R. V., 316
Cuperman, V., 2, 300

D

Davidson, G., 495, 499
Dianysian, R., 525
Dicharry, R. M., 170
Dunham, M. O., 140, 461

E

El-Gamal, A. A., 152
Ergezinger, S., 312
Equitz, W., 356

F

Fischer, T. R., 170, 175
Foster, J., 140
Furui, S., 465

G

Gersho, A., 2, 100, 108, 289, 300, 344, 495, 499, 517
Gibson, J. D., 160
Gish, H., 34
Goldberg, M., 368, 390
Gray, Jr., A. H., 208, 222
Gray, R. M., 9, 88, 140, 195, 208, 237, 258, 274, 333, 338, 485

H

Haoui, A., 297
Harashima, H., 386
Hemachandra, L. A., 152
Honda, M., 250

I

Itoh, A., 364

J

Jayant, N. S., 316, 352
Juang, B.-H., 222, 233, 438, 469

K

Kimball, O. A., 461
King, R. A., 72
Kopec, G. E., 434
Krasner, M. A., 461
Kroon, P., 320
Ku, T., 191
Kubula, G. F., 461

L

Levinson, S. E., 438
Linde, Y., 88
Lookabaugh, T., 195

M

Makhoul, J., 34, 246, 461
Martinez, H., 411
Messerschmitt, D. G., 297
Miyakawa, H., 386
Murakami, T., 364

N

Nasrabadi, N. M., 72
Ney, H., 457

P

Pan, K.-C., 442
Parthasarathy, S., 348
Price, P. J., 461

R

Rabiner, L. R., 438, 442, 469
Ramamoorthy, V., 352
Ramamurthi, B., 338, 344
Rebolledo, G., 208, 258
Reininger, H., 271
Rivera, C., 411
Rosenberg, A. E., 469
Rost, M. C., 160
Rothweiler, J., 521
Roucos, S., 34, 246, 461
Rutledge, C. W., 376

S

Sabin, M. J., 274
Sayood, K., 160
Schwartz, R. M., 246, 461
Sereno, D., 324
Shen, H.-H., 398
Shiraki, Y., 250
Shlien, S., 368
Shoham, Y., 289
Shore, J. E., 208, 415
Sloane, N. J. A., 118, 134
Smith, J. L., 513
Sondhi, M. M., 438
Soong, F. K., 316, 442, 469
Shperling, I., 152
Stanhope, T., 495
Sun, H., 390
Swaszek, P. F., 191

T

Tao, B. P. M., 485, 513
Thyagarajan, K. S., 348, 381, 402
Tremain, T. E., 328
Tsao, C., 237

W

Wei, V. K., 152
Welch, V. C., 328
Wellekens, C. J., 455
Wolf, D., 271
Wong, D. Y., 222, 233

Y

Yamazaki, E., 342

Subject Index

Acoustic pattern matchers, network-based connected digit recognition, 435
Adaptive bit allocation, subband speech coder and, 317–318
Adaptive coding, VQ and, 28
Adaptive discrete cosine transform coding
bit assignment, 388
color image coding, 388
complexity, 388
design method of vector quantizer, 387
experimental results, 388–389
structure of vector quantizer, 387
with VQ for color images, 386–389
Adaptive postfiltering, real-time vector APC speech coding with, 517–519
Adaptive transform coding using VQ, transform vector quantizer (TVQ), 77
Adaptive vector predictive coding, 7
Adaptive vector quantization
of color images, 372–373
image compression using, 368–374
of monochrome image, 369–372
Adaptive vector quantizer, 6–7, 23–24, 367
time-dependent vector quantization, 64–65
Adjacent points, codebook analysis, 114–115
Admissible/optimal polytopes, asymptotically optimal block quantization, 101–102
Algorithms
algorithm for quantizer design, 88–98
block quantizers, 88
classification algorithm for image vector quantization, 345
convergence of codebook design algorithm, 253–254
decoding algorithm, 134–138
design algorithm for VQ, 262
distortion measures, 88–89
efficient bit allocation, 291–295
for 800 bit/s vector quantization LPC vocoder, 226–227
encoding algorithm for PVQ, 178–180
fast nearest-neighbor search algorithms, 6
fast quantizing algorithm for, 134–138
FSVQ design algorithms, 144–146
guess by "splitting," 93–94
K-means algorithm for, 40
for linear predictive coding vector quantization, 226–227
Lloyd's example, 95–96
optimal quantization, 90–93
reproduction of alphabet and, 93
segment clustering, 247–248
simulated annealing algorithm, 154
speech example, 96–97
using generalized Lloyd algorithm, 237–245
vector quantization clustering algorithm, 356–362
for vector quantizer design, 160–167
Allocation problem, efficient bit allocation, 290
Arbitrary set of quantizers, efficient bit allocation, 289–296
Asymptotically optimal block quantization
admissible/optimal polytopes, 101–102
companding model of, 106
heuristic derivation of distortion integral, 103
infinite block length, 105–106
mathematical formula, 100–101, 104–105
minimization of distortion integral, 103–104
Asymptotic performance, pyramid vector quantizer (PVQ), 181–182
Asymptotic vector quantization, 49–51
experimental results, 51
space packing, 50–51

Backward adaptive prediction (APB), subband speech coder and, 317
Binary clustering, training and testing, 62
Binary codeword representation, pyramid vector quantizer (PVQ), 180–181
Binary search, codebook design, 56–58
Binary vector quantizer (BVQ), 81–83
block truncation coding (BTC) and, 82–83
facsimile coding and, 81–82
Bit allocation
adaptive bit allocation, 317–318
bit assignment, 388
CELP coder, 322
efficient bit allocation, 289–296
scalar vs. vector quantization of vector sources, 51–53, 56
Block quantizers, 88
BYBLOS
as BBN continuous speech recognition system, 461–464
grammatical knowledge source, 463
recognition performance, 463–464
recognition search strategy, 463
signal processing and analysis component, 462
system overview, 461–462
training/acquisition of phonetic coarticulation models, 462
word model generator, 462

Cascaded quantization
codebook design, 58–60
storage reduction, 59
Classified VQ, spatial vector quantizers (SVQ), 74–75
Codebook analysis
finding adjacent points, 114–115
remote hyperplanes, 114
vector quantizers, 114–115
Codebook design, 56–63
binary search, 56–58
cascaded quantization, 58–60
image coding and, 340–341
for image vector quantization, 346
product codes, 60–61
random codebooks, 61–62
rate-distortion speech coding, 216
training and testing, 62-63
VQ and, 39–40
Codebook generation
image sequence coding, 391
linear predictive coding matrix quantization, 235
VQ and, 391
Codebook replenishment VQ, spatial vector quantizers (SVQ), 75
Codebook robustness, training and testing, 62–63
Codebooks, 15–16
Codebook techniques, image sequence coding, 392–393
Code-excited linear predictive (CELP) coder
algorithm for, 328
bit allocation, 322
computational requirements, 329
error protection, 329–330
expandability, 328–329
for high-quality speech at 8 kbit/s, 324–326
long-term predictor, 321–322
performance characteristics, 330
postfiltering, 323
quantization procedures for, 320–323
short-term predictor, 320
system overview, 324–326
U.S. Standard 4800 BPS voice coder, 328–330
Coder design
vector distortion criteria for, 302
vector predictive coding system, 301–307
Codes, lattice quantizers used as, 134–138
Coding, vector quantizer design, 163–164
See also specific types of coding.
Coding quantizer output, rate-distribution theory, 44–45
Coding strategy, for image coding, 353–354
Color image
adaptive discrete cosine transform coding, 386–389
bit assignment, 388
color image coding, 388
complexity, 388
design method of vector quantizer, 387
experimental results, 388–389
image compression of, 372–373
structure of vector quantizer, 387
VQ and, 364–367
Color image VQ, 364–367
adaptive vector quantizer, 367
coder performance, 367
for color still pictures, 364–366
combined component VQ, 365–366
dynamic vector quantizer, 366–367
individual component VQ, 365
mean separated and normalized VQ, 364
multi-stage connection of DVQ/AVQ, 367
preprocessing color video signals, 366
Color still pictures, VQ and, 364–366
Combined component VQ, 365–366
Companding model, asymptotically optimal block quantization, 106
Complexity in pattern matching
reduction techniques, 4–6
VQ and, 3–4
Compression
image coding, 9
speech coding, 9
Compression system, discrete utterance recognition, 411–412
Computational considerations, discrete utterance recognition, 429–431
Computational costs, of VQ, 40–41
Computational requirements, code-excited linear predictive (CELP) coder, 329
Connected digit recognition
clustering, 457–458
deterministic single level recognition method, 458–459
recognition principle, 457
statistical models, 460
stochastic models, 459–460
using VQ, 457–460
Constant weight codes, good code design, 156
Constrained-entropy quantization, scalar quantization, 48–49
Costs
computational costs, 40–41
storage costs, 40–41

Decoding algorithm, for lattice quantizers, 134–138
Dependency characteristics, of VQ, 41
Deterministic single level recognition method, connected digit recognition, 458–459
Diagnostic acceptability measure (DAM), LPC vector quantization and, 229–231
Diagnostic rhyme test (DRT), LPC vector quantization and, 228–229, 236
Dimensionality, scalar vs. vector quantization of vector sources, 56
Discrete utterance recognition approach to, 416–419
compression system, 411–412
computational considerations, 429–431
multiple-speaker experiments, 428–429
pattern selection, 412–413
principles of, 415–416
recognition system, 413–414
source coding techniques for, 411–414
speaker-dependent performance on large data base, 419–422
speaker-dependent studies, 422–428
speaker-independent experiments, 428–429
without time alignment, 415–432
Distance measure, matrix quantizer, 349
Distortion, memoryless vector quantizers, 11–13
Distortion measures, 88–89
linear predictive coding matrix quantization, 235
linear predictive speech coding, 251–252
scalar vs. vector quantization of vector sources, 56
VQ and, 38–39
Distortion weighting, vector rotation, 54
Dual lattice, 134–135
Dynamic vector quantizer, 366–367

Edge classification
in image coding, 403
in image vector quantization, 345
Edge/shade classification, in image vector quantization, 344–345

Edge/shade segmented codebooks, image coding, 339–340
Efficient bit allocation
algorithm development for, 291–294
algorithmic variations, 294–295
allocation problem, 290
for arbitrary set of quantizers, 289–296
initial Lagrange multiplier, 295–296
main theorem for, 290–291
800 bit/s vector quantization LPC vocoder
algorithm complexity for, 226–227
DAM evaluation of, 229–231
DRT evaluation of, 228–229
gain quantization, 227
implementation of, 225–227
improved quantization, 222–223
parameter set reduction, 222
pattern matching, 226
pitch and voice coding, 227–228
theory/design of, 223–225
time redundancy reduction, 222
tree search, 225–226
unvoiced codebook experiment, 226
Embedded speech coding
multistage coders, 299
scalar embedded coders, 297
simulation results, 299
tree-expanded codebooks, 298–299
vector quantization approach, 297–299
vector quantizers, 297
VQ embedded coders, 297–298
Error protection, code-excited linear predictive (CELP) coder, 329–330
Expandability, code-excited linear predictive (CELP) coder, 328–329

Fast nearest-neighbor search algorithms, 6
Fast quantizing algorithm, for lattice quantizers, 134–138
Feedback vector quantizers, 19–22
FSVQ, 21–22
product/multistep FVQ, 21
vector predictive quantization, 21
Finite state/frame difference interpolative vector quantizer
components of, 398–399
designs for, 399–400
experimental results, 400–401
intra/inter frame decomposition, 398
for low-rate image sequence coding, 398–401
Finite-state vector quantizer (FSVQ), 21–22
FSVQ design algorithms, 144–146
Gauss-Markov sources, 146–148
labeled-state FSVQs, 143–144
labeled-transition FSVQs, 143–144
quantizer and, 142–143
sampled speech, 148–149
simulations of, 146–149
tracking finite-state source codes, 142
for waveform coding, 140–149
First layer of lattice, Voronoi regions and, 131–132
Frame repeat LPC vector quantization, linear predictive coding vector quantization, 233–234

Gain quantization, 800 bit/s vector quantization LPC vocoder, 227
Gain-shape vector quantizers, 5–6
memoryless vector quantizers, 18–19
Gauss-Markov sources
finite-state vector quantizer (FSVQ), 146–148
VQ and, 25–26
Good code design, 152–154
constant weight codes, 156
implementation issues, 157
simulated annealing algorithm for, 154
simulated annealing used for, 152–158
source codes, 155
spherical codes, 156–157
Grammatical knowledge source, BYBLOS, 463
Guess by "splitting," 93–94

Hardware realization
computer simulation results, 487–489
single codebook system, 489–492
two-codebook VQ SYSTEM DESIGN, 492–493
of waveform vector quantizers, 486–487
Heuristic derivation of distortion integral, asymptotically optimal block quantization, 103
Hidden Markov models (HMM)
continuous mixture density HMM, 438
diagonal vs. full covariance matrices, 440
energy/duration effects, 440
experimental evaluations, 440
internal vs. post-processor durational models, 441
mixture effects, 440
previous recognizers and, 441
recognizer duration, 439–440
speaker-independent isolated word recognition, 438–441
training HMM, 438–439
training set effects, 441
word clustering and, 440
Hierarchical VQ, spatial vector quantizers (SVQ), 75
High quality speech at 8 kbit/s, CELP coder for, 324–326
Human visual model
image coding, 352–354
matrix quantizer, 348–350

Image coding, 9
adaptive discrete cosine transform coding, 386–389
binary vector quantizer (BVQ), 81–83
coding strategy, 353–354
of color images, 364–367
edge classification, 403
edge/shade segmented codebooks, 339–340
experimental procedure, 403–404
finite state/frame difference interpolative vector quantizer, 398–401
human visual system model, 352–354
image compression, 333–336, 368–374
image sequence coding, 390–397
image vector quantization, 344–347
linear block prediction, 382–384
matrix quantizer, 348–350
methods for, 338–339
model-testing finite-state vector quantizer and, 352–353
predictive vector quantizer (PVQ), 76
region merging, 403
segmentation using region growing and, 402–404
spatial vector quantizers (SVQ), 73–76
subband vector quantizers (subband VQ), 83
training set and codebook design, 340–341
transform vector quantizer (TVQ), 77–81
using VQ, 338–341
vector quantization clustering algorithm, 356–362
vector quantizer of video signals, 342–343
VQ and, 28–29, 72–83, 364–367, 390–397
Image compression
coders used for, 334–335
of color images, 372–373
enhancements of, 336
experimental results, 335
of monochrome image, 369–372
performance characteristics, 335
using adaptive vector quantization, 368–374
using non-adaptive spatial VQ, 333–336
Image encoding applications, linear block prediction for, 382–384
Image sequence coding
codebook generation, 391
codebook techniques, 392–393
quantizer process, 391, 393–394
replenishment techniques, 391–397
training set generation, 391
using VQ, 390–397
vector formation, 391
Image/speech compression, for VQ, 9
Image transform coding application, vector quantizer design, 164–167
Image vector quantization
classification algorithm, 345
codebook design for, 346
edge-oriented classification in, 345
edge/shade classification in, 344–345
experimental results, 346
with perceptually-based cell classifier, 344–347
Imbedded VQ methods, 6–7
Implementation issues, good code design, 157
Individual component VQ, 365
Infinite block length, asymptotically optimal block quantization, 105–106
Initial Lagrange multiplier, efficient bit allocation, 295–296
Integer lattice, finding closest point of, 135–136
Interframe transform vector quantization (ITVQ), transform vector quantizer (TVQ), 79–81
Interframe VQ, spatial vector quantizers (SVQ), 75–76
Intra/inter frame decomposition, finite state/frame difference interpolative vector quantizer, 398

Joint segmentation and quantization, linear predictive speech coding, 252, 254–255

K-means initialization, training and testing, 62

Labeled-state FSVQs, 143–144
Labeled-transition FSVQs, 143–144
Laplacian source
geometric structure of, 176–177
memoryless source, 170–174
Large vocabulary word recognition
feature extraction and database, 465
preprocessing/DTW combination and, 468
preprocessing experiment using universal codebook, 467–468
preprocessing experiment using word-based codebooks, 465–466
VQ-based preprocessor using cepstral dynamic features, 465–468
Lattice quantizers, 122–133
comparison of, 132–133
decoding algorithm for, 134–138
fast quantizing algorithm for, 134–138
lattice quantizer design, 161–163
mean-square error of, 122–132
memoryless vector quantizers, 19
used as codes, 134–138
vector quantizers, 115–116, 161–163
Lattices, 122–133
first layer and Voronoi region, 131–132
lattice quantizers, 122–133
mean square error of, 122–132
root lattices, 123–125
Voronoi regions of, 122–133
Linear block prediction, 382–383
experimental results, 384
for image encoding applications, 382–384
with source classification, 383–384
vector DCPM, 383
Linear predictive coding (LPC)
analysis of for rate-distortion speech coding, 211–213
coding vector quantization, theory/design of, 223–225
implementation of, 225–227
tree search, 225–226
unvoiced codebook experiment, 226
Linear predictive coding matrix quantization
codebook generation, 235
distortion measure, 235
quantizer simulation, 235–236
speech segmentation, 234–235
Linear predictive coding vector quantization
algorithm complexity for, 226–227
DRT evaluation of, 236
frame repeat LPC vector quantization, 233–234
LPC matrix quantization, 234–236
pattern matching, 226
very low data rate speech compression with, 233–236
Linear predictive speech coding
convergence of codebook design algorithm, 253–254
distortion measure, 251–252
experimental results, 253–256
joint segmentation and quantization, 252, 254–255
subjective listening test, 255–256
variable-length segment quantization and, 250–257
variable-length SQ codebook design, 252–253
Lloyd algorithm, for memoryless vector quantizers, 15
Lloyd-Max quantization, scalar quantization, 47–48
Lloyd's example, 95–96
Long-term predictor, CELP coder, 321–322
Low-rate image sequence coding, finite state/frame difference interpolative vector quantizer, 398–401

Low-rate speech encoding
octave band SBC-VQ systems, 313
residually excited subband with classifier, 314
residual SBC-VQ systems with classifier, 314
using VQ and subband coding, 312–314

Markov source models
finite-state vector quantizer (FSVQ), 146–148
speech recognition and, 407–409
See also Hidden Markov models.
Matrix quantizer
distance measure, 349
experimental procedure, 349–350
with human visual model, 348–350
Matrix quantizer design for LPC speech
experimental results, 238–243
matrix quantization, 237–238
using generalized Lloyd algorithm, 237–245
Mean separated and normalized VQ, VQ and, 364
Mean/shape VQ, spatial vector quantizers (SVQ), 73–74
Mean-square error, of lattice quantizers, 122–132
Mean vector quantization, separating for memoryless vector quantizers, 19
Memoryless Gaussian/Gamma/Laplacian sources, vector quantizer design, 170–174
Memoryless vector quantizers, 11–15, 16–19
distortion, 11–13
gain/shape VQ, 18–19
lattice VQ, 19
Lloyd algorithm for, 15
multistep VQ, 17
product codes, 17–18
properties of, 13–14
quantization technique, 11
separating mean VQ, 19
tree-searched VQ, 16–17
Minimization of distortion integral, asymptotically optimal block quantization, 103–104
Minimum discrimination information
distortion measure, 208–218
model selection for, 208–218
Model selection, for minimum discrimination information, 208–218
Model-testing finite-state vector quantizer, image coding and, 352–353
Monochrome image, image compression of, 369–372
Multiple multiply-accumulate (MMAC), VLSI chip set, 526–527
Multiple rate low rate voice CODEC
approach to, 521
codebook development for VQ, 521–522
excitation coding, 523
simulation results, 523–524
variable rate vector coding, 522
Multiple-speaker experiments, discrete utterance recognition, 428–429
Multistage coders, embedded speech coding, 299
Multistage connection of DVQ/AVQ, 367
Multistage vector quantizers, 4–5
Multistep VQ, memoryless vector quantizers, 17

Network-based connected digit recognition
acoustic pattern matchers, 435
future of, 436
pronunciation networks, 434–435
pronunciation network search, 435–436
recognition experiments, 436
using VQ, 434–436
Next address selector (NAS), VLSI chip set, 527–528
Non-adaptive spatial VQ, image compression using, 333–336

Octave band SBC-VQ systems, low-rate speech encoding, 313
One-dimensional TVQ, 77–79
Optimal quantization, 90–93
vector quantizers, 111

Parameter set reduction, 800 bit/s vector quantization LPC vocoder, 222
Pattern matching
complexity issue, 3–4
800 bit/s vector quantization LPC vocoder, 226
vector PCM and, 2–3
Pattern selection, discrete utterance recognition, 412–413
Perceptually-based cell classifier, image vector quantization with, 344–347
Performance bounds for speech-like sources, VQ and, 263–264
Performance characteristics
code-excited linear predictive (CELP) coder, 330
image compression, 335
subband speech coder, 318–319
Performance evaluation
pyramid vector quantizer (PVQ), 185–188
vector quantizer design, 164, 171–174
Performance theories
asymptotic vector quantization, 49–51
rate-distortion theory, 44
scalar quantization, 47–49
Pitch and voice coding, 800 bit/s vector quantization LPC vocoder, 227–228
Polytopes
cells of, 120–121
comparison of, 122
crosspolytope, 120
Dirichlet's integral, 120
dodecahedron, 122
exception four-dimensional polytopes, 122
generalized octahedron, 120
icosahedron, 122
n-dimensional simplexes, 120
n-sphere, 120
regular polygons, 121
regular simplex, 120
second moment of, 119–122
truncated octahedron, 121
volume of, 120–121
Postfiltering, CELP coder, 323
Predictive vector quantizer (PVQ), 76
Preprocessing color video signals, VQ and, 366
Product code initialization, VQ and, 262
Product code PVQ, pyramid vector quantizer (PVQ), 182
Product codes
codebook design, 60–61
memoryless vector quantizers, 17–18
split-band codes, 61
split-vector codes, 61
Product code vector quantizers
design methods, 283–286
SGVQ design and, 278–283
shape-gain quantizers, 276–278
simulations of, 281–286
for waveform/voice coding, 274–288
Product/multistep feedback vector quantizer, 21
Pronunciation networks, network-based connected digit recognition, 434–436
Pyramid vector quantizer (PVQ), 175–189
applied to other sources, 183–185
asymptotic performance, 181–182
binary codeword representation, 180–181
encoding algorithm, 178–180
geometric structure of Laplacian source, 176–177
integer points on pyramid, 178
performance evaluation, 185–188
product code PVQ, 182
source coding theorem, 177–178
structural description of, 178–182

Quadrature minor filter (QMF), subband speech coder and, 317
Quantization
for CELP coder, 320–323
codes for Gaussian channel, 118–119
memoryless vector quantizers, 11
problem in vector quantizer design, 170–171
VQ and, 391, 393–394
for waveform coding, 140–149
Quantizers
finite-state vector quantizer (FSVQ), 142–143
performance bounds and VQ, 264
process in image sequence coding, 391, 393–394
quantizer-source mismatch, 173–174
simulation, linear predictive coding matrix quantization, 235–236
See also specific types of quantizers.

Random codebooks, codebook design, 61–62
Random quantizers, vector quantizers, 113–114
Rate-distortion speech coding
codebook design, 216
LPC analysis, 211–213
with minimum discrimination information distortion measure, 208–218
model selection for minimum discrimination information, 213–216
simulation of, 217
Rate-distribution theory, 44–45
coding quantizer output, 44–45
mathematical results, 45–47
Real-time implementation
hardware realization, 486–487
multiple rate low rate voice CODEC, 521–524
real-time speech compression, 495–498
real-time vector APC speech coding, 517–519
systolic architectures, 499–512
vector quantizer architecture for, 513–515
VLSI chip set, 525–528
Real-time speech compression
codebook search processor, 496–498
motivation for, 495–496
with VLSI VQ processor, 495–498
Real-time vector APC speech coding
with adaptive postfiltering, 517–519
coder performance, 519
coder structure, 517–518
encoding operation, 517–518
real-time implementation, 519
VQ of pitch and LPC predictors, 518
VQ of prediction residual, 518
Recognition
BYBLOS, 461–464
connected digit recognition, 457–460
deterministic single level recognition method, 458–459
discrete utterance recognition, 411–414, 415–432
hidden Markov models (HHM), 438–441
large vocabulary word recognition, 465–468
network-based connected digit recognition, 434–436
recognition experiments, 436
search strategy, 463
speaker-independent isolated word recognition, 438–441, 442–455
speaker recognition, 469–472
speech recognition, 407–409
Region merging, in image coding, 403
Remote hyperplanes, codebook analysis, 114
Replenishment techniques
codebook replenishment, 392
experimental results, 394–397
image sequence coding, 391–397
label replenishment, 392
VQ and, 391–397
Reproduction of alphabet, 93
Residually excited subband with classifier, low-rate speech encoding, 313–314
Residual SBC-VQ systems with classifier, low-rate speech encoding, 314
Root lattices, 123–125
definition of, 123, 134–135
Voronoi region of, 123–131

Sampled speech, finite-state vector quantizer (FSVQ), 148–149
Scalar coding, vs. vector coding, 9
Scalar embedded coders, embedded speech coding, 297
Scalar quantization, 47–49
constrained-entropy quantization, 48–49
Lloyd-Max quantization, 47–48
vector rotation and, 54
Scalar vs. vector quantization of vector sources, 51–56
bit allocation, 51–53, 56
dimensionality, 56
distortion measure, 56
vector rotation for correlated sources, 53–54
vector sources, 56
VQ comparisons, 54–56
Scalar waveform quantization, speech waveform coding, 65–66
Secondary structural decomposition, vector quantizers, 112
Segmentation using region growing, image coding and, 402–404
Segment clustering, segment vocoder at 150 B/S, 247–248

Segment network, segment vocoder at 150 B/S, 248
Segment quantization, time-dependent vector quantization, 64
Segment vocoder at 150 B/S
description of, 246–247
segment clustering, 247–248
segment network, 248
Selective frame transmission, time-dependent vector quantization, 64
Shannon theory, 9–10
Shape-gain vector quantizer (SGVQ), product code vector quantizers and, 276–283
Short-term predictor, CELP coder, 320
Signal processing and analysis component, BYBLOS, 462
Simulated annealing
algorithm for, 154
used for good code design, 152–158
Small code book design, vector quantizer design, 167–168
Source classification, linear block prediction with, 383–384
Source codes, good code design, 155
Source coding techniques, for discrete utterance recognition, 411–414
Source coding theorem, pyramid vector quantizer (PVQ), 177–178
Space packing, asymptotic vector quantization, 50–51
Spatial vector quantizers (SVQ), 73–76
classified VQ, 74–75
codebook replenishment VQ, 75
hierarchical VQ, 75
interframe VQ, 75–76
mean/shape VQ, 73–74
Speaker-dependent performance on large data base, discrete utterance recognition, 419–422
Speaker-dependent studies, discrete utterance recognition, 422–428
Speaker-independent experiments, discrete utterance recognition, 428–429
Speaker-independent isolated word recognition
database description, 446–448
hidden Markov models (HHM), 438–441
performance characteristics, 446–448, 451
pilot experiments, 448–451
VQ-based preprocessor for, 442–455
word-based VQ as preprocessor, 443–446
Speaker recognition
data base experiments, 470
experimental results, 470–471
LPC analysis, 470
speaker-based VQ codebook approach, 469–470
VQ approach to, 469–472
Speaker verification
approach to, 475–476
experimental results, 476–480
source model for, 474–475
text-dependent using VQ source coding, 473–482
Spectral template selection, speech recognition, 407–408
Speech coding, 9
code-excited linear predictive (CELP) coder, 320–323
efficient bit allocation, 289–296
800 bit/s vector quantization LPC vocoder, 222–231
embedded speech coding, 297–299
linear predictive coding vector quantization, 233–236
linear predictive speech coding, 250–257
low-rate speech encoding, 312–314
matrix quantizer design for LPC speech, 237–245
product code vector quantizers, 274–288
rate-distortion speech coding, 208–218
subband speech coder, 316–319
system for, 35–37
vector predictive coding of speech at 16 kbit/s, 300–310
vector quantization coding schemes, 271–273
vector quantizer architecture for, 513–515
VQ and, 34–68, 258–269
See also Speech waveform coding.
Speech example, 96–97
Speech recognition
experimental results, 408–409
Markov source models and, 407–409
spectral template selection, 407–408
stochastic modeling of phonetic elements, 408
VQ and, 407–409
Speech segmentation, linear predictive coding matrix quantization, 234–235
Speech/speaker independent codebook design, vector quantization coding schemes, 271–273
Speech/speech-like waveforms, VQ and, 258–269
Speech waveform coding
scalar waveform quantization, 65–66
vector waveform quantization, 66–67
VQ and, 26–27
Spherical codes, good code design, 156–157
Split-band codes, product codes, 61
Split-vector codes, product codes, 61
Stochastic modeling of phonetic elements, speech recognition, 408
Storage costs, of VQ, 40–41
Storage reduction, cascaded quantization, 59
Subband coding, VQ using, 7
Subband speech coder
with backward adaptive predictor, 316–319
with optimal time-frequency bit assignment, 316–319
performance characteristics, 318–319
QMF analysis and synthesis networks, 317
VQ of error power matrix, 318
Subband vector quantizers (subband VQ), 83
Subjective listening test, linear predictive speech coding, 255–256
Systolic architectures
applications for, 501–502
computational issues, 502–503
extensions of, 506–511
linear-array systolic VQ processor, 503–505
VLSI realizations, 511
for VQ, 499–512
VQ distortion measures, 502–503

Time alignment, discrete utterance recognition without, 415–432
Time-dependent vector quantization, 63–65
adaptive VQ, 64–65
segment quantization, 64
selective frame transmission, 64
Time redundancy reduction, 800 bit/s vector quantization LPC vocoder, 222
Tracking finite-state source codes, finite-state vector quantizer (FSVQ), 142
Training/acquisition of phonetic coarticulation models, BYBLOS, 462
Training set generation
image coding, 340–341
image sequence coding, 391
VQ and, 391
Training and testing
binary clustering, 62
codebook design, 62–63
codebook robustness, 62–63
K-means initialization, 62
Transform coding, vector rotation, 54
Transform vector quantizer (TVQ), 77–81
adaptive transform coding using VQ, 77
interframe transform vector quantization (ITVQ), 79–81
one-dimensional TVQ, 77–79
two-dimensional TVQ, 79
Tree-expanded codebooks, 4
embedded speech coding, 298–299
Tree search
800 bit/s vector quantization LPC vocoder, 225–226
memoryless vector quantizers, 16–17
Tree and trellis coding, vector waveform coding, 67
Trellis encoders, 22–23
Two-dimensional TVQ, 79

Unrestricted polar quantizers, asymptotic performance of, 191–194
Unvoiced codebook experiment, 800 bit/s vector quantization LPC vocoder, 226
U.S. Standard 4800 BPS voice coder
algorithm for, 328
code-excited linear predictive (CELP) coder, 328–330
computational requirements, 329
error protection, 329–330
expandability, 328–329
performance characteristics, 330

Variable-length segment quantization
codebook design, 252–253
linear predictive speech coding and, 250–257
Vector coding, vs. scalar coding, 9
Vector distortion criteria, vector predictive coding of speech at 16 kbit/s, 302
Vector DPCM, linear block prediction, 383
Vector formation
image sequence coding, 391
VQ and, 391
Vector linear predictor design, vector predictive coding of speech at 16 kbit/s, 302–304
Vector PCM, pattern-matching and, 2–3
Vector predictive coding of speech at 16 kbit/s, 300–310
AVPC system, 307–310
system design, 301–307
vector distortion criteria, 302
vector linear predictor design, 302–304
vector quantizer design, 304
Vector predictive coding system
vector distortion criteria, 302
vector linear predictor design, 302–304
vector quantizer design, 304–307
Vector predictive quantization, feedback vector quantizers, 21
Vector quantization-based preprocessor
using cepstral dynamic features, large vocabulary word recognition, 465–468
for speaker-independent isolated word recognition, 442–455
Vector quantization clustering algorithm
generalized Lloyd algorithm, 356
image coding application, 360–362
pairwise nearest neighbor clustering, 357–360
Vector quantization coding schemes, 271–272
experimental results, 272–273
speech/speaker independent codebook design in, 271–273
Vector quantization embedded coders, embedded speech coding, 297–298
Vector quantization of error power matrix, subband speech coder, 318
Vector quantization (VQ)
adaptive coding and, 28
adaptive transform coding using, 77
adaptive vector predictive coding, 7
adaptive vector quantizer, 23–24, 367
algorithm for quantizer design, 88–98
asymptotic VQ, 49–51
binary vector quantizer (BVQ), 81–83
block quantizers, 88
codebook design, 39–40, 56–63
codebook generation, 15–16, 391
coder performance, 367
of color images, 364–367, 386–389
for color still pictures, 364–366
combined component VQ, 365–366
complexity in pattern matching, 3–4
computational costs, 40–41
definition of, 72–73
dependency characteristics, 41
description of, 260–262
design algorithm, 262
distortion measures, 38–39, 88–89
dynamic vector quantizer, 366–367
experimental results, 264–267
fast nearest-neighbor search algorithms, 6
feedback vector quantizers, 19–22
FSVQ, 21–22
gain-shape vector quantizers, 5–6, 18–19
Gauss Markov sources and, 25–26
guess by "splitting," 93–94
image coding and, 28–29, 72–83
image sequence coding and, 390–397
image/speech compression for, 9
imbedded VQ methods, 6–7
individual component VQ, 365
K-means algorithm for, 40
lattice VQ, 19
Lloyd's example, 95–96
mean separated and normalized VQ, 364
memoryless vector quantizers, 11–15, 16–19
model for, 41–44
multi-stage connection of DVQ/AVQ, 367

multistage vector quantizers, 4–5, 17
network-based connected digit recognition using, 434–436
optimal quantization, 90–93
compared to other systems, 267–268
pattern matching, 2–4
performance bounds for speech-like sources, 263–264
performance theories, 44–51
predictive vector quantizer (PVQ), 76
preprocessing color video signals, 366
principle of, 390–391
problem formulation, 37–38
product code initialization, 262
product codes, 17–18
product/multistep FVQ, 21
quantization process, 391, 393–394
quantizer performance bounds, 264
replenishment techniques, 391–397
reproduction alphabet and, 93
scalar vs. vector sources, 51–56
separating mean VQ, 19
source coding for speaker verification, 473–482
spatial vector quantizers (SVQ), 73–76
speaker recognition approach to, 469–472
in speech coding, 34–38, 65–67, 96–97
speech recognition, 407–409
of speech/speech-like waveforms, 65–67, 258–269
speech waveform coding and, 26–27
storage costs, 40–41
subband coding using, 7
subband vector quantizers (subband VQ), 83
systolic architectures for, 499–512
techniques for reduced complexity, 4–6
time-dependent VQ, 63–65
training set generation, 391
transform vector quantizer (TVQ), 77–81
tree-search VQ, 16–17
tree-structured codebooks, 4
trellis encoders, 22–23
vector formation, 391
vector PCM, 2–3
vector predictive quantization, 21
vector quantizer design, 73
vector tree, 22–23
of video signals, 342–343
vocoding (LPC VQ) and, 27–28
Vector quantizer design, 73
algorithm for, 160–167
coding, 163–164
DPM/APC modules, 513–514
full-search system architecture, 514
image coding application, 515
image transform coding application, 164–167
lattice quantizer design, 161–163
for memoryless Gaussian/Gamma/Laplacian sources, 170–174
notation for, 160–161
performance evaluation, 164, 171–174
quantization problem, 170–171
quantizer-source mismatch, 173–174
small code book design, 167–168
for speech/image coding, 513–515
system function, 513
tree-search system architecture, 514–515
vector predictive coding of speech at 16 kbit/s, 304
VHSIC full search architecture, 515
Vector quantizers
codebook analysis, 114–115
embedded speech coding, 297
lattice quantizers, 115–116
optimal quantization, 111
prior research, 108–109
random quantizers, 113–114
secondary structural decomposition, 112
structure of, 108–116
Vector rotation
for correlated sources, 53–54
distortion weighting, 54
optimal scalar quantization, 54
transform coding, 54
Vector sources, scalar vs. vector quantization of vector sources, 56
Vector tree, 22–23
Vector waveform coding, tree and trellis coding, 67
Vector waveform quantization, speech waveform coding, 66–67
Very low data rate speech compression with, linear predictive coding vector quantization, 233–236
VLSI chip set
multiple multiply-accumulate (MMAC), 526–527
next address selector (NAS), 527–528
for real-time VQ of image sequences, 525–528
VQ encoding, 525–526
VQ encoding engine, 526
Vocoding (LPC VQ), VQ and, 27–28
Voronoi regions
first layer of lattice and, 131–132
of lattices, 122–133

Waveform coding
for finite-state vector quantizer (FSVQ), 140–149
product code vector quantizers for, 274–288
Word-based VQ as preprocessor, speaker-independent isolated word recognition, 443–446
Word model generator, BYBLOS, 462

Editor's Biography

Hüseyin Abut (S'70–M'74–SM'79) received the B.S. degree in electrical engineering from Robert College, Istanbul, Turkey, in 1968, and the M.S. and Ph.D. degrees from North Carolina State University, Raleigh, in 1970 and 1972, respectively.

Since 1981 he has been a professor in the Electrical and Computer Engineering Department at San Diego State University. Previously, he held positions in Marmara Research Institute, Gebze, Turkey (1972–73), Boğazici University, Istanbul, Turkey (1973–80), California State University at Long Beach (1979–80), National Semiconductor Corporation (1980–82), and Ecco Industries, Danvers, MA (1984–85). He was a visiting professor in Information Systems Laboratory, Stanford University (1979), in the Institut für Angewandte Physik, Frankfurt University, Federal Republic of Germany (Summer, 1982, 1985, and 1988), and in the Department of Computer Engineering, Ege Universitesi, Izmir, Turkey.

His current technical interests include digital speech and image encoding, speech/speaker recognition, and fast computational algorithms.